A

W9-DEW-424

A. Shasby

Modern Biology

Teacher's Edition

Albert Towle

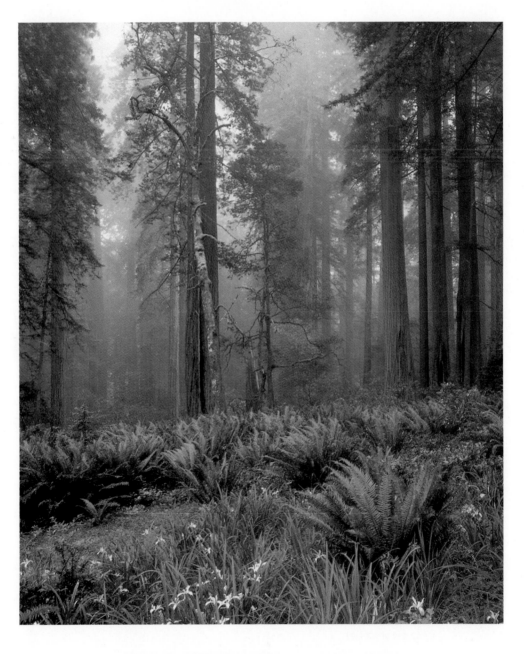

HOLT, RINEHART AND WINSTON

AUSTIN NEW YORK SAN DIEGO CHICAGO TORONTO MONTREAL

Acknowledgments

Daniel H. Franck, Ph.D.
Senior Editor, Project Director, National Science Consultant
Holt, Rinehart and Winston
New York, New York

The photograph on the title page shows redwood trees, *Sequoia sempervirens,* California. Photo by C. G. Schwartz, FPG

Developmental assistance by Ligature, Inc.

Copyright © 1989, 1985, 1981, 1977, 1973, 1969 by Holt, Rinehart and Winston, Inc.

All rights reserved. No part of this publication may be reproduced or transmitted in any form or by any means, electronic or mechanical, including photocopy, recording or any information storage and retrieval system without permission in writing from the publisher.

Requests for permission to make copies of part of the work should be mailed to: Permissions, Holt, Rinehart and Winston, 1627 Woodland Avenue, Austin, Texas 78741.

Printed in the United States of America
ISBN 0-03-013922-8
89012 039 987654321

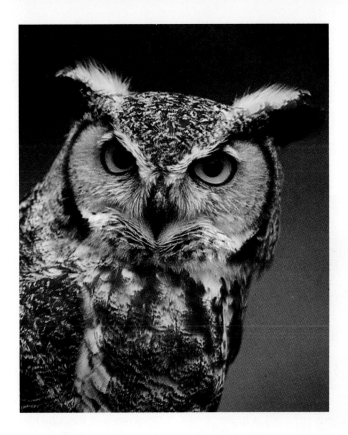

T*he photograph on the cover is of the great horned owl,* Bubo virginianus *ssp.* virginianus *(Family Strigidae, Order Strigiformes). The great horned owl is among the largest of owls, measuring up to 63 cm in length. It may have a wingspread of 140 cm.*

The great horned owl ranges throughout most of North and South America, from near the Arctic Circle in the north to below the 50th parallel in the south, and frequents a variety of habitats, from forest to open country, from desert to swamp. It can even be found from time to time in city parks. The great horned owl is common throughout its range, but because like most owls it is active by night, it is rarely seen.

The structural features common to owls are among the adaptations that have helped to make this bird successful. Like all owls the great horned owl is a predator; it locates prey by means of acute hearing and keen nighttime vision. It swoops down on its prey almost silently, the flapping of wings muffled by soft, fluffy feathers. A strong hooked beak and sharp talons enable it to capture and consume its prey.

Due to its exceptional size and strength, the great horned owl can capture a variety of prey, large and small—skunks, rabbits, chickens, grouse, frogs, lizards, and beetles. This adaptability in diet certainly contributes to the great horned owl's success over its wide range and various habitats.

Contents

Interleaved Chapter Planners

Lists of resources, planning information, and answers to reviews precede each chapter of the pupil text.

Modern Biology: *Research and Testing*

The *Modern Biology Program* was developed in conjunction with a thorough program of research and testing. The objective of this research and testing was to survey the wishes and concerns of American teachers of high school biology and to reflect those wishes and concerns in the various components of this program.

The great diversity of teaching styles in American high schools is a result of the diverse education of teachers and the variation in the political structures of educational units. Some schools have complete independence in establishing their curricula, while others are severely limited by state or local requirements. To develop a textbook and a biology program that meets the needs of such a diverse group, Holt, Rinehart and Winston embarked upon a series of activities to obtain information on the desires and the concerns of both teachers and students across the nation. The major components of these activities are described below.

Focus Groups

The scope, sequence, and design of *Modern Biology* were tested in focus groups throughout the country. Each focus group consisted of a moderator and about a dozen teachers from local high schools. The moderator showed the teachers various prototypes for the design, the table of contents, the writing style, and many other aspects of the book. The teachers responded, and representatives of Holt, Rinehart and Winston noted the teachers' concerns and then modified the components accordingly.

Focus-group testing was extensive and the testing was conducted in many areas of the country to determine the differences that resulted from the geographical location of the participants. As the focus-group testing proceeded, the design and the table of contents of the book became more and more accepted by members of the groups. A total consensus, of course, can never be obtained—biology teachers are a diverse group—but the factors that were tested in the focus groups finally met with general approval among the participants.

National Market Survey

Holt, Rinehart and Winston also conducted a national marketing survey and received completed surveys from over 3,000 high school biology teachers. The respondents represented users of all high school biology textbooks.

This national survey gave Holt, Rinehart and Winston a substantial look at the structure of the American high school biology classroom. Holt representatives learned about the time allotments for classroom and laboratory periods, the number of days the classes meet, the weekly time allotments for biology activities, district and state restraints and demands, student makeup, the division of students into ability levels, and many other aspects of classroom structure.

In addition, the survey yielded data on the time allocated to various subjects taught in the biology classroom. This data allowed Holt editors to determine which subjects in the biology curriculum took up the greatest amount of time during the school year. As a result, the editors reallocated book space to reflect the proportional amount of time spent on each subject.

The market survey also yielded data on what teachers liked and disliked about the presentation of subjects in various biology textbooks. Teachers were asked to describe what specific aspects of content they considered important in their biology curriculum, and to describe how well their current textbook treated each subject. The resulting data gave Holt editors a chance to ascertain the strengths and the weaknesses of current biology textbooks.

National Advisory Panel

Holt, Rinehart and Winston hosted a National Advisory Panel of nationally known scientists, science education professors, district science supervisors, and exemplary high school teachers to inform Holt of the current trends in biology education.

The present concerns of both scientists and science educators often foreshadow future trends in science education. The panel presented various viewpoints on the directions of both the biological sciences and the changing theories that address the best way to communicate science to students. The National Advisory Panel conducted spirited discussions on the merits of various approaches to teaching biology. These discussions had a profound impact on the structure of the *Modern Biology Program.*

Review Process

Holt, Rinehart and Winston then conducted an extensive series of content reviews of all aspects of the *Modern Biology Program.* Development of the manuscript for the text occurred in phases and at each phase the manuscript was reviewed for scientific and educational accuracy.

Each chapter of the manuscript was reviewed by at least two high school teachers and by one university professor who was a specialist in the subject of the chapter. Analysis of these reviews resulted in detailed modification of the style and the content of each chapter and insured that the most accurate and modern scientific thought about a subject was incorporated into the text.

As the manuscript proceeded to more advanced stages in the publishing process, teachers, scientists, and district science supervisors read the copy and suggested changes. These suggestions allowed editors of Holt, Rinehart and Winston to modify the manuscript at advanced stages of development.

Student Involvement

As the manuscript evolved, a number of high school students were asked to read and to comment on the manuscript and on the laboratory components of the program. Student responses allowed Holt editors to change the manuscript where needed and also to provide Holt editors with information on the readability of the manuscript. In addition, teachers acted as a conduit to Holt editors about the changing expectations and the abilities of their students, thus allowing the editors to create a text that meets students' needs.

Readability

Many districts and states require computerized readability tests to establish the reading level of a textbook. Any computerized readability test is highly suspect because it substitutes a mathematical analysis of sentence structure and word difficulty for the actual ease and effectiveness of communicating the written word.

However, to make readability analyses available to educators, Holt, Rinehart and Winston conducted computerized readability tests using the Bormuth, Dale-Chall, and Fry formulas. Results of these readability tests are available from Holt editors or from your local Holt, Rinehart and Winston representative. The readability tests were conducted on a chapter-by-chapter basis and the results allowed Holt editors to revise sections of the manuscript where it was needed.

Laboratory Field Testing

Of particular importance to Holt, Rinehart and Winston was the desire to make sure that the laboratory exercises worked in a high school laboratory setting. Because the availability of materials and resources varies widely between classrooms, laboratory exercises must be written in a way that could work in any reasonably equipped laboratory. Writers and editors also need to be aware of the time it takes to set up, conduct, and clean up a laboratory exercise. To address this need Holt field tested each laboratory exercise to determine its feasibility in the high school classroom. The responses provided by the field testers presented Holt editors with information about which laboratory exercises worked and which did not work. The scope and the sequence of the laboratory manual was revised in accordance with teacher and student responses.

Computerized Laboratory Field Testing

Computerized laboratory exercises were also field tested with both teachers and students after beta-testing had been completed. Beta-testing is a method by which computer professionals test the workability of a program under a variety of situations and eliminate any problems in the program. The responses from teachers and students enabled Holt editors to determine the success of the programs and to suggest ways to improve the programs.

Finally, Holt editors visited high school science departments to outline the kinds and the availability of computers in high schools and how computer time is allocated to biology students.

Introducing Modern Biology

Building on a long tradition of excellence, this edition of Modern Biology *preserves the best of past editions while setting a new standard for biology education. Totally redesigned and rewritten, it presents the most authoritative science content with a fresh and inviting voice.* Modern Biology *makes your job easier by engaging students' interest and setting them up for success. Each of its 10 units and 53 chapters has been carefully developed, using the following organizing principles.*

A thematic approach

Modern Biology presents the facts of biology within a pattern of unifying themes that helps students understand the larger significance of the details they are learning. Chapter 1 introduces students to seven key themes that reappear as strands throughout the book: (1) evolution, (2) reproduction and inheritance, (3) development, (4) structure and function, (5) energy relationships, (6) ecology, and (7) science and society.

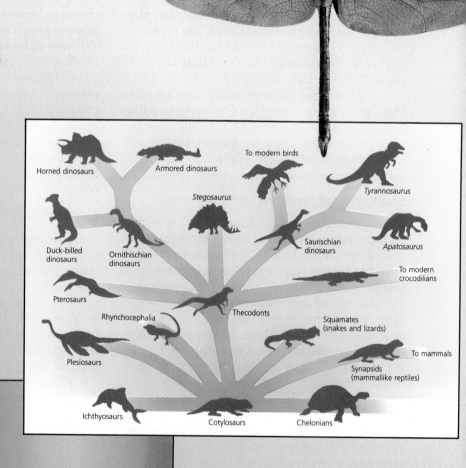

Horned dinosaurs

Armored dinosaurs

To modern birds

Tyrannosaurus

Stegosaurus

Duck-billed dinosaurs

Ornithischian dinosaurs

Saurischian dinosaurs

Apatosaurus

To modern crocodilians

Pterosaurs

Rhynchocephalia

Thecodonts

Squamates (snakes and lizards)

To mammals

Plesiosaurs

Synapsids (mammallike reptiles)

Ichthyosaurs

Cotylosaurs

Chelonians

A process perspective

Modern Biology also presents biology as a way of knowing. Chapter 2 uses a case study approach to introduce the basic processes biologists use to study the living world. In "Biology in Process" features throughout the book, students get an inside view of how biologists use these processes in their work.

A meaningful context

Modern Biology helps students think critically about concepts and relate them to the world in which they live. For example, the prologue and epilogue stimulate students' curiosity about biology and its significance for their lives now and as adults. Throughout the book questions lead students to evaluate and apply what they have learned.

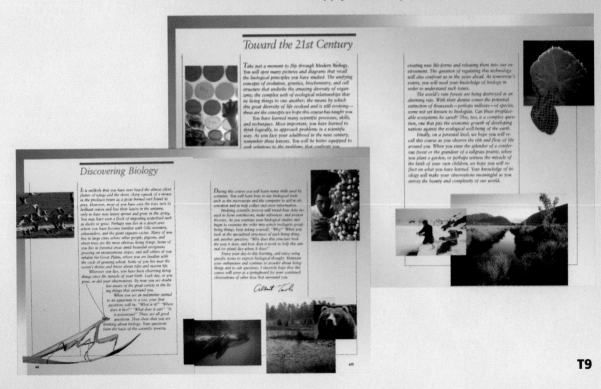

Openers set the stage for learning

Each unit and chapter opens with a blend of ideas and images that stimulate students' curiosity. These visual and verbal elements also work together to provide a preview of the key ideas and topics that lie ahead.

Unit Openers

Each two-page unit opener inspires wonder and reflection with a thought-provoking quotation and a mix of evocative illustrations and photographs. A list of chapter titles gives students an overview of the unit, and a brief introduction leads students to consider one or more of the central concepts of the unit.

Unit **4** Evolution

Chapters

14 | The Origin of Life
15 | Evolution: Evidence and Theory
16 | Evolution: Speciation
17 | Human Evolution
18 | Classification

"*Each organism instructs; its form and behavior embody general messages if only we can learn to read them. The language of this instruction is evolutionary theory.*"

Stephen Jay Gould

On the facing page you see a flower that looks like an insect and an insect that looks like a leaf. What biological processes have produced these oddities? What advantages do these organisms derive from their misleading appearances? You will learn answers to questions such as these as you study this unit on evolution, the process of change over time that modifies species in ways that enhance their survival.

Fossil of fish and palm leaf

Labyrinthodont, *Ichthyostega,* ancestral amphibian

202

Chapter **19** Viruses

Introduction

Viruses are organisms that straddle the threshold between living and nonliving things. They lack several of the essential features of all life-forms and reproduce by taking over the genetic machinery of host cells. The release of viruses from host cells results in the spread of many infectious diseases in both plants and animals.

Chapter Outline

19.1 Structure and Classification
• *Studying Viruses*
• *Structure*
• *Classification*

19.2 Reproduction and Evolution
• *The Lytic Cycle*
• *The Lysogenic Cycle*
• *Evolution*

Chapter Concept

As you read, look for ways in which the structure and reproduction of viruses make them highly effective agents of infection.

Artist's model of an influenza virus. Inset: Color-enhanced TEM of AIDS (HIV) virus inside a stricken T4 cell (×30,000).

For additional teaching information see pp. T194–T195.

282   Chapter 19

Chapter Openers

Each chapter opener draws students into the material through intriguing photographs and an introduction that relates the topic of the chapter to concepts that students already know. A Chapter Outline maps out the chapter, and a Chapter Concept focuses students' attention on a major concept.

Insect-mimic orchid, *Ophrys speculum*

Leaf mimic katydid from Brazil

203

Well-designed instruction helps students succeed

Every lesson in Modern Biology *is designed to help students succeed. The solid instructional design begins with clearly stated objectives that set goals for learning. Thematically organized text promotes comprehension, and carefully correlated reviews help students check their understanding.*

Section Objectives

Each lesson begins with a list of three to five objectives that preview the major concepts students will learn.

Headings

A numbered main heading signals the beginning of each lesson, and two levels of subheadings make visible the logical organization of topics.

Vocabulary

Each new biological term appears in boldfaced type and is clearly defined in context. Phonetic respellings help students pronounce difficult terms.

Section Objectives

- List the characteristics of body structure, fins, and integument of the yellow perch.

- Describe the internal anatomy of the yellow perch.

- Explain how gills function in respiration.

- Describe the spawning behavior of fishes.

Figure 36-12 The external features of the yellow perch, *Perca flavescens*, are representative of bony fishes. Note the growth rings on the scales shown in the inset. They indicate the fish's approximate age.

36.3 Morphology of a Bony Fish

The diversity of body shape in bony fishes reflects their varied ways of life. However, all have some characteristics in common. In this section we will examine the morphology of the yellow perch, a representative bony fish. This fish, *Perca flavescens*, grows up to 38 cm long and is common in the Great Lakes and other freshwater lakes of the eastern United States and Canada.

External Anatomy

Look at Figure 36-12 to see how the external appearance of the yellow perch shows several important adaptive features. These include the gill structures, specialized fins, and skin.

Body Structure

The yellow perch, like all bony fishes, has distinct head, trunk, and tail regions. On each side of the head is the **operculum**, a hard plate that opens at the rear and covers and protects the gills. The perch opens its mouth to take in water, and the water passes over the gills, where gases are exchanged. The water leaves through the operculum. Yellow perch have strong muscles along their dorsal backbone that thrust the tail from side to side. They can swim at speeds up to 16 km per hour.

External anatomy of a fish

Scales
Anterior dorsal fin
Lateral line
Posterior dorsal fin
Nostril
Eye
Mouth
Operculum
Pectoral fin
Pelvic fin
Anal fin
Caudal fin

546   *Chapter 36*

the olfactory lobes and the **cerebrum** *(suh-REE-brum)*, which respond mainly to smells. Smell is the most important sense in fish. At the posterior of the brain are the **cerebellum** *(SER-uh-BELL-um)*, which coordinates the muscles, and the **medulla oblongata** *(muh-DULL-uh muh-lawn-GAHT-uh)*, which regulates the internal organs. The major sense organs are connected directly with the brain via **cranial nerves**.

From the medulla the spinal cord extends the length of the body and carries nerve impulses to and from the brain. **Spinal nerves** connect the spinal cord with the internal organs, muscles, and sense organs. These nerves also carry impulses to the brain from the lateral line system.

Reproduction

The reproductive systems of most bony fishes resemble that of the yellow perch. The sexes are separate. Eggs are produced by ovaries in the female, and sperm is produced by the testes in the male; both are released through an opening just to the rear of the anus. Fertilization of the eggs of yellow perch takes place externally. Young fish hatch from fertilized eggs within a few hours in warm water or after many weeks in cold water. The adaptation of laying large numbers of eggs ensures that at least a few individual yellow perch will survive to become adult fish. Most eggs are not fertilized. Many others die or are eaten. The number of eggs bony fishes lay varies considerably. A brook trout lays from 80 to 5,000 eggs, but an ocean sunfish lays about 3 million.

Some fishes bear live young. Females receive sperm from males during mating, and fertilization is internal. These "live-bearers" include fishes common in aquariums, such as guppies, platys, mollies, and swordtails. Vertebra____ _____ are said to be viviparous.

Few generalizations can be made ab___ _____ spawning, behavior of bony fish. The f___ _____ eggs on the lake bottom, and the male fe_____ **milt**, a fluid containing sperm. Some spec__ build crude nests from plants, sticks, a__ migrate to warm, protected shallow wat__ eels from rivers in Europe migrate to the Atlantic and Pacific salmon migrate back They use their sense of smell to find the hatched. Here they spawn and die.

Section Review

1. Compare the atrium and the ventricle of a perch.
2. How do the caudal, dorsal, and pectoral fins function?
3. How do fish gills function in respiration?
4. Why do some fish lay a large number of eggs?
5. Bottom-dwelling fish often lack a gas bladder. Explain the adaptive advantage of this modification.

Figure 36-15 Among the many fish that fertilize their eggs externally is the white sturgeon, *Acipenser transmontanus.* Note that several of these white sturgeon eggs are hatching.

550 Chapter 36

Section Review

Five questions at the end of each lesson help students evaluate their progress. Question 1 checks students' understanding of new vocabulary. Questions 2-4 test their recall of facts. Question 5 requires students to think critically about a new concept.

Section Review

1. Compare the atrium and ventricle of a perch.
2. How do the caudal, dorsal, and pectoral fins function?
3. How do fish gills function in respiration?
4. Why do some fish lay a large number of eggs?
5. Bottom-dwelling fish often lack a gas bladder. Explain the adaptive advantage of this modification.

Chapter Review

A two-page Chapter Review presents four types of activities that review concepts and extend learning.

- **Vocabulary** consists of an alphabetical list of vocabulary terms and five questions that help build vocabulary skills.
- **Review** comprises ten multiple-choice and ten short answer questions that assess students' understanding of the chapter's major facts and concepts.

Chapter 36 Review

Vocabulary

anal fin (547)	deuterostome (537)	lobe-finned fish (545)	pharyngeal pouch (537)
appendicular skeleton (539)	dorsal fin (547)	lung (545)	pharynx (548)
artery (549)	dorsal nerve cord (537)	lungfish (545)	placoid scale (5__)
atrium (549)	endoskeleton (537)	medulla oblongata (550)	ray-finned fish (__)
axial skeleton (539)	esophagus (548)	milt (550)	skull (539)
capillary (549)	gas bladder (549)	notochord (537)	spawning (550)
caudal fin (547)	gill (540)	olfactory lobe (542)	spinal nerve (550)
cerebellum (550)	gill slit (537)	operculum (546)	vein (549)
cerebrum (550)	integument (547)	optic lobe (549)	ventricle (549)
chromatophore (547)	lateral line system (540)	pectoral fin (543)	vertebral column (537)
cranial nerve (550)		pelvic fin (547)	

In each set of terms below, choose the one that does not belong, and explain why.
1. notochord, vertebral column, dorsal nerve cord, gill slits
2. cerebrum, cerebellum, operculum, optic lobe

3. placoid scale, chromatophore, integument, milt
4. pharynx, atrium, ventricle, capillary
5. pectoral fin, gas bladder, operculum, lateral line system

Review

1. Chordates and echinoderms have a similar (a) body plan (b) embryonic development (c) habitat (d) feeding habit.
2. The skull functions to (a) protect the brain (b) anchor the axial skeleton (c) coordinate muscles (d) collect sensory information.
3. The density of water means that fish (a) must be heavy in order to float (b) do not need strong muscles to propel them (c) sink to the bottom when at rest (d) all of the above.
4. The lateral line system (a) keeps fish moving in a straight line (b) initiates migration (c) detects vibrations (d) acts as a camouflage.
5. Lamprey locate prey through a highly developed sense of (a) smell (b) hearing (c) sight (d) touch.
6. Sharks use claspers to (a) increase maneuverability (b) startle other fish (c) hold on

to the female while mating (d) hold on to prey while feeding.
7. To move up or down, a fish uses its (a) anal fin (b) dorsal fins (c) caudal fin (d) pelvic fins.
8. The esophagus of a fish (a) creates buoyancy (b) fertilizes eggs (c) carries food from the mouth to the stomach (d) holds wastes moving from the stomach to the anus.
9. A fish's ventricle (a) pumps blood through the body (b) collects blood returning to the heart (c) facilitates gas exchange through diffusion (d) carries oxygen through capillaries.
10. Some fish species (a) build nests for their eggs (b) return to their hatching grounds to spawn (c) lay huge numbers of eggs (d) all of the above.
11. Name the seven living vertebrate classes, and give a representative animal of each.

552 Chapter 36

12. Of what advantage is an endoskeleton in comparison with an exoskeleton?
13. Distinguish between the appendicular and axial skeletons.
14. How do researchers determine the habitat range of sharks?
15. Describe adaptations in rays and skates that allow a bottom-dwelling existence.

16. What is the advantage to lungfishes of having both lungs and gills?
17. How can scale pattern show a fish's age?
18. How do saltwater fish and freshwater fi__ maintain their salt balance?
19. Describe a fish's circulatory system.
20. Describe the function of the five lobes of a fish's brain.

Critical Thinking

1. Neoteny is a phenomenon in which larvae become sexually mature while retaining larval characteristics. The term has been used to explain in part the evolution of vertebrates from ancestral chordates. Use what you know about tunicates to support this hypothesis.
2. The shark has a large corkscrew-shaped structure in its intestine called a spiral valve. How might this organ function in digestion?
3. Many species of fishes that live deep in the ocean where little or no light reaches are luminescent. What might be the advantage of such an adaptation?
4. The four-eye butterfly fish shown here often swims backward. How might the fish's behavior and coloring work to its advantage?
5. Saltwater fish drink more water and produce much less urine than freshwater fish do. How do you account for this?

6. Cod and many other ocean fish lay eggs on the surface of the water. The male largemouth bass scoops out a nest in a lake or river bottom and waits for a female to deposit her eggs. What hypothesis would you make regarding the relative number of cod and bass eggs? Why?

Extension

1. The construction of the St. Lawrence Seaway allowed sea lampreys from the Atlantic Ocean to invade the Great Lakes. There they caused extensive damage to lake trout and other commercial fish species. Use your library to find out how the lamprey population was controlled and how successful this effort was. Present your findings to the class.
2. Read the article by Kenneth E. F. Watt entitled "Deep Questions about Shallow

Seas," in *Natural History,* July 1987, p. 96–97. Write a report explaining the diversity of fish species in shallow coastal waters.
3. Study the anatomy of any fish other than a yellow perch. Identify the operculum, backbone, and each fin. Label the internal parts, and note any structural difference from the yellow perch. What conclusion can you draw about habitat and behavior on the basis of the fish's structure?

Chordates and Fishes

- **Critical Thinking** questions require students to analyze and synthesize concepts and draw inferences from new information. One question promotes visual literacy by asking students to interpret a photo, illustration, or graph.
- **Extension** suggests additional readings, research projects, and hands-on activities.

An innovative visual program motivates, clarifies, and instructs

Modern Biology *features a state-of-the-art visual program that does more than just illustrate the text—it works in tandem with the text to teach and reinforce concepts. The work of award-winning photographers and some of the nation's top scientific and medical illustrators make it a program of beauty and eloquence.*

Closeups

Scanning electron micrographs (SEMs), illustrations paired with micrographs, and the innovative photography of Lennart Nilsson and others all give students a dramatic view of unseen structures in the living world.

Table 33-1   Common Insect Orders

Order	Number of Species	Examples	Characteristics
Orthoptera ("straight-wing")	23,000	Grasshoppers Crickets Katydids Cockroaches	Two pairs of straight wings; chewing mouthparts
Isoptera ("equal-wing")	1,800	Termites	Two pairs of membranous wings; chewing mouthparts
Dermaptera ("skin-wing")	1,100	Earwigs	One or two pairs of wings; chewing mouthparts; pincer-like appendages at tip of abdomen
Anoplura ("unarmed-tail")	200	Sucking lice	Wing... ...sucking ...
Hemiptera ("half-wing")	40,000	All true bugs	
Homoptera ("like-wing")	20,000	Aphids Mealy bugs Cicadas	
Ephemeroptera ("for-a-day-wing")	1,500	Mayflies	M... ...pre-w... ...unc-...ults

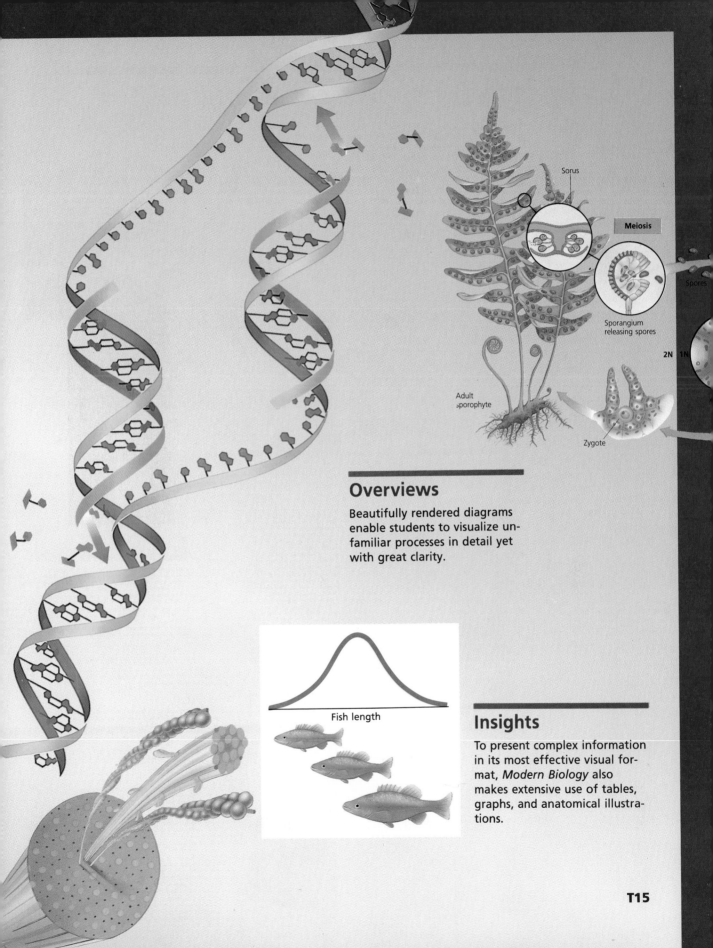

Sorus

Meiosis

Spores

Sporangium
releasing spores

2N 1N

Adult
sporophyte

Zygote

Overviews

Beautifully rendered diagrams
enable students to visualize unfamiliar processes in detail yet
with great clarity.

Fish length

Insights

To present complex information
in its most effective visual format, *Modern Biology* also
makes extensive use of tables,
graphs, and anatomical illustrations.

Special features enhance learning

Modern Biology broadens the scope of instruction through informative features and a useful reference section at the end of the book. While set off by their distinctive design, the features are nonetheless a integral part of each chapter.

Biotechnology

Sixteen chapters contain a "Biotechnology" feature that examines the tools, techniques, and knowledge biologists use to develop new products and technologies.

Biotechnology

New Foods from Plant Research

When early peoples began cultivating plants, they usually selected the most robust plants for food and seed. This selection process led to hardier crops with better yields. Unfortunately, it also led to decreased genetic variation. Furthermore, because only certain traits were selected for, many "wild" plants with potentially beneficial traits were not cultivated. Scientists are now rediscovering some of these wild varieties. By studying the adaptations of successful wild plants, biologists may discover ways to increase crops in areas of the world in which low food production leads to poor nutrition.

In this century the work of many agricultural researchers has been directed toward short-term solutions to problems rather than toward basic plant research. As a result scientists know a great deal about effective pesticides but comparatively little about plant genetics. Despite their use of expensive chemical pesticides and herbicides, farmers still lose up to one-third of their crops to pests and diseases. Some wild plants, however, are immune to pests and diseases.

Genetic engineering is leading the way in developing new plants and pest-resistant crops. Using the tools of genetic engineering, scientists can transfer genes from one plant to another. These scientists hope to develop plants with the ability, for instance, to fix atmospheric nitrogen, to resist frost, or to have immunity to pests or potent herbicides.

Scientists have already developed a hybrid potato plant that manufactures its own insect repellent. The repellent, a chemical called leptine, occurs naturally in wild potatoes. Researchers have fused cells of wild potat commercial duce a tas variety.

Once ties are de propagate cultures. B traits four develop h eties, sci vent disea potato far This famin blight tha the main i country. N also decrea chemical p

356   Chapter 24

The Importance o

from the inside out. Hagfish are extremely supple and can tie their bodies in knots to evade capture or to clean off the slime they secrete in self-defense. When not feeding, hagfish remain hidden in mud burrows on the ocean floor.

Answer to Biology in Process Answers might include using a yellow target from the start and then changing to a white target, but not with the abruptness of Clark's original experiment.

Chondrichthyes

Sharks, skates, and rays belong to the Class Chondrichthyes. Like the Agnatha, Chondrichthyes have skeletons of cartilage, not

Biology in Process

Thirty-five chapters include a feature called "Biology in Process," which explores ways in which biologists have used basic scientific processes to expand our knowledge of the living world.

Biology in Process

Understanding Sharks

People long feared all sharks without knowing a great deal about them. Scientists studied the anatomy of sharks by dissecting dead ones. Without observing live sharks, scientists were unable to answer some simple questions: How intelligent are sharks? How long do they live? How much food do they need?

Today scientists use a variety of sophisticated methods to collect information about the structure and the behavior of sharks. To study growth rates, researchers measure and tag individual sharks and then release them. When recaught, the sharks are again measured. To determine habitat range, scientists track sharks that have been fitted with transmitters that emit radio signals.

One American biologist, Eugenie Clark, is a world leader in shark study. She was one of the first scientists to construct a shark pen, which enabled her to observe and collect data from live sharks in captivity. In 1958 she conducted the now-classic experiments on the learning abilities of sharks. Up to that time sharks were considered somewhat stupid and poor subjects for the learning experiments that had been conducted on other animals. Clark hypothesized, however, that

542   Chapter 36

Laboratory

Structure and Function of Bird Feathers

Objective
To examine the structure and function of the three main types of bird feathers

Process Skills
Observing, inferring, hypothesizing

Materials
1 each of quill, contour, and down feathers; unlined paper; prepared slide of a contour feather; microscope; prepared slide of a down feather

Method
1. Examine a quill feather. Hold the base of the central shaft with one hand and gently bend the tip of the feather with your other hand. Be careful not to break the feather. Next hold the shaft and wave the feather in the air. Record your observations concerning the structure of the quill feather. Relate your observations to its possible function.
2. Examine the vane of the feather. Does the vane appear to be a solid structure? Write a description of its structure. Make a drawing of the quill feather. Label the shaft, vanes, and barbs.
3. Examine a contour feather and a down feather. Make a drawing of the contour feather and the down feather. Label your drawings as before. Do you notice a difference in the structure of the contour and the down feathers?
4. Examine a prepared slide of a contour feather under low power. Note the smaller barbs, called barbules, extending from each of the barbs.
5. How might you observe the region between the barbs? Locate the tiny hooks at the end of each barbule. Note the arrangement of the hooks on adjacent barbs. Why do you think the hooks are so small? Make a labeled drawing of the hooks.
6. Examine the prepared slide of the down feather under low power. Locate the barbs and barbules. Switch your microscope to high power, and make a labeled drawing of the down feather.

Conclusions
1. What is the function of the shaft? What is the function of the vanes and barbs?
2. How do hooks increase the strength and air resistance of a feather?
3. How is the structure of the quill feather related to its function of aiding flight?
4. Why are down feathers more effective at keeping a bird warm than the other two feather types are?

Inquiry
Each of the feather types you have examined has a specific structure and function. Review your observations, and try to think of features that account for the efficiency of the three types of feathers.

Birds **599**

Laboratory

The "Laboratory" at the end of each chapter provides a hands-on lab experience in which students explore not only the question at hand but also the processes they are using to answer it.

Think

Think

At least once in each chapter a feature called "Think" challenges students to think critically about a concept related to the topic they are studying.

The word *anthozoa* means "flower animal," a description that fits sea anemones like this one very well. Why then do you think that early biologists did not mistake sea anemones for plants as they did sponges?

...ugh the skeletons of sharks are made of

...issue made of cells surrounded by tough, ...es. Although these cartilaginous fish seem ...n bony fish, fossil evidence indicates that ...from bony fish and lost bone over time. ...rom the Agnatha by having movable jaws ...ed fins.

...roximately 275 species of sharks, skates, ...ter. Most are carnivores that typically use ...ans and lateral line systems to track prey. ...most rays is covered with **placoid scales**— ...that feel like sandpaper.

...redatory habits of some shark species have given ...on as killers. Yet fewer than 10 percent of shark ...wn to attack humans unprovoked. Whale sharks, ...e gentle 18-m giants that graze on plankton. ...n with a side-to-side motion of their asymmetric ...ehind their heads are paired **pectoral** *(PECK-tuh-* ...jut out from their bodies like the wings of a plane. ...pensate for the downward thrust of the tail fin.

Figure 36-8 What similarities can you note between a shark's tooth (top) and placoid scales (bottom)?

...lity of sharks

...dy Clark used ...n sharks, which ...d near the ...ine Laboratory, ...ta, Florida. A ...as lowered into ...eeding time. ...target caused ...also enabled ...et at food low- ...water on a string. ...ays the sharks ...nging the bell ...with food. ...d the routine. ...out at faster in- ...greater distances ...t. She altered the

volume of the bell sound. Each time the sharks adjusted to the change in routine.

However, Clark learned that sharks had a limited tolerance for change. One day she altered the color of the target to a bright yellow. The male shark started as usual for the target as it was lowered into

the water. Then about two feet away from the target he reacted to the change in color and did not hit the target.
■ How would you design an experiment to determine whether the shark reacted to the sudden change in color or whether the reaction was specifically to the color yellow?

Chordates and Fishes **543**

Writing About Biology

Once in each unit "Writing About Biology" exposes students to the thought-provoking perspectives of such gifted science writers as Rachel Carson, Annie Dillard, and Lewis Thomas. These excerpts represent a rich diversity of styles and modes of writing about biology, and each is accompanined by a brief writing assignment for students.

Writing About Biology

Pilgrim at Tinker Creek

This excerpt is from Annie Dillard's Pulitzer-prize-winning book, Pilgrim at Tinker Creek.

The plankton bloom is what interests me. The plankton animals are all those microscopic drifting animals that so staggeringly outnumber us. In the spring they are said to "bloom," like so many poppies. There may be five times as many of these teeming creatures in spring as in summer. Among them are the protozoans—amoebae and other rhizopods, and millions of various flagellates and ciliates, gelatinous moss animalcules or bryozoans; rotifers—which wheel around either free or in colonies; and all the diverse crustacean minutiae—copepods, ostracods, and cladocerans like the abundant daphnias. All these drifting animals multiply in sundry bizarre fashions, eat tiny plants or each other, die, and drop to the pond's bottom. Many of them have quite refined means of locomotion—they whirl, paddle, swim, slog, whip, and sinuate—but since they are so small, they are no match against even the least current in the water. Even such a sober limnologist as Robert E. Coker characterizes the movement of plankton as "milling around."

A cup of duck-pond water looks like a seething broth. If I carry the cup home and let the sludge settle, the animalcules sort themselves out, and I can concentrate them further by dividing them into two clear glass bowls. One bowl I paint all black except for a single circle where the light shines through; I leave the other bowl clear except for a single black circle against the light. Given a few hours, the light-loving creatures make their feeble way to the clear circle, and the shade-loving creatures to the black. Then, if I want to, I can harvest them with a pipette and examine them under a microscope.

I don't really look forward to these microscopic forays: I have been almost knocked [off] my kitchen chair on several occasions [when] I was following with strained eye[s the] career of a monostyla [...] enormous red [...] whipped into the [...] ing everything, an[d ...] huge flap[...] sions that [...] sweep my [...] the kitche[n ...] moral ex[...] microscop[...] forehead is [...] stant reminder [...] facts of creation [...] would just as soon forge[t ...] In the puddle or pond, in the city reservoir, ditch, or Atlantic Ocean, the rotifers . . . spin and munch, the daphnia . . . filter and are filtered, and the copepods . . . swarm hanging with clusters of eggs. These are real creatures with real organs leading real lives one by one. I can't pretend they're not there. If I have life, sense, energy, will, so does a rotifer. The monostyla goes to the dark spot on the bowl: To which circle am I heading? I can move around right smartly in a calm; but in a real wind, in a change of weather, in a riptide, am I really moving, or am I "milling around"?

Write

- Some people think that all animals have an equal right to life and that humans are not necessarily more important than other animals. What do you think? Write an essay in which you take a stand. Consider the ways that humans use other animals—as pets, as food, and as research subjects.

440 Excerpt from Pilgrim at Tinker Creek by Annie Dillard. Copyright © 1974 by Annie Dillard. Reprinted by permission of Harper & Row, Publishers, Inc.

Intra-Science: How the Sciences Work Together

The Question in Biology: Can We Save the Salmon?

In this unit you read about the diversity of behaviors exhibited by vertebrates. One example of vertebrate behavior is the annual migration of hundreds of thousands of salmon from the Atlantic and Pacific oceans to the inland riverbeds where they were hatched. For some this is a distance as long as 3200 km. In their hatching grounds, the salmon lay and fertilize their eggs.

At the time of migration, which varies by species, the estuaries, or mouths of rivers, on the Pacific and Atlantic coasts teem with salmon. The salmon remain in these waters while their systems adjust to the change from a saltwater to a freshwater environment. Sharks, and killer whales as well as commercial and sports fishers all catch great numbers of salmon in these estuaries.

The surviving fish swim upstream through the fast rivers, fighting against rushing currents and jumping over waterfalls in leaps as high as 3 m. Pacific coast salmon also try to elude the brown bears, which gather along the banks of the rivers.

About a century ago, the salmon were so plentiful that a Smithsonian Institution observer wrote: "Their numbers are beyond calculation and seem to be so great as to challenge human ingenuity to affect them in any way."

"Human ingenuity," however, has affected the salmon's numbers—in a negative way. Some observers have even predicted the imminent extinction of the Pacific coho salmon.

Many factors have contributed to this problem. While commercial overfishing has been partly responsible, activities on the land are the major contributors to the salmon's demise. Deforested land has eroded into streams and buried some of the salmon nurseries. Industrial pollutants have washed into rivers and poisoned some fish; massive dams, built to produce hydroelectric power, impede the fish's journey upstream.

The salmon, which is North America's most important food fish, needs to be protected by the application of scientific methods to ensure its survival. Preserving it will require the efforts of people working in many branches of science.

The Connection to Chemistry

In the course of its migration a salmon moves from saltwater to freshwater and its body undergoes changes to adjust to the new environment. The properties of the cell membranes in the gills change. The kidneys undergo physiological changes, and the sodium-potassium pump is modified to maintain homeostasis in fresh water. These processes allow the salmon's cells to remain in biological balance.

After it has adjusted to fresh water the salmon literally follows its nose to its spawning grounds. When a salmon hatches, the chemical makeup of the water around it is imprinted in its memory. As an adult, the salmon follows streams with a chemical makeup that most closely corresponds to this memory.

Scientists who work to replenish populations of salmon place scent chemicals in the water of fish hatcheries. Young salmon will remember this scent, and after they mature in ocean waters, they find their way back to the hatchery. There, workers retrieve them from the water and strip them ofeggs or milt. The sperm and eggs are then mixed, and another generation of fish are hatched.

The Connection to Physics

Even a salmon with its great jumping ability cannot leap over a dam 100 m or more high. However, the fish will keep trying until it dies. In some locations humans have been built to provide the fish with a route around high dams.

One type of fishway, similar to a ship lock, takes advantage of gravity. Water flows from a high lock into a lower lock where the fish have gathered. As the water level in the lower lock rises, so do the fish. In this way fish are gradually raised from one lock to the next until they reach the level of the river behind the dam.

Ladders, another type of fishway, consist of a series of pools, each one about a foot higher than the previous one. The salmon leap from pool to pool. This type of fishway requires careful control. If the water flows down the steps too quickly, the fish cannot jump against the flow. If the water flows too slowly, the fish may not try to leap to the next level because they always seek the maximum water flow.

Principles of physics have also been used to build artificial spawning channels. These channels provide the cold, oxygen-rich water that salmon need in order to spawn.

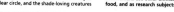

The Connection to Other Sciences

Bionic engineers solve many problems by studying natural systems. These scientists hope to learn how salmon manage to swim thousands of kilometers in the ocean, then find the mouth of their home river. As you know, the salmon relies on its sense of smell to navigate rivers. While still in the ocean, however, scientists think a salmon finds its way by using a kind of built-in magnetic compass and an ability to sense temperature differences and chemical variations.

The Connection to Careers

Salmon is a major food fish and thus provides jobs for the commercial fishers who catch it, the cannery workers who process it, the truckers who transport it, and the merchants who sell it. The fishways are designed by engineers and built by construction workers. The hatcheries employ fisheries biologists, chemists, and support staff. Ecologists and animal physiologists study the life cycle and adaptations of this unique migrating fish. For more information about related careers in the biological sciences, turn to page 842.

Intra-Science

Each unit ends with "Intra-Science," a feature that explores the ways in which biologists work together with physicists, chemists, and other scientists to find solutions to specific problems. Related careers are discussed in a special section.

632

Appendices

Four appendices provide useful information for student reference throughout the course:

- Five-Kingdom Classification of Organisms
- Safety
- Measurement
- Sources of Career Information

Glossary

An easy-to-use glossary provides definitions for all terms bold-faced in the text. Phonetic re-spellings also appear for all hard-to-pronounce terms.

Index

A comprehensive index provides quick access to all topics and individuals discussed in the text.

A comprehensive teacher's edition promotes success

This easy-to-use teacher's edition provides the tools and information you need for effective teaching. Important teacher information appears in three places: in the front matter, on annotated student pages, and in chapter planners that appear directly before each chapter.

Front Matter

The first 48 pages of the teacher's edition feature planning and pacing charts, suggestions for alternate approaches, and essays on topics related to biology teaching.

Annotations

The bulk of the teacher's edition consists of student pages annotated with answers to Section Reviews and other questions plus two types of additional teacher information. Annotations titled "Tip" provide teaching suggestions; those titled "Info" furnish background information that you may wish to share with students.

Info: Among mammals the gestation period of larger species generally is longer than that of smaller species. For example, the gestation period for rabbits is about 1 month; for elephants, over 20 months.

Tip: Suggest to students that they notice how the teeth of placental mammals are modified to perform specialized functions.

Chapter 33 | Insects

This chapter is devoted to familiarizing students with the structural physiological, and behavior characteristics of insects. Emphasis is placed on those characteristics that have contributed to the biological success of insects.

Correlation Chart

Section Number	Section Objectives	Pages	Features	Section Review
33.1	1. List the major reasons for the success of insects. 2. Give examples of three major kinds of variations among insects. 3. Name at least ten of the major orders of insects, and give an example of each. 4. Describe the relationship of insects to human society.	493–496		2, 3, 4 4 1, 5 1, 5
33.2	1. Summarize the function of each of the external parts of a grasshopper. 2. Summarize the function of each of the organ systems of a grasshopper. 3. Describe the processes of incomplete and complete metamorphosis, and explain how they contribute to the success of insects. 4. Give an example of three types of defense in insects.	497–502	Think, p. 502	
33.3	1. Describe the social organization and behavior of honeybees and the ways in which they behave. 2. Name three ways in which insects communicate, and give an example of each. 3. Summarize Karl von Frisch's discoveries about communication in honeybees.	503–507	Biology in Process, pp. 504–505 Biology Lab, p. 507	

Resources*

Print

Borror, Donald J., Dwight M. Delong, and Charles A. Triplehorn. *Introduction to the Study of Insects.* New York: CBS College Publishing, 1981.

Leahy, Christopher, *Peterson First Guides: Insects: Simplified Field Guide to Common Insects of North America.* Boston: Houghton-Mifflin, 1987.

Audiovisual/Software

Discover Entomology. Film or video. Entomological Society, 1987.

Insects: The Lovely and the Lethal. Britannica, 1981.

*Additional references are listed in Teacher's Program Guide.

33A

Section Teaching Guides

33.1 The Insect World

Approach

The main objective of this section is to familiarize students with the reasons for the great evolutionary success of insects and their major ecological importance. It builds on concepts of arthropod structure and function (Chapter 32), coevolution with flowering plants (Chapters 25), and diversity through adaptation (Chapters 15 and 16).

In presenting the text, guide students to develop an appreciation for the great positive significance of insects in ecosystems, helping them transcend stereotypical images of insects as pests. You may wish to use Table 33-1 as a tool to expand students' appreciation of insect diversity.

Concept Framework

Insects are one group of invertebrates about which all students will have some background knowledge. To tap this knowledge and provide a conceptual frame for approaching this section, engage students

33.3 Insect Behavior

Approach

The main objective of this section is to describe the social behavior of insects, particularly honeybees, and the ways in which insects communicate. It is important to convey to students that these genetically programmed behaviors are triggered by the release of hormones, which, in turn, are a result of gene expression. You might wish to review the concepts of protein production (Chapter 8) and gene expression (Chapter 12) to clarify this point. In addition, reinforce the idea that natural selection favored the evolution of social behavior and communication in insects.

Concept Framework

Anyone who has heard crickets chirping or has seen a line of ants marching toward a stale bread crust has witnessed a form of social behavior in insects. Students can apply their knowledge of social organization in humans to the study of insect behavior. For example, ask them to consider how division of labor and, consequently, communication allow human societies to operate efficiently.

Emphasis

In presenting the text stress the adaptive significance of communication and social behavior, as well as the difference between innate and learned behavior. These concepts will prepare students for the discussions of vertebrate communication and behavior in later chapters.

33.2 Aspects of Insect Biology

Approach

The main objective of this section is to describe how anatomy, patterns of development, and defensive strategies show adaptations that have led to the evolutionary success of insects. To help students appreciate the adaptive significance of such characteristics as wings and compound eyes, draw comparisons between insects and the arthropods discussed in Chapter 32. In discussing internal and external anatomy, stress the relationship between structure and function.

The information in this section on metamorphosis can be related back to the discussion of gene expression and morphogenesis in Chapter 12. Also identify metamorphosis as an adaptive strategy, a concept that students will also encounter when they read about echinoderms (Chapter 34) and amphibians (Chapter 37).

Concept Framework

Students will have some background knowledge...

Section Review Answers

(Continued from p. 502.)
3. the tympanum; on the first abdominal segment
4. Complete metamorphosis allows different developmental stages to become specialized for different functions, eliminates competition between larvae and adults for food and space, and increases the likelihood of survival during harsh weather.
5. The yellow jacket has warning coloration. Predators stay away from the fly because they identify it as a harmful insect. The fly displays Batesian mimicry.

Using the Features/The Dance Language of the Bees

The main process skills used in the feature are observing and experimenting. Karl von Frisch observed that when a bee returned to the hive after locating a food source, it performed a series of dancelike movements. By conducting a series of experiments in which he manipulated the placement of feeding stations, von Frisch determined that the bee's dance varies according to the distance of the food source from the hive.

In discussing the feature, point out to students that because two variables are observed together does not in itself establish a relationship between the two. Von Frisch at first thought that the type of food—pollen or sweetened water—determined the type of dance performed by the bee. Only through experimentation did he determine that distance of the food from the hive was the crucial factor.

Field Trip

Arrange for the class to visit a local apiary. Touring such a facility will enable students to observe directly the social behavior of bees. In addition, it will help students appreciate the commercial significance of bees to human society. Prior to the visit, prepare a list of specific points to notice or questions to answer about the facility. Some people have life-threatening allergies to bee stings. All students who plan to participate should provide parental permission and medical clearance.

Section Review Answers

(See p. 506.)

Laboratory Preparation

Background

In this laboratory exercise, students examine the external features of the paper wasp and compare them to diagrams of the grasshopper to verify that all insects share certain basic characteristics. The genus Polistes is a type of social wasp, with a body plan typical of the main characteristics of the insect body, like the honeybee described in the pupils' text. Students should notice that the female ovipositor in the worker has been modified into a stinger, which is

used in defending the nest from intruders. Students should know how the small sterile females become workers.

Preparation

Order the Polistes specimens ahead of time.

Safety

The preservative used on the specimen can be irritating to the skin. Caution students to avoid touching the specimen, and especially to avoid rubbing the eyes with contaminated hands.

Conclusions

1. A typical insect has three body segments: the head, the thorax, and the abdomen. The abdomen is divided into 11 segments. Insects also have three pairs of jointed legs that are attached to the thorax, one pair of compound eyes, and three simple eyes.
2. The ovipositor has been modified into a stinger.

Inquiry

Students should suggest that a number of representative insects be looked at to compare them morphologically to bees and wasps.

Chapter Review

Vocabulary

1. aorta; others are mouthparts
2. pheromone; others relate to metamorphosis
3. spiracle; others involved in digestion
4. behavioral variation; others are defenses involving coloration
5. nymph; others are insect body parts

Review

11. Small size may allow many species to inhabit same environment without competing.
12. different digestive enzymes or any other example involving adaptation of body systems, also different kinds of metamorphosis
13. three body parts, one pair of antennae, three pairs of jointed legs, and abdomen with 11 segments
14. by damaging crops, carrying disease, damaging wood in buildings

15. head, thorax, abdomen
16. In insects, oxygen travels directly to cells through air tubes called tracheae instead of being transported in the blood.
17. Camouflage allows insects to hide from predators by blending in with surroundings; warning coloration makes insects stand out in such a way that predators associate their appearance with their undesirability as food.
18. When levels of juvenile hormone are high, the release of molting hormone causes larva to molt. When levels of juvenile hormone drop, molting hormone causes the transition to pupa and then to adult.
19. workers, queen, drones
20. by rubbing scraper of one forewing against vein on other forewing; to attract mates and to protect territory

Critical Thinking

1. Insects with genes that confer resistance against insecticide survive and reproduce in large numbers; thus they proliferate through natural selection. This process occurs rapidly in insects because they have very short life spans.
2. Siphoning mouthparts suggest that the insect feeds on liquid food.
3. Odonata; two pairs of long, narrow membranous wings
4. A large organism probably could not transport oxygen quickly and efficiently enough through tracheae to supply all its cells; instead, the organism would require specialized organs (lungs) for gas exchange.
5. If a worker bee dies from stinging or through an attack by a predator, it has already completed most of its other functions. It would be a greater loss to the hive if death occurred early in the worker's life.
6. Answers will vary but may include: dormant stages in life cycle may have provided protection from coldest periods; ability to fly may have allowed migration to warmer areas; small size may have allowed them to inhabit protected environments; diverse food sources may not have been affected by colder climate; faster reproduction led to faster evolution.

33D

33C

Complete metamorphosis

Adult

Eggs laid by adult

Young larva hatching

Young larva

Older larva

Larva beginning to pupate

Pupa

Adult emerging from pupa

Chapter Planners

To help you in planning lessons, a four-page or six-page chapter planner appears just before the beginning of each chapter. These pages present

- A chart that correlates objectives with page numbers, questions, and features
- Lists of suggested readings and audiovisual and software resources
- Section teaching guides that include suggestions for approaching the lesson, activities, and tips for using the features
- A laboratory preparation guide
- Answers to the Chapter Review

A *full spectrum* of ancillaries supplements instruction

Modern Biology *features a complete package of ancillaries that provide abundant resources for laboratories, skill development, homework, and testing. For ease of use, each ancillary is clearly color-coded.*

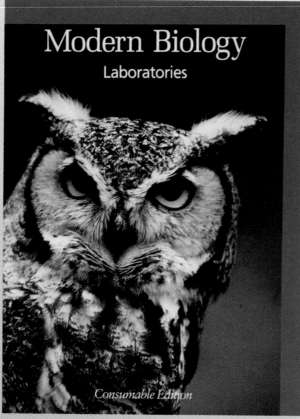

Components

- Laboratories (consumable)
- Laboratories (nonconsumable)
- Laboratories TE
- Process and Vocabulary Skills
- Process and Vocabulary Skills TE
- Tests/Alternate Tests
- Section Reviews
- Section Reviews TE
- Overhead Transparencies
- Overhead Teaching Masters
- Computerized Laboratories
- Computerized TestBank
- Teacher's Resource Organizer
- Teacher's Program Guide
- Posters

Teacher's Resource Organizer

The Teacher's Resource Organizer provides a unique storage system that enables you to keep blackline master versions of ancillaries at your fingertips.

HOLT,
RINEHART
AND WINSTON

1866

Teacher Resource Organizer

Computerized Components

The ancillary package includes a Computerized TestBank and Computerized Laboratories on photosynthesis, cell respiration, and genetics.

Modern Biology
Tests and Alternate Tests

Modern Biology
Section Reviews

Practice, Homework

HRW material copyrighted under notice appearing earlier in this work.

Investigation

43.1 *Daphnia Heartbeat*

How do chemicals affect daphnia heart rate?

Introduction

How do stimulants and depressants affect heart rate? When chemicals are ingested, they are carried through the body by the blood. While such chemicals sometimes act on the brain centers that control your heartbeat, they also affect the heart directly.

Daphnia, or water fleas, are small enough to be studied under a microscope and so transparent that we can see the heart beating. While daphnia are invertebrates, arthropods living in freshwater ponds, their hearts respond to chemical substances much as our own hearts do.

Prelab Preparation

Work in groups of 4 throughout this investigation.
What is your normal heart rate? How would you measure it?

1. Describe a simple technique for measuring your heart rate and record your measurement as "beats per minute."
2. How many times should you repeat this measurement? Why? Average your repeat measurements and record.

Look at the diagram of the water flea to familiarize yourself with the location of the internal organs. The water flea's heart rate is very fast (several times faster than yours). Do you think you could use the same technique for determining the heart rate of daphnia as you used to determine your heart rate? If not, what modification of your technique or what new technique would you propose?

3. Discuss your ideas with your lab team and record the technique that you have designed. Let your teacher check your technique before you begin. Remember that you want to keep your water flea alive and uninjured while you examine it and experiment with different chemical substances.

Procedure

A. Place the rubber bulb of a medicine dropper over the small end of the dropper as shown on the next page. Transfer a water flea from the culture jar to the center of a depression slide. Add 3 or 4 cotton fibers or bits of thread to the drop of water and cover with a coverslip. Examine the water flea under the low power of your microscope. Even though the heart is transparent, you will find it easily; it is constantly beating. CAUTION: Be careful not to crush the water flea with the coverslip or let the water on your slide dry up.

4. Follow the technique you designed in the Prelab Preparation and record in your data table the number of heartbeats per

Learning Objectives
- To compare the effects of different chemicals on the heartbeat of a daphnia.
- To relate changes in chemical concentration to the heart rate of a daphnia.

Process Objectives
- To experiment with a living animal under the microscope.
- To analyze team and class data.

Materials
For Group of 4
- Daphnia
- 2 Medicine droppers
- Depression slide
- Cotton fibers or thread
- Coverslip
- Compound microscope
- Stopwatch or clock with second hand
- Paper towel
- Dilute solutions of adrenaline (0.001%, 0.0001%, 0.00001%), ethyl alcohol (2%, 5%, 10%), coffee and/or cola drink (10%, 20%, 30%), or aspirin (0.5%, 1%, 2%)

Antennae
Compound eye
Heart
Brood pouch
Intestine

Average

Average

Investigation 43.1 **251**

Investigation 43.1 **253**

Laboratories

An innovative new laboratory manual integrates process and inquiry activities into a traditional laboratory program. The manual is available in either a consumable format with data sheets or a nonconsumable format without data sheets.

Process and Vocabulary Skills

One-page activities correlated to each chapter provide opportunities for students to expand their awareness of science processes and to build their vocabulary skills.

Name

Chapter 6

Vocabulary Ski

A. Compare and contrast each pa
of terms. Begin by describing th
similarities and then the differenc

B. Complete each sentence, using
terms from the vocabulary list.

Name   Class   Date

Chapter 32 **Process Skill**
Hypothesizing

A. How can we make hypotheses as precise as possible?

If observations are not carefully recorded and analyzed, they can lead to general hypotheses that are not very useful. For example, in exploring why a spider builds webs differently as it grows older, it is not very helpful to make general statements about the spider's ability to learn through experience. The statements must be broken down into precise, individual hypotheses to be tested. Is web-building an activity that is learned or innate? What changes does a spider undergo as it matures? How might these changes relate to the way in which a spider builds its web?

B. Spider Webs

The drawings depict three webs constructed by the same spider. Web A was built when the spider was a few days old. Web B was built at an age of one month, and Web C was built at an age of five months.

Answer each question.

1. Based on observations of the differences and similarities among the three webs, what three hypotheses can you formulate to explain the development of the webs?

a. _____

b. _____

Section Reviews

For each numbered section of a chapter, this ancillary supplies two pages of questions that can be used as quizzes, practices, or homework assignments.

Section **5.2** | *Parts of the Cell*

Read each statement. If the statement is true, write **T** in the space provided. If the statement is false, write **F** in the space provided.

_____ **1.** The ribosome is the site of protein synthesis within a cell.

_____ **2.** ATP is produced primarily in mitochondria.

_____ **3.** Most plant cells lack lysosomes.

_____ **4.** Primary cell walls continue to grow throughout the life of a plant cell.

_____ **5.** Chromatin carries a cell's genetic information.

Complete each statement by writing the correct term or phrase in the space provided.

6. Leucoplasts and chloroplasts are examples of organelles called _____ .

Tests and Alternate Tests

The testing ancillary provides two complete sets of chapter tests, unit tests, and final exams.

Chapter Test **30** | *Flatworms, Roundworms, and Rotifers*

1 in the diagram is a

_____ .

2 in the diagram is a

_____ .

Read each statement. If the statement is true, write **T** in the space provided. If the statement is false, write **F** in the space provided.

_____ **1.** A pseudocoelom is a body cavity in mesoderm.

_____ **2.** The mesoderm is the embryonic layer that forms between the ectoderm and the endoderm.

_____ **3.** Flatworms are members of the Phylum Nematoda.

_____ **4.** Planarians are free-living animals.

_____ **5.** During their larval stage, sheep liver flukes form cysts on blades of grass.

_____ **6.** Tapeworms absorb nutrients by means of a pharynx.

_____ **7.** Humans can best protect themselves from trichinosis by wearing shoes.

le layer of

d _____ .

_____ **8.** Cilia sweep foo~~___~~

_____ **9.** Tapeworms ha~~___~~ from digestion~~___~~

_____ **10.** A planarian's ~~___~~

Choose the one best response. Write the letter of that choice in the space provided.

_____ **11.** A single indivi~~___~~ organs is calle~~___~~

   a. host
   c. cyst

_____ **12.** Hookworm inf~~___~~ primarily by ca~~___~~

   a. interna~~___~~
   c. anemi~~___~~

_____ **13.** Schistosomiasis~~___~~

   a. flukes~~___~~
   c. tapewo~~___~~

_____ **14.** Which of the f~~___~~ food?

   a. phary~~___~~
   c. tegum~~___~~

21

Modern Biology

Section Reviews

HRW material copyrighted under notice appearing earlier in this work.

Planning and Pacing Guides

The success of a biology class depends upon many things. Of great importance is proper planning for the academic year. On the following pages are planning and pacing charts for three levels of student achievement. Though no classroom is composed of all of one kind of student, most biology classes can be thought of as basic, standard, or advanced.

Modern Biology is written at a readability level acceptable for 9th and 10th grade students. Teachers must be able to vary the length of time devoted to various subjects to bring the scope of the class in line with student abilities and with local curriculum requirements. One of the strong points of *Modern Biology* is its flexibility. The large number of chapters and the division of chapters into sections enable teachers to design a class around almost any type of class level or curriculum.

180-Day Curriculum and Extended Curriculum

For each of the planning and pacing charts on the following pages, recommendations are made for planning either a 180-day curriculum or an extended curriculum. By using these suggestions teachers can devise a curriculum suitable for their classes.

The 180-day curriculum is based on the average school year in which approximately 180 days are allotted for classroom study. Most teachers are required to prepare a curriculum in which all areas of biology are discussed within the 180-day period. In reality it is difficult to cover all of the required topics in depth within 180 days while taking into consideration the level of students in the classroom.

The extended curriculum is not time limited like the 180-day curriculum. The teachers and editors who designed the extended curriculum considered the time necessary for an in-depth coverage of the topic. As such, the total days allocated to an extended curriculum add up to greater than 180 days. The benefit of the extended curriculum guidelines is that they give the teacher an idea of the ideal time necessary to teach a subject. These guidelines set a general pace for a classroom in which the teacher is free to establish his or her own curriculum.

Basic Level Curriculum

The Basic Level curriculum is set up to aid teachers whose students are academically below the average American high school student in reading comprehension and general scholastic ability. Students must be able, however, to read and comprehend text that is written for 9th and 10th grade students. The enclosed planning and pacing guide is designed not to teach at a lower level, but to go more slowly over crucial topics and to avoid advanced topics more suitable to the more able student. By following the planning and pacing guide for Basic Level students, teachers will ensure that the most fundamental knowledge offered in the course is presented.

Standard Level Curriculum

The Standard Level curriculum is designed for average 9th and 10th grade students. These students are able to read the text with relative ease and comprehension. They should be able to master all but the most advanced topics.

Most but not all of the text should be required reading for students studying at this level. An average student should master the fundamentals of most of the text material in a year. Not all of the accessory material in the text is recommended, however, as these features are usually more advanced or challenging than the text.

Advanced Level Curriculum

The Advanced Level curriculum includes suggestions for teaching the most academically able students in the school. Advanced students are able to read the text quickly and comprehend it easily. These students will have the quickest grasp of the more abstract concepts of biology. In addition, advanced students will be able to master most of the features. Many advanced students should be encouraged to do the Extension exercises on the Chapter Review pages. These students should be assigned readings from the list found in the *Modern Biology Teacher's Program Guide*.

Basic Curriculum

Unit	1				2						3			4					5					6				
Chapter	1	2	3	4	5	6	7	8	9	10	11	12	13	14	15	16	17	18	19	20	21	22	23	24	25	26	27	28
Curriculum	■	■	■	■	■	■	■	■	■	■	■	·	·	■	■	·	·	■	■	■	■	■	■	■	■	■	■	·
Chapter Section Number	A	A	A	A	A	A	A	A	A	A	1,2	·	·	A	A	·	·	A	A	A	A	1	1	1	A	A	A	·
Laboratory	■	·	·			·		·			■	·	·		·	■	·	■	■			·	·	·	·	■	·	·
Biology in Process			·	·		·	■	·	■	·		·	·	■			·			·	■				·	■	·	·
Biotechnology		·			■								·							·	■	·	·					
Writing About Biology		·							·			·			■				·				·					
Think	■	·	·	·		■	·	·	·		■	·	·	·	■	·	·	■		■	·	·	·	·	·	·	■	·
Intra-Science					■											■							■				■	
Days (180-day curriculum)	4	4	4	5	5	3	5	3	5	5	3	·	·	3	5	·	·	3	4	4	4	3	3	3	4	4	4	·
Days (extended curriculum)	5	5	5	5	5	5	5	10	8	8	10	·	·	3	5	·	·	3	5	5	5	5	5	5	3	7	7	·

Unit		7							8						9							10				
Chapter	29	30	31	32	33	34	35	36	37	38	39	40	41	42	43	44	45	46	47	48	49	50	51	52	53
Curriculum	■	■	■	■	■	■	·	■	■	■	■	■	·	■	■	■	■	■	■	■	■	■	■	·	·
Chapter Section Number	A	A	A	A	A	1	·	A	A	A	A	A	·	A	A	A	A	A	A	A	A	A	A	·	·
Laboratory	■	·	·	■	·	·	·	■	■	■	■	·	·	■	·	■	■	·	·	■	·	·	■	·	·
Biology in Process	■	·			·	·		·	■		·		·		■	·	·				·	■	·		
Biotechnology			·	·						·		■				·	·	·	·						·
Writing About Biology	■											■					·								·
Think	■	·	·	·	■	·		■	·	·	·	■		·	■				■	·	■	·	·	·	·
Intra-Science							■						■						■					■	
Days (180-day curriculum)	4	4	4	4	4	3	·	5	4	4	4	4	·	5	5	5	5	5	5	5	4	4	3	·	·
Days (extended curriculum)	5	5	5	5	5	3	·	5	5	5	5	5	·	7	7	7	7	7	7	5	5	5	5	·	·

Legend:
- ■ recommended as part of curriculum
- · not recommended as part of curriculum
- (blank) not found in chapter
- A all sections

T27

Standard Curriculum

	1				2					3				4					5					6				
Chapter	1	2	3	4	5	6	7	8	9	10	11	12	13	14	15	16	17	18	19	20	21	22	23	24	25	26	27	28
Curriculum	■	■	■	■	■	■	■	■	■	■	■	·	■	■	■	■	■	■	■	■	■	■	■	■	■	■	■	■
Chapter Section Number	A	A	A	A	A	A	A	A	A	A	A	·	2	A	A	A	A	A	A	A	A	A	1	1	A	A	A	A
Laboratory	■	·	·			·		·		■	·	·	·			·		·	·			·		■	·			·
Biology in Process			·			·	·	■			·		·			·	·				·			■			·	·
Biotechnology		·											■								·			■				
Writing About Biology		■							·									·						■				
Think	■	·		·		·		·			·		·		■			·		·		·		■	·	·		·
Intra-Science				■									■					■						■				
Days (180-day curriculum)	4	5	5	5	5	4	8	5	5	5	5	·	3	4	4	4	3	3	4	4	4	3	3	3	4	4	4	3
Days (extended curriculum)	5	5	5	5	5	5	10	8	8	10	5	·	4	5	5	8	5	7	5	5	5	5	5	3	6	6	6	5

	7							8						9							10				
Chapter	29	30	31	32	33	34	35	36	37	38	39	40	41	42	43	44	45	46	47	48	49	50	51	52	53
Curriculum	■	■	■	■	■	■	·	■	■	■	■	■	·	■	■	■	■	■	■	■	■	■	■	■	·
Chapter Section Number	A	A	A	A	A	A	·	A	A	A	A	A	·	A	A	A	A	A	A	A	A	A	A	A	·
Laboratory	■	■	■	■	■	■	·	·	■	■	■	·	·	■	·	■	·	■	■	·	■	·	·	·	·
Biology in Process			·			·		·		·		·				·					·			·	
Biotechnology			■	·							■			■			■	·	■						·
Writing About Biology	■										■														
Think		·	■		■		·		■		·		■			·		■			·			·	·
Intra-Science							■						■												■
Days (180-day curriculum)	3	3	3	3	4	3	·	3	3	3	3	3	·	3	3	3	3	3	3	3	3	3	3	3	·
Days (extended curriculum)	5	5	5	5	6	5	·	5	5	5	5	5	·	7	7	7	7	7	7	5	5	5	5	4	·

Legend:

■ recommended as part of curriculum
· not recommended as part of curriculum
☐ not found in chapter
A all sections

T28

Advanced Curriculum

Unit	1				2					3				4				5					6				
Chapter	1	2	3	4	5	6	7	8	9	10	11	12	13	14	15	16	17	18	19	20	21	22	23	24	25	26	27	28
Curriculum																												
Chapter Section Number	A	A	A	A	A	A	A	A	A	A	A	A	A	A	A	A	A	A	A	A	A	A	A	A	A	A	A	A
Laboratory																												
Biology in Process																												
Biotechnology																												
Writing About Biology																												
Think																												
Intra-Science																												
Days (180-day curriculum)	3	3	4	4	4	4	4	4	4	4	4	4	3	3	4	4	4	3	4	4	4	4	4	3	4	4	4	3
Days (extended curriculum)	4	4	4	4	7	5	10	10	10	10	5	5	4	4	5	10	5	5	5	5	5	5	5	3	5	5	5	5

Unit	7							8						9							10				
Chapter	29	30	31	32	33	34	35	36	37	38	39	40	41	42	43	44	45	46	47	48	49	50	51	52	53
Curriculum																									
Chapter Section Number	A	A	A	A	A	A	A	A	A	A	A	A	A	A	A	A	A	A	A	A	A	A	A	A	A
Laboratory																									
Biology in Process																									
Biotechnology																									
Writing About Biology																									
Think																									
Intra-Science																									
Days (180-day curriculum)	3	3	3	3	3	3	4	3	3	3	3	3	4	3	3	4	4	4	4	2	2	2	2	2	2
Days (extended curriculum)	5	5	5	5	5	5	5	5	5	5	5	5	5	7	7	7	7	7	7	5	5	5	5	5	4

■ recommended as part of curriculum   ▪ not recommended as part of curriculum   ☐ not found in chapter   A all sections

Alternative Approaches

The diversity of high school biology classrooms makes it almost impossible to meet the needs of all teachers with a single textbook. Teachers vary widely in backgrounds, interests, and philosophies. In addition, local and state requirements often oblige teachers to follow a curriculum with which they are unfamiliar.

Modern Biology is often called a phylogenetic textbook. While it is true that organisms are discussed in a phylogenetic sequence, much of the book is structured not around taxa but around concepts such as the cell, ecology, or evolution. Although these concepts are unifying themes throughout the text, each of the ten units of the book can stand as an independent teaching tool. Consequently, once students master some basic concepts and vocabulary, teachers are free to use *Modern Biology* in many different ways.

On the following pages you will find suggestions for several approaches to using *Modern Biology*. The scope and sequence of *Modern Biology*, to our knowledge, reflects the wishes of the majority of teachers. From marketing surveys and long discussions with teachers, we found that most teachers prefer the phylogenetic approach because it places the diversity of life in an evolutionary context. The phylogenetic approach teaches students about whole organisms and allows teachers to emphasize the increasing complexity and adaptations of the earth's inhabitants. Other approaches, however, have strong adherents.

Systems Approach

The systems approach to teaching biology uses systems as a unifying concept. For example, a teacher might spend two weeks discussing digestion and the digestive systems of organisms, and another two weeks on excretion and the excretory system. The merit of this method lies in its synthesis and distillation of knowledge and in its emphasis on structure and physiology.

Modern Biology can be adapted to a systems approach because chapters are organized into sections that can be taught as independent units. By using the chart that follows, teachers can easily adapt their class to a systems approach.

Evolutionary Approach

Some teachers believe that biology must be understood as an outgrowth of the evolutionary process. *Modern Biology* is well suited to teaching an evolutionary approach. Students are introduced to the concepts of evolution, phylogeny, and adaptation on page 6, and most chapters reflect the evolutionary theme. With slight modifications in scope and sequence, *Modern Biology* can be easily used for the evolutionary approach.

Ecological Approach

Many teachers prefer to teach ecology first. In many locations this approach enables them to take advantage of fall weather for outdoor study. Teachers then discuss other topics in an ecological context, emphasizing an organism's role in its environment. The organization of *Modern Biology* into units makes it possible to use this approach. Unit 10, on ecology, can be taught first, after some initial preparation of vocabulary and introductory concepts.

Cellular/Molecular Approach

Teachers whose training has stressed physiological and molecular biology often prefer to introduce students to biochemistry and cell biology first and to reinforce these concepts throughout the school year. Because chemistry and cell biology are discussed early in the book, *Modern Biology* is well suited for teaching the cellular/molecular approach.

Process Approach

The process approach emphasizes the actual activities of science and tries to place scientific knowledge in the context of its discoveries. As such, teachers who prefer this approach are more inquiry and laboratory based. They try to get students to experience a phenomenon before they read about it in the text. Until this edition of *Modern Biology*, no standard high school biology text emphasized process skills. This text and its ancillaries rely on process and inquiry more than any other current text.

Systems Approach

Topic	Chapter	Basic	Standard	Advanced	Days (Standard) 180 days curriculum	extended study curr.
Biological Principles	1	A	A	A	4	5
	2	A	A	A	4	5
	3	A	A	A	5	5
	4	A	A	A	5	5
Human Body Systems	42	A	A	A	3	7
	43	A	A	A	3	7
	44	A	A	A	3	7
	45	A	A	A	3	7
	46	A	A	A	3	7
	47	A	A	A	3	7
	48	A	A	A	3	5
Invertebrate Systems	35	▪	A	A	5	5
	29	A	A	A		
	30	A	A	A		
	31 (choose relevant	A	A	A	20 days	25 days
	32 sections within	A	A	A		
	33 these chapters)	A	A	A		
	34	A	A	A		
Vertebrate Systems	41	▪	A	A	5	5
	36	A	A	A		
	37 (choose relevant	A	A	A	15 days	20 days
	38 sections within	A	A	A		
	39 these chapters)	A	A	A		
	40	A	A	A		
Microorganism Systems	19	A	A	A		
	20 (choose relevant	A	A	A	18 days	25 days
	21 sections within	A	A	A		
	22 these chapters)	A	A	A		
	23	A	A	A		
Plant Systems	24	A	A	A	4	3
	25	A	A	A	4	6
	26	A	A	A	4	6
	27	A	A	A	3	6
	28	▪	A	A	3	5
Cell Systems	5	A	A	A	5	5
	6	A	A	A	4	5
	7	A	A	A	8	10
	8	A	A	A	4	8
	9	A	A	A	5	8
Genetic Systems	10	A	A	A	3	10
	11	1, 2	A	A	2	5
	12	▪	▪	A	▪	▪
	13	▪	2	A	2	4
Evolution	14	A	A	A	2	5
	15	A	A	A	3	5
	16	▪	A	A	3	8
	17	▪	A	A	3	5
	18	A	A	A	2	4
Ecology	49	A	A	A	3	5
	50	A	A	A	3	5
	51	A	A	A	3	5
	52	▪	A	A	3	4
	53	▪	▪	A	▪	▪

Evolutionary Approach

Topic	Chapter	Basic	Standard	Advanced	Days (Standard) 180 days curriculum	Days (Standard) extended study curr.
Biological Principles	1	A	A	A	4	5
	2	A	A	A	5	5
	3	A	A	A	5	5
	4	A	A	A	5	5
DNA and Genetics	9	A	A	A	5	8
	10	A	A	A	5	10
	11	1, 2	A	A	5	5
	12	▪	▪	A	▪	▪
	13	▪	2	A	3	4
Principles of Evolution	14	A	A	A	4	5
	15	A	A	A	4	5
	16	A	A	A	4	8
	17	A	A	A	3	5
	18	A	A	A	3	4
Evolution of Microorganisms	19	A	A	A	4	5
	20	A	A	A	4	5
	21	A	A	A	4	5
	22	1	A	A	3	5
	23	1	1	A	3	5
Evolution of Invertebrates	29	1	A	A	3	5
	30	A	A	A	3	5
	31	A	A	A	3	5
	32	A	A	A	3	5
	33	A	A	A	4	5
	34	1	A	A	3	5
	35	▪	▪	A	▪	▪
Evolution of Vertebrates	36	A	A	A	3	5
	37	A	A	A	3	5
	38	A	A	A	3	5
	39	A	A	A	3	5
	40	A	A	A	3	5
	41	▪	▪	A	▪	▪
Evolution of Plants	24	A	A	A	3	3
	25	A	A	A	4	6
	26	A	A	A	4	6
	27	A	A	A	4	6
	28	▪	A	A	3	5
Ecology and Evolution	49	A	A	A	3	5
	50	A	A	A	3	5
	51	A	A	A	3	5
	52	▪	A	A	3	4
	53	▪	▪	A	▪	▪
The Cell	5	A	A	A	5	5
	6	A	A	A	4	5
	7	A	A	A	8	10
	8	A	A	A	5	8
Adaptations	42	A	A	A	3	7
	43	A	A	A	3	7
	44	A	A	A	3	7
	45	A	A	A	3	7
	46	A	A	A	3	7
	47	A	A	A	3	7
	48	A	A	A	3	5

Ecological Approach

Topic	Chapter	Basic	Standard	Advanced	Days (Standard) 180 days curriculum	Days (Standard) extended study curr.
Biological Principles	1	A	A	A	4	5
	2	A	A	A	4	5
	3	A	A	A	4	5
	4	A	A	A	4	5
Ecology	49	A	A	A	4	5
	50	A	A	A	4	5
	51	A	A	A	4	5
	52	A	A	A	4	5
	53	A	A	A	4	4
Plants	24	A	A	A	3	3
	25	A	A	A	3	6
	26	A	A	A	3	6
	27	A	A	A	3	6
	28	■	A	A	3	5
Microorganisms	19	A	A	A	3	5
	20	A	A	A	4	5
	21	A	A	A	4	5
	22	1	A	A	3	5
	23	1	1	A	3	5
Invertebrates	29	A	A	A	3	5
	30	A	A	A	3	5
	31	A	A	A	3	5
	32	A	A	A	3	5
	33	A	A	A	4	6
	34	1	A	A	3	5
	35	■	■	A	■	■
Vertebrates	36	A	A	A	3	5
	37	A	A	A	3	5
	38	A	A	A	3	5
	39	A	A	A	3	5
	40	A	A	A	3	5
	41	■	■	A	■	■
Genetics	10	A	A	A	5	10
	11	1, 2	A	A	5	5
	12	■	■	A	■	■
	13	■	2	A	3	4
Evolution	14	A	A	A	4	5
	15	A	A	A	4	5
	16	■	A	A	4	8
	17	■	A	A	3	5
	18	A	A	A	3	4
The Cell	5	A	A	A	4	5
	6	A	A	A	4	5
	7	A	A	A	8	10
	8	A	A	A	5	8
	9	A	A	A	5	8
Human Biology	42	A	A	A	3	7
	43	A	A	A	3	7
	44	A	A	A	3	7
	45	A	A	A	3	7
	46	A	A	A	3	7
	47	A	A	A	3	7
	48	A	A	A	3	5

Cellular/Molecular Approach

Topic	Chapter	Basic	Standard	Advanced	Days (Standard)	
					180 days curriculum	extended study curr.
Biological Principles	1	A	A	A	4	5
	2	A	A	A	4	5
	3	A	A	A	4	5
	4	A	A	A	4	5
The Cell	5	A	A	A	4	5
	6	A	A	A	8	5
	7	A	A	A	8	10
	8	A	A	A	4	8
	9	A	A	A	4	8
Genetics	10	A	A	A	4	10
	11	1, 2	A	A	4	5
	12	A	A	A	4	7
	13	A	2	A	3	4
Human Biology	42	A	A	A	3	7
	43	A	A	A	3	7
	44	A	A	A	3	7
	45	A	A	A	3	7
	46	A	A	A	3	7
	47	A	A	A	3	7
	48	A	A	A	3	5
Microorganisms	19	A	A	A	4	5
	20	A	A	A	4	5
	21	A	A	A	4	5
	22	1	A	A	3	5
	23	1	1	A	3	5
Invertebrates	29	A	A	A	3	5
	30	A	A	A	3	5
	31	A	A	A	3	5
	32	A	A	A	3	5
	33	A	A	A	4	6
	34	1	A	A	3	5
	35	■	■	A	■	■
Vertebrates	36	A	A	A	3	5
	37	A	A	A	3	5
	38	A	A	A	3	5
	39	A	A	A	3	5
	40	A	A	A	3	5
	41	■	■	A	■	■
Plants	24	A	A	A	3	3
	25	A	A	A	4	6
	26	A	A	A	4	6
	27	A	A	A	4	6
	28	■	A	A	3	5
Ecology	49	A	A	A	3	5
	50	A	A	A	3	5
	51	A	A	A	3	5
	52	■	A	A	3	4
	53	■	■	A	■	■
Evolution	14	A	A	A	4	5
	15	A	A	A	4	5
	16	■	A	A	4	8
	17	■	A	A	3	5
	18	A	A	A	3	4

Process Approach

Topic	Chapter	Basic	Standard	Advanced	Days (Standard) 180 days curriculum	Days (Standard) extended study curr.
Process and Principles	1	A	A	A	4	5
	2	A	A	A	5	5
	3	A	A	A	5	5
	4	A	A	A	5	5
The Cell	5	A	A	A	5	5
	6	A	A	A	4	5
	7	A	A	A	8	10
	8	A	A	A	5	8
	9	A	A	A	5	8
Genetics	10	A	A	A	5	10
	11	1, 2	A	A	5	5
	12	■	■	A	■	■
	13	■	2	A	3	4
Evolution	14	A	A	A	4	5
	15	A	A	A	4	5
	16	■	A	A	4	8
	17	■	A	A	3	5
	18	A	A	A	3	4
Ecology	49	A	A	A	3	5
	50	A	A	A	3	5
	51	A	A	A	3	5
	52	■	A	A	3	4
	53	■	■	A	■	■
Microorganisms	19	A	A	A	4	5
	20	A	A	A	4	5
	21	A	A	A	4	5
	22	1	A	A	3	5
	23	1	1	A	3	5
Plants	24	A	A	A	3	3
	25	A	A	A	4	6
	26	A	A	A	4	6
	27	A	A	A	4	6
	28	■	A	A	3	3
Invertebrates	29	A	A	A	3	5
	30	A	A	A	3	5
	31	A	A	A	3	5
	32	A	A	A	3	5
	33	A	A	A	4	6
	34	1	A	A	3	5
	35	■	■	A	■	■
Vertebrates	36	A	A	A	3	5
	37	A	A	A	3	5
	38	A	A	A	3	5
	39	A	A	A	3	5
	40	A	A	A	3	5
	41	■	■	A	■	■
Human Biology	42	A	A	A	3	7
	43	A	A	A	3	7
	44	A	A	A	3	7
	45	A	A	A	3	7
	46	A	A	A	3	7
	47	A	A	A	3	7
	48	A	A	A	3	5

*All levels to do all *Biology in Process*, *Biotechnology*, *Intrascience*, and *Think* features, plus laboratory exercises for all assigned chapters.

Developing Critical Thinking, Process, and Writing Skills

The high school biology classroom is no longer just a place where facts are disseminated and tested. The modern biology classroom is becoming a place where students not only learn the facts associated with the science of life but also learn the various higher level thinking skills that are a requirement for a successful adult life.

Modern biology education incorporates two related types of higher level skills, critical thinking and process. Loosely defined, critical thinking is the ability to analyze what is taught and to apply criteria of reliability to the things a person is taught. Additionally, the understanding of the processes that go into a scientific method is critical in making sound judgments about scientific knowledge.

Critical Thinking Skills

The various aspects of critical thinking have been debated for many years in our nation's high schools and universities. No absolutely precise definition of critical thinking is universally accepted. However, educators agree that critical thinking skills are decidedly different from the recall of facts and from the straight comprehension of verbal or written material.

Validating Facts

Critical thinking skills involve judging the validity of facts by collecting evidence that either supports the facts or refutes them. Sometimes judging the validity of facts involves observing phenomena. One of the functions of laboratories in biology classes is to allow students to observe organisms and to conduct experiments. These processes allow students to see firsthand if the facts presented in lectures and textbooks correspond to the reality of their experiences.

It is often impractical or impossible for students to check facts by observing or by experimenting. One aspect of critical thinking is for students to validate information by using other sources. Encourage students to learn how and when to consult other sources of information.

Critical thinking requires the knowledge of process skills, which are discussed below. Understanding that science is a process and that knowledge is gained by methods that employ process is essential for the validation of that knowledge. Students who are aware of scientific methods are better able to critically judge the validity of information and to determine the sources of error that lead to invalid information.

Making Decisions

Two of the essential results of highly developed critical thinking are the ability to make decisions and the ability to make value judgments. Making workable, ethical decisions means mastering the background knowledge of the subject in question. Students' ability to understand and to make judgments where values play a role can only be developed when they master critical thinking skills.

Biology classes are formed to disseminate information on the world of living things. Too often students leave biology classes with only a knowledge of facts and some ability to integrate those facts into abstract processes. The teacher who stresses original and critical thinking, however, brings a special gift to his or her students—the gift of analysis that leads to wise decision making.

Process Skills

Absolutely essential to the development of critical thinking skills is the knowledge of scientific processes—those processes scientists use to gain information. It is essential that teachers communicate to students that scientific processes are ways of going about learning things.

The best way to teach process skills is to place students in a laboratory setting, have them undertake a scientific investigation, and point out to them at key points in the experience the process skills they are using. Teaching process skills involves teaching students to look at and analyze their scientific activity.

The process skills discussed in Chapter 2 of this edition of *Modern Biology* are:

1. Observing and collecting data
2. Measuring
3. Organizing data
4. Classifying
5. Hypothesizing
6. Predicting
7. Inferring
8. Modeling
9. Communicating

These skills are expressed as verbs because they involve activity, and as students become aware of their scientific activity they will better understand process. When students use scientific processes, they find that separating the skills from one another and being aware of when they are using a particular skill will help them recognize and avoid scientific errors.

Using Process Skills

In the initial stages of exploring a problem, a scientist gathers data through observing some phenomenon and formulating a question based on those observations. In a rough, general way, this initial recognition of a question is based on the organization and analysis of data collected in observations. Often these beginning stages do not involve careful recorded observations or careful analysis.

Once a question is clearly formulated, further observations are made and analyzed. This may lead to revisions of the question until a hypothesis can be made. For example, the scientist may, on the basis of limited data, make an inference about the relationship between two species. After further observation the scientist may revise the inference and transform it into a hypothesis that can be tested through experimentation.

After formulating a hypothesis, but before performing an experiment, a scientist may make a prediction based on the hypothesis. Once the prediction is made, the scientist has a clear focus around which to design a formal scientific investigation—an experiment.

In most experiments the scientist begins by gathering data which is obtained by making observations and taking measurements. The scientist organizes the data by such methods as drawing graphs, making lists, creating tables, and developing models to more easily analyze the data. The scientist then performs the experiment.

While hypotheses form the basis of experimentation, an experiment cannot prove a hypothesis true beyond doubt. At best it can only support a hypothesis. However, if the results of an experiment discredit a hypothesis, the scientist must formulate a new hypothesis.

Modern Biology teaches process skills by introducing them in Chapter 2 for students to study. Then, the *Biology in Process* features found in many chapters reinforce students' knowledge and their recognition of process skills. In addition, the in-text laboratory exercises are process oriented, as are laboratory exercises in the ancillaries to the student's textbook.

Writing Skills

The development of critical thinking and process skills is closely tied to the ability to comprehend written material. Students need to locate and to organize information, make decisions, form hypotheses, test alternatives, and formulate new ideas based on information they have obtained from the text and from other scientific writings.

The reading skills important in science are best developed through reading materials required in the course. Using the basic text as a vehicle for learning and practicing these skills, as well as for a source of information, is a logical, practical means of accomplishing this goal.

Learning Parts of the Textbook

Very early in the course, it is a good idea to review with students the parts of the text and the function of each part. The easier it is for students to find information, the more they will get out of the text. Point out the special features in the front and back of the text: the introduction to the student, table of contents, appendix, glossary, and index. Show students how to use these aids to locate information and to define terms.

Encourage students to take advantage of such pedagogical aids as the chapter introductions and section objectives. These features help establish a conceptual framework that facilitates comprehension of the material presented in the text. In addition, section and chapter review questions test students' comprehension and encourage students to think critically about what they have read.

Studying Vocabulary

Scientific concepts are conveyed by words. In order to learn about biology, students need to understand the vocabulary used. Thus vocabulary development is an important component of science teaching. In *Modern Biology* new terms are printed in boldface and defined where they are introduced. These vocabulary terms are also listed at the end of each chapter, along with the page number where the word is defined. Review questions at the end of each section and chapter test students' knowledge and help them develop such vocabulary skills as studying etymology and comparing and contrasting meanings of scientific terms. Vocabulary words are also defined in the glossary for quick reference.

Interpreting Illustrations

Frequent in-text references to charts, diagrams, and photographs help to reinforce students' understanding of major concepts. Have students practice interpreting an illustration such as the one on page 6. Advise them to follow these steps:

1. Read the caption that accompanies the diagram.
2. Review the symbols or terms that are used in the diagram.
3. Think about what the diagram illustrates and why a diagram is an effective way to show it.
4. Read the text surrounding the diagram and see how it relates to the illustration.

Making Generalizations

A generalization is a statement or concept that includes the common characteristics of a group of observations or statements. The ability to make generalizations about new data or observations places this new information in context and enhances its meaning. Through analysis of information presented in the text and of their own observations and experiences, students should be able to form generalizations about biological concepts. Many of these generalizations will be related to the biological themes described in Chapter 1.

For example, the relationship between structure and function in living things is emphasized throughout *Modern Biology*. Making students aware of such a relationship will help them understand, for example, the significance of the structure of the cell membrane. It will also encourage students to look at familiar objects in new ways, as when they observe variations in the beaks and in the feet of birds.

Sequencing

Scientific processes, such as mitosis and meiosis or embryo development, occur in an exact, predictable sequence. Similarly, the presentation of facts and ideas in *Modern Biology* follows a logical sequence in which new information builds on concepts that have already been discussed. The ability to follow the sequence of facts, concepts, and experiments is an important aspect of reading comprehension. As students become familiar with the importance of sequencing in scientific study, they will be better able to predict outcomes and apply what they have learned to new situations.

Comparing and Contrasting

One way for students to improve reading comprehension is to compare and contrast the material being read with what they already know. If students know how to look for similarities and differences while reading, it will be easier to remember facts, to relate them to one another, and to apply them to new situations. For example, as students read about plant reproduction, they should look for key similarities and differences in the reproductive cycles of members of the different plant divisions. This process of comparing and contrasting sets of information will enhance students' appreciation of the various adaptations of plants.

Writing a Science Research Report

The Laboratory and Extension activities in *Modern Biology* offer students many opportunities to communicate their research findings in writing. A thorough research project includes clear statements of either the conclusions drawn from the research or the conditions and results of the experiment. The following guidelines will help students organize and document their research.

Writing a Library Research Report

The standard research paper includes the following components:

1. Title page
2. Introduction
3. Outline
4. Body of report

5. Student's conclusions
6. Footnotes
7. Bibliography

Students should clearly state or define their research topic. Sources might include reference texts, newspapers, and popular or scientific magazines. Encourage students to write letters to specific sources for additional information. The library houses many resources that can facilitate research, such as the card catalog, *Reader's Guide to Periodical Literature*, microfilm, data-base indexes, and so on.

Impress upon students the importance of accurately citing sources used in their research. For each source, students should record bibliographic information, which can appear as either an on-page or end-of-paper citation. Any direct quotations should be enclosed in quotation marks.

Once students have compiled their sources, they can construct a working outline of their report. You might want to approve their outline before they proceed with the body of the report. The outline should list the topics to be covered in a logical sequence, following standard outline form. Following approval of the outline, students can begin writing the body of the report. In their report students should not only summarize but offer their own conclusions about their research findings.

Writing a Laboratory Report

The finished laboratory report should include the following components:

1. Title page
2. Introduction
3. Library research
4. Hypothesis
5. Materials and methods
6. Observations and data collected
7. Conclusions
8. Footnotes
9. Bibliography

The report should clearly state a specific question and a hypothesis that the lab experiment will test. Students can also use the guidelines for compiling sources for a library research report to research information regarding the subject of the experiment. After compiling their sources, students should make an outline of the experimental design. Review the outline to be certain that the experiment will accomplish the purpose stated in the introduction, and that it is both practical and safe.

The reliability of a scientific experiment depends on whether other researchers can repeat the procedure and obtain the same results. Consequently, the methods and materials section of the lab report should contain all the information needed to repeat the experiment. Tell students to list all materials used and to explain the procedure in detail. This includes naming the organisms studied, describing the apparatus and the setup and conditions of the experiment, and identifying the control and the variables.

In the next section of the report, students should record all observations made and data obtained. When appropriate, data may be presented in the form of charts, graphs, tables, or diagrams. Finally, students should draw conclusions based on their analysis of the observations and data recorded above. Their conclusions should compare the results of the experiment with the original hypothesis, stating whether or not they agree and giving possible reasons for any discrepancies.

Safety in the Classroom

Safety is the responsibility of all school personnel—administrators, teachers, and students. To ensure students' safety, familiarize yourself with the federal, state, and local school safety regulations as well as the guidelines established by your school district or by your school. At the beginning of the year and periodically thereafter, conduct a thorough check to make sure that your classroom is in compliance with these regulations. Learn your school's procedure for handling accidents and emergencies. Most importantly, inform students of the safety regulations, review the regulations frequently in class, and set a good example by your own compliance and positive attitude about safety.

Classroom Layout and Safety Equipment

Make sure that your classroom is adequately ventilated. Maintain a well-ventilated, lockable storage space for chemicals. Post an inventory of the chemicals in the storage area and clearly label potentially hazardous chemicals. Periodically inspect the storage areas to prevent dangerous situations from developing.

Your classroom should have two fire exits. Review frequently with students instructions for what to do in case of fire. Make sure that a list of these instructions, including a map showing the fire exits, are prominently displayed in your classroom.

Make available and accessible to students the safety equipment listed below. Instruct students on the proper use of each piece of equipment.

- first-aid kit
- safety goggles
- fire blanket
- laboratory gloves
- eyewash facilities
- laboratory aprons

Equip your classroom with a dry-chemical fire extinguisher. Instruct students to immediately leave the scene of a fire and to never try to extinguish a fire without adult supervision.

Safety Symbols and Guidelines

The following safety symbols appear throughout *Modern Biology* in the laboratory activities at the end of each chapter. These safety symbols will help to remind students to follow the specific safety guidelines listed below. For additional information on classroom safety write the National Science Teachers Association, 1742 Connecticut Avenue, N.W., Washington, D.C., 20009.

ES follow safety guidelines when using electrical equipment

P follow safety guidelines with poisonous or volatile substances

CP follow safety guidelines with substances that can stain

EYE follow safety guidelines with substances that can harm eyes

FS follow safety guidelines when working around an open flame

HS follow safety guidelines when handling sharp instruments and glassware and avoid touching hot surfaces

CS follow safety guidelines when handling substances that can harm the skin

HT handle animals and other living organisms gently and with care

General Classroom Safety (EYE, CS, P)

DO enforce the following safety rules:

- wear safety goggles when there is potential danger to the face or eyes
- wash hands thoroughly at the end of each laboratory period
- clean up work areas after each activity or laboratory period

DO NOT allow students to:

- eat, drink, or engage in rough or boisterous play in the classroom
- conduct experiments without adult supervision
- touch face, mouth, or eyes when working with plants, rocks, soil, or chemicals

Fire Safety (FS, P, CP)

DO enforce the following safety rules:

- tie back long hair and confine loose clothing when using fire
- keep unlit matches away from open flames
- dispose of all extinguished matches properly
- extinguish all flames when not in use

DO NOT allow students to:
- use an open flame around volatile liquids
- reach across an open flame

Electrical Safety (ES, HS)

DO enforce the following safety rules:
- keep work areas dry
- make sure that electrical equipment is grounded and that electrical cords are in good condition
- turn off and unplug hot plates when not in use

DO NOT allow students to:
- touch electrical equipment with wet hands
- overload circuits

Chemical Safety (P, CS, CP, HS)

DO enforce the following safety rules:
- handle all chemicals with care
- avoid the buildup of harmful fumes
- label bottles and test tubes accurately
- store chemicals properly
- immediately wash away any chemicals spilled on skin or clothing with large amounts of water
- oversee the proper disposal of all chemicals

DO NOT allow students to:
- taste or inhale chemicals
- transfer chemicals with a mouth pipette
- pour water into an acid

Glassware Safety (HS, EYE)

DO enforce the following safety rules:
- handle all glassware with care
- take precautions with glass tubing especially when plugging it with a cork or when inserting or removing it from a rubber stopper
- provide a whisk broom and dustpan for sweeping up pieces of broken glass

DO NOT allow students to:
- drink from glassware used in the classroom
- touch heated glassware
- pick up broken glassware with bare hands

Equipment Safety (CS)

DO enforce the following safety guidelines:
- handle and store all fragile equipment with care
- if a thermometer breaks, dispose of mercury according to state or federal guidelines

- use sharp dissecting tools with care
- when cutting, always move the hand away from rather than toward the body

Specimen Safety (CS, EYE)

DO enforce the following safety rules:
- rinse preserved specimens before use
- provide adequate ventilation to dispel fumes from preservatives
- use gloves to avoid contact with preservatives
- refrigerate or preserve unpreserved specimens
- keep bacteria and fungi cultures in closed containers such as lidded petri dishes
- after using cultures of bacteria or fungi, kill them with a strong disinfectant

DO NOT allow students to:
- taste any of the materials used in biological investigations, including common food items
- culture any kind of harmful organism
- use unpreserved specimens that have decayed

Animal Safety (HT)

DO enforce the following safety rules:
- obtain specimens from a reputable source, such as a biological supply house, in accordance with state and local laws
- make sure that any mammals brought into the classroom are inoculated for rabies
- arrange for adequate care of all animals

DO NOT allow students to:
- handle animals roughly, tease them, or otherwise treat animals in an inhumane manner
- handle wild animals or material from wild animals such as fur or feathers

Field Trip and Activity Safety (CS, CP)

DO enforce the following safety rules:
- find out if any students have medical reasons to restrict their activities
- obtain consent from students' parents for trip
- visit the site first to note potential dangers
- discuss proper attire and conduct with students and arrange for adequate supervision
- warn students to avoid contact with poisonous plants, such as poison ivy or poison oak
- carry first-aid kit

Materials List

The materials needed for each in-text Laboratory are listed at the beginning of each laboratory exercise. A complete listing of materials is presented here for your convenience. This list is organized by chapter so that you can determine the materials you might have to order in advance. Refer to the Teacher's Guide for hints on preparing specimens and solutions and on ordering supplies. Remember to allow time in your planning to obtain the materials and the live specimens you cannot store ahead.

Quantities of various items are given in parentheses. Quantities are based on the quantity needed for one student or pair of students to do one Laboratory. Where no quanity is given for an item, it means that the quantity needed per student or pair of students is 1. Be sure to take into account the size of your class when assembling materials. No quantities are listed for readily available items, such as water.

See the particular Laboratory for more specific information. The page number where the Laboratory appears in the text is listed after the chapter number for your reference.

Chapter 1 (p. 13)
compound microscope; cotton batting; coverslip; medicine dropper, microscope slide; paper, unlined; pond water

Chapter 2 (p. 29)
paper; watch with second hand

Chapter 3 (p. 43)
ammonia, household; beakers, 30-mL (2); lemon juice; medicine dropper; milk, fresh skim; paper towels; pH paper, narrow-range; pH paper, wide-range; stirring rods (2)

Chapter 4 (p. 55)
beaker, 1-L; Benedict's solution; copper sulfate solution, 0.5%; cornstarch; food substance; glucose; hot plate; iodine-potassium iodide solution; lard; medicine droppers (4); milk, nonfat dry; sodium hydroxide solution, 10% aqueous; stirring rod, glass; Sudan III solution; test tube rack; test tube stoppers (2); test tubes (8); tongs; water, distilled

Chapter 5 (p. 77)
compound microscope; coverslip (2); Elodea plant; human cheek cells, prepared slide; methylene blue; microscope slides (2)

Chapter 6 (p. 89)
clock or watch; coverslip; fork, plastic; medicine droppers (2); microscope; microscope slide; onion, red; paper towel; petri dishes (2); potato, fresh; probe, blunt; scalpel; sodium solution, 10%

Chapter 7 (p. 109)
beakers, 250-mL; bulb, 200-watt; cylinder, graduated 250-mL; Elodea sprigs, 15-cm-long (4); pencil, glass-marking; pH paper; sodium bicarbonate; stirring rod, glass; water, distilled

Chapter 8 (p. 125)
beaker, 100-mL; copper sulfate solution, 0.5%; diphenylamine; egg white, raw digested and raw undigested; hot plate; medicine droppers (7); pencil, glass-marking; sodium hydroxide solution, 10% aqueous; test tube rack; test tube stoppers (2); test tubes (5); tongs; water, distilled

Chapter 9 (p. 139)
Allium root tip, longitudinal section; microscope

Chapter 10 (p. 159)
corn ears, R color alleles 3:1 (1), R and Su alleles 9:3:3:1 (1)

Chapter 11 (p. 177)
paper; pencil; phenylthiocarbamide paper

Chapter 12 (p. 187)
aerator; aquariums, 1-gallon (2); brine shrimp eggs; cardboard; compound microscope; dip net, fine mesh; dissecting microscope; medicine dropper; paper, unlined; pencil, glass-marking petri dish; slide, depression; sodium chloride solution, 1.5%; teaspoon; yeast, dry baker's

Chapter 13 (p. 197)
paper; pencil

Chapter 14 (p. 215)
aspartic acid; balance, metric; beaker, 500-mL; clock or timer; compound microscope; coverslip; graduated cylinder, 50-mL; flask, Erlenmeyer, 125-mL (2); glutamic acid; glycine; hot plate; medicine dropper; microscope slides; ring stand with clamp; sodium chloride solution, 1%; stirring rod, glass; tongs

Chapter 15 (p. 233)
Ascaris, preserved, 20 female and 20 male (40); graph paper; leaves (20); pan or tray; peanuts in the shell (20); ruler, metric

Chapter 16 (p. 251)
beads, 200 white and 200 black (400); containers (3)

Chapter 17 (p. 263)
paper; pencil

Chapter 18 (p. 275)
invertebrate specimens, living or preserved (11)

Chapter 19 (p. 291)
beaker, 100-mL; carborundum powder, 400-grit; cotton swabs dibasic potassium phosphate solution, 10 mL, 0.1 M; mortar and pestle; pencil, glass-marking; tobacco from several brands of cigarettes; tobacco plants (4)

Chapter 20 (p. 305)
bacteria, mixed broth culture of Bacillus subtilis, Micrococcus luteus, and Rhodospirillum rubrum; coverslips (2); inoculating loop or Pasteur pipette; paper towels; petroleum jelly; slides, depression (2); toothpicks (2)

Chapter 21 (p. 317)
compound microscope; coverslips (4); Euglena gracilis, prepared suspension; Euplotes culture; microscope slides (4); paper towel; Paramecium culture; pencil, glass-marking; pipettes (7); Stentor culture; vinegar, white; Vorticella culture; water, distilled

Chapter 22 (p. 331)
algae, mixed culture of Chlamydomonas, Eudorina, Gonium, Hydrodictyon, Pandorina, and Volvox; coverslips; medicine dropper; microscope; microscope slides; petroleum jelly; toothpicks

Chapter 23 (p. 345)
dissecting microscope; paper, unlined; petri dishes (2); probe; Rhizopus stolonifer culture, plus and minus strains; Rhizopus stolonifer culture, single-strain

Suppliers

A-V and Software

The codes for the audiovisual and software suppliers' names used in the Resources section of the Teacher's Guide *appear below in boldface type. The full name and mailing address of each company or distributor follow.*

AIMS
AIMS Media
6901 Woodley Ave.
Van Nuys, CA 91406-4878

Arthur Mokin
Arthur Mokin Productions, Inc.
2900 McBride La.
Santa Rose, CA 95401

Barr
Barr Films
3490 E. Foothill Blvd.
Pasadena, CA 91107

Benchmark
Benchmark Films, Inc.
145 Scarborough Rd.
Briarcliff Manor, NY 10510

Bio. Learning
Biological Learning
Distributed by Carolina Biological

Biology Media
Distributed by Carolina Biological

Britannica
Encyclopedia Britannica
425 N. Michigan Ave.
Chicago, IL 60611

Bullfrog
Bullfrog Films, Inc.
P.O. Box 149
Oley, PA 19547

Carolina Biological
Carolina Biological Supply Co.
2700 York Rd.
Burlington, NC 27215

Centre
Centre Productions, Inc.
1800 30th St., Suite 207
Boulder, CO 80301

Churchill
Churchill Films
662 N. Robertson Blvd.
Los Angeles, CA 90069

COMPress
Distributed by Carolina Biological

Conduit
Distributed by Carolina Biological

Coronet
Coronet Films and Video
108 Wilmot Rd.
Deerfield, IL 60015

Cross
Distributed by Carolina Biological

Educational Images
Educational Images, Ltd.
P.O. Box 3456
West Side Station
Elmira, NY 14905

Educational Media
Educational Media International
Distributed by Carolina Biological

Entomological Society
Entomological Society of America
Distributed by Carolina Biological

Fanlight
Fanlight Productions
47 Halifax St.
Jamaica Plain, MA 02130

Filmakers
Filmakers Library, Inc.
133 E. 58th St.
New York, NY 10022

Films
Films, Inc.
1213 Wilmette Ave.
Wilmette, IL 60091

Focal Point
Focal Point
Box 207
Pomfret, CT 06258

Focus
Distributed by Carolina Biological

Hagemann
Distributed by Carolina Biological

Hawkhill
Hawkhill Associates, Inc.
125 E. Gilman St.
Madison, WI 53703

Helix
Distributed by Carolina Biological

Humanities and Sciences
Films for Humanities and Sciences
Box 2053
Princeton, NJ 08540

Human Relations
Human Relations Media
175 Thompkins Ave.
Pleasantville, NY 10570

I.F.B.
International Film Bureau, Inc.
332 S. Michigan Ave.
Chicago, IL 60604

J & S Software
J & S Software, Inc.
135 Haven
Port Washington, NY 11050

Knowledge
Knowledge, Unlimited
Box 52
Madison, WI 53701

Lucerne
Lucerne Films, Inc.
37 Ground Pine Rd.
Morris Plains, NJ 07950

Marty Stouffer
Marty Stouffer Productions, Ltd.
300 S. Spring St.
Aspen, CO 81611

Media Guild
The Media Guild
11722 Sorrento Valley Rd.
San Diego, CA 92121

Nasco
Nasco
901 Janesville Ave.
Fort Atkinson, WI 53538

National Geographic
National Geographic Society
Educational Services
Seventeenth and M St., NW
Washington, DC 20036

New Dimension
New Dimension
85895 Lorane Hwy.
Eugene, OR 97405

Oxford
Oxford Scientific Films, Ltd.
Distributed by Carolina Biological

Pleasantville
Pleasantville Media
P.O. Box 415
Pleasantville, NY 10570

Pyramid
Pyramid Film and Video
Box 1048
Santa Monica, CA 90406-1048

Scholastic
Scholastic Software
2931 East McCarty St.
P.O. Box 7502
Jefferson City, MO 65102

Science & Mankind
Science & Mankind, Inc.
Communications Park
Box 2000
Mt. Kisco, NY 10549

Sunburst
Sunburst Communications, Inc.
P.O. Box 40
39 Washington Ave.
Pleasantville, NY 10570

Time-Life
Time-Life Films
1271 Avenue of the Americas
New York, NY 10020

U. of I.
University of Illinois
Film Center
1325 S. Oak
Champaign, IL 61820

Equipment and Materials

Biological equipment and materials can be obtained from a variety of sources. In addition to the suppliers listed below, many of the materials may also be purchased from other scientific supply houses.

AO Reichert, Scientific Instruments
P.O. Box 123
Buffalo, NY 14215

Bausch & Lomb
Scientific Optical Products
Division
1400 North Goodman St.
Rochester, NY 14602

Carolina Biological Supply Co.
2700 York Rd.
Burlington, NC 27215

Central Scientific Company
11222 Melrose Ave.
Franklin Park, IL 60131

Connecticut Valley Biological Supply
82 Valley Rd.
Southhampton, MA 01073

Delta Biologicals
P.O. Box 26666
Tucson, AZ 85726

Denoyer-Geppert Science Co.
5215 N. Ravenswood Ave.
Chicago, IL 60640

Difco Laboratories
920 Henry St.
Detroit, MI 48232

Edmund Scientific Co.
1010 E. Glouster Pike
Barrington, NJ 08007

Fisher Scientific Co.
Education Materials Division
4901 W. Lemoyne Ave.
Chicago, IL 60651

Frey Scientific Company
905 Hickory Lane
Mansfield, OH 44905

Hubbard Scientific Company
1946 Raymond
Northbrook, IL 60062

LaPine Scientific Co.
6001 S. Knox Ave.
Chicago, IL 60629

Nasco
901 Janesville Ave.
Fort Atkinson, WI 55538

Sargent-Welch Scientific Co.
7300 N. Linder Ave.
Skokie, IL 60076

Science Kit and Boreal Laboratories, Inc.
777 East Park Dr.
Tonawanda, NY 14150

VWR Scientific
P.O. Box 7900
San Fransisco, CA 94120
(or)
800 E. Fabyan Pkwy.
Batavia, IL 60510

Ward's Natural Science Establishment
P.O. Box 1712
Rochester, NY 14603
(or)
P.O. Box 1749
Monterey, CA 93940

Teaching the Challenged Student

Challenged children include those who have some physical, mental, or emotional disorder that impedes their functioning in the classroom. Federal law PL 94-142, enacted in 1975, requires all students with handicaps to be identified, tested, and evaluated by the appropriate agencies or school services. PL 94-142 also directs that challenged children be educated in the "least restrictive environment." In other words, students should be educated with nonhandicapped peers to the greatest extent possible.

Since the passage of PL 94-142, educators have worked to achieve greater integration of challenged children into regular school programs. As a result, at least two-thirds of exceptional children have been mainstreamed into regular classrooms.

A handicapped child may require special consideration in such areas as seating, communicating information, and achievement monitoring. At the same time, however, it is a good idea to accept the disability and to emphasize the similarities between the handicapped and nonhandicapped classmates as much as possible. Encourage other students in the class to do the same.

Prior to the beginning of the school year, learn all that you can about the handicapped student and the nature of the disability to determine what adjustments will be necessary. School psychologists or counselors, former teachers, and the student's parents can all provide valuable information. Although each student has individual needs that must be taken into account, the following guidelines should prove helpful.

Physical Impairments

Professionals classify physical impairments into two broad categories. Orthopedic impairments, caused by disorders such as cerebral palsy and muscular dystrophy, primarily affect the skeletal and muscular systems. Health impairments refer to heart conditions, diabetes, and other acute or chronic disorders that limit the student's strength and alertness.

If the student is confined to a wheelchair, check to make sure the classroom is as accessible as possible—that aisles are wide enough to accommodate the chair and that equipment is within reach. You may wish to enlist the aid of other students in the class to physically assist the disabled child as necessary. For students who cannot write for long periods of time, taped lectures, hand-held computers to facilitate note-taking, and oral exams may help.

Visual Impairments

Visually impaired students include both students with limited vision and the totally blind. For those students who have some vision, large-print editions of the text and the use of magnifiers or closed-circuit television make text materials and exams easily accessible. In the classroom, seat the child near the front of the room, away from glare if possible to maximize the view of both you and the chalkboard.

Students who are completely blind should be allowed to become familiar with the classroom layout before the first class session begins. Blind students can gain access to printed materials via Braille or taped versions. If available, blind students may also use computers equipped with voice synthesizers, or an Optacon, an electronic device that converts printed material into raised letters. In addition, science measurement kits with Braille markings are available.

Hearing Impairments

Hearing-impaired students may have either partial or total hearing loss. If you have a partially hearing-impaired student in your class, remember to always face the class while speaking. Try to speak in simple, direct language and to avoid digressions or sudden changes in topic. During class discussions, periodically summarize what students are saying and repeat students' questions before answering them, for the benefit of the hearing-impaired. Use visual media such as filmstrips, overhead projectors, and closed-captioned films when appropriate. You might arrange a buddy system in which another student provides copies of notes and assignments.

Students who are completely deaf may require a sign language interpreter. If so, let the student and the interpreter determine the most convenient seating arrangement. When asking the student a question, be sure to look at the student, not at the interpreter. If the student also has a speech impairment, group assignments for oral reports may be advisable.

Speech Impairments

Speech impairments include stuttering, voice disorders, and other disorders that interfere with a person's ability to communicate. Educators have had great success in mainstreaming speech-impaired children into regular classrooms.

Perhaps the most important quality to exhibit when teaching speech-impaired students is patience. For example, resist the temptation to finish sentences for a student who stutters. Also attend to such nonverbal cues as facial expression and body language. Making committee assignments for oral reports and using multisensory materials when possible will create a more comfortable learning environment for the speech-impaired student.

Learning Disabilities

Learning disabilities are believed to be caused by disorders of the central nervous system that obstruct a person's listening, reasoning, communication, or mathematical abilities. An estimated 2 percent of adolescents have some type of learning disability, making this the most common type of handicap. In teaching LD students, provide a supportive and structured environment in which rules and assignments are clearly stated. Use familiar words and short, simple sentences, repeating or rephrasing when necessary.

Depending on the severity of the disability, students may require extra time to complete exams or may need to tape-record lectures and answers to exam questions. For those who have difficulty organizing materials, you might provide chapter or lecture outlines for the student to fill in. Peer tutors who work with LD students on specific assignments and review materials have often proven effective.

If computers are available, they can be an extremely useful tool for reviewing material and exercising learning skills. With computer-assisted instruction, students receive simplified directions, can proceed in small, manageable steps, and receive immediate reinforcement and feedback.

Mild Mental Retardation

Mentally retarded individuals exhibit restrictions in such intellectual abilities as accumulation of knowledge, problem-solving, and abstract thinking. Students whose IQ measures between 50 and 75 are classified as mildly retarded. In many cases, mildly retarded students may be at least partially mainstreamed into regular classes.

Although the causes of learning disabilities and mental retardation differ, students with the two types of disability often exhibit similar difficulties in learning. Consequently the guidelines for teaching LD students also apply to mildly retarded students.

Behavior Disorders

Behavior disorders refer to emotional disturbances or behaviors that hinder a student's overall functioning. BD students may exhibit any of a variety of behaviors, ranging from extreme aggression to complete lack of interaction with others.

Behavioral psychologists disagree on the best way to deal with BD students. In general, however, a fair, consistent, yet flexible approach is recommended. Make sure to clearly state rules and expectations. Reinforce desirable behavior or even approximations of such behavior, and ignore or mildly admonish undesirable behavior. Because learning disabilities often accompany behavior disorders, you might also wish to refer to the guidelines for learning disabilities.

Behind the Scenes

The production of this edition of *Modern Biology* involved a total rewriting and redesigning of a textbook. The production of this text was undertaken not only by the author but by a series of highly trained editors, artists, designers, production managers, and other bookmaking professionals.

In any project of this magnitude, bookmakers must set guidelines for the style of the text. The scope and sequence of the text must conform to certain accepted grammatical and stylistic principles. Too often publishers take for granted that the users of their books will understand the decisions that went into making the text. By knowing certain "behind the scenes" decisions, teachers will be better informed about the special qualities of the textbook they have purchased.

Editors of this textbook generally used *Webster's Ninth New Collegiate Dictionary* as the reference source for words used in the text. For example, although a trend exists to use the word *genuses* in place of the more familiar *genera*, the editors of this text agreed with *Webster's* that the use of *genera* is still so widely accepted as to retain its usage. In other cases, however, contrary to *Webster's*, the editors chose what appears to be the modern consensus. Thus, in the case of *ameba*, which replaces the classic *amoeba*, the editors chose the more modern usage. Though not all teachers will agree with such decisions, editors know that such decisions must reflect some kind of consensus.

The decision to use such contractions as *can't* reflects an attempt to relax the writing of the book to make it more approachable for today's student. However, the editors decided to avoid colloquial phrasing whenever possible. In addition, much of the writing is reflective of modern scientific jargon. The trend toward a more scientific English is evident throughout this text. Students learn partly through style, and the style of *Modern Biology* is scientific.

Pronunciations in this textbook are taken directly from *Webster's Ninth New Collegiate Dictionary* and *Webster's Third New International Dictionary*. An example is the pronunciation of the word *cytoplasm*. Many people are more comfortable with the pronunciation *SITE-oe-PLAZZ-um*. However, the more relaxed *SITE-uh-PLAZZ-um* is used in the text to conform to dictionary usage. Most pronunciations in the text are taken directly from the dictionary.

Pronunciations are spelled phonetically according to standard phonetic usage. The decision to eschew nonphonetic spelling was based on teachers' wishes. Teachers often complained that students were unable to read the nonphonetic spellings.

Elements of style were based on *The Chicago Manual of Style* as the standard reference text. Decisions were made, however, specifically for *Modern Biology* users. In general introductory phrases are not set off by commas. Another general guideline was to write in a way that people speak, and not to be bound to grammatical rules that confuse the writing for the modern student. As such *Modern Biology* is written in a modern style, a scientific style best suited to today's high school student.

Holt, Rinehart and Winston is proud of this new edition of *Modern Biology* and wishes to keep the long tradition of this textbook alive. A textbook, however, is more than just an offering from a company. It is the result of a symbiotic relationship between educators, authors, and bookmaking professionals. We are eager to hear your comments, concerns, and criticisms. Please write to us concerning your feelings about the text and you will receive personal attention from our staff of editors.

Modern Biology

Albert Towle

Modern Biology

HOLT, RINEHART AND WINSTON
AUSTIN NEW YORK SAN DIEGO CHICAGO TORONTO MONTREAL

Acknowledgments

Reviewers and Consultants

Jocelyn Baker
Science Teacher
Windsor Forest High School
Savannah, GA

Wayne Becker, Ph.D.
Department of Botany
University of Wisconsin
Madison, WI

Gary Breckon, Ph.D.
Biology Department
University of Puerto Rico
Mayaguez, PR

Jeff Doering, Ph.D.
Associate Professor of Biology
Loyola University
Chicago, IL

Jody Lee Duek
Granada Hills High School
Granada Hills, CA

Susan Eichhorn
Department of Botany
University of Wisconsin
Madison, WI

John Einset, Ph.D.
EniChem Americas
Monmouth Junction, NJ

Barbara Foots
Science Director
Houston Independent School
District
Houston, TX

Jerry Franklin, Ph.D.
Bloedel Professor of Forest
Ecosystems Analysis
University of Washington
Seattle, WA

Joyce Gellhorn, Ph.D.
Boulder High School
Boulder, CO

Bob Gomprecht
Chairperson of Science Dept.
Fordham Prep School
Bronx, NY

Jerry Halpern
Columbia Grammar and
Preparatory School
New York, NY

Carl Helms, Ph.D.
Department of Biological Sciences
Clemson University
Clemson, SC

Garland E. Johnson
Science Coordinator
Fresno Unified School District
Fresno, CA

Jack Kelly
El Camino High School
Sacramento, CA

Ted King
Biology Teacher
R. Nelson Snider High School
Ft. Wayne, IN

David Largent, Ph.D.
Department of Biology
Humboldt State University
Arcata, CA

Ruben A. Leal
Biology and Physiology Teacher
J. W. Nixon High School
Laredo, TX

Welton L. Lee, Ph.D.
1235 44th Avenue
San Francisco, CA

Linda Barrett Phelps
Chemistry and Biology Teacher
Camden Central High School
Camden, TN

John Roller
Science Math Coordinator
Tulsa Public Schools
Tulsa, OK

John Smarelli, Jr., Ph.D.
Assistant Professor of Biology
Loyola University
Chicago, IL

Annice Steadman
Chairperson of Science Dept.
Little Rock Central High School
Little Rock, AK

Front cover: Great horned owl, *Bubo virginianus*.

Back cover: Otter Point, Acadia National Park, Maine.

Developmental assistance by Ligature, Inc.

Copyright © 1989, 1985, 1981, 1977, 1973, 1969 by Holt, Rinehart and Winston, Inc.

All rights reserved. No part of this publication may be reproduced or transmitted in any form or by any means, electronic or mechanical, including photocopy, recording, or any information storage and retrieval system, without permission in writing from the publisher.

Requests for permission to make copies of any part of the work should be mailed to: Permissions, Holt, Rinehart and Winston, Inc., 1627 Woodland Avenue, Austin, Texas 78741.

Printed in the United States of America

ISBN 0-03-013919-8

89012 039 987654321

James K. Stringfield, Ph.D.
Science Teacher
Harnett Central High School
Angier, NC

Aleta Sullivan
Science Educator
Hattiesburg High School
Hattiesburg, MS

Nina Surawicz, Ph.D.
Arizona State University
Tempe, AZ

David Sutton
Chairperson of Science Dept.
Wauwatosa West High School
Wauwatosa, WI

Lyn Swan
John Marshall High School
Los Angeles, CA

Bradley Switzer
Tates Creek Senior High School
Lexington, KY

Linda Thayer
Biology Teacher
Jeremiah Burke High School
Dorchester, MA

Salvatore Tocci
Chairperson of Science Dept.
East Hampton High School
East Hampton, NY

Betty Tumminello
Chairperson of Science Dept.
Pineville High School
Pineville, LA

Marge Tunnell
Biology Teacher
Irmo High School
Columbia, SC

Robert Vasconcellos
Science Instructor
Gunnarson High School
San Jose, CA

Contributors

Beth M. Atkin, Ph.D.
Research Associate
Department of Medicine
University of Chicago
Chicago, IL

Judy Berlfein
Del Mar, CA

Ronnee E. Bernberg
Biology Teacher
Lakeview High School
Chicago, IL

Suzanne Hagan
St. Louis, MO

Donald Igelsrud
LIFE Consultants
Calgary, Alberta, Canada

Glenn K. Leto
Biology Teacher
Barrington High School
Barrington, IL

Jon R. Luoma
Minneapolis, MN

Jennifer Matos, Ph.D.
Professor of Biology
Washington University
St. Louis, MO

Janice Moore
Assistant Professor of Biology
Colorado State University
Ft. Collins, CO

Randy Moore, Ph.D.
Professor of Biology
Baylor University
Waco, TX

Carl Reed
Biology Teacher
Barrington High School
Barrington, IL

Carol Savonen
Biologist, Science Writer
Leaburg, OR

Richard Strickland
Washington Sea Grant Program
University of Washington
Seattle, WA

Suzanne Winckler
Chatfield, MN

National Advisory Panel

Vernon Ahmadjian, Ph.D.
Professor of Botany
Clark University
Worcester, MA

Karen Arms, Ph.D.
Savannah, GA

Mildred Berry, Ph.D.
Science Supervisor
Dade County Public Schools
Miami, FL

Ray Evert, Ph.D.
Professor of Botany
University of Wisconsin
Madison, WI

Maura C. Flannery, Ph.D.
Associate Professor of Biology
St. Johns University
Jamaica, NY

Gerald Garner
Secondary Science Specialist
Los Angeles Unified School
   District
Van Nuys, CA

Mary Alice Garza-Gonzales
Biology Teacher
Jefferson High School
San Antonio, TX

Jerry Ivins, Ph.D.
Science Coordinator
Northwest Local School District
Cincinnati, OH

Donald Johnson
Science Department Chairperson
Mumford High School
Detroit, MI

Jane Matthews
Biology Teacher
B. C. Rain High School
Mobile, AL

Harold Slavkin, Ph.D.
Professor of Biochemistry
Laboratory for Dev. Biology
University of Southern California
Los Angeles, CA

Paul Sugg
Instr. Spec., Secondary Science
5000 Oakland Avenue
Dallas, TX

Herbert Vatali
Secondary Science Supervisor
Lynchburg City Schools
Lynchburg, VA

Dan Walker, Ph.D.
Professor of Science Education
Department of Biological Sciences
San Jose State University
San Jose, CA

Contents

Biology in Process

Each "Biology in Process" feature focuses on how one scientist or a team of scientists unraveled a complex scientific puzzle. By studying the processes these scientists used, you will become familiar with the way in which scientific inquiry is carried out.

Biotechnology

The "Biotechnology" feature focuses on new and innovative tools and techniques that scientists have developed. These new tools have enabled scientists to tackle problems that once were beyond the reach of existing technology and to explore new worlds that previously were unimagined.

Laboratory Investigations

The "Laboratory" in each chapter is designed to draw you into the process of science. Each lab will give you the opportunity to further explore some biological question using the same scientific processes that a biologist might employ.

Writing About Biology

The excerpts presented in "Writing About Biology" enable scientists to speak directly to the reader, not only about the scientists' ideas and methods but also about their feelings toward their work: their enthusiasm, their disappointments, their beliefs, and, sometimes, their triumphs.

Intra-Science

The "Intra-Science" feature focuses on issues of current interest in our world. Each two-page section considers one scientific question or process first from a biological perspective and then from the perspective of various other sciences. The interconnectedness of the sciences is reinforced by a concluding discussion of possible careers related to the issue at hand.

Discovering Biology

It is unlikely that you have ever heard the almost silent flutter of wings and the short, sharp squeak of a mouse in the predawn hours as a great horned owl found its prey. However, most of you have seen the trees turn to brilliant colors and lose their leaves in the autumn, only to have new leaves sprout and grow in the spring. You may have seen a flock of migrating waterfowl such as ducks or geese. Perhaps you live in a desert area where you have become familiar with Gila monsters, sidewinders, and the giant saguaro cactus. Many of you live in large cities where other people, pigeons, and street trees are the most obvious living things. Some of you live in forested areas amid beautiful evergreens growing on mountainous slopes, and still others of you inhabit the Great Plains, where you are familiar with the cycle of growing wheat. Some of you live near the ocean's shores and know about tides and marine life.

Wherever you live, you have been observing living things since the miracle of your birth. Each day, as you grew, so did your observations. By now you are doubtless aware of the great variety in the living things that surround you.

When you see an unfamiliar animal in an aquarium or a zoo, your first questions will be: "What is it?" "Where does it live?" "What does it eat?" "Is it poisonous?" These are all good questions. They show that you are thinking about biology. Your questions form the basis of the scientific process.

During this course you will learn many skills used by scientists. You will learn how to use biological tools such as the microscope and the computer to aid in observation and to help collect and store information.

Studying scientific process will reveal how data are used to form conclusions, make inferences, and project theories. As you continue your biological studies and begin to examine the order into which biologists group living things, keep asking yourself, "Why?" When you look at the specialized structures of each living thing, ask another question: "Why does this structure look the way it does, and how does it work to help this animal (or plant) live where it does?"

Enjoy your day-to-day learning, and enjoy using specific terms to express biological thought. Maintain your enthusiasm and continue to wonder about living things and to ask questions. I sincerely hope that this course will serve as a springboard for your continued observations of other lives that surround you.

Albert Towle

Exploring with Modern Biology

The first two pages of each unit are a doorway through which you can enter the study of a particular aspect of biology. The images and information on these pages provide a thought-provoking introduction to the topic of the unit.

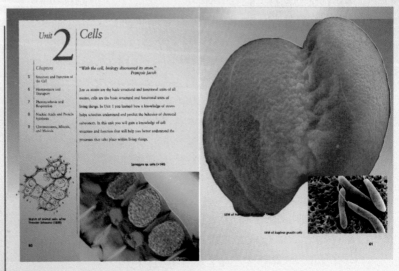

Look at the back of your hand. It's so familiar, you've probably never given it much thought. In reality, though, it's home to an array of unseen, unfelt life-forms—namely thousands of viruses, bacteria, and fungi, each one smaller than any skin cell. If all these expressions of life can occur on just your hand, imagine the forms of life in a square mile of rain forest, in the expanse of an entire ocean.

It is this amazing world of life that you are about to enter with Modern Biology as your guide. On these pages you'll see how the book has been designed to help you in your explorations.

The photos and text on the first page of each chapter help you identify the major themes to look for as you read.

A variety of images form an integral part of each chapter. Colorful photos present unfamiliar and hard-to-see aspects of biology, and scanning electron micrographs (SEMs) provide stunning three-dimensional views of the microscopic world. Dynamic diagrams help you visualize key processes and details.

As you make your way through a chapter, you will find certain mileposts to help guide you. Each major section is numbered and begins with a list of Section Objectives that help you anticipate its major concepts. The logically organized text has easy-to-find headings and subheadings and ends with Section Review questions that help you check your understanding.

5.1 Introduction to the Cell

A cell is the smallest unit that can carry on all the processes of life. Every living thing—from the tiniest bacterium floating in a drop of water to the largest elephant—is made up of one or more cells. A complete living thing that consists of only one cell is called a **unicellular organism**. Though small in size, unicellular organisms, such as the paramecium shown in Figure 5-1, are the most numerous of all life-forms. A living thing consisting of more than one cell is called a **multicellular organism**. Figure 5-2 shows that the buttercup, a multicellular organism, is made up of individual cells.

Discovery of the Cell

Only a few cells are large enough to be seen by the unaided eye. How, then, did scientists come to recognize that all forms of life are made up of cells? Our knowledge of cells began with the invention of the lens in the 1600s and grew along with advances in microscope technology. The light microscope allows us to view specimens magnified only up to about 2,000 times. At such magnifications plant and animal cells are visible, and many cell structures can be seen. The transmission electron microscope (TEM) is 1,000 times stronger than the light microscope. The scanning electron microscope (SEM) produces three-dimensional images. With these advances the smallest cells can be seen in clear detail.

Hooke and van Leeuwenhoek

Imagine the puzzlement of the English scientist Robert Hooke (1635–1703) when in 1665 he cut a thin slice of cork and observed it with a microscope. "I could exceedingly plainly perceive it to be all perforated and porous," he wrote, and further described it as consisting of "a great many little Boxes." Hooke used a microscope to observe the stems of elder trees, carrots, and ferns. He observed that each showed a similar formation. These "little Boxes" reminded him of the small rooms in which monks lived, so he called them cells.

What Hooke had observed were actually dead plant cells. The first person to observe living cells was a Dutch microscope maker, Anton van Leeuwenhoek (1632–1723), in 1675.

Section Objectives

- Outline the discoveries that led to the development of the cell theory.
- State the cell theory.
- Identify a limitation in the size of cells.
- Describe the relationship between cell shape and cell function.
- Distinguish between prokaryotes and eukaryotes.

shows how growth affects the ratio between surface area and volume. Volume increases by the cube, while surface area increases by the square. This suggests that as a cell grows large, its surface area becomes too small to maintain life functions.

Shape

We have said that most cells are roughly cuboidal or spherical. Many plant cells, for example, are shaped like triskaidekahedrons; that is, they are roughly round but have 13 flat sides. However, cells differ widely in shape. This diversity of form reflects a diversity of function. Look at the three human cells in Figure 5-4. The long extensions that reach out in various directions from the nerve cell body allow the cell to receive and transmit nerve impulses. In contrast, the flat shape of dead skin cells is well suited to their function of covering the body surface. White blood cells have the ability to change shape. This allows them to move through narrow openings and to surround and isolate bacteria that invade the body.

Internal Organization

Cells also differ in their internal organization. Look at the photomicrograph of an algal cell in Figure 5-5 (top). Note the large structure near the center of the cell. This structure is the **nucleus**, which directs the activities of the cell. It contains DNA, the hereditary information that passes from one cell to another during cell division. DNA directs the formation of the proteins that carry out cell activities. Look again at Figure 5-5 (top), and notice the other smaller structures in addition to the nucleus. These are the cell organelles, many of them enclosed by membranes. An **organelle** is a cell component that performs specific functions in the cell. Cells that contain a nucleus and membrane-bound organelles are called **eukaryotes** (yoo-KAR-ee-OTES). The cells of plants, animals, fungi, protozoa, and algae are all eukaryotes.

Note the bacterium in Figure 5-5 (bottom). Do you observe a nucleus or organelles? Cells that lack nuclei and membrane-bound organelles are called **prokaryotes** (proe-KAR-ee-OTES). Bacteria and related microorganisms are prokaryotes.

Figure 5-5   Two TEMs contrast the unicellular alga (top), a eukaryote, with the bacterium Bacillus licheniformis, a prokaryote (bottom, ×136,000). Note the nucleus (the area within the dark circle) in the eukaryote.

Section Review

1. Define *nucleus*, and describe the structure and function of the cell nucleus.
2. What is the cell theory?
3. What single factor limits the size that most cells are able to attain?
4. How would you determine whether a unicellular organism was a prokaryote or a eukaryote?
5. How do you account for the fact that three different scientists made major contributions to the cell theory within a 17-year period?

Structure and Function of the Cell   **65**

Adult male

Sperm

Adult medusa

Young medusa

Adult female

Egg

Blastula

Larva

Planula

Polyp

Biology in Process

Developing a Model of the Atom

By the late 1800s the electron had been identified as a subatomic particle. However, at the time scientists had little idea of where electrons were located in the atom or even of what the structure of the atom was.

The British scientist J. Thomson proposed a model of the atom in which electrons were embedded in the positive matter of the atom like "raisins in a plum pudding." This model was accepted until scientist Ernest Rutherford produced data that did not fit the "plum pudding" model.

In the early 1900s Rutherford, who was born in New Zealand, conducted several experiments in which he directed a beam of fast-moving positively charged particles called

34   Chapter 3

Biotechnology

Movement Along Microtubules

As early as 1879, scientists who observed the unicellular organism *Gromia* microscopically described an astonishing movement. Particles appeared to move along extensions from the organism at constantly varying speeds and in different directions. This behavior seemed to be at odds with the usual ways that particles move in organisms. How did these particles move? What was supplying the energy?

Robert Day Allen, a professor of biology at Dartmouth College, was one of the many

scientists who explored these questions. Allen pursued this mystery for 20 years until his death in 1986. He interpreted the old data and performed new experiments. He also ...

70   Chapter 5

Laboratory

Comparing Plant and Animal Cells

Objective
To examine similarities and differences in the morphology of plant and animal cells

Process Skills
Hypothesizing, classifying, observing

Materials
1 microscope slide, 1 coverslip, *Elodea* plant, compound microscope, prepared slide of human cheek cell

Method
1 Make a wet mount slide of a young *Elodea* leaf taken from near the growing tip of the plant.
2 Set the microscope on low power. Place the slide on the microscope stage, and focus on the top layer of cells. Make a drawing showing the shape of the *Elodea* cells. How would you describe the shape? Locate the cell wall and label it on your drawing.
3 Focus the microscope slightly above the cells. Then slowly focus downward through the top cell layer. This should make it possible to observe several layers of cells. What is the advantage of being able to see several cell layers at once?
4 Focus on the upper or lower edge of a cell. Note the green chloroplasts scattered across the surface. The chloroplasts may be moving in one direction along the cell membrane. This phenomenon is called cytoplasmic streaming. Describe it in words.
5 The central vacuole is like a water-filled balloon inside a box. You cannot see the vacuole with the microscope, but you can see evidence of its presence. Focus on either side of the cell. What evidence do you see of the presence of the central vacuole?
6 Note the cell nucleus. It looks like a small,

clear ball. Examine the nucleus under high power. Draw and label it.
7 Place a prepared cheek cell slide under the microscope. Locate the cells under low power and make a drawing showing their shape. Note the cell nucleus. Examine the nucleus under high power. Add the nucleus to your drawing and label it.

Conclusions
1 Compare the morphology of *Elodea* and cheek cells. Explain the reason for the differences between the two.
2 Compare the location of the nucleus in plant and animal cells. Account for the difference.

Inquiry
1 The cheek cell slides were stained so that they could be seen more clearly. Do you think that adding stain to a slide changes the cells in any way in addition to staining them? How could you test your hypothesis?
2 Methylene blue is often used to stain slides. What do you think would happen if you stained the *Elodea* leaf with methylene blue?

Structure and Function of the Cell   77

Biology is a body of knowledge, but also a way of knowing. Therefore, each chapter contains a special feature devoted to how scientists work. The feature takes one of two forms: "Biology in Process" reveals the mental detective work behind our biological knowledge; "Biotechnology" looks at the tools and techniques scientists have used to probe the unseen world.

The "Laboratory" at the end of each chapter gives you the chance to apply your knowledge to a scientific puzzle—first-hand experience with scientific processes.

Biology is a broad field, offering hundreds of opportunities for rewarding work. In each unit you'll find two windows on the workaday world of biology. One is called "Intra-Science." It poses a biological question and examines how biologists and also chemists, physicists, geologists, and other scientists must work together to find solutions. The second feature, called "Writing About Biology," showcases fine science writing that conveys information and often the powerful feelings scientists experience in their work.

Think

Pictured above is an airborne allergen called a pollen grain. How might such an allergen enter the body? How could the body rid itself of this allergen?

Why are some white cats deaf? Why would a plant smell like rotting meat? In each chapter you'll find a feature called "Think" that poses such questions. You'll be able to answer them using the knowledge you've gained from the chapter.

Intra-Science: *How the Sciences Work Together*

The Question in Biology: Can We Control Insects Safely?

Which would you rather eat—an apple riddled with worms, or an apple sprayed with poison? No doubt you would prefer an apple that had neither worms nor poison. But do you have a choice?

Of the 800,000 species of insects in the world, only about one percent are harmful pests. Left unchecked, however, these insects can cause famine by destroying food crops. This possibility becomes all too clear when you realize that pests ruin 500 million tons of grain and 130 million tons of potatoes every year in spite of pesticides and insecticides. In addition to economic damage caused by pests such as the cotton-attacking boll weevil, some insects are also carriers of disease. Mosquitoes can transmit organisms that cause malaria and yellow fever; the tsetse fly can carry protozoa that cause sleeping sickness; and the common household fly can carry organisms that cause dysentery and typhoid fever.

Obviously, insect control is necessary. A controversy exists, however, over the types of methods used. As in the example of the apple described above, it's clear that the so-called cure must not be worse than the insect problem it seeks to cure.

The war on harmful insects is fought on many fronts, among them, physical, chemical, and biological. The most successful strategy developed thus far is an integrated approach using elements of all three.

The Connection to Physical S...

Controlling the environment is one method of restraining insects. Draining swamps where mosquitoes breed and cleaning up rubbish and garbage heaps are effective means of combating mosquitoes and house flies. Controlling the temperature of food storage units discourages insects that feed on harvested crops. Above 65°C most insects cannot survive; below 5°C some live, but do not feed or reproduce.

Rotating crops reduces the numbers of insects that feed on only one crop. Agronomists also advise scheduling

532

Writing About Biology

The Tucson Zoo

This excerpt is from an essay in Lewis Thomas' book, The Medusa and the Snail.

Much of today's public anxiety about science is the apprehension that we may forever be overlooking the whole by an endless, obsessive preoccupation with the parts. I had a brief, personal experience of this misgiving one afternoon in Tucson, where I had time on my hands and visited the zoo, just outside the city. The designers there have cut a deep pathway between two small artificial ponds, walled by clear glass, so when you stand in the center of the path you can look into the depths of each pool, and at the same time you can regard the surface. In one pool, on the right side of the path, is a family of otters; on the other side, a family of beavers. Within just a few feet from your face, on either side, beavers and otters are at play, underwater and on the surface, swimming toward your face and then away, more filled with life than any creatures I have ever seen before. . . .

I was transfixed. As I now recall it, there was only one sensation in my head: pure elation mixed with amazement at such perfection. Swept off my feet, I floated from one side to the other, swiveling my brain, staring astounded at the beavers, then at the otters. I remember thinking, with what was left in charge of my consciousness, that I wanted never to know how they performed their mar-

vels; I wished for no news about the physiology of their breathing, the coordination of their muscles, their vision, their endocrine systems, their digestive tracts. . . . All I asked for was the full hairy complexity, then in front of my eyes, of whole, intact beavers and otters in motion.

It lasted, . . . for only a few minutes, and then I was back . . . wondering about the details by force of habit, but not, this time, the details of otters and beavers. Instead, me. Something worth remembering had happened in my mind. . . . I became a behavioral scientist, an experimental psychologist, an ethologist, and in the instant I lost all the wonder and the sense of being overwhelmed. But I came away from the zoo with something, a piece of news about myself: I am coded, somehow, for otters and beavers. I exhibit instinctive behavior in their presence, ... Beavers and otters possess a "releaser" for me. . . . What was released? Behavior. What behavior? Standing, swiveling flabbergasted, feeling exultation and a rush of friendship. I could not, as the result of the transaction, tell you anything more about beavers and otters than you already know. I learned nothing new about them. Only about me, and I suspect also about you, maybe about human beings at large. . . . Left to ourselves, . . . we hanker for friends.

• **Think** of a special animal you have observed and enjoyed—at home, at the zoo, on television. Write a journal entry describing your feelings about the animal.

24 From The Medusa and the Snail by Lewis Thomas. Copyright © 1974, 1975, 1976, 1977, 1978, 1979 by Lewis Thomas. All rights reserved. Reprinted by permission of Viking Penguin Inc.

Chapter 5 Review

Vocabulary

cell (63)	cytoplasmic streaming	microfilament (70)	organ system (76)
cell membrane (66)	(67)	microtubule (70)	plastid (72)
cell theory (64)	cytoskeleton (70)	middle lamella (73)	prokaryote (65)
cell wall (72)	endoplasmic reticulum	mitochondria (69)	ribosome (67)
chloroplast (74)	(68)	multicellular	selectively permeable
chromatin (72)	eukaryote (65)	organism (63)	membrane (66)
chromoplast (74)	flagella (71)	nuclear envelope (72)	spindle fibers (70)
chromosome (72)	fluid mosaic model	nucleolus (72)	tissue (75)
cilia (71)	(67)	nucleus (65)	unicellular organism
colonial organism	Golgi apparatus (68)	organ (75)	(63)
(76)	leucoplast (74)	organelle (65)	vacuole (72)
cytoplasm (66)	lysosome (70)		

1. Look up in a dictionary the meaning of the word parts *eu-*, *pro-*, and *kary-*. Explain why the words prokaryote and eukaryote are good terms for the organisms they describe. What do the terms suggest about the evolution of these organisms?
2. Describe the structure and function of a cell membrane, and compare it with the structure and function of a cell wall.
3. Look up in a dictionary the meaning of

the word part *cyto-*. Use what you have learned to explain the meanings of the terms cytoplasm and cytoskeleton.
4. Explain the relationship between microtubules, cilia, and flagella.
5. Choose the term that does not belong in the following group, and explain why: Golgi apparatus, endoplasmic reticulum, chromatin, and mitochondria.

Review

1. A prokaryote has (a) a cell nucleus (b) a cell membrane (c) organelles (d) all of the above.
2. The growth of cells is limited by the ratio between (a) volume and surface area (b) organelles and surface area (c) organelles and cytoplasm (d) nucleus and cytoplasm.
3. A cell membrane is composed of (a) lipids (b) proteins (c) nucleic acids (d) lipids and proteins.
4. The function of the Golgi apparatus is to (a) synthesize proteins (b) release energy (c) process and package proteins (d) synthesize lipids.
5. Mitochondria (a) transport materials (b) release energy (c) make proteins (d) control cell division.
6. Lysosomes function in cells to (a) recycle

cell parts (b) destroy disease-causing agents (c) shape developing body parts (d) all of the above.
7. The nucleolus is (a) the control center of the cell (b) the storehouse of genetic information (c) the site where ribosomes are synthesized (d) none of the above.
8. Plant cells differ from animal cells in having (a) fluid-filled vacuoles (b) cell walls surrounding the cell membrane (c) chromoplasts (d) all of the above.
9. Leucoplasts (a) synthesize proteins (b) store food (c) synthesize pigments (d) store pigments.
10. The stomach is an example of (a) a tissue (b) an organ (c) an organ system (d) none of the above.

11. How have microscopes been helpful in the study of cells? Explain.
12. What limits the size of cells?
13. Why is the cell membrane called a selectively permeable membrane?
14. Explain how the ribosomes, the endoplasmic reticulum, and the Golgi apparatus function together in protein synthesis.
15. Distinguish between the structure of rough ER and that of smooth ER.

16. Describe the inner structure of mitochondria, and explain their function.
17. How has AVEC technology enhanced our understanding of the structure and function of microtubules?
18. What is the difference between chromatin and chromosomes?
19. How are cell walls formed?
20. What is cell specialization? Give an example of cell specialization.

Critical Thinking

1. A mature human red blood "cell" is merely a membrane surrounding hemoglobin, the protein pigment that carries oxygen. Suggest why this "cell" has no organelles.
2. The coils of a radiator provide a large surface area from which heat is radiated into a room. Which cell organelles have a structure similar to that of a radiator? How is this structure related to their function?
3. Children who are born with Tay-Sachs disease are unable to carry out the normal breakdown of lipids in the brain because of a defect in a cell organelle. Based on this information, which cell organelle do you think may be defective in Tay-Sachs victims? Explain.
4. Livestock in the western United States often die after eating a locoweed, such as *Astragalus tnanus*. The chemical the plant

contains is also poisonous to plants. How does locoweed keep from poisoning itself?
5. Compare the two photos above. One shows cells from a grass stem, and the other shows cells from the trunk of an oak tree. Which photo do you think shows the tree cells? Why do you think so?
6. What characteristic of eukaryotic cells gives them a greater capacity for specialization than prokaryotic cells have? Explain your answer.

Extension

1. Read J. Stephenson's article, "Green Genes: Spice of Life," *Science Digest*, September 1986, p. 20. Prepare an oral report for the class on how genetic engineers use knowledge of membrane structure and function to engineer new plants.
2. Use the resources in your school or public library to learn more about the work of Schleiden, Schwann, or Virchow. Write a brief report summarizing the processes

that the researcher used to arrive at his conclusions about cells.
3. Place a sprig of the aquarium plant *Elodea* in water, and shine a bright light on it for an hour. Then examine a leaf under the microscope and make a drawing, noting the direction of chloroplast movement. How do you think the movement of cell organelles helps the cell to function?

78 Chapter 5

Structure and Function of the Cell 79

The two-page review at the end of each chapter helps you to survey the key terms and ideas you've learned.

Unit 1

Biological Principles

"Our life is a faint tracing on the surface of mystery, like the idle, curved tunnels of leaf miners on the face of a leaf. We must somehow take a wider view, look at the whole landscape, really see it, and describe what's going on here."

Annie Dillard

Answering questions about nature involves processes of investigation and discovery that are at the heart of biology. As biologists go about their work, they use scientific processes, special tools and techniques, and their knowledge of basic principles of biology and chemistry. This unit will introduce you to the themes, processes, tools, techniques, and principles of biology. Before you begin to study the unit, look at the illustrations on these two pages to see some of the types of questions that biologists seek to answer.

What developmental changes convert a larval dragonfly, *Anax junius,* into an adult?

How do a clownfish, *Amphiprion percula,* and a sea anemone depend on each other?

2

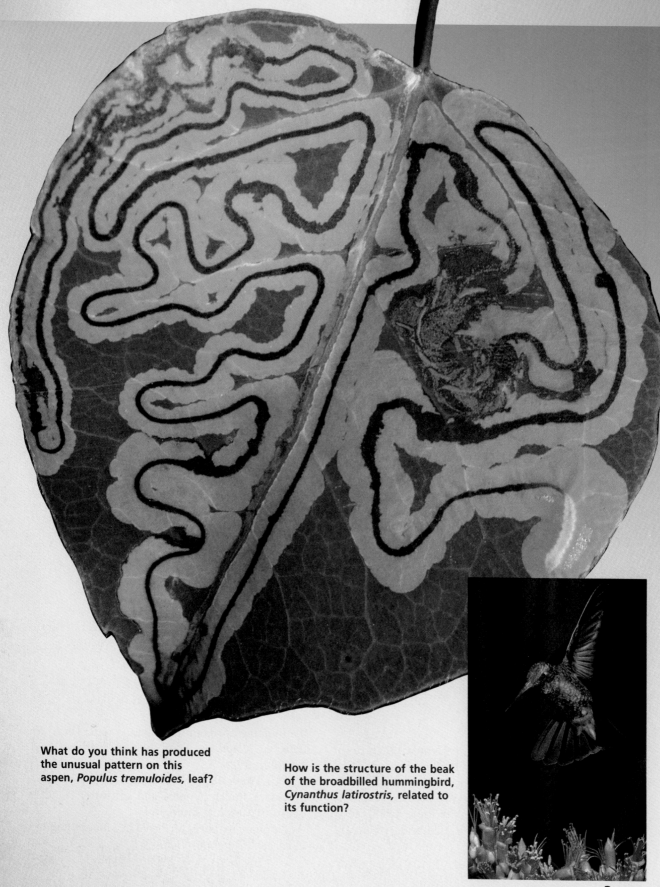

What do you think has produced the unusual pattern on this aspen, *Populus tremuloides,* leaf?

How is the structure of the beak of the broadbilled hummingbird, *Cynanthus latirostris,* related to its function?

Chapter 1 | *Biological Themes*

This chapter introduces the unifying themes of biology that allow scientists to study the diversity of life. Students learn about these recurring themes and the characteristics common to all life-forms.

Correlation Chart

Section Number	Section Objectives	Pages	Features	Section Review
1.1	1. List seven themes important in biology.	5–10	Biology in Process, pp. 8–9	1, 2, 3, 4, 5
	2. Discuss the relationship between adaptation and evolution.			2
	3. Explain how genes play a role in determining the developmental stages of an organism.		Think, p. 10	3
	4. Tell why the study of ecology is important in biology.			4
	5. Evaluate George Washington Carver's contributions to science and society.			
1.2	1. List seven characteristics of life.	11–13		2, 3, 4, 5
	2. State how living things are organized.		Biology Lab, p. 13	3
	3. Differentiate between a response and a behavior.			1

Resources*

Print

Gutteridge, Anne C. *Barnes and Noble Thesaurus of Biology: The Principles of Biology Explained and Illustrated.* New York: Barnes and Noble, 1983.

Smith, John Maynard. *The Problems of Biology.* New York: Oxford University Press, 1986.

Wilson, Edward O. *Biophilia.* Cambridge: Harvard University Press, 1984.

Audiovisual/Software

The Building of the Earth. Film, 55 min. Time-Life.

Living or Nonliving. Film, 16 min. National Geographic, 1986.

Tribbles. Apple II, 48K. Teaches scientific method. Includes students' notes and software. Conduit. Dist. by Carolina Biological.

*Additional references are listed in *Modern Biology, Teacher's Program Guide.*

Section Teaching Guides

1.1 Themes in Biology

Approach

The main objective of this section is to present seven themes important in the study of biology. The section will help students integrate knowledge from all later chapters and put details they learn about specific organisms into a larger context. In presenting the text, guide students to recognize how these themes unify the study of biology.

You may wish to point out to students that the diagrams in this section are representative of the types of illustrations they will be seeing in the rest of the book. By studying the relationships that are illustrated in the phylogenetic tree in Figure 1-2 and the salamander life-cycle in Figure 1-3, students will better understand similar diagrams that appear later in the book.

Concept Framework

Students will be familiar with the idea of themes mainly from studying literature. Tap their knowledge by assisting students in listing themes of familiar pieces of art or literature. Use these themes as a starting point to introduce students to the seven themes central to biology. Ask members of the class to jot down several reasons why each theme is important as they read the section.

Emphasis

Make certain that students can summarize each of the following biological themes: evolution, reproduction and inheritance, development, structure and function, energy relationships, ecology, and science and society.

Using the Features/Developing the Peanut as a Food Crop

The main process discussed in this feature is experimenting. The feature focuses on how George Washington Carver's scientific research on the peanut and other food crops benefited society. You may wish to suggest that students collect clippings from newspapers or magazines that report on the work of modern scientists. Discuss how the work of these scientists benefits society.

Biological Themes Mural

Divide students into seven groups. Assign each group one of the biological themes discussed in this section. Ask each group to design a large colored sketch or collage that illustrates its theme. Students may base their work on material from pages 6–10 of the text, and might add examples illustrating the themes from outside observations. Help students combine their sketches or collages to construct a mural that illustrates all seven biological themes. Arrange for students to display their mural in a hallway of the school.

Section Review Answers

(Continued from p. 10.)
5. Answers will vary. Possible questions include: Should people interfere to try to save particular species threatened with extinction, or is interfering taking too powerful a role in the fate of organisms? If people decide to interfere, what should they do—breed the organisms in captivity, change the environment, or move the species to another region?

1.2 The World of Biology

Approach

The main objective of this section is to give students information on the characteristics of all living things. The section builds on concepts presented in Section 1.1. In teaching this section guide students to realize that despite great differences between organisms, all organisms share certain general characteristics. These characteristics give scientists a working definition of *life*.

Concept Framework

Students will have some intuitive ideas about characteristics common to all organisms. Elicit some of these ideas and do the following demonstration to guide students into considering what constitutes life. Put a large drop of mercury in a petri dish. Use extreme caution when handling mercury. Mercury is poisonous and forms toxic vapors. Never flush mercury down the drain. Recycle or dispose of the mercury properly. Place the petri dish on an overhead projector. Ask students to observe the projection of the object in the petri dish and to decide whether the object is living or nonliving. Have them list features that make them think the object is living and several features that make them think the object is nonliving. You might push the mercury drop with a dissecting needle and use the needle to force the mercury to split into two or more drops.

Generate a discussion on the characteristics of living things based on students' observations. Talk about further tests students might do to find out whether the object is living or nonliving. At the end of the discussion, you may wish to tell students that the object was a drop of mercury. Then inform them that this section discusses characteristics shared by all organisms. As they read the text, ask them to consider whether the features that made the mercury drop appear to be alive are considered characteristics of life.

Emphasis

Make certain that students understand that nonliving things may exhibit one or two characteristics of life, but that living things exhibit all six characteristics. Students should be able to list and discuss the characteristics of life (cells, organization, energy use, response to the environment, growth, reproduction, and adaptation).

Evidences of Life

Display some items that were once part of a living organism, such as a piece of wood, a leather shoe, a seashell, and so on. Ask students to describe the types of evidence or clues that they might look for that could suggest that an item was once a living organism or part of a living organism. Assist students in generating a list of features to look for that might help establish that an object of unknown origin was once part of a living organism.

Section Review Answers

(See p. 12.)

Laboratory Preparation

Background

In this lab students examine a sample of pond water looking for evidence of life. Students relate the characteristics of living things as described in the text to the specimens they observe in the pond water. Stu-

dents discover that while some characteristics of living things are readily observable, others are not. Students speculate on what methods may be used to collect further evidence relating to the characteristics of living things.

Preparation

If pond water is not available, mixed cultures of protozoa can be ordered from a biological supply house. These organisms can be placed in tap water that has been set out overnight in open containers.

Conclusions

1. Answers will vary but may indicate that these organisms show cellular organization, locomotion, and response to stimuli.
2. The fact that these organisms move indicates that they use energy.
3. Reproduction, adaptation, and growth are difficult to observe.
4. Green coloration indicates a plantlike organism.

Inquiry

Students will think of investigations that provide ways of observing reproduction, growth, and adaptation. Long-range studies that measure population changes among a limited number of organisms are examples of such investigations.

Chapter Review

Vocabulary

1. Adaptations are inheritable traits that enable certain organisms to survive better in specific environments than organisms lacking those traits, leading to evolution; phylogeny is the evolutionary history of a group of organisms; evolution is the idea that species change over time.
2. Asexual reproduction does not involve the exchange of genetic material.
3. Morphology includes the internal and external structure of an organism or of any of its parts. Anatomy is internal morphology.
4. A behavior is a complex set of responses.
5. Cell division is the orderly formation of new cells from a parent cell. Cell enlargement is the increase in cell size. In multicellular organisms, both result in growth.

Review

11. Important themes in biology include evolution, reproduction and inheritance, development, structure and function, energy relationships, ecology, and science and society.

12. Organisms having favorable adaptations are better able to survive and reproduce than organisms lacking those adaptations. In time those organisms lacking favorable adaptations die out, leaving no offspring. Thus the species change over time.

13. At each of the developmental stages only some of the genes are expressed. As the salamander develops, different genes are expressed at different times.

14. Answers may vary. Accept examples such as the following: The shape of the owl's beak makes it functional for killing prey, or the structure of the starch molecule makes it functional as an energy-storing molecule in organisms.

15. Through the process of photosynthesis, producers use the energy of the sun to make food. Consumers eat the producers and gain energy in the food manufactured by the producers.

16. Scientists have learned that organisms are dependent on one another for food and other resources, that organisms are dependent on the environment for substances such as water, nutrients, and gases, and that the stability of the environment depends to some extent on the healthy functioning of organisms in that environment.

17. Answers will vary. Accept answers that show students recognize several different sides of the issue they discuss.

18. Carver did research on a soil-enriching crop, the peanut, worked for the increased production of the crop, and published information on improved farm techniques. Carver also inspired others to continue working in the field of agricultural research.

19. They are made of cells, are organized, use energy, respond to the environment, grow, and reproduce; species of organisms become adapted to different environments and evolve.

20. Organisms are organized on the molecular and cellular levels. They take in substances from the environment and organize them to produce complex chemicals. Their cells are organized internally, and in multicellular organisms cells are organized to form tissues, tissues are organized to form organs, and organs are organized to form systems.

Critical Thinking

1. Taxonomy deals with the classification of living things and the giving of names. It is necessary to describe and classify a new species before any other studies are made of it. This allows different scientists to compare their results.

2. Since ecology is the study of relationships of living things, ecologists must have a great deal of knowledge about individual organisms and their environments in order to study their interrelationships. This knowledge takes many years to obtain. Thus, ecology is a fairly recent study.

3. The tufted puffin in the photograph (*Lunda cirrhata*) belongs to one of several species that lives in arctic regions of the Pacific and along Alaska's coast. The puffin's thick body is adapted to the cold climate. Coloration of feathers and bills may serve to attract mates. Also, since such coloration varies among puffin species, it may be an adaptation related to species identification. The puffin's webbed feet are adapted for swimming.

4. The plant takes sunlight (solar energy) and transforms it into chemical energy. An animal takes the chemical energy of plants or other animals and transforms the energy for its own use.

Biological Themes

Introduction

Earth is probably the only planet in our solar system where life occurs. Scientists estimate that 40 million species of organisms exist on earth. Of these, only about a million have been named, and few have been studied in detail. Thus much of the world of biology is uncharted territory. However, this fascinating world is not without guideposts. In this chapter you will read about the unifying themes of biology that allow scientists to study the diversity of life.

Chapter Outline

1.1 Themes in Biology
- *Evolution*
- *Reproduction and Inheritance*
- *Development*
- *Structure and Function*
- *Energy Relationships*
- *Ecology*
- *Science and Society*

1.2 The World of Biology
- *Characteristics of Life*
- *The Living World*

Chapter Concept

As you read, notice the recurring themes of biology and the characteristics common to all organisms.

Waterfall with plants, Cascade Mountains, Oregon. Inset: Scientist in laboratory.

For additional teaching information see pp. T1A–T1D.

1.1 Themes in Biology

Life arose on earth over 3.5 billion years ago. The first **organism**—or living thing—was composed of a single cell. For millions of years single-celled organisms floating in the seas were the only life on earth. No fishes swam in the ocean, no birds flew through the air, no snakes slithered across the ground.

In time organisms diversified, became more complex, and came to inhabit almost every region of the earth. Today brilliant tropical birds glide through the trees of lush rain forests. Strange fish living in the dark depths of the ocean emit eerie red light that helps them locate their prey. Giant sequoia trees tower hundreds of feet above the ferns on forest floors. Colonies of green algae populate the tips of a polar bear's hollow hair.

These and millions of other types of organisms form the diverse citizenry of earth. **Biology**—the science of life—includes the study of every one of these organisms. It includes the study of the microscopic structure of a single organism as well as the study of the global interactions of millions of organisms. It includes the study of an organism's individual life history as well as the collective history of all organisms.

The science of biology is as varied as the organisms that are its subjects. However, biology is unified by certain themes. Seven major themes recur throughout this book: (1) evolution, (2) reproduction and inheritance, (3) development, (4) structure and function, (5) energy relationships, (6) ecology, and (7) science and society. The paragraphs below will give you a brief introduction to each theme. By reflecting on these themes as you study biology, you will recognize larger patterns in the study of life.

Section Objectives

- *List seven themes important in biology.*

- *Discuss the relationship between adaptation and evolution.*

- *Explain how genes play a role in determining the developmental stages of an organism.*

- *Tell why the study of ecology is important in biology.*

- *Evaluate George Washington Carver's contributions to science and society.*

Figure 1-1 Many organisms exhibit similarities, such as the similar patterns found on the zebra, *Equus* sp., and the zebra moth, *Heliconius chrithonius.* Yet these two organisms differ in many fundamental ways. The questions that biologists ask about similarities and differences of organisms are the basis of biological inquiry.

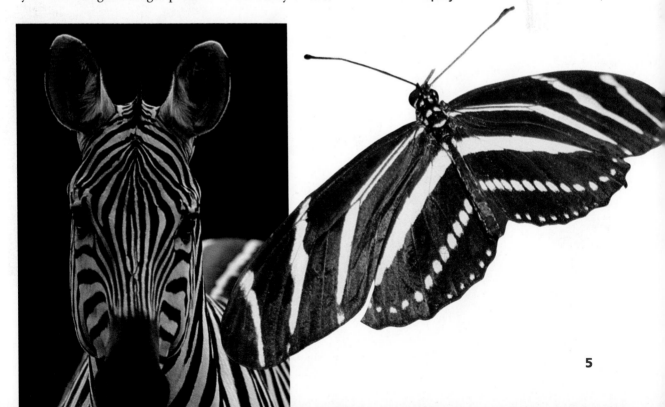

Info: The theory of evolution was first proposed by Charles Darwin (1809–1882) and Alfred Russel Wallace (1823–1913) in 1858. It has been modified, but the basic principles laid out by Darwin and Wallace still form the foundation of biological thought.

Evolution

Evolution—the theory that species change over time—is the unifying theme of biology. The theory of evolution helps explain how all the kinds of organisms came into existence. It helps us understand why organisms look the way they do and how organisms of the past are related to organisms alive today. It also helps explain relationships among various groups of living organisms.

Scientists suggest that evolution occurs by a process called natural selection. According to the theory of evolution by natural selection, organisms that have certain inheritable traits are better able to survive in specific environments than organisms that lack those traits. Such favorable traits are called **adaptations.** In most circumstances organisms having adaptations reproduce at a higher rate than organisms lacking the adaptations. In time organisms lacking the adaptations will die out. Individuals with adaptations live, and thus species change over time.

As you study biology, you will learn about evolution by examining adaptations of different organisms to various environments. For example, the cactus plant is adapted to a desert environment. Cactuses have thick, waxy coverings that reduce water loss. Their leaves are reduced to thin spines that further prevent evaporation and keep predators from eating the plant. In your study of biology, you will also learn about **phylogeny** *(fie-LAHJ-uh-nee)*—the evolutionary history of groups of organisms. The phylogeny of birds is shown in Figure 1-2.

Figure 1-2 Scientists developed phylogenetic trees, such as this phylogenetic tree of birds, to illustrate the probable evolutionary history of a group of organisms.

Reproduction and Inheritance

All organisms come from preexisting organisms by the process of reproduction. Reproduction involves the transfer of genetic information from parent to offspring. This information is contained in the structure of a molecule called **deoxyribonucleic acid,** or **DNA.** A short segment of DNA that contains a chemical message coding for the development of a certain trait is called a **gene.** The acquisition of traits from parents is **inheritance.**

Reproduction can be asexual or sexual. **Asexual reproduction** occurs when there is no exchange of genetic information between organisms. For example, when a bacterium reproduces asexually, it splits in two. The resulting bacteria usually contain identical genes. **Sexual reproduction** occurs when genetic information is exchanged between organisms. For example, when an egg of a female leopard frog is fertilized by sperm of the male leopard frog, the egg and sperm fuse, forming a zygote *(ZIE-GOTE)*. The zygote contains genetic information from both the female and the male. The zygote divides, and the resulting cells divide repeatedly and develop into a new individual that contains genes from both of its parents.

Development

All organisms are composed of and develop from cells. Some organisms are composed of only one cell. These organisms are called **unicellular organisms.** Most of the living things you see around you are composed of more than one cell. Such organisms are called **multicellular organisms.** Multicellular organisms usually arise from a zygote. The process by which a zygote becomes a mature individual is called **development.**

Consider the life cycle of a salamander, for example. An egg, a larva, and an adult salamander each represent a stage in this cycle. How can an egg, a larval salamander, and an adult salamander, whose cells contain the same genes, look so different

Figure 1-3   A strand of DNA (left, ×14,300) contains the genetic information of a cell. A single gene on this strand has been marked blue. The bacterial cell (right, ×17,650) is dividing, or undergoing asexual reproduction.

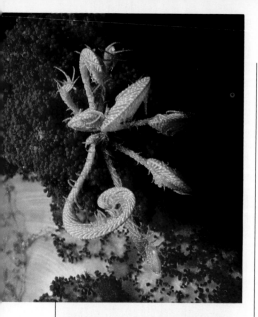

from one another? At each stage of the salamander's development, only some of the genes are active. When a gene is active, scientists say that it is expressed. In the salamander's egg some genes are expressed and others are not. The same is true for the larva and for the adult salamander. As any organism develops, different genes are expressed at different times.

Structure and Function

Biologists use the term **morphology** (*mor-FAHL-uh-jee*) to refer to the structure of an organism or of any of its parts. Morphology includes both internal and external structures. **Anatomy** is another term for internal morphology.

In biology structure is almost always related to function. A relationship between structure and function exists at all levels of biological organization. For example, a starch molecule consists of a long chain of sugar molecules. Its structure corresponds to its function—storage of sugar. The structure of the beak of the great horned owl shown on the cover of this book is related to the beak's function. The sharp, curved point allows the owl to tear the flesh of its prey.

Figure 1-4 How does the structure of this crinoid enable it to gather food?

Biology in Process

Developing the Peanut as a Food Crop

The development of new food crops is one of the most important ways that scientists work to benefit society. With his work on the peanut American scientist George Washington Carver made an enormous contribution to American agriculture.

During the 1800s and into the 1900s cotton was the chief cash crop grown in the southern part of the United States. However, cotton robs the soil of nutrients when it is planted in the same field year after year. Thus southern farmers, especially those who farmed small plots, made little money growing this crop.

George Washington Carver wanted to help the southern farmer. After receiving his master's degree in agriculture in 1896, Carver became head of the newly formed Agriculture Department at the Tuskegee Institute in Tuskegee, Alabama.

Carver set up a laboratory and experimental plots for agricultural research. He employed students to help carry out experiments on different crops and on products that could be made from those crops. Carver was especially interested in the peanut and the sweet potato, crops that harbor

Energy Relationships

All organisms use energy. How organisms get, use, and transfer energy is a major topic of study in biology. The sun supplies almost all of the energy for life on earth. Through the process of photosynthesis plants and certain unicellular organisms capture solar energy and convert it into a form of energy that can be used by living things.

Organisms that acquire energy by making their own food are called **autotrophs.** Plants and certain unicellular organisms are autotrophs. Organisms that gain energy by eating other organisms are called **heterotrophs.** Some unicellular organisms, as well as all animals and fungi, are heterotrophs. In studying biology you will learn about how energy flows from the sun to autotrophs and finally to heterotrophs.

How energy is produced, stored, and used within a single organism is also an important subject of study in biology. Most organisms on earth and every cell within these organisms use similar chemical processes to release energy from food. They all use this energy for growth and maintenance, as well as for carrying out essential life functions.

Answer to Biology in Process
Carver popularized a soil-enriching crop and came up with over 300 products that could be made from that crop.

bacteria on their roots that add nutrients to the soil. Carver discovered about 300 products that could be made from peanuts and over 100 products that could be made from sweet potatoes. These products included flour, cheese, milk, cosmetics, dyes, rubber, and peanut butter.

Carver's impressive list of products established the importance of the peanut and the sweet potato. Southern farmers began to grow these crops, which were especially suited to the warm weather and the sandy soil of that region.

In 1921 Carver went to Washington, D.C., to testify before Congress about the importance of the peanut. Because of Carver's testimony Congress acted to promote the growth of peanuts in the United States. In time the soil-enriching peanut became the second largest crop in the South. By 1950 it had become the sixth leading crop in the nation.

Carver continued his agricultural research. He published articles on practical matters such as improved farm techniques and food preservation. His work was especially appreciated during the Great Depression of the 1930s, when many people were out of work and had little money to buy food. Carver was also an influential teacher, inspiring young people to find ways to make science work for the betterment of all.

■ What are two ways in which Carver's work on the peanut benefited society?

Biological Themes **9**

Think

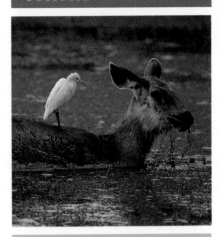

The egret, *Ardeola ibis,* and the sambar deer, *Cervus unicolor,* shown in the picture are dependent on each other. The egret feeds on organisms that live on the deer's skin and on insects stirred up when the deer passes. How might the deer be dependent on the egret? In what ways are both animals dependent on the environment?

Answer to Think
The deer benefits when the egret calls to warn of predators. Both animals depend on the environment for water and oxygen. The deer depends on plants for food.

Answers to Section Review
1. the internal or external structure of an organism or of any of its parts
2. Phylogeny reflects relationships between groups of organisms based on adaptations that each group has acquired during its evolutionary history.
3. Different genes are expressed at different developmental stages.
4. By studying ecology, biologists have come to recognize the interdependence of organisms.
(See pp. T1A–T1D.)

Ecology

The study of individual organisms is an important part of biology. However, to understand fully the biological world, scientists study interactions of organisms with each other and with the environment. This branch of biology is called **ecology.**

By studying ecology, biologists have come to recognize the interdependence of organisms. Even solitary spiders that spend most of their lives away from other members of their species depend on other animals for food. By studying ecology, biologists have also come to recognize the interdependence of organisms and the physical environment. To survive, all organisms need substances such as nutrients, water, and gases from the environment. The stability of the environment in turn depends to some extent on the healthy functioning of organisms in that environment. For example, the oxygen in the atmosphere is largely a product of plant life. When vast areas of forest are cut down, less oxygen is released. In time a great reduction of forests could alter the composition of the atmosphere. Such a change in the physical environment could eventually affect all life on earth.

Science and Society

Knowledge from biological science can be applied to specific problems in society to improve human life. For example, the development of the polio vaccine in the 1950s was a significant scientific breakthrough that had a large impact on society. By producing the polio vaccine in bulk and distributing it throughout the world, scientists, business leaders, and governments have worked together to reduce the threat of polio.

How biological knowledge should be used involves decisions based on **ethics**—the study of what is right and wrong and of our moral choices. **Bioethics** is the application of ethics to biological concerns. One biological issue involves genetic engineering—making new forms of life by placing genes from one organism into another. A second issue involves AIDS, a viral disruption of the immune system, which has profoundly affected our society. Understanding biology can help people responsibly answer ethical questions, such as, whether restrictions should be placed on genetic engineering, or whether AIDS test results should be publicized. Sound decisions will ensure a healthy future for society.

Section Review

1. What is *morphology?*
2. What is the relationship between adaptation and phylogeny?
3. Explain development in terms of gene expression.
4. Why is the study of ecology important in biology?
5. What bioethical questions does society face in deciding how to deal with species threatened with extinction?

1.2 The World of Biology

Our world abounds with a great variety of life. Giant tube worms thrive at the bottom of the ocean near bubbling volcanic vents. Red algae grow, covering the surface of arctic glaciers like a carpet. Bacteria find shelter even in the pores of your skin. Tube worms, algae, and bacteria and all other organisms—no matter how diverse they may be—share certain features characteristic of all living things.

Characteristics of Life

Although it is difficult to define life, biologists have formulated a list of characteristics by which we recognize living things. These characteristics are elaborated in the paragraphs below.

- **Cells**   All living things are composed of cells. In multicellular organisms some cells are highly specialized, each having a unique role in the function of the organism.
- **Organization**   All organisms are organized at both molecular and cellular levels. They take in substances from the environment and organize them in complex ways. Cells have an internal organization in which specific cell structures carry out particular functions. In multicellular organisms cells are organized into tissues, tissues are organized into organs, and organs are organized into systems.
- **Energy Use**   As you learned in Section 1.1, all organisms use energy for growth and maintenance.
- **Response to the Environment**   All organisms respond to the environment. A **response** is a reaction to a stimulus. For example, if a lion charges a gazelle, the gazelle may respond by running away. Responses may be simple or complex. A complex set of responses is called **behavior.** The annual migration of a bird is a behavioral response to changes in the seasons.

Section Objectives

- *List seven characteristics of life.*
- *State how living things are organized.*
- *Differentiate between a response and a behavior.*

Info: Underwater vent communities are fueled by sulfur gases from the vents. Bacteria use chemical energy from these gases to make food via a process called chemosynthesis. Organisms in the underwater vent communities differ from all other life-forms on earth. Nevertheless, scientists do not hesitate to classify them as living.

Figure 1-5   Life-forms vary greatly. The fluorescent coral (left) appears inanimate, while the gazelle, *Litocranius walleri* (right), is very swift. Yet both are classified as living things because they have the characteristics of life.

Figure 1-6 A lizard extending its frill, *Chlamydosaurus kingii* (left), and sky pointing by a blue-footed booby, *Sula nebouxii* (right), are two examples of behavior by animals.

Tip: Discuss the characteristics of life with the class. Ask students to speculate on whether and how these characteristics might be modified if new life-forms were found on other planets.

Answers to Section Review
1. A response is a reaction to a stimulus; a behavior is a complex set of responses.
2. Accept any of the following: cells, organization, energy use, response to the environment, growth, reproduction, and adaptation.
3. Organisms take in simple substances from the environment and use them to produce complex chemicals necessary for life. These substances are more highly organized than the substances that the organism uses to produce them. Cells have internal organization. In multicellular organisms, cells are organized into tissues, tissues are organized into organs, and organs are organized into systems. (See pp. T1A–T1D.)

- **Growth**   All living things grow. Growth occurs through cell division and cell enlargement. **Cell division** is the orderly formation of new cells from a parent cell. **Cell enlargement** is the increase in cell size. In multicellular organisms, cell division and cell enlargement together result in growth.
- **Reproduction**   All organisms have the ability to reproduce. Reproduction is not essential for the survival of an individual organism. It is essential for the continuity of life, because no organism lives forever.
- **Adaptation**   Organisms have adaptations. These adaptations are traits that give an organism an advantage in an environment. Variations of adaptations are essential for the continuity of life, because they confer an advantage on some members of a species in a changing environment.

The Living World

The world of biology is so rich that scientists have yet to explore all of it. New discoveries, like the algae population inside the polar bear's fur and the underwater vent community, are found every year. The wide variety of organisms in these communities shows us how much is still to be learned about organisms. Wherever you may live, the world around you teems with life. This book will help you appreciate this living world.

Section Review

1. What is the relationship between a *response* and a *behavior*?
2. List five characteristics of all organisms.
3. Describe three ways in which living things are organized.
4. Why is reproduction an important characteristic of life?
5. What kind of test might a scientist devise to figure out whether something is alive?

Laboratory

Life in a Drop of Water

Objective
To examine the variety of living organisms found in a drop of pond water

Classifying, inferring, communicating

Materials
Strands of cotton from a ball of cotton batting, microscope slide, medicine dropper, pond water, coverslip, compound microscope, unlined paper

Asterionella

Pediastrum

Stentor

Tubellaria

Meridion

Method
1 Place two or three strands of cotton on a microscope slide. Use a medicine dropper to apply one drop of pond water to the cotton strands on the slide. Place a coverslip over the drop of water, using the wet mount technique.
2 Place the slide on the microscope stage and observe the drop of water under low power. Record your observations. Center the microscope on an interesting organism, then shift to high power. Why is it necessary to begin your observation by using the low power objective of your microscope?
3 Try to distinguish between the nonliving objects and the living organisms you see in the water sample. What are some guidelines for distinguishing between living and nonliving matter? Make a sketch of each structure you identify as a living organism.
4 Under each sketch write a brief description of the organism's color, shape, and relative size. As you work through the following steps, record any other distinguishing physical features you observe, such as the presence of cilia.
5 Observe whether or not the organism moves under its own power. If it is motile, try to determine its means of locomotion. What

effect do the cotton strands have on the movement of the organisms through the water?
6 For each organism look for evidence that it responds to stimuli—such as light, obstacles, and movement—in its environment.
7 Look for evidence that the organism is capable of reproduction.
8 Compare your observations with those of your classmates.
9 Work as a class to compile a list of generalizations about organisms found in pond water.

Conclusions
1 What characteristics did you observe in living organisms in pond water?
2 From what observation can you infer that pond water organisms expend energy?
3 What characteristics of the pond water organisms are difficult to observe in this type of investigation?
4 What characteristics might be used to distinguish between plantlike and animallike organisms?

Inquiry
Think of an investigation that would reveal the characteristics of living organisms that you found difficult to observe in this investigation.

Chapter 1 Review

Vocabulary

adaptation (6)
anatomy (8)
asexual reproduction (7)
autotroph (9)
behavior (11)
bioethics (10)

biology (5)
cell division (12)
cell enlargement (12)
development (7)
deoxyribonucleic acid (DNA) (7)
ecology (10)
ethics (10)

evolution (6)
gene (7)
heterotroph (9)
inheritance (7)
morphology (8)
multicellular organism (7)

organism (5)
phylogeny (6)
response (11)
sexual reproduction (7)
unicellular organism (7)

1. What is the relationship between adaptation, phylogeny, and evolution?
2. What is the difference between asexual reproduction and sexual reproduction?
3. How is anatomy related to morphology?

4. What is the difference between a response and a behavior?
5. How do cell division and cell enlargement contribute to growth?

Review

1. Evolution helps explain how complex organisms came into existence, why organisms of the past differ from those alive today, and (a) how various groups of living organisms develop (b) how various groups of living organisms reproduce (c) how various groups of living organisms are related to each other (d) how various groups of living organisms obtain energy.
2. Reproduction involves the transfer of genetic information from (a) male to female (b) parents to offspring (c) unicellular organism to multicellular organism (d) egg to zygote.
3. When a gene is active, scientists say that it is (a) expressed (b) inherited (c) functional (d) interdependent.
4. In biology structure is almost always related to (a) evolution (b) adaptation (c) development (d) function.
5. The branch of science that includes the study of the interactions of organisms with one another and with the environment is called (a) biology (b) morphology (c) ecology (d) phylogeny.
6. All organisms are composed of (a) DNA (b) genes (c) cells (d) energy.

7. A response is a reaction to (a) a stimulus (b) a behavior (c) an adaptation (d) a developmental stage.
8. Growth occurs through (a) organization and reproduction (b) adaptation and evolution (c) stimulus and response (d) cell division and cell enlargement.
9. Reproduction is essential for the continuity of life because (a) organisms must reproduce to survive (b) organisms must reproduce to adapt to their environment (c) no organism lives forever (d) no organism is totally independent of other organisms.
10. The science of life is referred to as (a) phylogeny (b) biology (c) ecology (d) anatomy.
11. List seven themes that are important in biology.
12. What is the theory of evolution by natural selection?
13. How do genes play a role in the developmental stages of an organism such as a salamander?
14. Give an example of how structure is related to function in biology.
15. How does energy flow from the sun to producers and then to consumers?

16. What have scientists learned by studying ecology?
17. Explain the ethical implications involved in a biological issue, such as genetic engineering.
18. How did George Washington Carver work to benefit science and society?
19. List seven characteristics of all life-forms.
20. What do biologists mean when they say that organisms are organized?

Critical Thinking

1. One of the first branches of biology to be developed was taxonomy, the naming of organisms. Why might it be important for taxonomy to develop before the other branches of biology?
2. The branch of biology known as ecology has developed only fairly recently. Why might this be so? What sort of biological knowledge would be necessary before the study of ecology could begin?
3. Examine the photograph of the puffin at right. This animal, like all organisms, has adaptations that enable it to survive in its particular environment. Judging from the appearance of the puffin, what type of environment does it inhabit? Using the photograph, name two adaptations of the puffin that enable it to survive in its environment.
4. The first law of thermodynamics states that energy can be neither created nor destroyed during a process; only the form of

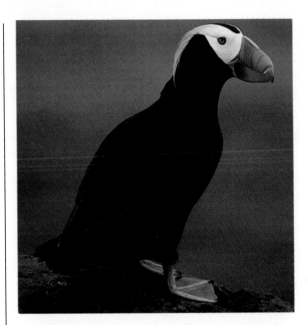

the energy can be changed. How does the biology of a plant or an animal reflect this law? Give an example.

Extension

1. Read the article by S. Weisburd entitled "Mussel Power from Methane," in *Science News*, September 27, 1986, p. 198. Make an oral report in which you include the following information: (1) the name of the new species of mussel that contains intracellular bacteria in its gills, (2) what the biologists must show in order to prove methane symbiosis, (3) what technique they used to show that the carbon in the methane consumed was identical to the carbon released as carbon dioxide, (4) what technique they used to show that the gill bacteria contained stacked membranes, and (5) why scientists did not find more examples of methane or hydrogen sulfide symbiosis earlier.
2. A hiker in the Wisconsin woods suddenly sees a brown bear. The bear runs in one direction and the hiker in the other. Later in the summer the hiker meets a mother bear with two cubs. The bear chases the hiker. Make a table in which you list the stimulus and response for the bear and the hiker in each situation. Why did the bear respond differently in the second situation?

Chapter 2 | Biological Processes

This chapter explains how scientists investigate the natural world. Students learn to identify the steps in the scientific method and the processes, tools, and techniques biologists use.

Correlation Chart

Section Number	Section Objectives	Pages	Features	Section Review
2.1	1. Define and give examples of observing, measuring, organizing and analyzing data, classifying, inferring, and modeling.	17–24	Think, p. 23	2, 5
	2. Explain the relationship between hypothesizing, predicting, and experimenting.			1, 3, 5
	3. Tell why communicating is an important process in science.			4
	4. List some similarities and some differences in the ways that scientists work.			5
2.2	1. Compare light microscopes with transmission and scanning electron microscopes in terms of magnification and resolution.	25–29	Biotechnology, pp. 26–27	2
	2. Define and discuss three techniques scientists use in the laboratory.		Biology Lab, p. 29	1, 5
	3. Explain why scientific sampling is an important technique in field studies.			3
	4. List ways scientists might use computers.			4
	5. Describe how a scanning tunneling microscope works and list its advantages over light and electron microscopes.			

Resources*

Print

Kavaler, Lucy. *A Matter of Degree*. New York: Harper & Row, 1981.

Kendig, Frank. *Life Spans*. New York: Holt, Rinehart, and Winston, 1980.

Pepi, David. *Thoreau's Method: A Handbook for Nature Study*. Englewood Cliffs, N.J.: Prentice-Hall, 1985.

Audiovisual/Software

Laboratory of Life. Film or video, 21 min. National Geographic, 1985.

Science Toolkit-Master Module. Disk. Apple 64K. User's manual and guide. Scholastic, 1986.

The Scientific Method. Film, 13 min. Britannica.

*Additional references are listed in *Modern Biology, Teacher's Program Guide*.

Section Teaching Guides

2.1 Scientific Processes and Methods

Approach

The main objective of this section is to define and present examples of processes and methods scientists use in their investigations of the natural world. The section prepares students for understanding specific examples of experiments and theories presented later in the book. The description of the various scientific processes will also prepare students for developing these skills in laboratory activities. In presenting this chapter, guide students to appreciate the varied nature of science. Point out that the methodology of science leads scientists to discover probabilistic truth rather than absolute truth.

Concept Framework

Students will probably be familiar with the scientific method that stresses controlled experiments. Use this knowledge as a springboard for discussing the steps in this scientific method, including posing a question or problem, formulating a hypothesis, designing an experiment, and drawing conclusions. Encourage students to realize that though many scientists follow this method, many areas of scientific investigation do not lend themselves to formal experimentation. Assist students in listing some of these areas. Tell students that they will be reading about different processes involved in the scientific method. As they read, encourage students to think about times they use these processes in their daily lives.

Emphasis

Make certain that students understand the tentative nature of scientific theories and can explain why hypotheses and theories can be refuted but cannot be proved true beyond all doubt.

Student Presentations

Ask each student to choose a scientist and to prepare an oral report focusing on the scientist's research and the scientific processes he or she used. Students can research the work of a modern scientist or a historic figure. They might even interview a scientist who works in the local community.

Section Review Answers

(Continued from p. 23.)

3. A hypothesis is a testable statement; a prediction is a statement of the results expected from testing a hypothesis; an experiment is the process of testing a hypothesis or a prediction by carrying out data-gathering procedures under controlled conditions.
4. Scientists often critique each other's work and benefit from each other's insight.
5. Accept answers that show students understand that scientific theories are tentative because they cannot be proved true beyond all doubt, and that scientists constantly weigh evidence, examine inferences, and modify theories.

2.2 Biological Tools and Techniques

Approach

Presenting tools and techniques useful in modern biology is the main objective of this section. The section builds on concepts discussed in 1.1 and will give students practical information for understanding the research of scientists presented throughout the rest of the book. In teaching this section, guide students to understand how scientists depend on certain tools and techniques to help them make observations, collect data, set up controlled conditions in the laboratory, and organize and analyze data.

Concept Framework

Most students have probably used a microscope to examine living specimens or prepared slides. If possible, set up stereoscopic microscopes and light microscopes. Allow students to examine specimens under both. Use this experience as a starting point to begin a discussion about the circumstances under which each type of microscope is useful. Tell them that as they read the section, they will be learning about different biological tools and techniques. Ask them to consider circumstances under which each tool or technique would be useful to a scientist.

Emphasis

Students should be able to describe and discuss the uses of the following tools or techniques: light microscopes, electron microscopes, cell cultures, fractionation, centrifugation, chromatography, field techniques, and computers.

Using the Features/Scanning Tunneling Microscope

The main tool discussed in this feature is the scanning tunneling microscope. It can magnify the surface of objects 100 million times and, can be used to examine living specimens. Direct students to study the diagram on p. 27 to learn how the microscope works. Point out that this tool has been used to observe the surfaces of DNA and of certain viruses. Lead an open discussion about how this microscope could be used to advance biological knowledge.

Field Trip

Take students to a local university or biological research company to see how an electron microscope works. If such facilities do not exist in your local area, you may wish to bring in books having electron micrographs of different objects to show the range of objects that can be observed with this technique.

Section Review Answers

(Continued from p. 28.)

2. A light microscope can magnify an object about 2,000 times and can be used for both living and preserved cells. A TEM can magnify an object up to 2,000,000 times and can be used to observe thin slices of nonliving specimens. An SEM can magnify an object up to 50,000 times and can be used to observe the surfaces of nonliving specimens.
3. A scientific sample can be used to represent an entire population.
4. Computers can be used to organize and analyze data and to make computer models.
5. Since the cells in a cell culture are identical, they eliminate the variable of differences in genetic makeup between individual cells or cell samples.

Laboratory Preparation

Background

In this exercise students investigate the scientific approach to problems and how scientists test their ideas. Students learn that scientific progress often consists of seeing old things in new ways, rather than seeing new things. William Harvey was the first anat-

omist to look at physiological phenomena in a quantitative way. By directing students toward thinking about circulation in a quantitative way, they should be able to reach Harvey's conclusions. Pulse rate is affected by a wide variety of factors and is an excellent example of the need for controls. Allow students to test any ideas they have that are safe and do not disturb other students. They may want to think about things that might affect their pulse rate, e.g., a pleasant or terrifying experience or fantasy. This lab offers an excellent opportunity for collecting and analyzing group data if there is time.

Preparation

Demonstrate to the students the correct way to find and count their pulse. Instruct students to place the index and middle fingers of their right hand on the thumb side of their left wrist. They should not begin counting until they can clearly feel the pulse. Remind students not to use their thumb to take a pulse.

Conclusions

1. Answers will vary, depending on the activity chosen and the hypothesis.
2. The control was the pulse rate recorded while sitting.

Inquiry

Some possible sources of error in this investigation include inaccurate timing and the effects of unperceived independent variables, such as food intake, medication, room temperature, and time of day.

Chapter Review

Vocabulary

1. To measure a scientist must observe; but observing can be done without making measurements.
2. Collecting data involves the gathering and recording of specific information; organizing data involves placing the data in some kind of logical framework; analyzing data involves determining whether the data are reliable and whether they support or refute a specific prediction or hypothesis.
3. Hypothesizing involves formulating a testable statement; experimenting involves carrying out a data-gathering procedure to test a hypothesis or

prediction; inferring involves drawing conclusions based on a set of facts, such as those gathered in an experiment or premises.

4. Classifying involves grouping objects, organisms or phenomena into an established organizational scheme; modeling involves constructing a representation of an object, system or process that helps show relationships between large amounts of complex data.

5. A TEM transmits electrons through a thinly-sliced specimen, producing magnification of up to 2,000,000×; an SEM scans the surface of a specimen with a beam of electrons, producing magnification of up to 50,000×.

Review

11. Accept descriptions based on a stated hypothesis or prediction and that have a control group and an experimental group which are identical except for one factor, identified as the independent variable and in which one thing is measured, identified as the dependent variable. Students might use the experiment described on page 20 as their example.

12. Accept any reasonable answers. Students might use the example of a testable inference based on the experiment discussed on page 21. They might use the example of the fossil horses on page 21 as an untestable inference.

13. A model helps show relationships between large amounts of complex data.

14. Scientists critique each other's work, benefit from the insight of other scientists, and work directly with scientists from other institutions or other countries.

15. Differences include that each scientist devises his or her own scientific method combining some or all of the scientific processes, taylored to a particular research project. Similarities include that all their answers to scientific questions are tentative.

16. A specimen is illuminated by a source of light; the light passes through the objective lens and the ocular lens, both of which together produce a magnified image of the specimen.

17. Advantages include that electron microscopes magnify to a great degree and SEMs can show great surface detail (they are more powerful than light microscopes); disadvantages include that living organisms do not survive the preparation techniques necessary for viewing specimens under electron microscopes.

18. Accept answers that include descriptions of any three of the following: cell culture; fractionation; centrifugation; chromatography. Descriptions are given on pages 27 and 28.

19. Accept any reasonable answer. Students might mention the use of computers to organize data, analyze data, and model.

20. With the scanning tunneling microscope a biologist can view living specimens; the magnification is on an atomic level.

Critical Thinking

1. The test of the truth of an observation in science is its repeatability. The more times an observation can be repeated, the greater the probability that it is true.

2. Such details tell other scientists how the experiment was done, and the conditions under which the conclusions are valid. Also, they allow other scientists to try to repeat the experiment.

3. The metric system is convenient because its units are in multiples of ten. Also, it is used and understood all over the world.

4. Answers will vary, but students might suggest that the cells in a culture may not behave exactly as they do in the body.

5. The SEM shows a three-dimensional image and reveals details about surface features of the organism. The TEM shows a much greater magnification and gives a better indication of internal organization.

Chapter 2 | Biological Processes

Introduction

The science of biology has been built on the observations of many scientists over thousands of years. Their curiosity about the natural world has led to a body of knowledge that grows larger every day. In this chapter you will learn about the processes, tools, and techniques biologists use to add to the body of knowledge we call biology.

Chapter Outline

Chapter Concept

As you read this chapter, notice how scientific processes, tools, and techniques are used by biologists in their scientific methods.

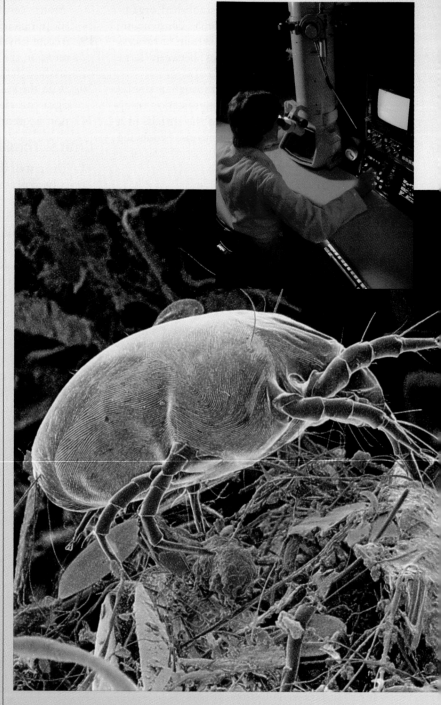

House dust mite, *Dermato phaggides,* on a piece of dust (×380). Inset: Electron microscope.

For additional teaching information see pp. T2A–T2D.

2.1 Scientific Processes and Methods

Science is a body of knowledge. Science is also a way of learning about the natural world. One of the best introductions to science comes from watching scientists in action. Dr. Luis Baptista is a biologist who studies birds that live in the rain forests of South America. These equatorial forests are home to thousands of bird species, many of which have not even been named, much less studied in detail. Simply finding out what's there is one of Dr. Baptista's primary aims. To do this he does a series of field studies. A field study involves spending several weeks or months at one site, observing and collecting birds.

Each day Dr. Baptista puts in long hours of intense observation. Carrying a portable tape recorder to record unusual bird calls, he spends the day hiking through the rain forest, visiting different habitats, such as a river edge, a rocky ledge, or a clearing, to observe and identify birds. He may also spend part of the day identifying birds that have been caught in collecting nets. In the evening, when most birds are inactive, he records information on what he observed that day. Such records of his observations enable Dr. Baptista to communicate with his colleagues about what he observed. They also can help other biologists classify the birds and learn more about their behavior.

Dr. Baptista is just one of thousands working to add to our knowledge of the rain forest. Not all of these biologists are field biologists—scientists who work outdoors. Some biologists work in laboratories doing research on plants or animals brought back from the rain forest. Others use information collected by field and laboratory scientists to develop computer models of interactions between organisms and their resources in the complex environment of the rain forest.

The ecology of the rain forest is just one of many broad topics that biologists study. Whatever they study, all biologists, in fact all scientists, use certain processes to obtain knowledge about nature. Understanding scientific processes will help you further appreciate how scientists work.

Scientific Processes

All areas of science involve posing questions about nature and discovering facts that can help answer those questions. The work of Dr. Baptista centers on the question, "What species of birds live in a particular area of the South American rain forest?" His field study is an attempt to discover facts that can help answer that question. As you read about the scientific processes, note how each process contributes to the formulation of questions or the discovery of facts that can help answer those questions.

Section Objectives

- *Define and give examples of observing, measuring, organizing and analyzing data, classifying, inferring, and modeling.*

- *Explain the relationship between hypothesizing, predicting, and experimenting.*

- *Tell why communicating is an important process in science.*

- *List some similarities and some differences in the ways that scientists work.*

Figure 2-1 Studying the tropical rain forest requires the use of many scientific processes.

Figure 2-2 The green-headed tanager, *Tangara seledon* (left), can be found in a habitat such as the tropical rain forest (right).

Figure 2-3 Biologists use graphs to organize quantitative information in a visual format.

Observing and Collecting Data

All scientific understanding of the natural world ultimately is based on observations. **Observing** is the use of one or more of the five senses to perceive objects or events. For example, in the rain forest Dr. Baptista uses his senses of sight and hearing to perceive the location of birds in different habitats. Scientists use various tools to aid in making observations. Dr. Baptista uses binoculars to help him see birds at a distance, for example. Scientists use other tools such as microscopes to help them perceive objects or events that would otherwise go unobserved.

While making observations, scientists often collect data. **Collecting data** is the gathering and recording of specific information based on observations. For example, Dr. Baptista might tape a bird call and then record the time, location, habitat, and weather conditions where he made the specific observation.

Measuring

Observations are most useful when they involve quantitative data—data that can be measured in numbers. **Measuring** is the process of determining the dimensions of an object, the number of objects in a group, the duration of an event, or other characteristics in precise units. For example, when Dr. Baptista counts the number of birds of different species collected in a particular habitat, he is making measurements that may be useful in understanding the behavior of the species. Most scientific measurements throughout the world are made using the metric system. Refer to Measurement on p. 840 to review this system.

Organizing Data

Data are of little use unless they are organized. **Organizing data** involves placing observations and measurements in some kind of logical order, such as in a graph, chart, table, or map. Dr. Baptista might organize his quantitative data as shown in Figure 2-3. This bar graph shows the number of birds of various species collected at a particular location on a specific date.

Tip: Discuss the value of quantifying data. Quantitative data can be compared by scientists repeating other's experiments.

Classifying

Classifying is the process of grouping objects, organisms, or phenomena into an established organizational scheme, or developing new organizational schemes. In biology classifying is most often used to organize living things into groups that share morphological traits. Dr. Baptista might collect many birds in the rain forest. Some he will be able to classify immediately because their anatomy, behavior, and habitats are similar to those of known species. Others may not have been studied by scientists before. To classify these birds, biologists will study the anatomy, behavior, and habitats of the unidentified birds to find out how the birds might be related to known species.

Hypothesizing

The scientific processes discussed so far often lead to the formulation of questions. For example, after collecting and organizing his data, Dr. Baptista might notice that one species of bird is usually found in a riverbank habitat early in the day and in a thicket habitat during midday. He might wonder why this would be so. To help answer this question, he might think of possible explanations for this behavior. For example, insects the birds feed on could be more prevalent near the riverbank early in the day and more prevalent in the thicket during midday. Or predators might be less common near the riverbank early in the day and less common in the thicket during midday.

Dr. Baptista might put a possible explanation into the form of a testable statement—or **hypothesis.** His hypothesis might be: The birds avoid predation by moving from the riverbank to the thicket during the middle of the day.

Hypothesizing, the process of forming testable statements about observable phenomena, is often one of the first steps in a scientific investigation. A statement is testable if evidence can be collected that either supports the hypothesis or refutes it. For example, the above hypothesis could be shown to be false if the predators of the birds were observed to be equally common in both habitats in the early morning and in the middle of the day.

Although a hypothesis may be refuted, it can never be proved true beyond all doubt. It can only be supported by evidence. At any time new data might be collected that indicate a previously accepted hypothesis does not hold true in all instances. Good scientists look for ways to refine and revise their hypotheses as they uncover new evidence.

Predicting

To test a hypothesis, a scientist usually makes a prediction that follows from the hypothesis. **Predicting** is stating in advance the result that will be obtained from testing a hypothesis. A prediction often takes the form of an "if–then" statement. The biologist might make the following prediction based on the hypothesis that

Figure 2-4 After observing this great egret, *Egretta alba,* in its water biome, a biologist might formulate questions about the behavior of the bird and then form a hypothesis that answers those questions.

is discussed on the last page: *If* it is true that the birds move from riverbank habitats to thicket habitats to avoid predators, *then* I will observe fewer predators in the riverbank habitat in the morning than during midday, and I will observe fewer predators in the thicket habitat during midday than in the morning. Showing this prediction to be false refutes the hypothesis; showing this prediction to be true supports the hypothesis.

Experimenting

Some hypotheses or predictions can best be tested through careful observations in a natural setting, for example during a field study. Others can best be tested through experiments. **Experimenting** is the process of testing a hypothesis or prediction by carrying out data-gathering procedures under controlled conditions. Such conditions eliminate extraneous influences and allow close observations to be made.

Most experiments in biology are controlled experiments. A **controlled experiment** is based on a comparison of a control group or phase with an experimental group or phase. The control group and the experimental group are designed to be identical except for one factor. This factor is called the **independent variable** or the manipulated variable. During the course of a controlled experiment, a scientist observes or measures one main factor in both the control group and the experimental group. This factor is called the **dependent variable** or the responding variable.

The following example shows how a controlled experiment works. Suppose that a biologist collects cuttings from a species of vine that grows up trunks of small trees in the rain forest. The biologist might hypothesize that the vine depends on the tree only for support. From this hypothesis the biologist might make the following prediction: If the vine depends on the tree only for support, then a vine transplanted onto an artificial tree should grow as rapidly as a vine transplanted onto a real tree.

To test this prediction, the biologist might design the controlled experiment shown in Figure 2-6. The control group consists of vine cuttings transplanted onto small trees similar to those in the rain forest. The experimental group consists of vine cuttings transplanted onto plastic models of the trees in the control group. The biologist conducting the experiment would treat the vines in the control group and those in the experimental group identically, using the same type of soil, keeping them at the same temperature, exposing them to the same amount of light, and watering them with the same amount of water. The type of tree, real or artificial, is the independent variable; the rate of growth of the vines is the dependent variable.

Analyzing Data

After a scientist has collected and organized data from a field study or an experiment, the data must be analyzed. **Analyzing data** is the process of determining whether data are reliable and

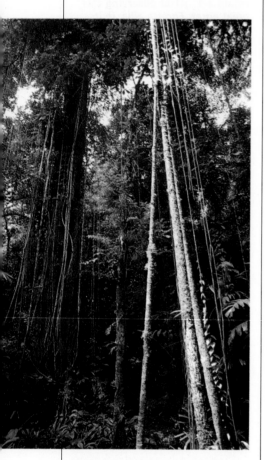

Figure 2-5 The vine growing up this tree in a tropical rain forest is the control in an experiment conducted to see if the vine depends on the tree only for support.

Info: Biologists often restrict the term *hypothesis* to a general testable statement that indicates cause-and-effect relationships.

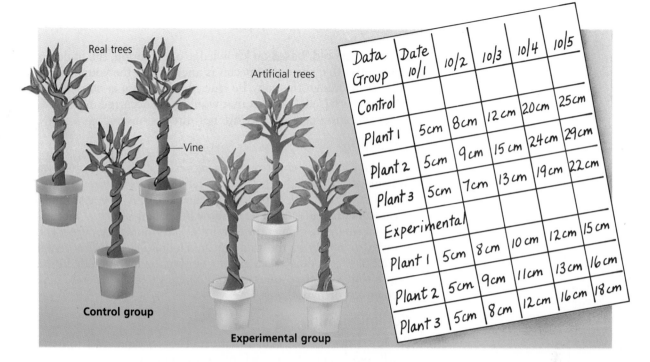

Real trees

Artificial trees

Vine

Control group

Experimental group

Data Group	Date 10/1	10/2	10/3	10/4	10/5
Control					
Plant 1	5cm	8cm	12cm	20cm	25cm
Plant 2	5cm	9cm	15cm	24cm	29cm
Plant 3	5cm	7cm	13cm	19cm	22cm
Experimental					
Plant 1	5cm	8cm	10cm	12cm	15cm
Plant 2	5cm	9cm	11cm	13cm	16cm
Plant 3	5cm	8cm	12cm	16cm	18cm

whether they support or refute a given prediction or hypothesis. Scientists analyze data in many ways, including using statistics, interpreting graphs, determining relationships between variables, comparing the data to those obtained from other studies, and determining possible sources of experimental error.

For example, suppose that the biologist who conducted the experiment on vines obtained the data charted in Figure 2-6. In analyzing the data, the biologist might first average the heights of the plants in each group for every set of measurements. Figure 2-7 shows a graph of these averages. It indicates that the plants in the experimental group grew slower than did the plants in the control group. Further analysis would determine whether the data refute the hypothesis and whether the data are reliable.

Inferring

Inferring is the process of drawing conclusions on the basis of facts or premises instead of direct perception. Facts might include data gathered during a field study or an experiment; premises might include conclusions drawn from previous knowledge or from past experience. Based on the data gathered in the experiment discussed above, the biologist might infer that all vines obtain nutrients from the trees on which they grow. Such an inference may lead this biologist to gather more information. Based on this new information, new predictions or hypotheses about the relationship between vines and trees may be made.

Some inferences, such as the one discussed above, are theoretically testable. However, other inferences are not testable. For example, suppose a biologist notices that two species of fossil horses show many similarities in structure. Further analysis might indicate that one fossil is 30 million years old and the other is 20

Figure 2-6 The control and experimental groups initially differ only in the independent variable.

Figure 2-7 After four days the control and experimental groups differ in the dependent variable.

million years old. Based on knowledge of evolution, the biologist might infer that the older species is ancestral to the younger species. Even though this may be true, the inference is not testable because it is based on events that would have occurred millions of years ago and that therefore are not directly observable.

Modeling

Modeling involves constructing a representation of an object, a system, or a process that helps show relationships between data. A model may be visual, verbal, or mathematical. A biologist might create a mathematical model to show how environmental factors, such as sunlight, temperature, rainfall, and humidity, affect the growth of plants in the rain forest. Figure 2-8 shows a visual model of the feeding relationships between many animals in the rain forest. Scientists sometimes use models to help generate new predictions or hypotheses.

Communicating

Scientists do not work in isolation. They critique each other's work and benefit from each other's insight. In addition, many scientists work with scientists from other institutions or other countries. For these reasons **communicating**—sharing information—is essential to their work. When Dr. Baptista, whose work we examined earlier, shares information on the birds of the rain forest by making his notes and collections available to other scientists, he is communicating. He also might communicate by publishing his observations in a scientific journal or by giving presentations at scientific meetings or conferences. Then other scientists can use his data to come up with new hypotheses about the rain forest.

Figure 2-8   This model represents how various plants are food for animals. These animals, in turn, are food for other animals.

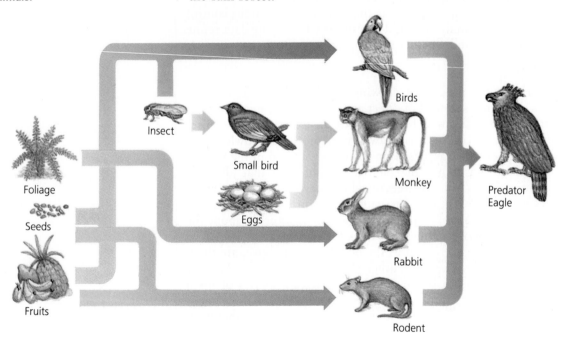

Scientific Methods

Science does not prescribe one single method for asking and seeking answers to questions. Scientists combine some or all the processes you have read about in a logical way to devise a **scientific method** best suited for a particular research project.

For example, in his field study of birds of the rain forest, Dr. Baptista might use a scientific method that combines observing, collecting data, measuring, organizing and analyzing data, and communicating. The biologist doing the experiment on vines might use a scientific method that combines all the science processes. A biologist making a mathematical model of environmental factors affecting plant growth in the rain forest might use a scientific method that combines organizing and analyzing data, predicting, inferring, modeling, and communicating.

Some scientists use less complex methods. Ethologists, scientists who study behavior, study animals in their natural habitats. By observing over long periods, ethologists can uncover individual and societal patterns of behavior. The processes of observing, organizing data, and communicating are integrated into one kind of scientific method.

Though scientists use different scientific methods, they share certain ways of reasoning that are essential to science. All answers to scientific questions are tentative. Hypotheses, posed as possible answers to scientific questions, can be refuted, but they cannot be supported beyond exception.

If a hypothesis continues to be supported by evidence gathered by many different scientists, it may come to be called a theory. A **theory** is the most probable explanation for a set of data, based on the best available evidence. A good theory can be used not only to explain existing data but also to make predictions about new data that may be found. However, if new data or a reanalysis of existing data indicates that a theory is unacceptable, the theory may have to be modified or discarded. The tentative nature of scientific theories leads scientists to constantly weigh evidence, examine inferences, and modify hypotheses.

Section Review

1. How does a *hypothesis* differ from a *theory*?
2. Define each of the following processes: observing, measuring, organizing and analyzing data, classifying, inferring, and modeling.
3. What is the relationship between hypothesizing, predicting, and experimenting in science?
4. Why is communicating an important process in science?
5. It is sometimes said that science is based on *probabilistic* rather than *absolute* truth. Based on what you have read about scientific processes and methods, what do you think this statement means?

Think

Suppose a biologist does a month-long field study on eagles nesting in the mountains. What scientific processes might the biologist use while working in the field? What processes might the biologist use on returning to the laboratory?

Answer to Think
Scientific processes used in the field might include observing, collecting data, measuring, classifying, and organizing data; processes used on returning might include analyzing data, modeling, inferring, and communicating.

Answers to Section Review
1. A hypothesis is a testable statement; a theory is a statement backed by a large amount of evidence gathered by many scientists over a long period of time.
2. Accept student wording of the definitions as given in the glossary. (See pp. T2A–T2D.)

Writing About Biology

The Tucson Zoo

This excerpt is from an essay in Lewis Thomas' book, The Medusa and the Snail.

Much of today's public anxiety about science is the apprehension that we may forever be overlooking the whole by an endless, obsessive preoccupation with the parts. I had a brief, personal experience of this misgiving one afternoon in Tucson, where I had time on my hands and visited the zoo, just outside the city. The designers there have cut a deep pathway between two small artificial ponds, walled by clear glass, so when you stand in the center of the path you can look into the depths of each pool, and at the same time you can regard the surface. In one pool, on the right side of the path, is a family of otters; on the other side, a family of beavers. Within just a few feet from your face, on either side, beavers and otters are at play, underwater and on the surface, swimming toward your face and then away, more filled with life than any creatures I have ever seen before. . . .

I was transfixed. As I now recall it, there was only one sensation in my head: pure elation mixed with amazement at such perfection. Swept off my feet, I floated from one side to the other, swiveling my brain, staring astounded at the beavers, then at the otters. . . . I remember thinking, with what was left in charge of my consciousness, that I wanted . . . never to know how they performed their marvels; I wished for no news about the physiology of their breathing, the coordination of their muscles, their vision, their endocrine systems, their digestive tracts. . . . All I asked for was the full hairy complexity, then in front of my eyes, of whole, intact beavers and otters in motion.

It lasted, . . . for only a few minutes, and then I was back . . . wondering about the details by force of habit, but not, this time, the details of otters and beavers. Instead, me. Something worth remembering had happened in my mind.

. . . I became a behavioral scientist, an experimental psychologist, an ethologist, and in the instant I lost all the wonder and the sense of being overwhelmed. But I came away from the zoo with something, a piece of news about myself: I am coded, somehow, for otters and beavers. I exhibit instinctive behavior in their presence, ... Beavers and otters possess a "releaser" for me, . . . What was released? Behavior. What behavior? Standing, swiveling flabbergasted, feeling exultation and a rush of friendship. I could not, as the result of the transaction, tell you anything more about beavers and otters than you already know. I learned nothing new about them. Only about me, and I suspect also about you, maybe about human beings at large. . . . Left to ourselves, . . . we hanker for friends.

- **Think of a special animal you have observed and enjoyed—at home, at the zoo, on television. Write a journal entry describing your feelings about the animal.**

From The Medusa and the Snail *by Lewis Thomas. Copyright © 1974, 1975, 1976, 1977, 1978, 1979 by Lewis Thomas. All rights reserved. Reprinted by permission of Viking Penguin Inc.*

2.2 Biological Tools and Techniques

To study the living world, biologists need to observe cells and cell parts, analyze chemical reactions, and determine relationships between organisms. Biological tools and techniques aid them in their study. As you read this section, notice the uses of tools and techniques in various scientific processes.

Microscopes

Among the most widely used tools in biology are microscopes. A **microscope** is an instrument that can form an enlarged image of an object. Biologists use microscopes to study organisms or objects too small to be seen with the naked eye. The **magnification** of a microscope refers to its power to increase an object's apparent size. The **resolution** of a microscope refers to its power to show detail clearly. Several kinds of microscopes are discussed below. They vary in magnification, resolution, and usage.

Light Microscope

In a light microscope, a beam of light passing through one or more lenses produces an enlarged image of the object or specimen being viewed. Use Figure 2-9 to follow the path of light through a typical light microscope. Notice that the specimen is illuminated by an electric light bulb. The light passes through two lenses. The lens closest to the specimen is called the **objective lens.** Most microscopes have a series of interchangeable objective lenses that can be rotated into place. The objective lens directs light to the **ocular** *(AHK-yuh-lur)* **lens** in the eyepiece. The magnification of a light microscope is determined by the power of its lenses. The objective lens of a typical light microscope can magnify an object so that it appears to be 40 times (×40) its actual size. The ocular lens can magnify an object ten times (×10). When an object is viewed through both lenses, the magnification is 400 times (×400). The most powerful light microscopes can magnify an object about 2,000 times (×2,000). Scientists use light microscopes to observe living organisms and preserved cells.

Section Objectives

- *Compare light microscopes with transmission and scanning electron microscopes in terms of magnification and resolution.*

- *Define and discuss three techniques scientists use in the laboratory.*

- *Explain why scientific sampling is an important technique in field studies.*

- *List ways scientists might use computers.*

- *Describe how a scanning tunneling microscope works and list its advantages over light and electron microscopes.*

Figure 2-9   The lenses in a light microscope (right) can magnify objects up to 2,000 times. The *Paramecium aurelia* (left) was viewed at a magnification of 100.

Ocular lens

Coarse focus

Fine focus

Objective lens

Stage

Specimen

Light bulb

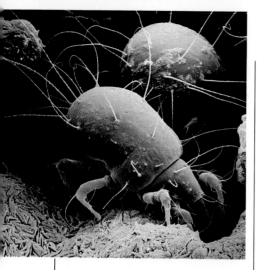

Info: Electron beams do not produce visible images. However, electrons projected onto a plate produce a visible image much as light reacts with camera film.

Electron Microscopes

In an electron microscope, a beam of electrons focused by magnets produces an enlarged image of an object. Electron microscopes are more powerful than light microscopes. However, an electron microscope cannot be used to observe living organisms, which cannot survive the necessary preparation techniques.

A **transmission electron microscope**, called a TEM, transmits a beam of electrons through a very thinly sliced specimen. Magnetic lenses enlarge the image and focus it on a fluorescent screen or photographic plate, producing a visible image. Transmission electron microscopes can magnify objects up to 2,000,000 times.

A **scanning electron microscope**, called an SEM, scans the surface of an object with a beam of electrons. In preparation for scanning the surface of the object is sprayed with a fine metallic mist. The electron beams bounce off the surface of the object and are projected onto a fluorescent screen or photographic plate. This produces a three-dimensional image of the object. Scanning electron microscopes can magnify objects up to 50,000 times. However, the importance of the SEM is its ability to produce a clear, striking image of the surfaces of an object.

Figure 2-10 The term *SEM* refers both to the scanning electron microscope and to the image it produces, which is a scanning electron micrograph. This SEM shows a soft-bodied mite, *Acarus siro* (×37).

Biotechnology

Scanning Tunneling Microscope

The important role of microscopy in science was recognized in 1986 when the Nobel prize in physics went to three scientists for their microscope designs. One winner, German scientist Ernst Ruska, invented the first electron microscope in 1931.

The other two winners were Gerd Binnig of West Germany and Heinrich Rohrer of Switzerland. Binnig and Rohrer designed a scanning tunneling microscope in 1981. The microscope they invented maps the surfaces of objects atom by atom.

The scanning tunneling microscope has revealed the surface structure of gold, silicon, nickel, and graphite on an atomic scale. Researchers have used this type of microscope to scan the surfaces of DNA and of certain viruses.

Unlike other microscopes, the scanning tunneling microscope doesn't rely on light or electrons from an external source to produce an image. Because of this it requires no lenses.

Instead, the scanning tunneling microscope makes use of the electrons that leak—or tunnel—from the surface of all matter. These electrons form

Laboratory Techniques

The study of living organisms hinges on gaining a deeper knowledge of cells and how they work. Laboratory techniques enable biologists to experiment with cells and analyze substances cells produce. Scientists can then model the workings of cells and the organisms they compose. Several laboratory techniques are discussed below.

Cell Cultures

The ability to grow cells in the laboratory has allowed biologists to study cell growth, interaction between cells, and cells' response to substances in the environment. A population of identical cells grown in the laboratory is referred to as a **cell culture**. A biologist can isolate one type of cell, such as a heart muscle cell, and grow a population of these cells in a cell culture. The biologist then can use the cell culture in controlled experiments in order to study the effects of certain nutrients on these cells. The use of a cell culture allows biologists to determine how specific types of cells react without having the data obscured by the presence of other types of cells.

Computer screen

Scanning needle

Electronic cloud

Specimen

clouds just above any surface, be it soft or hard, metal or nonmetal, living or nonliving. The density of the electron clouds decreases with distance from the surface of the matter.

In a scanning tunneling microscope, an electric needle with a point about the size of an atom scans the surface of a specimen, detecting electrons tunneling from the specimen's surface. As the scanning needle moves across the specimen, it follows the contours of the surface by maintaining a constant distance from the specimen. This precision is possible because the needle is able to detect and move in response to slight differences in the density of the electron cloud.

The motion of the needle is analyzed by a computer, which projects a three-dimensional map of the surface of the specimen on a fluorescent screen. In this way the scanning tunneling microscope can achieve magnifications of 100 million times the size of the specimen.

The scanning tunneling microscope can indicate the atomic composition of a specimen as well as the arrangement of atoms on the surface of the specimen. Unlike the scanning electron microscope, which requires specimens to be killed and coated with metal for viewing, the scanning tunneling microscope can be used with living specimens. For this reason the scanning tunneling microscope promises to become an important tool in biology.

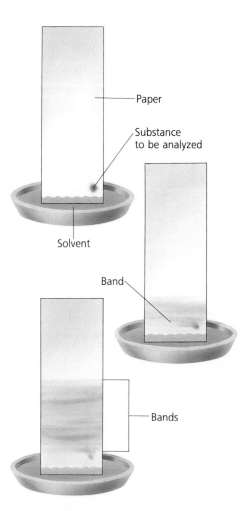

Paper

Substance
to be analyzed

Solvent

Band

Bands

Figure 2-11 In paper chromatography a paper with a substance to be analyzed is hung with its edge in a liquid. As the liquid moves up the paper, components of the substance separate into bands based on their molecular weight and stability.

Tip: Discuss whether or not the biology class is a scientific sample of the school.

Answers to Section Review
1. Fractionation releases components of cells by grinding them in a blender; centrifugation splits cells by spinning fractionated cells at high speed.
(See pp. T2A–T2D.)

Fractionation, Centrifugation, and Chromatography

Cell **fractionation** is a technique used to release the components of a cell. Cells can be fractionated in a blender. The grinding action breaks open cells, releasing the cell parts.

Centrifugation is a technique used to separate cell parts by type by spinning fractionated cells at high speeds. The spinning action causes heavier components to settle to the bottom of a tube while the lighter components remain near the top.

Chromatography is used to separate chemical components of a substance based on the varying rates of movement up a piece of paper or down a column of beads. Figure 2-11 shows how paper chromatography works. A piece of paper with a dot of the substance to be analyzed is hung in a liquid. As the liquid moves up the paper, components of the substance are deposited as separate bands on the paper. The bands can then be identified.

Field Techniques

Working in a natural environment such as a rain forest, a biologist has little control over variables. However, certain field techniques aid the biologist in collecting accurate data.

Scientific sampling is a technique in which a small sample is used to represent an entire population. A biologist studying one region of the rain forest might identify and count the plants in one acre. From this sample the biologist could estimate the total number of plants of different species in the region. However, the biologist's estimate will be accurate only if the acre is representative of the entire region being studied.

Field biologists often collect organisms. Such collections are most useful when the biologist keeps careful records of location and conditions. Scientists can use such information to classify the organisms and learn about their behavior.

Computers

Many biologists collect data that can best be organized and analyzed through the use of computers. A biologist studying the rain forest might collect data on factors affecting plant growth. Using a computer, the biologist can construct a model showing the effect of environmental factors on plant growth.

Section Review

1. Compare cell *fractionation* and *centrifugation*.
2. How do microscopes differ in magnification and use?
3. Why is scientific sampling necessary in field research?
4. How might computers be used in biology?
5. How might a cell culture allow for greater control of variables in an experiment on cell growth?

Laboratory

Scientific Methods

Objective
To understand how scientists approach and solve problems

Process Skills
Hypothesizing, experimenting, collecting and analyzing data

Materials
Watch with second hand, paper

Method
1 One of the most important characteristics of modern science is its quantitative approach to solving problems. One of the first scientists to use quantitative methods was William Harvey, who discovered that blood circulated through the body. At the time Harvey began his work, anatomists believed that the liver produced blood from the food that the body consumed. The blood was then carried by veins to the heart, purified in the lungs, and then pumped to the various organs of the body, where it was consumed. Harvey measured that the left ventricle of the heart held roughly 100 mL of blood. He also measured that the heart beats an average of 64 times per minute. Assuming that 1 mL of blood weighs 1 g, how much blood would the body need to produce per hour to replace the blood consumed by the organs? Roughly how much food, do you think, would the body need to consume to produce this much blood? Harvey hypothesized that the same blood must circulate continuously throughout the body. Keeping Harvey's measurements in mind, can you determine how he formed this hypothesis?

2 While sitting quietly at your desk, find the pulse in your wrist and count the beats for one minute. The number of beats per minute is your pulse rate. Record your result.

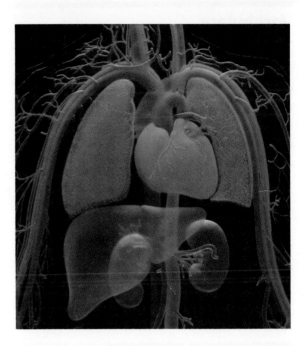

3 Repeat step 2 twice. Calculate your average pulse rate and record your result.

4 How do you think standing or holding your breath will affect your pulse rate? Choose one of these activities and formulate a hypothesis about its effect on pulse rate. What is the independent variable? What is the dependent variable?

5 Again record your pulse rate for one minute, this time under the influence of the independent variable. Record your result.

6 Repeat step 5 twice. Calculate your average pulse rate, and record your result.

7 Compare your data from step 3 with your data from step 6.

Conclusions
1 How do your results in step 7 compare with the hypothesis you made in step 4?
2 What measurement did you use as a control in this investigation?

Inquiry
What are some possible sources of error in this experiment?

Chapter 2 Review

Vocabulary

analyzing data (20)
cell culture (27)
centrifugation (28)
chromatography (28)
classifying (19)
collecting data (18)
communicating (22)
controlled experiment (20)

dependent variable (20)
experimenting (20)
fractionation (28)
hypothesis (19)
hypothesizing (19)
independent variable (20)
inferring (21)
magnification (25)

measuring (18)
microscope (25)
modeling (22)
objective lens (25)
observing (18)
ocular lens (25)
organizing data (18)
predicting (19)
resolution (25)

scanning electron microscope (26)
scientific method (23)
scientific sampling (28)
theory (23)
transmission electron microscope (26)

Explain the difference between each set of related terms.
1. observing, measuring
2. collecting data, organizing data, analyzing data
3. hypothesizing, experimenting, inferring
4. classifying, modeling
5. transmission electron microscope, scanning electron microscope

Review

1. Scientific knowledge is ultimately based on (a) observations (b) hypotheses (c) experiments (d) models.
2. Data that is quantitative can be (a) described in words (b) measured in numbers (c) recorded on a tape recorder (d) seen through a microscope.
3. A hypothesis is a testable statement that (a) can help a biologist classify organisms (b) can be supported in all circumstances (c) can be demonstrated false through experimentation (d) is the first step in every research project.
4. "If–then" statements are used in (a) collecting data (b) organizing data (c) predicting (d) inferring.
5. Scientists analyze data to (a) gather and record specific information based on observations (b) place observations and measurements in a logical order (c) group objects, organisms, or phenomena into an established organizational scheme (d) determine whether data are reliable and whether they support or refute a given hypothesis.
6. A theory is defined as the most probable explanation for a set of data based on (a) a testable hypothesis (b) dependent and independent variables (c) a visual, verbal, or mathematical model (d) the best available evidence.
7. The resolution of a microscope refers to (a) its power to increase an object's apparent size (b) its power to show detail clearly (c) its series of interchangeable objective lenses (d) its power to scan the surface of an object.
8. Cell cultures allow a biologist to (a) observe cells in a controlled environment (b) release cell parts (c) isolate cell parts (d) separate dissolved substances.
9. In scientific sampling researchers use a sample to represent (a) a control group (b) an experimental group (c) an entire population (d) an entire field study.
10. The scanning tunneling microscope magnifies objects up to (a) 1,000 times (b) 100,000 times (c) 500,000 times (d) 100,000,000 times.
11. Describe a controlled experiment.

12. Give an example of a testable inference and an untestable inference.
13. What is the purpose of a scientific model?
14. How do scientists benefit by communicating with each other?
15. What are some similarities and differences in the ways that scientists work?
16. How does a light microscope work?

17. What are some advantages and disadvantages of electron microscopes in biology?
18. Describe three techniques a biologist might use in the laboratory.
19. How might a biologist use a computer?
20. For a biologist, what advantages does a scanning tunneling microscope have over a scanning electron microscope?

Critical Thinking

1. Scientists often try to repeat, or replicate, the results of other scientists' experiments. Why do you think this is an important part of the scientific process?
2. One of the most important parts of any scientific publication is the part called "Methods and Materials," in which the scientist describes the procedure used in the experiment. Why do you think such details are so important?
3. Why do scientists prefer to use the metric system of measurement?
4. Cell culture has the advantage of being controllable. What might be one disadvantage of using a cell culture in a scientific experiment?

5. Look at the photographs above. The TEM (left, ×33,000) is of *Staphylococcus aureus*. The SEM (right, ×2,500) is also of *S. aureus*. Compare and contrast what each electron micrograph tells you about this organism.

Extension

1. Read the article by S. Weisburd entitled "Extinction Wars," in *Science News*, Vol. 129, No. 5, February 1, 1986, pp. 75–77. Make an oral report to the class in which you include answers to the following questions: (1) Where did R. G. Prinn get his "data" for the acid rain hypothesis? (2) What is the scientific evidence that mass extinctions at the Cretaceous–Tertiary (K–T) boundary were neither purely catastrophic nor entirely gradual? (3) What scientific evidence is there that the oceans experienced large and rapid drops in temperature millions of years ago? (4) Why will the cause of mass extinction millions of years ago continue to remain in doubt?

2. Go to your local supermarket or grocery store and observe the fresh, frozen, and canned vegetable sections and the people in them. Note ten customers in each area and estimate their age. Make a bar graph of your data. Record the ages of individuals on the *x*-axis of the graph. Group the ages as follows: (1) under 20, (2) 20–40, (3) 40–60, and (4) over 60. Record the number of individuals in each age group in the *y*-axis. Remember that you will need a separate bar for each of the three vegetable sections. Can you draw any conclusions as to whether fresh vegetables are bought by a certain age group? What are some of the possible sources of error in this experiment?

Chapter 3 | Chemistry

This chapter presents an overview of the basic principles of chemistry. Students learn how the laws of chemistry determine the basic characteristics shared by all organisms and govern all biological phenomena.

Correlation Chart

Section Number	Section Objectives	Pages	Features	Section Review
3.1	1. Draw a model of the structure of the atom.	33–37	Biology in Process, pp. 34–35	2
	2. Define *element, isotope,* and *radioisotope*.		Think, p. 36	1, 5
	3. Define *compound* and list two properties of compounds.			3
	4. Contrast two types of bonds.			4
	5. Describe the experimental data on which Rutherford formulated his model of the atom.			
3.2	1. Compare and contrast mixtures and compounds.	38–40		1
	2. List some characteristics that distinguish solutions from other mixtures.			2
	3. Define *suspension* and give an example of one.			3
	4. Compare the sol state and the gel state of a colloid.			5
	5. Contrast properties of acids and bases.			4
3.3	1. Contrast potential energy with kinetic energy.	41–43	Biology Lab, p. 43	2
	2. Describe the relationship between the states of matter and the kinetic energy of molecules, atoms, or ions that compose matter.			3
	3. Describe how chemical reactions are illustrated by chemical equations.			1
	4. Explain how every chemical reaction involves either a net absorption or a net release of energy.			4, 5

Resources*

Print

McQuarrie, Donald, and Peter Rock. *Descriptive Chemistry*. San Francisco: W. H. Freeman, 1984.

Porterfield, William W. *Inorganic Chemistry: A Unified Approach*. Reading, Mass.: Addison-Wesley, 1984.

Audiovisual/Software

An Introduction to Chemistry. Filmstrips and cassettes. National Geographic, 1984.

Elements, Compounds and Mixtures, 2nd ed. Film or video, 20 min. Coronet, 1985.

*Additional references are listed in *Modern Biology, Teacher's Program Guide*.

Section Teaching Guides

3.1 Composition of Matter

Approach

The main objective of this section is to describe the basic properties of atoms and of the compounds they form. The section will prepare students for understanding biochemistry (Chapter 4), as well as chemical activity within organisms. In presenting the text, guide students to appreciate how the knowledge of the fundamental properties of matter can help them to understand many biological phenomena.

Concept Framework

Students will be familiar with the atom as the basic unit of matter. Show students a model of the atom. Ask them to identify protons, neutrons, and electrons. Tap this knowledge to introduce students to the basic properties of matter. Point out that electrons play a pivotal role in chemical bonding. Tell students that as they read, they should note the types of bonds that form between atoms as a result of the particular electron arrangements of the atoms.

Emphasis

Students should be able to differentiate between the following pairs of chemical terms: *atoms* and *molecules; elements* and *compounds; isotopes* and *radioisotopes; ionic bonds* and *covalent bonds.*

Using the Features/Developing a Model of the Atom

The process skills discussed in this feature are experimenting, inferring, and modeling. The feature describes Rutherford's experiments that led to his formulation of the atomic model. Encourage students to study the Rutherford model shown on p. 35 as they read about the inferences Rutherford made in formulating this model.

Skits on Chemical Bonding

Divide students into two groups. One group should do a skit on ionic bonding; the other, on covalent bonding. Each student should portray a proton, a neutron, or an electron. Have students arrange themselves in the configuration of two or more atoms. Students might identify the atomic particles they represent by wearing construction paper signs. Ask each group to choose a director who will guide the "atoms" to form ionic bonds and covalent bonds.

Section Review Answers

(Continued from p. 37.)
4. An ionic bond forms as a result of the electrical attraction between ions with opposite charges. A covalent bond is formed when pairs of electrons are shared between two or more atoms.
5. Radioisotopes can be injected into the blood, and scientists can observe their movement by detecting the radiation.

3.2 Mixtures

Approach

A description of the properties of different kinds of mixtures is the main objective of this section. The section builds on concepts introduced in Section 3.1. In teaching this section, guide students to understand how mixtures and compounds differ.

Concept Framework

Students will be familiar with different kinds of mixtures. To tap this knowledge, assist the class in making a list of familiar mixtures. Guide them to realize that a solution, such as sugar water, is a mixture. As they read, encourage them to consider why a solution is defined as a mixture.

Emphasis

Make sure students can differentiate between solutions, suspensions, and colloids and that they realize that acidity or alkalinity is a property of solutions.

Making Mixtures

Divide the class into three groups. Have one group make a solution; have the second make a suspension; have the third group make a colloid. Ask each group to demonstrate to the class, how to make the mixture. Ask students to include a description of how their mixture may change over the length of a day, and list some biological examples of the mixture. You might have students observe their mixtures the following day.

Section Review Answers

(Continued from p. 40.)
3. Blood in an undisturbed test tube will separate

into layers, with the blood cells settling to the bottom of the test tube.

4. An acid contains more hydronium ions than hydroxide ions; a base contains more hydroxide ions than hydronium ions.

5. Nutrients and waste products would not be able to move through the thick medium.

3.3 Energy

Approach

The main objective of this section is to define and discuss different forms of energy. The section prepares students for concepts presented in Chapter 4 (Biochemistry) and in Chapter 7 (Photosynthesis and Respiration). In presenting the text, guide students to understand that how energy is stored, obtained, and released is of fundamental importance for organisms.

Concept Framework

Energy is a familiar concept to students. Ask students to find pictures or think of words that connote energy. Use students' knowledge of energy to introduce them to the scientific definition of the term *energy*, explaining that it differs somewhat from the common definition of the term. Direct students to think about the scientific definition as they read the section.

Emphasis

Students should be able to distinguish between potential energy and kinetic energy and to explain the role of energy in chemical reactions.

Chemical Reaction Demonstration

Most changes that accompany chemical reactions cannot be directly observed, but must be inferred. Set up the following demonstration and discuss how the results can be inferred. Place a small piece of moist steel wool in the bottom of a test tube. Attach this test tube and an empty test tube to a test-tube stand and invert the test tubes in a shallow pan of water. Mark the level of the water in each test tube. Observe the test tubes for three days. Students should note that the steel wool rusts and the level in that test tube rises. Point out that the change in the steel wool is the result of a chemical reaction. Guide them to infer that the reaction somehow caused some of the air in the test tube to be used up, causing the water level to rise.

Section Review Answers

(Continued from p. 42.)

3. Atoms or molecules in gases have the greatest kinetic energy; atoms or molecules in liquids have greater kinetic energy than those in solids.

4. In an endothermic reaction more energy is required to sustain the reaction than is released by the reaction. In an exothermic reaction more energy is released than is required to start the reaction.

5. When oil is burned, energy is released quickly. However, the body breaks down oil in many steps of intermediary metabolism. Thus, energy is released slowly, preventing cell damage.

Laboratory Preparation

Background

In this lab students study the effect of acids and bases on organic molecules. Hydrogen bonding determines the precise shape of biological molecules. Since the number of hydrogen ions, or pH, greatly affects hydrogen bonding, pH is an important determinant of biological activity. Acid rain is so dangerous because it affects basic biological processes. Most organisms have buffers in their cells and fluids which limit the effect of changes in the number of hydrogen and hydroxyl ions by reacting with them. This exercise uses pH paper; expect variation in color.

Preparation

Use fresh skim milk. Cut the pH paper into squares the width of the strip and make enough so that students may make more than one measurement if they question their results.

Safety

Ammonia is a potentially dangerous reagent. Caution students to handle it with care and avoid breathing fumes. If spilled on skin, flush affected area with water immediately.

Conclusions

1. No. The milk was acidic when lemon juice was added and basic when ammonia was added.

2. The curdling was a manifestation of the protein precipitating out of solution because the acid

lemon juice changed the shape of the protein molecules. By contrast the ammonia, a base, dissolved some of the protein in the milk.

Inquiry

Students will design an experiment that compares the amount of a particular acid that must be added to milk and to water in order to change the pH. They will then offer hypotheses to explain the results.

Chapter Review

Vocabulary

1. All three subatomic particles differ in electrical charge and location in the atom: an electron is negatively charged and circles the nucleus; a proton is positively charged and is part of the nucleus; a neutron has no charge and is part of the nucleus.
2. An element is a substance that cannot be broken down chemically into simpler kinds of matter. A compound is composed of two or more elements joined by chemical bonds.
3. A solution is a mixture in which substances are uniformly dissolved in another substance. A suspension is a mixture in which particles spread through a liquid or a gas but settle over time. A colloid is a mixture with particles intermediate in size between particles in a solution and a suspension.
4. An acid has a greater concentration of hydronium ions than hydroxide ions, whereas a base has a greater concentration of hydroxide ions than hydronium ions.
5. An endothermic reaction involves a net absorption of energy. An exothermic reaction involves a net release of energy.

Review

11. In a radioisotope the nucleus emits detectable radiation. Scientists use radioisotopes to trace uptake of certain chemicals by mixing radioisotopes with nonradioactive isotopes and observing the movement of radioisotopes through the organism.
12. In compounds elements combine in definite proportions; substances forming mixtures combine in any proportion. In compounds the elements

do not retain their original properties; in mixtures the substances do retain their original properties.
13. An ion is an atom or molecule with an electrical charge.
14. In a covalent bond two atoms share one or more pairs of electrons.
15. Rutherford inferred that the atom was composed mainly of empty space, based on the observation that most alpha particles passed through the gold sheets. He inferred that the atom had a small nucleus made up of a concentration of positively charged particles, based on his observation that a few of the alpha particles bounced back.
16. In a sol the particles spread out, and the colloid is in a liquid state. In contrast, in a gel the particles form a loose network, and the colloid is in a semisolid state.
17. An acid solution contains more hydronium ions than hydroxide ions, while a base solution contains more hydroxide ions than hydronium ions. The control of pH is important for digestion. Glands in the stomach and small intestine secrete substances to appropriately adjust the pH of the food.
18. Potential energy is stored energy; kinetic energy is energy of motion.
19. $2H_2 + O_2 \rightarrow 2H_2O$
   reactants $\rightarrow$ products
20. because most reactions in the cell require more energy than they release

Critical Thinking

1. six
2. fifteen
3. sixteen
4. nineteen
5. Na^+, Cl^-, H_3O^+, OH^-
6. Salt and sugar will dissolve in the water, forming a solution. Sand will form a suspension. After an hour, the sand will settle to the bottom of the container.
7. The flowers contain a pigment that changes from blue to pink as the soil pH changes from acidic to alkaline.
8. the kinetic energy of falling water

Chapter 3 | Chemistry

Introduction

The earth supports a variety of living organisms. However, as different as they appear, all organisms share certain characteristics, which you learned about in Chapter 1. All these characteristics are governed by the laws of chemistry. In this chapter you will learn some of the fundamental principles of chemistry. They will give you a better understanding of living organisms and how they function.

Chapter Outline

Chapter Concept

As you read, notice how a knowledge of chemistry helps you understand and explain biological phenomena.

Raindrops on a tamarack tree, *Larix laricina*. Inset: *Paramecium bursaria* surrounded by phytoplankton (×50).

For additional teaching information see pp. T3A–T3D.

3.1 Composition of Matter

Everything in the universe is made up of matter. This includes all living things, from amebas to whales, from bacteria to giant sequoias. **Matter** is anything that occupies space and has mass. **Mass** is the quantity of matter an object has. The pull of gravity on the mass of an object gives the object the property of weight.

Changes in matter are essential to all life processes. For example, when an organism takes in food, it alters the physical and chemical properties of the food. Such changes allow the organism to make use of the nutrients in the food. By learning how changes in matter occur you will gain an understanding of organisms and how they adapt to the environment.

Atoms

The **atom** is the fundamental unit of matter. The nature and properties of atoms determine the structure and behavior of matter. Atoms are about 10 nm in diameter. Thus they can be observed only with a very powerful electron microscope, and even then the internal structure cannot be observed. However, through experimentation, scientists have developed models that describe the structure and behavior of the atom.

Figure 3-1 shows a simplified model of the atom. The various components that make up an atom are called subatomic particles. The central core, or **nucleus,** consists of two kinds of subatomic particles. One, the **proton,** has a positive electrical charge. The other, the **neutron,** has no electrical charge. Most of the mass of the atom is concentrated in the nucleus. **Electrons** are particles with a negative electrical charge that move about the nucleus at tremendous speeds. An atom has an equal number of protons and electrons. The electrical charges of these particles offset one another, so that the net electrical charge of the atom is zero.

Section Objectives

- *Draw a model of the structure of the atom.*

- *Define* element, isotope, *and* radioisotope.

- *Define* compound, *and list two properties of compounds.*

- *Contrast two types of chemical bonds.*

- *Describe the experimental data on which Rutherford formulated his model of the atom.*

Tip: If students have not seen a three-dimensional model of an atom, they may have difficulty understanding that the energy levels are not planes. Explain that the levels may be compared to balls that fit inside each other. The electrons at a particular level would appear to move over the surface of the ball constituting that level.

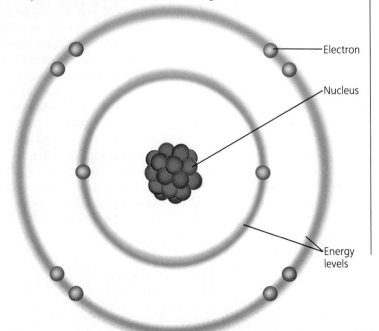

Figure 3-1   The electrons in this model of a neon atom are distributed into two energy levels. The innermost level holds a maximum of two electrons. The second level has a maximum of eight electrons.

Electron

Nucleus

Energy levels

1	Atomic number
H	Symbol
1	Mass number

Hydrogen

6	7	8
C	N	O
12	14	16
Carbon	Nitrogen	Oxygen

10	11	17
Ne	Na	Cl
20	23	35
Neon	Sodium	Chlorine

Figure 3-2 The chemical symbol, atomic number, and mass number of six common elements are shown here. What is the mass number of sodium? How many neutrons does sodium have?

Elements

Elements are substances that cannot be broken down chemically into simpler kinds of matter. Each element has a unique chemical symbol. More than 100 elements have been identified. Ninety-two elements occur in nature. The rest have been synthesized in physics laboratories.

Each atom of any element has the same number of protons. The number of protons in an atom is called the **atomic number** of the element. For example, since each atom of the element carbon has six protons, the atomic number of carbon is 6.

Isotopes

Although in any given element the number of protons in each atom is the same, the number of neutrons can vary. Atoms of the same element that differ in the number of neutrons they contain are called **isotopes**. Most elements have two or more isotopes. The **mass number** is the sum of the protons and neutrons in an atom. For each element shown in Figure 3-2 the mass number for the most common isotope is given just below the chemical symbol.

Biology in Process

Developing a Model of the Atom

By the late 1800s the electron had been identified as a subatomic particle. However, at the time scientists had little idea of where electrons were located in the atom or even of what the structure of the atom was.

The British scientist J. J. Thomson proposed a model of the atom in which electrons were embedded in positive matter of the atom like "raisins in a plum pudding." This model was accepted until scientist Ernest Rutherford produced data that did not fit the "plum pudding" model.

In the early 1900s Rutherford, who was born in New Zealand, conducted several experiments in which he directed a beam of fast-moving positively charged particles called alpha particles toward thin sheets of gold. He used both photographic plates and a Geiger counter to detect the paths of the alpha particles after they hit the gold sheets.

Most of the alpha particles passed directly through the gold as would be expected from the plum pudding model. However, Rutherford was shocked to observe that a few of the particles bounded back. Rutherford used this observa-

Radioisotopes

Some isotopes have unstable nuclei—that is, their nuclei tend to release particles or radiant energy or both. Such isotopes are called radioactive isotopes or **radioisotopes.** When the nucleus of a radioisotope emits radiation, the number of protons or neutrons in the nucleus is reduced.

Radioisotopes are useful in certain types of biological experiments because the radiation can be detected. A botanist, for example, can add a radioisotope of phosphorus to the soil. The botanist can trace the movement of phosphorus through the plant by measuring the radiation present in various parts of the plant at various times.

Compounds

Compounds consist of atoms of two or more elements that are joined by chemical bonds. In a compound the proportions of individual atoms are fixed. For example, in the compound water the atoms of hydrogen (H) and oxygen (O) are always in a proportion

Figure 3-3 By studying the proportion of certain radioactive isotopes in a fossil, scientists can determine its age. A commonly used isotope for dating organisms is carbon-14.

Nucleus

Electron

tion to formulate a new model of the atom, and he presented this model in 1911.

Study the picture of the Rutherford atom as you read about Rutherford's formulation of this model. Rutherford reasoned that because most of the alpha particles passed through the gold sheets undeflected, the atoms, of which the gold sheets were composed, must consist mainly of empty space.

From the scattering of the alpha particles Rutherford inferred that only a concentrated mass of positively charged material could repel the alpha particles, which were also positively charged. Only a large, concentrated mass could deflect them at the large angles he detected in his experiment. From these observations, he inferred the existence of the nucleus—a dense, central, positively charged core.

Rutherford then speculated that electrons were arranged around this positive core in the way that rings are arranged around Saturn. He also realized, however, that compared to the entire atom, the nucleus must be relatively very small.

Since Rutherford's time scientists have modified the model of the atom. Experiments have revealed more about the nature of the nucleus and about the arrangement of electrons around the nucleus. Yet Rutherford's experiments and model were a crucial stage in the process of modeling the atom.

■ How did Rutherford use inferring and modeling to formulate his concept of the atom?

Info: Elements rarely occur in nature in pure form as single atoms. Instead, atoms tend to bond, forming compounds. Atoms in which the outermost energy levels are full are the least likely to form bonds with other atoms.

Answer to Think
This binding helps prevent water from leeching these ions from the soil, thus keeping a supply of the ions available to the plant at all times.

Think

Many minerals needed by plants can be washed away by rain. Calcium, potassium, and sodium compounds are nutrients essential for the growth of plants. These nutrients occur as positively charged ions, which allows them to bind to negatively charged particles in the soil. Why is it important that calcium, potassium, and sodium ions can bind to soil?

of 2 to 1. A chemical formula shows the kind and proportion of atoms that form a particular compound. The chemical formula for water is H_2O. A compound differs in physical and chemical properties from the elements that compose it. For example, in nature the element oxygen is usually found as a gas (O_2) and the element hydrogen is usually found as a gas (H_2). However, when oxygen gas and hydrogen gas combine, they form the compound water (H_2O), a liquid.

How elements combine to form compounds depends on the number and arrangement of the electrons in the atoms. Experiments show that higher-energy electrons are located at a greater distance from the nucleus than lower-energy electrons are. Thus scientists have developed a model of the atom in which electrons occupy discrete energy levels at varying distances from the nucleus. Figure 3-1 illustrates this model.

According to the model each energy level can hold only a certain number of electrons. For example, the level nearest the nucleus can hold only two electrons. The next level can hold up to eight electrons. Most elements are most stable when the outermost energy level contains eight electrons. An important exception relevant to biology is hydrogen, which is most stable when its outer energy level contains two electrons. Aside from helium, neon, and a few other elements, most elements do not have such a stable configuration of electrons in their atoms. Hence most elements undergo reactions that cause their atoms to become stable. These reactions involve forming bonds with other atoms.

Ionic Bonds

Figure 3-4 shows how sodium and chlorine bond to form table salt. Note that the sodium atom has one electron in its outer energy level and that the chlorine atom has seven electrons in its outer energy level. Remember that most atoms are stable when their outer energy level contains eight electrons.

The diagram shows that when sodium and chlorine interact, the outer electron of the sodium atom moves to the chlorine atom. Afterward sodium and chlorine each have eight electrons in their outer energy level. In this process each atom becomes an **ion**—an atom or a polyatomic particle with an electrical charge. Sodium has lost an electron, and thus it becomes a positively charged ion, abbreviated as Na^+. Chlorine has gained an electron. Thus it becomes a negatively charged ion, chloride, abbreviated as Cl^-.

Because positive and negative charges attract each other, the sodium ion and the chloride ion attract each other and form a bond based on this attraction. Such a bond is an **ionic bond.** The resultant compound, in this case NaCl, is an ionic compound.

Ions are important in a number of biological processes. The elements that plants absorb from the soil—such as magnesium, nitrogen, iron, phosphorus, potassium, and calcium—are all in ionic form. Sodium and potassium ions play a major role in the transmission of nerve impulses in the human body.

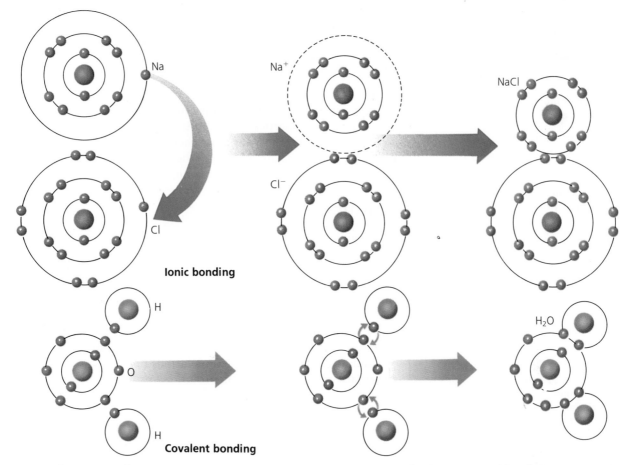

Ionic bonding

Covalent bonding

Covalent Bonds

A **covalent bond** forms when two atoms share one or more pairs of electrons. Water is made up of one oxygen atom and two hydrogen atoms held together by covalent bonds. Since hydrogen has one electron, it needs another to give it the stable arrangement of two electrons in its outer energy level. Since oxygen has six electrons in its outer energy level, it needs two more electrons to give it the stable arrangement of eight electrons. The hydrogen and oxygen atoms gain stability by sharing pairs of electrons.

A **molecule** is a group of atoms held together by covalent bonds. Many substances important in biology are molecules, including carbon dioxide (CO_2) and water. A molecule can be represented either by a chemical formula, such as H_2O, or by a structural formula, such as H—O—H. The structural formula shows the arrangement of atoms in the molecule.

Section Review

1. Define *element, isotope,* and *radioisotope.*
2. How are subatomic particles arranged in the atom?
3. What are two properties of compounds?
4. How does an ionic bond differ from a covalent bond?
5. How would medical researchers use radioactive substances to trace the movement of substances in the body?

Figure 3-4   Table salt forms through the ionic bonding of sodium ions and chlorine ions (top). Water forms through the covalent bonding of hydrogen atoms and oxygen atoms (bottom).

Answers to Section Review
1. Elements are substances that cannot be broken down chemically to form simpler kinds of matter. Isotopes are atoms of the same element that differ in the number of neutrons they contain. Radioisotopes have nuclei that emit radiation.
2. Protons and neutrons form the nucleus at the center of the atom, and electrons orbit the nucleus at various distances, depending on their energy level.
3. The proportion of elements that form the compound is fixed; a compound differs in physical and chemical properties from the elements that comprise it.
(See pp. T3A–T3D.)

Info: Solutions have two distinguishing characteristics: Because the solute is uniformly distributed in the solvent, every part of a given solution has the same chemical and physical properties. Particles of a solute are ions, atoms, or small molecules, which cannot be removed from a solution by mechanical means because they are so small.

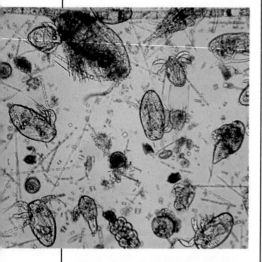

Figure 3-5 Marine microorganisms, such as these zooplanktons and phytoplanktons, live in aqueous solutions.

3.2 Mixtures

In a compound the combined substances do not retain their original chemical properties. In a **mixture,** however, the combined substances do retain their original chemical properties. For example, if sand and salt are mixed, neither changes chemically. How are the different types of mixtures important in biology?

Solutions

A **solution** is a mixture in which one or more substances are uniformly distributed in another substance. The **solute** is the substance being dissolved in the solution; the particles that compose a solute may be ions, atoms, or small molecules. The **solvent** is the substance in which the solute is dissolved. When sugar, a solute, and water, a solvent, are mixed, a solution of sugar water results. Though the sugar dissolves in the water, neither the sugar molecules nor the water molecules are altered chemically.

Solutions can be composed of various proportions of a given solute in a given solvent. Thus solutions can vary in concentration. The **concentration** of a solution is a measurement of the amount of solute dissolved in a fixed amount of solution. The more solute dissolved, the greater the concentration of the solution. For example, sugar can be added to a glass of water, spoonful by spoonful, increasing the concentration of the solution. At some point, however, sugar will no longer dissolve in the solution. The solution is then said to be saturated.

Solutions can be mixtures of liquids, solids, or gases. Brass, for example, is a solid solution, formed by cooling a mixture of liquid zinc and liquid copper. Most biologically important solutions, however, are those in which gases, liquids, or solids are dissolved in water. **Aqueous solutions**—solutions in which water is the solvent—are universally important to living organisms. Marine microorganisms spend their lives immersed in the sea, an aqueous solution. Most nutrients that plants need are in aqueous solution in moist soil. Plasma, the liquid part of the blood, is an aqueous solution containing dissolved nutrients and gases. Body cells exist in an aqueous solution of intercellular fluid.

Suspensions

A **suspension** is a mixture in which particles spread through a liquid or a gas but settle out over time. Waves on a beach can form a suspension of sand and water. The whipping of the wind can form a suspension of dust particles in the air. The particles in a suspension are larger than solute particles in a solution.

One suspension of importance in biology is blood. Red corpuscles and white blood cells are suspended in the plasma. The motion of blood coursing through blood vessels keeps these blood

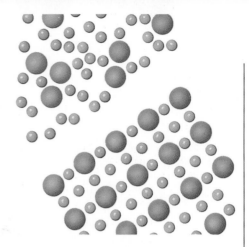

cells distributed throughout the plasma. However, blood in an undisturbed test tube soon becomes separated into layers, with the blood cells settling to the bottom of the test tube.

Colloids

Colloids *(KAHL-OIDZ),* also called colloidal suspensions, are mixtures in which, like solutions, particles do not settle out over time. However, particles in a colloid are intermediate in size between particles in a solution and those in a suspension.

Colloids can exist in a sol state or a gel state. For example, a colloid is formed when warm water and gelatin powder are mixed. The colloid is first a liquid called a **sol.** When the mixture cools, the gelatin molecules become arranged in a way that forms a loose network, and the colloid becomes a **gel,** a semisolid colloid. A colloid can change from a sol to a gel and back to a sol.

Cytoplasm, the material inside of cells, is an example of a colloid important in biology. Cytoplasm in the gel state helps maintain cell shape. For example, in an ameba the cytoplasm away from the cell membrane is in a gel state and helps maintain the shape of the organism. Cytoplasm near the membrane is often in a sol state. The change from sol to gel and back causes locomotion in amebas, as shown in Figure 3-6.

Acids and Bases

The health and functioning of a living thing depend on the chemical makeup both of the organism's internal environment and of the external environment. One of the most important aspects of an environment is the degree of its acidity or alkalinity.

Acidity or alkalinity is a measure of the relative amounts of hydronium ions (H_3O^+) and hydroxide ions (OH^-) dissolved in a solution. In water these ions form naturally: water molecules break up, or dissociate, to form hydrogen ions (H^+) and hydroxide ions (OH^-). The hydrogen ions spontaneously recombine with water molecules to form hydronium ions (H_3O^+). Pure water contains equal numbers of hydronium ions and hydroxide ions; thus water is a neutral solution.

Figure 3-6 An ameba (left) moves as the cytoplasm near the membrane changes from a gel to a sol. The sol changes back to a gel as the molecules become arranged in a loose network (right).

Tip: Have students test several soil samples in class. Point out that farmers frequently test their soil to determine its acidity. They use fertilizers to adjust the chemical levels of soil.

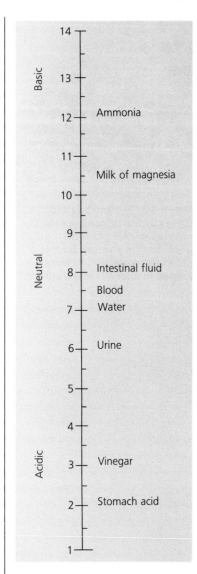

Figure 3-7 The pH scale indicates the relative concentrations of hydroxide ions and hydronium ions in a solution. Some litmus paper turns red in acidic solutions and blue in alkaline solutions.

Answers to Section Review
1. In mixtures two or more substances combine in indefinite proportions and in such a way that the substances continue to retain their original properties.
2. a mixture in which one or more substances are dissolved in another substance
(See pp. T3A–T3D.)

Acids

If hydrogen chloride (HCl) is dissolved in water, some of its molecules dissociate to form hydrogen ions (H^+) and chloride ions (Cl^-). The hydrogen ions combine with water molecules to form hydronium ions. Such an aqueous solution then contains many more hydronium ions than hydroxide ions and is therefore defined as an **acid solution.** Acid solutions are often just called acids. Acids tend to have a sour taste; they produce a tingling or burning sensation if they come into contact with the skin. In concentrated forms they are highly corrosive. Citric acid, found in oranges, lemons, and grapefruit, is an example of a common acid of biological significance.

Bases

If sodium hydroxide (NaOH) is dissolved in water, some of its molecules dissociate to form sodium ions (Na^+) and hydroxide ions (OH^-). Such a solution contains many more hydroxide ions than hydronium ions and is therefore defined as a **base solution.** Base solutions are often called bases. The adjective *alkaline* refers to bases. Bases have a bitter taste; they tend to feel slippery. Ammonium hydroxide (NH_4OH), a waste product given off by many living organisms, is an example of a base found in nature.

pH

Scientists have developed a scale for measuring the relative concentrations of hydronium ions and hydroxide ions in a solution. It is called the **pH scale** and ranges from 1 to 14, as shown in Figure 3-7. A solution with a pH of 1 is very acidic; a solution with a pH of 7 is neutral; a solution with a pH of 14 is very basic. The pH of a solution can be measured electronically or with litmus paper, a treated paper that changes color at various pH levels.

The control of pH is important for living systems. For example, the chemicals that digest nutrients in the stomach can work only in an acidic environment. Glands in the stomach secrete acids that lower the pH of the food as it enters the stomach. These acids have a pH of about 2. The chemicals that digest nutrients in the intestine, on the other hand, can function only in a more basic environment. Glands that empty into the intestine secrete bases to raise the pH of the fluid that enters the intestines. These bases have a pH of about 8.

Section Review

1. How do mixtures differ from compounds?
2. What is a solution?
3. Why is blood considered an example of a suspension?
4. How do the compositions of an acid and a base differ?
5. If all the cytoplasm in a cell changed permanently to a gel, the cell would die. Explain why this is true.

3.3 Energy

One characteristic of living organisms is that they use energy. To understand how organisms function, you need to learn some basic facts about energy.

Forms of Energy

Scientists define **energy** as the ability to do work. Work is defined as the movement of a mass. Energy occurs in various forms, including light energy, heat energy, electrical energy, chemical energy, and mechanical energy. Energy can be converted from one form to another. For example, when plants manufacture glucose during photosynthesis, they transform light energy from the sun into chemical energy.

Energy of any form can be classified as either potential or kinetic. **Potential energy** is stored energy or energy of position. A boulder at rest at the top of a hill or the water at the top of a waterfall has potential energy. If the boulder begins to roll down the hill or the water passes over the waterfall, their potential energy will change to **kinetic energy**—energy of motion.

All the atoms and molecules in any substance are in constant motion; thus they have kinetic energy. The kinetic energy of the atoms or molecules of a substance determines its state: solid, liquid, or gas. Particles of a solid are tightly bound together in a definite shape. Though the particles vibrate, they have relatively little kinetic energy. Particles of a liquid are not as tightly bound as those in a solid. They have greater kinetic energy and move more freely, giving a liquid its ability to flow and to conform to the shape of any container. Particles of a gas have even greater kinetic energy than particles of a liquid or a solid. Gas particles move rapidly and fill the volume of any container. With the addition of sufficient energy matter changes from solid to liquid to gas.

Section Objectives

- *Contrast potential energy with kinetic energy.*

- *Describe the relationship between the states of matter and the kinetic energy of molecules, atoms, or ions that compose matter.*

- *Describe how chemical reactions are illustrated by chemical equations.*

- *Explain how every chemical reaction involves either a net absorption or a net release of energy.*

Tip: Demonstrate how the addition of energy can change the state of matter. Heat energy changes ice to water; additional heat changes the water to steam.

Figure 3-8 The energy in the boulder at the top of the incline is potential energy. The energy in the waterfall is kinetic energy. If the boulder moves, its energy will become kinetic energy.

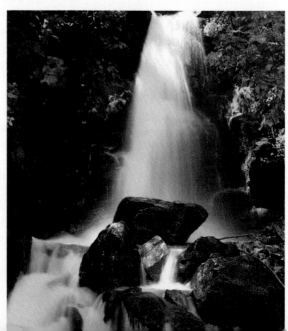

Table 3-1
Chemical Equations

Reactants	Products
$2Na + Cl_2 \longrightarrow$	$2NaCl$
$2H_2 + O_2 \longrightarrow$	$2H_2O$
$6CO_2 + 6H_2O \longrightarrow$	$C_6H_{12}O_6 + 6O_2$

Info: In most chemical reactions energy transformations occur. For example, the human body changes the potential energy of food into the kinetic energy required for muscle movement and tissue growth.

Figure 3-9 A cell in the corn root tip must be supplied continuously with nutrients to get the energy it needs.

Answers to Section Review
1. A chemical reaction is the process of breaking chemical bonds, forming new bonds, or both. Accept any of the chemical equations given in the chapter.
2. Potential energy is stored energy or the energy of position. Kinetic energy is energy of motion.
(See pp. T3A–T3D.)

Chemical Reactions

A **chemical reaction** is the process of breaking chemical bonds or of forming new bonds or both. In a chemical reaction the atoms of the reactants combine to form the products of the reaction. For example, the reactants Na^+ and Cl^- combine in a chemical reaction to form the product NaCl.

Chemical equations show how reactants change during chemical reactions. The reactants are shown on the left side of the equation; the products of the reaction are shown on the right side. The number of each kind of atom must be the same on either side of the arrow. Identify the products and reactants in the chemical equations shown in Table 3-1.

Energy and Chemical Reactions

For most chemical reactions to occur, energy must be added to the reactants. The **activation energy** is the amount of energy required for the reaction to begin. Touching a burning match to a piece of paper supplies the activation energy necessary for the molecules in the paper to react with oxygen—that is, to burn. Once started, every chemical reaction involves either a net release of energy or a net absorption of energy. Chemical reactions that involve a net release of energy are called **exothermic** (*ECK-suh-THUR-mick*) **reactions.** Reactions that involve a net absorption of energy are called **endothermic** (*EN-duh-THUR-mick*) **reactions.** Many reactions that occur in cells are endothermic. Because these reactions use energy, cells need to be continually replenished with nutrients that can be broken down to supply energy.

Exothermic reactions often release energy quickly. A rapid release of energy in living things, however, would damage the organism's cells. Therefore living things have **intermediary metabolism,** a series of chemical reactions by which energy is released slowly in controlled amounts that will not damage cells. When sugar is broken down in a cell, its chemical energy is released in a complex series of chemical reactions. This slow release of energy prevents damage to the cell.

Section Review

1. Define *chemical reaction,* and write a chemical equation for a common chemical reaction.
2. How does potential energy differ from kinetic energy?
3. How does the kinetic energy of the atoms or molecules of a substance determine its state?
4. State the difference between endothermic and exothermic reactions.
5. Burning oil causes an exothermic reaction in which energy is released as a hot flame. The body can also use oil as a source of energy. How do these two processes differ?

Laboratory

pH and Living Systems

Objective
To study the effect of acids and bases on organic molecules

Process Skills
Experimenting, observing

Materials
Small squares of wide-range pH paper, paper towels, medicine dropper, household ammonia, lemon juice, fresh skim milk, 2 30-mL beakers, small squares of narrow-range pH paper, 2 stirring rods

Method
1 Line up four squares of wide-range pH paper 1 cm apart on a paper towel.
2 Use a medicine dropper to put one drop of water on a pH paper square. Determine the pH and then record it.
3 Use a clean dropper to place a small drop of ammonia on another square of the pH paper. Determine the pH of the ammonia and record your result.
4 Repeat step 3 with lemon juice and then repeat it again with milk.
5 Compare your results for the four substances that you have tested. Which substance is the most acidic? Which substance is the most basic? Is one of these substances neither acidic nor basic?
6 Molecules that make up, or are produced by, living organisms usually can function only within a very narrow pH range. To show how a biological molecule is affected by pH, place 100 drops of milk in a beaker. Measure the pH of the milk by transferring a drop of milk from the stirring rod to a square of narrow-range pH paper placed on a paper towel. Record the result. Why should you remeasure the pH of the milk with the narrow-range pH paper?

7 Add a drop of lemon juice to the milk in the beaker. Stir the mixture thoroughly. Measure the pH of the solution with the narrow-range pH paper.
8 Repeat step 7 until you notice an obvious change in the appearance of the milk. What is the pH of the milk at the point of change? The milk changes appearance because the acid in the lemon juice has changed the shape of the proteins in the milk, causing the proteins to precipitate out of solution. What evidence do you see of this?
9 Repeat step 7 until you no longer note a change in the pH of the milk. Record the pH at this point. Compare this figure with your initial result in step 6.
10 Repeat steps 6 and 7 using a clean beaker and fresh milk, and substitute ammonia for the lemon juice. Add ammonia to the milk until the change in pH of the milk is equal to the change in pH you measured in step 9. Does the appearance of the milk change after this point?

Conclusions
1 Was the pH of the milk when you stopped adding lemon juice the same as when you stopped adding ammonia?
2 Why do you think lemon juice curdled the milk but ammonia did not?

Inquiry
Design an experiment investigating whether milk or water is more easily made acidic.

Chapter 3 Review

Vocabulary

acid solution (40)	compound (35)	intermediary metabolism (42)	neutron (33)
activation energy (42)	concentration (38)	ion (36)	nucleus (33)
aqueous solution (38)	covalent bond (37)	ionic bond (36)	pH scale (40)
atom (33)	electron (33)	isotope (34)	potential energy (41)
atomic number (34)	element (34)	kinetic energy (41)	proton (33)
base solution (40)	endothermic reaction (42)	mass (33)	radioisotope (35)
chemical equation (42)	energy (41)	mass number (34)	sol (39)
chemical reaction (42)	exothermic reaction (42)	matter (33)	solute (38)
colloid (39)	gel (39)	mixture (38)	solution (38)
		molecule (37)	solvent (38)
			suspension (38)

Explain the difference between the terms in each of the following sets.
1. electron, neutron, proton
2. element, compound
3. solution, suspension, colloid
4. acid, base
5. endothermic reaction, exothermic reaction

Review

1. The nucleus of the atom is made up of (a) protons and neutrons (b) protons and electrons (c) elements and compounds (d) sols and gels.
2. Isotopes are atoms of the same element that differ in their number of (a) ions (b) molecules (c) formulas (d) neutrons.
3. How elements bond to form compounds depends on the (a) activation energy of the compound (b) dissociation of the ions in the compound (c) number and arrangement of electrons in the component atoms (d) model of the atom.
4. The results of Rutherford's experiment were surprising because after hitting the gold foil, some alpha particles (a) passed through the foil (b) bounced back from the foil (c) became radioactive (d) were stripped from the atoms.
5. When sugar dissolves in water, (a) ionic bonds break (b) covalent bonds break (c) the sugar molecules and water molecules are altered (d) the sugar molecules and water molecules are not altered.
6. Particles in a suspension (a) will settle over time (b) will not settle over time (c) are always acidic (d) are neutral.
7. Particles in a colloid are larger than those in a (a) solution (b) suspension (c) sol (d) gel.
8. Atoms in a solid have (a) more kinetic energy than those in a gas (b) less kinetic energy than those in a gas (c) more kinetic energy than those in a liquid (d) less potential energy than those in a gas.
9. The amount of energy required for a chemical reaction to get started is called (a) potential energy (b) kinetic energy (c) electrical energy (d) activation energy.
10. Intermediary metabolism (a) releases more energy than it uses (b) releases less energy than it uses (c) releases energy slowly in controlled amounts that will not damage cells (d) releases smoke, heat, and light as products of the reactions.
11. How do radioisotopes differ from nonradioactive isotopes, and how are radioisotopes useful in certain experiments?
12. Contrast the properties of compounds with those of mixtures.

13. What is an ion?
14. How are electrons distributed in a covalent bond?
15. What inferences did Rutherford use to formulate his model of the atom?
16. How does the sol phase of a colloid differ from its gel phase?
17. Define the terms *acid solution* and *base solution,* and give an example of how the control of pH is important in biology.

18. What is the difference between potential energy and kinetic energy?
19. Write the chemical equation for the formation of water from hydrogen and oxygen, and identify both the reactants and the products.
20. Most reactions in the cell are endothermic. How does this explain why cells need a continual supply of nutrients?

Critical Thinking

1. A carbon atom has six electrons. How many protons does it have?
2. Phosphorus-32 is a radioisotope of phosphorus commonly used by scientists studying DNA. Ordinary phosphorus has 15 protons and a mass number of 31. How many protons does phosphorus-32 have?
3. Common sulfur has 16 electrons and a mass number of 32. How many neutrons does it have?
4. Sulfur-35 is a radioisotope of sulfur used by scientists studying gene expression. How many neutrons does it have? (See information in question 3.)
5. If a scientist adds NaCl and HCl to pure water, what ions will be in the resulting solution?
6. A scientist mixes salt, sugar, and sand in water. Will the result be a suspension, a solution, or both? What will the result be after the water sits for an hour?

7. The plant in the photograph has blue flowers, as shown, when the soil is acidic but pinkish flowers when the soil is alkaline. Explain.
8. A dam that is located on a fast-flowing mountain stream generates electricity. What is the source of energy for the electricity?

Extension

1. Read the article by J. Raloff entitled "Hybrid Grass Roots Out Soil Salinity" in *Science News,* June 15, 1985, p. 374. Make an oral report in which you describe the effect of salt ions on mechanisms by which the hybrid of sorghum and sudangrass removes salt from the soil.
2. Dissolve table salt or sugar in water at room temperature until it is saturated.

The solution will be saturated when excess salt or sugar stays at the bottom of the container. Remove the clear solution at the top with a spoon. Place the remaining saturated solution in the refrigerator and leave it for two hours. Observe the results. What do your observations tell you about the effect of temperature on molecules in solution inside a living cell?

Chapter 4 | Biochemistry

This chapter introduces the basic biochemical compounds necessary to life on earth. Students concentrate on the structure and function of the four major types of organic compounds: carbohydrates, proteins, lipids, and nucleic acids.

Correlation Chart

Section Number	Section Objectives	Pages	Features	Section Review
4.1	1. Define organic, inorganic, and polar compounds.	47–48		1
	2. Summarize the way in which polarity in the water molecule gives water its unique properties and makes it essential for life.			2, 5
	3. Describe why carbon readily combines with other elements, and discuss its significance in living things.			3
	4. Compare condensation reactions and hydrolysis.			4
4.2	1. Define *monosaccharide, disaccharide,* and *polysaccharide,* and discuss their significance to organisms.	49–55	Think, p. 50	1
	2. Relate the structure of lipids to their functions.			3
	3. Compare the structures of the various types of proteins.		Biology in Process, pp. 52–53	2, 5
	4. List two essential functions of nucleic acids.			2
	5. Outline Sanger's experiment for deciphering the structure of insulin.		Biology Lab, p. 55	

Resources*

Print

Herriott, Jon. *Papers in Biochemistry*. Reading, Mass.: Addison-Wesley, 1984.

Wilabraham, Antony C., and Michael S. Matta. *Introduction to Organic Chemistry and Biological Chemistry*. Menlo Park, California: Benjamin/Cummings, 1984.

Audiovisual/Software

Biology Programs for the Apple II. 8 diskettes. J & S Software.

Functional Chemistry in Living Cells. Filmstrips or slides with cassettes. Human Relations, 1984.

*Additional references are listed in *Modern Biology, Teacher's Program Guide*.

Section Teaching Guides

4.1 Compounds Important to Life

Approach

The main objective of this section is to present the properties of water and of carbon compounds, which are essential to living things. The section builds upon concepts of basic chemistry (Chapter 3). It also provides students with information they need to understand cell structure and function (Chapter 5) and homeostasis and transport (Chapter 6). In presenting the text, guide students to relate the structures of water and carbon compounds to the essential functions of living cells.

Concept Framework

Students already realize the importance of water for living things. Ask students to think of other substances organisms need to stay alive. Write the chemical formulas for these substances on the chalkboard. Elicit the fact that most of these substances contain carbon. As they read the section, encourage students to note the properties of water and the properties of carbon that make these substances essential for living organisms.

Emphasis

Students should understand the importance of the polarity, adhesion, and cohesion of water and realize the significance of the bonding properties of carbon.

Demonstration

Demonstrate the capillarity of water by pouring water into a graduated cylinder or a glass tube. Let students observe how the surface of the water curves and is higher at the sides than in the middle of the tube. This curvature is called a meniscus. Point out that the attraction of water molecules to the surface of the glass causes the level of the water at the sides of the tube to rise.

Section Review Answers

(Continued from p. 48.)
3. Carbon has four electrons in its outer energy level; as a result, it readily bonds to itself and to other elements.

4. Polymers link up in a condensation reaction in which a water molecule is given off; polymers break down by hydrolysis, in which a water molecule is split and the hydrogen and hydroxide ions each combine with a monomer.
5. In most environments the temperature remains below the boiling point of water. Thus organisms are rarely exposed to conditions in which their contents, which are mainly water, would be destroyed by boiling away.

4.2 Organic Compounds

Approach

The main objective of this section is to present the structure and function of the four main types of organic compounds—carbohydrates, lipids, proteins, and nucleic acids. It builds upon students' knowledge of basic chemistry (Chapter 3) and prepares them to understand homeostasis and transport (Chapter 6), photosynthesis and respiration (Chapter 7), and protein synthesis (Chapter 8). In presenting the text, guide students to understand that the functions of organic compounds in living things are directly related to the structures of these organic compounds.

Concept Framework

Students are familiar with structures composed of small, repeating units. Help students form a conceptual framework for understanding the structure of the organic compounds discussed in this section, by helping them make a list of these structures. The list might include buildings made of bricks, jewelry made of beads, mosaics made of tiles, or any other examples. Tell students that the organic compounds they will be studying in this section can also be thought of as structures composed of small, repeating units. Ask students to consider how this would allow organisms to make a wide variety of substances from only four organic compounds.

Emphasis

Make certain students understand the structure of each of the following basic units and how the units are arranged to form organic compounds: monosaccharides forming polysaccharides; fatty acids forming lipids; amino acids forming proteins; nucleotides forming nucleic acids.

Constructing Models

Divide the class into four groups. Assign each group to make a model of one of the organic compounds. The model should clearly show the units of which the compound is made. Refer students to the illustrations in this section to help them design their models. They might compose models using plastic-foam balls, building blocks, papier mâché, or chemistry model kits.

Using the Features/Analyzing the Structure of a Protein

The main process illustrated in this feature is experimenting. The feature outlines Sanger's experiments to discover the structure of insulin. Refer students to the diagram on page 53 for a visual model of some of the steps in his experiment. The picture on page 52 shows a computer graphic of the structure of insulin. The graphic shows the two polypeptide chains, one in yellow and green, the other in blue. The ribbons represent the sulfur bonds that hold the two chains together and the white balls represent the zinc atoms that form the center of the molecule.

Section Review Answers

(Continued from p. 54.)

2. Amino acids have a central carbon to which is attached a hydrogen, an amino group, a carboxyl group, and a variable R group. Proteins are formed from one or more chains of amino acids.
3. The carboxyl end is hydrophilic; the hydrocarbon end is hydrophobic.
4. DNA stores information relating to the control and replication of the cell; RNA stores information relating to the manufacture of proteins.
5. According to the model presented in the chapter, an enzyme hooks up with a specific molecule—the substrate that has a specific shape. If the shape of the enzyme were altered, it would no longer hook up with the correct substrate and would not be effective as an enzyme.

Laboratory Preparation

Background

In this lab, students test for carbohydrates, proteins and lipids, and consider the function of these nutri-

ents by evaluating an unidentified food. Standard methods are used and can be modified to suit the school's equipment. The foods were chosen because of their light colors which do not interfere with the colors produced by the tests. Benedict's test is specific for glucose so other monosaccharides should not be used.

Preparation

Do not reveal that the food used is lowfat cottage cheese until students have completed the investigation and named the predominant nutrient in the unknown substance. Benedict's solution can be purchased from a biological supply house. To prepare iodine–potassium iodide solution, add 10 g of potassium iodide to 50 mL of distilled water. Stir until dissolved. Add 5 g of iodine and stir again. Add enough distilled water to bring the total volume to 100 mL.

The 10% aqueous sodium hydroxide and the 0.5% copper sulfate can be purchased from a biological supply house. Results are better, however, when the chemicals are prepared and used separately. Sudan III can be purchased in solid form from a biological supply house. To prepare the solution, make a 2% stock solution with absolute alcohol. Just before using, mix this solution with an equal amount of 45% alcohol. The colors obtained for corn starch, milk, and lard will be blue-black, purple, and pink respectively. The unknown substance will probably produce pale versions of these colors because it contains a small amount of these nutrients.

Safety

Benedict's solution stains hands and clothing. Pour into dropper bottles for student use. If iodine solution comes into contact with skin, flush affected areas immediately with water. Store in brown bottles away from bright light.

Sodium hydroxide is a corrosive. Flush affected areas immediately with water.

Conclusions

1. Protein is the main nutrient in the unknown substance. The protein test should be the only one in which the color of the unknown substance is close to the color of the known nutrient.
2. The four known substances—iodine–potassium iodide solution, aqueous sodium hydroxide, copper sulfate, and Sudan III—are the controls.

Inquiry

Possible sources of error in the glucose test are not boiling the substances long enough, not allowing sufficient cooling time, not measuring accurately when preparing solutions, having impurities in food substances, and using unclean equipment.

Chapter Review

Vocabulary

1. Monomers are the units, consisting of a simple molecule, that make up polymers.
2. Amino acids are the monomers that compose proteins. Amino acids are joined by peptide bonds to form proteins.
3. Monosaccharides are simple sugars; disaccharides are double sugars, which are made up of two monosaccharides; polysaccharides are large, complex polymers composed of three or more monosaccharides.
4. Hydrophilic means water-loving and refers to a compound or the end of a compound that is polar; hydrophobic means water-hating and refers to a compound or the end of a compound that is nonpolar.
5. RNA is a type of nucleic acid; nucleic acids are composed of monomers called nucleotides.

Review

11. Because a charge is distributed unevenly over the surface of a water molecule, water is a good solvent for ionic compounds and for other polar compounds. Water molecules also adhere to solid surfaces, so water tends to spread through pores and move upward against the force of gravity. Because water molecules adhere, or bond to each other, water heats up and cools down slowly, thus protecting organisms from extreme fluctuations in the temperature of the environment.
12. Both involve water. In a condensation reaction, two monomers bond together releasing a molecule of water; in hydrolysis, a molecule of water is split and two monomers disassociate.
13. Disaccharides are sugars composed of two monosaccharides. Sucrose and lactose are examples of disaccharides.

14. The sequence of amino acids determines the shape of a protein.
15. Accept diagrams that show steps similar to those shown in Figure 4-8.
16. Lipids have more carbon-hydrogen bonds than most other organic molecules. These bonds store more energy than carbon-oxygen bonds.
17. The carboxyl end is polar and thus is hydrophilic; the hydrocarbon end is nonpolar and thus is hydrophobic.
18. Triglycerides are composed of three fatty acids joined to a molecule of glycerol. They include fats and oils. Waxes consist of a long-chain fatty acid joined to a long-chain alcohol. Steroids are lipids composed of four carbon rings.
19. DNA stores information relating to the replication of the cell; RNA stores information relating to the manufacture of proteins.
20. Sanger knew that an amino group of one amino acid attaches to a carboxyl group of another when two amino acids bind. Thus he knew that every polypeptide chain would have one free amino group and one free carboxyl group. Using a chemical that bonded with free amino groups, Sanger could determine how many free amino groups insulin had, and therefore how many polypeptide chains insulin had.

Critical Thinking

1. The rounded shape is due to the attraction of the water molecules for each other.
2. Because water molecules are so cohesive, a large amount of heat energy must be used to overcome the bonding between the water molecules before they can evaporate.
3. The oil would provide the animal with greater flexibility than the fat would. A cold-weather animal must carry a great deal of insulation. However, fat might make it difficult for the animal to move.
4. The enzyme breaks some of the peptide bonds of the proteins and thus weakens the muscle fiber of the meat.

Chapter 4 | *Biochemistry*

Introduction

Carbon, hydrogen, oxygen, and nitrogen make up 99 percent of all living things. These elements join in various combinations and arrangements to form four major types of compounds: carbohydrates, proteins, lipids, and nucleic acids. In this chapter you will learn about the structure of these four types of compounds and about the function of the compounds in living organisms.

Chapter Outline

4.1 Compounds Important to Life
- *Water*
- *Carbon Compounds*

4.2 Organic Compounds
- *Carbohydrates*
- *Lipids*
- *Proteins*
- *Nucleic Acids*

Chapter Concept

As you read, notice how structure relates to function in each of the four organic compounds.

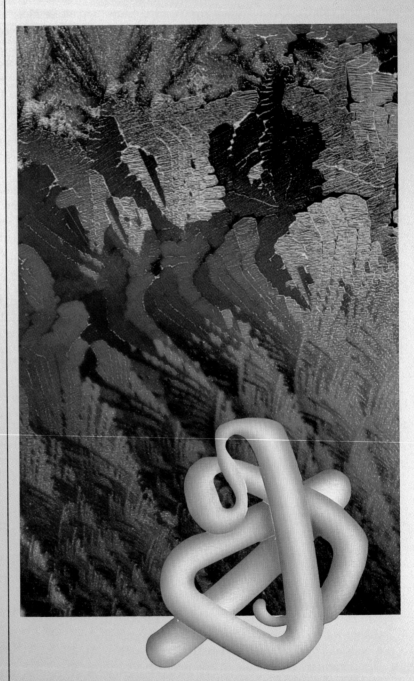

Crystallized amino acid. Inset: Three-dimensional model of a protein molecule.

For additional teaching information see pp. T4A–T4D.

4.1 Compounds Important to Life

Biologists classify compounds into organic compounds and inorganic compounds. **Organic compounds** are derived from living things and contain carbon. **Inorganic compounds** are generally derived from nonliving things. Both organic and inorganic compounds are important to life. In this section you will learn about water and carbon compounds, both of which are essential to living things.

Water

Water, an inorganic molecule, has unique properties that make it one of the most important compounds for living things. In the water molecule (H_2O) the hydrogen and oxygen atoms bond so that the electrical charge is unevenly distributed. The area of the molecule containing oxygen has a slightly negative charge, and the areas containing hydrogen have slightly positive ones. Because of this uneven charge, water is called a **polar compound.**

Because water is a polar compound, it is an effective solvent. For example, when ionic compounds interact with water, the compounds often break apart, or dissociate into ions. This splitting or breaking up of an ionic compound frees ions to participate in biological reactions. Most polar molecules also tend to dissolve in water.

Water molecules and the molecules of solid surfaces are attracted to each other. This phenomenon, called adhesion, gives water its property of **capillarity**—the ability to spread through fine pores or to move upward through narrow tubes against the force of gravity.

Section Objectives

- *Define* organic, inorganic, *and* polar compounds.

- *Summarize the way in which polarity in the water molecule gives water its unique properties and makes it essential for life.*

- *Describe why carbon readily combines with other elements, and discuss its significance in living things.*

- *Compare condensation reactions and hydrolysis.*

Tip: Run a thin stream of water from a sink. Rub a glass rod with a silk cloth and place it next to the stream. The positive charges on the rod will attract the oxygen in the water molecule and bend the stream toward the rod.

Figure 4-1 Attraction between water molecules and the molecules of a solid surface can result in capillarity, which allows water to flow against the force of gravity.

Capillary movement

Capillary movement

Water molecule

47

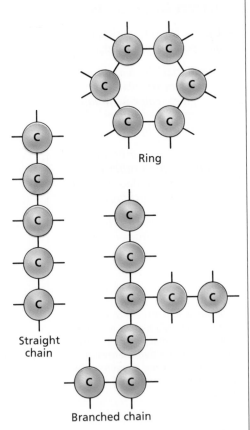

Ring

Straight
chain

Branched chain

Figure 4-2 A carbon atom, with only four electrons in its outer energy level, readily bonds with other carbon atoms in various formations.

Info: Carbon makes up approximately 18 percent of the human body, yet it makes up only 3 percent of the earth's crust.

Answers to Section Review
1. Organic compounds contain carbon and are derived from living things. Most inorganic compounds are derived from nonliving things.
2. polarity, which enables it to be an effective solvent; capillarity, which allows it to flow against the force of gravity; and cohesion, which allows it to heat up and cool down slowly
(See pages T4A–T4D.)

The uneven electrical charge in the water molecule causes the molecules to bond with one another. This phenomenon, called **cohesion,** explains why water heats up and cools down slowly in comparison with other substances. When water is heated, most of the heat energy goes into breaking the bonds between the molecules. Only after the bonds have been broken will the heat energy increase the motion of the molecules and thus raise the temperature of the water. In organisms, the cohesive property of water helps keep cells at an even temperature despite temperature fluctuations in the environment.

Carbon Compounds

As you learned in Chapter 3, the carbon atom has four electrons in its outer energy level. Remember that atoms become stable when their outer energy level contains eight electrons. Carbon, therefore, readily forms four covalent bonds with other elements. As shown in Figure 4-2 carbon also bonds to other carbon molecules, forming straight chains, branched chains, or rings. Because carbon can bond in such a variety of ways, organic compounds exhibit great variability. This variability is responsible for the great variation among living things. Among the most familiar organic compounds are sugars, fats, and proteins. Each of these kinds of organic compounds results from the unique bonding properties of carbon.

Many carbon compounds that are essential for life are **polymers,** compounds consisting of repeated linked units. Each unit in a polymer is called a **monomer** and consists of a simple molecule. Large polymers are called **macromolecules.**

Monomers link together, often forming polymers through a chemical reaction called a condensation reaction. In a **condensation reaction,** also called dihydration synthesis, the reactants give off a hydrogen ion (H^+) and a hydroxide ion (OH^-), which in turn combine to produce a water molecule (H_2O). Conversely, the breakdown of some polymers into monomers occurs through a reversal of this process, called hydrolysis. In **hydrolysis** *(hie-DRAHL-uh-suss)* a water molecule splits into a hydrogen ion and a hydroxide ion. As the polymer breaks apart, the hydrogen and hydroxide ions each combine with one of the monomers.

Section Review

1. Distinguish between *organic* and *inorganic compounds.*
2. What properties make water essential for living things?
3. What property allows carbon compounds to exist in a number of forms?
4. How does the reactive process by which polymers bond differ from the process by which they dissociate?
5. The cohesion between water molecules gives water a high boiling point. Why might this be important for living things?

4.2 Organic Compounds

For all living organisms four types of organic compounds are essential: carbohydrates, lipids, proteins, and nucleic acids. As you read, study the structure of each organic compound shown in Figure 4-3 and note its function in the cell.

Carbohydrates

Carbohydrates are organic compounds composed of carbon, hydrogen, and oxygen in a ratio of two hydrogen atoms to one oxygen atom. The number of carbon atoms varies. Carbohydrates exist as monosaccharides, disaccharides, or polysaccharides.

Monosaccharides

A **monosaccharide** (*MAHN-uh-SACK-uh-ride*)—or simple sugar—contains carbon, hydrogen, and oxygen in a ratio of one to two to one. The most common monosaccharides are glucose, fructose, and galactose. Glucose is manufactured by plants during photosynthesis. It is the main source of energy for both plants and animals and is metabolized during the process called cellular respiration. Fructose is found in fruits and is the sweetest of the monosaccharides. Galactose is found in milk and is usually in

Section Objectives

- *Define* monosaccharide, disaccharide, *and* polysaccharide, *and discuss their significance to organisms.*

- *Relate the structure of lipids to their functions.*

- *Compare the structures of the various types of proteins.*

- *List two essential functions of nucleic acids.*

- *Outline Sanger's experiment for deciphering the structure of insulin.*

Figure 4-3 The four major kinds of organic compounds are proteins, carbohydrates, nucleic acids, and lipids.

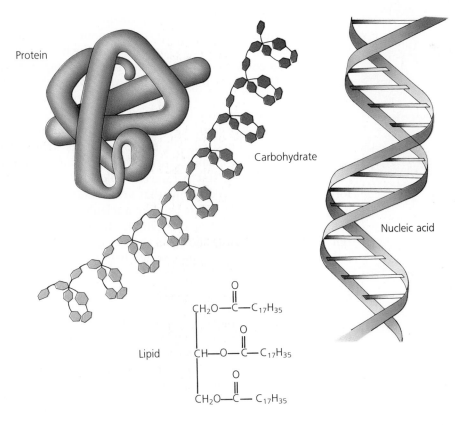

Protein

Carbohydrate

Nucleic acid

Lipid

$$CH_2O-\overset{\overset{\textstyle O}{\|}}{C}-C_{17}H_{35}$$

$$CH-O-\overset{\overset{\textstyle O}{\|}}{C}-C_{17}H_{35}$$

$$CH_2O-\overset{\overset{\textstyle O}{\|}}{C}-C_{17}H_{35}$$

Figure 4-4 The monosaccharides—glucose, galactose, and fructose—all have the same molecular formula although their structures vary.

Glucose

Galactose

Fructose

Answer to Think
Birds need a large amount of energy for long migratory flights. If energy were stored as carbohydrates, the bird would be too heavy to fly.

Think

Why would it be to the advantage of an animal, such as a migratory bird, to store energy in the form of lipids rather than in the form of carbohydrates?

combination with glucose or fructose. Glucose, fructose, and galactose all have the identical molecular formula, $C_6H_{12}O_6$, but not the same structural formulas. Therefore glucose, fructose, and galactose are **isomers**—compounds that differ in structure but not in molecular composition. Look at the isomers in Figure 4-4.

Disaccharides and Polysaccharides

In organisms two monosaccharides combine in a condensation reaction forming a double sugar—or **disaccharide** (*die-SACK-uh-ride*). Sucrose, a diasaccharide found in sugarcane and sugar beets, is composed of fructose and glucose. Lactose, a diasaccharide found in milk, is composed of glucose and galactose.

A **polysaccharide** (*PAHL-ih-SACK-uh-ride*) is a complex molecule composed of three or more monosaccharides. Like disaccharides, polysaccharides form through the linking of monosaccharides in condensation reactions. The most important polysaccharides found in living organisms are glycogen, starch, and cellulose.

Animals store glucose in the form of a polysaccharide called glycogen (*GLIE-kuh-jun*). Glycogen, sometimes called animal starch, consists of hundreds of glucose molecules strung together in a highly branched chain. Plants store glucose in the form of a polysaccharide called starch. Starch molecules have two basic forms—highly-branched chains that are similar to glycogen and long unbranched chains that coil like a telephone cord. Cellulose is a polysaccharide that gives strength and rigidity to the plant cell. In a cellulose molecule thousands of glucose monomers are linked in long, straight chains, which help support plants.

Lipids

A **lipid** is a fatty compound made up of carbon, hydrogen, and oxygen. A lipid molecule has a larger number of carbon and hydrogen atoms and a smaller number of oxygen atoms than carbohydrates do. Fats, oil, and waxes are lipids. Unlike the other organic compounds, lipids do not dissolve in water.

Info: There are 13 possible isomers of any 6-carbon sugar, called a hexose.

The cell membrane is composed mostly of lipids. The insolubility of the lipids in water allows the membrane to form a barrier between the aqueous environments inside and outside the cell. Lipids also store energy efficiently. The lipid molecule has a relatively large number of carbon–hydrogen bonds, which store more energy than the carbon–oxygen bonds more common in the molecules of other organic compounds.

Fatty Acids

Fatty acids are the monomers that make up most lipids. Figure 4-5 shows that a **fatty acid** is composed of a long, straight hydrocarbon chain with a carboxyl (COOH) group attached at one end.

The contrasting properties of the two ends of the fatty acid molecule account for the characteristics of lipids. The carboxyl end of the fatty acid molecule is polar. Thus it is attracted to water molecules, which are also polar. Because of this attraction the carboxyl end of the fatty acid molecule is said to be **hydrophilic** (*HIE-druh-FIL-ick*), which means that it is "water loving."

In contrast, the hydrocarbon end of the fatty acid molecule is nonpolar. It tends not to interact with water molecules and in effect repels water. The hydrocarbon end of the fatty acid molecule is said to be **hydrophobic** (*HIE-druh-FOE-bick*), or "water fearing." In cell membranes the hydrophilic ends of the fatty acid molecules are oriented to the aqueous side of the membrane and the hydrophobic ends are oriented to the center of the membrane.

Triglycerides, Waxes, and Steroids

Three kinds of lipids exist: triglycerides, waxes, and steroids. A **triglyceride** (*trie-GLISS-uh-ride*) is a type of lipid in which the macromolecule is composed of three molecules of fatty acids joined to one molecule of glycerol. Each fatty acid molecule combines with the glycerol molecule through a condensation reaction. There are two main types of triglycerides. Triglycerides that are liquid at room temperature are called **oils.** The triglycerides found in plants are usually oils. These are often found in seeds, where they serve as an energy source for sprouting plants. Triglycerides that are solid at room temperature are called **fats.** The triglycerides found in animals are usually fats.

A **wax** is a type of lipid in which the molecule consists of a long fatty acid chain joined to a long alcohol chain. These long chains make waxes highly waterproof. In plants, wax forms a protective coating on the outer surfaces. In animals, wax forms protective layers, too. For example, earwax forms a barrier that keeps microorganisms from entering the middle ear.

In contrast with most other lipid polymers, which are composed of fatty acids, a **steroid** (*STIR-OID*) is a lipid in which the molecule is composed of four carbon rings. Steroids are considered lipids because they do not dissolve in water. Steroids are found in substances as varied as hormones, nerve tissue, toad venoms, and plant poisons.

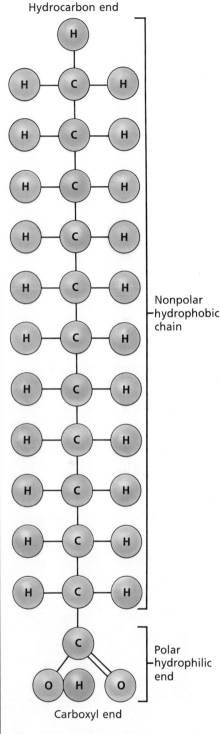

Hydrocarbon end

Nonpolar hydrophobic chain

Polar hydrophilic end

Carboxyl end

Figure 4-5 A fatty acid consists of a chain of carbon atoms. At the "head" end of each fatty acid is a carboxyl group; this end is polar. The hydrocarbon, nonpolar, end is the "tail."

Glycine

Serine

Figure 4-6 Glycine and serine contain the same elements but in different amounts. How does the structure of glycine differ from that of serine?

Proteins

Proteins are organic compounds composed mainly of hydrogen, oxygen, carbon, and nitrogen. Formed from the linkage of monomers called amino acids, which you will read about below, **proteins** are structural and functional compounds of universal importance in cells and thus in organisms. The skin and muscles of animals are made up mostly of proteins, as are the chemicals that are essential for the life processes in both plants and animals. Each species of organism has thousands of unique proteins.

Amino Acids

The 20 **amino acids**—monomers that form proteins—share the same basic structure. As Figure 4-6 shows, each amino acid contains a central carbon atom to which four other atoms or groups of atoms bond covalently. A single hydrogen atom, shown in orange, bonds at one site. A carboxyl group (COOH), shown in green, bonds at a second site. An amine group (NH_2), shown in blue, bonds at the third site. A group of atoms called the R group, shown in yellow, bonds at the fourth site. The differences between kinds of amino acids result from different R groups.

Biology in Process

Analyzing the Structure of a Protein

In 1943 the British scientist Frederick Sanger set out to analyze the sequence of amino acids in a protein. Sanger chose to study a relatively small protein, insulin—a protein hormone that helps control the amount of blood sugar.

First Sanger broke insulin into its constituent amino acids. Using chromatography, he identified the kinds and amounts of the amino acids. He used this information about insulin to verify his work as he proceeded to decipher the sequence of amino acids in the insulin molecule.

When amino acids bond together, the amino group of one amino acid attaches to the carboxyl group of the next amino acid. Thus at one end of any polypeptide chain is a free amino group, and at the other end is a free carboxyl group. Sanger used this knowledge to figure the number of polypeptide chains in insulin. He used a chemical that bonded only with free amino groups to show that insulin was composed of two free amino groups and thus consisted of two polypeptide chains.

Next Sanger had to determine the order of the amino acids in each of the two chains.

Answer to Biology in Process: He used chromatography to identify the kinds and amounts of the amino acids and then deciphered the sequence of the amino acids.

Dipeptides and Polypeptides

Two amino acids bond together to form a **dipeptide** (die-PEP-TIDE). Bonding occurs by a condensation reaction in which the amino group of one amino acid releases a hydrogen ion (H^+) and the carboxyl group of the second amino acid releases a hydroxide ion (OH^-). Then the nitrogen atom from the amino group and the carbon atom from the carboxyl group bond covalently, linking the two amino acids. A covalent bond between a nitrogen atom and a carbon atom is called a **peptide bond**.

A long chain of amino acids is called a **polypeptide**. Proteins are composed of two or more polypeptides. The amino acid sequence of the proteins determines how each protein bends and folds on itself and how the proteins intertwine.

Enzymes

Intermediary metabolism in living things usually involves catalysts—substances that speed up chemical reactions without being affected by the reactions themselves. **Enzymes**—proteins that act as catalysts in intermediary metabolism—are essential for the functioning of any cell. An enzyme speeds up a chemical reaction

Figure 4-7 When amino acids such as glycine and alanine bond, a dipeptide forms. The covalent bond that joins a nitrogen atom and a carbon atom is a peptide bond.

The diagram shows the steps in the complex procedure he followed. In step 1 he used chemicals that broke the bonds between specific amino acids to split the chains into smaller fragments. Then he treated these fragments with chemicals that broke off amino acids one at a time. Using chromatography he then identified the individual amino acids. Sanger checked his results with the results of his first experiment, which had revealed the composition of insulin.

Finally Sanger had to determine how the fragments fit together. In step 2 he repeated the fragmentation of insulin, using a different chemical, which cleaved the polypeptide chains at different sites. He analyzed the amino acid sequences in this set of fragments and compared them with the sequences from the first set of fragments. By comparing the two sets of fragments Sanger could determine the order of the amino acids in insulin (step 3). Sanger's painstaking labor paid off. In 1953 he deciphered the amino acid sequence of insulin. In 1958 he won the Nobel Prize in chemistry for his achievement.

■ What was the process Sanger used to discover the sequence of amino acids in insulin?

Substrate

Products of the reaction

Enzyme

Figure 4-8 An enzyme, which is the essential catalyst in cell metabolism, binds in an exact fit to a specific site on a molecule, or substrate.

Phosphate group

Sugar

Nitrogen base

Figure 4-9 Each nucleotide in RNA and DNA consists of a phosphate group, a five-carbon sugar, and a ring-shaped nitrogen base.

Answers to Section Review
1. A monosaccharide is a simple sugar containing carbon, hydrogen, and oxygen in a ratio of one to two to one. A disaccharide is a double sugar formed from two monosaccharides. A polysaccharide is a sugar composed of three or more monosaccharides.
(See pages T4A–T4D.)

by lowering the activation energy, that is, the amount of energy needed for a reaction to occur.

Scientists have proposed several models of enzyme action. Figure 4-8 illustrates one model. Notice how the shape of a particular enzyme allows the enzyme to hook up with a specific molecule. This molecule, called the **substrate,** is the reactant in a chemical reaction that is catalyzed by the enzyme. The linkage of the enzyme and the substrate probably weakens some chemical bonds in the substrate. This weakening might explain how the enzyme lowers the activation energy for a chemical reaction. Once the reaction has occurred, the products are released. The enzyme itself is not altered, thus allowing the cell to reuse the enzyme thousands of times without having to synthesize new enzymes.

Nucleic Acids

Nucleic acids are complex organic molecules that store important information in the cell. The two most important types of nucleic acids are DNA and RNA. Deoxyribonucleic *(DEE-AHK-sih-RIE-boe-noo-KLEE-ick)* acid, or DNA, stores information that is essential for almost all cell activities, including cell division. **Ribonucleic** *(RIE-boe-noo-KLEE-ick)* acid, or **RNA,** stores and transfers information that is essential for the manufacturing of proteins.

Both DNA and RNA are composed of thousands of monomers called **nucleotides** *(NOO-klee-uh-TIDEZ).* As shown in Figure 4-9, each nucleotide is made up of three main components: a phosphate group, a five-carbon sugar, and a ring-shaped nitrogen base. You will learn more about these important compounds in Chapter 8.

Section Review

1. Define *monosaccharide, disaccharide,* and *polysaccharide.*
2. Describe the structure of amino acids and proteins.
3. How do the two ends of a fatty acid differ?
4. What are two types of nucleic acids and their functions?
5. High temperatures can weaken bonds between different parts of a protein molecule, thus changing its shape. How might this change alter the effectiveness of an enzyme?

Laboratory

Testing Food for Nutrients

Objective
To evaluate the nutrient content of an unidentified food

Process Skills
Experimenting, observing

Materials
1-L beaker, hot plate, glucose, unknown substance, 8 test tubes, 4 medicine droppers, distilled water, Benedict's solution, glass stirring rod, tongs, test tube rack, cornstarch, iodine–potassium iodide solution, nonfat dry milk, 10% aqueous sodium hydroxide solution, 0.5% copper sulfate solution, test tube stopper, lard, Sudan III solution

Method

1 Fill a 1-L beaker half full with water and heat it on the hot plate.
2 Place pea-sized portions of glucose and the unknown substance in separate test tubes. What type of nutrient is glucose? Use a dropper to add about 2.5 cm of distilled water and 10 drops of Benedict's solution to each test tube. Mix with a stirring rod.
3 When the water boils, use tongs to place the test tubes in the water bath. Leave the test tubes in the water bath for 10 minutes.
4 Remove the test tubes with tongs and place the tubes in a test tube rack. When the tubes cool, an orange or red precipitate will form if large amounts of glucose are present. Small amounts of glucose will form a yellow or green precipitate. Record your observations.
5 You will use chemical reagents to test the unknown for specific nutrients. By comparing the color change a reagent produces in the unknown with the change it produces in the known nutrient, you can estimate the amount of that nutrient. Use small samples. Record all observations.

6 Place cornstarch in a clean test tube and some of the unknown substance in another. What type of nutrient is cornstarch?
7 Use a clean dropper to add 10 drops of iodine–potassium iodide solution to each test tube. Observe the results.
8 Place nonfat dry milk in a clean test tube and some of the unknown substance in another. What is the main nutrient in milk?
9 With a clean dropper slowly add an amount of sodium hydroxide solution about equal to the amount of the milk sample, and mix carefully. Then add 10 drops of copper sulfate solution one drop at a time. Stopper and gently shake the tube between drops.
10 Repeat step 9 with the unknown substance.
11 Place a small piece of lard in a clean test tube and some of the unknown substance in another. What type of nutrient is lard?
12 Use a clean dropper to add 10 drops of Sudan III solution to each test tube. Mix with a stirring rod, and observe.

Conclusions
1 What is the main nutrient in the unknown?
2 What are the controls in this investigation?

Inquiry
What are sources of error in this investigation?

Chapter 4 Review

Vocabulary

amino acid (52)	fatty acid (51)	monosaccharide (49)	polypeptide (53)
capillarity (47)	hydrolysis (48)	nucleic acid (54)	polysaccharide (50)
carbohydrate (49)	hydrophilic (51)	nucleotide (54)	protein (52)
cohesion (48)	hydrophobic (51)	oil (51)	ribonucleic acid
condensation reaction	inorganic compound	organic compound	(RNA) (54)
(48)	(47)	(47)	steroid (51)
dipeptide (53)	isomer (50)	peptide bond (53)	substrate (54)
disaccharide (50)	lipid (50)	polar compound (47)	triglyceride (51)
enzyme (53)	macromolecule (48)	polymer (48)	wax (51)
fat (51)	monomer (48)		

Explain the relationship between the terms in each of the following sets.
1. monomer, polymer
2. amino acid, peptide bond, protein
3. monosaccharide, disaccharide, polysaccharide
4. hydrophilic, hydrophobic
5. nucleotide, nucleic acid, RNA

Review

1. Most organic compounds contain carbon and are (a) hydrophobic (b) not important to life (c) made by living organisms (d) not made by organisms.
2. The distinguishing feature of a polar compound is its (a) even distribution of electrical charge (b) uneven distribution of electrical charge (c) even temperature (d) uneven temperature.
3. Because the carbon atom has four electrons in its outer energy level carbon (a) participates in condensation reactions (b) participates in hydrolysis (c) bonds easily to itself and to other elements (d) adheres to solid surfaces.
4. The main source from which plants and animals get their energy is the monosaccharide (a) glucose (b) fructose (c) galactose (d) glycogen.
5. Plants store glucose in a polysaccharide called (a) cellulose (b) maltose (c) lactose (d) starch.
6. When two amino acids bond, (a) the carboxyl group of one amino acid joins to the amine group of the other in a condensation reaction (b) the carboxyl group of one amino acid joins to the amino group of the other during hydrolysis (c) a dipeptide is formed through intermediary metabolism (d) a polypeptide is formed through intermediary metabolism.
7. Enzymes speed up chemical reactions in the cell by (a) raising the activation energy (b) lowering the activation energy (c) releasing the products of the reaction (d) emerging from the reaction unaltered.
8. Lipids are distinguished from other organic molecules because they (a) contain carbon, hydrogen, and oxygen in a ratio of 1:2:1 (b) do not dissolve in water (c) store energy (d) form structural and functional compounds of universal importance in cells.
9. Steroids differ from other lipid polymers in that steroids (a) do not occur in varied substances (b) are not hydrophilic (c) are not hydrophobic (d) are not composed of fatty acid monomers.
10. A nucleotide is composed of a phosphate group, a five-carbon sugar, and a (a) substrate (b) triglyceride (c) nitrogen base (d) macromolecule.

11. How does the polarity in the water molecule give water unique properties that make it essential to life?
12. Compare and contrast a condensation reaction with hydrolysis.
13. Define *disaccharide* and give examples.
14. What factor determines the shape of a protein?
15. Diagram a model of one of the ways scientists think that enzymes might work.
16. How do lipids store energy efficiently?
17. How does the carboxyl end of the fatty acid molecule differ from the hydrocarbon end of the molecule?
18. Compare and contrast triglycerides, waxes, and steroids.
19. What are two nucleic acids and their functions?
20. How did Sanger use his knowledge of the structure of amino acids to discover the number of polypeptide chains in insulin?

Critical Thinking

1. Examine the shape of the water droplets on the leaf in the photograph at the right. What property of water molecules might give them that shape?
2. When we sweat, our bodies are cooled as the water evaporates. Explain how the polarity of the water molecules makes this an efficient way of cooling.
3. As you learned in the chapter, the triglycerides in animals are usually fats, and those in plants are usually oils. However, in many animals of the Arctic and Antarctic, the triglycerides are proportionately oils. What adaptive advantage would the storage of body fat as oil instead of fat be to animals that live in freezing climates?

4. Meat tenderizers contain papain, an enzyme that digests proteins. These tenderizers are derived from unripe papayas. How might a tenderizer make meat tender?

Extension

1. Read the article by Lee Torrey entitled "Endurance Training: Pushing the Limit," in *Science Digest*, October 1985, pp. 62–67. Give an oral report in which you explain what carbohydrate loading is and why athletes use it. What are the possible dangers in some carbohydrate-loading plans?
2. Cut fibrous meat into four one-inch cubes. Sprinkle three of the cubes with equal amounts of meat tenderizer. Place one cube in the refrigerator, leave one at room temperature, and place the other in an incubator at 32°C. For the fourth cube, place the same amount of meat tenderizer and a few tablespoons of water in a container, boil for three minutes, and place on the meat. After three hours observe the texture of all four meat cubes. What do you conclude about the effect of temperature on the enzyme in meat tenderizer? Write a report in which you explain the method used, give the temperatures, and compare your observations of the meat before and after the incubation with the enzyme.

Intra-Science: *How the Sciences Work Together*

The Question in Biology: Can Alaskan Wildlife and a Pipeline Coexist?

In this unit you read about science and society. The construction of the Alaska pipeline provides an example of how they work together. The Alaskan wilderness is an expanse of towering mountains, rushing rivers, and permanently frozen ground. You might not think that a single construction project could disrupt the ecological balance of such a vast region. But the Alaskan wilderness is far more vulnerable than it might seem.

In 1968 geologists discovered one of the world's largest oil fields near Prudhoe Bay on the coast of northern Alaska. The problem was that during most of the year oil tankers could not reach Prudhoe Bay, which is located near the polar ice pack. Thus to gain access to the oil, engineers designed and built a 1,300-km pipeline to Valdez, an icefree port on Alaska's southern coast, where tankers could pick up the petroleum and transport it to market.

The challenge was to build the pipeline without damaging the wilderness environment and harming the plant and animal life there. Most of the terrain the pipeline would cross was tundra—plains where fragile vegetation grows sparsely on the permafrost. Huge herds of caribou have migration routes that cross the area. The region also contains nesting areas of the peregrine falcon, an endangered species, as well as many rivers where salmon spawn. As you can see, the building of the Alaska pipeline required more than engineering. It also demanded the ingenuity of scientists concerned with soil, plants, animals, and human beings.

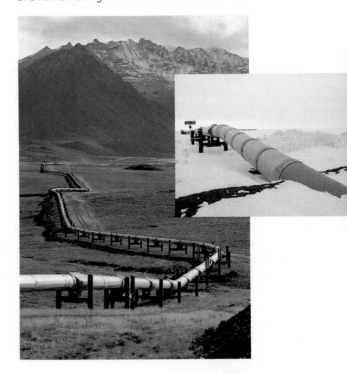

The Connection to Physics

The physical phenomena of freezing and melting were of great concern to pipeline engineers. For about half its route, the pipeline could not be buried without melting the permafrost. When oil under high pressure flows through a pipe, its motion against the pipe produces friction. This friction in turn generates heat. As a result, temperatures inside the pipe could be high enough to melt the surrounding permafrost. Not only would the pipeline sink in the thawed mud, but the resulting soil erosion would cause major damage to the terrain.

Wherever the permafrost was unstable because of its high ice content, workers insulated the pipeline and elevated it on supports sunk more than 7 m into the ground. Each vertical support contains a tube through which flows liquid ammonia, a refrigerant that prevents the transfer of heat to the permafrost.

The pipeline also had to cross 20 large rivers and about 300 streams. At most crossings, workers coated the pipeline with a thick layer of concrete so that it could not float and then buried it at least 1.5 m under the river. Workers constructed bridges across the Yukon and other extremely wide rivers and then attached the pipeline to the bridges.

Three mountain ranges also stood in the route of the pipeline. As a result, the pipe had to cross elevations as high as 1,440 m. Forty huge pumps were installed to keep about 1 million barrels of petroleum flowing through the pipeline each day. The pipeline also has 60 valves that can be closed in case of a break in the line.

The Connection to Chemistry

Materials science is a branch of applied chemistry. The construction of the pipeline involved the use of many different types of materials. The properties of those materials are determined by their chemical composition. For example, the pipe had to have special properties to withstand the thermal shock of being welded in extreme cold. These properties were obtained by using steel that had a small grain size because of its low content of carbon and oxygen.

The designers of the pipeline also took precautions to assure that the oil would not solidify if it were to stop flowing for some reason. Thus they insulated the above-ground pipes with fiberglass sheathing and polyurethane panels.

The Connection to Other Sciences

Many design measures were taken to protect the wildlife of the region. For example, wherever biologists suspected that caribou might be afraid to pass under the elevated sections of pipeline, short stretches of the pipe were buried. In these places special refrigeration devices were installed to keep the permafrost frozen. Similarly, designers changed the location of a pumping station to save the nest of a peregrine falcon. And where pipeline was laid at river crossings, workers took precautions to avoid stirring up silt that could drift into areas where salmon spawn.

Wherever archaeological sites lay in the path of the pipeline, archaeologists excavated the sites and removed artifacts. Anthropologists also helped arbitrate the land claims of the native Alaskans whose property was affected.

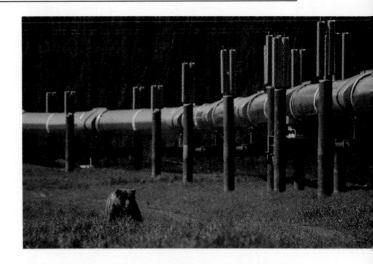

The Connection to Careers

The Arctic remains one of the last great wildernesses. Alaska and the neighboring Canadian Northwest Territories still have vast, untapped supplies of natural gas. Gaining access to the resources of the north requires the skills of surveyors, welders, heavy-equipment operators, engineers, and truck drivers. As you have seen, however, it also takes the knowledge of geologists, chemists, biologists, and anthropologists to ensure that these resources are obtained with as little damage as possible to the environment and its inhabitants. For more information on related careers in the biological sciences, turn to page 842.

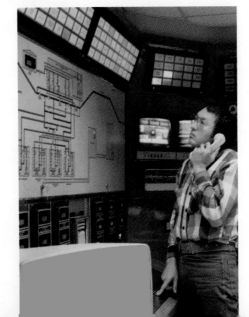

Unit 2 | Cells

Chapters

"With the cell, biology discovered its atom."
François Jacob

Just as atoms are the basic structural and functional units of all matter, cells are the basic structural and functional units of living things. In Unit 1 you learned how a knowledge of atoms helps scientists understand and predict the behavior of chemical substances. In this unit you will gain a knowledge of cell structure and function that will help you better understand the processes that take place within living things.

Spirogyra sp. cells (×140)

Sketch of animal cells, after Theodor Schwann (1839)

SEM of human cell dividing (×1,800)

SEM of *Euglena gracilis* cells

Chapter 5 | Structure and Function of the Cell

This chapter explains the characteristics of cells, the structures and functions of cell parts, and the organization of specialized cells into multicellular organisms. It discusses how cell structures relate to their specialized functions.

Correlation Chart

Section Number	Section Objectives	Pages	Features	Section Review
5.1	1. Outline the discoveries that led to the development of the cell theory.	63–65		5
	2. State the cell theory.			2
	3. Identify a limitation in the size of cells.			3
	4. Describe the relationship between cell shape and cell function.			1
	5. Distinguish between prokaryotes and eukaryotes.			4
5.2	1. Describe the structure, composition, and function of the cell membrane.	66–74	Biotechnology, pp. 70–71	2
	2. Name the major organelles found in the cell, and describe their functions.		Think, p. 74	1, 5
	3. Describe the structure and functions of the cell nucleus.			4
	4. Compare and contrast animal and plant cells.			3, 5
	5. Describe how AVEC microscopy is useful in observing microtubules.			
5.3	1. Distinguish between tissues, organs, and organ systems.	75–77	Biology Lab, p. 77	1, 2, 3
	2. Describe the way in which cells are organized in multicellular organisms.			4, 5

Resources*

Print

de Duve, Christian. *A Guided Tour of the Living Cell,* Vols. 1 & 2. New York: Scientific American Books, 1985.

Sheeler, P. and D. E. Bianchi. *Cell Biology: Structure, Biochemistry, and Function.* New York: Wiley, 1983.

Audiovisual/Software

Cell Structure and Function. Filmstrip and cassette. Biology Media.

The Architecture of Cells: Special Structure, Special Function. 3 filmstrips and cassettes, 15 min. each. Human Relations, 1983.

*Additional references are listed in *Modern Biology, Teacher's Program Guide.*

Section Teaching Guides

5.1 Introduction to the Cell

Approach

The main objective of this section is to introduce students to the study of cells, a topic explored in greater detail in Chapters 6–9. It builds on concepts of biology as process (Chapter 2) and the relationship between structure and function (Chapter 4). In presenting the text, help students to develop an awareness of the enormous diversity of cells in shape, size, and internal organization. Recognizing the differences among cells will provide a basis for the discussion of specialization of function that appears in Sections 5.2 and 5.3.

Concept Framework

Students all know that there is a basic difference between living things and nonliving things. To build on that knowledge, engage students in a discussion of the smallest units, or "building blocks" of living and nonliving things. Elicit the fact that in all living things the smallest unit of organization is the cell. You may wish to end the discussion by asking students to consider which differences among two living organisms are most likely to result from differences among cells.

Emphasis

Make certain that students understand that individual cells are limited in the size they can attain; large organisms are built of a multiplicity of cells.

Time Line

Ask the class to jointly create a time line of the history of cell study. Ask them to record the dates new technologies were introduced and the dates major theories were first presented.

Section Review Answers

(Continued from p. 65.)
4. Look at the cell through a microscope. If the cell has a nucleus and membrane-bound organelles, it is a eukaryote; if it does not, it is a prokaryote.
5. By the mid-nineteenth century, the light microscope had been developed and was widely available. All three scientists must have utilized the achievements of scientists who preceded them, and thus the historical time was right to interpret the data that had accumulated and those they had gathered in their own experiments.

5.2 Parts of the Cell

Approach

The main objective of this section is to demonstrate the close relationship between the structure of a cell part and its function. The text builds on the discussion of cell diversity in Section 1 of this chapter. You may wish to extend Table 5-1 to include structure, component parts, and function. Have students fill in these parts of the table as they read through the section.

Concept Framework

The parts of the cell carry out specialized functions. The concept of division of labor is a familiar one to students, particularly when presented through an analogy such as a factory or a city. Choose one analogy, and ask students to list specialized parts such as the factory wall or lunchroom. End the discussion by asking students to consider why it is more efficient to organize a factory into specialized parts and why cells may be able to function more efficiently with specialized parts.

Emphasis

Make certain that students understand that an organism that functions efficiently has an evolutionary advantage over one less efficient.

Field Trip

Visit a university or hospital laboratory that has a TEM or SEM. Arrange to have students view both animal and plant cells.

Using the Features/Movement Along Microtubules

The main technique used in Allen's observations of microtubules is a form of video-enhanced contrast microscopy, called AVEC. This technique enabled Allen to view particles moving along a microtubule. To help students understand the usefulness of this technique, ask them to experiment with the contrast knobs of their television sets or computer screens. If possible demonstrate this in the classroom.

Section Review Answers

(Continued from p. 74.)

3. Both are containers, though the lysosome also serves a digestive function. The lysosome is found primarily in animal cells; it holds digestive enzymes. The vacuole is more common in plants; it stores food and waste products.
4. The nucleus is the control center of the cell. It contains the nucleolus, which is the site where ribosomes are synthesized.
5. The cell walls provide support and stability to plants, which do not move. All but a very few animals move about to feed. The cell walls would interfere with their mobility.

5.3 Multicellular Organization

Approach

This section takes the concept of cell specialization discussed in Section 5.2 a step further by applying it to groups of cells. It prepares students for a discussion of organ systems in Units 6–9. In presenting the text, use the same analogy of a division of labor that was used in Section 5.2. Explain that just as parts of cells perform specialized functions, cells are also organized into groups that perform like functions.

Concept Framework

Most students are familiar with some of the organs of the body. Build on this knowledge by asking students to name organs of the body and then to describe the function of each. Draw a comparison between the specialized functions of organs in the body to the functions in cells that are specific to cells in a particular tissue or organ system.

Emphasis

Make certain that students understand that individual cells of a multicellular organism are unable to survive independently of the total organism.

Section Review Answers

(Continued from p. 76.)

3. An example might be the vascular tissue in the root, the root itself, and the transport system.
4. Only to some extent: a few cells in a colony reproduce; most are modified for movement.

5. Scientists consider it to be multicellular but lacking in cell walls. Any responsible answer, however, is acceptable.

Laboratory Preparation

Background

The purpose of this lab is to allow students to observe some of the fundamental differences between plant and animal cells. Students will discover that important plant cell structures, such as cell walls, a central vacuole, and chloroplasts, are absent in animal cells. The central vacuole has several functions. By taking up space, it reduces the area of cell activity and thus reduces the energy requirements of the cell. The central vacuole also stores waste products. The turgidity caused by the vacuole pressing the cell contents against the cell wall provides support to plants. Most animals have an external or internal skeletal system to support them, and so there is no necessity for animal cells to contain a central vacuole and cell walls.

Preparation

Methylene blue solution can be purchased premixed or prepared by adding 0.5 g of methylene blue powder to 100 mL of distilled water and mixing thoroughly. Let solution stand for 24 hours.

Safety

Methylene blue stains skin and clothing. Pour into dropper bottles for student use.

Conclusions

1. *Elodea* cells resemble elongated cubes while cheek cells are spherical or irregular in shape. Plant cells keep their elongated shape because of the rigid cell wall.
2. The nuclei of plant cells are usually found at the side of the cell. The nuclei of animal cells are found near the center. Plant cell nuclei are pressed against the side by the large central vacuole.

Inquiry

1. Students will devise ways to observe the changes that occur when stain is added.
2. Students should stain *Elodea* and then compare the stained cells with unstained *Elodea* cells.

Chapter Review

Vocabulary

1. *Eu-* means "true," *pro-* means "before," and *karyo-* means "kernel" or "nucleus." Eukaryotes are cells that have a true nucleus, enclosed in a membrane. Prokaryotes are cells that may have evolved before the true nucleus formed. The meaning of the word reflects the fact that prokaryotic cells are older than eukaryotic cells.

2. The cell membrane is made up of two layers of lipid and protein molecules and holds the contents of the cell together while regulating the movement of particles through its surface into and out of the cell. The cell wall is constructed between cell membranes of adjoining plant cells and gives support and strength to the cell.

3. *Cyto-* means cell. Thus the cytoplasm is the plasm, or fluid, of the cell, and the cytoskeleton is the framework of the cell.

4. Microtubules are hollow protein tubes that form the internal structure of cilia and flagella.

5. Chromatin is the term that does not belong. The other three terms refer to membrane-bound organelles.

Review

11. Each new advancement in the technology of microscopes has enabled scientists to observe more of the structure and workings of the cell.

12. The size is limited by the ratio of its volume to its surface area.

13. A cell membrane is called a selectively permeable membrane because the structure of the membrane allows some materials to pass through it but does not allow others to do so.

14. Proteins are synthesized on the ribosomes, then some pass from rough ER to smooth ER, where they are packaged, and finally go to the Golgi apparatus for further processing and packaging. Then they go to the cell membrane for export outside the cell, or to another cell organelle for use within the cell.

15. Rough ER is covered with many ribosomes, while smooth ER has few or no ribosomes.

16. Mitochondria have an inner membrane that has many folds, called cristae. This increases the amount of surface area available for the work of the mitochondria, which is to break down or-ganic molecules and make ATP for the cell's energy needs.

17. AVEC technology has enabled scientists to observe the interaction between microtubules and particles moving along them.

18. Chromatin is fine threads of genetic material. When a cell is dividing, the chromatin coils up and is more easily visible under a microscope. It is then called chromosomes.

19. First the middle lamella forms between two cells. Then the cell on each side of the middle lamella forms a primary wall. Some plants finally form a secondary wall that provides extra stiffness.

20. Some cells perform specific functions on behalf of the whole organism. An example is a skin cell of an animal.

Critical Thinking

1. None of the functions of organelles is needed for this "cell" to carry out its function. It limits the cell because it cannot divide. It lacks mitochondria and consequently cannot burn sugar the way other cells do.

2. The ER and the cristae inside the mitochondria have shapes similar to radiator coils. The large surface area made possible by the coil arrangement is essential to the function of both the organelles and the radiator.

3. Lysosomes are defective; they are organelles that contain enzymes that digest lipids.

4. The poisonous chemical is stored in a vacuole, isolating it from the rest of the cell.

5. The tree trunk cells (right photo) have secondary thickenings.

6. The presence of organelles makes possible a greater specialization of function.

Structure and Function of the Cell

For additional teaching information see pp. T5A–T5D.

Introduction

Living things, like nonliving things, are composed of chemicals—chemical compounds and ultimately chemical elements such as carbon, hydrogen, oxygen, and nitrogen. It is the organization of these elements into cells that distinguishes living things from all other matter. In this chapter you will learn about the characteristics of cells and the organization of specialized cells into larger organisms. You will also learn how structural variations within cells relate to specialized functions.

Chapter Outline

5.1 Introduction to the Cell
- *Discovery of the Cell*
- *Cell Diversity*

5.2 Parts of the Cell
- *Cell Membrane*
- *Cytoplasm and Organelles*
- *Cell Nucleus*
- *Plant Cells*

5.3 Multicellular Organization
- *Tissues, Organs, and Organ Systems*
- *Organisms*

Chapter Concept

As you read, look for examples of how cell structures vary with their functions.

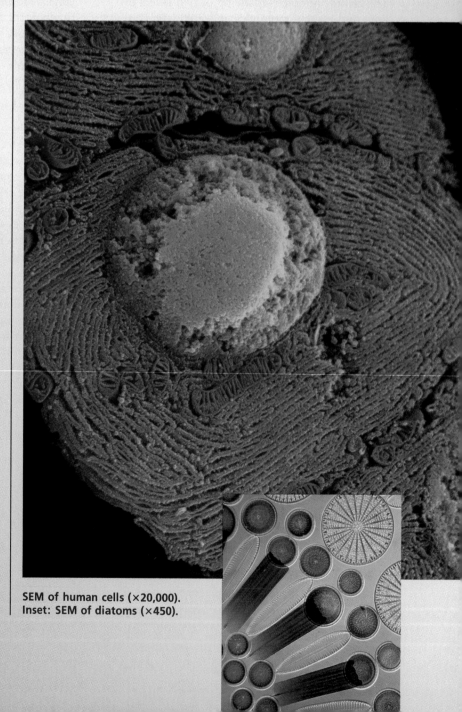

SEM of human cells (×20,000).
Inset: SEM of diatoms (×450).

5.1 Introduction to the Cell

A **cell** is the smallest unit that can carry on all the processes of life. Every living thing—from the tiniest bacterium floating in a drop of water to the largest elephant—is made up of one or more cells. A complete living thing that consists of only one cell is called a **unicellular organism.** Though small in size, unicellular organisms, such as the paramecium shown in Figure 5-1, are the most numerous of all life-forms. A living thing consisting of more than one cell is called a **multicellular organism.** Figure 5-2 shows that the buttercup, a multicellular organism, is made up of individual cells.

Discovery of the Cell

Only a few cells are large enough to be seen by the unaided eye. How, then, did scientists come to recognize that all forms of life are made up of cells? Our knowledge of cells began with the invention of the lens in the 1600s and grew along with advances in microscope technology. The light microscope allows us to view specimens magnified only up to about 2,000 times. At such magnifications plant and animal cells are visible, and many cell structures can be seen. The transmission electron microscope (TEM) is 1,000 times stronger than the light microscope. The scanning electron microscope (SEM) produces three-dimensional images. With these advances the smallest cells can be seen in clear detail.

Hooke and van Leeuwenhoek

Imagine the puzzlement of the English scientist Robert Hooke (1635–1703) when in 1665 he cut a thin slice of cork and observed it with a microscope. "I could exceedingly plainly perceive it to be all perforated and porous," he wrote, and further described it as consisting of "a great many little Boxes." Hooke used a microscope to observe the stems of elder trees, carrots, and ferns. He observed that each showed a similar formation. These "little Boxes" reminded him of the small rooms in which monks lived, so he called them cells.

What Hooke had observed were actually dead plant cells. The first person to observe living cells was a Dutch microscope maker, Anton van Leeuwenhoek (1632–1723), in 1675.

Section Objectives

- *Outline the discoveries that led to the development of the cell theory.*
- *State the cell theory.*
- *Identify a limitation in the size of cells.*
- *Describe the relationship between cell shape and cell function.*
- *Distinguish between prokaryotes and eukaryotes.*

Info: Specimens for the TEM must be sliced very thin and placed in a vacuum. Therefore only dead, preserved cells can be observed.

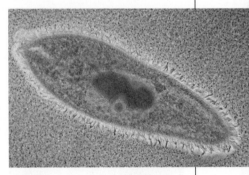

Figure 5-1   All the life activities of a paramecium, *Paramecium caudatum,* **shown here in color-enhanced magnification, occur within a single cell (×140).**

 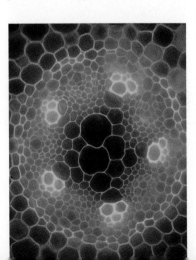

Figure 5-2   A buttercup, *Ranunculus* **sp. (left), a multicellular organism, is composed of individual cells like those in the root tip (right, ×60).**

Figure 5-3   A 1-mm cube has a surface area of 6 mm^2 and a volume of 1 mm^3, giving a surface-to-volume (S/V) ratio of 6 : 1. For a 2-mm cube, surface area and volume are 24 mm^2 and 8 mm^3 respectively, a 3 : 1 S/V ratio. For a 3-mm cube, surface area and volume are 54 mm^2 and 27 mm^3 respectively, and the S/V is 2 : 1.

Figure 5-4   Color-enhanced SEMs of human cells with structure-related functions include nerve (left), skin (center, ×750), and white blood (right, ×250).

The Cell Theory

About 150 years passed before scientists began to organize the observations of Hooke and van Leeuwenhoek into a unified theory. In 1838 a German botanist, Matthias Schleiden (1804–1881), concluded that all plants were composed of cells. A year later the German zoologist Theodor Schwann (1810–1882) came to the same conclusion about animals. In 1855 a German physician, Rudolph Virchow (1821–1902), while studying how disease affects living things, determined that cells come only from other cells. The observations of these three scientists, taken together, are known as the **cell theory**. This theory has three parts:

1. All living things are composed of one or more cells.
2. Cells are organisms' basic units of structure and function.
3. Cells come only from existing cells.

Cell Diversity

Although all living things are composed of cells, not all cells are alike. Even cells within the same organism may show enormous diversity in size, shape, and internal organization.

Size

Cells range in size from 2 m long, as in the thin nerve cells that extend down a giraffe's leg, to .2 micrometers (μm) in diameter, as in some bacteria. Most plant and animal cells are about 10 to 50 μm in diameter and are visible only with a microscope.

Many cells are roughly cuboidal or spherical. A cell with such a shape is limited in size by the ratio between its volume and its outer surface area. The food, oxygen, and other materials a cell requires must enter through its surface. Likewise, waste products must leave through its surface. The larger the cell, the larger the surface area required to maintain it. As a cell grows, its volume increases more rapidly than its surface area does. Figure 5-3

Tip: Remind students that because cells are pictured on a flat page, it is easy to mistakenly think of all cells as flat.

shows how growth affects the ratio between surface area and volume. Volume increases by the cube, while surface area increases by the square. This suggests that as a cell grows large, its surface area becomes too small to maintain life functions.

Shape

We have said that most cells are roughly cuboidal or spherical. Many plant cells, for example, are shaped like triskaidekahedrons; that is, they are roughly round but have 13 flat sides. However, cells differ widely in shape. This diversity of form reflects a diversity of function. Look at the three human cells in Figure 5-4. The long extensions that reach out in various directions from the nerve cell body allow the cell to receive and transmit nerve impulses. In contrast, the flat shape of dead skin cells is well suited to their function of covering the body surface. White blood cells have the ability to change shape. This allows them to move through narrow openings and to surround and isolate bacteria that invade the body.

Internal Organization

Cells also differ in their internal organization. Look at the photomicrograph of an algal cell in Figure 5-5 (top). Note the large structure near the center of the cell. This structure is the **nucleus,** which directs the activities of the cell. It contains DNA, the hereditary information that passes from one cell to another during cell division. DNA directs the formation of the proteins that carry out cell activities. Look again at Figure 5-5 (top), and notice the other smaller structures in addition to the nucleus. These are the cell organelles, many of them enclosed by membranes. An **organelle** is a cell component that performs specific functions in the cell. Cells that contain a nucleus and membrane-bound organelles are called **eukaryotes** *(yoo-KAR-ee-OTES).* The cells of plants, animals, fungi, protozoa, and algae are all eukaryotes.

Note the bacterium in Figure 5-5 (bottom). Do you observe a nucleus or organelles? Cells that lack nuclei and membrane-bound organelles are called **prokaryotes** *(proe-KAR-ee-OTES).* Bacteria and related microorganisms are prokaryotes.

Figure 5-5 Two TEMs contrast the unicellular alga (top), a eukaryote, with the bacterium *Bacillus licheniformis*, a prokaryote (bottom, ×136,000). Note the nucleus (the area within the dark circle) in the eukaryote.

Section Review

1. Define *nucleus,* and describe the structure and function of the cell nucleus.
2. What is the cell theory?
3. What single factor limits the size that most cells are able to attain?
4. How would you determine whether a unicellular organism was a prokaryote or a eukaryote?
5. How do you account for the fact that three different scientists made major contributions to the cell theory within a 17-year period?

Answers to Section Review
1. It directs the activities of the cell and contains DNA.
2. It states that all living things are made up of cells, that cells are the basic structural and functional units in organisms, and that cells come only from other cells.
3. the ratio between the volume and the outer surface area of the cell
(See pp. T5A–T5D.)

Tip: Point out that the wall forming at the center of the bacterium in Figure 5-5 indicates that the organism is beginning to divide.

Section Objectives

- *Describe the structure, composition, and function of the cell membrane.*

- *Name the major organelles found in the cell and describe their functions.*

- *Describe the structure and functions of the cell nucleus.*

- *Compare and contrast animal and plant cells.*

- *Describe how AVEC microscopy is useful in observing microtubules.*

5.2 Parts of the Cell

Each living cell carries out the tasks of taking in food, transforming food into energy, getting rid of wastes, and reproducing. In this section you will learn about the structures of eukaryotic cells that help perform these functions. Prokaryotic cells will be discussed in Chapter 20.

Although there is no typical eukaryotic cell, most have three main components:

1. The **cell membrane** is the outer boundary of the cell and separates the cell from its surroundings and other cells.
2. The **cytoplasm** *(SITE-uh-PLAZZ-um)* lies inside the cell membrane, contains water and salts, and surrounds organelles.
3. The nucleus contains DNA and directs the activities of the cell.

Cell Membrane

The cell membrane, sometimes called the plasma membrane, is the structure that separates the cell from its external environment. It gives shape and flexibility to the cell. The cell membrane is a complex barrier that keeps out some molecules but allows others to permeate, or pass, into the inside of the cell. The cell membrane is therefore called a **selectively permeable membrane.** Some scientists use the term *semipermeable membrane.* The membrane is about 7.5 to 10 nanometers (nm) thick.

The cell membrane is composed of two layers of molecules. Each layer is made up of a sheet of lipids. Recall from Chapter 4 that a lipid molecule has a carboxyl group as a "head" and a

Figure 5-6   A typical animal cell contains many organelles within a cell membrane.

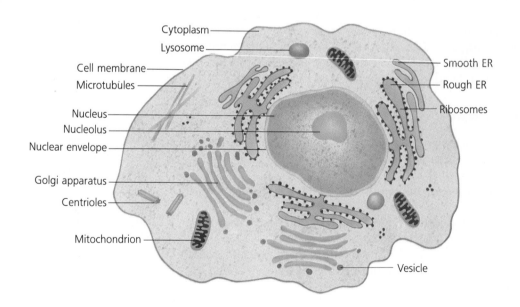

Cytoplasm
Lysosome
Cell membrane
Microtubules
Nucleus
Nucleolus
Nuclear envelope
Golgi apparatus
Centrioles
Mitochondrion
Smooth ER
Rough ER
Ribosomes
Vesicle

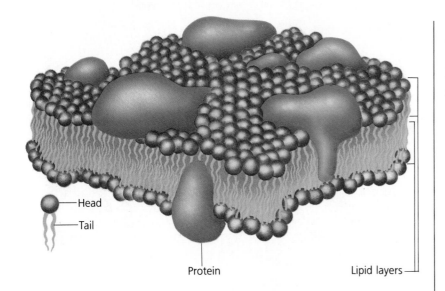

- Head
- Tail
- Protein
- Lipid layers

Figure 5-7 Lipids and proteins can move in relation to one another in the fluid mosaic model of the cell membrane. Why does this two-layer model look as though it is composed of three layers?

hydrocarbon chain as a "tail." As Figure 5-7 shows, the lipid molecules in membranes are arranged so that their heads form the outside of the membrane and their tails form the inside of the membrane. In addition, protein molecules are embedded in the lipid layers. Some are associated with only one layer, while others penetrate the entire membrane. You will learn more about how this structure affects transport of materials in Chapter 6.

The lipid molecules that form the membrane are fluid. They can move about relative to one another in a fluid manner. Some of the proteins are also free to move about, so that the mosaic, or pattern, of lipids and proteins changes. Because of these characteristics, scientists call their model of the dynamic cell membrane the **fluid mosaic model.**

Cytoplasm and Organelles

The cytoplasm is the jellylike material found inside the cell membrane. It surrounds organelles and contains water, salts, and organic molecules. The cytoplasm is in constant motion as particles and organelles move around. This motion, called **cytoplasmic streaming,** is most apparent in some plant cells and in protozoa.

Cytoplasm surrounds organelles. The word *organelle* means "little organ." Like the organs of the body, each organelle·performs a specific activity.

Ribosomes

The most numerous of the cell's organelles are **ribosomes** *(RIE-buh-SOMEZ)*, the organelles where proteins are made. As you can see in Figure 5-8, each ribosome is a spherical structure of about 15 to 20 nm in diameter. It is composed of three nucleic acid molecules and over 50 proteins. The ribosome is the site of protein

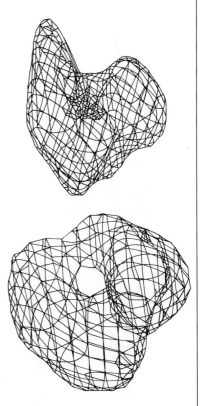

Figure 5-8 Ribosomes are formed from RNA and proteins. They function as the template for protein synthesis. Note their three-dimensional structure in top (top) and side (bottom) views.

Figure 5-9 Rough ER cells of the human pancreas are covered with ribosomes. These cells produce and export enzymes used in the digestive process.

synthesis, a process that you will learn more about in Chapter 8. The distribution of the ribosomes within the cell depends on how the proteins they produce will be used. Proteins used within the cell are produced by ribosomes that float freely in the cytoplasm. Proteins that will be exported for use outside the cell are produced by ribosomes attached to a system of membranes called the endoplasmic reticulum.

Endoplasmic Reticulum

The **endoplasmic reticulum** *(EN-duh-PLAZZ-mick ree-TICK-yuh-lum)* is a membrane system of folded sacs and tunnels. It is called ER for short. Endoplasmic reticulum that is covered with ribosomes, as shown in Figure 5-9, is known as rough ER; ER with few or no ribosomes is called smooth ER. The relative amounts of rough and smooth ER vary from one type of cell to another. Cells that make a lot of proteins that will be exported contain a lot of rough ER. For example, rough ER predominates in those cells of the pancreas that produce and export digestive enzymes. Smooth ER functions primarily as an intracellular highway, a path along which molecules move from one part of the cell to another. In addition, smooth ER can serve as a storage area for proteins that will later be exported from the cell.

Golgi Apparatus

The **Golgi apparatus** is the processing, packaging, and secreting organelle of the cell. It consists of a stack of membranes or sacs filled with fluid and dissolved or suspended substances. To understand how the Golgi apparatus works, follow the process in Figure 5-10. The Golgi apparatus operates like a production line in a factory, where a product is assembled at one end, then packaged,

Figure 5-10 An ingenious system moves a protein from its point of synthesis to the cell exterior.

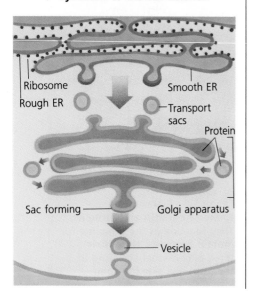

Ribosome
Rough ER
Smooth ER
Transport sacs
Protein
Sac forming
Golgi apparatus
Vesicle

and finally put into a mail bag at the other end. After protein is synthesized on a ribosome, the protein passes into the interior of the ER membrane. Then it moves through the interior of the membrane to an area of smooth ER. There the protein is enclosed in a vesicle, or membranous pouch, that buds off from smooth ER. The vesicle migrates to and fuses with a Golgi sac. The contents of the vesicle are then modified as they pass from sac to sac. Finally a new vesicle is formed from the last Golgi sac, which moves to the cell membrane and discharges its contents outside the cell.

Mitochondria

Scattered throughout the cytoplasm are large organelles that can grow, divide, or fuse with one another. These are the **mitochondria** *(MITE-uh-KAHN-dree-uh)*, the respiration centers of the cell. Chemical reactions within mitochondria release the energy from the nutrients taken into the cell. As Figure 5-12 shows, a mitochondrion has two membranes. The smooth outer membrane serves as the boundary between the mitochondrion and the cytoplasm. The inner membrane has many long folds, called cristae, that enlarge the internal surface area, providing more space for the reactions of cell respiration that take place there.

The mitochondrion is the place where adenosine triphosphate (ATP) is formed. You will learn more about this process in Chapter 7. ATP provides the chemical energy that drives the chemical reactions of the cell. Mitochondria are therefore most numerous in cells that use a lot of energy, such as muscle cells. A liver cell, which uses energy to carry out a host of biochemical activities, may contain as many as 2,500 mitochondria.

Mitochondria have their own DNA, and new mitochondria arise only when existing ones divide. These facts have led to the theory, discussed further in Chapter 14, that mitochondria developed from prokaryotic cells that came to live inside eukaryotic cells. The prokaryotes, according to the theory, may have sought protection by living inside the eukaryotes, supplying energy for the eukaryotes in turn.

Figure 5-11 TEM of a Golgi apparatus from a rat liver cell (top, ×34,000) is shown in greater detail in an artist's rendering (bottom).

Figure 5-12 TEM of a mitochondrion in a bat pancreas (left, ×79,000) is diagrammed in detail at right.

Inner membrane

Outer membrane

Lysosomes

Organelles that contain digestive enzymes are called **lysosomes** (*LIE-suh-SOMEZ*). These small, spherical organelles are surrounded by a single membrane. They exist primarily in animal and fungal cells and contain some 40 kinds of enzymes. These enzymes digest food particles, disease-causing bacteria captured by the white blood cells, and worn-out and broken parts of cells. Lysosomes also play a role in early development. For example, the embryonic human hand begins as a solid stump. Then lysosomal enzymes selectively destroy tissue and form fingers.

Microtubules and Microfilaments

Long, slender protein tubes called **microtubules** and fine protein threads called **microfilaments** help shape and support cells. Collectively, they form the **cytoskeleton,** the framework of the cell. They are assembled as needed and are then broken down and reassembled to form new structures. Specialized microtubules, called **spindle fibers,** aid in movement of chromosomes during cell division. Microfilaments, which lie just under the cell membrane, help move cellular materials and play a role in cytoplasmic streaming.

Biotechnology

Movement Along Microtubules

As early as 1879, scientists who observed the unicellular organism *Gromia* microscopically described an astonishing movement. Particles appeared to move along extensions from the organism at constantly varying speeds and in different directions.

This behavior seemed to be at odds with the usual ways that particles move in organisms. How did these particles move? What was supplying the energy?

Robert Day Allen, a professor of biology at Dartmouth College, was one of the many scientists who explored these questions. Allen pursued this mystery for 20 years until his death in 1986. He interpreted the old data and performed new experiments. He also applied new techniques he had developed in which a video screen was used to increase the contrast between the light and dark parts of a microscopic image. This technique was called AVEC, for Allen video-enhanced contrast microscopy. With AVEC, Allen was able to conclude that microtubules were the highways of both intra- and extracellular traffic.

Microtubules were not even known until 1963, and at

Cilia and Flagella

Cilia *(SILL-ee-uh)* and **flagella** *(fluh-JELL-uh)* are two kinds of structures that extend out from the surface of the cell and function in movement. Cilia are short, hairlike extensions that occur in large numbers on the surface of some cells. A unicellular paramecium, for example, moves forward by beating the 17,000 cilia that cover its surface. In many animals cilia also line the surface of organs, where they function to move fluids and mucus. The cilia in the human respiratory tract, for example, move mucus that traps dust and bacteria.

Flagella are whiplike, longer than cilia, and occur singly or in pairs. By whipping back and forth, flagella swiftly move unicellular organisms, such as *Euglena*, or cells in multicellular organisms. A sperm cell is rapidly propelled toward the egg by the power of its single flagellum.

Cilia and flagella are internally similar. The cross section in Figure 5-13 reveals that each is composed of nine pairs of microtubules located around its circumference and one pair of microtubules running down its center. Biologists refer to this as the "9 + 2" structure. Each cilium or flagellum is anchored to the cell by a cylindrical structure called a basal body.

Figure 5-13 This TEM of the cross section of a single cilium shows the typical 9 + 2 structure (×70,000).

first their fragility and their size—only 25 nm in diameter—created problems in viewing them. Researchers overcame the fragility by growing *Allogramia,* another unicellular organism, on microscope slides. The slides were treated with a positively charged substance that caused the organisms to stick to the glass. Then, using AVEC microscopy, Allen observed *Allogromia* microtubules and recorded the interaction of particles with the microtubules.

Allen turned his attention next to the giant nerve cell of the squid (right). Allen invited several squid experts to his laboratory. First they observed the squid nerve cell through a microscope. No movement was visible. Then Allen applied AVEC techniques and switched the nerve cell image to a television monitor.

"Suddenly the screen was teeming with tiny particles moving in both directions," Allen wrote later. Excitement rippled through the group as scientists saw the particles move along threadlike channels, the microtubules.

Using AVEC technology, Allen saw the mitochondria lurch forward in a jerky motion. He saw small Golgi sacs move swiftly "like dense schools of small fish" toward the surface of the cell.

Allen still had not explained how the particles moved along the microtubules and why they moved in opposite directions. Other researchers determined that microtubules had "side arms," or cross bridges, that connected particles to the microtubules. Allen and other biologists hypothesized that these side arms were made up of enzymes that provided the chemical energy needed to move the particles.

Figure 5-14 This SEM of a yeast cell prepared by the freeze-fracture technique shows the nuclear pores. In the freeze-fracture technique, cells are quick frozen and fractured along a line that splits membranes down the middle of the lipid layer, exposing the inner membrane.

Cell Nucleus

The nucleus is the site where nucleic acids are synthesized and it therefore directs the activities of the cell. It is surrounded by a double membrane called the **nuclear envelope.** Substances enter and leave the nucleus through holes in this envelope called nuclear pores. The pores form channels that allow large molecules, such as nucleic acids, to pass back and forth from the interior of the nucleus to the cytoplasm.

Inside the nuclear envelope is a dense, protein-rich substance called nucleoplasm. The nucleoplasm contains fine strands of **chromatin,** a combination of DNA and proteins. When a cell is dividing, a strand of chromatin coils up and condenses and can be seen under a microscope as a **chromosome.** Chromosomes thus are mostly DNA, the genetic information of the cell.

Most nuclei also contain at least one spherical body called the **nucleolus.** Ribosomes are synthesized and partially assembled in the nucleolus. After assembly they migrate to the cytoplasm.

Plant Cells

Most of the parts of the cell just described are common to all eukaryotic cells. Fungal cells and cells of protozoa and algae have some modifications, which will be discussed in the chapters on those organisms. Plant cells possess some structures in addition to those just described. In plants, a **cell wall** surrounds the cell membrane and helps support and protect the cell. The cell wall is one characteristic that distinguishes a plant cell from an animal cell. Two others are fluid-filled cavities or sacs called **vacuoles** and large organelles called **plastids,** which store food or pigments.

Cell Wall

A cell wall is the rigid covering of a plant cell. It is made primarily of long chains of cellulose embedded in hardening compounds such as pectin and lignin. Pores in the cell wall allow ions and molecules to pass to and from the cell membrane. Cell walls are of two types. Primary cell walls are formed during cell growth. Secondary cell walls are formed after growth has ceased. Each type of wall aids in strengthening the cell.

When a plant cell divides, a middle lamella is first formed between the new cells. The **middle lamella** is an intercellular glue made of pectin, the substance that makes jelly gel. Next, the primary wall is formed between the middle lamella and the cell membrane. Its cellulose fibers are laid down in layers. Each layer has all of the fibers oriented in one direction. The growing cell is thus able to expand sequentially at a right angle to the most recent fiber layer. This structure functions both to protect the cell and simultaneously to allow the cell to grow.

After growth ceases, a secondary cell wall is formed. This wall is made of cellulose and lignin fibers interwoven so that no further expansion of the cell is possible. The secondary wall functions to strengthen the mature cell. A compound called lignin in secondary walls makes them woody. When you pick up a piece of wood, you hold in your hand secondary cell walls.

Vacuoles

Vacuoles are organelles found only in plant cells. Like lysosomes in animal cells, vacuoles store enzymes and waste products. In a mature plant cell, a vacuole typically takes up as much as 90 percent of the volume, crowding the other structures into the

Figure 5-15 Most organelles of a plant cell (below left) are also found in animal cells. Which organelles are not common to both? The elements that make up a cell wall appear at the right.

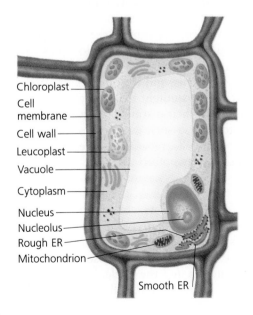

Chloroplast
Cell membrane
Cell wall
Leucoplast
Vacuole
Cytoplasm
Nucleus
Nucleolus
Rough ER
Mitochondrion
Smooth ER

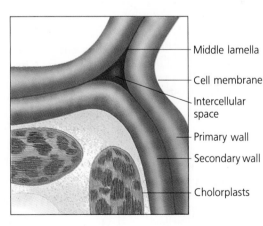

Middle lamella
Cell membrane
Intercellular space
Primary wall
Secondary wall
Cholorplasts

Table 5-1 Comparison of Prokaryotes and Eukaryotes

Feature	Prokaryotes	Eukaryotes	
		Animals	Plants
Average size (diameter)	1–10 μm	10–20 μm	30–50 μm
Cell membrane	Yes	Yes	Yes
Cell wall	Yes	No	Yes
Nucleus	No	Yes	Yes
Ribosomes	Yes	Yes	Yes
Endoplasmic reticulum	No	Yes	Yes
Golgi complex	No	Yes	Yes
Mitochondria	No	Yes	Yes
Vacuoles	No	Small or absent	Usually a single large one per cell
Lysosomes	No	Often present	Rare
Microtubules	No	Yes	Yes
Cilia and flagella	Yes	Often present	Mostly absent except for sperm

Think

In many plant cells the vacuole takes up a large space. As the above photograph of a spinach leaf cell shows, this crowds the other organelles to the edge of the cell. In addition to providing a storage area for nutrients and wastes, how else might the large area occupied by a vacuole function to benefit the cell?

Answers to Section Review
1. the cell membrane, the cytoplasm, and the nucleus
2. The protein molecules form patterns, or mosaics, in and on the surface of the membrane. The lipid molecules and some protein molecules move relative to one another. As the molecules change positions, the patterns also change.
(See pp. T5A–T5D.)

space that's left. Some of the waste products stored by vacuoles are toxic materials that need to be kept away from the rest of the cell. Some acacia trees store cyanides in their vacuoles as a poisonous defense against plant-eating animals.

Plastids

Plants depend on light as a source of energy. Solar energy must be converted into chemical energy and then be stored. In plants plastids carry out these functions. **Chloroplasts** contain a green pigment called chlorophyll that absorbs sunlight as the first step in the conversion process. You will learn more about chloroplasts in Chapter 7. **Chromoplasts** synthesize and store pigments such as orange carotenes, yellow xanthophylls, and various red pigments, some of which function in trapping sunlight for energy. Chromoplasts are what give certain plants their distinctive colors. **Leucoplasts** (LOO-kuh-PLASTS) store food such as starches, proteins, and lipids. Present in most plant cells, leucoplasts are especially prominent in storage organs such as potato tubers.

Section Review

1. Name the three main parts of the eukaryotic cell.
2. Summarize the fluid mosaic model of the cell membrane.
3. Compare and contrast the vacuole and the lysosome.
4. Distinguish between the nucleus and the nucleolus.
5. Why do you think plant, and not animal, cells have walls?

5.3 Multicellular Organization

Biologists suggest that the earliest cells on earth were simple pro-karyotic cells similar to some present-day bacteria. Like most bacteria they lacked the internal structures to synthesize their own food; they depended on organic nutrients in the environment. As they reproduced and their numbers increased, however, they began to compete for limited natural resources. With increasing competition cells with adaptations evolved. Some of these cells were eukaryotic. These unicellular eukaryotes may have existed in temporary groups, or colonies, with other cells, with some cells acting as special energy producers. Eventually, such temporary colonies may have evolved into multicellular organisms in which groups of cells became specialized for various functions. You will learn more about the evolution of prokaryotes and eukaryotes in Chapter 14.

Tissues, Organs, and Organ Systems

You have learned how the organelles within a cell perform specific functions that contribute to cell maintenance. Division of labor is also carried out by the individual cells that compose a multicellular organism. Each cell depends on other cells to perform one or more of the functions that keep the entire organism alive. The division of labor among cell types is called cell specialization.

In most multicellular organisms, cells are organized into tissues. A **tissue** is a group of similar cells that carry out a common function. Epithelial tissue, for example, consists of sheets of closely packed cells that form surface coverings, such as the outermost living layer of the skin and the lining of the nose. Cells that pull against one another make up muscle tissue. Cells that are specialized for transmitting messages make up nervous tissue.

Several types of tissue that interact to perform a specific function form an **organ**. The stomach, for example, is an organ. In the

Section Objectives

- *Distinguish between tissues, organs, and organ systems.*

- *Describe the way in which cells are organized in multicellular organisms.*

Info: The cells of multicellular organisms have spaces between them that are filled with a watery material called interstitial fluid.

Figure 5-16 Spongy tissue (left) is the foundation for the tiny branching tubes and air sacs that form the lungs, an organ (center). The lungs, in turn, are part of the respiratory system (right).

Tissue

Organ

Organ System

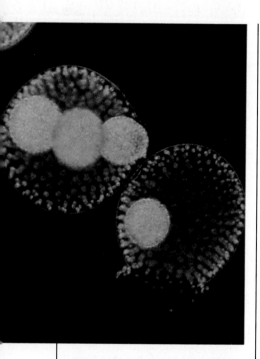

stomach, muscle tissue causes movement, epithelial tissue secretes enzymes, and nervous tissue transmits messages back and forth between the stomach and the brain.

An **organ system** is made up of a group of organs that work together to perform a set of related tasks. For example, the mouth, esophagus, stomach, intestines, and several other organs make up the digestive system. Each of these organs performs a specific aspect of the complex process we refer to as digestion.

Plants have tissue systems and organs. These are organized somewhat differently from those in animals. A dermal tissue system forms the outer layer of a plant. A ground tissue system makes up the bulk of roots and stems. A vascular tissue system is the water transport system of the plant. The three plant organs are roots, stems, and leaves.

Organisms

In a unicellular organism one cell carries out all the functions of life. Some life-forms, called colonial organisms, appear to be neither unicellular nor multicellular. A **colonial organism** is a group of more or less similar cells that live together in closely connected groups. An example is *Volvox*, a green alga that can be seen in Figure 5-17.

The organization of *Volvox* may give a clue to how multicellular organisms evolved. In *Volvox* each cell maintains its own individual existence. Sometimes, however, individual cells carry out specialized functions, such as reproduction or movement, for the whole colony. Each hollow *Volvox* sphere contains 500 to 60,000 cells. The spheres move in the water by the combined beating of the flagella of the specialized outer cells. A few of the cells are specialized for reproduction. They produce daughter colonies, which you can see as the darker spheres floating within the main sphere.

The same principle of specialization found in *Volvox* is found within all multicellular organisms. However, in multicellular organisms cells are so specialized that they are dependent on the function of other cells of the organism.

Section Review

1. Explain how a cell, a tissue, and an organ are related.
2. Give an example of an organ system, and name some of the parts that form it.
3. Name the plant tissue systems and organs.
4. To what extent are the individual cells within a *Volvox* colony independent of one another?
5. Green algae, such as *Codium*, enlarge by dividing the nucleus but do not form cell walls between the parent and the daughter cells. Would you call *Codium* a unicellular or a multicellular organism? Explain your answer.

Figure 5-17 New *Volvox* colonies are formed in the interior of other colonies and are released when the old colony ruptures.

Tip: Students are usually surprised to learn that the skin—with its complex system of cell layers, nerve endings, glands, and fat tissue—is considered an organ.

Answers to Section Review
1. A tissue is composed of one or more types of cells that are specialized to perform a particular function. An organ is composed of one or more types of tissue that perform a specialized function.
2. One possible example would be the digestive system, made up of the mouth, esophagus, stomach, and intestines.
(See pp. T5A–T5D.)

Laboratory

Comparing Plant and Animal Cells

Objective

To examine similarities and differences in the morphology of plant and animal cells

Process Skills

Hypothesizing, classifying, observing

Materials

1 microscope slide, 1 coverslip, *Elodea* plant, compound microscope, prepared slide of human cheek cell

Method

1 Make a wet mount slide of a young *Elodea* leaf taken from near the growing tip of the plant.
2 Set the microscope on low power. Place the slide on the microscope stage, and focus on the top layer of cells. Make a drawing showing the shape of the *Elodea* cells. How would you describe the shape? Locate the cell wall and label it on your drawing.
3 Focus the microscope slightly above the cells. Then slowly focus downward through the top cell layer. This should make it possible to observe several layers of cells. What is the advantage of being able to see several cell layers at once?
4 Focus on the upper or lower edge of a cell. Note the green chloroplasts scattered across the surface. The chloroplasts may be moving in one direction along the cell membrane. This phenomenon is called cytoplasmic streaming. Describe it in words.
5 The central vacuole is like a water-filled balloon inside a box. You cannot see the vacuole with the microscope, but you can see evidence of its presence. Focus on either side of the cell. What evidence do you see of the presence of the central vacuole?
6 Note the cell nucleus. It looks like a small,

clear ball. Examine the nucleus under high power. Draw and label it.
7 Place a prepared cheek cell slide under the microscope. Locate the cells under the microscope under low power and make a drawing showing their shape. Note the cell nucleus. Examine the nucleus under high power. Add the nucleus to your drawing and label it.

Conclusions

1 Compare the morphology of *Elodea* and cheek cells. Explain the reason for the differences between the two.
2 Compare the location of the nucleus in plant and animal cells. Account for the difference.

Inquiry

1 The cheek cell slides were stained so that they could be seen more clearly. Do you think that adding stain to a slide changes the cells in any way in addition to staining them? How could you test your hypothesis?
2 Methylene blue is often used to stain slides. What do you think would happen if you stained the *Elodea* leaf with methylene blue?

Chapter 5 Review

Vocabulary

cell (63)
cell membrane (66)
cell theory (64)
cell wall (72)
chloroplast (74)
chromatin (72)
chromoplast (74)
chromosome (72)
cilia (71)
colonial organism (76)
cytoplasm (66)

cytoplasmic streaming (67)
cytoskeleton (70)
endoplasmic reticulum (68)
eukaryote (65)
flagella (71)
fluid mosaic model (67)
Golgi apparatus (68)
leucoplast (74)
lysosome (70)

microfilament (70)
microtubule (70)
middle lamella (73)
mitochondria (69)
multicellular organism (63)
nuclear envelope (72)
nucleolus (72)
nucleus (65)
organ (75)
organelle (65)

organ system (76)
plastid (72)
prokaryote (65)
ribosome (67)
selectively permeable membrane (66)
spindle fibers (70)
tissue (75)
unicellular organism (63)
vacuole (72)

1. Look up in a dictionary the meaning of the word parts *eu-*, *pro-*, and *kary-*. Explain why the words *prokaryote* and *eukaryote* are good terms for the organisms they describe. What do the terms suggest about the evolution of these organisms?
2. Describe the structure and function of a cell membrane, and compare it with the structure and function of a cell wall.
3. Look up in a dictionary the meaning of the word part *cyto-*. Use what you have learned to explain the meaning of the terms *cytoplasm* and *cytoskeleton*.
4. Explain the relationship between microtubules, cilia, and flagella.
5. Choose the term that does not belong in the following group, and explain why: Golgi apparatus, endoplasmic reticulum, chromatin, and mitochondria.

Review

1. A prokaryote has (a) a cell nucleus (b) a cell membrane (c) organelles (d) all of the above.
2. The growth of cells is limited by the ratio between (a) volume and surface area (b) organelles and surface area (c) organelles and cytoplasm (d) nucleus and cytoplasm.
3. A cell membrane is composed of (a) lipids (b) proteins (c) nucleic acids (d) lipids and proteins.
4. The function of the Golgi apparatus is to (a) synthesize proteins (b) release energy (c) process and package proteins (d) synthesize lipids.
5. Mitochondria (a) transport materials (b) release energy (c) make proteins (d) control cell division.
6. Lysosomes function in cells to (a) recycle cell parts (b) destroy disease-causing agents (c) shape developing body parts (d) all of the above.
7. The nucleolus is (a) the control center of the cell (b) the storehouse of genetic information (c) the site where ribosomes are synthesized (d) none of the above.
8. Plant cells differ from animal cells in having (a) fluid-filled vacuoles (b) cell walls surrounding the cell membrane (c) chromoplasts (d) all of the above.
9. Leucoplasts (a) synthesize proteins (b) store food (c) synthesize pigments (d) store pigments.
10. The stomach is an example of (a) a tissue (b) an organ (c) an organ system (d) none of the above.

11. How have microscopes been helpful in the study of cells? Explain.
12. What limits the size of cells?
13. Why is the cell membrane called a selectively permeable membrane?
14. Explain how the ribosomes, the endoplasmic reticulum, and the Golgi apparatus function together in protein synthesis.
15. Distinguish between the structure of rough ER and that of smooth ER.

16. Describe the inner structure of mitochondria, and explain their function.
17. How has AVEC technology enhanced our understanding of the structure and function of microtubules?
18. What is the difference between chromatin and chromosomes?
19. How are cell walls formed?
20. What is cell specialization? Give an example of cell specialization.

Critical Thinking

1. A mature human red blood "cell" is merely a membrane surrounding hemoglobin, the protein pigment that carries oxygen. Suggest why this "cell" has no organelles.
2. The coils of a radiator provide a large surface area from which heat is radiated into a room. Which cell organelles have a structure similar to that of a radiator? How is this structure related to their function?
3. Children who are born with Tay-Sachs disease are unable to carry out the normal breakdown of lipids in the brain because of a defect in a cell organelle. Based on this information, which cell organelle do you think may be defective in Tay-Sachs victims? Explain.
4. Livestock in the western United States often die after eating a locoweed, such as *Astragalus toanus*. The chemical the plant

contains is also poisonous to plants. How does locoweed keep from poisoning itself?
5. Compare the two photos above. One shows cells from a grass stem, and the other shows cells from the trunk of an oak tree. Which photo do you think shows the tree cells? Why do you think so?
6. What characteristic of eukaryotic cells gives them a greater capacity for specialization than prokaryotic cells have? Explain your answer.

Extension

1. Read J. Stephenson's article, "Green Genes: Spice of Life," *Science Digest,* September 1986, p. 20. Prepare an oral report for the class on how genetic engineers use knowledge of membrane structure and function to engineer new plants.
2. Use the resources in your school or public library to learn more about the work of Schleiden, Schwann, or Virchow. Write a brief report summarizing the processes

that the researcher used to arrive at his conclusions about cells.
3. Place a sprig of the aquarium plant *Elodea* in water, and shine a bright light on it for an hour. Then examine a leaf under the microscope and make a drawing, noting the direction of chloroplast movement. How do you think the movement of cell organelles helps the cell to function?

Chapter 6 | *Homeostasis and Transport*

This chapter teaches students how the cell membrane plays a role in maintaining homeostasis between the cell and its environment. Students learn about the processes of diffusion, osmosis, and other kinds of transport.

Correlation Chart

Section Number	Section Objectives	Pages	Features	Section Review
6.1	1. Define homeostasis.	81–83	Think, p. 82	2
	2. Describe the movement of molecules to achieve equilibrium.			2, 3, 5
	3. Explain the process by which water molecules move across a membrane.			1, 2, 3, 4
	4. Differentiate between diffusion and osmosis.			1
6.2	1. Describe how carrier molecules transport materials across the cell membrane.	84–89	Biology in Process, pp. 86–87	2, 3, 5
	2. Distinguish between facilitated diffusion and active transport.			1, 3
	3. Explain how protein channels function in cell transport.		Biology Lab, p. 89	2
	4. Compare and contrast endocytosis and exocytosis.			4
	5. Describe the process that led researchers to hypothesize that communication between cells occurred.			

Resources*

Print

Robertson, R.N. *The Lively Membranes.* New York: Cambridge University Press, 1983.

Wolfe, S. L. *Biology of the Cell,* 2nd ed. Belmont, Calif.: Wadsworth, 1981.

Audiovisual/Software

Biology: The Cell. Apple 48K. 2 disks. Scholastic, 1986.

Cell Biology: Life Functions. Film or video, 20 min. Coronet.

Diffusion and Osmosis. Film or video, 11 min. Coronet.

Diffusion and Osmosis. Film, 14 min. Britannica.

*Additional references are listed in *Modern Biology, Teacher's Program Guide.*

Section Teaching Guides

6.1 Diffusion and Osmosis

Approach

The main objective of this section is to demonstrate how the cell membrane functions to maintain equilibrium between the cell and its environment. It builds on information about the structure of the cell presented in Chapter 5 and on the definitions of molecule, solution, solute, and solvent given in Chapter 3. In presenting the text, help students develop an awareness of the cell as a dynamic, changing organism that is constantly adjusting to changes in the surrounding environment.

Concept Framework

All students will be familiar with the concepts of diffusion and osmosis even though the terms may be new to them. Review these concepts by citing common examples, such as the diffusion of smoke through the air or food coloring in a glass of water. To relate the concept to plant and animal cells, ask students why placing celery stalks or carrot sticks in water keeps them crisp while salting the vegetables or placing them in salty water wilts them. You may wish to end the discussion by asking the students to consider as they read the text how the structure of the cell membrane facilitates these changes in the plant.

Emphasis

Make certain that students understand that substances such as organic and inorganic compounds are constantly moving in and out of the cell through the cell membrane.

Constructing Diagrams

Divide the class into teams of three students each. Direct each team to develop diagrams or drawings to illustrate the conditions of solute and cell concentrations shown on Table 6-1 and to present their drawings to the class.

Section Review Answers

(Continued from p. 83.)
4. The cell is hypertonic relative to its environment, because turgor pressure is caused by water moving from the outside into the cell.

5. It is dangerous because the humans' cells will lose water. If too much water is lost, the cell will shrivel and die.

6.2 Other Kinds of Transport

Approach

The main objective of this section is to explain how the cell membrane functions to regulate the passage of materials in and out of the cell in ways other than simple diffusion. The text builds on the discussion of electrons, energy levels, and ionic bonds in Chapter 3. An understanding of active transport as exemplified by the sodium-potassium pump will help students deal with the difficult concept of chemiosmosis that occurs in Chapter 7. In presenting the text, make sure that students understand that active transport requires the expenditure of energy. You may wish to direct students to create a chart of the various types of cell transport, with one column identifying the transport mechanism as either active or passive, and another column giving examples. Have students fill in the chart as they read.

Concept Framework

Most students will be familiar with the concepts of active and passive transport if presented to them in the form of analogies. For active transport, use the analogy of a cyclist exerting energy to pedal up a hill. For passive transport, ask students to think of a cyclist coasting down a hill. Point out that chemical energy is required for active transport of materials through a cell membrane, while no energy is required for passive transport. You may wish to end the discussion by asking students to consider while they read the text the sources of the energy that make active transport possible across a cell membrane.

Emphasis

Make certain that students understand that all forms of transport across the cell membrane work either for or against a gradient, such as a concentration gradient or an electric gradient.

Guest Speaker

Invite a cell biologist from a university or research laboratory to speak on current research on cell membranes. Ask the speaker to bring TEM or SEM slides of cell membranes to show the class.

Research Report

Invite interested students to research the technique of freeze-fracturing and its usefulness to the study of cell membranes. Ask the students to present a report to the class, and to include, if possible, illustrations taken from or based on articles in specialized books or in science journals.

Using the Features/Connections Between Cells

The main process skills used in this feature are observing and experimenting. Werner Loewenstein and Yoshinobu Kanno observed that electrically charged ions passed from one cell to another. The two scientists then devised two separate series of experiments to first, determine the size range of molecules that could pass through a gap junction, and then, how various ions affected the permeability of the junction. Point out that experimenting also includes the process skills of collecting data, inferring, and hypothesizing. Discuss with students other possible experimental uses of tracer molecules, such as tracing the path of carbon dioxide when taken up by plants.

Section Review Answers

(See p. 88.)

Laboratory Preparation

Background

In this lab students observe the osmotic forces that are vital in the support and movement of plant structures. Students should be aware that these forces produce four or five atmospheres of pressure, as compared to the two to four atmospheres of pressure commonly put into tires. From this comparison, students should understand why cell membranes are easily broken by osmotic pressure and why strong cell walls are required. The size of the potato sticks is important. If they are too thin, osmotic pressure will not support the stick in the air. If they are too thick, the sticks will not bend when osmotic pressure is lost. Students will not see rapid changes in the purple onion cells. In order to observe rapid changes, cells

with a high ratio of surface area to volume, e.g., *Nitella,* should be used.

Preparation

To prepare 10% salt solution, place 10 g of NaCl in a graduated cylinder and add distilled water to the 100 mL mark.

Safety

Remind students to handle scalpels with care.

Conclusions

1. The salt solution caused water to leave the cells because the concentration of solute outside the cell was greater than the concentration inside the cell. Water moves from an area of lower solute concentration to an area of higher solute concentration.
2. Water caused more turgor pressure. Water moved into the cell, swelling the cytoplasm and increasing its pressure against the cell wall.

Inquiry

Students could set a standard for measuring potato stick bend, such as measuring the length of the stick and its length when picked up in the center with a fork. Students could then test various dilutions of the salt solution to find a solution that produces the same amount of bend as that of a fresh-cut stick.

Chapter Review

Vocabulary

1. Diffusion is the movement of molecules from an area of higher concentration to an area of lower concentration. Facilitated diffusion also refers to the movement of molecules from high to low areas of concentration, but carrier molecules speed up the process of diffusion.
2. A selectively permeable membrane is one that allows some substances to pass through but not others. The cell membrane is semipermeable.
3. As turgor pressure inside a plant cell decreases, the cell contents shrink. This is the condition known as plasmolysis.
4. A contractile vacuole is an organelle that collects water and then by contracting, forces the water outside the organism.

5. Pinocytosis means "drinking cell" and is named for the process during which the cell surrounds and takes in liquids or small particles. Phagocytosis means "eating cell" or "destroying cell" and is named for the similar endocytosis process during which a cell surrounds and ingests solid substances. *Cyto-* means "cell".

Review

11. Equilibrium is when molecules moving randomly through a space are evenly distributed.
12. No, because some molecules are too large or of an incompatible type and some membranes are selectively permeable, allowing some substances and not others to pass through the membrane.
13. The organism may burst, because although water continues to flow into it from the surrounding environment, it has no means of getting rid of the excess water.
14. Direction depends on concentrations of water and solutes in the solutions outside and inside the cell.
15. It is a boundary layer that prevents the flow of molecules from aquatic environment to aquatic environment.
16. The carrier molecule binds to a particular substance, allows it to be transported across the membrane, and releases the substance.
17. facilitated transport
18. ATP is the energy source used by the cell to work the sodium-potassium pump.
19. Endocytosis is the process by which solid substances that are too large to cross through the cell are brought into the cell. Exocytosis is the process by which large substances are moved out of the cell.
20. Researchers observed that electrically charged ions moved from cell to cell without moving outside the cells. The hypothesis was confirmed when the researchers observed that flourescent-tagged tracer molecules of various sizes also moved from cell to cell.

Critical Thinking

1. It is like turgor pressure in that the cell membrane is kept taut by the outward motion of the water molecules. It is different in that the cell membrane is more permeable to water than the balloon is to air. Furthermore, the air is forced into the balloon by mechanical pressure, whereas the water is drawn into the cell by the presence of a higher solute concentration than that on the outside.
2. The organism has a selectively permeable membrane through which unwanted water will enter, and it will use a contractile vacuole to pump it out. The sump pump uses electric energy, and the vacuole used energy from ATP.
3. It might suggest that the two layers that form the outside and the inside of the cell membrane are essentially the same. It also underscores the dynamic, flexible nature of the cell membrane.

Chapter 6 | Homeostasis and Transport

Introduction

A cell exists in a constantly changing environment. By adjusting to changes, a cell maintains a balance with its environment. In the previous chapter you learned that the cell membrane is the boundary of the cell. In this chapter you will learn how the cell membrane functions to regulate the passage of materials into and out of the cell, maintaining a balance between the cell and its environment.

Chapter Outline

6.1 Diffusion and Osmosis
- *Diffusion*
- *Osmosis*

6.2 Other Kinds of Transport
- *Carrier Transport*
- *Gated Channels*
- *Endocytosis and Exocytosis*

Chapter Concept

As you read, look for the ways in which a cell regulates the movement of materials and maintains homeostasis.

TEM of endocytosis (×80,000). Inset: Artist's representation of atoms in motion.

For additional teaching information see pp. T6A–T6D.

6.1 Diffusion and Osmosis

To remain alive and function optimally, cells, tissues, organs, and organisms must maintain a biological balance with their immediate environment. Cells maintain this biological balance, or **homeostasis,** by controlling and regulating what gets into and out of the cell. Though cells can exist within a range of environmental states, there is a limit to how much they can adjust.

Diffusion

To understand how cells maintain homeostasis, you must understand how molecules move. Molecules are in constant motion. They travel in straight lines until they hit something, then rebound, still traveling in a straight line but in a different direction. They rebound again, and so on. As the molecules move, they tend to move into those areas where the molecules are less concentrated. This tendency results in an overall direction of movement among the molecules. **Diffusion** is the process by which molecules spread from an area of greater concentration to an area of lesser concentration.

Equilibrium

An example of diffusion occurs when you put a cube of sugar in water. The sugar molecules diffuse away from the cube into the water. At first the molecules are more concentrated near the sugar cube and less concentrated farther from the cube. This difference in concentrations of a substance across space is called a **concentration gradient.** Eventually the sugar molecules diffuse throughout the water so that they are equally distributed. When the concentration of the molecules of a substance is the same throughout a space, a state called **equilibrium** exists. Once equilibrium is established the random movement of molecules continues, and equilibrium is maintained.

Section Objectives

- *Define homeostasis.*

- *Describe the movement of molecules to achieve equilibrium.*

- *Explain the process by which water molecules move across a membrane.*

- *Differentiate between diffusion and osmosis.*

Tip: To help students understand Brownian motion, put a drop of India ink on a slide, add a cover-slip, and have students observe it under the microscope at high power.

Figure 6-1 Ink dropped into a beaker of water (left) diffuses from areas of high concentration to areas of lower concentration. Over time the ink diffuses evenly throughout the water (center and right). Eventually equilibrium is established.

The normal red blood corpuscles in the top photo are in a plasma solution. Now look at the red blood corpuscles in the bottom photo. Do you think they are in a salty solution or in distilled water? Why?

Answer to Think
They are in distilled water. The corpuscles are swollen because water has diffused into them. This suggests they are in a hypotonic solution. The concentration of solute molecules inside the cell membrane was higher than the concentration of solute molecules in the solution outside the cell membrane.

Diffusion Across Membranes

Some kinds of molecules can also diffuse across a membrane. They do this by moving between the molecules that make up the membrane. If a substance can pass through a membrane, the membrane is **permeable** to it. As with all diffusion, molecules diffuse from an area of greater concentration on one side of the membrane to one of lesser concentration on the other.

Not all molecules can diffuse through all membranes. The ability of a molecule to pass through a membrane depends on the size and type of molecule and the molecular structure of the membrane. Membranes that let only some substances through are called selectively permeable membranes, or semipermeable membranes. Cell membranes are semipermeable; thus they regulate what gets into and out of a cell. Oxygen, for example, enters by diffusion, but large molecules like starch do not.

Osmosis

Recall from Chapter 3 that a solution is composed of a solute dissolved in a solvent. In the example cited on the previous page sugar is the solute and water is the solvent. In organisms the solvent is water and the solutes are inorganic and organic compounds. In the sugar example the solute diffused. However, it is also possible for the solvent to diffuse. The process by which water molecules diffuse through a membrane from an area of greater concentration to an area of lesser concentration is called **osmosis** *(ahz-MOE-suss)*. Note that in biology osmosis refers only to the diffusion of water.

Direction of Osmosis

What determines the direction in which the water molecules diffuse across a cell membrane? The direction depends on the concentrations of water and of solutes dissolved in the solutions. Study Table 6-1. If the concentration of solute molecules in the

Table 6-1   Direction of Osmosis

Conditions	Environment solution is	Cell solution is	Water will move
If solute concentration in the environment is lower than in the cell	Hypotonic	Hypertonic	Into the cell
If solute concentration in the environment is higher than in the cell	Hypertonic	Hypotonic	Out of the cell
If solute concentration in the environment is equal to that in the cell	Isotonic	Isotonic	Water will not move

environment outside the cell is lower than that in the cell, the solution outside is **hypotonic** relative to its environment. The prefix *hypo-* means "lower." Water will move into the cell until equilibrium is established. When the concentration of solute molecules outside the cell is greater than that inside, the solution outside is **hypertonic** relative to its environment. The prefix *hyper-* means "higher." Water will diffuse out of the cell until equilibrium is established. When the concentration of solutes outside and inside the cell are equal, the solution is **isotonic** relative to its environment. The prefix *iso-* means "equal." Water will not diffuse in either direction.

Notice that the prefixes *hypo-*, *hyper-*, and *iso-* refer to the relative concentration of solute. Remember that a hypotonic solution exists in relation to a coexisting hypertonic solution. A hypertonic solution exists in relation to a coexisting hypotonic solution. An isotonic solution has an adjacent isotonic solution.

Role of Osmosis

A plant cell swells as it fills with water. It does not swell endlessly because it is restricted by cell walls. As water diffuses into the cell, pressure—called **turgor pressure**—increases. The pressure forces the cytoplasm and cell membrane against the cell wall and the cell becomes rigid. Such a cell has high turgor pressure.

Turgor pressure is maintained while the cells are in a hypotonic environment. If the environment loses water or gains solutes, the concentration of solutes outside the cell may become greater than that inside. Water then leaves the cell, turgor pressure is lost, and the cell wilts. This condition is called **plasmolysis** *(plazz-MAHL-uh-suss)*. The reverse condition, in which cells take in so much water that they burst, is called **cytolysis.**

Unicellular freshwater organisms exist in a hypotonic environment and are thus hypertonic relative to their environment. Water will therefore diffuse into the cell. Since these organisms need to maintain a relatively lower concentration of water inside the cell to function normally, they must rid themselves of the excess water that enters the cell by osmosis. Some organisms do this with **contractile vacuoles,** organelles that remove water. The contractile vacuoles collect the excess water and then contract, squeezing the water out of the cell.

Section Review

1. Distinguish between *diffusion* and *osmosis.*
2. How does the cell membrane help maintain homeostasis?
3. What is one factor that determines the direction in which molecules move across a membrane?
4. When the turgor pressure of a plant cell increases, does that indicate that the cell is hypotonic, hypertonic, or isotonic relative to its environment? Explain your answer.
5. Why would drinking sea water be dangerous for humans?

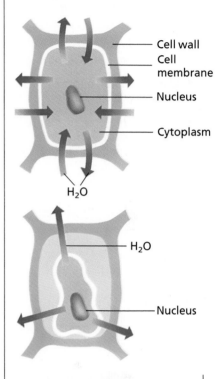

Figure 6-2   The plant cell at the top is in osmotic balance since conditions are such that water is leaving and entering the cell in equal amounts. In the plant cell at the bottom, a hypertonic solution outside the cell has caused plasmolysis.

Cell wall
Cell membrane
Nucleus
Cytoplasm
H_2O

H_2O
Nucleus

Answers to Section Review
1. Diffusion is the movement of molecules from an area of higher concentration to an area of lower concentration. Osmosis is the diffusion of water molecules through a membrane.
2. The membrane allows certain molecules to pass through the membrane and restricts the passage of others.
3. the concentration gradient across the membrane
(See pp. T6A–T6D.)

Section Objectives

- *Describe how carrier molecules transport materials across the cell membrane.*

- *Distinguish between facilitated diffusion and active transport.*

- *Explain how protein channels function in transport.*

- *Compare and contrast endocytosis and exocytosis.*

- *Describe the process that led researchers to hypothesize that communication between cells occurred.*

6.2 Other Kinds of Transport

Diffusion and osmosis are forms of passive transport. **Passive transport** is the movement of any substance across a membrane without the use of chemical energy. **Active transport** is the movement of any substance across a cell membrane with the use of chemical energy. Materials cross the cell membrane by either passive or active transport, depending on the size and chemical makeup of the material. The structure of the cell membrane also plays an important role in both types of transport.

As you recall from Chapter 5, the cell membrane consists of two layers, each containing lipid molecules. The carboxyl, hydrophilic ends of the lipid molecules are on the outside surface of the membrane. The glycerol, hydrophobic ends of the lipid molecules form the inside. This structure ensures that molecules that dissolve in water do not automatically pass through the membranes. A layer of lipids separates the two aquatic environments, that of the cell and that of the surrounding environment. The lipid layer thus forms a chemical barrier to the free movement of molecules into and out of the cell. However, many mechanisms serve to ensure the passage of biologically important materials across a membrane.

Carrier Transport

The cell membrane has proteins associated with the lipid layers. Some of these proteins extend across the membrane and others are embedded in it. Each of these two types of proteins aids in the transport of molecules across the membrane. Proteins that function in transport are called **carrier molecules.** These proteins are also called permeases. The carrier molecules and solute molecules move across the membrane. Generally, each carrier molecule is specialized to allow the movement of only one type of molecule across a membrane.

Figure 6-3 Unlike active transport, none of the several types of passive transport requires the expenditure of chemical energy. Gated channels are explained on p. 86.

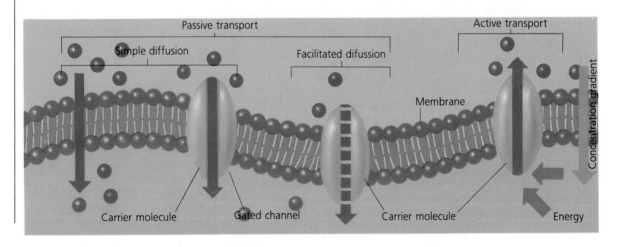

Passive transport — Simple diffusion — Facilitated difussion — Membrane — Active transport — Concentration gradient — Carrier molecule — Gated channel — Carrier molecule — Energy

Facilitated Diffusion

Some carrier molecules transport solute molecules through a membrane without expending energy. This passive form of carrier transport is called **facilitated diffusion.** In this type of diffusion carrier molecules speed up the diffusion. Glucose molecules, for example, are too large to diffuse quickly across a cell membrane. Sometimes the cell needs glucose in large quantities as a source of energy. If the glucose level within the cell falls too low, carrier molecules accelerate the transport of glucose molecules across the membrane. Facilitated diffusion also operates when the level of glucose is too high inside the cell and must be lowered. The carrier molecules speed the rate of diffusion out of the cell.

Active Transport

Carrier molecules also function in the active transport of solute molecules across cell membranes. One common active transport system is the **sodium–potassium pump,** a chemical mechanism that moves sodium ions (Na^+) out of the cell and forces potassium ions (K^+) in. Movement in both cases is against the respective concentration gradients. The process depends on the ability of the protein carrier molecule to change shape. Three sodium ions fit into the carrier molecule. A chemical reaction using high-energy ATP molecules changes the shape of the carrier molecule, which causes the sodium to be released outside the cell. The carrier molecule can accept two potassium ions. The loss of phosphate from ATP causes a second shape change and the release of the potassium inside the cell.

The pump moves three sodium ions in one direction for every two potassium ions moved in the opposite direction. The change in their distribution on either side of the membrane results in a buildup of a positive charge outside the cell and creates a gradient of electrical charge across the membrane.

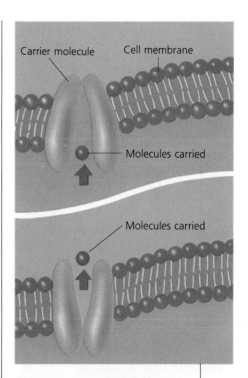

Figure 6-4 In facilitated diffusion carrier molecules speed up diffusion without expending energy.

Figure 6-5 The sodium–potassium pump moves sodium ions out of the cell and potassium ions into the cell. In both cases the pump works against the concentration gradients.

85

Info: Active transport can move potassium ions into a cell even when the potassium concentration within the cell is as high as 1,000 times that of the extracellular environment.

The sodium–potassium pump is crucial to many physiological processes in animals, including the conduction of nerve impulses and the contraction of muscles. One-third of the energy expended by an animal at rest goes to power this pump.

The sodium–potassium pump is only one example of an active transport system. Other systems work in similar ways to provide the energy for transporting sugars, amino acids, and other important metabolic materials across cell membranes.

Gated Channels

Another form of passive transport occurs when proteins in the membrane form a gated channel across the lipid membrane. A **gated channel** is a protein-controlled passage that permits the cell membrane to be permeable as needed. Molecules that for some reason cannot cross the membrane can diffuse through these channels. Some channels are permanently open. Others open in response to environmental stimuli. For example when an impulse travels down a nerve, stimulates a muscle, and causes it to contract, at least four types of gated channels open and close in the neuromuscular junction in less than one second.

Biology in Process

Connections Between Cells

For 150 years scientists had thought of cells as separate, autonomous units. The development of the electron microscope seemed to confirm this view. The electron microscope allowed researchers to see that cell membranes enclosed cells. Then in 1962 two teams of U.S. researchers independently made startling discoveries. They detected ions that moved from one cell interior to another without passing through the cell membrane and without moving outside the cells. How did this movement occur?

Werner Loewenstein and Yoshinobu Kanno asked this question while they were engaged in a study of cells from the salivary gland of the fruit fly larva. As part of their study they caused the nucleus of one cell to carry an electric charge. To their surprise they found that an electric charge was soon evident in adjacent cells but not in the external environment. This discovery indicated to Loewenstein and Kanno that ions must have passed from one cell to another. At about the same time two other researchers, Stephen W. Kuffler and David D. Potter, made a similar observation while studying leech cells.

Endocytosis and Exocytosis

Some molecules, such as food particles and waste materials, are too large to pass through the cell membrane. However, the processes called endocytosis and exocytosis enable these large molecules and particles to enter and exit the cell without passing through the cell membrane. These processes are part of the cell's maintenance of homeostasis. Cells use endocytosis and exocytosis to stay in balance with their environment by taking in nutrients, exporting proteins, and excreting wastes. Such transfers are made possible by the dynamic nature of the cell membrane.

Endocytosis

Endocytosis *(EN-duh-sie-TOE-suss)* is the process by which cells engulf substances that are much too large to enter the cell by passing through the cell membrane. These large materials outside the cell are enclosed by a portion of the cell, which folds into itself and forms a pouch. The pouch, called a vesicle, then pinches off from the cell membrane and enters the cytoplasm. Once inside the cytoplasm the contents of the vesicle are digested by cellular enzymes.

Answer to Biology in Process
The ions could only have passed into the cells through junctions between cells in contact with one another.

Nonjunctional membrane

Junctional membrane

Junctional seals

Electrical measurements and detailed examinations of electron photomicrographs suggested to the researchers that a passage existed between the cell membranes of the two cells. Such passages have since become known as "gap junctions" (left).

Loewenstein and Kanno then decided to determine whether or not particles larger than ions could pass through a gap junction. The researchers injected fluorescent tracer molecules into the interior of one cell. These tracer molecules later showed up in adjacent cells as well. By using tracer molecules of different sizes, the researchers determined the size range of molecules that were able to pass through a gap junction.

The team then set out to find the role that various ions played in affecting the permeability of the gap junctions.

The researchers injected calcium ions in the cytoplasm. They determined that an increase in calcium ions lowered the permeability of the gap junctions.

■ What was the significance of the observation that the tracer molecules injected into the cell pass only to adjacent cells?

Endocytosis

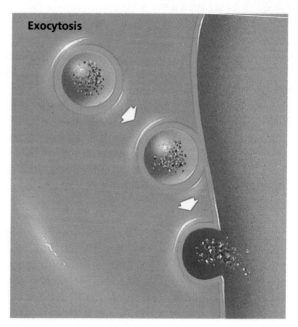

Exocytosis

Figure 6-6 In endocytosis (left) cells engulf fluids or food particles that are too large to enter the cell through the cell wall. In exocytosis (right) cells excrete wastes or export proteins that cannot exit through the cell membrane.

Answers to Section Review
1. Facilitated diffusion does not require cell energy. Active transport does require cell energy.
2. The two lipid bilayers are embedded with protein molecules that can act as carrier molecules. Substances bind to the carrier molecules on one side of the membrane and are released on the other side.
3. Sodium ions are moved out and potassium ions are brought into the cell against the concentration gradient.
4. Endocytosis is the ingestion of solid molecules by a cell. The cell membrane engulfs the particle, and the membrane pinches off to form a sac inside the cell.
5. Carrier molecules speed up the passage of glucose into the muscle cells, which have been depleted during exercise.

Biologists distinguish two kinds of endocytosis on the basis of the materials enclosed. **Pinocytosis** *(PIN-uh-sie-TOE-suss)* is the movement of solutes or fluids into the cell, while **phagocytosis** *(FAG-uh-sie-TOE-suss)* is the movement of large food particles or whole microorganisms into the cell. Amebas feed by phagocytosis. In mammals white blood cells called phagocytes engulf and digest bacteria. Review the process of endocytosis by studying Figure 6-6.

Exocytosis

Maintaining homeostasis within a cell often involves exporting large molecules as well as importing them. **Exocytosis** *(ECK-soe-sie-TOE-suss)*, also shown in Figure 6-6, is the passage of large molecules out of a cell. For example, cells manufacture proteins necessary for various physiological reactions outside the cell. Recall that ribosomes make the proteins and the Golgi apparatus processes and packages them into vesicles. The packaged proteins then move to the cell membrane. The vesicle membrane fuses with the cell membrane and dumps its contents out of the cell.

Section Review

1. Explain the difference between *facilitated diffusion* and *active transport.*
2. How does the structure of the cell membrane facilitate transport by carrier molecules?
3. How does the sodium–potassium pump work?
4. Describe the process of endocytosis.
5. How might carrier molecules help maintain homeostasis during exercise, when extra energy is needed quickly?

Tip: Make sure students understand that the main difference between pinocytosis and phagocytosis is in the size of the ingested particle.

Laboratory

Osmosis and Turgor Pressure

Objective
To examine the osmotic behavior of cells and to understand the concept of turgor pressure

Process Skills
Observing, experimenting

Materials
Fresh potato, scalpel, plastic fork, 2 petri dishes, 10% salt solution, clock or watch, piece of red onion, 2 medicine droppers, microscope slide, coverslip, blunt probe, microscope, paper towel

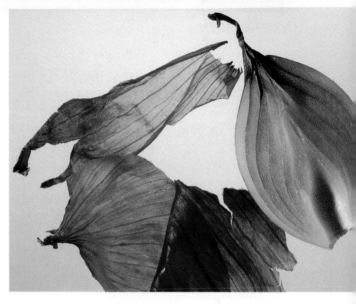

Method

1 In this investigation you will study potato and onion cells in order to observe the relationship between osmosis and turgor pressure. Cut a potato in half lengthwise. Cut a 3-mm-thick slice along the length of one of the halves. Cut four sticks 3 mm wide along the length of the slice.

2 Holding a fork in one hand, place a potato stick across the middle of the tines so that equal amounts overhang each side. Observe how much the potato stick bends. What does the amount of bend tell you about the turgor pressure inside the potato cells? How might this observation serve as a control for later observations?

3 Fill a petri dish about half full with water. Place an equal amount of 10% salt solution in another petri dish. Add two potato sticks to each dish.

4 Once each minute, use the fork to pick up each stick by its center to test it for bend. Test each stick until you no longer notice any difference in how much it bends. Compare your observations about the sticks placed in water and the sticks placed in salt solution with your observations in step 2. What inference can you draw from your observations? How might you test your inference directly?

5 Take a small piece of onion and peel off a sheet of the purple skin. With a scalpel cut a piece of skin about 6 mm².

6 Use a medicine dropper to place a drop of 10% salt solution on a microscope slide. Place the piece of onion skin over the drop and cover the slide with a coverslip.

7 Examine the slide under low power of a microscope. What do you observe? How do these observations relate to those in step 4?

8 Use a clean dropper to add water to one side of the coverslip and paper towel to absorb the salt solution on the other side. What happens to the onion cells? What can you infer about the movement of water between cells and their external environment?

Conclusions

1 What effect did the salt solution have on the potato and onion cells?

2 Which caused greater turgor pressure inside the cell—water or salt solution?

Inquiry
How can you use what you have learned about the flexibility of the potato sticks to determine the salt concentration naturally present in the cells of a freshly cut stick?

Chapter 6 Review

Vocabulary

active transport (84)
carrier molecules (84)
concentration gradient (81)
contractile vacuole (83)
cytolysis (83)

diffusion (81)
endocytosis (87)
equilibrium (81)
exocytosis (88)
facilitated diffusion (85)
gated channels (86)

homeostasis (81)
hypertonic (83)
hypotonic (83)
isotonic (83)
osmosis (82)
passive transport (84)
permeable (82)

phagocytosis (88)
pinocytosis (88)
plasmolysis (83)
sodium–potassium pump (85)
turgor pressure (83)

1. Distinguish between diffusion and facilitated diffusion.
2. Explain what is meant by a selectively permeable membrane. Give an example of a biological system that employs a semipermeable membrane.
3. How is plasmolysis related to turgor pressure?
4. What is a contractile vacuole, and how does it function?
5. Use a dictionary to find the meaning of the word parts *pino-*, *phago-*, and *cyto-*. Why are the terms *pinocytosis* and *phagocytosis* good names for the processes they describe?

Review

1. Homeostasis is (a) the movement of materials into and out of a cell (b) the transport of proteins against the concentration gradient (c) a state of biological balance (d) the process of molecules spreading out evenly across an environment.
2. The part of the cell that functions to maintain homeostasis relative to the cell's environment is the (a) cytoplasm (b) Golgi apparatus (c) nucleus (d) cell membrane.
3. Gap junctions aid the (a) movement of molecules against a concentration gradient (b) movement of particles in response to random movement of molecules (c) movement of molecules from one cell to another (d) paired movement of sodium and potassium ions.
4. Glucose enters a cell by (a) diffusion (b) pinocytosis (c) phagocytosis (d) active transport.
5. When the solution inside the cell is hypertonic relative to its environment, (a) water will move from inside the cell to outside the cell (b) water will move from outside the cell to inside the cell (c) the water balance will be maintained (d) the cell will burst.
6. When the turgor pressure of a cell is high, a plant (a) wilts (b) dies (c) is rigid (d) explodes.
7. Plasmolysis refers to a cell's (a) taking in too much water (b) shriveling (c) consuming dangerous bacteria (d) maintaining a fluid balance.
8. Gated channels function to (a) export proteins out of the cell (b) transport proteins against the concentration gradient (c) transport molecules by diffusion through the channel (d) import proteins into the cell.
9. The cell must expend energy to transport substances using (a) a protein channel (b) facilitated diffusion (c) a sodium–potassium pump (d) osmosis.
10. As part of the immune system, a white blood cell engulfs, digests, and destroys an invading bacterium through the process of (a) facilitated diffusion (b) phagocytosis (c) pinocytosis (d) osmosis.
11. Explain equilibrium.

12. Can all molecules diffuse through all membranes? Explain your answer.
13. What would happen to a freshwater unicellular organism if its contractile vacuole stopped functioning? Explain your answer.
14. What determines the direction of movement of water across a cell membrane?
15. How does the lipid layer of a membrane form a barrier to molecules?
16. How does a carrier molecule transport materials across a cell membrane?
17. What form of transport is used by the cell membrane to speed up the intake of glucose when needed?
18. What energy source does the cell use to work the sodium–potassium pump?
19. Distinguish between endocytosis and exocytosis.
20. What did researchers observe that led them to hypothesize that a connection existed between cells?

Critical Thinking

1. In an inflated ballon there is a higher concentration of air molecules inside than there is in the air outside. Because of their constant random motion these molecules push against the balloon and keep it taut. In what way is this pressure similar to turgor pressure? In what way is it different?
2. Sometimes water seeps through the concrete wall of a basement after a heavy rain, and the homeowner removes it with a sump pump. How can this situation be compared with the action of a unicellular organism that lives in a pond?
3. When a cell like the one pictured on the right takes in substances through endocytosis, the cell membrane forms an inside-

out pouch. The outside of the cell membrane becomes the inside of the vesicle. What might this suggest about the structure of the cell membrane?

Extension

1. Read the article by J. Silburner entitled "Sugar/Water Switch Allows Dry Life," in *Science Digest*, September 15, 1984, p. 167. Make an oral report in which you explain how the use of trehalose helps plants and animals adapt to their environment. Also talk about how trehalose might be useful to humans.
2. Kidney dialysis is the artificial filtering of a patient's blood to remove wastes when the patient's kidneys can no longer function well. Selectively permeable membranes are used in kidney dialysis and in other areas of medicine. Consult your local hospital, and obtain information on how kidney dialysis is performed and on whom. You may wish to read the article entitled "The Clean Machine: The Kidney," in *Current Health*, February 1985, pp. 28–29.
3. Examine a number of food items at a grocery store. List at least four foods in which salt is used as a preservative. Keeping in mind what you have learned about osmosis, explain why salt is used so commmonly to preserve food.

Chapter 7 | Photosynthesis and Respiration

This chapter discusses the necessity of energy for carrying on life processes and how energy is trapped, transferred, and released. Students focus on photosynthesis and respiration, and they learn how cell structure is related to these activities.

Correlation Chart

Section Number	Section Objectives	Pages	Features	Section Review
7.1	1. Compare and contrast the functions of photosynthesis and respiration.	93–95		2, 5
	2. Distinguish between autotrophs and heterotrophs.			3
	3. Name the parts of and describe the structure of ATP.			2, 4, 5
	4. Summarize the process of phosphorylation.			1, 5
7.2	1. State the reactants and the products of photosynthesis.	96–102	Biology in Process, pp. 100–101	2
	2. Explain how the structure of the chloroplast facilitates photosynthesis.			3, 5
	3. Distinguish between the main events of the light reactions and the dark reactions.			4
	4. Name three carbon-fixing methods used by plants.			1
	5. Summarize Calvin's investigations of the events of dark reactions.			4
7.3	1. Distinguish between anaerobic and aerobic respiration.	103–109	Think, p. 105 Biology Lab, p. 109	1, 5
	2. Name the end products of glycolysis.			2
	3. Describe the process of fermentation.			3
	4. List the major events of aerobic respiration.			4

Resources*

Print

Gouindjee, ed. *Photosynthesis*. 2 vols. New York: Academic Press, 1982.

Nadakavukaren, Mathew, and Derek McCracken. *Botony: An Introduction to Plant Biology*. St. Paul: West, 1985.

Audiovisual/Software

How Green Plants Make and Use Food, revised. Film or video, 12 min. Coronet, 1986.

Photosynthesis: Unlocking the Power of the Sun. Apple 128K. IBM PC 128K. 2 disks. Scholastic, 1986.

*Additional references are listed in *Modern Biology, Teacher's Program Guide*.

Section Teaching Guides

7.1 The Need for Energy

Approach

The main objective of this section is to indicate that all living organisms require energy to function. The text assumes knowledge of the definitions of *compound, molecule, ion,* and *chemical reaction* that were presented in Chapter 3. You may also wish to ask students to review the explanations of an endothermic reaction and of activation energy that were given in Chapter 3. For a prior discussion of water and carbon molecules and of enzymes, refer students to Chapter 4.

This section serves as an introduction to the chapter as a whole. This section presents an explanation of the complementary processes by which cells trap energy—photosynthesis—and release the energy for use by the cell—cellular respiration. This section introduces and defines the molecule ATP, which will be an important focus of Sections 2 and 3. This section also describes the process of phosphorylation; an understanding of this important reaction will help students comprehend the more complex processes of photosynthesis and respiration ahead. You may wish to use the schematic diagram in Figure 7-1 to preview the chapter.

A second objective of the section is to introduce the idea of a food chain and to distinguish between autotrophs and heterotrophs. This theme will reoccur in Unit 8 and again in Unit 10.

In presenting the text help students to recognize the cell's fundamental need for energy to carry on all life processes and to maintain homeostasis and balance with its environment. Explain to students that there are several kinds of energy—physical, kinetic, electrical, and chemical. In this section the references are to chemical energy. You may wish to preview all the illustrations in the section with the students before asking them to read the text. Use Figure 7-2 to help students grasp the relationship between the trapping and release of energy. Point out that the end products of photosynthesis become the raw materials of respiration, and vice versa.

Concept Framework

Most students will know that their bodies need energy to grow and to function. However, students may have difficulty relating the formation and breakdown of ATP to their own bodies. To provide a starting point for introducing students to ATP, ask students for examples of how they use energy. Most will first think of some intense physical activity such as running a race. Then ask students where their bodies got the energy, and if the answer is food, ask where that energy came from, and so on, leading the discussion back to plants and to sunlight. Use the analogy of ATP as the currency of energy transactions in living organisms. Like currency, ATP can be used in small or large amounts; it can be spent as soon as it is earned or stored until it is needed. Ask students to consider the biological pathways by which energy transactions are made as they read.

Assist students in working through some simple food chains. You might begin by naming a mammal high on a food chain and by asking students to name its food source, and the source of its food source, and so on down the chain.

Emphasis

Make certain that students understand that although not all living organisms are capable of trapping energy from the sun, all organisms carry on some form of cellular respiration. Only autotrophs carry on photosynthesis or a similar form of trapping and storing energy, but both autotrophs and heterotrophs must process the stored energy to release it to do work.

Compiling a Chart

Have students list the activities of a living organism that require energy. After compiling the list on the board, ask students to organize the activities in a chart form, grouping similar activities under functions such as respiration, nutrition, and so on.

Debate

Divide the class into two teams, one called the "Autotrophs" and the other the "Heterotrophs." Ask them to debate the question: Which is worse, a world without plants or a world without animals? Remind students that not only do animals depend on plants for food energy, but plants depend on animals for the recycling of CO_2 back into the atmosphere.

Designing Closed Systems

Explain to students that the earth operates as a closed system in the recycling of O_2 and CO_2. Ask students to design other closed systems that sustain life and that allow for a continual recycling of necessary materials, such as an aquarium or a spaceship. Be sure that all students include an energy source in their planning. Have students present selected designs to the class.

Section Review Answers

(Continued from p. 95.)
4. An ATP molecule is composed of adenine, which contains nitrogen; ribose, which is a five-carbon sugar; and three phosphate groups.
5. Plants are called producers because they produce glucose; animals are called consumers because they consume the glucose produced by plants.

7.2 Photosynthesis

Approach

The main objective of this section is to explain how plants trap and store energy. The text builds on the discussion of the structure of the chloroplast presented in Chapter 5. The text also uses information about the movement of ions across a membrane given in Chapter 6. In presenting the text, preview with students how this section is organized. Explain to students that in most cases the text gives an overview of the process, then reviews the details of the reactions that take place, and finally concludes with a summary that states the products of the reactions. Point out that the illustrations should be studied closely along with the text. Explain that a biological pathway is a series of reactions that are repeated over and over. Some pathways, such as the electron transport chain, will be introduced in this section and will reappear in the next section on respiration. Encourage students to focus on the biological pathways the reactions take, and not to get lost in the details of the processes.

Concept Framework

Most students will know that plants cannot survive without sunlight. Build on that knowledge by asking students if they know what the sunlight contributes to the plants. Use the students' answers to make a transition from Section 7.1 to Section 7.2. Elicit from students the information that sunlight is a form of energy. Ask students to describe observations they

have made that demonstrate the energy of sunlight. Some students will probably mention the heat of sunlight or the way some plants move so that their leaves receive the maximum sunlight. Explain that this section will deal with the energy transactions that take place when the sunlight reaches the plant. You may wish to end the discussion by asking students to keep something in mind as they read. Have them consider which transactions are common to all photosynthetic autotrophs and which transactions are specific to particular types of plants.

Emphasis

Make certain that students understand that photosynthesis is the process by which plants and many microorganisms trap and store energy from sunlight; some other forms of life use different energy sources. During photosynthesis one form of energy—light energy—is transformed into another form—chemical energy.

Using a Prism

Have students use a prism to separate sunlight onto a white piece of paper. Suggest that they use colored pens or crayons to draw and label the parts of the spectrum. Interested students may wish to turn their drawings into an illustrated chart similar to Figure 7-7 by adding wavelengths and drawings of the organisms that use various wavelengths. Students should be encouraged to use their library resources to research this information.

Debate

Divide the class into two-person debate teams. Have half of the teams take the pro position and half the con position in debating the statement: Developing a close relationship to your house plants and talking into their leaves helps them to grow. If necessary, provide further stimulus to the debate by suggesting that in talking to plants many people will also be breathing carbon dioxide on them.

Student Report

Have interested students research the life and work of the British scientist Peter Mitchell. Ask them to consider what his contributions have been to photosynthetic studies and how the scientific world reacted when he first presented his hypothesis about chemiosmosis. Have selected students present the best reports to the entire class.

Field Trip

Visit a plant nursery, botanical garden, or avid amateur gardener to observe plants that use C_4 and CAM pathways for photosynthesis. Have the students prepare questions beforehand on the environment in the wild that makes these alternate pathways suitable for these plants.

Using the Features/Calvin: The Dark Reactions

The main process skills used this feature are experimenting and interpreting data. Remind students that experimenting incorporates other process skills, chiefly observing, inferring, and hypothesizing. After an experiment is over, the data obtained has to be interpreted. Explain that Calvin first designed an experiment in which algal cells absorbed radioactive carbon along with light during photosynthesis. Calvins' experiment resulted in data that indicated whether various compounds were formed early or late in the dark reactions cycle. Point out that Calvin had to analyze the data in order to form a theory establishing the sequence of events that occurred during the dark reactions. To help students understand the part radioactive isotopes played in the experiment, discuss other uses of radioactive isotopes, such as in medical diagnosis. To explain how Calvin analyzed the data, provide students with any set of sequential data in a jumbled order, such as the growth measurements of a baby over a number of months. Be sure that some sets of data have identical height measurements and different weight measurements. Have students place the measurements in a possible sequence and formulate a hypothesis based on the results.

Section Review Answers

(See p. 102.)

7.3 Respiration

Approach

The main objective of this section is to describe how the energy stored in glucose during photosynthesis is released in the cell during cellular respiration. The text includes a discussion on the differences between fermentation and aerobic respiration, and the uses of each by organisms. The text builds on the discussion of carbohydrates in Chapter 4 and of the structure of the mitochondria in Chapter 5. Because all living organisms carry on cellular respiration, the concepts presented here are important to all fields of biology, and particularly to Units 5-10.

In presenting the text, review again with students the way in which the section is organized, beginning with an overview, followed by details of the reactions, followed by a summary of the results. As you did for Section 7.2, encourage students to study the illustrations as they read the text and to look for broad biochemical pathways that are used. You may wish to preview the text by reading through Figure 7-12 with students.

Concept Framework

Most students apply the term *respiration* to multicellular organisms and to the process of breathing in oxygen and breathing out carbon dioxide. Make certain that all students understand that, in the context of a single cell, the word *respiration* applies to the breakdown of energy; oxygen may or may not be present. To avoid this confusion and to provide a smooth transition for studying this section, emphasize the energy-releasing aspect of the process of respiration. You may wish to end the discussion by asking the question: Where does the energy come from that is required to do the work of cellular respiration? Suggest to students that they look for the answer to this question as they read the text.

Emphasis

Make certain that students understand that all cells require respiration and that both fermentation and anaerobic respiration begin with the same first step—glycolysis. The objective of all forms of respiration is the same: to break down the stored energy into a form that is usable to the cell in order for it to carry on cellular activity.

Guest Speaker

Invite a local nutritionist or a dietician to speak to the class on the relative value of various foods in both a normal diet and in diets specialized for people engaged in certain activities or for people suffering from special health problems. Have students prepare for the speaker's visit by framing questions on the nutritional values of fermented products such as yeasts and yogurts.

Student Reports on Yeast Fermentation

Have interested students research the work of Louis Pasteur on yeast fermentation. Ask them to answer the question: Why is Pasteur credited with saving the French wine industry? Ask selected students to present their reports to the class.

Field Trip

Arrange a visit to a sports clinic or hospital physical therapy unit. Ask your guide to demonstrate the energy requirements of various forms of strenuous physical activity. Initiate a discussion of how and why oxygen debts are likely to occur.

Section Review Answers

(Continued from p. 108.)

4. The Krebs cycle begins with a four-carbon compound combining with acetyl-CoA to form citric acid. The four-carbon molecule is ultimately reformed and thus can be used to start the cycle over again.

5. Organisms using fermentation probably appeared on earth first, at a time when there was no oxygen gas on earth.

Laboratory Preparation

Background

In this lab students observe that the rate of photosynthesis increases as the intensity of light increases. Students should be familiar with the summary equation for photosynthesis because they indirectly measure the rate of photosynthesis by noting CO_2 levels. Since pH levels of the water indicate CO_2 levels, students also should be familiar with the pH measurements.

Preparation

Elodea is available from pet stores or can be collected locally from ponds or slow-moving streams.

Conclusions

1. In solutions B and C, the pH increased. In solutions A and D, the pH did not change. The change in the pH of the solutions B and C indicates that the CO_2 was being used by the *Elodea* for photosynthesis.

2. Solution D served as a control to demonstrate that *Elodea*, alone, did not change the pH.

3. The rate of photosynthesis increases as the intensity of light increases.

Inquiry

No. The plant eventually would use up all of the CO_2 in the beaker, and photosynthesis would stop.

Chapter Review

Vocabulary

1. C_4; the others are components of aerobic respiration
2. stomata; the others are accessory pigments
3. heterotroph; the others are types of biochemical pathways
4. chemiosmosis; the others are ways plants fix carbon
5. stomata; the others are parts of the chloroplast

Review

11. Autotrophs can build organic molecules directly from inorganic compounds. Some heterotrophs eat autotrophs and other heterotrophs eat heterotrophs. In both cases the heterotrophs are using energy first converted by autotrophs.

12. Like currency, ATP can be stored until needed, then "cashed in." ATP can also be compared to currency because it can be broken down into small units or accumulated in large amounts. Also, like currency, ATP is valuable.

13. A biochemical pathway is a series of reactions. The electron transport chain and the Calvin cycle are both examples of biochemical pathways that occur during photosynthesis.

14. Both in photosynthesis and in respiration, chemiosmosis occurs when an accumulation of H^+ builds up on one side of a membrane. As electrons pass down a transport chain, hydrogen ions move across the membranes of thylakoids (during photosynthesis) or mitochondria (during respiration). This results in an electrochemical gradient, which in turn initiates the passage of chemicals through the membrane in an attempt to restore equilibrium. This movement of molecules releases stored electrochemical energy, which is used to synthesize ATP.

15. Cactuses and similar plants that live in extremely dry climates carry on carbon fixation at night. This means that their stomata need not be open during the day, when they would be subject to much greater water loss.

16. Yeast cells are used in wine-making because they carry on alcoholic fermentation, producing the alcohol in wine.

17. The pyruvic acid that is the end product of glycolysis is oxidized and combined with coenzyme A to form acetyl-CoA. This compound enters the Krebs cycle and combines with a four-carbon molecule to form citric acid.

18. Both electron transport chains occur within a membrane—the photosynthesis chain in the thylakoid membrane and the respiration chain in the mitochondrion membrane. Both chains function in the same way—as the hydrogen electrons pass down the energy gradient, hydrogen ions accumulate on one side of the membrane, causing an electrochemical gradient. The movement of the ions through the membrane powers the formation of ATP.

19. The maximum net yield depends in part on the type of cell. Only 36 ATP are formed in brain cells and skeletal muscle cells, for example. Actual yields may also be lower because hydrogen has leaked through the membrane or has been used for some other purpose in the cell. All ATP formed is also not available; some is used for other purposes, including starting reactions.

20. Calvin learned the sequence in which the compounds were formed during photosynthesis.

Critical Thinking

1. Vitamin B_1 is important in the human diet because without it the pyruvic acid from glycolysis would not be converted to acetyl-CoA, and would be unavailable for the Kreb's cycle. As a result the cells would have less energy.

2. The inefficiency of the electron transport system during the cold winter months would have the advantage of providing the animal with more heat at a time when it was needed.

3. The alcohol content does not increase because the yeast has fermented all the available sugar by the time the alcohol level reaches 12 percent.

4. The appearance of oxygen in the earth's atmosphere allowed the development of the Kreb's cycle, which uses fuel more efficiently. Therefore, larger, more complex organisms were able to develop. The source of the oxygen was probably photosynthesis in blue-green bacteria and plants.

Chapter 7 | *Photosynthesis and Respiration*

Introduction

In Chapter 3 you learned that many life processes involve endothermic reactions—that is, reactions that require the addition of energy. Where does this energy come from? For most living things the answer is ultimately the sun. In this chapter we will look closely at the processes through which cells trap energy, transfer it, and release it to power cellular activities.

Chapter Outline

Chapter Concept

As you read about complex biochemical pathways, keep in mind that glucose is made in photosynthesis and broken down in respiration.

Leaf of a tulip tree, *Liriodendron tulipifera*. Inset (left): TEM of mitochondrion in a bat pancreas (×6,000). Inset (right): Color-enhanced TEM of a chloroplast in the moss *Physcomitrella potens* (×5,700).

For additional teaching information see pp. T7A–T7F.

7.1 The Need for Energy

All living things require energy to survive. Whether they are unicellular or multicellular, all organisms need energy to power the many endothermic reactions that are essential to cell function. Cells are constantly active—building, repairing, growing, and reproducing. Where do organisms get the energy to support these activities? The simplest answer is from their food. However, this chapter will answer the question much more fully by introducing you to two fundamental biological processes—photosynthesis and respiration. **Photosynthesis** is the process that converts the radiant energy of sunlight into chemical energy. **Respiration** is the process that releases chemical energy for use by the cell. Figure 7-1 provides an overview of these processes.

Energy for Life

Scientists divide living things into two groups according to the way in which they get their food. Plants and other organisms that meet their energy needs by building organic molecules from inorganic substances are called **autotrophs** *(AWT-uh-TROFES)*. Organisms that do not make their own food but depend directly or indirectly on autotrophs for food are **heterotrophs** *(HET-uh-ruh-TROFES)*. A cow, for example, is a heterotroph; it depends directly on autotrophic plants for food. A human who eats the meat of the cow is also a heterotroph. Because the energy in the meat was originally derived from plants, the human is indirectly dependent on plants for food. Ultimately all life on earth depends on autotrophs.

Photosynthesis is the major way that autotrophs produce chemical energy. Photosynthesis is carried out in plants and many microorganisms. In photosynthesis the radiant energy from sunlight is used to produce glucose ($C_6H_{12}O_6$) from carbon dioxide (CO_2) and water (H_2O). During the process oxygen (O_2) is released into the atmosphere.

Respiration is the process by which glucose molecules are broken down and their stored energy released. Like photosynthesis, respiration occurs in a controlled series of reactions. All organisms use some form of respiration to obtain energy.

In any biological system materials are constantly recycled. During photosynthesis plants take in CO_2 and release O_2. When

Section Objectives

- *Compare and contrast the functions of photosynthesis and respiration.*
- *Distinguish between autotrophs and heterotrophs.*
- *Name the parts of and describe the structure of ATP.*
- *Summarize the process of phosphorylation.*

Info: Some organisms produce energy through chemical means rather than photosynthesis. For a discussion of bacteria that are chemoautotrophic see Chapter 20.

Figure 7-1 Energy for most living organisms comes from the sun and is converted to usable form through photosynthesis and respiration.

Figure 7-2 Autotrophs produce glucose and oxygen by photosynthesis. Both autotrophs and heterotrophs produce carbon dioxide and water by cellular respiration.

Figure 7-3 The energy in ATP is stored in the bonds linking the phosphate groups. The energy is released when the bonds are broken.

animals eat plants, for example, they convert glucose to energy through respiration. This energy-releasing process requires O_2 and releases CO_2. Thus the alternation of photosynthesis and respiration results in the continual recycling of CO_2 and O_2 in the environment. However, a constant input of energy is needed to keep the cycle going. For almost all living things the ultimate source of this energy is sunlight.

Adenosine Triphosphate

Both photosynthesis and respiration are processes that involve a series of biochemical reactions. Such a series of reactions is called a **biochemical pathway.** You will encounter several biochemical pathways in this chapter. Usable energy produced by one reaction may be stored and used in a later reaction. In most cases this usable energy is stored in a molecule called **adenosine triphosphate** *(uh-DEN-uh-SEEN trie-FAHS-fate),* or **ATP.**

The ATP molecule, shown in Figure 7-3, has three parts: adenine, a nitrogen-containing molecule; ribose, a five-carbon sugar; and three phosphate groups. The adenine bonds to ribose, forming the compound adenosine. The phosphate groups attach in sequence to the adenosine. **Adenosine monophosphate,** or **AMP,** has one attached phosphate group; **adenosine diphosphate,** or **ADP,** has two; and adenosine triphosphate, or ATP, has three.

ATP stores energy in the bonds between the phosphate groups. A molecule of ATP is often written as A–P~P~P, with *A* representing adenosine, and each *P* representing a phosphate group. Each wavy line (~) represents a bond that, when broken, releases energy.

ATP–ADP Cycle

The phosphate bonds of ATP must be broken before cells can use the energy stored in them. When a cell needs energy, an enzyme called ATPase breaks the bond between the second and third

Cellular respiration

ATP · ATP synthetase · ATP · ATPase · ATP

ADP + P · Energy · ADP + P · Energy · ADP + P

Energy for cell activities · Energy for cell activities

phosphate groups. The terminal phosphate group is then removed, ADP is formed, and energy is released:

$$A–P\sim P\sim P \longrightarrow A–P\sim P + P + energy$$

Sometimes a similar reaction breaks the bond between the first and second phosphate groups:

$$A–P\sim P \longrightarrow A–P + P + energy$$

This reaction releases energy and converts ADP to AMP. Thus cells obtain energy by splitting phosphate groups from ATP.

The breakdown of ATP to ADP may result in free phosphate ions. Usually, however, the breakdown of ATP is accompanied by the transfer of a phosphate group to another molecule, a process called **phosphorylation** (*FAHS-FOR-uh-LAY-shun*). The phosphorylated molecule gains both the phosphate and the energy.

The formation of ATP is the reverse of its breakdown. ADP is phosphorylated to form ATP. An enzyme called **ATP synthetase** (*AY-TEE-PEE SIN-thuh-TAZE*) catalyzes the synthesis of ATP. Adding a phosphate group to ADP requires energy:

$$A–P\sim P + P + energy \longrightarrow A–P\sim P\sim P$$

Photosynthesis, respiration, the formation of ATP, and the breakdown of ATP form a fundamental biological cycle. Photosynthesis stores energy in glucose molecules; respiration releases that energy. The energy freed forms ATP as another storage molecule. Energy from the breakdown of ATP fuels cell activity.

Section Review

1. What is *ATP synthetase*?
2. What is the importance of ATP to cells?
3. Name one example of an autotroph and of a heterotroph. How does each obtain its energy?
4. Name the component parts of a molecule of ATP.
5. Why are plants called producers and animals consumers?

Figure 7-4 When phosphate bonds break, ATP becomes ADP and energy is released. ATP is re-formed when ADP and P bond.

Tip: Make sure students realize that the energy to start the cycle such as that depicted in Figure 7-4 is fueled by energy from the preceding cycle and that energy from the current cycle will provide the energy to start the succeeding cycle.

Answers to Section Review
1. an enzyme that catalyzes the synthesis of ATP
2. The energy of ATP is used to power chemical reactions.
3. An autotroph is any photosynthetic organism; a heterotroph, any organism that consumes an autotroph or another heterotroph. Most autotrophs acquire their energy from the sun, and heterotrophs acquire their energy either directly or indirectly from autotrophs. (See pp. T7A–T7F.)

Section Objectives

- *State the reactants in and the products of photosynthesis.*

- *Explain how the structure of the chloroplast facilitates photosynthesis.*

- *Distinguish between the main events of the light reactions and those of the dark reactions.*

- *Name three carbon-fixing methods used by plants.*

- *Summarize Calvin's investigations of the events of dark reactions.*

Figure 7-5 Photosynthesis includes light reactions, during which ATP and NADPH + H⁺ are formed. During the dark reactions, glucose is formed.

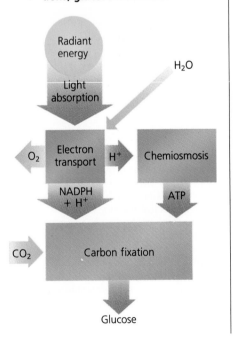

7.2 Photosynthesis

As you have read, photosynthesis is a process that converts solar energy into energy that is stored in chemical bonds. Just how does this transformation of radiant energy occur during photosynthesis? Here is a general equation for photosynthesis:

$$6\ CO_2 + 6\ H_2O + \text{light energy} \longrightarrow C_6H_{12}O_6 + 6\ O_2$$

While the general equation summarizes the process, photosynthesis is actually a complex biochemical pathway. Photosynthesis consists of two interdependent pathways called the light reactions and the dark reactions. Look at Figure 7-5, which summarizes these reactions. During the **light reactions** the energy in sunlight is trapped, O_2 is released, and both ATP and the hydrogen-carrier molecule, which is written as NADPH + H⁺, are formed. During the **dark reactions** the ATP, NADPH + H⁺ react with CO_2 from the atmosphere and form glucose. The process results in the transformation of light energy from the sun into energy stored in the bonds of the glucose molecule.

As we study these complex reactions, it will be helpful to keep certain points in mind. We will be tracing the path of electrons as they pass from molecule to molecule. The function of these molecules is to shuttle electrons and hydrogen ions as they move from one reaction to another.

Structure of a Chloroplast

Both the light reactions and the dark reactions take place in chloroplasts. As you learned in Chapter 5, chloroplasts are membrane-bound organelles that contain both the pigment chlorophyll and the enzymes necessary for photosynthesis. All photosynthetic cells contain at least one chloroplast, but many contain more than 50.

To understand the structure of the chloroplast, study Figure 7-6. Note that the chloroplast has a double membrane. The interior membrane is organized into flattened sacs of photosynthetic membranes called **thylakoids** *(THIE-luh-KOIDZ)*. The thylakoid membrane has an internal reservoir called a **lumen.** Stacks of thylakoids are called **grana.** Each granum may contain up to several dozen thylakoids. The thylakoids are embedded in a protein-rich solution that is called the **stroma.** The light reactions of photosynthesis take place in the thylakoid membranes; the dark reactions take place in the stroma.

Chlorophyll

Photosynthesis requires the presence of pigments that are acted upon by sunlight. Sunlight consists of particles of energy that move in waves of different wavelengths. A wavelength is the distance between the crests of successive waves. The shorter the wavelength, the more energy the light has. For example, violet

Lumen

ATP synthetase

Thylakoid membrane

Granum

stroma

light, which has a short wavelength, has more energy than red light, in which the wavelength is long. **Pigments** are light-absorbing compounds. Pigments appear colored because they absorb light of certain wavelengths and reflect that of others. Chlorophyll is a pigment that absorbs red and blue light and reflects green light. Plants containing chlorophyll appear green.

Photosynthetic pigments in plants absorb energy from sunlight. Although plants contain several kinds of pigments, the primary pigment in all plants is chlorophyll. There are five types of chlorophyll: a, b, c, d, and bacteriochlorophyll. These occur in different organisms and absorb different wavelengths of light.

Chlorophylls a and b are the most common of the photosynthetic pigments. Chlorophyll a is found in all plants and algae. Chlorophyll b occurs in plants, green algae, and euglenoids.

Carotenoids and Phycobilins

Chloroplasts contain other pigments called accessory pigments. Accessory pigments trap wavelengths of light that cannot be absorbed by chlorophyll and then transfer the energy to chlorophyll molecules for use in photosynthesis. Accessory pigments are

Figure 7-6 Chloroplasts are the site of light and dark reactions. Light reactions occur in the thylakoid membrane; dark reactions, in the stroma (beige).

Info: Very short wavelengths, such as ultraviolet rays and X rays, are so strong that they can destroy plant cells. Very long wavelengths, such as infrared rays, are not strong enough to power the chemical changes necessary in photosynthesis.

important because they enable plants to use a greater amount of the sun's energy than is available to chlorophyll alone.

Carotenoids are yellow, orange, and brown pigments that are present in most plants. The most common carotenoids are **carotenes** and **xanthophylls** *(ZAN-thuh-FILLZ)*. **Phycobilins** *(FIE-koe-BIE-lunz)* are accessory pigments found in red algae and blue-green bacteria. Phycobilins enable red algae to live far below the ocean surface because phycobilins absorb the green, violet, and blue light waves that penetrate deep into the water. Review the absorption spectra of some of the pigments, graphed in Figure 7-7.

Light Reactions

The pigments that are in the chloroplasts intercept light and begin the light reactions of photosynthesis. The light reactions involve many kinds of molecules. As you will learn, an electron freed from chlorophyll is passed from molecule to molecule in a sequence of reactions called the **electron transport chain.**

Reactions

The light reactions occur in two photosystems, photosystem I and photosystem II. A **photosystem** is a unit of several hundred chlorophyll molecules and associated acceptor molecules. The two photosystems have different forms of chlorophyll a. Photosynthesis begins when pigments trap light energy. The light energy boosts the energy level of the electrons in the chlorophyll molecules to such a high level that the electrons can escape the chlorophyll. Electrons with a high level of energy are called "excited."

Follow the path of the electrons shown in Figure 7-8. In photosystem I the excited electrons pass through a series of molecules until they reach a molecule called $NADP^+$. Two electrons and two hydrogen ions then attach to the $NADP^+$ and form $NADPH + H^+$. $NADPH + H^+$ is one product of the light reactions.

Notice that the chlorophyll molecule in photosystem I has now lost electrons. How are they replaced? Electrons from photosystem II take the place of the electrons lost in photosystem I. In photosystem II electrons are excited by light and are passed from acceptor molecule to acceptor molecule. The final acceptor molecule is the chlorophyll molecule of photosystem I.

Now notice that the chlorophyll in photosystem II has lost electrons. Where do the electrons come from that are needed to reestablish the chlorophyll in photosystem II? They come from water. When light interacts with chlorophyll in photosystem II, a water molecule is split into two hydrogen ions and one atom of oxygen. The hydrogen ions move through the thylakoid membrane and into the lumen. The oxygen atom unites with another oxygen atom, and O_2 is released into the atmosphere.

The photosystems and their acceptor molecules are embedded in the thylakoid membrane. As electrons move along the electron transport chain, hydrogen ions are transported across the

Absorbance

400   500   600   700
Wavelength (nm)

Chlorophyll a
Chlorophyll b
Beta-carotene

Figure 7-7 How do these pigments differ in their capacity for light absorption?

Thylakoid membrane

Lumen

ATP formation

$2 H^+ + \frac{1}{2} O_2$

H_2O

$2 H^+$

e^-

e^-

Photosystem I

Photosystem II

Stroma

e^-

e^-

Electron path

Sunlight

e^-

Sunlight

$2H^+$

$2H^+$

$NADP^+$ $NADPH + H^+$

thylakoid membrane. These ions are added to the H$^+$ already in the lumen from the splitting of water.

As you will recall from Chapter 6, the concentration of ions inside the membrane creates an electrochemical gradient. This gradiant sets the stage for **chemiosmosis** (*KEM-ee-oz-MOE-suss*), or the passage of chemicals through a membrane. This movement of molecules generates the energy for the synthesis of ATP. As the trapped hydrogen ions move out of the lumen through the membrane channels, stored electrochemical energy is released. This energy is used by ATP synthetase to make ATP molecules. Thus ATP, NADPH + H$^+$, and O$_2$ are products of the light reactions.

Products

Let's review the light reactions of photosynthesis.

1. Electrons are excited when light energy strikes chlorophyll.
2. Electrons are passed along the electron transport chain. Electrons, NADP$^+$, and H$^+$ unite to form NADPH + H$^+$.
3. Light hits a second chlorophyll molecule. Electrons replace those lost in photosystem II. Water is split, and O$_2$ is released into the atmosphere. Electrons from H$_2$O replace those lost in photosystem I. Hydrogen ions are added to the lumen.
4. The electron transport chain carries more hydrogen ions across the thylakoid membrane into the lumen.
5. The hydrogen ions move back across the thylakoid membrane and generate the energy for the formation of ATP.

The end products of the light reactions, then, are NADPH + H$^+$, O$_2$, and ATP. O$_2$ is released into the atmosphere; NADPH + H$^+$ and ATP are released into the stroma of chloroplasts and are used to drive the dark reactions of photosynthesis.

Figure 7-8 Electrons lost by photosystem I are replaced by electrons from photosystem II, which in turn are replaced by electrons released from the breakdown of H$_2$O.

Stroma

Thylakoid membrane

ATP synthetase

H$^+$

ADP + P

ATP

Figure 7-9 Energy for the manufacture of ATP is generated during chemiosmosis as hydrogen ions move through the thylakoid membrane.

Tip: Ask students how the structure of the membrane, as discussed in Chapter 6, contributes to chemiosmosis.

Dark Reactions

The dark reactions make up the second phase of photosynthesis. The term *dark reactions* is misleading. Although these reactions are light-independent, meaning they can occur in the dark, they also occur in the light. They result in the formation of glucose. First CO_2 is incorporated, or "fixed," into organic molecules.

The dark reactions are often called the **Calvin cycle**, after Melvin Calvin (1911–), the American scientist who figured out the pathway. They are also called the C_3 cycle, because their first stable products are molecules that contain three carbon atoms.

Reactions

The Calvin cycle occurs in the stroma. The cycle consists of four major steps: CO_2 is fixed by ribulose bisphosphate *(RIE-byuh-LOSE BIS-FAHS-FATE)*, RuBP; phosphoglyceraldehyde (PGAL) is formed; glucose is formed; and RuBP is regenerated. Look at Figure 7-10 as you read.

1. CO_2 enters the plant from the atmosphere. RuBP, a five-carbon sugar molecule, binds to CO_2. This binding process is

Answer to Biology in Process
The methanol inactivated the enzymes, so that the process of photosynthesis was stopped immediately at precisely the time the cell was killed.

Biology in Process

Calvin: The Dark Reactions

In 1945 a group of botanists at the University of California at Berkeley began investigating the photosynthetic pathways. Melvin Calvin and his co-workers designed an experiment to trace the path of CO_2 during the dark reactions. This experimental design utilized a new and important tool—a radioactive isotope of carbon designated carbon-14 (^{14}C).

For this experiment Calvin chose a unicellular green alga called *Chlorella* (left, ×400). By growing *Chlorella* under controlled conditions Calvin established a steady rate of photosynthesis. Calvin then exposed algal cells to light in the presence of radioactive CO_2 ($^{14}CO_2$). The photosynthesizing cells absorbed the $^{14}CO_2$. As the $^{14}CO_2$ was used in the dark reactions, it became incorporated into the molecules of the reactants. Such radioactive compounds are said to be labeled.

In order to determine the sequence of the chemical steps occurring in the dark reactions, Calvin had to isolate the products at different points in the photosynthetic pathway. To do this, he built a device that enabled him to expose *Chlorella* to $^{14}CO_2$ for a set period of time and then kill the cells by called **CO_2 fixation.** An enzyme in the stroma called RuBP carboxylase catalyzes this fixation reaction. RuBP carboxylase makes up about 25 percent of the total protein in the chloroplast and is the most abundant protein on earth. The addition of CO_2 to RuBP forms an unstable six-carbon sugar molecule that immediately splits into two molecules of phosphoglyceric acid, or PGA, a three-carbon molecule. Thus the fixation of one molecule of CO_2 by RuBP results in the production of two molecules of PGA.

2. PGAL is formed through the following two-step process: First, a PGA molecule binds with a phosphate group supplied by ATP:

$$PGA + A–P{\sim}P{\sim}P \longrightarrow PGA{\sim}P + A–P{\sim}P$$

Then the molecule reacts with hydrogen from $NADPH + H^+$, breaking the phosphate bond and forming the three-carbon molecule PGAL:

$$PGA{\sim}P + NADPH + H^+ \longrightarrow PGAL + NADP^+ + P$$

3. Some of the PGAL is used to make glucose. Two PGAL molecules combine to form a molecule of glucose ($C_6H_{12}O_6$).

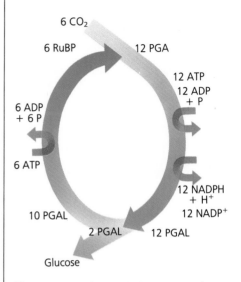

Figure 7-10 Glucose is formed in most autotrophs through a process called the Calvin cycle.

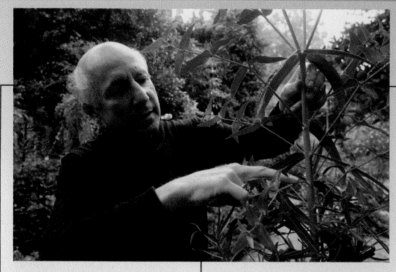

dropping them into boiling methanol. The alcohol also inactivated the enzymes that catalyze parts of the photosynthetic reactions.

By varying the time between exposure and killing of the cells Calvin obtained photosynthetic products from different points in the sequence.

After a short time only a few compounds were labeled. After a longer time many compounds were labeled.

Calvin then isolated the photosynthetic compounds by using paper chromatography. Once the products were isolated, the sequence in which they occurred could be determined. Calvin reasoned that the compounds that were labeled in the shortest time period were the first compounds in the cycle to have incorporated $^{14}CO_2$. Compounds that appeared after a longer time were formed later in the sequence. In this way Calvin recreated the steps of the dark reactions.

The cycle discovered by this technique was renamed the Calvin cycle. In 1961 Calvin was awarded a Nobel Prize for his research.

■ Why did Calvin plunge the algal cells into methanol immediately after exposing the cells to radioactive CO_2?

Radiant energy

Electron transport chain

Chemiosmosis

NADPH + H⁺   ATP

Calvin cycle

Glucose

Figure 7-11 This illustration summarizes the process of photosynthesis: from light energy to glucose formation.

Answers to Section Review
1. an abbreviation for crassulacean acid metabolism
2. Photosynthesis requires CO_2 and H_2O and produces glucose ($C_6H_{12}O_2$) and O_2.
3. chlorophyll a and b
4. No. The dark reactions can and usually do take place in sunlight; they are not dependent on sunlight.
5. The flattened sacs of membranes and the stacking of disks in grana provide a great deal of surface area on which the events of photosynthesis can take place.

4. Most of the PGAL is used to regenerate more RuBP, using energy supplied by ATP. The RuBP is needed to continue the cycle. Therefore, six turns of the cycle are required to produce one molecule of glucose.

Products

Glucose is the product of the dark reactions of photosynthesis. It is an energy-storage molecule and is either used quickly by plants for doing work or stored as starch. Look at Figure 7-11, which summarizes photosynthesis.

C₄ and CAM Pathways

The Calvin cycle is the most common method of fixing carbon. Two other pathways are C_4 and CAM photosynthesis. These pathways have adaptive value to the plants that use them.

Plants adapted to hot, dry climates use the C_4 pathway. In C_4 photosynthesis, a molecule of CO_2 binds, in a special leaf cell, to a three-carbon compound rather than to the five-carbon RuBP. This fixation results in a four-carbon compound, hence the name C_4 photosynthesis. The compound is then transported to an adjacent cell. When released inside the cell, the CO_2 is trapped. Here it is fixed by RuBP and travels through the Calvin cycle.

The advantage of C_4 photosynthesis is that C_4 plants make efficient use of CO_2 by fixing carbon up to four times as fast as C_3 plants do. This allows them to grow in higher temperatures and at much faster rates than C_3 plants can. Although C_4 photosynthesis requires more energy than C_3 photosynthesis, C_4 plants grow in abundant light where the extra energy is readily available.

Cactuses and other plants adapted to dry habitats have crassulacean acid metabolism, or **CAM**. In CAM photosynthesis the plant absorbs CO_2 at night and fixes it in the form of a four-carbon compound. During the day CO_2 is released and refixed by RuBP. It then enters the Calvin cycle. Whereas in C_4 plants carbon fixation and the Calvin cycle occur in different parts of the leaves, in CAM plants these processes occur at different times of the day.

How does CAM photosynthesis help plants survive? Plant pores called **stomata** (*STOE-muh-tuh*) regulate the movement of CO_2 and water vapor in and out of leaves. For carbon fixation to occur the stomata must be open. However, plants lose water when their stomata are open. Since heat and sunlight increase water loss, plants lose less water by opening their stomata only at night.

Section Review

1. What is *CAM?*
2. What are the reactants in and products of photosynthesis?
3. What two types of chlorophyll are most common?
4. Must the dark reactions take place in the dark? Explain.
5. What is the adaptive significance of the shape and arrangement of thylakoid membranes?

7.3 Respiration

Autotrophs produce their own food, and heterotrophs eat plants or other heterotrophs. Before food can be used to perform work, however, its energy must be released. We'll now examine the energy-releasing process called respiration.

Two main types of respiration exist in living things. Both begin with **glycolysis** *(glie-KAHL-uh-suss)*, a process by which one glucose molecule is broken down into two pyruvic acid molecules. In one kind of respiration, pyruvic acid is then broken down without the use of oxygen by a process called **fermentation.** The second kind of respiration, called **aerobic** *(AR-OE-bick)* **respiration,** occurs when pyruvic acid is metabolized using oxygen.

Figure 7-12 shows a simplified version of the biochemical pathways involved in respiration. As you read further, compare these pathways with those of photosynthesis.

Glycolysis

Glycolysis occurs in the cytoplasm. It does not require oxygen. Each of its four stages is catalyzed by a specific enzyme.

1. Glucose is a stable energy-storage molecule. For its energy to be released, the glucose must be converted into a reactive compound. This conversion occurs through the phosphorylation of a molecule of glucose by two molecules of ATP.
2. The six-carbon product of glucose phosphorylation, called glucose-6-phosphate, ultimately breaks into two three-carbon molecules of PGAL.

Section Objectives

- *Distinguish between fermentation and aerobic respiration.*
- *Name the end products of glycolysis.*
- *Describe the process of fermentation.*
- *List the major events of aerobic respiration.*

Figure 7-12 The glucose formed during photosynthesis is broken down within heterotrophs through the process of respiration.

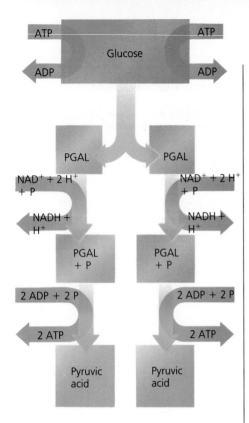

Figure 7-13 During glycolysis, the first phase in anaerobic and aerobic respiration, one glucose molecule produces four ATP molecules.

3. A phosphate group is added to each PGAL molecule and hydrogen atoms are removed. The hydrogen is picked up by a molecule called NAD^+. Since there are two PGAL molecules, two molecules of NADH + H^+ are formed.
4. As the phosphate bonds formed in stage 3 are broken, ATP is generated. Two ATP molecules and one molecule of NADH + H^+ are generated as each of the original PGAL molecules is converted to pyruvic acid. Therefore, a total of four ATP molecules are produced in the fourth stage of glycolysis.

Briefly review the results of glycolysis, illustrated in Figure 7-13. One glucose molecule is broken down into two pyruvic acid molecules. The reaction series produces two molecules of NADH + H^+ and four molecules of ATP. Since two of the ATP molecules are used to initiate glycolysis, there is a net gain of two molecules of ATP. These two molecules represent only 2 percent of the total energy contained in a molecule of glucose.

Fermentation

Fermentation is the breakdown of pyruvic acid without the use of oxygen. Some scientists refer to the combined processes of glycolysis and fermentation as **anaerobic** *(AN-uh-ROE-bick)* **respiration**. The metabolism of pyruvic acid during fermentation does not produce any ATP. What, then, is the function of fermentation? Recall that NAD^+ is needed for glycolysis to proceed. However, when pyruvic acid is formed, NAD^+ is used to form NADH + H^+. The function of fermentation is to break down pyruvic acid and regenerate NAD^+ for reuse in glycolysis, where ATP is formed. Fermentation occurs in two forms: lactic acid fermentation and alcoholic fermentation.

Lactic Acid Fermentation

Lactic acid fermentation occurs in animal cells and in some unicellular organisms when O_2 is in short supply. Lactic acid forms when pyruvic acid accepts hydrogen from NADH + H^+:

$$\text{Pyruvic acid} + \text{NADH} + H^+ \longrightarrow \text{lactic acid} + NAD^+$$

As O_2 is consumed in aerobic respiration, oxygen becomes scarce. When no oxygen is available, NAD^+ + H^+ donates its hydrogen atoms to pyruvic acid, and lactic acid soon accumulates in tissues.

Figure 7-14 Fermentation frees NAD^+ molecules so that glycolysis can continue within a cell.

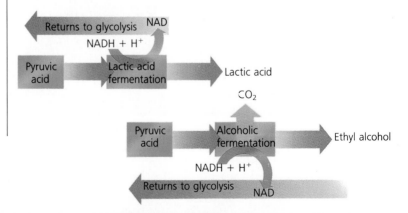

Lactic acid makes muscles feel tired and sore—you've probably experienced muscle fatigue after strenuous exercise. Lactic acid is transported to the liver where it is converted to glucose.

Alcoholic Fermentation

Alcoholic fermentation is a process that occurs in some plant cells and some unicellular organisms, such as yeasts. In alcoholic fermentation pyruvic acid is converted to ethyl alcohol:

Pyruvic acid $+$ NADH $+$ H$^+$ $\longrightarrow$
$$\text{ethyl alcohol} + CO_2 + NAD^+$$

Just as lactic acid accumulates during lactic acid fermentation, ethyl alcohol accumulates during alcoholic fermentation. The alcohol in wine and beer is produced by fermentation carried out by some microorganisms.

Lactic acid and alcoholic fermentation do not supply cells with any additional ATP molecules. Most of the energy remains in the fermentation products, lactic acid or ethyl alcohol.

Aerobic Respiration

The result of glycolysis and anaerobic respiration is shown by the reaction:

$$C_6H_{12}O_6 + 6\,O_2 \longrightarrow 6\,H_2O + 6\,CO_2 + 38\,\text{ATP}$$

Most organisms carry on aerobic respiration, which releases a great deal more energy from a glucose molecule than anaerobic respiration does. Glycolysis occurs in the cytoplasm, but aerobic respiration occurs in mitochondria. As you will recall from Chapter 5, the mitochondrion has an outer and an inner membrane. Look at Figure 7-15. The enzymes associated with the series of reactions known as the Krebs cycle occur on the inner membrane and in the matrix, the dense solution enclosed by the inner membrane. The inner membrane, like the thylakoid membrane in a chloroplast, also contains the molecules of an electron transport chain and the ATP synthetase enzymes.

Think

Bread rises as a result of yeast fermentation. How might fermentation cause the bread to rise? Why is sugar often added to the bread dough?

Answer to Think
Yeast fermentation produces bubbles of CO_2 that become trapped in bread dough, causing it to rise. Sugar is added to the dough to provide the glucose for fermentation.

Info: Athletes who require quick bursts of energy, such as sprinters, have an above-average percentage of white fibers in their muscle tissue. These fibers are fueled by anaerobic respiration.

Figure 7-15 Mitochondria, such as the one shown at the left, are the site of aerobic respiration. The Krebs cycle occurs in the inner membrane.

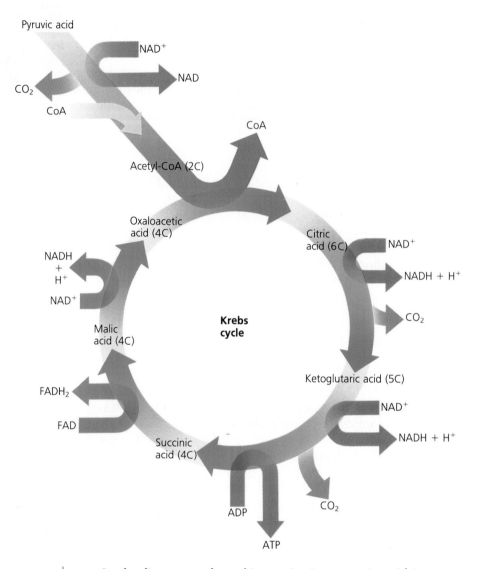

Pyruvic acid

NAD⁺

NAD

CO_2

CoA

CoA

Acetyl-CoA (2C)

Oxaloacetic
acid (4C)

Citric
acid (6C)

NAD^+

$NADH + H^+$

CO_2

NADH
+
H⁺

NAD⁺

**Krebs
cycle**

Malic
acid (4C)

Ketoglutaric acid (5C)

NAD^+

$NADH + H^+$

FADH₂

FAD

Succinic
acid (4C)

CO_2

ADP

ATP

Figure 7-16 Study this illustration of the Krebs cycle to understand how the chemical energy in pyruvic acid is converted to the chemical energy in ATP.

Tip: Tell students that they can obtain a clearer idea of the aerobic respiration process by overlapping Figures 7-16 and 7-13 at the *Pyruvic acid* labels. Be sure they understand why they need two copies of Figure 7-16 to have a more complete picture.

In the first step of aerobic respiration pyruvic acid is converted to a molecule called acetyl-CoA. In a series of reactions a three-carbon pyruvic acid molecule first loses CO_2 and becomes a two-carbon acetyl group. Study Figure 7-16, which illustrates the reactions. This acetyl group combines with coenzyme A, abbreviated CoA, forming **acetyl-CoA**. During this transformation pyruvic acid loses hydrogen. This happens when NAD^+ accepts the hydrogen, and a molecule of NADH forms:

Pyruvic acid + CoA + NAD^+ ⟶
acetyl-CoA + CO_2 + NADH + H^+

Acetyl-CoA now enters a biochemical pathway called the Krebs cycle. Remember that each molecule of glucose produces two molecules of pyruvic acid. Therefore, for each molecule of glucose, two molecules of acetyl-CoA enter the Krebs cycle.

The Krebs Cycle

The **Krebs cycle** is the central biochemical pathway of aerobic respiration. It is named after its discoverer, Sir Hans Krebs (1900–1981). Because citric acid is formed in the process, it is also known as the citric acid cycle. It begins when the two-carbon molecule acetyl-CoA combines with a four-carbon molecule to form the six-carbon molecule citric acid. Two carbon atoms are removed and form two CO_2 molecules, which are released from the cell. A four-carbon compound then combines with another acetyl-CoA molecule to start the cycle again.

The citric acid is immediately oxidized by enzymes in the mitochondria. Two carbon atoms are removed in the form of CO_2, and eight hydrogen atoms are released. These hydrogen atoms are picked up by two molecules—NAD^+ and a related molecule called FAD. Thus for each molecule of acetyl-CoA the cycle produces three molecules of $NADH + H^+$, one molecule of $FADH_2$, and one molecule of ATP. Since one molecule of glucose produces two molecules of pyruvic acid, each of which in turn becomes a molecule of acetyl-CoA, the breakdown of one molecule of glucose via the Krebs cycle produces six $NADH + H^+$, two $FADH_2$, and two ATP molecules.

Electron Transport Chain

Glycolysis, the conversion of pyruvic acid to acetyl-CoA, and the Krebs cycle complete the breakdown of glucose. Up to this point, 4 new ATP molecules have been made. As you can see in Figures 7-13 and 7-16, 12 carrier molecules—10 $NADH + H^+$ and 2 $FADH_2$—have been produced. These molecules carry electrons to an electron transport chain, where additional ATP is produced.

You already know how an electron transport chain works. Here the series of electron transport molecules is located in the inner membrane of the mitochondria. Figure 7-17 shows how electrons move down an electron transport chain, with oxygen as the final acceptor. Water is formed in the process.

As in the light reactions, ATP is generated by chemiosmosis. As electrons move down the energy gradient in the inner membrane of the mitochondrion, hydrogen ions are pumped across the membrane into the mitochondrial matrix. An electrochemical gradient forms, and hydrogen ions diffuse through channels that contain ATP synthetase. The energy generated by the movement of hydrogen ions is used by ATP synthetase to make ATP.

Let us review how many molecules of ATP are produced. Ten $NADH + H^+$ and 2 $FADH_2$ molecules are produced by aerobic respiration. Each molecule of $NADH + H^+$ yields 3 molecules of ATP. $FADH_2$ yields 2 molecules of ATP. Thus, the electron transport chain yields a maximum of 34 molecules of ATP:

$$10 \ NADH + H^+ = 30 \ ATP$$
$$2 \ FADH_2 = 4 \ ATP$$

Figure 7-17 The electron transport chain during aerobic respiration operates along the inner membrane of mitochondria.

Inner membrane

Matrix

Matrix

$NADH + H^+$ e^-

$2 H^+$
$FADH_2$

$2 H^+$

$2 H^+$

$2 H^+$

e^-

$\frac{1}{2} O_2$
$+$
$2 H^+$

H_2O

$2 H^+$

ATP synthetase   Inner membrane

H^+ in matrix

ADP + P

ATP

Glycolysis

Glucose

2 PGAL

2 ATP

2 ADP

2 pyruvate

2 acetyl CoA

Krebs cycle

2 ATP

2 NADH + H⁺

2 NADH + H⁺

2 ATP

6 NADH + H⁺

2 FADH₂

Electron transport

34 ATP

Figure 7-18 A single molecule of glucose can yield up to 38 ATP molecules, 34 generated by the electron transport chain and 2 each by glycolysis and the Krebs cycle.

Table 7-1 Summary of ATP Yield from Aerobic Respiration

Glycolysis	= 2 ATP
Krebs cycle	= 2 ATP
Electron transport	= 34 ATP
	38 ATP

Answers to Section Review
1. Anaerobic respiration is the breakdown of glucose without the use of oxygen. Aerobic respiration is the breakdown of glucose with the use of oxygen.
2. NAD⁺ bonds to hydrogen. Without a molecule to accept the hydrogen, glycolysis would stop.
3. Muscle soreness following exercise is probably a result of an oxygen debt in the muscles. Once oxygen is returned, as through heavy breathing, the lactic acid is converted back into pyruvic acid and the soreness disappears.
(See pp. T7A–T7F.)

Why is O_2 so important in aerobic respiration? At the end of the electron transport chain, O_2 is the final hydrogen acceptor. Water forms in the process:

$$\tfrac{1}{2}\,O_2 + 2\,H^+ \longrightarrow H_2O$$

Energy Yield

What is the maximum number of ATP molecules produced in the aerobic breakdown of glucose? Remember that the Krebs cycle and glycolysis yield two ATP molecules each, and the electron transport chain generates up to 34 ATP molecules. Thus, aerobic respiration produces a maximum of 38 ATP molecules.

The actual number of ATP molecules produced may vary for several reasons. For example, in brain tissue the electron transport chain yields 32 ATP molecules, for a maximum of 36 ATP molecules. In addition, some H^+ in chemiosmosis may leak through the membrane, or be used for other purposes, thus lowering the ATP yield.

Aerobic respiration is generally 19 times more efficient than anaerobic respiration. Nevertheless, the ATP produced represents only about one-half the total energy stored in a molecule of glucose. The remaining energy is unavailable for use by organisms. However, the conversion of glucose to ATP provides the cell with enough energy for cellular activities.

Section Review

1. What do the terms *anaerobic respiration* and *aerobic respiration* mean?
2. Why is NAD^+ called a hydrogen acceptor?
3. What is the relation between the sore muscles that result from strenuous exercise and lactic acid fermentation?
4. With what product does the Krebs cycle begin and end?
5. In ancient times the earth's atmosphere contained no oxygen gas. What form of respiration, do you think, was used by the earliest forms of life? Explain your answer.

Laboratory

Light Intensity and Photosynthesis

Objective
To measure the effect of light intensity on the rate of photosynthesis

Process Skills
Experimenting, observing, inferring

Materials
Glass-marking pencil, 4 250-mL beakers, 250-mL graduated cylinder, distilled water, pH paper, sodium bicarbonate, glass stirring rod, 4 15-cm-long sprigs of *Elodea*, electric light with 200-watt bulb

Method
1 Use a glass-marking pencil to label four beakers *A, B, C,* and *D.* Using a graduated cylinder, pour 200 ml of distilled water into each beaker.
2 In this experiment you will determine the relative rates of photosynthesis in sprigs of *Elodea* by observing changes in pH. Make a table to record your pH data. Head the columns *Distilled Water, Sodium Bicarbonate Solution,* and *Solution with* Elodea. Label the rows *A, B, C,* and *D.*
3 Using pH paper, test the pH of the water in each beaker. Record the results in your table. Is the water acidic, neutral, or basic?
4 Add 500 mg of sodium bicarbonate to beakers A, B, and C, and stir the solutions with a glass stirring rod until the chemical is dissolved. When you add sodium bicarbonate to water, carbon dioxide (CO_2) will form. How could you use the concentration of CO_2 to measure the rate of photosynthesis?
5 Test the pH of solutions A, B, and C. Record your results in the table. Are the solutions acidic, neutral, or basic? How does the pH measurement of these solutions differ from the previous pH measurement? What can

you infer about the relationship between the pH of a solution and its CO_2 content?
6 Obtain four 15-cm-long sprigs of freshly cut *Elodea* from your teacher. Place one sprig of *Elodea* in each beaker.
7 Place beaker A in darkness, beaker B in room light, and beakers C and D 0.5 m from an electric light lit by a 200-watt bulb.
8 After 20 minutes, use pH paper to test the pH of the solution in each beaker. Why is it important to measure the pH of each solution after a set amount of time?

Conclusions
1 How did the pH of the solution in each of the beakers change after the *Elodea* was added? What is the relationship between change in pH and rate of photosynthesis?
2 What purpose did solution D serve in this experiment? Did its pH change after *Elodea* was added?
3 According to your experiment, how is photosynthesis affected by light intensity?

Inquiry
If beaker C is covered with an airtight lid and kept in bright light, will the pH rise indefinitely? Why or why not?

Chapter 7 Review

Vocabulary

acetyl-CoA (106)
adenosine
  diphosphate (ADP)
  (94)
adenosine
  monophosphate
  (AMP) (94)
adenosine
  triphosphate (ATP)
  (94)
aerobic respiration
  (103)

alcoholic fermentation
  (105)
anaerobic respiration
  (103)
ATP synthetase (95)
autotroph (93)
biochemical pathway
  (94)
C_4 (102)
Calvin cycle (100)
CAM (102)
carotene (98)
carotenoid (98)

chemiosmosis (99)
CO_2 fixation (101)
dark reactions (96)
electron transport
  chain (98)
fermentation (103)
glycolysis (103)
granum (96)
heterotroph (93)
Krebs cycle (107)
lactic acid
  fermentation (104)

light reactions (96)
lumen (96)
phosphorylation (95)
photosynthesis (93)
photosystem (98)
phycobilin (98)
pigment (97)
respiration (93)
stomata (102)
stroma (96)
thylakoid (96)
xanthophyll (98)

From each set of terms below, choose the term that does not belong and explain why it does not belong.
1. ATP, C_4, acetyl-CoA, glycolysis
2. xanthophyll, carotene, phycobilin, stomata
3. photosynthesis, fermentation, respiration, heterotrophs
4. chemiosmosis, Calvin cycle, C_4, CAM
5. grana, stroma, stomata, thylakoid

Review

1. When cells break down glucose, they release (a) pigments (b) CO_2 (c) O_2 (d) light.
2. Phosphorylation occurs when (a) a phosphate group is added to a molecule (b) a phosphate group is removed from a molecule (c) water is split (d) O_2 is formed.
3. ATP synthetase is an important factor in photosynthesis and respiration because it (a) accepts hydrogen (b) replaces oxygen (c) adds phosphorus (d) catalyzes ATP formation.
4. The products of photosynthesis are (a) glucose and O_2 (b) CO_2 and H_2O (c) ATP and H_2O (d) RuBP and H^+.
5. Accessory pigments (a) add color to plants but do not trap light energy (b) trap light energy that chlorophyll cannot absorb (c) occur only in algae (d) transfer energy from chlorophyll to other pigments.
6. A reactant used in the Calvin cycle is (a) H_2O (b) glucose (c) CO_2 (d) citric acid.
7. C_4 photosynthesis enables plants in hot, dry climates to fix carbon (a) only at night (b) at a faster rate (c) with less energy (d) with minimum H_2O.
8. Glycolysis takes place in the (a) stroma (b) mitochondrion (c) cristae (d) cytoplasm.
9. Before aerobic respiration can proceed, pyruvic acid must be converted to (a) citric acid (b) glucose (c) acetyl-CoA (d) alcohol.
10. The maximum number of ATP molecules produced by aerobic respiration is (a) 38 (b) 36 (c) 34 (d) 4.
11. Why is it said that all life depends on autotrophs?
12. Explain the phrase "ATP is the currency of energy."

13. What is meant by the term *biochemical pathway?* Give an example of one that occurs during photosynthesis.
14. What role does chemiosmosis play in the formation of ATP?
15. Explain how CAM photosynthesis is advantageous for cactuses.
16. Why are yeast cells used in the production of wine?
17. Summarize the events that occur from the end of glycolysis through the first reaction of the Krebs cycle.
18. Compare the electron transport chain in photosynthesis and in respiration.
19. What factors determine the total net yield in aerobic respiration?
20. What did Calvin learn by labeling photosynthetic cells with radioactive CO_2?

Critical Thinking

1. The enzyme complex that removes carbon dioxide from pyruvic acid requires vitamin B_1, or thiamine. Why is this vitamin so important in the human diet?
2. The generation of ATP by the electron transport system is not 100 percent efficient, and some energy is lost as heat. In most animals the electron transport chain becomes even less efficient during the cold winter months. In what way might this decreased efficiency be an advantage?
3. The alcohol content of most wines is about 12 percent. The alcohol is produced by the fermentation of the natural sugar in grape juice by yeast. What might explain the alcohol content's not being higher?
4. What effect did the appearance of oxygen in the earth's atmosphere have on evolution? What was the most likely source of this oxygen?

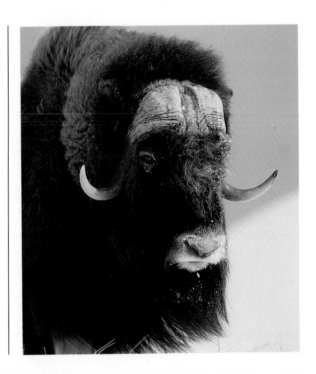

Extension

1. Read "How Purple Was My Valley," in *Discover*, November 1987, pp. 14–15. According to Goldsworthy, what adaptive advantage does a chlorophyll that uses red and blue light give a plant? Why might humans have something in common with *Halobacterium halobium?* Write a report answering these questions.
2. Suppose that tomorrow the sun's rays are blocked by dust or smoke clouds from a massive volcanic eruption. What short-term and long-term effects might the lack of sunlight have on the earth? How might the clouds affect the energy cycle?
3. With algae obtained from a pond or from your local aquarium store, make two identical cultures. Examine the cultures under the microscope, and note the number and types of algae. Place one culture in the light and the other in total darkness. Observe the cultures every day for the next ten days. Make drawings of the organisms that you see. What do your observations tell you about the role of light?

Chapter 8 | Nucleic Acids and Protein Synthesis

This chapter describes the structures and functions of DNA and RNA, as well as the processes of replication and transcription. Students view the translation process from DNA to RNA to protein synthesis.

Correlation Chart

Section Number	Section Objectives	Pages	Features	Section Review
8.1	1. Explain the principal function of DNA in cells.	113–117	Biotechnology, pp. 116–117	3, 5
	2. Describe the basic structure of a DNA molecule.			1, 2
	3. Summarize the process of replication.			3, 4
	4. List some major scientific discoveries that led to the description of DNA structure by Watson and Crick.			
8.2	1. Compare and contrast the structure of RNA with that of DNA.	118–119		2
	2. Name the three types of RNA, and describe the structure of each.			3
	3. Summarize the process of transcription.			1, 3, 4
8.3	1. List two characteristics of the genetic code.	120–123	Think, p. 120	4
	2. Distinguish between a codon and an anticodon, and state where each is found.			2
	3. Summarize the process of translation.			1, 3, 5

Resources*

Print

Creighton, Thomas E. *Proteins: Structures and Molecular Principles*. New York: Freeman, 1984.

Freifelder, David. *Essentials of Molecular Biology*. Boston: Jones and Bartlett, 1985.

Herriott, Jon, et al. *Papers in Biochemistry*. Reading, Mass.: Addison-Wesley, 1984.

Audiovisual/Software

The Chemistry of the Cell I: The Structure of Proteins and Nucleic Acids. Film, 21 min. U. of I.

Nucleic Acids. Apple II, 48K. With study sheet, teacher's guide and disk. Bio. Learning. Dist. by Carolina Biological.

Protein Synthesis. Apple II, 48K. Includes disk, backup, and complete documentation. Helix. Dist. by Carolina Biological.

*Additional references listed in *Modern Biology, Teacher's Program Guide*.

Section Teaching Guides

8.1 DNA

Approach

The main objective of this section is to lead students to understand that DNA controls cellular activities by directing the production of proteins in living organisms. The discussion of DNA structure builds upon the concepts of organic chemistry (Chapter 4). In presenting the text, help students to appreciate how the structure of DNA enables it to store information as well as participate in replication.

Concept Framework

Students read in Chapter 5 that cells take many forms and perform a variety of functions. Build on this concept to provide students with a perspective on the role of DNA in cells by asking students to list some of the familiar forms and functions of cells. Emphasize the diversity of structures and functions. Have students speculate about what ultimately controls this diversity. Elicit the observation that DNA stores all of the information needed by cells. Ask students as they read to consider how DNA stores information.

Emphasis

Make certain that students understand the importance of accuracy in DNA replication. You might review the concept of base pairing.

Guest Speaker

Invite a geneticist from a local university to talk with the class about DNA replication. Ask the speaker to describe the experiments that revealed the details of the replication process.

Using the Features/Discovering the Structure of DNA

The main process skill in this feature concerns communication among members of the scientific community. A scientific discovery often draws upon a large body of previous work. For example, James Watson and Francis Crick's discovery of DNA structure drew upon research conducted by Lawrence Bragg, Frank Griffith, and Erwin Chargaff. Point out to students that, because the results of earlier research had been published, Watson and Crick were

able to synthesize information and concepts from different sources in developing their model of DNA.

Section Review Answers

(See p. 117.)

8.2 RNA

Approach

The main objective of this section is to describe the structure of RNA. By comparing the structure of RNA with that of DNA, this section reinforces the concept of structure-function relationships. It also builds on the concepts of organic chemistry and prepares students for the discussion of protein synthesis in Section 8.3. In presenting the text, explain that protein synthesis is the primary function of RNA.

Concept Framework

Students are familiar with some template systems such as keys and locks, peg and hole games, and so on. Elicit from students other examples of template systems. Use this knowledge to engage students in a discussion of DNA and RNA synthesis. Then ask students to compare the processes of replication and transcription. You might end the discussion by asking students as they read to consider what triggers the exit of replicated RNA from the nucleus and its entry into the cytoplasm.

Emphasis

Students should see how transcription maintains an organism's genetic code by creating RNA with a base sequence that is complementary to DNA.

Section Review Answers

(See p. 119.)

8.3 Protein Synthesis

Approach

The primary objective of this section is to familiarize students with the process of protein synthesis. Emphasize that the specific base sequences of mRNA "code for" amino acids, amino acids then bond together to form polypeptides, and one or more polypeptides compose proteins. In presenting the text, relate protein synthesis to the two previous sections by reminding students that DNA makes RNA and RNA makes proteins.

Concept Framework

All students have heard that protein is a human nutritional requirement. Tap this knowledge to provide a conceptual perspective for this section by engaging students in a discussion of why proteins are important in human nutrition. Elicit the suggestion that proteins can be broken up and reassembled to serve different functions and that protein synthesis is the process by which such reassembling occurs. Encourage students to study the processes in protein synthesis as they read this section.

Emphasis

Make sure that students comprehend that 64 possible codons code for 20 different amino acids as well as give a few start and stop signals. Stress the variation that is possible in combining the four nucleotides into three-unit codons. You may wish to compare the genetic code to letters of an alphabet and codons to words, in order to help students grasp the information-transmission aspects of protein synthesis.

Protein Synthesis Analogies

Assist students in developing analogies of some processes that resemble protein synthesis. Such processes should be sequential, involve several sources of information, and result in a product. You may wish to have students vote on the process they think is most analogous to protein synthesis and then name the points of analogy.

Section Review Answers

(Continued from p. 123.)

3. The cloverlike structure of transfer RNA allows an amino acid attached at one end to be positioned properly for insertion in a polypeptide when the anticodon attaches to a codon in messenger RNA.
4. The genetic code is formed by the arrangement of nitrogen bases in a particular sequence of DNA and it is universal because every organism possesses the bases in the same order.
5. Students should reason that not all genes are always active because not all proteins are needed at all times or in all cells. Some students may recognize that cell diversity and specialization arises from the fact that different genes are "turned on" in different cells.

Laboratory Preparation

Background

In this lab students learn how scientists identify chemical substances found in living organisms. Students should be reminded that organisms store heredity material in DNA and RNA. Also, students should be familiar with the fact that the biuret reagent reacts with protein and breaks it down into amino acids. Students observe a color change when all the protein is broken down.

Preparation

To prepare diphenylamine, add 15 g diphenylamine to 1 L glacial acetic acid. Add 15 mL concentrated H_2SO_4. Store in a dark bottle. On the day of lab, add 1 mL acetaldehyde solution. To prepare acetaldehyde solution, add 1 mL acetaldehyde to 500 mL distilled water. DNA and RNA can be purchased from a biological supply house. To prepare DNA solution, add 0.1 g DNA for every 200 mL distilled water. Add one drop glacial acetic acid to every 200 mL DNA solution. If necessary, use heat to dissolve DNA. Store in freezer. Prepare RNA solution in the same manner. Hydrolyzing the polypeptides in raw egg white with pepsin forms digested egg white. To prepare digested egg white, add 0.5 g pepsin to 50 mL distilled water and stir until dissolved. 24 hours before the investigation, add 50 mL pepsin solution to each 100 mL raw egg white. Pepsin can be purchased from a biological supply house. 10% aqueous sodium hydroxide and 0.5% copper sulfate can be purchased mixed together from a biological supply house. However, results are better when chemicals are prepared and used separately.

Conclusions

1. The undigested protein required more copper sulfate solution for identification. Some of the protein had been broken down into amino acids; therefore, there was less protein present. The smaller amount of protein required less reagent to produce a color change.
2. Test tube A contained DNA. Test tube B contained RNA.

Inquiry

Students should question whether heat would change the structure of protein and cause it to react differently to the reagents. They will form a hypothesis and test it.

Chapter Review

Vocabulary

1. Translation refers to a process in which one "language" (sequence of bases in mRNA) is converted to another (sequence of amino acids in a protein).
2. Messenger RNA carries the "message" of information for protein synthesis from DNA in the nucleus to ribosomes in the cytoplasm.
3. Transcription refers to a process in which the sequence of bases in DNA is "transcribed" into a complementary sequence of bases in RNA.
4. The prefix "deoxy-" indicates that a molecule of deoxyribose has one less oxygen atom than a molecule of ribose.
5. A hydrogen bond is a chemical bond in which atoms share a hydrogen nucleus.

Review

11. The double-stranded structure of DNA allows it to replicate itself to pass on information to new cells; its sequence of four different bases allows it to encode information.
12. Enzymes separate the two strands of DNA, and each strand serves as a template upon which nucleotides are assembled into complementary strands with the aid of other enzymes.
13. Franklin's images indicated that DNA is in the form of a helix with the sugar-phosphate backbone on the outside. They also indicated that a molecule of DNA must consist of more than one strand of nucleotides.
14. Messenger RNA transmits information from DNA for use as a template in building proteins; transfer RNA carries amino acids to the ribosomes; ribosomal RNA forms part of the structure of ribosomes and functions in protein synthesis in ways not yet fully understood.
15. RNA is single-stranded, DNA is double-stranded; RNA has ribose, DNA has deoxyribose; RNA has uracil, DNA has thymine.
16. Enzymes separate the two strands of DNA and catalyze the formation of bonds between sugars and phosphates to form a new strand of nucleotides.

17. The sequence of bases in an anticodon is complementary to that in its corresponding codon.
18. They serve as "start" and "stop" signals.
19. Each gene codes for one polypeptide, and many proteins consist of more than one polypeptide.
20. The ribosome attaches to mRNA and matches up the anticodons of tRNA with the codons of mRNA in such a way that a polypeptide is formed.

Critical Thinking

1. The four codons for alanine differ only in the base found in the third position. This pattern tends to minimize the errors that occur in translation because of the substitution of bases in the third position.
2. methionine, serine, asparagine, lysine, cysteine, serine, stop
3. TAC AGA TTG TTT GTC CGA ATT
4. If the units consisted of two bases, each of the four bases could combine with another base of the same type or with any of the other three bases, allowing for $4 \times 4 = 16$ different combinations—not enough to code for the 20 existing amino acids.
5. In uracil a CH_3 group is attached to one of the carbons, which in thymine has a H atom attached. This difference does not affect bonding because the hydrogen bonds with adenine form on the other side of the molecule.

Chapter 8 Nucleic Acids and Protein Synthesis

Introduction

Every cell, no matter how simple or complex, must store and transmit the information needed to manufacture proteins. DNA stores the information for protein synthesis. RNA transmits this information for use in building proteins. In this chapter you will learn how the functions of DNA and RNA are related to their molecular structures.

Chapter Outline

8.1 DNA
- *Structure of DNA*
- *Replication of DNA*

8.2 RNA
- *Structure of RNA*
- *Transcription*

8.3 Protein Synthesis
- *Protein Structure*
- *Codons*
- *Translation*

Chapter Concept

As you read, pay attention to the way that DNA and RNA store and transmit information and synthesize proteins. Keep in mind that DNA makes RNA, which in turn makes proteins.

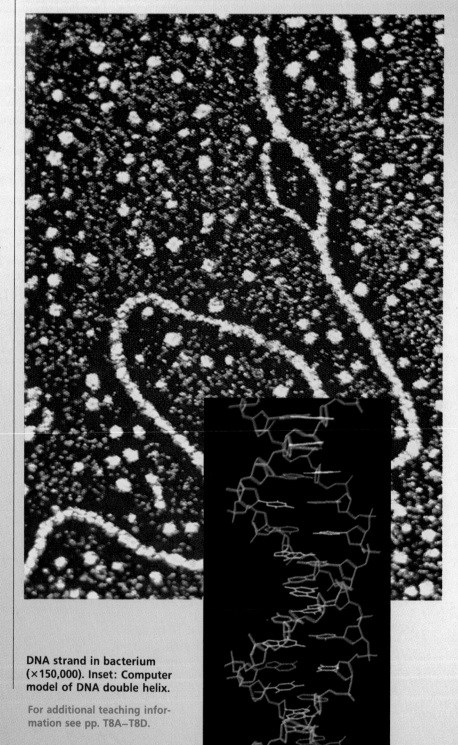

DNA strand in bacterium (×150,000). Inset: Computer model of DNA double helix.

For additional teaching information see pp. T8A–T8D.

8.1 DNA

What enables cells to have so many different forms and perform so many different functions? Ultimately the source of this amazing diversity is DNA, or deoxyribonucleic acid. In simple terms, DNA controls the production of proteins within the cell. These proteins in turn form the structural units of cells and control all chemical processes within the cell.

Every new cell that develops in your body needs an exact copy of the DNA from its parent cell. Likewise, humans and all other organisms must be able to pass copies of their DNA on to their offspring in order to continue the species. As you will see in this chapter, the structure of the DNA molecule is related to its two primary functions—to store and use information to direct the activities of the cell and to copy itself exactly for new cells that are created.

Structure of DNA

As you recall from Chapter 4, the nucleic acids DNA and RNA are complex organic molecules. DNA and RNA are polymers—that is, they are composed of repeating subunits, or monomers. The repeating subunits in both DNA and RNA are **nucleotides.**

Nucleotides

The DNA molecule consists of two long strands, each of which is a chain of nucleotide monomers. Each nucleotide has three parts: (1) a five-carbon sugar molecule called **deoxyribose** *(dee-AHK-sih-RIE-BOSE)*, (2) a phosphate group, and (3) a nitrogen base.

While the sugar molecule and phosphate group are the same in every nucleotide, the nitrogen base may be any one of four different kinds. These bases are **adenine** *(AD-'n-EEN)*, **guanine** *(GWAHN-EEN)*, **thymine** *(THIE-MEEN)*, and **cytosine** *(SITE-uh-SEEN)*. Study the structure of each in Figure 8-1. The first two bases—adenine and guanine—belong to a class of organic molecules known as **purines** *(PYOOHR-EENZ)*, which have a double ring of carbon and nitrogen atoms. The second two bases—thymine and cytosine—are **pyrimidines** *(pie-RIM-uh-DEENZ)*. These molecules have a single ring of carbon and nitrogen atoms.

Section Objectives

- *Explain the principal function of DNA in cells.*
- *Describe the basic structure of a DNA molecule.*
- *Summarize the process of replication.*
- *List some major scientific discoveries that led to the description of DNA structure by Watson and Crick.*

Figure 8-1 The nitrogen base in a nucleotide is either a purine or a pyrimidine.

Purines: Adenine, Guanine
Pyrimidines: Thymine, Cytosine

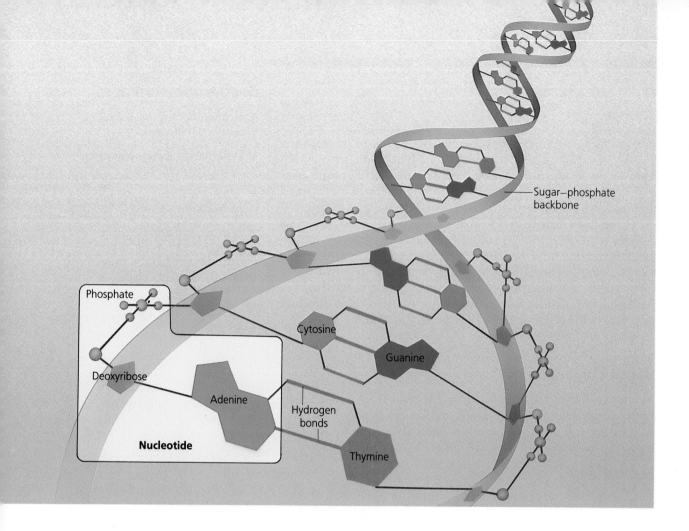

Sugar–phosphate backbone

Phosphate

Cytosine

Deoxyribose

Guanine

Adenine

Hydrogen bonds

Nucleotide

Thymine

Figure 8-2 The DNA structure resembles a twisted ladder and is described as a double helix.

Info: Other theories besides the double helix theory were presented. At one time Linus Pauling was working on a three-stranded theory that was thought of highly.

The Double Helix

Each nucleotide—deoxyribose, phosphate, and nitrogen base—bonds to other nucleotides to form a long strand. Two of these strands bonded together form a molecule of DNA. The two strands twist around a central axis to form a spiral structure called a **double helix.** The double helix structure was first described in 1953 by James Watson (1928–) and Francis Crick (1916–). Their discovery is one of the most significant of the twentieth century.

Figure 8-2 shows the details of the structure of DNA. As you can see, the DNA molecule looks something like a twisted ladder. The sides of the ladder are formed by alternating sugar and phosphate units, and the rungs consist of bonded pairs of nitrogen bases. Notice that the rungs are of uniform length because in each case one base is a double-ringed purine and the other is a single-ringed pyrimidine. Note also that the DNA molecule has a right-hand twist, with each full turn consisting of ten base pairs.

The two strands of DNA are held together by hydrogen bonds between the bases. A **hydrogen bond** is a type of chemical bond in which atoms share a hydrogen nucleus—that is, one proton. In DNA, hydrogen bonds form between purines and pyrimidines. More specifically, adenine always bonds with thymine, and cytosine always bonds with guanine. This occurs because the structure of adenine complements that of thymine in such a way that hydrogen bonds are established. The same kind of complementarity exists between cytosine and guanine. However, adenine and thymine form only two hydrogen bonds, while cytosine and guanine form three.

The fact that adenine always bonds to thymine and cytosine always bonds to guanine has structural and functional consequences. The sequential arrangement of nitrogen bases along one strand is the exact complement of the sequential arrangement of bases on the adjacent strand. Because of this characteristic, the two strands are said to be complementary strands.

Replication of DNA

The scientific world was stunned by Watson and Crick's proposed structure for DNA because it immediately suggested how DNA might be copied—and copied exactly. If the two complementary strands of DNA were to unwind, then each strand could serve as the template, or mold, on which a new complementary strand could be built. This process would produce two new DNA molecules that are exact replicas of the original DNA. Subsequent research has shown that this is exactly what happens.

Process of Replication

The process of the duplication of a DNA molecule is called **replication.** Refer to Figure 8-3 as you read about the process of replication. Replication begins when an enzyme called DNA helicase attaches to a DNA molecule, moves along the molecule, and "unzips" the two strands of DNA. DNA helicase acts by breaking the hydrogen bonds between the nitrogen bases.

After the DNA strands are separated, the now unpaired bases in each strand react with the complementary bases of nucleotides that are floating freely in the nucleus. These complementary bases bond with the bases in the DNA strands by forming hydrogen bonds. As each new set of hydrogen bonds links a pair of bases, an enzyme called DNA polymerase catalyzes the formation of the sugar-to-phosphate bonds that connect one nucleotide to the next one. This process results in two new DNA molecules, each of which consists of one "old" strand of DNA and one "new" strand of DNA.

The process of replication maintains the complementary nature of the DNA molecule. Since adenine (A) can pair only with thymine (T) and cytosine (C) can pair only with guanine (G), the sequence of nucleotides in each new strand exactly matches that in the original molecule. Suppose, for instance, that the sequence of nucleotides in one strand of the original molecule was A-T-T-C-C-G. The complementary strand in the original molecule must have had the sequence T-A-A-G-G-C. Any new strand formed

Figure 8-3 Replication results in two new DNA molecules, each with one old and one new strand.

Tip: When giving students a sequence of bases to practice replication, it is good to give four or five of the same base in a row so the student can more readily see that the replicated strand is the same.

using the first strand as a template will also follow the sequence T-A-A-G-G-C. Similarly, any new strand formed using the original T-A-A-G-G-C strand as a template must follow the sequence A-T-T-C-C-G.

Replication of DNA doesn't begin at one end of the molecule and proceed to the other. Rather, copying of DNA occurs simultaneously at many points on the molecule. Enzymes join the individual segments of DNA to each other as they are copied. Because of the size of the DNA molecule, if replication didn't occur in this way, it would take 16 days to copy just one DNA molecule of a fruit fly. Actually, the replication of this fruit fly DNA takes only about three minutes, because about 6,000 sites are being copied simultaneously.

Accuracy and Repair

The entire process of DNA replication occurs with great accuracy. The cell has a built-in "proofreading" function, performed by enzymes or proteins depending on the organism. The degree of accuracy is on the order of one error per billion nucleotides.

DNA also has a mechanism for repair. In human cells, body heat, radiation, chemicals, and other factors can damage DNA.

Biology in Process

Discovering the Structure of DNA

In 1912 British physicist Lawrence Bragg figured out how to use X rays to reveal the structure of crystals. In 1928 British bacteriologist Frank Griffith determined that traits could be transmitted between strains of bacteria. These discoveries seem to have little to do with DNA, but they launched lines of research that converged in the discovery of DNA structure in the early 1950s.

Although the names of James Watson and Francis Crick are forever associated with the structure of DNA, they drew upon the work of many other researchers in developing their model. Truly in the words of Isaac Newton, they "stood upon the shoulders of giants."

When Watson and Crick began their work together, they already knew a great deal about DNA. Researchers following up on the work of Griffith had demonstrated that DNA is the genetic material. Others had revealed many details of the chemical composition of DNA. Of special note was a discovery by American biochemist Erwin Chargaff. He had shown that in DNA the amount of cytosine equals that of guanine, and the amount of thymine equals that of adenine.

Answer to Biology in Process: Watson and Crick communicated between themselves to synthesize information provided from the work of others in DNA research.

Body heat alone continuously breaks up bonds between deoxyribose and purines. A group of 20 or more repair enzymes recognizes and removes damaged nucleotides and replaces them with new ones, thus ensuring the accurate replication of new DNA.

In studying the structure and replication of DNA, make a note of one thing. The sequence of bases along a strand is, as you will learn, not random. Inherent in the sequence of nitrogen bases is all the information determining the structure and function of an organism.

Section Review

1. What is the difference between a *purine* and a *pyrimidine?*
2. How are the two strands of a DNA molecule bonded together to form a double helix?
3. Why is the process of DNA replication necessary?
4. What important roles do enzymes play during the replication process?
5. Prokaryotic cells lack a nucleus; eukaryotic cells have a nucleus. Suggest an adaptive advantage for confining DNA in a nucleus.

Answers to Section Review
1. A purine has a double ring of carbon and nitrogen atoms; a pyrimidine has a single ring of carbon and nitrogen atoms.
2. by hydrogen bonds between complementary bases
3. Replication is necessary to produce DNA for new cells produced by cell division.
4. Enzymes separate the strands of DNA and catalyze the formation of sugar-to-phosphate bonds that form new strands of DNA.
5. It is probably advantageous for a cell to have DNA and the enzymes involved in replication and transcription confined to a small area rather than dispersed throughout the cell.

The most direct information about DNA structure available to Watson and Crick came from Bragg's technique of X-ray crystallography. Maurice Wilkins and Rosalind Franklin, researchers at King's College in London, had used X-ray crystallography to produce images of DNA like the one shown at left. By 1952 Franklin's interpretation of these images indicated that DNA is a helix with the sugar–phosphate backbone on the outside. Franklin's images also showed that the diameter of the helix was so large that it must consist of more than one strand of nucleotides.

The main questions left for Watson and Crick (above) to answer were how many strands of nucleotides make up DNA and how those strands are bonded together. Watson and Crick succeeded in answering those questions by combining available information with a technique of model building, which they knew that American chemist Linus Pauling had used to determine aspects of protein structure.

The chain of events that led to the discovery of DNA structure exemplifies the way in which a scientific discovery often draws upon the work done by many scientists over many years. In this particular case, Lawrence Bragg was no doubt pleased that the technique he had developed 40 years earlier played a major role in this landmark discovery, which Watson and Crick made in the laboratory he directed at Cambridge University.
■ In what ways was the process of communicating vital to the success of Watson and Crick?

Section Objectives

- *Compare and contrast the structure of RNA with that of DNA.*

- *Name the three types of RNA, and describe the structure of each.*

- *Summarize the process of transcription.*

Info: Note that only one strand of DNA is transcribed in any one segment of DNA.

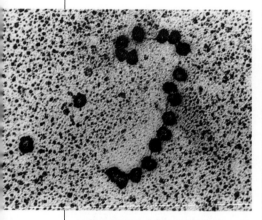

Figure 8-4 Ribosomes, shown in this SEM, are composed of rRNA and proteins.

8.2 RNA

DNA stores and transmits the information needed to make proteins but it does not actually use that information to synthesize proteins. That is the primary function of RNA, or ribonucleic acid. In this section we will examine the basic structure of RNA and the process by which it is produced from DNA. In Section 8.3 we will study the role of RNA in the synthesis of proteins.

Structure of RNA

Like DNA, RNA is a polymer consisting of nucleotide monomers or subunits. However, RNA differs from DNA in three important ways. First, an RNA molecule consists of only one strand of nucleotides instead of the two strands of the DNA molecule. Second, RNA has **ribose** as its five-carbon sugar rather than deoxyribose. Finally, RNA has the nitrogen base **uracil** instead of thymine.

In addition, RNA exists in the following three structural forms. Each plays a different role in protein synthesis.

- **Messenger RNA (mRNA)** is a single, uncoiled strand that transmits information from DNA for use during protein synthesis. It serves as a template, or pattern, for the assembly of amino acids during protein synthesis.
- **Transfer RNA (tRNA)** is a single strand of RNA folded back on itself in hairpin fashion, allowing some complementary bases to pair. tRNA exists in 20 or more varieties, each with the ability to bond to only one specific type of amino acid.
- **Ribosomal RNA (rRNA)**, which is RNA in globular form, is the major constituent of the ribosomes. The precise function of rRNA in protein synthesis is not yet fully understood.

Transcription

RNA is produced from DNA through a process called **transcription**. RNA molecules are transcribed according to the information encoded in the base sequence of DNA. As you read about transcription, follow the process in Figure 8-5.

An enzyme called RNA polymerase first binds to a DNA molecule and then causes the separation of the complementary strands of DNA. The enzyme next directs the formation of hydrogen bonds between the bases of a DNA strand and complementary bases of RNA nucleotides that are floating in the nucleus. RNA polymerase then moves along the section of DNA, establishing the sugar-to-phosphate bonds between the RNA nucleotides in much the same way as in DNA replication. When RNA polymerase reaches the sequence of bases on the DNA that acts as a termination signal, the enzyme triggers the release of the newly made RNA.

Tip: To help explain the term *transcription,* relate it to the word *transcript.* In a sense the DNA code is "written down" on the mRNA strand.

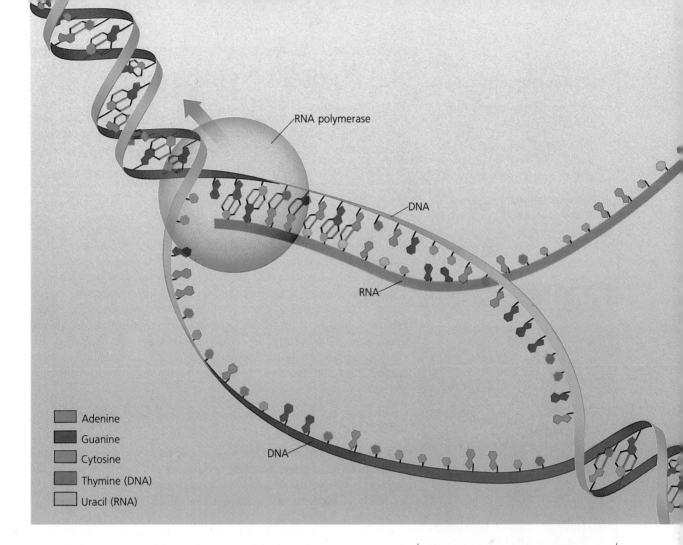

Adenine

Guanine

Cytosine

Thymine (DNA)

Uracil (RNA)

RNA polymerase

DNA

RNA

DNA

All three types of RNA are transcribed in this manner. Each type of RNA then leaves the nucleus and travels to the cytoplasm, where it is involved in the synthesis of protein.

Notice that the transcription process creates RNA with a base sequence complementary to DNA. As you learned in Section 8.1, genetic information resides in the sequence of bases on a DNA molecule. During transcription this genetic code of DNA becomes inherent in the sequence of bases in RNA.

Section Review

1. What is *transcription*?
2. In what three ways does the structure of RNA differ from that of DNA?
3. What are the three types of RNA?
4. How is the sequence of bases in DNA reflected in the sequence of bases in the RNA made through the transcription process?
5. Why is the presence of nuclear pores advantageous for the movement of RNA?

Figure 8-5   During transcription a DNA strand is linked to complementary bases of RNA nucleotides by hydrogen bonds.

Answers to Section Review
1. the process through which RNA is produced from DNA
2. RNA is single stranded; it has ribose instead of deoxyribose; and it has uracil instead of thymine.
3. messenger RNA, transfer RNA, ribosomal RNA
4. The bases in the RNA are complementary to the DNA the RNA was transcribed from.
5. Nuclear pores allow RNA to move from the nucleus to the cytoplasm, where protein synthesis takes place.

Nucleic Acids and Protein Synthesis   **119**

Think

These organisms have the same genetic code. What does this suggest about evolution of life on earth?

8.3 Protein Synthesis

Now that you know about the structures of DNA and RNA, let's take a look at the process of protein synthesis itself. **Protein synthesis** is the formation of proteins using information coded on DNA and carried out by RNA.

Protein Structure

Depending on its complexity, an organism may have from several hundred to several thousand different proteins as part of its makeup. Each protein molecule is made up of one or more polymers called polypeptides, each of which consists of a specific sequence of amino acids linked together by peptide bonds. As you know from Chapter 4, there are 20 different amino acids. A protein may consist of hundreds or thousands of amino acids, and all these amino acids must be arranged in a particular sequence for the protein to function properly. In fact, all structural and functional characteristics of a protein are determined by its amino acid sequence.

Codons

The assembly of amino acids into a specific protein is the task accomplished during protein synthesis. Before you study the sequence of events that occurs in protein synthesis, you must understand how the information necessary for creating proteins is encoded in nucleic acids.

The **genetic code** is the system that contains information needed by cells for proper functioning. This information is built into the arrangement of the nitrogen bases in a particular sequence of DNA. Since DNA makes RNA, which makes proteins, the DNA ultimately contains the information needed to put the amino acids together in the proper sequence.

As you learned in Section 8.2, mRNA is transcribed using DNA as a template. The genetic code inherent in the DNA is thus reflected in the sequence of bases in mRNA. A specific group of three sequential bases of mRNA is called a **codon.** In effect, each codon codes for, or recognizes, a specific amino acid using tRNA as an intermediary. As you will see, each codon attracts a group of bases on tRNA, and each tRNA has a specific amino acid attached to it.

Table 8-1 lists all of the 64 possible codons. You will note that each amino acid listed in the table has several codons. These codons often differ from each other only by the base in the third position. A few codons do not encode amino acids at all. Instead, they are "start" signals that engage a ribosome to start reading an mRNA molecule or "stop" signals that cause the ribosomes to stop reading mRNA.

Tip: Tell students that during translation the genetic code is "explained" or "translated" for the cell by changing it to a sequence of amino acids.

Info: Once the peptide bond forms, the tRNA molecules leave the amino acid and can be reused to transfer the same amino acid to another part of the mRNA strand.

Table 8-1 Codons in mRNA

First Base	Second Base				Third Base
	U	C	A	G	
U	UUU } Phenyl- UUC } alanine UUA } UUG } Leucine	UCU UCC UCA } Serine UCG	UAU } Tyrosine UAC } UAA } UAG } Stop	UGU } Cysteine UGC } UGA Stop UGG Tryptophan	U C A G
C	CUU CUC } Leucine CUA CUG	CCU CCC } Proline CCA CCG	CAU } Histine CAC } CAA } CAG } Glutamine	CGU CGC } Arginine CGA CGG	U C A G
A	AUU } AUC } Isoleucine AUA } AUG Methionine	ACU ACC } Threonine ACA ACG	AAU } Asparagine AAC } AAA } AAG } Lysine	AGU } Serine AGC } AGA } Arginine AGG }	U C A G
G	GUU GUC } Valine GUA GUG	GCU GCC } Alanine GCA GCG	GAU } Aspartic GAC } acid GAA } Glutamic GAG } acid	GGU GGC } Glycine GGA GGG	U C A G

The coding for amino acids shown in Table 8-1 is universal. For example, UUU is the codon for the amino acid phenylalanine in bacteria, mice, and humans. AUG is the universal "start" code. Every organism has the same genetic code.

Translation

The process of assembling protein molecules from information encoded in mRNA is called **translation.** The process of translation begins when mRNA moves out of the nucleus by passing through the nuclear pores. The mRNA then migrates to a group of ribosomes, where the actual synthesis of proteins takes place.

Amino acids floating freely in the cytoplasm are transported to the ribosomes by tRNA. Notice in Figure 8-6 that each kind of tRNA has a region that bonds to a specific amino acid. Notice also that the opposite loop of the cloverlike structure of tRNA bears a sequence of three bases called an **anticodon.** The tRNA anticodon bases are complementary to—and thus pair with—the codon of mRNA.

The assembly of a polypeptide begins when a ribosome attaches at an AUG codon on the mRNA. The AUG codon pairs with an anticodon UAC on a specific tRNA. Since the tRNA with this anticodon carries methionine, the first amino acid in the chain is always methionine. However, this does not mean that all polypeptides begin with methionine, because this first methionine is later removed from the chain.

Figure 8-6 Each kind of tRNA transports a specific amino acid to the ribosomes.

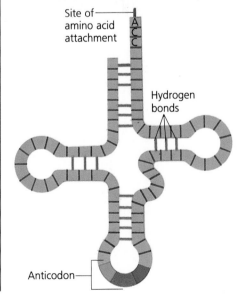

Site of amino acid attachment

Hydrogen bonds

Anticodon

Nucleus

Pore

Amino acid

Transfer RNA

Amino acid

Anticodon

Ribosome

Codon

Messenger RNA

A U G A C A G U U C U G

As you can see in Figure 8-7, usually several ribosomes are simultaneously translating the same mRNA. As a ribosome moves along the strand of mRNA, each codon is sequentially paired with its anticodon and the specified amino acid is added to the chain. An enzyme in the ribosome catalyzes a reaction that binds each new amino acid to the chain. In this reaction a peptide bond is formed with the last amino acid in the chain. The process continues, with each amino acid being linked to the chain in turn. Eventually the ribosome reaches a "stop" codon. This codon brings the

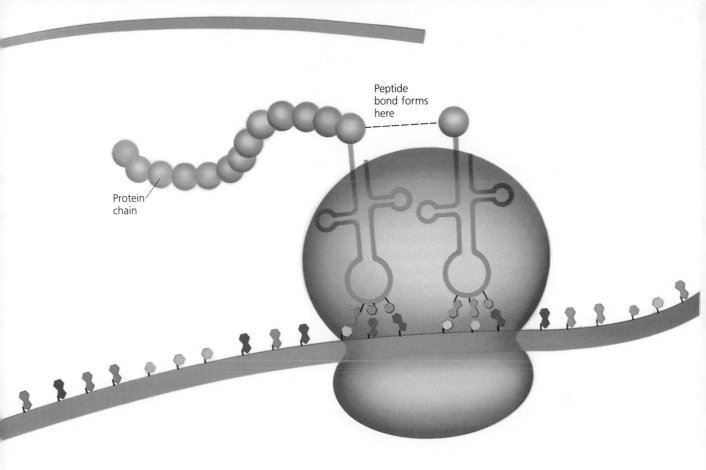

Peptide
bond forms
here

Protein
chain

process to an end. At this point the mRNA is released, and the polypeptide is complete.

As you have seen, the manufacture of specific proteins is a function of the order of bases in mRNA. In turn, the base sequence of mRNA is complementary to that of a specific region of DNA. The region of DNA that directs the formation of a polypeptide is called a **gene.** Since most proteins consist of more than one polypeptide, information from several genes may be required to produce a particular protein.

Section Review

1. What is *translation?*
2. Distinguish a codon from an anticodon, and explain the significance of each.
3. How does the physical structure of transfer RNA relate to its function?
4. Explain how the genetic code is formed and why the genetic code is said to be universal.
5. Do you think that all of an organism's genes are always involved in the continuing process of making proteins? Defend your answer.

Figure 8-7   When a codon is sequentially paired with its anticodon, a specified amino acid is added to the chain.

Answers to Section Review
1. the process by which protein molecules are assembled from information encoded in mRNA
2. A codon is a series of three bases in mRNA that codes for a specific amino acid; an anticodon is a series of three bases in tRNA that is complementary to a codon and pairs with it during translation.
(See pp. T8A–T8D.)

Nucleic Acids and Protein Synthesis   **123**

Writing About Biology

The Double Helix

The following excerpts are from The Double Helix, James Watson's account of the discovery of the structure of DNA.

The α-helix had not been found by staring at X-ray pictures; the essential trick, instead, was to ask which atoms like to sit next to each other. In place of pencil and paper, the main working tools were a set of molecular models superficially resembling the toys of preschool children. . . .

I went ahead spending most evenings at the films, vaguely dreaming that at any moment the answer would suddenly hit me. . . .

Not until the middle of the next week, however, did a nontrivial idea emerge. It came while I was drawing the fused rings of adenine on paper. Suddenly I realized the potentially profound implications of a DNA structure in which the adenine residue formed hydrogen bonds similar to those found in crystals of pure adenine. If DNA was like this, each adenine residue would form two hydrogen bonds to an adenine residue related to it by a 180-degree rotation. Most important, two symmetrical hydrogen bonds could also hold together pairs of guanine, cytosine, or thymine.

I thus started wondering whether each DNA molecule consisted of two chains with identical base sequences held together by hydrogen bonds between pairs of identical bases. There was the complication, however, that such a structure could not have a regular backbone since the purines (adenine and guanine) and the pyrimidines (thymine and cytosine) have different shapes.

Despite the messy backbone, my pulse began to race. . . . The existence of two intertwined chains with identical base sequences could not be a chance matter. Instead it would strongly suggest that one chain in each molecule had at some earlier stage served as the template for the synthesis of the other chain. . . .

[One day elapsed during which American crystallographer Jerry Donahue convinced Watson that his model was incorrect.]

When I got to our still empty office the following morning, I quickly cleared away the papers from my desk top so that I would have a large, flat surface on which to form pairs of bases held together by hydrogen bonds. Though I initially went back to my like-with-like prejudices, I saw all too well that they led nowhere. When Jerry came in I looked up, saw that it was not Francis, and began shifting the bases in and out of various other pairing possibilities. Suddenly I became aware that an adenine-thymine pair held together by two hydrogen bonds was identical in shape to a guanine-cytosine pair held together by at least two hydrogen bonds. All the hydrogen bonds seemed to form naturally; no fudging was required to make the two types of base pairs identical in shape. Quickly I called Jerry over to ask him whether this time he had any objection to my new base pairs. When he said no, my morale skyrocketed. . . .

Upon his arrival Francis did not get more than halfway through the door before I let loose that the answer to everything was in our hands. . . .

Write

- James Watson used time away from his laboratory and a set of models similar to preschool toys to help him solve the puzzle of DNA. In an essay discuss how play and relaxation help promote clear thinking and problem solving.

124 *James D. Watson, excerpted from* The Double Helix. *Copyright © 1968 James D. Watson. Reprinted with permission of Atheneum Publishers, an imprint of Macmillan Publishing Company.*

Laboratory

Testing for Proteins and Nucleic Acids

Objective
To learn chemical tests for detecting proteins and nucleic acids

Process Skills
Experimenting, inferring, hypothesizing

Materials
100-mL beaker, hot plate, glass-marking pencil, 5 test tubes, test-tube holder, 7 medicine droppers, diphenylamine, solution A, solution B, distilled water, tongs, undigested raw egg white, digested raw egg white, 10% aqueous sodium hydroxide solution, 0.5% copper sulfate solution, test-tube stopper

Method

1 Prepare a hot water bath by pouring 50 mL of water into a 100-mL beaker and boiling the water on a hot plate.
2 While the water is heating, label three test tubes *A*, *B*, and *C*. Place these in a test-tube holder, and put 40 drops of diphenylamine in each tube. Diphenylamine, a reagent that tests for the presence of nucleic acids, stains DNA purple and RNA green.
3 Using a clean dropper, add 20 drops of solution A to test tube A.
4 Using a clean dropper, add 20 drops of solution B to test tube B.
5 Using a clean dropper, add 20 drops of distilled water to test tube C. What is the function of test tube C?
6 Use tongs to place the test tubes in the boiling water. Heat for 25 minutes. Using tongs, remove the test tubes from the water bath, and allow them to cool for 10 minutes.
7 While the tubes are heating, label two test tubes *D* and *E*. Using a clean dropper, put 40 drops of undigested raw egg white in tube D.
8 Using a clean dropper, place 40 drops of digested raw egg white in tube E. Egg white is mostly protein. In a digested protein, the peptide bonds that hold the polypeptide chain together have been broken.

9 Using a clean dropper, slowly add 40 drops of sodium hydroxide solution to tube D and mix carefully. Then add copper sulfate solution one drop at a time. Stopper and gently shake the tube between drops. A color change indicates the presence of undigested protein. Record the number of drops you added to produce a color change.
10 Repeat step 9 using test tube E.
11 After tubes A, B, and C have been heated and cooled, observe the three test tubes for color changes. Record your observations.

Conclusions
1 Which form of protein required the greater amount of copper sulfate solution for identification? Why do you think this was so?
2 Which test tube contained DNA? Which contained RNA?

Inquiry
If you had heated the undigested protein for 25 minutes, would it have affected your test results? Form a hypothesis, and test it.

Vocabulary

adenine (113)	gene (123)	protein synthesis (120)	thymine (113)
anticodon (121)	genetic code (120)	purine (113)	transcription (118)
codon (120)	guanine (113)	pyrimidine (113)	transfer RNA (tRNA) (118)
cytosine (113)	hydrogen bond (114)	replication (115)	translation (121)
deoxyribose (113)	messenger RNA (mRNA) (118)	ribose (118)	uracil (118)
double helix (114)	nucleotide (113)	ribosomal RNA (rRNA) (118)	

Explain why each of the terms below is an appropriate name for the process or thing to which it refers.

1. translation
2. messenger RNA
3. transcription
4. deoxyribose
5. hydrogen bond

Review

1. DNA (a) encodes the information needed for making proteins within the cell (b) directs RNA to make lipids (c) directs RNA to produce glucose (d) produces carbohydrates.
2. The physical structure of the genetic code is DNA's sequence of (a) sugars (b) nitrogen bases (c) phosphates (d) hydrogen bonds.
3. Proteins are composed of (a) nucleotides (b) nitrogen bases (c) deoxyribose (d) amino acids.
4. The general shape of a DNA molecule is a (a) sphere (b) folded chain (c) helix (d) branched chain.
5. RNA (a) is double-stranded like DNA (b) has the same four nitrogen bases as DNA (c) contains ribose (d) contains thymine.
6. The pattern of base pairing in DNA can be summarized by the rule that (a) purines always pair with purines (b) any purine can pair with any pyrimidine (c) pyrimidines only pair with pyrimidines (d) adenine pairs with thymine, and cytosine pairs with guanine.
7. DNA replication (a) involves RNA polymerase (b) occurs at one site (c) requires a DNA molecule to start the process (d) needs several days' time to occur.
8. Transporting amino acids to ribosomes for assembly into needed proteins is the function of (a) DNA (b) mRNA (c) rRNA (d) tRNA.
9. Transcription occurs (a) at ribosomes (b) in the cytoplasm of the cell (c) in the nucleus of the cell (d) as the last step in protein synthesis.
10. The number of bases in a codon is (a) 2 (b) 3 (c) 4 (d) 5.
11. What are some ways in which the structure of a DNA molecule is related to its function?
12. How is a DNA molecule duplicated in the process of replication?
13. What information did Rosalind Franklin's work with X-ray crystallography provide to Watson and Crick?
14. Name the three types of RNA present in cells, and describe the function of each.
15. Name at least three major structural differences between RNA and DNA.
16. What roles do enzymes play in the transcription process?
17. What is the relationship between codons and anticodons?

18. What functions are carried out by those few codons that do not code for amino acids?
19. Why may the production of a particular protein require information from more than one gene?
20. What is the role of the ribosome in protein synthesis?

Critical Thinking

1. Use Table 8-1 to find the codons that compose the amino acid alanine. How many codons code for alanine? How do they differ from one another? What significance might this pattern of differences have?
2. Refer again to Table 8-1. Using the table, determine the sequence of amino acids that is specified by the following list of codons: AUG UCU AAC AAA CAG GCU UAA.
3. Using the list of codons in question number 2, figure out the base sequence of DNA that codes for them. List the order of DNA bases that must have been transcribed to establish this sequence of codons.
4. Suggest a reason why a system composed of three bases per codon is required to code for amino acids. Would a system with two bases per codon work to code amino acids?

Uracil

5. The diagram above shows the chemical structure of uracil. Compare this diagram with Figure 8-1. How does the chemical structure of uracil differ from the chemical structure of thymine? How is it possible that both thymine and uracil can bond with adenine despite this difference in structure? (You may wish to look again at the patterns of hydrogen bonding in Figure 8-2 to help you answer this last question.)

Extension

1. Prepare a report on the contributions of one of the following scientists to the discovery of the structure of DNA: Erwin Chargaff, Francis Crick, Rosalind Franklin, Linus Pauling, James Watson, or Maurice Wilkins. James Watson's book *The Double Helix* is a good source of information on Crick and Watson. *The Eighth Day of Creation* by Horace Freeland Judson covers all of these scientists.
2. Read the article "New Assay Identifies Southpaw DNA," in *Science News*, November 14, 1987, p. 308. What effects do researchers think this "left-handed" DNA may have in cells?
3. On a piece of cardboard or construction paper, draw the DNA bases. Cut them out and label each one. Demonstrate to your classmates how hydrogen bonds are established, and why only complementary bases pair.
4. Read the article by Harold Vaimus entitled "Reverse Transcription" in *Scientific American*, September 1987, pp. 56–64. Explain the process of reverse transcription, telling in what organisms reverse transcription takes place and what purpose reverse transcription serves in those organisms.

Chapter 9 | Chromosomes, Mitosis, and Meiosis

This chapter explains the processes of cell reproduction that contribute to the growth and development of organisms and the creation of new individuals. Students concentrate on how the cell structures change to assist in these processes.

Correlation Chart

Section Number	Section Objectives	Pages	Features	Section Review
9.1	1. Identify the structures of a chromosome. 2. Summarize how DNA becomes a chromosome. 3. Distinguish between a diploid cell and a haploid cell.	129–130	Think, p. 130	1, 5 2, 3 4
9.2	1. Define *cell cycle* and tell what occurs in a cell during each phase of this cycle.	131–135	Biology in Process, pp. 134–135	1, 2, 5
	2. Describe the changes that take place in a cell during each of the three phases of interphase. 3. Name the structures that appear in a cell during mitosis, state when they appear, and explain their functions. 4. Describe the process by which Walther Flemming developed a model of cell division.			2, 4 3, 4
9.3	1. Define *meiosis* and state its purpose. 2. Summarize the events that occur in a cell during meiosis I and meiosis II. 3. Describe the formation of egg and sperm in animals.	136–139	 Biology Lab, p. 139	1 1, 2 3, 5
	4. Explain why sexually reproducing organisms have a survival advantage over organisms that reproduce only by mitosis.			4

Resources*

Print

Baserga, Renato. *The Biology of Cell Reproduction.* Cambridge, Mass.: Harvard University Press, 1985.

Edmunds, Leland N., Jr., ed. *Cell Cycle Clocks.* New York: Marcel Dekker, 1984.

Audiovisual/Software

Cell Division and the Life Cycle. 2 filmstrips and cassettes, 15 min. each. Human Relations, 1985.

Mitosis and Meiosis: How Cells Divide. Video or 4 filmstrips with cassettes, 10–13 min. each. Teacher's guide. Science and Mankind.

*Additional references are listed in *Modern Biology, Teacher's Program Guide.*

Section Teaching Guides

9.1 Chromosomes

Approach

The main objective of this section is to describe what happens to genetic material during the first stages of cell division. The information in this section builds on students' knowledge of cell structure and function (Chapter 5) and their knowledge of nucleic acids (Chapter 8). In presenting the material, guide students to realize that the passing of genetic information contained in the chromosomes is of crucial importance in reproduction.

Concept Framework

Students will be familiar with the term *DNA* and will have read about its structure in Chapter 8. Tap this knowledge to build a conceptual framework for the material presented in this section by reviewing the structure and function of DNA with students. You may wish to show them a photograph of a cell treated with stain so that the chromatin is visible in the nucleus. Tell students that they will be reading about the changes that take place in this genetic material during cell division. Suggest that students try to visualize these changes as they read about them in this section.

Emphasis

Make sure that students can describe the structure of chromatin and chromosomes and that they understand the functional difference between haploid and diploid cells.

Cell Division Demonstration

Set up a series of microscopes showing cell division in cells with different numbers of chromosomes. Let students identify and count the number of chromosomes in each cell. After students have finished looking at the slides give them a key that lists the cell type and the number of chromosomes for each type of cell so that they can check their work.

Section Review Answers

(Continued from p. 130.)
3. When DNA becomes visible as a rod-shaped structure during the beginning of cell division, it is referred to as a *chromosome*.

4. A diploid cell contains both members of a homologous pair of chromosomes; a haploid cell contains only one member of each pair.
5. The organisms would be different because traits and appearances are determined by heredity material carried on chromosomes, not by the quantity of chromosomes.

9.2 Mitosis

Approach

The main objective of this section is to explain the cell cycle and to describe the process of mitosis. This section builds on the concept of cell surface to volume ratio (Chapter 5). In presenting the text, guide students to realize that the processes of mitosis and cytokinesis together result in two cells that are genetically identical to the parent cell. They should also realize that most cells spend most of their lifetimes in interphase.

Concept Framework

Students will be familiar with processes that take place in stages, such as growth. To build a conceptual framework for this section, make an analogy between these familiar processes and mitosis. You may wish to give a definition of *mitosis* at this time. Tell students as they read to outline the stages of mitosis.

Emphasis

Students should be able to describe the phases of the following events in the cell cycle: interphase, mitosis, and cytokinesis.

Using the Features/A Model of Cell Division

The main processes presented in this feature are observing, hypothesizing, inferring, and modeling. This feature describes how Walther Flemming developed a method of staining chromosomes so that he could observe different stages of mitosis. From his observations he constructed a model of cell division.

Modeling the Cell Cycle

Divide students into three groups. Assign each group one of the stages of the cell cycle and ask students to draw and label pictures showing what happens to a cell during that stage. Direct students to compose captions for their drawings. Let them use their drawings to construct a bulletin-board display of the cell cycle.

Section Review Answers

(Continued from p. 135.)

1. Cytokinesis is the division of the cytoplasm of a parent cell and all of its contents into two daughter cells; mitosis is strictly the division of the cell nucleus.
2. the S phase
3. Polar fibers extend from centriole to centriole; kinetochore fibers extend from the centromere of a chromosome to a centriole.
4. In prophase spindle fibers appear and polar fibers begin their attachments. In metaphase kinetochore fibers extend from centriole to centromere; they move the chromosomes to the equator of the cell and hold them there. In anaphase polar fibers are forced toward the opposite sides of the cell, and kinetochore fibers disappear at the place of attachment to the centriole; this results in the moving of the chromatids to opposite sides of the cell. In telophase spindle fibers disappear.
5. Cytokinesis is less complex than mitosis.

9.3 Meiosis

Approach

The main objective of this section is to describe the phases of meiosis, which lead to the formation of egg and sperm cells and to introduce students to the adaptive advantages of sexual reproduction. This section builds on concepts developed in 9.1 and 9.2 and helps prepare students for understanding the reproductive cycle set out in Chapter 47. In presenting the material, guide students to understand that meiosis leads to the formation of egg and sperm cells essential for sexual reproduction.

Concept Framework

Students will be familiar with the idea of cell division from having studied the previous sections of this chapter. Build a conceptual frame for this chapter by asking students to list and describe the phases of mitosis. Tell them that in meiosis the nucleus divides twice. Encourage students as they read to notice other differences between meiosis and mitosis.

Emphasis

Make sure that students can differentiate between the following: meiosis I and meiosis II, egg cells and sperms cells, asexual reproduction and sexual reproduction.

Phases of Meiosis and Mitosis Charts

Ask students to make charts comparing mitosis and meiosis and showing cell during the different phases of these processes. Direct students to focus on the main event that occurs during each phase. The chart should point out differences in number and arrangement of chromatids during the different phases of meiosis and mitosis.

Section Review Answers

(Continued from p. 138.)

2. to exchange portions of homologous chromatids, which results in the exchange of genes
3. An ootid receives almost all the cytoplasm that belonged to the parent cell whereas the polar bodies receive little or no cytoplasm. Thus the polar bodies disintegrate.
4. by leading to genetic variation between parents and offspring, which gives some offspring a more favorable chance of surviving in a changing environment
5. Homologous chromosomes are sorted independently during anaphase I. This occurs because tetrads line up and separate randomly. Because chromosomes are sorted independently genetic recombination can occur.

Laboratory Preparation

Background

This lab procedure acquaints students with the division of plant cells. By examining many cells and determining their stage of division, students discover that mitosis is a continuous process. The students examine cells of an onion root tip, collecting data in order to estimate the length of the major stages of mitosis. Students should be aware of the differences between these stages before they begin to examine the cells. A major difficulty in determining the stage of mitosis arises when only part of the cell is visible because of the orientation, for example, when you have a polar view from the end of the spindle. Students should be aware of this problem before they begin to collect data.

Preparation

Prepare slides showing a longitudinal section of an *Allium* root tip.

Conclusions

1. The various stages of mitosis are not always easy to classify; they blend into one another. By making this observation, students can infer that the events are continuous.
2. Answers will vary somewhat. The approximate relative lengths of the stages of mitosis are: prophase, 85%; metaphase, 8%; anaphase, 3%; and telophase, 4%.

Inquiry

Because several longitudinal sections are made from each root tip, the data from the entire class will provide a better representation of what is occurring in the whole volume of dividing tissue than would the data collected by an individual.

Chapter Review

Vocabulary

1. Mitosis results in two diploid cells each identical to the parent cell; meiosis results in four unique haploid cells.
2. In meiosis I homologous chromosomes are separated into separate cells. In meiosis II the chromatids of each chromosome are segregated into separate cells.
3. Synapsis is the pairing of homologous chromosomes preceding the formation of tetrads. A tetrad is a group of two homologous chromosomes forming after synapsis.
4. Asexual reproduction is the production of offspring from one parent that does not involve the union of gametes. Sexual reproduction is the production of offspring involving the fusion of gametes.
5. Crossing-over is the exchange of chromatid portions, resulting in the exchange of genes. Any process resulting in an exchange of genes is genetic recombination.

Review

11. Each daughter cell must receive the same complement of DNA as the parent cell.
12. Chromosomes always occur in pairs in sexually reproducing organisms.
13. When a cell grows its surface area expands more slowly than its volume. The volume eventually becomes so much larger than its surface area that sufficient nutrients might not enter the cell and wastes might not leave the cell quickly enough for the cell to survive. Cell division gives the new cells a smaller surface to volume ratio, which may allow them to survive.
14. Differentiating between these distinct events enables biologists to communicate clearly about cell division.
15. No. A cell either remains in interphase or it divides, forming two new cells. The two new cells may then go through the cell cycle.
16. Cells that stop growing and dividing remain in the G_1 phase. Such cells, include nerve cells and muscle cells.
17. A cell plate helps plant cell that has undergone mitosis complete cytokinesis. It forms a barrier across the middle of the cell. On both sides of the plate a new cell wall then forms.
18. Flemming discovered a dye that made cell structures visible, and he developed a model of cell division.
19. The process of mitosis that creates new unicellular organisms by asexual reproduction is called binary fission.
20. Multicellular organisms reproduce asexually by cloning or by budding.

Critical Thinking

1. Meiosis. Pollen cells are gametes and thus form by meiosis.
2. The DNA might have replicated and the chromatids split into separate chromosomes during anaphase of mitosis, but cytokinesis probably did not occur.
3. Mitochondria are in the cytoplasm. Since the female contributes more cytoplasm, the female contributes more mitochondria.
4. 36 chromosomes

Chapter 9 | Chromosomes, Mitosis, and Meiosis

Introduction

In the previous chapters in this unit you have learned about the inner workings of the cell and about the structure of DNA. In this chapter you will learn about the ways in which cells divide. Cell division perpetuates life and allows for the growth and reproduction of organisms.

Chapter Outline

9.1 Chromosomes
- *Chromosome Structure*
- *Chromosome Numbers*

9.2 Mitosis
- *The Cell Cycle*
- *Interphase*
- *Phases of Mitosis*
- *Cytokinesis*

9.3 Meiosis
- *Phases of Meiosis*
- *Formation of Egg and Sperm*
- *Asexual and Sexual Reproduction*

Chapter Concept

As you read about cell processes, pay attention to the way that intercellular structures change during these processes.

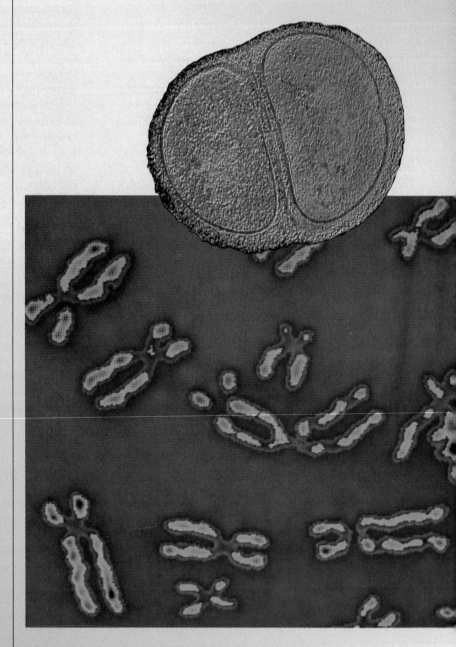

Human chromosomes (×55,000). Inset: Cell division occurring in the bacterium *Staphylococcus epidermidis* TEM (×17,500).

For additional teaching information see pp. T9A–T9D.

9.1 Chromosomes

Cell division is the process by which cells reproduce. A unicellular organism divides and forms two new individuals. Multicellular organisms result from divisions of the zygote, a single cell. Even the zygote is produced by the union of two other cells, usually a sperm cell and an egg cell, and each of these cells arose from other cells. Insightful nineteenth-century biologists recognized a pattern: all cells come from preexisting cells. How cells divide is the subject of this chapter.

Information in the cell is contained in genes. Genes code for the proteins that carry out cellular functions. When a cell divides, genes are separated into daughter cells in an orderly, predictable process. This process involves the formation of chromosomes.

Chromosome Structure

Recall from Chapter 5 that densely stained areas exist inside the nuclear membrane of a nondividing cell. The substance that takes up stain in these areas is chromatin *(KROE-muh-tin)*. Experimental evidence has shown that chromatin is composed of DNA and proteins. The DNA in chromatin exists as thin, uncoiled strands.

When cell division begins, DNA replicates. Then the replicated strands of DNA begin to coil, as shown in Figure 9-1. As coiling proceeds, DNA becomes tightly wrapped around globules of **histones,** a group of special protein molecules. DNA and histones in this coiled form can be seen as a rod-shaped structure called a chromosome. Note that only the physical arrangement of the DNA strand is changed in its transition from chromatin to chromosome. The chemical makeup of the DNA remains unchanged.

Each chromosome consists of two identical parts, each called a **chromatid** *(KROE-muh-tid)*. The two chromatids are often called sister chromatids. The point at which each pair of chromatids is attached is called the **centromere** *(SEN-truh-mir)*.

Section Objectives

- *Identify the structures of a chromosome.*
- *Summarize how DNA becomes a chromosome.*
- *Explain the difference between a diploid cell and a haploid cell.*

Info: Although the nucleus controls the activities of a cell, it is the chromosomes that carry the messages for those activities to the cells.

Figure 9-1 During the initial phases of cell division the DNA within cells coils and becomes organized into chromosomes, each of which is composed of two chromatids.

| DNA double helix | DNA histones | One coil within supercoil | Further coiling within supercoil | Supercoil within chromosome | Chromosomes |

Tip: Speculate with students on what type of organism might develop if in sexual reproduction the gametes were 2N instead of 1N. Conclude the discussion with the observation that the haploid condition resolves the problem.

Answer to Think
They differ because each chromosome has different kinds and amounts of DNA and because each controls different aspects of the organism.

Think

These are human chromosomes as seen under a scanning electron microscope. Why do chromosomes differ in size and shape?

Answers to Section Review
1. DNA in a coiled, rod-shaped form is referred to as a chromosome. Each chromosome consists of two identical parts called chromatids, which are joined at a centromere.
2. The DNA exists as thin, uncoiled strands in chromatin.
(See pp.T9A–T9D.)

Table 9-1 Chromosome Numbers of Various Species

Species	Number	Species	Number
Alligator	32	Fruit fly	8
Ameba	50	Garden pea	14
Brown bat	44	Goldfish	94
Bullfrog	26	Grasshopper	24
Carrot	18	Horse	64
Cat	32	Human	46
Chicken	78	Lettuce	18
Chimpanzee	48	Onion	16
Corn	20	Redwood	22
Earthworm	36	Sand dollar	52

Chromosome Numbers

Every species has a characteristic number of chromosomes in each cell. The number varies considerably among species. A certain nematode worm species, for example, has only 2 chromosomes in each cell, while certain protozoa have over 300 chromosomes in each cell. A potato, a plum, and a chimpanzee each have 48 chromosomes in each. As you might guess from this information and from the chromosome counts given in Table 9-1, chromosome number is not an indication of the complexity of the species.

In all sexually reproducing organisms chromosomes occur in pairs. The two members of each pair are called **homologous chromosomes** or homologues. Each chromosome of a pair has the same size and shape as its homologue. Viewed with a microscope, the two homologous chromosomes are indistinguishable. However, the pair is structurally different from all other homologous pairs in the cell.

A cell that contains both chromosomes of a homologous pair is termed **diploid.** The notation 2N refers to diploid cells. For example, the diploid, or 2N, number of a human body cell is 46, and the cell contains 23 homologous pairs of chromosomes. A cell that has only one chromosome of each homologous pair is termed **haploid.** The notation 1N refers to haploid cells. For example, the haploid, or 1N, number of a human egg or sperm cell is 23, and there are no homologous chromosomes in either cell.

Section Review

1. Explain how a chromatid differs from a chromosome.
2. What is the structure of the DNA in chromatin?
3. When do we refer to DNA as a chromosome?
4. What is the difference between a diploid cell and a haploid cell?
5. Why might two organisms that contain the same number of chromosomes not possess similar traits and appearance?

9.2 Mitosis

Mitosis *(mie-TOE-suss)* is the division of the cell nucleus in which the chromosomes in the parent cell divide into two identical sets. In body cells, also called somatic cells, mitosis is the process by which the number of cells is increased without changing the information contained in the DNA or the amount of DNA in those cells. In unicellular organisms, such as a yeast cell, mitosis is a means of reproduction. All offspring have the same DNA and hence contain the same genetic information.

What triggers a cell to divide? Cell biologists have observed that the size of a cell seems to be a factor that stimulates its division. You may recall from Chapter 5 that when a cell grows, its surface area expands more slowly than its volume does. Thus its volume eventually becomes disproportionately large in comparison with its surface area. The surface area is not large enough to allow the necessary amounts of nutrients to enter the cell and to allow wastes to pass out. Therefore as the surface-to-volume ratio becomes relatively small, the cell may divide. Many other mechanisms probably play a role in initiating cell division.

The Cell Cycle

The process of mitosis is part of the cell cycle. The **cell cycle** is the sequence of events that occurs in a cell from mitosis to mitosis. Figure 9-2 represents five events of the cell cycle and the amount of time usually required for each. The five events of the cell cycle consist of the three phases of interphase, mitosis, and cytokinesis. Although biologists give these events names, the cell cycle is a continuous process. **Interphase** is a period of cell growth and development that precedes mitosis and follows cytokinesis. Mitosis is the division of the cell nucleus. The process of **cytokinesis** is the division of the cytoplasm of a parent cell and its contents into two daughter cells.

Section Objectives

- *Define* cell cycle, *and tell what occurs in a cell during each phase of this cycle.*

- *Describe the changes that take place during each of the three phases of interphase.*

- *Name the structures that appear in a cell during mitosis, state when they appear, and explain their functions.*

- *Describe the process by which Walther Flemming developed a model of cell division.*

Tip: Demonstrate to students how the surface-to-volume ratio changes when a cell divides. Have students figure the surface area of a cube of, say, 10 cm. Dissect the cube twice and figure the combined surface area of the eight 5-cm cubes that result. Point out that while the total volume of the smaller cubes remains the same as that of the large cube, the surface area has doubled.

Figure 9-2   From mitosis to mitosis a cell typically passes through a five-event cycle. Note that there are three phases in interphase.

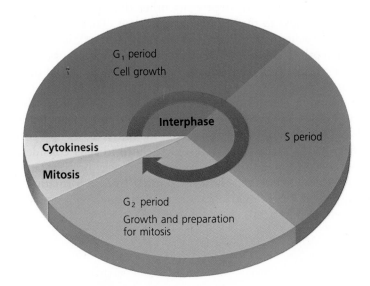

131

Interphase

Interphase, like cell division, is a continuous process. Biologists have differentiated three distinct phases in interphase. Look back at Figure 9-2 to see the three phases of interphase, G_1, S, and G_2. The G_1 phase is the first period of interphase. In G_1 the cell doubles in size, and enzymes and organelles, such as mitochondria and ribosomes, roughly double in number. In the S phase the DNA that makes up the chromatin replicates. In the G_2 phase the cell undergoes rapid growth that prepares it for mitosis, synthesizing necessary enzymes and structures.

As you can see from Figure 9-2, the three phases of interphase constitute the largest part of the cell life cycle. In other words, cells spend most of their lifetime in interphase. However, the actual life spans of eukaryote cells vary. In ideal conditions a yeast cell lives for about two hours before it divides and forms two new cells. An ameba in a laboratory culture may live for a few days before it divides. In animals some cells of the early embryo may divide every 15 to 20 minutes.

In multicellular organisms many cells remain in interphase permanently. Cells that stop growing and dividing remain in the G_1 phase. The muscle and nerve cells of animals are incapable of dividing and so remain in the G_1 phase throughout their existence. In dividing cells interphase immediately precedes mitosis.

Phases of Mitosis

The fourth event in the cell cycle is mitosis, a continuous process that scientists divide into four phases: prophase, metaphase, anaphase, and telophase. As you read, keep in mind that mitosis produces two identical nuclei with the same number of chromosomes as the parent cell.

Prophase is the first phase of mitosis. Prophase can be subdivided into three steps. During early prophase, as shown in Figure 9-3, the chromatin coils and forms chromosomes. Simultaneously both the nucleolus and the nuclear membrane break down and disappear. In organisms other than plants two dark spots appear next to the disappearing nucleus. These small, dark cylindrical bodies, called **centrioles** (*SEN-tree-OLEZ*), move away from each other, going toward opposite ends, or poles, of the cell.

The middle step of prophase is marked by the initial development of spindle fibers—microtubules of protein. There are two types of spindle fibers—polar and kinetochore fibers. **Polar fibers** extend across the cell from centriole to centriole. **Kinetochore** (*kuh-NET-uh-kor*) **fibers** extend from the centromeres of a chromosome to the centrioles. Each kinetochore fiber is considered part of a centromere.

The last step of prophase is marked by the formation of **asters,** protein fibers that radiate from each centriole. Asters are not found in plant cells.

Early prophase

Middle prophase

Late prophase

Figure 9-3 Prophase, the first phase of mitosis, is characterized by the formation of chromosomes and the development of spindle fibers and asters.

Metaphase is the second phase of mitosis, as seen in Figure 9-4. During metaphase kinetochore fibers move the chromosome to the center, or equator, of the cell. Once in place at the equator of the cell, the chromosome is held in the center by the kinetochore fibers. As you can see from Figure 9-4, metaphase is characterized by the arrangement of all chromosomes along the equator of the cell.

Anaphase is the third phase of mitosis. First, the centromere of each pair of chromatids divides. Then the chromatids separate and move toward opposite poles of the cell. The movement of chromosomes is rapid and dramatic.

How do chromosomes move? Spindle fibers—both polar fibers and kinetochore fibers—play important roles in the movement of chromosomes. A polar fiber that arises from one centriole has a molecular structure that is oriented in an opposite manner from the fiber that arises from the other centriole. These polar fibers are connected to each other by bridges of protein. With the aid of ATP these polar fibers push against each other and force each other toward opposite sides of the cell. In addition, the kinetochore fibers are broken down and shortened at their place of attachment to the centriole. The overall effect is that the chromatids move to opposite sides of the cell.

Telophase is the final phase of mitosis. At the beginning of telophase the two identical sets of chromatids are clustered at opposite sides of the cell. During telophase the centrioles and spin-

Info: Through mitotic regeneration some cells are able to generate lost body parts, such as the arm of a starfish. Human beings are able to regenerate skin and muscle cells but not whole body parts.

Figure 9-4   Scientists divide mitosis into four phases: prophase, metaphase, anaphase, and telophase.

Prophase   Metaphase   Anaphase   Telophase

Animal cell

Figure 9-5 In animal cells the cell membrane narrows at the center and eventually divides in two. This process is called cytokinesis.

dle fibers disappear. The chromatids unwind and elongate into the threadlike structures of DNA that are now called chromatin. A nuclear membrane then forms again around each mass of chromatin, and finally a nucleolus appears. Cytokinesis then occurs.

Cytokinesis

The process of cytokinesis usually takes place immediately after mitosis. Cytokinesis occurs when cytoplasm from the original cell splits and forms two new cells. Each newly formed cell houses one of the two nuclei formed during mitosis. In dividing the cytoplasm cytokinesis also separates the other structures distributed throughout the cytoplasm, such as ribosomes, Golgi bodies, and mitochondria. The two new cells that have formed during cytokinesis are generally equal in size.

Figure 9-5 demonstrates the way in which cytokinesis occurs in animal cells. Notice how it differs from the same process in plants, which is shown in Figure 9-6. In animal cells cytokinesis begins during early anaphase. The cell membrane pinches in and, after telophase, divides the cell at its center, forming two daughter

Biology in Process

A Model of Cell Division

In the mid-1800s, biologists who wanted to learn how cells divide were hindered by the fairly transparent nature of cells. Therefore, developing a technique to make the substructures of cells visible became a priority among these biologists. Only if scientists could distinguish the inner parts of a cell could they then identify and observe the structures of cell division and so come to understand the cell division process.

Biologists hypothesized that cell structures would react to dye as other materials do—each structure, according to its individual properties, absorbing a certain amount or kind of dye. They predicted that the application of dye to a cell would thus make transparent cell parts visible.

Biologists began testing their hypothesis by applying various dyes to various types of cells and observing the visual effects of each dye. The first biologists who applied dyes to cells achieved mixed results. For example, one biologist found that a dye obtained from logwood stained a nucleus black. This dying procedure distinguished the nucleus from the cytoplasm but masked the inner structures of the nucleus.

cells. In plant cells vesicles formed by the Golgi bodies fuse at the equator and form the **cell plate,** a membrane across the middle of the cell. A new cell wall then forms on both sides of the cell plate. One cell is thus separated into two.

Section Review

1. Define the term *cytokinesis*, and explain how cytokinesis differs from mitosis.
2. During which of the three phases of interphase—G_1, S, or G_2—does DNA replicate?
3. What is the distinction between polar fibers and kinetochore fibers?
4. What is the role of spindle fibers in each phase of mitosis? At what points do the spindle fibers appear and disappear? What changes do the fibers undergo? What functions do they perform in each phase?
5. Why might a cell require so much less time to complete the process of cytokinesis than it does to complete the process of mitosis?

Plant cell

Figure 9-6   In plants a cell plate forms and eventually divides the cell during cytokinesis into two daughter cells.

Finally, in 1879 a German anatomist named Walther Flemming noted that by applying synthetic red dyes to cells he could observe their substructures. In fact, certain threadlike substructures absorbed so much dye that Flemming called these darkly stained structures *chromatin,* from the Greek word for "color."

This was the result scientists had predicted could be achieved with the application of dye. However, the stain killed the cell to which it was applied. Therefore, Flemming could not simply stain cells to observe them dividing as scientists had hoped. Nevertheless,

using the technique of cell staining, Flemming attempted to infer how cells divide.

He devised a method of creating a kind of moving picture. He based this method on the assumption that he would be viewing cells in various phases of division. Flemming

inferred that he would capture the equivalent of still photos of cells in these various phases if he drew a picture of each cell as he saw it. Therefore Flemming drew each cell after he stained it. Then he sorted these drawings until he created a model of the cell division process. Flemming published this model in 1882. He called this process *mitosis,* which means "thread," because he observed that the threadlike chromatin plays a central role in the process.

■ What assumption did Flemming make that could have rendered his model inaccurate had he been incorrect?

Chromosomes, Mitosis, and Meiosis   **135**

9.3 Meiosis

Remember that mitosis produces daughter cells having the same number of chromosomes as the parent cell. **Meiosis** *(mie-OE-suss)* is the process of nuclear division that reduces the number of chromosomes by half. This process of cell division is involved in sexual reproduction. In animals meiosis often results in haploid egg cells and haploid sperm cells. In plants meiosis results in haploid spores that later lead to the production of egg and sperm cells. In both plants and animals the two 1N cells fuse during fertilization to form a 2N zygote that develops into a new organism.

Phases of Meiosis

In meiosis two nuclear divisions take place instead of one as in mitosis. In **meiosis I,** the first division, the homologous chromosomes are separated into separate cells. In **meiosis II,** the second division, the chromatids of each chromosome are segregated into separate cells. The result is that all daughter cells have one-half the number of chromosomes of the parent cell.

Meiosis I

Meiosis I, like mitosis, follows a period of interphase. As you read the following description of the four phases of meiosis I, refer to Figure 9-7.

- Prophase I: DNA strands coil, shorten, and thicken, and are referred to as chromosomes. As in the prophase of mitosis, spindle fibers appear. Then the nuclear membrane and the nucleolus disappear. A second step, which does not take place during mitosis, occurs: Every chromosome lines up next to its homologue. This pairing of homologous chromosomes is called **synapsis.** These homologous chromosomes twist around each other to form a **tetrad,** which is a group of two chromosomes. The word *tetrad* means "four" and refers to the four chromatids that compose the two chromosomes. As tetrads form, portions of chromatids may be exchanged, either between the two

Figure 9-7 Notice that during prophase of meiosis I homologues line up next to each other, forming tetrads. Two daughter cells are formed by the end of telophase I.

Prophase I

Metaphase I

Anaphase I

Telophase I

homologues or between sister chromatids. This phenomenon, called **crossing-over,** results in the exchange of genes.

- Metaphase I: The tetrads are moved by spindle fibers to the equator of the cell. The homologous pairs of chromosomes remain together.
- Anaphase I: The homologous pairs of chromosomes separate, as shown in Figure 9-7. As in the anaphase phase in mitosis, one chromosome of each pair is pulled by action of the spindle fibers to one pole of the cell, and the other is pulled to the opposite pole. Each chromosome is still composed of two chromatids joined by a centromere.
- Telophase I: The cytoplasm divides, forming two daughter cells. This is the final stage of meiosis I.

During meiosis I the parent cell produces two daughter cells, each with one member of each pair of homologous chromosomes. Chromosome number has been halved, but each chromosome, having been replicated earlier, has twice the original amount of DNA.

Meiosis II

Meiosis II occurs in each cell formed during meiosis I and is not preceded by DNA replication. Refer to the diagrams in Figure 9-9 as you read about the phases of meiosis II.

- Prophase II: The chromosomes of some organisms—particularly plants—coil again, having uncoiled between telophase I and prophase II. New spindle fibers form. In animals chromosomes generally do not uncoil after meiosis I.
- Metaphase II: The chromosomes are moved to the cell equator. Each chromosome is made of two sister chromatids that are joined by a centromere that is attached to a spindle fiber.
- Anaphase II: The centromeres joining the chromatids duplicate, freeing each sister chromatid from the other. Then each sister chromatid is moved toward an opposite pole.
- Telophase II: As the spindle dissolves, a nuclear membrane forms around the chromosomes in each daughter cell. Meiosis II is complete, and cytokinesis occurs. The two nuclear divisions of meiosis result in four daughter cells from a single parent cell, each with half the number of chromosomes of the parent.

Figure 9-8 Crossing-over occurs when chromosomes that make up a tetrad exchange portions of their chromatids. This results in an exchange of genes.

Tip: Emphasize that mitosis concerns itself basically with an increase in the number of cells in order to enlarge the organism. Meiosis, on the other hand, is concerned basically with the production of new organisms.

Figure 9-9 By the end of meiosis II each 2N parent cell has divided into four haploid daughter cells.

Prophase II

Metaphase II

Anaphase II

Telophase II

Formation of sperm

Sperm

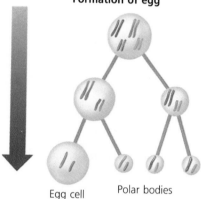

Formation of egg

Egg cell   Polar bodies

Figure 9-10 In the formation of male gametes a parent cell divides by meiosis. This leads to the development of four sperm. In the formation of female gametes a parent cell divides by meiosis. This leads to the development of an egg and three polar bodies.

Info: With 23 pairs of chromosomes in the human body there are more than 8 million different combinations possible for every human cell formed by meiosis.

Answers to Section Review
1. Meiosis I is the division in which homologous chromosomes are separated into separate cells. Meiosis II is the division in which chromatids of each chromosome are segregated into separate cells.
(See pp. T9A–T9D.)

Formation of Egg and Sperm

In animals 1N daughter cells produced by meiosis often differentiate to form **gametes** *(GAM-eets),* sexual reproductive cells. In many animals male gametes, called sperm, are produced in the male reproductive organs. Each new cell becomes a sperm cell. Female gametes, or eggs, are formed in the female reproductive organs. The cytoplasm of the parent cell that undergoes meiosis often divides unequally, with the **ootid**—the egg cell—receiving almost all of the cytoplasm. The other three cells formed during meiosis I and meiosis II—called **polar bodies**—receive little or no cytoplasm and eventually disintegrate. Thus the result of each meiotic division is one haploid egg cell.

Asexual and Sexual Reproduction

Asexual reproduction is the production of offspring from one parent, without the union of gametes. In unicellular organisms new cells are created by binary fission, or mitotic division: a normal cell cycle. In multicellular organisms new cells are formed by various means, such as cloning or budding. Since asexual reproduction occurs only by mitosis, it results in offspring that are genetically identical with the parent.

Sexual reproduction is the production of offspring through meiosis and subsequent fusion of gametes. Such offspring are genetically different from both parents. This is because in meiosis genes are combined in new ways through **genetic recombination.** Genetic recombination results when (1) crossing-over occurs between homologues or between chromatids, (2) homologous pairs separate independently in meiosis I, or (3) sister chromatids separate independently in meiosis II.

Genetic recombination introduces variation between parents and their offspring and hence among the different offspring. This variation gives some offspring a better chance of surviving in a changing environment. For example, if disease strikes a crop of grain, a few plants may have variations that make them resistant to the disease. While many individuals will die, these few plants will survive and reproduce, keeping the species alive. Thus organisms that reproduce sexually have a survival advantage over organisms that reproduce asexually.

Section Review

1. Define *meiosis I* and *meiosis II.*
2. What is the function of crossing-over?
3. What is the difference between an ootid and polar bodies?
4. How does genetic recombination help a species survive?
5. During which phase of meiosis I do you think genetic recombination occurs besides during crossing-over in prophase? How might genetic recombination occur in this stage?

Laboratory

Observing Mitosis

Objective
To examine dividing cells in order to determine the relative length of each stage of mitosis

Process Skills
Classifying; collecting, organizing, and analyzing data

Materials
Microscope, prepared slide of a longitudinal section of *Allium* (onion) root tip

Method
1 Review with your teacher the visible characteristics of each stage of mitosis. Make a list of these characteristics to aid you in your observations. Remember: Interphase is not a phase of mitosis.
2 Using low power on your microscope, focus on the apical meristem of the onion root tip. This is the area just behind the root cap. Why is this a good place to look for cells undergoing mitosis?
3 Examine the apical meristem carefully and choose a sample of about 50 cells to classify. Look for a group of cells that seems to have been actively dividing. The cells will appear to be in rows, so it should be easy to keep track of them.
4 Prepare a chart for entering your data. Head five columns *Stage, Tally Marks, Count, Percent,* and *Time.* Under the column labeled *Stage* list *Prophase, Metaphase, Anaphase,* and *Telophase.*
5 For each of the cells in your sample identify the stage of mitosis, and place a mark in the *Tally Marks* column next to the appropriate stage. Count the tallies for each stage, and fill in the *Count* column.
6 Calculate the percentage of cells found in each stage, and enter the figures under *Percent.* The percentage of cells found in each

stage is a measure of how long each stage lasts. Why is this so?
7 In *Allium* interphase is about 15 hours long, while mitosis normally takes 83 minutes at room temperature. With this information as your basis, how would you fill in the column labeled *Time?* Make the necessary calculations, and record your results in the column.
8 Make another, similar chart, but leave out the *Tally Marks* column. Collect and record the count for each stage of mitosis for the entire class. Fill in the percent and time information using the data collected by the entire class.

Conclusions
1 How can you determine that mitosis is continuous by examining fixed cells?
2 What are the relative lengths, in percentages, of the stages of mitosis in onion root tip?

Inquiry
In addition to the fact that it is a larger sample, why might the class data be more representative of what is happening in the onion root tip than individual data?

Chromosomes, Mitosis, and Meiosis **139**

Chapter 9 Review

Vocabulary

anaphase (133)

asexual reproduction (138)

aster (132)

cell cycle (131)

cell plate (135)

centriole (132)

centromere (129)

chromatid (129)

crossing-over (137)

cytokinesis (131)

diploid (130)

G₁ phase (132)

G₂ phase (132)

gamete (138)

genetic recombination (138)

haploid (130)

histone (129)

homologous chromosome (130)

interphase (131)

kinetochore fiber (132)

meiosis (136)

meiosis I (136)

meiosis II (136)

metaphase (133)

mitosis (131)

ootid (138)

polar body (138)

polar fiber (132)

prophase (132)

S phase (132)

sexual reproduction (138)

synapsis (136)

telophase (133)

tetrad (136)

Explain the difference between the terms in each of the following sets.
1. mitosis, meiosis
2. meiosis I, meiosis II
3. synapsis, tetrad
4. asexual reproduction, sexual reproduction
5. crossing-over, genetic recombination

Review

1. All cells come from (a) sperm cells (b) zygotes (c) egg cells (d) preexisting cells.
2. Chromatin is (a) a dark stain (b) a dense substance within the nuclear membrane of a nondividing cell (c) one of the two identical parts that make up a chromosome (d) the point at which each pair of chromatids is joined.
3. Every species has (a) diploid and haploid cells (b) a distinctive number of chromosomes per cell (c) at least eight chromosomes per cell (d) a number of chromosomes that varies with the complexity of the organism.
4. Mitosis (a) can increase the number of body cells without changing the information contained in the DNA of those cells (b) is a means of reproducing sexually (c) is never triggered by cell size (d) results in daughter cells that are genetically different from each other.
5. Interphase is (a) a time when cells rest between divisions (b) the phase when the cytoplasm divides (c) a time of cell growth and development (d) a small part of the life cycle of the cell.
6. Cytokinesis (a) differs in animals and plants (b) does not occur in plants (c) immediately precedes mitosis (d) is a process of nuclear division.
7. Polar fibers are (a) the final product of meiosis (b) the final product of cytokinesis (c) the microtubules of protein that extend across a cell from centriole to centriole (d) the dark, cylindrical bodies that move toward opposite ends, or poles, of a cell.
8. When all chromosomes are arranged along the equator of a cell, the cell is in (a) prophase (b) metaphase (c) anaphase (d) telophase.
9. Meiosis I produces (a) a parent and two daughter cells (b) four daughter cells (c) two diploid cells (d) two haploid cells.
10. The ootid receives (a) almost all the cytoplasm from the original parent cell (b) little cytoplasm (c) the same amount of cytoplasm as other gametes (d) no cytoplasm.
11. Why do cells divide equally and in an orderly process in mitosis?

12. In what type of organisms do chromosomes always occur in pairs?
13. Explain why size might trigger cell division.
14. The cell cycle is a continuous process. Why have biologists differentiated five distinct events in this cycle?
15. Does the same cell go through the cell cycle more than once? Why?
16. Some cells remain in the G_1 phase of interphase. Describe the character of these cells, and give an example.
17. What function does a cell plate perform? How?
18. What are the two most significant contributions Walther Flemming made to the study of cell biology?
19. What does the term *binary fission* refer to? In what type of organisms does this process occur?
20. What are two of the ways by which multicellular organisms may reproduce asexually?

Critical Thinking

1. The photograph at the right shows cell division in a lily anther. The daughter cells will form pollen grains. Do you think that the photograph shows mitosis or meiosis? Support your answer.
2. Some human liver cells have 92 chromosomes. Explain how such cells probably arose.
3. In the female meiosis results in an ootid and three polar bodies. The ootid receives almost all the cytoplasm from the meiotic cell. In sperm formation the cytoplasm is divided equally among four sperm. Which parent contributes more mitochondria to the new embryo?
4. Suppose that in a certain plant with a haploid number of 12, DNA replication took place and the chromatids in anaphase of meiosis II separated but that cytokinesis did not occur. How many

chromosomes would the zygote have if the resulting egg was fertilized by a normal sperm?

Extension

1. Read the article by J. A. Miller entitled "Mouse Procreation is Mom-Pop Affair," in *Science News*, October 6, 1984, p. 213. Give an oral report in which you include a discussion of the role of sex-related conditioning of the maternal and paternal genes in embryo development.
2. Make a bar graph showing the stages in meiosis, including interphase, on the *x*-axis and the N number of human chromosomes on the *y*-axis.
3. A chemical called colchicine blocks the process of cytokinesis. Use your school library or public library to research colchicine. Present a report to your class explaining how colchicine prevents cytokinesis and describing how colchicine is used in horticulture.

Intra-Science: *How the Sciences Work Together*

The Question in Biology: How Feasible Is Desalination?

Over 97 percent of the earth's water is sea water and is therefore too salty to drink. In the past 40 years, scientists have developed processes for removing the salt from water. Through desalination, sea water can be turned into drinking water.

In the 1960s numerous desalination plants were built in the Middle East, Central America, South America, and the United States. Many of these have been shut down, but there are still major desalination plants in the Middle East. The expense of desalination limits the process to areas where salt water is abundant and fresh water is scarce. In such areas, desalinated water supports people, livestock, and agriculture through irrigation.

The Connection to Physics

The most common methods of desalination involve changing states of matter. In distillation, water is heated in a container until it turns into a vapor. The vapor is collected and cooled until it becomes liquid again, while the salt and other minerals remain in the original container.

You may have observed distillation when heating liquid in a covered pot. After you remove the pot lid, steam will eventually condense back into water inside the lid. People have been distilling water for centuries. Sailors routinely boil sea water to create drinking water. Some people collect dew, which is condensed water vapor from the air.

In a process called multistage-flash distillation, heated salt water is pumped into successive low-pressure chambers, where it "flashes" into steam. The steam collects on cool condensing coils and drips into special containers as it turns into water. The salt remains in the low-pressure chamber.

Freezing water will force salt molecules to separate from water molecules. In this type of desalination, water is frozen in a vacuum chamber. Because ice is lighter than water, the ice floats above the salty brine, or "slurry." In a second chamber, any remaining salt is scraped or washed off of the surface of the ice crystals. The ice is then melted, yielding pure water.

The Connection to Chemistry

Other methods of desalination are based on an understanding of the chemical changes of the water and salt molecules. When salt (NaCl) is dissolved in water, its ionic bonds dissociate and the components separate into a positively charged sodium ion and a negatively charged chloride ion.

In the desalination process called electrodialysis, sea water is pumped into a chamber divided into sections by semipermeable membranes. One type of membrane will permit only negatively charged ions to pass through; the other will allow only positively charged ions to pass. Positive and negative electrodes are placed at opposite ends of the chamber. When an electric charge is passed through the chamber, the sodium ions move toward the negative electrode and the chloride ions move toward the positive electrode, leaving salt-free water in the center of the chamber.

Electrodialysis is expensive because it requires the use of electricity. The process is used more frequently on brackish inland ground water than on sea water, because more electricity is required to separate the ions in the saltier water.

Another method of desalination used on inland ground water is reverse osmosis. As you know, the process of osmosis would normally cause fresh water to flow into a container of salt water through a membrane. However, by applying pressure greater than the natural osmotic force to the salt water, water is forced into the freshwater chamber while salts are concentrated in the high-pressure chamber. Water contains a number of minerals and chemicals in addition to salt. Today, desalination often takes place at sophisticated water filtration and purification plants, where a variety of substances are removed from sea water or inland water.

The Connection to Other Sciences

Geologic factors contribute to the mineral and salt content of sea water or inland ground water. For a desalination plant to be cost effective, it must process large quantities of water using a method that is suited to its salt content. Desalination plants are often associated with electric power plants. The large amounts of water processed in desalination can be used to generate power. Falling water provides kinetic energy that can be harnessed by generators.

Desalination may be intimately tied to a country's economy. By the late 1940s, Kuwait was aggressively building desalination plants to process both sea water and brackish ground water. By pairing desalination plants and electric power plants, Kuwait could afford to expand the water treatment plants to meet growing needs in the country. From 1977 to 1982, for example, the consumption of desalinated water in Kuwait increased by 62 percent.

The Connection to Careers

We need fresh water to drink and to irrrigate crops. While desalination is too expensive to be feasible in many areas of the world today, it has provided fresh water to people in places where cost is less of a factor. There are many opportunities to work in the area of water filtration and purification. In some cases, these processes include desalination. People who work in the field of water treatment have backgrounds in chemistry, geology, hydrology (the study of water), and biology.

Engineers are involved in researching and designing new, more cost-effective ways to create fresh water from salt water. Also, manufacturers and power plant operators must develop plans for treating the water they use in their facilities, so that the water is relatively salt- and mineral-free when it is returned to the ocean, rivers, and lakes. For more information about related careers in the biological sciences, turn to page 842.

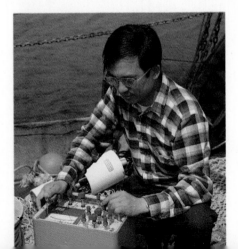

Genetics

"The multiplicity of shape, color, and behavior in individuals and in species is produced by the coupling of genes, as Mendel guessed. . . . But the question is not how the genes are arranged; the modern question is, How do they act? The genes are made of nucleic acid. That is where the action is."

Jacob Bronowski

The striking coloration of Indian corn is inherited just like your eye color or the shape of a bird's beak. Scientists working in the field of genetics study the patterns of inheritance of such traits as well as the biochemical mechanisms that underlie those patterns. In this unit you will learn about these patterns and mechanisms as well as about technologies that alter these mechanisms to produce desired traits.

Salivary chromosomes of fruit fly, *Drosophila melanogaster*

Indian corn, *Zea mays*

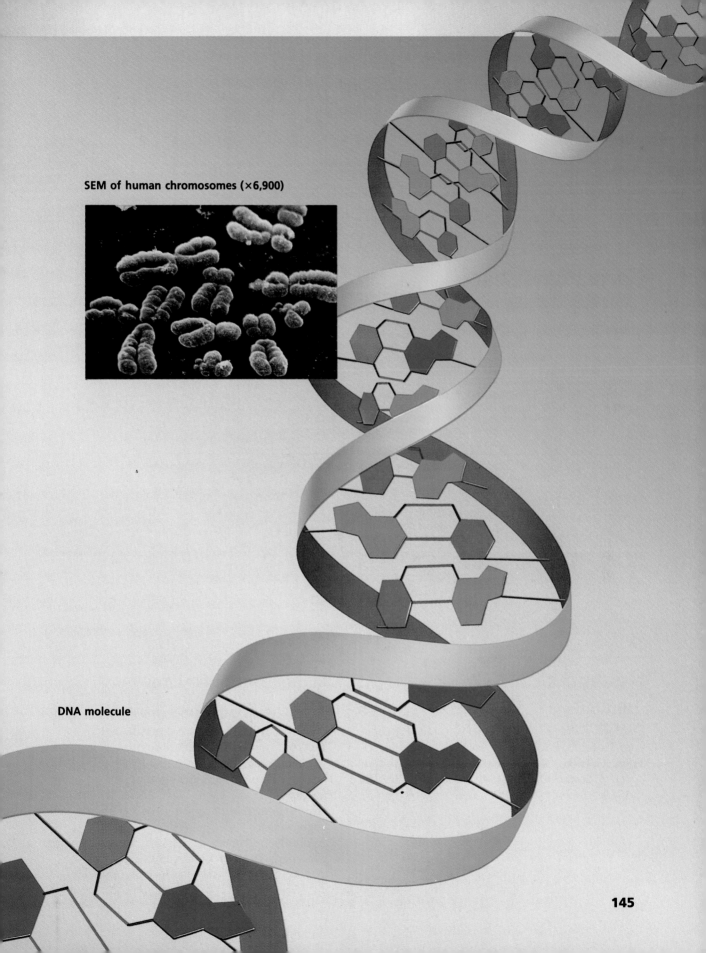

SEM of human chromosomes (×6,900)

DNA molecule

145

Chapter 10 | *Fundamentals of Genetics*

This chapter teaches students how to predict which parental traits are likely to appear in offspring. Students apply the principles of dominance, recessiveness, segregation, and independent assortment.

Correlation Chart

Section Number	Section Objectives	Pages	Features	Section Review
10.1	1. Describe Mendel's experiments, results, and conclusions.	147–152	Biology in Process, pp. 150–151	1, 2, 3
	2. State and explain Mendel's three principles.			3
	3. Give examples of how scientific knowledge of genes and chromosomes supports Mendel's principles.			4, 5
	4. Summarize the observations that led Sutton to propose the chromosome theory.			
10.2	1. Define *probability*, and explain how it is used to predict the results of genetic crosses.	153–159	Think, p. 154	2
	2. Use a Punnett square to predict the results of genetic crosses.		Biology Lab, p. 159	4, 5
	3. Explain how a testcross reveals the genotype of an individual whose phenotype is dominant.			1, 3
	4. Distinguish monohybrid crosses from dihybrid crosses.			5

Resources*

Print

Gardner, Eldon J., and D. Peter Snustad. *Principles of Genetics,* 7th ed. New York: Wiley, 1984.

Edlin, Gordon. *Genetic Principles: Human and Social Consequences.* Boston: Jones and Bartlett, 1984.

Audiovisual/Software

The Fascinating Story of Cell Growth. Apple 128K. IBM PC 128K. 2 Disks. Scholastic, 1986.

The Gene. Video or 2 filmstrips with cassettes, 16–19 min. each. Teacher's guide. Hawkhill, 1984.

Heredity. Film. 15 min. Britannica.

Heredity I: Traits, Genes, and Alleles. Filmstrip or transparencies with cassette. Teacher's guide. Biology Media.

*Additional references are listed in *Modern Biology, Teacher's Program Guide.*

Section Teaching Guides

10.1 The Legacy of Gregor Mendel

Approach

This section focuses on the scientific processes Gregor Mendel used to formulate the principles of inheritance. In presenting the text, point out Mendel's use of such process skills as observing, collecting and interpreting data, inferring, and predicting to formulate his theories of inheritance. You might review the processes of protein synthesis, mitosis, and meiosis (Chapters 8 and 9). Note that Mendel predicted the outcome of chromosome behavior even though his work predated knowledge of DNA and chromosomes.

Concept Framework

All students are aware that certain traits appear frequently within families while other traits are more rare. To prepare students for understanding Mendel's research, ask them what inferences they can draw from their observations of the appearance of certain traits from generation to generation and how these inferences might be tested quantitatively.

Emphasis

Make certain that students understand the connection between Mendel's principles of inheritance and the actual mechanisms of meiosis and mitosis.

Using the Features/Sutton and the Chromosome Theory

The main process skills used in the feature on Walter Sutton and the chromosome theory are observing and inferring. Specifically, Sutton combined his observations of chromosome behavior with his previous knowledge of inheritance and inferred that chromosomes carried the "factors" that Mendel had referred to years earlier.

Section Review Answers

(Continued from p. 152.)

4. Each contrasting form of a gene is an allele.
5. Mendel would not have observed independent assortment occurring consistently. Since independent assortment would not have appeared to be operating for all traits, Mendel would have been less likely to arrive at his principle of independent assortment.

10.2 Genetic Crosses

Approach

The main objective of this section is to teach students how to use Mendel's principles to predict the outcome of genetic crosses. This knowledge will prepare students for the material presented in Chapters 11 and 13 on human genetics and applied genetics. This section also provides background for the discussion of allele frequency and genetic equilibrium in Chapter 16. Encourage students to use the process skills of predicting, inferring, and interpreting data to solve the genetics problems included in this section.

Concept Framework

Students will have encountered the use of probability in activities such as predicting heads and tails. Having students verbalize this knowledge can help prepare them for understanding and solving genetics problems. In addition, ask students to consider the limitations as well as the uses of probability statistics as a tool for making predictions.

Emphasis

Make sure that students can use Punnett squares to predict the outcomes of the genetic crosses described in the text.

Field Trip

Plan a visit to a botanic garden, dog breeder, or similar facility where students can observe the inheritance of selected traits from one generation to the next. Prior to the trip, prepare a list of problems in which students predict the results of genetic crosses.

Section Review Answers

(See p. 158.)

Laboratory Preparation

Background

In this lab students collect and analyze data from genetic corn ears reflecting the results of simple monohybrid and dihybrid crosses. The observed data is related to expectations predicted through the use of Punnett squares to arrive at conclusions regarding parental genotypes and phenotypes. The students are also asked to relate their conclusions to the Mendelian principles of segregation and independent assortment.

Preparation

Order the two types of corn ears from a biological supply house. Ear A should have a ratio of 3 purple grains : 1 yellow grain. Ear B should have a ratio of 9 purple, smooth : 3 purple, wrinkled : 3 yellow, smooth : 1 yellow, wrinkled. When ordering, be aware that *smooth* is described in supply catalogs as "starchy" and *wrinkled* is described as "sweet."

Conclusions

1. The parental ears are both heterozygous purple (Pp). A comparison of the observed ratios with expected ratios is the evidence.
2. The observed ratios should be similar to students' predicted ratios.

Inquiry

The phenotypic ratio will be the same as when purple and smooth are dominant, except the opposite alleles will be expressed. Students should refer to the Punnett square they constructed in Step 8.

Chapter Review

Vocabulary

1. Cross-pollination may take place when pollen is transferred from the anthers of one pure plant to the stigma of another pure plant, whereas self-pollination may take place when pollen is transferred from anthers to stigma in the same flower or in flowers of the same plant.
2. The alleles for the trait that is expressed is dominant while the allele for the trait that is masked is recessive.
3. Each contrasting form of a gene is an allele.
4. When both alleles of a pair are the same, an organism is homozygous for that trait. When the two alleles in the pair are not the same, the organism is heterozygous for that trait.
5. A genetic cross between individuals that vary only in a single trait is a monohybrid cross. A genetic

cross between individuals that vary in two traits is a dihybrid cross.

Review

1. The gardener should cross each plant with a short plant. If a tall plant is homozygous, the offspring of such a cross will all be tall. If a tall plant is heterozygous, the offspring of such a cross will be half tall and half short. Such a cross is called a testcross.

	t	t			t	t
T	Tt	Tt		T	Tt	Tt
T	Tt	Tt		t	tt	tt

2. The gametes are B and b. The phenotypic ratio is 100 percent black. The geotypic ratio is 100 percent Bb.

	b	b
B	Bb	Bb
B	Bb	Bb

3. The phenotype of both the male and female should be smooth coat. The genotype would be rr.

	r	r
r	rr	rr
r	rr	rr

4. The phenotypic ratio will be 1 red : 1 pink. The genotypic ratio is 1 RR : 1 RR′.

	R	R′
R	RR	RR′
R	RR	RR′

5. The most probable genotype of the two parents is AaIi. The abnormal ratio is no doubt due to the small number of seeds planted. The explanation could be checked by planting more seeds and/or by doing a testcross for both parents. For a testcross the double recessive aaii should be used.

	AI	Ai	aI	ai
AI	AAII	AAIi	AaII	AaIi
Ai	AAIi	AAii	AaIi	Aaii
aI	AaII	AaIi	aaII	aaIi
ai	AaIi	Aaii	aaIi	aaii

6.
AAIi	×	AAII	AAIi	×	AaII
AAII	×	AAII	AAII	×	AaII
AaII	×	AAII	AaII	×	AaII
AaIi	×	AAII	AaIi	×	AaII
AAIi	×	AAIi	AAIi	×	AaIi
AaII	×	AAIi	AAII	×	AaIi
AaII	×	AAIi	AaII	×	AaIi
AaIi	×	AAIi	AaIi	×	AaIi

7. The phenotypic ratio for this cross is 9 tall, round : 3 tall, wrinkled : 3 short, round ; 1 short, wrinkled.

	TR	Tr	tR	tr
TR	TTRR	TTRr	TtRR	TtRr
Tr	TTRr	TTrr	TtRr	Ttrr
tR	TrRR	TtRr	ttRR	ttRr
tr	TtRr	Ttrr	ttRr	ttrr

8. The phenotypic ratio is 6 red, tall : 6 pink, tall : 2 red, short : 2 pink, short. The genotypic ratio is 4 RR′Tt : 4 R′RTt : 2 RRTT : 2 R′RTT : 2 RRtt : 2 RR′tt.

	RT	Rt	R′T	R′t
RT	RRTT	RRTt	RR′TT	RR′Tt
Rt	RRTt	RRtt	RR′Tt	RR′tt
RT	RRTT	RRTt	RR′TT	RR′Tt
Rt	RRTt	RRtt	RR′Tt	RR′tt

9. The most probable explanation for your results is that the alleles for red flower color and white flower color are codominant. You could prove this hypothesis by self-pollinating the orange plants. You should get a phenotypic ratio of 1 red : 2 orange : 1 yellow which would reflect a genotypic ratio of 1 RR : 2 RR′ : 1 R′R′. The Punnett square is as follows:

	R	R′
R	RR	RR′
R′	RR′	R′R′

10. Yes, you could get rabbits with brown curly hair in the first generation if the male rabbit was heterozygous for both curly and brown hair. The other genotypes would not produce this. The Punnett squares of the four possible crosses are as follows:

1. BbSs × bbss

	bs	bs	bs	bs
BS	BbSs	BbSs	BbSs	BbSs
Bs	Bbss	Bbss	Bbss	Bbss
bS	bbSs	bbSs	bbSs	bbSs
bs	bbss	bbss	bbss	bbss

Genotype: 4 BbSs : 4 Bbss : 4 bbSs : 4 bbss
Phenotype: 4 black, straight : 4 black, curly : 4 brown, straight : 4 brown, curly

2. BBSS × bbss

	bs
BS	BbSs

Genotype: 100 percent BbSs
Phenotype: 100 percent black, straight

3. BbSS × bbss

	bs
BS	BbSs
BS	BbSs
bS	bbSs
bS	bbSs

Genotype: 1 BbSs : bbSs Phenotype: 1 black, straight : 1 brown, straight

4. BbSS × bbss

	bs
BS	BbSs
BS	BbSs
Bs	Bbss
Bs	Bbss

Genotype: 1 BbSs : 1 Bbss Phenotype: 1 black, straight : 1 black, curly

Critical Thinking

1. The chances are 1 in 100,000, or $10 \times 10 \times 10 \times 10 \times 10$, or 10^5.

2.

	F	F
F	FF	FF
f	Ff	Ff

	F	F
F	FF	FF
f	Ff	Ff

	F	f
F	FF	Ff
f	Ff	ff

	F	f
F	FF	Ff
f	Ff	ff

Eight will be carriers; two will have the disease.

Chapter 10

Fundamentals of Genetics

Introduction

Parents pass their traits on to their offspring. However, not all of the parents' traits will actually appear in the offspring. The white tiger shown here had yellow parents. In this chapter you will learn how to predict which traits are likely to appear in the offspring. Such predictions are based on the work of Gregor Mendel, an Austrian monk and a pioneer in the field of genetics.

Chapter Outline

Chapter Concept

As you read, notice how Mendel's experimental design helped him develop his principles. Then apply these principles to help predict the outcome of genetic crosses.

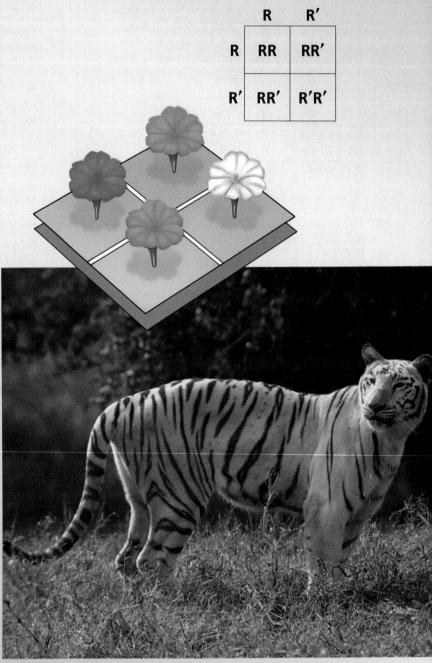

	R	R'
R	RR	RR'
R'	RR'	R'R'

White tiger, *Panthera tigris*, India. Inset: Japanese four o'clock, *Mirabilis japonica*, Punnett square.

For additional teaching information see pp. T10A–T10D.

10.1 The Legacy of Gregor Mendel

Section Objectives

- *Describe Mendel's experiments, results, and conclusions.*
- *State and explain Mendel's three principles.*
- *Give examples of how scientific knowledge of genes and chromosomes supports Mendel's principles.*
- *Summarize the observations that led Sutton to propose the chromosome theory.*

In 1843 at the age of 21 Gregor Mendel entered a monastery in Brünn, Austria, where he was given the task of tending the garden. In 1851 he was sent to the University of Vienna to study science and mathematics. His studies of mathematics included training in the new field of statistics, which would later prove invaluable to his research.

When Mendel returned to the monastery, he used his knowledge of statistics to analyze biological phenomena. He observed that some tall plants produced tall offspring, while other tall plants produced short offspring. Likewise some of the plants with yellow seeds yielded offspring with yellow seeds, but others yielded offspring with green seeds. The phenomenon Mendel was observing is called **inheritance,** the passing of traits by heredity. **Heredity** is the transmission of traits from parents to their offspring.

To see whether there was any predictable pattern to inheritance, Mendel decided to study the hereditary patterns of pea plants. Mendel studied seven characteristics of pea plants, each of which occurred in two contrasting traits: seed texture (round or wrinkled), seed color (yellow or green), seed coat color (colored or white), pod appearance (inflated or constricted), pod color (green or yellow), position of flowers along the stem (axial or terminal), and stem length (tall or short). Mendel hoped to discover if there were predictable patterns in the inheritance of traits—and, if so, what these patterns were. Mendel used statistical methods to analyze the results of thousands of crosses between individual plants. In nine years he bred thousands of pea plants. The patterns he found form the basis of our knowledge of genetics.

Mendel's Experiments

An important aspect of Mendel's experimental design was his decision to study each characteristic and its two contrasting traits individually. He began by growing plants that were pure for each trait. Plants that are **pure** for a trait always produce offspring with that trait. Pea plants pure for the trait of yellow pods always produce offspring with yellow pods. A **strain** is the term used to denote all plants pure for a specific trait. Mendel produced strains by allowing the plants to go through a natural process called self-pollination.

Self-pollination is a reproductive process in which fertilization occurs within a single plant. The reproductive structures of seed plants are located inside the flowers. In pea plants each flower has both sperm-producing and egg-producing structures. The anthers produce pollen grains that contain sperm. Eggs are

Figure 10-1 Mendel transferred pollen from an anther to a stigma of the same pea plant in order to produce purebred strains by self-pollination.

Petal
Stigma
Anther

Tip: Emphasize to the students that *characteristics* have contrasting *traits.*

P₁

Green-podded plant ✕ Yellow-podded plant

F₁

✕

All green-podded plants

F₂

3 green-podded plants   1 yellow-podded plant

Figure 10-2 Pure green-podded plants crossed with pure yellow-podded plants produce only green-podded plants. Yet when this second generation is permitted to self-pollinate, some yellow-podded plants appear.

Info: People have used principles of heredity since ancient times to improve plants and animals, but before the investigations of Gregor Mendel no one had developed a theory of how these principles actually worked.

produced in structures called ovules. The ovule is contained in an ovary, the tip of which is called the stigma. **Pollination** is the transfer of pollen from anthers to stigma. In self-pollination pollen is transferred from anthers to stigma in the same flower or in flowers of the same plant. By allowing the plants to self-pollinate, Mendel obtained 14 strains, one for each of the 14 traits. He called each strain a **parental generation,** or **P₁ generation** for short.

Mendel then applied **cross-pollination** to these strains by transferring pollen from the anthers of a plant pure for one trait to the stigma of another plant pure for the contrasting trait. For example, if he wanted to cross a plant pure for the trait of yellow pods with one pure for the trait of green pods, he dusted the pollen from a yellow-podded plant onto the stigma of a green-podded plant. Then he allowed the seeds to grow.

Next he recorded how many of each type of offspring each P₁ plant produced. Mendel called the offspring of the P₁ generation the **first filial generation,** or **F₁ generation.** He then self-pollinated flowers from the F₁ generation and collected the seeds. Mendel called the plants in this generation the **second filial generation,** or **F₂ generation.** Following this process, Mendel performed thousands of crosses and documented the results of each.

Mendel's Results and Conclusions

In one of his experiments Mendel crossed two plants, one pure for green pods and one pure for yellow pods. The resulting seeds produced an F₁ generation with only green-podded plants. No yellow pods developed even though one parent had been pure for yellow pods. Only one of the two traits found in the P₁ generation appeared in the F₁ generation. If Mendel had concluded his research at this point, he might have interpreted the results to mean that crosses between green-podded plants and yellow-podded plants always produced offspring with green pods. However, Mendel self-pollinated the F₁ plants and then planted seeds from the resulting F₂ generation. When they grew, he observed that about three-fourths of the F₂ plants had green pods and one-fourth had yellow pods. The apparently lost trait had reappeared.

Study Table 10-1 to see the results of Mendel's crosses. Whenever Mendel crossed strains, one of the P₁ traits vanished from all of the F₁ plants. In every case that trait reappeared in a ratio of about 3 : 1 in the F₂ generation. This pattern emerged in thousands of pollinations. Mendel called the trait that appeared in the F₁ generation the dominant trait. The trait that appeared to be lost in the F₁ generation but that reappeared in the F₂ generation he called the recessive trait.

After studying the results of these crosses, Mendel concluded that patterns of inheritance were governed by three principles: (1) the principle of dominance and recessiveness, (2) the principle of segregation, and (3) the principle of independent assortment.

Table 10-1   Mendel's Crosses and Results

Trait	P_1 cross	F_1 generation	F_2 generation	Actual ratio	Probability ratio
Position of flowers along stem	Axial × terminal	Axial	651 axial 207 terminal	3.14 : 1	3 : 1
Height of plant	Tall × short	Tall	787 tall 277 short	2.84 : 1	3 : 1
Pod appearance	Inflated × constricted	Inflated	882 inflated 299 constricted	2.95 : 1	3 : 1
Pod color	Green × yellow	Green	428 green 152 yellow	2.82 : 1	3 : 1
Seed texture	Round × wrinkled	Round	5,474 round 1,850 wrinkled	2.96 : 1	3 : 1
Seed color	Yellow × green	Yellow	6,022 yellow 2,001 green	3.01 : 1	3 : 1
Seed coat color	Colored × white	Colored	705 colored 224 white	3.15 : 1	3 : 1

Principle of Dominance and Recessiveness

Mendel suggested that something within a plant controlled the expression of a characteristic. He named this "something" a factor. Each characteristic he was studying had two traits; for example, the characteristic of pod color could be expressed as either green or yellow. Thus he inferred that each characteristic was the result of the interaction of a pair of factors.

Mendel looked next at how factors interacted by studying the F_1 generation of plants. He observed that only one trait of each characteristic appeared in the F_1 generation. In attempting to explain this phenomenon, Mendel arrived at the **principle of dominance and recessiveness:** one factor in a pair may mask the other factor, preventing it from having an effect. Mendel called the one factor **dominant** since it masked or dominated the other factor for this trait. He called the other factor **recessive** since it was masked in the presence of a dominant factor for that characteristic. Interpreting the data in light of this principle, Mendel then concluded that the factors that produced traits visible in the F_1 generation were dominant over the factors that produced traits that didn't appear in that generation.

Biology in Process

Sutton and the Chromosome Theory

Mendel's work elegantly illustrated that traits are passed from parent to offspring in predictable patterns. However, Mendel could not identify the cellular mechanism by which the information was communicated from generation to generation. He lacked information about cell structure that might have enabled him to discover the mechanics of inheritance.

In the 1870s and 1880s, biologists provided the first accurate descriptions of the processes by which genetic information is communicated from generation to generation: mitosis and meiosis. The discovery of these processes led to the realization that a connection exists between Mendel's work and the modern science of genetics. In 1902, a young biologist named Walter S. Sutton bridged the two bodies of work by proposing that Mendel's "factors" are carried on chromosomes.

Two processes that Sutton used in his studies were observing and inferring. Through microscopic observations of grasshopper sperm, Sutton discovered that before meiosis occurs, the chromosomes lined up in pairs that were alike in size and shape. From these observa-

Principle of Segregation

Mendel then asked why the traits that disappeared in the F_1 generation reappeared in the F_2 generation. Mendel reasoned that if each parent had two factors, each offspring must have two factors. Each parent could not be passing two factors or the offspring would have four. Mendel concluded that each reproductive cell received only one factor for each characteristic. He then stated the **principle of segregation:** the two factors for a characteristic segregate, or separate, during the formation of eggs and sperm.

Principle of Independent Assortment

Mendel also crossed plants that differed in two characteristics, such as in height and seed color. The data from these more complex crosses showed that traits produced by dominant factors did not necessarily appear together. A green seed pod produced by a dominant factor could appear in a short pea plant, produced by recessive factors. Mendel concluded that the factors for different characteristics were not connected. He stated the **principle of independent assortment:** factors for different characteristics are distributed to reproductive cells independently.

Answer to Biology in Process
Sutton based his chromosome theory on the following inferences:
(1) chromosomes, like Mendel's factors, occur in pairs and (2) the sperm and the egg are the only possible carriers of hereditary factors.

tions, Sutton inferred that chromosomes, like Mendel's factors, occurred in pairs.

Sutton next observed that pairs of chromosomes separated during meiosis, one member of each pair going to each gamete. He also noted that pairs generally separated independently. This is what Mendel proposed in his laws of segregation and independent assortment. Finally, Sutton saw that a sperm cell and an egg each carry half the number of chromosomes found in body cells. Through fertilization they produced a new organism with a full set of chromosomes.

Sutton drew upon his knowledge of reproductive biology to interpret these observations. He knew that the egg and sperm provided the only physical link between generations. Since these cells were the sole link, Sutton inferred that they were the only possible carriers of hereditary factors.

Therefore, Sutton concluded that Mendel's hereditary "factors" were carried on chromosomes. Sutton hypothesized that hereditary information was transmitted from one generation to the next generation via chromosomes. This hypothesis corresponds well with Mendel's principles and is supported by both Mendel's and Sutton's observations. It explains how offspring receive one set of traits from each parent, and it accounts for the independent transmission of the separate traits.

■ What were the inferences Sutton made on which he based his chromosome theory?

Figure 10-3 The arrangement of chromosomes (left or right) during the first division of meiosis determines the type of gametes produced. DNA replication and the second meiotic division are omitted for simplicity.

Answers to Section Review
1. Self-pollination is a reproductive process in which pollination occurs within the same plant.
2. Mendel studied each characteristic and its contrasting trait individually.
3. The principle of dominance and recessiveness maintains that one factor in a pair will mask the other factor, preventing it from being expressed. The principle of segregation maintains that a pair of factors separates during the formation of eggs and sperm. The principle of independent assortment maintains that each pair of factors is distributed to reproductive cells independently of each other pair. (See pp. T10A–T10D.)

Chromosomes and Genes

Most of Mendel's findings agree with what biologists now know of the structure and function of genes and chromosomes. However, when Mendel presented his findings in a paper he wrote in 1865, scientists had not yet discovered chromosomes, nor did they possess any knowledge of meiosis. Lacking a way to explain Mendel's findings, the scientific community did not understand or care about them. Also, scientists of that time had not yet used statistics to understand biological phenomena. Mendel's pairing of statistics and natural science seemed odd to them.

Not until the year 1900 did biologists realize the connection between Mendel's obscure paper and ongoing work on inheritance. During that year Mendel's paper was rediscovered three times by biologists working independently. Shortly thereafter biologist Walter Sutton proposed the chromosome theory, which linked observations of meiosis with Mendel's principles. Now not only are Mendel's principles accepted, but his quantitative methods are also used extensively by biologists.

Instead of just studying factors, biologists today also study genes. A **gene** is the segment of DNA on a chromosome that controls a particular hereditary trait. Because chromosomes occur in pairs, genes occur in pairs. Each contrasting form of a gene, called an **allele,** is the equivalent of Mendel's "factor." Scientists use letters to represent alleles, capital letters to refer to dominant alleles and lowercase letters to refer to recessive alleles. For example, the allele for the dominant trait of tallness in pea plants is represented by T, and the allele for the recessive trait of shortness by t. The allele for the dominant trait of green pod color is represented by G; the allele for yellow pod color by g.

Current knowledge of chromosomes and meiosis also supports the principle of segregation. When gametes are formed, they receive one chromosome from each homologous pair of chromosomes. This means that when the gametes combine in fertilization, the offspring receives one allele from each parent. This fact—that when chromosomes segregate independently, any alleles on these chromosomes also segregate independently—supports the principle of independent assortment. However, though these principles are correct in general, some variations in hereditary patterns are not explained by Mendel's principles.

Section Review

1. Define *self-pollination.*
2. Name an important aspect of Mendel's experimental design.
3. Name and briefly explain Mendel's three principles.
4. What is the relationship between genes and alleles?
5. How might Mendel's conclusions have differed if he had studied two traits determined by alleles carried on the same chromosome?

Tip: Review with students the material in Chapter 9 that outlines the steps in meiosis.

10.2 Genetic Crosses

Geneticists still rely on Mendel's principles to predict the likely outcome of genetic crosses. In this section you will learn how to use Mendel's principles to determine the likely outcome of several examples of two kinds of genetic crosses. You will acquire the knowledge to predict the probable genetic makeup of offspring resulting from specified crosses. You will also learn how to interpret this genetic information to determine the probable appearance of these offspring.

Genotype and Phenotype

The genetic makeup of an organism is its **genotype.** The external appearance of an organism is its **phenotype.** For example, the genotype of a pure tall plant is TT. It consists of two dominant alleles for height—T and T. The plant's phenotype, or appearance, is tall. The genotype of an F_1 pea plant is Tt if one parent has contributed one allele for tallness and one for shortness. The phenotype of that plant is tall, since the dominant gene is expressed and the recessive gene is not expressed. Thus a phenotype of tallness may be produced by either of two genotypes—TT or Tt.

When both alleles of a pair are the same, an organism is said to be **homozygous** for that characteristic. An organism may be homozygous dominant or homozygous recessive. A pea plant homozygous dominant for height will have the genotype TT. A pea plant homozygous recessive for height will have the genotype tt. When the two alleles in the pair are not the same, for example, when the genotype is Tt, the organism is **heterozygous** for that characteristic.

Mendel studied traits that had only two alleles. Biologists now know that some traits are determined by many alleles. When three or more alleles control a trait, it is said to have **multiple alleles.** The trait of blood type in humans is determined by multiple alleles. The alleles called I^A, I^B, and i^O all affect the characteristic of blood type. Among the possible genotypes for blood type are $I^A i^O$, $I^B i^O$, and $i^O i^O$. However, no matter how many alleles determine a trait, an individual can only have two alleles for a given characteristic, one inherited from each parent.

Section Objectives

- *Define* probability, *and explain how it is used to predict the results of genetic crosses.*

- *Use a Punnett square to predict the results of monohybrid and dihybrid genetic crosses.*

- *Explain how a testcross reveals the genotype of an individual whose phenotype is dominant.*

- *Distinguish monohybrid crosses from dihybrid crosses.*

Figure 10-4   The difference in coloration in these cottontail rabbits (*Sylvilagus* sp.) is a phenotypic difference. A difference in phenotype indicates a difference in genotype.

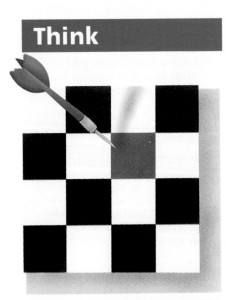

If you were blindfolded and threw a dart and hit the board above, what is the probability that you would hit the red square? If you threw two darts, what is the probability that you would hit the red square at least once? Express your answers in fractions.

Answer to Think
In this case, number of possibilities would be substituted for number of events. With one throw you would have one possibility of hitting the red square and 16 possible outcomes. Therefore, your probability of hitting the red square would be 1 in 16 or 1/16. With two throws your chances would be 2 in 16 or 1/8.

Probability

Probability is the likelihood that a specific event will occur. It is expressed as a decimal, a percentage, or a fraction. It is determined by the following formula:

$$\text{Probability} = \frac{\text{number of one kind of event}}{\text{number of all events}}$$

For example, in Mendel's experiments the dominant trait of yellow seed color appeared in the F_2 generation 6,022 times. The recessive trait of green seed color appeared 2,001 times. The total number of individuals was thus 8,023. Using the probability formula above, the probability that the dominant trait will appear in a similar cross is 6,022/8,023, or 0.75. Expressed as a percentage, the probability is 75 percent. Expressed as a fraction, the probability is 3/4. The probability that the recessive trait will appear in an F_2 generation 2,001/8,023, or 0.25. Expressed as a percentage, the probability is 25 percent. Expressed as a fraction, the probability is 1/4. Probability tells us that there are three chances in four that an offspring of two heterozygous individuals will have the dominant trait and one chance in four that it will have the recessive trait.

In genetics, as in other systems based on probability, the expected ratios occur only when there are many trials. For example, many coin tosses should yield a result of heads 50 percent of the time and tails 50 percent of the time. However, if you tossed a coin only a few times, you probably wouldn't get this result. Only after many, many tries would you be likely to get the ratio of heads predicted on the basis of probability—0.5, or 50 percent, heads and 0.5, or 50 percent, tails.

Predicting Results of Monohybrid Crosses

A cross between individuals that involves one pair of contrasting traits is called a **monohybrid cross**. A cross between a pea plant that is pure for tallness and one that is pure for shortness is an example of a monohybrid cross. There is a simple way to establish the probabilities of the results of a monohybrid cross. Biologists use a diagram called a **Punnett square**, named for its inventor, R. C. Punnett, to aid them in predicting probabilities.

Example 1: Homozygous × Homozygous

Let's use a Punnett square to predict the results of a monohybrid cross between a homozygous tall pea plant (TT) and a homozygous short pea plant (tt). Draw a square and divide it into four parts. Represent the alleles contributed by the homozygous dominant parent by writing two *T*'s, one next to each of the boxes on the left side of the square. Represent the alleles contributed by the homozygous recessive parent by writing two *t*'s, one above each

	t	t
T	Tt	Tt
T	Tt	Tt

Homozygous × homozygous

box across the top of the square. Fill in each box with the letters that are above it and beside it outside the square. The Punnett square should contain the same results as the square in Figure 10-5. These results indicate the probable genotypes of such a cross. One hundred percent of the offspring would be Tt.

What phenotypes will result from such a cross? According to Mendel's principle of dominance and recessiveness, a dominant allele masks the expression of a recessive allele. Since all offspring are Tt, each has a dominant and a recessive allele for height. Therefore in all offspring the dominant allele for height will be expressed, and 100 percent of the offspring will be tall.

Example 2: Homozygous × Heterozygous

Now let's predict the outcome of a cross between a guinea pig that is homozygous for the dominant trait of rough coat (RR) and a guinea pig that is heterozygous for this trait (Rr). The lower case r stands for the recessive allele. Genotype rr results in a smooth coat. Set up and complete the square. Compare your results with those shown in Figure 10-6. The probable genotypic ratio is 2RR : 2Rr. In other words, you could expect that 50 percent of the offspring of this pair would be homozygous dominant for rough coat and the other 50 percent would be heterozygous for this trait. The probable phenotypic ratio is four out of four rough coated, or 100 percent rough coated.

Figure 10-5 A tall homozygous and a short homozygous pea plant will produce only tall heterozygous offspring.

Figure 10-6 A guinea pig homozygous for rough coat and one heterozygous for rough coat produce all rough-coated individuals, half of which are homozygous.

	R	r
R	RR	Rr
R	RR	Rr

Homozygous × heterozygous

155

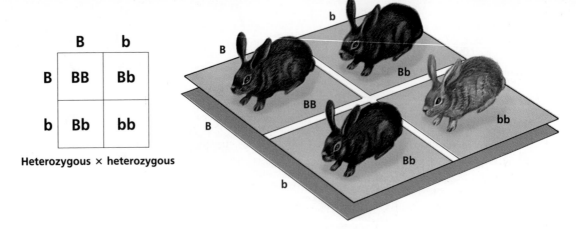

	B	b
B	BB	Bb
b	Bb	bb

Heterozygous × heterozygous

Figure 10-7 Crossing two rabbits both heterozygous for black coat tends to produce two heterozygous black offspring, plus one purebred black and one purebred brown individual.

Tip: Tell students that a testcross can be used in a dihybrid cross if the organism is mated with a double-recessive individual. A test organism homozygous for both sets of genes will produce only one type of offspring. If the test organism is heterozygous for both sets of genes, four types of offspring will appear.

	b	b
B	Bb	Bb
B	Bb	Bb

	b	b
B	Bb	Bb
b	bb	bb

Figure 10-8 A testcross is used to determine the genotype of an unknown individual.

Example 3: Heterozygous × Heterozygous

In rabbits the allele for black coat (B) is dominant over the allele for brown coat (b). To predict the results of crossing two parents that are both heterozygous (Bb) for coat color, set up and complete a Punnett square. Your completed square should match the square shown in Figure 10-7. As you can see, one-fourth of the offspring would have genotype BB, one half would have the genotype Bb, and one-fourth would have the genotype bb. Since B is dominant over b, three-fourths of all offspring from this cross would have the black phenotype, and one-fourth would have the brown phenotype. The probable genotypic ratio of a monohybrid cross of two heterozygous individuals is 1BB : 2Bb : 1bb. The probable phenotypic ratio is 3 black : 1 brown.

Example 4: Testcross

In rabbits both BB and Bb result in a black coat. How might you determine whether a black rabbit was homozygous (BB) or heterozygous (Bb)? You could perform a **testcross,** the procedure in which an individual of unknown genotype is crossed with a homozygous recessive individual. A testcross can determine the genotype of any individual whose phenotype is dominant.

If the rabbit in question were homozygous, what type of offspring would a testcross produce? To predict this, set up a Punnett square. Write the alleles for one parent, B and B, next to the boxes along the left side of the square and the alleles for the other parent, b and b, above the boxes of the square. Now complete the square, and compare your results with those in Figure 10-8. Notice that all genotypes produce dominant phenotypes. Therefore, if all the offspring of the testcross have the dominant phenotype, then the individual in question must be homozygous dominant.

Now suppose the rabbit in question were heterozygous. Set up another Punnett square, writing the alleles for one parent, B and b, next to the boxes along the left side of the square and the alleles of the other parent, b and b, above the boxes. Complete the square. Figure 10-8 illustrates the results. The probable genotypic ratio is 2Bb : 2bb. The probable phenotypic ratio is 2 black coated : 2 brown coated. Thus, if about half of the offspring are brown, the individual in question must be heterozygous.

Example 5: Codominance

Mendel restricted his studies to traits for which one allele dominated completely over the other. However, this is not true for all traits. **Codominance,** or incomplete dominance, is the phenomenon that occurs when two or more alleles influence the phenotype. This results in an trait intermediate between the dominant and recessive traits. In Japanese four o'clocks, for example, both the allele for red flowers (R) and the allele for white flowers (R′) influence the phenotype. Neither allele is dominant. When these plants self-pollinate, red-flowered plants produce only red-flowered offspring, and white-flowered plants produce only white-flowered offspring. However, when red and white four o'clocks are crossed, the F_1 offspring all have pink flowers. One hundred percent of the offspring of this cross possess the (RR′) genotype which results in a pink phenotype.

What would be the result of crossing two pink-flowered (RR′) Japanese four o'clocks? You can solve the problem by filling in a Punnett square, as shown in Figure 10-9. The Punnett square method yields the following probable genotypic ratio: 1RR : 2RR′ : 1R′R′. Given that neither the allele for red flowers (R) nor the allele for white flowers (R′) is completely dominant, the probable phenotypic ratio must be 1 red, 2 pink, and 1 white.

Predicting Results of Dihybrid Crosses

A **dihybrid cross** is a cross between individuals that involves two pairs of contrasting traits. Predicting the results of a dihybrid cross is more complicated than predicting the results of monohybrid cross. There are more possible combinations of alleles to work out. For example, to predict the results of a cross involving both seed texture and seed color, you have to consider how the four alleles from each parent can combine.

Example 6: Homozygous × Homozygous

Suppose that you want to predict the results of a cross between a pea plant that is homozygous for round, yellow seeds and one that is homozygous for wrinkled, green seeds. In pea plants the allele for round seeds (R) is dominant over the allele for wrinkled seeds (r), and the allele for yellow seeds (Y) is dominant over the allele for green seeds (y). Therefore you will predict the results of a cross between a parent of the genotype RRYY and a parent of the genotype rryy. Use a Punnett square to do this. This Punnett square will need to contain 16 boxes. Now list the independently sorted alleles from one parent—RY, RY, RY, and RY—along the left side of the Punnett square and the independently sorted alleles from the other parent—ry, ry, ry, and ry—along the top of the square. Then fill in each box with the letters that are above it outside the square and those that are to the left of it outside the square. Compare your results with those shown in Figure 10-10.

	R	R′
R	RR	RR′
R′	RR′	R′R′

Codominance

Figure 10-9 Crossing pink four o'clocks produces red, white, and pink offspring.

	ry	ry	ry	ry
RY	RrYy	RrYy	RrYy	RrYy
RY	RrYy	RrYy	RrYy	RrYy
RY	RrYy	RrYy	RrYy	RrYy
RY	RrYy	RrYy	RrYy	RrYy

Homozygous × homozygous

Figure 10-10 This particular dihybrid cross produces only one type of offspring.

	SB	Sb	sB	sb
SB	SSBB	SSBb	SsBB	SsBb
Sb	SSBb	SSbb	SsBb	Ssbb
sB	SsBB	SsBb	ssBB	ssBb
sb	SsBb	Ssbb	ssBb	ssbb

Heterozygous × heterozygous

	SB	Sb	sB	sb
SB	SSBB	SSBb	SsBB	SsBb
Sb	SSBb	SSbb	SsBb	Ssbb
sB	SsBB	SsBb	ssBB	ssBb
sb	SsBb	Ssbb	ssBb	ssbb

Figure 10-11 A dihybrid cross of two heterozygous individuals produces nine genotypes and four phenotypes.

As illustrated in this figure, the genotype of all of the offspring of this cross will be heterozygous dihybrids, RrYy. Therefore the phenotype of all the offspring will have round, yellow seeds.

Example 7: Heterozygous × Heterozygous

In guinea pigs the allele for short hair (S) is dominant over the allele for long hair (s). The allele for black hair (B) is dominant over the allele for brown hair (b). Use a Punnett square to predict the probable offspring of a cross between two individuals heterozygous for both characteristics (SsBb). As shown in Figure 10-11 the offspring are likely to possess nine different genotypes that will result in the following four phenotypes:

- $9/16$ of the guinea pigs will have short, black hair.
- $3/16$ of the guinea pigs will have short, brown hair.
- $3/16$ of the guinea pigs will have long, black hair.
- $1/16$ of the guinea pigs will have long, brown hair.

The results from this dihybrid cross support the principle of independent assortment. In the next chapter you will learn about variations to this principle.

Section Review

1. Explain the terms *genotype* and *phenotype*.
2. What is the formula used to determine probablility? In what ways can probability be expressed?
3. Would a testcross be needed to determine the genotype of a Japanese four o'clock with pink flowers? Why or why not?
4. Explain how you would use a Punnett square to predict the probable outcome of a monohybrid cross.
5. The offspring of two short-tailed cats have a 25 percent chance of having no tail, a 25 percent chance of having a long tail, and a 50 percent chance of having a short tail. What can you hypothesize based on this information about the genotypes of the parents?

Answers to Section Review
1. An organism's genetic makeup is its genotype. An organism's external appearance is its phenotype. The phenotype of an organism is the expression of its genotype.
2. Probability is equal to the number of one kind of event divided by the number of all events. Probability can be expressed as a decimal, a percentage, or a fraction.
3. No. A Japanese four o'clock with a pink phenotype possesses one allele for red flowers and one for white flowers because red alleles and white alleles in this plant are codominant.
4. You would divide a square into four boxes. You would place symbols representing the alleles of one parent along the left hand side and symbols representing the alleles of the other parent across the top. To complete the square, you would fill in each box with the two symbols that converge on that square. The contents of each box would then reveal possible genotypes and phenotypes in predictable proportions.
5. The two short-tailed parents both have an allele for a long tail and an allele for no tail. These alleles are codominant. When both occur together, they are expressed as a short-tailed phenotype.

Laboratory

Genetic Inheritance in Corn

Objective
To analyze Mendelian factors in monohybrid and dihybrid crosses

Process Skills
Collecting data, inferring, predicting

Materials
Corn ears A and B

Method

1 In the first part of this investigation you will work with the offspring of a monohybrid cross. Working with a partner, count the number of purple and yellow kernels on corn ear A. One team member should call off the colors while the other records them. Purple results from the dominant allele (P), and yellow from the recessive allele (p).

2 Combine your data with that of two other teams. Use these data for all calculations. Why would you want to combine data?

3 Find the ratio of purple to yellow kernels. Record this ratio.

4 Draw the framework of a Punnett square for a monohybrid cross.

5 Divide the larger number in your ratio by the smaller. The answer, rounded to the nearest whole number, should show the relative proportion of P and p. Based on this information, label the boxes in the Punnett square either *purple* or *yellow*. What is the genotype of the yellow kernels? Mark this genotype in the appropriate box.

6 Label each of the remaining boxes with a P. How do you know that it is correct to assume that all other kernels had at least one allele for purple color?

7 Draw an inference about the parental genotypes. Label the top and left sides of the Punnett square with the parental alleles. Fill in the remaining genotypes.

8 Corn ear B contains the offspring of a dihybrid cross involving both color and texture. Purple (P) is dominant to yellow (p), and smooth texture (S) is dominant to wrinkled (s). Both parental plants are heterozygous for both traits. Draw a Punnett square, complete it, and predict the expected ratio of phenotypes. How many boxes will you need?

9 Count the kernels on corn ear B that represent each of the four phenotypes. Combine your data with those of two other teams. Calculate the phenotypic ratio. Compare the observed and expected ratios.

Conclusions

1 What are the genotypes and phenotypes of the parental plants for corn ear A? What is the evidence for this?

2 Does the observed ratio of kernels on corn ear B support your prediction for the expected ratio in a dihybrid cross?

Inquiry
Without making a Punnett square, predict the expected phenotypic ratio of corn ear B if yellow and wrinkled were dominant characteristics.

Vocabulary

allele (152)	heredity (147)	pollination (148)	probability (154)
codominance (157)	heterozygous (153)	principle of	Punnett square (154)
cross-pollination (148)	homozygous (153)	dominance and recessiveness (150)	pure (147)
dihybrid cross (157)	inheritance (147)		recessive (150)
dominant (150)	monohybrid cross (154)	principle of independent assortment (151)	second filial, or F_2, generation (148)
first filial, or F_1, generation (148)	multiple alleles (153)		self-pollination (147)
gene (152)	parental, or P_1, generation, (148)	principle of segregation (151)	strain (147)
genotype (153)	phenotype (153)		testcross (156)

Explain the difference between the terms in each of the following sets.

1. cross-pollination, self-pollination
2. dominant, recessive
3. allele, gene
4. heterozygous, homozygous
5. monohybrid cross, dihybrid cross

Review

1. A gardener has two tall pea plants. How can the gardener determine whether the two plants are homozygous or heterozygous for the gene determining tallness? Show the two Punnett squares as evidence for your conclusion. What is this type of cross called?

2. In rabbits the allele for black coat color (B) is dominant over the allele for brown coat color (b). What would be the results of a cross between an animal homozygous for black coat color (BB) and one homozygous for brown coat color (bb)? Show the gametes, Punnett square, and phenotypic and genotypic ratios.

3. In guinea pigs the allele for rough coat (R) is dominant over the allele for smooth coat (r). In order for all the offspring to be smooth coated, what should the phenotype and genotype of the male and female parents be? Show the Punnett square to support your answer.

4. In Japanese four o'clocks the alleles for red flower color (R) and for white flower color (R') are codominant. Thus a plant with the phenotype RR' will have pink flowers. Show the Punnett square and genotypic and phenotypic ratios for a cross between a plant with red flowers and one with pink flowers.

5. In peas the allele for axial flowers (A) is dominant over the allele for terminal flowers (a). The allele for inflated seed pods (I) is dominant over that for constricted pods (i). A cross between two peas with axial flowers and inflated pods gives the following offspring: 20 with axial flowers and inflated pods, 7 with axial flowers and constricted pods, and 5 with terminal flowers and inflated pods. What is the most probable genotype for the two parents? Explain the results and show the Punnett square. How can the explanation be checked?

6. For question 5 give all the possible crosses between two peas with axial flowers and inflated pods.

7. What would be the results of a cross between two pea plants that were heterozygous for both tall plants and round seeds? Show the Punnett square, and state the genotypic and phenotypic ratios.

8. In Japanese four o'clocks the alleles for red flower color (R) and the allele for white flower color (R') are codominant. Show the Punnett square and the genotypic and pheontypic ratios of the cross of the genotype: RRTt × RR'Tt. Assume tall (T) is dominant over short (t).

9. Suppose that you are an explorer who has found a new species of plant. Some of the plants have red flowers and some have yellow flowers. You cross a red-flowered plant with a yellow-flowered plant and all the offspring have orange flowers. What is the most probable explanation? How would you prove it? Support your answer with a Punnett square.

10. In rabbits the allele for black coat color (B) is dominant over the allele for brown coat color (b). The allele for straight hair (S) is dominant over that for curly hair (s). Someone gives you a male rabbit with black straight hair and a female with brown curly hair. If you bred these two, would you expect to get any with brown curly hair in the first generation? Explain your answer with four Punnett squares.

Critical Thinking

1. The law of independent assortment applies not only to genetics but also to games of pure chance. Suppose that you have 10 Ping-Pong balls in a jar, each marked with a number from 1 to 10. If you mix them thoroughly and take out one, the chances are 1 in 10 that you will get a particular number, for example 5. If you have two such jars, the chances of drawing a 5 from each jar is 1 in 100 (1 in 10 for each jar, or a total of 1 in 100 for drawing 5, 5):

$$\frac{1}{10} \times \frac{1}{10} = \frac{1}{100}$$

Suppose that you have 5 such jars. What is the chance of drawing the number 31456?

2. Cystic fibrosis is a genetically transmitted condition in which the body produces excess mucus that clogs the lungs and intestines. Persons who are heterozygous (Ff)

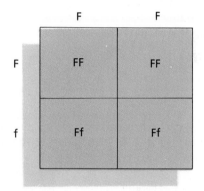

for cystic fibrosis are carriers. Persons who are homozygous recessive (ff) have the disease. The Punnett square above shows the cross between a normal adult and a carrier. Show the Punnett squares for the crosses between each of these four F_1 individuals and another who is a carrier. How many individuals in the F_2 generation will be carriers? How many will be affected by the disease?

Extension

1. Read the article by Julie Ann Miller entitled "Mendel's Peas: A Matter of Genius or Guile?" in *Science News,* February 18, 1984, pp. 108–109. Give an oral report in which you emphasize the problems of analyzing data.

2. Toss a penny and a nickel together 32 times, and record the results as they fall: number of heads, nickel; number of tails, nickel; number of heads, penny; and number of tails, penny. How does this illustrate independent assortment?

Chapter 11 | *Inheritance Patterns and Human Genetics*

This chapter explains principles of inheritance. It examines chromosome behavior, gene mutation, sex-linked traits, and human genetics. Students become aware of how chromosomes control certain disorders in humans.

Correlation Chart

Section Number	Section Objectives	Pages	Features	Section Review
11.1	1. Distinguish between a point mutation and a frame shift mutation.	163–164		1, 2, 5
	2. Name three types of chromosome mutations.			1, 2, 3
	3. Identify two external causes of mutation.			4
11.2	1. Explain how sex chromosomes determine sex inheritance in *Drosophila*.	165–169	Think, p. 168	1, 2
	2. Explain why sex-linked traits appear more often in males than in females.			5
	3. Describe the evidence that indicates that all chromosomes have linkage groups.			4
	4. Define *crossing-over,* and tell how it creates genetic variation.			3
	5. Summarize the procedure involved in constructing a chromosome map.			4
11.3	1. Name three methods of studying human genetics.	170–177	Biology in Process, pp. 172–173	2
	2. Name two disorders that are controlled by a single allele.			3
	3. Distinguish between polygenic traits and traits controlled by multiple alleles.		Biology Lab, p. 177	5
	4. Identify the methods of detecting genetic disorders during pregnancy.			4
	5. Describe the procedure scientists used to identify the location of a gene for manic-depressive illness.			

Resources*

Print

Barrett, James T., ed. *Contemporary Classics in the Life Sciences*. Vol. 1, *Cell Biology*. Philadelphia: ISI Press, 1986.

Kenney, Martin. *Biotechnology: The University-Industrial Complex*. New Haven, Conn.: Yale University Press, 1986.

Audiovisual/Software

Chemistry of Heredity I: Identification of Genetic Material and DNA Structure. Film, 15 min. Britannica.

The Gene. Two filmstrips with cassettes, 16-19 min. each. Hawkhill.

Heredity. Film, 15 min. Britannica.

*Additional references are listed in *Modern Biology, Teacher's Program Guide*.

Section Teaching Guides

11.1 Mutation

Approach

The main objective of this section is to describe the different types of mutations and their causes. The first type, chromosome mutation, occurs during cell division; consequently the discussion of deletion, inversion, translocation, and nondisjunction draws on the discussion of mitosis and meiosis in Chapter 9. The material on gene mutation builds on the concepts of DNA replication and protein synthesis in Chapter 8.

In presenting the text, note that mutations cause a change in phenotype because a change in DNA means that different proteins will be produced. Also point out that although many mutations are harmful, beneficial mutations are an important source of genetic variation.

Concept Framework

Students have probably heard that cigarettes, asbestos, ultraviolet light, and other substances cause cancer. Ask students to think about how these substances actually produce such drastic changes in the body. Use this knowledge to introduce students to the fact that some cancers result from changes in a cell's genetic structure. As they read about the different types of mutations, encourage students to think about how these molecular changes can have far-reaching effects on the body, regardless of what caused the changes.

Emphasis

Make certain that students can distinguish between chromosome and gene mutations, and that they understand how each alters the genetic code.

Guest Speaker

Invite a chemist or other researcher who can explain how the Ames test is conducted and how pharmaceutical companies and other industries use the test to determine possible mutagenic effects of new products.

Section Review Answers

(Continued from p. 164.)
3. Deletion is the loss of part of a chromosome.

Inversion occurs when a broken chromosome piece reattaches itself in the reverse order. In translocation the broken piece reattaches itself to a nonhomologous chromosome.
4. Mutagenic substances include ultraviolet light, cigarette tars, asbestos, some viruses, and atomic radiation.
5. One answer is that more than one codon may code for an amino acid. Thus a change in one nitrogen base may not affect the code. Alternately, the substituted amino acid may not alter the structure or chemical properties of the protein.

11.2 Genetic Patterns

Approach

This section focuses on Thomas Hunt Morgan's research with *Drosophila*, and how it led to scientists' understanding of inheritance patterns. To understand the role of chromosomes in inheritance, students will need to understand the principles of mitosis and meiosis discussed in Chapter 9. In addition, the description of Gregor Mendel's work in Chapter 10 provides important background for this section. Morgan's experiments yielded concrete evidence in support of Mendel's laws of inheritance—dominance and recessiveness, segregation, and independent assortment. In your discussion, note Morgan's use of scientific process skills to arrive at his conclusions.

Concept Framework

Students will be able to make many inferences about inheritance patterns simply by observing the traits of friends and family members. They know, for example, that the population is approximately half male and half female. Tap this knowledge to introduce the concept of inheritance patterns. As they read the section, ask students to consider whether their own observations of inheritance patterns are supported by the principles of inheritance described in the text.

Emphasis

Students should be able to construct Punnett squares that show how sex, sex-linked traits, and linked traits are inherited. They should also understand how calculating crossing-over frequency within a linkage group enables geneticists to construct chromosome maps.

Section Review Answers

(Continued from p. 168.)

3. by redistributing genetic material between chromosome pairs so that traits occur in different combinations
4. by determining the frequency of crossing-over and constructing a chromosome map
5. The square would show a cross between a white-eyed male (X^rY) and a heterozygous red-eyed (X^RX^r) female.

11.3 Human Genetics

Approach

The main objective of this section is to describe some of the ways geneticists have applied knowledge of inheritance patterns to issues in human genetics. The section builds on the concepts of Mendelian genetics (Chapter 10) and inheritance patterns (Section 11.2), and prepares students for Chapter 13, on applied genetics. In presenting the text, refer students to the pedigree charts in Figures 11-8, 11-11, and the feature so that they can see how traits are passed on from generation to generation.

Concept Framework

Students know that some disorders or unusual traits run in families, perhaps in their own. Use this knowledge as a framework to prepare students for understanding the material presented in this section. As students read, ask them to look for information in the text that might enable them to predict which family traits might appear in future generations.

Emphasis

Students should be able to predict the inheritance patterns of single-allele, polygenic, multiple-allele, and sex-linked traits.

Using the Features/Genetic Markers and Manic-Depressive Illness

The main process skills used in this feature are hypothesizing, collecting and interpreting data, and observing. Researchers studying the Amish community in Pennsylvania began with the hypothesis that manic-depressive illness is at least partly caused by genetic factors. The analysis of data on the inheritance of the disease through several generations supported this hypothesis. The identification of a genetic marker on chromosome 11 of many manic-depressives provided further evidence.

In discussing the feature, point out that although no definite conclusions have been reached, the results of the Amish study have generated additional hypotheses. As scientists devise new experiments to test these hypotheses, they will come closer to determining the cause of manic-depressive illness. You might also ask students to consider the implications of this research in terms of society's perception of mental illness.

Field Trip

Arrange for the class to visit the prenatal care unit of a nearby hospital so that students can see the equipment used in fetal examinations.

Section Review Answers

(See p. 176.)

Laboratory Preparation

Background

In this lab students survey a variety of human hereditary traits to produce a personalized phenotypic profile. Students combine information regarding the genetic nature of each trait with each profile, allowing them to speculate about their genotype for each trait. Students collect data reflecting the frequency of each trait within the class and comment on the relationships, if any, between the frequency of a trait, its evolutionary advantage, and its genetic nature.

Preparation

No advance preparation is necessary for this laboratory activity.

Conclusions

1. Answers to the question will vary. Point out to students that the same combination of phenotypes would not necessarily reflect the same combination of genotypes. The same phenotype could result from either a heterozygous or a homozygous dominant genotype.
2. Yes. Although having five fingers is a recessive trait, it is much more common than having six fingers, a trait expressed by a dominant allele.

Inquiry

Students will make suggestions about the adaptative advantage of each trait.

Chapter Review

Vocabulary

1. One map unit is the distance that equals one percentage point of crossing-over.
2. The addition or deletion of a nitrogen base is called a frameshift mutation.
3. A genetic marker is a short DNA segment that can be used to identify a disease.
4. Monosomy involves having only one of a particular chromosome; trisomy involves having three of a particular chromosome.
5. In genetic screening, karyotypes of the prospective parents are prepared, and their blood is tested for enzymes that could suggest genetic abnormalities. In genetic counseling people are informed about genetic problems that could affect potential offspring.

Review

11. Mutations may simply occur by chance, without external cause. They may also be caused by environmental factors called mutagens.
12. when a replicated chromosome pair fails to separate during cell division
13. Male fruit flies have the genotype XY, and females are XX. In birds males are XX, and females are XY. In grasshoppers males are XO, and females are XX.
14. In a cross between a heterozygous red-eyed female ($X^R X^r$), and a red-eyed male ($X^R Y$), all white-eyed offspring were male ($X^r Y$).
15. because they have the same genetic makeup; differences between them may be the result of environmental factors
16. Geneticists have identified a DNA strand called a genetic marker that appears in 96 percent of HD patients.
17. The sickle-cell allele results from a point mutation that causes a base substitution, thus altering the structure of the hemoglobin molecule.
18. genotypes $I^A I^A$ and $I^A i^o$
19. hair, skin, and eye color

20. In the Amish families they tested, scientists identified an abnormal segment of DNA on chromosome 11 of manic-depressives.

Critical Thinking

1.

2. The sequence should be:
TAC, GAG, ATC, GGG, TAT, T.
3. The phenotypic ratio is 3 gray-bodied, long-winged : 1 black-bodied, short-winged. The genotypic ratio is: 1 GGLL : 2 GgLl : 1 ggll.

	GL	gl
GL	GGLL	GgLl
gl	GgLl	ggll

4. The normal sequence of the codon is: GAG (mRNA codon). The mutant is GUG.
5. A sample should include people of both sexes, all races, different ages, and as many ethnic groups as possible.
6. Exercise at high altitudes requires a great deal of oxygen, yet oxygen levels are low at high altitudes. This shortage of oxygen could cause symptoms in heterozygous individuals since they do have some sickle cells.

Chapter 11

Inheritance Patterns and Human Genetics

Introduction

The DNA of chromosomes contains information that regulates inheritance patterns. What exactly are these patterns? Why do some traits run in families, while others seem to appear out of nowhere? In this chapter you will learn how biologists use their knowledge of chromosome behavior to study how inheritance patterns operate.

Chapter Outline

11.1 Mutation
- *Chromosome Mutations*
- *Gene Mutations*
- *Mutagens*

11.2 Genetic Patterns
- *Sex Determination*
- *Sex Linkage*
- *Linkage Groups*
- *Crossing-Over*
- *Chromosome Mapping*

11.3 Human Genetics
- *Studying Human Genetics*
- *Human Genetic Traits*
- *Detecting Genetic Disorders*

Chapter Concept

As you read, pay attention to the scientific processes that were used to discover inheritance patterns and to the way these processes are applied to issues of human genetics.

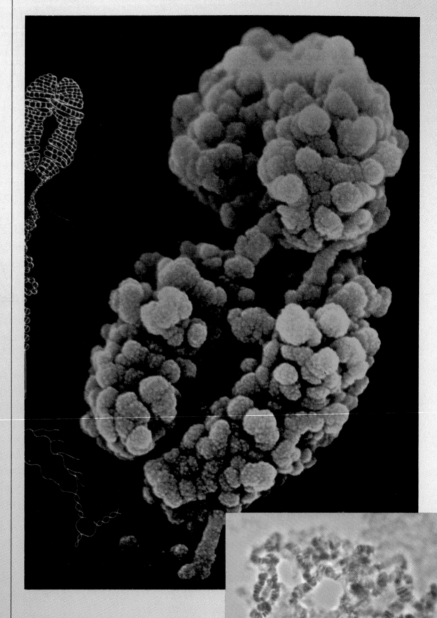

Color-enhanced SEM of human chromosome pair (×125,000). Inset: Chromosomes in *Drosophila* salivary gland (×2,048).

For additional teaching information see pp. T11A–T11D.

11.1 Mutation

As you learned in Chapter 8, DNA ultimately directs the production of the proteins that affect an organism's metabolism and development. Therefore any change in DNA will affect the production of proteins and in turn result in a new phenotype. A change in DNA is called a **mutation**. Mutations can involve entire chromosomes or specific genes. They may take place in any cell. **Germ cell mutations** occur in sex cells, such as eggs and sperm. They do not affect the organism itself but are passed on to offspring. **Somatic mutations** take place in body cells. They are passed on to daughter cells through mitosis.

The majority of mutations are harmful. Most of these are called lethal mutations because they prevent the individual from developing much beyond the zygote stage. However, some mutations result in phenotypes that have an evolutionary advantage and are therefore beneficial. These mutations ultimately provide the variation upon which natural selection acts.

Chromosome Mutations

Chromosome mutations often occur during cell division. These mutations are either changes in the structure of a chromosome or loss of an entire chromosome. **Deletion** occurs when a piece of a chromosome breaks off. All of the information on that piece is lost. **Inversion** occurs when a piece breaks from a chromosome and reattaches itself to the chromosome in the reverse orientation. If a broken piece attaches to a nonhomologous chromosome, the result is a **translocation**. Figure 11-1 shows these mutations.

Section Objectives

- *Distinguish between a point mutation and a frame shift mutation.*

- *Name three types of chromosome mutation.*

- *Identify two external causes of mutation.*

Tip: Why does an X-ray technician stay behind a lead-lined screen while taking X rays of patients? (Overexposure to X rays can result in cell mutations in any body organ and can lead to cancer.)

Figure 11-1 In chromosome mutations a piece of chromosome breaks off. In some cases the piece may reattach in a reversed position, and in others it may attach to a nonhomologous chromosome.

Deletion   Inversion   Translocation

Base substitution

DNA ➡ mRNA ➡ Amino acid

Normal

G
A ➡ G
A   A ➡ Glutamic
A   A   acid

Substitution

G   G
T ➡ U ➡ Valine
A   A

Frameshift

DNA ➡ mRNA ➡ Amino acid

Normal

T   T
G   G
G ➡ G ➡ Trypto-
G   G   phane
A ➡ A ➡ Glutamic
G   G   acid

Frameshift

T
G   G
G ➡ G ➡ Glycine
G   G
A   A
G   G

Figure 11-2 In a point mutation one nitrogen base changes to produce a different amino acid. A frameshift mutation also disturbs this process of protein synthesis.

Info: Atmospheric ozone acts as a barrier to solar ultraviolet radiation. Therefore the destruction of the ozone layer by fluorocarbon compounds may have serious genetic ramifications.

Answers to Section Review
1. a change in DNA
2. a somatic mutation; a germ cell mutation
(See pp. T11A–T11D.)

Another kind of chromosomal mutation, called **nondisjunction,** occurs when a replicated chromosome pair fails to separate during cell division. When nondisjunction occurs, one daughter cell receives an extra copy of a chromosome, and the other daughter cell lacks that chromosome entirely. Some human disorders resulting from nondisjunction are discussed in Section 11.3.

Gene Mutations

Gene mutations may involve a single nitrogen base within a codon or larger segments of DNA. A codon, as you recall from Chapter 8, consists of three nitrogen bases that cause a specific amino acid to be inserted into a polypeptide during protein synthesis. The substitution, addition, or removal of a single nitrogen base is called a **point mutation.** As Figure 11-2 shows, if one nitrogen base is substituted for another, the altered code may signal the insertion of a different amino acid into the protein chain. The protein will thus have a different chemical structure. The addition or deletion of a nitrogen base is a point mutation called a **frameshift mutation.** An addition or deletion causes the genetic message to be read out of sequence. The result is usually the inability of the gene to code for the proper amino acids. The bottom of Figure 11-2 illustrates the effects of a frameshift mutation.

Mutagens

What causes mutations? Some mutations are chemical mishaps with no apparent external cause. As in all complex systems, sometimes by chance things go wrong. Often, however, mutations are caused by **mutagens**—environmental factors that damage DNA. For example, overexposure to sunlight can cause skin cancer, for ultraviolet light has a mutagenic effect on the DNA in skin cells. Other mutagens, some of which cause cancer, include cigarette tars, asbestos, and certain viruses. Researchers often use a procedure called the **Ames test** to identify mutagenic substances.

Radiation causes mutations in both germ cells and somatic cells. The children of many survivors of the atomic bomb attacks on Hiroshima and Nagasaki in 1945 were stillborn or suffered birth defects. Radiation increased somatic mutations in the bone marrow cells of many survivors, resulting in a higher incidence of leukemia, a form of cancer.

Section Review

1. What is a *mutation?*
2. Which type of mutation affects the organism but not its offspring? Which type affects offspring?
3. List and describe three types of chromosome mutation.
4. List three examples of mutagenic substances.
5. Why might base substitutions frequently go unnoticed?

11.2 Genetic Patterns

In the early 1900s geneticist Thomas Hunt Morgan of Columbia University began a series of breeding experiments with *Drosophila melanogaster*, the common fruit fly. Many characteristics of *Drosophila* make them ideal subjects. They are easy to maintain in the laboratory, have a generation time of only 10 to 15 days, produce hundreds of individuals from each mating, and have easily distinguishable physical characteristics. Also, *Drosophila* has only eight chromosomes, making it easier to determine the relationship between chromosomes and inheritance patterns.

Sex Determination

Morgan observed that of *Drosophila*'s four pairs of homologous chromosomes, one pair was different in males and females. In females the chromosomes of this pair were identical. In males the chromosomes were different. One chromosome looked like those of the pair of females, but the other was shorter and hook-shaped. Morgan called the large chromosomes X chromosomes. He called the short, hook-shaped chromosome the Y chromosome. Morgan correctly hypothesized that X and Y chromosomes are **sex chromosomes,** chromosomes that determine an individual's sex. All the other chromosomes, those not included in sex determination, are called **autosomes.**

Recall from Chapter 9 that homologous chromosomes segregate during meiosis I. The sex chromosomes pair and segregate in the same manner as other chromosomes. Female gametes each receive one X chromosome. Male gametes receive either an X or a Y chromosome. If an egg is fertilized by a sperm containing an X chromosome, the resulting zygote will be XX, and the new individual will be female. If an egg is fertilized by a sperm containing a Y chromosome, the resulting zygote will be XY, and the new individual will be male.

As shown in Figure 11-3, these combinations of chromosomes make it probable that 50 percent of the offspring of any mating will be male and 50 percent female. In most organisms the males are XY, and the females are XX. In birds, however, the males are XX, and the females are XY. In grasshoppers, the males, with only an X chromosome, are XO, and the females are XX.

Sex Linkage

The discovery of sex chromosomes led Morgan to hypothesize that some traits were always associated with one sex or the other. His experiments confirmed this hypothesis. Although most fruit flies have red eyes, Morgan once observed a male fruit fly with white eyes. When he crossed a female with this male, the results followed Mendel's prediction: the F_1 generation all had red eyes.

Section Objectives

- *Explain how sex chromosomes determine sex inheritance in* Drosophila.

- *Explain why sex-linked traits appear more often in males than in females.*

- *Describe the evidence that indicates that all chromosomes have linkage groups.*

- *Define* crossing-over, *and tell how it creates genetic variation.*

- *Summarize the procedure involved in constructing a chromosome map.*

Figure 11-3 A female fruit fly has XX sex chromosomes, and a male, XY sex chromosomes. As this Punnett square shows, half of the offspring will probably be female, and half will be male.

	X	Y
X	XX	XY
X	XX	XY

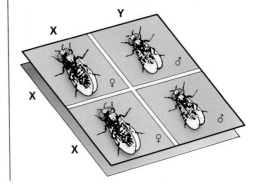

Inheritance Patterns and Human Genetics **165**

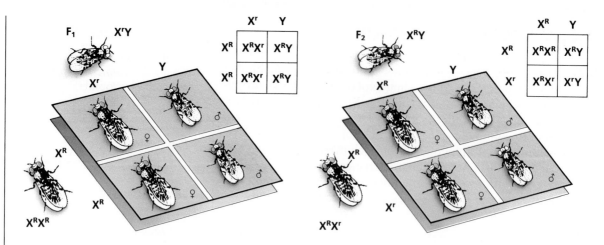

	X^r	Y
X^R	$X^R X^r$	$X^R Y$
X^R	$X^R X^r$	$X^R Y$

	X^R	Y
X^R	$X^R X^R$	$X^R Y$
X^r	$X^R X^r$	$X^r Y$

Figure 11-4 Eye color is a sex-linked trait in fruit flies carried on the X chromosome—R for dominant red and r for white. Determine the sex and eye color of the F_1 and F_2 generations.

Info: Dominance assures an organism's adaptability since a trait usually dominates because of success through natural selection.

Figure 11-5 When traits are linked, their probable distribution is the same as that for body color and wing length in the fruit fly, as shown here.

	GL	gl
GL	GGLL	GgLl
gl	GgLl	ggll

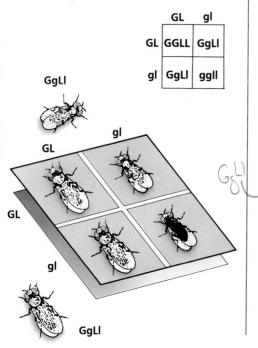

Morgan next crossed members of the F_1 generation and produced the expected ratio of three red-eyed flies to one white-eyed. However, all the white-eyed flies were male. Why weren't any females white-eyed? Morgan hypothesized that the allele for red eyes was carried on the X chromosome. Figure 11-4 shows his reasoning. Assume the X chromosome carries the gene for eye color, either X^R or X^r. If an $X^R X^R$ female is crossed with an $X^r Y$ male, all of the F_1 females will be $X^R X^r$, and all of the F_1 males $X^R Y$. In the F_2 generation half the females will have the $X^R X^R$ genotype, and half the $X^R X^r$ genotype. All will have red eyes. Half the F_2 males will be $X^R Y$, or red-eyed; half will be $X^r Y$, or white-eyed. Morgan concluded that eye color in *Drosophila* was a **sex-linked trait**—one determined by alleles on the sex chromosomes.

Linkage Groups

Genes are found on chromosomes. Since there are thousands more traits than chromosomes, each chromosome must carry many genes. The group of genes located on one chromosome is a **linkage group**. These genes are usually inherited together.

Morgan demonstrated the existence of linkage groups by studying *Drosophila*. In *Drosophila* the allele for gray body (G) is dominant to that for black body (g), and the allele for long wings (L) is dominant to that for short wings (l). Mendelian genetics predicts that when strains are crossed, the F_2 generation will have a phenotypic ratio of $9:3:3:1$. When Morgan performed the cross GGLL × ggll, the F_2 generation showed the $3:1$ phenotypic ratio seen in Figure 11-5, not the Mendelian ratio.

Morgan correctly hypothesized that the genes for body color and wing size are in the same linkage group and are inherited together. By doing many genetic crosses within a species it is possible to determine the number of linkage groups for that species. The number always equals the number of chromosomes. *Drosophila*, with 4 pairs of chromosomes, has 4 linkage groups. A human, with 23 pairs of chromosomes, has 23 linkage groups.

Crossing-Over

Morgan noticed that sometimes not all the alleles in a linkage group were inherited together. For example, in *Drosophila,* the allele G for gray bodies is dominant to g for black bodies, and R for red eyes is dominant to r for purple eyes. Morgan crossed a ggrr fly with a GgRr fly. In the F_2 generation 151 offspring had a gray body and red eyes, and 131 had a black body and purple eyes, almost a 1:1 ratio. Morgan expected this result from his studies of linkage groups. However, 8 offspring had gray bodies and purple eyes, and 10 had black bodies and red eyes.

Morgan realized that these results could have occurred only if the alleles somehow had changed or become rearranged. Mutations could have caused the changes, but mutations usually occur in only one individual out of tens of thousands. The alleles must therefore have become rearranged. Morgan inferred that this rearrangement had taken place through crossing-over, the exchange of alleles between homologous chromosomes. Figure 11-6 shows how crossing-over accounts for the individuals in this cross that are not predicted by Mendelian inheritance laws.

Chromosome Mapping

Crossing-over occurs more often between some alleles than between others. Since most alleles occupy a fixed place on a chromosome, it follows that the smaller the distance between two alleles, the greater the likelihood that these alleles will not be separated by crossing-over. Alfred H. Sturtevant, Morgan's student, used crossing-over data to construct a **chromosome map**—a diagram of allele positions on a particular chromosome.

Tip: Students may be confused by the concept of crossing-over. Use different colors for alleles in a graphic presentation of this phenomenon.

Figure 11-6 When traits are not distributed according to the expected ratio in offspring, crossing-over of alleles may have occurred.

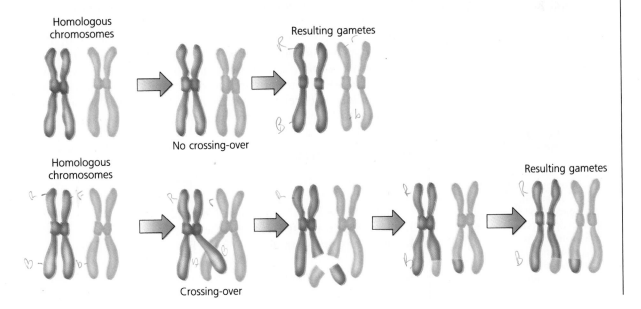

Homologous chromosomes
Resulting gametes
No crossing-over

Homologous chromosomes
Resulting gametes
Crossing-over

Figure 11-7 Geneticists study the rate of crossover between alleles to construct a chromosome map, which shows the relative position of an allele on a chromosome.

Think

In *Drosophila* the alleles for red eyes (R) and brown body (B) are dominant to the alleles for white eyes (r) and yellow body (b). Suppose Morgan crossed a heterozygous red-eyed, brown-bodied female (RrBb) with a white-eyed, yellow-bodied male (rrbb). Of 500 offspring, 7 have red eyes and yellow bodies, and 8 have white eyes and brown bodies. How many map units apart are the two alleles?

Answer to Think
Since there was a 3 percent crossover (15 out of 500 offspring), the two alleles are 3 map units apart.

Answers to Section Review
1. Sex chromosomes determine the sex of an organism; autosomes are nonsex chromosomes.
2. An individual has a 50 percent chance of inheriting either an X or a Y chromosome from the male. This chromosome determines the sex of the individual.
(See pp. T11A–T11D.)

To construct a chromosome map, researchers compare the frequency of crossover for three characteristics, such as eye color, body color, and wing shape. Suppose, for example, that in crosses of *Drosophila,* wing shape and eye color were linked 90 percent of the time. Then the alleles for the two characteristics would be separated during crossing-over 10 percent of the time. If wing shape and body color were linked 93 percent of the time, they would be separated through crossing-over 7 percent of the time. If eye color and body color were linked 83 percent of the time, they would be separated through crossing-over 17 percent of the time.

Geneticists use such data when constructing a chromosome map. Two alleles that are separated by crossing-over 1 percent of the time are considered to be 1 **map unit** apart. In our example the alleles for wing shape and eye color are 10 map units apart, the alleles for wing shape and body color are 7 map units apart, and the alleles for eye color and body color are 17 map units apart. With this data it is possible to establish the relative position of these three alleles on a chromosome. By doing many such comparisons it is possible to determine the actual position of the alleles on a chromosome. Figure 11-7 shows the position of three alleles on *Drosophila* chromosome number 2. It also shows how data on crossover rates are used to determine the position of the three chromosomes relative to each other. Scientists have used this procedure to map the chromosomes of numerous species.

Section Review

1. Distinguish between *sex chromosomes* and *autosomes*.
2. How does sex chromosome inheritance result in an approximate 50 : 50 ratio of males to females?
3. How does crossing-over increase genetic variability?
4. How do geneticists determine the position of an allele on a chromosome?
5. Construct a Punnett square to show how a white-eyed female *Drosophila* could be produced.

Writing About Biology

Inborn Errors

The following selection is an excerpt from the book The Gene Doctors *by Yvonne Baskin.*

In recent decades researchers have come to realize that genes have a much wider influence on our health than causing . . . individually uncommon inborn disorders and birth defects. The interplay of genes and environmental factors—from viruses and cigarette smoke to radiation, asbestos, and diet—is responsible for some of our most common afflictions: diabetes, heart disease, cancer, mental illness, and many of the chronic and debilitating diseases of later life.

No one has yet isolated the gene, or set of genes, involved in most of these diseases, or pinpointed all the environmental factors that trigger or modulate them. But, as is the case for cleft lip, the evidence for genetic involvement is clear in the statistics and the frequency with which certain disorders tend to crop up in families.

If one twin develops insulin-dependent diabetes, for instance, the other has a fifty-fifty chance of getting it, too. In the case of non-insulin-dependent diabetes, the risk goes up to more than 90 percent that if one identical twin develops the disease, both twins will suffer it. The odds are 45 percent that both twins will develop schizophrenia if one twin has it. Certain types of cancer and heart disease, which account for almost two thirds of the deaths in the United States, have long been known to cluster in families.

The interaction of genes and environment in disease should come as no surprise. The same subtle mix of factors shapes most of the normal characteristics of living things. A grove of cedars growing near the sea, constantly buffeted by ocean breezes, may grow shrublike and crooked, branched only on the leeward side. Inland, out of the wind, genetically identical cedars will grow into straight and substantial trees, evenly branched all around. A child may inherit the tendency to grow to a taller-than-average height and all the genes needed to carry out that tendency. Yet poor nutrition, infectious diseases, or certain drugs can thwart this inborn programming.

In the past, even when afflictions seemed to run in the family, it was often impossible to tell whether it was because of physical heredity or the life-style and fortunes of the clan. Environmental factors acting alone can often produce results that look just like those produced by abnormal genes: Retardation can be caused by injuries or oxygen deprivation at birth instead of faulty genes or chromosomes; anemia by iron deficiency rather than defective hemoglobin; depression or psychosis by grief, stress, or drugs rather than inborn chemistry.

Statistics and studies of twins and adoptees have given scientists their first clues, confirming a role for genes in many of our most common diseases. Now the techniques of molecular biology are allowing them to begin sorting out the interwoven roles of genes and environmental factors and to devise interventions at the proper level, whether this means changing family-life-styles or replacing faulty proteins—or, eventually, the genes that produced them.

Write
- **Write a paragraph explaining why replacing faulty genes would be more effective than traditional methods of medical treatment.**

From The Gene Doctors: Medical Genetics at the Frontier *by Yvonne Baskin. Copyright © 1984 by Yvonne Baskin. Reprinted by permission of William Morrow & Co.*

- *Name three methods of studying human genetics.*

- *Name two disorders that are controlled by a single allele.*

- *Distinguish between polygenic traits and traits controlled by multiple alleles.*

- *Identify the methods of detecting genetic disorders during pregnancy.*

- *Describe the procedure scientists used to identify the location of a gene for manic-depressive illness.*

Table 11-1   Inheritance Patterns of Human Traits

Single Allele, Dominant
- Huntington disease
- Achondroplasia (dwarfism)
- Cataracts
- Polydactyly (extra fingers or toes)

Single Allele, Recessive
- Albinism
- Cystic fibrosis
- Phenylketonuria
- Hereditary deafness

Sex-linked
- Color blindness
- Hemophilia
- Muscular dystrophy
- Icthyosis simplex (scaly skin)

Polygenic
- Skin, hair, and eye color
- Foot size
- Nose length
- Height

Multiple Alleles
- ABO blood groups
- Rh blood factor

11.3 Human Genetics

The 23 pairs of human chromosomes contain 35,000 to 40,000 genes, about 7 times more than are found in *Drosophila*. Consequently human inheritance patterns are significantly more complicated. Nevertheless, the principles you read about in Chapter 10 and in this chapter apply to human beings. In this section we will study some human genetic patterns. Most of the examples are diseases or disorders, but keep in mind that the vast majority of genes control beneficial traits. Geneticists concentrate on disease-carrying genes because they are easily traced through generations and because they are of great concern to people. Table 11-1 lists some well-known human traits and their pattern of inheritance.

Studying Human Genetics

The study of human genetics poses difficulties that Morgan didn't face in studying *Drosophila*. For example, it may take 75 years to produce three generations of humans, and each pair of humans will produce only a few individuals. In addition, ethical concerns prevent scientists from using the same procedures to study humans that they use to study other organisms. How then do scientists gather evidence about human inheritance?

Population Sampling

One tool for studying human genetics is **population sampling,** in which researchers select a small number of individuals that represent the whole population. Selecting members of the sample according to carefully formulated statistical rules helps ensure that the study will yield accurate results.

For instance, geneticists used population sampling to estimate the percentage of people in the United States who could taste the chemical phenylthiocarbamide (PTC). "Tasters" detect a bitter taste to PTC, while "nontasters" do not detect any taste. By testing a few thousand randomly selected individuals the geneticists have estimated with a high degree of certainty that in the entire United States 65 percent of the people are tasters and 35 percent are nontasters.

Twin Studies

Geneticists also study identical twins to distinguish between genetic and environmental influences on specific traits. Identical twins are those that develop from the union of a single egg and sperm. Because identical twins have the same genetic inheritance, their differences may result from environmental influences such as home life or education. Genetic and environmental effects are even more easily seen if the twins are raised by different families. Twin studies are controversial among scientists, however, and their results must be analyzed with care.

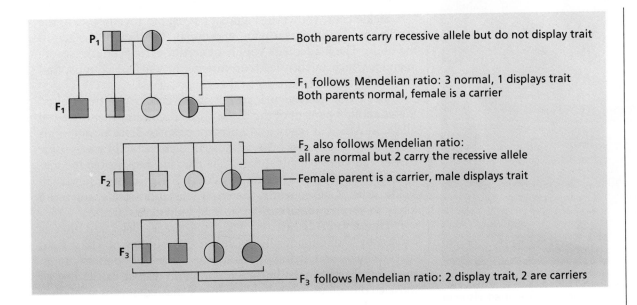

Both parents carry recessive allele but do not display trait

F_1 follows Mendelian ratio: 3 normal, 1 displays trait
Both parents normal, female is a carrier

F_2 also follows Mendelian ratio:
all are normal but 2 carry the recessive allele

Female parent is a carrier, male displays trait

F_3 follows Mendelian ratio: 2 display trait, 2 are carriers

Pedigree Studies

Geneticists may also trace a **pedigree,** a family record that shows how a trait is inherited over several generations. Analysis of inheritance patterns produces a pedigree like the one shown in Figure 11-8. A pedigree often reveals a **carrier,** someone who is heterozygous for a trait. Carriers do not have the trait themselves but may pass it on to their offspring.

Human Genetic Traits

Studies of human genetics, including population sampling, twin studies, and pedigree analysis, have identified a number of patterns of inheritance. More modern genetic techniques, which are discussed in Chapter 12, have added to our knowledge of human genetics.

Single-Allele Traits

Geneticists have identified about 200 human traits that are coded for by single, dominant alleles. About 250 other traits are coded for by homozygous recessive alleles. Some examples of codominance also exist. Examples of single-allele disorders include sickle-cell disease and Huntington disease.

    Sickle-cell disease, often called sickle-cell anemia, is a single-allele disorder that operates in a codominant system. The dominant allele A produces normal hemoglobin that results in round erythrocytes. The codominant allele A′ codes for abnormal hemoglobin and results in sickle-shaped erythrocytes. AA individuals have normal hemoglobin and normal erythrocytes. Heterozygous individuals (AA′) have both normal red blood cells and sickle-shaped cells. A′A′ individuals only have sickle cells.

Figure 11-8 By studying inheritance patterns of a certain trait over several generations of a family, geneticists sometimes can determine who is a carrier of the trait.

Tip: You may wish to discuss random selection methodology, perhaps introducing students to the statistical *p* for probability value with its associated confidence levels.

Info: Sickling results when bonding patterns in hemoglobin are changed as the amino acid sequence changes.

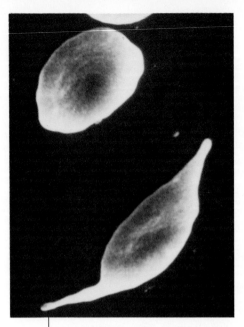

Sickle cells, shown in Figure 11-9, clump together, clog the capillaries, causing great pain, and impair the flow of oxygen to the body. The inadequate supply of erythrocytes produces severe anemia, which in turn leads to fatigue, headaches, cramps, and eventually to the failure of vital organs.

Sickle-cell disease primarily affects Blacks. In the United States about 9 percent of the Black population carry the sickle-cell allele. In Africa as many as 40 percent of some Black populations carry it. In these areas the sickle-cell allele actually gives some carriers an adaptive advantage. The parasite responsible for malaria, a leading cause of death in western Africa, cannot survive in sickle cells. Consequently, heterozygous individuals are more likely to survive malaria than normal individuals are.

Huntington disease, or HD, is a single-allele trait that is caused by a dominant allele. The first symptoms of HD—mild forgetfulness and irritability—appear when the victims are in their thirties or forties. In time Huntington disease causes loss of muscle control, uncontrollable physical spasms, severe mental illness, and eventually death.

Because the HD allele is dominant, any person who inherits the allele will develop the disease. Unfortunately, most carriers do

Figure 11-9 Sickle-cell disease is a genetically transmitted disorder that causes erythrocytes to become deformed and to clump together.

Biology in Process

Genetic Markers and Manic-Depressive Illness

Scientists long suspected that some forms of mental illness had a biochemical basis, but they lacked evidence to back up their hypothesis. In 1987, however, scientists announced that they had found a link between a genetic abnormality and a mental disorder.

Between 1 million and 2 million Americans have some type of manic-depressive illness. Manic-depressives suffer severe mood swings, ranging from elation to despair.

Researchers studying Amish families near Lancaster, Pennsylvania, observed a high incidence of manic-

depression. Of the 81 members of three families being studied, 14 had manic-depressive illness and 5 had other forms of depression.

The Amish are genetically isolated—that is, they seldom marry out of their community. They also keep detailed family records, which enabled geneticists to trace the disease through several generations. In addition, alcoholism, drug abuse, violence, and other behaviors that might mask the effects of manic-depressive illness are rare among the Amish.

The scientists obtained blood samples from all family members. They then isolated not know they have the disease until after they have had children. Thus the disease is unknowingly passed on from one generation to the next. Recently, however, geneticists discovered a **genetic marker**—a short section of DNA that indicates the presence of an allele that codes for a trait—for the HD allele. People who possess this genetic marker have a 96 percent chance of developing HD. By conducting a simple test on sample cells, geneticists are now able to inform a person of the presence of the marker before he or she has children.

Polygenic Traits

In humans characteristics are usually controlled by a single gene, but sometimes a characteristic may be controlled by several genes. A trait that is controlled by two or more genes is called a **polygenic trait**. Skin color, for example, is determined by the additive effect of from four to seven genes. Each contributes a certain amount of the protein melanin, a color pigment. The more melanin produced, the darker the skin; the less melanin produced, the lighter the skin. A similar mechanism determines eye color. Light blue eyes have little melanin. Very dark brown eyes have a great deal of melanin.

Answer to Biology in Process
It facilitated tracing inheritance of the disease and reduced the possibility of other genetic causes.

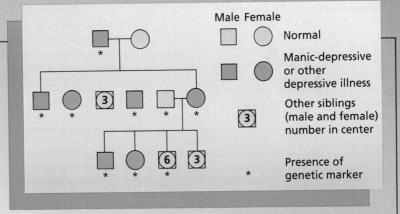

sections of DNA from the white blood cells. When they compared the DNA of normal individuals with that of manic-depressives, they saw that a gene segment on chromosome 11 of the manic-depressives was different.

Researchers have not yet located a specific gene for manic-depressive illness. However, they know that a gene on chromosome 11 is involved in the production of dopamine, a chemical in the brain that affects mood. Some scientists hypothesize that a defect in this gene may create a predisposition to the disease or cause the disease.

Some studies of the inheritance of manic-depression reveal no genetic abnormality. Also, of the Amish with abnormal DNA 37 percent did not have the disease. Researchers concluded that several genes and environmental factors contribute to the development of manic-depression.

An understanding of the genetic mechanisms of illness should enable scientists to predict which individuals are at risk and should help scientists develop more effective treatments.
■ How did the genetic isolation of the Amish help scientists collect and interpret data?

Inheritance Patterns and Human Genetics   **173**

Table 11-2 Possible Genotypes and Phenotypes

Genotype	Blood Type
I^A I^A	A
I^A i^O	A
I^B I^B	B
I^B i^O	B
I^A I^B	AB
i^O i^O	O

Multiple-Allele Traits

Some human traits are controlled by multiple alleles, three or more alleles of the same gene that code for a single trait. For example, in humans there are three alleles that code for blood type: I^A, I^B, and i^O. Any individual has two of these alleles, which together make up the gene for blood type. The alleles I^A and I^B are codominant, and both are dominant to the i^O allele. Study Table 11-2 to see which combination of alleles determines each blood type.

The kind of antigen or antigens produced by a specific combination of alleles determines blood type. An **antigen** is a substance that causes the body to produce an antibody. The presence of antigen A or B determines the type of blood a person can receive in a transfusion. Mixing incompatible types of blood causes the cells to clump together, or agglutinate. For example, type O blood has no antigens and contains A and B antibodies. Therefore people with type O blood can receive blood only from donors with type O blood.

Sex-Linked Traits

Certain human traits are sex linked; that is, the alleles for these traits appear only on the X chromosome. Males have only one X chromosome. Since no complementary portion of the Y chromosome exists, any single recessive allele on an X chromosome will be expressed because it cannot be masked by a dominant allele. Males are therefore far more likely than females to express recessive sex-linked alleles.

Color blindness is a recessive sex-linked disorder in which an individual cannot distinguish between certain colors. Many kinds of color blindness exist, the most common being the inability to distinguish red from green. About 8 percent of males are color blind. The condition is rare in females. Figure 11-10 presents test patterns for different types of color blindness.

Figure 11-10 Color blindness, a recessive sex-linked trait, affects about 8 percent of males. The patterns test for different types of color blindness: red–green (left and center) and blue–yellow (right). An individual who cannot see the characters in the pattern may have that type of color blindness.

Another recessive sex-linked trait that is found primarily in males is **hemophilia.** The blood of hemophiliacs lacks a protein that is essential for clotting. Though very different in their manifestations, both color blindness and hemophilia are inherited through the same pattern. Study Figure 11-11 to see how a recessive, sex-linked allele is inherited.

Sex-Influenced Traits

The presence of male or female sex hormones influences the expression of certain human traits, called **sex-influenced traits.** The alleles that code for most of these traits are on the autosomes. Males and females have the same alleles, but the trait is expressed in only one sex. Pattern baldness is a familiar sex-influenced trait. The allele for baldness, B, is dominant in males but recessive in females. Both men and women with BB alleles will eventually lose their hair. A heterozygous woman (BB′) will not lose her hair, but a heterozygous man (BB′) will. These differences result from the presence of particular sex hormones.

Nondisjunction

Some genetic disorders result from nondisjunction. Remember from Section 11.1 that nondisjunction is the failure of chromatids to separate during cell division. When this occurs during meiosis II, the result may be a sperm or an egg cell with either an extra or a missing chromosome. In humans, for example, if nondisjunction occurs during sperm formation, one sperm cell may have 22 chromosomes, and another may have 24. If one of these gametes combines with a normal egg, the zygote will have either 45 or 47 chromosomes. A zygote with 45 chromosomes has only one of a particular chromosome, a condition called **monosomy.** A zygote with 47 chromosomes has three of a particular chromosome, a condition called **trisomy.**

An extra chromosome 21, for example, results in **Down syndrome,** a disorder characterized by mental retardation, a fold of skin above the eyes, and weak muscles. Nondisjunction also affects sex chromosomes. Klinefelter syndrome results from the trisomic genotype XXY. Klinefelter individuals may be mentally retarded and have low fertility. Turner syndrome is a monosomic condition with the genotype XO. An XO female is characterized by immature physical development, sterility, and a webbed neck.

Detecting Genetic Disorders

A person with a family history of genetic disorders may wish to undergo genetic screening before having children. **Genetic screening** is an examination of a person's genetic makeup. It involves making a karyotype. A **karyotype** is a picture of an individual's chromosomes. The individual's blood is also tested for the presence or absence of certain enzymes. These procedures can reveal potential genetic disorders that might be passed on to children.

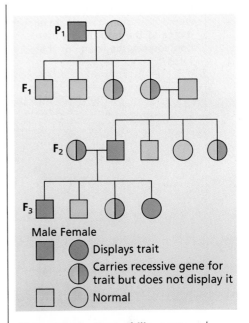

Figure 11-11 Hemophilia, a sex-linked trait, is expressed primarily in males. Most females with the trait are heterozygous and therefore are carriers.

Figure 11-12 By using ultra-sound physicians can create an image of the fetus in the uterus. Can you see the image of the fetus?

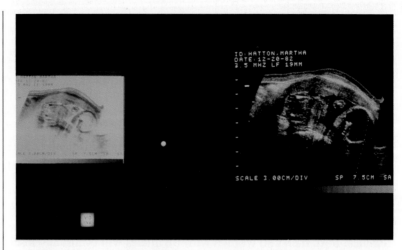

Figure 11-12 By using ultra-sound physicians can create an image of the fetus in the uterus. Can you see the image of the fetus?

To prepare a karyotype, a scientist removes chromosomes from a cell, then stains and photographs them. Each chromosome is cut from the enlarged photo and matched with a photograph of its homologue. Scientists can then spot chromosomal abnormalities. Couples at genetic risk may wish to undergo **genetic counseling,** a type of counseling that informs them about problems that could affect their offspring.

Physicians can also diagnose more than 200 genetic disorders in the fetus through procedures such as amniocentesis, fetoscopy, ultrasound, and chorion villi sampling. In **amniocentesis** the physician uses a needle and syringe to remove a small amount of amniotic fluid from the placenta, the sac that surrounds the fetus. The physician makes a karyotype and analyzes cells that have sloughed off from the fetus and float in the amniotic fluid. In **fetoscopy** the physician inserts a tiny camera through a small incision in the uterus. The physician can observe the fetus's development and take skin and blood samples for analysis. Physicians may also analyze fetal development by using **ultrasound,** a technique in which high-frequency sound waves bounce off the fetus, forming an image. Figure 11-12 shows such an image, called a sonogram. In **chorion villi sampling** the physician analyzes a sample of the chorion villi, which grow between the mother's uterus and the placenta. Since the villi have the same genetic makeup as the fetus, the physician is able to detect genetic disorders.

Section Review

1. Distinguish between *sex-linked* and *sex-influenced traits*.
2. What methods do geneticists use to study human genetics?
3. How does an individual who is heterozygous for the sickle-cell gene benefit in equatorial regions?
4. List three genetic tests that can be done on the fetus during pregnancy.
5. If a couple has four children with different blood types, what are the genotypes of the parents' blood types?

Answers to Section Review
1. Sex-linked traits are carried on the X chromosome. Sex-influenced traits are carried on autosomes, and their expression is controlled by sex hormones.
2. population sampling, twin studies, and pedigree studies
3. Being heterozygous improves the chances of surviving malaria, a leading cause of death in equatorial areas.
4. amniocentesis, fetoscopy, ultrasound, chorion villi sampling
5. One parent is $I^A i^O$, and the other is $I^B i^O$.

Laboratory

Exploring Hereditary Traits

Objective
To determine your possible genotypes for some inherited traits

Process Skills
Observing, communicating, inferring

Materials
Pencil, paper, phenylthiocarbamide paper

Method
1 Make a table with three columns and nine rows. Head the columns *Trait, Phenotype,* and *Possible Genotypes.* Down the first column list the ten traits shown below.
2 In the second column of the table, you will write *yes* or *no* depending on whether or not you possess each trait. Work with a partner. For traits that you cannot observe directly, ask your partner for help. Each of the traits listed is controlled by a dominant allele.

- *Tongue-rolling* (R): Stick out your tongue and try to roll up the sides so that the tongue forms a U-shape.
- *Free earlobe* (F): The lobe of the ear hangs freely below the point of attachment to the head.
- *Widow's peak* (W): The midpoint of the hairline along the front of the forehead points downward.
- *Straight thumb* (N): When extended from the palm of the hand, the top segment of the thumb forms a straight line with the bottom segment. (See figure.)
- *Straight little finger* (S): The last segment of the little finger forms a straight line with the rest of the finger rather than bending toward the ring finger.
- *Left-over-right thumb crossing* (L): When the hands are folded in a natural fashion, the left thumb crosses over the right.

- *Chin cleft* (C): The center of the chin has an indentation resembling a deep dimple.
- *Mid-digital hair* (H): Hair is present on the middle section of any of the fingers.
- *Six-fingers* (F): Six fingers are present on either hand (or were present at birth).

3 In the third column of the table, record your possible genotypes for each trait, using the appropriate symbols.
4 Compare your traits with those of your classmates. Note the trait or traits that are most prevalent among your classmates.

Conclusions
1 Do any two people in the class have exactly the same combination of phenotypes for the traits studied?
2 Is there evidence that a trait shared by most of the population is not controlled by a dominant allele?

Inquiry
Review all the traits studied. Then discuss with the class the possible adaptive advantage of each of the traits.

Chapter 11 Review

Vocabulary

Ames test (164)	frameshift mutation (164)	karyotype (175)	sex chromosome (165)
amniocentesis (176)	genetic counseling (176)	linkage group (166)	sex-influenced trait (175)
antigen (174)		map unit (168)	
autosome (165)	genetic marker (173)	monosomy (175)	sex-linked trait (166)
carrier (171)	genetic screening (175)	mutagen (164)	sickle-cell disease (171)
chorion villi sampling (176)	germ cell mutation (163)	mutation (163)	somatic mutation (163)
		nondisjunction (164)	
chromosome map (167)	hemophilia (175)	pedigree (171)	translocation (163)
	Huntington disease (172)	point mutation (164)	trisomy (175)
deletion (163)		polygenic trait (173)	
Down syndrome (175)	inversion (163)	population sampling (170)	ultrasound (176)
fetoscopy (176)			

1. What is a map unit?
2. What is a frameshift mutation?
3. Define a genetic marker.

4. Distinguish monosomy from trisomy.
5. Distinguish between genetic screening and genetic counseling.

Review

1. The addition or removal of a single nitrogen base results in (a) nondisjunction (b) trisomy or monosomy (c) a frameshift mutation (d) a translocation.
2. A point mutation involves substitution, addition, or deletion of a (a) codon (b) gene (c) DNA strand (d) nitrogen base.
3. The advantages of using *Drosophila* in genetic studies do not include individuals having (a) a short life cycle (b) many individuals produced in a mating (c) easily distinguishable characteristics (d) many chromosomes.
4. If a heterozygous red-eyed female *Drosophila* is crossed with a red-eyed male, the eyes of male offspring will be (a) all red (b) three-fourths red, one-fourth white (c) half red, half white (d) one-fourth red, three-fourths white.
5. The first chromosome map was constructed by (a) Alfred H. Sturtevant (b) Thomas Hunt Morgan (c) Gregor Mendel (d) Walter Sutton.

6. Geneticists may use pedigrees to determine (a) genetic and environmental effects on a trait (b) whether someone is a carrier for a disorder (c) the frequency of a gene in a population at large (d) the position of a gene on a chromosome.
7. Baldness is a (a) sex-linked trait (b) polygenic trait (c) sex-influenced trait (d) multiple-allele trait.
8. Klinefelter syndrome is an example of (a) monosomy (b) trisomy (c) a point mutation (d) a translocation.
9. The test that can yield information on the fetus is (a) fetoscopy (b) chorion villi sampling (c) ultrasound (d) all of the above.
10. Karyotypes enable scientists to identify (a) chromosomal abnormalities (b) blood type (c) point mutations (d) carriers for recessive traits.
11. What causes mutations?
12. What is nondisjunction?
13. Identify the genotypes of male and female fruit flies, birds, and grasshoppers.

14. What evidence led Morgan to hypothesize that the gene for eye color in *Drosophila* was carried on the X chromosome?
15. Why are identical twins useful in studies in human genetics?
16. How do geneticists detect the allele for Huntington disease?
17. How is the sickle-cell allele produced?
18. Identify the possible genotypes that could produce blood type A.
19. Give an example of a polygenic trait.
20. What evidence from the study of Amish families indicates that manic-depressive illness has a genetic basis?

Critical Thinking

1. The diagram at the right shows the banding pattern of human chromosomes 8 and 14. These chromosomes sometimes break at the dotted lines, and translocation occurs. Draw the mutant chromosomes that would result from such a translocation. Note: This translocation has been found in Burkitt lymphoma, a cancer of the immune system.

2. The mutagens known as substituted acridines are planar molecules with three-ring structures. They are about the same size as a DNA base pair and can slip between DNA base pairs, causing a frameshift mutation. Suppose one strand of DNA read TAC, GAA, TCG, GGT, ATT before mutation. If acridine caused a *G* to be inserted between two *A*s, what would the new sequence be? The commas are only to help identify groups of three.

3. In *Drosophila* body color and wing length are on the same chromosome. Gray body (G) is dominant to black body (g) and long wings (L) are dominant to short wings (l). Draw a Punnett square for the cross GgLl × GgLl, and show the phenotypic and genotypic ratios. Assume both dominants are on the same chromosome.

4. Sickle-cell disease is caused by a point

Chromosome 8   Chromosome 4

mutation in the hemoglobin gene, resulting in a valine being substituted for a glutamic acid. Show the sequence of the codon before and after the point mutation. Refer to Chapter 8 for the code.

5. The levels of enzymes in normal humans vary. Suppose you are a geneticist who wants to determine the normal level of a certain enzyme in human white blood cells. What types of people should you include in your sample?

6. Generally individuals who are heterozygous for sickle-cell disease have no symptoms of the disease. However, they should avoid extreme oxygen stress such as vigorous sports at high altitudes. Explain why this would be necessary.

Extension

1. Read the article by Joe Levine entitled "Do They Really Want to Know?" in *Time,* October 20, 1986, p. 80. Discuss the pros and cons of testing for Huntington disease.

2. Using the notation given in this chapter, make a pedigree of your own family. If there is a history of disease, such as cancer or diabetes, be sure to note that.

Chapter 12 | *Gene Expression*

This chapter describes how genetic information is implemented at the cellular level. Students learn how genes control the synthesis of proteins and how this in turn affects the development of the individual organism.

Correlation Chart

Section Number	Section Objectives	Pages	Features	Section Review
12.1	1. Define the *lac* operon, and explain how it is turned on and off.	181–183		1, 2
	2. Describe control of a eukaryotic enhancer sequence.			3, 5
	3. Explain how mRNA cleavage controls gene expression in eukaryotes.			4
12.2	1. Describe *Drosophila* development and the influence of homeotic genes.	184–187	Biology in Process, pp. 184–185	2, 5
	2. Explain why scientists think that cancer represents out-of-control cell division.			1, 3
	3. Describe the ways in which viruses and other carcinogens can cause cancer.		Think, p. 186	4
	4. Summarize the experimental procedure used to identify oncogenes.		Biology Lab, p. 187	

Resources*

Print

Friedberg, Errol C., ed. *Cancer Biology: Readings from Scientific American*. New York: Freeman, 1986.

Prescott, David M., and Abraham S. Flexer. *Cancer: The Misguided Cell,* 2nd ed. Sunderland, Mass.: Sinauer, 1986.

Scott, Andrew. *Pirates of the Cell: The Story of Viruses from Molecule to Microbe*. New York: Basil Blackwell, 1985.

Audiovisual/Software

The Fascinating Story of Cell Growth. Apple 128K. IBM PC 128K. 2 disks. Scholastic, 1986.

Origin of Cellular Life. Still-frame video. 46 min. Biology Media.

Regulation of Gene Expression in Eucaryotes. Still-frame video. 32 min. Carolina Biological.

*Additional references listed in *Modern Biology, Teacher's Program Guide.*

Section Teaching Guides

12.1 Control of Gene Expression

Approach

The main objective of this section is to describe the mechanisms by which genes are expressed in both prokaryotes and eukaryotes. Since gene expression essentially results in the initiation of protein synthesis, this section draws heavily on the material presented in Chapter 8. In particular, you may wish to review the processes of transcription and translation. The description of the *lac* operon also builds on the concepts of saccharide structure and enzyme activity in Chapter 4, on biochemistry. To clarify the differences between gene expression in prokaryotes and eukaryotes, it may be helpful to review the differences between prokaryotic and eukaryotic cell structure (Chapter 5).

Concept Framework

Students know that complex organisms such as humans contain many different kinds of cells that perform a variety of specialized functions. They also know from Chapter 9 that the chromosomes in each cell contain all of the DNA necessary to carry out the cellular activity of the entire body. To provide a framework for understanding the concept of gene expression, ask students what would happen if all the genes in every cell were continuously expressed. Elicit the fact that cell specialization could not exist if there were no control of gene expression.

Emphasis

Make certain that students understand the functions of the inducer, promoter, operator, and repressor in prokaryotic gene expression. They should also understand the significance of DNA coiling, enhancer control, and mRNA cleavage in eukaryotic gene expression.

Modeling Components of the *Lac* Operon

To help students visualize the turning on and turning off of genes, demonstrate the components of the *lac* operon by plugging in a lamp and turning it on and off. Ask students to identify which items of the lamp represent the inducer, promoter, operator, and repressor.

Section Review Answers

(Continued from p. 183.)

2. Lactose binds with the repressor molecule, thus freeing the operator region so that transcription can take place.
3. the extent to which DNA is uncoiled
4. A gene can be expressed only after introns are removed and the exons are spliced together.
5. the operator region

12.2 Morphogenesis

Approach

The main objective of this section is to describe the processes by which gene expression results in normal or abnormal development of an organism. This section builds on the concepts of cell growth and specialization in Chapter 5 and of protein synthesis in Chapter 8. The discussion of cancer and its causes also relates back to the information on mutagens in Chapter 11. An understanding of the process of morphogenesis will prepare students for the discussions of metamorphosis in insects (Chapter 33), echinoderms, (Chapter 34), and amphibians (Chapter 37). In presenting the text, reinforce the idea that all cellular activity, whether normal or abnormal, is ultimately the result of gene expression.

Concept Framework

Students know that certain changes take place in the body at certain times of life. For example, they know that growth occurs throughout childhood and that puberty takes place during adolescence. Tap this knowledge in order to provide a framework for understanding the effects of gene expression, by asking students how the body "knows" when to grow and when to stop growing. Elicit the fact that growth and development, like other life processes, are regulated by sequential gene expression.

Emphasis

Make certain that students understand the relationship between morphogenesis and the development of cancer. In other words, stress the idea that gene expression is responsible for both normal growth and development and cancerous growth.

Using the Features/Oncogene Proteins and Normal Proteins

The main process skills used in this feature are interpreting data and inferring. Michael Waterfield used a computer to compare the amino acid sequence of a protein coded for by an oncogene with the amino acid sequences of normal proteins. By analyzing his data, Waterfield determined that the oncogene protein closely resembled a protein normally expressed only in blood platelets. The discovery that normal and cancerous cells can produce similar proteins led Waterfield to infer that the development of cancer might be caused by a malfunction in the signal that triggers protein synthesis. In discussing the feature, point out that Waterfield's work represents a small but significant development in the effort to understand oncogenes. As scientists learn more about the molecular basis of cancer, they will be better equipped to design research that will lead to its cure.

Guest Speaker

Invite a geneticist or other researcher from a local university who can expand on the discussion of oncogenes and describe the most recent findings about their role in causing cancer.

Section Review Answers

(Continued from p. 186.)
3. Cancer is caused by a disorder in gene expression that results in uncontrolled cell division.
4. genetic predisposition for cancer, amount of carcinogen in exposure, and number of exposures
5. Probably nothing would happen, because the genes that code for the hormones that cause metamorphosis would not be expressed.

Laboratory Preparation

Background

In this lab students observe the changes that occur during the development of organisms. Some students are aware of developmental changes from watching pets mature or siblings change in morphology as they grow. Students study brine shrimp in order to understand that morphological change is due to different developmental stages. Any of the lepidopterans are also useful in a discussion of developmental changes.

Preparation

To prepare 1 L of 1.5% salt solution, add 15 g NaCl to 985 mL distilled water. Brine shrimp eggs are available from pet stores or from a biological supply house. Fill the aquariums just before the students begin the experiment. Replace the salt water in the aquariums once a week. This investigation will take portions of 12 lab sessions.

Conclusions

1. Different genes are activated at different stages of development, resulting in the observed sequence of changes.
2. The genetic information coding for the morphologies of previous stages is still in the cells, but it has been deactivated and is no longer available for transcription.

Inquiry

Students will form a hypothesis and test it. A reasonable hypothesis would state that a lower temperature would tend to slow development because the chemical reactions involved in growth would slow down.

Chapter Review

Vocabulary

1. In the *lac* operon the operator is attached to the repressor, thus preventing transcription until lactose enters the cell. The promoter produces RNA polymerase when the operon is turned on.
2. Heterochromatin is the tightly coiled form of DNA and contains inactive genes. Euchromatin is the uncoiled form of DNA and the site of active transcription of DNA into mRNA.
3. An exon is a segment of DNA that actively codes for a protein. An intron is an inactive segment of DNA. After being transcribed into mRNA, the introns must be removed in order for protein synthesis to occur.
4. A benign tumor remains self-contained and is not life threatening. A malignant tumor metastasizes and forms new tumors in other locations, resulting in a form of cancer.
5. A sarcoma is a tumor that grows in bone or muscle tissue. A carcinoma is a tumor that grows in skin or nerve tissue.

Review

11. Jacob and Monod studied the *lac* operon in *E. coli.*

12. The *lac* operon shuts off when lactose is no longer present in the cell and the repressor molecule attaches to the operator.

13. After it binds to a receptor protein, estrogen enters the nucleus of the cell, where it is broken down by enzymes.

14. Introns are removed from an mRNA strand after the mRNA migrates to the cytoplasm.

15. Cell differentiation is the change in structure or morphology of a cell in relation to other cells. Morphogenesis is the change in form of an organism resulting from cell differentiation.

16. If an imaginal disk such as a leg disk is transplanted to a wing disk area, a leg will grow where the wing should be.

17. A blastula is the hollow ball of cells formed in many organisms from a zygote after it undergoes repeated cycles of cell growth. A gastrula is formed when one surface of a blastula forms a pocket. The cells of a gastrula differentiate further, forming the tissues of the embryo and adult organism.

18. When a tumor metastasizes, it breaks away from the original cell mass and forms new tumors in other parts of the body.

19. Tobacco, asbestos, ultraviolet lights and X rays are carcinogens.

20. Waterfield discovered that the amino acid sequence for the protein making up an oncogene was nearly identical with that of a normal protein.

Critical Thinking

1. Protein synthesis requires amino acids and energy. Therefore, it is more efficient to make digestive enzymes when they can be used on the lactose.

2. The results are probably correct because different genes, and therefore different mRNA, are expressed in the two tissues.

3. The genes that control the enzymes are switched off or have very low activity when the diet lacks adequate protein. The genes are switched on by a high-protein diet but the former lack of protein temporarily results in inadequate protein metabolism.

4. The abnormal fly is a homeotic mutant, possessing two thoraxes.

5. Because there are many different kinds of cancer, with different causes, the same treatment will not work in all cases. Also, even people who have the same kind of cancer may respond differently to the same treatment because of their body chemistry, and so on.

6. The expression of the gene that codes for production of dehydrogenase is controlled by the amount of alcohol present.

Chapter 12 | Gene Expression

Introduction

As you learned in Chapter 8, the information for protein synthesis is encoded in the DNA of a cell. Yet not all the cells of a single organism require identical amounts or types of proteins at all times. In this chapter you will learn how genes control the production of proteins in prokaryotes and eukaryotes and how this production affects the development of the organism. The study of gene expression is one of the most active areas of modern biological research.

Chapter Outline

12.1 Control of Gene Expression
- *Gene Expression in Prokaryotes*
- *Gene Expression in Eukaryotes*

12.2 Morphogenesis
- *Homeotic Genes*
- *Cancer*

Chapter Concept

As you read, notice how gene control results in changes in the organism's morphology and physiology.

Adult form of green darner, *Anax junius.* Below: Larval form of green darner.

For additional teaching information see pp. T12A–T12D.

12.1 Control of Gene Expression

Genes control the synthesis of proteins, which are the basis of the structure and physiology of a cell. **Gene expression** is the activation of a gene that results in the formation of a protein. By controlling gene expression cells regulate which proteins are active at any given time.

Gene Expression in Prokaryotes

Scientists first studied gene expression in prokaryotes. Much of our initial knowledge of gene expression comes from the work of French scientists François Jacob (1920–) and Jacques Monod (1910–1976). Jacob and Monod determined how genes control the metabolism of the disaccharide lactose in *Escherichia coli*.

Lactose provides energy for the intestinal bacterium *E. coli*. However, enzymes must both alter the cell's permeability to lactose and cleave lactose into two monosaccharides—glucose and galactose. These enzymes are proteins made up of a number of polypeptides. The DNA segment on the *E. coli* chromosome that codes for these polypeptides is called the *lac* **operon,** shown in Figure 12-1. An **operon** consists of three segments: a **promoter,** an **operator,** and **structural genes.** The activation of these segments on the operon controls gene expression.

Section Objectives

- *Define the* lac *operon, and explain how it is turned on and off.*

- *Describe control of a eukaryotic enhancer sequence.*

- *Explain how mRNA cleavage controls gene expression in eukaryotes.*

Figure 12-1 During the metabolism of lactose, the *lac* operon turns on. Lactose removes the repressor, allowing RNA polymerase to begin transcription of mRNA.

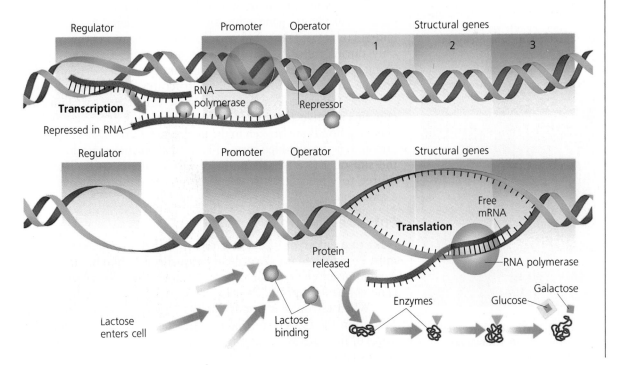

When lactose is not present in the cell, one repressor attaches to the operator region of the *lac* operon while other repressor molecules float freely in the cell. A **repressor** is a molecule that inhibits a gene from being expressed. A **regulator gene,** which lies close to the *lac* operon, codes for the repressor. The presence of a repressor molecule prevents the action of the enzyme RNA polymerase. Since RNA polymerase is needed to make the mRNA that will produce the enzymes that break down lactose, no enzymes will be produced while a repressor is attached to the operon.

An **inducer** is a molecule that initiates gene expression. In this example lactose is the inducer. It binds to the repressor on the operator portion and removes it. Lactose also binds to all other repressor molecules in the cell, so that none can interfere with the promoter. RNA polymerase can then begin to act. It moves from the promoter section of the operon, past the now-free operator region, and into the structural genes of the operon. Here RNA polymerase transcribes the DNA code into mRNA, which then migrates to the ribosomes and forms three enzymes: beta-galactosidase, permease, and transacetylase. These enzymes cleave lactose into glucose and galactose and regulate cell permeability to lactose. These actions decrease the amount of lactose in the cellular environment. When the level of lactose in the cellular environment is low, the repressor rebinds to the operator, and the *lac* operon is shut off.

Gene Expression in Eukaryotes

Gene expression in eukaryotes is different from that in prokaryotes. In eukaryotes gene expression is partly related to the coiling and uncoiling of DNA. Eukaryotic DNA coils into chromosomes only during mitosis or meiosis. At other times uncoiled DNA, along with histones, exists as chromatin. DNA wraps around the histone complex to form beadlike structures called **nucleosomes.**

The tightly coiled form of DNA, called **heterochromatin,** contains inactive genes, those that are not being expressed. When chemical bonds are broken in the nucleosomes, the DNA begins to uncoil. It is the uncoiled form of DNA, called **euchromatin,** that is the site of active transcription of DNA into mRNA. Thus the degree to which DNA is uncoiled indicates the degree of gene expression.

Enhancer Control

Next to a eukaryotic gene is a region called an **enhancer,** which must be activated for the gene in the euchromatin to be expressed. Activation of enhancers has been studied in the expression of the gene controlling production of estrogen, a female sex hormone.

When estrogen is present in the cell, it binds to a receptor protein in the cytoplasm. This complex then travels through the pores of the nuclear membrane. Once inside the nucleus the receptor–estrogen complex attaches to the enhancer portion of the

Tip: Have students review the discussion of the process by which DNA becomes a chromosome in Chapter 9.

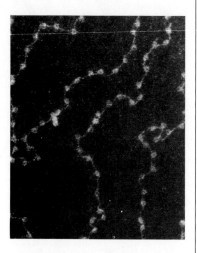

Figure 12-2 The knoblike areas along these DNA strands are nucleosomes.

gene. RNA polymerase can then synthesize the RNA that makes the protein that breaks down estrogen. The presence of the protein reduces the amount of estrogen, and the lack of estrogen blocks the action of RNA polymerase.

mRNA Cleavage

The cleavage of mRNA after it has been formed on a DNA segment also controls gene expression. In eukaryotes a gene consists of two interspersed regions—introns and exons. **Introns** are the inert segments of DNA, which do not code for proteins. **Exons** are the segments that function as the codons during protein synthesis.

Both introns and exons are transcribed into mRNA. When mRNA migrates from the nucleus to the cytoplasm, it is cleaved by enzymes, and all of the introns drop out of the nucleotide sequence. The exons remaining are then spliced together. A cap and a tail, each made up of a nitrogen base, attach to each end. The shortened mRNA strand then migrates to the ribosomes, where protein synthesis occurs. The expression of genes is thus regulated through chromosome uncoiling, control of the enhancer region, and cleavage of mRNA that has been formed by the gene.

Section Review

1. What is a *lac operon?*
2. How does lactose affect the functioning of the *lac* operon?
3. What dictates the degree of DNA transcription in eukaryotes?
4. How does mRNA cleavage control gene expression in eukaryotes?
5. What region of a prokaryotic gene would you say is analogous to the enhancer region of a eukaryotic gene?

Info: When introns are dropped out of the nucleotide sequence, their excision must be precise. Some introns may contain up to 10,000 nucleotides, and the shift of a single nucleotide can produce defective protein.

Figure 12-3 How does gene expression control the change from juvenile to adult?

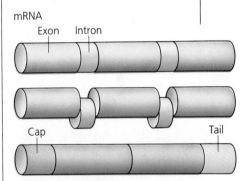

mRNA

Exon   Intron

Cap   Tail

Figure 12-4 When mRNA migrates to the cytoplasm, introns (red) are excised, and exons (blue) are spliced together. A nitrogen base attaches to each end of the gene, forming a cap and a tail.

Answers to Section Review
1. the segment of DNA that controls the production of enzymes that break down lactose
(See pp. T12A–T12D.)

Gene Expression **183**

Answer to Biology in Process
Waterfield had to determine the amino acid sequence of both types of protein. He then compared the two amino acid sequences.

12.2 Morphogenesis

The development of form in an organism is called **morphogenesis.** In Section 12.1 you learned how genes are turned on and off in prokaryotes and eukaryotes. How is the expression of genes manifested in the morphogenesis of these organisms? Let's study some cases of gene expression to better understand the normal and abnormal development of organisms.

Homeotic Genes

All multicellular, sexually reproducing organisms begin life as zygotes. A zygote undergoes repeated cycles of cell growth and division and in many organisms eventually forms into a hollow ball of cells called a **blastula. Cell differentiation,** the change in the morphology or physiology of a cell in relation to its neighboring cells, begins in the blastula. One surface of the blastula eventually forms a pocket, and the ball of cells is then called a **gastrula.** Cells in the gastrula differentiate further and form the tissues of the embryo and the adult organism.

Biology in Process

Oncogene Proteins and Normal Proteins

In 1983 Michael Waterfield sat down at his computer terminal in the Imperial Cancer Research Fund in London. An American scientist, Thomas Deuel, had sent him a protein sample coded for by an oncogene. The protein was one that is found in cancerous cells but not normal cells.

Waterfield knew that this protein, like every other protein, consisted of a series of amino acids. He had determined the exact order of its 104 amino acids, but he wondered whether this sequence was similar to that of any known normal protein.

Waterfield entered into the computer the sequence of the 104 amino acids in the protein. Then he directed the computer to compare this sequence with the amino acid sequences of other proteins, which had previously been stored in the memory of the computer.

Waterfield observed the results and analyzed his data. He noticed that the amino acid sequence of the protein coded for by the oncogene was nearly identical with that of a protein made by normal cells and found in human blood platelets (left, false-color SEM, ×4,400).

The normal protein, which is stored in the platelets, is re-

One of the most thoroughly studied examples of morphogenesis is that of the fruit fly, *Drosophila*. In *Drosophila* the gastrula changes into an elongated larva. Genes control the morphogenesis of a specific region in the larva into a specific part in the adult *Drosophila*. Genes that control the development of specific adult structures are called **homeotic** *(HOE-mee-AHT-ick)* **genes.** In the fruit fly these homeotic genes are located on the third chromosome of the genome.

Homeotic genes instruct cells of the larva to form tissues called **imaginal** *(im-AJ-un-'l)* **disks,** which become prominent in the pupa stage, also called the imago. A *Drosophila* larva has nine pairs of imaginal disks and a single genital disk. Hormones cause each disk to develop into a specific structure characteristic of the adult. Figure 12-5 illustrates the correspondence between the structural characteristics of the adult and the imaginal disks that will develop into those characteristics.

Under normal conditions an imaginal disk always develops into the same specific adult structure. For instance, a leg disk will always produce a leg. In fact, if a developmental biologist transplants a leg imaginal disk to the wing disk area, a leg will form on the adult where a wing should be.

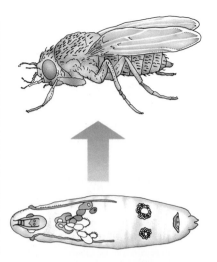

Figure 12-5 Each imaginal disk in the fruit fly larva develops into a specific structure in the adult.

leased into the bloodstream when the body is injured. The protein contacts cells near the wound and probably stimulates them to undergo cell division. This process is necessary for normal healing to take place.

Under normal conditions the gene for this protein is expressed only in platelets, not in other types of cells. Furthermore, this protein is normally stored inside platelets and released only when injury occurs.

However, if this gene is expressed in a nonplatelet cell, it may cause the cell to divide. Continued production of the protein might result in uncontrolled cell division, which in turn would lead to cancer.

Waterfield thus demonstrated that similar proteins could be produced by normal cells and cancerous cells. He inferred from this evidence that the signal that triggers the manufacture of protein plays a

key role in turning a normal cell into a cancerous cell.

Today scientists all over the world are working to determine the origins and workings of all sorts of oncogenes. To accomplish this goal scientists continue to use computers to assemble or manage data. However, their rare flashes of insight don't come from any machine. Instead, these leaps of knowledge occur as scientists like Waterfield follow an organized scientific method.

■ What data did Waterfield need in order to prove that the oncogene protein and the blood platelet protein were similar?

Cancer

Cancer, the uncontrolled growth of one or more cells, is morphogenesis gone wild. Cancer is caused by disorders in gene expression that result in cell division disorders.

Tumors

A **tumor** is an abnormal mass of cells that results from ungoverned cell division. A tumor may be either benign or malignant. The cells of a **benign tumor** remain within the mass. Generally, benign tumors pose no threat to life. The cells of a **malignant tumor** undergo **metastasis** *(muh-TASS-tuh-suss),* or break away and cause new tumors to form in other locations. The diseases caused by malignant tumors are collectively referred to as cancer.

Malignant tumors can be categorized according to the types of tissue they affect. **Carcinomas** grow in the skin and nerves. **Sarcomas** grow in bone and muscle. **Lymphomas** are solid tumors that grow in the tissues that form blood cells. **Leukemia** is an abnormal growth of immature white blood cells.

Causes of Cancer

The term **carcinogen** *(kahr-SIN-uh-jun)* refers to any substance that causes cancer. Whether a person actually develops cancer seems to depend on many factors, including genetic predisposition, the number of exposures, and the amount of carcinogen in each exposure. Well-known carcinogens include tobacco, asbestos, and ultraviolet light. Viruses also cause cancer in plants and animals and have been implicated in various kinds of leukemia.

Oncogenes

Scientists have determined that some genes, when expressed, can cause a normal cell to become cancerous. **Oncogene** is the term for such genes, which are the most recently identified sources of cancer. In some cases oncogenes may be genes that have a normal function in cell metabolism. However, as you have read, the control of gene expression is a complex process. If the control of gene expression or an oncogene goes awry, the gene may be expressed continually, resulting in the uncontrolled division of a cell. Carcinogens may in some way interfere with oncogene control and thus cause cancer.

Section Review

1. List the four types of malignant tumors.
2. How is morphogenesis regulated in *Drosophila?*
3. How is cancer related to gene expression?
4. Name three factors that determine whether cancer will develop after exposure to a carcinogen.
5. What do you infer would happen if an imaginal disk were transplanted into an adult fruit fly?

Answer to Think
Both observations strongly suggest that the onset of cancer is linked with some type of error in gene expression that results from mutation. However, scientists do not yet know the exact cause-and-effect relationship between gene expression and cancer.

Think

Many carcinogens, such as X rays, ultraviolet light, asbestos, and tobacco, are also known mutagens. In addition, chromosomal mutations such as nondisjunction and translocation have frequently been observed in cancerous cells. What does this suggest about the relationship between cancer and gene expression?

Answers to Section Review
1. carcinoma, sarcoma, lymphoma, leukemia
2. Homeotic genes instruct cells to form imaginal disks. Hormones then cause each disk to develop into an adult structure.
3. Cancer is caused by a disorder in gene expression that results in uncontrolled cell division.
4. genetic predisposition for cancer, amount of carcinogen in exposure, and number of exposures
(See pp. T12A–T12D.)

Laboratory

Gene Expression and the Development of Brine Shrimp

Objective
To infer the role of sequential gene expression by observing brine shrimp metamorphosis

Process Skills
Observing, inferring, hypothesizing

Materials
Glass-marking pencil, 2 1-gallon aquariums filled with 1.5% salt solution, aerator, teaspoon, brine shrimp eggs, medicine dropper, depression slide, compound microscope, unlined paper, dry baker's yeast, cardboard, fine-mesh dip net, petri dish, dissecting microscope

Method
1 Use a glass-marking pencil to label an aquarium filled with salt water *Aquarium 1*. Attach an aerator and begin pumping air through the water.
2 Add $\frac{1}{2}$ teaspoon of brine shrimp eggs to the aquarium. What is the function of the aerator in the aquarium?
3 Twenty-four hours after adding the brine shrimp eggs to the salt water, remove a few eggs with a clean dropper and place them on a depression slide. Examine the eggs under a compound microscope set to low power.
4 If the eggs have not begun to hatch, wait another 24 hours and repeat the procedure. What do you see when the eggs begin to hatch? Draw a sketch of your observations.
5 During the next class period, label a second aquarium of salt water *Aquarium 2*. Transfer the aerator to the second aquarium and begin pumping air through the water. Leave the aerator in place during the rest of the investigation.
6 Add $\frac{1}{2}$ teaspoon of baker's yeast to the water in aquarium 2.
7 Cover one end of aquarium 1 with cardboard. The brine shrimp will swim toward the light end of the aquarium. Use a dip net to remove as many of the brine shrimp as possible. Transfer the brine shrimp to aquarium 2. Why did you add the baker's yeast to aquarium 2?
8 Once a week for the next three weeks repeat steps 6 and 7.
9 Every other day for the next three weeks, transfer a few organisms from aquarium 2 to a depression slide and examine the slide under the microscope. Make a drawing of the organisms you observe.
10 As the brine shrimp grow larger you will have to place them in petri dishes and examine them under a dissecting microscope.

Conclusions
1 Based on your observations, what controls the sequence of changes that occur as the brine shrimp develop?
2 What happens to gene sequences that code for the morphology of previous stages in a brine shrimp's development?

Inquiry
Would lowering the temperature of the water affect the development of brine shrimp? Form a hypothesis and test it.

Chapter 12 Review

Vocabulary

benign tumor (186)
blastula (184)
cancer (186)
carcinogen (186)
carcinoma (186)
cell differentiation (184)
enhancer (182)
euchromatin (182)

exon (183)
gastrula (184)
gene expression (181)
heterochromatin (182)
homeotic gene (185)
imaginal disk (185)
inducer (182)
intron (183)
lac operon (181)

leukemia (186)
lymphoma (186)
malignant tumor (186)
metastasis (186)
morphogenesis (184)
nucleosome (182)
oncogene (186)

operator (181)
operon (181)
promoter (181)
regulator gene (182)
repressor (182)
sarcoma (186)
structural gene (181)
tumor (186)

Explain the difference between the terms in each of the following sets.
1. operator, promotor
2. heterochromatin, euchromatin
3. exon, intron
4. benign, malignant
5. sarcoma, carcinoma

Review

1. Gene expression occurs through the process of DNA (a) replication (b) translation (c) transcription (d) synthesis.
2. The repressor molecule is coded for by the (a) structural gene (b) regulator gene (c) imaginal disk (d) nucleosome.
3. The site of active DNA transcription in eukaryotes is called (a) an operon (b) an enhancer (c) heterochromatin (d) euchromatin.
4. When present in a cell, lactose acts as (a) an inducer (b) a repressor (c) a promoter (d) a regulator.
5. In eukaryotes introns and exons are interspersed regions of (a) chromatin (b) DNA (c) tRNA (d) histones.
6. Lactose is to estrogen as inducer is to (a) repressor (b) operon (c) enhancer (d) regulator.
7. The process of cell differentiation begins in the (a) zygote (b) blastula (c) gastrula (d) larva.
8. Prokaryotes differ from eukaryotes in that prokaryotes have (a) introns and exons (b) no nucleus (c) no chromosomes (d) an enhancer region.
9. Imaginal disks become prominent during the (a) zygote stage (b) gastrula stage (c) larva stage (d) pupa stage.
10. A malignant tumor that grows only in bone or muscle tissue is a (a) sarcoma (b) lymphoma (c) carcinoma (d) leukemia.
11. In which organism did Jacob and Monod study the *lac* operon?
12. What causes the *lac* operon to shut off?
13. What happens to estrogen after it binds to a receptor protein?
14. At what stage are introns removed from an mRNA strand?
15. Distinguish between cell differentiation and morphogenesis.
16. Describe evidence that the morphological destiny of an imaginal disk is unrelated to its location.
17. Describe the characteristics of a blastula and a gastrula.
18. What happens when a malignant tumor undergoes metastasis?
19. Name three commonly known carcinogenic substances.
20. What key discovery made by Michael Waterfield suggests that the appearance of a cancerous cell is probably related to gene expression?

Critical Thinking

1. Of what advantage is it to the bacterial cell to make digestive enzymes for lactose only when lactose is present in the cellular environment?

2. A molecular biologist isolates mRNA from the brain and liver of a mouse and finds that the types of mRNA are different. Can these results be correct, or has the biologist made an error? Explain your answer.

3. Kwashiorkor is a disease in children caused by a diet high in carbohydrates but low in protein. When children with kwashiorkor suddenly are put on a diet rich in protein, they may become very ill with ammonia poisoning, even dying from it. The high level of ammonia in their blood is due to the inadequate metabolism of protein. What does this tell you about the enzymes that metabolize protein?

4. Examine the photograph at the right of an abnormal *Drosophila*. What type of mutation, do you think, is present in this abnormal fly?

5. In one person cancer may be cured through the use of a certain drug or

by means of radiation treatment, whereas in another person cancer may be unaffected by these treatment methods. Suggest the most likely explanation for this difference in response.

6. *Drosophila* feed on fermenting fruit, which often contains a large amount of alcohol. If *Drosophila* are fed a diet that has a high alcohol content, there is an increase in the amount of dehydrogenase, an enzyme that metabolizes alcohol, in the digestive tract. What does this increase tell you about the enzyme?

Extension

1. Read the article by M. D. Lemonick entitled "Of Fireflies and Tobacco Plants," in *Time*, November 17, 1986, p. 87. Prepare an oral report including information on what luciferase is, what a "reporter" gene is, and how the system described in the article will help scientists understand gene expression.

2. Construct models showing gene expression in prokaryotes and eukaryotes. Your model of the prokaryote should include promoter, operator, structural genes; repressor; regulator gene; RNA polymerase; inducer; mRNA; and ribosomes. Your model of the eukaryote should show DNA, heterochromatin, euchromatin, enhancer, receptor protein for the gene

you choose, RNA polymerase, mRNA with introns and exons, and ribosomes. The gene for estrogen is a good example of a eukaryote to use for your model.

3. Construct a chart in which you compare the qualities in bacteria and fruit flies that make them ideal experimental tools for the scientist. Make sure to include similarities as well as differences between the two species.

4. Use the library to research the latest information on oncogenes and their protein products. Make an oral report to the class in which you show how your findings support the theory that cancer arises from out-of-control cell division or from differentiation.

Chapter 13 | *Applied Genetics*

This chapter discusses the various ways science applies knowledge about hereditary patterns and chemical ways of altering genes to develop new or improved plants, animals, and other products that benefit society.

Correlation Chart

Section Number	Section Objectives	Pages	Features	Section Review
13.1	1. Describe the process of mass selection. 2. Compare the advantages and disadvantages of inbreeding. 3. Explain how hybridization can produce new varieties of organisms.	191–192	Think, p. 192	1, 5 1, 2 1, 3, 4
13.2	1. Explain how scientists can alter an organism's genetic makeup by inducing mutations and polyploidy. 2. List the steps involved in engineering bacteria able to manufacture human interferon. 3. Describe the process of gene sequencing and name some of its uses. 4. Summarize the process of manufacturing artificial chromosomes.	193–197	Biotechnology, pp. 194–195 Biology Lab, p. 197	2 1, 3, 5 4

Resources*

Print

Drlica, Karl. *Understanding DNA and Gene Cloning: A Guide for the Curious.* New York: Wiley, 1984.

Tyckoson, David A. *Test Tube Babies: In Vitro Fertilization and Embryo Transfer.* (From the *Oryx Science Bibliographies* Series, Vol. 4.) Phoenix, Ariz.: Oryx Press, 1986.

Wiegele, Thomas C. *The Social Impacts of Biotechnology: An Annotated Bibliography of Recent Works.* DeKalb, Ill.: APLS, 1986.

Audiovisual/Software

Children of the Future. Film, 55 min. Britannica, 1987.

Genetic Engineering. Apple II, 48K. Operating instructions, glossary, and software on disk with backup. Helix. Dist. by Carolina Biological.

Significant Hazards? The Somerville DNA Debate. Film or video, 24 min. Fanlight, 1984.

*Additional references are listed in *Modern Biology, Teacher's Program Guide.*

Section Teaching Guides

13.2 Controlled Breeding

Approach

The main objective of this section is to describe the techniques by which people can manipulate the hereditary characteristics of offspring of a specific organism. The three primary methods of controlled breeding—mass selection, inbreeding, and hybridization—are all based on the principles of heredity first outlined by Gregor Mendel. Therefore a review of Mendel's laws will be a useful preparation for presenting this section to students. The discussion also builds on the concepts of inheritance patterns described in Chapter 11.

In this section students will learn that breeders can manipulate hereditary characteristics by selecting for breeding those individuals with desirable traits. Later in the book they will read about other applications of controlled breeding techniques to produce plants that are useful to humans (Chapter 24). Tell students that, as they will learn when they study Chapter 15, organisms produced by controlled breeding are also subject to the forces of natural selection, which favors organisms whose traits have the greatest likelihood of reproductive success.

Concept Framework

Students are familiar with shows that exhibit top individuals of various breeds of dogs, cats, cattle, flowers, or other organisms. To provide a framework for understanding the concepts of controlled breeding, ask students to think about what constitutes a champion breed. Ask how they think plant and animal breeders obtain their prize specimens. Tell them to look for examples in the text that show that controlled breeding is based on principles of genetics.

Emphasis

Make certain that students can identify the major controlled breeding techniques and describe the advantages and disadvantages of each. Students should also be able to explain how each technique involves the application of principles of heredity, including segregation, independent assortment, and dominance and recessiveness.

Devising a Breeding Experiment

Tell students to devise a hypothetical controlled breeding experiment in which they try to develop a new variety of plant or animal. For example, they might want to produce a sweet-tasting lime or a dog that does not bark. Have students prepare a chart or diagram that depicts the method of breeding to be used (mass selection, inbreeding, or hybridization), the parent organisms to be bred, and the desired offspring. Arrange the diagrams in a display, and have students critique each other's ideas and methods.

Section Review Answers

(See p. 639.)

13.2 Manipulating Genes

Approach

This section focuses on the methods by which geneticists can alter the genetic makeup of an organism by manipulating the genes of the organism. The description of the first method, inducing mutations, draws on the discussion of mutation in Chapter 11. The process of inducing polyploidy builds on the concepts of plant cell structure (Chapter 5) and meiosis and mitosis (Chapter 9).

The process of genetic engineering is a complex procedure that involves the direct manipulation of genes. In discussing genetic engineering and gene sequencing, you might wish to review the structure of DNA and genes (Chapter 8), and the structure of chromosomes (Chapter 9). Convey to students the idea that the recent advances in genetic engineering and gene sequencing techniques are based on knowledge of these structures.

Concept Framework

Students know from previous chapters that in all organisms, the coded instructions carried in DNA control all aspects of the organism's metabolism. Use this knowledge as a starting point for a discussion by asking students to consider how this fact opens up the possibility for the direct manipulation of genetic material. As students read this section, ask them to consider the fact that because the genetic code is, for the most part, universal, the capability exists for the genes of one organism to be expressed in another organism.

Emphasis

Students should be able to explain how the processes of inducing mutation, inducing polyploidy, genetic engineering, and gene sequencing enable scientists to manipulate the genetic makeup of an organism.

Using the Features/Artificial Chromosomes

The biotechnology feature on artificial chromosomes describes techniques that enable scientists to construct entire chromosomes, rather than individual genes. Scientists constructed artificial yeast cell chromosomes by isolating those regions that govern the essential activities of the chromosome and splicing those regions into a bacterial plasmid. The plasmid was then cut between the two telomeres, creating a linear chromosome, and reinserted into the yeast cell.

Researchers are learning more about the mechanisms of chromosomal segregation and replication by studying these artificial chromosomes. For example, they have determined that chromosome length affects the rate of segregation error. Encourage students to think about how such knowledge might be used to determine the cause of Down syndrome and other chromosomal abnormalities.

Guest Speaker

Invite a geneticist to speak to the class about gene therapy. The discussion should include a description of the process involved in gene therapy, as well as an explanation of how the procedure can be used to treat hereditary disorders.

Section Review Answers

(See p. 644.)

Laboratory Preparation

Background

This laboratory exercise introduces students to methods utilized in modern molecular biology. Students should be familiar with three fundamental concepts: alleles, linkage, and crossing-over. Alleles are the different forms of a gene, consisting of a different sequence of DNA bases, and found in the same location of different homologous chromosomes. A gene can have many different alleles. Linkage means DNA sequences are found on the same chromosome. Linked sequences become separated by crossing-over. Students use a DNA probe (a restriction enzyme) to remove a small piece of DNA from a region five cross-over units from the gene that causes a kidney disease. Because the DNA sequences in this region are highly variable, several forms or alleles exist, and the students can use these to identify chromosomes that carry the disease-causing gene. Students use the DNA probe data to determine which markers are carried by members of a family with an inherited kidney disease. Students use cross-over information to predict the likelihood that members of the third generation of the family carry the disease.

Preparation

No advance preparation is required for this lab.

Conclusions

1. Allele B is probably on the same chromosome as the APKD gene because the marker appears twice in generation II in association with the disease while E appears only once in association with the disease.
2. The APKD gene would have been acquired by crossing-over. The probability is 5%.
3. The probability is 95% since E is linked to the APKD gene in the mother.
4. Each has a 5% chance of having the disease because the disease would have to be acquired by crossing-over.

Inquiry

If the grandfather were the APKD gene bearer, the pedigree shown for generation II would still be possible. However, the situation in generation III would change. The probability of III-1 or III-3 acquiring the disease would be 95%. The probability of III-2 acquiring the disease would be 5%.

Chapter Review

Vocabulary

1. Luther Burbank practiced mass selection.
2. Inbreeding is the mating of two similar, closely related organisms. Hybridization is the mating of two different but related organisms.

3. *Poly-* means many; *-ploid* means multiplication of chromosome number.
4. Restriction enzymes are used in gene splicing.
5. A plasmid is the small ring of DNA found in *E. coli.*

Chapter Review

11. The selection and cultivation of specific plants and animals for food was the earliest form of applied genetics.
12. The seedless grape and spineless cactus are two plant varieties that Luther Burbank developed.
13. Inbreeding of dachshunds eventually produced offspring with weak spines.
14. Hybrids are often sterile because they receive different chromosomes from each parent. Consequently the chromosomes are unable to match up with a homologue and cell division cannot take place.
15. H. J. Muller induced mutations in *Drosophila* by exposing them to X rays, thus producing new traits in the fruit flies.
16. Polyploid organisms have multiple sets of chromosomes. They are usually larger and hardier than normal organisms.
17. Scientists determine the base sequence of a DNA segment by analyzing the DNA segment with a DNA sequencer.
18. Restriction enzymes cut apart DNA molecules at specific points, thus making it possible to isolate a particular gene. Restriction enzymes also break open plasmids so that new genetic material can be spliced onto them.
19. Gene cloning is an efficient way to produce interferon because the bacterium that carries the gene for interferon reproduces rapidly and thus can produce large amounts of interferon in a short amount of time.
20. Scientists splice a gene that codes for a pigment-inhibiting protein into the artificial chromosome. They can tell whether abnormal segregation has taken place by observing the color of the yeast cells.

Critical Thinking

1. The advantage of using mass selection is that it selects for a total phenotype instead of merely considering one gene. Thus, the plant will be better adapted overall than a plant created using genetic engineering. A disadvantage of mass selection is that it takes longer to develop a new variety when using that method than when genetic engineering is used.
2. Breeding parents with their offspring, as in this example, is inbreeding. The resulting offspring would be likely to possess many positive attributes because it received genes from both parents that result in the expression of favorable traits.
3. Federal regulations were enacted because of concerns that genetically engineered organisms might cause disease or have other harmful effects on the environment.
4. Triploid bananas are not able to reproduce naturally because they possess an uneven number of chromosome pairs.
5. You would use restriction enzymes *Bam, Hi,* and *Bal*I. The gene would be 1,071 base pairs long. To determine whether the procedure was successful, you would reinsert the plasmid into the bacterial cell, clone the bacteria, and expose the new bacterial colony to tetracycline. If the bacteria are not resistant to the tetracycline, the procedure has been successful.

Chapter 13 | *Applied Genetics*

Introduction

Scientists have used a knowledge of genetics to develop new sources, and improve existing sources, of food, medicines and vaccines, and other products essential to life. In this chapter you will learn how scientists apply genetic principles to benefit human society. You will learn about traditional practices and about new technologies, the potential of which have yet to be realized.

Chapter Outline

13.1 Controlled Breeding
- *Mass Selection*
- *Inbreeding*
- *Hybridization*

13.2 Manipulating Genes
- *Inducing Mutations*
- *Inducing Polyploidy*
- *Genetic Engineering*
- *Gene Sequencing*

Chapter Concept

As you read, note the ways in which scientists have applied their knowledge of both the chemical aspects of genetics and inheritance patterns.

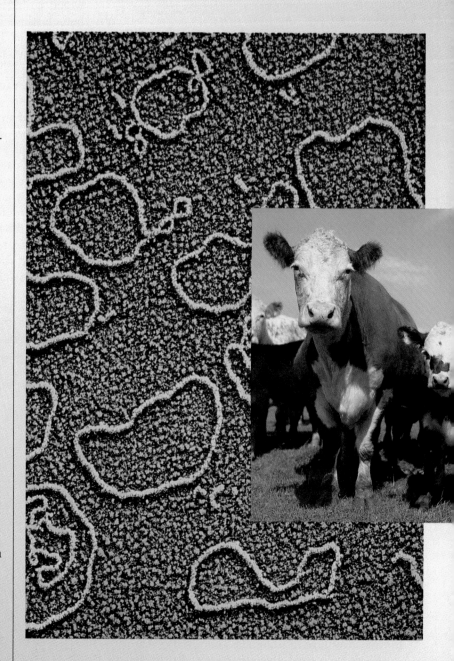

Color-enhanced TEM of *Escherichia coli* plasmids (×14,285). Inset: Hereford cow and crossbred calf, Iowa.

For additional teaching information see pp. T13A–T13D.

13.1 Controlled Breeding

Archeological evidence shows that humans began cultivating plants and animals for food about 10,000 years ago. Their selection of specific organisms for breeding was a simple form of applied genetics. **Applied genetics** is the manipulation of the hereditary characteristics of an organism to improve or create specific traits in offspring.

Controlled breeding involves manipulating the hereditary characteristics of offspring by selecting parents with specific phenotypic traits. This process enables a breeder to develop new strains of a species or to maintain existing strains. For example, the controlled breeding and cultivation of corn plants have produced varieties that yield more than ten times as much grain as varieties used early in this century. Applying controlled breeding to plants and animals has traditionally been based upon three techniques: mass selection, inbreeding, and hybridization.

Mass Selection

Mass selection is the process of choosing a few individuals to act as parents from a large pool of individuals. For example, a farmer may breed only the two or three cows in a herd that produce the most milk. Because these individuals produce the most milk, they are the most likely to produce offspring that will also produce large amounts of milk.

Luther Burbank (1849–1926), an American plant breeder, achieved great success with mass selection around the turn of the century. Burbank grew large numbers of fruits, flowers, vegetables, and grains, and selected only those with desirable traits for further breeding. He developed over 800 new plant varieties, including the seedless grape and the spineless cactus.

Figure 13-1 Luther Burbank developed daisies and potatoes by the process of mass selection.

Think

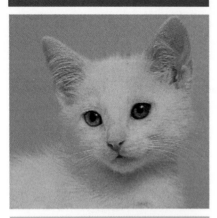

Many white cats with blue eyes are deaf. Suggest a mechanism by which the selection for this phenotype also results in deafness.

Answer to Think
Recessive alleles controlling deafness are probably found close to the alleles for blue eyes on the chromosome.

Tip: To help students understand why hybrids are usually sterile, have them (1) illustrate how hybrids end up with nonhomologous chromosomes and (2) explain what is missing for the process of meiosis to occur. Also, point out that because mitosis in hybrids is normal, they are able to grow.

Answers to Section Review
1. selection of parents with specific traits in order to manipulate the hereditary characteristics of offspring
2. because it increases the incidence of harmful homozygous recessive traits
3. Two different but related varieties of plants or animals are crossed.
4. because hybridization reverses the negative effects of inbreeding by increasing the number of heterozygous genes
5. Today people have a greater understanding of inheritance patterns and can breed organisms more selectively.

Inbreeding

A breeder who wants to maintain or to intensify desirable traits in a plant or animal species may practice a type of controlled breeding referred to as **inbreeding.** Inbreeding is based on the assumption that individuals that possess similar phenotypes also will possess similar genotypes. When phenotypically similar individuals are bred, offspring are more likely to be similar to their parents. In addition, offspring are less likely to possess new, undesired characteristics.

Most modern domestic breeds of dogs and cats were developed through inbreeding. Unfortunately, inbreeding can eventually produce weaker organisms because it increases the incidence of harmful homozygous recessive traits. For example, dachshunds were initially bred to dig badgers and rodents out of long narrow holes. However, while their long spinal column is adapted for this task, the spinal column is also prone to slipped or stressed vertebral disks.

Hybridization

Hybridization is another type of controlled breeding. In **hybridization** two different but related species or varieties of plants or animals are crossed. The products of this type of crossbreeding are called **hybrids.** Hybrids possess a different genotype and usually a phenotype different from that of either parent. When two different but closely linked species are crossed, the chromosome incompatibility during meiosis causes most of their hybrid offspring to be sterile.

Some hybrid individuals grow faster and larger and are healthier than either parent. Such an individual is said to display **hybrid vigor.** For example, hybridization has produced large tomatoes that resist wilting in hot weather. In addition, hybrid cattle may be resistant to some diseases that attacked their parents. Hybridization can reverse the damaging effects of inbreeding, which results in hybrid vigor in the offspring. This occurs because hybridization increases the number of heterozygous genes in an organism, thus reducing the likelihood that a harmful, recessive allele will be expressed.

Section Review

1. What is *controlled breeding?*
2. Why does inbreeding of a plant or animal species often produce weaker offspring?
3. What method does a plant or animal breeder use to produce hybrid organisms?
4. Why do hybrid organisms often display hybrid vigor?
5. Do you think mass selection is practiced less frequently today than it was in Burbank's time? Why?

13.2 Manipulating Genes

Controlled breeding is used to create the characteristics of progeny by selecting parents according to phenotype. Other techniques in applied genetics manipulate the genes themselves. Among these techniques are induction of mutations, induction of polyploidy, and use of genetic engineering and gene splicing.

Inducing Mutations

As you know, mutations are changes in the DNA of an organism. Mutations introduce new alleles to the genetic makeup of an organism. Mutations occur at a very low rate in nature. However, in 1927 the American geneticist H. J. Muller (1890–1967) found that he could induce a much greater rate of mutation in *Drosophila* by treating the flies with X rays. He then selected the resulting mutants for controlled breeding. Breeders have since used mutagens to create new traits in many organisms. People can then use those organisms whose mutations are beneficial to humans. For example, X rays induce mutations in bacteria. These mutations are then selected and propagated if they are useful.

Inducing Polyploidy

Breeders can also introduce desirable traits into a species by inducing **polyploidy**, a condition in which cells contain multiple, complete sets of chromosomes. Although polyploidy is rare and usually lethal in animals, it often occurs naturally in plants. As you can see in Figure 13-2, polyploid individuals are often larger or hardier than their parents.

Plant breeders artificially induce polyploidy by administering **colchicine**, a chemical that prohibits the formation of the cell plate during cell division. Colchicine is usually applied by placing the roots of the plant in a colchicine solution. Chromosomes in the cells of the plant go through the phases of cell division, but the cell plate does not form. Therefore, two complete sets of chromosomes exist. When the plant is removed from the colchicine solution the cells will again form cell plates during mitosis. Thus all resulting cells will contain an extra set of chromosomes.

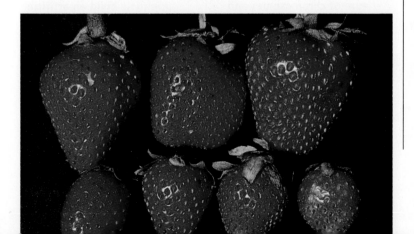

Figure 13-2 Note the significant size difference between the polyploid strawberries in the top row and the normal strawberries in the bottom row.

Genetic Engineering

The understanding of the molecular basis of inheritance has led to a new form of applied genetics called **genetic engineering,** in which scientists directly manipulate genes. Genetic engineering frequently involves the use of **recombinant DNA,** which is composed of DNA segments from at least two different organisms.

Genetic engineering has many commercial applications. An example is the use of recombinant DNA technology to make **interferon,** a virus-destroying protein produced naturally by the human body. The production of synthetic interferon involves (1) isolating the human gene that codes for interferon production, (2) splicing this gene into a strand of bacterial DNA, (3) inserting recombinant DNA into a bacterium, and (4) cloning the bacterium and collecting the product.

Isolation of a Gene

The first step in this process is isolating the human interferon gene. Genetic engineers use **restriction enzymes,** proteins that cut a DNA molecule into pieces. The restriction enzyme *Eco*RI cuts DNA wherever the sequence C-T-T-A-A-G occurs, as shown in

Biotechnology

Artificial Chromosomes

Until recently genetic engineering was restricted to the manipulation of individual genes. However, thanks to gene splicing techniques perfected in the 1980s, scientists can now construct entire chromosomes. Scientists hope to learn more about the nature of chromosomes and the mechanisms of cell division by studying these artificial chromosomes.

To construct an artificial chromosome, researchers must isolate three regions of a chromosome: the region where replication begins, the centromere, and the telomeres, or terminal ends of the chromosome. These regions govern the essential activities of the chromosome. Researchers have had the greatest success constructing yeast chromosomes because they are relatively small and have these regions clearly defined.

Scientists build an artificial chromosome step by step. They isolate each chromosomal region from a yeast cell and splice it into a bacterial plasmid. Eventually the plasmid contains the region where replication begins, the centromere, and the telomeres. Restriction enzymes are used to cut the

Figure 13-3. Other restriction enzymes cut DNA at different nucleotide sequences. By using the proper restriction enzyme scientists can cut the human interferon gene out of its chromosome. Once the gene for interferon is removed, it is separated from the rest of the DNA and then inserted into a strand of bacterial DNA.

Gene Splicing

Gene splicing is the process by which a gene from one organism is placed into the DNA of another organism. In our example, the human interferon gene is placed into the DNA of *E. coli,* the common bacterium of the human intestine. In addition to a single, circular chromosome, *E. coli* contains a single, small ring of DNA called a **plasmid.** Human DNA is inserted into this plasmid.

The plasmid ring is removed from the bacterium and then opened with a restriction enzyme. The human interferon gene and the bacterial plasmid have "sticky ends"—unpaired bases at each end of the DNA segment, where they were cleaved by restriction enzymes. As the human DNA is spliced into the plasmid DNA, the unpaired bases of each bond readily. As shown in Figure 13-4 on the following page, a newly formed plasmid contains both human and bacterial DNA.

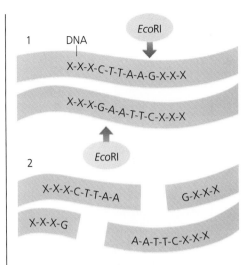

Figure 13-3   Restriction enzyme *Eco*RI cleaves double-stranded DNA at the sequence C-T-T-A-A-G, freeing two single-stranded complementary ends.

plasmid between the two telomeres, which are fused, thus producing a single linear chromosome. The newly created chromosome is then reinserted into the yeast cell.

Scientists also insert a gene that when expressed inhibits the production of red pigment in the yeast cell. This gene acts as a kind of genetic marker for the presence of the artificial chromosome. A yeast cell that contains one artificial chromosome appears pink (p. 194). If replication fails to occur or a homologue is lost during mitosis, the daughter cell that inherits the chromosome will be pink and the one that does not will be red. If the chromosomes fail to segregate, the daughter cell that inherits both chromosomes will be white and the other will be red. Such abnormalities will show up in the yeast colony as uneven bands or sectors of color (above).

In the first yeast cells that contained an artificial chromosome, segregation occurred randomly. That is, a daughter cell was equally likely to inherit one, two, or no chromosomes. After determining that the chromosomal regions functioned properly, researchers inferred that the length of the chromosome affected segregation. When they increased the chromosome length from 11,000 to 55,000 base pairs, the rate of segregation error dropped to 1.5 percent.

Researchers are also using artificial chromosomes to study meiosis. They hope this research will provide answers to such questions as what causes Down syndrome and other human disorders related to abnormal chromosome segregation.

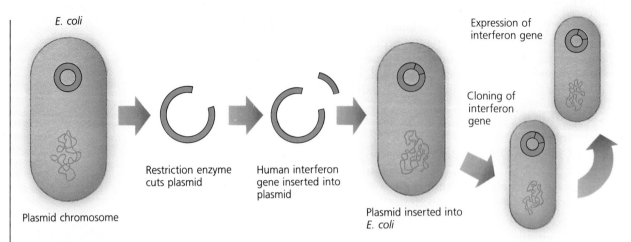

E. coli

Expression of
interferon gene

Cloning of
interferon
gene

Plasmid chromosome

Restriction enzyme
cuts plasmid

Human interferon
gene inserted into
plasmid

Plasmid inserted into
E. coli

Figure 13-4   A plasmid with both human and bacterial DNA results when a human interferon gene is spliced into an E. coli plasmid.

Info: Humulin NPH insulin, a breakthrough in the treatment of diabetes mellitus, is made through gene splicing. This insulin is derived from humans instead of from animals. Humulin insulin has two advantages: It is less likely to cause allergic reactions in diabetics and it does not have to be refrigerated.

Answers to Section Review
1. DNA segments from at least two different organisms
2. because polyploid organisms are often larger and hardier than normal organisms
3. through gene cloning
4. Scientists obtain a sample of DNA and then run it through a DNA sequencer, which determines the sequence of base pairs in the original DNA fragment.
5. The small size of E. coli makes the growing of large amounts of the bacteria in cultures easy, and its circular shape facilitates gene splicing.

Insertion, Cloning, and Collecting

Once a DNA fragment is incorporated into a plasmid, the plasmid is inserted into another bacterium, which is then placed in a culture medium, where it divides. Each time a bacterium divides a new copy of the plasmid DNA, which includes the human DNA gene, is created. This process by which the human gene is replicated is called **gene cloning**. Because *E. coli* can divide every 20 minutes, gene cloning is an efficient way to produce many copies of a specific genetic sequence. The gene for human interferon is thus expressed in bacterial cultures and the resulting interferon protein is collected and eventually used by physicians.

Gene Sequencing

A technique called gene sequencing has greatly expanded the potential for manipulating genes. **Gene sequencing** is the process of determining the exact order of bases in a fragment of DNA. Scientists place a sample of DNA, which has been cut from a chromosome by restricting enzymes, into a special laboratory instrument called a DNA sequencer. This machine almost automatically determines the sequence of the original DNA fragment. Knowing the sequence of a DNA segment helps scientists learn the exact structure of the protein coded for by a certain gene. Such knowledge can be used by scientists in developing cures for hereditary diseases.

Section Review

1. What is *recombinant DNA?*
2. Why do scientists induce polyploidy in plants?
3. How is recombinant DNA expressed?
4. Summarize the process of gene sequencing.
5. What characteristics of the *E. coli* plasmid do you think make it a useful organism for use in recombinant DNA technology?

Laboratory

Molecular Genetic Diagnosis

Objective
To understand how diseases can be diagnosed by using genetic markers

Process Skills
Inferring, predicting, analyzing data

Materials
Pencil and paper

Method
1 Adult polycystic kidney disease (APKD) is controlled by a dominant gene. The gene that causes APKD is a mutation of one of the genes for kidney function. Persons with APKD usually do not show symptoms until middle age. Recently it has become possible to genetically diagnose the disease by sampling a small segment of DNA, called a marker, from a site near this gene.
2 This DNA segment comes in various lengths. Markers of different lengths can be thought of as different alleles. The markers can be identified by a technique called gel electrophoresis, which separates markers by determining how far each type of DNA can travel in an electric field.
3 The gene and the marker are 5 map units apart. Although the gene and the marker are linked, crossing-over can separate them. What is the probability that the two will become separated by crossing-over?
4 The figure (top) shows the pedigree of an APKD family. Generations are indicated by Roman numerals; individuals by Arabic numerals. Circles indicate females; squares males. Solid symbols indicate persons with APKD. The third generation is all children.
5 The markers shown in the figure (bottom) are labeled according to length—ranging from A, the longest, to H, the shortest. Why does each person have two markers?

6 Make a copy of the pedigree.
7 Under the symbol for each individual, write the markers identified for that person. Why are no third-generation individuals shown as having the disease?

Conclusions
1 The grandmother (I-2) has APKD. Based on the expression of the disease in generation II, which marker is probably linked with the APKD gene? Explain your answer.
2 If APKD is linked with the B marker, how would II-3 have acquired the APKD gene? What is the probability of this event?
3 What is the probability that III-2 has APKD?
4 What is the probability that III-1 or III-3 has the disease?

Inquiry
Suppose that in generation I the grandfather was the individual with APKD. Would disease in the second generation be possible as shown? How would the change affect members of the third generation?

Chapter 13 Review

Vocabulary

applied genetics (191)
colchicine (193)
controlled breeding (191)
gene cloning (196)
gene sequencing (196)

gene splicing (195)
genetic engineering (194)
hybrid (192)
hybridization (192)

hybrid vigor (192)
inbreeding (192)
interferon (194)
mass selection (191)
plasmid (195)

polyploidy (193)
recombinant DNA (194)
restriction enzyme (194)

1. Name the type of controlled breeding practiced by Luther Burbank.
2. Distinguish between inbreeding and hybridization.
3. Look up the meanings of the word roots *poly-* and *-ploid* and tell why polyploidy is appropriately named.
4. Which type of genetic engineering involves the use of restriction enzymes?
5. What is a plasmid?

Review

1. Controlled breeding involves control of (a) the genetic makeup of an organism (b) parent selection (c) the breeding environment (d) the number of offspring that will be produced.
2. Humans began cultivating plants and animals for food about (a) 2,000 years ago (b) 1,000 years ago (c) 10,000 years ago (d) 5,000 years ago.
3. Luther Burbank developed the seedless grape through the use of the controlled breeding process of (a) mass selection (b) inbreeding (c) hybridization (d) inducing mutation.
4. The purpose of using the inbreeding technique is to (a) develop new species (b) develop new traits within a species (c) maintain or intensify desirable traits (d) eliminate undesirable traits.
5. Compared to its parents, a hybrid individual usually has the (a) same phenotype and same genotype (b) different phenotype and different genotype (c) same phenotype and different genotype (d) different phenotype and same genotype.
6. H. J. Muller induced mutations in *Drosophila,* common fruitflies, by exposing them to (a) ultraviolet radiation (b) colchicine (c) *Eco*RI (d) X rays.
7. Polyploid individuals contain multiple sets of (a) all chromosomes (b) sex chromosomes only (c) nuclei (d) cell plates.
8. Interferon is a protein that (a) cleaves DNA molecules together (b) destroys viruses (c) induces polyploidy (d) inhibits blood clotting.
9. Recombinant DNA technology is used to produce the substance (a) insulin (b) interferon (c) human growth hormone (d) human sex hormones.
10. *E. coli* containing recombinant DNA can reproduce as often as (a) every 20 minutes (b) every 30 minutes (c) every hour (d) every 2 hours.
11. What was the earliest form of applied genetics practiced by humans?
12. Name two plant varieties developed by Luther Burbank.
13. What weakness developed when dachshunds were inbred?
14. Why are hybrid organisms often unable to reproduce?
15. What was H. J. Muller's contribution to the field of applied genetics?
16. How do polyploid organisms differ from normal organisms?
17. How do scientists determine the exact

base sequence of a DNA segment?

18. What role do restriction enzymes play in the creation of recombinant DNA?

19. Why is gene cloning an efficient method of producing a substance such as interferon artificially?

20. How do scientists observe the inheritance patterns of artificial chromosomes in yeast cells?

Critical Thinking

1. List an advantage and a disadvantage of developing new plant and animal varieties using mass selection instead of genetic engineering techniques.

2. Male championship horses are often bred with their female offspring. This usually produces horses that possess many of the same exceptional traits as the paternal parent. What method of controlled breeding is being used here? Why might the resulting offspring inherit positive traits?

3. The United States government has stringent regulations requiring researchers to confine genetically engineered organisms to the laboratory. What concerns do you think might have led to the enactment of these regulations?

4. Bananas sold commercially are of the triploid variety—that is, they possess three sets of chromosomes. However, these bananas are infertile and must be produced through hybridization. Why are they unable to reproduce?

5. Examine the restriction map of the *E. coli* plasmid pBR 322, shown at the right. This is a commonly used plasmid com-

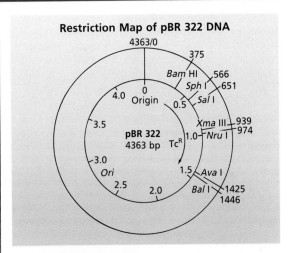

Restriction Map of pBR 322 DNA

posed of 4,363 base pairs. The map shows sites at which certain restriction enzymes cut the DNA of the plasmid. For example, *Sph*I cuts the plasmid at base pair 566. Suppose you want to isolate from the plasmid the gene that codes for resistance to the antibiotic tetracycline, which is indicated as *Tc^r*. What restriction enzymes would you use? How many base pairs long is the *Tc^r* gene? How might you check to be certain that your procedure was successful?

Extension

1. Read the article by Mary Murray entitled "Battling Illness with Body Proteins" in *Science News*, January 17, 1987, pp. 42–45. Present an oral report in which you include answers to the following questions: (1) What four gene-cloned medical products are now approved by the FDA? (2) What does ANF do? (3) What does tPA do? (4) What does TNF do? and (5) How is gene-cloned human growth hormone superior to growth hormone extracted from cadavers?

2. Recent practical applications of genetic engineering are often in the news. Look in issues of *Time* and *Newsweek* from the last several months for recent medical products produced with the help of genetic engineering, and present an oral report on these developments.

Intra-Science: *How the Sciences Work Together*

The Question in Biology: Can We Stop the Killer Bees?

In this unit you studied the application of genetics to human welfare. A related problem concerns *Apis mellifera,* the European honeybee, an important agricultural commodity in the United States. Not only do members of the species provide honey and wax, they also fertilize nearly 100 important crops, including alfalfa, apples, and almonds. In recent years, however, a strain of unusually aggressive honeybees, *Apis mellifera var. scutellata,* has entered the Western Hemisphere. Swarms of these killer bees, as they are called, are migrating toward North America. These deadly insects could kill thousands of animals and people, endanger the honeybee population, and cause billions of dollars worth of damage to crops.

The problem dates to 1956, when scientists imported a number of African bees to Brazil for breeding experiments. In 1957, swarms of bees, including 26 queens, escaped from a Brazilian apiary and began mating with wild European bees. The two varieties of honeybees are genetically similar. However, studies show that Africanized bees can be aroused to anger much more quickly than European bees. In addition, they sting 8 to 10 times as often, and take a half hour to become calm, 8 times longer than the European bees. Unlike European bees, the Africanized bees attack in swarms without provocation. Compounding the problem is the fact that these traits are passed on to the offspring produced by Africanized bees and European bees. Consequently, most mixed varieties are just as much of a menace as the pure African strain.

Since 1957, the rapidly multiplying killer bees have been traveling north at the rate of about 480 km per year. Isolated colonies were found in southern California in 1985. Experts predict that entire swarms of the bees could reach the United States by 1989. In the meantime, entomologists, geneticists, computer scientists, chemists, and meteorologists are combining their efforts to find a way to control the killer bees.

The Connection to Computer Science

The initial challenge facing scientists is determining whether a honeybee is actually a killer bee. Because European and Africanized bees are virtually identical, entomologists have worked with computer scientists to develop techniques for analyzing characteristics of unidentified bees. Technicians remove and then measure the length of the wings of a specimen. They also calculate the bee's net body weight after removing the contents of the bee's digestive system. These data, along with about two dozen other measurements, are fed into a computer for analysis. From the results, scientists have been able to determine the probability that a bee is either European or Africanized.

The Connection to Chemistry

Because European and Africanized bees are so similar genetically, any pesticide capable of killing one will probably kill the other as well. Consequently chemists are trying to develop other means of both controlling killer bees and protecting domestic European colonies. One plan, called drone trapping, involves setting traps baited with synthetic "queen substance." Queen substance is a chemical secreted by queen bees to attract drones, or male bees. The traps are used to capture both European and Africanized wild bees.

Chemists are also experimenting with poison baits. Wild bees consume the bait and carry it back to their hives. One type of poison sterilizes the queen bee; another type kills her.

The Connection to Other Sciences

Geneticists know that Africanized drones carry out their afternoon mating flights about 25 minutes later than European drones. Through controlled breeding, they have produced late-flying European bees. Populating an area with these bees could increase the likelihood that both African and European queens will be inseminated by European drones. Either way the result would be reduced numbers of pure-bred Africanized bees.

In addition to protecting areas already inhabited by killer bees, scientists are also trying to predict which areas are likely targets for future invasion. Africanized bees survive year around in areas where the mean high winter temperature is at least 15°C. Meteorologists have provided long-range forecasts of temperatures and wind patterns that can be used to predict where and when the bees will swarm.

The Connection to Careers

As you have seen, efforts to prevent killer bees from invading the United States involve biologists, chemists, and computer scientists. In addition, the employees of the U.S. Department of Agriculture will be needed to inspect northbound ships and trucks and to perform on-site inspections of managed bee colonies. Should the threat of the Africanized bees become imminent, health care professionals will likely be called upon to advise the public on how to protect and defend themselves from the killer bees. For more information on related careers in the biological sciences, turn to page 842.

201

Unit 4 Evolution

"Each organism instructs; its form and behavior embody general messages if only we can learn to read them. The language of this instruction is evolutionary theory."

Stephen Jay Gould

On the facing page you see a flower that looks like an insect and an insect that looks like a leaf. What biological processes have produced these oddities? What advantages do these organisms derive from their misleading appearances? You will learn answers to questions such as these as you study this unit on evolution, the process of change over time that modifies species in ways that enhance their survival.

Labyrinthodont, *Ichthyostega*, ancestral amphibian

Fossil of fish and palm leaf

Insect-mimic orchid, *Ophrys speculum*

Leaf-mimic katydid from Brazil

Chapter 14 | *The Origin of Life*

This chapter introduces students to processes scientists have used in answering questions about the origin of life on earth. Included in this study are the scientific processes of observing, hypothesizing, and experimenting.

Correlation Chart

Section Number	Section Objectives	Pages	Features	Section Review
14.1	1. Define *spontaneous generation* and list some observations people used as evidence for this theory about how life began.	205–207		1
	2. Summarize the results of experiments by Francesco Redi and by Lazzaro Spallanzani to test the theory of spontaneous generation.			2, 3
	3. Describe how Louis Pasteur's experiment disproved the spontaneous generation of microorganisms under existing conditions.			4, 5
14.2	1. Outline the modern scientific understanding of the formation of the earth.	208–211		1, 2, 3
	2. List four developments scientists hypothesize were necessary before life could have originated on earth.			2, 3, 4
	3. Summarize Oparin's theory of the origin of life and describe an experiment performed to test the theory.			4
	4. Compare and contrast coacervates and microspheres with living cells.			5
14.3	1. List three characteristics that describe the organisms scientists think were the first forms of life on earth.	212–215	Biology in Process, pp. 212–213	2
	2. Describe how prokaryotes and eukaryotes might have evolved.		Think, p. 214	1, 3, 4
	3. Name two ways that oxygen in the atmosphere influenced the evolution of life.		Biology Lab, p. 215	4, 5
	4. Summarize the endosymbiont hypothesis, and describe some of the evidence that supports the hypothesis.			1, 3

Resources*

Print

Margulis, Lynn and Dorion Sagan. *Microcosmos: Four Billion Years of Evolution from Our Microbial Ancestors.* New York: Summit (Simon & Schuster), 1986.

Audiovisual/Software

From Protistans to First Multicellular Animals. Film, 18 min. Benchmark, 1986.

*Additional references are listed in *Modern Biology, Teacher's Program Guide.*

Section Teaching Guides

14.1 Spontaneous Generation

Approach

The main objective of this section is to present the theory of spontaneous generation and to show how it was eventually discredited through the experiments of Francesco Redi, Lazzaro Spallanzani, and Louis Pasteur. The section builds upon the concept of scientific processes (Chapter 2), illustrating how a theory is evaluated and modified through observation and experimentation. In presenting the text, guide students to develop an appreciation for the complex nature of scientific inquiry and to gain historical perspective on a great scientific debate about the origin of life.

Concept Framework

Most students are probably aware of some testable theories of how life originated on earth as well as untestable theories on this topic. Ask students to identify testable and nontestable theories of life origin. Point out that this chapter addresses the question of how life originated in a scientific manner, posing only testable answers. Suggest that students make note of answers posed by different scientists as they read this section.

Emphasis

Students should be able to identify the control group and the experimental groups in each of the following experiments: Redi's experiment on flies and meat; Spallanzani's experiment on broth in sealed flasks; and Pasteur's experiment on broth in curve-necked flasks.

Debate/Simulation

Encourage students to imagine they are scientists in the late 1700s. Divide students into three groups: the first group should support the theory of spontaneous generation; the second group should oppose this theory; the third group should serve as a panel of judges. Ask the first two groups to develop their arguments and bring in evidence to present in a debate. Encourage students to confine their evidence to that which could be gathered in the 1700s. The judges should decide the winner based only on the arguments presented in class.

Section Review Answers

(Continued from p. 207.)
5. The broth would have become contaminated with microorganisms, supporting Pasteur's conclusion. The only source of contamination would have been contact between the broth and the trapped particles when the flask was tipped.

14.2 Origin of Life on Earth

Approach

The main objective of this section is to familiarize students with four developments scientists infer must have occurred for life to begin on earth. The chapter builds upon concepts of basic chemistry (Chapter 3), biochemistry (Chapter 4), and cell processes (Chapter 7). In teaching the text, guide students to understand the complex conditions required for the origin of life.

Concept Framework

Many students are aware that earth is the only planet in our solar system known to support life. Encourage them to compare characteristics such as temperature, relative distance from the sun, and surface features of earth and other planets. Ask them to relate these characteristics to the presence and absence of life. As they read this section, encourage students to note how conditions on earth changed over time.

Emphasis

Students should understand why the following developments are important in the origin of life: formation of simple and complex organic compounds; concentration and enclosure of these organic compounds; evolution of growth, metabolism, and reproduction.

Guest Speaker

Invite an astronomy professor from a local college or university to talk to the class about theories concerning the origin of the universe or the origin of our solar system.

Section Review Answers

(Continued from p. 211.)
4. Oparin hypothesized that hot gases in the atmosphere would react and form simple organic compounds. Miller and Urey tested the theory by

sending electric current through the heated gases, producing some simple organic compounds Oparin hypothesized would form.

5. Accept charts that reflect features discussed on pp. 210–211.

14.3 The First Forms of Life

Approach

The main objective of this section is to present current theories on the evolution of prokaryotes and eukaryotes. The section reinforces knowledge of cell structure and function (Chapter 5) and cell processes (Chapter 7). Guide students to gain a general understanding of evolution as a change in a type of organism over time.

Concept Framework

Most students will have familiarity with certain unicellular organisms such as bacteria. To provide a framework for understanding the information presented in this section, remind them that bacteria are prokaryotes. Assist the class in making a list of the characteristics of prokaryotes and eukaryotes. As students read this section, ask them to note evidence that suggests how prokaryotes and eukaryotes might have evolved.

Emphasis

Make sure students understand that scientists infer that prokaryotes preceded eukaryotes because, in general, simpler organisms have evolved before more complex organisms.

Time Line

Have students develop a time line of the evolution of life beginning with the formation of the earth and extending to the present. Begin by adding the evolution of prokaryotic and eukaryotic cells to the time line. Students can add other organisms as they read later chapters.

Using the Features/The Evolution of Eukaryotes

The main process skills used in this feature are observing and hypothesizing. Dr. Lynn Margulis has compiled evidence to support the endosymbiont hypothesis. Ask students if they think this hypothesis is directly testable or not and why.

Section Review Answers

(Continued from p. 214.)

2. Scientists infer that the first organisms were the following: simple and thus prokaryotic; anaerobic because no oxygen was available; heterotrophic because organic molecules would have been available in the water.
3. They might have evolved through an ability to make their own food, an advantage when food became scarce.
4. The initial function of aerobic respiration was to bind oxygen with other compounds to prevent damage to coenzymes.
5. The ozone layer protects cells from damage from ultraviolet light. Destruction of the ozone layer could destroy life on earth.

Laboratory Preparation

Background

In this laboratory experiment, students will form microspheres by heating certain amino acids. They will observe the microspheres and compare their characteristics with those of living cells. Students will note shared characteristics. For example, the contents of both microspheres and living cells differ from any surrounding liquid. They will both shrink if placed in a hypertonic solution, or swell if placed in a hypotonic solution. Like bacteria cells, microspheres can be either gram-positive or gram-negative, depending on their amino acid makeup. Fission can be induced in acidic microspheres by raising the pH. Current hypotheses about prebiotic earth consider these characteristics as indications that the evolution of life could have begun with something similar to simple microspheres.

Preparation

Order enough amino acids ahead of time to supply all lab stations with 1 g each of aspartic acid, glutamic acid, and glycene. Prepare 1% NaCl solution by adding 5 g of sodium chloride to 500 mL of distilled water.

Safety

Caution students not to ingest the amino acid powders, and to wash their hands after handling them.

Conclusions

1. Microspheres and living cells both have membranelike boundaries.
2. Microspheres do not have a nucleus or organelles, and living cells do.

Inquiry

1. Students should form a hypothesis about how different amounts of heat might affect the results of this experiment, and suggest ways to test this hypothesis.
2. Students should suggest similar experiments to the one they have conducted but with different amino acids.

Chapter Review

Vocabulary

1. *Bio-* refers to life; *-genesis* refers to the origins or coming into being of something; the theory that life comes from life is an explanation of how life comes into being.
2. Fossils show traces or remains of living organisms. Scientists use fossils to date when the first organisms might have appeared on earth.
3. Coacervates and microspheres are microscopic droplets that have been concentrated and enclosed within a boundary. They are similar to living cells in that they can selectively take up substances from their surroundings; their contents become organized with different types of molecules collected in different areas; and certain chemical reactions can take place more easily within them than they can in water; they grow and appear to reproduce. They look similar to prokaryotic cells.
4. "Spontaneous" means that something happens quickly and naturally; "generation" refers to bringing into existence. This theory assumed that living organisms formed in a matter of minutes, hours, or days from nonliving matter.
5. *Endo-* means "within" and *sym-* means "together." "Endosymbiosis" is a good name for the theory because it refers to the idea that bacteria and cyanobacteria living within a prokaryotic cell may have evolved into mitochondria and plastids of eukaryotic cells.

Review

11. anaerobic, heterotrophic prokaryotes
12. when nutrients the heterotrophs used for food became scarce due to competition
13. by chemosynthesis
14. by preventing the gas from destroying essential chemicals in the organisms' cells
15. The ozone layer absorbed damaging energy from ultraviolet light.
16. Fish appeared in ponds that had gone completely dry during the previous season; flies swarmed from rotten meat; broth became cloudy with bacteria.
17. If living organisms can come only from other living organisms, how could life on earth have begun?
18. to provide energy to drive the chemical reactions in the chamber; lightning
19. because many of the organisms were autotrophs not heterotrophs (they were photosynthetic)
20. cases of gene transmission not explainable by nuclear chromosomes alone

Critical Thinking

1. Answers should include some way to identify the fish eggs which give rise to the fish in the pond. Conduct an experiment to compare two artificial mud-bottomed "ponds": one with mud made from sterile soil; the other with mud from a pond that had fish in it.
2. Answers should include an evaluation of conditions which would probably have to be met for life to form, including an atmosphere containing carbon, temperatures between the boiling and freezing point of water, some form of energy to drive the chemical reactions to form life.
3. by using other sources of energy, including heat, sunlight, ultraviolet radiation, mechanical energy, simulating an earthquake or volcanic eruption, or other sources
4. the taking in of organic molecules as food from the oceans by the first heterotrophs; the generation of oxygen gas by photosynthetic organisms; the uptake of oxygen gas by aerobic organisms; the colonization of land by living organisms
5. Prokaryotes first appeared at about 3:00 A.M., eukaryotes at about 9:00 A.M.

Chapter 14 | The Origin of Life

Introduction

How did life on earth begin? Over the centuries scientists have sought the answer to this question by engaging in the various processes of science. In this chapter you will learn how scientists have used observing, hypothesizing, experimenting, and other processes in studying the origin of life.

Chapter Outline

Chapter Concept

As you read, look for examples of the many processes of science involved in answering questions about how life began on earth.

Lake of Kilauea Volcano, Hawaii. Inset: The Andromeda Galaxy.

For additional teaching information see pp. T14A–T14D.

14.1 Spontaneous Generation

Can mud produce live fish? Will rotting meat give rise to flies? Does the air generate microorganisms? In the past, people asked such questions in pondering a deep mystery of science: How did life arise on earth? Then, as today, people attempted to unravel this mystery by relying primarily on their observations. Fish appeared in ponds that had gone completely dry during the previous season. That mud might have given rise to such fish seemed a reasonable explanation for what people observed. Flies swarming from rotten meat and clear broth clouding up with bacteria were other phenomena people attempted to explain. Their observations led them to the conclusion that living organisms could arise from nonliving matter. This process, known as **spontaneous generation,** was one way people explained how new life formed.

In attempting to learn more about the process of spontaneous generation, scientists performed controlled experiments. As you read about each of the experiments described below, refer to the figures that show the experimental setup.

Redi's Experiment

In times past, as now, flies were such pesky creatures that most people were too busy brushing them off to bother studying them. In fact, not until the 1600s did anyone notice that flies had different developmental forms. In the middle of that century, the Italian scientist Francesco Redi (1626–1697) observed that tiny white wormlike maggots turned into sturdy oval cases from which flies eventually emerged. He also observed that maggots seemed to appear where adult flies had previously landed. These observations led him to question the commonly held hyphothesis that flies were generated spontaneously from rotten meat.

Figure 14-1 shows an experiment Redi set up in 1668 to test the spontaneous generation of flies. The control group consisted

Section Objectives

- *Define* spontaneous generation *and list some observations people used as evidence for this theory about how life began.*

- *Summarize the results of experiments by Francesco Redi and by Lazzaro Spallanzani to test the theory of spontaneous generation.*

- *Describe how Louis Pasteur's experiment disproved the spontaneous generation of microorganisms under existing conditions.*

Figure 14-1   In Redi's experiment maggots were found only in the control jars because only there could flies reach the meat.

Redi's Experiment

Control group

Experimental group

Netting

Meat — Maggots

Meat
No maggots

Spallanzani's Experiment

Control group   Experimental group

Broth is boiled.   Broth is boiled.

Flask is open.   Flask is sealed.

Broth becomes cloudy.   Broth remains clear.

Figure 14-2 Spallanzani's flask was tightly sealed after boiling. The contents remained uncontaminated.

Tip: Ask students to think of a possible source of error in Spallanzani's experiment. Answers might include not sealing the flask before microorganisms entered it.

of uncovered jars that contained meat. The experimental group consisted of similar jars covered with netting. The netting allowed air to enter, while preventing flies from landing on the meat. After a period of time, maggots swarmed on the meat in the open jars. The net-covered jars remained free of maggots despite the flies buzzing all around. The results of Redi's experiment convinced most people that flies come from eggs laid by other flies; flies do not arise spontaneously from rotting meat.

Spallanzani's Experiment

At about the same time that Redi carried out his experiment, other scientists began using a new tool—the microscope. Their observations with the microscope revealed that the world was teeming with tiny creatures. These microorganisms seemed so simple in structure, so numerous, and so widespread that many people concluded that microorganisms must arise spontaneously from a "vital force" in the air.

In the 1700s another Italian scientist, Lazzaro Spallanzani, (1729–1799), designed an experiment, shown in Figure 14-2, to test spontaneous generation of microorganisms. Spallanzani hypothesized that microorganisms formed not from air, but from other microorganisms. He assumed that boiling broth in a flask would kill all the microorganisms in the flask—those in the broth, those on the inside of the glass, and those in the air in the flask. Spallanzani knew that boiled broth was clear. He also knew that a flask of boiled broth left unsealed would soon become cloudy, contaminated with microorganisms. Spallanzani sealed his experimental flask of boiled broth by melting the glass top shut. He found that the broth in the sealed flask remained clear.

Spallanzani concluded that boiled broth became contaminated only when microorganisms from the air entered the flask. However, Spallanzani's opponents objected to his method and disagreed with his conclusions. They claimed Spallanzani boiled his flasks so long that he destroyed the "vital force" in the air inside the flask. Air lacking this "vital force" could not generate life, they claimed. Scientists continued to debate the spontaneous generation of microorganisms into the 1800s.

Pasteur's Experiment

By the mid-1800s the spontaneous generation controversy had grown so fierce that the Paris Academy of Science offered a prize to anyone who could clear up the issue once and for all. The winner of the prize turned out to be the French scientist Louis Pasteur (1822–1895).

Refer to Figure 14-3 to see how Pasteur set up his prize-winning experiment. To answer objections to Spallanzani's experiment, Pasteur made a curve-necked flask that allowed air to enter. Broth boiled in this flask remained clear and uncontami-

Pasteur's Experiment

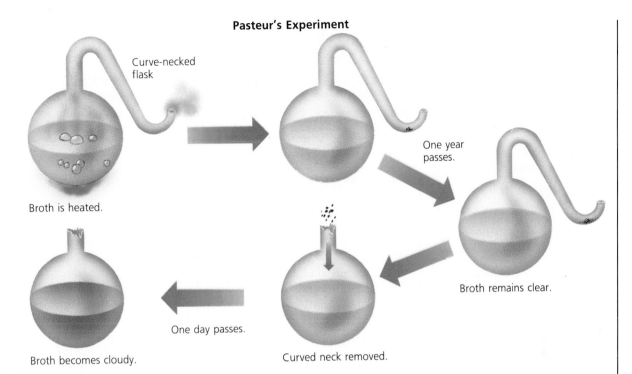

Curve-necked flask

Broth is heated.

One year passes.

Broth remains clear.

Curved neck removed.

One day passes.

Broth becomes cloudy.

nated for more than a year, even though it was exposed to air. The curve in the neck of the flask trapped particles in the air, preventing them from reaching the broth. When Pasteur removed the curved neck, the broth clouded with microorganisms within a day. From such experiments, Pasteur concluded that "productions of infusions [liquids contaminated with microorganisms], previously heated, have no other origin than the solid particles which the air always transports."

With this experiment Pasteur claimed to have "driven partisans of the doctrine of spontaneous generation into the corner." The theory died a sudden death. **Biogenesis** (*BIE-oe-JEN-uh-suss*), the theory that living organisms come only from other living organisms, became a cornerstone of biology.

Section Review

1. What is the theory of spontaneous generation, and how did people use the theory to explain what they observed?
2. How did Redi's experiment help disprove spontaneous generation of flies?
3. Why did proponents of the theory of spontaneous generation object to Spallanzani's experiment?
4. Describe Pasteur's experiment and his results.
5. What would have happened if Pasteur had tipped one of his flasks so that the broth in the flask came into contact with the inside of the curved neck? Would these results have supported his conclusion? Why or why not?

Info: Point out that *abiogenesis* is another term for spontaneous generation. The *a-* prefix means "without."

Figure 14-3 In Pasteur's experiment broth became contaminated only when the curved neck was removed from a flask.

Answers to Section Review
1. The theory states that living organisms can arise from nonliving matter. Accept any of the examples given on p. 205.
2. Maggots did not appear on meat when flies were kept away.
3. They claimed Spallanzani had destroyed the "vital force" necessary to spontaneous generation by heating the air in the flask for so long a time.
4. Pasteur used a curve-necked flask that allowed air, but not dust and other particles, to reach the sterile broth. The broth remained uncontaminated, proving that particles in the air, not the air itself, contaminated the broth.
(See pp. T14A–T14D.)

Figure 14-4   It took more than a billion years for the modern earth to form from a swirling mass of gases.

14.2 Origin of Life on Earth

Pasteur's experiment left scientists with a dilemma. If living organisms in the existing environment can arise only from other living organisms, how could life on earth have begun? To answer this question, we first need to take a look at the early history of the earth.

Formation of the Earth

Use Figure 14-4 to trace the steps in the early history of earth. Using evidence from computer models of the sun, scientists hypothesize that about 5 billion years ago the solar system was a swirling mass of gas and dust. Within a few million years, most of this material had collapsed inward and formed the sun. The remaining dust and gas ringed the young sun as an enormous disk. This material is thought to have collected in clumps that formed the planets, including the earth.

According to evidence from meteors assumed to be the same age as earth, our planet is dated at about 4.6 billion years of age. Beginning as a ball of very hot rock, the young planet was quite different from today's earth. Meteorites bombarded its surface. Volcanic eruptions constantly shook the planet. Lava released hot gases that formed the early atmosphere. In time the temperature at

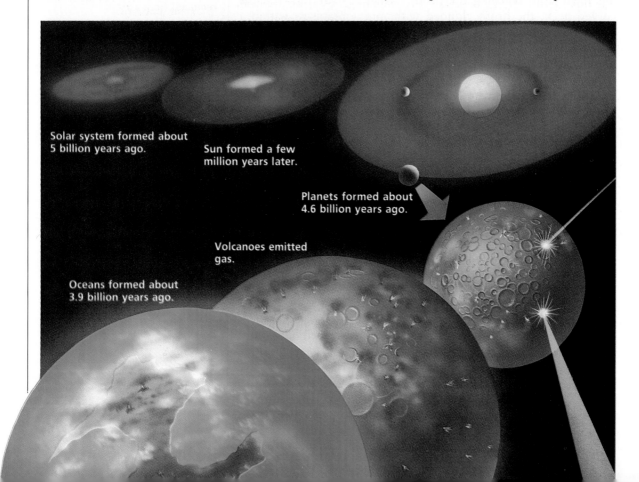

Solar system formed about 5 billion years ago.

Sun formed a few million years later.

Planets formed about 4.6 billion years ago.

Volcanoes emitted gas.

Oceans formed about 3.9 billion years ago.

the earth's surface dropped below the boiling point of water. The water vapor in the air condensed and fell to earth as rain. In this way the oceans formed about 3.8 billion to 3.9 billion years ago.

The Appearance of Life on Earth

The above account of the earth's formation and early history is the theory that best fits the available scientific evidence. However, no rocks have survived to provide direct evidence about the earth's earliest eras. The oldest rocks on earth are only about 3.9 billion years old.

Though we know little about the earth's infancy, we do have records of early forms of life on earth. Some rocks up to 3.5 billion years old contain **fossils**—remains or traces of once-living organisms. These first fossils are of ancient prokaryotic cells. (Remember that prokaryotes lack a membrane-bound nucleus and membrane-bound organelles.) These microscopic fossils, or **microfossils**, are probably not the remains of the first cells that lived on earth. Instead these fossilized cells probably descended from a long line of unpreserved prokaryotes.

For life to come into being, scientists agree that four developments must have occurred. These include the following: the formation of simple organic compounds important to life; the formation of complex organic compounds, such as proteins; the concentration and enclosure of these organic compounds; and the linking of chemical reactions involved in growth, metabolism, and reproduction. Let's look at each of these developments.

Formation of Simple Organic Compounds

As you know, organic compounds—those containing carbon—are essential to life. Carbon and all the other elements found in organic compounds existed on earth when the planet came into being. In time the main elements of life—carbon, hydrogen, nitrogen, and oxygen—came to form gases in the atmosphere.

How these elements present in the atmosphere could have formed simple organic compounds important to life is a challenging scientific puzzle. One of the most popular hypotheses put forth to solve this puzzle was developed by the Soviet scientist Alexander I. Oparin in 1923. Oparin (1894–1980) suggested that the atmosphere of the primitive earth contained ammonia (NH_3), hydrogen gas (H_2), water vapor (H_2O), and compounds made of hydrogen and carbon, such as methane (CH_4). At temperatures well above the boiling point of water, the gases in the atmosphere might have formed simple organic compounds, such as amino acids. According to Oparin, when the earth cooled and water vapor condensed to form lakes and seas, these simple organic compounds would have collected in the water. Over time the compounds entered complex chemical reactions. Oparin hypothesized that proteins and other organic compounds important to life would have resulted from these reactions.

Figure 14-5 These microfossils show spheroid cells and filaments (top) and a closeup of a filament (bottom).

Info: Only 50 km below its surface, the earth is still hot. A small portion of the earth is still molten.

Glycine

$$NH_2 \quad O$$
$$H_2C-C$$
$$OH$$

Alanine

$$NH_2 \quad O$$
$$H_3C-CH-C$$
$$OH$$

Figure 14-6 Miller and Urey's device testing Oparin's hypothesis produced organic compounds, including the amino acids shown.

Labels on figure: CH$_4$ H$_2$ NH$_3$ enter chamber; Spark; Condensor; Water vapor forms H$_2$O; Organic compounds

Oparin developed his hypothesis by reading the writings of hundreds of scientists working on a wide range of problems. His hypothesis rests on the results of experiments done by these scientists. However, Oparin did not run experiments to test his hypothesis. That work was left for later scientists.

In 1953 the American scientists Stanley L. Miller (1930–) and Harold C. Urey (1893–1981) set up an experiment using Oparin's hypothesis as a starting point. Their setup is illustrated in Figure 14-6. They made a chamber containing the gases Oparin assumed were present in the young earth's atmosphere. As the gases circulated in the chamber, sparks, representing lightning, supplied energy to drive chemical reactions. The experiment generated organic compounds including the amino acids—the building blocks of proteins—listed in the figure.

Since the 1950s, scientists have continued to explore the origin of simple organic compounds. Their experiments have produced a variety of compounds, including various amino acids, ATP, and the nucleic acids in DNA. Such results suggest many ways that vital organic molecules might have formed on the young earth.

Formation of Complex Organic Compounds

How could simple organic compounds have formed complex organic compounds essential to life? This question has been partially answered for one group of organic compounds—proteins. Remember that proteins are chains of amino acids. Some scientists have performed experiments that suggest that amino acids might have formed chains spontaneously in the early atmosphere. Other scientists have shown that amino acids will link up when heated in the absence of oxygen gas. In either or both of these ways, amino acids could have combined and formed complex proteins. Similar mechanisms might have led to the formation of carbohydrates, lipids, and nucleic acids.

Concentration and Enclosure of Organic Compounds

The next step in the origin of life, according to most scientists, was the concentration and enclosure of complex organic compounds. Experiments have shown that proteins clump together to form microscopic droplets. Coacervates and microspheres are two types of these droplets. **Coacervates** *(koe-AS-ur-VATES)* are actually collections of droplets, made of molecules of different types, that have irregular shapes. **Microspheres,** on the other hand, are round and usually form from only one type of molecule.

Coacervates and microspheres bear a resemblance to living cells. Coacervates, microspheres, and cells are all set off from the environment by a membranelike boundary. All three can selectively take up certain substances from their surroundings. Internally, different types of molecules cluster in different areas of these structures. Also, certain chemical reactions can take place more easily inside these structures than they can in water. Coacervates can grow; microspheres can bud, forming smaller microspheres. Under the microscope both of these structures sometimes look so much like living cells that scientists have sometimes mistaken them for new species of bacteria.

One process that must have occurred on the young earth was the enclosure of nucleic acids in membranes. Once DNA was separated from the environment by some kind of boundary, it would be protected and might be able to carry out the precise reactions of replication. Such a process has been recreated in the laboratory. Figure 14-7 shows a mixture of lipids and DNA exposed to conditions of wetting and drying that may simulate those in ancient tidepools. When the mixture was hydrated, the DNA and lipids intertwined with one another. When it was dried, the lipids sandwiched the DNA molecules. When rehydrated, the lipid membranes dispersed again, but now the DNA was enclosed in lipid vessels. In this way nucleic acids might have been concentrated and enclosed in membranes on the young earth.

Evolution of Growth, Metabolism, and Reproduction

As lifelike as coacervates and microspheres appear, they lack the complexity of living cells. Coacervates and microspheres can neither maintain stable growth nor reproduce. They also cannot metabolize—that is, they are unable to obtain energy from substances in the environment. But over time coacervates and microspheres might have developed these capabilities, which are necessary for life. So endowed, these nonliving structures might have eventually crossed over the border into the living world.

In summary, evidence suggests that, sometime between 4.6 billion and 3.5 billion years ago, life arose on earth, generated from nonliving matter. However, such spontaneous generation could not occur on earth today. The atmosphere of the young earth lacked oxygen gas. Though this substance is essential to most forms of life today, it would have destroyed organic compounds in the hot atmosphere of the young earth by the chemical process called oxidation. In fact, the oxygen gas in our atmosphere was almost certainly generated by living organisms. These living organisms helped change conditions on earth. Thus, the unique circumstances under which life spontaneously arose are no longer present on earth.

Section Review

1. What is a *fossil*?
2. How do scientists think the earth formed, and what do they think the earth was like during its first billion years?
3. According to scientists what four developments must have occurred before life came into being?
4. What was Oparin's hypothesis, and how was it tested?
5. Make a chart comparing and contrasting coacervates and microspheres with living cells.

Tip: Discuss with students why budding of microspheres is not true reproduction. (Genetic material is not transferred.)

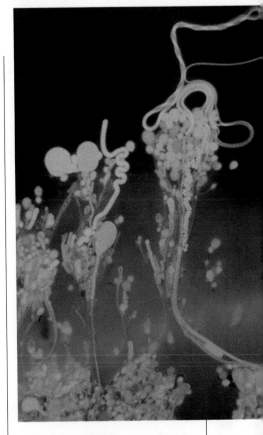

Figure 14-7 Membrane vessels and twisted multilayer membranes have been formed in the laboratory under conditions that may have existed on the young earth. The membranes in this photograph have formed boundary layers that have encapsulated DNA.

Answers to Section Review
1. the remains or traces of a once-living organism
2. from a cloud of gas and dust orbiting the sun; a ball of very hot rock
3. the formation of both simple and complex organic compounds; the concentration and enclosure of organic molecules; the linking of chemical reactions involved in cell growth, metabolism, and reproduction
(See pp. T14A–T14D.)

14.3 The First Forms of Life

A remote and desolate corner of Australia was nicknamed "the North Pole" by prospectors of the 1800s. However, this region was a gold mine for twentieth-century scientists. There they found the 3.5-billion-year-old microfossils mentioned in Section 14.2. As you know, these fossils were probably not the earth's first organisms. Scientists hypothesize that the first cells were anaerobic, heterotrophic prokaryotes.

The First Prokaryotes

Prokaryotes were probably the first organisms on earth. As you have read, the atmosphere of the young earth lacked oxygen gas. This suggests that the first prokaryotes must have been anaerobic. They probably derived their energy from the metabolic pathway of glycolysis. Remember that glycolysis does not require oxygen.

Bathed in the earth's warm seas, the first prokaryotes probably just floated into the organic molecules they used for food. Because they gathered food from their surroundings rather than

Biology in Process

The Evolution of Eukaryotes

Look at the photographs of a chloroplast (far right) and a blue-green bacterium (near right). Then study the photographs of a mitochondrion (top left) and a bacterium (bottom left). Do you notice similarities in the size and appearance of these pairs of structures? If so, you have made an observation that led scientists in the late 1800s to propose the endosymbiont hypothesis of eukaryote evolution.

In the early 1900s this hypothesis fell out of favor. Genetic studies suggested that all cell traits are carried on chromosomes in the nucleus. The endosymbiont hypothesis implied that some traits, for example mitochondria and chloroplasts, would be inherited independently of nuclear chromosomes. At first scientists found no supporting evidence.

In the late 1950s Lynn Margulis, at that time a graduate student in genetics, learned of cases of gene transmission that could not be explained by nuclear chromosomes alone. She inferred that these unusual patterns of heredity meant that some DNA existed outside the nucleus.

Margulis's hunch proved correct. In 1962 scientists discovered DNA in the chloro-making it, they were heterotrophic. In time, however, the heterotrophic prokaryotes would have multiplied, increasing competition for nutrients. Organisms that could make their own food would have had a distinct advantage. These conditions may have led to the development of the earth's first autotrophs. These autotrophs probably made glucose through chemosynthesis rather than through photosynthesis. Next, photosynthesizing prokaryotes probably evolved. The cell structure and organization of the 3.5-billion-year-old microfossils indicates that these organisms were among the first photosynthesizing prokaryotes.

The Changing Earth

Changes in types and numbers of fossils indicate that major changes in earth's environment occurred about 2.8 billion years ago. At that time photosynthesizing organisms that gave off oxygen gas as a waste product first arose. The generation of oxygen by prokaryotes changed the world. Oxygen could destroy some essential coenzymes, but organisms that bonded oxygen gas to other compounds prevented it from doing damage. This bonding is one of the first steps in aerobic respiration. So it seems that

Answer to Biology in Process
This fact helps support the hypothesis. Since the association has been shown to be possible, it could have arisen before.

 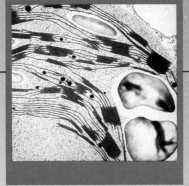

plasts of algae. How did DNA get into the chloroplasts? To Margulis, the existence of DNA in these complex organelles suggested that the endosymbiont hypothesis might be valid. After completing her Ph.D. in genetics in 1965, she set out to make a case for the hypothesis.

Over the last twenty-five years Margulis has compiled compelling evidence to support the hypothesis that mitochondria evolved from bacteria and

that chloroplasts evolved from blue-green bacteria. Some of Margulis's evidence rests on observations like those you have made: The cell structures exhibit a startling similarity to the free-living organisms. Other evidence includes the fact that DNA and ribosomes have been found in both mitochondria and chloroplasts. The ribosomes of these organelles are very similar to the DNA and ribosomes of bacteria and cyano-

bacteria. Margulis's arguments have convinced many scientists of the endosymbiotic origins of chloroplasts and mitochondria.

Margulis continues to compile evidence for an extension of the endosymbiont hypothesis: that centrioles evolved from swimming bacteria. Remember that these structures seem to help guide the movement of chromosomes during cell division.

■ Some prokaryotes live in close association with eukaryotes today. Could this fact be used as evidence for the endosymbiont hypothesis? Why or why not?

Think

Some environments still support anaerobic, heterotrophic prokaryotes. Which of these two environments do you think would be able to support such organisms? Why?

Answer to Think
The environment pictured in the top photo would support such organisms, because it is oxygen poor.

Answers to Section Review
1. Blue-green bacteria and bacteria lived in symbiosis within other prokaryotes and eventually evolved into the chloroplasts and mitochondria of eukaryotes.
(See pp. T14A–T14D.)

aerobic respiration may have first been an adaptation to prevent the destruction of essential organic compounds by oxygen.

As you read in Chapter 7, aerobic respiration can liberate much more energy from a glucose molecule than anaerobic respiration can. Therefore aerobic respiration probably gave some organisms an advantage and in time they became dominant.

The release of oxygen through photosynthesis had another effect. Ultraviolet light from the sun is hazardous. It can damage DNA. Before oxygen built up in the atmosphere, the only way organisms could shield themselves from these deadly rays was by living in water. Water absorbs energy from ultraviolet light.

When oxygen was released into the atmosphere, it provided a second protective shield from ultraviolet radiation. Some oxygen gas (O_2) absorbs energy from ultraviolet light and splits into single oxygen atoms (O). Then oxygen gas reacts with the single atoms of oxygen and forms a layer of ozone (O_3) that encircles the globe. This process, which continually occurs in the atmosphere, prevents harmful radiation from reaching the earth.

The First Eukaryotes

Prokaryotes are so much simpler than eukaryotes that most scientists infer that prokaryotes evolved first. The fossil record supports this inference. Scientists do not know exactly how eukaryotes evolved. Certain prokaryotes, specifically bacteria and blue-green bacteria, may have come to live inside other prokaryotes, thus gaining protection. The bacteria in turn may have provided extra energy for the host prokaryotes. The blue-green bacteria may have provided extra food. Such a relationship, in which different organisms live in close association, is called symbiosis.

Eventually the bacteria and cyanobacteria might have evolved into mitochondria and chloroplasts. In addition, a nucleus could have formed when membranes enclosed DNA. Remember that a nucleus and organelles are distinguishing features of eukaryotes. The idea that a eukaryote's mitochondria evolved from bacteria and its chloroplasts evolved from cyanobacteria is the basis of the theory of **endosymbiosis,** also called the endosymbiont hypothesis. Fossil evidence indicates that eukaryotes had evolved by 1 billion years ago.

Section Review

1. Explain the endosymbiont hypothesis of eukaryote evolution.
2. Why do scientists think the first forms of life on earth were anaerobic, heterotrophic prokaryotes?
3. How might autotrophic prokaryotes have evolved?
4. What do scientists speculate was the initial function of aerobic respiration?
5. Some forms of air pollution damage the earth's ozone layer. How might such damage affect life?

Laboratory

Making Microspheres

Objective
To make microspheres from amino acids and to compare their structure to living cells

Process Skills
Observing, hypothesizing

Materials
500-mL beaker, hot plate, 2 125-mL Erlenmeyer flasks, ring stand with clamp, balance, aspartic acid, glutamic acid, glycene, glass stirring rod, tongs, clock or timer, 1% NaCl solution, 50-mL graduated cylinder, dropper, microscope slide, coverslip, microscope

Method

1 Fill a 500-mL beaker one-half full with water and heat it on a hot plate. Leave space on the hot plate for a 125-mL Erlenmeyer flask to be added later.

2 While waiting for the water to boil, clamp a 125-ml Erlenmeyer flask to a ring stand. Add 1 g each of aspartic acid, glutamic acid, and glycene to the flask and combine these dry powders with a stirring rod.

3 When the water in the beaker begins to boil, move the ring stand carefully so the flask of amino acids sits in the hot-water bath.

4 When the amino acids have heated for 20 minutes, measure 10 mL of NaCl solution in a graduated cylinder and pour the solution into a second Erlenmeyer flask. Place the second flask on the hot plate beside the hot-water bath.

5 When the NaCl solution begins to boil, use tongs to remove the flask containing the solution from the hot plate. Then, still holding the flask with tongs, slowly add the NaCl solution to the hot amino acids while stirring.

6 Let this solution boil for 30 seconds. Why do you think you heated the solutions?

7 Remove the solution from the water bath, and allow it to cool for 10 minutes.

8 Use a dropper to place a drop of the solution on a microscope slide and cover the drop with a coverslip.

9 Place the slide on the microscope stage. Examine the slide under low power for tiny spherical structures. Then examine the structures under high power. These tiny sphere-shaped objects are microspheres. How do you think the microspheres were formed?

Conclusions

1 In what ways do microspheres visually resemble living cells?

2 In what ways do microspheres differ visually from living cells?

Inquiry

1 What do you think would happen if you added too much or too little heat? How can you test for the right amount of heat to use?

2 Do you think your microsphere experiment would have worked if you had substituted other amino acids? How can you test your hypothesis?

Prepare the 1% salt solution by adding 5 g of sodium chloride to 500 mL of distilled water.

Chapter 14 Review

Vocabulary

biogenesis (207)
coacervate (210)

endosymbiosis (214)
fossil (209)

microfossil (209)
microsphere (210)

spontaneous
 generation (205)

1. Using a dictionary, find the meaning of the word parts *bio-* and *-genesis* and relate them to the theory of biogenesis.
2. How do fossils give evidence about the origin of life?
3. Define coacervates and microspheres; compare them with living cells.

4. Why is "spontaneous generation" a good name for the theory it describes?
5. Use a dictionary to find the meaning of the prefixes *endo-* and *sym-*. Why is "endosymbiosis" a good name for the theory it describes?

Review

1. In the 1600s and 1700s people used the theory of spontaneous generation to explain (a) how new life started (b) how simple organic compounds formed (c) how coacervates and microspheres originated (d) how eukaryotes evolved.
2. Redi's experiment was important because it showed that (a) maggots give rise to microorganisms (b) flies swarm on rotting meat (c) flies do not form from rotting meat (d) air contains a "vital force."
3. People objected to Spallanzani's experiment because (a) he used an open jar of meat as a control (b) he heated his flasks of broth for a long time (c) he drew the necks of his flasks into a curved shape (d) his flasks contained microorganisms.
4. The neck of Pasteur's flasks (a) allowed both air and particles to enter the flask (b) allowed air to enter but kept particles out (c) allowed particles to enter but kept air out (d) kept both air and particles out.
5. Scientists think the earth formed (a) from the sun (b) from water vapor (c) from volcanoes (d) from gas and dust.
6. The oldest fossil is about (a) 4.6 billion years old (b) 3.5 billion years old (c) 3.5 million years old (d) 2 million years old.
7. The elements that formed simple organic molecules essential to life probably originally came from (a) the atmosphere (b) the moon (c) the sun (d) the oceans.

8. Miller and Urey's experiment (a) proved Oparin's hypothesis about the origin of life (b) disproved the hypothesis (c) provided evidence for the hypothesis (d) did not apply to the hypothesis.
9. Coacervates and microspheres are (a) collections of organic molecules, concentrated and enclosed within a boundary (b) the first forms of life (c) the oldest microfossils (d) new forms of bacteria.
10. The generation of organisms from nonliving material does not occur today mainly because (a) the presence of oxygen in the atmosphere prevents organic compounds from spontaneously forming (b) coacervates and microspheres take up all the extra nutrients on earth (c) heterotrophs take up all the extra nutrients on earth (d) there is not enough energy to drive the chemical reactions to form complex organic compounds necessary for life.
11. What kinds of organisms are thought to have been the first life-forms on earth?
12. Which environmental factors probably favored the evolution of autotrophs?
13. By what process did the first autotrophs probably make glucose?
14. How might the bonding of oxygen gas have served evolving organisms?
15. How did the formation of the ozone layer permit organisms to colonize land?
16. What were some observations that might

have supported the theory of spontaneous generation?

17. Why did the theory of biogenesis pose a dilemma regarding the origin of life?

18. In the Miller–Urey experiment, what was the function of the electrical sparks and what did they represent?

19. Why do scientists assume that the 3.5-billion-year-old microfossils found in Australia were probably not representatives of the first organisms on earth?

20. What genetic phenomena originally led Lynn Margulis to become interested in the endosymbiont hypothesis?

Critical Thinking

1. People once believed fish could form from the mud of a temporarily dry pond. How could you demonstrate that this conclusion is false?

2. Do you think life could have originated on other planets? Justify your answer.

3. According to a recent hypothesis lightning may not have existed on the young earth. How could you modify the Miller–Urey experiment to reflect this new idea? What sources of energy could you use to replace the "lightning"?

4. Some scientists hypothesize that life changed the conditions on earth and in the earth's atmosphere just as conditions on the earth and in its atmosphere change life. What examples from this chapter might back up this hypothesis?

5. The diagram on the right compresses the history of the earth into a 12-hour clock to help you understand the relative time of different events. Note that the forma-

tion of the earth occurred at midnight on the clock. The oceans formed at about 2:00 A.M., while the oldest human fossils date from just before noon. Based on fossil evidence, about what time on the clock did the first prokaryotes appear? About what time did the first eukaryotes appear?

Extension

1. Scientists studying the atmosphere have observed a large hole in the ozone layer above the continent of Antarctica. Use current science magazines to find out about this phenomenon. Present a report to your class summarizing different hypotheses about the causes of this hole.

2. Read the chapter entitled "Life Emerges from the Soup" in *Thread of Life* by Roger Lewin, Smithsonian Books (1982) pp. 90–105. What is a stromatolite? How did the organisms that formed stromato-

lites differ from the photosynthesizing organisms most common today?

3. Make a time line to show the important events in the history of earth described in this chapter. Start your time line with the formation of the earth. Be sure to indicate dates for the formation of the oceans, the age of the oldest microfossils, the first photosynthetic organisms to generate oxygen gas, and the origin of eukaryotic cells. You may wish to continue this time line as you study the rest of the unit.

Chapter 15 | Evolution: Evidence and Theory

This chapter describes the theory of evolution. Students evaluate whether the evidence presented supports the theory of evolution by natural selection.

Correlation Chart

Section Number	Section Objectives	Pages	Features	Section Review
15.1	1. State what is meant by evolution.	219–224	Think, p. 223	1, 2, 3, 4
	2. Describe different types of fossils, and explain how the age of a fossil can be determined.			2, 5
	3. Outline how the fossil record suggests that species have changed over time.			2, 5
	4. Summarize the evidence of evolution provided by living organisms.			1, 3, 4
15.2	1. Compare Lamarck's theory of evolution with Darwin's theory of evolution.	225–230	Biology in Process, pp. 228–229	5
	2. List some of the evidence that led Darwin to the idea of how species might change over time.			3, 4
	3. Define *natural selections* and outline the reasoning that led Darwin to conclude evolution could occur by means of natural selection.			1, 2, 3, 4
	4. Describe Kettlewell's experiment with the two varieties of moths, summarize his results, and explain how he interpreted these results.			
15.3	1. Explain why adaptive radiation is a form of divergent evolution.	231–233	Biology Lab, p. 233	1, 2, 5
	2. Compare and contrast divergent evolution with convergent evolution.			2, 3
	3. Define and give an example of coevolution.			4

Resources*
Print

Scott, Andres. *The Creation of Life: Past, Future, Alien.* New York: Basil Blackwell, 1986.

Audiovisual/Software

Biovideo: The Evidence for Evolution. Film, 30 min. Dist. by Carolina Biological.

Stephen Jay Gould: His View of Life. Film and video, 57 min. Time-Life, 1984.

*Additional references are listed in *Modern Biology, Teacher's Program Guide.*

Section Teaching Guides

15.1 Evidence of Evolution

Approach

Presenting evidence of evolution is the main objective of this section. The evidence hinges upon concepts of basic chemistry (Chapter 3), biochemistry (Chapter 4), and gene expression (Chapter 12). This section prepares students to understand evidence of evolution of specific groups of organisms presented in subsequent chapters. In presenting this section, guide students to evaluate each piece of proposed evidence for its scientific validity.

Concept Framework

All students encounter situations in which they need to evaluate indirect evidence and make inferences. Elicit examples of such situations. Use the examples as an introduction to scientific methods of evaluating indirect evidence and making inferences. Encourage students to note the use of indirect evidence and inferences as they read this section.

Emphasis

Students should understand the significance of the following types of evidence of evolution: fossil evidence, evidence of common ancestry, homologous structures, vestigial organs, biochemical evidence, and embryological evidence. They should also be able to give examples of each.

Exhibit

Bring in different kinds of fossils for students to examine. Encourage them to note similarities and differences between fossil organisms and organisms alive today.

Section Review Answers

(Continued from p. 224.)

4. Biochemical studies have demonstrated the universality of compounds essential to metabolism and of blood proteins. This universality is considered evidence of common ancestry.
5. Both require evidence be collected and used to formulate a theory of past events. The study of evolution differs from the study of history in that experiments demonstrate the evolution of rapidly reproducing species.

15.2 Darwin's Theory of Evolution

Approach

This section focuses on Darwin's theory of evolution by natural selection in both its historic and scientific context. It includes examples of scientific processes and methods presented in Chapter 2 and provides background information on evolution that will aid understanding of subsequent chapters. In presenting the text, guide students to appreciate the explanatory power of this theory.

Concept Framework

Charles Darwin is a familiar name to most students. However, many students' familiarity is limited to Darwin's association with the theory that links human evolution to primates. To provide a conceptual framework for presenting the full breadth of Darwin's work, ask students to list Darwin's accomplishments. Point out that Darwin's book *The Origin of Species* is still widely read today. Tell students that they will be learning more about Darwin's life and work as they read this section.

Emphasis

Students should realize that Darwin's contribution hinged on his development of the idea of natural selection as the mechanism of evolution.

Dramatic Presentation

Encourage students to produce a play or video presentation of the life of Charles Darwin. Assign one group of students to write the script. Assign another group to act out the parts. Assign a third to find or make costumes and props. Presentations should include Darwin's childhood, student days, voyage on the *Beagle,* and development and public presentation of his theory. Students may wish to refer to *The Autobiography of Charles Darwin and Selected Letters* edited by Francis Darwin or any of Darwin's writings.

Using the Features/Observing Natural Selection

Observing, experimenting, and interpreting data are the main science processes discussed in this feature. Help students identify the conflicting observations that scientists have made in the same or similar natural settings. Relate differences in observations to conflicts in conclusions. Lead a discussion on how scientists might resolve unanswered questions about peppered moths.

Section Review Answers

(Continued from p. 230.)

4. evidence of common ancestry; homologous structures; vestigial organs; and embryological evidence; changes in species due to domestic breeding; fossil evidence
5. Lamarck could have inferred the ancestral salamander's legs became smaller through disuse and the salamanders passed on the acquired trait of small legs and eventually no legs to their offspring. Darwin would have inferred that salamanders having shorter legs had a selective advantage. These salamanders would be more likely to reproduce and pass the trait to their offspring. In time the longer-legged salamanders would die out and eventually salamanders with no legs would evolve.

15.3 Patterns of Evolution

Approach

The main objective of this section is to define and give examples of patterns of evolution. The section builds on Darwin's theories, presented in Section 15.2. The section also prepares students to recognize the importance of adaptation to the environment discussed in subsequent chapters. In presenting the text, guide students to appreciate the variety of adaptations that enable particular organisms to survive in specific environments.

Concept Framework

Most students will be aware that differences among organisms is due in part to the organisms' adaptations to different environments. Discuss the crucial role adaptation plays in the theory of evolution by natural selection. Tell students to watch for examples of adaptations as they read this section.

Emphasis

Make sure that students can give at least one example for each of the following patterns of evolution: adaptive radiation, divergent evolution, convergent evolution, and coevolution.

Field Trip

Take students to a local nature preserve, zoo, or botanic garden. Arrange for a staff naturalist to discuss adaptations of various plants or animals to the natural environment.

Section Review Answers

(See p. 232.)

Laboratory Preparation

Background

This laboratory exercise demonstrates that variation is found throughout the living world. Ask students to think about the ways in which natural selection could act on variation in different traits. Students should understand that not all variation is inherited, or subject to evolutionary change, and they should begin to contemplate factors that maintain variation in populations. In a broader context, students realize that a report of an "average" trait—be it bean size or the "average" human height—may not adequately represent the spectrum of individuals encountered in a given population.

Preparation

Order the preserved *Ascaris* specimens ahead of time (enough for each lab station to have 20 males and 20 females). *Ascaris* males are smaller that females and have a distinct curve at the posterior end.

Safety

Be sure that students work with the preserved *Ascarus* specimens in a well-ventilated area. The preservative used on the specimens may irritate skin. Caution students to wear gloves or to wash hands thoroughly after handling the specimens.

Conclusions

1. Yes. The graphs suggest bell-shaped curves.
2. The combined graph does not show that female and male *Ascaris* differ in size.

Inquiry

1. Possible sources of error are careless measuring, inaccurate measuring tools, and samples not large enough to reflect the overall population.

2. Students should recognize that the validity of the results would be affected because the graphs would merely show variation in growth at different ages, not variation within a species.

Chapter Review

Vocabulary

1. Casts, imprints, and molds are all fossils. An imprint is a thin film of carbon left after all other elements have evaporated. A mold is a hollow impression left after remains of an organism have disintegrated. A cast forms when sediments fill a fossil mold.
2. An adaptation is an inherited trait that increases an organism's chance of survival in a particular environment. Natural selection is the process by which individuals that have advantageous adaptations are most likely to survive and reproduce than those that do not.
3. Homologous structures are body parts of different species that are similar in structure and develop from the same part of the embryo. Vestigial organs serve no apparent function in one species but are homologous to useful structures in other species.
4. Evolution is the theory that species change over time. Coevolution is the joint evolution of two or more species in close interaction.
5. In convergent evolution unrelated species become more similar due to their adapting to the same environment. In divergent evolution related species become increasingly dissimilar by adapting to different environments.

Review

11. The two concepts are that species change over time and that today's species have arisen by descent and modification from ancestral forms.
12. Students should list the following: fossil evidence, evidence of common ancestry, homologous structures, vestigial organs, embryological evidence, biochemical evidence.
13. The skeletal and muscle structure of the 23 species of Hawaiian honeycreepers indicates the birds are closely related.
14. Salamanders would pass the trait of short legs, acquired through disuse during their lifetime, to their offspring. Thus the legless salamander would eventually evolve through the inheritance of acquired characteristics.
15. Members of a species with favorable variations would be more likely to survive and reproduce than others. If such variations could be inherited, over many generations organisms with unfavorable variations would die. In this way species would evolve.
16. Wallace sent Darwin an essay in which he put forth a theory identical to Darwin's. This spurred Darwin to let his theory be presented and to publish *The Origin of Species*.
17. He cited fossil evidence, homologous structures, vestigial organs, embryological evidence, and evidence from breeding. The theory was scientifically accepted due to the strength of Darwin's evidence and logic.
18. Recent experiments indicate that moths rest in shadows on the underside of branches rather than on trunks. In this position dark moths are less visible than light moths.
19. In adaptive radiation species many diverge from a single ancestral form.
20. Coevolution is the joint change of two or more species in close interaction. Coevolution occurs with predators and prey, with parasites and hosts, with plant-eating animals and plants, and with flowers and animals that feed on nectar or pollen.

Critical Thinking

1. This could increase the variation in populations and might speed up evolution, for if the environment changed some organisms would survive.
2. The same DNA in different species could be explained by descent from a common ancestor.
3. With this varied background Darwin made generalizations that provided the basis for his theory and made observations that provided evidence to support his theory.
4. The panda's "thumb" is an adaptation that would have decreased competition for food for the panda because pandas having the "thumb" could eat bamboo instead of competing for prey.
5. Hunting at night, the fox would rely more on hearing and less on sight to find its prey.

Chapter 15 | Evolution: Evidence and Theory

Introduction

Life on earth began billions of years ago with simple unicellular organisms. Today organisms exhibit an enormous range of differences as well as some surprising similarities. In this chapter you will learn about the theory of evolution. Evolution can help explain similarities and differences between species and also how lifeforms change.

Chapter Outline

15.1 Evidence of Evolution
- *Evidence from Fossils*
- *Evidence from Living Organisms*

15.2 Darwin's Theory of Evolution
- *Evolutionary Theory Before Darwin*
- *Darwin's Background*
- *The Voyage of the* Beagle
- *Darwin in England*
- *Evolution by Natural Selection*
- The Origin of Species

15.3 Patterns of Evolution
- *Divergent and Convergent Evolution*
- *Coevolution*

Chapter Concept

As you read, evaluate whether the evidence presented in the chapter supports the theory of evolution.

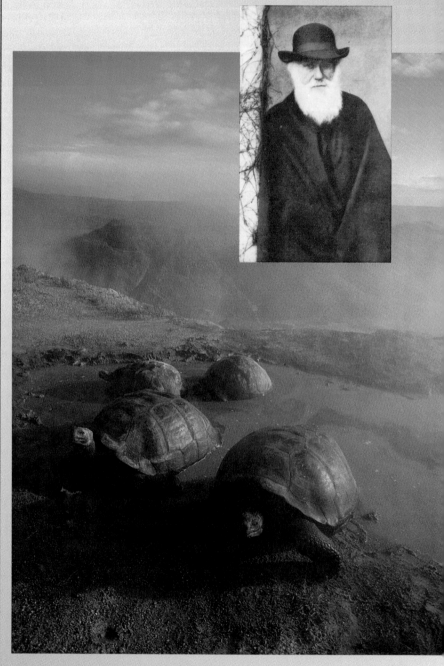

Giant tortoises, *Geochelone elephantopus,* on the Galápagos Islands. Inset: Charles Darwin.

For additional teaching information see pp. T15A–T15D.

15.1 Evidence of Evolution

The Australian microfossils you read about in Chapter 14 are the oldest evidence of life on earth. As you remember, these were the remains of 3.5-billion-year-old prokaryotes. The first life-forms were probably older, perhaps by millions of years. Biologists theorize that all organisms that exist today descended from those first cells. **Evolution** is the theory that species change over time. According to this theory, today's species descended from more ancient forms of life by structural and physiological modifications. If the theory is true, then the world we observe should contain evidence supporting the idea that species change. Let us examine the natural world to see if such evidence exists.

Evidence from Fossils

If today's species have come from more ancient forms, then we should be able to find remains of those species that no longer exist. Scientists have found such evidence in the form of fossils. Fossils—traces of once-living organisms—are found most commonly in layers of sedimentary rock. This type of rock begins to form when water and wind form layers of sand and silt. These actions can bury remains of an organism so quickly that bacteria and other decomposers are sealed off, preventing decomposition. After time the sedimentary layers become rock.

The most common fossils found in sedimentary rocks are from the hard parts of organisms, including shells, bones, teeth, and woody stems. Sometimes minerals replace the original remains, often molecule by molecule. Such replacement preserves the microscopic structure of the organism.

Fossils can form in other ways. For instance, insect fossils have been found trapped in hardened resin. Such fossils show details as small as the insect's tiny leg hairs. In another instance, woolly mammoths that were frozen in Arctic ice have been dug up with their skin, bones, and muscles perfectly preserved.

Fossils are not always the body parts of an organism. An **imprint** is a type of fossil in which a film of carbon remains after the other elements of an organism have decayed. A **mold** is a type of fossil in which an impression of the shape or track of an organism has survived. Figure 15-1 shows how a fossil **cast** forms when sediments fill in the cavity left by a fossil mold.

Dating Fossils

If scientists suggest that fossils are the remains of organisms that lived long ago, they should be able to prove it. Scientists have therefore developed ways to date, or determine the age of, fossils.

The relative age of fossils is determined from their position in sedimentary rock. In undisturbed sedimentary rock the bottom layers are the oldest and the top layers are the youngest. Fossils

Figure 15-1 A fossil mold is formed when a trilobite buried by sediments dissolves. The mold is then filled with sediments, forming a cast.

Shell

Mold

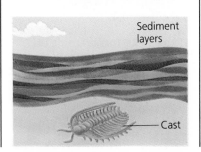
Sediment layers

Cast

found in the lower layers are older. Those found in the upper layers are younger. Thus a fossil's position in sedimentary rock beds gives its age relative to other fossils.

The absolute age of a fossil is determined by dating the fossil with radioactive isotopes. Radioactive isotopes have unstable nuclei that break down, or decay, and form other elements. These isotopes decay at a constant, known rate. The period of time it takes for one-half of the radioactive material to decay is called the **half-life** of the isotope. Remember that isotopes of an element differ in number of neutrons. Carbon-12, the most common and stable isotope of carbon, has 6 protons and 6 neutrons and therefore has an atomic mass of 12. Carbon-14, an unstable, radioactive isotope of carbon has 6 protons and 8 neutrons and an atomic mass of 14. Carbon-14 decays to nitrogen-14. The half-life of carbon-14 is about 5,700 years.

The ratio of carbon-14 to carbon-12 in the atmosphere is assumed to be constant over time. Organisms take up the two isotopes of carbon in about the same ratio that the isotopes are found in the atmosphere. Thus, when an organism dies, the ratio of carbon-14 to carbon-12 in its remains will be the same as that in the atmosphere. As time goes by, the amount of carbon-14 will decrease as it changes into nitrogen-14. However, carbon-12 does not decay. By comparing the ratio of carbon-14 to carbon-12 in a fossil with the ratio of these isotopes in the atmosphere, scientists can date fossils that are up to 50,000 years old.

Scientists use other radioactive isotopes, such as potassium-40, to date older fossils. Potassium-40 has a half-life of 1.28 billion years. Uranium-238 is another radioactive isotope with a long half-life that can be used to date older fossils. Uranium-238 is used to determine the age of cores of silt taken from seabeds.

The Fossil Record

By dating fossils and examining geologic strata, scientists have been able to put together a time scale for the history of life on earth. Table 15-1, the geologic time scale, outlines this history. Study the time scale carefully to learn when various kinds of organisms first appeared.

Fossil evidence indicates that over time organisms of increasing complexity appeared on the earth. Bacteria and blue-green bacteria are the first fossils that were preserved from the Precambrian era. During the beginning of the Paleozoic era, complex multicellular invertebrates dominated life in the oceans. By the end of the Paleozoic era, plants and animals had colonized the land surface of the earth.

The fossil record contains many examples that could be interpreted to mean that species evolved from more ancient organisms. Examine the fossil evidence shown in Figure 15-2. Compare the foot structure of the different species of fossil horses. The similarities and differences between the feet of horses from different epochs strongly suggest that horses changed over time.

Equus

Merychippus

Mesohippus

Hyracotherium (Eohippus)

Millions of years ago

3

7

25

40

60

Figure 15-2   Fossil records of horse foot structure show structural differences over time.

Table 15-1 Geologic Time Scale

End Date (millions of years ago)	Era	Period	Epoch	Organisms
0.1	Cenozoic	Quaternary	Recent	Modern humans appear
			Pleistocene	Woolly mammoth
2.5		Tertiary	Pliocene	Large carnivores; apes
7			Miocene	Land mammals diversify
26			Oligocene	Primitive apes; horses
38			Eocene	Small horses
53			Paleocene	First carnivores; primates
65	Mesozoic	Cretaceous		Dinosaurs die out; flowering plants appear
135		Jurassic		Age of the dinosaurs; birds arise
195		Triassic		First dinosaurs; mammals; forests of conifers
225	Paleozoic	Permian		First seed plants
280		Carboniferous		First reptiles appear
345		Devonian		First insects and amphibians appear
395		Silurian		Fishes dominant; first modern vascular plants (end)
430		Ordovician		Modern groups of algae and fungi appear
500		Cambrian		First fish; plants invade land; many invertebrates and marine plants
600		Precambrian		First eukaryotes; blue-green bacteria and bacteria abound

Evidence from Living Organisms

By examining fossils and by determining their relative and absolute ages, scientists have collected evidence that supports the theory that species changed over time. Further evidence is derived from living organisms. In order to determine if species change, scientists compare common ancestry, structure, biochemistry, and development of organisms alive today. As you read this section, study this evidence and critically evaluate whether it indicates that species may have arisen by descent and modification from ancestral species.

Evidence of Common Ancestry

If species change over time, then scientists should be able to cite examples showing that a group of living species may have come from a common ancestor. Let us examine one of many cases for which this seems to be true. Gracing the islands of Hawaii is a family of birds commonly called the Hawaiian honeycreepers. Figure 15-3 shows 7 of the 23 Hawaiian honeycreeper species, all of which are found only in Hawaii. All Hawaiian honeycreepers have similarities in skeletal and muscle structure that indicate they are closely related. However, each of the Hawaiian honeycreeper species has a bill specialized for eating certain foods. Scientists suggest that all 23 honeycreeper species apparently arose from a single finch species that migrated to Hawaii.

Homologous Structures

If a bat, a human, an alligator, and a penguin all evolved from a common ancestor, then they should share common anatomical traits. In fact, they do. Compare the forelimbs of the human, the

Tip: Remind students that as they study and evaluate the evidence presented in this chapter, they are inferring.

Figure 15-3   Like all species of Hawaiian honeycreepers, the seven shown below can be identified by their beaks.

Ciridops (ula-ai-hawane) eats fruits, seeds, some insects

Himatione (apapane) eats primarily nectar, some insects

Loxops (akepa) eats insects, some nectar

Pseudonestor (parrot-billed koa finch) eats insects

Drepanis (mamo) eats primarily nectar, some insects

Hemignathus (akialoa) eats insects, nectar

Vestiaria (iiwi) eats primarily nectar, some insects

bat, the penguin, and the alligator, shown in Figure 15-4. Find the humerus, radius, ulna, and carpals in each forelimb. Though the limbs look strikingly different on the outside and though they vary greatly in function, they are very similar in skeletal structure. More significantly, they are derived from the same structures in the embryo. Structures that are embryologically similar are called **homologous structures.** Though these animals look different, a comparison of homologous structures indicates that they are quite similar. Does this suggest that these animals may have evolved from a common ancestor?

Vestigial Organs

Some organisms have structures or organs that seem to serve no useful function. For example, humans have a tailbone at the end of the spine that is of no apparent use. Some snakes have tiny pelvic bones and limb bones, and some cave-dwelling salamanders have eyes even though members of the species are completely blind. Such seemingly functionless parts are called **vestigial** *(veh-STIJ-ul)* **organs** or structures. Vestigial organs are often homologous to organs that are useful in other species. The vestigial tailbone in humans is homologous to the functional tail of other primates. Thus vestigial structures can be viewed as evidence for evolution: organisms having vestigial structures probably share a common ancestry with organisms in which the homologous structure is functional.

Biochemistry

Biochemistry also reveals similarities between organisms of different species. For example, the metabolism of vastly different organisms is based on the same complex biochemical compounds. The protein cytochrome c, essential for aerobic respiration, is one such universal compound. The universality of cytochrome c is evidence that all aerobic organisms probably descended from a common ancestor that used this compound for respiration. Certain blood proteins found in almost all organisms give additional evidence

Think

The wings of moths and the wings of birds serve similar functions but differ in structure. Thus they are called *analogous* structures. Do you think analogous structures provide evidence of evolution?

Figure 15-4   The forelimbs of the penguin, the alligator, the bat, and the human are all derived from the same embryological structure.

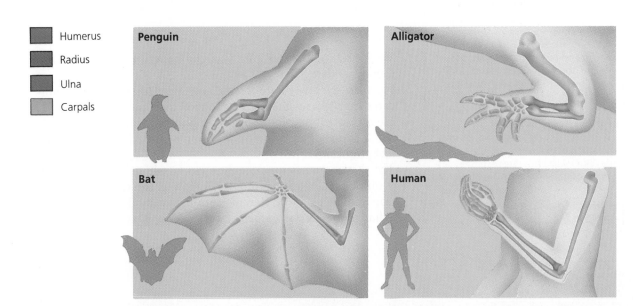

Humerus
Radius
Ulna
Carpals

Penguin

Alligator

Bat

Human

Fish   Rabbit   Gorilla

Figure 15-5 Although all three adult organisms look quite different, the developing embryos of the rabbit and gorilla look similar.

Answers to Section Review
1. Homologous structures are body parts that are similar in structure among different species and develop from the same part of the embryo. Vestigial structures are functionless structures that are homologous to useful structures in other species. Both represent similarities between organisms that can be explained by descent from a common ancestor.
2. by considering the relative position of the fossils in layers of rock and by using radioactive elements to measure fossil age; the fossil record shows organisms of increasing complexity over time
3. Embryos of different species may have almost identical patterns of development. This suggests they share common genes for embryonic development and hence a common ancestor.
(See pp. T15A–T15D.)

that these organisms descended from a common ancestor. Such biochemical compounds, including cytochrome c and blood proteins, are so complex it is unlikely that almost identical compounds would have evolved independently in widely different organisms.

Further studies of cytochrome c in different species reveal variations in the amino acid sequence of this molecule. For example, the cytochrome c of monkeys and cows is more similar than the cytochrome c of monkeys and fish. Such similarities and differences suggest that monkeys and cows are more closely related than are monkeys and fish. Scientists have similarly compared the biochemistry of universal blood proteins. Their studies reveal evidence of degrees of relatedness between different species. This evidence implies that some species share a more recent common ancestor than other species do. From such evidence scientists have inferred the evolutionary relationships between different species of organisms.

Embryological Development

The similarities described above are not the only ones scientists have noticed among organisms of different species. Figure 15-5 shows that embryos of certain species develop almost identically, especially in the early stages. Based on your knowledge of gene expression from Chapter 11 you can infer that such similarities indicate the species have genetic similarities. Again, these similarities can be considered evidence that the organisms shown probably descended from a common ancestor.

The similarities between living species—in ancestry, in homologous and vestigial structures, in embryological development, and in biochemical compounds—all could be explained as extremely remarkable coincidences. However, a far more probable explanation of these similarities is that species have arisen by descent and modification from more ancient forms. Additionally, the fossil record contributes compelling evidence that species have changed over time. The fossil evidence and evidence from living organisms strongly suggest that species evolve.

Section Review

1. What are homologous and vestigial structures, and how do they provide evidence of evolution?
2. How do scientists date fossils, and what fossil evidence indicates that species may have changed over time?
3. How does embryology suggest common ancestry?
4. What evidence of common ancestry is provided by biochemical studies?
5. The American scientist Stephen Jay Gould calls the study of evolution "a historical inquiry." How is the study of evolution similar to and different from the study of history? What role does the gathering of evidence play in both?

15.2 Darwin's Theory of Evolution

Scientists at the beginning of the 1800s knew of some kinds of fossils, and they were very aware of homologous and vestigial structures. Many scientists suspected that some kind of evolution had given rise to the living things around them. However, they had no unifying theory to explain how evolution might have occurred. Two scientists led the way in the search for a mechanism of evolution. The first was Jean Lamarck. The second was one of the greatest figures in biology, Charles Darwin.

Evolutionary Theory Before Darwin

The first systematic presentation of evolution was put forth by the French scientist Jean Baptiste de Lamarck (1774–1829) in 1809. Lamarck described a mechanism by which he believed evolution could occur. This mechanism, known as "the inheritance of acquired characteristics," is illustrated in Figure 15-6.

Assume that there were salamanders living in some grasslands. Suppose, Lamarck argued, that these salamanders had a hard time walking because their short legs couldn't trample the tall grasses or reach the ground. Suppose that these salamanders began to slither on their bellies to move from place to place. Because they didn't use their legs, the leg muscles wasted away from disuse and the legs thus became small. Lamarck's theory said that the salamanders passed this acquired trait to their offspring. In time the salamander's legs were used so rarely that they disappeared. Thus, Lamarck argued, legless salamanders evolved from salamanders by inheriting the acquired characteristic of having no legs.

Lamarck presented no experimental evidence or observation and his theory fell out of scientific favor. The next significant idea came from the British scientist Charles Darwin.

Darwin's Background

Charles Darwin (1809–1882), like many people of genius, did not at first appear to have extraordinary talents. From a young age

Section Objectives

- *Compare Lamarck's theory of evolution with Darwin's theory of evolution.*

- *List some of the evidence that led Darwin to the idea of how species might change over time.*

- *Define natural selection, and outline the reasoning that led Darwin to conclude evolution could occur by means of natural selection.*

- *Describe Kettlewell's experiment with the two varieties of moths, summarize his results, and explain how he interpreted these results.*

Figure 15-6 According to Lamarck's theory the legs of salamanders could disappear because of disuse. Thus legless salamanders would have evolved from salamanders with legs.

Darwin disliked school and preferred observing birds and collecting insects to studying. He was sent to medical school in Scotland when he was 16. Young Darwin found medicine "intolerably dull." He was much more interested in attending natural history lectures.

Seeing that Darwin lacked enthusiasm for becoming a doctor, his father suggested he study for the clergy. Darwin was agreeable to the idea and enrolled in the university at Cambridge, England, in 1827. Here again, Darwin admitted, "My time was wasted, as far as the academical studies were concerned."

However, Darwin found that his friendship with John S. Henslow, professor of botany, made life in Cambridge extremely worthwhile. Through long talks with Henslow, Darwin's knowledge of the natural world increased. Henslow encouraged Darwin in his studies of natural history. In 1831 Henslow recommended that Darwin be chosen for the position of naturalist on the ship the HMS *Beagle*.

The Voyage of the *Beagle*

The *Beagle* was chartered for a five-year mapping and collecting expedition to South America and the South Pacific. Darwin's job as ship naturalist was to collect specimens, make observations, and keep careful records of anything he observed that he thought significant.

At the beginning of the voyage Darwin read a geology book given to him by Henslow. This book, *Principles of Geology* by Charles Lyell, spurred his interest in the study of landforms. In Chile Darwin observed the results of an earthquake: the land had been lifted by several feet. In the Andes he observed fossil shells of marine organisms in rock beds at about 4,300 m. He came to agree with Lyell that over millions of years earthquakes and other geologic processes could change the geology of the land. Because the land changed, new habitats would form. Darwin realized that animals would have to adapt to these changes.

During the *Beagle*'s five-year trip the captain often dropped Darwin off at one port and picked him up months later at another. One reason that Darwin was so eager to study life on land was that he suffered from terrible seasickness and couldn't wait to get off the *Beagle*. During his time on land Darwin trekked hundreds of miles through unmapped regions. He observed thousands of species of organisms and collected many different types of fossils. On the long sea voyages he used his time to catalog his specimens and write his notes.

Darwin in England

When Darwin returned to England in October 1836, his collections from the voyage were praised by the scientific community. Darwin sent many specimens to experts for study. A bird special-

Figure 15-7 Darwin spent five years on the *Beagle* as ship naturalist, collecting new species and recording his observations.

Writing About Biology

Wyoming Reverie

This excerpt is from The Dinosaur Heresies *by Robert Bakker.*

From my Field Book, 1981 . . . July 3, 6:35 A.M. Como Bluff, Wyoming. 7,020 feet above sea level. No human being or human structure visible. Air clear, dry, cool. . . .

I'm still excited by the first dinosaur of the summer. I sit here on the crest of a little sandstone hogback, remnant of a stream that flowed a hundred million years ago, and look down on my crew's work of the last four days. It's becoming a sizeable hole, a proper dinosaur dig, twenty-five feet across, dug by pickax, army-surplus trenching shovel, ice-pick, and fingernails.

I saw my first dinosaur in . . . the American Museum of Natural History in New York, at the age of nine. But those skeletons seemed tamed by civilization, mounted as they were on steel and plaster. . . . A dinosaur in the rock is different. This one before me is huge, and its six-foot long thigh bone, which would dwarf any elephant's, lies half exposed to the Wyoming sunrise. Its coal black form is clearly etched against the surrounding pale rock by thousands of careful chisel marks. . . .

From where I sit on the quarry's rim I can see the dinosaur's great trochanters, the attachment site of the immense hip muscles, and the bone surface pitted and rough where tendons and ligaments were anchored to the femur. A hundred thousand millenia ago, those tendons and muscles were full of dinosaur blood coursing through capillary beds, bringing oxygen to the cells that powered the stride of this ten-ton giant. Muscles pulsed in cycles of contraction and release, and the hind limb, fully twelve feet long from hip to toenails, swung through its stroke covering six feet with every pace.

Broken chips of bone lie under my boots, wretched fragments from now unidentifiable bones which had eroded long before we found the site. . . . Some of the broken bits are incredibly delicate bubbles of bone, a frothy texture of holes and vesicles that housed the living substance of the animal's cells. These bits crumble into shards if I rub them too hard, but in life the brittle bone crystals were embedded in a fabric of tough connective tissue, collagen fibers whose great tensile strength combined with the hardness of the bone crystals to produce a living bony architecture capable of resisting enormous loads of both compression and tension. Collagen has long since rotted away, along with all the muscle fibers and blood vessels. But the fossilized bone faithfully preserves the canals left by every capillary that made its passage through it to serve the dinosaur in life. . . .

Reverie is normal in Wyoming at sunrise. I suppose a no-nonsense laboratory scientist might think I was wasting my time communing with the spirit of the fossil beast. But scientists need reverie. We need long walks and quiet times at the quarry to let the whole pattern of fossil history sink into our consciousness.

Write

- Write a short essay explaining why you think scientists may need reverie. In what way might it help them find answers to their questions? Does Bakker's line of reasoning here have anything to do with the concern that Lewis Thomas expressed in the beginning of his essay on p. 28?

From The Dinosaur Heresies *by Robert Bakker. Copyright © 1986 by Robert Bakker. Reprinted by permission of William Morrow & Co.*

Large ground finch

Warbler finch

Woodpecker finch

Figure 15-8 The large ground finch, *Geospiza magnirostris* (top), warbler finch, *Certhidea olivacea* (middle), and woodpecker finch, *Camarynchus pallidus* (bottom), have beaks specialized for their food source.

ist, or ornithologist, studied Darwin's bird collections from the Galápagos Islands, located about 1,000 km west of South America. He reported that Darwin had collected 13 similar but separate species of finches. Each finch species had a distinctive bill specialized for a particular food source. Other experts studied Darwin's fossils and classified them as remains of extinct mammals. The fossils included rodents the size of hippopotamuses.

The similarities of the Galápagos finches led Darwin to infer that the finches shared a common ancestor. The similarities between the fossil mammals Darwin collected and modern mammals led him to believe that species change over time.

In 1837 Darwin began his first notebook on evolution. For several years Darwin filled his notebooks with facts that could be used to support the theory of evolution. He found evidence from his study of the fossil record: he observed that fossils of similar relative ages are more closely related than those of widely different relative ages. Comparing homologous structures, vestigial organs, and embryological development of living species gave him additional evidence of evolution. He consulted animal and plant breeders about changes in domestic species. He ran his own breeding experiments and also did experiments on seed dispersal.

Biology in Process

Observing Natural Selection

Direct evidence of natural selection began to come to light before Darwin published *The Origin of Species*. This evidence involved two varieties of a moth species called the peppered moths—one light and one dark. Find the light and dark peppered moths on the lichen-covered tree trunk in the picture. In Darwin's day the light variety was found throughout the countryside, where such lichen-covered trees were common.

The dark variety of peppered moth was first observed near English cities in 1848. This variety was hardly noticeable on the dark, soot-covered tree trunks near industrial areas. Over the next hundred years scientists observed a greater number of dark moths relative to light moths near the cities.

In the 1950s H. B. Kettlewell, an English scientist, tested the hypothesis that the difference in numbers of the light and dark peppered moths was due to natural selection. Kettlewell marked, released, and recaptured the peppered moths in the country and near a city. Kettlewell assumed that the relative numbers of moths of the two varieties he recaptured indicated their relative survival

Evolution by Natural Selection

The central question still remained: if evolution occurred, by what means did it occur? In 1838 Darwin read a book called *Essay on the Principle of Population* by a British economist, Thomas Malthus (1776–1834). Malthus stated that a human population growing unchecked would double every 25 years. Resources such as food, air, and water cannot increase at the same rate, Malthus argued. Thus human beings are involved in an intense "struggle for existence," competing for the limited resources. This idea helped Darwin uncover the mechanism he needed.

Combining the idea of competition with his other observations, Darwin explained how evolution could occur. First, he stated that variation exists among individuals of a species. Second, he stated that scarcity of resources in a burgeoning population would lead to competition between individuals of the same species because all use the same limited resources. Such competition would lead to the death of some individuals, while others would survive. From this reasoning Darwin concluded that individuals having advantageous variations are more likely to survive and reproduce than those without the advantageous variations.

Tip: Show students illustrations of all 13 varieties of bills of Darwin's finches. Discuss the distinctive bill features and speculate on an adaptive advantage of each.

Answer to Biology in Process
Some possible sources of experimental error include the different proportions of the two moth varieties released in the two environments and the relatively small number of moths recaptured. Changes to reduce the experimental error might include releasing the same number of each type of moth in both environments, improving the recapturing method in order to recover more specimens, and subjecting the data to rigorous statistical analysis.

Kettlewell's data for 1953

Location		Color of peppered moth Light	Dark
Unpolluted countryside (Dorset)	Number released	469	473
	Number recaptured	62	30
Polluted city (Birmingham)	Number released	137	447
	Number recaptured	64	154

rates. His results are shown in the bar graph. Kettlewell interpreted the data to mean that dark moths survived better near cities and light moths survived better in the country. Thus the experiment demonstrated natural selection had occurred in these species of peppered moths.

Kettlewell inferred that the adaptation of color conferred a selective advantage to the moths. He hypothesized that birds could see light moths more easily on sooty trunks near cities and dark moths more easily on lichen-covered trunks in the countryside. Therefore, near cities more light moths would serve as prey, and more dark moths would survive and reproduce. In the countryside the opposite result would occur.

Recent experiments, however, have raised questions about this explanation. They indicate that the peppered moths probably rest not on tree trunks, but in shadows on the underside of tree branches. In this position dark moths are less visible than light moths whether the trees are soot-covered or lichen-covered. Scientists now question whether color is the main adaptation conferring a selective advantage to the light moths. Further experiments may help scientists better understand the selective mechanism in this case.

■ What are some possible sources of experimental error in Kettlewell's experiment? How could you change the experiment in order to reduce the chance of introducing experimental errors?

Figure 15-9 The willow ptarmigan, *Lagopus lagopus*, is brown (top) through most of the year but turns white (bottom), in winter as a means of camouflage.

Answers to Section Review
1. the idea that individuals having variations that give them an advantage over others are most likely to survive and reproduce.
2. The book presented evidence that the earth's geologic features can best be explained by gradual changes occurring over millions of years. Darwin recognized that if the land had changed, animals would have had to adapt to their new habitat.
3. the limited distribution of the closely related Galapagos finches and the comparison of the fossil mammals to modern mammals (See pp. T15A–T15D.)

Darwin coined the term **natural selection** to describe the process by which organisms with favorable variations survive and reproduce at a higher rate. An inherited variation that increases an organism's chance of survival in a particular environment is called an **adaptation.** Over many generations, an adaptation could spread throughout the entire species. In this way, according to Darwin, evolution by natural selection would occur.

As an example Darwin noted that the ptarmigan turns white in winter. This color change, he inferred, helped protect it from predators, which would have a hard time spotting the bird in snow. Ptarmigans that didn't change color in winter would be spotted easily and eaten. In this way, Darwin implied, ptarmigans that turned white in winter would be more likely to survive, reproduce, and pass this adaptation to future generations.

The Origin of Species

Darwin compiled evidence for evolution by natural selection for about 20 years. Between 1842 and 1844 he wrote a 230-page essay summarizing his theory and the evidence for it. In the 1850s he began working on a detailed, multivolume book to present his theory to the scientific community.

Darwin might never have completed the book if another British scientist, Alfred Russel Wallace (1823–1913), had not come up with the same idea in 1858. While living in the Malay Archipelago in the Pacific Ocean, Wallace formulated his theory and wrote it in an essay, which he sent to Darwin. Darwin's fellow scientists persuaded him to let them present his theory and Wallace's essay jointly at a scientific meeting. The presentation "excited very little attention," according to the modest Darwin.

However, the publication of Darwin's book *The Origin of Species* in 1859 changed biology forever. The first printing of the book sold out in one day. Darwin clearly and logically presented the idea that natural selection is the mechanism of evolution. In Darwin's own lifetime many scientists became convinced that evolution occurs by means of natural selection. Today this theory is the unifying one for all biology.

Section Review

1. What is *natural selection?*
2. How might reading Lyell's book on geology have influenced Darwin's thinking about evolution?
3. What evidence from Darwin's voyage on the *Beagle* originally led him to the idea that species undergo change?
4. What types of evidence did Darwin present to back up the theory of evolution by natural selection?
5. Use the example of the evolution of the legless salamander to compare and contrast Darwin's theory of evolution to Lamarck's theory of evolution.

Info: Darwin and Mendel were contemporaries, but Mendel's work was so little known during his lifetime that Darwin never heard of him.

15.3 Patterns of Evolution

As you will read in Chapter 16, natural selection can ultimately lead to the formation of new species. Sometimes many species evolve from a single ancestral species. As you know, similarities in skeletal and muscular structure of Hawaiian honeycreepers led scientists to conclude that the 23 species of honeycreepers evolved from one ancestral species. Such an evolutionary pattern, in which many related species evolved from a single ancestral species, is called **adaptive radiation.** Adaptive radiation most commonly occurs when a species of organisms successfully invades an isolated region where few competing species exist. If new habitats are available, new species will evolve.

Divergent and Convergent Evolution

Adaptive radiation is one example of divergent evolution. **Divergent evolution** is the process of two or more related species becoming more and more dissimilar. The red fox and the kit fox provide an example of two species that have undergone divergent evolution. The red fox lives in mixed farmlands and forests, where its red color helps it blend in with surrounding trees. Compare the color of the red fox to that of the kit fox, shown in Figure 15-10. The kit fox lives on the plains and in the deserts, where its sandy color helps conceal it from prey and predators. Notice also that the ears of the kit fox are larger than those of the red fox. The kit fox's large ears are an adaptation to its desert environment. The enlarged surface area of its ears helps the fox get rid of excess body heat. Similarities in structure indicate that the red fox and the kit fox had a common ancestor. As they adapted to different environments, the appearance of the two species diverged.

In **convergent evolution,** on the other hand, unrelated species become more and more similar in appearance as they adapt to the same kind of environment. The two unrelated types of plants

Section Objectives

- *Explain why adaptive radiation is a form of divergent evolution.*

- *Compare and contrast divergent evolution with convergent evolution.*

- *Define and give an example of coevolution.*

Tip: Ask students to think of a possible negative effect of convergent evolution. (It could lead to greater competition if species came to occupy the same environment.)

Figure 15-10 The color of the red fox, *Vulpes vulpes* (left), helps it blend in with the trees in its habitat, while the sandy color of the kit fox, *Vulpes macrotis* (right), conceals it from predators in the desert.

Figure 15-11 The cactus, *Cereus gigantea* (left), and the spurge, *Euphorbia* sp. (right), both have fleshy stems that store water and protect them from predators.

shown in Figure 15-11 have adapted to desert environments. Notice the resemblance of the cactus, which grows in the American desert, to the euphorbia, which grows in the African deserts. Both have fleshy stems armed with spines. These adaptations help the plants store water and ward off predators.

Coevolution

Coevolution is the joint change of two or more species in close interaction. Predators and their prey sometimes coevolve; parasites and their hosts often coevolve; plant-eating animals and the plants upon which they feed also coevolve. One example of coevolution is between plants and the animals that pollinate them.

In tropical regions bats visit flowers to eat nectar. The fur on the bat's face and neck picks up pollen, which the bat transfers to the next flower it visits. Bats that feed at flowers have a slender muzzle and a long tongue with a brushed tip. These adaptations aid the bat in feeding. Flowers that have coevolved with bats are light in color. Therefore, bats, which are active at night, can easily locate them. The flowers also have a fruity odor attractive to bats.

Divergent and convergent evolution and coevolution are different ways organisms adapt to the environment. These are examples of how the diversity of life on earth is due to the ever-changing interaction between a species and its environment.

Section Review

1. Define and give an example of *adaptive radiation*.
2. How do the red and the kit fox show divergent evolution?
3. What is convergent evolution, and how are cactuses and euphorbs an example of this evolutionary pattern?
4. How do bats and bat-pollinated flowers provide an example of coevolution?
5. Adaptive radiation most commonly occurs on newly formed remote islands. Why do you think this is true?

Answers to Section Review
1. Adaptive radiation is the evolutionary pattern in which many related species evolve from a single ancestral species. Hawaiian honeycreepers are an example.
2. Similarities in structure indicate a common ancestor. However, differences in color and in ear size show that they have become dissimilar in adapting to different environments.
3. a pattern in which unrelated species become more and more similar as they adapt to the same environment; cactuses and euphorbs are unrelated plants that have both adapted to a desert environment, exhibiting spines and fleshy stems.
4. Bats have a slender muzzle and a long tongue with a brushed tip that aid in feeding on nectar. Their hairy faces and necks pick up pollen, thus aiding pollination. Bat-pollinated flowers are light in color, an adaptation that allows bats to see the flowers easily. The flowers have a fruity odor attractive to bats.
5. Newly formed, remote islands are isolated environments where little competition exists. In such an environment many different habitats are available for colonization. If members of the same species can adapt to the different habitats, a number of different species can evolve from a single ancestral species.

Laboratory

Variation and Natural Selection

Objective
To observe variation within a species

Process Skills
Measuring, collecting and interpreting data

Materials
Metric ruler, 20 leaves from a nearby tree, 20 peanuts in the shell, 20 preserved female *Ascaris*, 20 preserved male *Ascaris*, graph paper, pan or tray

Humane Alternative
Measure male and female human finger lengths to avoid using preserved organisms.

Method
1 Measure the length in millimeters of each leaf blade from its union with the petiole to the tip of the blade. Record each length.
2 Make a line graph for leaf length. Label the horizontal axis *Length*. Label the vertical axis *Number*.
3 Count the number of leaves that fall within each 5-mm interval. Transfer the length and number data to the graph.
4 Measure and record the width in millimeters of the 20 leaves.
5 Make a line graph for leaf width. Label the horizontal axis *Width*. Label the vertical axis *Number*. Repeat step 3.
6 Measure and record the length in millimeters of 20 petioles.
7 Make a line graph for petiole length. Label the graph as in step 2. Repeat step 3.
8 Measure and record the length in millimeters of 20 peanut shells.
9 Make a line graph for peanut shell length. Label the graph as in step 2. Count the number of peanuts that fall within each 5-mm interval. Transfer the data to the graph.

10 Measure the length in millimeters of 20 female *Ascaris* and record the length to the nearest 5 mm. Label the graph as in step 2 and transfer the data to the graph.
11 Measure the length in millimeters of 20 male *Ascaris* and record the length to the nearest 5 mm. Label the graph as in step 2 and transfer the data to the graph.
12 Using your data on female and male *Ascaris* lengths, make a graph combining all 40 measurements.

Conclusions
1 Do your graphs tend to have a similar shape? If so, how would you describe the shape of your graphs?
2 Look at your graph on which the lengths of female and male *Ascaris* were combined. What information was lost?

Inquiry
1 What are possible sources of error in the data you have collected?
2 Suppose that some of the leaf and *Ascaris* specimens you measured had not reached maturity while others had. Would this affect the validity of your results? If so, how?

Chapter 15 Review

Vocabulary

adaptation (230)
adaptive radiation
  (231)
cast (219)
coevolution (232)

convergent evolution
  (231)
divergent evolution
  (231)
evolution (219)

half-life (220)
homologous
  structures (223)
imprint (219)
mold (219)

natural selection
  (230)
vestigial organ (223)

Explain the difference between the terms in each of the following sets.
1. cast, imprint, mold
2. adaptation and natural selection
3. homologous structures and vestigial organs
4. evolution and coevolution
5. convergent and divergent evolution

Review

1. Most fossils are found in (a) ice (b) resin (c) sedimentary rocks (d) sand.
2. Radioactive isotopes such as carbon-14 can be used to date fossils because (a) the isotopes vary in number of protons (b) the isotopes decay at a relatively constant rate (c) the position of the fossil in undisturbed sedimentary rock indicates its age (d) the geologic time scale outlines the history of life on earth.
3. Changes in the foot of different species of fossil horses indicate that (a) horses have changed over time (b) horses have homologous structures (c) horses have vestigial organs (d) horses have a long half-life.
4. The fact that the metabolism of different species of organisms is based on the same biochemical compounds is evidence that (a) species change over time (b) organisms descended from a common ancestor (c) species have different genes (d) cytochrome c is essential for anaerobic respiration.
5. The first systematic presentation of evolution was proposed by (a) Erasmus Darwin (b) Charles Darwin (c) Alfred Russel Wallace (d) Jean Baptiste de Lamarck.
6. In his essay on population Malthus argued that a human population would double every 25 years if (a) species inherited acquired characteristics (b) the earth did not change over millions of years (c) resources were unlimited (d) species did not undergo adaptive radiation.
7. According to the theory of natural selection, individuals having variations that give them an advantage in obtaining limited resources will likely (a) be preserved as fossils (b) survive and reproduce (c) diverge markedly (d) evolve.
8. Adaptive radiation is most likely to occur (a) in isolated regions (b) in mixed farmlands and forests (c) on the plains and in deserts (d) in tropical regions.
9. When related species become dissimilar as they evolve, they undergo (a) convergent evolution (b) coevolution (c) adaptive radiation (d) divergent evolution.
10. As environments change, organisms must (a) diverge evolutionarily (b) converge evolutionarily (c) coevolve (d) adapt or move to new habitats.
11. What two concepts formed Darwin's basis for the theory of evolution by natural selection?
12. Name five types of evidence of evolution.
13. How do Hawaiian honeycreepers provide evidence of evolution?
14. How did Lamarck think legless salamanders evolved?
15. How could natural selection result in evolution of new species, according to Darwin?
16. What role did Wallace play in Darwin's presentation of the theory of evolution?

17. What evidence did Darwin use to prove that evolution had occurred, and why was his theory accepted by scientists?
18. Kettlewell assumed that color was the adaptation that conferred a selective advantage to peppered moths in their respective environments. What evidence causes scientists to question this assumption?
19. Why is adaptive radiation considered a form of divergent evolution?
20. Define *coevolution,* and give four examples of types of coevolution.

Critical Thinking

1. Scientists have recently found that organisms can exchange DNA without sexual reproduction. What role do you think this ability may have played in evolution?
2. Biologists have discovered that many genes are the same in different species. How does this fact provide evidence for evolution?
3. In addition to formulating the theory of evolution by natural selection, Charles Darwin also did detailed scientific investigations of many different types of organisms. For example, Darwin published studies of coral reefs, barnacles, orchids, insectivorous plants, and fungi. How might the information Darwin gathered by doing this type of research have contributed to the formulation of the theory of evolution?
4. The ancestor of the modern panda was a meat-eating hunter. Its paws were therefore adapted for running after prey, not for making fine movements. Yet the mod-

ern panda appears to have a "thumb." The panda, a vegetarian, uses this "thumb" to help it strip leaves off bamboo stalks. The panda then eats the shoots. As the diagram above shows, the panda's "thumb" is actually a small bone in its wrist, not a separate digit. Explain how natural selection could have led to this unique "thumb".

5. In the desert rodents are more active in the cool of night. The kit fox mentioned in the text preys on such animals. What role other than cooling does this suggest for the enlarged ears of the kit fox?

Extension

1. Read the article by Robert T. Bakker, "Evolution by Revolution," *Science '85,* Vol. 6, No. 9, November 1985, pp. 72–80. It had been assumed that evolution was a gradual process. New fossil findings have led some scientists to suggest that evolution may have occurred in spurts. Report on the concept of punctuated equilibrium.
2. Though he never became as famous as Charles Darwin, Alfred Russel Wallace made many important contributions to the field of biology in the late 1800s and early 1900s. Use biographies, autobiographies, and an encyclopedia to find information on the life of Alfred Russel Wallace. Write a report highlighting Wallace's contributions to biology.
3. Make an instant fossil. Take 2 or 3 leaves, a coin, or even an apple and drop dampened plaster of Paris around the object. Let the plaster of Paris dry and then break it open with a hammer. Point out the mold.

Chapter 16 | *Evolution: Speciation*

This chapter explains the process of speciation. Students learn about evolution by examining the steps in the process and by outlining the relationship between adaptation and extinction.

Correlation Chart

Section Number	Section Objectives	Pages	Features	Section Review
16.1	1. Compare the morphological concept of species with the biological concept of species.	237–241	Biology in Process, pp. 238–239	2, 5
	2. Describe how scientists study variation of a given trait in a population.		Think, p. 240	1, 3, 4
	3. Define genetic equilibrium.			1, 3, 4
	4. Explain the conditions of the Hardy-Weinberg principle.			4
	5. Contrast Gould's method of classifying *Cerion* with that of nineteenth century scientists.			2
16.2	1. Summarize how mutation and migration each disrupt genetic equilibrium.	242–246	Think, p. 244	2
	2. Define genetic drift and give an example of how genetic equilibrium could be disrupted through genetic drift.			1
	3. Describe how natural selection differs from mutation, migration, and genetic drift in terms of its ability to cause significant changes in a gene pool.			1, 2, 3, 4
	4. Contrast the effects of stabilizing, directional, and disruptive selection on variation in a trait over time.			3
16.3	1. Define geographic isolation, and explain how it can lead to speciation.	247–251	Biology Lab, p. 251	2
	2. Give two examples of ways reproductive isolation can lead to speciation.			3
	3. Summarize the theory of punctuated equilibrium.			1
	4. Outline the role of adaptation in the survival or extinction of a species.			4, 5

Resources*

Print

Harrison, James, ed. *Nature's Secret World*. New York: Arco, 1984.

Wynne-Edwards, V.C. *Evolution through Group Selection*. Boston: Blackwell Scientific, 1986.

Audiovisual/Software

Natural Selection. Video, 20 min. Teacher's guide. Humanities and Sciences, 1985.

*Additional references are listed in *Modern Biology, Teacher's Program Guide*.

Section Teaching Guides

16.1 Genetic Equilibrium

Approach

The main objective of this section is to help students understand the genetic basis of evolution. It builds on concepts of the theory of evolution (Chapter 15) and will help students understand examples of evolution discussed in later chapters. In presenting the text, guide students to realize that any change in a species from one generation to the next involves genetic change. Reinforce this idea when discussing variation in a population and genetic equilibrium.

Concept Framework

Students will have learned about alleles and genes in Chapter 10. Review the definitions of these terms and engage students in a discussion of the relationship between evolution and genetics. Elicit the idea that traits are passed from generation to generation through genes and that for species to change over time, changes in the species's genes must occur. Ask students to formulate a definition of evolution that includes terms from genetics as they read the section.

Emphasis

Students should understand the integral role of genes in the following concepts: the concept of species; the concept of variation in a population; the concept of genetic equilibrium.

Using the Features/Classifying *Cerion*

The main process skills illustrated in the feature are observing and classifying. The feature discusses the observations and reasoning Gould used in classifying certain species of *Cerion*. Refer students to the map on page 239 to help them understand the distribution of the two snail species. Have students locate the ocean coasts and interior coasts of the islands.

Field Trip

Take the class on a field trip to a seed company or to a plant breeding laboratory. Before the trip arrange with your guide to include discussions of why genetic stability is important in seed production or plant breeding and of the techniques used to ensure genetic stability in his or her presentation. Afterward ask students how these techniques help prevent muta-tion, migration, nonrandom mating, and natural selection, thus upholding the Hardy–Weinberg conditions.

Section Review Answers

(See p. 241.)

16.2 Disruption of Genetic Equilibrium

Approach

This section's main objective is to outline the conditions under which genetic equilibrium is disrupted and thus evolution occurs. It builds on the concept of natural selection (Chapter 15) and on concepts presented in Section 1 of this chapter. In presenting the text, guide students to realize that many factors can influence the evolution of particular organisms.

Concept Framework

The concept of natural selection, presented in Chapter 15, will be familiar to students. Engage students in a discussion of how natural selection leads to adaptation. Ask them to give a definition of natural selection that incorporates the idea of genetic change in a population. Encourage them to think of other factors that could lead to genetic change. Tell students that they will be reading about all of these factors in this section.

Emphasis

Students should be able to compare the ability of the following factors to cause changes in a gene pool: mutation, migration, genetic drift, and natural selection.

Modeling: Disruptive Selection

This activity models disruptive selection. Make a template of a moth by taping together two triangles with 3cm sides. Use the template to make three types of moths: one out of black paper; one out of newspaper; one having one black-paper wing and one newspaper wing. Make sixteen or more of each type of moth. Clear an area of the classroom and tape down a five-foot square of black paper and a five-foot square of newspaper. Scatter equal numbers of each of the three types of moths over the two squares. Divide the class into two groups, one to collect moths on the black paper and one to collect moths on the newsprint. Give each student 30 seconds to collect moths. Record the numbers of each type collected on black paper and on newsprint. Ask which type of moth was easier to find, those that contrasted with the background or those that matched. Ask students to state the analogy to the peppered moths discussed in Chapter 15. Elicit the idea that the activity models disruptive selection.

Section Review Answers

(Continued from p. 246.)
4. the preferential choice of a mate based on the presence of a specific trait
5. Number of male peacocks on the y-axis; variation in circles (no circles, colored circles, circles that look like eyes) on the x-axis. Graph should show a normal bell curve and bell curves shifted toward the right.

16.3 Formation of Species

Approach

The main objective of this section is to describe how geographic and reproductive isolation lead to the formation of new species. It prepares students to understand factors involved in both speciation and extinction, which will be discussed in many of the chapters to come. In presenting the text, guide students to appreciate that the debate on rates of speciation is an indication of healthy ferment in the scientific community; it shows that the theory of evolution has led biologists to new hypotheses that are testable through scientific processes.

Concept Framework

Students will be familiar with the definition of evolution. Encourage students to distinguish between evolution, change in species over time, and speciation—the formation of new species. Ask them to try to identify the key factor in the formation of new species as they read the section.

Emphasis

Make sure students understand how geographic isolation and reproductive isolation are essential for the formation of new species.

Guest Speaker

Invite a speaker who can discuss endangered species in your local region to visit the class. Suggestions include a naturalist or a representative from the Sierra Club or the Audubon Society.

Section Review Answers

(See p. 250.)

Laboratory Preparation

Background

In this lab exercise, students investigate what happens to gene frequencies in populations with and without selection pressure. Students will observe that without selection, gene frequencies in a large, randomly mating population will stay the same; the population is in Hardy–Weinberg equilibrium. With selection (in this case, strong selection) students observe that in a large, randomly mating population, allele frequencies will change. They also learn that it is difficult to completely eliminate an allele from a population.

Preparation

If time does not permit students to complete the entire investigation, divide the class into two groups. One group can do Steps 1 through 5, and the second group can do Steps 6 through 12.

Conclusions

1. The recessive allele decreases in frequency in steps 6 through 12.
2. The answer is "No" because the decrease in the frequency of the b allele began to slow in later generations.

Inquiry

Students can test their hypotheses by continuing to form new generations in which the homozygous recessive individuals cannot reproduce.

Chapter Review

Vocabulary

1. Members of separate biological species cannot interbreed or produce fertile offspring; members of separate morphological species differ in internal or external anatomy.

2. In directional selection one extreme is at a selective advantage; in disruptive selection the extreme forms of a trait are at a selective advantage; in stabilizing selection the average form of a trait is at a selective advantage.

3. Genetic drift is a change in gene frequencies due to chance. Genetic equilibrium is a lack of change in gene frequencies from one generation to the next.

4. A species is a group of organisms whose members are *capable* of breeding with one another in nature and producing fertile offspring. A population is any *actual* interbreeding group of organisms living in a particular location at the same time.

5. Geographic isolation results when segments of a population are physically separated. Reproductive isolation results when biological barriers arise between members of the same population in the same geographic area.

Review

11. This concept doesn't account for the reproductive compatibility of morphologically different organisms.

12. stabilizing selection

13. The Hardy–Weinberg principle applies when all of the following conditions are met: no mutations occur; individuals neither enter nor leave the population through migration; the population is large; individuals mate randomly; natural selection does not occur.

14. The morphological similarities between the species indicate that they might have shared a common ancestor. The species could have diverged after populations became geographically isolated; for example, geologic activity might have caused sudden geographic isolation by forming the deep channel that now separates the two landmasses.

15. Migration can cause the introduction of new alleles when individuals immigrate to a new area, the loss of alleles when individuals emigrate, or the shift in proportion of alleles when individuals immigrate or emigrate.

16. It causes the most significant changes in allele frequencies in a population.

17. Sexual selection is considered a type of natural selection.

18. During the ice age, fish in Death Valley all lived in one huge lake. The lake evaporated at the end of the ice age, leaving geographically isolated ponds. The fishes eventually adapted to their respective ponds, diverged genetically, and were no longer able to interbreed, thus forming new species.

19. The wood frog and the leopard frog breed at different times and thus are reproductively isolated. In the ancestral frog species, both frogs that bred earlier than average and frogs that bred later than average may have been selected for, while frogs that bred in between those times may have been selected against.

20. The conversion of the prairie into farmland and grazing land and hence the decline in population of the prairie dog is one factor in the potential extinction of the black-footed ferret. The ferret's inability to adapt to these changes—for example, by finding another food source—is another factor in its potential extinction.

Critical Thinking

1. Accept any answers that the students can support with facts. Examples might include insects or rodents that have evolved in response to insecticides or other poisons.

2. The horse and the donkey can mate but they do not produce fertile offspring. Thus, by definition, the horse and donkey would not belong to the same biological species.

3. It might have led to genetic drift and possibly to nonrandom mating and natural selection.

4. In the left map, the continents were connected, so organisms could have crossed back and forth between the continents. In the right map many continents have drifted apart, leading to geographic isolation of former breeding populations. This would result in increased speciation.

5. The birds could not leave the island by flying, so they were easy prey for human hunters. If dodos had been able to fly, some might have flown away and escaped.

Chapter 16 | Evolution: Speciation

Introduction

No one knows exactly how many species of organisms exist on earth today. In a tropical rain forest alone there are many thousands of species of plants, insects, birds, and mammals, each with its own adaptations. How did all of these species evolve? In this chapter you will trace the process of speciation and gain a better understanding of evolution.

Chapter Concept

As you read, follow the steps that lead to the evolution of a new species.

Poppies in bloom, Antelope Valley, California. Inset: Speciation among red-tailed monkeys, *Cercopithecus ascanius.*

For additional teaching information see pages T16A–T16D.

16.1 Genetic Equilibrium

A **species** is a group of individuals that look similar and whose members are capable of producing fertile offspring in the natural environment. In this section you will learn about this modern definition of species. You will also learn how to view populations of species as genetic entities and how changes in genes in a population can result in evolution.

The Concept of Species

The word *species* is derived from a Latin word meaning "kind" or "appearance." You are probably already familiar with this definition. For instance, you might realize that all gorillas are of the same species. You would base your belief on the fact that individuals in this group look alike. We use the concept of species because we naturally tend to group together things that look alike.

The Morphological Concept of Species

Long ago scientists used to classify organisms solely on the basis of similarities and differences in internal and external structures—that is, on their **morphology.** When differences in appearance are used as the major criterion for classifying an organism, the organism is said to have been classified by the **morphological species concept.** Because morphological characteristics are easy to observe, the morphological species concept is extremely useful. Using this concept, scientists can readily communicate about the characteristics, behavior, and relationships of organisms.

However, using the morphological species concept has limitations. For example, it sometimes happens that organisms that appear different enough to be called different species will interbreed and produce fertile offspring. Figure 16-1 shows two vari-

Section Objectives

- *Compare the morphological concept of species with the biological concept of species.*

- *Describe how scientists study variation of a given trait in a population.*

- *Define genetic equilibrium.*

- *Explain the conditions of the Hardy–Weinberg principle.*

- *Contrast Gould's method of classifying* Cerion *with that of nineteenth-century scientists.*

Figure 16-1   Interbreeding of the red-shafted flicker and the yellow-shafted flicker produces a fertile hybrid with the coloration shown.

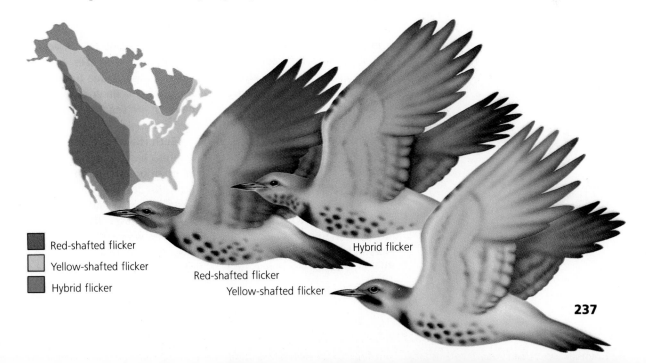

Red-shafted flicker

Yellow-shafted flicker

Hybrid flicker

Hybrid flicker

Red-shafted flicker

Yellow-shafted flicker

eties of birds, a yellow-shafted flicker and a red-shafted flicker. The yellow-shafted flicker lives in the eastern part of North America, and the red-shafted flicker lives in the western part. However, where their ranges overlap, the birds interbreed and produce fertile offspring. The offspring of two morphologically dissimilar organisms is called a **hybrid**. The hybrid flickers have a different coloration from that of either the yellow-shafted or the red-shafted flickers.

Using just the morphological concept of species, scientists might have classified the yellow-shafted, red-shafted, and hybrid flickers each as separate species. The morphological concept of species is limited because it does not account for the reproductive compatibility of morphologically different organisms.

The Biological Concept of Species

Because of limitations of the morphological concept of species, scientists expanded their definition of species to include reproductive compatibility. When organisms are classified solely by whether or not they naturally breed with one another and produce fertile offspring, they are said to have been classified by the **biological species concept**.

Biology in Process

Classifying Cerion

The snails shown are members of a group of Bahamian land snails of the genus *Cerion* (SIR-EE-on). Scientists have observed variation in the snails' shell size, structure, and color. Nineteenth-century scientists, using the morphological species concept, classified up to 600 species.

This long list of snail species discouraged scientific study of *Cerion*. Under the morphological species concept members of different species could be distinguished only by minute differences in their shells. The American scientist Stephen Jay Gould

studied *Cerion* by using both the biological and the morphological concepts of species.

Gould observed *Cerion* in the northern Bahamas, a group of islands formed from two submerged landmasses, or banks. Little Bahama Bank and Great Bahama Bank, shown on the map, are separated by a deep channel.

Each island of the northern Bahamas has one coast facing the open ocean and one coast facing the interior of its bank. Gould observed that *Cerion* on the ocean coasts of both banks had thick, ribbed shells and a uniform color. Those on the interior coasts of both banks had

Using both the biological and morphological concepts of species, modern scientists have arrived at the definition of species stated at the beginning of this chapter, that is, a group of individuals that look similar and whose members are capable of producing fertile offspring in the natural environment. In reality scientists today classify organisms by using both morphology and reproductive compatibility, as well as many other characteristics that you will learn about in Chapter 18.

The modern concept of species is important to the study of evolution because it acknowledges the fact that individuals pass on genes to the next generation. Scientists use this concept of species not only to classify organisms but also to study how species change over time.

Variation of Traits in a Population

To study how species change, scientists must first understand populations. A **population** is made up of all the members of the same species that live in a particular location at the same time. For example, all the fish of a single species that live in a pond make up a population.

Answer to Biology in Process
Gould observed the geographic distribution of varieties of *Cerion* on each island and found that the two coastal varieties could interbreed and produce fertile hybrids. Gould also observed reproductive anatomy in order to determine whether *Cerion* from different islands could interbreed.

thin shells, few ribbings, and a mottled color.

On each island where the ranges of the ocean and interior varieties of *Cerion* overlapped, Gould observed *Cerion* with various combinations of ribbing and mottling. This indicated that the two varieties on each island interbred and produced fertile hybrids. Thus Gould concluded that all varieties on a single island were members of the same species.

Detailed study of the snail's reproductive anatomy revealed that all varieties of *Cerion* on each island of Little Bahama Bank had the capacity to interbreed with one another.

Likewise the *Cerion* on all the islands of the Great Bahama Bank had the capacity to interbreed. Thus Gould hypothesized that *Cerion* on Little Bahama Bank were members of the same species, as were those on Great Bahama Bank. However, as a result of differences in reproductive anatomy *Cerion* of Little Bahama Bank were probably unable to interbreed with those of Great Bahama Bank. Thus *Cerion* from the two banks were members of two separate species.

By using both the biological and the morphological concepts of species, Gould was able to untangle the web of detail left by previous researchers of *Cerion*. Further study may help scientists learn how the species evolved.

■ How did Gould use observation in determining the classification of *Cerion*?

Evolution: Speciation   **239**

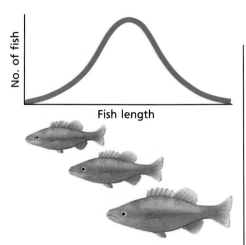

No. of fish

Fish length

Figure 16-2 A bell curve illustrates how most members of a population are grouped in an average range for a given trait while only a few are at the extreme ends of the range.

Think

Organisms often show variation caused by environmental factors. For example, most plants grown in nitrogen-poor soil will be smaller than plants of the same species grown in nitrogen-rich soil. What are some other examples of variation in a species due to environmental factors?

Answer to Think
Accept any examples students back up with facts. For example, variation in skin wrinkling among humans of the same age might be due to differences in exposure to the sun and wind.

Tip: Point out that in a population in genetic equilibrium the phenotypic ratio can vary although the allele frequency does not. This is because allele combinations can vary.

Within a population, individuals vary in many observable traits. For example, fish in a pond vary in length, weight, and color. Scientists study variation in a certain trait by measuring that trait in a large sample. Figure 16-2 shows the results of such measurements. It is a graph of the range of lengths in a population of mature fish. The shape of the graph looks like a bell, and the graph is called a **bell curve.** The bell curve shows that a few fish are extremely short and a few are extremely long, but most are of average length. In nature many traits in populations show variation according to the bell curve pattern.

What causes variation in traits? Sometimes variations are due to environmental factors. Often variations are due to heredity, and this is the kind we are going to discuss. Each individual inherits a different combination of genes from its parents. As you remember from Chapter 10, a difference in **genotype**—the genetic makeup of an individual—usually results in a difference in **phenotype**—the outward appearance of the organism.

As you learned in Chapter 10, variations in genotype arise through mutation, recombination, and crossing-over. Mutation changes individual genes. Genetic recombination, the reassortment of genes during meiosis, results in new gene combinations. Crossing-over, the interchange of chromatid portions of homologous chromosomes during meiosis, also leads to new genotypes. All of these mechanisms cause the variations in traits in a population.

Allele Frequencies and Genetic Equilibrium

From the above discussion you can see that the variation of traits in a population is often controlled by genes. It follows, then, that changes in the genes cause changes in the corresponding traits. Often the most meaningful way to understand how species change over time—that is, how they evolve—is to think of a population as simply a collection of genes. The collection of genes for all the traits in a population is called a **gene pool.** The gene pool of a population thus contains all the alleles for all the genes. An **allele frequency** is the percentage of a specific allele of a gene in the gene pool.

Consider the population of Japanese four o'clock flowers shown in Figure 16-3. As you learned in Chapter 10, these flowers show codominance for color. This means that Japanese four o'clock flowers with red flowers are homozygous (RR) for the red trait, the plants with white flowers are homozygous (R′R′) for the white trait, and the plants with pink flowers are heterozygous (RR′).

Because Japanese four o'clocks exhibit codominance for color, you can infer the genotype by observing the phenotype. Then, when you know all the genotypes, you can calculate the allele frequencies. Compare the parent generation and the off-

First generation

Phenotype frequency		Genotype frequency
White	0	$R = 0.75$
Pink	0.5	$R' = 0.25$
Red	0.5	

RR   RR   RR′   RR′   RR   RR′   RR′   RR

Second generation

Phenotype frequency		Genotype frequency
White	0.125	$R = 0.75$
Pink	0.25	$R' = 0.25$
Red	0.625	

RR   RR′   R′R′   RR   RR   RR′   RR   RR

spring of Japanese four o'clocks in Figure 16-3. Notice that the two generations differ in numbers of individuals of each of the three phenotypes. Nevertheless, the allele frequencies for R and R′ are the same for both generations. This means that the allele frequencies of these Japanese four o'clock plants has not changed from generation to generation. A population in which allele frequencies do not change from generation to generation is said to be in **genetic equilibrium.**

Figure 16-3 Though these Japanese four o'clocks differ phenotypically from generation to generation, the allele frequencies remain the same.

The Hardy–Weinberg Principle

In 1908 a British mathematician, Godfrey Hardy (1877–1947), and a German physician, Wilhelm Weinberg (1862–1937), independently outlined the conditions necessary for genetic equilibrium. The **Hardy–Weinberg principle** states that a population will remain in genetic equilibrium if, and only if, all of the following conditions are met: (1) no mutations occur; (2) individuals neither enter nor leave the population through migration; (3) the population is large; (4) individuals mate randomly; (5) and natural selection does not occur. However, if even one of these conditions does not hold true, allele frequencies of the population may change. When the allele frequency changes, the range of traits in a population will change. In other words, evolution will occur.

Section Review

1. What is the definition of *allele frequency,* and how does it relate to evolution?
2. Compare the morphological concept of species with the biological concept. How are both concepts important?
3. What is genetic equilibrium?
4. What is the Hardy–Weinberg principle, and under what conditions does this principle apply?
5. In what way is the understanding of the biological concept of species important for understanding evolution?

Answers to Section Review
1. The percentage of an allele of a gene in a gene pool. Allele frequency changes when a population is undergoing evolution.
2. The morphological concept concerns differences in the internal and external anatomy. The biological concept concerns the ability of organisms to interbreed. Both concepts help define *species.*
3. the state in which the allele frequencies of a population do not change from generation to generation
4. The principle that states the conditions under which a population will remain in genetic equilibrium. The conditions are that no mutations occur, individuals neither enter nor leave the population through migration, the population is large, individuals mate randomly, and natural selection does not occur.
5. The concept shows how changes in the breeding patterns of organisms can lead to evolution.

- *Summarize how mutation and migration each disrupt genetic equilibrium.*

- *Define genetic drift, and give an example of how genetic equilibrium could be disrupted through genetic drift.*

- *Describe how natural selection differs from mutation, migration, and genetic drift in terms of its ability to cause significant changes in a gene pool.*

- *Contrast the effects of stabilizing, directional, and disruptive selection on variation in a trait over time.*

Figure 16-4 Most koalas, *Phascolarctos cinereus,* are gray (top). Why might some be white (bottom)?

16.2 Disruption of Genetic Equilibrium

Evolution occurs when genetic equilibrium is disrupted. Review the Hardy–Weinberg conditions presented in Section 16.1. In Section 16.2 you will learn how changes in the allele frequencies of a population occur when these conditions are not met. You will learn also that natural selection is the most significant factor causing changes in the gene pool.

Mutation

Mutation is a physical change in a gene or chromosome. Remember from Chapter 11 that mutations include point mutations, such as insertions or deletions of nucleotides in genes, and chromosome mutations, such as the gain or loss of one or more chromosomes.

Mutations affect genetic equilibrium by producing totally new alleles for a trait. In addition mutations can change the frequency of the alleles already present in the gene pool. For example, the mutation that causes the sickle-cell trait occurs spontaneously in about 5 out of 100 million people. This means that an allele for normal hemoglobin will spontaneously undergo mutation into the allele for the abnormal hemoglobin in 5 out of 100 million people. The occurrence of mutation in the sickle-cell allele, and in most other alleles, is too low to cause major changes in the allele frequencies in a population.

Migration

Migration—the movement of individual organisms into or out of a population—can alter allele frequencies in a population. Consider a population of squirrels, for example. This population is characterized by particular allele frequencies for certain traits. Suppose individuals from another population of squirrels move into the area and interbreed with the original population. The second group may introduce alleles that were absent in the gene pool of the first group. Also, the second group will probably have allele frequencies for certain traits that differ from those of the original population. Thus because of immigration—the movement of new individuals into a population—allele frequencies of this population of squirrels will be altered, and genetic equilibrium will be disrupted. Emigration—the departure of individuals from a population—also disrupts genetic equilibrium. The movement of genes into or out of a population through migration is called **gene flow**. Gene flow does not always involve the movement of individuals from place to place. It may involve the movement of sperm, for example, when sperm in plant pollen is transported by wind.

Genetic Drift

Two of the necessary conditions for maintaining genetic equilibrium as stated by the Hardy–Weinberg principle are the presence of a large population and random mating. The Hardy–Weinberg principle is based on the laws of probability, which do not necessarily hold for small and medium-sized populations. In small populations chance can significantly affect allele frequencies from one generation to the next. **Genetic drift** is the phenomenon by which allele frequencies in a population change as a result of random events or chance.

Several species of giant land turtles called Galápagos tortoises live on the Galápagos Islands. Before the 1900s thousands of these tortoises inhabited the Galápagos island of Española. Settlers introduced goats and pigs to the island, and these animals competed with the tortoises for food. As a result tortoise populations dwindled. By 1970 the total population of this species of Galápagos tortoises had shrunk to fewer than 20 members. In such a small population it is possible that one allele of a particular gene will be found in only one tortoise. If that tortoise failed to breed, the allele would be eliminated from the gene pool forever. Elimination of the allele would change allele frequencies for that trait. Thus in the small population of Galápagos tortoises genetic equilibrium could be disrupted by chance. Also, when a population is too small, one individual may not be able to mate as frequently as another individual. This is an example of nonrandom mating, which may also affect allele frequency. All examples of genetic drift occur in small to medium-sized populations. In these populations the effect of nonrandom mating may cause major changes in allele frequencies.

While mutation, migration, and genetic drift can change the gene pool, they usually do not cause significant changes that lead to the formation of new species. Natural selection does cause significant changes leading to the formation of new species.

Figure 16-5 One tortoise subspecies living on the Galápagos Islands is *Geochelone elephantopus* ssp. *vandenburghi*.

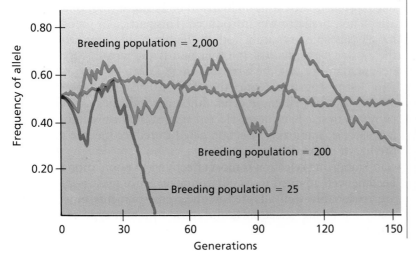

Figure 16-6 Genetic drift is significant only in small and medium-sized populations. In a small population a particular allele may disappear completely over a few generations. In a larger population a particular allele may vary widely in frequency due to chance but still be present in enough individuals to be maintained in the population. In a much larger population the frequency of a particular allele may vary slightly due to chance but will remain relatively stable over generations.

Answer to Think
Answers will vary but may include the following: seed eating, shape of bill, large eyes, coloration, perching feet, dark streak from corner of eye.

Natural Selection

The Hardy–Weinberg principle states that for a population to remain in genetic equilibrium natural selection must not occur. In fact natural selection does occur. It is an ongoing process in nature, and it is the single most significant factor disrupting genetic equilibrium. As you learned in Chapter 15, natural selection results in higher reproductive rates for individuals with certain phenotypes, and, hence, certain genotypes. As a result of natural selection some members of a population are more likely to contribute their genes to the next generation than others are. Thus by the action of natural selection allele frequencies change from one generation to the next. Each of four types of natural selection—stabilizing, directional, disruptive, and sexual—causes changes in the gene pool of a population.

Stabilizing Selection

Stabilizing selection is a type of natural selection in which individuals with the average form of a trait have an advantage in terms of survival and reproduction. The extreme forms of the trait, on the other hand, confer a disadvantage to the organism. For example, in some species of lizards conditions may exist where average size is an advantage in terms of survival and reproduction. Lizards that are larger than average will be more easily spotted, captured, and eaten by predators than average-sized lizards will. On the other hand, lizards that are smaller than average might not be able to run fast enough to escape predators and therefore will also be at a disadvantage. Thus conditions are such that average-sized lizards would be most likely to survive and therefore to reproduce.

The graph on the upper left in Figure 16-7 illustrates the effects of stabilizing selection on the trait of body size in a population of lizards. The red line shows the initial variation in lizard size as a standard bell curve. Remember that the pattern of variability for many traits may be described as a bell curve. The blue line represents the variation in lizard body size several generations after a new predator was introduced. This particular predator was most easily able to capture both the highly visible large lizards and the slower small lizards. In other words the introduction of the new predator caused the large and small lizards in this population to be selected against. Notice that the blue bell curve is not as wide as the original bell curve. This difference indicates that the two extremes in size have been selected against by the predator. Thus, in this case, stabilizing selection acted to reduce the size range in the lizard population.

Stabilizing selection is most effective in a population that has become well adapted to its environment. Since most populations are reasonably well adapted to their environments most of the time, stabilizing selection is the most common type of natural selection.

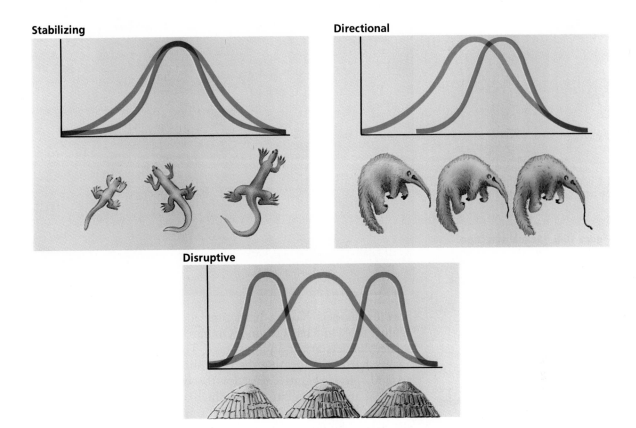

Stabilizing

Directional

Disruptive

Directional Selection

Directional selection is a type of natural selection in which individuals with one of the extreme forms of a trait have an advantage in terms of survival and reproduction. The graph on the upper right in Figure 16-7 shows the effects of directional selection on the trait of tongue length in anteaters. Anteaters feed by breaking open termite nests, extending a sticky tongue into the nest, and lapping up the termites. Suppose that the area was invaded by a new species of termite that built very deep nests. Anteaters with long tongues could more effectively prey on these termites than could anteaters with short tongues. In other words anteaters with long tongues would likely have a selective advantage. Thus directional selection acts to favor one extreme of a trait.

Disruptive Selection

Disruptive selection is a type of natural selection in which individuals with either of the extreme forms of a trait have an advantage in terms of survival and reproduction. The average form of the trait, on the other hand, confers a selective disadvantage to the organism. The bottom graph in Figure 16-7 also shows the effects of disruptive selection on the trait of shell color in marine animals called the limpets. The shell color of limpets varies from pure white to dark tan. On rocks covered with goose barnacles, which are white, white-shelled limpets are at an advantage. Birds that

Figure 16-7 Stabilizing, directional, and disruptive selection can all be illustrated as changes (blue) of the standard bell curve (red).

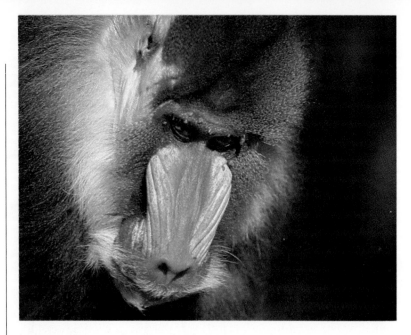

Figure 16-8 Sexual selection causes the vivid coloration that distinguishes the male mandrill baboon, *Papio sphinx,* from the female of the species.

Answers to Section Review
1. a change in the gene frequencies of a population due to chance
2. Mutation changes allele frequencies when new alleles are introduced. Migration changes allele frequencies when individuals enter or leave a population.
3. In stabilizing selection an organism with the average form of a trait has an advantage over one with an extreme form. In some species of lizards average-size lizards have an advantage over both larger- and smaller-than-average-size lizards. Directional selection shifts variation of a trait toward one extreme. Anteaters with longer tongues can prey on termites more effectively. In disruptive selection an organism with an extreme form of a trait has an advantage while one with the average form has a disadvantage. White limpets on rocks with goose barnacles are at an advantage because they are camouflaged against their background. Dark limpets on dark, bare rocks also have a selective advantage through camouflage. However, limpets of intermediate color on either background are more easily seen and preyed on by birds. (See pp. T16A–T16D.)

prey on limpets have a hard time identifying the white shells against the white background. On bare, dark-colored rocks dark-shelled limpets are at an advantage. The limpet-eating birds also have a hard time locating the dark shells against the dark background. However, it is easy for the birds to spot limpets with shells of an intermediate color, which are visible against both the white and the dark backgrounds. In summary, disruptive selection eliminates the intermediate form and favors the extreme forms.

Sexual Selection

Sexual selection is the preferential choice of a mate based on the presence of a specific trait. For example, in the tropical beetle, females preferentially mate with males having an elongated snout. Therefore male beetles with a longer snout are more likely to reproduce and pass their genes on to the next generation. Thus the longer snout is sexually selected for in the male of this species. Sexual selection may be stabilizing, directional, or disruptive.

Natural selection is the most significant factor that disrupts genetic equilibrium and causes changes in the gene pool of a population. Such changes, by definition, lead to evolution in a species.

Section Review

1. What is *genetic drift?*
2. Explain how mutation and migration disrupt genetic equilibrium.
3. Compare and contrast stabilizing, directional, and disruptive selection, and give an example of each.
4. What is sexual selection?
5. Scientists infer from observation that the eyespots on the tail of the male peacock evolved through sexual selection. Draw graphs that show possible mechanisms for the sexual selection of the male peacock's tail.

Info: The peacock's tail, the lion's mane, and the deer's antlers are all considered results of sexual selection.

16.3 Formation of Species

Disruption of genetic equilibrium, as discussed in Section 16.2, leads to changes in the gene pool of a population. As you have read, it is natural selection that causes the most significant genetic changes in a population. However, such changes do not necessarily lead to **speciation**—the formation of new species. In this section you will learn how speciation occurs.

Isolated Populations

Remember that a species is defined as a group of organisms that look similar and have the ability to interbreed and produce fertile offspring in the natural environment. For a new species to arise, either interbreeding or the production of fertile offspring must somehow cease among members of a formerly successful breeding population. For this to occur, populations or segments of a population must somehow become isolated.

Two forms of isolation prevent interbreeding or cause infertility among members of the same species. These forms of isolation are geographic isolation and reproductive isolation.

Geographic Isolation

Geographic isolation is the physical separation of members of a population. Populations may be physically separated when their original habitat becomes divided, as, for example, when new land or water barriers form. Also, when part of a population colonizes a new, remote area such as an island, the colonizers may become geographically isolated from other populations of the species. For example, when a group of American finches colonized the Hawaiian islands, the group became geographically isolated from the parent population. These finches eventually gave rise to the 23 species of Hawaiian honeycreepers discussed in Chapter 15.

Geographic isolation of a population may occur as a result of physical changes in an environment. When a river changes course or even when a highway is built across a field, populations may become geographically isolated.

Let's study a specific example in which geographic isolation may have led to speciation. The desert of Death Valley, California, has a number of isolated ponds formed by springs. Each pond contains a species of fish that lives only in that pond. Scientists suggest that these species arose through geographic isolation.

How did these fish become isolated in Death Valley? Geologic evidence from a study of wave patterns in sedimentary rocks indicates that most of Death Valley was covered by a huge lake during the last ice age. When the ice age ended, the region became dry. Only small, spring-fed ponds remained. Members of a fish species that previously formed a single population in the lake may have become isolated in different ponds.

Section Objectives

- *Define geographic isolation, and explain how it can lead to speciation.*

- *Give two examples of ways reproductive isolation can lead to speciation.*

- *Summarize the theory of punctuated equilibrium.*

- *Outline the role of adaptation in the survival or extinction of a species.*

Info: Fossil evidence suggests that the camel originated in the western United States, crossed by a land bridge to Asia, and became isolated there when the land bridge disappeared.

Figure 16-9 Two types of pupfish that live in ponds in Death Valley are *Cyprinodon salinus* (top) and *Cyprinodon milleri* (bottom).

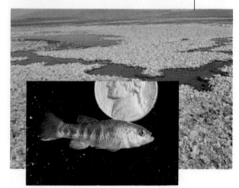

The environments of the isolated ponds differed enough that natural selection and perhaps genetic drift acted on the separate populations. Eventually the fish in the different ponds may have diverged so much genetically that they could no longer interbreed even if brought together. In this way geographic isolation of fishes in Death Valley probably led to the formation of new species. Geographic isolation, plus reproductive isolation, probably is the usual cause of the formation of new species.

Reproductive Isolation

Sometimes groups of organisms within a population become isolated genetically without prior geographic isolation. When barriers to successful breeding arise among population groups in the same area, the result is reproductive isolation. **Reproductive isolation** is the inability of formerly interbreeding organisms to produce offspring.

Reproductive isolation can arise through disruptive selection. Remember that in disruptive selection the two extremes of a specific trait in a given population are selected for. The wood frog and the leopard frog, shown in Figure 16-10, have become reproductively isolated, possibly as a result of disruptive selection. Though the wood frog and the leopard frog sometimes interbreed in captivity, they do not interbreed where their ranges overlap in the wild. The wood frog usually breeds in early April, and the leopard frog usually breeds in mid-April.

This reproductive isolation may have resulted from disruptive selection. In the ancestral frog species frogs that bred earlier and frogs that bred later may have both been selected for, while frogs that bred between these times may have been selected against, perhaps because some predator was especially active during that time. The two groups of frogs may have become reproductively isolated because of differences in breeding times. Probably it was in part through such reproductive isolation that speciation occurred in these frogs. Eventually different selection pressures led to the type of morphological variations seen in Figure 16-10.

Figure 16-10 As the graph shows, peak mating activity in frog species can vary widely. Such variance has led to reproductive isolation in the wood frog, *Rana sylvatica* (top), and the leopard frog, *Rana pipiens* (bottom).

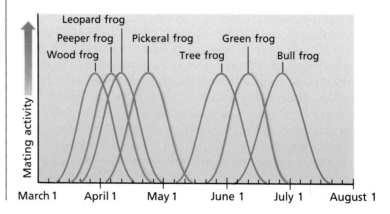

Rates of Speciation

How fast do new species form? Because their generation times are short, new species of unicellular organisms may evolve in years, months, or even days. For plants and animals Darwin theorized that new species formed gradually over millions of years. Yet scientists know of examples in which species arose in only thousands rather than millions of years. For example, archaeological evidence indicates that settlers from Polynesia introduced banana trees to the Hawaiian islands about a thousand years ago. Today several species of moths, unique to the Hawaiian islands, feed on bananas. These moth species are closely related to other plant-eating moths in Hawaii. Thus scientists suggest that the banana-eating moths arose from other plant-eating moths, undergoing adaptive radiation in less than the thousand years that banana trees have existed in Hawaii.

Evidence from the fossil record has led some scientists to propose that speciation need not occur gradually but can occur in spurts. According to the theory of **punctuated equilibrium,** all populations of a species may exist for a relatively long time at or close to genetic equilibrium. Then this equilibrium may be interrupted by a brief period of rapid genetic change in which speciation occurs.

Some scientists argue that if new species evolved gradually, the fossil record should show many examples of transitional forms—species with characteristics intermediate between those of ancestral species and new species. However, for most organisms such transitional forms are absent from the fossil record. Instead the fossil record shows that most species remained the same for hundreds of thousands or for millions of years. Then new, related species suddenly appeared. Whether new species form gradually or rapidly is still a point of debate among scientists. However, scientists agree that natural selection, whether gradual or rapid, is the most important factor in speciation.

Extinction

Just as new species form through natural selection, species also die off—or become **extinct.** What causes extinction? For a species to continue to exist, some members must have traits that allow them to survive and pass their genes on to the next generation. If the environment changes, for instance, the species will become extinct unless some members have adaptations that allow them to survive and reproduce successfully under the new environmental conditions. Changes in climate and competition among species are examples of environmental changes to which species must adapt in order to survive.

Environmental changes caused by human beings have led to the extinction of hundreds of organisms in the past few centuries. Most of these changes involve the destruction of habitats. For

Hedylepta blackburni

Figure 16-11   Moths that feed on Hawaiian banana trees are believed to be an example of a rapidly evolving species.

Tip: Point out that environmental changes can include changes in geologic features, changes in flora and fauna, and even sunspot activity.

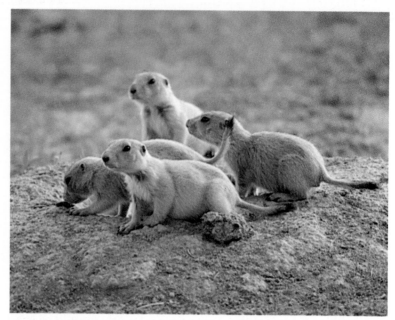

Figure 16-12 The black-footed ferret, *Mustela nigripes* **(left), may be on the way to extinction because humans have encroached on the habitat of its prey, the prairie dog,** *Cynomys ludovicianus* **(right).**

Answers to Section Review
1. A species may exist for a long time at or close to genetic equilibrium followed by the sudden interruption of this equilibrium by a brief period of rapid genetic change.
2. When segments of a population become physically isolated, each segment adapts to its own environment. Eventually the populations diverge so much genetically that members of these segments can no longer interbreed.
3. Organisms having two extreme forms of a trait may not be able to interbreed.
4. The inability to adapt to changes in the environment can lead to extinction of the species.
5. Accept either answer as long as the student supports the answer with facts.

example, the conversion of the prairies of central North America into farmland and grazing ranges caused a decline in the large population of prairie dogs in the region. In turn the black-footed ferret, which preys solely on the prairie dog, has also greatly declined in numbers. This weasellike animal may soon become extinct, because members of the species do not have variations that result in reproductive success in this changing environment.

The example of the ferret shows how species depend on others for survival. The decline in population of one species has led to the near extinction of another. Extinction is a natural process. However, the rapid rate at which species are becoming extinct as a result of the destruction of habitats by human beings may endanger the survival of many life-forms. Over the billions of years of evolution, species of organisms have evolved with unique adaptations to problems presented by a diverse and changing environment. Once a species has become extinct, the unique solution to life we call a species will be lost forever.

Section Review

1. What is the theory of *punctuated equilibrium?*
2. How can geographic isolation lead to speciation?
3. Explain how disruptive selection can lead to reproductive isolation.
4. What is the relationship between adaptation to change and extinction?
5. Do you think that the potential extinction of the black-footed ferret is mainly a natural process or is strictly the result of human intervention? Support your opinion with facts.

Laboratory

Factors That Influence Natural Selection

Objective
To demonstrate the effect of natural selection on genotypic frequencies

Process Skills
Modeling, predicting, hypothesizing

Materials
200 black and 200 white beads, 3 containers labeled *Parental*, *Next Generation*, and *Dead*

Method
1 Put 200 black beads and 200 white beads into the container labeled *Parental*. The black beads represent the dominant allele for black coat (B); the white beads represent the recessive allele for white coat (b) in a hypothetical animal. Assume that the box contains gametes from a parental population of 50 BB, 100 Bb, and 50 bb.
2 Simulate random mating by removing two beads from the container. Record the genotype and the phenotype of the zygote produced. Place the alleles in the container labeled *Next Generation*.
3 Repeat step 2 fifty times.
4 Predict whether the allelic and phenotypic frequencies of all the new individuals produced will be similar to the beginning frequencies.
5 Calculate the frequencies of the alleles and phenotypes produced. Compare your results with your prediction.
6 Put all the beads back into the container labeled *Parental*.
7 Simulate random mating by removing two beads at a time from the container. Record the genotype and phenotype of each zygote produced. Assume that individuals with the white coat phenotype are incapable of reproducing and place them in the container

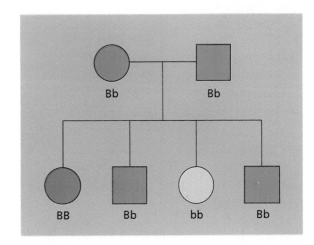

labeled *Dead*. Which genotype will these have? Place alleles with the black coat phenotype in the container labeled *Next Generation*. Which genotypes will these have?
8 Repeat step 7 until the parental pool is depleted.
9 Transfer the beads in the container labeled *Next Generation* to the *Parental* container. Repeat step 7 until the parental pool is depleted.
10 Repeat step 9 two more times.
11 Predict whether the allelic frequencies of each generation will change.
12 Calculate the frequencies of the final genotypes produced and compare the results with your prediction.

Conclusions
1 Compare the frequency of recessive alleles produced by steps 1 through 5 with that produced by steps 6 through 12.
2 Did the frequency of the b allele change uniformly through all generations? If not, what happened?

Inquiry
If you continued performing steps 6 through 12 of this investigation, would you eventually succeed in completely eliminating the b allele? Form a hypothesis and test it.

Chapter 16 Review

Vocabulary

allele frequency (240)
bell curve (240)
biological species
    concept (238)
directional selection
    (245)
disruptive selection
    (245)
extinct (249)
gene flow (242)

gene pool (240)
genetic drift (243)
genetic equilibrium
    (241)
genotype (240)
geographic isolation
    (247)
Hardy–Weinberg
    principle (241)

hybrid (238)
migration (242)
morphological species
    concept (237)
morphology (237)
mutation (242)
phenotype (240)
population (239)

punctuated
    equilibrium (249)
reproductive
    isolation (248)
sexual selection
    (246)
speciation (247)
species (237)
stabilizing selection
    (244)

Explain the difference between the terms in each of the following sets.
1. biological species, morphological species
2. directional, disruptive, and stabilizing selection
3. genetic drift, genetic equilibrium
4. population, species
5. geographic isolation, reproductive isolation

Review

1. A morphological species is characterized by (a) similarities of appearance in its members (b) gene flow (c) the ability of members of different species to interbreed (d) the inability of members of different species to interbreed.
2. Organisms classified as separate species according to the biological species concept (a) can usually interbreed (b) cannot usually be identified on the basis of morphology (c) cannot usually interbreed (d) are formed through mutation.
3. If a population is in genetic equilibrium, (a) evolution has occurred (b) speciation has occurred (c) allele frequencies change from one generation to the next (d) allele frequencies remain the same from one generation to the next.
4. Mutations affect genetic equilibrium by (a) changing the position of a gene (b) introducing new alleles (c) causing immigration (d) causing emigration.
5. Genetic drift is any change in gene frequencies that is due to (a) migration (b) crossing-over (c) extinction (d) chance.

6. Mutation, migration, and genetic drift (a) tend to lead to the formation of new species (b) tend to lead to disruptive selection (c) do not alone cause speciation (d) cause sexual selection.
7. Directional selection, disruptive selection, and stabilizing selection are all examples of (a) genetic equilibrium (b) natural selection (c) random changes in a gene pool (d) speciation.
8. The most common way for new species to form is through (a) mutation (b) disruptive selection (c) geographic and reproductive isolation (d) genetic equilibrium.
9. According to the theory of punctuated equilibrium (a) speciation occurs gradually (b) speciation occurs rapidly (c) speciation occurs when populations are isolated (d) speciation never occurs.
10. Extinction of a species is most often due to (a) geographic isolation of the species (b) reproductive isolation of the species (c) the lack of successful variations (d) random changes in the gene pool of the species.

11. What are the limitations of the morphological concept of species?
12. What type of selection is shown when the bell curve narrows over time?
13. Under what conditions does the Hardy–Weinberg principle apply?
14. How might geographic isolation have led to the evolution of the two *Cerion* species classified by Gould?
15. How can migration alter allele frequencies for certain traits?
16. Why is natural selection considered the most important factor in evolution?

17. What is the relationship between natural selection and sexual selection?
18. How do scientist hypothesize that geographic isolation led to the formation of new species of fishes in California's Death Valley?
19. Describe how the wood frog and the leopard frog are reproductively isolated. How might this isolation have arisen as a result of disruptive selection?
20. What factors have contributed to the potential extinction of the black-footed ferret?

Critical Thinking

1. Give three examples showing that evolution is still occurring today. Refer to magazines such as *National Geographic* and *National Wildlife*.
2. The biological concept of species states that a species is a group of organisms that can interbreed and produce fertile offspring in nature. A mule is the sterile offspring of a horse and a donkey. By the definition above do a horse and a donkey belong to the same species?
3. In the late 1800s hunting reduced the population of a species of giant elephant seals to about 20 individuals. How might such a reduction in population have disrupted genetic equilibrium?
4. Look at the maps on the right showing the earth as it was 550 million years ago

and as it is today. Notice the difference in the placement of the continents. How could this change have affected the evolution of organisms in these areas?
5. A species of flightless bird called the dodo once inhabited Mauritius Island in the Indian Ocean. The animals thrived until humans came and killed them all. How did the dodo's inability to fly affect its extinction?

Extension

1. Read the article by Roger Lewin entitled "I Want a Girl Just Like the Girl...," in *Discover*, November 1986, pp. 65–66. How can the choice of mates resembling ourselves or our parents affect human evolution?
2. Visit an area where plants or animals are bred. Possible places include farms, zoos, plant arboretums, seed companies, and nurseries. Find out how the breeders manipulate the genetic makeup of the plants or animals, and prepare an oral report on how this manipulation speeds up or slows down evolution. If you cannot visit one of the suggested locations, look in the *Readers' Guide to Periodical Literature* for articles on plant and animal breeding.
3. Do research and make a list of five or more examples of sexual selection in animals.

Chapter 17 | *Human Evolution*

This chapter presents the latest theories about human origin and hominid evolution. Students learn about the many scientific processes paleoanthropologists use in this study, including careful observation, drawing inferences, and hypothesizing.

Correlation Chart

Section Number	Section Objectives	Pages	Features	Section Review
17.1	1. Describe how scientists gather evidence of human ancestry.	255–257	Think, p. 256	1
	2. List some traits shared by all primates.			2
	3. Name two distinguishing characteristics of anthropoids.			3
	4. Give examples of traits unique to humans.			4, 5
17.2	1. List three evolutionary trends characteristic of the hominid line.	258–263	Biology in Process, pp. 260–261	2, 3
	2. Contrast the age and physical features of the remains of different australopithecine species.			1, 2
	3. Compare *Homo habilis, H. erectus,* and *H. sapiens.*		Biology Lab, p. 263	1, 3
	4. Outline two theories of hominid evolution.			4, 5
	5. Explain how the discovery of the black skull has caused paleoanthropologists to reevaluate their theories of hominid evolution.			

Resources*

Print

Kornberg, Warren, ed. *Human Evolution* (from NSF *Mosaic* Reader Series). Wayne, N.J.: Avery, 1982.

Lewin, Roger. *Human Evolution: An Illustrated Introduction.* New York: Freeman, 1984.

Audiovisual/Software

Cave Dwellers of the Old Stone Age. Film, 18 min. Britannica.

How Far from the Apes? A New Debate on the Ancestry of Man. Filmstrip. Teacher's guide. Knowledge, 1984.

The Story of Evolution. Video of filmstrip with cassette, 16 min. Teacher's guide. Hawkhill, 1985.

*Additional references are listed in *Modern Biology, Teacher's Program Guide.*

Section Teaching Guides

17.1 The Study of Human Origins

Approach

The main objective of this section is to present distinguishing characteristics of different groups of primates. This section builds upon concepts of adaptation and evolution discussed in Chapters 15 and 16. In presenting the text, guide students to realize how knowing the characteristics of different groups of primates helps paleoanthropologists in their studies of human origins.

Concept Framework

Students will probably be familiar with the theory that humans and primates share a common ancestor. Build on this knowledge by helping students list some of the traits that humans and primates share. Encourage them to add to this list as they read the section and also make a separate list of traits that are uniquely human.

Emphasis

Make certain that students can list traits characteristic of all primates and name distinguishing traits of the following groups of primates: anthropoids, apes, and humans.

Simulation

To help students realize the advantages of the opposable thumb, have students restrict use of their thumbs. Provide each student with masking tape to secure their thumbs against their other fingers. Carry on normal class activities for about ten minutes. Then let students discuss the activities that they found impossible to do without the use of their thumb. Talk about the advantages having the opposable thumb gives to primates.

Section Review Answers

(Continued from p. 257.)

3. Features that distinguish anthropoids from other primates include a well-developed collarbone, rotating shoulder joints, partially rotating elbow joints, an opposable thumb, and a relatively large cranial capacity.

4. bipedalism, a round jaw with relatively unspecialized teeth, and an enlarged frontal portion of the brain, which contributes to language ability

5. Since trees were farther apart, primates couldn't easily swing from branches. This could have led to adaptations for bipedalism. Primates that walked upright would have had a selective advantage by standing upright to locate predators.

17.2 Hominid Evolution

Approach

Tracing the evolution of hominids is the main objective of this section. The section illustrates processes of science (Chapter 2) and builds upon concepts of adaptation and evolution (Chapters 15 and 16). In presenting the text, guide students to be aware of the time span involved in hominid evolution by referring to the time line on pages 258 and 259. Point out that the span of about five million years is a relatively rapid evolutionary rate for mammals.

Concept Framework

Students will have general familiarity with anthropological activity such as the search for bones, tools, and other evidence of human ancestors. Build on this knowledge by discussing some of the inferences about hominids scientists might make from examining such evidence. Ask students to evaluate the evidence used to develop theories of hominid evolution as they read this section.

Emphasis

Students should be able to list the hominids in order from the most ancient to the most recent and to give characteristics of the following hominids: *A. afarensis*, *A. africanus*, *A. robustus*, *A. boisei*, *H. habilis*, *H. erectus*, Neanderthals, and Cro-Magnons.

Using the Features/The Black Skull

The main process skills discussed in this feature are observing and inferring. The feature discusses the discovery of the black skull by Richard Leakey and Alan Walker and outlines why this discovery has caused scientists to reevaluate former theories of hominid evolution. Suggest that students review information presented on p. 262 before reading the feature.

Student Presentations

Divide the class into seven groups. Assign each group one of the hominids discussed in the section. You may wish to assign *A. robustus* and *A. boisei* to a single group. Ask each group to prepare a presentation on the assigned hominid. The presentation should include a map of where the hominid remains have been found and the following information: who found the remains and when, what the remains consist of, and the phylogenetic status of the hominid in the most current theories of hominid evolution. Refer students to *The General Science Index* for the latest articles on these hominids.

Section Review Answers

(Continued from p. 262.)

3. From *H. habilis,* the most ancient species, to *H. sapiens,* the most modern, brain size became large, height probably increased, and culture became more sophisticated.
4. One theory says that *A. afarensis* was the common ancestor to the australopithecines and modern humans. The other theory says that *A. afarensis* is not a valid species and that *Australopithecus* and *Homo* diverged about 6 million years ago with the remains of the common ancestor yet to be found.
5. Accept any answer students can justify, including the following: evidence of warfare among the two species might include sites of mass hominid killing, evidence of warlike culture in art and weapons, or evidence of hominid migration to escape war. Evidence of interbreeding might include fossil remains showing characteristics of both Neanderthals and Cro-Magnons, or fossil remains of the two species buried at the same site.

Laboratory Preparation

Background

In this lab students gain an understanding of the assumptions derived from the use of comparisons of amino acid sequences as a measure of evolutionary relationships. Students should understand that proteins are three-dimensional and depend on shape to interact properly. They will then comprehend the assumption that some amino acids in the sequence are less important for function than others. If different amino acids make the same contribution to shape, or if a slight change in shape does not interfere with function, different amino acids will appear at that location as a result of random genetic change, i.e., mutation and recombination. Students need to be aware that it is easier to compare the amino acid sequences of adjacent organisms than to compare each one with the human sequence. Remind students to compare each animal to human cytochrome C.

Preparation

No outside preparation is required for this lab.

Conclusions

1. No. The values listed are based on comparison with humans. The values give no indication of the number of amino acid sequences the chicken and the horse share with each other.
2. Proteins can be said to behave like a clock because they change gradually over time. The number of differences among them might be considered a measure of time.

Inquiry

No. Students should understand that differences that occur in the amino acid sequences of hemoglobin over time are not necessarily related to differences that occur in cytochrome C over time. However, in the case of humans and chimpanzees, even though we do not know the human and chimpanzee gene sequences, antigen-antibody reactions between the two species quantitatively demonstrate the closeness of their relationship.

Chapter Review

Vocabulary

1. *Bi-* means "two"; *-ped-* means "feet"; *-ism* means "the condition of." *Bipedalism* means "the condition of having two feet," which implies walking on two rather than four feet.
2. *Hom-* means "man"; *anthropo-* means "resembling a human being."
3. The thumb can be positioned opposite the four other fingers for grasping.
4. *Cranial* refers to *cranium,* which originally comes from a Greek root meaning "head." The cranium

is made up of the bones that surround the brain. Cranial capacity suggests the amount of space that the cranium encloses, which would be the limiting factor for the size of the brain.

5. *Paleo-* means "ancient"; *anthropo-* refers to human beings; and *-logy* means "the study of." An anthropologist may or may not study fossil bones, which are the main subject of study for *paleo*anthropologists.

Review

11. Anthropoid traits include a well-developed collarbone, rotating elbow and shoulder joints, opposable thumb and/or big toe, and large cranial capacity.

12. Upright walking allows movement in plains and allows for locating predators and prey.

13. bipedalism and the development of language for communication

14. A human jaw is rounder than the jaw of an ape; humans lack spaces between their teeth, which are adaptations apes have for stripping tree branches of leaves; the human teeth are smaller and less specialized than those of apes.

15. The size of the brain case of a fossil skull indicates the cranial capacity.

16. Most hominid fossil remains have been found on the continent of Africa.

17. Evidence indicates that Neanderthals lived in caves and stone shelters; remains of stone tools for scraping hides for clothing have been found.

18. Scientists have found sophisticated tools, including harpoons, scrapers, drills, fishhooks, and needles, along with Cro-Magnon remains; Cro-Magnons also decorated the walls of caves with paintings of the animals they hunted.

19. Accept answers that include information in Figure 17-8.

20. Previous theories of hominid evolution assumed that *A. boisei* descended from *A. robustus* because scientists had inferred that the trait of robustness evolved over time in australopithecines. Because of its massiveness, the black skull has been classified as *A. boisei;* it is dated as older than any other *A. boisei* or *A. robustus* remains; it showed that some early australopithecines were robust. Therefore *A. boisei,* previously considered to be a descendant of *A. robustus,* could not have been a descendant, because the black skull is older than any other

A. robustus; likewise, *A. boisei* cannot be considered an ancestor of *A. robustus* because the black skull is more robust, therefore more advanced. For these reasons *A. robustus* and *A. boisei* must have evolved as two separate lines.

Critical Thinking

1. Accept any inference that students can back up with specific observations. For example, the spacing of the baboon teeth suggests that the baboons clean leaves off branches as apes do, and that baboons use their pointed teeth to pierce prey.

2. Apes would not easily be able to use spoken language because the voice box is not well developed. The lack of development in the facial muscles prevents apes from using spoken language because they cannot move the lips and tongue with precision, as is necessary to produce the sounds in spoken language.

3. Accept answers students can back up with a logical reason. For example, the use of foresight in tool making is a sign of increased intelligence and learned behavior, which accompanies cultural development. Cultural development gave hominids more flexible behavior, enhancing their ability to adapt to environmental changes.

4. The reindeer remains suggest that Cro-Magnons hunted this animal. The remains of this animal in southern Europe suggest that during the time of the Cro-Magnons, the climate of Europe was colder than it is today. Accept any other reasonable answer.

Chapter 17 | *Human Evolution*

Introduction

Twelve years after *The Origin of Species* was published, Charles Darwin wrote another book, on the subject of human evolution. In this book, *The Descent of Man*, Darwin cited evidence that humans are most closely related to the group of animals that includes monkeys and apes. Darwin's work was a landmark in the scientific study of human origins. In this chapter you will read about some of the latest theories of human origins and evolution.

Chapter Outline

17.1 The Study of Human Origins
- *Classification of Humans*
- *Primate Characteristics*

17.2 Hominid Evolution
- Australopithecus
- Homo habilis
- Homo erectus
- Homo sapiens
- *Theories of Hominid Evolution*

Chapter Concept

As you read, look for the many scientific processes involved in studying human evolution.

Homo sapiens skull, 300,000 years old. Inset: Face of a chimpanzee, *Pan troglodytes*.

For additional teaching information see pp. T17A–T17D.

17.1 The Study of Human Origins

Scientists who study human evolution are called **paleoanthropologists** *(PAY-lee-oe-AN-thruh-PAHL-uh-jists)*. Paleoanthropologists unearth fossils and observe them carefully. Because most fossil skeletons are far from complete, these scientists must make inferences from subtle clues. For example, a curved piece of skull bone allows them to infer the size of the brain; wear on a fossil tooth may give some indication of diet; markings on fossil limb bones often show where muscles were attached and what mode of locomotion was used.

Based on their observations and inferences, paleoanthropologists construct models that shed light on the morphology and habits of early humans and their ancestors. They use these models to formulate theories about human evolution. Because human fossils are few and fragmentary, each new fossil find is a potential source of significant information. Thus paleoanthropology is a rapidly changing field that is characterized by exciting new theories and vigorous debates.

Classification of Humans

Before paleoanthropologists can decide how to classify fossils, they must understand how humans are related to other animals. Evidence from fossils, comparative morphology, and biochemistry indicates that humans belong to the order of mammals called primates. **Primates** include tree shrews, lemurs, tarsiers, monkeys, and apes, as well as humans.

Primate Characteristics

Fossil evidence and evidence from comparative morphology indicate that the majority of extinct primate species lived in trees. Most modern primate species are also tree dwelling, or arboreal. Scientists infer from this that many of the characteristics that primates share may have evolved as adaptations to arboreal life. For example, most primates have highly movable fingers and toes, which have flattened nails rather than claws. These characteristics allow arboreal primates to grasp branches and cling to tree trunks securely.

Good vision is another adaptive trait of tree dwellers. Cones, color-sensitive cells in the eyes, give primates color vision. This allows primates to locate ripe fruit in trees. Front-facing eyes and a reduced snout enable primates to integrate images from both eyes simultaneously and thus perceive depth accurately. Therefore arboreal primates can gauge distances as they leap or swing from branch to branch.

Info: The word *paleoanthropologist* is composed of three Greek roots: *paleo-* which means "ancient"; *anthropo-* which refers to human beings; and *-ology* meaning "the study of."

Figure 17-1 The movable fingers and toes, front-facing eyes, and small snout of the tarsier, *Tarsius syrichta,* are adaptations for living in trees.

Figure 17-2 The gibbon, *Hylobates* sp. (top), and the gorilla, *Gorilla gorilla* (bottom), are anthropoids.

Think

Chimpanzees sometimes strip blades of grass to use for pulling termites out of their nests. How does this compare with the way humans make and use tools?

Tip: Have students tape both thumbs to their forefingers and try to carry out a series of routine tasks.

The ability to hold the body in a vertical position is another adaptation for life in the trees. As primates evolved, the sense of smell became less acute. Therefore primates probably became more dependent on their eyesight. Being able to hold themselves erect while feeding probably allowed them to easily spot and flee from approaching predators.

Characteristics of Anthropoids

Monkeys, apes, and humans form a subgroup of primates called the **anthropoids** *(AN-thruh-POIDZ)*. The well-developed collarbone, rotating shoulder joints, and partially rotating elbow joints give anthropoids skeletal strength and flexibility. The **opposable thumb**—a thumb that can be positioned opposite the other fingers—gives anthropoids precision grip. Most anthropoids also have an opposable big toe, which gives them increased grasping ability.

Compared with other primates, anthropoids have a large brain for their body size. The fossil record shows that as primates evolved, brain size increased. Scientists estimate brain size by measuring the brain case of fossil skulls. The size of the brain case is called **cranial capacity**. A large cranial capacity relative to body size is a distinguishing characteristic of anthropoids. Large cranial capacity is not necessarily a sign of intelligence, although in some primate lineages a larger brain may correspond to greater mental abilities.

In comparison with other primates, humans and apes—including gibbons, orangutans, gorillas, and chimpanzees—have a large cranial capacity relative to body size. Humans and apes have also become increasingly capable of sitting, standing, or walking erectly. The lower vertebrae that form the tail in monkeys are reduced in apes and humans, leaving them without a tail.

Characteristics of Humans

Bipedalism—upright walking on two legs—is a uniquely human trait. Other primates occasionally walk on two legs, but usually they swing from branches or walk on all fours. Differences in the pelvic bones of apes and humans, shown in Figure 17-3, correlate with differences in modes of locomotion. The broad shape and muscular attachments of the human pelvis are adaptations for bipedalism. The pelvis supports the internal organs during upright walking. Another adaptation for bipedalism is the shape of the human foot. The bones of the big toe are aligned with those of the other toes. This causes body weight to be distributed evenly during upright walking.

Figure 17-3 shows that the human jaw and teeth differ from those of apes. These differences reflect differences in diet. The ape's U-shaped jaw has wide spaces between the teeth that allow it to pull branches through the teeth and strip off the leaves. The human jaw is rounder than the ape jaw. The teeth are smaller and less specialized, an adaptation to the varied diet of humans.

Answer to Think: Humans make tools deliberately, using foresight and planning. Chimpanzee behavior does not indicate a plan to use tools. In many cases it is a rudimentary form of tool use.

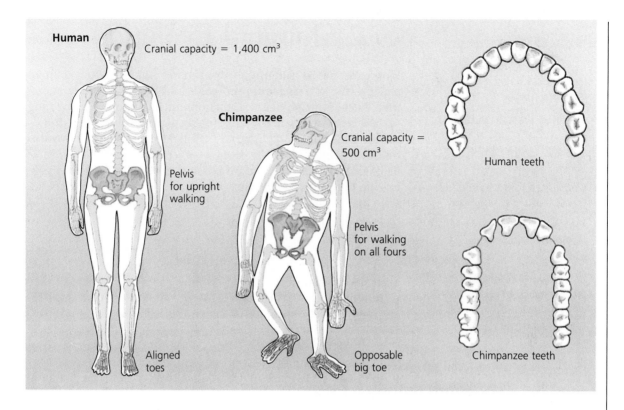

Human

Cranial capacity = 1,400 cm^3

Chimpanzee

Cranial capacity = 500 cm^3

Pelvis for upright walking

Pelvis for walking on all fours

Aligned toes

Opposable big toe

Human teeth

Chimpanzee teeth

Humans have an average cranial capacity of 1,400 cm^3. The human cranial capacity is larger than that of any other primate. The high forehead of humans is indicative of the large size of the frontal part of the brain. This part of the brain is important for a unique human trait—the ability to communicate through verbal language. Apes communicate through sounds and gestures; they can also be taught to use certain forms of sign language. However, apes living in the wild have not developed any complex, flexible set of signals that can compare to those that make up the languages of humans.

Section Review

1. What do paleoanthropologists study, and how do they gather their information?
2. Name several characteristics of primates, and explain how these characteristics are adaptations to life in the trees.
3. What features distinguish anthropoids from the other primates?
4. What are three specifically human traits?
5. Evidence from fossil pollen found with human remains indicates that primates ancestral to humans probably inhabited the edges of the forests, where trees were scattered rather than close together. How might this type of environment have influenced the evolution of bipedalism?

Figure 17-3 Human physical characteristics, compared with those of the chimpanzee, include a larger cranial capacity, a broader pelvis, and a big toe aligned with the other toes. The human jaw is rounder and the teeth are less specialized. Can you find other comparisons suggested by the drawings?

Answers to Section Review
1. Paleoanthropologists unearth fossil fragments, observe them, make inferences, compare them to living primates, and construct models to shed light on the morphology and behavior of early humans and their ancestors.
2. Primates have: movable digits with nails rather than claws, which allow them to grasp branches and tree trunks; color vision and good depth perception, which allow them to locate fruit in trees and gauge distances as they swing through trees; and the ability to hold the body erect, which helps them spot and flee from predators. (See pp. T17A–T17D.)

Section Objectives

- *List three evolutionary trends characteristic of the hominid line.*

- *Contrast the age and physical features of the remains of different australopithecine species.*

- *Compare* Homo habilis, H. erectus, *and* H. sapiens.

- *Outline two theories of hominid evolution.*

- *Explain how the discovery of the black skull has caused paleoanthropologists to reevaluate their theories of hominid evolution.*

17.2 Hominid Evolution

Hominids are a subgroup of primates that includes human beings—the species *Homo sapiens*—and their immediate ancestors. Paleoanthropologists study hominid evolution by examining fossils, from which they have inferred the evolutionary trends toward bipedalism and increased cranial capacity. They have also found evidence of the evolution of culture—behavior that is dependent on learning and on passing knowledge from one generation to the next. Note the evidence of these trends as you read.

Australopithecus

The earliest genus of hominids is called *Australopithecus (aw-STRAY-loe-PITH-uh-kuss),* meaning "southern ape." The first australopithecine fossils were dubbed "southern" because they were found in South Africa. However, remains of *Australopithecus* have since been found in eastern Africa as well.

Australopithecus afarensis

The oldest australopithecine skeleton unearthed to date was found in eastern Africa in 1974. This fossil is between 3 million and 3.5 million years old. The shape of the pelvic bone indicates that the remains are those of a female. Scientists have nicknamed her Lucy. They classify Lucy and morphologically similar skeletons as *Australopithecus afarensis.* This species existed between 3 million and 4 million years ago.

A. afarensis was shorter than modern humans, less than 1.5 m tall. Fossil footprints and the shapes and sizes of pelvic and leg bones indicate that the species was bipedal. From the fossil skulls scientists infer that *A. afarensis*'s cranial capacity was between 380 and 450 cm^3, less than one-third that of modern humans.

Figure 17-4 Various hominids have existed over the past 4 million years. Today only *Homo sapiens sapiens* remains. Keep in mind that males and females evolved simultaneously.

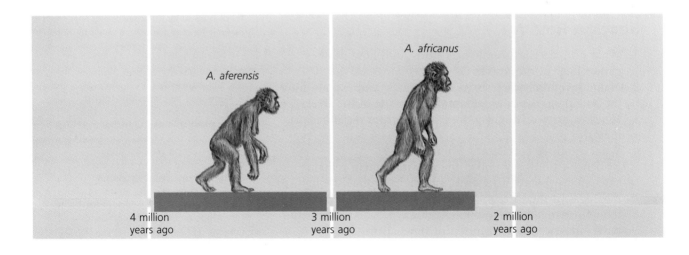

A. aferensis

A. africanus

4 million years ago

3 million years ago

2 million years ago

Australopithecus africanus

The next oldest species of australopithecines, *Australopithecus africanus*, lived 2.2 million to 3 million years ago. The first fossils of this species were found in 1924 in South Africa. *A. africanus* was slightly taller and heavier than *A. afarensis*. Its cranial capacity was slightly larger, between 430 and 550 cm³.

Australopithecus robustus and Australopithecus boisei

Two other australopithecine species probably arose after *A. africanus*. *A. robustus* fossils have been found in southern Africa; *A. boisei* fossils have been found in eastern Africa. Most of their remains are between 1 million and 2 million years old. Both species had heavier skulls and larger back teeth than *A. africanus*. Their cranial capacity was between 450 and 600 cm³.

Homo habilis

In the early 1960s paleoanthropologists working in East Africa found a hominid skull with a relatively large brain case. This fossil was the first evidence of such an early hominid species having a cranial capacity as great as 600 to 800 cm³. These remains were found along with stone tools that were presumably made and used by the hominids. This finding led to the naming of a new species, *Homo habilis*, meaning "handy human." Remains of *H. habilis* were later found in southern as well as eastern Africa. These fossils are between 1.6 million and 2 million years old.

Some studies of the insides of *H. habilis* skulls indicate that one region of the brain essential to speech was developed in this species. Tool marks on animal bones found near the hominid fossils indicate that *H. habilis* ate meat. Yet some fossils indicate that *H. habilis* was no taller than earlier australopithecines.

Figure 17-5 The skull of *Australopithecus africanus* (top) is more apelike than that of *Homo habilis* (bottom).

A. robustus   A. boisel   H. habilis   H. erectus   H. sapiens   H. sapiens

Neanderthal   Cro-Magnon

1 million years ago   Present

Figure 17-6 This reconstruction of *Homo erectus* indicates a thick skull, a prominent brow, a low forehead, and a small chin.

Homo erectus

In 1891 paleoanthropologists working on the Pacific island of Java discovered remains of a skull and thigh bone that were older and more distinctly human than any fossils found previously. They named this species *Homo erectus*, "upright human." *H. erectus* fossil remains have since been found in many other parts of the world. They range between 0.5 million and 1.6 million years old.

Compared with modern humans, *H. erectus* had a thick skull, large brow ridges, a low forehead, and a very small chin. Cranial capacity ranged between 700 and 1,250 cm^3. An almost complete *H. erectus* skeleton was found in East Africa in 1984. These remains are of a 12-year-old male who stood about 1.7 m tall. This indicates that an adult *H. erectus* might easily have been as tall as a modern human.

Although *H. erectus* ate mainly plant foods, members of this species were also successful hunters. They used stone tools, such as the hand ax, possibly for removing thick animal hides. Traces of charred bones indicate that *H. erectus* built fires for cooking and probably also for warmth.

Biology in Process

The Black Skull

In Kenya during the summer of 1985 a team of paleoanthropologists led by Richard Leakey and Alan Walker found a hominid skull estimated to be about 2.6 million years old. They named it the black skull because of its distinctive dark color. It is also referred to as KNM-WT 17000, its catalog number at the Kenya National Museum.

To which hominid species does the black skull belong? Because of its extreme thickness scientists first classified it as *A. boisei*. However, the morphology of this massive skull has caused scientists to question both in-terpretations of hominid evolution presented on p. 262.

First, the black skull is almost a million years older than other *A. boisei* remains. So *A. boisei* could not have been the descendant of *A. robustus,* as postulated.

Second, evidence suggests that australopithecine species became more robust over time. Scientists had inferred that robustness was a recently acquired trait in *Australopithecus.* However, the discovery of the black skull led scientists to reexamine this inference and the implications they had based on it. If robustness was an advanced trait, the older, more

Homo sapiens

Two skulls between 250,000 and 350,000 years old were found in Germany and Great Britain in the 1930s. These skulls are evidence of hominids transitional between *H. erectus* and *H. sapiens*. They have both the large brow ridges of *H. erectus* and the very large cranial capacity of *H. sapiens*.

Neanderthals

Hominid fossil skeletons from 35,000 to 130,000 years ago have been found in Europe, Asia, and Africa. They belong to a group called Neanderthals *(nee-AN-dur-THAWLZ)*. Neanderthals were early *H. sapiens*. According to the fossil evidence, Neanderthals had heavy bones, thick brow ridges, and a small chin. However, skull sizes indicate that Neanderthals had a cranial capacity averaging 1,450 cm³. This is slightly larger than the cranial capacity of modern humans. Fossils of leg and trunk bones indicate that the Neanderthals were about 1.5 m, or 5 ft, tall and stocky. Neanderthals were adapted to the cold climate of northern Europe. They lived in caves and stone shelters. They carefully shaped stone tools for scraping hides to make clothing.

Answer to Biology in Process
The black skull is older than any other *A. robustus*. The black skull is also more advanced than *A. robustus*. Hence *A. robustus* and *A. boisei* must have evolved as two separate lines.

robust species, *A. boisei*, could not have been the ancestor of *A. robustus*.

Faced with the new evidence that *A. boisei* could be neither an ancestor nor a descendant of *A. robustus*, some scientists have offered revised theories of hominid evolution. The diagram illustrates one of the newest theories.

Notice that these scientists suggest that *A. boisei* and *A. robustus* evolved along separate lines. They further suggest that several different species of australopithecines arose within a 500,000-year span. This is considered a rapid adaptive radiation for primates. What changes in the environment could have led to such rapid speciation in australopithecines?

Someday the answer to this question and others may be found. Until then paleoanthropologists will continue observing, collecting data, formulating inferences, re-evaluating theories, and communicating their ideas to other scientists. Using these varied processes, scientists may come closer to solving the mystery of human ancestry.

■ Why has the discovery of the black skull led scientists to re-evaluate their theories of hominid evolution?

Figure 17-7 Paleoanthropologists offer at least two different views of human evolution.

Answers to Section Review
1. *A. afarensis, A. africanus, A. robustus, A. boisei, H. habilis, H. erectus,* and *H. sapiens.*
2. Basically, they became increasingly larger in size and in cranial capacity over time. Accept answers that include more detail.
(See pp. T17A–T17D.)

Cro-Magnons

Hominid fossil skulls from about 35,000 years ago were first found in caves in southwestern France in 1868. Since then the fossil remains of these people, referred to as Cro-Magnons, have been found elsewhere in Europe as well as in Africa, Asia, and even Australia. Cro-Magnons had a cranial capacity of about 1,400 cm^3. They are distinguished from Neanderthals by their high forehead, prominent chin, and lack of brow ridges. Taller than Neanderthals, they stood about 1.8 m, or almost 6 feet, tall. Cro-Magnons were *H. sapiens* and are regarded by scientists as modern humans.

Cro-Magnons had a sophisticated culture. They made many types of tools, including a wide variety of blades, harpoons, scrapers, and even drills, fishhooks, and needles. They decorated the walls of caves with colorful paintings of the animals they hunted.

Theories of Hominid Evolution

Paleoanthropologists differ in their interpretations of how humans evolved. This is because scientific theories are based on the best available evidence and hominid fossil evidence is fragmentary. Compare the two diagrams of hominid evolution in Figure 17-7. The top diagram suggests that *A. afarensis* gave rise to two lines of hominids about 3 million years ago. One line led to the later australopithecines. The other led to hominids who were directly ancestral to modern humans.

The bottom diagram shows an alternate interpretation. According to this theory skeletal remains classified as *A. afarensis* are placed into a preaustralopithecine genus. This theory suggests that *Homo* and two lines of *Australopithecus* diverged between 4.5 million and 6.0 million years ago.

The debate between paleoanthropologists illustrates that this science is an ongoing process. Precise observation and analysis of new evidence and reevaluation of theories will present a more accurate and detailed picture of human evolution.

Section Review

1. Name seven hominid species.
2. How did the australopithecines differ from one another in terms of changes in their physical features over time?
3. Compare *Homo habilis, H. erectus,* and *H. sapiens* with regard to evolutionary trends.
4. Compare two theories of hominid evolution.
5. According to fossil evidence Neanderthals died out about 35,000 years ago. Some scientists hypothesize that the Neanderthals were killed off by the Cro-Magnons. Others hypothesize that the two groups interbred. What evidence would you look for to evaluate these theories?

Amino Acid Sequences and Evolutionary Relationships

Objective
To infer evolutionary relationships among species by comparing the amino acid sequences of the same protein in different organisms

Process Skills
Classifying, inferring

Materials
Pencil, paper

Method

1. The greater the time organisms have been diverging from a common ancestor, the greater the difference that can be expected in their amino acid sequences. This statement is based on the assumption that the rate of change of a specific amino acid sequence is the same in different organisms. Think of other methods used to determine evolutionary relationships. How is this biochemical method different?

2. Cytochrome c, a protein found in the mitochondria of many organisms, consists of a chain of about 100 amino acids. The figure shows the corresponding parts of the cytochrome c amino acid sequence of nine vertebrates. The numbers along the side of the figure refer to the position of these sequences in the chain. The letters identify the specific amino acids in the chain.

3. Make a table to record your data. Head columns *Species* and *Number of Differences*. For each vertebrate count the amino acids in the sequence that differ from the human sequence, and list them in your table.

4. List the eight vertebrate sequences in descending order according to their degree of evolutionary closeness to humans. Which organism of all the vertebrates listed is most closely related to humans?

	Horse	Chicken	Tuna	Frog	Human	Shark	Turtle	Monkey	Rabbit
42	Q	Q	Q	Q	Q	Q	Q	Q	Q
43	A	A	A	A	A	A	A	A	A
44	P	E	E	A	P	Q	E	P	Y
46	F	F	Y	F	Y	F	F	Y	P
47	T	S	S	S	S	S	S	S	S
49	T	T	T	T	T	T	T	T	T
50	D	D	D	D	A	D	E	A	D
53	K	K	K	K	K	K	K	K	K
54	N	N	S	N	N	S	N	N	N
55	K	K	K	K	K	K	K	K	K
56	G	G	G	G	G	G	G	G	G
57	I	I	I	I	I	I	I	I	I
58	T	T	U	T	I	T	T	I	T
60	K	G	N	G	G	Q	G	G	G
61	E	E	N	E	E	Q	E	E	E
62	E	D	D	D	D	E	E	D	D
63	T	T	T	T	T	T	T	T	T
64	L	L	L	L	L	L	L	L	L
65	M	M	M	M	M	R	M	M	M
66	E	E	E	E	E	I	E	E	E
100	K	D	S	S	K	K	D	K	K
101	A	A	A	A	A	T	A	A	A
102	T	T	T	G	T	A	T	A	T
103	N	S	S	S	N	A	S	N	N
104	E	K	–	K	E	S	K	E	E

Conclusions

1. Examine the list you prepared in step 4. The values listed for the chicken and the horse differ by only one. Can you infer from this that the chicken and the horse are closely related to each other? Why or why not?

2. Why can it be said that proteins behave like molecular clocks?

Inquiry
Other proteins, such as hemoglobin, can be used to establish degrees of evolutionary relatedness among organisms. Would you expect to find the same number of differences in the cytochrome c and hemoglobin amino acid chains when comparing organisms? Why or why not?

Vocabulary

anthropoid (256)
bipedalism (256)

cranial capacity (256)
hominid (258)

opposable thumb
 (256)

paleoanthropologist
 (255)
primate (255)

1. Using a dictionary, find the meanings of the word parts that make up the word *bipedalism*. Why is "bipedalism" a good name for the mode of locomotion it describes?
2. What is the meaning of the root *hom-* in *hominid*? What is the meaning of the root *anthropo-* in *anthropoid*?
3. Why is the primate thumb described as *opposable*?

4. Using a dictionary, look up the meaning of *cranial*. Why is "cranial capacity" an accurate way to describe the size of the brain case?
5. Use a dictionary to find the meaning of the word parts that make up *paleoanthropology*. How might the work of a paleoanthropologist differ from that of an anthropologist?

Review

1. Paleoanthropologists reconstruct the bodies of human ancestors on the basis of (a) fingerprint patterns (b) comparisons of living species (c) written descriptions in ancient sources (d) fossil fragments.
2. The earliest primates probably lived in (a) the grasslands (b) the forests (c) the deserts (d) the mountains.
3. Humans are well adapted to upright walking because they have (a) good color vision (b) a broad pelvis (c) a big toe that is opposable to the other toes (d) the ability to use language.
4. One difference between *A. africanus* and the two species *A. robustus* and *A. boisei* is that (a) *A. africanus* had a less massive skull (b) *A. africanus* had a more massive skull (c) *A. africanus* had bigger teeth (d) *A. africanus* made tools.
5. *H. habilis* was so named because evidence indicates this species (a) was bipedal (b) used language (c) made tools (d) ate meat.
6. One feature that distinguishes *H. erectus* from *H. sapiens* is that *H. erectus* (a) had a large chin (b) had thinner bones (c) had a low forehead (d) used fire.
7. Hominid fossils that range in age between 250,000 and 350,000 years of age show some features of *H. erectus* and some features of (a) *A. boisei* (b) *H. habilis* (c) Cro-Magnons (d) *H. sapiens*.
8. Cro-Magnons are regarded as (a) a transitional species between *H. erectus* and *H. sapiens* (b) a transitional species between australopithecines and ancestral humans (c) ancestors to Neanderthals (d) identical to modern humans.
9. Among the hominids, fossil evidence indicates that there was an evolutionary trend toward bipedalism, increased cranial capacity, and (a) the development of culture (b) the development of vegetarianism (c) the development of robustness (d) the development of brow ridges.
10. Because of the black skull's extreme thickness, scientists classify it as (a) *A. robustus* (b) *A. boisei* (c) *A. afarensis* (d) *A. africanus*.
11. What are some traits of anthropoids?
12. How is upright walking an adaptive trait for humans?
13. What are two behavioral traits that are unique to humans?
14. How do the jaw and teeth of humans differ from those of apes?

15. How do scientists use fossil evidence to draw conclusions about the cranial capacity of hominids?
16. On which continent have most fossil remains of hominids been found?
17. What evidence do paleoanthropologists have that Neanderthals were well adapted to the cold climate of northern Europe?

18. Why do scientists hypothesize that Cro-Magnons had a sophisticated culture?
19. Diagram two theories of hominid evolution.
20. How did the discovery of the black skull show that previous theories of hominid evolution must have been in error?

Critical Thinking

1. The photo at the right shows the size and arrangement of the teeth of a baboon. What can you infer about the baboon's diet from observing the size and spacing of its teeth?
2. In comparison with humans, apes do not have a well-developed voice box or facial muscles. How might the ape's anatomy affect its ability to use spoken language?
3. According to the fossil evidence *H. habilis* made stone tools deliberately. They traveled long distances to collect specific types of rocks and minerals; then they shaped the rocks by chipping them at the edges. What does this reveal about *H. habilis*'s ability to use foresight? Why might this ability be significant?

4. Cro-Magnon remains have been found with reindeer bones in certain areas of southern Europe. What does this fact suggest about the diet of Cro-Magnons and the environment in which they lived?

Extension

1. Read the article by Pat Shipman entitled "When Skeletons Speak, It's Not Just Dem Bones Talkin'," in *Discover*, April 1986, pp. 94–99. How does Lucy's pelvis differ from that of modern females? What does this indicate about childbirth, brain size, and bipedalism in *A. afarensis*?
2. Visit a local zoo to observe the behavior of monkeys, apes, or other primates. Pay close attention to the facial expressions of the animals. Notice the ways in which the animals interact. Take notes on what you observe. What similarities and differences do you see between the behavior of the primates you observed and that of humans? If it is not possible to visit a zoo, study pictures of primates in books or magazine articles about primates. What inferences can you make about primate behavior from the pictures?
3. Some of the fossils you have read about in this chapter were discovered by Richard Leakey. Leakey comes from a family of famous anthropologists. Both his father, Louis Leakey, and his mother, Mary Leakey, have made important contributions to the study of human evolution. Use an encyclopedia or other reference to find biographical information about the Leakeys. Write a report about the Leakeys and about the discoveries that members of this family have made.

Chapter 18 | *Classification*

This chapter introduces taxonomy and explains how scientists use it to bring order to the biological world. Students learn how the study of evolution influences and is influenced by taxonomy.

Correlation Chart

Section Number	Section Objectives	Pages	Features	Section Review
18.1	1. Summarize the importance of taxonomy for biologists. 2. Describe the system of binomial nomenclature. 3. List the categories of the classification system used today.	267–268		1, 2 3, 5 4
18.2	1. List the kinds of evidence used by modern taxonomists in classifying organisms. 2. Define *phylogeny* and explain its role in classifying organisms. 3. Explain how family trees illustrate phylogeny. 4. Define the role of biosystematics in inferring evolutionary relationships.	269–271	Think, p. 271	2, 5 1 3 4
18.3	1. Describe the five-kingdom system of classification. 2. List the characteristics that distinguish monera from protists. 3. Describe how RNA-sequencing studies may lead to a revision of the five-kingdom classification system.	272–275	Biology in Process, pp. 272–273 Biology Lab, p. 275	1, 2, 4 3

Resources*

Print

Gutnik, Martin J. *The Science of Classification.* Danbury, Conn.: Watts, 1980.

Savage, R.J.G. *Mammal Evolution: An Illustrated Guide.* New York: Facts on File, 1986.

Audiovisual/Software

Classifying Living Things. Film, 21 min. Britannica, 1986.

General Biology Series. 30 charts, 44" x 36" each. Denoyer-Geppert Science Co.

Life Science. Apple. Curriculum Data Base. 2 disks. User's handbook. Activity units. Scholastic.

*Additional references are listed in *Modern Biology, Teacher's Program Guide.*

Section Teaching Guides

18.1 History of Taxonomy

Approach

The main objective of this section is to familiarize students with the fundamental goals and nomenclature of biological classification. It builds upon concepts of diversity through adaptation (Chapters 15 and 16) and prepares students for future concept presentation of organismic diversity (Chapters 20–23, 25, and 29–41). In presenting the text, guide students to develop an appreciation for the specificity of binomial nomenclature. You may wish to use Table 18.1 to help students investigate the classifications of specific organisms.

Concept Framework

All students will have been exposed to classification systems such as alphabetization or the Dewey Decimal System. Build on this knowledge to present a conceptual framework for introducing taxonomy. Ask students to provide examples of the levels of categories used to classify items in familiar systems such as alphabetization, grocery store shelving, numbers for players on a sports team, and so on. End the discussion by asking students to speculate on possible categories of classification for living things.

Emphasis

Make sure that students understand that taxonomy is a system created by humans to organize living things based on scientific comparisons. Actual classifications may or may not reflect actual evolutionary relationships.

Observing and Classifying

Bring to class six to ten different items in one of the following categories: leaves, nuts, seashells, or any other group of diverse items that, together, are commonly called by a single name. Divide the class into several groups. Give students time to study and take notes on each of the items, then have each group classify the items into three categories of its choosing. Discuss the process with the class, pointing out the legitimate sources of differences among the students' classification schemes. You may wish to have students explain their classification schemes.

Section Review Answers

(Continued from p. 268.)
5. Examples might include the sea horse and the prairie dog.

18.2 Modern Taxonomy

Approach

The major objective of this section is to familiarize students with the observations and inferences that form the basis of modern taxonomy. It builds upon concepts of biochemical and genetic diversity (Chapters 7–10 and 12) and diversity through adaptation (Chapters 15 and 16). In presenting the text, help students understand that taxonomists consider many different kinds of evidence in classifying organisms and inferring phylogenies.

Concept Framework

Students have learned that genetic recombination results in variation. Tap this knowledge to provide a conceptual frame for approaching this section, have students discuss the similarities and differences between two groups of organisms, such as plants and animals or elephants and bears. Elicit the generalization that there are similarities among the variations. As they read this section ask students to consider the similarities and differences among the wide variety of life they see in their own environments.

Emphasis

Make sure that students understand that taxonomy reflects a scientific process of collecting evidence, observing and organizing observations, and sometimes inferring evolutionary relationships.

Guest Speaker

Invite a taxonomist or natural historian to speak to the class about the process of identifying, naming, and classifying a new species.

Using the Features/Classifying Archaebacteria

The main process skills used in the feature are observation and inference. Carl Woese observed substantial differences among groups of bacteria and inferred new phylogenetic relationships based on his observations. Remind students that these skills are essential in scientific experimentation.

Section Review Answers

(See p. 271.)

18.3 Five-Kingdom System

Approach

The main objective of this section is to introduce students to the range of diversity among living things. It builds upon concepts of diversity through adaptation (Chapters 15 and 16), and prepares for future concept development in organismal diversity. In presenting the text, guide students to an appreciation of the habitats occupied by living things. Encourage them to consider how the environment influences the modes of nutrition and reproduction exhibited by members of the different kingdoms.

Concept Framework

Students have learned that plants are autotrophic and animals are heterotrophic. Build on this knowledge to provide insight into one group of classification criteria. Engage students in a discussion of how autotrophs and heterotrophs interact with their environments. Tell students as they read to consider the fact that biomes, large regions of the biosphere, are technically defined in terms of plant life.

Emphasis

Make sure that students understand that the five-kingdom system is generally accepted today, but that scientific trends may lead future taxonomists to use systems of fewer or more kingdoms.

Field Trip

Visit a zoo, nature preserve, or park and have students observe the diversity of organisms in a particular area. Have them focus on classifying organisms into kingdoms.

Section Review Answers

(See p. 275.)

Laboratory Preparation

Background

In this lab students learn organized ways to approach the complexity and diversity of the natural world and to retrieve information efficiently through the concept of flow-chart thinking. Students observe traits of animals and make independent choices to identify the animals. The key illustrates levels of classification schemes. Students realize that the key, while useful, does not necessarily reflect phylogeny.

Preparation

All of the organisms are available from a biological supply house and may be obtained live. Possible specimens include: sea star (Class Asteroidea), sea urchin (Class Echinoidea), sea cucumber (Class Holothuroidea), *Dipylidium Caninum* (Class Cestoidea), *Ascaris Lambricoides* (Phylum Nematoda), leech (Class Hirundinea), *Nereis* (Class Polychaeta), *Lambricus*, (Class Oligochaeta), *Limulus* (Subphylum Chelicerata), grasshopper (Subphylum Uniramia), crayfish (Subphylum Crustacea).

Safety

If live specimens are used, remind students to handle them in a humane manner.

Conclusions

1. Specimen names will vary according to how many specimens the student has time to key.
2. Characteristics need not be opposite ones as shown by key step 3. What is necessary is that the step be constructed so that *a* and *b* include all organisms to be classified.
3. Answers will vary. Some closely related species may be difficult to key on the basis of physical characteristics.

Inquiry

Students should suggest ways to repeat the investigation using a trichotomous key, such as by organizing the information in the form of a branching chart.

Chapter Review

Vocabulary

1. A common name is the vernacular name given to an organism.
2. The word *phylum* is used for organisms other than plants; *division* is used for plants.
3. *Phylo* is from the Greek word meaning "tribe, race, or clan"; *geny* is from the Greek word for

being born. Phylogeny describes the evolution of a genetically related group of organisms.

4. Biosystematics is the study of reproductive mechanisms and gene flow.
5. Most dictionary definitions of biological species mention that the organisms within a species are presumed to be related and that they can interbred to produce fertile offspring. Linnaeus did not presume relationships among organisms nor did he include interbreeding in his definition.

Review

11. Both divided organisms into plants and animals; they further grouped organisms on the basis of similar features.
12. Scientific names prevent confusion and allow scientists to communicate accurately.
13. The second largest level of classification is the phylum; for plants it is called the division.
14. Plants are autotrophic; fungi are heterotrophic.
15. The major type of evidence taxonomists use is morphological.
16. One example might be the populations of maggots in Wisconsin that have separated into two populations living on either apple or cherry trees. As the trees bear fruit at different times, these populations may become reproductively isolated.
17. The Kingdom Monera contains the greatest numbers of living things on earth; almost every moneran species has billions of individuals.
18. A scientific name can reveal one or more characteristics of the organism, the geographic range or geographic location of the organism, or something about the person who discovered the organism.
19. Kingdom Monera includes two types of bacteria with strikingly different genetic material and cell chemistry.
20. Taxonomists can use embryological evidence to determine the tissues in the embryo that gave rise to a given structure. Using that information, they can determine whether two organisms may be closely related.

Critical Thinking

1. Students might take into consideration the age of the book, record, or song; characteristics or categories such as genre, author or composer, performer, and so on. It is important that the same criteria be applied to each individual item classified at each level.
2. The microscope revealed the vast diversity of microorganisms and revealed details about organisms that had not been apparent.
3. Some taxonomists compare nucleotide sequences in DNA and RNA to determine relationships between organisms.
4. Scientists have not collected every known organism and therefore presume that many unknown species exist. One example of a new and poorly investigated habitat is a hot-water vent beneath the ocean.
5. They might have had to use an inconsistent system of long descriptive names and short names that varied from region to region or country to country.
6. Euglena have a single cell, chloroplasts for photosynthesis, and flagella for movement. The presence of chloroplast indicates it is alga, but if chloroplasts are undeveloped it may be classified as a protozoan.

Introduction

Scientists discover and name thousands of new species every year. The science of identifying, classifying, and naming organisms is called taxonomy. In their work taxonomists may also make inferences about the evolutionary relationships among species. In this chapter you will learn about the modern taxonomic system by which all living things are classified.

Chapter Outline

Chapter Concept

As you read, consider how scientists try to develop workable classification systems to organize their knowledge.

Bioluminescent ctenophore.

Inset: Oak, *Quercus* sp., leaves.

For additional teaching information see pp. T18A–T18D.

18.1 History of Taxonomy

Taxonomy (*tacks-AHN-uh-mee*) is the science of grouping organisms according to their presumed natural relationships. Organisms were first classified more than 2,000 years ago by the Greek philosopher Aristotle, who classified living things as either plants or animals. He divided animals into land, water, and air dwellers. He divided plants into three subcategories on the basis of stem differences. Aristotle's system was favored by scientists until the eighteenth century. Then, in a period of expanding scientific exploration in which many new organisms were found, biologists began to recognize that Aristotle's categories did not accommodate all the variations among living things. Driven by their need to communicate about these organisms, biologists devised a number of classification systems, most of which proved to be somewhat unworkable.

These early scientists used either common names or long scientific descriptions, both of which caused confusion. **Common names,** such as *robin* and *fir tree,* are the everyday names given to organisms. Common names may not describe organisms accurately. For example, a jellyfish is not a fish. Sometimes the same common name is used for different species. For instance, a maple tree might actually be a sugar maple, a silver maple, or a red maple. Also, some organisms have more than one common name. In the United States the tree shown in Figure 18-1 is called a buttonwood tree, a plane tree, or a sycamore, depending on the region. To avoid this kind of confusion, some early scientists used long Latin descriptions to name organisms.

Binomial Nomenclature

Understandably, scientists tried to devise simpler naming systems. The system used today, called binomial nomenclature, is based on the work of the Swedish naturalist Carolus Linnaeus.

Linnaeus (1707–1778) developed a system of grouping organisms into hierarchical categories. He placed structurally similar organisms into a group he called a species, similar species into a larger group called a **genus,** and similar genera into a **family.** Similar families were placed in an **order;** similar orders, in a **class;** and similar classes, in a **phylum.** Finally, Linnaeus placed phyla into either the plant or animal **kingdom.**

Rather than list all seven categories in naming an organism, Linnaeus chose to use only the species name—that is, the genus name and the epithet denoting the species. For example, Linnaeus classified human beings into the genus *Homo.* The epithet that denotes the human species is *sapiens.* Therefore humans belong to the species *Homo sapiens.* Linnaeus's system is called **binomial nomenclature** *(bie-NOE-mee-ul NOE-mun-CLAY-chur).* Binomial means "two names."

Section Objectives

- *Summarize the importance of taxonomy for biologists.*
- *Describe the system of binomial nomenclature.*
- *List the categories of the classification system used today.*

Tip: As an exercise in classifying, have students select a category (kingdom) of things and devise their own classification system. You may wish to suggest a general category such as fasteners (screws, nuts, tape, twist ties, paper clips, buttons) and have students establish the subcategories.

Figure 18-1 Although called by different names in different parts of the country, this tree is known as *Platanus occidentalis* to scientists throughout the world.

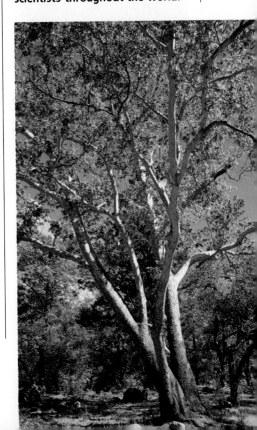

Table 18-1   Classification of Three Organisms

Common Name	Kingdom	Phylum/ Division	Class	Order	Family	Genus	Specific Epithet
Bacterium	Monera	Schizophyta	Scotobacteria	Spirochaetales	Spirochaetaceae	*Cristispira*	*petinis*
Box elder	Plantae	Anthophyta	Dicotyledonae	Sapindales	Aceraceae	*Acer*	*negundo*
Human being	Animalia	Chordata	Mammalia	Primates	Hominidae	*Homo*	*sapiens*

Info: The scientific classification for the grizzly bear is *Ursus horribilis,* which stands for "horrible bear."

Figure 18-2   The twinflower, *Linnaea borealis,* is a member of the cornflower family.

Answers to Section Review
1. Taxonomy is the science of identifying, classifying, and naming living things; it helps scientists study and communicate about a variety of species.
2. the structural similarities or differences among organisms
3. genus name and specific epithet
4. kingdom, phylum or division, class, order, family, genus, species
(See pp. T18A–T18D.)

The binomial name of a species is called its **scientific name.** For example, *Rana pipiens* is the scientific name of the leopard frog. Usually the scientific name that a taxonomist chooses either describes the organism or the range of the organism, or honors another scientist or a friend. For instance, the Latin name *Trifolium agrarium,* meaning "three-leaved" and "found in fields," refers to a clover that is common in meadows. Figure 18-2 shows *Linnaea borealis,* a species that grows in northern regions. The flower was Linnaeus's favorite; *borealis* means "northern."

Levels of Classification

Today taxonomists use Linnaeus's seven levels of classification. Botanists use the term **division** in place of phylum when classifying plants. Study the sample classifications in Table 18-1.

Often taxonomists classify species into subspecies, varieties, or strains, thus denoting variation within a species. The subspecies of a species are morphologically different and are often geographically separated. For example, the northern timber wolf is a subspecies of *Canis lupus* called *Canis lupus* ssp. *occidentalis.* The notation *ssp.* means "subspecies."

The varieties of a species are morphologically different and are often not geographically separated. Some have been produced by humans. For instance, peaches and nectarines are varieties of *Prunus persica.* Botanists refer to the peach tree as *Prunus persica* var. *persica,* and to the nectarine as *Prunus persica* var. *nectarina.* *Var.* means "variety."

A strain is a biochemically dissimilar group within a species. Strain is usually used in reference to microorganisms. A strain is often represented by letters or numbers. For instance, two strains of the intestinal bacterium *Serrata marcescens* are called *S. marcescens* D1 and *S. marcescens* 933.

Section Review

1. What is *taxonomy,* and why is it important?
2. What criterion did Linnaeus use to classify organisms?
3. What do the two parts of a scientific name denote?
4. List the levels of classification from the largest group down.
5. Give several examples of common names that might cause confusion if scientists were to use them.

18.2 Modern Taxonomy

In this section we will focus first on the evidence used to identify organisms that have already been classified or to classify new organisms as they are discovered. You will learn how studying morphology, embryology, chromosomes, biochemistry, and physiology enables scientists to identify and classify organisms. You will also learn that many modern taxonomists are not content just to identify and classify species. They are also interested in inferring the **phylogeny** *(fie-LAHJ-uh-nee)*, or the evolutionary history, of a species.

Evidence Used in Classification

Most taxonomists base their classifications primarily on comparative morphology. They study an organism's morphological features to determine its taxonomic category. When taxonomists attempt to either identify or classify a plant, for example, they look first at the structure of the plant to determine its similarity to known species. Are its flower petals separate, for example, or are they fused into a tube? Are its flowers purple or white or pink? Does the plant have one leaf per stem or several? By looking at a series of such features taxonomists can usually determine the species of an organism—or can determine the species to which it is most similar. Taxonomists usually rely on special books or manuals to aid them. Identification often involves the use of a **dichotomous key**—a written set of choices that leads to the name of the organism.

However, morphological evidence in itself may not be sufficient for accurately identifying or classifying an organism. Therefore taxonomists also compare the embryology, chromosomes, or biochemistry of organisms.

Embryological evidence is often used in classification, especially at the level of the larger classification categories. For example, the wings of a bird and the wings of an insect arise from different tissues in the embryo. The wing structures in the two organisms are therefore analogous and not homologous. This indicates that the organisms are not closely related. However, embryological evidence can also indicate that seemingly dissimilar species may well be related. For example, the bones of the forelimb in the lizard are embryologically similar to those in the cat. This suggests that the two organisms may share a similar phylogenetic origin.

Taxonomists find other evidence for identification and classification by comparing the chromosomes of different organisms. For example, taxonomists studying two similar mushrooms can prepare a karyotype for each organism. By counting chromosomes and comparing chromosome shapes these scientists may determine whether two mushroom species are closely related.

Section Objectives

- *List the kinds of evidence used by modern taxonomists in classifying organisms.*
- *Define phylogeny, and explain its role in taxonomy.*
- *Explain how family trees illustrate phylogeny.*
- *Define the role of biosystematics in inferring evolutionary relationships.*

Info: The coelacanth *Latimeria chalumnae*, once thought to be extinct, was rediscovered deep in the Indian Ocean. Morphologically this fish is considered a possible link in the transition to ambulatory terrestrial life.

Figure 18-3 Through the study of morphological differences taxonomists have classified the brown bear (top) as *Ursus arctos* and the polar bear (bottom) as *Ursus maritimus.*

Taxonomists can use biochemical comparisons of organisms to further refine their classifications. Recall from Chapter 8 that the sequence of nucleotide bases in genes is reflected in the amino acid sequence of a protein. Comparing the proteins, amino acid sequences, DNA, and RNA in different organisms can help researchers determine the degree of genetic similarity among various organisms.

Physiological studies are especially useful in the classification of bacteria. For example, members of the genera *Enterobacter* and *Escherichia* can both ferment the sugar lactose. However, members of *Enterobacter* can use citric acid as their sole source of carbon, while members of *Escherichia* cannot. Species of these bacterial genera look similar under the microscope, but they are classified into different genera on the basis of their physiological differences.

Inferring Phylogeny

A major task of some modern taxonomists is inferring the phylogeny of a group of species. Taxonomists are able to infer the

Figure 18-4 A phylogenetic tree shows the evolutionary relationship of species. The finches closest to the ends of the branches are the most recently evolved species.

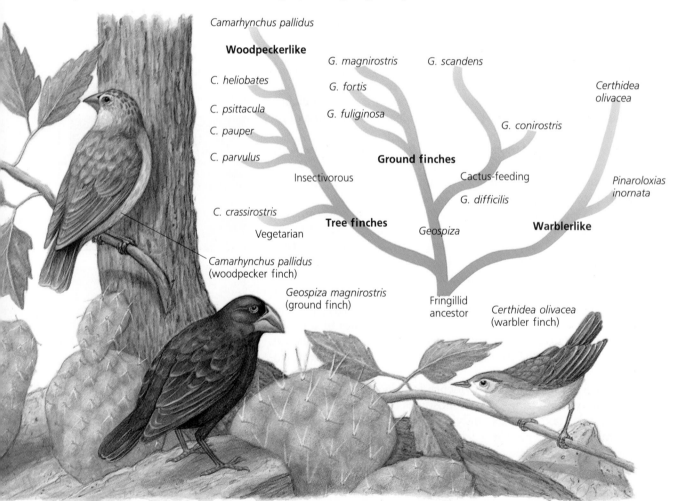

Camarhynchus pallidus

Woodpeckerlike

C. heliobates

C. psittacula

C. pauper

C. parvulus

C. crassirostris

Vegetarian

Tree finches

Camarhynchus pallidus
(woodpecker finch)

Geospiza magnirostris
(ground finch)

G. magnirostris   G. scandens

G. fortis

G. fuliginosa

Certhidea
olivacea

G. conirostris

Ground finches

Insectivorous   Cactus-feeding

G. difficilis

Geospiza   **Warblerlike**

Pinaroloxias
inornata

Fringillid
ancestor   Certhidea olivacea
(warbler finch)

probable evolutionary relationships among species that have been classified on the basis of the kinds of evidence listed on the last two pages. Look at the Galápagos finches in Figure 18-4 as an example. As you read in Chapter 15, the finches that Darwin observed on the Galápagos Islands were classified based on similarities and differences of their bills. Taxonomists infer that the group of species having the most similar bills evolved along the same evolutionary lines.

Taxonomists commonly illustrate the inferred phylogeny of related organisms with a diagram called a phylogenetic tree. A **phylogenetic tree** is a visual model of the inferred evolutionary relationships among organisms. Figure 18-4 shows a phylogenetic tree of Galápagos finches. The species that are illustrated at the tips of the branches represent those organisms that have evolved most recently. The main branches and the trunk of the tree represent the organisms from which the recently evolved organisms arose. Branches in close proximity imply that a close evolutionary relationship exists between the organisms on those branches.

Biosystematics

As you learned in Chapter 16, many taxonomists use reproductive compatibility to infer evolutionary relationships. **Biosystematics** is the study of reproductive compatibility and gene flow. Biosystematists assess the genetic variation in populations and among species. Using information from their genetic studies, biosystematists make inferences about the ancestral history of a species.

Essentially, biosystematists study speciation, the evolution of one species into two. They collect data on variations in a population of organisms. The biosystematists then analyze the data to construct a model of the genetics of the population. As an example, in Wisconsin a single population of fly maggots has begun to feed, mature, and then reproduce near either apple trees or cherry trees. Because the two kinds of trees bear fruit at different times of the year, the apple-eating maggots and the cherry-eating maggots may eventually become reproductively isolated from one another through the mechanism discussed in Chapter 16. A biosystematist might infer that in this case speciation is occurring.

Section Review

1. What is *phylogeny*?
2. List some types of evidence used by modern taxonomists to classify organisms.
3. What do branches that are close together on a phylogenetic tree indicate?
4. How does the science of biosystematics aid in the classification of species?
5. Why is it important for taxonomists to continually collect new specimens?

Think

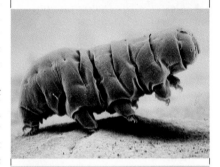

The organism shown above is called a tardigrade. What characteristics do you think might be considered important in classifying this organism?

Answer to Think
Answers will vary but may include limb structure, organ structure, movement, and the morphology of the filaments.

Tip: You can use a comparative anatomy text to illustrate species relationships. For example, compare foreleg development in amphibians, lizards, whales, and humans and the adaptation of the wing in birds.

Answers to Section Review
1. the evolutionary history of a species
2. Possibilities include evidence from the morphology, embryology, biochemistry, and chromosomes.
3. that the organisms on those branches are closely related
4. by using information from genetic studies to make inferences about the ancestral history of a species
5. New specimens might be new species. Also, new specimens can be used to study variation within a species.

- *Describe the five-kingdom system of classification.*

- *List the characteristics that distinguish monera from protists.*

- *Describe how RNA-sequencing studies may lead to a revision of the five-kingdom classification system.*

Answer to Biology in Process
Although challenged at first, Woese continued to publish evidence, which led many to support him and to search for more evidence.

18.3 Five-Kingdom System

Aristotle divided organisms into either the plant or the animal kingdom, but today scientists recognize that some organisms belong to neither. Most modern taxonomists use a five-kingdom system. As you read, keep in mind that this system is only one of many current attempts to organize the diversity of life-forms.

Taxonomists place an organism into one of the following kingdoms: Monera *(moe-NIR-uh)*, Protista *(proe-TISS-tuh)*, Fungi *(FUN-jie)*, Plantae *(PLAN-tee)*, and Animalia *(AN-uh-MAY-lee-uh)*. The criteria used to define kingdoms include cell structure, tissue structure, nutritional requirements, and developmental patterns.

Monera

The Kingdom Monera is made up of prokaryotic organisms, which lack nuclei and other membrane-bound organelles. Many monerans live in aquatic and terrestrial habitats and obtain nutrients primarily through absorption. Other monerans are autotro-

Biology in Process

Classifying Archaebacteria

While there is great diversity in the Kingdom Monera, all of the prokaryotic unicellular microorganisms in the kingdom have been considered basically similar. Taxonomists have also thought that these organisms must be closely related evolutionarily. However, by studying the genetic traits of prokaryotes, one scientist has noted that two different types of monerans may exist.

Using new ribosomal RNA-sequencing techniques, Carl R. Woese

began to collect genetic data on various organisms in 1969. Woese classified his data into "rRNA dictionaries." In these dictionaries he described a species as a group in which each individual had the same sequence of six or more nucleotides. By comparing the data contained in rRNA dictionaries of different species, he could infer their phylogenetic relationships.

By 1981 Woese had compared the rRNA dictionaries of almost 200 species of prokaryotes and eukaryotes. He found that most prokaryotes formed a large, distinctive phylogenetic group, which conphic, producing food either by photosynthesis or chemosynthesis. Most monerans reproduce asexually, although members of some species exchange genetic material. Scientists have identified only about 5,000 moneran species—a small number in comparison with other kingdoms. Nevertheless, the Kingdom Monera includes the greatest number of living things on earth, since almost every moneran species has billions of individuals. Familiar monerans include *Escherichia coli*, the bacteria that are normally present in the intestines of humans, and *Clostridium tetani*, the bacteria that cause tetanus.

Info: Taxonomists are continuously evaluating species with the aid of advanced technology. Thus they are finding new ways to classify the populations of the five kingdoms. It is conceivable that with the accumulation of enough evidence a sixth kingdom, viruses, may eventually be established.

Protista

The Kingdom Protista is made up of eukaryotic organisms that lack specialized tissue systems. Remember that eukaryotes do have true nuclei and membrane-bound organelles. Protists may be unicellular or multicellular. They live in aquatic or moist habitats and obtain their food by ingestion, absorption, or photosynthesis. They reproduce sexually and asexually. More than 50,000 species have been placed in the Kingdom Protista. Examples of protists include the algae and the protozoa.

firmed existing classification schemes. However, Woese also identified a group of prokaryotes that were as different from other prokaryotes as they were from eukaryotes. He called this group archaebacteria to distinguish them from the group called eubacteria.

Woese also noticed that the habitats of modern archaebacteria are oxygen free, very salty, or boiling hot and acidic. Some archaebacteria photosynthesize with a pigment called bacterial rhodopsin, which is chemically similar to a visual pigment in animals.

Woese proposed splitting the Kingdom Monera into two kingdoms to reflect the differences he observed. Woese thought that archaebacteria should be taxonomically separated from eubacteria. Furthermore, he suggested that archaebacteria, eubacteria, and eukaryotes arose in three distinct groups from a common ancestor. The identity of that ancestor is still in question, but Woese provided a compelling hypothesis for its existence.

When Woese first communicated his findings, many people challenged his ideas. However, as Woese and his colleagues published further evidence of genetic similarities between archaebacteria and eukaryotes, other investigators began to look for evidence to support Woese's hypothesis.
■ How did Woese's process of communicating with other scientists affect the acceptance of his hypothesis?

Table 18-2 The Five-Kingdom System

Kingdom	Monera	Protista	Fungi	Plantae	Animalia
Type of cells	Prokaryotic	Eukaryotic	Eukaryotic	Eukaryotic	Eukaryotic
Unicellular or multicellular	Most are unicellular	Most are unicellular	Most are multi-cellular	Multicellular	Multicellular
Mode of nutrition	Heterotrophy and autotrophy	Heterotrophy and autotrophy	Heterotrophy	Autotrophy	Heterotrophy
Examples					

Tip: To emphasize how the plant and animal kingdoms have come to dominate the earth, have students list the kingdoms with their respective numbers of species.

Answers to Section Review
1. Monera, Protista, Fungi, Plantae, and Animalia
2. cell structure, tissue structure, nutritional requirements, and developmental patterns
3. Monerans are prokaryotic.
4. Fungi are heterotrophic; plants are autotrophic.
5. Possibilities include conducting genetic and biochemical studies comparing members of the kingdoms, searching for fossil evidence for similarities among protists and other organisms, and studying other kingdoms to identify characteristics their members may share with protists.

Fungi

The Kingdom Fungi is made up of heterotrophic unicellular and multicellular eukaryotic organisms. Fungi absorb nutrients rather than ingesting them. Most species are terrestrial. They reproduce sexually and asexually. The 100,000 fungi species include mushrooms, yeasts, puffballs, rusts, and bread molds.

Plantae

The Kingdom Plantae consists of the plants, all of which are multicellular and autotrophic. Most plants are terrestrial. They reproduce sexually and asexually. The more than 350,000 species of plants include mosses, ferns, conifers, and flowering plants.

Animalia

The Kingdom Animalia is made up of eukaryotic, multicellular, heterotrophic organisms that obtain nutrients by ingesting food. Animals are terrestrial and aquatic. Most reproduce sexually; a few can reproduce asexually. There are more than a million animal species, including organisms as diverse as spiders and elephants. Table 18-2 summarizes the five-kingdom system.

Section Review

1. Name the five kingdoms.
2. What are the main criteria used to define kingdoms?
3. How are monerans different from all other organisms?
4. What is the primary difference between fungi and plants?
5. Suggest some possible techniques that could be used to reclassify the organisms in the Kingdom Protista.

Laboratory

Using a Dichotomous Key to Identify Organisms

Objective
To correctly identify unknown organisms using a dichotomous key

Process Skills
Observing, classifying

Materials
11 unidentified invertebrate specimens, living or preserved

Method
1 In this investigation you will use a dichotomous key to identify and classify a group of unknown invertebrate organisms. Pick a specimen and study it briefly. Then go to step 1 of the key. Note that only one of each pair of statements applies to the organism you are identifying. Pick either 1a or 1b—whichever describes your specimen. Then follow the directions to the next step.
2 Continue in this manner until you arrive at the name of the specimen.
3 Choose another specimen and repeat the process.
4 Continue working through the key with new specimens until your teacher says to stop.

Key
1a	Body has radial symmetry (Phylum Echinodermata)	Go to 2
1b	Body has lateral symmetry	Go to 4
2a	Radiating arms present	Class Asteroidea
2b	No arms present	Go to 3
3a	Covered with spines	Class Echinoidea
3b	Elongated body	Class Holothuroidea
4a	Body flat	Class Cestoidea
4b	Body not flat	Go to 5
5a	Body divided into repeating units	Go to 6
5b	Body not divided as above	Phylum Nematoda

6a	Hard external covering (Phylum Arthropoda)	Go to 9
6b	Soft external covering (Phylum Annelida)	Go to 7
7a	Bristles absent	Class Hirudinea
7b	Bristles present	Go to 8
8a	Leglike appendages present	Class Polychaeta
8b	Leglike appendages absent	Class Oligochaeta
9a	Feelerlike antennae on head absent	Subphylum Chelicerata
9b	Feelerlike antennae on head present	Go to 10
10a	Unbranched antennae	Subphylum Uniramia
10b	Branched antennae	Subphylum Crustacea

Conclusions
1 Name the specimens you have identified.
2 When designing a dichotomous key, must the two parts of each step describe opposite characteristics? Can you think of a more important criterion for deciding how to construct the parts of a step? Explain your answer.
3 Are there some species of organisms that cannot be identified with a dichotomous key? Explain your answer.

Inquiry
In this investigation you used a dichotomous key to identify some unknown organisms. How might you repeat the investigation using a trichotomous key?

Chapter 18 Review

Vocabulary

binomial
    nomenclature (267)
biosystematics (271)
class (267)
common name (267)

dichotomous key
    (269)
division (268)
family (267)
genus (267)

kingdom (267)
order (267)
phylogenetic tree
    (271)

phylogeny (269)
phylum (267)
scientific name (268)
taxonomy (267)

1. Distinguish between a common name and a scientific name.
2. What is the difference between a phylum and a division?
3. Using a dictionary, find the meaning of the word parts *phylo-* and *-geny* in *phylogeny*. Explain why this term is appropri-ate for discussing common ancestry of organisms.
4. Define *biosystematics*.
5. Look up the word *species* in a dictionary. How does the dictionary definition compare with Linnaeus's use of the term? Suggest explanations for the differences.

Review

1. A scientific name includes information about (a) species and phylum (b) division and genus (c) genus and order (d) genus and species.
2. Aristotle classified plants on the basis of differences in their (a) stems (b) flowers (c) leaves (d) roots.
3. A group of related classes of organisms make up a (a) genus (b) order (c) phylum (d) kingdom.
4. Linnaeus defined a species as a group of organisms that (a) reproduced asexually (b) reproduced sexually (c) shared many common characteristics (d) shared a common ancestor.
5. Biosystematists study the process of (a) speciation (b) taxonomy (c) genetics (d) development.
6. A unique characteristic of the Kingdom Monera is that all of its members are (a) prokaryotes (b) heterotrophs (c) eubacteria (d) archaebacteria.
7. In a phylogenetic tree, close branches indicate (a) close evolutionary relationships (b) the most recently evolved species (c) common habitats (d) common aspects of behavior.
8. The Kingdom Protista includes (a) bacteria (b) plants (c) algae (d) mushrooms.
9. Some protists are similar to plants in that they (a) carry on photosynthesis (b) have advanced tissues (c) may ingest nutrients (d) are very small.
10. Taxonomists can use data from RNA-sequencing techniques to (a) predict future changes in species (b) determine when a species diverged from a common ancestor (c) identify genetic similarities among species (d) explain the origin of life.
11. How were Aristotle's and Linnaeus's classification systems for organisms similar?
12. Why are scientific names important in scientific work?
13. What is the second largest level in today's classification system?
14. What are the differences between plants and fungi?
15. What is the major type of evidence that scientists use to identify and classify organisms?
16. Give an example that shows how speciation may be occurring within a population.
17. Which of the five kingdoms contains the greatest numbers of living things on earth?
18. Name three things you might learn about an organism by investigating the meaning of its scientific name.

19. Why do some taxonomists propose that the Kingdom Monera be split into two separate kingdoms?

20. How might a taxonomist use embryological evidence in classifying an organism?

Critical Thinking

1. Develop a hierarchical classification system with at least three levels for your books or record albums or for your favorite songs. Explain the criteria you use to group individual items at each of the different levels.
2. Immediately after the microscope was invented, taxonomists began to use it in their work. What reasons might they have had for doing so?
3. Recently scientists found ways to "map" DNA molecules down to specific genes. New technology allows them to create "DNA fingerprints," genetic profiles of individuals. Crime specialists think the DNA fingerprints will make the positive identification of criminals an exact science. How is the creation of genetic profiles similar to one method used by some taxonomists?
4. Biologists think that there are probably millions of undescribed and unclassified species on earth. Why do you think so many species might still be undescribed or unclassified today?

5. If no one had devised a system of binomial nomenclature, what problems would scientists have had today in communicating about organisms?
6. Examine the photograph above of a member of the Division Euglenophyta. What characteristics do you see that might make this organism difficult to classify further?

Extension

1. Collect half a cup of water from a shallow pond. Using a microscope, study several samples from the water. Draw the organisms you find, and classify them as best you can into kingdom and phylum.
2. A number of years ago scientists found a living fish, called the coelacanth, which is one of a class of fishes long thought to be extinct. Write a research report describing how taxonomists decided where to classify this discovery.
3. Read the article entitled "Artful Adapter," in *Science News*, August 2, 1986, pp. 72–

73. Explain why the chambered nautilus is now classified as a cephalopod, along with the squid and the octopus, even though it has an external shell like the gastropods.

4. Visit the zoo and list the scientific names of all the animals you see, or use your library to research ten organisms. Record the scientific and common names of these organisms. For each animal list a trait that led taxonomists to classify the organism in its genus or family.

Intra-Science: *How the Sciences Work Together*

The Question in Biology: Is There Life on Other Planets?

People have wondered for centuries whether or not life exists outside the earth. Despite the advances in space technology in the past 30 years, scientists have not yet found conclusive evidence that life exists on other planets. At the same time, however, there is no evidence that it doesn't. Consider the following: Earth is one of nine planets orbiting the sun. There are about 100 billion suns in our galaxy. There may be over 100 billion galaxies. With potentially countless trillions of planets in the universe it is very possible that many have the conditions that could support life.

The branch of biology concerned with the possibility of alien life forms is called exobiology. Exobiologists study conditions on other planets to determine whether life could exist there. Their studies include an examination of the conditions that might have existed on ancient Earth. As you learned in Chapter 14, the atmosphere of ancient Earth included gases such as methane, ammonia, nitrogen, carbon dioxide, hydrogen, and water vapor. These gases contained carbon, hydrogen, oxygen, and nitrogen, the elements that eventually combined to form organic compounds.

As a result of such studies, some scientists speculate that life may have evolved on planets with atmospheres similar to Jupiter's. The gases that make up the atmosphere of a Jupiter-like planet—ammonia, helium, hydrogen, methane, water, and hydrogen sulfide—contain all of the elements necessary to form cells. In addition, lightning in a Jupiter-like atmosphere could provide the energy necessary to synthesize amino acids, the building blocks of protein. However, to prove whether life exists on Jupiter-like planets anywhere in the universe, biologists will have to rely on the efforts of other scientists, particularly astronomers and physicists.

The Connection to Astronomy

Although most scientists believe the universe contains many planetary systems, no planets have actually been observed outside of our own solar system. In 1987, however, a team of Canadian astronomers presented the most convincing evidence yet of the existence of another planetary system. Over a period of six years, they observed a slight wiggling motion in seven nearby stars.

The pattern of movement suggests that the wiggling may be caused by the gravitational pull of orbiting planets. If additional research confirms that planets are circling these stars, astronomers will have discovered the first planetary systems outside our solar system. Positive identification of such systems will in turn facilitate the search for extraterrestrial life.

The Connection to Physics

Given present technology and the laws of physics it seems improbable that human beings will soon venture to other solar systems. It takes four years for light from earth to travel to the nearest star, Proxima Centauri, and four more years to return. Light travels at about 300,000 km per second. Even if engineers could design a spaceship that could approach the speed of light, the energy required to power such a craft would surpass the amount of energy consumed by the entire world in one year. Consequently, at present, our strongest chance of communicating with life forms in other solar systems lies in electromagnetic communication.

The first radio search for extraterrestrial life was conducted in 1960, without success. So far the most extensive search has been conducted by Harvard physicist Paul Horowitz and The Planetary Society. Their project, called META (Megachannel Extraterrestrial Assay), involves an ongoing spot search for radio signals emanating from stars in the northern sky.

In 1987 NASA requested federal funding for a massive two-part search for extraterrestrial intelligence. The first part of this project would use radio telescopes around the world—including the world's largest, a 1,000-foot dish in Puerto Rico—to listen for signals from

stars within 100 light years of the earth. At the same time, scientists would use a new computer system called a multichannel spectrum analyzer to scan the entire sky at a wide range of frequencies. Such a system can monitor 8 million channels simultaneously. If approved the entire search will take six years to complete.

The Connection to Other Sciences

The search for life on other planets is based on the assumption that life can only exist under conditions similar to those on earth. Chemists are currently investigating the possibility that carbon-based life might evolve in liquid methane or some other medium besides water. They are also considering the possibility that life forms might be based on elements other than carbon, such as silicon. If such life forms do exist, their biochemistry and biology would be far different from that of earth's carbon-based organisms.

The Connection to Careers

As technological advances have expanded our capabilities for space exploration, scientists have begun to take the search for alien life much more seriously. Not only NASA but other research facilities and universities are accelerating their efforts to learn about the universe. The field of space exploration attracts people from all areas of science, including biologists, astronomers, physicists, chemists, and geologists. In addition, engineers, computer scientists, and even anthropologists are contributing to the study of life on other planets. For more information on related careers in the biological sciences, turn to page 842.

Unit 5 | Microorganisms

"Nature is to be found in her entirety nowhere more than in her smallest creatures."

Pliny the Elder

When you think about living things, you may tend to think only of plants and animals. In this unit, however, you will learn about the great diversity of living things that are neither plants nor animals—bacteria, protozoa, algae, and fungi. You will also learn about viruses, which stand at the boundary between the living and the nonliving worlds.

Mushrooms, *Hygrophorus miniatus*

Three protozoa (left to right): *Stentor, Didinium, Vorticella*

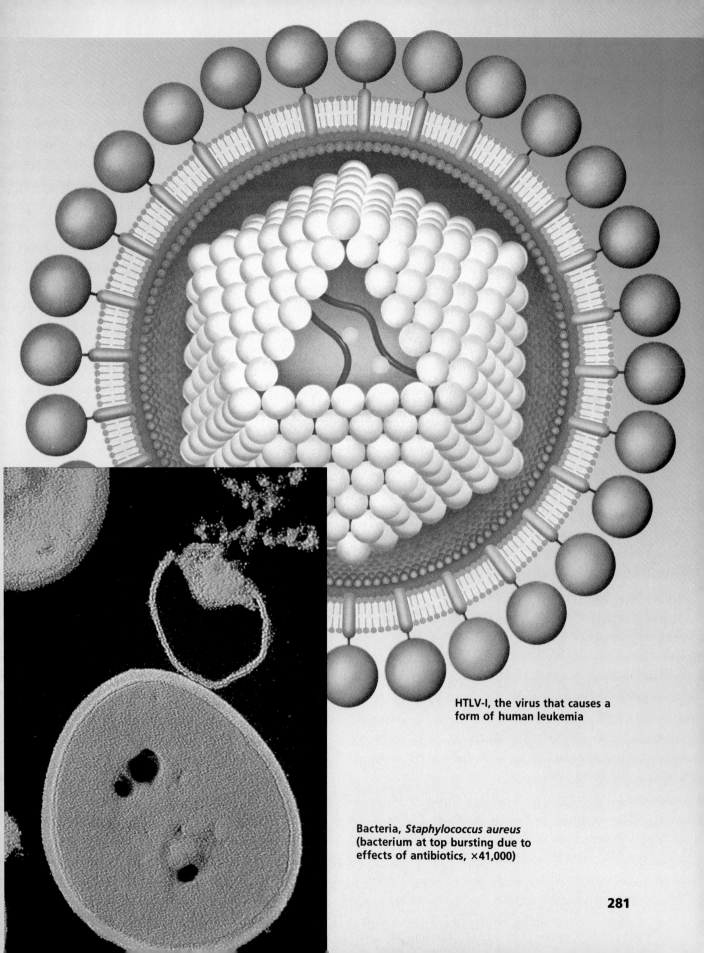

HTLV-I, the virus that causes a
form of human leukemia

Bacteria, *Staphylococcus aureus*
(bacterium at top bursting due to
effects of antibiotics, ×41,000)

Chapter 19 | Viruses

This chapter examines the structure of various types of viruses. Special attention is paid to how viruses reproduce and why they are such potent agents of infection.

Correlation Chart

Section Number	Section Objectives	Pages	Features	Section Review
19.1	1. Compare and contrast viruses and cells.	283–287		1, 2
	2. Describe the structure of three types of disease-causing viruses.			3
	3. Distinguish between DNA and RNA viruses.			3, 4
	4. Name and describe two disease-causing particles that are smaller than viruses.		Biology in Process, pp. 286–287	5
	5. State how Pasteur developed a rabies vaccine.			
19.2	1. Describe the structure of a bacteriophage.	288–291		2
	2. Name the five phases of the lytic cycle.		Biology Lab, p. 291	1, 3
	3. Explain how transduction occurs.			5
	4. State a possible theory for the evolution of viruses.			4

Resources*

Print

Barrett, James T., ed. *Contemporary Classics in the Life Sciences,* Vol. 1: *Cell Biology.* Philadelphia: ISI Press, 1986.

Mahy, B.W.J., ed. *Virology: A Practical Approach.* Washington, D.C.: IRL Press, 1985.

Scott, Andrew. *Pirates of the Cell: The Story of Viruses from Molecule to Microbe.* New York: Basil Blackwell, 1985.

Audiovisual/Software

The Living Cell: An Introduction. Film, 20 min. Britannica.

Viruses: The Mysterious Enemy. 2 filmstrips with cassettes. Human Relations.

Viruses, Retroviruses and AIDS: An Interview with Dr. Luc Montagnier. Video, 28 min. Carolina Biological, 1987.

*Additional references are listed in *Modern Biology, Teacher's Program Guide.*

Section Teaching Guides

19.1 Structure and Classification

Approach

The main objective of this section is to relate the structure of viruses to their highly effective success as agents of infection. The section builds on the discussion of characteristics of living things in Chapter 2. In addition, the text builds on information about lipids, proteins, and nucleic acids presented in Chapter 4. The comparison of viruses with cells makes use of the discussion of cell structure in Chapter 5. In presenting the text, help students appreciate the difficulties in classifying viruses as either living or nonliving. By recognizing the inability of viruses to reproduce in the same way as living cells, students are prepared to question how viruses do in fact reproduce and how their patterns of reproduction effect the transmission of diseases. This information will also serve as a basis for understanding the text on AIDS in Chapter 44 and the Unit 9 Writing About Biology feature.

Concept Framework

Most students will know that viruses are responsible for such familiar diseases as the common cold and the flu. You can tap this knowledge to create a starting point for a discussion of viruses. Engage students in a discussion of an infectious viral disease such as the common cold. Ask students what causes this disease, how the infectious agent enters the body, and what are the best treatments. Elicit the information that doctors have no cure for the cold, and that most treatments such as resting or drinking plenty of fluids are ways of helping the body deal with the effects of the virus rather than ways of killing the virus. You may wish to end the discussion by raising the question of whether a virus is living or nonliving, and whether it is possible to speak in terms of killing a nonliving thing.

Emphasis

Make certain that students understand that the characteristics that make viruses difficult to classify as living or nonliving also make them highly effective agents of disease.

Guest Speaker

Invite a guest speaker from a public health clinic, hospital, or a research laboratory to talk about the techniques biologists use to study viruses. Suggest to students that they ask the speaker what precautions he or she takes when studying viruses.

Using the Features/Louis Pasteur and the Rabies Treatment

The main process skills used in "Louis Pasteur and the Rabies Treatment" are hypothesizing and experimenting. Pasteur hypothesized that by inoculating an animal with a series of increasingly viral samples of a virus, he could make the animal immune to that virus. He first tested his hypothesis on dogs using the rabies virus. To reinforce students' understanding of these skills, ask students to suggest a hypothesis on how the common cold is transmitted and to design an experiment to test their hypothesis.

Section Review Answers

(Continued from p. 287.)
5. Viroids and prions are much smaller than viruses, so more powerful electron microscopes are required to view them. Also, because viroids and prions each lack certain features of viruses, some types of viral studies might not be useful for studying them.

19.2 Reproduction and Evolution

Approach

The main objective of this section is to demonstrate how the structure of a bacteriophage relates to its method of reproduction. It builds on the explanations of DNA and RNA given in Chapter 9. In presenting the text, emphasize that viruses of differing morphologies reproduce in similar ways. Tell students that the text focuses on the example of a bacteriophage because bacterial viruses are the most studied and hence the best understood of all the viruses. You may wish to use Figure 19-4 as your main tool in organizing the information in the section. Have the students closely follow the steps illustrated in Figure 19-4 as they read.

Concept Framework

Most students will be familiar with some viral infections such as cold sores that disappear and then reappear months or even years after their first appearance. Other examples are shingles and some sexually transmitted diseases such as genital herpes. Discussion of such disorders will provide an opportunity to correct any misunderstandings students may hold about these disorders. Explain that the reappearance of the disease need not occur because of a reinfection. Relate the discussion to the reproductive process in viruses. Ask students as they read this text to consider what causes dormant viruses to become reactivated.

Emphasis

Make certain that students understand that viruses can only multiply by using the genetic machinery of their host cells.

Section Review Answers

(Continued from p. 290.)

2. The bacteriophage has a protein collar beneath the polyhedral coat and a sheath that runs from the collar to the base. Fibers emerge from the base plate. Polyhedral viruses do not have these characteristics.
3. The phases are adsorption, entry, replication, assembly, and release.
4. Because viruses are obligate intracellular parasites and cannot exist without cells, they probably formed spontaneously from nonliving organic material or evolved from existing cells.
5. When the genetic code of a cell is altered through the process of transducton, it will have a new phenotype.

Laboratory Preparation

Background

In this lab, students observe the effects of the tobacco mosaic virus, TMV, on healthy plants. Students obtain TMV from cigarette tobacco, which demonstrates that viruses are common components of the environment. Students should conduct this investigation away from gardens or houseplants because TMV spreads from plant to plant.

Preparation

Start tobacco seedlings about two weeks prior to the investigation. To prepare 0.1 M dibasic potassium phosphate solution, add 6.8 g KH_2PO_4 to 500 mL of distilled water.

Safety

Remind students to thoroughly wash their hands and all laboratory equipment used in step 2.

Conclusions

1. Plant A will remain healthy and show ordinary growth over the course of the investigation. The leaves of plants B, C, and D will become mottled with light green and yellow blotches. Eventually, the leaves will drop off and the plants die.
2. The leaves of plant C will become infected more quickly than the leaves of plant D. The virus will also spread more quickly among the leaves of plant C. This is because the leaves near the top of the plant are younger and more susceptible to the virus than the leaves of plant D.
3. Healthy plants may be infected by TMV in tobacco smoke. As a result, smoking in greenhouses is usually prohibited.

Inquiry

Answers will vary. Some possible sources of error include not using the same amount of mixture or carborundum powder for each plant, not applying the mixture evenly, or having more TMV in one part of the mixture than in another.

Chapter Review

Vocabulary

1. *Obligate* means "restricted to only one life style"; *intracellular* means "occurring within a cell"; *parasite* means "an organism that lives in or on another organism." Taken together as a single phrase, *obligate intracellular parasite* means that viruses can exist only by living within a cell.
2. In electrophoresis electrical changes are used to identify the protein components of specific viruses and to provide information about the structure of viruses. Serology is the study of antibodies that develop in an animal's bloodstream in re-

sponse to a virus; it is used to identify the virus present and thus to diagnose disease. Tissue culture requires the growing of a group of like tissue cells in the controlled conditions of a laboratory and then the injection of the cells with a specific virus. This enables the scientist to closely study the structure of the virus.

3. A phage is another word for a bacteriophage, which is any virus that infects bacteria. Prophage is a segment of the DNA of a virus that has attached to the DNA of the host cell.

4. Transduction is the viral transport of DNA from one cell to another.

5. *Lyses* means "to break open." *Lytic* cycle refers to a cycle or series of phases that begins and ends with the breaking open of a cell. *Lysogenic* means "having the capacity to become lytic, or to break open."

Review

11. Five different techniques used to study viruses include the use of the electron microscope, the infection of a tissue culture with a specific strain of virus, serology, electrophoresis, and nucleic acid sequencing.

12. The AIDS virus has a capsid that consists of a lipid layer in which glycoprotein molecules are embedded. The lipid layer in turn surrounds two protein layers.

13. Retroviruses reproduce by making DNA with the help of an enzyme they contain, reverse transcriptase. The DNA in turn makes new RNA, which makes the parts that are assembled into new viruses.

14. Kuru is caused by prions, contracted by handling infected human brains. The effect is a degenerative nerve disease.

15. The tail fibers allow the virus to attach itself to the host cell. The tail fibers contain proteins that have an affinity with the cell wall of specific bacteria.

16. The viral DNA inactivates the DNA of the host cell. It then uses the genetic machinery of the host cells to make the proteins and DNA needed to assemble new bacteriophages.

17. A virulent virus is one that causes disease by destroying its host cell. A temperate virus may invade a host cell but it does not immediately destroy the host cell.

18. A temperate phage might become virulent and enter a lytic cycle by some external stimuli, such as exposure to radiation or to certain chemicals.

19. The short generation time of a virus means that thousands of new individuals are created before the immune system manages to destroy most of them. Those that have not been destroyed have developed resistance and have reproduced new resistant viruses.

20. Pasteur cultivated the rabies virus in the spinal cord of a rabbit. As the cord dried, the virulence of the virus decreased. He was able to isolate virus samples of varying degrees of virulence.

Critical Thinking

1. AIDS is caused by the retrovirus HIV that must use reverse transcriptase to reproduce. By blocking the reverse transcriptase, the drug stops the virus from reproducing.

2. Although the research is not complete, the virus probably stays in the liver cells in a latent stage and may be activated by certain mutagens.

3. Infection by a second phage could kill the cell by lysis or cause abnormal DNA replication. Death of the *E. coli* would stop phage reproduction.

4. Shingles is caused by a temperate phage that has become virulent and entered the lytic cycle. The phage may have been dormant for many years after it initially caused chicken pox.

5. The jump in the graph corresponds to the moment in the lytic cycle when the new viruses break out of the cell.

Chapter 19 | *Viruses*

Introduction

Viruses are organisms that straddle the threshold between living and nonliving things. They lack several of the essential features of all life-forms and reproduce by taking over the genetic machinery of host cells. The release of viruses from host cells results in the spread of many infectious diseases in both plants and animals.

Chapter Outline

19.1 Structure and Classification
- *Studying Viruses*
- *Structure*
- *Classification*

19.2 Reproduction and Evolution
- *The Lytic Cycle*
- *The Lysogenic Cycle*
- *Evolution*

Chapter Concept

As you read, look for ways in which the structure and reproduction of viruses make them highly effective agents of infection.

Artist's model of an influenza virus. Inset: Color-enhanced TEM of AIDS (HIV) virus inside a stricken T4 cell (×30,000).

For additional teaching information see pp. T19A–T19D.

19.1 Structure and Classification

A **virus** is a biological particle composed of genetic material and protein. A typical virus consists of either RNA or DNA encased in a protein coat called a **capsid**. When a virus causes a disease, the virus is said to be **virulent** *(VIR-yuh-lunt)*. If the virus does not cause disease immediately, it is said to be **temperate**.

Viruses are constructed of compounds usually associated with cells, but they are not considered living organisms. They have some, but not all, of the characteristics of life listed in Chapter 2. They have no nucleus, cytoplasm, organelles, or cell membrane. They do not reproduce by either mitosis or meiosis, nor are they capable of carrying out cellular functions. Study Table 19-1, in which viruses are compared and contrasted with cells.

Because they are not cells, viruses can only reproduce by invading a host cell and using the enzymes and organelles of the host cell to make more viruses. They are therefore **obligate intracellular parasites**, which means they require a host cell to reproduce. Outside a host cell a virus is a lifeless particle with no control of its movements. It is spread on the wind, in water, in food, or via blood or other bodily secretions.

Studying Viruses

Scientists knew as early as 1800 that some factor existed that was smaller than bacteria and that could transmit disease. However, they were unaware of the nature of the factor until 1935, when Wendell Stanley (1904–1971) first isolated the virus that causes tobacco mosaic disease. The electron microscope was the first tool that permitted scientists actually to look at viruses. Recent advances such as tissue culture, serology, electrophoresis, and nucleic acid sequencing help virologists learn more about viruses.

Tissue culture is the growing of living cells in a controlled medium. This technique is the foundation for viral research. By

Section Objectives

- *Compare and contrast viruses and cells.*

- *Describe the structure of three types of disease-causing viruses.*

- *Distinguish between DNA and RNA viruses.*

- *Name and describe two disease-causing particles that are smaller than viruses.*

- *State how Pasteur developed a rabies vaccine.*

Table 19-1 Comparing Viruses and Cells

	Virus	Cell
Structural parts	Protein, nucleic acid core	Nucleus, cytoplasm, organelles, membrane
Nucleic acid	Either DNA or RNA	Both DNA and RNA
Reproduction	Requires a host cell	By mitosis and meiosis
Cellular respiration	No	Yes
Crystallization	Yes	No

infecting a tissue culture of living cells with a strain of a virus, researchers can grow viruses and examine them.

Researchers grow large quantities of a virus in tissue cultures then subject the viruses to serological and electrophoretic studies. **Serology** *(suh-RAHL-uh-jee)* is the study of biological fluids. In particular, serologists often determine an organism's antibody responses to viruses. **Electrophoresis** *(ih-LECK-truh-fuh-REE-suss)* is a process that separates molecules, especially proteins, on the basis of their specific electrical charges. Electrophoresis is used to separate and examine the protein components of viruses. In addition, recent advances in DNA and RNA sequencing have enabled virologists to determine the sequence of bases in the viral nucleic acids. Using all these techniques and others, virologists have determined the structure of many viruses.

Structure

The structure of the polio virus shown in Figure 19-1 is typical of many viruses. The virus particle is about 20 to 30 nm in diameter. The capsid is shaped like an **icosahedron** *(ie-KOE-suh-HEE-drun)*—a polyhedron with 20 triangular faces. The capsid is made of protein subunits that fit together like the pieces of leather on a soccer ball. The capsid surrounds a single strand of RNA.

Most icosahedral viruses are between 15 and 200 nm in diameter. The approximately 200 kinds of viruses that cause the common cold are mostly icosahedral viruses about the size and shape of the polio virus. Some viruses, such as the virus that causes tobacco mosaic disease, a disease of tobacco plants, are rod shaped when viewed under the electron microscope. These viruses have a helical strand of nucleic acid that runs the length of the

Figure 19-1 Many viruses, including the polio virus, are shaped like polyhedrons and typically are covered by a protein coat called a capsid.

Table 19-2 Some Common Viruses That Cause Diseases

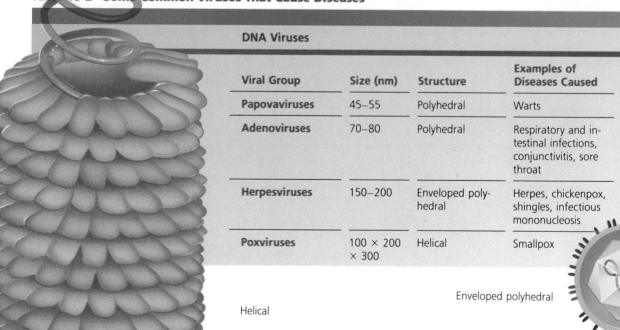

	DNA Viruses		
Viral Group	**Size (nm)**	**Structure**	**Examples of Diseases Caused**
Papovaviruses	45–55	Polyhedral	Warts
Adenoviruses	70–80	Polyhedral	Respiratory and intestinal infections, conjunctivitis, sore throat
Herpesviruses	150–200	Enveloped polyhedral	Herpes, chickenpox, shingles, infectious mononucleosis
Poxviruses	100 × 200 × 300	Helical	Smallpox

Helical

Enveloped polyhedral

virus and is surrounded by a helically arranged protein coat. The protein coat makes up about 95 percent of the mass of this type of virus. Rabies and mumps are caused by helical viruses.

The virus that causes acquired immune deficiency syndrome, or AIDS, is even more complex. The AIDS virus is shown on the right. This virus, called the HIV, has two single strands of RNA in its core. These strands are surrounded by two layers of protein. A layer of lipids surrounds these inner protein layers. **Glycoprotein** molecules, proteins with sugar chains attached, are embedded in the lipid layer and form the capsid of the virus.

Classification

Because viruses are not considered living things, they are not classified by the system of nomenclature discussed in Chapter 18. Viruses instead are classified as DNA viruses or RNA viruses, depending on the type of nucleic acid in the capsid. As shown in Table 19-2, viruses contain either RNA or DNA, never both.

DNA and RNA Viruses

DNA and RNA viruses differ in the manner in which they alter the machinery of a host cell. Once inside the host cell, a DNA virus may directly produce new RNA that then makes more viral proteins. Alternatively the virus DNA may join to the DNA of the host cell and then direct the synthesis of new viruses.

RNA viruses perform in another way. Some RNA viruses enter the cell and make new proteins directly. They do so by releasing the RNA, which then migrates directly to the cytoplasm, where it uses the host ribosomes to make proteins. The polio virus, which is an RNA virus, acts in this manner.

Tip: Give students diagrams of several different viruses and ask them to rank the viruses in order of complexity. Have students defend their choices.

Tip: AIDS is discussed in Chapter 44.

Info: The NIH uses the term HIV, not HTLV-III or LAV, for the AIDS virus.

Info: Of the influenza viruses listed in Table 19-2 type A is responsible for most flu epidemics, type B is found primarily in children, and type C is relatively mild and rare.

AIDS virus

RNA Viruses

Viral Group	Size (nm)	Structure	Examples of Diseases Caused
Picornaviruses	20–30	Polyhedral	Poliomyelitis, infectious hepatitis, common cold
Myxoviruses	80–120	Enveloped helical	Influenza A, B, C
Rhabdovirus	70 × 180	Enveloped helical	Rabies
Retroviruses	100	Glycoprotein enveloped complex	AIDS (depressed immune system)

Polyhedral

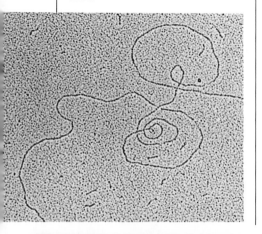

Figure 19-2 Viroids, which are smaller than viruses, cause diseases in some plants. They consist of a short, single strand of RNA.

Other RNA viruses, called retroviruses, act differently. In addition to RNA, **retroviruses** contain an enzyme called reverse transcriptase. **Reverse transcriptase** is an enzyme that makes DNA from RNA. In normal cells, you recall, DNA makes RNA. The RNA in turn makes proteins. In retroviruses RNA makes DNA with the aid of reverse transcriptase. The DNA then makes the new RNA. This RNA in turn makes the proteins that become part of the new viruses. The AIDS virus is a retrovirus.

Viroids and Prions

In the last three decades scientists have isolated certain disease-causing particles that are smaller and simpler than viruses. One of these is the viroid. A **viroid** *(VIE-ROID)* is a short, single strand of RNA with no surrounding capsid. Look at the viroid pictured in Figure 19-2. The strand of RNA shown there looks like a tangled piece of string. It does not contain enough RNA to make proteins. Scientists suggest that this RNA strand somehow interferes with normal cell functions and causes the production of new viroid strands by using the host cell's enzymes. Diseases caused by viroids include those that harm potato, coconut, chrysanthemum, and citrus crops, as well as other plants.

Biology in Process

Louis Pasteur and the Rabies Treatment

Louis Pasteur was almost 60 years old and semi-paralyzed when in 1881 he began his search for a treatment to keep people who had contracted rabies from dying. To discover a treatment for this dread disease, Pasteur first had to find a way of growing a weakened strain of the germ. When injected, a weakened strain would help the body's disease-fighting mechanisms combat the rabies virus (right, ×65,400). Then he had to test the treatment before releasing it for widespread use.

Pasteur knew that rabies attacked nerve tissue; the mad behavior exhibited by rabid dogs was proof. Therefore, he reasoned, if he inoculated the brain of a healthy dog with the rabies virus, he could grow the virus in the animal. Pasteur did this and succeeded in culturing the virus.

Pasteur noted that all the dogs he had inoculated contracted the disease and that some managed to recover. Those that did recover became resistant to subsequent injections. This reaction, because it showed that this virus behaved like other microorganisms, suggested to Pasteur that a single infection could provide future immunity.

A **prion** (*PREE-AHN*) is a glycoprotein particle containing a polypeptide of about 250 amino acids. Even without nucleic acids prions are capable of reproducing in some mammalian cells. Prions are implicated in diseases with long incubation periods. For instance, prions cause scrapie, a slow degeneration of the nervous system in sheep and goats. Another prion disease is kuru, a degenerative nerve disease that can be contracted by touching the brains of deceased individuals. It occurs among some New Guinea highland tribes whose funeral rites include touching the brains of deceased ancestors.

Section Review

1. What is a *capsid*?
2. What are the similarities between viruses and cells? What are the differences?
3. Compare and contrast the structures of a polio virus and a tobacco mosaic virus.
4. What is reverse transcriptase? What is its function?
5. Why might the study of viroids and prions lag behind the study of viruses?

Answers to Section Review
1. A capsid is a protein coat that encloses the genetic material, either RNA or DNA, of a virus.
2. Both viruses and cells are composed of the same types of compounds. Viruses lack a nucleus, cytoplasm and organelles, and a cell membrane; cannot reproduce on their own; and do not carry on cellular respiration.
3. Both have a protein coat that surrounds a nucleic acid core. The polio virus is icosahedral in shape, and the core is a strand of RNA. The tobacco mosaic virus is rod shaped, and the protein is arranged around a spiral length of DNA.
4. an enzyme present in retroviruses that enables the RNA of the viruses to make DNA
(See pp. T19A–T19D.)

Pasteur next turned his attention to cultivating a weakened strain of the virus to use in his treatment. This time Pasteur injected the virus into a spinal cord removed from a rabbit. The spinal cord was then slowly dried. As the cord dried, the virulence of the virus decreased. By varying the time of drying Pasteur was able to produce virus samples of varying degrees of strength.

By 1884 Pasteur had developed a rabies treatment that consisted of a series of shots of increasingly virulent virus. He inoculated dogs with a series of virus samples. The first preparation contained dead viruses, which are harmless to the animal. Each subsequent shot was more virulent than the last. Finally he was able to prevent the expression of rabies in dogs.

Though this treatment worked well in animals, Pasteur was reluctant to try it on humans. "What is possible on the dog may not be so on man," he worried. In July 1885, however, two physicians pleaded with Pasteur to treat nine-year-old Joseph Meister, who had been bitten by a rabid dog. The physicians convinced Pasteur that without treatment Joseph was "doomed to inevitable death." The treatment was administered, and Joseph survived.

By the time of Pasteur's death in 1895, 20,000 people had been treated with "the Pasteur Treatment." The death rate was less than one-half of one percent.

■ Could Pasteur have tested his treatment in any way other than by using it on a human victim of rabies?

Section Objectives

- *Describe the structure of a bacteriophage.*

- *Name the five phases of the lytic cycle.*

- *Explain how the process of transduction occurs.*

- *State a possible theory for the evolution of viruses.*

Tip: Have some students construct a model of a T phage from a round-headed bolt, one or two nuts, and stiff wire for legs. Have them explain the model and tell where the nucleic acids would be located.

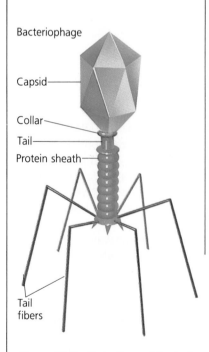

Bacteriophage

Capsid—

Collar—

Tail—

Protein sheath—

Tail fibers

Figure 19-3 Note the polyhedral structure of a T4 bacteriophage (above). The false-color TEM (right) shows phages that infect *E. coli* (×60,000).

19.2 Reproduction and Evolution

Scientists first learned about virus reproduction by studying **bacteriophages,** viruses that infect bacteria. Phages *(FAY-juz),* as they are called, can be easily studied because their bacterial hosts multiply quickly in cell cultures.

The most commonly studied phages are those of the T group. They are named T1, T2, T3, and so forth. The T phages infect the bacterium *Escherichia coli,* the common bacterium of the human digestive tract. The T-even phages (T2, T4, T6) are virulent. They are capable of destroying *E. coli* cells.

Examine the structure of the T4 phage in Figure 19-3. Notice that its morphology is different from that of the viruses described in Section 19.1. DNA in the viral core is surrounded by a protein coat that forms a polyhedron. Beneath the head is a collar of protein and a sheath that rests on a base plate. Tail fibers emerge from the base plate. As you read about the reproduction of the T4 phage, notice how the structure of the virus suits its function.

The Lytic Cycle

The **lytic cycle** is a fundamental reproductive process in viruses. The term *lyse* means to "break open," a reference to the liberation of the new viral particles from the host cell. The T4 phage reproduces by the lytic cycle and thus can serve as an example of viral reproduction. The lytic cycle has five phases, each of which is continuous with the others. The phases are adsorption, entry, replication, assembly, and release. Study Figure 19-4 as you read.

1. During adsorption the virus attaches itself to a specific host cell. The tail fibers of the virus contain proteins that have a chemical affinity with the bacterial cell wall. In fact, specific areas of the wall, called **receptor sites,** are the places where the virus attaches itself.
2. During entry the T4 phage releases an enzyme that weakens a spot in the cell wall of the host. Then, much like a hypodermic needle, the T4 presses its sheath against the cell and injects its

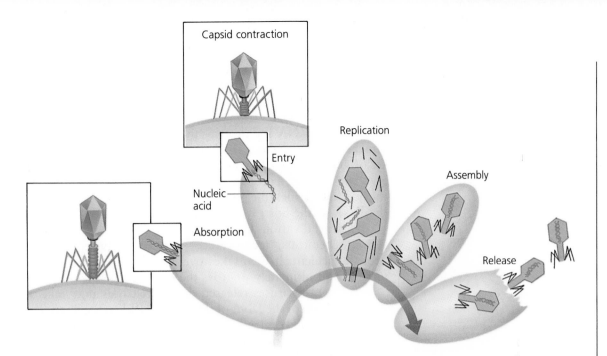

Capsid contraction

Entry

Nucleic acid

Replication

Assembly

Absorption

Release

DNA into the host cell through the weak spot in the wall. The empty capsid remains on the outside of the cell. In contrast, many viruses enter their host cell intact. Once inside, the capsid dissolves and the genetic material is released. This process is called uncoating.

3. During replication the viral DNA takes complete control of cell activity. It inactivates the *E. coli* DNA. The genes contained in the DNA of the viral genome then take over. They direct the cell to make viral DNA and the viral proteins that make up the structural portions of the phage. This happens when viral DNA makes RNA from nucleotides in the host cell by using the enzymes of the host cell.

4. During assembly proteins coded for by phage DNA act as enzymes that put new virus particles together. The entire metabolic activity of the cell is thus directed toward assembling new T4 phages. The result is a cell stuffed with new viruses.

5. During release the T4 phages release an enzyme that digests the bacterial cell wall from within. The disintegration of the infected host cell, called **lysis** (*LIE-SISS*), allows new viruses to leave the cell. The new virus particles can then infect other cells, and the process can start again.

The Lysogenic Cycle

During the lytic cycle viruses enter the cell, use its components to make new viruses, and destroy the cell in one continuous process, which usually takes a day or two. However, some temperate viruses can infect a cell without causing its immediate destruction. Temperate viruses undergo a kind of life cycle called a **lysogenic cycle,** which has been most thoroughly studied in bacteriophages. Study this cycle in Figure 19-6 on the following page as you read.

Info: Herpes zoster virus resides in cells deep in the spinal cord of humans. When the body is under stress, the viruses may migrate to mucous membrane cells and go through a lytic cycle to cause the disease shingles.

Figure 19-4 During a lytic cycle a phage goes through five phases during which it invades a host cell and eventually destroys the cell.

Figure 19-5 The dark spots are phages that have been inside this *E. coli* cell 30 minutes. Notice that the heads of the phages point toward the center of the cell and the tails toward the membrane (×30,000).

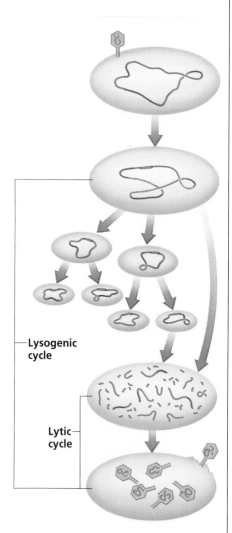

Lysogenic
cycle

Lytic
cycle

Figure 19-6 In contrast to a lytic cycle in which a phage quickly destroys a cell, in a lysogenic cycle a phage enters a cell and remains inactive until an external stimulus causes the phage to become virulent.

Answers to Section Review
1. A prophage is the DNA of a temperate phage that attaches to the host genome. The prophage is replicated when the host DNA replicates, and replication is continued within the host DNA when the host cell divides. At any stage the prophage can become virulent and produce new viruses.
(See T19A–T19D.)

A temperate phage enters a bacterium in much the same way the T4 bacteriophage does, by attachment of tail fibers and injection of the DNA into the host cell. At this point, however, the lysogenic cycle differs from the lytic cycle. Instead of immediately creating new RNA and proteins the DNA of the temperate phage attaches itself to the host DNA. It becomes, in effect, an additional set of genes. The phage, now represented only by a short DNA segment, is called a **prophage.** When host DNA replicates or when the host cell divides, the prophage acts just like an inert segment of the DNA of the host. It causes no harm to the cell.

However, various external stimuli—exposure to radiation or certain chemicals, for instance—can cause the prophage to become virulent. It then takes over the host cell, produces new viruses, and ultimately destroys the cell.

Temperate viruses released during cell lysis may take with them a portion of the DNA of the host cell. When the phage enters a new host, it may introduce genes from the former host into the new host. In this process, called **transduction,** a virus transfers DNA from cell to cell and thus causes a change in the genetic code of bacterial cells. This results in genetic recombination and hence phenotypic variation in the new host bacterium.

Evolution

How and when did viruses evolve? No fossil evidence of viruses has been found. However, scientists form inferences about the evolution of viruses. Because they are obligate intracellular parasites, viruses probably did not arise until cells had evolved, since their existence requires cells. If this is so, then viruses probably either formed spontaneously from existing nonliving organic material or evolved as simplifications of previously existing cells.

Whatever their origin, existing viruses often evolve very rapidly by natural selection. Suppose that cold viruses invade a human body. The human immune system may destroy most of these viruses. The few that remain will have been naturally selected and will be resistant to immediate attack by the immune system. These resistant viruses enter cells and produce hundreds or thousands of viruses in a few days. The immune system will respond to repel the viruses eventually but not until many new ones have been formed. The short generation time of a virus means that natural selection acts quickly to create new viral types that are capable of withstanding destruction in the next host.

Section Review

1. What is a *prophage,* and how does it function?
2. How does a bacteriophage differ from a polyhedral virus?
3. What are the five phases of the lytic cycle?
4. Why do scientists think that viruses evolved after cells?
5. What is the significance of transduction?

Laboratory

Tobacco Mosaic Virus

Objective
To study the effect of the tobacco mosaic virus on plants

Process Skills
Experimenting, observing

Materials
4 tobacco plants, glass-marking pencil, tobacco from several brands of cigarettes, mortar and pestle, 10 mL 0.1 M dibasic potassium phosphate solution, 100-mL beaker, cotton swabs, 400-grit carborundum powder

Method

1 The tobacco mosaic virus, TMV, infects tobacco and other crops. Use a glass-marking pencil to label four pots containing tobacco plants A, B, C, and D.

2 Place pinches of tobacco from several different brands of cigarettes in a mortar. Add 5 mL of potassium phosphate solution and grind the mixture with a pestle. Pour the mixture into a beaker. Wash your hands and all laboratory equipment used in this step. Given that you will use this mixture to infect plants with TMV, what assumption are you making about the dried tobacco?

3 Moisten a swab with potassium phosphate solution, and sprinkle it with carborundum powder. Apply the mixture by scraping the swab over several leaves of plant A.

4 Rinse the leaves with water and place plant A away from the other plants. What role does plant A play in this investigation?

5 Moisten a swab with the mixture from step 2, and sprinkle it with carborundum powder. Apply the mixture to the leaves of plant B.

6 Prepare two more swabs as in step 5. Apply the mixture to the upper leaves of plant C and the lower leaves of plant D. Why is the carborundum powder added to each swab?

What do you think treating different areas of the plants will tell you about TMV?

7 Gently rinse the leaves of plants B, C, and D with water. Why should these plants be kept away from plant A?

8 Water all of the plants daily. Observe the plants for about two weeks. Record your observations.

Conclusions

1 Compare the leaves of plant A with the leaves of plants B, C, and D. What difference, if any, do you see? What are the characteristics of leaves infected with tobacco mosaic virus?

2 Compare plants C and D. Did the virus affect one plant more than the other? Explain your answer.

3 Greenhouse operators generally do not allow smoking in their greenhouses. Aside from health and safety issues, how might your results support this rule?

Inquiry
What are some possible sources of error in this investigation?

Chapter 19 Review

Vocabulary

bacteriophage (288)	lysogenic cycle (289)	receptor site (288)	tissue culture (283)
capsid (283)	lytic cycle (288)	retrovirus (286)	transduction (290)
electrophoresis (284)	obligate intracellular	reverse transcriptase	viroid (286)
glycoprotein (285)	parasite (283)	(286)	virulent (283)
icosahedron (284)	prion (287)	serology (284)	virus (283)
lysis (289)	prophage (290)	temperate (283)	

1. In a dictionary look up the words *obligate*, *intracellular*, and *parasite*. How do these words apply to viruses?
2. How do electrophoresis, serology, and tissue culture each contribute to the study of viruses?
3. Distinguish a phage from a prophage.
4. What is transduction?
5. Use what you know about the meaning of the word *lyses* to explain the meaning of *lysogenic* and of *lytic* cycles. If necessary use your dictionary to help you.

Review

1. Viruses and cells are alike in that both (a) have membranes (b) reproduce (c) respire (d) synthesize proteins.
2. Electrophoresis is used in the study of viruses to (a) diagnose and classify them (b) determine their nucleic acid sequence (c) identify their protein components (d) observe their structure.
3. *Icosahedron* refers to the structure of a virus's (a) nucleic acid (b) capsid (c) phage (d) lipid layer.
4. Viruses are classified primarily according to (a) disease caused (b) structure (c) size (d) nucleic acid.
5. Reverse transcriptase functions to (a) make DNA from RNA (b) unite viral DNA with the host DNA (c) release viral RNA to make proteins (d) transfer host DNA from one cell to another.
6. Viroids differ from viruses in (a) their larger size (b) the absence of a capsid (c) the absence of nucleic acids (d) their longer incubation period.
7. The sheath of a bacteriophage is used to (a) attach the phage to the bacterial wall (b) enclose the nucleic acid (c) transmit the nucleic acid to the bacterium (d) take over the genetic machinery of the cell.
8. *Lyses* refers to the disintegration of (a) the bacteriophage capsid on entering the cell (b) the DNA of the host cell (c) a T4 phage enzyme (d) the host cell.
9. Temperate phages reproduce by (a) immediately taking over the genetic machinery of the host cell (b) producing DNA that in turn produces viral DNA (c) attaching themselves to the ribosomes of the host cell (d) attaching themselves to the DNA of the host cell.
10. Transduction occurs when (a) a T4 phage moves from outside to inside a cell (b) a virus transfers DNA from one cell to another (c) a phage uses protein from one host cell and DNA from another (d) a virus transfers from the lysogenic cycle to the lytic cycle.
11. Name five techniques for studying viruses.
12. Describe the capsid of the AIDS virus.
13. How do retroviruses reproduce?
14. What causes the disease kuru, and how does kuru affect the body?
15. What is the function of the tail fibers of a bacteriophage?
16. Once a bacteriophage is inside a host cell, how are the components of new phages produced?

17. Distinguish between a virulent virus and a temperate virus.
18. What causes a temperate phage to become virulent?
19. How does the short generation time of most viruses work to their advantage?
20. How did Pasteur develop a weakened strain of rabies to use as a treatment?

Critical Thinking

1. Some people suggest that the drug AZT (azidothymidine) can help patients with AIDS. This drug blocks the enzyme reverse transcriptase. Explain how AZT might help patients.
2. People who have had a hepatitis-B viral infection have a greater chance of developing liver cancer later in life, especially if they are exposed to aflatoxin. Aflatoxin is a mold toxin found in some foods, such as contaminated peanut butter. What does this relationship suggest to you about the role of the hepatitis virus in causing cancer to develop?
3. Lambda (λ) is a bacteriophage that attacks *E. coli* and may enter a lysogenic life cycle. Once it enters the lysogenic life cycle, the phage inhibits the entry of more phages. What is the evolutionary advantage of the phage's "immunizing" its host against further lambda infection?

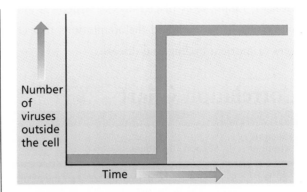

4. Shingles is a disease caused by the same herpes virus that causes chicken pox. How do you account for the fact that shingles often appears years after the initial chicken pox attack?
5. Look at the graph above. Discuss how the sharp jump in the number of viruses outside the cell corresponds to the phases of the lytic cycle.

Extension

1. Read the article by J. Raloff entitled "Virus Allows Wasps to Kill Crop Pests," in *Science News,* July 13, 1985, page 22. Give an oral report in which you discuss the role of the virus in helping the wasp and that of the wasp in helping the virus.
2. Look in the *Reader's Guide to Periodical Literature* for three articles on AIDS and its causes. Give a report to the class in which you discuss what you find.
3. Call your local hospital or family doctor and ask how viral diseases are diagnosed. If possible, visit the clinical laboratory of a hospital to see how the tests are done.
4. Research and write a report on some viral disease, such as polio or smallpox, that is now preventable. In your report discuss the process scientists followed in identifying the cause of the disease, isolating the virus, formulating a vaccine, and testing the vaccine.
5. Research and write a report on present-day scientists, such as those at the Center for Disease Control in Atlanta, who are searching for preventions or cures for viral diseases. In your report discuss the processes these scientists use to find preventions and cures and the precautions they take to ensure that they do not become contaminated with the viruses that they work with.

Chapter 20 | *Bacteria*

This chapter discusses the diversity and importance of the Kingdom Monera. Emphasis is placed on the structures and function of the four phyla of Monera and the adaptations that contribute to their roles in nutrient cycling and disease.

Correlation Chart

Section Number	Section Objectives	Pages	Features	Section Review
20.1	1. Explain the differences between the members of the Phylum Archaebacteria and the members of the Phylum Schizophyta.	295–299		2, 5
	2. Describe the four classes in the Phylum Schizophyta.			3
	3. Distinguish between Gram-positive and Gram-negative bacteria.			3
	4. Describe the Phyla Cyanophyta and Prochlorophyta.			4
20.2	1. List the distinguishing features of moneran cells.	300–305	Think, p. 301	1, 4
	2. Describe autotrophy and heterotrophy in monerans.			2, 3
	3. Describe the way in which organisms can survive in vent communities deep in the ocean.		Biology in Process, pp. 302–303	
	4. Distinguish between endotoxins and exotoxins.		Biology Lab, p. 305	5
	5. Explain the basic effect of antibiotics.			5

Resources*

Print

Hauser, Juliana T. *Techniques for Studying Bacteria and Fungi.* Burlington, N.C.: Carolina Biological Supply Co., 1986.

Hobby, Gladys L. *Penicillin: Meeting the Challenge.* New Haven: Yale University Press, 1985.

Singleton, Paul, and Diana Sainsbury. *Introduction to Bacteria.* New York: Wiley-Interscience, 1981.

Audiovisual/Software

Bacteria. Film, 19 min. Britannica.

Bacteria: Friend and Foe. Film, 11 min. Britannica.

Bacteria and Other Microorganisms Set. Overhead transparencies. Hagemann. Dist. by Carolina Biological.

Introduction to the Bacteria. Still-frame video, film-strip, or transparencies with cassette, 20 min. Carolina Biological.

*Additional references are listed in *Modern Biology, Teacher's Program Guide.*

Section Teaching Guides

20.1 Evolution and Classification

Approach

The primary objective of this section is to identify the criteria used by scientists to classify the major groups of monerans. It builds on concepts of basic chemistry and cell biology (Chapters 3 through 5), the origins of life (Chapter 14), diversity through adaptation (Chapters 15 and 16), and classification (Chapter 18). In presenting the text, help students appreciate that bacteria were probably the first organisms on earth. Review the differences between prokaryotic and eukaryotic cells with them if necessary (Chapter 5). You may also wish to use Figure 20-1 to help students visualize the three main shapes in which bacteria are found.

Concept Framework

Students have studied the possible conditions that existed on the early earth (see Chapter 14). To reinforce their understanding and to provide a conceptual framework for studying bacteria, engage students in a discussion of the traits an organism would have to have had to survive in the early atmosphere. Elicit the observation that the environment would have been inhospitable to many of today's organisms. You may wish to end the discussion by asking students to think about the genetic mechanisms that would have been most efficient in the environment of the early earth.

Emphasis

Make certain students understand that bacteria are extremely diverse. Point out that monerans vary in shape, metabolism, cell wall composition, and abilities to move. There are very few environments on earth that do not support some kind of bacterial life.

Comparing Autotrophs

Divide the class into three groups and assign each group one of the following types of organisms: blue-green bacteria (cyanophytes); green algae (chlorophytes); and plants. Ask each group to prepare a simple chart showing the basic characteristics of the organisms assigned. You may wish to refer students to Chapter 23, which describes chlorophytes and to Chapter 25, which describes plants. Let each group present their chart to the class. Then ask students to identify three similarities and three differences among these autotrophic organisms.

Section Review Answers

(Continued from p. 298.)
4. Population explosions occur when the nutrients on which bacteria thrive, such as those found in sewage, are abundant.
5. Archaebacteria live in habitats that are believed to resemble the conditions that existed on the early earth.

20.2 Biology of Monerans

Approach

The main objective of this section is to familiarize students with the key features of moneran physiology. It builds on the concepts of basic chemistry and cell biology (Chapters 3 through 5), metabolism (Chapter 7), and DNA replication (Chapter 8). In presenting the material, highlight the relationship of structure to function. You may wish to use Figures 20-5 and 20-6 to demonstrate features of moneran physiology. You might also wish to emphasize features of bacterial nutrition and respiration in preparation for covering pathogenic bacteria in Chapter 44, on the immune system.

Concept Framework

Most students have heard of tetanus, strep throat, food poisoning, and other bacterial diseases. Tap this knowledge to provide a conceptual starting point for studying this section by engaging students in a discussion of where in the human body pathogenic bacteria can live. Elicit the observation that many disease-causing bacteria can live in the lungs and airways of the body, while others live in the digestive tract. You may wish to end the discussion by asking students to consider the characteristics of these and other possible bacterial habitats as they study various aspects of moneran physiology.

Emphasis

Make sure that students understand the diversity in bacterial nutritional requirements and respiration. Point out that such variation permits bacteria to thrive in a greater range of environments than members of any other kingdom.

Display

Ask students to bring in items or pictures of items used in the home that help prevent the spread of bacterial infection. Such items might include disposable cups, disposable tissues, disinfectants, antibacterial soaps, and so on. Encourage students to write signs that can be placed next to the items that explain how using the items reduces the risk of infection by interrupting the mode of transmission. Set up a display of the items with their signs for other students to view.

Using the Features/Deep-Sea Hot Springs

The main process skills used in this feature are observation and inference. Scientists observed features of the vent communities that led them to infer that bacteria metabolized hydrogen sulfide and thus provided usable energy for other vent organisms. By demonstrating in the laboratory that bacteria from the vents could indeed convert hydrogen sulfide into carbohydrates, researchers confirmed the validity of this inference. You may wish to point out to students that scientists' growing understanding of the vent community will help us to understand other communities involving chemoautotrophic bacteria, which may lead to new food and energy sources.

Section Review Answers

(Continued from p. 304.)

3. In nitrogen fixation bacteria convert gaseous nitrogen to ammonia. Plants can use the nitrogen in ammonia, but not gaseous nitrogen. Thus nitrogen fixation makes an essential nutrient available to plants, which are in turn eaten by animals.
4. The pili of two cells form a bridge through which a portion of DNA from one bacterium can enter another bacterium.
5. The penicillin produced by fungi in the natural environment probably serves to reduce competition from bacteria.

Laboratory Preparation

Background

In this lab students observe living bacteria large enough to be seen with the 40K objective. Students can examine bacteria in saliva, yogurt, and other common sources, as well. Students use sterile techniques to prevent contamination of cultures. Students do not need to take any special precautions with these bacteria, but they should follow the same procedures required for other biological materials, i.e., protozoa. The most common problem in making hanging drop preparations is using too large a drop. Small drops stay in place when the slide is turned over; large drops contact the edge of the depression and coverslip and are spread by the surface tension between them.

Preparation

Cultures of *Bacillus subtilis*, *Micrococcus luteus*, and *Rhodospirillum rubrum* are available from a biological supply house. To prepare nutrient broth, bring 350 mL distilled water to a boil in a 500 mL beaker. Add 3 g dehydrated nutrient broth and stir until dissolved. Pour broth in test tubes and sterilize tubes for 15 minutes at 15 pounds of pressure. Transfer bacteria to tubes of sterile nutrient broth by using an inoculating loop and following sterile techniques. Store at room temperature for two to three days to allow bacteria to grow.

Conclusions

1. The rod-shaped *Bacillus subtilis* and spiral *Rhodospirillum rubrum* bacteria move in this culture by means of flagellation.
2. *Micrococcus* appears in clusters and *Bacillus* and *Rhodospirillum* appear singly.

Inquiry

The hanging drop technique allows the students to observe the organisms more accurately because the cover slip does not crush the organisms and the organisms are able to breathe.

Chapter Review

Vocabulary

1. chemoautotrophs; because they make their own energy instead of using food produced by other organisms
2. methanogens; because they are archaebacteria
3. archaebacteria; because they are not involved in pathogenic relationships
4. endospores; because they protect genetic material

and they are not involved in the exchange of genetic material

5. obligate anaerobes; because they cannot live in the presence of oxygen

Review

11. The Phylum Schizophyta is divided into the Classes Eubacteria, Actinomycota, Rickettsiae, and Spirochaeta.

12. Gram-negative bacterial cells have a lipopolysaccharide covering and stain pink; Gram-positive bacteria have no such covering and stain purple. The Gram-positive bacteria are more susceptible to antibiotics.

13. Actinomycota are Gram-positive bacteria that form colonies of filaments. Some are decomposers, some cause disease, some produce antibiotics. Rickettsiae are Gram-negative, parasitic bacteria. Spirochetes are spiral-shaped bacteria that usually move with flagella.

14. Blue-green bacteria are similar to plants in that they are photosynthetic, capture light with chlorophyll a, use water during photosynthesis, and produce oxygen. Colonies of blue-green algae also show specialization of cells, which is a trait more common in multicellular organisms.

15. The capsule is a protective layer of polysaccharides that surrounds the cell wall.

16. By decomposing dead organic matter, saprophytic bacteria release carbon and other elements for use by other organisms.

17. Photoautotrophs use light energy to create their food; chemoautotrophs use chemical compounds.

18. In conjugation a portion of the DNA from one bacterium may pass across a bridge formed by the pili, line up with the homologous portion of the DNA of a second bacterium, and be integrated into that DNA.

19. An exotoxin is generally more deadly.

20. hydrogen sulfide

Critical Thinking

1. Fossil bacteria are not only very small and therefore easily missed, but their cells are far more fragile—and less often preserved—than those of bones and tree limbs.

2. A deep puncture wound is more likely to develop tetanus because oxygen is less likely to be present deep under the skin.

3. Gram-positive bacteria cannot survive without their rigid walls. Gram-negative bacteria do not have such walls and are therefore less likely to be affected by penicillin.

4. Many bacteria do not produce useful substances yet are not normally harmful. Because they consume excess food in the intestines, these bacteria prevent the establishment of foreign—and perhaps dangerous—bacteria in the intestinal tract.

5. The bacteria are stained pink and must therefore be Gram-negative. Since most Gram-negative bacteria produce endotoxins, it is likely that these do also.

6. In vent communities the supply of bacteria is virtually unlimited, and predation is limited. These conditions make a very productive community possible.

Chapter 20 | Bacteria

Introduction

Monerans, or bacteria, are the most numerous organisms on earth. You probably think of them first as agents of disease. Some bacteria do cause deadly diseases, such as bubonic plague, typhus, and gangrene. However, many others are essential to all other living things. For example, many monerans are decomposers, that is, they break down organic matter into substances that other organisms can use. In this chapter, we will study the structure and physiology of monerans, as well as ways in which monerans affect humans.

Chapter Outline

20.1 Evolution and Classification
- *Evolution*
- *Classification*

20.2 Biology of Monerans
- *Structure and Movement*
- *Nutrition*
- *Respiration*
- *Reproduction*
- *Production of Toxins*
- *Antibiotics*

Chapter Concept

As you read, look for morphological and physiological variations among monerans.

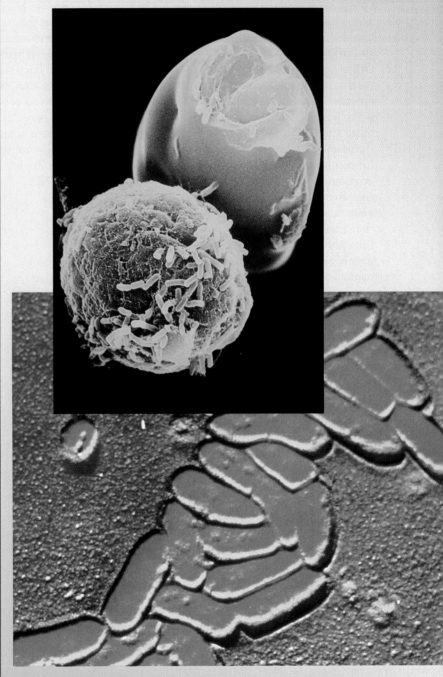

Aggregate of the bacillus *Erysipelothrix rhusiopathiae* (×2,900). Inset: Color-enhanced SEM of the bacillus *Agrobacterium tumifaciens* on tobacco cells (×1,070).

For additional teaching information see pages T20A–T20D.

20.1 Evolution and Classification

Bacteria is the common name given to most organisms in the Kingdom Monera. All monerans are prokaryotic—they are characterized by the absence of a nucleus and membrane-bound organelles. They are also microscopic. For example, the smallest cells known to exist are monerans called mycoplasmas, which measure 0.20 to 0.25 μm in diameter. The average moneran is about 1 μm long; some, however, may be as long as 500 μm.

Evolution

Fossil monerans have been found in Australian deposits more than 3.5 billion years old. Monerans have evolved into many forms and now live in nearly every environment, many of which cannot support other living things.

The classification of monerans into an evolutionary hierarchy is not easy. Many of them look alike; it can be difficult to separate them into species based solely on structure. However, by relying on physiology in addition to morphology, scientists have determined a probable phylogenetic classification of monerans.

Classification

The Kingdom Monera is subdivided into four phyla: Archaebacteria *(AHR-kee-oe-back-TIR-ee-uh)*, Schizophyta *(skuh-ZAHF-uh-tuh)*, Cyanophyta *(SIE-uh-NAHF-uh-tuh)*, and Prochlorophyta *(PROE-klor-AHF-uh-tuh)*. These organisms differ in both morphology and physiology.

As you can see in Figure 20-1, monerans are usually spherical, rod-shaped, or spiral. Spherical monerans are called *cocci*; rod-shaped monerans, *bacilli*; and spiral monerans, *spirilli*.

Section Objectives

- *Explain the differences between the members of the Phylum Archaebacteria and the members of the Phylum Schizophyta.*

- *Describe the four classes in Phylum Schizophyta.*

- *Distinguish between Gram-positive and Gram-negative bacteria.*

- *Describe the Phyla Cyanophyta and Prochlorophyta.*

Figure 20-1 The most common shapes among monerans are represented here by *Staphylococcus aureus*, sphere-shaped (×3,945); *Bacillus cereus*, rod-shaped (×8,000); and *Leptospirilla ichterohaemorrhagiae*, spiral-shaped (×4,400).

Some monerans cluster together; others form chains or filaments. The prefix *staphylo-* is used to describe cells that form clusters. Therefore a cluster of spherical moneran cells is called a *staphylococcus*. The prefix *strepto-* is used to describe cells that form filaments. Therefore a filament of rod-shaped moneran cells is called a *streptobacillus*.

Archaebacteria

Members of the Phylum Archaebacteria are adapted to the harsh environments in which they live. Although **archaebacteria** are similar in many ways to other monerans, they also have unique characteristics that have led scientists to place them in their own phylum. For example, the transfer RNAs and cell walls of archaebacteria differ from those of other bacteria. Archaebacteria include methanogens, extreme halophiles, and thermoacidophiles.

The **methanogens** live only in the absence of free oxygen. These anaerobic monerans are called methanogens because they use carbon dioxide and hydrogen to form methane and water. Methanogens live in the digestive tracts of sheep and cattle and at the bottoms of bogs, lakes, and sewage treatment ponds.

The **extreme halophiles** *(HAL-uh-FILEZ)* live only in areas of high salt concentration, such as the Great Salt Lake and the Dead Sea. The **thermoacidophiles** *(ther-moe-uh-SID-uh-FILEZ)* live only in places that are very acidic and where temperatures are very hot, often reaching 90° C. For example, members of the genus *Sulfolobus* live in acidic hot springs where the temperature averages 80° C and the pH is less than 2. Nevertheless, these monerans maintain an internal pH that is nearly neutral. Members of the genus *Thermoplasma* are found in smoldering coal tailings where the temperatures are often higher than 90° C.

Some other kinds of archaebacteria live around volcanic vents that are found miles below the ocean's surface. These monerans use the sulfur gases that bubble from the vents as their energy source. Scientists think that the environments in which present-day archaebacteria live may be similar to those that existed during the formative stages of life on earth.

Schizophyta

The largest moneran phylum is Schizophyta, all the members of which are commonly referred to as bacteria. Schizophyta is divided into four classes:

- Class Eubacteria *(YOO-back-TIR-ee-uh)* contains the largest number of and many of the most familiar bacteria.
- Class Actinomycota *(ACK-tuh-noe-MIE-KOE-tuh)* contains rod-shaped organisms that form branched filaments.
- Class Rickettsiae *(rick-ET-see-ee)* contains mostly nonmotile intercellular parasites.
- Class Spirochaeta *(SPIE-ruh-KEE-tuh)* contains large spiral-shaped organisms.

Figure 20-2 Some archaebacteria thrive in hot springs such as this one in Yellowstone National Park.

Tip: Refer students to Chapter 14 for a review of the conditions under which life began.

Most members of the Class Eubacteria are free-living soil and water bacteria. **Eubacteria** tend to live in less harsh environments than archaebacteria. Eubacteria are further classified by their reaction to a Gram stain. A **Gram stain** is a test that uses a series of dyes to stain bacterial walls. **Gram-negative bacteria** have an outer covering of lipopolysaccharides and stain pink. **Gram-positive bacteria** lack this covering and stain purple. Gram-positive eubacteria are susceptible to antibiotics. Gram-negative eubacteria are difficult to treat with antibiotics. Mycoplasmas, the smallest monerans, are eubacteria.

Members of the Class Actinomycota, called **actinomycotes,** are Gram-positive bacteria that form colonies of branching, multicellular filaments. Some actinomycotes decompose dead plants and animals; some cause diseases, such as diphtheria and tuberculosis. Others, especially members of the species *Streptomyces,* are the source of many antibiotics.

Members of the Class Rickettsiae, called **rickettsiae,** are parasitic Gram-negative bacteria that can reproduce only in certain cells of a specific host. Insects often carry these bacteria and transmit them to mammals. Typhus, for example, is a rickettsial disease transmitted by lice.

Members of the Class Spirochaeta, called **spirochetes,** are spiral-shaped or curved bacteria. Most spirochetes use flagella to move. One species of spirochete causes the sexually transmitted disease syphilis. Another causes Lyme disease, an ailment with symptoms similar to those of arthritis that is transmitted by ticks.

Cyanophyta and Prochlorophyta

Members of the Phylum Cyanophyta are called **blue-green bacteria.** These bacteria have some traits that are similar to those of plants and plantlike protists. Blue-green bacteria are photosynthetic. They also capture solar energy with chlorophyll a, one of the pigments found in plants, use water during photosynthesis, and produce oxygen.

Figure 20-3 The four classes in the Phylum Schizophyta are shown in these color-enhanced photos (left to right): *Klebsiella pneumoniae,* **a eubacterium (×1,970);** *Corynebacterium diphtheriae,* **an actinomycote (×12,000);** *Chlamydia* **sp., a rickettsia (×17,600); and** *Treponema pallidum,* **a spirochete (×700).**

Figure 20-4 The Phylum Cyanophyta includes such diverse organisms as (left to right) *Chroococcus* sp. (from a peat bog, ×400), *Lyngbia spirillans* (on agar, ×10), and *Anabaena* sp. with *Microcystis* sp. (×60).

Info: Blooms result in depletion of oxygen in polluted lakes, thus contributing to the death of many aquatic organisms.

Answers to Section Review
1. Streptococcus is a filament of spherical bacteria; staphylococcus is a cluster of spherical bacteria.
2. Methanogens live in the digestive tracts of sheep and cattle and at the bottom of bogs, lakes, and sewage-treatment ponds; extreme halophiles live in areas where salt is highly concentrated; thermoacidophiles live in highly acidic areas where the heat is extreme.
3. Gram-negative bacteria, of which rickettsiae are an example, have a lipopolysaccharide covering. Gram-positive bacteria, of which actinomycetes are an example, lack a lipopolysaccharide covering. (See pp. T20A–T20D.)

Most scientists now classify cyanophytes as monerans because, like other prokaryotes, they lack membrane-bound nuclei. In addition, their cell walls are chemically similar to those of other prokaryotes. Unlike other prokaryotes, however, blue-green bacteria are usually encased in a jellylike substance and often clump together in colonies. For example, *Anabaena* is a single filament, while *Nostoc* is a cluster of cells.

Some blue-green bacteria are also distinguished from other prokaryotes by having specialized cells and therefore a division of labor within the colony. The members of some genera produce cells called heterocysts when their nitrogen supplies are depleted. **Heterocysts** are specialized cells that convert nitrogen from the atmosphere into a form that the organism can use in cellular metabolism.

A rapid increase in the population of a blue-green bacteria, which may discolor a lake, river, or part of the ocean, is called a **bloom**. Since blue-green bacteria thrive on the phosphates and nitrates found in sewage, these population explosions often occur in polluted water.

Members of the Phylum Prochlorophyta are photosynthetic bacteria that live symbiotically with the marine vertebrates known as tunicates. Members of the genus *Prochloron* contain photosynthetic pigments unlike those in blue-green bacteria but similar to the chloroplasts of eukaryotes.

Section Review

1. Differentiate between *streptococcus* and *staphylococcus*.
2. List the habitats of the three types of archaebacteria.
3. Explain the structural difference between a Gram-positive bacterium and a Gram-negative bacterium. Give an example of each.
4. When do blue-green bacteria form blooms?
5. Why do scientists think archaebacteria were the first form of life on earth?

A New Class of Antibiotics

This excerpt is from an article by Lawrence K. Altman that appeared in the New York Times *of June 26, 1987.*

Swedish scientists have developed a new class of antibiotics that promise one day to assist in the treatment of infections caused by some of the most dangerous and resistant strains of bacteria. . . .

The antibiotics are synthetic chemicals that scientists at the Astra Pharmaceutical Company in Sweden have tested in laboratory experiments against a large group of bacteria that cause many infections of the urinary tract and complicate surgery and hospital stays.

The new antibiotics are unusual, the scientists said in an article in a professional journal, because they have been designed to be absorbed by the bacteria and then to attack it from the inside.

The new antibiotics have not yet been tested on humans and face many scientific and governmental regulatory hurdles before they can become licensed drugs. That process usually takes several years. . . .

The new class of antibiotics is designed to kill only Gram-negative bacteria—so named because they do not retain a crystal violet chemical stain used as a standard laboratory test in everyday medical practice. The test was developed by a Danish physician, Dr. Hans Gram.

Some antibiotics are effective against some Gram-positive and some Gram-negative bacteria while others, such as penicillin, are effective primarily against Gram-positive but not against Gram-negative microbes.

Infections caused by Gram-negative bacteria have become an increasingly important problem in medical practice, in part because they have developed resistance to antibiotics. With the emergence of such resistant organisms, scientists have striven to discover new antibiotics that are effective against the resistant organisms. . . .

The new drugs were specifically designed to combat Gram-negative bacteria. Thus because they "cannot previously have been encountered by the target organism," the factors in bacteria that determine resistance "are unlikely to be present in the population," Dr. Higgins said.

The drugs are a combination of two substances that stop the manufacture of lipopolysaccharide, a key component of the outer wall of Gram-negative bacteria.

To sabotage that process, researchers had to find a way to get their drug inside their bacteria, which normally resist its entry. The Swedish team did it by attaching molecules of the drug to substances called peptides that are readily received by the cell and thus trick it into accepting the drug as well.

Once the combination enters the bacteria, the natural breakdown of the peptides leads to release of the drug that retards an enzyme that is vital to the production of lipopolysaccharide, killing the microbes.

Write

- The new antibiotics described in the article seem to be needed as soon as possible. However, it may be several years before they are approved because of "scientific and governmental regulatory hurdles." Write a newspaper editorial in which you discuss your stand on this issue. Should governmental regulations be less stringent? Model your writing on actual editorials.

Copyright © 1987 by The New York Times Company. Reprinted by permission.

Section Objectives

- *List the distinguishing features of moneran cells.*

- *Describe autotrophy and heterotrophy in monerans.*

- *Describe the way in which organisms can survive in vent communities deep in the ocean.*

- *Distinguish between endotoxins and exotoxins.*

- *Explain the basic effect of antibiotics.*

20.2 Biology of Monerans

The structures and physiology of monerans are diverse. These variations reflect the numerous habitats to which monerans are adapted.

Structure and Movement

Study the electron micrograph of the moneran in Figure 20-5. Monerans are prokaryotic; their DNA is located in the cytoplasm. The DNA is arranged in a single, circular chromosome. In addition, monerans do not have membrane-bound organelles. Some monerans have **plasmids,** smaller circular strands of DNA that are capable of replicating independently. All monerans except the mycoplasmas have walls that differ structurally and chemically from those of eukaryotes. Some monerans have rigid cell walls; others have flexible ones.

Many monerans can produce **capsules,** protective layers of polysaccharides around their cell walls. Under natural conditions many prokaryotes also produce a net of polysaccharides called the **glycocalyx** *(GLIE-koe-KAY-licks)* that helps them stick to the surface of rocks, teeth, and host cells. Some monerans attach themselves to objects with protein strands called **pili.**

Under adverse conditions, many monerans encase their DNA and some of their cytoplasm in a tough envelope. This structure, called an **endospore,** can lie dormant for years. When favorable conditions return, the endospore coating breaks and the cell becomes active. Anthrax, an often fatal disease of sheep and cattle, can survive in endospore form for 60 years.

Monerans have different ways of moving. Many move by rotating stiff flagella that are bent into **S**-curves. When the flagella are rotated in one direction, the monerans move in a straight line; when rotated in the opposite direction, they tumble. Spirochetes use filaments that wrap around the cell to move.

Figure 20-5 *Enterobacter* is a typical moneran. In the color-enhanced TEM (left, ×23,520), the DNA appears green. The diagram (right) shows structural features.

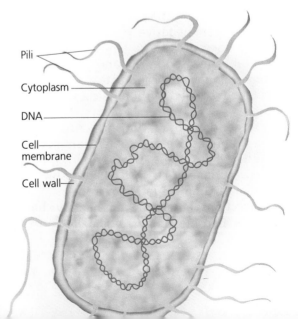

Pili

Cytoplasm

DNA

Cell membrane

Cell wall

Nutrition

Monerans may be heterotrophic or autotrophic. Most monerans are **heterotrophs**—that is, they use food produced by other organisms. A heterotroph that feeds on dead or decaying organic matter is called a **saprophyte** *(SAP-ruh-FITE)*. By decomposing matter, saprophytic bacteria release carbon and other elements for use by other organisms. Saprophytes are therefore essential to the recycling of nutrients in every ecosystem.

Monerans that produce their own food from inorganic material are **autotrophs.** Those that use sunlight as a source of energy are called **photoautotrophs.** Photoautotrophs have different chlorophylls that enable them to absorb light of various wavelengths. Some have the same pigments found in plants; others have unique pigments that absorb longer wavelengths.

Some autotrophic monerans, called **chemoautotrophs,** use the energy of chemical reactions instead of sunlight to synthesize food. Some chemoautotrophs can also fix nitrogen. **Nitrogen fixation** is the process by which gaseous nitrogen (N_2) is converted into ammonia compounds. Plants need nitrogen to synthesize proteins, but they cannot use gaseous nitrogen. They can, however, use the nitrogen in NH_4OH. Because most of the nitrogen on earth is gaseous and because only monerans can convert gaseous nitrogen into forms that plants can use, plants—and therefore most life-forms—depend on these nitrogen-fixing monerans.

Respiration

Many monerans are **obligate anaerobes,** that is, they cannot survive in the presence of oxygen. Among these are the methanogens. **Facultative anaerobes** can live with or without oxygen. *Escherichia coli,* which is common in the human digestive tract, is a facultative anaerobe. Monerans that cannot survive without oxygen are called **obligate aerobes.** The obligate aerobe that causes tuberculosis, *Mycobacterium tuberculosis,* lives in the lungs.

Think

Many organisms find the odor of decaying organic matter repulsive. What adaptive advantage might this ability give living organisms?

Answer to Think
Many infectious bacteria can be found in decaying matter. The perception that such substances are foul would allow organisms to avoid them, making it less likely that the organisms would become infected.

Tip: Suggest that students review the role of oxygen in respiration explained in Chapter 7 and speculate about the details of moneran respiration.

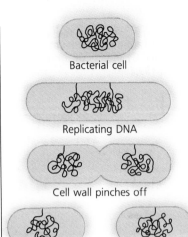
Bacterial cell

Replicating DNA

Cell wall pinches off

Two identical cells

Figure 20-7 Binary fission is the most common means of bacterial reproduction.

Reproduction

Some monerans reproduce rapidly. Under optimal conditions, *E. coli* cells, for example, can divide every 15 minutes. In one week, a moneran can produce millions of cells. Heat, cold, predation, and lack of food inhibit the rate of reproduction.

Binary Fission

Monerans usually reproduce by splitting in two, an asexual process called **binary fission.** First the DNA in the cell replicates. Then the plasma membrane and cell wall grow inward, forming two daughter cells with identical genetic material.

Conjugation

Occasionally the genetic materials of two monerans of the same species are recombined in a process called **conjugation.** A portion of the DNA of one moneran passes across a bridge, formed by the pili, into another moneran. This piece of DNA then lines up with the homologous piece of DNA in the recipient cell. The homologous portion is destroyed and the new DNA is substituted. This exchange of genetic information increases genetic variability.

Biology in Process

Deep-Sea Hot Springs

The Mid-Atlantic Ridge is a zone of earthquakes and volcanoes. In the 1970s scientists observed the ridge from the submarine research vessel *Alvin.* They discovered vents through which heat and gases escaped from deep below the ridge. To their surprise, they also found that the areas around the vents were teeming with life. In fact, these communities are some of the most productive communities known to exist.

The lack of sunlight in the vent communities prevents photosynthetic organisms from living there. Yet the researchers found crabs, clams, giant tube worms, and fish living in the total darkness around the vents. What were the organisms using for food?

The researchers analyzed water samples from the vents, which turned out to be rich in hydrogen sulfide, a compound that is toxic to most organisms. The researchers knew that on land certain chemoautotrophic bacteria were able to convert this compound into energy. It seemed possible that chemoautotrophs performed the same function in the deep-sea vent communities.

Next researchers isolated bacteria from the vent waters. In the laboratory, they demon-

Production of Toxins

Any organism that causes a disease is called a **pathogen.** Diseases that are caused by pathogenic bacteria include bubonic plague, leprosy, tuberculosis, and typhus. Most pathogenic bacteria enter the human body through the respiratory or gastrointestinal tract. You will learn more about these diseases in Chapter 44.

In most cases bacterial diseases are caused by the toxins that are produced by bacteria. A **toxin** is a poisonous substance that disrupts the metabolism of the infected organism. Bacteria produce two types of toxins, endotoxins and exotoxins. **Endotc ins** are found in the cell walls of most Gram-negative bacteria. Endotoxins all cause the same symptoms: fever, weakness, and damage to the circulatory system. Typhoid fever is one example of a disease that is caused by endotoxins. **Exotoxins,** which are products of the metabolism of some bacteria, are secreted into the area surrounding the bacteria. Exotoxins are the most potent poisons known. One nanogram of the tetanus exotoxin, for example, is sufficient to kill a guinea pig. Exotoxins cause diseases such as diphtheria and botulism.

Answer to Biology in Process Researchers observed that plant life, the usual basic source of energy, could not exist in the total darkness of the vent community. However, the vents were rich in hydrogen sulfide, which researchers knew could be metabolized by bacteria found on land. This led researchers to hypothesize that in vent communities bacteria were the source of energy.

strated that the bacteria were indeed able to metabolize hydrogen sulfide. Researchers then hypothesized that the vent bacteria produced carbohydrates that were the main source of energy for the other vent organisms.

To support their hypothesis, researchers continued to observe the organisms that lived near the vents. They found that most of the larger animals did consume bacteria by filter-feeding and therefore probably derived nutrients from them. However, one of the dominant vent animals— the tube worm *Riftia pachyptila*—had no ability to filter

out the bacteria. Instead *Riftia* absorbed nutrients through an enclosed sac called a trophosome, or feeding body.

When researchers examined *Riftia* under a microscope, however, they found vast colonies of chemoautotrophic bacteria living in the trophosome. Further investigations showed that the bacteria and the tube worm have an important sym-

biotic relationship. *Riftia* receives nutrition from the chemoautotrophic bacteria. In turn, it provides the bacteria with compounds they need in their metabolism, such as carbon dioxide and oxygen.
■ What observations led researchers to believe that the vent bacteria were the basic source of energy in the vent communities?

Table 20-1 Some Bacterial Diseases

Bacterium	Disease	Transmission	Symptoms
Clostridium tetani	Tetanus	Endospores enter wound	Neurotoxins are produced that cause muscle spasms; death results from spasms of respiratory muscles
Clostridium perfringens	Gas, gangrene	Endospores enter wound	Gases are produced that cause soft tissues to swell and die
Rickettsia rickettsii	Rocky Mountain Spotted Fever	Carried by ticks	Chills, fever, pain in legs and joints
Rickettsia typhi	Endemic murine typhus	Carried by rodents and fleas	Rashes, prolonged high fever
Salmonella sp.	Food poisoning	Contaminated or inadequately cooked food	Nausea, abdominal cramps, diarrhea
Yersinia pestis	Bubonic plague	Carried by fleas on rodents	Enlarged lymph nodes, hemorrhaging, fever, possibly death
Streptococcus sp.	Strep throat	Airborne	Fever, exhaustion, sore throat, enlarged lymph nodes
Mycobacterium tuberculosis	Tuberculosis	Airborne	In active disease, coughing, weight loss, possibly death
Corynebacterium diphtheriae	Diphtheria	Airborne	Fever, sore throat, swelling of neck, possibly death

Info: Many bacteria contain R (resistance) factors, plasmids including genes that code for enzymes that inactivate specific drugs. Various R factors can be recombined in conjugation.

Answers to Section Review
1. *Glyco-* is from the Greek word for "sweet"; *calyx* is from the Greek word for "chalice" or "cup". The glycocalyx is a sticky polysaccharide covering.
2. A saprophyte is a heterotroph that feeds on dead or decaying organic matter. A chemoautotroph synthesizes inorganic compounds by using chemicals instead of sunlight.
(See pp. T20A–T20D.)

Antibiotics

Antibiotics are chemicals that are capable of inhibiting the growth of some bacteria. Many of these antibiotics are produced by living organisms. The first antibiotic, penicillin, a product of the fungus *Penicillium,* was discovered by Alexander Fleming (1881–1955), a British bacteriologist.

Antibiotics are useful and necessary. However, antibiotics often destroy not only pathogens but also useful bacteria. Many pathogenic bacteria have also become resistant to antibiotics. When a population of bacteria is first exposed to antibiotics, most members of the population die. However, some may have mutations that allow them to survive the antibiotic. These individuals then reproduce, passing their resistance on to their offspring.

Section Review

1. Look up the word *glycocalyx* in the dictionary. Why is this term appropriate for the structure it describes?
2. How does a chemoautotroph differ from a saprophyte?
3. Describe nitrogen fixation and its importance.
4. How does the pilus function in conjugation?
5. What is the role of penicillin in a natural environment?

Laboratory

Observing Living Bacteria

Objective
To observe the shapes and the motility of living bacteria

Process Skills
Observing, classifying, inferring

Materials
2 coverslips; paper towels; 2 toothpicks; vaseline; inoculating loop or Pasteur pipette; mixed broth culture of *Bacillus subtilis, Micrococcus luteus,* and *Rhodospirillum rubrum;* 2 depression slides; a local source of living bacteria

Method
1 The hanging drop technique permits observation of microscopic living organisms without flattening these organisms between the microscope slide and the coverslip. This technique also permits gas exchange between the drop of water and the atmosphere. To make a hanging drop preparation, begin by placing a coverslip on a clean piece of paper towel. Use a toothpick to put a very small dot of vaseline near each corner of the coverslip.
2 Use an inoculating loop or Pasteur pipette to remove a small drop of bacteria from the culture, and place the drop in the center of the coverslip.
3 Take a depression slide and, with the depression facing down, place it on top of the coverslip. Be sure that you position the drop in the center of the depression. The vaseline will affix the coverslip to the slide.
4 Gently lift the slide off the paper. Turn the slide over quickly so that the drop will be disturbed as little as possible. If the drop was too large, or if too much vaseline was used, the drop may spread out and become dispersed over the edges of the depression. If this happens, make a new preparation.

5 Begin your examination of the drop under a microscope set to low power. Center the drop in the field of view and then change to high power.
6 Each of the three types of bacteria in the culture has a different cell shape. Identify and name each shape.
7 Observe the various methods of locomotion.
8 Repeat steps 1 through 5 using another bacterial preparation supplied by your teacher. It may be necessary to dilute the drop of culture in water if the population is too dense to allow clear observation. Identify the shape and method of locomotion of each species of bacteria.

Conclusions
1 Of the three species of bacteria observed in step 6, which move? What is their method of locomotion?
2 Does each bacterium appear singly or as part of a cluster?

Inquiry
Why are the advantages provided by the hanging drop technique important for making accurate observations of living bacteria?

Vocabulary

actinomycotes (297)	conjugation (302)	Gram-positive	pathogen (303)
antibiotic (304)	endospore (300)	bacterium (297)	photoautotroph (301)
archaebacterium	endotoxin (303)	Gram stain (297)	pilus (300)
(296)	eubacterium (297)	heterocyst (298)	plasmid (300)
autotroph (301)	exotoxin (303)	heterotroph (301)	rickettsiae (297)
binary fission (302)	extreme halophile	methanogen (296)	saprophyte (301)
bloom (298)	(296)	nitrogen fixation	spirochete (297)
blue-green bacterium	facultative anaerobe	(301)	thermoacidophile
(297)	(301)	obligate aerobe (301)	(296)
capsule (300)	glycocalyx (300)	obligate anaerobe	toxin (303)
chemoautotroph	Gram-negative	(301)	
(301)	bacterium (297)		

In each of the following sets, choose the term that does not belong and explain why.
1. heterotroph, saprophyte, chemoautotroph
2. methanogen, spirochete, rickettsiae
3. archaebacterium, exotoxin, pathogen
4. pilus, conjugation, endospore
5. blue-green bacterium, obligate anaerobe, bloom

Review

1. The smallest cell known to exist is a (a) spirochete (b) mycoplasma (c) archae-bacterium (d) eubacterium.
2. Rod-shaped monerans are called (a) cocci (b) bacilli (c) halophile (d) spirilli.
3. A thermoacidophile is a(n) (a) spirochete (b) cyanophyte (c) archaebacterium (d) actinomycete.
4. Gram-positive bacteria stain (a) blue (b) pink (c) red (d) purple.
5. Blooms that occur in lakes are the result of (a) antibiotics (b) pathogens (c) rickett-siae (d) population explosions.
6. Bacterial DNA is (a) circular (b) encased in a capsule (c) linear (d) found in the nucleus.
7. The glycocalyx helps monerans to (a) survive unfavorable environmental conditions (b) stick to surfaces (c) metabolize gaseous nitrogen (d) ingest food.
8. In nitrogen fixation, gaseous nitrogen is converted to (a) carbon (b) glucose (c) ammonia (d) methane.
9. An organism that must have oxygen to survive is (a) an obligate aerobe (b) a facultative anaerobe (c) a facultative aerobe (d) an obligate anaerobe.
10. Genetic recombination in monerans can occur during the process of (a) conjugation (b) heterocyst formation (c) binary fission (d) endospore production.
11. What are the four classes in Phylum Schizophyta?
12. Explain the major difference between Gram-positive and Gram-negative bacteria. Which of these two groups is more susceptible to antibiotics?
13. Describe the main characteristics of actinomycotes, rickettsiae, and spirochetes.
14. List three ways in which blue-green bacteria resemble plants.
15. Describe the capsule of a moneran, and state its function.
16. Explain how saprophytic monerans contribute to the recycling of nutrients in the environment.

17. How do photoautotrophs differ from chemoautotrophs?
18. Describe one way in which some monerans can exchange genetic information.
19. Which is generally more deadly, an endotoxin or an exotoxin?
20. What compound is metabolized by bacteria in deep-sea vents?

Critical Thinking

1. Scientists have only recently discovered fossil bacteria. Explain why this discovery may have taken so long.
2. *Clostridium tetani,* the bacteria that cause tetanus, are obligate anaerobes. From this information would you infer that a deep puncture wound or a surface cut would be more likely to become infected by tetanus bacteria? Explain the reason for your inference.
3. Penicillin works by interfering with the ability of bacteria to build a cell wall. Given this fact, explain why Gram-positive bacteria are more susceptible to the effects of penicillin than Gram-negative bacteria are.
4. Some of the bacteria that are normally found in the human intestinal tract are beneficial. For example, *E. coli* produces Vitamin K. However, *E. coli* can also cause diarrhea under exceptional circumstances and can cause serious infections if it invades other parts of the body. Other bacteria found in the digestive tract do not produce substances the body can use, nor do they produce substances that are harmful. What role might they play?

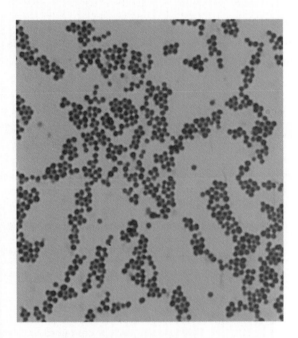

5. Examine the photograph above of bacteria that have been treated with a Gram stain. Would you hypothesize that these bacteria produce endotoxins?
6. Vent communities have some of the densest and most productive populations known to exist. What might explain this?

Extension

1. Read the articles entitled "Missing Mycoplasmas" and "Sexual Mycoplasmas" in *Science News,* September 1986, p. 184. Give an oral report on what you learn, mentioning three of the diseases caused by mycoplasmas, the reason scientists don't know how many diseases are caused by mycoplasmas, the research that must be done to determine whether *M. genitalium* causes gonococcal urethritis, and the way in which biotechnology may help in this research.
2. Research and write a report on the use of bacteria in either the processing of several foods or in the treatment of sewage.
3. Visit or telephone your local public health department, and ask for information on bacterial diseases that have been reported in your area in recent weeks.

Chapter 21 | *Protozoa*

This chapter investigates the four phyla of protozoa. Students learn about their structure and methods of locomotion, feeding, and reproduction. Included in this study are descriptions of diseases caused by various protozoa.

Correlation Chart

Section Number	Section Objectives	Pages	Features	Section Review
21.1	1. Explain how protozoa may have evolved from prokaryotes.	309–310		2
	2. List the four phyla of protozoa and their characteristics of movement.			3
	3. Discuss the ecology of protozoa and its relation to their physiology.			1, 4, 5
	4. Describe why euglenoids are not classified with protozoa.			
21.2	1. Compare and contrast methods of movement in the four phyla of protozoa.	311–317	Think, p. 313	1, 2
	2. Describe binary fission in amebas and ciliates.			4
	3. Describe conjugation in ciliates.		Biotechnology, pp. 314–315	3
	4. List and explain two diseases caused by zoomastiginoids.			
	5. List and explain two diseases caused by sporozoans.		Biology Lab, p. 317	5

Resources*

Print

Begon, Michael, and Martin Mortimer. *Population Ecology: A Unified Study of Animals and Plants,* 2nd ed. Boston: Blackwell Scientific (Dist. by Sinauer), 1986.

Jahn, T.L., E. Bouce, and F. Jahn. *How to Know the Protozoa.* 2nd ed. Dubuque, Iowa: William C. Brown Group, 1979.

Teasdale, Jim. *Microbes.* Morristown, N.J.: Silver Burdett, 1985.

Audiovisual/Software

Living Things in a Drop of Water, 2nd ed. Film, 12 min. Britannica, 1985.

Locomotion. Film, 30 min. Carolina Biological, 1986.

Protozoa. Still-frame video or filmstrip with cassette, 15 min. Carolina Biological.

Protists: Form, Function, and Ecology. Film, 23 min. Britannica, 1985.

The Microorganism Simulator. Apple II 48K. Disk. Focus, 1987.

*Additional references are listed in *Modern Biology, Teacher's Program Guide.*

Section Teaching Guides

21.1 Overview of Protozoa

Approach

The main objective of this section is to familiarize students with the distinguishing features of the so-called animal-like protists. It builds upon concepts of metabolism (Chapter 7) and classification (Chapter 18). In presenting the section, help students develop an appreciation for the diversity among the simplest eukaryotes. You may wish to use Figure 21-2 to emphasize the variation in locomotion exhibited by some protozoa.

Concept Framework

Students have studied the differences between prokaryote and eukaryote cells (Chapter 5) and the endosymbiosis hypothesis (Chapter 14). They have also studied monerans (Chapter 20). To tap this knowledge and highlight the differences among protozoans, engage students in a discussion of what prokaryotic traits might, when coordinated, benefit eukaryotic organisms. You may wish to end the discussion by asking students why modes of movement might serve as a key to classifying protozoa.

Emphasis

Make certain that students understand that protozoa are called the animal-like protists because they are heterotrophic and mobile. They are distinguished from the autotrophic protists (algae) and from the funguslike protists based on reproductive criteria.

Class Discussion

Describe ciliar and flagellar movement to students, stressing undulation. Ask students to explain how undulations can cause an organism to move forward. Encourage them to interact, make drawings and models, and/or experiment to reach an explanation.

Section Review Answers

(See p. 310.)

21.2 Protozoan Phyla

Approach

The main objective of this section is to encourage students to appreciate diversity among protozoans. It builds upon concepts of cellular differentiation (Chapter 5) and diversity through adaptation (Chapters 15 and 16). In presenting the text, guide students to develop an appreciation of the diversity in protozoan locomotion. To stress the criteria scientists use in grouping protozoa together, encourage students to look for similarities among the members of the four protozoan phyla that are used as examples in the text.

Concept Framework

Students have studied endocytosis and exocytosis (Chapter 6). To build on this knowledge and apply their previous learning to actual organisms, engage students in a review of the key types of endocytosis (pinocytosis and phagocytosis) and discuss the adaptive advantages of being able to engulf food particles. You may wish to end the discussion by asking students to speculate on how amebas might resemble cells of members of the Kingdom Animalia.

Emphasis

Make certain students understand that parasitic protozoa must be carried from host to host in blood or other bodily fluids, whether or not the protozoan has a flagellum. In this sense the protozoa are completely dependent upon host organisms.

Guest Speaker

Invite a protozoologist or health professional from a nearby university or hospital to speak to the class about parasitic protozoa that can cause disease in humans. Have the speaker address modes of disease spread as well as specific organisms and the symptoms and treatments of disease.

Using the Features/Controlling Chagas Disease

The main technique used in the feature concerns internationally coordinated disease control. WHO has developed a three-part campaign intended to control Chagas disease, or South American trypanosomiasis. The campaign involves using pesticides and simple housing renovations to eliminate kissing bugs, carriers of the protozoan that causes the disease. It also involves programs designed to reduce the spread of the disease among humans. This type of coordinated campaign could be used in controlling other protozoan diseases in developing countries. Make sure students understand that such methods of disease control must be feasible in the area or country where they are applied. People must be willing to comply with health-oriented programs.

Section Review Answers

(See p. 316.)

Laboratory Preparation

Background

In this lab students study the cellular specialization and adaptation of a variety of ciliates. Students observe the ability of these cells to develop different kinds of locomotion, to distinguish between nutritive and non-nutritive materials, and to develop a wide range of life-styles. Students can best observe ciliate behavior under natural conditions rather than in agents designed to slow them down. Wheat grains that have been heated can be added to protozoan cultures as a source of nutrients. Subcultures can also be made in spring water or aged tap water. Congo Red–stained yeast results may vary.

Preparation

Cultures of ciliated protozoa are available from biological supply houses. To prepare Congo Red, add a pinch of Congo Red and a teaspoon of dry yeast to a tablespoon of distilled water in a small flask. Cover and let mixture sit overnight. Bring the mixture to a gentle boil for five minutes. Allow the mixture to cool. Add just enough of the yeast mixture and activated charcoal to a dense *Euglena* culture to tint the culture without obscuring the *Euglena*. To save time have each student prepare two wet mounts and exchange them with another student who has prepared two others.

Conclusions

1. The cilia are organized differently in different organisms.
2. *Paramecium* and *Stentor* can swim freely. *Euplotes* crawls rather than swims. *Vorticella* is sessile; it does not move.
3. Freshwater protozoa contain a higher concentration of solutes than their environment, making them hypertonic to their environment. They would burst if water were not removed from them continuously.

Inquiry

Students will devise a method of measuring the rate of contractile vaculoes, such as counting with a stopwatch and then testing the animals in water of different temperatures.

Chapter Review

Vocabulary

1. *Pseudopodium* means "false foot." It is appropriate for the footlike projections used in movement by sarcodines.
2. Ectoplasm is the colloid sol found just inside the cell membrane of sarcodines; endoplasm is found in the interior of the cell.
3. A food vacuole encloses food particles and carries them through the cytoplasm, where they are digested and absorbed. A contractile vacuole regulates the amount of water in the organism.
4. *Zoo* means "life" or "alive," and in biological nomenclature tends to be used to mean "animal." *Plankton* is from a Greek root meaning "drifting." The image of "drifting life" accurately describes organisms that float in ocean water.
5. The response is called avoidance behavior.

Review

11. Marine sarcodines of the genus *Foraminifera* have calcium carbonate shells with spikelike protrusions and the genus *Radiolaria* have supportive silicon dioxide inside their shells. Heliozoans, a type of freshwater sarcodines, have thin, extended projections on their shells.
12. In ameboid movement the endoplasm begins to push forward, facilitated by the slimy ectoplasm. It forms a pseudopodium. As the pseudopodium moves forward, old pseudopodia are retracted into the cytoplasm, so the organism moves forward.
13. A contractile vacuole removes excess water from the cell. This maintains homeostasis, and the organism is able to survive in aquatic environments, where water is constantly diffusing into the cell.
14. The sarcodine surrounds the food particle with pseudopodia. Then part of the cell membrane

forms a food vacuole around the food particle.

15. The macronucleus disintegrates and does not play a direct role in conjugation; the micronucleus divides both meiotically and mitotically, eventually producing a micronucleus that will fuse with one from another organism, thereby permitting genetic recombination.

16. The ciliate digests food by means of food vacuoles formed in the gullet.

17. Humans contract giardiasis by drinking water contaminated with *Giardia,* which is carried by muskrats and beavers.

18. Sporozoans have no means of locomotion.

19. The diagram should show *Plasmodium* gametes forming a zygote in the *Anopheles* mosquito gut and should show the zygote forming and releasing spores that are carried to the mosquito's mouth. The spores should then be shown reproducing in a human liver, being released, and infecting red blood cells. The spores develop into sexual forms and are again ingested by mosquitos that bite the infected human.

20. Malaria can cause anemia and jaundice, enlargement of the liver and spleen, and sometimes kidney failure.

Critical Thinking

1. *Euglena* is a good choice for use in the laboratory because *Euglena* can be autotrophic and it would not need to be fed.

2. The adaptations that justify classifying paramecia as being more complex than amebas is that they have many more organelles, including specialized ones such as the oral groove, gullet, and so on.

3. These shells have spines that provide protection against predators.

4. The proximity of food is the stimulus that might cause pseudopodia to form in the unicellular animals.

5. Movement by means of cilia is more efficient than movement by means of pseudopodia because pseudopodia are not specialized for movement. Cilia are more specialized for movement in that they move in synchronized strokes.

Chapter 21 | Protozoa

Introduction

The Kingdom Protista is made up of widely diverse organisms. In this chapter you will learn about the four phyla of protozoa, the so-called animallike protists. They are all heterotrophic and live in both aquatic and terrestrial environments. In Chapter 22 the autotrophic plantlike protists will be covered. Funguslike protists, which are also heterotrophic, are discussed in Chapter 23.

Chapter Outline

Chapter Concept

As you read, examine how different protozoa are structurally adapted for either free-living or parasitism.

Stained paramecia, *Paramecium caudatum*, (×100). Inset: Color-enhanced shell of a radiolarian (×700).

For additional teaching information see pp. T21A–T21D.

21.1 Overview of Protozoa

The approximately 56,000 species of organisms called protozoa are heterotrophic members of the Kingdom Protista. Protozoa live either freely or as parasites. Most are unicellular, though some form colonies in which cell specialization occurs. Most protozoa are microscopic, but some grow as large as 5 cm in diameter.

Unlike algal protists, protozoa do not make their own food through photosynthesis. Protozoa differ from funguslike protists in their methods of reproduction. Like all other protists, protozoa are eukaryotic.

Evolution

The first cells were prokaryotic and originated over 3.5 billion years ago. The first eukaryotes found in the fossil record are 1.45 billion years old. Eukaryotes may have arisen much earlier, but because their cells are generally soft, it is not likely that many were preserved in rocks. However, evidence from comparative anatomy and cell physiology strongly suggests that eukaryotes evolved from prokaryotes. According to the theory of **endosymbiosis,** prokaryotic parasites once lived inside other prokaryotic cells. Then the parasitic prokaryotes lost the ability to live independently of their hosts and evolved into various cell organelles. Structural and biochemical evidence suggests that mitochondria arose from parasitic bacteria and that chloroplasts arose from parasitic bluegreen bacteria. The nucleus, however, probably did not arise from an endosymbiont. It probably came to exist as an organelle when DNA was enclosed within a double membrane.

Classification

Biologists classify protozoa into four phyla: Sarcodina (SAHR-kuh-DINE-uh), Ciliophora (SILL-ee-AHF-ur-uh), Zoomastigina (ZOE-uh-MASS-tuh-JINE-uh), and Sporozoa (SPOR-uh-ZOE-uh). These are shown in Figure 21-2 on the following page. The main criterion for classification is mechanism of locomotion.

- Members of the Phylum Sarcodina move by using cytoplasmic projections called pseudopodia.
- Members of the Phylum Ciliophora move by the use of cilia.
- Members of the Phylum Zoomastigina move by use of flagella.
- Members of the Phylum Sporozoa are immobile and parasitic.

Some biologists suggest that organisms called euglenoids should also be considered protozoa. Although *Euglena* and related organisms look and behave much like protozoa, they photosynthesize under certain circumstances. In other words, they can be autotrophic as well as heterotrophic. For this reason scientists often consider them more like algae than like protozoa.

Section Objectives

- *Explain how protozoa may have evolved from prokaryotes.*

- *List the four phyla of protozoa and their characteristics of movement.*

- *Discuss the ecology of protozoa and its relation to their physiology.*

- *Describe why euglenoids are not classified with protozoa.*

Tip: Try wrapping a handful of soil in a tissue, placing it in a funnel, and flooding the funnel with water. Wait 24 hours and draw off some fluid to collect soil protists for observation.

Info: Of the approximately 800 species of euglenoids, about a third have chloroplasts. However, they lack some vitamins and amino acids that most algae manufacture on their own.

Figure 21-1 These foraminiferan fossils are similar to marine foraminiferans living today.

Figure 21-2 The four protozoan phyla are represented by the following organisms, clockwise from upper left: Ciliophora, *Vorticella*; Sporozoa, *Laptotheca*; Zoomastigina, *Trypanosoma lewisi*; Sarcodina, *Ameba*.

Answers to Section Review
1. An eyespot is a specialized, pigmented region found in many protozoa. It senses the quantity and quality of light. By sensing light the organism can move to conditions necessary for its survival.
2. When parasitic prokaryotes lost the ability to live independently of their prokaryotic hosts, they evolved into the cell organelles common in eukaryotes.
3. Sarcodina move by pseudopodia; Ciliophora move with cilia; Zoomastigina move with flagella; Sporozoa do not move.
4. CO_2 and O_2 diffuse across the semipermeable cell membrane.
5. Protozoa recycle nutrients because they are heterotrophic; they are a major component of zooplankton, an important food source for marine organisms; protozoa in the soil are probably eaten by certain soil animals such as earthworms.

Ecology and Physiology

Some species of modern-day protozoa are free-living; others are parasitic. Free-living protozoa live in all habitats in which water is available at some time of the year. Many free-living protozoa make up **zooplankton,** unicellular heterotrophic organisms that drift in water. Other free-living protozoa live in the soil. Parasitic protozoa live in the tissues and bloodstream of their hosts. Many parasitic protozoa cause diseases, which are among the most severe in humans and other animals.

Whether free-living or parasitic, most protozoa lack a protective outer covering. Therefore a semipermeable cell membrane serves as the regulatory boundary between the cytoplasm and the environment. Potassium, calcium, oxygen gas, and carbon dioxide diffuse across the cell membrane.

Most protozoa have physiological mechanisms for monitoring and responding to the environment. For example, many free-living forms have **eyespots,** localized regions of pigments that detect changes in the quantity and quality of light. Protozoa also sense touch and chemical changes in their environments. For instance, when they encounter a noxious chemical, some protozoa back up and try to bypass the chemical.

Section Review

1. What is an *eyespot*?
2. Briefly summarize the theory of endosymbiosis.
3. List the four phyla of protozoa, and describe how members of each move.
4. How do O_2 and CO_2 enter and leave protozoa?
5. What might be some possible ecological roles of protozoa?

21.2 Protozoan Phyla

As you read about the diversity of protozoa in the four phyla, keep in mind that each species is adapted to either a free-living or a parasitic life-style. All are heterotrophic; most are unicellular. All have cellular mechanisms that monitor the environment.

Sarcodina

Biologists have identified about 40,000 species of the Phylum Sarcodina. The most familiar sarcodines are members of the freshwater genus *Ameba*. Amebas are bottom-dwelling scavengers; they feed on decaying organic matter in rivers, streams, and lakes.

Most sarcodines have flexible cell membranes. Many do not have any added protective covering. However, some, such as the marine sarcodines of the genus *Foraminifera*, have calcium carbonate shells with spikelike protrusions. Others, such as the marine sarcodines of the genus *Radiolaria*, have supportive silicon dioxide inside their shells. Freshwater sarcodines called heliozoans have a shell with thin, extended projections.

Movement, Structure, and Nutrition

Sarcodines move by means of pseudopodia. **Pseudopodia** *(SOO-duh-POE-dee-uh)* are cytoplasmic extensions that function in movement. The cytoplasm of a sarcodine is made up of two regions—the ectoplasm and the endoplasm. The **ectoplasm** is the thin, slippery colloidal sol directly inside the cell membrane. The **endoplasm** is the colloidal sol and gel found in the interior of the cell. When movement begins, the endoplasm pushes outward, facilitated by the slippery ectoplasm, and becomes distinguishable as a pseudopodium. At the same time, previously formed pseudopodia are retracted. Hence the sarcodine moves forward by **ameboid movement,** as shown in Figure 21-3. Ameboid movement is a form of **cytoplasmic streaming,** the internal flowing of the contents of a cell.

Section Objectives

- *Compare and contrast methods of movement in the four phyla of protozoa.*

- *Describe binary fission in amebas and ciliates.*

- *Describe the process of conjugation in ciliates.*

- *List and explain two diseases that are caused by zoomastiginoids.*

- *List and explain two diseases caused by sporozoans.*

Info: The white cliffs of Dover, England, are actually the chalky deposits of ancient foraminiferan calcium carbonate shells.

Figure 21-3 When sarcodines move, the endoplasm and ectoplasm push out to form pseudopodia. Meanwhile, previously formed pseudopodia retract.

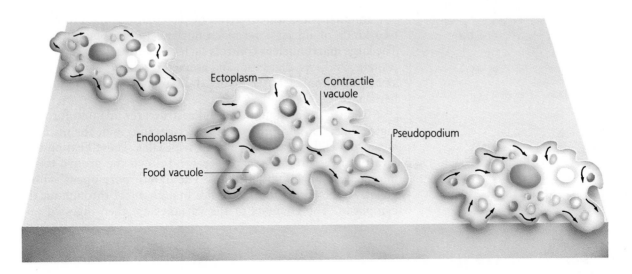

Tip: Discuss with students the various forms of protist movement: flagellar, ameboid, and ciliary.

Figure 21-4 A paramecium moves by means of cilia that beat in waves. Water currents created by cilia that line the oral groove move food down the groove to the mouth pore.

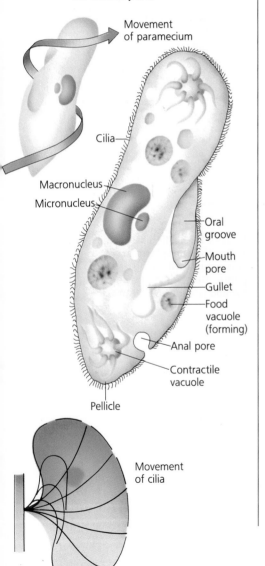

Movement of paramecium

Cilia

Macronucleus

Micronucleus

Oral groove

Mouth pore

Gullet

Food vacuole (forming)

Anal pore

Contractile vacuole

Pellicle

Movement of cilia

Most freshwater sarcodines are internally similar to the ameba in Figure 21-3. Notice the **contractile vacuole,** an organelle that excretes water. Freshwater organisms are usually hypertonic relative to their environment, and water diffuses into them. To maintain homeostasis, many freshwater unicellular organisms have contractile vacuoles that excrete excess water.

Sarcodines absorb many nutrients by diffusion from the surrounding water and ingest nutrients by **phagocytosis,** the engulfing of food. When a sarcodine contacts food, it surrounds the food with pseudopodia. A portion of the cell membrane then pinches together and surrounds the food in a **food vacuole.** Enzymes from the cytoplasm enter the vacuole and digest the food. Any undigested food leaves the cell in a reverse process that is known as exocytosis.

Reproduction

Sarcodines reproduce by binary fission, an asexual, mitotic division that produces identical offspring. Under ideal conditions it takes less than an hour for an ameba to undergo binary fission. When conditions are unfavorable, most sarcodines form **cysts,** protective outer walls. When conditions improve, the cell breaks out of the cyst and resumes normal activities.

Ciliophora

The 8,000 species of Phylum Ciliophora move by means of **cilia,** short, hairlike projections that line the cell membrane and beat in synchronized strokes. Ciliates live in marine and freshwater habitats. Members of the freshwater genus *Paramecium,* shown in Figure 21-4, are the most thoroughly studied of the ciliates.

Structure, Nutrition, and Responses

A paramecium never changes shape like an ameba because it is surrounded by a rigid protein covering, the **pellicle.** The pellicle is covered with thousands of cilia arranged in rows. The cilia beat in waves. Each wave passes slantwise across the long axis of the body of the paramecium, causing it to rotate as it moves forward.

Internally a distinctive trait of the ciliates is the presence of two kinds of nuclei—the macronucleus and the micronucleus. The large **macronucleus** controls such cell activities as respiration, protein synthesis, digestion, and asexual reproduction. The much smaller **micronucleus** is involved in sexual reproduction and heredity.

Ciliates have numerous cellular structures adapted for feeding on bacteria and other protists. A paramecium, for instance, has a funnellike **oral groove** lined with cilia. These beating cilia create water currents that sweep food down the groove to the **mouth pore** of the paramecium. The mouth pore opens onto a **gullet,** which forms food vacuoles that circulate throughout the cytoplasm. The contents of the vacuoles are then digested and

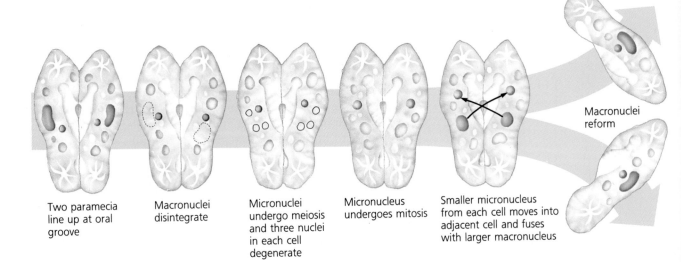

| Two paramecia line up at oral groove | Macronuclei disintegrate | Micronuclei undergo meiosis and three nuclei in each cell degenerate | Micronucleus undergoes mitosis | Smaller micronucleus from each cell moves into adjacent cell and fuses with larger macronucleus | Macronuclei reform |

absorbed. The indigestible matter remaining in the vacuole moves to the **anal pore,** an opening where waste is eliminated.

Most ciliates exhibit **avoidance behavior,** a reaction away from a potentially harmful situation. A paramecium will back up and move away when it encounters an obstacle in its environment.

Reproduction

Asexual reproduction in ciliates occurs primarily by binary fission. In this process only the micronucleus divides by mitosis. The macronucleus, which contains up to 500 times more DNA than the micronucleus, simply elongates and splits into two parts, each part going to a daughter cell.

Sexual reproduction occurs in many ciliates by a process known as **conjugation.** Conjugation involves individuals from two mating strains, such as the two paramecia in Figure 21-5. The two paramecia lie next to each other. Then the macronucleus of each individual disintegrates. Each diploid micronucleus then undergoes meiosis, producing four haploid micronuclei. In each cell three of these disappear; the fourth moves to the oral groove. Here it undergoes mitosis, producing two haploid micronuclei of unequal size. The smaller micronucleus from one paramecium then exchanges places with the smaller micronucleus from the other paramecium. Each small micronucleus then fuses with each larger micronucleus, forming diploid micronuclei. The two paramecia separate, and macronuclei form again.

Zoomastigina

The 2,500 species of Phylum Zoomastigina, also called Mastigophora, are characterized by the presence of one or more long **flagella.** The undulations of whiplike flagella push or pull the protozoan through the water. Many zoomastiginoids are free-living, moving through lakes or ponds and feeding on small organisms. Some zoomastiginoids are parasites. Among these are members of the genera *Trypanosoma, Leishmania,* and *Giardia.*

Figure 21-5 Paramecia reproduce by conjugation, a form of sexual reproduction in which genetic material is exchanged by two mating organisms.

Think

Based on what you learned in Chapter 10, what is the adaptive significance of conjugation?

Answer to Think
Conjugation results in recombination of genes, creating phenotypic variation in offspring that can be acted on by natural selection.

Trypanosoma

Several members of the genus *Trypanosoma* are powerful agents of disease. Trypanosomes are slender, elongate, flattened protozoa with one posterior flagellum. They live in the blood of their hosts, which include humans and other animals, and are carried from host to host by bloodsucking invertebrates such as flies.

African **trypanosomiasis,** or sleeping sickness, is a trypanosome disease that occurs in two forms—Gambian, caused by *T. gambiense,* and Rhodesian, caused by *T. rhodesiense.* Both are transmitted by the tsetse fly, which lives only in Africa. Trypanosomiasis is characterized by fever and swollen lymph nodes. In later stages the parasites invade the brain, causing uncontrolled sleepiness. *T. gambiense* may take several years to invade the brain. *T. rhodesiense* affects the host more quickly, and Rhodesian trypanosomiasis is usually fatal.

Leishmania and Giardia

Zoomastiginoids of the genera *Leishmania* and *Giardia* cause serious diseases in humans. *Leishmania,* carried by sand flies, cause leishmaniasis, which afflicts millions of people in Africa,

Biotechnology

Controlling Chagas Disease

Chagas disease, or South American trypanosomiasis, is caused by *Trypanosoma cruzi.* The disease is usually transmitted to humans through the bites of conenose bugs. These insects, also called kissing bugs, live in the crevices of mud walls and thatch roofs. Mud and thatch dwellings are common in rural areas of Latin America. The disease can also be transmitted through contaminated blood in a transfusion.

From southern Mexico to northern Argentina 35 million people are exposed to South American trypanosomiasis. Of these, 10 to 15 million become infected.

The acute phase of the disease is most commonly found in children. Symptoms include facial swelling, called chagoma, around the area of the insect bite; fever; and swollen lymph nodes. The bite frequently occurs on the eyelid, causing it to swell shut.

The chronic phase of the disease, which is more prevalent in men, causes severe heart damage. Often after years of carrying the disease, the victims suddenly die of heart failure.

The World Health Organization (WHO) works throughout Latin America to combat Chagas

Asia, and Latin America. The disease is characterized by disfiguring skin sores and may be fatal.

Giardia, carried by muskrats and beavers, cause giardiasis, a disease characterized by fatigue, diarrhea, cramps, and weight loss. Drinking water that is contaminated with *Giardia* will not affect muskrats and beavers, but it can harm humans.

Sporozoa

All 6,000 species in the Phylum Sporozoa have adults with no means of locomotion. All are parasitic. They are carried in the blood and other bodily fluids of their hosts, absorbing nutrients from the fluid in which they float. Several sporozoans, including *Toxoplasma* and *Plasmodium,* cause diseases in humans.

Toxoplasma

The sporozoan *Toxoplasma* is a parasite found in animals. In humans it causes toxoplasmosis, a disease that has few or no symptoms in adults with healthy immune systems but that is often fatal to newborns. One species of *Toxoplasma* causes coccidiosis, a deadly disease affecting birds and young cattle.

disease. WHO has a three-part agenda: (1) controlling the kissing bug by means of insecticides, (2) modifying houses to make them unsuitable for the bugs, and (3) screening blood at blood banks to halt the spread of the disease.

Over 15 years ago WHO began helping countries establish national insect control programs. It is only through coordinated efforts, such as those in Argentina, Brazil, and Venezuela, that a dent can be made in kissing bug populations.

Programs in areas in which Chagas disease is endemic call for spraying houses with insecticides every three to six months, a successful but expensive approach. In 1982, for example, Argentina spent $14 million on spraying.

WHO also conducts pilot projects throughout Latin America to help rural people modify their houses so as to eliminate the kissing bugs. Modification may include plastering mud walls, replacing thatch with metal roofing, and putting in cement floors. For four years after being modified one group of Venezuelan houses remained free of kissing bugs without repeated applications of insecticides.

Since blood transfusions are another way Chagas disease can be spread, many affected countries are initiating the testing of blood already in blood banks and of blood donors to ensure they are not carrying the disease.

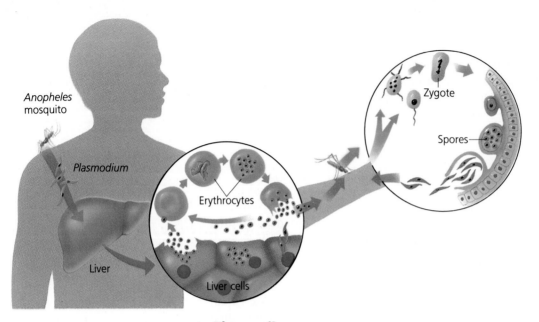

Anopheles
mosquito

Plasmodium

Liver

Erythrocytes

Liver cells

Zygote

Spores

Figure 21-6 The life cycle of
***Plasmodium*, which causes ma-**
laria, involves more than one
host and asexual and sexual re-
production stages.

Answers to Section Review
1. Ectoplasm is a sol found directly
inside the cell membrane; endoplasm is a sol and gel found in the
interior of the cell.
2. by means of pseudopodia
3. The micronucleus undergoes meiosis, producing haploid micronuclei,
one of which produces more micronuclei mitotically. Eventually the
micronuclei of different organisms
are exchanged and fused.
4. two genetically identical offspring
5. by interrupting the life cycle of
Plasmodium through the use of insecticides to control *Anopheles*
mosquitoes

Plasmodium

Members of the genus *Plasmodium* cause malaria, a disease that kills about 1 million people a year and that is most prevalent in the tropics. Like most sporozoans *Plasmodium* has a complex life cycle with more than one host. As Figure 21-6 shows, when a female *Anopheles* mosquito that is carrying *Plasmodium* spores bites a person, the spores enter the person's bloodstream. They travel to the liver, where they reproduce asexually. The new spores infect erythrocytes and continue asexual reproduction. Every two to three days spores burst out of the erythrocytes and release toxins into the blood. Some of the spores develop into plus and minus cells.

When a female *Anopheles* bites the infected person, it ingests these cells. In the digestive system of the mosquito the cells develop into gametes that combine to form a zygote. The nucleus of the zygote divides internally many times and forms internal spores. When the zygote membrane bursts, the spores migrate to the salivary glands of the mosquito. If the mosquito then bites another person, the cycle begins again.

The destruction of red corpuscles and the release of toxins into the blood cause the fever, anemia, and other symptoms of malaria. The disease causes the liver and spleen to become enlarged, and in the most acute cases the kidneys fail.

Section Review

1. What is the difference between ectoplasm and endoplasm?
2. How do sarcodines move?
3. How does the micronucleus behave in ciliate conjugation?
4. What is the result of binary fission?
5. How might health workers attempt to control malaria?

Laboratory

Observing Ciliates

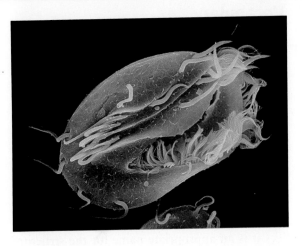

Objective
To observe some ciliate characteristics

Process Skills
Observing, inferring

Materials
4 microscope slides; glass-marking pencil; 7 pipettes; cultures of *Euplotes, Paramecium, Stentor,* and *Vorticella;* 4 coverslips; compound microscope; distilled water; paper towel; white vinegar; prepared suspension of *Euglena gracilis*

Method
1 Label four microscope slides as follows: *E* for *Euplotes, P* for *Paramecium, S* for *Stentor,* and *V* for *Vorticella.*
2 Using a pipette, place a small drop from the bottom of each culture on the appropriate slide. Use a different pipette for each culture. Be sure that each slide contains several protozoa. You should be able to see the organisms without a microscope.
3 Examine each of these organisms under low power. Note the cilia. Notice how each organism moves. Are any of these organisms colonial?
4 Search for contractile vacuoles. What is the function of contractile vacuoles? Do all contract at the same rate?
5 With a clean pipette, place a drop of distilled water at the edge of one of the slides. Does the organism's rate of contraction change? To answer the question, you may have to slow down the movement of the protozoa. To do this remove some fluid from beneath the coverslip by placing a piece of paper towel near the edge of the coverslip opposite the side where you added distilled water.
6 Some protozoa have special structures called trichocysts that they can discharge from

their bodies for defense or for attachment to objects. While observing each ciliate under the microscope, use a clean pipette to place a drop of vinegar at the edge of the coverslip. Which of the ciliates have trichocysts?
7 Prepare another wet mount using a drop of *Stentor* culture and a drop of the prepared *Euglena gracilis* suspension. Examine the wet mount under low power. Congo Red–stained yeast and very fine activated charcoal have been added to the suspension. Can the *Stentor* discriminate between the charcoal, which it cannot use for food, and the yeast, which is a food material?
8 Congo Red dye is red at pH 5 and blue at pH 3. Watch the color of the yeast as it is eaten and note any change. Why does the color change from red to blue?

Conclusions
1 Do all of the organisms have the same kind of cilia?
2 Which organisms are motile and which are sessile?
3 Why do freshwater protozoa need to excrete water continually?

Inquiry
Devise an experiment that will test whether water temperature has an effect on the excretion rate of contractile vacuoles.

Vocabulary

ameboid movement (311)	contractile vacuole (312)	eyespot (310)	oral groove (312)
anal pore (313)	cyst (312)	flagellum (313)	pellicle (312)
avoidance behavior (313)	cytoplasmic streaming (311)	food vacuole (312)	phagocytosis (312)
cilia (312)	ectoplasm (311)	gullet (312)	pseudopodium (311)
conjugation (313)	endoplasm (311)	macronucleus (312)	trypanosomiasis (314)
	endosymbiosis (309)	micronucleus (312)	zooplankton (310)
		mouth pore (312)	

1. Look up the meaning of the word *pseudopodium* in a dictionary, and explain why it is an appropriate name for the structure it represents.
2. Distinguish between the two regions of cytoplasm that are known as ectoplasm and endoplasm.
3. Compare and contrast the functions of a food vacuole and a contractile vacuole in a sarcodine.
4. Look up the word *zooplankton* in a dictionary, and explain the relationships between its word roots.
5. What is the name for the response that protozoa exhibit when trying to escape predators?

Review

1. The theory of endosymbiosis suggests that cell organelles of eukaryotic organisms evolved from (a) free-living protozoa (b) trypanosomas (c) parasitic prokaryotes (d) euglenoids.
2. A key characteristic of protozoan habitats is that at some point during the life cycle they must include (a) algae (b) water (c) blood (d) soil.
3. Many protozoa monitor light intensity with (a) pseudopodia (b) eyespots (c) cilia (d) contractile vacuoles.
4. Pseudopodium consist primarily of (a) a pellicle (b) endoplasm (c) ectoplasm (d) cyst cells.
5. The pellicle is a characteristic of the (a) zoomastiginoids (b) sarcodines (c) ciliates (d) sporozoans.
6. Ciliates form food vacuoles in a (a) mouth pore (b) contractile vacuole (c) oral groove (d) gullet.
7. Flagella are characteristic of the Phylum (a) Zoomastigina (b) Sarcodina (c) Sporozoa (d) Ciliophora.
8. African trypanosomiasis or sleeping sickness is transmitted by (a) tsetse flies (b) *Anopheles* mosquitoes (c) kissing bugs (d) muskrats.
9. Members of the genus *Plasmodium* cause (a) toxoplasmosis (b) malaria (c) giardiasis (d) coccidiosis.
10. Chagas disease is caused by (a) *Trypanosoma cruzi* (b) *Trypanosoma rhodesiense* (c) *Trypanosoma conorhini* (d) *Trypanosoma gambiense*.
11. Describe the range of variation in types of protective covering among members of the Phylum Sarcodina.
12. Describe the process called ameboid movement.
13. What adaptive significance does the contractile vacuole have in the freshwater sarcodine?
14. Describe the process of phagocytosis in a sarcodine.
15. What is the behavior of the macronucleus and the micronucleus of the two individuals in ciliate conjugation?

16. How does a ciliate digest food?
17. How is the disease known as giardiasis spread to humans?
18. What is the characteristic of the Phylum Sporozoa that distinguishes it from the three other protozoan phyla?

19. Draw a simplified life cycle of a malaria-causing *Plasmodium,* showing the protozoan activity in both the mosquito and the human hosts.
20. What are the symptoms of malaria?

Critical Thinking

1. *Euglena* are often used in experiments in high school laboratories. In view of the fact that *Euglena* can be autotrophic, why is this organism a good choice for use in the laboratory?
2. Many scientists suggest that paramecia are more complex than amebas. What adaptations in paramecia would justify such a classification?
3. Look at the photograph at the right, which shows the shells of several radiolarians, protozoa found in the ocean. What is the function of such shells?
4. Many protozoa feed by means of pseudopodia. What stimulus might cause these cytoplasmic extensions to form in the unicellular animals?

5. Why is movement by means of cilia more efficient than movement by means of pseudopodia?

Extension

1. Read the article by Dean Baker entitled "Giardia!" in *National Wildlife,* August/September 1985, pp. 19–21. What health problems do *Giardia* cause in humans? How do *Giardia* enter the human body?
2. Write a brief research paper on the protozoa found as zooplankton in the ocean. Include in your report information about the types of protozoa found in zooplankton and about the species of fish that depend on these protozoa as a food source.
3. Collect water from at least three sources—ponds, lakes, tap, or ditches—and look at the samples under a microscope. Count the different kinds of protozoa in each sample, and draw a picture of all the types you see.
4. Obtain a culture of paramecia or *Euglena*

from a biological supply house. Place the protozoa in a depression slide, and test their reactions to different chemicals. Dip a thread into a salt solution, a sugar solution, and ammonia and then hold the thread in the culture. See what reaction takes place. Next cover part of the culture with a piece of cardboard, and observe the reaction. Write a report describing these experiments. Include a title, description of the procedure used, and the results obtained.
5. In chart form compare conjugation and binary fission as forms of reproduction. Include headings such as: protozoa with these forms of reproduction, sexual or asexual, genetic variation, and so forth.

Chapter 22 | Algae

This chapter examines the different types of algae. Students learn that algae are characterized by widely diverse modes of reproduction that are adaptations to their environments. Students view these adaptations as possible phylogenetic clues.

Correlation Chart

Section Number	Section Objectives	Pages	Features	Section Review
22.1	1. Explain why algae are classified in the Kingdom Protista. 2. Describe the three parts of a thallus. 3. Distinguish unicellular, colonial, and filamentous algae. 4. List the six divisions of algae and the characteristics of each. 5. Describe the uses of aquaculture.	321–327	Think, p. 322 Biotechnology, pp. 324–325	1 3 1, 2, 5 4, 5
22.2	1. Describe asexual and sexual reproduction in unicellular algae. 2. Explain conjugation in filamentous algae. 3. Distinguish between antheridia and oogonia. 4. Describe the two main stages in alternation of generations.	328–331	Biology Lab, p. 331	2, 3, 5 4 1

Resources*

Print

Bold, Harold C., and Michael J. Wynne. *Introduction to Algae: Structure and Reproduction*, 2nd ed. Englewood Cliffs, N.J.: Prentice-Hall, 1985.

Van Iterson, Jr., G., L.E. Den Dooren De Jong, and A.J. Kluyver. *Martinus Willem Beijerinck: His Life and His Work*. Madison, Wisc.: Science Tech., 1983.

Audiovisual/Software

Making Seaweed Worth Eating. Still-frame video or transparencies with cassette, 22 min. Teacher's guide. Biology Media.

Mariculture of Porphyra. Still-frame video or transparencies with cassette, 21 min. Teacher's guide. Biology Media.

Simple Plants: The Algae. Film, 18 min. Britannica.

*Additional references are listed in *Modern Biology, Teacher's Program Guide*.

Section Teaching Guides

22.1 Structure and Classification

Approach

The main objective of this section is to familiarize students with the similarities and differences among algal protists. It builds on concepts of basic chemistry (Chapters 3 and 4), cell structure (Chapter 5), metabolism (Chapter 7), and classification (Chapter 18), and prepares for future concept presentation of plant diversity (Chapter 25). In presenting the text, guide students to develop an appreciation for variations apparent among algae. Make sure that students also gain an appreciation for the varying complexity of unicellular, filamentous, colonial, and thalloid algae. You may wish to use Table 22-1 to summarize the key criteria used in classifying members of each algal division.

Concept Framework

Students are aware of the diversity among mammals, one group in the Kingdom Animalia. To build on this knowledge and provide students with a perspective with which to approach the study of diversity among algae, have the class generate a list of five to ten extremely diverse mammals. Then ask students to identify the traits the mammals on the list have in common. Elicit the generalization that the classification of diverse organisms into one group often rests on several fundamental characteristics shared by members of the group. You may wish to end the discussion by asking students to consider, as they read the text, the usefulness of the criteria scientists have chosen to consider in classifying algae.

Emphasis

Make certain that students understand that all organisms referred to as algae are autotrophic, unlike the protozoa and funguslike protists. This is why algae are sometimes referred to as the plantlike protists.

Classification Exercise

Show the class slides or photographs of *Ulva* (Division Chlorophyta), *Macrocystis* or *Fucus* (Division Phaeophyta), and *Porphyra* or *Chondrus* (Division Rhodophyta). You may use other multicellular algae from these three divisions if appropriate. Have students generate a list of the similarities and differences they are able to observe from the illustrations among these multicellular algae. As students read the text, ask them to take note of key features of thalli of multicellular algae, along with differences in photosynthetic pigments and storage substances among the divisions.

Using the Features/Harvest from the Sea

The main techniques used in "Harvest from the Sea" are related to aquaculture. In 1943 a British researcher helped devise a means of "seeding" nets that would produce edible algae. This technique could be managed by humans to improve yields from oceanic crops. In addition, through genetic engineering, algae are being developed that can tolerate more extreme environments than they presently do, and may have more nutritive value. Such aquacultural techniques can be used in coastal regions around the world to produce additional food for humans and other animals.

Section Review Answers

(See p. 327.)

22.2 Comparative Reproduction

Approach

The main objective of this section is to provide students with a foundation for understanding increasingly complex and diverse reproductive mechanisms. The material in the section builds upon concepts of mitosis and meiosis (Chapter 9) and prepares for future concept presentations of reproduction in fungi (Chapter 23) and of alternation of generations in plants (Chapters 25 and 27). In presenting the text, guide students to an understanding of the key reproductive differences between unicellular and multicellular algae. You may wish to ask students to consider whether multicellularity actually necessitates more complex reproductive mechanisms.

Concept Framework

Most students will have observed growths of algae (or funguslike protists) in stagnant puddles or ponds.

To build on this experience and to provide a starting point for discussing algal reproductive diversity, engage students in a discussion of the possible environmental conditions that support rapid population growth. Elicit the observation that nutrient-rich environments will support rapid population growth. You may wish to ask students to speculate on the importance of water and the nutrients it contains in the reproductive processes in various groups of algae.

Emphasis

Make certain that students understand the adaptive advantages of asexual and sexual reproduction, as well as the possible advantages of being able to reproduce in both ways under different environmental conditions. You can use this emphasis to lead into the discussion of alternation of generations in *Ulva*.

Class Exhibit

Have students draw a generalized diagram of alternation of generations to keep on the wall and use in discussing fungi and plants.

Section Review Answers

1. In *Oedogonium,* a filamentous green alga, the oogonium is the female gametangium and produces eggs. The antheridium is the male gametangium and produces sperm.
2. In asexual reproduction in unicellular algae, as well as in sexual reproduction in *Oedogonium* and *Ulva,* the haploid zoospores grow into asexual, haploid new individuals.
3. At a certain critical point, asexually reproducing diatoms shed their small shells and grow to full size, then generate another shell.
4. *Spirogyra* reproduces sexually through conjugation, in which a gamete in one filament moves to another filament through a conjugation tube.
5. Sexual reproduction increases genetic variability in a species, giving the species an advantage in changing environments.

Laboratory Preparation

Background

In this lab students examine a variety of algae that demonstrate early evidence of specialization and division of labor among cells. Students should be aware that the algae are all members of the same order, the Volvocales, and represent an attempt by nature to develop motile, multicellular plants. By using the key, students can identify the organisms and indicate which algae show evidence of specialization. Students may see reproductive activity and, consequently, evidence of isogamy, heterogamy, or oogamy. Students can observe flagella if the diaphragm is closed down to provide contrast. Students should be reminded that these organisms vary in size, and it is important not to crush *Volvox* while examining other forms.

Preparation

Algae can be collected locally from streams, ponds, and fish hatcheries. Thoroughly rinse the algae to remove debris and small organisms, using tap water that has been standing at room temperature for at least 24 hours. Loosely place algae in jars containing aged tap water. Add a 0.1% solution of commercial 1-10-5 fertilizer to the jars. Store uncovered in a cool room beneath fluorescent lights.

Conclusions

1. Flagella become coordinated, and cytoplasmic connections appear between cells, indicating that communication is occurring.
2. Yes. Cells begin to differ in size as the colony size increases.

Inquiry

Students will design a method to study the evolutionary relationships among algae, such as by studying the amino acid sequences of specific proteins present in each genus.

Chapter Review

Vocabulary

1. Pyrenoid; laminarin and carageenan are components of algal cell walls.
2. Sporangium; the oogonium and antheridium produce gametes.
3. Valve; the blade is part of the thallus in multicellular algae.
4. Blade; it is a structural feature of brown algae, while red tied and bioluminescence are features related to dinoflagellates.

5. Valve; the holdfast and stipe are part of a multicellular thallus.

Review

11. Members of the Division Pyrrophyta have members that are bioluminescent.
12. Algal classification involves consideration color, food storage, and cell-wall structure.
13. Members of the Division Euglenophyta do not have cell walls.
14. The blade is specialized for photosynthesis.
15. Laminarin is an oily, not starchy, carbohydrate.
16. Colonial algae are groups of unicellular organisms that act in a coordinated manner. Filamentous algae are linear arangements of single cells.
17. Accessory pigments include chlorophylls a, b, c, and d.
18. Environmental stress usually triggers sexual reproduction in algae.
19. *Euglena* can be heterotrophic, like protozoa; they have a contractile vacuole, as do protozoa, which rids the organism of excess water, and they have no cell wall.
20. The similar reproductive patterns in green algae and plants suggest a phylogenetic relationship.

Critical Thinking

1. Living things exhibit a continuous variation and often do not "naturally" fall into neat groups.
2. May, June, July, and August. During these warmer months the number of *Gonyaulax* increase. Shellfish feed by filtering these organisms and thus may accumulate the toxin and become poisonous.
3. This common feature is another example of convergent evolution. The two genera are not closely related.
4. Because it lacks a cell wall, *Euglena* has no mechanical support to help prevent the cell membrane from bursting when excess water enters the cell. Therefore the contractile vacuole must pump out the excess water.
5. You would be suprised to find such a specimen because the function of an eyespot is to help direct the organism to light for photosynthesis.
6. In many species of *Spirogyra* sexual reproduction is directly related to the ratio betwen the cell's surface area and its volume. This is somewhat influenced by environmental conditions such as temperature.

7. Some possible ways the diatom "knows" to divide sexually are by the ratio of its surface to its volume or by the ratio of the size of its nucleus to the size of its cytoplasm.
8. The cells of one filament all have the same genetic makeup. The cells of two different filaments may have come from different zygospores and thus have different genetic compositions. Thus conjugation between them will yield more genetic variation.

Chapter 22 | *Algae*

Introduction

Algae are diverse organisms ranging from microscopic, flagellated single cells to giant kelps anchored to rocks by huge rootlike structures. Algae belong to the Kingdom Protista, along with the protozoa. Unlike protozoa, however, algae can make their own food by photosynthesis. In this chapter you will explore the different types of algae. You will also see that algae are characterized by widely diverse structures and methods of reproduction.

Chapter Outline

22.1 Structure and Classification
- *Structure*
- *Classification*

22.2 Comparative Reproduction
- *Unicellular Reproduction*
- *Multicellular Reproduction*

Chapter Concept

As you read, notice that algae are actually a widely diverse group of organisms that vary in structure, biochemistry, and reproduction.

Kelp (*Macrocystis porifera*), California. Inset: Color-enhanced SEM of a dinoflagellate (×100).

For additional teaching information see pp. T22A–T22D.

22.1 Structure and Classification

Algae are a diverse group of eukaryotic, plantlike organisms. They are classified into six divisions of the Kingdom Protista. Algae are autotrophic protists—that is, they have chloroplasts and produce their own food by photosynthesis. In the past some biologists classified algae as plants. Today they do not. One reason is that algae and plants have different methods of reproduction. Algae have gametes that are formed in and protected by **unicellular gametangia** (GAM-uh-TAN-jee-uh), or single-celled gamete holders. Plants have gametes formed in multicellular gametangia.

Although algae form a very diverse group of protists, all algae have several features in common. For example, algal cells often have **pyrenoids** (PIE-ruh-NOIDZ), organelles that synthesize and store starch. In addition, almost all algae are aquatic, and even terrestrial forms require H_2O for reproduction. Many aquatic algae also possess flagella.

Structure

The body of an alga is called a **thallus** (THAL-us). The thallus can be unicellular, colonial, filamentous, or thalloid. Colonial algae are groups of independent cells that move and function as a unit. Filamentous algae consist of cells in a linear arrangement. Thalloid algae are organisms in which cells divide in many directions to create a body that is multicellular and often modified into rootlike, stemlike, or leaflike parts.

Unicellular algae are mostly aquatic. Organisms thus adapted are called plankton. Photosynthetic microorganisms are called **phytoplankton** (FIE-toe-PLANK-tun). Phytoplankton provide food for numerous aquatic organisms. They also generate a great amount of the oxygen we breathe.

Colonial algae are groups of individual algal cells. In colonies the algal cells act in a coordinated manner. Certain cells may become specialized for certain functions. This division of labor allows colonial algae to move, feed, and reproduce efficiently.

A filamentous alga is composed of a row of cells. Some filamentous algae have structures that anchor them to the ocean bottom, where they can exploit an environment inhospitable to many other organisms. Some have branching filaments.

Multicellular thalloid algae are not organized into specialized tissues but can often be very large and outwardly complex. Some algae grow as a complex thallus and are referred to as seaweeds. For example *Ulva*, commonly known as sea lettuce, has a leaflike thallus that may be several centimeters wide but only two cells thick. *Macrocystis*, called the giant kelp, has a rubbery leaflike portion, stemlike areas, and enlarged air bladders.

Section Objectives

- *Explain why algae are classified in the Kingdom Protista.*
- *Describe the three parts of a thallus.*
- *Distinguish unicellular, colonial, and filamentous algae.*
- *List the six divisions of algae and the characteristics of each.*
- *Describe the uses of aquaculture.*

Tip: Place a bowl of tap water aside for several days. Then pour out the water and have students feel the bottom of the bowl for a slimy residue. If possible, examine a specimen from the bottom of the bowl under the microscope.

Figure 22-1 *Volvox* is a colonial green alga that exhibits division of labor between cells.

Think

Many multicellular algae have gas-filled structures called bladders in or beneath their leaflike regions. What sort of adaptation might this be?

Answer to Think
The gas-filled bladders are adaptations for flotation. They keep the leaflike portions of the thallus on top of the water, where sunlight is available for photosynthesis.

Classification

Biologists classify algae into six divisions, based on color, food-storage substances, and the composition of cell walls. Members of each division have distinctive colors, depending on the photosynthetic pigments in their cells. These pigments absorb light. All algae contain the pigment chlorophyll a. However, different divisions of algae also contain other forms of chlorophyll, such as b, c, or d, each of which absorbs a different wavelength of light. Members of different algal divisions have different accessory pigments as well. Algae also vary in methods of reproduction, as you will learn in Section 22.2. As you read the descriptions of the divisions, refer to Table 22-1.

Chlorophyta

Members of the Division Chlorophyta *(kloe-RAHF-uh-tuh)*, the green algae, are a diverse group of organisms of over 7,000 species. They can be unicellular, colonial, filamentous, or thalloid. Most green algae are aquatic, although many inhabit moist terrestrial environments such as soil, rock surfaces, and tree trunks.

Three observations have led biologists to conclude that green algae are the ancestors of plants. First, both green algae and plants have chloroplasts that contain chlorophylls a and b. Second, both green algae and plants store food as starch. Finally, both green algae and plants have cell walls made of cellulose.

Table 22-1 Six Divisions of Algae

Division	Cell Types	Photosynthetic Pigments	Form of Food Storage	Cell Wall Composition
Chlorophyta (*Chloro* = "green"): 7,000 species	Both unicellular and multicellular	Chlorophylls a and b, xanthophylls, carotenes	Starch	Polysaccharides, sometimes cellulose
Phaeophyta (*Phaeos* = "brown"): 1,500 species	Mostly multicellular	Chlorophylls a and c, fucoxanthin (a carotene)	Laminarin (oil)	Cellulose with alginic acids
Rhodophyta (*Rhodo* = "red"): 4,000 species	Mostly multicellular	Chlorophyll a and d, phycobilins	Starch	Cellulose or pectin, many with calcium carbonate
Chrysophyta (*Chryso* = "gold"): 10,000 species	Mostly unicellular	Chlorophylls a and c, carotenes, including fucoxanthin	Chrysolaminarin (oily carbohydrate)	Cellulose, some with silica; possibly no cell wall
Pyrrophyta (*Pyrro* = "fire"): 1,000 species	All unicellular	Chlorophylls a and c, carotenes	Starch	Cellulose
Euglenophyta (*Euglena* = "true eye"): 800 species	Mostly unicellular	Chlorophylls a and b, carotenes in genera with chloroplasts	Paramylon (a starch)	No cell wall; protein-rich pellicle

Figure 22-2 *Chlamydomonas* (left) and desmid (right) illustrate the diversity of shape in green algae.

To understand some of the diversity in unicellular green algae, examine Figure 22-2. *Chlamydomonas* is common in soils and in freshwater ponds and streams. It has a single cup-shaped chloroplast. Each chloroplast contains a pyrenoid, where starch is made. Two anterior flagella enable the organism to swim. An eyespot, which is an area sensitive to light, enables the alga to move either toward or away from light. Desmids are unusual unicellular algae that live primarily in fresh water. In fact, the presence of desmids often indicates degree of water pollution. As you can see, each cell is divided into halves, which are called semicells.

Unlike unicellular algae, colonial algae have some characteristics of multicellular organisms. *Gonium* is perhaps the simplest colonial green alga, consisting of a colony that is one cell thick and shaped in a rectangle. A *Volvox* colony, on the other hand, is round and much larger, containing up to 60,000 cells and exhibiting a remarkable division of labor. Intercellular communication allows the coordination of the many cells in the colony. *Volvox* cells are connected by fine cytoplasmic strands that enable adjacent cells to chemically communicate with each other.

Spirogyra is a filamentous green alga with unusual spiral chloroplasts that stretch from one end of the cell to the other. Examine the *Spirogyra* in Figure 22-3. Can you locate the dotlike pyrenoids associated with the chloroplasts? *Oedogonium* is another common freshwater filamentous green alga. Members of this genus have netlike chloroplasts.

Ulva has a leaflike, photosynthetic body and commonly grows on rocks and pilings. Its thallus collapses during low tide to prevent water loss in the intertidal zone, the area between high and low tides.

Figure 22-3 Notice how the chloroplasts in *Spirogyra* form spirals (×50).

Phaeophyta

Members of the Division Phaeophyta *(fee-AHF-uh-tuh)*, the brown algae, are multicellular and usually large. Most of the approximately 1,500 species are marine organisms. Their brown color results from the accessory pigment fucoxanthin. The food they produce is stored as **laminarin** *(lam-uh-NAR-in)*, a carbohydrate with glucose units linked differently from those in starch.

The large brown alga shown in Figure 22-4 is *Macrocystis*. Members of the genus live in the intertidal zone. Individuals may be more than 100 m long. The thallus is composed of a holdfast, a stipe, and blades. The **holdfast** anchors the thallus to a rock. The **stipe** is the stemlike region. Each leaflike **blade** is a region modified for photosynthesis. The cell walls of these brown algae contain alginic acid, a source of commercially important alginates. Alginates are polysaccharides used to make gels for ice cream and other foods.

Rhodophyta

Members of the Division Rhodophyta *(roe-DAHF-uh-tuh)* are called red algae. Most of the approximately 4,000 species are

Figure 22-4 The structure of these brown algae allows them to exist in the intertidal zone.

Biotechnology

Harvest from the Sea

Think for a moment of the number of people inhabiting this planet. As our population grows, so does our demand for food. How can the needs of the human population be met? One solution may be found in the world's oceans.

Although oceans cover three-fourths of the earth's surface, humans obtain 96 percent of their food from the land. The origin of agriculture dates back to around 10,000 B.C. In some coastal areas aquaculture, or farming of the oceans, is almost as ancient.

As early as 1670 the Japanese were harvesting *Porphyra*

tenera, a red alga commonly called nori, from Tokyo Bay. The technique they used was simple. Every fall farmers placed twigs or pieces of bamboo in the bay. Spores landed on the twigs. As the spores divided and the algae grew, the twigs were moved closer to land, where they were more accessible. By winter the algae could be harvested.

For nearly three centuries that same technique was used. In 1943, however, an English botanist, Elizabeth Drew, made a discovery. Drew learned that *Porphyra* releases spores as it dies. During spring and summer the spores enter an asexual marine and multicellular. The multicellular forms are generally less than 1 m long. A few unicellular species inhabit land and freshwater environments. Red algae can survive at greater depths than any other algae can; they commonly grow at depths of 150 m and have been discovered at depths of 268 m. How can they photosynthesize at such depths? Examine Figure 22-5. Red algae contain chlorophylls a and d as well as accessory pigments called phycobilins. Phycobilins absorb the violet, blue, and green light that penetrates the depths at which these algae grow.

The cell walls of red algae contain cellulose and are sometimes coated with a sticky substance called **carageenan** (*KAR-uh-GEE-nun*). Carageenan, which is a polysaccharide, is used to produce cosmetics, gelatin capsules, and some cheeses. Coralline algae, a group of red algae that deposits calcium carbonate in their cell walls, are important components of coral reefs.

Chrysophyta

Members of the Division Chrysophyta (*kruh-SAHF-uh-tuh*) are called golden-brown algae. There are over 10,000 species of golden-brown algae, the majority of which are commonly called diatoms. Chrysophytes contain chlorophylls a and c and the

Figure 22-5   Red algae, such as *Spermothamnion turneri*, are able to synthesize food at a water depth of more than 250 m.

phase of growth. By understanding the alga's life cycle, Drew enabled farmers to begin collecting *Porphyra* spores. Special synthetic nets for spore collecting were developed, as were hatcheries where those nets could be "seeded." By 1960 a method for freezing seeded nets was developed, and unproductive nets could be replaced with fresh ones. As a result nori production increased tremendously.

More recently, through genetic manipulation and selection experiments, scientists have developed new strains of algae. Some of these strains are more resistant to disease and pests; others grow faster and more abundantly. Thus previously unsuitable waters can be used for aquaculture.

Algae harvested from the sea have many uses. In some areas of the world, particularly the Orient, seaweeds are regularly eaten fresh, boiled, stewed, fried, or even cooked into bread. Algae contain protein, iodine, vitamins, and mineral salts.

Algae are also used as fish fodder and as food for cattle, pigs, and chickens, which can digest algal carbohydrates.

Advances in aquaculture and genetic engineering will continue to offer possible new sources of food. Algae and other marine organisms may well provide for our ever-growing human population.

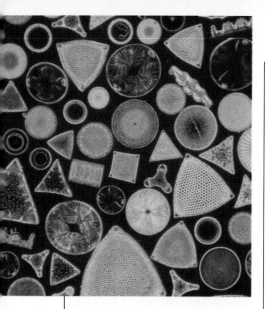

Figure 22-6 The walls of these various diatoms contain silicon dioxide, which makes the shells rigid.

accessory pigment fucoxanthin. Because of the pigment similarities between golden-brown algae and brown algae, scientists suggest that the two divisions have a close evolutionary relationship. Chrysophytes store food in the form of oil, not starch.

Diatoms are unicellular or colonial, nonflagellated, photosynthetic algae with silica-impregnated shells. They inhabit both freshwater and marine environments and are an essential component of phytoplankton. Diatoms are so abundant in some marine environments that they are responsible for the bulk of worldwide photosynthesis.

Diatoms have highly ornamented double walls containing silicon dioxide. The two halves of the wall fit together like the two parts of a box. Each half is called a **valve.** Figure 22-6 shows two types of diatoms. **Centric** *(SEN-trick)* **diatoms** have circular or triangular valves and are most abundant in marine waters. **Pennate** *(PEN-ate)* **diatoms** have rectangular valves and are most abundant in freshwater ponds and lakes. Some pennate diatoms move by secreting threads that attach to the surface of the water. When these threads contract, they pull the diatom forward.

Because their cell walls contain silicon dioxide, diatom shells do not decompose. Rather, the rigid shells of dead diatoms sink and eventually form a layer of material called **diatomaceous** *(DIE-uh-tuh-MAY-shuss)* **earth.** Diatomaceous earth is slightly abrasive; it is an ingredient of many commercial products, such as detergents, paint removers, fertilizers, and insulators.

Pyrrophyta

The approximately 1,100 species of the Division Pyrrophyta *(puh-RAHF-uh-tuh)* are called fire algae, or dinoflagellates. Most dinoflagellates are marine and photosynthetic. Dinoflagellates are an important component of oceanic phytoplankton.

The cell walls of dinoflagellates are made of cellulose. The cellulose forms plates that look like armor when seen under a microscope. The majority of cells are shaped like tops, but many have spinelike projections. Most dinoflagellates have two flagella that each fit into a groove in the cell wall. One groove runs vertically and is called a sulcus. The other groove runs horizontally and is called the girdle. Look at this arrangement in Figure 22-7. Movement of the flagella causes the dinoflagellate to spin like a top as it propels itself through the water.

Figure 22-7 The surface of dinoflagellates is characterized by ornamented silicon dioxide walls.

Noctiluca exemplifies another characteristic of many dino-flagellates—the ability to produce light. Organisms that produce light are said to display **bioluminescence** *(BIE-uh-LOO-muh-NESS-ens)*. If you have ever seen a sparkling light display in the sea, dinoflagellates were probably responsible.

Dinoflagellates are also responsible for a phenomenon known as red tide. **Red tides** are discolorations of sections of the ocean caused by population explosions of dinoflagellates such as *Gonyaulax*. During these population explosions, called **algal blooms,** the water appears red because of the pigments in the algae. *Gonyaulax* produces toxins that can cause respiratory paralysis in vertebrates. For example, if people eat mussels that feed on these toxic dinoflagellates, they may suffer a severe neurotoxic reaction called mussel poisoning, which has the potential to be fatal.

Euglenophyta

The approximately 1,000 species of the Division Euglenophyta *(YOO-gluh-NAHF-uh-tuh)* are unicellular algae that have many features in common with green algae. However, they also have many characteristics similar to those of protozoa. Euglenoids contain chlorophylls a and b and store food as starch, but they do not have cell walls. Unlike other algae, euglenoids are not completely autotrophic. If a euglenoid is placed in the dark, it will become heterotrophic. For these reasons some scientists classify euglenoids as protozoa.

The most familiar genus of euglenoids is *Euglena,* members of which are abundant in freshwater ponds and lakes. Examine the *Euglena* shown in Figure 22-8. *Euglena gracilis* is generally elongate and fairly flexible. Like all species of *Euglena*, it is able to change shape because of the presence of a pellicle, a flexible proteinaceous covering. *Euglena gracilis* contains a structure called a reservoir with small openings that lead to the outside. A contractile vacuole is located in the reservoir. The contractile vacuole functions to rid the organism of excess water. Growing out of the reservoir and extending far beyond the cell is the long flagellum. A second flagellum is contained within the reservoir. Only the emergent flagellum moves the euglenoid about. A red-orange eyespot functions as a light detector and guides the photosynthetic alga toward bright areas.

Section Review

1. What is a *unicellular gametangium?*
2. Give an example of a unicellular, a colonial, and a filamentous green alga.
3. Name the three parts of a kelp thallus.
4. Which divisions of algae store food as starch?
5. In what way do unicellular algae differ from members of the Kingdom Monera?

Figure 22-8 *Euglena,* the most familiar type of euglenoid, is propelled by a long flagellum. An eyespot guides the alga toward light.

- Long flagellum
- Short flagellum
- Eyespot
- Reservoir
- Contractile vacuole
- Nucleus
- Cell membrane
- Chloroplast

Answers to Section Review
1. a single-celled gamete holder
2. *Chlamydomonas* and desmids are unicellular. *Volvox* and *Gonium* are colonial. Filamentous green algae include *Spyrogyra* and *Oedogonium*.
3. holdfast, stipe, blade
4. Chlorophyta, Rhodophyta, Pyrrophyta, Euglenophyta
5. Algae are eukaryotic, while monerans are prokaryotic. This means that unicellular algae have nuclei and membrane-bound organelles.

Section Objectives

- *Describe asexual and sexual reproduction in unicellular algae.*

- *Explain conjugation in filamentous algae.*

- *Distinguish between antheridia and oogonia.*

- *Describe the two main stages in alternation of generations.*

Tip: What advantage does a flagellated gamete have over a nonflagellated gamete? (Active motility is more likely to result in successful fusion with another gamete than is a passive method.)

22.2 Comparative Reproduction

Some species of algae reproduce only asexually, thereby generating new organisms that are genetically identical with parent organisms. Other algae can reproduce either asexually or sexually. In these algae sexual reproduction is often triggered during periods of environmental stress.

Unicellular Reproduction

Earlier in this chapter we examined *Chlamydomonas*, a typical unicellular green alga. Members of this genus have both asexual and sexual reproduction. During asexual reproduction *Chlamydomonas* first absorbs its flagella. The haploid cell then divides mitotically one to three times, and two to eight haploid, flagellated daughter cells called **zoospores** *(ZOE-oe-SPORZ)* develop within the parent cell. These motile, asexual reproductive cells break out of the parent cell, disperse, land, and eventually grow to full size.

Sexual reproduction also begins when haploid cells divide mitotically to produce either plus or minus gametes. The plus and minus terminology is used when gametes look similar but differ in their chemical composition. A plus gamete and a minus gamete come into contact with one another and shed their cell walls. They fuse and form a diploid zygote, which develops a thick protective wall. A zygote in such a resting state is called a **zygospore** *(ZIE-goe-SPOR)*. The zygospore can withstand unfavorable environmental conditions. When conditions are favorable, the zygospore breaks out of the thick wall. It then divides by meiosis and forms typical haploid *Chlamydomonas* cells.

Figure 22-9 *Chlamydomonas* reproduces both asexually (blue arrows) and sexually (red arrows).

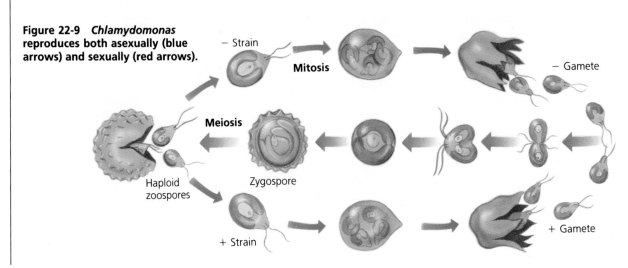

− Strain

Mitosis

− Gamete

Meiosis

− Gamete

Haploid zoospores

Zygospore

+ Strain

+ Gamete

Diatoms also can reproduce sexually or asexually. During asexual reproduction the two valves of the diatom shell split apart. Each valve then grows another valve within itself. This form of reproduction creates increasingly smaller diatoms. However, when it reaches a certain size, a diatom sheds its shell, grows to full size, regenerates its shell, and begins the cycle again.

In sexual reproduction a diploid diatom undergoes meiosis to produce a gamete. Plus and minus gametes unite to form a zygote that will grow into a mature diatom.

Multicellular Reproduction

Reproduction of multicellular algae varies widely among the divisions. Reproduction in red and brown algae is very complex, with that of the red algae often involving three states in a sexual life cycle. Examination of three less complex algal life cycles will enable you to understand something of the reproductive variation in algae.

Conjugation: *Spirogyra*

Spirogyra, a filamentous green alga, reproduces sexually by a process called conjugation, in which one gamete moves to the other through a conjugation tube between adjacent filaments. Study the process in Figure 22-10. First two filaments align side by side. The walls between adjacent cells then dissolve and a conjugation tube forms between the cells. One cell is considered to be a plus gamete. Its contents move through the conjugation tube, enter the adjacent cell, and fuse with the minus gamete. After fertilization the resulting zygote develops a thick wall, falls from the parent filament, and becomes a resting spore. It later produces a new filament.

Egg and Sperm: *Oedogonium*

Oedogonium is another filamentous green alga. As shown in Figure 22-11, *Oedogonium* has cells specialized for producing gametes. Recall that the modified cells that produce and hold the

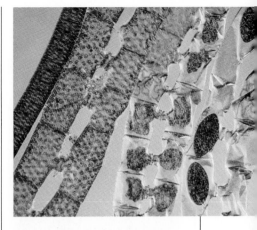

Figure 22-10 *Spirogyra* reproduces sexually by a process called conjugation. Fusion occurs when a plus gamete moves through a conjugation tube to join a minus gamete.

Info: Adverse environmental conditions trigger sexual reproduction in some algae. Offspring produced sexually differ from both parents, and the resulting variation of offspring gives an advantage to the species in a changing environment.

Figure 22-11 *Oedogonium* reproduces sexually by producing male and female gametes. The sperm, released into surrounding water, swim to the egg.

Diploid sporophyte

Sporangium

Meiosis

Haploid zoospores

Haploid gametophytes

− Gamete

Mitosis

+ Gamete

Diploid zygote

Diploid sporophyte

Figure 22-12 Reproduction in *Ulva* occurs through alternation of gametophyte and sporophyte generations.

gametes are called unicellular gametangia. The male unicellular gametangium is called an **antheridium** *(AN-thuh-RID-ee-um)*. It produces sperm. The female unicellular gametangium, the **oogonium** *(OE-uh-GOE-nee-um)*, produces an egg. Flagellated sperm are released from the antheridium into the surrounding water, swim to an oogonium, and enter through small pores. After fertilization the resulting zygote is released from the oogonium and forms a thick-walled, resting spore. The diploid spore then undergoes meiosis, forming four haploid zoospores that are released into the water. Each zoospore settles and divides. One of the new cells will become an anchoring holdfast; the other will divide and form a new filament.

Alternation of Generations: *Ulva*

The last form of multicellular algal reproduction that we'll consider is the complex cycle known as alternation of generations. This cycle is characterized by two distinct multicellular phases: a haploid, gamete-producing phase called the **gametophyte** *(guh-MEET-uh-FITE)*, and a diploid, spore-producing phase called the **sporophyte** *(SPOR-uh-FITE)*.

Examine Figure 22-12, which shows alternation of generations in *Ulva*. The *Ulva* illustrated at the top of the diagram is a diploid sporophyte. The sporophyte forms reproductive cells called **sporangia** *(spo-RAN-jee-uh)* that produce haploid zoospores by meiosis. These zoospores divide mitotically, forming more motile spores. The spores settle through the water, land on rocks, and grow into multicellular, haploid gametophytes. Note that the gametophyte looks exactly like the sporophyte. The gametophyte produces gametangia and then produces plus and minus gametes that unite and form zygotes. As Figure 22-12 shows, the diploid zygote completes the cycle by dividing mitotically into a new diploid sporophyte.

It is of exceptional significance that this life cycle occurs in a green alga. Plants, which presumably evolved from green algae, also have an alternation of generations as their sexual life cycle. As you will learn in Chapter 25, the plant life cycle differs from that of *Ulva* in two respects. The sporophyte and gametophyte do not look alike, and gametes are formed in multicellular rather than unicellular gametangia.

Section Review

1. Explain the difference between an *oogonium* and an *antheridium*.
2. What is the role of a zoospore in algal reproduction?
3. Since diatoms grow smaller whenever they reproduce asexually, why don't they gradually disappear?
4. How does *Spirogyra* reproduce sexually?
5. What is a possible adaptive advantage of sexual reproduction in the diatom?

Answers to Section Review (See pp. T22A–T22D.)

Laboratory

Colonial Organization in Algae

Objective
To examine colonial algae in order to explore the features of colonial organization

Process Skills
Classifying, inferring

Materials
Mixed culture of the following algae: *Chlamydomonas, Eudorina, Gonium, Pandorina, Hydrodictyon,* and *Volvox;* microscope slides; coverslips; medicine dropper; toothpicks; petroleum jelly; microscope

Method
1 Make a wet mount from the mixed algae culture. Be careful not to press down on the algae with the coverslip.
2 If possible, observe the algae through the scanning lens of a microscope. Otherwise use low power. Describe how you might use the microscope light to aid your observations in this experiment.
3 Use the key below to identify each of the algae represented. Choose one organism to identify and start at 1. Decide which of the two choices fits the organism. Go to the next number indicated and make the next decision. Continue until you have identified the organism. Identify the remaining organisms.

1a	Single cells	*Chlamydomonas*
1b	Colony of cells	Go to 2
2a	Colony flattened or netlike	Go to 3
2b	Colony round	Go to 4
3a	Horseshoe-shaped colony	*Hydrodictyon*
3b	Netlike colony	*Gonium*
4a	More than 100 cells in colony	*Volvox*
4b	Less than 100 cells in colony	Go to 5
5a	Cells close together	*Pandorina*
5b	Cells apart from each other	*Eudorina*

4 Draw each type of algae you have identified. What features should you include?

5 All of these algae are closely related. Do you think they all belong to the same division?
6 What examples of specialization, division of labor, and interdependence of cells can you observe? Why are these observations important? What do they tell you about the algae?
7 Reproduction rates change as colony size increases. The simplest type of reproduction, called isogamy, produces motile gametes of the same size regardless of sex. Heterogamy is the type of reproduction that produces motile gametes of different sizes. The most specialized reproduction produces motile male and nonmotile female gametes. It occurs in *Volvox* and is called oogamy.

Conclusions
1 What changes in behavior occur as colony size increases?
2 Did you observe any indirect indication of cellular specialization occurring as colony size increased? Be specific.

Inquiry
How might you study the evolutionary relationships among these algae?

Chapter 22 Review

Vocabulary

algal bloom (327)	diatomaceous earth (326)	phytoplankton (321)	thallus (321)
antheridium (330)	gametophyte (330)	pyrenoid (321)	unicellular gametangia (321)
bioluminescence (327)	holdfast (324)	red tide (327)	valve (326)
blade (324)	laminarin (324)	sporangium (330)	zoospore (328)
carageenan (325)	oogonium (330)	sporophyte (330)	zygospore (328)
centric diatom (326)	pennate diatom (326)	stipe (324)	

In the following groups of terms, choose the term that does not belong and explain why it does not.

1. laminarin, carageenan, pyrenoid
2. oogonium, antheridium, sporangium
3. blade, thallus, valve
4. red tide, blade, bioluminescence
5. valve, holdfast, stipe

Review

1. *Chlamydomonas* and *Euglena* both have (a) a stipe (b) a valve (c) an eyespot (d) an antheridium.
2. In *Ulva* the sporophyte produces (a) sporangia (b) antheridia (c) gametangia (d) carageenan.
3. Diatoms are in the Division (a) Rhodophyta (b) Euglenophyta (c) Chlorophyta (d) Chrysophyta.
4. Red algae contain an accessory pigment known as (a) phycobilin (b) fucoxanthin (c) carotene (d) xanthophyll.
5. An algal bloom that produces red tides in the sea is a population explosion among (a) diatoms (b) red algae (c) desmids (d) dinoflagellates.
6. Motile reproductive cells in algae are called (a) sporangia (b) zoospores (c) zygospores (d) sporophytes.
7. A pyrenoid is (a) a molecule in the cell wall (b) an adaptation for anchorage (c) a food-storage structure (d) a unicellular green alga.
8. Biologists think that the probable ancestors of plants were from the (a) Division Chrysophyta (b) Division Euglenophyta (c) Division Phaeophyta (d) Division Chlorophyta.
9. Conjugation occurs in (a) some unicellular algae (b) some filamentous algae (c) some colonial algae (d) some multicellular algae.
10. Farming the ocean is called (a) aquaculture (b) domestication (c) collecting (d) agriculture.
11. Which algal division has members that are bioluminescent?
12. Name the characteristics considered in the classification of algae.
13. Which division of algae does not have a cell wall?
14. What is the adaptive advantage of a blade?
15. How is laminarin different from other storage molecules?
16. Explain the difference between colonial and filamentous algae.
17. List the types of chlorophyll found in the algae.
18. Under what conditions do most algae undergo sexual, as opposed to asexual, reproduction?
19. In what ways are *Euglena* species like protozoa?
20. What is the evolutionary significance of the alternation of generations found in green algae?

Critical Thinking

1. A few years ago many botanists classified algae, including the blue-green bacteria, as plants. *Euglena* was classified as a photosynthetic protozoan because it was free-swimming. What does this tell you about how scientists go about classifying living things?

2. Along the Pacific Coast of North America there is a saying that it is unsafe to collect and eat shellfish during the months that lack an *r* in their spellings. Name these months, and give a possible scientific explanation for the belief. Keep in mind that shellfish are filter feeders.

3. The horsetail, or *Equisetum,* has silicon dioxide (SiO_2) in its outer cells. Diatoms also contain SiO_2. Explain whether the presence of SiO_2 in both genera indicates a close relationship. (Hint: Consider the types of evolution described in Chapter 15.)

4. How does the absence of a cell wall in *Euglena* make the function of the contractile vacuole critically important?

5. Suppose that you discover a new species of unicellular algae that has chlorophyll and an eyespot but no flagella or cilia. Why would you be surprised at the lack of locomotion?

6. What conditions might trigger sexual reproduction in *Spirogyra?* (Hint: Recall

from Chapter 5 what can trigger the division of cells.)

7. How does a diatom "know" it's gotten small enough and must divide sexually?

8. Examine the drawing of lateral conjugation in *Spirogyra,* in which adjacent cells of the same filament have conjugated. Explain whether this type of conjugation offers more or less genetic recombination than does sclariform conjugation, in which conjugation is between the cells of two filaments.

Extension

1. Read the article by S. Weisburd entitled "The World's Deepest-Dwelling Plant," in *Science News,* January 5, 1985, p. 4. Make an oral report. Be sure to discuss the significance of the red pigment in the algae and the lack of calcium carbonate at the top and bottom of the cells.

2. Go to your local grocery store or supermarket, and find at least three items that contain carageenan, a polysaccharide derived from red algae.

3. Leave a jar of pure clean water near a window or light source. Do not cover the jar. Examine it once a week for signs of green algae. Examine the algae under a microscope. How can algae grow in clean water?

4. Gather samples of soil, dead leaves, old grass, and fresh leaves if possible. Place each sample in a jar or tube of clean water, and leave in light. Examine the samples under a microscope once a week, and make drawings of the types of organisms you find. What does this tell you about the different habitats of algae?

Chapter 23 | *Fungi*

This chapter reviews the origin and evolution of fungi and four phyla of protists with funguslike characteristics. Students identify adaptations that contribute to the role fungi play in the ecosystem.

Correlation Chart

Section Number	Section Objectives	Pages	Features	Section Review
23.1	1. Describe the origin and evolution of fungi.	335–342	Biotechnology, pp. 340–341	5
	2. Distinguish between a hypha and a mycelium.			1
	3. List the four divisions of fungi and their key reproductive characteristics.		Think, p. 336	2, 3, 5
	4. Explain how geneticists use yeasts in laboratory research.			
	5. Define *mycorrhizae* and *lichen* and distinguish between them.			4
23.2	1. List the four phyla of funguslike protists and their distinguishing characteristics.	343–345	Biology Lab, p. 345	1, 3, 5
	2. Describe the life cycle of a plasmodial slime mold.			2
	3. Explain how oomycetes differ from other funguslike protists.			4

Resources*

Print

Arora, David. *Mushrooms Demystified: A Comprehensive Guide to the Fleshy Fungi*, 2nd ed. Berkeley, Calif.: Ten Speed Press, 1986.

Strobel, F. A., and G. N. Lanier. "Dutch Elm Disease." *Scientific American* 245 (August 1981): 56–66.

Teasdale, Jim. *Microbes*. Morristown, N.J.: Silver Burdett, 1985.

Audiovisual/Software

Fungi. Film, 18 min. New York: Benchmark, 1983.

The Impact of Fungi on Man and His Environment. Still-frame video, filmstrip, or transparencies with cassette, 21 min. Teacher's guide. Carolina Biological.

*Additional references are listed in *Modern Biology, Teacher's Program Guide*.

Section Teaching Guides

23.1 Biology of Fungi

Approach

The primary objective of this section is to familiarize students with the distinguishing characteristics of members of the Kingdom Fungi. This section builds upon concepts of cell structure (Chapter 5), metabolism (Chapter 7), mitosis and meiosis (Chapter 9), and classification (Chapter 18). In presenting the text, guide students to develop an understanding of the variation present in fungal morphology and reproductive mechanisms. You may wish to use Table 23-1 to highlight the key similarities in classification criteria for members of the four fungal divisions.

Concept Framework

Most students are aware that some fungi are poisonous. Build on this awareness and provide a conceptual framework for the understanding of fungal metabolism by engaging students in a discussion of possible sources of toxins in mushrooms. Lead students to consider whether or not the toxicity could be related to parasitic or saprophytic life forms. You may wish to end the discussion by asking students to look for other potential sources of toxicity as they study the fungal divisions.

Emphasis

Make sure that students understand that fungi are known to enter into several important ecological relationships. Some are parasitic or saprophytic. Some fungi live symbiotically with roots of plants, creating mycorrhizae; others live symbiotically with lichens. Review the distinctions among these ecological relationships with students if necessary.

Describing Lichen Thalli

Bring in examples or pictures of the four forms in which lichen thalli occur. You might also wish to provide pictures of algae and fungi for comparison. Ask students to describe the outward appearance of each form, the microscopic arrangement of cells that results in this appearance, and the possible transfer of nutrients between the fungal partner and the algal partner. Students may also wish to indicate if a particular thallus type is likely to occur in a particular environment.

Using the Features/Yeast Genetics

The main tools used in this feature relate to genetic engineering. Yeasts can reproduce sexually or asexually under different environmental conditions. By manipulating the yeast's environment, geneticists can encourage the production of hundreds of clones exhibiting particular traits. These clones can then be used to study traits more closely. Yeasts represent the simplest eukaryotes that bear a genetic resemblance to humans, so genetic research with yeasts can yield important information relating to human genetic mechanisms.

Section Review Answers

(Continued from p. 342.)

5. Zygomycetes probably evolved before ascomycetes, because the former are coenocytic.

23.2 Funguslike Protists

Approach

The main objective of this section is to familiarize students with the physiological characteristics of members of the four funguslike divisions of the Kingdom Protista. It builds on concepts of cell structure (Chapter 5), metabolism (Chapter 7), mitosis and meiosis (Chapter 9), and classification (Chapter 18). In presenting the text, point out the differences between the four divisions of protists, particularly in reproductive mechanisms. You may wish to use Table 23-2 to summarize key funguslike characteristics for students.

Concept Framework

Students have probably heard of the Irish potato famine that caused many people to leave Ireland in the mid-1800s. To expand on their knowledge, ask students to speculate about why it was so difficult to eliminate the oomycete that caused the famine. Lead the class to the observation that oomycetes are able to reproduce rapidly, like other protists. You may wish to end the discussion by asking students to consider what damp weather might have to do with the persistence of the blight-causing oomycete in crops.

Emphasis

Be certain that students understand that the funguslike protists are heterotrophic, like protozoa. While the slime molds in their feeding stages resemble amebas and some invertebrate animals, they also resemble fungi and plants in their reproductive cycles. For example, slime molds produce spores in sporangia, as plants do in alternation of generations. Chytrids and oomycetes bear certain resemblances to protozoa (i.e., the presence of flagella).

Observing Slime Molds

Show the class several 35 mm slides of a plasmodium of a slime mold, then several slides of an ameba engulfing a food particle. Have students list at least three animal-like qualities and three funguslike qualities of the slime mold. Then have students explain why slime and amebas are both classified in the Kingdom Protista.

Section Review Answers

(See p. 344.)

Laboratory Preparation

Background

This lab procedure familiarizes students with the basic fungal morphology. Fungi are made up of hyphae, a mass of which form a mycelium. Students learn that fungi reproduce both sexually and asexually and that when sexual reproduction occurs, it involves two different mating types. This is an appropriate lab to review why sexual reproduction is so ubiquitous. Students observe that organisms require certain environment conditions for growth, i.e., *Rhizopus* does not grow on dry bread. Additionally, students learn that most fungi are saprophytic and are important decomposers.

Preparation

Rhizopus cultures are available from a biological supply house. If you are growing your own cultures, start the cultures five to seven days before the lab session is scheduled.

Conclusions

1. No. There is no visible difference between the plus and minus strains.
2. The zygospores are diploid because they are formed from two gametes, one from each of the two *Rhizopus* strains.

Inquiry

Students will devise an experiment using controls and testing reactions to such environmental factors as moisture and dryness.

Chapter Review

Vocabulary Answers

1. *Myco-* means "fungus," and *-rhizo* means "root." The word *mycorrhizae* appropriately describes the mutualistic relationship between fungi and plant roots.
2. *Spor-* means "seed," *-angio* means "vessel," and *-phore* means "bearer" or "producer." Since a sporangiophore bears sporangia, or "seed holders," this name is almost appropriate though spores, not seeds are borne.
3. *Coeno-* means "in common," and *-cyte* means "hollow." No septa are present in a coenocytic hypha, so one could say that all the nuclei in the hyphae "share" the hollow space.
4. *Hypha* means "web." While intertwined hyphae form a weblike mycelium, this word does not seem altogether appropriate for the coenocytic or septate single hypha.
5. *Conidio-* comes from Greek roots meaning "dust" or "ashes" (*konis*) and "small one" (*idium*). The suffix *-phore* means "bearer." Conidiophores bear asexual spores in ascomycetes. The spores, called conidia, resemble fine dust particles arranged in chains or clumps.

Review

11. In yeasts and other ascomycetes, diploid cells produce four haploid daughter cells by meiosis.
12. They are 500 million years old, from the Ordovician Period.
13. In the coenocytic hypha the nuclei do not form individual cells and do not act in a coordinated manner as they do in pseudoplasmodia.
14. The fact that yeasts reproduce rapidly and can survive as either diploid or haploid individuals, depending upon environmental conditions, makes yeasts useful in studying dominant and recessive mutations as well as other features of the cell cycle.

15. Fungi digest organic materials and thereby return organic molecules and other nutrients to the ecosystem, where they can be used by fungi and other organisms.

16. Mutualism is a type of symbiotic relationship in which both parties benefit. Mycorrhizae, the symbiotic relationship between fungi and plant roots, is mutualistic because the fungus increases the absorptive surface available to the plant, while the plant sugars made during photosynthesis provide nutrients for the fungus.

17. Rusts and smuts are parasitic basidiomycetes.

18. Deuteromycetes do not have sexual reproductive phases or sexual life cycles that have been discovered. Therefore they cannot be classified on the basis of the features of their sexual forms as other fungi are.

19. Myxomycota are terrestrial and feed on decaying organic matter, so would most likely be found on a deciduous forest floor, in a swamp, or anywhere on land where organic matter is dead and disintegrating.

20. Two hyphae make contact with each other and fuse, producing a gametangium. Nuclei from the two hyphae fuse and form a zygote. The zygote then develops a thick outer covering and becomes a zygospore. When environmental conditions are favorable, the zygospore germinates, sending up new hyphae.

Critical Thinking

1. It is another example of convergent evolution.

2. The advantage is that they are not eaten. The disadvantage is that mammals don't carry them and their spores to a new area (while eating).

3. Moldy bread sometimes contained *Penicillium* which produced the antibiotic penicillin.

4. These fungi may have more space for reproduction and growth.

5. In humans as a pathogen; it is the cause of aspergillosis, a sometimes fatal lung infection.

6. The yeast sticks to its body; as it flies from fruit to fruit, it spreads the yeast, which can ferment the sugars in the fruit.

7. The soil serves as a source of fungi needed to form mycorrhizae.

8. The fruiting bodies resemble fungal fruiting bodies rather than other protist forms.

Chapter 23 | *Fungi*

Introduction

The Kingdom Fungi is made up of diverse organisms adapted to absorb food from living or dead organisms. Members of this kingdom range from unicellular yeasts to giant bracket fungi. This chapter also discusses four phyla of protists that have funguslike traits but are distinct enough from fungi to be classified in the Kingdom Protista. These phyla are discussed in this chapter because their life cycles are unlike those of other protists.

Chapter Outline

Chapter Concept

As you read, identify the importance of fungi and funguslike protists to humans.

Slime mold in the plasmodial stage. Inset: Mushrooms, *Marasmius* sp., in a Costa Rican rain forest.

23.1 Biology of Fungi

Members of the Kingdom Fungi are a diverse group of over 65,000 species living in many different environments. Molds, mildews, rusts, smuts, mushrooms, and yeasts are all fungi. Most fungi are microscopic. Some microscopic fungi, such as the pathogens responsible for athlete's foot, ringworm, and chestnut blight, are harmful. Many microscopic fungi, such as yeasts, are beneficial. Yeasts cause bread to rise and are among the most widely used research organisms. Not all fungi are entirely microscopic. Some fungi produce fruiting bodies that are visible to the unaided eye, such as puffballs, mushrooms, and morels.

Most fungi are saprophytic or parasitic, and a few species are predatory. A saprophyte is an organism that feeds on dead organic material. As fungi digest this material, they recycle the nutrient molecules in it. Organisms that break down organic materials are decomposers. For example, fungi release carbon, nitrogen, and phosphorus molecules when they digest dead plants and animals. These molecules are used by the fungus or by other organisms. Without decomposers, nutrients would not be reused and life could not continue on earth.

A parasite derives its nutrients from a living host organism at the host's expense. Parasitic fungi harm host organisms and cause many plant and animal diseases. For example, the fatal Dutch elm disease is caused by the fungus *Ceratocystis ulmi*. Other fungi, such as wheat rusts, attack crops and cause serious economic loss. Fungi that attack animals cause a variety of diseases, such as histoplasmosis, a serious lung disease.

A predatory fungus is one that captures prey for food. *Pleurotus ostreatus* is an example of a predatory fungus. Also called the oyster fungus, *P. ostreatus* obtains nitrogen by capturing roundworms. After secreting a substance that makes the roundworm sluggish the fungal cells engulf the worm, penetrate its body, and absorb its contents.

Evolution

The oldest fossils that resemble modern fungi are found in Precambrian rocks that are about 900 million years old. The earliest fossils that are distinctly fungi are about 500 million years old, from the Ordovician period. By about 300 million years ago, in the late Carboniferous period, all modern divisions of fungi had evolved. Since most present-day fungi are terrestrial, scientists speculate that fungi underwent adaptive radiation shortly after plants and animals colonized the land. Scientists reason that fungi, like all other eukaryotes, arose from prokaryotes. Fungi are heterotrophic—they cannot make their own food—and therefore probably arose from other heterotrophs. Some scientists, however, theorize that some fungi evolved from red algae.

Section Objectives

- *Describe the origin and evolution of fungi.*
- *Distinguish between a hypha and a mycelium.*
- *List the four divisions of fungi and their key reproductive characteristics.*
- *Explain how geneticists use yeasts in laboratory research.*
- *Define* mycorrhiza *and* lichen *and distinguish between them.*

Tip: To help students appreciate the importance of the fungi and protists covered in this chapter, discuss the impact of a fungus such as *Penicillium* and a protist such as *Phytophthora infestans.*

Figure 23-1   This bracket fungus is a saprophyte that aids in the breakdown of organic material.

Info: Fungi easily enter plants through their stomata. Research has shown that some rusts of a nanometric size are so sensitive that they enter the plants by determining the elevations of the guard cells. Scientists therefore are working to develop plants with lower or higher guard cells.

Figure 23-2 A mushroom, such as *Amanita parcivolvata* shown at the top, has the main parts illustrated below. Note the presence of two nuclei in a single cell.

It seems possible that the first true fungi were unicellular and gave rise to coenocytic fungi. **Coenocytic** *(SEE-nuh-SIT-ick)* refers to filaments without internal cross walls. A coenocytic filament contains many nuclei that move through the cytoplasm. Coenocytes may have given rise to club fungi. Club fungi, in turn, may have given rise to sac fungi. Club fungi and sac fungi have bodies that are divided into distinct cells by perforated cross walls.

Characteristics and Classification

The vegetative filament of a fungus is called a **hypha** *(HIE-fuh)*. Hyphae may or may not be divided into cells by perforated cross walls, each of which is called a **septum**. A hypha grows at its tip, where new membrane material is added by the action of Golgi bodies. A mat of interwoven hyphae is known as a **mycelium** *(mie-SEE-lee-um)*. Branches form in areas away from the tip.

Hyphae cell walls are made of **chitin,** a complex polysaccharide also found in the external skeleton of insects and other invertebrates. Fungi differ from plants, which have cellulose in their cell walls. Fungi, like animals, store food as glycogen, while plants store food as starch.

Fungi reproduce both sexually and asexually. Sexual reproduction in many fungi begins when cells or gametes from two mating strains undergo cytoplasmic fusion. The nuclei, however, may not immediately fuse, but will divide independently. When genetically different nuclei coexist within a hypha, the hypha is said to be **heterokaryotic.** A hypha with genetically similar nuclei is called **homokaryotic.**

Fungi reproduce asexually by producing spores, usually on special branches or modified cells of the hyphae. The small, lightweight spores are often released into the air. They may be carried long distances before they land and germinate into new hyphae. Fungi also reproduce asexually by the fragmentation, or breaking apart, of the hyphae.

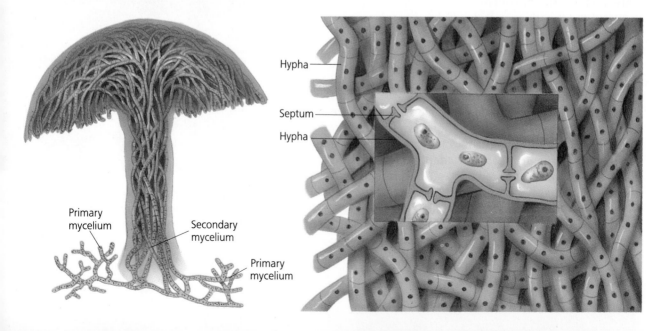

Primary mycelium

Secondary mycelium

Primary mycelium

Hypha

Septum

Hypha

Table 23-1   Classification of Fungi

Division	Body	Asexual Reproduction	Sexual Reproduction	Habitat	Examples
Zygomycota (600 species)	Coenocytic hyphae	Spores from sporangia	Conjugation; zygospores	Mostly terrestrial	*Pilobus* spp., *Rhizopus* spp.
Basidiomycota (25,000 species)	Septate hyphae	Fragmentation	Basidia produce basidiospores	Mostly terrestrial	Rusts, smuts, mushrooms
Ascomycota (30,000 species)	Septate or unicellular hyphae	Conidia; budding	Asci produce ascospores	Terrestrial and aquatic	Yeasts, molds, mildews, morels
Deuteromycota (10,000 species)	Septate hyphae	Conidia	Unknown	Terrestrial	*Cladosporium* spp., *Sporothrix* spp.

Members of the Kingdom Fungi are classified into four divisions based primarily on the structure of hyphae or on the type of reproduction. As you read about the four divisions of the fungi kingdom, refer to Table 23-1.

Zygomycota

The approximately 600 species of the Division Zygomycota *(ZIE-goe-mie-KOE-tuh)* are mostly terrestrial organisms. They are commonly found in soil and dung. Zygomycetes are coenocytic—that is, their hyphae lack septa. However, their reproductive structures are blocked off from the rest of the hyphae by unperforated cross walls.

The Division Zygomycota includes *Rhizopus stolonifer,* the common black bread mold. *Rhizopus* appears as a fuzzy growth on bread and consists of three different types of hyphae. **Rhizoids** are anchoring hyphae that penetrate the bread, produce digestive enzymes, and absorb nutrients. The hyphae that grow across the surface of the bread are called **stolons.** Specialized upright hyphae called **sporangiophores** *(spuh-RAN-ji-oe-forz)* produce sporangia at their tips.

Asexual reproduction in zygomycetes occurs when hormonal action causes upright sporangiophores to form. Sporangia form at the tips of the sporangiophores. The sporangia produce spores that are dispersed by wind.

Sexual reproduction in *Rhizopus* is called conjugation. Conjugation in fungi is a sexual process that occurs when two filaments line up next to each other. A similar kind of conjugation occurs in filamentous algae such as *Spirogyra.* In *Rhizopus* hyphae of two mating strains come close together. Each hypha encloses haploid (1N) nuclei. Hormones cause short branches to form on each hypha and grow outward until they touch. Then septa form near the tip of each branch. Each resulting cell is a gametangium that contains several nuclei. The gametangia fuse;

Spore

Sporangiophore

Two (1N) mating
strains meet

\+ \-

Gametangia

Zygote (2N)

Zygospore

Meiosis

Zygospore (1N)
germination

Figure 23-3 *Rhizopus* **goes
through a sexual reproduction
process called conjugation in
which two mating strains join.
Meiosis occurs when the zygospore germinates.**

then the nuclei fuse in pairs. Each pair contains one nucleus from each mating strain. Thus the zygote contains many diploid (2N) nuclei. Figure 23-3 shows the life cycle of *Rhizopus*.

The wall surrounding the zygote then thickens. The resulting protective, temporarily dormant structure is called a zygospore. Meiosis occurs when the zygospore germinates. Germination usu- ally takes place after one to three months, depending on environ- mental conditions.

Reproducing sexually and asexually provides an adaptive advantage. During periods when the environment is favorable, the rapid, asexual formation of spores ensures the quick spread of the species. In periods of environmental stress, sexual reproduction ensures genetic recombination before the hyphae die.

Basidiomycota

Scientists place about 25,000 species in the Division Basidiomy- cota *(buh-*SID-*ee-oe-mie-KOE-tuh)*, or club fungi. Mushrooms, toadstools, and puffballs are classified as basidiomycetes, as are rusts and smuts.

Under proper conditions, such as after a warm rain in the spring, underground hyphae grow upward, intertwine, and produce a **basidiocarp,** the reproductive body of a basidiomycete. A mushroom is a basidiocarp with a flattened **cap** attached to a stem that is called a **stalk.**

On the underside of the cap are radiating rows of **gills.** A specialized club-shaped reproductive cell called a **basidium** *(buh-*SID-*ee-um)* forms on the gills. Each gill contains thousands of basidia. In each basidium two nuclei become isolated by a com- plete septum. The nuclei fuse and form a diploid zygote. Meiosis then results in four nuclei that are pushed into cytoplasmic exten- sions to form **basidiospores.** At maturity basidiospores are released and germinate into new homokaryotic hyphae.

As the homokaryotic hyphae grow, septa form so that each cell contains one nucleus. These homokaryotic, septate hyphae are called the **primary mycelium.** These hyphae grow and fuse with hyphae from another mating strain. When separate mating strains fuse, a **secondary mycelium** results. Hyphae of these myce- lia are heterokaryotic, containing one nucleus from each mating strain in each cell. Secondary mycelium intertwines and forms a basidiocarp. A mushroom, for example, is a basidiocarp.

Rusts are a group of parasitic basidiomycetes that cause dis- eases in plants. Rusts attack many plants but are of particular concern to us when they infect cereal crops such as wheat and oats. Infected plants look rusty because they are covered with reddish brown fungal spores. Rusts do not form basidiocarps. Instead basidia develop on the surface of infected plants.

Smuts, like rusts, are plant pathogens that do not form basid- iocarps. The approximately 1,000 species of smuts cover infected organisms with dark masses of spores.

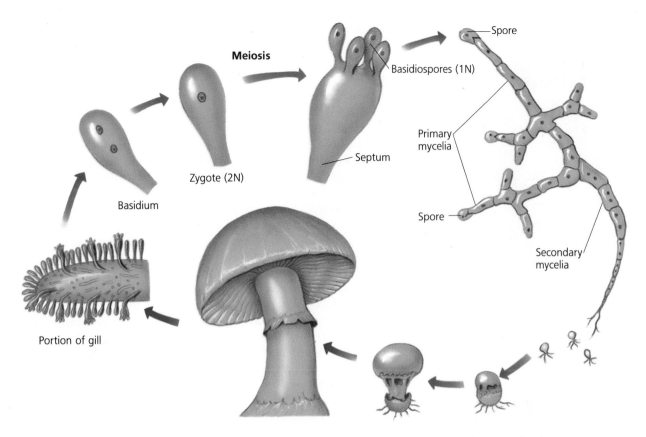

Meiosis

Zygote (2N)

Basidium

Basidiospores (1N)

Spore

Septum

Primary mycelia

Spore

Secondary mycelia

Portion of gill

Ascomycota

The Division Ascomycota *(AS-koe-mie-KOE-tuh)*, or sac fungi, contains over 30,000 species. This is the largest division of fungi, and some biologists suspect it may have been the most recent group to evolve. Sac fungi live in a variety of habitats, including salt and fresh water. In terrestrial environments ascomycetes live in soil, dung, and rotting logs. Common ascomycetes are morels, powdery mildews, yeasts, and cup fungi.

Sexual reproduction begins when the hyphae of two mating strains form either male or female gametangia. The female gametangium is called an **ascogonium** *(AS-kuh-GOE-nee-um)*, and the male gametangium is called an antheridium. The gametangia fuse, and male nuclei move into the ascogonium. Male and female nuclei pair but do not fuse. These cells divide, and the resulting heterokaryotic hyphae grow and intertwine, forming an ascocarp. The **ascocarp** is the reproductive body of an ascomycete. It has various forms, depending on the species.

Sacs called **asci** *(AS-ie)* form on the surface of the ascocarp near the tips of some hyphae. Each ascus *(AS-kuss)* encloses two nuclei. The nuclei fuse in the ascus, and the resulting diploid nucleus immediately undergoes meiosis, producing four haploid nuclei. Each nucleus usually then divides once by mitosis, so each ascus finally contains eight haploid **ascospores**. When an ascus ruptures, it releases ascospores, usually into the air. These land on the ground and germinate, forming new hyphae.

Figure 23-4 The reproductive body of a basidiomycete, a mushroom, develops basidia on the gills lining the underside of its cap. New mushrooms form from a secondary mycelium.

Figure 23-5 The morel, *Morchella deliciosa*, shown here is an edible ascomycete, or sac fungus.

Think

Conidiophores often produce thousands of conidia (SEM, ×500). What is the adaptive advantage of producing large numbers of asexual spores?

Asexual reproduction in ascomycetes involves the production of spores, each of which is called a **conidium** *(koe-NID-ee-um)*. Many conidia are formed on the ends of specialized branches called **conidiophores.**

In unicellular ascomycetes called yeasts asexual reproduction occurs by bud formation. Sexual reproduction involves the formation of a diploid zygote. The zygote results from the fusion of two ascospores. The zygote either may function as an ascus, producing eight ascospores, or may produce diploid cells by budding. If the zygote undergoes meiosis and produces ascospores, the haploid ascospores can reproduce in one of two ways. They can either produce new haploid cells by budding, or they can fuse in sexual reproduction.

Humans are interested in the approximately 600 species of yeasts because of their ability to break down carbohydrates. This ability makes yeast important to the food industry. When carbohydrates are thus broken down, they form alcohol and carbon dioxide. This process causes bread dough to rise and grapes to ferment. Domesticated yeast, usually a strain of *Saccharomyces cerevisiae*, is used in most modern brewing processes. Yeasts are also important tools in genetic research.

Biotechnology

Yeast Genetics

For many years scientists have worked with *Escherichia coli*, a bacterium, to explore gene structure and function. Now the unicellular ascomycetes known as yeasts rival *E. coli* as the favored organisms in genetic research.

The molecular and chromosomal processes in the yeast life cycle are easier to unravel than the same processes in higher organisms. For example, yeast cells have about 10,000 genes, compared with nearly 400,000 in humans. Scientists have found that while yeasts are structurally simpler than multicellular organisms, many features of their biochemistry closely resemble those of plants and animals. Most important, yeasts are eukaryotes with true cellular organelles. *E. coli* is prokaryotic and does not offer the same insights into plant and animal genetics.

Geneticists began to work with yeasts in the 1950s, when it was found that yeasts could live both as diploid and haploid organisms. When food is plentiful, yeasts live as diploid individuals and reproduce asexually by budding (SEM, left, ×2,885). The diploid mother cell produces new diploid cells through mitosis. When food is

Deuteromycota

Deuteromycetes, sometimes called the imperfect fungi, or **Fungi Imperfecti,** are a group of about 10,000 species of fungi in which a sexual reproductive phase has not been discovered. In effect, the Division Deuteromycota (*DOO-tuh-roe-mie-KOE-tuh*) is a taxonomic "holding tank." Biologists can classify these organisms on the basis of their type of asexual reproduction. However, they will not assign deuteromycetes to another division unless they can identify a sexual reproductive phase. The group contains forms that probably have evolved from several different strains. Today it appears that most imperfect fungi have characteristics similar to those of ascomycetes. Deuteromycetes are the organisms that cause ringworm and athlete's foot.

Aspergillus and *Penicillium* are two deuteromycetes that have recently been reclassified into the Division Ascomycota. Each of these organisms is valuable to humans. Humans use *Aspergillus* to ferment soy beans and make soy sauce. *Penicillium* produces the antibiotic penicillin and gives flavor to Roquefort and Camembert cheeses. These organisms have been renamed as a result of their reclassification.

Mitosis   Meiosis

scarce, the diploid yeast cells undergo meiosis, and each produces four haploid cells. The haploid cells stay in the mother cell wall until food is again plentiful, when they are released.

Haploid yeast cells can produce new haploid cells by mitosis or can reproduce sexually. Separate strains secrete hormones, fuse, and form new diploid individuals.

By controlling laboratory food supplies researchers can influence the mode of yeast reproduction. First they used yeasts to study recessive mutations. In diploid research animals a recessive trait will be expressed in only about one-fourth of the offspring. However, two of the haploid offspring of diploid yeast will express a recessive trait. Since yeast cells reproduce almost once every two hours, many mutations can be studied in a very short time.

Data on recessive mutations gives clues to the identities of a number of functioning genes. By the mid-1970s research with yeast had provided models for identifying human genes that influence cancer, Down syndrome, and other genetic disorders. In the early 1980s a procedure called transformation was perfected. Transformation uses recombinant DNA technology to study specific mutations in yeast cells, which gives researchers control over specific mutations and promises to contribute to AIDS research. It has already helped produce a vaccine against hepatitis B.

Answers to Section Review
1. The primary mycelium is homo-karyotic; the secondary mycelium is heterokaryotic.
2. No sexual life cycle or phase has been discovered in deuteromycetes.
3. The drawing should include the stalk, cap, and gills as well as ba-sidia indicated on the gills.
4. In mycorrhizae, fungi benefit from plant sugars, and plants bene-fit from nutrients absorbed from the fungi. The association benefits both organisms, so it is mutualistic. (See pp. TA23A–T23D.)

Mycorrhizae and Lichens

Mycorrhizae and lichens represent two types of symbiotic relationships between fungi and autotrophic organisms. These relationships are of great ecological significance: they are instances of **mutualism,** a type of symbiosis in which both organisms benefit.

A **mycorrhiza** (MIE-kuh-RIE-zuh) is a symbiotic association between fungi and plant roots. They occur in more than 80 percent of plants. Mycorrhizae help plants absorb water and nutrients, such as phosphorus and potassium, by forming extensive networks of fungal hyphae in the soil. The fungus increases the surface area in the soil available for absorption. The digestive action of fungal enzymes provides nutrients that can be readily absorbed by the plant. The fungi absorb some of the sugars created by the plant during photosynthesis.

Fungal hyphae penetrate roots. Mycorrhizae that contain fungi that penetrate the root cortex cells are called endomycorrhizae, the most common type. The fungus is usually a zygomycete. Mycorrhizae with fungal hyphae that do not penetrate cortex cells are called ectomycorrhizae. They are less common than endomycorrhizae, and the fungus is usually a basidiomycete. Ectomycorrhizae are most common in coniferous forests, especially those with nutrient-poor soils. Mycorrhizae help plants absorb nutrients in such environments and also offer the plant some protection from harsh conditions such as cold and drought.

About 20,000 species of fungi form lichens. A **lichen** (LIE-kun) is a symbiotic association between a fungus, usually an ascomycete, and a green alga or cyanobacterium. Although the relationship is mutualistic, it may also be thought of as controlled parasitism. The fungal hyphae penetrate the cells of the photosynthetic partner and absorb food.

Biologists classify lichens into groups according to the nature of their thallus, or body. Crustose lichens grow as a layer on the surfaces of rocks or trees. Foliose lichens, such as the one you see in Figure 23-6, are loosely attached to the substrate and have thin, leafy thalli. Shrubby lichens with upright growths are called two-fold lichens. Long, thin growths hanging from tree branches are foliose lichens.

Figure 23-6   What kind of symbi-otic relationship results in this foliose lichen?

Section Review

1. Explain how a *secondary mycelium* differs from a *primary mycelium*.
2. What is the main characteristic that distinguishes the deuteromycetes from other fungi?
3. Draw a generalized mushroom showing the three main features of the basidiocarp and the location of basidia.
4. Explain why mycorrhizae are said to be mutualistic.
5. Given the morphological evidence, what evolutionary relationship might exist between zygomycetes and ascomycetes?

23.2 Funguslike Protists

The organisms discussed in this section are all protists, not fungi. For many years scientists classified them as fungi because some phases of their life cycles looked funguslike and because they stored food as glycogen. However, many taxonomists now consider these morphological similarities to be superficial. They place these organisms into the four protist phyla shown and described in Table 23-2.

Acrasiomycota

Approximately 65 species of cellular slime molds are placed in the Phylum Acrasiomycota *(uh-KRAZZ-ee-oe-mie-KOE-tuh)*. They live on land or in fresh water. The feeding stage of a cellular slime mold is a **myxameba** *(MICKS-uh-MEE-buh)*, a uninucleate cell. Myxamebas live on the forest floor or on decaying plants, moving and feeding like amebas. During periods of environmental stress some myxamebas secrete a compound called acrasin. Other myxamebas in the area move against the concentration of acrasin until they clump together, forming a pseudoplasmodium. A **pseudoplasmodium** *(SOO-duh-plazz-MOE-dee-um)* is a group of cells that act together as a unit to form sporangia that produce spores.

The acrasiomycetes are simple both morphologically and physiologically. Because of this organisms such as *Dictyostelium* have been used to study physiology and morphogenesis. Since they are eukaryotic, the cellular slime molds have much more complex mechanisms of gene expression than prokaryotes do. Organisms such as *Dictyostelium* can be easily grown in the laboratory, where scientists monitor the chemical and morphological responses of the organisms under varying environmental conditions. The simplicity of the organisms is relative, however, because their complex physiology is still very far from being completely understood.

Figure 23-7 *Dictyostelium* is a cellular slime mold widely used in scientific research.

Table 23-2 Classification of Funguslike Protists

Phylum	Body	Reproduction	Habitat
Acrasiomycota (65 species)	Uninucleate myxameba	Sporangia on stalks	Mostly terrestrial
Myxomycota (450 species)	Multinucleate plasmodium	Sporangia on stalks	Mostly terrestrial
Chytridiomycota (750 species)	Motile cells	Gametes and zoospores	Mostly aquatic
Oomycota (475 species)	Coenocytic hyphae	Gametes and zoospores	Aquatic and terrestrial

Figure 23-8 A plasmodial slime mold (top), a chytrid (center), and an oomycete (bottom) are examples of funguslike protists.

Tip: Contrast the simple cell division of the ameba with the sporangia developed by a plasmodium.

Answers to Section Review
1. A plasmodium is multinucleate cytoplasm; a pseudoplasmodium is made up of individual cells.
2. Sporangia, produced during periods of environmental stress, produce spores that later produce myxamebas or swarm cells.
3. chytridiomycota
4. Oogonia form eggs, and antheridia form sperm.
5. because their pseudoplasmodia are made up of coordinated single cells rather than multinucleate cytoplasm

Myxomycota

Taxonomists place the plasmodial slime molds in the Phylum Myxomycota (*MICKS-uh-mie-KOE-tuh*), which contains over 450 species. The feeding stage is called a **plasmodium**—multinucleate cytoplasm surrounded by a membrane that moves as a mass, feeding on organic matter.

During periods of stress the plasmodium becomes stationary and produces sporangia on stalks. The sporangia produce spores. Spores may stay dormant for several years; under proper conditions they release either myxamebas or flagellated swarm cells. These fuse, and the nucleus of the resulting zygote divides repeatedly by mitosis; but the cells do not undergo cytokinesis. The result is the multinucleate cytoplasm of the plasmodium.

Chytridiomycota

Biologists place approximately 750 protist species in the Phylum Chytridiomycota (*kih-TRID-ee-oe-mie-KOE-tuh*). The chytrids are primarily aquatic protists characterized by gametes and zoospores with a single, posterior flagellum. Most chytrids are unicellular. Some chytrids are coenocytic and have hyphae that anchor the organism. Many chytrids are parasites on algae, plants, and insects, while others are saprophytes.

Oomycota

The Phylum Oomycota (*OE-uh-mie-KOE-tuh*) contains over 475 aquatic and terrestrial species, including water molds, white rusts, and downy mildews. Oomycetes are parasitic or saprophytic. One oomycete, *Phytophthora infestans,* the late blight fungus, causes the potato rot that led to the devastating Irish potato famine of 1845–1847.

Oomycetes are coenocytic with branched hyphae. Their cell walls are composed of cellulose, not chitin. Oomycetes form male and female gametangia. The female gametangium, or oogonium, forms eggs; the male gametangium, the antheridium, forms sperm. A zygote resulting from the fusion of egg and sperm develops into a thick-walled, diploid **oospore** (*OE-uh-SPOR*). Oospores are released into water or moist areas and germinate into coenocytic hyphae. In asexual reproduction oomycetes produce flagellated zoospores, which then produce new hyphae.

Section Review

1. How does a *plasmodium* differ from a *pseudoplasmodium?*
2. Explain the role of sporangia in the myxomycete life cycle.
3. Which phylum contains mostly aquatic organisms?
4. How do oogonia and antheridia function in oomycetes?
5. Why are acrasiomycetes called cellular slime molds?

Info: The late blight that destroyed the potato crop in Ireland in the 1800s was responsible for a million deaths and the migration of 1.5 million people, primarily to the United States and Canada.

Laboratory

Fungal Morphology

Objective
To study the vegetative and reproductive morphology of *Rhizopus stolonifer*, the common black bread mold

Process Skills
Observing, inferring

Materials
Petri dish containing a single-strain culture of *Rhizopus stolonifer*, petri dish containing plus and minus strains of *Rhizopus stolonifer*, dissecting microscope, probe, unlined paper

Method

1 Remove the cover from a petri dish containing a single-strain culture of *Rhizopus stolonifer*. Do you recognize a familiar odor? What does it smell like?

2 Place the culture dish under a dissecting microscope and observe the mycelium. Note the threadlike stolons that extend horizontally along the surface of the agar.

3 Use a probe to tease apart the hyphae. Note the rhizoids that grow down into the agar. What is the function of the rhizoids?

4 Replace the cover on the petri dish. Carefully turn the petri dish upside down and put the dish back under the dissecting microscope. Focus through the agar to get a better view of the rhizoids.

5 Turn the petri dish right side up. Remove the cover and replace the petri dish under the dissecting microscope. Note the hyphae that extend from the mycelium. These are the sporangiophores. What is their function?

6 Observe the round objects at the ends of the sporangiophores. These are the sporangia. The sporangia produce asexual, haploid spores. Can you think of an adaptive advantage to spores that are produced above the mycelium?

7 Make a drawing of what you see under the dissecting microscope. Label a stolon, rhizoid, sporangiophore, and sporangia.

8 Place a petri dish containing plus and minus strains of *Rhizopus stolonifer* under the dissecting microscope. To what do the plus and minus designations refer?

9 To look for evidence of mating between the two different mating strains, search for zygospores that have formed where the two strains have united. Describe the zygospores. Make a drawing of the zygospores, and show as much detail as you can.

10 Examine your culture carefully. Can you see the stages leading up to the formation of zygospores? Based on your observations, make a drawing that models the mating between the two different strains.

Conclusions

1 Did you observe differences between the plus and minus strains of the mold?

2 Are the zygospores haploid or diploid? Explain your answer.

Inquiry
Design an experiment to determine the optimum conditions for growth of *Rhizopus*.

Vocabulary

ascocarp (339)	coenocytic (336)	mutualism (342)	rhizoid (337)
ascogonium (339)	conidiophore (340)	mycelium (336)	secondary mycelium
ascospore (339)	conidium (340)	mycorrhiza (342)	(338)
ascus (339)	Fungi Imperfecti (341)	myxameba (343)	septum (336)
basidiocarp (338)	gill (338)	oospore (344)	sporangiophore
basidiospore (338)	heterokaryotic (336)	plasmodium (344)	(337)
basidium (338)	homokaryotic (336)	primary mycelium	stalk (338)
cap (338)	hypha (336)	(338)	stolon (337)
chitin (336)	lichen (342)	pseudoplasmodium	
		(343)	

Look up the meaning of the word roots of the following terms. Tell whether the words are appropriate, and explain why.
1. mycorrhizae *(myco-, -rhizo)*
2. sporangiophore *(spor-, -angio, -phore)*
3. coenocyte *(coeno-, -cyte)*
4. hypha
5. conidiophore *(conidio-, -phore)*

Review

1. Hyphae of secondary mycelium are (a) coenocytic (b) homokaryotic (c) heterokaryotic (d) plasmodial.
2. The common edible mushroom is classified in the Division (a) Basidiomycota (b) Ascomycota (c) Oomycota (d) Zygomycota.
3. The female gametangium in oomycetes is called an (a) ascogonium (b) oogonium (c) antheridium (d) oospore.
4. Fungi that feed on decaying organic matter are referred to as (a) saprophytic (b) parasitic (c) mutualistic (d) symbiotic.
5. *Rhizopus* grows most commonly on (a) soil (b) fruit (c) bread (d) decaying logs.
6. The cross walls that separate cells in hyphae are known as (a) rhizoids (b) gills (c) asci (d) septa.
7. Lichens are symbiotic associations of fungi and (a) roots (b) roundworms (c) water molds (d) algae.
8. The gills of a mushroom contain (a) basidia (b) conidiophores (c) basidiocarps (d) ascocarps.
9. The body of a lichen is referred to as (a) a stipe (b) a hypha (c) a thallus (d) an ascus.
10. Myxameba clump together and form (a) plasmodia (b) mycorrhizae (c) hyphae (d) pseudoplasmodia.
11. How is the yeast life cycle similar to that of other ascomycetes?
12. How old are the first fossils distinctly identified as fungi?
13. How does a coenocytic hypha differ from a pseudoplasmodium?
14. What aspect of the yeast life cycle makes yeasts useful in genetic research?
15. Explain how fungi contribute to nutrient cycling in the environment.
16. Describe mutualism, and give an example of such a relationship involving fungi.
17. What are rust and smuts?
18. Why are deuteromycetes more difficult to classify than other fungi?
19. Where might you look to locate a plasmodial slime mold?
20. Describe the process of conjugation in a zygomycete.

Critical Thinking

1. The cell walls of fungi and the exoskeletons of insects and crustaceans contain chitin. Tell whether this is of phylogenetic significance, and explain why.
2. Many fungi are fatally poisonous to mammals. What is the adaptive advantage of this to the fungi? What is the disadvantage?
3. Long before antibiotics were discovered, it was the practice to place moldy bread on wounds. Explain why this might have helped the wounds to heal.
4. Fungi such as *Penicillium* engage in a kind of "chemical warfare" against other microorganisms by producing chemicals that diffuse outward and kill nearby organisms. Suggest an adaptive advantage of this characteristic of these fungi.
5. Most fungi grow best at a temperature of around 15° to 21° C. *Aspergillus fumagatus*, however, can grow well at 37° C. Knowing this, where would you expect *A. fumagatus* to grow?
6. *Drosophila*, the common fruit fly, eats juices from ripe fruit and yeast. How

might the fruit fly help speed the natural decay of fruit?
7. When transplanting a wild plant to garden soil it is very important to include some of the soil from the original habitat. Give a possible explanation, based on the information from this chapter.
8. Examine the photo above of the fruiting bodies of the slime mold *Leocarpus fragilis*. How does the appearance of *Leocarpus fragilis* make it difficult to classify this organism?

Extension

1. Read the article entitled "Fungus Degrades Toxic Chemicals" by J. Dusheck in *Science News*, June 22, 1985, p. 391. Make an oral report, discussing what is unusual about the metabolism of the fungus, what persistent poisons it degrades, and how radioactive carbon is used in the experiment.
2. Use the library to research ways in which medical science has been influenced by fungi. Investigate the discovery of penicillin and other drugs derived from fungi. Investigate the role played by fungi in various diseases.
3. Obtain basidiospores of edible mushrooms and collaborate with the other members of the class in growing the mushrooms under various conditions of

light and temperature. If time permits, experiment with various food sources. Keep records of your results and report your findings to the class.
4. Prepare a yeast culture by adding yeast to a mixture of nine parts of water to one part of molasses. Allow the culture to ferment and examine it under a microscope. Make a drawing of several yeast cells and identify any visible cell parts. Look for instances of budding and add them to your drawing.
5. Research ways in which lichens are valuable to other life forms, including human beings. In what way do lichens help to cultivate new soil? How do they contribute nitrogen to the atmosphere?

Intra-Science: *How the Sciences Work Together*

The Process in Biology: Ocean Aquaculture

As you learned in Chapter 22, 96 percent of the world's food supply comes from the land. Yet standing amid the vast corn fields of Iowa or wheat fields of Kansas, you can easily lose sight of the fact that just three percent of the land on this planet can be cultivated. The rest is too rugged or has climatic conditions too harsh for growing crops. Nearly all the food for today's world population of five billion comes from this small amount of land.

The world's oceans are a relatively untapped source of additional food for the world's growing population with advantages the land doesn't offer. For example, the oceans are vast, covering about 70 percent of the earth's surface. Furthermore, the climatic changes that produce droughts, floods, frosts on land never affect the ocean; conditions underwater are far more constant. Finally, while agriculture is limited to areas where topsoil is

abundant, food can theoretically be harvested in every region of the ocean, from the surface to the bottom. The cultivation of food crops in the ocean is called ocean aquaculture; harvesting food from different regions in the ocean is called ocean polyculture.

Over 80 percent of today's ocean aquaculture takes place along Asian coastlines. A growing amount, however is taking place in other areas of the world. To date most aquaculturists have cultivated animals. Seven percent of the fish and shellfish harvested in the world currently comes from aquaculture facilities. In the U.S. scallops, salmon, and oysters are among the major cultivated

species. Oysters are also cultivated in Europe and Asia. Free-swimming oyster larvae will attach themselves to twigs or bamboo sticks placed in the water.

The Japanese have grown an edible red alga known as nori for more than 300 years. To grow nori, Japanese aquaculturists place bamboo sticks in shallow, muddy water, and wait for algal spores to settle on them. Then the sticks are moved to river estuaries where the algae grow in the changing mixture of salt water and fresh water. Large mowers are used in some coastal areas today to harvest brown algae known as kelp. However, it does not appear that it will be economically feasible to increase the harvest of algae for food in the near future. Much more energy must go into such cultivating and harvesting than is provided by the algae.

The Connection to Physics

Many of the most productive areas from which marine animals and algae can be harvested are intertidal zones. A key feature of these zones is constant wave action. In general, physicists define a wave as a disturbance that travels in a space or another medium. Most importantly, waves transport energy without moving matter.

Waves protect organisms such as kelps and mussels from predators that cannot tolerate rough seas. Waves

improve the flow of nutrients in the intertidal areas. Wave action also keeps the photosynthesizing parts of marine protists exposed to the sun by keeping them in constant motion. Finally, the action of waves continually dislodges some organisms, making their habitats available for new organisms such as mussels.

The Connection to Chemistry

Algae produce such chemical substances as algin, agar, and carageenin, which give ice cream, cosmetics, soaps, film, and paints a smooth texture and the ability to stay moist. You may know that agar is also used as a growth medium in scientific laboratories. Seaweeds contain iodine, an important element of human nutrition. Iodine is added to table salt since it is difficult to predict how much will be available in other foods. Iodine compounds are also used as antiseptics. Finally, iodine can be used in some chemical analyses because it produces a deep blue color when it contacts starch.

Although algae are also high in carbohydrates, humans do not have the enzymes to digest the materials in algal cell walls. Algae are harvested today primarily for their industrial value. Algae and inedible fish parts can be used as fertilizers. This is because the compounds of potassium, calcium, and nitrogen they contain are important plant nutrients.

The Connection to Other Sciences

In some parts of the world, agriculture and aquaculture support one another. For example, after their grain crops are harvested, Indonesian farmers may use the same fields to raise crustaceans that thrive on decaying vegetation. Carp can be fattened in rice paddies replenished by summer rains.

Aquaculture is limited by ecological factors. Although bacteria, algae, and small animals provide food for most of the ocean's fish, humans cannot consume them in similar quantities. The food chain from producer to consumer begins with marine phytoplankton and ends when humans eat fish or shellfish.

The Connection to Careers

Aquaculture is a developing field. In many coastal areas, of course, fishing has been an important source of income for people for years. Today, however, biochemists and other researchers explore ways in which additional nutritional value can be obtained from the available food.

Professional and technical jobs in aquaculture require a bachelor's degree in biology. Courses in managing fish and mollusk hatcheries are available at many two-year and four-year colleges in the United States. Most fishery workers enter the field through on-the-job experience. Some then go into business for themselves, harvesting and marketing fish and shellfish. For more information on related careers in the biological sciences, turn to page 842.

Unit **6** | *Plants*

"Plants are the strong scaffoldings on which the original symbiosis of photosynthesis has been raised from ooze and stretched as a canopy across rock bottom."

David Rains Wallace

Eating an insect may strike you as an unusually active thing for a plant to do, but plants actually interact in many dynamic ways with the environment. Consider, for example, the crucial role that plants play in transforming sunlight into usable forms of energy upon which almost all living things depend. What other kinds of interactions are suggested by the illustrations on these pages?

Maple, *Acer* sp., samara

Tuliptree, *Liriodendron tulipifera*

350

Venus's-flytrap, *Dionaea muscipula*

Orchid, *Cattleya* sp.

351

Chapter 24 | *The Importance of Plants*

This chapter explains the importance of plants as food producers. Students learn about the agricultural, medicinal, and commercial uses of plants. Plant ecology and the coevolution of some plants and animals are discussed.

Correlation Chart

Section Number	Section Objectives	Pages	Features	Section Review
24.1	1. Summarize briefly the history of the cultivation of plants by humans.	353–358		5
	2. Define cereals, legumes, root crops, and fruits, and distinguish between them.			3
	3. List the lessons learned from the Green Revolution.			2
	4. Define *famine,* and suggest two cultivation practices that can be used to fight famine.		Technology Feature, pp. 356–357	1
	5. List several medicinal and commercial uses of plants.			4
24.2	1. Describe what is meant by plant ecology.	359–361	Think, p. 360	1
	2. Give two examples of how plants recycle nutrients in the environment.			2
	3. Identify the major threat to the tropical rain forests.		Biology Lab, p. 361	4
	4. Explain the coevolution of flowers and pollinators.			3, 5

Resources*

Print

Dirzo, Rodolfo, and Jose Sarukhan, eds. *Perspectives on Plant Population Ecology.* Sunderland, Mass.: Sinauer, 1984.

Heiser, Charles B., Jr. *Of Plants and People.* Norman, Okla.: University of Oklahoma Press, 1985.

Audiovisual/Software

All Things Plant. Film, 18 min. Discussion guide. Barr, 1986.

Jungles: The Green Oceans. Film, 23 min. Britannica.

Seas of Grass. Film or video, 55 min. Time-Life, 1984.

The Tropical Rain Forest. Film, 17 min. Britannica.

*Additional references are listed in *Modern Biology, Teacher's Program Guide.*

Section Teaching Guides

24.1 Plants and People

Approach

The primary objective of this section is to provide students with a perspective on the interactions of plants and humans. It builds on concepts of biochemistry (Chapter 4) and prepares for future concept presentations of plant biology (Chapters 25 through 28) and ecology (Chapters 49 through 53). In presenting the text, guide students to develop an appreciation for the numerous ways people use and depend upon plants. Remind students that plants, as autotrophs, are essential to heterotrophic nutrition. You may wish to use Figure 24-2 to point out and emphasize the vast amounts of land devoted to food plant cultivation.

Concept Framework

All students have eaten fruits and vegetables. Use their experience as a conceptual starting point for studying this section by engaging them in a discussion of the nutritional properties of cereals, legumes, root crops, fruits, and other plants such as leafy green vegetables and stems such as celery and asparagus. Elicit the observation that many of the nutrients humans need to function on a daily basis come from plants. You may wish to conclude this discussion by suggesting that students think of examples from their own experience as they read about the various types of food plants.

Emphasis

Make certain that students understand the profound influence humans have had on plant crops. By selecting plants for nutritious, medicinal, or other qualities, people have turned many natural areas into cultivated fields. When over-cultivation of these areas depletes nutrients in the soil or drought occurs, crops fail and famine may result.

Classifying Food Plants

Bring a diverse group of food plants to class, including: cereal products such as oatmeal, breads, raw rice, and corn; legumes such as lentils, peas, and soybeans or tofu; root, stem, and leaf crops such as carrots, broccoli, lettuce, and onions; and fruits such as squash, tomatoes, pears, pineapples, and walnuts.

Divide students into three or four groups and give each group a set of foods containing at least one representative of each major category. Give the groups ten or fifteen minutes to classify their foods and list at least one nutritional benefit derived from eating each. The groups may report this information to the class or create a wall display of their findings.

Using the Features/New Food from Plant Research

The principal technique discussed in the feature is genetic engineering. Scientists have used recombinant DNA technology to create new varieties of plants that can withstand environmental variability. In some cases, the scientists are crossing crop plants with "wild" plants that are sturdier, grow more quickly or slowly, and so on. Through such procedures, researchers can increase crop yields, develop food plants more resistant to bacterial or fungal infections, and create plants with high nutritional value. For example, scientists developed a hybrid of wheat and rye called triticale in the late 1930s that has a high protein content. You may wish to point out to students that in developing food crops, people may have initially overlooked some of the beneficial traits of the "wild" variety of a plant. In restoring such traits to crop plants, today's scientists can sometimes correct "mistakes" of the past.

Section Review Answers

(Continued from p. 357.)

4. Plant fibers are used in cloth, paper, and rope; lumber is used in building; some plants are used in medicines; rubber is obtained from rubber trees; and guayule produces a substance that can be substituted for rubber.

5. If the world population continues to grow at its current rate, food production and distribution systems will have to improve to keep pace with increased demands, which will necessitate some type of Green Revolution.

24.2 Plants and the Environment

Approach

The main objective of this section is to identify and describe the role plants play in ecosystems. The section builds on concepts of chemistry and biochemistry (Chapters 3 and 4), metabolism (Chapter 7), and coevolution (Chapter 16). In presenting the material, lead students to appreciate the interactive nature of plant-habitat and plant-animal relationships. You may wish to particularly emphasize the critical role of plants in the cyclical transfer of nutrients in the biosphere.

Concept Framework

Students have learned about coevolution in Chapters 16 and 18. Build upon this base of knowledge to provide a conceptual framework for understanding plant-environment interactions. Engage students in a discussion of the mechanism of coevolution as they understand it. Elicit the generalization that plant-animal interactions can influence the reproductive success of both groups. You may wish to end the discussion by asking students as they read this section to think of other environmental factors that may have influenced plant evolution.

Emphasis

Make sure that students can relate the human uses of plants studied in Section 24.1 to the interactive concepts presented in this section. For example, you might point out that excessive farming can strip soil of its nutrients, eventually resulting in barren land, or that extensive deforestation, even in areas with healthy soil, permanently changes an ecosystem by eliminating habitats.

Field Trip

Take the class to any area near the school where several different types of plants are growing (such as a park, garden, nature preserve, or even along a residential street). Ask students to speculate about what the plants are taking from and contributing to this particular environment. Then ask students to speculate about the adaptations each plant exhibits that reflect its interaction with animals and other environmental factors. Sum up the experience by reinforcing the concept of the cyclical nature of plant ecology.

Section Review Answers

(Continued from p. 360.)

5. A brightly colored insect that is distasteful is more easily remembered by a predator than one that is not. Therefore the brightly colored species will have a better chance of survival.

Laboratory Preparation

Background

In this lab, students study the morphology and internal anatomy of several economically important food plants. A fresh asparagus tip is used to illustrate the structure of a terminal bud and the elongation that occurs below the apical meristem. A brussels sprout lateral bud is used to show the origin of branches. Students should be made aware of the inhibiting effect of auxin on the lateral buds (apical dominance). This can be seen if the lateral buds at the tip of the asparagus are compared with those farther down the stem.

Students make paper-thin cross sections of the asparagus stem with a scalpel and stain it with toluidine blue, a stable stain that produces "stained glass window images" and can be used with a wide variety of fresh biological preparations. Students learn to identify the vascular system and fibers. Students examine and describe a snow pea pod and a pea. The snow pea pod can be used to introduce the concepts of flowers, fruits, and seeds. With this basic structural background, students survey some of the economic uses of plants. Students may consider what their classroom would look like if all the plant products were removed.

Preparation

Toluidine blue stain can be purchased from a biological supply house as a 1% solution. To further dilute it to a 0.05% solution, add 1 part of the 1% solution to 19 parts of distilled water. In step 6, some of the colors produced by the toluidine blue stain might be difficult to predict. In general, vascular cells will stain red.

Safety

Remind students to handle the scalpels with care.

Conclusions

1. The bud contains, in embryonic form, a stalk and leaves.
2. The plant fibers have commercial value. They are used as paper.

Inquiry

Students will think of various uses of plants such as medicine, clothing, and shelter.

Chapter Review

Vocabulary

1. Plant ecology is the study of the interactions of plants and their environment.
2. Cereals include rye, corn, wheat, millet, sorghum, corn, barley, and oats.
3. Root crops include carrots, potatoes, yams, cassava, beets, parsnips, turnips, rutabaga, radishes, manioc, and taro.
4. Agroforestry is the planting of crops near trees; mixed cropping is the planting of two or more crops in the same area.
5. A famine is a long, widespread food shortage that occurs when food production and distribution cannot meet the need for food.

Review

11. The grains of grasses, or cereals, are the major source of food in the world.
12. Deforestation can result in wastelands and the destruction of biological communities.
13. Answers may vary. Wood provides the cellulose that is used in cellophane and rayon; cotton, flax, and hemp fibers are woven into cloth; rubber is used in a variety of products; jojoba extract is used in cosmetics.
14. Agroforestry is the planting of crops near trees. The trees help prevent soil erosion and shelter the crops.
15. Scientists study wild varieties to find beneficial adaptations, such as pest or drought resistance, that can be bred into cultivated varieties.
16. The Jimson weed produces steroid compounds that taste unpleasant to many animals. However, some animals that have coevolved with the Jimson weed have developed a tolerance for the steroids and therefore are able to continue to eat the plant.
17. The potential for famine can be decreased if farmers use plants and cultivation technologies that are compatible with their region. Examples include planting cassava or millet in the desert, or using agroforestry and mixed cropping in the tropics.
18. People act as selecting agents when they propagate only those plants that have characteristics they find valuable, such as robustness or nutritional value.
19. Digitalis, aspirin (or salicylic acid), quinine, morphine, and codeine are all medicines that are extracted from plants.
20. Colorful flowers attract pollinators. Plants that are more frequently pollinated are more likely to reproduce than plants that are not.

Critical Thinking

1. The soybean has a high protein content and is able to grow well in nitrogen-poor soil because it can fix nitrogen. These characteristics make it an ideal crop in a variety of countries.
2. The drawing shows an example of mixed cropping, or companion planting. Corn needs a lot of nitrogen to grow, some of which is fixed by the roots of the beans. The corn provides support for the beans to climb on, giving them maximum exposure to light.
3. The fish help to disperse the seeds of the fruits.
4. Many diseases are carried by insects; malaria is carried by mosquitos. Pesticides that were used in some of the new techniques of the Green Revolution were not specific but killed or weakened many insects, including mosquitos, in addition to the targeted species. Those that survived, however, formed the basis of an expanding pesticide-resistant population. In many cases the new farming techniques provided these populations with new habitats, such as irrigation ditches, in which to reproduce.

Chapter 24 | The Importance of Plants

Introduction

Plants are the producers of food. They form the base of the ecological food chain on which humans ultimately depend for survival. People also need plants for medicines, clothing, and building materials. In this chapter you will learn why plants are important to humans. You will also learn about the ecological roles of plants and the coevolution of some plants and animals. In the rest of the chapters in this unit you will examine specifics of plant biology.

Outline

24.1 Plants and People
- *Plants for Food*
- *Food Production*
- *Other Uses of Plants*

24.2 Plants and the Environment
- *Plant Ecology*
- *Coevolution of Plants and Animals*

Chapter Concept

As you read, pay attention to the relationship between plants and humans, and consider how an understanding of plant biology will lead to a better understanding of the interaction of science and society.

Potatoes, *Solanum tuberosum,* in test tubes at an agricultural research lab. Inset: Tissue culture of carrots, *Daucus carota,* for cloning.

For additional teaching information see pp. T24A–T24D.

24.1 Plants and People

Throughout human history people have used plants. For example, on the evidence of fossilized pollen found at Neanderthal burial sites, scientists believe that flowers have been placed on graves for more than 35,000 years. We still use plants in such ritualistic ways, but the relation between plants and people is primarily that of producer and consumer. Of the 350,000 species of plants people now use about 1,000 for food.

Plants for Food

The cultivation of plants probably began about 11,000 years ago in a region of the Middle East known as the Fertile Crescent. Wheat, barley, lentils, and peas were the first plants to be domesticated. Olives, dates, and grapes were later cultivated.

In cultivating plants, people propagated those individuals with characteristics that they found most valuable. This form of natural selection—with people acting as selecting agents—resulted in the evolution of food plants. For example, the wild wheat stalk breaks easily in the wind, an adaptation that increases the dispersal of ripe grains. Early people selected seeds from those plants with stalks that did not break easily. When these seeds were grown, the grains could be harvested before they fell off the plant.

Cereals

Cereals are members of the family Gramineae, the grass family. The grains of cereals are rich in carbohydrates. Cereals are the basis of human nutrition. Over half of the world's cultivated land is devoted to crops such as rice, wheat, corn, oats, sorghum, rye, and millet.

Section Objectives

- *Summarize briefly the history of the cultivation of plants by humans.*

- *Define cereals, legumes, root crops, and fruits, and distinguish between them.*

- *List the lessons learned from the Green Revolution.*

- *Define* famine, *and suggest two cultivation practices that can be used to fight famine.*

- *List several medicinal and commercial uses of plants.*

Figure 24-1 The first plants to be cultivated were cereals. Cultivation began about 11,000 years ago in the Fertile Crescent region of the Middle East. The inset map shows the present-day national boundaries in this region.

Wheat
Corn

Barley
Millet

Rice Wheat

Corn

•Yumen

Corn
Beijing

Wheat Wheat Corn
Wheat
Corn
Barley
Rice •**Shanghai**

Rice
Wheat

Millet
Wheat
Rice
Rice

Corn

Rice

Rice

•**Canton**
Rice •**Victoria (Hong Kong)**

Intensively cultivated land

Other cultivated land

Figure 24-2   Rice is one of the main crops grown in China.

Rice, *Oryza sativa,* is a source of food for more than 60 percent of the world's population. It thrives in regions with abundant rainfall. Figure 24-2 shows the distribution of rice and other cereal crops in China. Wheat, *Triticum officinale,* grows well in moderate to cold climates, including parts of the United States, the Soviet Union, and Canada.

Legumes

Legumes *(LEG-YOOMZ)* are members of the pea family. They bear protein-rich seeds in long pods. Peas, *Pisum sativum;* peanuts, *Arachis hypogaea;* and soybeans, *Glycine max,* are legumes. Legumes often form symbiotic associations with nitrogen-fixing bacteria, which enable them to grow and form protein-rich seeds in nitrogen-poor soil. About 45 percent of a soybean is protein.

Root Crops

Root crops are plants that are grown for the nutrients stored in their roots or modified underground stems. Root crops—such as carrots, *Daucus carota,* and potatoes, *Solanum tuberosum*—are excellent sources of carbohydrates, vitamins, and minerals.

Tip: Have students add iodine to carrot, potato, and turnip slices to see where starch is stored.

Fruits

Fruits are modified flower parts that enclose a seed or seeds. Grapes, *Vitis* sp., and tomatoes, *Lycopersicon esculentum,* are fleshy fruits called berries. Apples, *Malus* sp., are moist fruits called pomes; walnuts, *Juglans* sp., are dry fruits called nuts. Grains are the fruits of cereals.

Food Production

Scientists estimate that by the year 2000 there will be 6.2 billion people on the earth. To ensure that food will be available to such a large population, there must be more efficient national and international food production and distribution systems. Economists can help improve these systems. Biologists can also play a role by developing new food plants.

The Green Revolution

Between 1950 and 1970 researchers concentrated on improving the productivity of land already under cultivation. Their work led to a great increase in worldwide food production known as the **Green Revolution.** For example, Norman Borlaug, of the International Rice Institute in the Philippines, developed strong-stemmed varieties of rice that produced heavy heads of grain. These varieties were also adapted to tropical and subtropical regions. Borlaug later received the Nobel Peace Prize for his work.

Although the Green Revolution brought about many advances, the resulting distribution of new grains and agricultural techniques brought about some problems. New techniques were often too expensive to use in developing regions. Many farmers could not afford the fertilizers needed to grow the new varieties of plants. Moreover, tropical soils, which are low in essential nutrients, were often not suitable for raising some of the new crops. Researchers today are working on more effective ways to use the grains and techniques developed in the Green Revolution.

Famine

Long, widespread food shortages, called **famines,** occur when food production and distribution cannot meet the need for food. When severe drought struck Ethiopia and several other northern and central African nations in the early 1980s, many people starved.

Some agricultural methods may help reduce the possibility of famine. For example, some root crops, such as cassava, *Manihot* sp., and millet, *Panicum miliaceum,* can be grown in dry areas without expensive fertilizers or irrigation systems. **Agroforestry,** the planting of crops alongside trees in the tropics, maintains the forest, protects the soil from erosion, and shields young crops. In **mixed cropping,** the planting of two or more crops in one field, if one crop fails, the second or third may still produce.

Info: Another agricultural technique is intensive farming, in which crops are crowded together very tightly but one complements the other by taking different minerals from the soil.

Figure 24-3 Progress in agriculture is represented by research at the Rice Institute of Banos University in the Philippines (top) and mixed cropping, including plants for fodder, a root crop, and a cereal, in Devon, England (bottom).

Other Uses of Plants

Plants play an important role in many areas of our lives. Like the first medicines, many of today's medicines are made from plants. **Digitalis** *(DIJ-uh-TAL-uss)*, which is used in the treatment of heart disease, is obtained from the leaves of the foxglove plant, *Digitalis purpurea.* Periwinkles, *Vinca* sp., are the source of two drugs that are effective against Hodgkin's disease and some leukemias. Two powerful pain relievers, morphine and codeine, are extracted from the opium poppy, *Papaver somniferum.* The chief ingredient of aspirin, **salicylic** *(SAL-uh-SILL-ick)* **acid,** was originally obtained from the bark of the willow tree. The willow is in the genus *Salix,* hence the name salicylic acid. The bark of the cinchona tree, *Cinchona* sp., contains **quinine** *(KWIE-nine)*, a drug used to treat malaria.

People also use many plant fibers. Cotton, *Gossypium hirsutum,* is the most widely used plant fiber in the world. It is most often woven into cloth. Linen is made from the fibers in the stem of the flax plant, *Linen* sp., and the fibers of the hemp plant, *Cannabis sativa,* are made into ropes. The papyrus reed, *Cyperus papyrus,* has been made into paper as well as reed boats.

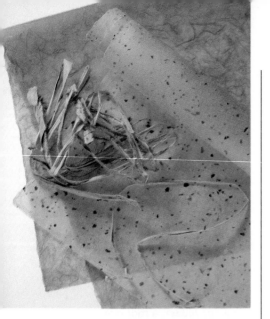

Figure 24-4 Kozo, a fiber that comes from mulberry trees grown in Japan, is the chief ingredient in a fine, very lightweight paper used in printing.

Biotechnology

New Foods from Plant Research

When early peoples began cultivating plants, they usually selected the most robust plants for food and seed. This selection process led to hardier crops with better yields. Unfortunately, it also led to decreased genetic variation. Furthermore, because only certain traits were selected for, many "wild" plants with potentially beneficial traits were not cultivated. Scientists are now rediscovering some of these wild varieties. By studying the adaptations of successful wild plants, biologists may discover ways to increase crops in areas of the world in which low food production leads to poor nutrition.

In this century the work of many agricultural researchers has been directed toward short-term solutions to problems rather than toward basic plant research. As a result scientists know a great deal about effective pesticides but comparatively little about plant genetics. Despite their use of expensive chemical pesticides and herbicides, farmers still lose up to one-third of their

Linseed oil, which is used in wood finish and paint, is processed from flax seeds. Cellophane and rayon are made from tree cellulose. Building materials can be obtained from conifers, cork trees, and bamboo plants. In some parts of the world, people use palm fronds to cover and insulate their homes.

Other useful products come from plants found in deserts. The oil of the jojoba *(huh-HOE-buh)* shrub, *Simmondsia californica,* is used in cosmetics, waxes, and industrial oils. The guayule *(gwie-YOO-lee), Parthenium argentatum,* produces a substitute for the natural rubber found in tropical trees such as the Brazilian rubber tree, *Hevea brasiliensis.*

Section Review

1. Explain *mixed cropping,* and suggest how it might ensure successful crops.
2. Name two problems associated with the Green Revolution.
3. Describe legumes, root crops, and fruits.
4. List three nonnutritional uses of plants.
5. Is the world facing the need for a new Green Revolution? Defend your answer.

Answers to Section Review
1. Mixed cropping is the planting of two or more crops in one area. If one crop fails because of pests or disease, the others may still be harvested.
2. The Green Revolution provided tropical regions with plants that were not suitable for such environments. Farmers in these countries could not afford the fertilizers needed to grow the new varieties.
3. Legumes are members of the pea family; they bear protein-rich seeds in pods. Root crops are modified underground stems or roots that store nutrients. Fruits are modified flower parts that enclose a seed or number of seeds.
(See pp. T19A–T19D.)

crops to pests and diseases. Some wild plants, however, are immune to pests and diseases.

Genetic engineering is leading the way in developing new plants and pest-resistant crops. Using the tools of genetic engineering, scientists can transfer genes from one plant to another. These scientists hope to develop plants with the ability, for instance, to fix atmospheric nitrogen, to resist frost, or to have immunity to pests or potent herbicides.

Scientists have already developed a hybrid potato plant that manufactures its own insect repellent. The repellent, a chemical called leptine, occurs naturally in wild potatoes. Researchers have fused cells of wild potatoes with those of commercial potatoes to produce a tasty, insect-resistant variety.

Once such resistant varieties are developed, they can be propagated by cloning in tissue cultures. By using the beneficial traits found in wild species to develop hardier domestic varieties, scientists might help prevent disasters such as the Irish potato famine of 1845–1847. This famine resulted from a blight that destroyed much of the main food source of the country. New varieties might also decrease the need for chemical pesticides.

Writing About Biology

Earth's Green Mantle

This excerpt is from the first chapter of Rachel Carson's book, Silent Spring.

The earth's vegetation is part of a web of life in which there are intimate and essential relations between plants and the earth, between plants and other plants, between plants and animals. Sometimes we have no choice but to disturb these relationships, but we should do so thoughtfully, with full awareness that what we do may have consequences remote in time and place. But no such humility marks the booming "weed killer" business of the present day, in which soaring sales and expanding uses mark the production of plant-killing chemicals.

One of the most tragic examples of our unthinking bludgeoning of the landscape is to be seen in the sagebrush lands of the West, where a vast campaign is on to destroy the sage and to substitute grasslands. If ever an enterprise needed to be illuminated with a sense of the history and meaning of the landscape, it is this. For here the natural landscape is eloquent of the interplay of forces that have created it. It is spread before us like the pages of an open book in which we can read why the land is what it is, and why we should preserve its integrity. But the pages lie unread.

The land of the sage is the land of the high western plains and the lower slopes of the mountains that rise above them, a land born of the great uplift of the Rocky Mountain system many millions of years ago. It is a place of harsh extremes of climate: of long winters when blizzards drive down from the mountains and snow lies deep on the plains, of summers whose heat is relieved by only scanty rains, with drought biting deep into the soil, and drying winds stealing moisture from leaf and stem.

As the landscape evolved, there must have been a long period of trial and error in which plants attempted the colonization of this high and windswept land. One after another must have failed. At last one group of plants evolved which combined all the qualities needed to survive. The sage—low-growing and shrubby—could hold its place on the mountain slopes and on the plains, and within its small gray leaves it could hold moisture enough to defy the thieving winds. It was no accident, but rather the result of long ages of experimentation by nature, that the great plains of the West became the land of the sage. . . .

So in a land which nature found suited to grass growing mixed with and under the shelter of sage, it is now proposed to eliminate the sage and create unbroken grassland. Few seem to have asked whether grasslands are a stable desirable goal in this region. Certainly nature's own answer was otherwise. The annual precipitation in this land where the rains seldom fall is not enough to support good sod-forming grass; it favors rather the perennial bunchgrass that grows in the shelter of the sage.

Write

- Rachel Carson is writing about proposals that land management agencies create more grazing land for cattle interests. Write a short essay describing how Rachel Carson feels about people altering the natural plant life of a region. Discuss the writing techniques Rachel Carson uses to persuade readers to agree with her viewpoint. How much of her argument is based on fact, and how much is opinion?

From Silent Spring *by Rachel Carson. Copyright © 1962 by Rachel Carson. Reprinted by permission of Houghton Mifflin Company.*

24.2 Plants and the Environment

Plants use solar energy in the process of photosynthesis to convert water and carbon dioxide into glucose and oxygen. The biosphere—the portion of the earth's surface on which all life exists—depends on photosynthesis.

Plant Ecology

The study of the interaction of plants and the environment is called **plant ecology.** In addition to creating food, plants cycle chemical elements and nutrients. For example, plants release oxygen through their leaves into the atmosphere. They also absorb elements released from weathering rocks, such as phosphorus, potassium, iron, and magnesium, through their roots. The organisms that eat plants, called consumers, incorporate these elements into their own bodies. Eventually these same nutrients are returned to the earth when the consumers are decomposed by organisms such as bacteria and fungi.

Some plant ecologists apply their knowledge of ecology to solving problems caused in some way by the interaction of plants and people. For example, the demands of a growing human population for food and living space have led to the destruction of vast tracts of the world's tropical rain forests. Erosion soon follows the mass cutting of trees, called **deforestation.** The nutrients in the thin topsoil are easily washed away by rains. Wastelands then develop where lush vegetation and abundant animal life once flourished. Plant ecologists are now studying ways to reduce this destruction and to solve problems caused in other regions by the actions of people.

Tip: How do you explain the fact that some of the carbon in your body right now was floating in the air as CO_2 last year? (The CO_2 was used by plants during photosynthesis. The carbon became part of the plants, which were consumed directly by humans, or by animals, which were then consumed by humans.)

Info: About 2 percent of the rain forest, or an area approximately the size of Iowa, is currently being destroyed each year.

Figure 24-5   A tropical rain forest supports hundreds of plant and animal species (left). Because the top layer of soil is very thin, trees grow buttresses for added support (right.)

Figure 24-6 Some orchid species resemble a fly (left). The tobacco hornworm (right) has coevolved as a predator on plants.

Think

The flower of this plant, *Stapelia* sp., smells like rotten meat. Suggest an adaptive significance for this trait.

Answer to Think
The flowers attract carrion flies and beetles that act as pollinators.

Answers to Section Review
1. the study of the interaction of plants and their environment
2. Plants release oxygen during photosynthesis and help recycle nutrients by absorbing chemicals from the soil. When the plants are eaten, the nutrients are passed along.
3. In some orchid species the flowers have coevolved with their pollinators, which are flies, so that the flowers resemble a female fly. The male fly lands on the flower because he thinks he has located a mate. Pollen sticks to his body and is transferred to the next orchid he visits.
4. erosion
(See pp. T19A—T19D.)

Coevolution of Plants and Animals

The study of plants and the environment must include an understanding of the interdependence of plants and animals. Some plant and animal species may adapt successively over time in ways that affect their interactions. That is, the organisms may coevolve. These changes may be mutually beneficial adaptations, as in some pollination relationships, or antagonistic interactions, as in producer–consumer relationships.

Many flowering plants have coevolved with their pollinators, organisms that carry pollen from one plant to another while gathering food. In some flowers size, shape, color, or odor make them attractive to their pollinators. For example, in some orchid species the flowers have coevolved with their fly pollinators and now look like the female fly. A male lands on a flower as if he had located a mate. The pollen he touches sticks to his body and is transferred to the next orchid he lands on.

In one example of a producer–consumer relationship, a plant produces toxic chemicals that discourage animals from eating the plant. Jimson weed, a plant in the nightshade family, produces steroid compounds that are distasteful to many animals. However, some animals coevolved with the Jimson weed and developed a tolerance for the steroids. Some insects are actually stimulated by the steroids and continue to eat the plant. In another example, poison ivy leaves are often red, a warning that the plant is toxic. Again, some vertebrates are immune to the toxins. Often families of insects that have coevolved with toxic plants and are immune to their toxins have bright warning coloration that warns predators that they are toxic.

Section Review

1. What is *plant ecology?*
2. How do plants provide essential nutrients in ecosystems?
3. Discuss how the process of coevolution operates in the pollination relationship between orchids and the flies that pollinate them.
4. What phenomenon usually occurs after deforestation?
5. What advantage does an insect that is toxic gain by being brightly colored?

Laboratory

Food Plant Morphology

To study the external morphology and internal anatomy of several important food plants

Process Skills
Observing, inferring

Materials
Fresh stalk of asparagus, scalpel, culture dish, 0.05% aqueous toluidine blue stain, forceps, paper napkin, ruler, scissors, fresh brussels sprout, 2 microscope slides, 2 coverslips, compound microscope, pod of snow peas

Method

1 Obtain a stalk of asparagus. Can you find the leaves? What is the structure at the very top of the plant?

2 Use a scalpel to cut the stem about 2.5 cm below the top of the plant. Then cut several thin cross sections of the stem.

3 Fill a culture dish one-half full with toluidine blue stain. Use forceps to place the asparagus cross sections in the solution. Place 1 cm² of paper napkin in the stain along with the asparagus. Set the culture dish aside.

4 Cut the top part of the asparagus in half lengthwise. This part of the asparagus is the apical meristem, or growing tip, where most cell division takes place. Examine the tip. Suggest how the asparagus grows in length at the cellular level. Examine the leaves on the inside carefully. What is beneath them?

5 Cut the brussels sprout in half lengthwise. How is it similar to the asparagus? Brussels sprouts are buds. Note the way these buds are placed along the stem. Try to infer what develops from these buds.

6 Return to the asparagus sections in the dish of toluidine blue stain. Using forceps, remove one of the asparagus cross sections from the stain and make a wet mount of it.

Examine the slide with a microscope set to low power. Toluidine blue stains in several colors depending on the structure stained. Plant cells containing lignin, which strengthens cell walls, stain green or blue-green. Other cell walls stain red or purple. The large, round cells are the plant's vascular system, the cells through which water and nutrients pass. Notice the small circles. These are plant fibers seen in cross section.

7 Make a wet mount of the square of stained napkin. The plant fibers you see here will be seen in longitudinal section.

8 Examine a snow pea pod. The snow pea pod is a fruit. Open the pod and break open a snow pea. Describe what you see inside.

Conclusions

1 The longitudinal sections of the asparagus and the brussels sprout look similar from the inside. Yet one is a stalk with leaves and the other is a bud. Can you explain this?

2 What plant part did you examine that has commercial value other than as food?

Inquiry

Discuss with classmates ways other than those covered in which plants are useful to humans.

Chapter 24 Review

Vocabulary

agroforestry (355)
cereal (353)
deforestation (359)

digitalis (356)
famine (355)
fruit (355)

Green Revolution
  (355)
legume (354)
mixed cropping (355)

plant ecology (359)
quinine (356)
root crop (354)
salicylic acid (356)

1. Define plant ecology.
2. List three cereals.
3. List three root crops.

4. Describe the difference between agroforestry and mixed cropping.
5. Define famine.

Review

1. Some species have evolved so that they produce toxic chemicals as a defense in response to (a) large numbers of pollinators (b) predation (c) nutrients in the soil (d) deforestation.
2. Some plants have coevolved with their pollinators and now produce (a) spines (b) grains (c) toxic chemicals (d) modified flowers.
3. Edible underground structures that store nutrients are called (a) root crops (b) legumes (c) orchards (d) grains.
4. A mutually beneficial or antagonistic adaptive change in two organisms is called (a) predation (b) coevolution (c) pollination (d) famine.
5. Organisms that use plants for food are referred to as (a) predators (b) consumers (c) producers (d) pollinators.
6. The foxglove plant aids in the treatment of heart disease because it is the source of a medication called (a) quinine (b) morphine (c) digitalis (d) salicylic acid.
7. Fruits enclose (a) seeds (b) grains (c) roots (d) berries.
8. Many of the cultivation techniques that were introduced in developing areas to increase productivity during the Green Revolution did not succeed because they were (a) too complicated (b) too expensive (c) too time-consuming (d) too practical.
9. Legumes can produce seeds that are protein-rich even in poor soil because of their symbiotic associations with bacteria that (a) break down carbohydrates (b) absorb oxygen (c) fix nitrogen (d) produce carbon.
10. The cultivation of plants by humans probably began in (a) the Middle East (b) South America (c) North America (d) northern Europe.
11. What group of plants is the major source of food for the world today?
12. List two possible long-term effects that can be caused by the deforestation of a tropical rain forest.
13. Give three examples of plants that are used commercially by humans, and describe how they are used.
14. Describe the technique of agroforestry and list the advantages of its use.
15. Why have scientists begun to study wild varieties of cultivated plants?
16. Describe how the Jimson weed operates as an antagonistic producer in the producer–consumer relationship.
17. Give two examples of ways in which the potential for famine in the world can be decreased.
18. Describe the way in which people act as selecting agents in the evolution of food plants.
19. Name three medicines that were developed from plants.
20. What adaptive advantage might colorful flowers have for plants?

Critical Thinking

1. The soybean is the third largest crop grown in the United States. In Asia, soybeans are a staple food. They are eaten fresh, dried, fermented, and processed into tofu. Some soybeans are used for animal feed or for oil. What aspects of the biology of soybeans might explain their importance to various countries of the world?

2. The drawing to the right shows pole beans that have grown up around corn stalks. What is this an example of? What advantage might each plant gain from this arrangement?

3. During the rainy season in the Brazilian rain forest the rivers flood the land. Many fish from these rivers then swim among the land plants and eat their fruit. How might this intermingling of fish and plants help the plants?

4. Many of the new crops that were planted as part of the Green Revolution required heavy applications of pesticides. In many countries the Green Revolution was followed by an increase in such diseases as malaria. Why might this have happened?

Extension

1. Read the article by Kenneth R. Sheets entitled "A Bountiful Harvest That's Hard to Swallow," in *U.S. News and World Report,* November 11, 1987, pp. 82–83. Prepare an oral report on the adverse consequences of the Green Revolution. Include in your presentation two reasons for the success of the revolution, two ways in which overabundance has affected the American economy, and the names of two nations that traditionally have imported American grain but that are now self-sufficient.

2. Using your library, research the role that plants played in the European discovery and early development of America. Write a report on what you find. Include answers to the following questions in your report: What plant product was Columbus interested in when he sailed to America? What roles did tobacco, cotton, and spruce timber play in the colonial economy?

3. Make a list of all the news items about plants or plant products that appear in your daily newspaper over one week. Don't forget that petroleum, plastics, natural gas, and coal are derived from plants.

4. Choose one food crop—potatoes, sorghum, tea, barley, tomatoes, or sugar cane—and write a report on its history, cultivation, uses, and current production.

5. Using your library, research a famine that has occurred within the last 200 years. Give an oral report in which you describe the causes of the famine, the events that brought about its end, and any changes that were made to prevent future famines in the region.

6. Read the article by Suzanne W. T. Batra entitled "Deceit and Corruption in the Blueberry Patch," in *Natural History,* August 1987, pp. 57–59. Write a report on the article, which includes an example of coevolution, that of the fungus and blueberry.

Chapter 25 | *Plant Evolution and Classification*

This chapter provides students with an overview of plants, concentrating on features that set plants apart from other living things. Students look for characteristics that indicate possible relationships among divisions.

Correlation Chart

Section Number	Section Objectives	Pages	Features	Section Review
25.1	1. Compare and contrast green algae and plants.	365–368	Biology in Process, pp. 366–367	2
	2. Name three adaptations of plants to life on land.			3, 5
	3. Compare vascular and nonvascular plants.			1
	4. Define and describe alternation of generations.			4
	5. Summarize what paleobotanists have inferred regarding plant evolutionary history.			
25.2	1. Name the three types of plants that make up the Division Bryophyta.	369–370		
	2. List the distinguishing characteristics shared by nonvascular plants.			1, 2, 3, 4, 5
	3. Compare sporophytes and gametophytes in bryophytes.			
25.3	1. Describe the two main functions of vascular tissues.	371–377	Think, p. 375 Biology Lab, p. 377	1
	2. Distinguish between gymnosperms and angiosperms.			2, 3
	3. Summarize the adaptive advantages of seeds.			5
	4. Distinguish between monocots and dicots.			4

Resources*

Print

Margulis, Lynn and Karlene V. Schwartz. *Five Kingdoms: An Illustrated Guide to the Phyla of Life on Earth.* New York: W.H. Freeman, 1982.

Wexler, Jerome. *From Spore to Spore: Ferns and How They Grow.* New York: VanNostrand Reinhold, 1984.

Audiovisual/Software

Plants of the Galápagos. Still-frame video, filmstrip, or transparencies with cassette, 15 min. Teacher's guide. Carolina Biological.

The Frozen World. Film or video. 55 min. Time-Life, 1984.

*Additional references are listed in *Modern Biology, Teacher's Program Guide.*

Section Teaching Guides

25.1 Overview of Plants

Approach

The main objective of this section is to give students an overview of the Kingdom Plantae. The section surveys nonvascular and vascular plants, with information including plants' adaptations to land, evolution, classification, and alternation of generations. Building on the concept of evolution (Chapters 15 and 16) and on knowledge of meiosis and mitosis (Chapter 9) and algae (Chapter 22), this section provides students with background information relevant to concepts presented in subsequent chapters in Unit 6. In presenting the text, guide students to realize that adapting to land was the single most important step in the evolution of plants. You may refer students to the Geologic Time Scale in Chapter 15 to review the outline of evolution of life on earth.

Concept Framework

Students will be familiar with the general characteristics of many common plants. To tap into this knowledge, assist students in listing names and characteristics of plants on the chalkboard. Elicit representatives from all major categories of plants. End the discussion by asking students to think about features that all plants share.

Emphasis

Make sure that students understand the distinctions between the following categories of plants: nonvascular plants; vascular, seedless plants; vascular, seed plants. Stress the adaptive advantages of a vascular system and seeds in helping them appreciate these distinctions.

Field Trip

Take students on a plant-collecting trip. Divide students into three groups. One group should collect five examples of nonvascular plants. The second group should collect five examples of vascular, seedless plants. The third should collect five examples of seed plants. Instruct students to note conditions favorable for the growth of each particular plant. As much as possible, collections should reflect the range of diversity of each category. Let each group give a presentation on their collection.

Using the Features/Reconstructing the Earliest Plant Fossils

This feature emphasizes the process skills of observing, modeling, and inferring. It focuses on how the botanist Harlan Banks determined the sequence of development in plant evolution. When discussing this feature, stress Bank's technique of observing cross sections of fossil plants under the microscope and modeling the plants. Help students identify the inferences Banks made in his research.

Section Review Answers

(Continued from p. 368.)
5. It gave protection from desiccation, allowing plants to survive outside the aquatic environment and to begin to colonize dry land.

25.2 Nonvascular Plants

Approach

This section's main objective is to familiarize students with the nonvascular plants that compose the Division Bryophyta. It will help students understand later chapters on ecology by introducing the role bryophytes play in succession and stability of certain ecosystems. In presenting the text, guide students to realize that bryophytes depend on free water in the environment to function and reproduce successfully.

Concept Framework

Most students will have seen green moss covering the forest floor. If possible obtain a moss sample and allow students to look at the parts of an individual moss plant under a hand lens or stereoscopic microscope. Point out that mossy carpets students are familiar with are made up of thousands of such individual moss plants. Students will be less familiar with hornworts and liverworts. Tell students that as they read this section they should watch for the similarities and differences among the mosses, hornworts, and liverworts.

Emphasis

Students should be able to list adaptations that allow mosses, hornworts, and liverworts to obtain enough water from the environment without the aid of vascular tissue.

Exhibit

Make a terrarium with moist soil to grow liverworts. Allow students to examine the liverworts and to record their observations over a few weeks. Encourage them to watch for the growth of reproductive structures.

Section Review Answers

(See p. 370.)

25.3 Vascular Plants

Approach

The main objective of this section is to introduce students to the major characteristics of vascular plants including both the seedless and the seed plants. This introduction will help students understand the details of plant morphology presented in later chapters of this unit. As you present this text, guide students to develop appreciation for the great diversity in the vascular plants.

Concept Framework

Students already know that vascular plants have vascular tissue. Introduce the function of this tissue by making an analogy with pipes in a building. Just as pipes carry water from a water main in the ground to the top of a tall building, vascular tissue carries water from the roots of a plant to the tips of its leaves. Discuss the importance of vascular tissue in the evolution of plants. Elicit the fact that vascular plants, unlike nonvascular plants, can colonize environments that do not have a ready supply of water. End the discussion by reviewing the divisions of vascular plants presented in Table 25-1. This will aid student understanding of this section.

Emphasis

Make certain students can list the major features of each of the following divisions: Psilophyta, Lycophyta, Sphenophyta, Pterophyta, Cycadophyta, Ginkgophyta, Gnetophyta, Coniferophyta, and Anthophyta.

Posters

Divide the class into three groups. Ask students in each group to work together to research the life cycle of one of the three main groups of plants and to present the information in an instructional poster. One group should work on nonvascular plants. Another group should work on vascular, seedless plants. The third should work on vascular, seed plants. Allow students time to present their posters to the class.

Section Review Answers

(See p. 376.)

Laboratory Preparation

Background

This lab procedure enables students to observe the variations and patterns in leaf venation of major groups of plants. Students will observe the stained lignin of vascular leaves and nonvascular leaflike organs under the dissecting microscope. Students will sketch and compare the various vascular tissue patterns. Students should be familiar with the basic structural differences between vascular and nonvascular plants and structural differences between gymnosperms and angiosperms.

Preparation

This investigation requires fresh leaves and is best done when the leaves can be picked rather than purchased from a greenhouse. Try to include both palmately veined and pinnately veined dicot leaves. Clear and stain leaves five days in advance. Place leaves in 5% sodium hydroxide for five days. Wash the leaves with water and place them in a 50% chlorox solution for 10 to 15 minutes. Wash leaves three times in water and place them in 70% ethyl alcohol for 30 minutes. Remove the leaves and place them in 1% safranin O for 25 minutes until stained. Place the leaves in 70% ethyl alcohol for 2 minutes, then in 90% ethyl alcohol for 2 minutes, then in 100% ethyl alcohol for 15 minutes. Place the leaves in a solution of equal parts of ethyl alcohol and xylene for 15 minutes. Finally, place the leaves in xylene for 10 minutes.

Conclusions

1. Both have dichotomous venation.
2. Dicots usually have a main vein with smaller veins branching from it. Monocots usually have veins of similar size.

3. Monocots differ in the arrangement and size of the veins. Some have veins that fork.
4. Some dicots have compound leaves that grow in two rows on each side of a central stalk (pinnate), and some have compound leaves that grow from the tip of the stalk (palmate).
5. The leaves on the left are dicots; on the right are a ginkgo and a monocot.

Inquiry

1. Students will devise ways to test whether venation is consistent within leaves of a single tree.
2. Students will devise ways to test whether venation patterns of all maples are alike.

Chapter Review

Vocabulary

1. The rhizoid is an anchor but does not conduct water; a rhizome is a stem and conducts water and nutrients.
2. The gametophyte is a haploid plant that produces gametes; the sporophyte is a diploid plant that produces spores.
3. An angiosperm is a flowering plant with seeds protected inside fruit; a gymnosperm bears naked seeds on cones.
4. A monocot has one cotyledon in the seed; a dicot has two.
5. Xylem conducts water and minerals; phloem transports sugars.

Review

11. at the end of the Silurian period; 395 million years ago
12. the green algae of the Division Chlorophyta, because they have the same photosynthetic pigments, have cellulose cell walls, and store food as starch
13. The gametophyte is haploid, while the sporophyte is diploid.
14. The cycle is the same, but in nonvascular plants the gametophyte is the dominant generation; in vascular plants, the sporophyte is dominant.
15. The details of anatomy in the fossils were too tiny to be observed without microtechniques.
16. The lack of vascular tissues for transport and support limits size.

17. They require water to transport sperm for fertilization.
18. tissue systems that transport water and nutrients to all parts of the plant
19. true roots, stems, and leaves
20. Ferns are seedless vascular plants; flowering plants are seed plants.

Critical Thinking

1. The cuticle will be thicker on the upper side of the water lily leaf, because it is exposed to air and would dry out with a thin cuticle. The cuticle on the underside of the leaf would be lying on the water, so would not have to be as thick to protect the leaf from water loss.
2. First, fossils of trees are larger than those of small plants and are therefore easier to find without specialized equipment. Second, the woody thickening of cells in tree trunks decays less before fossilization than do the softer parts of smaller vascular and nonvascular plants.
3. This moss grows near the mouths of caves (but inside the caves), in dark holes, deep woods, and sometimes in basements. Its special shape is adapted to focus as much light as possible on its chloroplasts when very little light is available.
4. Cactuses have very tough, thick cuticles that reduce the amount of moisture lost from the leaves. The plants also have fleshy stems that store water for long dry periods. Vascular tissues help transport water and nutrients throughout the plant. Also, the waxy cuticle protects the vascular plant from losing too much water by evaporation.
5. Bryophytes are not likely to survive in an open area that does not have a regular water supply, because they can only reproduce under moist conditions. However fields contain microhabitats; any microhabitat suggested by a student should be plausible.
6. As an example, as a bee goes from flower to flower collecting nectar and other nutrients, it will carry pollen from plant to plant. This increases the opportunity for pollination and ensures greater frequency of fertilization.

Chapter 25

Plant Evolution and Classification

Introduction

Kingdom Plantae includes over 350,000 species of multicellular, land-dwelling, photosynthesizing organisms. This chapter will provide you with an overview of plants, from the seemingly most simple to the most complex. It will also concentrate on features that set plants apart from other living things. You will learn more about the details of plant biology as you read the rest of the chapters in this unit.

Chapter Outline

25.1 Overview of Plants
- *Adaptations to Land*
- *Evolution*
- *Classification*
- *Reproduction*

25.2 Nonvascular Plants
- *Musci*
- *Hepaticae and Anthocerotae*

25.3 Vascular Plants
- *Seedless Vascular Plants*
- *Seed Plants*

Chapter Concept

As you read, look for characteristics of plant divisions that reflect their relative level of complexity and reveal their possible phylogenetic relationships.

Carboniferous forest. Inset: orchid.

For additional teaching information see pp. T25A–T25D.

25.1 Overview of Plants

Section Objectives

- *Compare and contrast green algae and plants.*
- *Name three adaptations of plants to life on land.*
- *Compare vascular and non-vascular plants.*
- *Define and describe alternation of generations.*
- *Summarize what paleobotanists have inferred regarding plant evolutionary history.*

At the start of the Silurian period, about 430 million years ago, organisms lived only in aquatic environments. Fishes and mollusks were diversifying, and vast colonies of marine algae thrived near turbulent shores. Corals flourished in warm, shallow waters, but the land was almost as barren as the face of the moon.

At the end of the Silurian, about 395 million years ago, evolutionary events occurred that were to have a huge biological impact. Club-shaped plants, no bigger than your little finger, began to grow and survive in the wet mud at the edge of aquatic habitats. Plants had arrived on land.

Adaptations to Land

Biologists infer that land plants evolved from green algae of the Division Chlorophyta that lived in ancient oceans or ponds. The strongest evidence for this inference lies in observed similarities between modern green algae and plants. Both have the same photosynthetic pigments, chlorophyll *a* and *b;* both store food in the form of starch; and both have cellulose cell walls.

Moving from water to land offered organisms some distinct competitive advantages: greater availability of sunlight for photosynthesis, increased levels of carbon dioxide, and decreased vulnerability to predation. However, the land environment also presented challenges. For example, plants on land were exposed to the dangers of drying out, or desiccating, through evaporation. One early adaptation to life on land was the **cuticle** *(KYOOT-ih-kul)*, a waxy protective covering. Cuticles prevent desiccation and were naturally selected in early land dwellers.

Gametes also were susceptible to desiccation. Successful land plants developed multicellular structures that helped protect gametes—eggs and sperm—during development. The presence of these **multicellular gametangia** *(GAM-uh-TAN-jee-uh)*, or gamete "holders," is the distinguishing feature of plants.

Certain plants evolved with specialized structures that absorbed water from the soil. Some of these plants also evolved complex tissues, called **vascular** *(VASS-kyuh-lur)* **tissues,** that transported water and food from one part of the plant to another. Vascular tissues also helped support the plants, an important function since plants living on land are not supported by the buoyancy of water.

As evolution proceeded, other adaptations occurred. Adaptations that increased the ability of the stems to absorb light were selected for, as were underground stems that anchored plants to the land and absorbed nutrients and water. Many plants evolved with tissues that stored the food made in photosynthesis. Fossil evidence indicates that light-absorbing stems gradually evolved into true leaves. Underground stems gradually evolved into true

Info: The protective cuticle restricted the exchange of oxygen and carbon dioxide with the environment. Small pores in the cuticle were a later adaptation.

Figure 25-1 *Fritschiella,* **a green alga, has some characteristics that are similar to plants.**

roots. Resistant reproductive cells called **spores** evolved with the first land plants and allowed widespread dispersal of species. Much later some plants developed seeds, which protected and nourished embryos. Some developed woody tissues and grew to great heights, giving them an advantage in gathering light.

Evolution

Much of what we know about plant phylogeny has come from the study of the fossil record. The earliest known plant fossil is of the genus *Cooksonia*. This genus appears in rock dating from the late Silurian period, about 400 million years ago. *Cooksonia* grew to only about 10 cm high and had forked, leafless stems with reproductive structures at the tips. It also had vascular tissue, which suggests that it lived on land. Other leafless plants, such as *Rhynia* and *Zosterophyllum*, appear in fossils from the early Devonian period, about 390 million years ago. Mid-Devonian fossils from about 345 million years ago include plants with true leaves.

Fossils from the late Devonian period indicate that by 340 million years ago trees had evolved. Giant horsetails, giant club mosses, and tree ferns dominated the forests of the late Devonian

Figure 25-2 *Cooksonia,* the earliest known plant, had reproductive structures at the tips of leafless stems.

Biology in Process

Reconstructing the Earliest Plant Fossils

How did modern plants become such a complex and diverse group? What did early land plants look like? Where should we look for evidence of plant evolution?

Major construction projects can sometimes yield more than rubble. For example, as roads were being built in New York, layers of Devonian rock were exposed and yielded hundreds of microfossils (top left) and fossilized plants (right). Harlan Banks of Cornell University visited these sites and collected fossils for analysis.

Harlan Banks is a paleobotanist studying the plant life of the past. Luckily for Banks, northern New York is a gold mine of 350- to 400-million-year-old plant fossils. Banks scours fresh road cuts for rock layers that contain black marks, iron stains, or other clues that might lead him to fossils.

Banks wants to know about the internal tissues of early plants in order to model the sequence of evolutionary events in land plants. He therefore uses painstaking techniques to prepare fossils for microscopic observation. To see the structural detail of an ancient plant, Banks may cut a fossil-containing rock into 6,000 paper-thin slices by using a

Answer to Biology in Process: Banks used inferring to correlate his models of early plants with data from other studies and to determine the sequence of developments in plant evolution.

and Carboniferous periods. Ancestors of today's pine trees appeared in the late Carboniferous period, about 280 million years ago. Flowering plants, the dominant group of plants on the earth today, did not evolve until the early Cretaceous period, only 135 million years ago.

Classification

Taxonomists divide land plants into two groups based on whether or not they have vascular tissue. The **nonvascular plants,** members of the Division Bryophyta *(brie-AHF-uh-tuh),* have neither vascular tissue nor true roots, stems, or leaves. Members of the nine divisions of **vascular plants** have water-conducting tissues and most have true roots, stems, and leaves.

Study the classification of plants in Table 25-1 on the following page. You can see that vascular plants are further divided into two groups, the seedless plants and the seed plants. Seedless plants include ferns and the three divisions of plants called fern allies. Seed plants are classified into four divisions of **gymnosperms** *(JIM-nuh-SPERMZ)* and one division of **angiosperms** *(AN-jee-uh-SPERMZ),* or flowering plants.

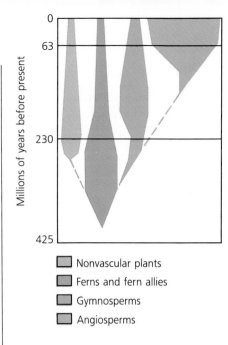

Millions of years before present

- Nonvascular plants
- Ferns and fern allies
- Gymnosperms
- Angiosperms

Figure 25-3 The width of each branch is proportionate to the number of species in existence at a given time.

diamond-edged saw. He polishes each slice until it is thin enough to let light shine through. By observing each successive slice, Banks can reconstruct three-dimensional details and make models of the plants as they may have appeared.

Banks correlates his models of early plants with data from geographic studies and radioactive dating procedures. He then arranges his fossils in order of increasing complexity, modeling the progression of adaptations he observes.

From models such as these, paleobotanists can infer the sequence of developments that occurred over time in plant evolution. Then, they can formulate hypotheses concerning the rate of such developments.

With today's techniques for observing details in fossils, paleobotanists can distinguish fossils of the earliest land plants from their waterborne ancestors. Using the microscope, scientists search for xylem, phloem, and waxy coverings.

Studies of fossils from the early Devonian period have revealed details of the development of roots, intricate branching patterns, spore-bearing organs, and woody tissues. From further fossil evidence paleobotanists infer that leaves and other characteristics emerged afterward. Banks and his colleagues have concluded that after only 15 million years on land, plants had already developed many of the features of present-day plants.

■ In which instances did Banks use the process of inferring?

Table 25-1   Land Plant Classification

Nonvascular plants	Vascular plants: Seedless	Vascular plants: Seed

Division Bryophyta: Mosses, liverworts, hornworts
 	
    Class Musci
    Class Hepaticae
    Class Anthocerotae
	Division Psilophyta: Whisk ferns
	Division Lycophyta: Clubmosses
	Division Sphenophyta: Horsetails
	Division Pterophyta: Ferns
		Gymnosperms
		Division Cycadophyta: Cycads
		Division Ginkgophyta: Ginkgos
		Division Coniferophyta: Conifers
		Division Gnetophyta: Gnetophytes
		Angiosperms
		Division Anthophyta: Flowering plants
		    Class Monocotyledonae
		    Class Dicotyledonae

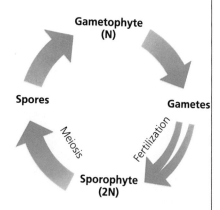

Figure 25-4   In the alternation of generations, plants alternate between a multicellular haploid stage and a multicellular diploid stage.

Answers to Section Review
1. Nonvascular plants do not have specialized tissue systems that move food, water, and minerals from one part of the plant to another, while vascular plants have these complex conducting systems.
2. Both have the same photosynthetic pigments; both have cellulose walls; and both store food as starch.
3. Spores are resistant to drying out and aid in dispersal.
4. The gametophyte produces gametes, and the sporophyte produces spores.
(See pp. T25A–T25D.)

Reproduction

All plants have a sexual life cycle called **alternation of generations,** in which a haploid (1N) phase alternates with a diploid (2N) phase. As Figure 25-4 shows, a haploid multicellular **gametophyte** *(guh-MEET-uh-FITE)* produces gametes that join in fertilization to form a zygote. The zygote (2N) develops by mitosis into a diploid **sporophyte** *(SPOR-uh-FITE)*. The sporophyte then produces a **sporangium** *(spor-AN-jee-um)*, a spore-bearing sac. Some cells of the sporangium undergo meiosis and produce haploid (1N) spores. These spores are released and then develop into new 1N gametophytes. In all land plants the gametophyte looks different from the sporophyte. This type of life cycle is called a heteromorphic alternation of generations.

In nonvascular plants the gametophyte is the larger and more long-lived phase of the sexual life cycle. For example, the familiar, leafy, highly visible part of the moss plant is the gametophyte. In contrast, vascular plants have dominant sporophytes. In seed-bearing plants the gametophyte is not a separate plant as it usually is in mosses and ferns. An apple tree is an example of a sporophyte. Apple gametophytes are microscopic parts of the apple flower.

Section Review

1. What is the main difference between *vascular plants* and *nonvascular plants?*
2. What are the similarities between today's land plants and the green algae?
3. How has the evolution of spores contributed to the chances of success for land plants?
4. What is the principal difference between a gametophyte and a sporophyte?
5. Why did the cuticle represent an adaptive advantage for early land plants?

25.2 Nonvascular Plants

The nonvascular plants include about 23,000 species of mosses, liverworts, and hornworts that form the Division Bryophyta. Structural evidence suggests that bryophytes, though lacking vascular tissue, probably arose from a line of plants that had vascular tissues. For example, bryophyte capsules have nonfunctioning stomata that resemble the stomata found in vascular plants, which you will learn more about in Chapter 26.

Without complex support and vascular systems, bryophytes seldom grow large. Most are small and lie close to the soil or to the bark of trees, where they can absorb water rapidly.

Bryophytes need water to reproduce because the sperm must swim through water to an egg. These plants live in damp areas but can also live in deserts and other dry environments, reproducing during periods of moisture. They can survive in extreme conditions by occupying small areas that are protected from the surrounding climate. For example, you might find mosses growing in the cool, moist cracks of desert rocks.

Musci

Almost every land environment in the world is home to at least one of the more than 14,000 species of mosses of the Class Musci. The thick green carpets you see on shady forest floors actually consist of thousands of tiny moss gametophytes. Each gametophyte is attached to the soil by rootlike structures called **rhizoids** *(RIE-zoidz)*. Unlike roots, rhizoids do not conduct water. However, they do serve as anchors for the plants. Mosses also have leaflike structures only one cell thick that are adapted for photosynthesis and absorb moisture rapidly. Moss gametophytes can be found in sizes ranging from 1 cm to 1 m, but most moss plants are less than 15 cm in length.

Periodically moss gametophytes (1N) produce eggs and sperm. Sperm from one gametophyte fertilizes an egg from another gametophyte. The resulting zygote produces a stalklike sporophyte. The sporophyte forms a sporangium called a **capsule,** which matures and releases spores. The spores develop into tiny, green-branched filaments that grow into mature gametophytes.

Some mosses serve the important environmental function of acting as pioneer species in barren areas. By gradually breaking up the surface of rocks, for example, mosses create a layer of soil in which other plants can begin to grow. In areas devastated by fire, volcanic action, or human activities, pioneer mosses can help trigger the development of new biological communities. Their rhizoids also help prevent erosion by holding soil in place.

Sphagnum is a genus of moss that is a major component of peat bogs in the northern parts of the world. People in northern European and Asian countries mine and dry peat to use as fuel. In

Tip: Tell students that the sexual organs of bryophytes must be close to the ground because that is where available water is most likely to be found.

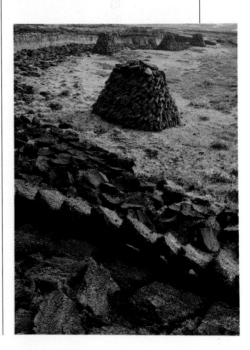

Figure 25-5 Hand-cut sections are removed from peat bogs, dried in the sun, and used for home heating on the Isle of Lewis, Scotland.

Figure 25-6 The conspicuous, leafy forms of moss plants, *Mnium hornum* (left), and liverworts, *Marchantia* sp. (right), are the gametophyte generation.

Answers to Section Review
1. a sporangium produced by a moss gametophyte
2. They do not have vascular tissue.
3. By breaking up rock surfaces, mosses create a layer of soil in which other plants can grow.
4. Because the plant has no vascular transport system, the liverwort must absorb as much water as possible over as much of its surface as possible. The more rapidly and efficiently the plant can absorb water, the healthier and more able to reproduce it will be.
5. Bryophytes have multicellular gametangia.

Ireland and other countries, this mining is so extensive that it threatens the survival of the bog communities. *Sphagnum* mosses give off an acid that slows down decomposition in the bogs. Archeologists have recovered intact human remains that have been buried in bogs for thousands of years.

Hepaticae and Anthocerotae

Liverworts are bryophytes of the Class Hepaticae. Hornworts are bryophytes placed in the Class Anthocerotae. Combined, both classes have fewer species than does Musci, the class of mosses.

Some liverworts have a thalloid form—that is, a flat body that has distinguishable upper and lower surfaces. Other liverworts have a leafy form, with thin, transparent, leaflike structures arranged along a stemlike axis. Both kinds of liverworts lie close to the ground, an adaptation that allows the plant to absorb soil water readily. In some species the sporophyte is topped by a parasol-shaped structure that holds sporangia.

Hornworts resemble liverworts, but the structural details of their sporangia are different. Like liverworts, hornworts grow in moist, shaded habitats. They share an unusual characteristic with algae: each of their cells has a single large chloroplast rather than numerous small ones.

Section Review

1. What is a *capsule?*
2. How are bryophytes different from all other plant divisions?
3. What role do mosses play in the early development of biological communities?
4. Some liverworts have leaflike structures that absorb water rapidly. In what way is this an important feature for a nonvascular plant?
5. Why are bryophytes classified along with plants instead of with algae?

25.3 Vascular Plants

Why are vascular plants so numerous? Biologists suggest that certain specializations permit vascular plants to live in more kinds of environments than nonvascular plants can. For example, the strong stems of vascular plants enable them to grow tall. The height of vascular plants allows them to rise above other plants, thereby getting an unimpeded source of sunshine. Without the hindrance of shade from other plants, vascular plants are able to make and distribute food efficiently.

Two types of tissue make up the vascular system. **Xylem** *(ZIE-lum)* carries water and dissolved minerals from the roots to the stems and leaves. **Phloem** *(FLOE-um)* transports the sugar made by photosynthesis from the leaves to the rest of the plant. Phloem also carries stored food from the roots to other plant parts. Both types of vascular tissue are composed of cells with thick walls, which also help support the plant stem.

Seedless Vascular Plants

Seedless vascular plants were probably much more common during the Devonian period than they are today. The four divisions of modern seedless plants are Psilophyta *(sie-LAHF-uh-tuh)*, Lycophyta *(lie-KAHF-uh-tuh)*, Sphenophyta *(sfee-NAHF-uh-tuh)*, and Pterophyta *(ter-AHF-uh-tuh)*. Members of the first three divisions are fern allies, while members of the last are ferns.

Psilophyta, Lycophyta, and Sphenophyta

The Division Psilophyta is represented by whisk ferns of the genus *Psilotum*. Despite their name, members of this genus are not ferns at all. They have no roots or leaves and produce sporangia on the ends of short branches. These features suggest that *Psilotum* resembles early land plants. Whisk ferns live in tropical and subtropical climates.

The Division Lycophyta includes three related genera: the spikemosses, club mosses, and quillworts. Spikemosses are vascular plants in the genus *Selaginella*. Club mosses, which make up the genus *Lycopodium*, are small evergreen plants that often look like tiny pine trees. Club mosses have numerous leaves with a

Section Objectives

- *Describe the two main functions of vascular tissues.*
- *Distinguish between gymnosperms and angiosperms.*
- *Summarize the adaptive advantages of seeds.*
- *Distinguish between monocots and dicots.*

Figure 25-7 The whisk fern, *Psilotum* sp. (left), a club moss, *Lycopodium* (center), and the spikemoss, *Selaginella* sp. (right), illustrate the diversity of seedless vascular plants.

Figure 25-8 The quillwort, *Isoetes storkii* (left), bears spores on all leaves, while the horsetail, *Equisetum tamateia* (right), has a spore-bearing, whitish shoot as well as green nonfertile shoots.

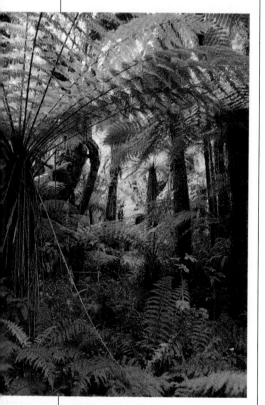

Figure 25-9 Tree ferns, *Cyathea smithii,* are the largest living ferns. Which generation is represented here?

tough cuticle that helps retain water. The small quillworts, members of the genus *Isoetes,* live submerged in ponds and lakes. You can see in Figure 25-8 that their tubular leaves resemble small feather quills. Members of all three of these genera bear spores near the bases of their leaves. However, some species of *Lycopodium* have their sporangia-bearing leaves grouped into a conelike structure called a **strobilus.**

The Division Sphenophyta includes plants of the genus *Equisetum,* or horsetails. Horsetails have jointed stems and scalelike leaves. The outer cells of these gritty-feeling plants contain silicon dioxide, the major component of sand. Native Americans and early settlers used horsetails as pot scrubbers. This practice led to another common name for the genus, scouring rushes.

Horsetails thrive near streams and ponds and also grow hardily in the tropics. Some may even grow in dry sand along roadbeds or railroad tracks. As you can see in Figure 25-8, horsetails may have two types of shoots—vegetative and fertile. The vegetative shoots of horsetails have scalelike leaves that are arranged in a whorl and lie flat on the stem. At the top of each fertile shoot is a spore-bearing strobilus. The spores blow away in the wind. They land and develop into small gametophytes.

Pterophyta

Ferns probably originated in the early Carboniferous period, about 335 million years ago. Ancestral ferns, along with giant types of club mosses and horsetails, composed the vast coal-forming forests illustrated at the beginning of this chapter.

The 12,000 species of modern ferns are members of the Division Pterophyta. Ferns are a diverse group. Some are tiny aquatic plants. Others live high in jungle trees or in the cracks of mountaintop rocks. Ferns also grow above the Arctic Circle and in desert regions. The largest living ferns are tree ferns, shown in Figure 25-9. These ferns can reach 25 m in height, and some have leaves 5 m long.

Most ferns have an underground stem called a **rhizome.** The rhizome contains vascular tissues and produces roots that absorb water and minerals from the soil. The leaves of ferns arise as tiny coiled leaves called **fiddleheads.** Fiddleheads uncoil and develop into mature leaves called **fronds.** Even the fronds of tropical tree ferns begin as fiddleheads.

Ferns produce sporangia and spores on the lower surface of their leaves. If you turn a fern leaf over, you can see the usually circular sori *(SOE-ree).* These are clusters of sporangia where spores are formed. When the spores are mature, the sporangium breaks and the spores are scattered.

Seed Plants

Spores of land plants are unicellular reproductive structures formed by meiosis in a sporangium. A **seed,** however, is a multi-cellular structure that contains an embryo. You will learn more about the nature of a seed in Chapter 27.

Seeds are an evolutionary success story. Seeds increase the chances that a plant will have reproductive success. Inside the tough protective coat of a seed is an embryo and a food supply. When conditions are too hot or too cold, or too wet or too dry, the seed remains inactive. When conditions favor growth, the seed germinates—that is, the embryo begins to grow into a young plant called a seedling.

Seeds helped the seed plants survive while many seedless land plants died out in the dry climate of the late Mesozoic era. Today there are five divisions of seed plants. Seven hundred species of living gymnosperms compose four divisions: Cycadophyta, Ginkgophyta, Coniferophyta, and Gnetophyta. Gymnosperms produce "naked" seeds; they are not enclosed in fruits. The other division is Anthophyta, the angiosperms. Angiosperm seeds are enclosed in fruits.

Cycadophyta and Ginkgophyta

Cycads *(SIE-kadz)* are gymnosperms of the Division Cycadophyta. They flourished during the Mesozoic era, the age of the dinosaurs. Only about 100 species of cycads survive today, most of which are native to the tropics. Like palm trees, most cycads have their leaves at the top of a short, thick trunk. Cycad plants are either male or female and bear large cones like those shown in Figure 25-10.

Like the cycads, ginkgos *(GINK-oze)* flourished in the Mesozoic era. Ginkgos are gymnosperms of the species *Ginkgo biloba,* the only remaining species of the Division Ginkgophyta. Ginkgos are deciduous trees that grow well in cities, since they can tolerate air pollution. Female trees are rarely planted, however, because they produce messy seeds that have a strong, disagreeable odor. Notice the plum-shaped fleshy seeds on the ginkgo shown in Figure 25-10.

Figure 25-10 The cycad, *Encephalartos princeps* (below), and the ginkgo, *Ginkgo biloba* (bottom), are both gymnosperms with male and female individuals.

Fir needles and cones

Pine needles and cones

Yew needles and cones

Figure 25-11 The needles and cones of conifers come in many shapes and sizes: fir, *Abies* sp. (top), pine, *Pinus* sp. (middle), and yew, *Taxus* sp. (bottom).

Figure 25-12 *Welwitschia mirabilis* leaves direct excess water to the root. The underground root is the largest part of the plant.

Coniferophyta

The conifers *(KAHN-uh-ferz)*, which are gymnosperms of the Division Coniferophyta, include pine trees, cedars, junipers, redwoods, yews, firs, and spruces. Conifers are woody plants with needles or scalelike leaves. Each individual bears both male and female cones. The small male cones, which are strobili, appear in clusters. Most release clouds of dustlike pollen and then fall off their branches. After pollination, the complex, woody female cones close up tightly. This protects the developing seeds, which ripen and then are released.

The oldest living organisms in the world are conifers called bristlecone pines, which are found in the mountains of the Great Basin in California and Nevada. Some of these trees have survived for 4,900 years. Their age has been determined by biologists who specialize in analyzing the rings in the wood of trees and has been supported by radioactive dating techniques. The tallest trees in the world are also conifers. Redwoods growing along the Pacific coast of northern California may grow as high as 110 m, the height of a 30-story skyscraper. The tree most widely used for lumber and papermaking, the Douglas fir, is also a conifer.

Gnetophyta

An odd group of cone-bearing gymnosperms called gnetophytes *(NEE-tuh-FITES)* have vascular systems more like those of angiosperms than of gymnosperms. Botanists classify the 70 species of the Division Gnetophyta into three genera: *Ephedra*, *Gnetum*, and *Welwitschia*.

Ephedra (ih-FED-ruh) is an American desert genus, with clustered stems that look like horsetails. *Gnetum (NEE-tum)* is a genus of tropical vines or trees, with broad leaves that resemble those of flowering plants.

Welwitschia mirabilis (well-WIH-chee-uh mir-AH-buh-lus) is the only species in the genus. It grows in the Namib desert of southwestern Africa. The Namib lies near the Atlantic Ocean and receives more moisture than most deserts. Some nights a thick fog rolls in over the desert. *W. mirabilis* is modified to take advantage of the fog. This plant has a single, huge, deeply penetrating root and only two leaves. The leaves extend out from the flat, tablelike stem. The leaves grow at their bases, and would be quite long if not for the wear and tear the tips take in the desert wind. *Welwitschia* leaves absorb the dew that condenses from the surrounding fog.

Anthophyta: Diversity

Today, about 275,000 species of the Division Anthophyta dominate the plant world. Angiosperms are flowering seed plants characterized primarily by the presence of a fruit. You will learn about the fruit in Chapter 27.

Angiosperms grow in many forms. Some are delicate herbs with showy blossoms, such as violets or orchids. In plant biology, the term *herb* refers to a nonwoody plant. Others, such as rosebushes, are woody shrubs. Some, such as the ivy you see growing on buildings, are vines. The oaks, aspens, and birches of the forest are all flowering plants, although you may never have noticed their blossoms. Grasses are also angiosperms, but you must look closely to see their tiny, highly modified flowers.

Angiosperms first appear in the fossil record at the beginning of the Cretaceous period, about 135 million years ago. At that time gymnosperms dominated the plant world. By 100 million years ago, however, flowering plants had probably begun to outnumber gymnosperms.

What led to the relatively sudden success of this new kind of plant? One factor is that angiosperm seeds are themselves enclosed in protective fruits. The fruits of flowering plants protect the seeds and aid in their dispersal. Additionally, angiosperms have highly efficient vascular systems.

Another important factor in the success of angiosperms is that they evolved along with insects. The fossil record indicates that this coevolution, or mutual adaptation, of flowering plants and insects began in the Cretaceous period (135 million to 65 million years ago). Many angiosperms can reproduce only when insects carry pollen from one plant to another. The insects visit the flowers for nourishing nectar and pollen. In doing so they pollinate plants.

Some plants have evolved remarkable means of attracting insects. For example, some flowers produce scent only at night, when the moths that pollinate them are active. Some species produce small amounts of nectar so that insects must visit many plants. You will learn more about pollination in Chapter 27.

Anthophyta: Classification

The Division Anthophyta has two classes—Monocotyledoneae and Dicotyledoneae. One feature that distinguishes these two groups is the number of **cotyledons**, or seed leaves. **Monocots** have a single cotyledon in their embryos, while **dicots** have two. Mature monocot leaves have parallel venation—arrangement of

Think

Through the coevolution of plants and insects, many unique methods of attracting insects to flowers have evolved. What do you think is the value of the bright yellow nectar guides on the purple irises shown here?

Answer to Think
The pollen guides lead insects to the center of the flower, where they will find the nectar.

Dicot
_____

Bean seed   Periwinkle   Maple leaf

Monocot
_____

Corn grain   Iris   Corn leaf

Figure 25-13 Monocots and dicots each have their own characteristic number of cotyledons, flower symmetry, and leaf venation.

Answers to Section Review
1. Xylem and phloem are the two types of tissue that make up the vascular system. Xylem transports water; phloem transports food. Both lend support to stems.
2. Gymnosperm seeds are not protected by a fruit.
3. *Ephedra, Gnetum,* and *Welwitschia*
4. Monocots have one seed leaf, leaves with parallel veins, and flowers with three-part symmetry. Dicots have two seed leaves, leaf veins in a netlike pattern, and flowers with four- or five-part symmetry.
5. Seeds increase the chances that a plant will successfully reproduce. The embryo and a food supply are protected inside the seed's tough coat. There they lie dormant until favorable environmental conditions permit germination.

veins—while dicot leaves have net venation. You can also recognize monocots by the three-part symmetry of their flowers. In contrast, most dicot flower parts occur in fours or fives, as you can see in Figure 25-13.

The approximately 90,000 monocot species include lilies, sedges, irises, onions, orchids, corn, bamboo, and leeks. A small number of woody trees, including coconuts and other palms, are also monocots. Much of the world's food supply comes from the grains of monocot grasses, such as wheat, oats, corn, rye, and rice.

Dicots include the majority of the flowering plants, such as magnolias, mustards, and maples as well as cactuses and most forest trees. In fact, with about 185,000 species, dicots are the most abundant and diverse group of all the plants that have adapted to life on land.

Section Review

1. What are *xylem* and *phloem*, and what functions do they perform?
2. What is the main difference between gymnosperms and angiosperms?
3. What are the three genera that compose the Division Gnetophyta?
4. What are the three differences that distinguish monocots and dicots?
5. What reproductive advantages do seed plants have over seedless vascular plants and nonvascular plants?

Laboratory

Patterns in Leaf Venation

Objective
To observe the variations in patterns of leaf venation in major groups of plants

Process Skills
Inferring, classifying

Materials
Stained leaflike organs of moss and liverwort; leaves of ferns, conifers, ginkgos, and 3 kinds each of monocots and dicots; petri dish; dissecting microscope; unlined paper

Method

1 Your teacher has stained several types of leaves with a red dye called safranin in order to make the venation pattern of the leaves stand out. Safranin stains lignin, which is a major hardening component of leaf vascular tissue. Place stained moss and liverwort "leaves" in a petri dish and add enough water to cover them.

2 Observe the "leaves" under the dissecting microscope. Record your observations by sketching each leaf. Do you see any red-stained venation in either the moss or the liverwort leaves? Explain why or why not.

3 Repeat the procedure with the fern leaf. Make a careful sketch of the leaf, noting the venation pattern.

4 Repeat the procedure with the conifer and ginkgo leaves. To what major group do conifer and ginkgos belong? Continue sketching and identifying vascular tissue.

5 Repeat the procedure with monocot leaves. To what major group do monocot leaves belong? Observe similarities and differences among monocot leaves. Continue sketching and identifying vascular tissue.

6 Repeat the procedure with dicot leaves. To what major group do dicot leaves belong? Observe similarities and differences among dicot leaves. Continue sketching and identifying vascular tissue.

Conclusions

1 What kind of venation does a ginkgo have? In what way is it similar to fern venation?

2 How does monocot venation differ from dicot venation?

3 Compare the venation in the different types of monocot leaves. In what ways do the veins differ?

4 Compare the venation in the different types of dicot leaves. How do the veins differ?

5 Look at the photographs of leaf venation on this page. Can you tell, by observing patterns in venation, what major group of plants each of these four leaves came from?

Inquiry

1 Do you think venation is consistent among all the leaves on a single tree? How could you tell?

2 Describe a method that provides an answer to the question: Are the venation patterns of all maples alike?

Include both palmately veined and pinnately veined dicot leaves.

Chapter 25 Review

Vocabulary

alternation of generations (368)	fiddlehead (373)	nonvascular plant (367)	spore (366)
angiosperm (367)	frond (373)	phloem (371)	sporophyte (368)
capsule (369)	gametophyte (368)	rhizoid (369)	strobilus (372)
cotyledon (375)	gymnosperm (367)	rhizome (373)	vascular plant (367)
cuticle (365)	monocot (375)	seed (373)	vascular tissue (365)
dicot (375)	multicellular gametangia (365)	sporangium (368)	xylem (371)

Explain the difference between each of the following pairs of terms.
1. rhizoid and rhizome
2. gametophyte and sporophyte
3. angiosperm and gymnosperm
4. monocot and dicot
5. xylem and phloem

Review

1. The Division Bryophyta includes (a) mosses and club mosses (b) hornworts and liverworts (c) liverworts and cycads (d) conifers and ginkgos.
2. Mosses help start new biological communities by (a) detecting air pollution (b) forming new soil (c) producing spores (d) slowing decomposition.
3. Phloem conducts (a) air (b) sugars (c) chlorophyll (d) soil.
4. The presence of silicon dioxide in the outer cells of the plant is a characteristic of (a) ferns (b) mosses (c) horsetails (d) club mosses.
5. Flowering plants are in the Division (a) Psilophyta (b) Anthophyta (c) Gnetophyta (d) Sphenophyta.
6. Seedless vascular plants include all of the following *except* (a) ferns (b) horsetails (c) cycads (d) club mosses.
7. Gymnosperms include all of the following *except* (a) cycads (b) gnetophytes (c) ginkgos (d) quillworts.
8. The great success of angiosperms is due in part to (a) highly efficient vascular systems (b) seeds protected inside fruits (c) coevolution with insects (d) all of the above.
9. A monocot has (a) parallel-veined leaves (b) two seed leaves (c) four-part flowers (d) five-part flowers.
10. The distinguishing feature of all plants is the presence of (a) vascular tissue (b) pollen grains (c) multicellular gametangia (d) rhizoids.
11. When did the first land plants appear?
12. Which division of organisms is believed to represent the ancestors of land plants? Why?
13. What is the genetic difference between the gametophyte and the sporophyte?
14. How is alternation of generations different in nonvascular plants and vascular plants?
15. Why were microtechniques for analyzing fossils necessary before the details of the evolution of land plants could be revealed?
16. What factors limit the size of nonvascular plants?
17. Why must nonvascular plants live in damp environments?
18. What is the principal distinguishing feature of vascular plants?
19. List the structural features of vascular plants that nonvascular plants lack.
20. Why are ferns and flowering plants placed in different divisions?

Critical Thinking

1. The water lily, illustrated at the right, is a vascular plant that lives in aquatic habitats. Do you think its cuticle would be thicker on the upper surface of the leaf or on the lower surface? Explain.

2. Fossils of trees are easier to find than fossils of small plants. Give two possible explanations for this fact.

3. Luminous moss *(Schistostega pennata)* has cells shaped like lenses, with chloroplasts spread out behind the curving cell membrane. In what sort of environment would you expect to find this moss?

4. Cactuses are vascular plants that are adapted to dry environments. What specific adaptations might have been selected for over the course of their evolution?

5. Explain whether or not you would expect to find mosses growing in open fields.

6. While gymnosperm cones are either male or female, angiosperm flowers usually contain both male and female structures. How might this characteristic benefit angiosperms?

Extension

1. Read the article by E. W. Stiles, "Fruit for All Seasons," *Natural History,* August 1984, p. 42. Make an oral report on plants having seeds that are dispersed by birds. Concentrate on the relationship that appears to exist between the fruiting times of plants and the paths followed by migrating birds. What are the adaptive advantages to plants when their fruits undergo double color changes? What advantages do plants with "foliar fruit flags" have? Offer a possible evolutionary scenario that describes how natural selection may have influenced such plants.

2. Go to your local florist or garden shop and get a handful of *Sphagnum* moss. Let it dry overnight in a warm place. Weigh the moss to the nearest 0.1 g, and record the weight. Then place the moss in water, and weight it down. Let the moss sit overnight. Remove the moss, let the excess water drain off, and weigh the moss again. Calculate the percentage increase in weight as follows:
   (a) Wet weight − dry weight = weight due to water
   (b) Weight due to water/dry weight × 100 = percentage increase due to water

3. Collect leaves from five different types of plants growing in deep shade (such as under a porch, in a forest, or in the shadow of a tall building). Also collect leaves from five different types of plants growing in full sun. Examine the leaves with a hand lens or dissecting microscope, and create a chart using the following columns: (top) sun, shade; (side) leaf hairs on top, leaf hairs on bottom, length of leaf, width of leaf, and thick/thin. Make drawings of the hairs from each leaf. Do the leaf hairs from the different types of plants look different? Do the leaves from the various plants differ in thickness? Are there more hairs on the leaves from the sun or on the leaves from the shade? What else can you conclude?

Chapter 26 | *Plant Structure and Function*

This chapter describes plant tissue systems and how they are organized for specialized functions. Students learn how these tissues are arranged differently in different plants in response to environmental conditions.

Correlation Chart

Section Number	Section Objectives	Pages	Features	Section Review
26.1	1. Describe the three plant tissue systems and the kinds of cells that occur in each.	381–382		1, 3, 5
	2. Describe the two main types of meristem tissue.			4, 5
	3. Differentiate between primary and secondary growth.			2
26.2	1. Explain the difference between a taproot system and a fibrous root system.	383–386		2
	2. Distinguish between primary growth and secondary growth in roots.			3, 4, 5
	3. Describe primary root tissues.			1, 4, 5
	4. List the three principal functions of roots.			3
26.3	1. Describe differences between monocot and dicot stems.	387–392	Think, p. 388	2
	2. List the secondary tissues in the dicot stem.		Biology in Process, pp. 390–391	1, 2
	3. Explain the formation of an annual ring.			1
	4. Explain the pressure-flow hypothesis, and tell what substances are transported in phloem.			3, 4
	5. Describe the transpiration-cohesion theory.			5
26.4	1. Identify the main external features of leaves.	393–397	Biology Lab, p. 397	1, 5
	2. Explain the importance of stomata.			3, 4
	3. Describe the three main regions of leaf tissue.			2, 5

Resources*

Print

Selsam, Millicent E. *Tree Flowers*. New York: Morrow, 1984.

Wilkens, Malcom B. *Advanced Plant Physiology*. London: Pitman, 1984.

Audiovisual/Software

The Primary Plant Body. Still-frame video, filmstrip, or transparencies with cassette, 36 min. Teacher's guide. Biology Media.

The Secondary Plant Body. Still-frame video, filmstrip, or transparencies with cassette, 20 min. Teacher's guide. Biology Media, 1987.

*Additional references are listed in *Modern Biology, Teacher's Program Guide*.

Section Teaching Guides

26.1 Plant Tissues

Approach

The main objective of this section is to provide students with a general appreciation for the three basic plant tissue systems and meristems. It builds upon concepts of structure-function relationships (Chapter 1), cell structure (Chapter 5), diversity through adaptation (Chapters 15 and 16), and plant evolution and diversity (Chapter 25). It also prepares students for future concept presentation of plant reproduction (Chapter 27) and plant responses (Chapter 28).

In presenting the material, help students to retain a perspective on the plant as an entire organism, which will be useful as they study the following sections on roots, stems, and leaves. You may wish to refer to Figure 26-2 as a tool in assisting students to maintain this whole-plant perspective. Guide students to understand that all flowering plants, gymnosperms, and ferns contain ground tissues, vascular tissues, dermal tissues, and meristems. You may also want to remind them that nonvascular plants lack xylem and phloem, and that the presence of xylem and phloem has great adaptive significance (see Chapter 25).

Concept Framework

Most students have observed tree growth over time, and have probably noticed that trees increase in girth as well as in height. You can build on these observations to provide students with a conceptual starting point for studying this section. Engage students in a discussion of the types of tissues that might be "added" to the tree as the tree grows in each direction. Elicit the observation that as a tree grows larger, additions to the tissue systems are required by the tree to maintain its metabolic and storage functions.

You might want to point out to students that many plants continue growing throughout their lives (indeterminate growth), unlike most members of other kingdoms. You might also bring up some of the issues related to surface-to-volume ratios in large trees such as giant redwoods and Douglas firs. You may wish to end the discussion by asking students if they have observed any nonwoody (herbaceous) plants that also grow in girth as well as height.

Emphasis

Make certain students understand that primary growth is produced by apical meristems, while secondary growth is produced by lateral meristems. Stress that most monocots do not have lateral meristems and therefore do not have secondary growth. Point out to students that in plants with lateral meristems, primary growth precedes secondary growth. Emphasize that both kinds of growth then continue to take place at the same time in a given plant.

Reports on Local Plant Life

Divide the class into two groups. Have students in the first group research and prepare individual reports on a different herbaceous plant common in your area. Have students in the second group prepare similar reports on woody plants common to your area. If appropriate, students in either group may choose to report on plants that were once plentiful in your region but have disappeared, or nonnative plants introduced to the area (such as kudzu) that have pushed native plants out of their former habitats. Have each group present an oral report to the class and prepare a wall display conveying the results of its research.

Section Review Answers

1. *Dermal* means "skin." Epidermis and cork are analogous to the skin.
2. Mostly they do not. Cork is a secondary tissue, and most monocots do not have secondary growth.
3. Tracheids are long cells with tapered ends, and transport between tracheids takes place through pits in the cell walls. Vessel elements are wider than tracheids and have end walls with holes through which substances can pass.
4. apical
5. No, sclerenchyma occurs only in nongrowing regions of the plant.

26.2 Roots

Approach

The main objective of this section is to show how primary and secondary root tissues develop and function in the plant. The section builds on concepts of structure and function (Chapter 1), cell structure (Chapter 5), diffusion and osmosis (Chapter 6), and diversity through adaptation (Chapter 15 and 16). It also prepares students for future concept presentation of plant reproduction (Chapter 27).

In presenting the material, assist students in relating structures to functions, both in terms of tissue systems within the root and the relationship of the root to the entire plant. Help students identify representative ground, vascular, and dermal tissues in the root. You may wish to use Table 26-1 as a tool to help students see how the absorptive functions of roots affect the rest of the plant body. When you discuss the absorptive function of roots, remind students of the nutrients necessary for photosynthesis and cellular respiration (see Chapter 7).

Concept Framework

All students know that most roots grow in the soil. This knowledge can help students evaluate concepts covered in this section. Engage students in a discussion of how different soil qualities (such as soil density, mineral content, water content, and organic content) might affect the growth and functioning of roots. Elicit as many observations and generalizations as possible about the ways in which soil quality might affect the availability of essential water and nutrients to the plant. If possible, contribute examples from your own community of different soil compositions and the types of vegetation supported in different areas. You may wish to end the discussion by asking students to consider the habitat in which a plant lives when studying the functions of its roots, stems, and leaves.

Emphasis

Make certain that students are able to track the development of different root tissue systems in primary and secondary growth. Try to help them see that the proportion of the root occupied by each tissue system is related to the functions served by the particular root at the particular age in question. You may wish to ask students to speculate on the tissue content of a root specialized for storage, such as a potato tuber, as opposed to the tissue content of a tap root or the air roots of mangroves.

Guest Speaker

Invite someone to speak to the class who is employed in the field of forestry, in the lumber business or by a paper company, or in some form of agriculture. Ask the person to address the class concerning the importance of considering soil quality and content in planning, managing, and conserving large plant-producing tracts of land. If possible, have the speaker use local or familiar examples of soils that support different types of plant life.

Section Review Answers

(See p. 386.)

26.3 Stems

Approach

The major objective in this section is to relate stem tissue variations to the functions served by stems in monocots and dicots. The section builds on concepts of structure and function (Chapter 1), cell structure (Chapter 5), transport between cells (Chapter 6), metabolism (Chapter 7), diversity through adaptation (Chapters 15 and 16), and plant evolution and diversity (Chapter 25).

In presenting the material in this section, guide students to develop an appreciation of the principal transport function of the stem, and help them view primary and secondary growth in the stem from a functional perspective. Suggest to students that they try to visualize the inside of a stem as they study various types of growth. You may wish to refer students to Figures 26-9 (page 388) and 26-10 (page 389) to stress the changes that occur in stem tissue systems as a plant grows.

Concept Framework

All students should know that wood comes from trees. Use this knowledge as a starting point for discussing secondary growth in stems. Engage students in a discussion of all the qualities of wood products they have noticed in their environments. For example, some students may point out that wooden furniture often has a visible grain, and that wood used in household construction may have knots. Some may have seen furniture with veneer. Students will also point out that there are woods of varying hardness, and so on.

Write these examples of variation in wood on the chalkboard and elicit the observation that part of the variety in wood products reflects the variety of secondary growth in different genera of trees. Provide students with examples of specific differences in woods from various trees, and encourage students to contribute whatever they know about specific woods. You may wish to end the discussion by asking students to study this section and then to look for stem structures, adaptations, and growth marks in their own environments.

Emphasis

Make certain students understand that the mechanisms of movement of food, water, and minerals through vascular systems in stems is still not thoroughly understood by researchers. Help them develop an appreciation for the physical and chemical questions of pressure and ion movement that remain to be answered. You may wish to review the material presented in Chapter 6 on cell transport to highlight the functions of the cell that are important in these mechanisms.

Characteristics of Stem Growth

Have interested students collect several twigs or woody stems that are several internodes long. Have students research the growth of stems, twigs, and leaves in the species of plant their samples represent. Then ask students to present a report to the class in which they explain what can be inferred from a stem—in terms of the plant on which it occurs, the habitat in which the plant is usually found, and environmental influences on growth—by observing its external characteristics.

Using the Features/Using Aphids to Study Phloem

The primary process skill used in this feature involves data collection. Scientists have collected data through various means to try to support or refute hypotheses concerning the flow of sap through phloem. Scientists can measure the rate of flow of sap and analyze its content by using aphids, which have a stylus that can be inserted in the phloem without extensively damaging the vascular tissue. Such data collection not only contributes to the results of single experiments, but it provides a base of knowledge from which other inferences may be drawn and other hypotheses proposed. You may wish to point out to students that relevant data can be found in a variety of ways and gathered from many sources. You might also remind them that data once considered relevant in a single experiment may later be found to be more relevant to other hypotheses.

Section Review Answers

(Continued from p. 392.)

5. Plants in cool, humid environments probably lose less of their total water content through transpiration than plants in hot, dry environments do. This would mean that water is perhaps not "pulled" up the xylem as rapidly in cool, humid environments as it is in hot, dry ones.

26.4 Leaves

Approach

The main objective of this section is to acquaint students with the relationships between structures and functions in leaves. The section builds on concepts of structure and function (Chapter 1), basic chemistry and cell biology (Chapters 3 through 6), metabolism (Chapter 7), diversity through adaptation (Chapters 15 and 16), and plant evolution and diversity (Chapter 25). In presenting the text, help students develop an appreciation for the significance of photosynthesis in the environment. You may also want to expand your presentation to include a review of the cellular site of photosynthesis (review sections of Chapter 7), trace the movement of sugar and water through the rest of the plant, discuss in detail the photosynthetic and storage cells in the mesophyll, and/or discuss the gas-exchange processes that occur elsewhere in the leaf. You may wish to use Figure 26-15 on page 394 as a tool to help students identify the three tissue regions in the leaf.

Concept Framework

Students know that the leaves of deciduous plants in North America change color and fall off of the plants in autumn. They also know that even nondeciduous trees lose leaves periodically. Use this knowledge to relate leaves to the plant as a whole. Engage students in a discussion of the possible adaptive advantages to the plant of seasonal and/or periodic leaf loss. As they read this section, ask students to consider possible adaptive advantages of the color change in leaves in the fall.

As students speculate about the reason for color changes in leaves, lead them to infer that the change in the balance of pigments in leaf chlorophyll that causes the color change may be related to environmental and internal changes in the plant as a whole. You may wish to end this part of the discussion by asking students to consider what substances might build up in leaves as a result of photosynthesis, influencing the proportion of different pigments in the chloroplasts.

As students speculate about reasons for leaf fall (abscission), lead them to infer that waste-loaded leaves that are unable to produce adequate nutrients, are best eliminated and replaced by protective dermal tissue. You may wish to describe the process of leaf abscission. (At the base of the petiole, an abscission zone consisting of structurally weak cells undergoes periodic hormonal changes that cause cell walls to dissolve. After the leaf and petiole have fallen off, a second layer of cells grows over the former area of attachment and becomes the leaf scar.) You may wish to conclude this part of the discussion by asking students to consider possible advantages of plant dormancy as they study leaf structure and function.

Emphasis

Make certain students understand the relationship between stomatal functioning, gas exchange, and photosynthesis. If appropriate, review the sections of Chapter 7 that explain the role of carbon dioxide in photosynthesis. You may also wish to review concepts of basic chemistry, diffusion, and osmosis that relate to stomatal opening and closing.

Guest Speaker

Invite a botanist to speak to the class about modifications in leaves. The speaker should explain, for example, the adaptive advantages of specialized leaves such as tendrils, spines, storage leaves (such as those of onions), window leaves, succulent reproductive leaves, bracts, and insect-trapping leaves. Make sure the speaker relates modifications to habitats as well as to tissue-system distribution.

Section Review Answers

(See p. 396.)

Laboratory Preparation

Background

In this lab students observe the structure and function of plant root tips, shoot tips, and leaves. By looking at prepared slides under the microscope and at living plant material, students become familiar with the microstructure of root tips, shoot tips, and leaves. Students will make stomatal prints by painting clear nail polish on the underside of a leaf, peeling it off, and examining the imprint of guard cells under the microscope. Students need to be familiar with and make use of labeled text photomicrographs of root tips, shoot tips, and leaves.

Preparation

Prepare longitudinal section slides of a corn root tip and longitudinal section slides of a coleus shoot tip. Prepare cross-sectional slides of a leaf.

Conclusions

1. Typical answers will suggest that root tips have hairs for absorbing water, and shoot tips conduct photosynthesis.
2. Most photosynthesis occurs in the upper half of the leaf where most of the chloroplasts are found. This part of the leaf also receives the most direct sunlight.

Inquiry

Students should state a hypothesis about water absorption and suggest an experiment that will test it, such as removing the root hairs from one plant and comparing its ability to absorb water with that of another plant that has root hairs.

Chapter Review

Vocabulary

1. Epidermis; it is not a secondary growth tissue.
2. Mesophyll; it is not a vascular tissue cell.
3. Vascular cambium; it is not related to primary growth.
4. Sclerenchyma; it is not related to water absorption and transport.
5. Cork; it is not found in leaves.

Review

11. Growth produced by the lateral meristems, vascular cambium and cork cambium causes plants to grow wider.
12. Heartwood is the dark wood in the center of a stem, where xylem is clogged with waste products. Sapwood is nearer the outside of the stem and still transports water and nutrients.

13. Transpiration causes a low pressure to build up in the xylem. Water moves to this area of low pressure; it is a cohesive fluid—its molecules stick together.
14. Sugar is a major component of phloem sap, along with P-protein.
15. Modified leaves include spines, tendrils, and the leaves of insectivorous plant.
16. A sweet potato is a root; a white potato is a stem. Both are specialized for food storage.
17. When guard cells fill with water, they pull apart, thus creating an opening through which gases can pass.
18. In a monocot vascular bundles are scattered throughout the stem; in a dicot vascular bundles form a ring in the stem.
19. Primary growth originates in apical meristems and occurs in all plants; secondary growth originates in lateral meristems and does not occur in monocots.
20. Herbaceous plants are usually annuals, such as zinnias, and do not develop vascular cambium or secondary xylem and phloem. Woody plants, usually perennials such as forsythia, develop vascular cambium and secondary growth.

Critical Thinking

1. It suggests that each cell has the genetic potential to form a new plant.
2. When the tiny roots and root hairs are destroyed by moving the plant, the plant cannot absorb as much water and minerals.
3. The thin annual rings are a result of poor growth. The most probable cause of this is a drought. It is also possible that the tree was shaded by another, bigger tree or suffered a severe insect attack that slowed its growth.
4. The petioles are twisted and shortened or lengthened in order to bring the leaves into a flat plane in which no leaf shades another.

Chapter 26 | Plant Structure and Function

Introduction

Plants have adapted to a vast range of environments over the course of their evolution. Most plants begin life by sending down roots and sending up shoots. As plants grow, their cells become specialized for various functions. The tissues made up of these specialized cells are arranged differently in different groups of plants. Such variations directly reflect the forces of natural selection that still operate where the plants live.

Chapter Outline

26.1 Plant Tissues
- *Tissue Systems*
- *Meristems*

26.2 Roots
- *Root Structure*
- *The Function of Roots*

26.3 Stems
- *Stem Structure*
- *The Function of Stems*

26.4 Leaves
- *Leaf Structure*
- *The Function of Leaves*

Chapter Concept

As you read, look for ways in which plant tissue arrangements reflect adaptations to various environmental conditions.

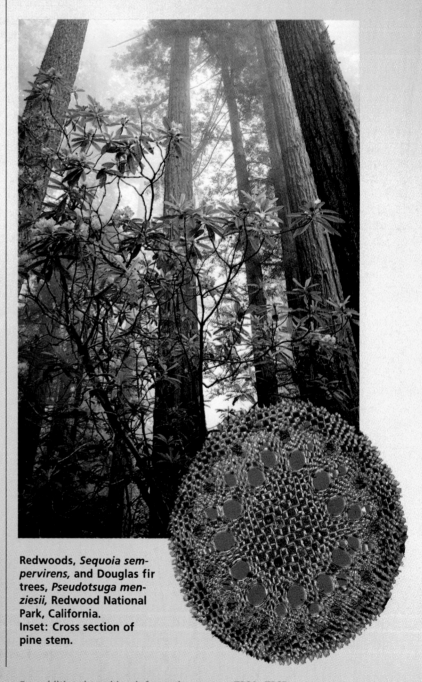

Redwoods, *Sequoia sempervirens*, and Douglas fir trees, *Pseudotsuga menziesii*, Redwood National Park, California.
Inset: Cross section of pine stem.

For additional teaching information see pp. T26A–T26F.

26.1 Plant Tissues

Like all organisms plants are composed of cells. Groups of cells that perform a common function make up a **tissue**. Tissues are arranged into tissue systems. Three types of tissue systems are found in plants: ground system, dermal system, and vascular system. These systems are then organized into one of the three plant organs: **roots, stems,** and **leaves**. All of these levels of organization reflect adaptations to the environment.

Tissue Systems

The **ground system** is made up of three kinds of tissue. **Parenchyma** *(puh-REN-kuh-muh)* is living tissue that consists of large, loosely packed, roughly cube-shaped cells with thin cell walls. Parenchyma cells are involved in many functions, including photosynthesis, food storage, and wound healing. In plants that are not woody, they make up over 80 percent of the cells. **Collenchyma** *(koe-LEN-kuh-muh)* is living tissue made up of elongated cells with unevenly thickened, flexible walls. Collenchyma cells are specialized for supporting growing regions of the plant. **Sclerenchyma** *(skluh-REN-kuh-muh)* tissue, which at maturity is made up of the walls of dead cells, supports and strengthens the plant. Sclerenchyma will not stretch and thus is found primarily in regions that have ceased growing. Some sclerenchyma walls are long and thin and are called **fibers.**

Ground tissues are modified for adaptive advantage. Cactuses, for example, have large amounts of parenchyma tissue, which store water in dry environments. Herbs that must be flexible to withstand wind possess a lot of collenchyma tissue. Sclerenchyma tissue is found in plant parts in which hardness is an advantage, such as the shells of nuts or the spines of cactuses.

The **vascular system** includes xylem and phloem tissues. Xylem conducts water and minerals. Phloem conducts food. In angiosperms xylem tissue is made up of the walls of dead cells. In angiosperms xylem has two components—tracheids and vessel elements. Look at Figure 26-1. A **tracheid** *(TRAY-kee-id)* is a long, thick wall with tapering ends. Water moves from tracheid to tracheid through **pits**, which are thin, porous areas of the wall. A **vessel element** is a wall that has slanting ends with holes through which water can pass. Vessel elements are arranged end to end, thus forming long, hollow tubes called vessels. Gymnosperm xylem contains only tracheids.

Unlike xylem, phloem tissue is alive at maturity. The conducting cell of angiosperm phloem is called a **sieve tube member.** Sieve tube members are stacked end to end in a long tube called a sieve tube. Compounds move from cell to cell through regions at the ends of the sieve tube members called **sieve plates.** Each sieve tube member lies next to a specialized neighboring **companion**

Section Objectives

- *Describe the three plant tissue systems and the kinds of cells that occur in each.*

- *Describe the two main types of meristem tissue.*

- *Differentiate between primary and secondary growth.*

Tip: Explain to students that although it is convenient to classify plant parts into roots, stems and leaves, it is important to keep in mind that a plant functions as a whole unit.

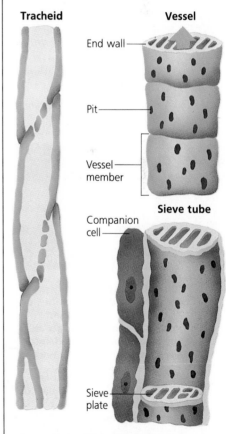

Figure 26-1 Water is conducted in angiosperms through tracheids and vessels. Food is conducted through sieve tubes, which are accompanied by companion cells.

Leaves

Twig

Bud

Apical meristem

Stem

Shoot system

Roots

Root system

Figure 26-2 Apical meristems, the growing points of plants, occur at the tips of stems and roots. Nodes are the points at which leaves are attached. Internodes are areas between nodes.

Info: The cells of the apical meristem divide so rapidly that they can go through a complete cell cycle in as little as 12 hours.

Answers to Section Review
(See pp. T26A–T26F.)

cell. Botanists suggest that companion cells help control the movement of sugars through sieve tubes.

Like ground tissue vascular tissue is modified for adaptive advantage. For example, xylem forms the wood of trees, providing strength while conducting water. In aquatic plants such as duckweeds xylem is not used for either support or water conduction and may be nearly absent from the mature plant.

The **dermal tissue system** forms the outside covering of plants. The dermal tissue is composed of the **epidermis** (*EP-uh-DER-muss*), the outer layer of cells, which is covered with a waxy cuticle. The epidermis functions to protect the plant, reduce water loss, and aid in gas exchange. Openings in the epidermis are called **stomata** (*STOE-muh-tuh*). Stomata are structures that regulate the passage of gases into and out of the plant. In mature stems and roots the protective epidermis is eventually replaced by **cork**, another dermal tissue. Cork cells are packed closely together and their walls contain a protective fatty substance called suberin.

Dermal tissue is also modified in various plants. For example, plants in dry environments have stomata that are sunk deep in the epidermis, preventing water loss. Stomata, usually found on the lower sides of leaves, are found on the upper side in water lilies, allowing free passages of gases to and from the air.

Meristems

Plants possess **meristems** (*MER-ih-stemz*), growing regions where cells divide. Plants grow in length by **apical** (*AY-pih-kul*) **meristems** at the tips of stems and roots. Look at Figure 26-2.

Plants also have **lateral meristems,** which grow and increase their circumference. One is the **vascular cambium,** which produces additional vascular tissues. Another is the **cork cambium,** which produces cork. The tissues produced by apical meristems are called **primary growth** tissues. Those produced by the lateral meristems are called **secondary growth** tissues.

Adaptations occur in meristems and their growth patterns. Annual plants are adapted for rapid growth and reproduction. They do not have a vascular cambium, so their tissues remain nonwoody. They are called **herbaceous plants.** Trees are perennial plants, which are adapted for growth year after year. They develop much secondary xylem. They are called **woody plants.**

Section Review

1. Look up the word *dermal* in a dictionary. Why is this an appropriate term for epidermis and cork?
2. Do monocots have cork? Explain your answer.
3. How do tracheids and vessel elements differ?
4. What kind of meristem do monocots and dicots share?
5. Would you expect to find sclerenchyma cells near meristems? Why or why not?

26.2 Roots

The tissue systems you read about previously are arranged into organs: roots, stems, and leaves. In this section you will study roots. During growth most plants develop a **primary root,** the first root to grow out of a seed. If, as the root system develops, the primary root grows downward and remains the largest root, the organ is called a **taproot.** Many plants, such as carrots, have taproots that are adapted to store food. Taproots of large trees are adapted for growing hundreds of feet down, supporting the plant while reaching water deep in the ground.

In some plants the primary root does not grow large, and numerous small roots develop and remain near the top of the soil. The resulting organ is called a **fibrous root** system. Fibrous roots spread out and anchor the plant. Many monocots, such as grasses and orchids, have fibrous root systems, which are adapted for gathering water close to the surface of the soil.

Other root systems are modified for specialized functions. The prop roots of the corn plant support the stem. Another variation of the root system occurs in mangroves, which are adapted for growing in the shifting soils of tidal areas. When the area is covered with water, no oxygen is available to the roots. Some mangrove root branches are modified for growing up into the air and absorbing oxygen when the tide recedes.

Section Objectives

- *Explain the difference between a taproot system and a fibrous root system.*

- *Distinguish between primary growth and secondary growth in roots.*

- *Describe primary root tissues.*

- *List the three principal functions of roots.*

Figure 26-3 Four plants exhibiting four types of roots are (clockwise from top left) the carrot, *Daucus carota,* (taproot), the orchid (fibrous root), corn, *Zea mays,* (prop root), and the mangrove, *Avicennia* sp., (air root).

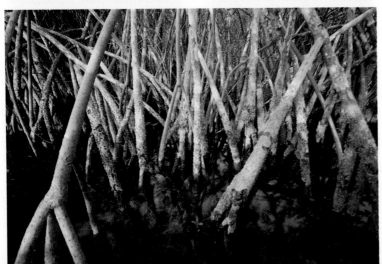

Root Structure

The root apical meristem produces cells that differentiate into ground tissue, vascular tissue, and dermal tissue. These then differentiate into cells that form the mature root.

The Root Tip

Study Figure 26-4. Notice that the root tip is organized into three parts. The apical meristem is the region of cell division. Cells formed here later grow longer in the region of elongation. Here a root increases in length. Behind this region is the region of maturation, where cells mature and differentiate.

The cells of the meristem also produce parenchyma cells that come to lie in front of the tip. These cells make up the **root cap,** a shield covering the apical meristem. The root cap produces a slimy substance that functions like a lubricating oil, allowing the root tip to move more easily through the soil. Cells lost from the root cap are constantly replaced by the apical meristem.

Primary Growth

Growth, elongation, and maturation in the root tip establish the three tissue systems in the root. Dermal tissue matures to form an epidermis, the outermost cylinder of the root. In young roots epidermal cells often form extensions called **root hairs,** which are specialized for absorbing water and minerals. The root hairs increase the surface area of the root and thus increase its ability to absorb water and nutrients.

Ground tissue matures into the cortex and the endodermis. The **cortex** is located just inside the epidermis, as you can see from Figure 26-6. The largest region of the young root, it is made up of loosely packed parenchyma cells. The innermost cylinder of the cortex is the **endodermis,** a specialized layer of cells that regulates substances entering the center of the root. Each cell of the endodermis is belted by a narrow band of suberin called the **Casparian** *(kass-PAR-ee-un)* **strip.** Substances absorbed by root hairs move freely around the walls of the loosely packed cortex cells and toward the root center. When these substances reach the endodermis, the Casparian strip blocks their passage through the intercellular spaces. Substances must therefore pass through the semipermeable cell membranes of the endodermal cells. The cell membrane of the endodermal cells thus regulates the transport of substances into the interior vascular tissues.

Vascular tissue matures to form the **vascular cylinder,** the innermost part of the root, which contains xylem and phloem. In most plants xylem makes up the central core of the vascular cylinder. Xylem usually has a fluted structure, as shown in Figure 26-6. Phloem lies between the xylem lobes. The outermost rings of cells in the central vascular area are called the **pericycle.** All lateral roots are formed from divisions of these cells. The lateral root forces its way through the cortex and emerges.

Figure 26-4 The three parts of a root tip, such as the one from a willow tree shown here, are the regions of division, elongation, and maturation (×150).

Figure 26-5 A germinating kernel of field corn, *Zea mays*, has already established primary and secondary roots complete with root hairs.

Epidermis

Vascular cylinder

Cortex

Developing
lateral root

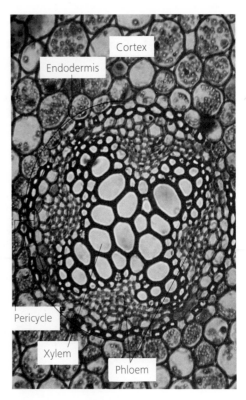

Cortex

Endodermis

Pericycle

Xylem

Phloem

The structure of roots is an adaptation for the function of absorbing materials from the soil. The vascular tissues are located in the center of the root, protected by cylinders of both epidermis and cortex. Lateral roots are initiated inside the root and are thus protected during their earliest growth stages. In addition the Casparian strip regulates the flow of material to the interior tissues.

Figure 26-6 Cross section of a buttercup root, *Ranunculus* sp., shows the process of primary growth.

Secondary Growth

Some roots have secondary growth. Secondary growth begins when a vascular cambium forms between primary xylem and primary phloem. As shown in Figure 26-7, vascular cambium produces secondary xylem toward the inside of the root and secondary phloem toward the outside. The secondary root tissues gradually crush the endodermis, cortex, and epidermis, all of which are eventually replaced by cork. Secondary growth in roots is an adaptation for strength and support of the plant stem.

The Function of Roots

In addition to anchoring a plant in the soil, roots serve two other primary functions. Roots absorb water and minerals, and they store food made during photosynthesis.

Absorption of Water and Inorganic Nutrients

Root hairs contain many dissolved substances such as minerals, sugars, and amino acids. Water in the soil, on the other hand, contains fewer dissolved substances. As a result of the concentra-

Figure 26-7 Secondary growth in a fir tree root causes the formation of woody tissue (×50).

Table 26-1 Some Essential Plant Nutrients and Their Role in Plant Growth

Element	Role in Plant
Macronutrients	
Nitrogen	Synthesis of proteins, nucleic acids, chlorophyll, coenzymes
Potassium	Synthesis of amino acids; helps open and close stomata in leaves
Calcium	Part of cell walls and membranes; influences cell permeability
Phosphorus	Found in nucleic acids, coenzymes, ATP
Sulfur	Found in proteins, coenzymes
Magnesium	Found in chlorophyll, enzymes
Micronutrients	
Iron	Found in enzymes, chlorophyll
Manganese	Found in enzymes
Zinc	Found in enzymes
Copper	Found in enzymes
Chlorine	Acts in photosynthesis
Boron	Exact role unknown

tion gradient between root hairs and soil water, water moves into the root hairs by osmosis. See Chapter 6 for a review of the process of osmosis. The rate of water absorption is influenced by the amount of water lost through **transpiration,** the evaporation of water from stems and leaves.

Roots absorb a variety of inorganic nutrients, usually by active transport. Plant cells use some nutrients, such as nitrogen and potassium, in large amounts. These elements are called **macronutrients.** They use other nutrients, such as manganese, in smaller amounts. These are called **micronutrients.** Too much or too little of any macronutrient or micronutrient will harm or kill a plant. Nutrient requirements vary among species.

Storage of Food

Roots are also adapted for food storage. Phloem carries the sugars made in leaves to roots and stems for storage. In roots these sugars are converted to starch and stored in areas containing parenchyma cells. You may be familiar with storage roots such as those of beets, carrots, turnips, and sweet potatoes.

Section Review

1. Name the three regions of the root tip.
2. How does a primary root become a taproot?
3. Explain how root hairs increase the ability of a plant to absorb water from the ground.
4. Name two areas of the root where you would probably find parenchyma cells.
5. Compare and contrast endodermal cells and epidermal cells.

Answers to Section Review
1. division, elongation, and maturation
2. The primary root continues to grow and becomes the largest root in the system.
3. They increase the surface area of the root.
4. root cap and cortex
5. Endodermal cells are the innermost cells of the cortex, have a line of fatty suberin called the Casparian strip in their cell walls, and regulate the water entering the vascular cylinder. Epidermal cells are outside the cortex, are covered with a waxy cuticle, and function in the absorption of water.

26.3 Stems

Look at Figure 26-8. Like roots, stems are adapted to different environments. Stems of some plants grow only along the surface of the soil, as you can see by examining the strawberry stolon. Cactuses have green, fleshy stems that both store water and carry on photosynthesis. Stems such as the fleshy underground tuber of a white potato are modified for storing food. Gladiolus and crocus plants have underground stems called corms that are specialized for food storage. All of these differences in stem shape and growth represent adaptations to the environment.

Stem Structure

Did you know that a sign nailed 2 m high on a tree will remain 2 m high, even though the tree may grow much taller? That is because stems grow in length only at their tips, where apical meristem produces new primary tissues. Stems grow in circumference through growth of lateral meristems.

Primary and secondary growth also take place in woody twigs tipped with buds. Buds are lateral branches that include the apical meristem and modified leaves called bud scales. In spring the bud opens and the bud scales fall off the twig. The scales leave scars on the surface of the twig.

Section Objectives

- *Describe differences between monocot and dicot stems.*

- *List the secondary tissues in the dicot stem.*

- *Explain the formation of an annual ring.*

- *Explain the pressure–flow hypothesis, and tell what substances are transported in phloem.*

- *Describe the transpiration–cohesion theory.*

Tip: Ask students to state the main difference between how plants grow and how animals grow. (In animals, cell division takes place everywhere; in plants, cell division is limited to certain areas.)

Figure 26-8 Various stem adaptations are illustrated by (clockwise from left) the strawberry, *Fragaria* sp., plant (stolon), the cactus, *Cereus* sp., (green, fleshy stem), a plant with a typical woody stem, and a potato, *Solanum tuberosum*, plant (tuber).

Bud

Leaf scar

Bud scales

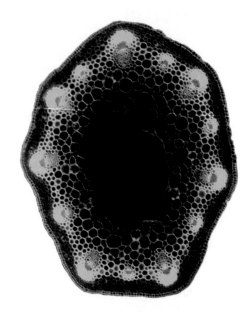

Figure 26-9 Cross-sectional photographs contrast a typical monocot stem (×10) and a typical dicot stem (×10).

Think

Which of the three basic ground tissues would you expect to find the most of in this euphorb stem? Why?

Primary Growth

Apical meristems give rise to cells that form the primary dermal, ground, and vascular tissues of a stem. Locate each of these tissues in Figure 26-9. As you can see, the dermal tissue is represented by the epidermis, the outer layer of the stem. Its primary functions are to protect the plant and to reduce the loss of water to the atmosphere.

The ground tissue consists mostly of parenchyma cells. In gymnosperms and dicots, ground tissue forms a cortex and a pith. The cortex lies immediately inside the epidermis, as it does in the root. It frequently contains some flexible collenchyma tissue and may be photosynthetic. The **pith** is located in the center of the stem. Monocot stems are not so precisely divided into pith and cortex as dicot stems are.

Vascular tissue formed near the apical meristem occurs in bundles, long strands that lie embedded in the cortex. Look at the **vascular bundles** shown in Figure 26-9. Each bundle contains primary xylem and primary phloem. Xylem is located nearer the inside of the stem, and phloem is located nearer the outside. This arrangement of xylem and phloem differs from that found in a root, where both types of vascular tissue are located in the center.

Compare the arrangement of vascular bundles in monocots and dicots. In a monocot stem vascular bundles are scattered throughout the ground tissue. In dicot stems the vascular bundles occur in a single ring. Most monocots have no secondary growth and thus retain the scattered arrangement of vascular bundles throughout their lives. However, in many dicots the vascular bundles are eventually replaced by secondary tissues produced by the lateral meristems.

Answer to Think: Euphorb stems consist mainly of parenchyma, which is specialized for water storage.

Secondary Growth

The vascular cambium in stems arises first between the xylem and phloem in a vascular bundle. Eventually the vascular cambium forms a cylinder. Stems increase in thickness as a result of divisions of cells in the vascular cambium.

Examine Figure 26-10. The vascular cambium produces secondary xylem to the inside and secondary phloem to the outside. It usually produces much more secondary xylem, called wood.

In addition to the vessel elements and tracheids, discussed in Section 26.1, wood contains sheets of parenchyma cells called **rays.** As you can see in Figure 26-10, rays form a flat tissue that radiates from the center of the trunk. Rays store food and also transport food and water laterally across the xylem.

When water is plentiful during spring, the vascular cambium forms new secondary xylem with cells that are wide and thin walled. This wood is called **springwood.** In hot, dry summers the vascular cambium produces **summerwood,** which has smaller cells with thicker walls. The visible difference in density between spring wood and summer wood is called an **annual ring.** Look again at Figure 26-10. The circles that look like bull's-eyes are the annual rings of the tree. Since one such ring is formed each year, you can estimate the age of a plant simply by counting its annual rings. How old is the tree in the figure?

Some vessels of the secondary xylem eventually become clogged and thus useless in transporting water. They also give the xylem a dark color. This dark wood in the center of a tree is called heartwood. The functional, lighter-colored wood nearer the outside of the trunk is sapwood.

The secondary phloem produced toward the outside of the wood is part of the bark. **Bark** is the protective outside covering of woody plants. It consists of cork, cork cambium, cortex, and secondary phloem. The cork cambium produces cork toward the outside and living ground tissues toward the inside. Since cork cells quickly die and cannot expand, the continued production of secondary xylem and phloem causes the outer layers of cork to split. The bark of familiar trees such as oaks and maples thus appears rough and irregular in texture.

The Function of Stems

Whatever their sizes and shapes, most stems have three functions: support, transport, and storage. You have already learned about the support function. Let's examine transport and storage.

Transportation of Food

Sugars are the food that moves in the phloem of plants. Movement of these sugars occurs from **source,** where it is made, to **sink,** where it is stored or used. This movement is explained by the **pressure–flow hypothesis.**

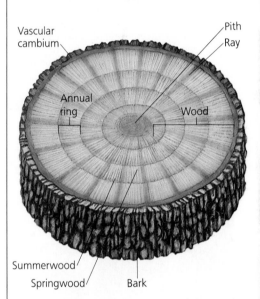

Figure 26-10   A cross section of the stem of a white pine, *Pinus strobus,* shows the rays and annual rings.

Sugars are manufactured in photosynthetic cells and pumped by active transport into sieve tubes. This increases the concentration of dissolved substances in the sieve tube, and water enters the sieve tube by osmosis. The sieve tube becomes filled with water. Thus a positive pressure builds up at the source end of the sieve tube. This is the "pressure" part of the hypothesis.

At the sink end of the sieve tube this process is reversed. Sugars are pumped out, and water leaves the sieve tube by osmosis. Therefore pressure is reduced at the sink. The difference in pressure causes sugars to move from source to sink. In other words, the difference causes sugars to "flow."

Transportation of Water and Minerals

Water, which is required by all living cells, is constantly evaporating from stems and leaves. In addition stems and leaves require inorganic nutrients. Water and inorganic nutrients enter a plant through the roots. How do these substances travel from the roots to the top of a tree?

At one time scientists thought that the water absorbed by the roots created a pressure that "pushed" other water up the plant. This sometimes happens. When it does, the pressure of water

Answer to Biology in Process
Data indicating rapid phloem sap movement from source to sink have provided support for the notion of pressure in the sieve tube.

Biology in Process

Using Aphids to Study Phloem

What substances move through the phloem? Although this question seems like a rather simple one, its exact answer eluded biologists for many years.

Researchers hypothesized that phloem carried sugars made in photosynthesis. The first experiments designed to test this hypothesis involved cutting into the phloem and collecting the sap that came out. These experiments indicated that sugars were indeed a major component of the phloem sap.

However, the experiments also revealed that the phloem sap contained a slimy substance called P-protein. Furthermore, some of the P-protein remained on the sieve plate and formed a slime plug that blocked the movement of materials out of the sieve tube member. On the basis of this evidence some investigators inferred that slime plugs occurred in undisturbed phloem, and they encouraged a revision of the pressure–flow hypothesis. Other scientists inferred that the formation of slime plugs was a form of wound healing. What researchers needed was a way to ensure that the phloem sap they studied was not contaminated with wound-healing substances that causes water to be forced out of the leaves. This process, called guttation, usually occurs at night. Guttation cannot account for most of the movement of water through a plant, however. Scientists now think that water may be "pulled" up a plant. They suggest that water forms a continuous column from roots, through stems, to leaves. This is possible because water is a cohesive fluid; its molecules are strongly attracted to each other.

The movement of this column of water is the basis for the **transpiration–cohesion theory**. Figure 26-12 on the following page shows how water that evaporates from leaves through stomata is eventually replaced by water from the xylem. When a water molecule from the xylem replaces water that has evaporated from the leaf, a low pressure is formed at the top of the water column. Vessel elements and tracheids do not collapse under this low pressure because their walls are thick. Water moves from areas of high pressure to areas of low pressure. This, combined with the strong cohesion of water, pulls the entire water column up from the roots through the plant.

The speed of water movement through a stem changes throughout the day. At midday evaporation is greatest, and water moves rapidly from the roots through the plant. This rapid move-

Figure 26-11 The rate of water uptake peaks about six hours after the rate of transpiration peaks.

were released when they cut into the phloem.

In the early 1950s two insect physiologists suggested a novel research tool—aphids. They observed that aphids obtained their food from phloem by inserting a "tap" into individual sieve tube members.

The aphid's tap (left) is an extremely small, snoutlike mouthpart called a stylet. Since the contents of the phloem are under positive pressure, sugars in the phloem surge into the aphid. The phloem sap often passes right through the insect and forms a droplet of sugary liquid called honeydew on the aphid's body.

Once an aphid has inserted its stylet into a sieve tube member, researchers anesthetize it with a gentle stream of carbon dioxide. They then sever the aphid from the stylet, leaving the stylet in the sieve tube. About 1 mm³ of sap per hour can be collected from a stylet for up to five days.

Studies using aphids' stylets, in addition to studies involving radioactive isotopes, have provided evidence that slime plugs do not occur in undisturbed plants. P-proteins are

not present in the phloem sap collected from aphid stylets. The phloem sap instead contains a fluid that is approximately 20 percent sugar. Amino acids make up approximately 0.1 percent of the sap.

By measuring the rate of flow through stylets placed at various points on the plant, researchers can construct a model of substance movement. They have found that, in some plants, the movement of phloem sap from source to sink occurs at the rate of nearly 100 cm per hour.

■ How have scientists used the process of collecting data in studying phloem?

Transpiration

Absorption

Figure 26-12 The diagrams above show how water moves through xylem from areas of high pressure to areas of low pressure.

Answers to Section Review
1. Summerwood is made up of small cells with thick cell walls. Springwood is made up of large cells with thin cell walls.
2. Dicot vascular bundles occur in a single ring. Monocot vascular bundles are scattered.
3. Sugars are pumped into sieve tubes by active transport. Water follows by osmosis, resulting in high pressure in the source end of the sieve tube. At the sink end of the sieve tube the sugars are pumped out and water again follows by osmosis. The difference in pressure at the source and sink causes the sugars to flow through the phloem.
4. Tubers are short, thickened stem areas used for food storage. Thickened stems for storing water and nutrients are found in plants that grow in dry areas.
(See pp. T26A–T26F.)

ment creates a low pressure in the xylem, so strong that tree trunks actually shrink slightly in diameter during the day. At night there is little transpiration. Water does not move in the xylem, and the trunk enlarges slightly in diameter.

Storage in Stems

Plant stems are adapted for food storage in some species. As you read earlier, white potatoes are tubers, stems that are specialized for storing food. The "eyes" of white potatoes are actually buds that will grow into other stems if placed in the proper environment.

Many plant stems are full of parenchyma cells, which indicates that they store water and food for use by the rest of the plant. Succulents have particularly fleshy stems that retain water and nutrients in areas where water is often unavailable.

Section Review

1. Explain the difference between summerwood and springwood.
2. How does the arrangement of vascular tissues in a stem differ between monocots and dicots?
3. Describe how food moves through stems from source to sink.
4. Describe some adaptations of stems for food storage.
5. How might the rate of transport of water differ between plants in cool, humid environments and those in hot, dry environments?

26.4 Leaves

Being specialized for capturing sunlight for photosynthesis, most leaves are thin and flat. Although this structure may be typical, it is certainly not universal. Leaves are the most variable organ of plants. This variability represents adaptations to environmental conditions.

Look at the leaves in Figure 26-13. The coiled structure in the figure is a tendril, a kind of modified leaf. The tendrils of plants such as peas and morning glories wrap themselves around objects as the plant grows toward a light source.

Leaves of desert plants are often modified into protective spines. Since spines are relatively small, they reduce the surface area from which water can evaporate. In cactuses, little photosynthesis occurs in the spines; most photosynthesis takes place in the stems.

An uncommon modification of leaves occurs in insectivorous plants such as the pitcher plant. In insectivorous plants leaves often function as animal traps. These plants usually grow in soil that is relatively poor in nitrogen and phosphorus. The insects contain a large amount of these nutrients, and the plant receives essential macronutrients when it digests them.

Figure 26-13 **Plants with diverse leaf adaptations include (left to right) the pea plant, *Pisum sativum* (tendrils), the cactus, *Opuntia* sp. (spines), and the pitcher plant, *Sarracenia* sp. (tubular leaves).**

Leaf Structure

Leaves can be round, straplike, needlelike, or heart shaped, and their edges range from smooth to serrate. Regardless of their external structure, all leaves are made up of ground, dermal, and vascular tissues. Secondary tissues rarely occur in leaves.

The broad, flat portion of a typical leaf is the **blade;** this is the site of most of the photosynthesis in the plant. The blade is attached to the stem by a stalklike **petiole.** The maple leaf shown in Figure 26-14 is a **simple leaf;** it has only one blade and one petiole. In **compound leaves,** such as those of the date palm, the blade is divided into leaflets. In some species the leaflets are divided. The result is a doubly compound leaf.

Venation is the arrangement of veins in a leaf. Leaves of most monocots, such as grasses, have parallel venation, meaning that their main veins are parallel to each other. Leaves of most dicots, such as sycamores, have net venation, meaning that smaller veins branch out from a large central vein.

Figure 26-15 shows the internal structure of a leaf. Leaves consist of cells from three tissue systems. The dermal tissue system is represented by the epidermis. In most leaves the epidermis is a single layer of cells coated with an impermeable cuticle. Water, oxygen, and carbon dioxide enter and exit from the leaf through stomata in the epidermis.

The number of stomata per unit area of leaf varies in different plants. For example, submerged leaves of aquatic plants do not have stomata. However, leaves of corn plants, which grow best in hot weather, have approximately 7,500 stomata per square centimeter of leaf epidermis.

The number of stomata also varies on the upper and lower sides of leaves. Most plant leaves have more stomata on their

Figure 26-14 Silhouetted above are the leaves of the sugar maple tree, *Acer saccharum* (top), and the golden-rain tree, *Loelreuteria paniculata* (bottom). The first is a simple leaf; the second is compound.

Figure 26-15 Cells from three tissue systems are represented in the internal structure of a leaf. Can you name the systems?

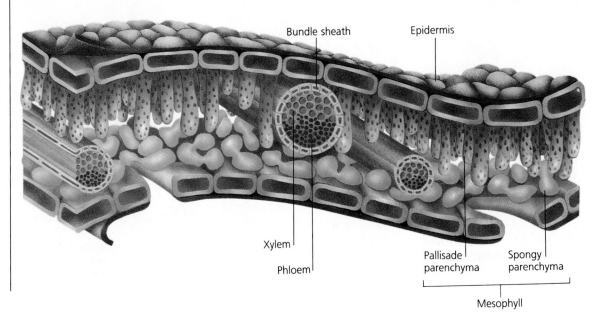

Bundle sheath Epidermis

Xylem

Phloem

Pallisade parenchyma Spongy parenchyma

Mesophyll

shaded lower epidermis than on their exposed upper epidermis. For instance, oak leaves have no stomata on their upper surfaces but have thousands of stomata per square centimeter on their lower surfaces. Regardless of their exact distribution, stomata are functional: they regulate gas exchange.

Photosynthesis occurs in the leaf **mesophyll** *(MEZZ-oe-FILL)*, a ground tissue that is made up of chloroplast-rich parenchyma cells. In most plants, the mesophyll is organized into palisade parenchyma and spongy parenchyma. **Palisade parenchyma** occurs directly below the upper epidermis; this is where most of a photosynthesis in the leaf occurs. Palisade cells are shaped like columns and are packed tightly together, increasing the exposure of chloroplasts to sunlight. Between the palisade parenchyma and the lower epidermis is **spongy parenchyma**. It consists of irregularly shaped cells surrounded by numerous air spaces, which allow oxygen and carbon dioxide to diffuse into and out of a leaf.

The vascular tissue system is represented by vascular bundles. In leaves they are called veins. They are continuous with the vascular tissue of the stem and the petiole and lie embedded in the mesophyll. Veins are often surrounded by a tight-fitting **bundle sheath** made up of parenchyma cells.

Tip: Review with students the process of photosynthesis.

The Function of Leaves

Leaves are the primary site of photosynthesis in plants. As such, they must absorb light energy and take in carbon dioxide from the environment. Palisade cells in leaves use light energy, carbon dioxide, and water to make sugars. Sugars manufactured in leaves can then be used as an energy source for growth or be transported to roots or fruits and stored until needed.

The temperature of a leaf increases when it absorbs light, just as your skin can get hot on a bright summer day. Transpiration cools leaves while pulling water and dissolved minerals from roots through stems. In corn more than 98 percent of the water that is absorbed by the roots is lost through transpiration.

Modifications for Capturing Light

Leaves absorb light, which provides the energy for photosynthesis. In addition, the chloroplasts in photosynthetic leaf cells are typically arranged in a thin, uniform layer near the lighted side of the leaf. This minimizes the shading of one chloroplast by another.

In many plants leaves are modified for specialized environmental conditions. In deserts, where light is sometimes very intense, leaves often have thick cuticles or hairs. These reflect or block the excess light and reduce water loss. Another adaptation related to light intensity occurs in the so-called window plants. The leaves of these plants actually grow underground. All of the light for photosynthesis comes through a tiny "window" of transparent tissue that protrudes above soil level.

Figure 26-16 The leaves of a stone plant, *Lithops* sp., are modified so that the plant mimics a small stone.

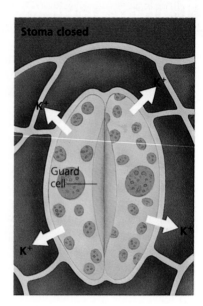

Stoma open

Guard cell

Stoma closed

Guard cell

Figure 26-17 SEMs of the leaf of a tobacco plant, *Nicotiana tabacum,* show open (left, ×2,000) and closed (right, ×2,000) stomata.

Answers to Section Review
1. A simple leaf has one blade and one petiole. In a compound leaf the blade is divided into leaflets.
2. The epidermis has an impermeable cuticle and contains stomata that help regulate gas exchange. The mesophyll carries out photosynthesis. The vascular tissues transport water and food.
3. probably the sugar maple leaf, because it has to transpire more to keep cool
4. During the day, light causes the pumping of potassium ions into the guard cells. Water follows by osmosis. The water-filled guard cells bulge apart, creating an opening through which gases can pass. At night potassium ions are pumped out, causing water to leave by osmosis. The guard cells shrink and the stoma closes.
5. The greater the surface area exposed to the sun, the greater the amount of light that will be captured for photosynthesis.

Gas Exchange

Plants obtain carbon dioxide for photosynthesis from the air. Carbon dioxide diffuses into leaves through stomata. A stoma is bordered by two kidney-shaped **guard cells,** each of which has flexible cell walls that allow it to bend.

The opening and closing of a stoma is regulated by turgor pressure in guard cells. Turgor pressure is related to the amount of water in the cells. The stomata of most plants open during the day and close at night. Look at Figure 26-17. During the day light causes nearby epidermal cells to pump potassium ions (K^+) into the guard cells. Water then moves into the cells by osmosis. This influx of water increases the turgor pressure in guard cells and causes them to expand. The cells expand lengthwise, the result being that they bow apart and form a pore. At night the potassium ions are pumped out of the epidermal cells. Water then leaves the guard cells by osmosis; the guard cells shrink, and the pore closes. The opening and closing of stomata reflects one of the major ways in which plants survive diverse environments.

Section Review

1. What is the difference between a simple leaf and a compound leaf?
2. Describe the basic functions of the three leaf tissues.
3. Which leaf would have more stomata, that of a water lily or a sugar maple? Why?
4. Explain the function of the guard cells in regulating stomatal opening and closing.
5. How might the shape of a leaf affect its ability to capture sunlight?

Laboratory

Plant Structure and Function

Objective
To examine the structure and function of plant root tips, shoot tips, and leaves

Process Skills
Observing, inferring, hypothesizing

Materials
Prepared longitudinal section slide of a corn root tip, microscope, live coleus plant, prepared longitudinal section slide of a coleus shoot tip, prepared cross-sectional slide of a leaf, live leaf, clear nail polish, forceps, microscope slide

Method
1 Examine a corn root tip slide under low power. Note the translucent, loosely arranged cells at the extreme tip of the root. These cells make up the root cap.
2 Note the dense, opaque region just above the root cap. This is the apical meristem. New root cells arise in this region. Just behind this region is the zone of elongation. Compare the cells in the meristem with those in the zone of elongation. Find the root hairs. Examine a root hair under high power.
3 Examine a coleus plant. How are the leaves and buds arranged on each shoot?
4 Examine a prepared slide of a coleus shoot tip under low power. Find the paired rudimentary leaves. What other structure do they resemble? Can you relate the dense cluster of cells just below the rudimentary leaves to another structure you have viewed?
5 Examine a prepared slide of a leaf under low power. Look for openings on the lower surface of the leaf. These openings are called stomata. How does their location relate to their function? Examine one of the openings under high power.
6 Switch to low power and locate a vascular bundle, the cuticle, and the epidermis.

7 Locate the spongy and palisade parenchyma near the center of the cross section. Focus the microscope on the palisade parenchyma, and then switch to high power. You will notice a grouping of small, green, oval bodies. These structures are chloroplasts.
8 Paint a light coat of clear nail polish on the underside of a live leaf. Allow a few minutes for the nail polish to dry, and then apply a second coat. How is this process like the formation of some types of fossils?
9 When the nail polish is dry, use forceps to peel the polish from the leaf. Place the impression on a microscope slide with the smooth side facing up. Examine the impression under low power and low light. Locate the guard cells on the impression. What is the function of these cells?

Conclusions
1 How do root tips differ from shoot tips?
2 Based on your observations, in what part of the leaf does the most photosynthesis occur?

Inquiry
How important are root hairs in the ability of the plant to absorb water? How can you test your hypothesis?

Chapter 26 Review

Vocabulary

annual ring (389)	guard cell (396)	primary root (383)	summerwood (389)
apical meristem (382)	herbaceous plant (382)	ray (389)	taproot (383)
bark (389)	lateral meristem (382)	root (381)	tissue (381)
blade (394)	leaf (381)	root cap (384)	tracheid (381)
bundle sheath (395)	macronutrient (386)	root hair (384)	transpiration (386)
Casparian strip (384)	meristem (382)	sclerenchyma (381)	transpiration–cohesion theory (391)
collenchyma (381)	mesophyll (395)	secondary growth (382)	vascular bundle (388)
companion cell (381)	micronutrient (386)	sieve plate (381)	vascular cambium (382)
compound leaf (394)	palisade parenchyma (395)	sieve tube member (381)	vascular cyclinder (384)
cork (382)	parenchyma (381)	simple leaf (394)	vascular system (381)
cork cambium (382)	pericycle (384)	sink (389)	venation (394)
cortex (384)	petiole (394)	source (389)	vessel element (381)
dermal tissue system (382)	pit (381)	spongy parenchyma (395)	woody plant (382)
endodermis (384)	pith (388)	springwood (389)	
epidermis (382)	pressure–flow hypothesis (389)	stem (381)	
fiber (381)	primary growth (382)	stoma (382)	
fibrous root (383)			
ground system (381)			

Choose the term that does not belong in each of the following sets, and explain why.
1. wood, bark, epidermis, ray
2. sieve tube member, mesophyll, tracheid, companion cell
3. apical meristem, primary xylem, vascular cambium, endodermis
4. root hair, transpiration, vessel element, sclerenchyma
5. blade, cork, petiole, vein

Review

1. Bark does *not* contain (a)secondary phloem (b) secondary xylem (c) cork cambium (d) cortex.
2. Dark wood in the center of a tree is (a) springwood (b) sapwood (c) heartwood (d) summerwood.
3. Mesophyll is the site of (a) water absorption (b) food storage (c) photosynthesis (d) secondary growth.
4. Water moves between tracheids through (a) end walls (b) stomata (c) vessel elements (d) pits.
5. Stomata open and close as the result of turgor pressure in the (a) guard cells (b) sieve tube members (c) companion cells (d) bundle sheath.
6. The root apical meristem is protected by the (a) suberin (b) cuticle (c) root cap (d) cortex.
7. Collenchyma and sclerenchyma are (a) vascular tissues (b) meristems (c) dermal tissues (d) ground tissues.
8. Most monocot stems do *not* have (a) primary growth (b) secondary growth (c) xylem (d) phloem.
9. The Casparian strip occurs on (a) epidermal cells (b) endodermal cells (c) pallisade parenchyma cells (d) vessel elements.

10. The main site of photosynthesis in cacti is the (a) root (b) spine (c) flower (d) stem.
11. What causes a plant to grow wider?
12. Explain the difference between heartwood and sapwood.
13. Briefly describe the transpiration–cohesion theory of water movement in plants.
14. What substances are transported through phloem?
15. List three modified types of leaves.

16. How is a sweet potato different from a white potato? How are they similar?
17. Explain how guard cells regulate the opening and closing of stomata.
18. Describe the different arrangements of vascular bundles in monocots and dicots.
19. Explain the difference between primary growth and secondary growth.
20. Distinguish between herbaceous and woody plants, and give an example of each.

Critical Thinking

1. A pith cell from a plant can be isolated in tissue culture and, if given the proper plant hormones, will form new plants. What does this tell you about the growth potential of these cells?
2. When transplanting a plant, it is important not to remove any more soil than necessary from around the roots. From your knowledge of the function of roots and root hairs, why is this so important?
3. Suppose you examined a tree stump and noticed that 50 rings in from the surface the annual rings became thinner and closer together than the others. What would you conclude about the climate in that area 50 years ago?

4. Examine the photograph above. Explain how this leaf arrangement is adaptive for plants that grow in the shady areas that receive little sunlight.

Extension

1. Read the article by Thomas Pawlick entitled "What's Killing Canada's Sugar Maples?" in *International Wildlife,* January/February 1985, pp. 30–40. Report to the class on the possible role of roots in mineral uptake.
2. Measure transpiration in a common herbaceous plant. Buy two small flower or tomato plants from a local nursery. Make sure the plants are well watered and then cover the pots (including the tops of the pots but not the plants themselves) with a plastic bag or plastic wrap and secure tightly. Weigh the plants to the nearest 0.1 g and record the weight. Turn a fan on one plant, and set another in sunlight for two hours. Weigh the plants again to the nearest 0.1 g and calculate the percentage of water loss resulting from transpiration. Use the following formula:

Original weight − weight after 2 hr = weight change
Weight change ÷ original weight × 100 = percentage weight loss

If no difference is noticed at first, continue the experiment for another two hours or overnight.

Chapter 27 | *Plant Reproduction*

This chapter investigates sexual and asexual plant reproduction. Students study plant adaptations of structure, function, and process that help protect and disperse offspring.

Correlation Chart

Section Number	Section Objectives	Pages	Features	Section Review
27.1	1. Describe the life cycle of mosses. 2. Summarize the life cycle of ferns. 3. Describe the life cycle of gymnosperms.	401–404		1, 2, 3 2, 3, 4 2, 3, 4, 5
27.2	1. Name the four main parts of a flower, and state the function of each. 2. Compare pollen formation and ovule formation. 3. Relate flower structure to method of pollination. 4. Describe fertilization in flowering plants. 5. Summarize the evidence for the adaptive value of nectar guides.	405–410	Think, p. 407 Biology in Process, pp. 408–409	2 1, 4 1, 5 3
27.3	1. Distinguish between simple, aggregate, and multiple fruits. 2. List the similarities and differences between monocot and dicot embryos. 3. Describe three adaptations that promote the dispersal of fruits and seeds. 4. Describe the conditions necessary for seed germination.	411–414		1, 2 3 5 4
27.4	1. State the advantages and disadvantages of asexual reproduction. 2. Compare asexual reproduction with sexual reproduction. 3. Explain the difference between natural and artificial propagation, and give an example of each.	415–417	Biology Lab, p. 417	3 2, 3 1, 2, 4, 5

Resources*

Print

Dowden, Anne Ophelia. *From Flower to Fruit*. New York: Crowell, 1984.

Gibbons, Bob. *How Flowers Work: A Guide to Plant Biology*. Poole, England: Blanford Press (dist. by Sterling), 1984.

Audiovisual/Software

Plant Life Cycles: Alternation of Generations in Plants. Still-frame video or transparencies with cassette, 19 min. Biology Media.

Sexual Encounters of the Floral Kind. Video, 60 min. Oxford Scientific Films. Dist. by Carolina Biological, 1986.

*Additional references are listed in *Modern Biology, Teacher's Program Guide*.

Section Teaching Guides

27.1 Alternation of Generations

Approach

The main objective of this section is to reinforce the concept of alternation of generations by reviewing the life cycles of nonflowering plants. It builds upon the concepts of cell structure and function (Chapter 5), mitosis and meiosis (Chapter 9), diversity through adaptation (Chapters 15 and 16), and plant evolution and diversity (Chapter 25). In presenting the material in the section, assist students in developing a clear understanding of the key characteristics of the sporophyte and gametophyte phases in alternation of generations. Also make sure that students understand the differences between meiosis and mitosis, since such understanding will permit them to comprehend how the gametophyte and sporophyte generations come about. You may wish to briefly review alternation of generations in algae and the alternating diploid-haploid life cycles of some sexually reproducing fungi. You may also wish to encourage students to compare Figures 27-1, 27-2, and 27-4 to see that the pattern of alternation of generations stays the same, although the appearance and dominance of the gametophyte and sporophyte may differ drastically in various groups of plants.

Concept Framework

Most students have seen a plant grow, and many have planted seeds. To use their observations of plants as a starting point for discussing alternation of generations and various specific life cycles, engage students in a discussion of the appearance of common gametophytes and sporophytes. (You may refer them to the discussion of alternation of generations in Chapter 25 for an overview of the two phases.) Most students will describe the sporophyte of flowering plants and conifers, but will have some difficulty describing a seed-plant gametophyte or a moss sporophyte. Elicit from students the inference that a dominant sporophyte may serve to protect the seed-plant gametophyte. You might end the discussion by asking students to consider the possible adaptive advantages of the dominant gametophyte generation in mosses as they read the text.

Emphasis

Make certain students understand that the seed plants have an adaptive advantage in that their gametes are enclosed in a protective structure. This means that the gametes do not have to stay in direct contact with water to remain viable. Also stress the significance of heterospory in increasing variation among populations of seed plants.

Mapping Plant Life

Provide students with a map of an area that shows locations with varying average rainfall and/or varying average temperature by week or month. Ask students to mark the map to indicate where one would expect to find mosses, ferns, and conifers based upon what is known about their reproductive requirements. You might ask students to expand the map to indicate where flowering plants would thrive. You could also plan a follow-up field trip to see if vegetation is located in similar environments where students predicted it would be.

Section Review Answers

(See p. 404.)

27.2 The Life Cycle of Flowering Plants

Approach

The primary objective of this section is to provide students with an understanding of the basic mechanisms of sexual reproduction in flowering plants. The section builds on concepts of structure and function relationships (Chapter 1), mitosis and meiosis (Chapter 9), diversity through adaptation (Chapters 15 and 16), plant evolution and diversity (Chapter 25), and plant tissues (Chapter 26). It prepares students for future concept development concerning animal reproduction, in that the intricate processes of pollen and ovule formation in plants provide students with an example of the genetic events involved in the production of gametes.

In presenting the material in the section, you may wish to draw a generalized diagram of alternation of generations on the board. Ask students to identify at what stages in the life cycle certain events, such as pollen and egg formation, occur. Alternatively or in addition to this exercise, you could draw a generalized flower so that students can identify where in the plant as a whole these events take place. You may wish to use Figures 27-8, 27-9, and 27-10 as tools to assist students in seeing discrete processes in the context of the general reproductive process.

Concept Framework

All students have seen and smelled flowers. To build on such experiences and to provide a conceptual framework for approaching this section, engage students in a discussion of the possible adaptive advantages of having bright colors and attractive odors. Elicit the generalization that such colors and odors serve to attract pollinators (or, perhaps, to warn certain predators away). Point out that without successful pollination, in which pollen is transferred to a stigma from an anther, reproduction in flowering plants would be impossible. You may wish to end the discussion by asking students to consider as they read the text what structural adaptations over time possibly led to the present-day form of a typical flower.

Emphasis

Make certain students understand the essential processes of fertilization and meiosis in the context of alternation of generations. Reinforce the fact that meiosis produces haploid individuals, while fertilization results in diploid individuals. The processes of meiosis and fertilization produce the alternation of haploid gametophytes and diploid sporophytes common in plant life cycles.

Guest Speaker

Invite a local university professor, farm manager, agricultural extension agent, or nursery operator to speak to the class about breeding plant hybrids, such as new brands of seed corn or new varieties of roses. Ask the speaker to explain the methods used in hybridization and to identify the point in the life cycle of a flowering plant at which interventions are made to produce new varieties. Also ask the speaker to explain why hybridization is a desirable or beneficial technique in the particular case being discussed with the class.

Using the Features/Nectar Guides and Pollination

The principal process skill exemplified in this feature is experimentation. In order to test the hypothesis that insect pollinators follow markings on a flower that lead to nectar, Manning and Free designed experiments using various models constructed to resemble flowers. The experimental results supported the hypothesis and also revealed that the bees learned to follow the lines, rather than following the lines because of some unidentified instinct. The experimental results also suggested that additional observations of pollinators in nature should provide information to scientists relating to the learning behavior of pollinators.

Students may be curious about how Manning and Free trained their experimental bees. Explain to students that by using honey as bait, the researchers enticed the insects to land on the models. They then observed the landing patterns exhibited by the bees on different models with and without honey as bait. Point out to students that in a sense, the initial training experiments served as controls in this experimental process.

Section Review Answers

(Continued from p. 410.)
4. polar nuclei (dentral cell), egg, and two synergids, and three antipodals
5. Cross-pollination increases genetic variation and thereby increases the chances that members of a species will survive in a changing environment.

27.3 Fruits and Seeds

Approach

The main objective of this section is to provide students with an understanding of the structures that ensure sexual reproduction in seed plants. The section builds on concepts of structure and function relationships (Chapter 1), plant evolution and diversity (Chapter 25), and plant tissues (Chapter 26). In presenting the material, make sure that students comprehend how fruits and seeds fit into the overall life cycle of an individual plant. Also stress that the adaptive advantages of various modifications for seed dispersal contribute to the reproductive success of a species.

Concept Framework

Many students are aware that there is a difference between fruits and vegetables, but few will know exactly what it is. To provide a conceptual starting point for discussing the importance of fruits and seeds, engage students in a discussion focused on defining the terms *fruit* and *vegetable*. (Webster's defines *vegetable* as "a usually herbaceous plant (as the cabbage, bean, or potato grown for an edible part that is usually eaten as part of a meal . . .".) Elicit the suggestion that a fruit contains seeds but a vegetable does not. You may wish to end the discussion by asking students to read the section and to look for the similarities, in terms of tissue content, between fruits and so-called vegetables.

Emphasis

Make certain that students understand that the germination of a seed is essential in the reproductive cycle of flowering plants. While seeds may stay dormant for many years and still germinate successfully under favorable conditions, some seeds deprived of the necessary conditions for germination eventually fail to produce new plants. The extinction of the bird known as the dodo, for example, has resulted in the complete absence of reproduction among the tropical trees whose seeds must pass through a dodo's crop before germination can take place. Also remind students that the gymnosperms have naked seeds; that is, they lack fruits.

Botany Spelling Bee

Conduct a botany spelling bee, using vocabulary words from this chapter. You may also use words from Chapter 26 if desired. You may wish to have students elect to either spell the word or to define the word when their turn comes.

Section Review Answers

(Continued from p. 414.)
5. Through dormancy plants may survive conditions that are not favorable for germination and growth. Dormant seeds can be carried great distances by wind or other means before they germinate. Dormancy ensures that the seed is mature before growth begins.

27.4 Asexual Reproduction

Approach

The primary objective of this section is to help students gain an appreciation for the adaptive advantages of asexual reproduction in plants. The section builds on concepts of mitosis and meiosis (Chapter 9), plant evolution and diversity (Chapter 25), and plant tissues (Chapter 26). In presenting the material to students, use as many familiar examples as possible to help them understand variation in vegetative reproduction. You may also wish to artificially propagate several plants for students to study as they read this section. Figure 27-15 illustrates grafting, a process that may be difficult to duplicate in a classroom setting.

Concept Framework

Most students will be familiar with the common Idaho potato, and will be aware that the "eyes" have something to do with reproduction. Build upon students' familiarity with the potato, in order to introduce them to the possible cellular mechanisms of vegetative propagation in the tuber. Point out to them that a piece of potato containing an "eye" can develop into a new plant. (Each "eye" is actually a node formed in a spiral around a modified stem, and consists of an axillary—or lateral—bud.) Elicit from students the suggestion that the eye contains an apical meristem that produces primary growth. Students may also suggest that the nutrients stored in the section of tuber are useful in the development of the new plant. Ask students to consider as they read the text, why it is usually better to artificially propagate a plant during the plant's dormant phase than during its growing period.

Emphasis

Make certain students understand that vegetative propagation results in the rapid production of genetically identical clones. While the clones might not survive in the company of plants with greater genetic variability, they are well suited to disturbed areas where a supportive environment might exist temporarily. Point out to students that plants known for vegetative propagation are also able to reproduce sexually.

Field Trip

Visit a local nursery or greenhouse and investigate as many means of artificial propagation as possible. Arrange to have someone speak to the class and demonstrate such techniques as planting leaf and root cuttings, tip and air layering, propagation from specialized roots and stems, and various types of grafting. You may wish to extend this activity into the classroom by providing students with an opportunity to propagate cuttings of their own.

Guest Speaker

Invite a local university professor or horticulturist to speak to the class about mericloning and tissue culture, two laboratory techniques for artificial propagation. The speaker should point out reasons for attempting to keep the propagation process free of bacteria and fungi, and should give examples that illustrate the benefits of these techniques.

Section Review Answers

(See p. 416.)

Laboratory Preparation

Background

This lab procedure allows students to observe both the differences and the similarities between monocot and dicot seeds and seedlings. Similarities include: both have seed coats, both have an embryo that can absorb starch stored in the seed, although the absorptive structure may be different, and both have a radicle that will grow to produce a primary root (which will degenerate in the case of monocots).

Preparation

Prepare bean and corn seeds by soaking them in water for 24 hours. To prepare the seedlings, plant corn and bean seeds in a planting medium, such as vermiculite, about ten days before the investigation.

Safety

Remind students to handle scalpels with care.

Conclusions

1. The corn seed is a monocot because it has one cotyledon. The bean seed is a dicot because it has two cotyledons.
2. Students should name the following: a monocot has one cotyledon, and the hypocotyl does not elongate; a dicot seed has two cotyledons, and the hypocotyl elongates.
3. The dicot cotyledons emerge from the soil and shrivel. The monocot remains below the ground.

Inquiry

Students should form a hypothesis and devise a means of testing it, such as by testing for the presence of starch in monocot and dicot cotyledons with iodine solution and observing the results.

Chapter Review

Vocabulary

1. The combining form -cotyl means "cup-shaped hollow"; epi- means "on"; and hypo- means "under." Epicotyl is an appropriate term because it is the part that lies on the cotyledons, while the hypocotyl lies below.
2. A sporangium encloses spores; a seed encloses a diploid embryo.
3. Both are found in females, are related to reproduction, and are found in flowering plants.
4. In double fertilization one sperm unites with the egg in the ovule, producing the diploid embryo; the other sperm joins with the two polar nuclei, producing the triploid endosperm nucleus.
5. The fruit is a mature ovary. Fruits contain seeds, which are fertilized ovules.

Review

11. The three factors essential for successful wind pollination are: the plant must release a great deal of pollen, there must be good air circulation to carry the pollen to other plants, and plants must be relatively close.
12. The megaspore mother cell in the ovule undergoes meiosis and produces megaspores.
13. A drupe, such as an olive or a peach, has a seed with a woody coating. A berry, such as a tomato or a grape, does not have a seed with a woody coating. Pomes, such as apples and pears, have a leathery interior coating.
14. The diagram should show the following: microspore mother cell, meiosis, four haploid microspores, the tube cell, the generative cell, and the pollen grain.
15. Gymnosperms do not undergo double fertilization as do angiosperms.
16. Seeds can be dispersed by wind, as when maple seed are blown out of trees. They can be dispersed by animals, as when a bird eats a fruit and the seed passes through the bird's digestive system. They can be dispersed by water, as when a coconut floats to a new site and germinates.
17. In monocot germination the hypocotyl is

straight and protected by the coleoptile, and the cotyledon remains underground as the shoot grows. In dicot germination the hypocotyl is often hook shaped and pulls the epicotyl and cotyledons out of the soil as it grows upward. The cotyledons eventually fall off the plant.

18. Seed dormancy is a feature of angiosperm reproduction that increases chances for success by ensuring that seeds germinate only when conditions are favorable for growth.

19. Bees follow the fanned lines of the nectar guide to the nectar source. The bees learn to follow these guides once they have learned that nectar is available if they follow a particular guide.

20. The principal difference is that nonvascular plants have a dominant gametophyte and vascular plants have a dominant sporophyte.

Critical Thinking

1. The gametes would have 12 chromosomes.
2. Such features would be disadvantageous because the large petals and sepals could block the wind currents needed to spread the pollen.
3. When germination begins the endosperm can quickly be converted to sugars. If the endosperm were sugary, the seed would be eaten before germination.
4. Humans and other animals take advantage of the rich store of starch, fat, and protein stored up for the newly emerging embryo after germination.
5. The new plant has an independent supply of energy and protein that can be used until it can take root.
6. The shoots would produce undesirable flowers.

Chapter 27 | Plant Reproduction

Introduction

You know that all plants, from liverworts to apple trees, reproduce by alternation of generations, a form of sexual reproduction. Plants may also reproduce asexually, resulting in offspring that are genetically identical to their parents.

Chapter Outline

Chapter Concept

As you read, note the many plant adaptations that help ensure the protection and dispersal of offspring.

Milkweed, *Asclepias* sp. Inset (top): spore caps of sphagnum moss. Inset (bottom): *Dryopteris borre.*

For additional teaching information see pp. T27A–T27D.

27.1 Alternation of Generations

Recall from Chapter 25 that the life cycle of a plant involves alternation of generations. All plants have a diploid (2N) sporophyte generation that alternates with a haploid (1N) gametophye generation. In mosses and other bryophytes the small sporophyte is attached to and nutritionally dependent on the larger gametophyte. However, the evolution of fern allies, ferns, gymnosperms, and angiosperms involved a reversal of this relationship. In these plants the sporophyte is the larger generation.

Seed plants benefit from having a reduced gametophyte. While the sperm of nonseed plants must swim through water to reach eggs in a gametophyte, the gametes of seed plants are protected by being enclosed in structures produced by the much larger sporophyte. As a result seed plants are able to live in a much wider variety of habitats than plants whose reproduction depends on the presence of moisture. In Sections 27.1 and 27.2 we will explore similarities and differences in sexual reproduction in various plants.

The Life Cycle of Mosses

Look at the life cycle of a moss, as illustrated in Figure 27-1, and notice that the larger generation is the haploid gametophyte. During certain times of the year most moss gametophytes produce two types of multicellular reproductive structures near their growing tips. The **antheridium** *(AN-thur-ID-ee-um)* is a male reproduc-

Section Objectives

- *Describe the life cycle of mosses.*
- *Summarize the life cycle of ferns.*
- *Describe the life cycle of gymnosperms.*

Tip: Review with students the major groups of vascular plants and whether they are classified as seed plants or nonseed plants.

Figure 27-1 In the moss life cycle the sporophyte is the physically smaller generation. The gametophyte is the larger, photosynthetic generation.

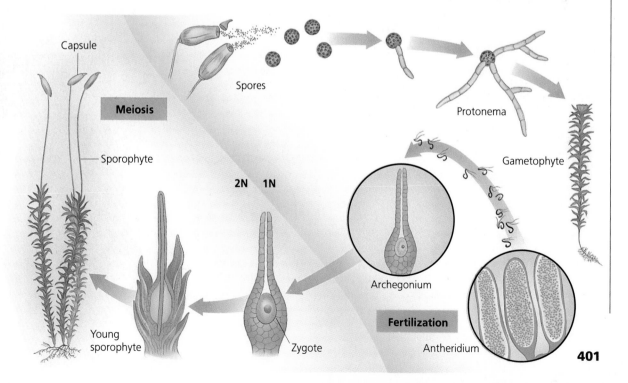

Capsule

Spores

Meiosis

Sporophyte

Protonema

Gametophyte

2N   1N

Archegonium

Fertilization

Young sporophyte

Zygote

Antheridium

tive structure that produces hundreds of flagellated sperm by mitosis. The **archegonium** (AR-kuh-GOE-nee-um) is a female reproductive structure that produces a single egg by mitosis.

During moist periods sperm break out of an antheridium and swim to an archegonium. They then enter the long neck, travel to the egg, and fertilize it, forming a diploid zygote. The zygote is the first cell of the sporophyte generation. It develops into an embryo. Repeated mitotic divisions of the embryo form a multicellular sporophyte.

The sporophyte at first is a stalk that remains attached to and dependent on the gametophyte for nourishment. Soon, cells at the tip of the stalk divide and form a sporangium called a capsule. Cells within the capsule undergo meiosis to form haploid spores. When the spores are mature the capsule splits open, and the spores are released. They travel on currents of wind. Those that land in favorable environments germinate and produce a slender thread called a **protonema.** The protonema is an early stage of gametophyte development. As the protonema grows, it soon begins to look like a mature moss.

All the moss spores look alike, and all produce similar gametophytes. This phenomenon is known as **homospory.** A life cycle such as that of mosses is thus called a homosporous alternation of generations.

The Life Cycle of Ferns

Compare the life cycle of a moss with that of a typical fern, shown in Figure 27-2. As you can see, the life cycle of a fern is also a homosporous alternation of generations, but the sporophyte, not the gametophyte, is the larger generation.

Figure 27-2 The fern life cycle, like that of a moss, includes a sexually reproducing generation that requires moisture for the sperm to swim to the egg.

Sorus

Meiosis

Spores

Mature gametophyte

Sporangium releasing spores

2N 1N

Sperm

Antheridium

Fertilization

Adult sporophyte

Zygote

Archegonium

Fern gametophytes are heart shaped and about 5 to 15 mm in diameter. They are anchored to the soil by rootlike rhizoids on their lower surfaces. Antheridia and archegonia form on the upper surface of the gametophytes. When water covers the mature gametophyte, the antheridia release spiral-shaped sperm, which swim to the archegonia. Here one sperm fuses with the egg and forms a zygote. The zygote, the first cell of the sporophyte generation, rapidly undergoes mitosis, grows into an embryo, and develops into a young fern. As the sporophyte grows it crushes the gametophyte. The mature fern has an underground rhizome that produces fronds.

In most ferns, cells on the underside of the frond differentiate from surrounding leaf cells and develop into multicellular sporangia. Locate a sporangium in Figure 27-3. In many ferns the sporangia are grouped together in clusters called **sori**. Cells inside a sporangium undergo meiosis and form haploid spores. At maturity the sporangium opens. The spores are released and then carried by air currents to different locations. They land, and if conditions are proper, they germinate and form gametophytes. Notice that all of the spores look alike.

Figure 27-3 Sori are clusters of sporangia found on the underside of fern leaves.

The Life Cycle of Conifers

The conifer life cycle, shown in Figure 27-4, illustrates the major ways in which alternation of generations in seed plants differs from that in mosses and ferns. Conifers and all other seed plants produce two types of spores. This phenomenon is called **heterospory,** and plants with this life cycle have a heterosporous alternation of generations. As you read about the conifer life cycle, refer to Figure 27-4. Notice that even though the life cycle of the

Figure 27-4 Conifers, like all seed plants, have a life cycle that includes the production of two types of spores.

403

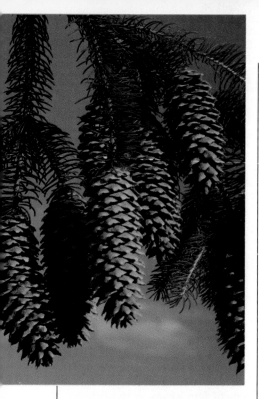

Figure 27-5 The cones of a spruce tree, *Picea* sp., are typical of gymnosperms.

Answers to Section Review
1. An antheridium is a male gametangium that produces sperm. An archegonium is a female gametangium that produces an egg.
2. The student should include sporophyte, gametophyte, meiosis, gametes, zygote, and fertilization at the minimum.
3. The spores of conifers, which are seed plants, are of two different types—megaspores and microspores—while bryophyte and fern spores are all alike.
4. Gymnosperms have seeds and ferns do not; ferns are homosporous and gymnosperms are heterosporous; ferns require free water for reproduction and gymnosperms do not. Other answers may be acceptable.
5. By eliminating the need for sperm to swim to the egg, the microgametophyte permits reproduction in a dry environment.

Info: The female cone of conifers secretes a sticky fluid that traps pollen grains in order to draw them toward the female gametophyte.

conifer is more complex than that of mosses and ferns, the sequence of phases in the conifer life cycle is the same.

The sporophyte of the pine life cycle is the tree. The entire life cycle of the pine takes two years to complete. During the reproductive season the pine tree produces cones. These cones are of two types, male and female. Cells inside female cones produce sporangia called megasporangia. The megasporangium of seed plants is called an **ovule.** Cells inside the male cones produce sporangia called microsporangia.

As you learned in Chapter 25, the cells in all sporangia divide by meiosis and produce haploid spores. Megasporangia cells produce large spores known as **megaspores.** Megaspores produce megagametophytes. Microsporangia cells produce smaller spores known as **microspores.** Microspores produce microgametophytes.

In seed plants the megagametophyte never breaks out of the megaspore wall. Likewise, microspores produce microgametophytes that form inside the microspore wall. The microgametophyte of seed plants is called a **pollen grain.**

The pollen grain is released from the microsporangium and at the same time from the male cone. It travels on the wind and lands on the female cone. It filters down between the scales until it nears the megagametophyte, which never leaves the cone. By this time the megagametophyte has produced archegonia and eggs. The pollen grain splits open, and a **pollen tube** begins to grow toward the egg. Sperm forms and travels down the pollen tube, fertilizes the egg, and a zygote is formed. A zygote develops into an embryo. A fertilized ovule is called a **seed.**

At the beginning of this section you read that seed plants live in a wider range of habitats than plants whose sperms must swim freely through water to reach an egg. One reason for this becomes obvious by examining the life cycle of the pine. Sperm of seed plants are protected in the microgametophyte and thus do not have to swim to the egg. Reproduction in seed plants can therefore take place independently of seasonal rains or other periods of moisture.

Section Review

1. Distinguish between antheridium and archegonium and state the function of each.
2. Draw a generalized diagram of alternation of generations, showing haploid and diploid phases.
3. How are the spores of conifers different from the spores of bryophytes and the spores of seedless vascular plants?
4. List three differences between the life cycles of ferns and gymnosperms.
5. In what way did the enclosure of the microgametophyte contribute to the great evolutionary success of gymnosperms and angiosperms?

27.2 The Life Cycle of Flowering Plants

Undoubtedly you have admired flowers for their beauty, their bright colors, attractive shapes, and pleasing aromas. Each flower is actually a set of adaptations that helps to ensure successful reproduction. In fact, the very traits that make flowers attractive to people are adaptations that attract animals that carry pollen grains from flower to flower.

Not all flowers are colorful. As you might guess, insects and animals seldom visit flowers that are not colorful. Instead the wind carries the pollen from plant to plant.

Parts of a Flower

Botanists have determined from an examination of plant development that a flower is a highly modified branch. The parts of the flower are modified leaves, usually found in four successive whorls or rings. All these modified leaves are formed on the swollen tip of the branch, which is called the **receptacle.**

Sepals make up the bottom whorl. They surround and protect the developing flower before it blooms. You can see the leaf-like sepals in Figure 27-6. All the sepals are collectively called the calyx. **Petals** make up the next whorl and are collectively called the corolla. Most insect-pollinated flowers have brightly colored petals. Petals and sepals of wind-pollinated plants are usually reduced or absent. Because they are not directly involved in reproduction, sepals and petals are called nonessential flower parts.

Section Objectives

- *Name the four main parts of a flower, and state the function of each.*

- *Compare pollen formation and ovule formation.*

- *Relate flower structure to method of pollination.*

- *Describe fertilization in flowering plants.*

- *Summarize the evidence for the adaptive value of nectar guides.*

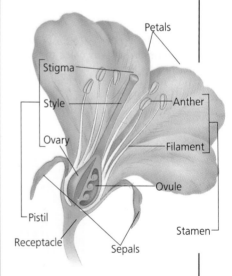

Figure 27-6   The various parts of a flower can be seen in the hare-bell, *Campanula rotundifolia* **(left), and the rock nettle,** *Urtica* **sp. (right). These parts are labeled in the illustration below.**

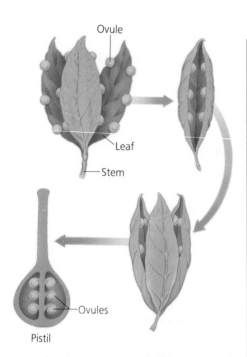

Ovule

Leaf

Stem

Ovules

Pistil

Figure 27-7 A pistil is composed of ovule-bearing leaves that have evolutionarily folded and fused.

Figure 27-8 The process of pollen formation takes place in the anther and results in the formation of a two-celled pollen grain.

Info: The wall of the pollen grain contains a hardening substance called sporopollenin, which resists desiccation.

The two innermost whorls of flower parts are directly involved in sexual reproduction and are called essential flower parts. The male reproductive structures are **stamens,** and the female reproductive structures are **carpels.** Each stamen has two parts. The **anther** is the microsporangium that produces microspores, then pollen grains. The stalklike **filament** supports the anther. In most flowers the carpels are fused to each other, forming a structure called the **pistil.** Many species have flowers with three, four, or five stamens but only one pistil. Not all species of flowers have both stamens and carpels. Flowers with both are called perfect. Those flowers that lack either one of these structures are called imperfect. Two types of imperfect flowers exist. When male flowers and female flowers are on the same plant, the species is called dioecious. If the flowers are on different plants, the species is called monoecious.

Figure 27-7 shows how carpels are leaves that have folded and fused to form a pistil. This fusion is evident when you make a cross section of the **ovary,** the base of the pistil. The ovary is often composed of chambers, each of which often represents one folded carpel. A stalklike **style** arises from the ovary. The style has a **stigma** at its tip. In most species the stigma is sticky or has hairs, enabling it to trap pollen grains.

Pollen Grain Formation

Figure 27-8 shows the process of pollen formation. Each anther contains four pollen sacs that contain diploid microspore mother cells. Each microspore mother cell undergoes meiosis and produces four haploid microspores. The nucleus of each microspore

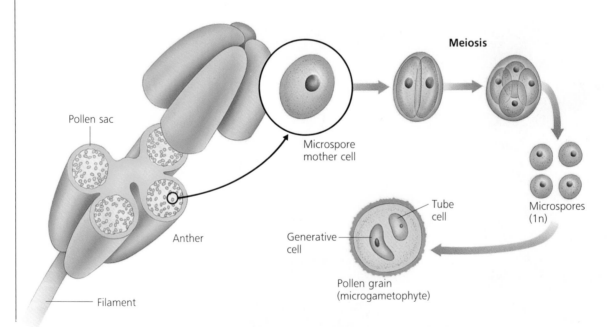

Pollen sac

Microspore mother cell

Meiosis

Microspores (1n)

Tube cell

Generative cell

Anther

Pollen grain (microgametophyte)

Filament

Answer to Think: The rough-shaped pollen grains are dispersed by animals. The protrusions on their surfaces are adapted to catch in the fur of animals or on the limbs of insects. The smooth pollen grains are dispersed by wind. Their smooth surfaces offer little resistance to wind currents.

Tip: Make sure students understand that although all seed plants have ovules, only angiosperms have ovaries.

then divides by mitosis and produces two haploid cells. The two cells remain inside the microspore, which produces a thick wall that surrounds them both. The resulting haploid, two-celled structure is the male gametophyte, the pollen grain. The larger of the two cells is the **tube cell,** which later forms the pollen tube. The smaller cell, called the **generative cell,** will form two sperm.

Ovule Formation

Ovules are megasporangia that are formed in ovaries. Figure 27-9 shows that an ovule contains a large diploid megaspore mother cell. This cell undergoes meiosis to produce four haploid megaspores. Three of these megaspores degenerate and die. The remaining megaspore then enlarges and undergoes three consecutive mitotic divisions to produce eight haploid nuclei. Cytokinesis, however, does not occur. By the time these nuclei have formed, the ovule has formed one or two outer layers called **integuments** *(in-TEG-yoo-ments).* These integuments do not completely encase the ovule. At one end of the ovule is a small pore called a **micropyle** *(MIE-kruh-PILE),* through which the pollen tube will enter and grow.

What about the eight haploid nuclei? As you can see in the figure, they are initially arranged in two groups of four, with one group at each end of the cell. Here's what happens next:

- One nucleus from each group migrates to the center of the cell. These two nuclei are called **polar nuclei.**
- Cell walls form around each of the remaining six nuclei. The three cells farthest from the micropyle are called antipodals and later may function in nutrition.
- One of the three cells nearest the micropyle enlarges and becomes the egg. The remaining two cells are called helpers or synergids.

This structure, which contains seven cells and eight nuclei, is the mature megagametophyte. It is referred to as the **embryo sac.**

Think

Compare these two pollen grains. Which do you think might be carried by animals? Which is carried by the wind? Why do you think so?

Figure 27-9 In the ovule a mother cell undergoes a series of divisions resulting in the formation of seven cells, one of which is the egg.

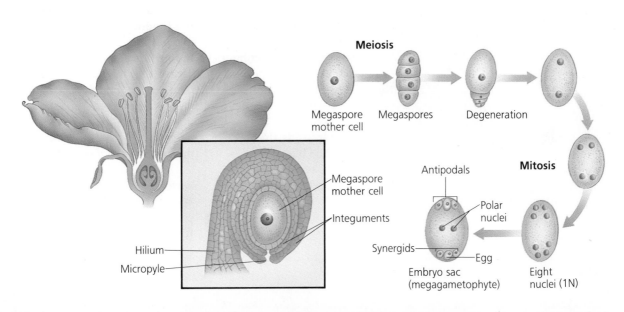

Meiosis

Megaspore mother cell   Megaspores   Degeneration

Mitosis

Antipodals   Polar nuclei

Synergids   Egg

Embryo sac (megagametophyte)   Eight nuclei (1N)

Megaspore mother cell

Integuments

Hilium

Micropyle

Pollination

Before sperm fertilize the eggs contained in embryo sacs, pollen must first be transferred from an anther to a stigma of a flower of the same species. This process is called **pollination.** The transfer of pollen from one plant to another plant of the same species is known as **cross-pollination.** The transfer of pollen from one flower to another flower on the same plant is known as **self-pollination.**

In many plants such as oak trees, pollen is released into the air and carried away in wind currents. Successful wind pollination depends on: (1) the release of large amounts of pollen, (2) ample circulation of air to carry pollen and, (3) individuals being located relatively close to one another.

Most plants with colorful flowers are pollinated by animals. The bright petals and distinctive scents attract animals that feed on pollen and **nectar,** a nourishing solution of sugars and amino acids. When animals gather nectar, the pollen attaches to their bodies. The animals deposit pollen on other flowers as they collect more nectar. Pollinators include bees, hummingbirds, moths, mosquitos, butterflies, beetles, and bats.

Biology in Process

Nectar Guides and Pollination

For many years biologists have wondered if floral nectar guides play a role in insect pollination. Could the lines that appear to lead to the nectar-containing structures in flowers be an adaptation that attracts pollinators?

Researchers A. S. Manning and John Free hypothesized that lines or rows of dots guide insect pollinators to the nectar inside the flower. The biologists created a variety of models with which they could test this hypothesis. They began by training bees to land on flat, colored shapes, using honey as bait. Then they experimented with models that did not contain honey in order to learn whether a patterned flower would be more attractive to the bees. Manning and Free used round, blue models with and without yellow lines leading toward the center. They observed that the bees did not land on one model more often than on the other. Thus they inferred that line patterns alone did not make flowers more attractive to insects.

In their next experiment each model had six "petals" and contained honey in the center. One model had no markings; the other had lines

Fertilization

Fertilization is not the same as pollination. Pollination is the transfer of pollen from an anther to a stigma, while fertilization is the union of gametes. In order for fertilization to occur, the pollen grain must travel to the egg, and sperm must form. The sequence of events leading to fertilization is shown in Figure 27-10 on the following page.

When a pollen grain germinates, the tube cell forms a pollen tube that grows through the stigma and style toward the ovule. The haploid generative cell divides mitotically and forms two sperm. The pollen tube grows through the micropyle and into the embryo sac, forming a passageway through which the sperm enter the embryo sac. One of the sperm fuses with the egg and forms a diploid zygote. This zygote will eventually develop into the embryo.

The second sperm fuses with the two polar nuclei in the embryo sac, producing a triploid (3N) nucleus. This nucleus then divides by mitosis. Repeated nuclear division and, later, cytokinesis, results in a tissue called **endosperm** *(EN-duh-SPERM)*. The endosperm is a nutritive tissue.

Answer to Biology in Process
Natural observations could include watching insects on their first and subsequent visits to a variety of nectar-containing flowers. Flowers with and without lines could also be tested to find out if the insects can see patterns humans cannot see. Insects could be observed for response to other patterns, perhaps even patterns such as those on screens and flyswatters.

leading toward the center of the "flower." Manning and Free observed the spots where bees landed on each model. They found that bees first landed on the edges of both models. However, when the bees returned to the models, they landed more frequently in the center of the model with lines. From these observations, Manning and Free inferred that

bees could learn to follow "guides" to the region of the flower containing nectar.

In a third experiment Manning and Free first observed bees landing on a honeyless model with a "fan" of lines that radiated from a single point. Bees landed on the point only 8.3 percent of the time, and on the lines themselves almost 70 percent of the time.

The bees were then trained with another fan model in which honey was placed at points of intersection. Later, the bees were returned to the model without the honey. The experiment showed that the bees had learned to follow the lines to the point where nectar might be available. Almost 35 percent landed first on the intersection point, and only 57 percent landed first on the lines themselves.

■ Experiments such as these provide evidence that insects follow nectar guides to nectar. What kinds of observations in nature might also support this hypothesis?

Plant Reproduction **409**

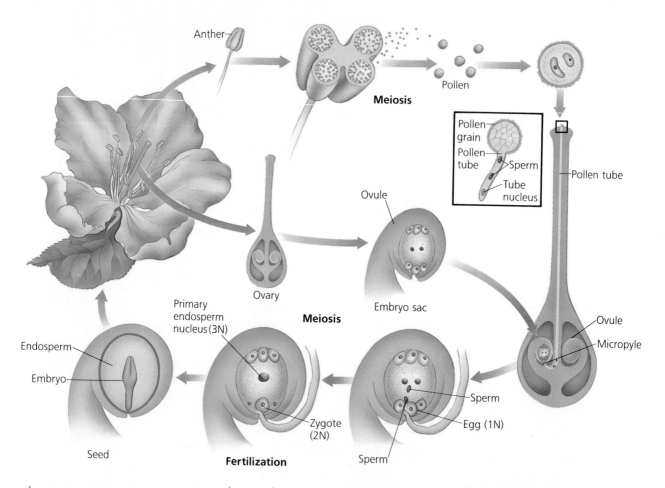

Anther

Meiosis

Pollen

Pollen grain
Pollen tube
Sperm
Tube nucleus

Pollen tube

Ovule

Ovary

Ovule

Embryo sac

Meiosis

Micropyle

Primary endosperm nucleus (3N)

Sperm

Endosperm

Egg (1N)

Embryo

Zygote (2N)

Sperm

Seed

Fertilization

Figure 27-10 Following pollination, a pollen tube and sperm must be produced before fertilization can take place.

Answers to Section Review
1. Self-pollination is the transfer of pollen from one flower to another on the same plant. Cross-pollination is the transfer of pollen from the flower of one plant to the flower of another plant.
2. The drawing should show four rings with sepals forming the outermost ring and, moving inward, petals, stamens, and pistil.
3. The tube cell produces the pollen tube, and the generative cell produces sperm. Sperm travel down the pollen tube, reach the ovary, and fertilize an ovule.
(See pp. T27A–T27D.)

As you can see, two types of cell fusion take place in the embryo sac: one that produces the zygote and one that ultimately produces the endosperm. This process, called **double fertilization,** is unique to angiosperms. Double fertilization enhances the survival of angiosperms by providing the angiosperm embryo with the endosperm to use as a food supply. You can review the angiosperm life cycle in Figure 27-10.

Section Review

1. What is the difference between self-pollination and cross-pollination?
2. Draw a generalized flower, showing the sepals, petals, stamens, and a pistil in relation to each other. Be sure to label each structure.
3. Explain the roles of the tube cell and the generative cell in fertilization.
4. Name the seven cells of the embryo sac.
5. What are some possible adaptive advantages associated with cross-pollination?

27.3 Fruits and Seeds

Fertilization initiates the development of fruits and seeds. A seed is a fertilized ovule. Ovules are found in an ovary. A mature ovary is called a **fruit.** From these definitions you can easily see that many foods that are commonly called vegetables, such as green peppers, okra, cucumbers, and tomatoes, are actually fruits because they contain seeds.

Structure of Seeds

You learned in Chapter 25 that angiosperms are classified as either monocots or dicots on the basis of the number of cotyledons present in the embryo. However, the number of cotyledons is just one of the differences between monocots and dicots. To understand the other differences, let's examine a bean seed, which is a dicot and a corn grain, which is a monocot.

Look at the bean seed shown in Figure 27-11. Bean seeds are usually kidney shaped. The seed is enclosed by a **seed coat,** which develops from the ovule wall and the integuments of the ovule. Along the concave edge of the seed is the **hilum,** a scar marking the location where the seed was attached to the ovary wall. Near the hilum is the micropyle, the opening through which the pollen tube grew to the embryo sac.

Most of the internal part of the seed is occupied by two large, fleshy cotyledons. Between the two cotyledons is the rest of the embryo having three parts. The **radicle** is the embryonic root. The **hypocotyl** *(HIE-puh-KAHT-'l)* is a stemlike area between the cotyledons and radicle. Finally, the **epicotyl** *(EP-uh-KAHT-'l)* is the region above the cotyledons. The epicotyl plus any embryonic leaves is called a **plumule** *(PLOO-myool).*

Now examine the corn grain shown in the figure. The grain of corn is actually a fruit. The wall of the fruit is, however, very thin. Technically corn is a fruit, but the seed occupies almost the entire grain. In corn the umbrella-shaped cotyledon is pressed close to the starchy endosperm. It does not store food as bean cotyledons do, rather, its function is to absorb nutrients from the starchy endosperm and transfer them to the growing embryo. The embryonic shoot of corn is covered with a cylindrical sheath called a **coleoptile** *(KOE-lee-AHP-t'l),* which protects the shoot during germination. Similarly, the embryonic root is covered with a protective sheath called a **coleorhiza** *(KOE-lee-uh-RIE-zuh).*

Types of Fruits

Fruits protect seeds, aid in their dispersal, and may provide a source of energy for the early growth of the seedling. Scientists distinguish between three major kinds of fruit: simple fruit, aggregate fruit, and multiple fruit. Similarly, there are several variations

Section Objectives

- *Distinguish between simple, aggregate, and multiple fruits.*

- *List the similarities and differences between monocot and dicot embryos.*

- *Describe three adaptations that promote the dispersal of fruits and seeds.*

- *Describe the conditions necessary for seed germination.*

Monocot grain

- Endosperm
- Cotyledon
- Plumule
- Epicotyl
- Hypocotyl
- Radicle

Dicot seed

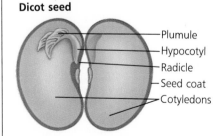

- Plumule
- Hypocotyl
- Radicle
- Seed coat
- Cotyledons

Figure 27-11 In beans the cotyledon is the site of food storage. In corn food is stored in the endosperm and absorbed through the cotyledon.

Table 27-1 Variation in Fruits

Simple fruits
Dry dehiscent fruits have papery or leathery walls and open along seams and usually contain several seeds. ■ **Follicles** come from one pistil and split along one seam (columbine, milkweed). ■ **Legumes** come from one pistil and split along two seams (peas, honey locust). ■ **Capsules** come from more than one pistil and split in a variety of ways (poppy).
Dry indehiscent fruits have paper or leathery walls and do not open along seams and usually contain just one seed. ■ **Grains** have a thin ovary wall fused to the seed coat (corn, wheat). ■ **Nuts** have a woody ovary wall fastened to a single seed (walnut, butternut). ■ **Achenes** have separate seed coats and ovary walls (buttercup, knotweed). ■ **Samaras** have ovary walls that form winglike structures (maple, ash).
Fleshy fruits have soft walls; most are indehiscent. ■ **Drupes** have woody interior coating forming a stone or pit; they are formed from one to many single-seeded carpels (peach, plum). ■ **Berries** lack woody interior coatings; the ovary is fleshy and often juicy; they are formed from one to several many-seeded carpels (tomato, grape). ■ **Pomes** have fleshy floral tubes and a leathery interior (apple, pear).
Aggregate fruits
Aggregate fruits develop from flowers having several separate carpels (raspberry, strawberry).
Multiple fruits
Multiple fruits develop from a cluster of flowers (pineapple, fig).

of each type. The kinds of fruits and their variations are described in Table 27-1. Simple fruits derive from one carpel or several united carpels. Aggregate fruits develop from flowers having several separate carpels. Multiple fruits are a grouping of many fruits derived from a cluster of flowers.

Dispersal of Fruits and Seeds

Fruits and seeds are dispersed by animals, wind, and water. You may have walked through a field or park and unwittingly collected stickers or burrs on your shoes or socks. These burrs are actually fruits. The flavor, fragrance, and bright color of many fruits attract animals. When animals eat such fruits, the seeds often pass unharmed through their digestive systems.

Fruits and seeds of some plants are adapted for wind dispersal. For example, fruits of orchids are extremely small and can be carried away by even the smallest gusts of wind. The fruits of maple and ash trees, called samaras, have winglike appendages that cause the fruit to flutter through the air.

Fruits and seeds of plants growing near water often contain air chambers that allow them to float. Fruits of coconuts, for example, may float thousands of kilometers on ocean currents. When they finally wash ashore, they can germinate and produce new palm trees.

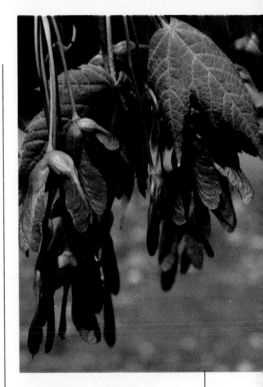

Seed Germination

Seeds are the dormant stage of a plant's life cycle. **Dormancy** is a state of decreased metabolism that allows an organism to survive conditions unfavorable for growth, such as prolonged drought or intense cold.

Conditions Needed for Germination

What kinds of environmental signals break dormancy and begin germination? Environmental factors such as water, oxygen, temperature, and light can trigger germination. Seeds are very dry and must absorb water to germinate. Water softens the seed coat and activates enzymes that begin turning the endosperm of the seed into simple sugars that provide energy for the growth of the embryo. The germinating embryo needs oxygen for cellular respiration. Also, most seeds will germinate only if the temperature is above 10° C. Light also plays a role in seed germination. You will learn about its role in Chapter 28.

Seeds of some plants germinate only under drastic conditions. For example, some seeds germinate only after being scorched by fire or after passing through the digestive systems of animals. Some seeds, such as those of apples, must be exposed to freezing temperatures before they will germinate.

Seeds of many desert plants contain chemicals that inhibit germination. These chemicals are washed from seeds by heavy rain but not by light showers. This trait ensures that the seeds will germinate only when enough water is available for growth.

Figure 27-12 The samara of the red maple, *Acer rubrum*, has winglike parts that enhance its ability to travel from the source.

Tip: Ask students to explain the adaptive advantage of an apple seed germinating only after prolonged exposure to freezing cold. (This ensures that the seed will not germinate in the summer or fall and expose the new seedling to killing frost.)

Info: A germinating seed must take in much water. This hydration increases enzyme activity and results in a marked rise in metabolism. The higher metabolic rate makes it possible for the embryo to resume cell division.

Process of Germination

Let's return to the bean seeds we studied earlier. The first visible change associated with seed germination is emergence of the radicle. The shoot begins growing soon after the radicle ruptures the seed coat. The shoots of dicot seedlings emerge differently from those of monocot seedlings. As you can see in Figure 27-13 on the following page, the bean hypocotyl soon curves and becomes shaped like a hook. Once the curved portion of the hook breaks through the soil, it straightens and pulls the cotyledons and plumule into the air. The embryonic leaves in the plumule emerge, unfold from between the cotyledons, synthesize chlorophyll, and begin to photosynthesize. The seedling then begins to grow rapidly at the apical meristem end of the epicotyl. After food moves into the plant, the shrunken cotyledons fall off.

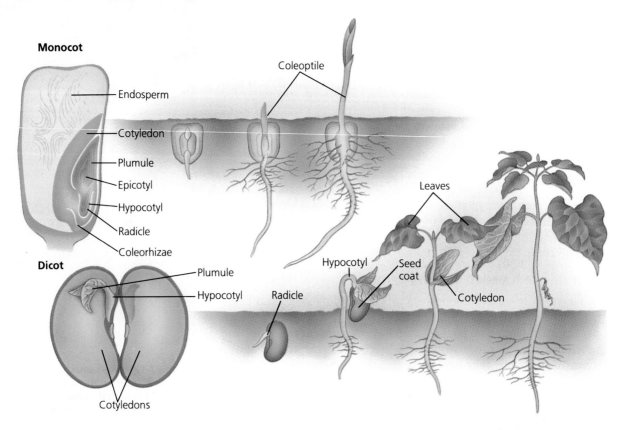

Monocot
- Endosperm
- Cotyledon
- Plumule
- Epicotyl
- Hypocotyl
- Radicle
- Coleorhizae

Coleoptile

Dicot
- Plumule
- Hypocotyl
- Cotyledons

Leaves

Hypocotyl
Seed coat
Radicle
Cotyledon

Figure 27-13 During germination, monocot and dicot cotyledons perform different functions.

Answers to Section Review
1. Simple fruits come from one carpel or several fused carpels; Aggregate fruits come from flowers with several separate carpels. Multiple fruits come from a cluster of flowers.
2. Dry fruits, such as peanuts and sunflower seeds, have papery or leathery walls. Fleshy fruits, such as raspberries and avocados, have soft walls and juicy, succulent tissue.
3. The diagrams should also include the epicotyls, hypocotyls, radicles, plumules, coleoptile, coleorhiza, and hilum.
4. Students should answer with three of the following: water, oxygen, proper temperature, and light. (See pp. T27A–T27D.)

In contrast the cotyledon of the corn grain remains underground and transfers sugars from the endosperm to the growing parts of the embryo. Unlike the bean shoot, the corn shoot is not hook shaped. It is straight and protected by a coleoptile that pushes through the soil. When the coleoptile breaks through the soil surface, leaves of the plumule quickly unfurl.

In beans the radicle becomes the primary root. It eventually produces lateral roots that help absorb water and help anchor the young plant in the soil. In corn the radical gives rise to the many lateral roots that make up the diffuse root system characteristic of grasses. In both beans and corn, leaves eventually emerge, and the plants begin to carry on photosynthesis. If it is not eaten or cut down, the seedling matures. Eventually the seedling becomes an established plant.

Section Review

1. Explain how carpel arrangement is related to different types of fruits.
2. Describe two main types of simple fruits, and give an example of each.
3. Diagram a dicot seed and a monocot seed, showing endosperm, cotyledons, and the parts of the embryo.
4. What are three main requirements for germination?
5. List three possible evolutionary advantages of dormancy.

27.4 Asexual Reproduction

Sexual reproduction in plants involves the fusion of gametes and produces genetic variation among offspring. Genetic variation is advantageous to plants in several ways. It increases the probability of producing individuals adapted to new environments that may arise through climatic or other changes. Variation makes it less likely that changes in the environment will eliminate an entire species. Variability brought about by sexual reproduction allows some offspring to succeed in new environments. Adaptations to new environments account for much of the diversity in the plant life we see.

Despite the benefits of sexual reproduction, many new individuals do not arise from seeds. Instead some plants arise through asexual reproduction, the production of a new individual without the union of gametes. In asexual reproduction all of the offspring are genetically identical with the one parent. Offspring produced by asexual reproduction are called **clones.**

Asexual reproduction gives a species several adaptive advantages. Clones can be produced rapidly. This feature of asexual reproduction is frequently observed in weeds. A **weed** is a plant specialized for quick growth and reproduction in disturbed areas. In weeds, genetically identical plants rapidly colonize newly created space. Often we think of weeds as just nuisance plants. In fact, they are a highly specialized group of organisms adapted for very specific ecological conditions. The American writer Ralph Waldo Emerson once said, "A weed is a plant whose virtues have not yet been discovered."

Asexual reproduction occurs in plants naturally or artificially. When it occurs naturally, asexual reproduction is called vegetative propagation. When humans manipulate the process of vegetative propagation in plants, asexual reproduction is known as artificial propagation.

Vegetative Propagation

The naturally occurring production of new plants from nonreproductive plant parts is known as **vegetative propagation.** Examine the leaves of *Kalanchoe* shown in Figure 27-14. Buds at the edges of these leaves produce plantlets that when separated from the leaf, become new individuals. Plants such as ferns and crabgrass reproduce asexually with underground rhizomes. Leaves and roots form at each node of the rhizome. The node is a place where leaves attach. Eventually these structures break off and become independent individuals.

Other plants, such as poplars and black locust trees, produce buds on their underground roots that become new individuals. Strawberries produce stolons, aboveground runners that break off from the parent plant and produce new individuals at their nodes.

Section Objectives

- *State the advantages and disadvantages of asexual reproduction.*

- *Compare asexual reproduction with sexual reproduction.*

- *Explain the difference between natural and artificial propogation, and give an example of each.*

Figure 27-14   Vegetative propagation occurs in the leaves of *Kalanchoe.*

Figure 27-15 Grafting usually involves the insertion of a branch of one plant into the body of another.

Answers to Section Review
1. Artificial propagation is asexual plant propagation conducted by humans. Vegetative propagation is naturally occurring asexual plant reproduction.
2. plantlets produced from buds at the edges of leaves; new individuals produced by ferns at nodes on their rhizomes; buds produced on the underground roots of poplars; and new individuals produced at the nodes of stolons of strawberries
3. Sexual reproduction results in genetic variation. This variation increases the likelihood that at least some individuals will survive if environmental conditions change.
4. A portion of one plant is united with the root or stem of another plant, resulting in a new combination of tissues.
5. The seedless fruits may have begun as mutants that were then artificially propagated by humans.

Artificial Propagation

Many economically important plants are propagated artificially in order to rapidly produce large crops. **Artificial propagation** is the human application of vegetative propagation. For example, if you owned a potato farm and found that one of your plants produced large numbers of huge potatoes, you'd like more of these plants since they would increase the size of your crop. You could easily achieve this by planting a piece of a huge potato in which there was at least one eye. The eye will grow into a new individual, genetically identical with its high-yielding parent.

Plants such as bananas and ivy are propagated from cuttings that grow into new individuals. Many trees can also be propagated by grafting, as shown in Figure 27-15. **Grafting** is a technique in which a portion of one individual is inserted into the root or shoot of another individual. This technique allows growers to design their own plants. For example, the drought-resistant roots of one plant can be grafted to shoots that produce large, sweet fruit. Plants such as seedless navel oranges can be propagated only with cuttings or through grafting.

Section Review

1. Distinguish between artificial and natural propagation.
2. Give three examples of vegetative propagation.
3. What are the advantages of sexual reproduction when compared with asexual reproduction?
4. Describe the process of grafting.
5. Explain how fruits such as seedless oranges and seedless green grapes might have come into being.

Laboratory

Examining the Seeds of Flowering Plants

Objective
To examine the structure of monocot and dicot seeds and seedlings

Process Skills
Observing, classifying, hypothesizing

Materials
Hand lens, soaked bean seed, dissecting tray, scalpel, spoon, bean seedling, soaked corn seed, potted corn seedling

Method

1 Use a hand lens to examine the outside of a bean seed. Closely observe the concave surface of the seed. Are there any marks or discolored areas?

2 Place the seed in a dissecting tray. Use the scalpel to carefully remove the seed coat. How is the interior of the seed divided? These structures are cotyledons, or seed leaves. Gently force the cotyledons apart. What is inside the cotyledons? Observe this structure closely using the hand lens. Note the end of the fingerlike projection that is nearest the center of the seed. What is the function of this structure? The fingerlike projection up to the point of attachment to the cotyledon is the hypocotyl.

3 Use a spoon to remove a bean seedling from the soil in which it is growing. Shake the soil loose from the root. What has happened to the hypocotyl? What has happened to the cotyledons?

4 Examine the exterior of a corn seed. On one of the flat sides of the seed is a whitish disk. This is the embryo. How does the structure resemble what you found inside the cotyledons of the bean seed?

5 Lay the seed down with the embryo side up. On the dissecting tray cut the seed from top to bottom through the middle of the embryo. Inspect the inside of the seed with the hand lens. What structure makes up the largest part of the embryo? Compare this structure to the embryonic plant you observed in step 2. Next observe the structures surrounding the embryo. How do these structures compare with the hypocotyl you observed in step 2? Do you think these structures function in similar ways?

6 Use the spoon to gently remove the corn seedling from its soil and shake the soil loose from the root. Compare the hypocotyl of the corn seed with that of the bean seed.

Conclusions

1 Which seed is a monocot and which is a dicot? Explain your answer.

2 What differences did you observe between the monocot and dicot seeds?

3 What difference did you observe in the position of the cotyledons as the bean and corn seedlings grew?

Inquiry
Starch for nourishing the embryo is stored in both monocot and dicot seeds. Where is most of the starch found in each type of plant? How would you test for this?

Chapter 27 Review

Vocabulary

anther (406)	embryo sac (407)	microspore (404)	receptacle (405)
antheridium (401)	endosperm (409)	nectar (408)	seed (404)
archegonium (402)	epicotyl (411)	ovary (406)	seed coat (411)
artificial propagation (416)	filament (406)	ovule (404)	self-pollination (408)
	fruit (411)	petal (405)	sepal (405)
carpel (406)	generative cell (407)	pistil (406)	sorus (403)
clone (415)	grafting (416)	plumule (411)	stamen (406)
coleoptile (411)	heterospory (403)	polar nucleus (407)	stigma (406)
coleorhiza (411)	hilum (411)	pollen grain (404)	style (406)
cross-pollination (408)	homospory (402)	pollen tube (404)	tube cell (407)
	hypocotyl (411)	pollination (408)	vegetative propagation (415)
dormancy (413)	integuments (407)	protonema (402)	
double fertilization (410)	megaspore (404)	radicle (411)	weed (415)
	micropyle (407)		

1. Using a dictionary, find the meaning of the word roots in *epicotyl* and *hypocotyl*. Explain why these terms are appropriate for the structures they describe.
2. Explain the difference between a sporangium and a seed.
3. List three similarities between an ovule and an ovary.
4. What are the stages involved in the process of double fertilization?
5. How is fruit related to a seed?

Review

1. The integuments of the ovule are interrupted by (a) a microspore (b) a micropyle (c) a hilum (d) a stamen.
2. The ovary contains (a) anthers (b) sepals (c) ovules (d) filaments.
3. Bryophyte sperm are produced in (a) antheridia (b) anthers (c) archegonia (d) sori.
4. Pollination occurs when (a) a sperm fuses with an egg (b) insects ingest nectar (c) a spore leaves a sporangium (d) pollen lands on a stigma.
5. Attaching part of one plant to another plant is a process known as (a) grafting (b) cloning (c) cutting (d) natural propagation.
6. The epicotyl in a seed is part of the (a) hilum (b) hypocotyl (c) plumule (d) cotyledon.
7. In pollen, the generative cell forms (a) polar nuclei (b) sperm (c) epicotyl tissue (d) endosperm tissue.
8. Pollen sacs are filled with (a) megaspores (b) microspores (c) megaspore mother cells (d) sperm.
9. The fern rhizome produces (a) gametes (b) rhizoids (c) fronds (d) antheridia.
10. The petals of a flower are collectively called a (a) pistil (b) calyx (b) sepals (d) corolla.
11. What three factors are essential for successful wind pollination?
12. How are megaspores formed?
13. Describe three types of fleshy fruits, and give examples of each.
14. Draw a diagram showing the events and the plant structures involved in the formation of pollen.

15. How do gymnosperm life cycles differ from angiosperm life cycles?
16. Name three methods of seed dispersal, and give an example of each.
17. In what ways does germination differ between monocots, such as corn, and dicots, such as beans?
18. Why is seed dormancy a significant evolutionary development?
19. How do bees use nectar guides to find a nectar source?
20. What is the basic difference between alternation of generations in nonvascular and vascular plants?

Critical Thinking

1. Suppose that a type of moss normally has a 2N chromosome number of 12. If meiosis does not occur in the formation of spores but healthy spores form and grow into new gametophytes, what will the chromosome number of the gametes be?
2. Flowers that are pollinated by wind have smaller petals and sepals than those pollinated by animals do. How might big petals and sepals actually be an adaptive disadvantage to them?
3. Why is it advantageous for a plant to have a starchy endosperm as opposed to a sugary endosperm?
4. Keeping in mind their original function, why are fruits and seeds used as a food source by humans and other animals?
5. The nuts of many trees often germinate on the surface of ground that is hard or covered with grass. The new root may grow an inch or two on the surface before it reaches a soft spot and enters the soil. How does the store of nutrients in the nut make this growth possible?

6. Look at the picture above. Most roses are propagated by grafting a new variety onto a stronger wild or an old garden rootstock. When pruning roses in the early spring, it is important to cut off any new shoots that grow below the graft. Give a reason for pruning them.

Extension

1. Read the article by Julie Ann Miller entitled "Somaclonal Variation," in *Science News*, August 24, 1985, pp. 120–121. Give an oral report in which you explain how scientists use somaclonal variants to engineer new, more useful plants.
2. Examine a head of lettuce, a carrot, a radish, a tomato, a green pepper, a potato, a cucumber, and some green peas or beans. These are all commonly called vegetables, but botanically speaking, some are fruits. Determine which are fruits. Discuss the structure of each in terms of plant anatomy.
3. Read the article by Willem Meijer entitled *Saving the World's Largest Flower*, in *National Geographic*, July 1985, pp. 136–140. Write a report about the unusual reproductive process that the *Raffesia arnoldii* undergoes.

Chapter 28 | *Plant Responses*

This chapter details ways in which plants respond to hormones and external stimuli, including seasonal changes. Students observe that these responses reflect adaptation to environmental conditions.

Correlation Chart

Section Number	Section Objectives	Pages	Features	Section Review
28.1	1. List six types of plant hormones and give some functions of each.	421–424	Biology in Process, pp. 422–423	1, 2, 4
	2. Give examples of the adaptive advantages of hormonal responses.			5
	3. Describe three agricultural applications of plant hormones.			3
	4. Summarize the experiments that led to the discovery of auxins.			
28.2	1. Explain the difference between tropisms and nastic movements.	425–426	Think, p. 426	1
	2. List the environmental stimuli to which plants respond and the tropism named for each.			2, 3, 4
	3. Describe common nastic movements, and explain their survival values.			5
28.3	1. Define *critical length,* and explain the role it plays in flowering responses.	427–429	Biology Lab, p. 429	1, 5
	2. Describe vernalization, and give examples of human applications of this phenomenon.			2
	3. Relate abscission zones to dormancy.			4
	4. Describe the role of phytochrome in plant responses.			3

Resources*

Print

Raven, Peter H., et al. *Biology of Plants,* 4th ed. New York: Worth, 1986.

Wilson, Brayton F. *The Growing Tree,* revised ed. Amherst, Mass.: University of Massachusetts Press, 1984.

Audiovisual/Software

Growth of Flowers, 2nd ed. Film or video, 11 min. Teacher's guide. Coronet, 1984.

How Plants Grow: The Inside Story. Apple 128K. IBM 128K. Disk. Scholastic, 1986.

Plant Growth Regulators. Filmstrip or video, 12 min. Teacher's guide. Carolina Biological.

*Additional references are listed in *Modern Biology, Teacher's Program Guide.*

Section Teaching Guides

28.1 Plant Hormones

Approach

The main objective of this section is to examine the effects of the plant hormones. It builds on concepts of structure and function relationships (Chapter 1), basic chemistry and cell biology (Chapters 3 through 5), and plant tissues (Chapter 26). In presenting the text, encourage students to examine the structural diagrams of the hormones in order to more fully understand their effects. Stress that hormones can have different effects in different concentrations.

Concept Framework

Students have learned from reading Chapter 26 that primary growth occurs through cell division, elongation, and maturation in the apical meristems. To reinforce this understanding and to provide a starting point for the discussion of hormones, ask students to speculate about what causes and controls the growth in the meristems. Elicit the observation that something must cause cells to begin to elongate and later stop elongation. Point out to students that auxin, one of the major plant hormones, is primarily responsible for regulating growth. You may wish to end the discussion by asking students to look for evidence as they read the section whether auxin is also mainly responsible for secondary growth in the lateral meristems.

Emphasis

Make certain students understand that many of the processes of growth and reproduction they have studied are controlled or even initiated by plant hormones. You may wish to review certain aspects of growth and fruit production with students as they study this section.

Guest Speaker

Invite a college professor, member of the local forestry or parks department, or a horticulturist to address the class on the topic of hormones as biological controls and herbicides. Ask the speaker to explain how hormones applied in particular doses and/or at particular times during the growth of a plant can limit its growth or even kill it.

Using the Features/The Discovery of Auxin

The main process skill discussed in this feature is experimentation. Charles Darwin and his son Francis designed several experiments that showed that a certain "influence" caused unequal growth in relation to a plant's curvature toward light. Plant physiologist Frits Went later identified this "influence" as a chemical now called auxin. The feature shows that careful experimentation can provide data that supports a hypothesis or suggests new hypotheses that merit investigation.

Section Review Answers

(See p. 424.)

28.2 Plant Movements

Approach

The main objective of this section is to distinguish among the main types of plant movements. It builds on concepts of basic chemistry and cell biology (Chapters 3 through 5) and plant tissues (Chapter 26). In presenting the material, make sure that students understand the distinction between tropisms and nastic movements.

Concept Framework

Most students have seen plants bending or turning toward the sun but do not know what causes these movements. To use such observations as a starting point for studying plant movements, engage students in a discussion of other plant movements they have studied. If students have difficulty starting the discussion, point out that growth produces movement. Elicit the inference that plant movements are often influenced by hormones. You may wish to end the discussion by asking students as they read the text to consider the interactions of different hormones that might create certain types of movement.

Emphasis

Make certain students understand that tropisms and nastic movements are responses to environmental stimuli including light, touch, gravity, and chemical contact. Plant movements are adaptive in that they adjust the pattern of development to suit environmental conditions.

Comparing Water and Hormone Transport

You may wish to have students discuss the mechanism of turgor as it results in the "movement" of a recently watered plant. Have them compare the movement of water in the vascular system to the transport and action of hormones. Ask students if target tissues are involved in both instances of transport.

Section Review Answers

(See p. 426.)

28.3 Photoperiodism and Vernalization

Approach

The main objective of this section is to investigate seasonal responses of plants. It builds on concepts of basic chemistry, biochemistry, and cell biology (Chapters 3 through 5), plant tissues (Chapter 26), and plant reproduction (Chapter 27). In presenting the text, guide students to develop an understanding of the critical length, or photoperiod, concept in plant flowering.

Concept Framework

Students are aware of the fact that flowers bloom and trees bud in your area at certain times of the year. Tap this knowledge to present a discussion of day lengths at times of year that seem particularly significant. Elicit from students the generalization that as days grow shorter, fewer and fewer plants blossom or bud. On the other hand, many regions of the country have autumn-blooming plants such as snapdragons or cylamen. You may wish to end the discussion by asking students to look for what adaptive advantages there might be in blooming later in the year than most other plants in a region as they read the text.

Emphasis

Make certain students understand the adaptive significance of dormancy. You may wish to review the material in Chapter 27 on seeds and seed dispersal to reinforce the advantages of dormancy with environmental conditions that are not favorable for germination.

Predicting Plant Flowering

Provide students with data on annual day lengths and average temperatures for your area. Ask them to predict what kinds of plants would most likely flower during early spring, mid-summer, and mid-autumn. Follow up on their predictions by providing information on the actual plants flowering in the area at different times of the year.

Section Review Answers

(See p. 428.)

Laboratory Preparation

Background

In this lab procedure, students set up a controlled experiment, collect data, and formulate a hypothesis about the effects of the plant growth hormone gibberellin on bean seedlings. Students calculate the percentage increase in height of plants treated with the hormone and compare this percentage to that of control plants. Students should be aware of the function of experimental control, treatment, hypothesis, and sources of error in the scientific method.

Preparation

Soak the bean seeds overnight. Place the soaked bean seeds between layers of damp paper towels. Keep the paper towels wet for three days. Prepare the potting mixture by mixing equal parts of sand, vermiculite, and potting soil. To prepare the gibberellic acid solution, add 100 mg gibberellic acid powder to 5 mL ethyl alcohol; mix into 1 L distilled water; add 1 mL wetting agent. Pour solution in dropper bottles for student use. To prepare the control solution, add 1 mL of wetting agent to 1 L distilled water. This investigation requires one complete lab period plus brief parts of five additional periods.

Safety

Ethyl alcohol is flammable. If spilled, flush affected area immediately with water.

Conclusions

1. Answers will vary. In general the plants treated with gibberellic acid will show a greater percentage increase than the untreated seedlings.

2. Bean seedlings treated with gibberellic acid generally grow faster than those that are treated only with water.

Inquiry

As possible sources of error, students may suggest individual genetic differences between seeds; differences in the environment of the two pots; disease, insect predation, accidents, or damage to individual plants; error in measuring, labeling, or calculating; incorrect preparation of the gibberellin solution.

Chapter Review

Vocabulary

1. The words relate to responses to touch, gravity, and chemicals, respectively.
2. A hormone is also called a plant growth regulator; a target tissue is the tissue in which a hormone produces its effect.
3. Vernalization is an adaptation in which a seed or plant must be exposed to cold before germinating or flowering; dormancy is an adaptation in which a seed or plant undergoes a period of reduced metabolism before germinating or growing; photoperiodism is an adaptation of certain plants for flowering in response to specific day lengths.
4. A negative tropism is a response away from a stimulus; a positive tropism is a response toward a stimulus; and a nastic movement is a response that is independent of the direction of the stimulus.
5. A critical length is a day length above or below which a plant will flower. Long-day plants flower when days are longer than their critical length, and short-day plants flower when days are shorter than their critical length.

Review

11. Auxin produced by the shoot tip inhibits the growth of buds beneath the shoot tip. This results in upward growth at the top of the plant, with little growth in the branches directly under the shoot tip.
12. Bolting is a process in which the length of the internodes increases as a result of the presence of gibberellin.

13. Oligosaccharins influence growth, development, and tissue repair. They may also be involved in defense against disease.
14. The critical length is the period of daylight above or below which a photoperiodic plant will flower. Photoperiodic plants flower in response to days longer or shorter than their critical lengths.
15. Ethylene can be used to bring about ripening in fruits that are harvested before they are mature.
16. The abscission process allows the plant to preserve nutrients by transporting nutrients from the doomed leaves into the stem for storage.
17. By coiling around nearby objects as it grows, a plant may continue to grow upward and keep a maximum of its surface exposed to the sun.
18. The influence of a hormone is affected by the degree of its concentration and by the target tissue it reaches.
19. The weed killer 2,4-D kills dicots but does not harm monocots, such as corn.
20. Hydrotropism and chemotropism are both responses to chemical substances. However, chemotropism occurs between parts of the same plant, while hydrotropism is the response of a plant part to environmental water.

Critical Thinking

1. The plants near the road have probably been sprayed with a weed killer such as 2,4-D.
2. The hormone can be transported to nearby plants by air currents and thus acts as a chemical messenger within a population.
3. It is important that 2,4-D not contaminate groundwater supplies because these are the source of irrigation water. Such polluted water could harm dicots.
4. The base of the leaves may be phototropic due to auxin distribution.
5. It provides a shade-free environment for new leaves that emerge in spring.
6. If the trees were sensitive to temperature as well as to day length, they might start to grow after a brief warm spell in early spring. If this were followed by a severe cold period, as often happens, the plants would die.

Chapter 28 | *Plant Responses*

Introduction

Plants reproduce and grow in response to many environmental factors, such as light, moisture, and temperature. Chemicals inside the plant called hormones are intricately involved in this growth and development. Hormones and other chemicals influence a variety of plant movements and often the timing of flowering and reproduction. These chemical regulators have adaptive significance for plant survival.

Chapter Outline

28.1 Plant Hormones
- *Auxin*
- *Cytokinins*
- *Gibberellins*
- *Other Hormones*

28.2 Plant Movements
- *Tropisms*
- *Nastic Movements*

28.3 Photoperiodism and Vernalization
- *Flowering*
- *Vernalization and Dormancy*

Chapter Concept

As you read, note the ways in which plants respond to external stimuli and to hormones. Note that these are adaptations to environmental conditions.

Forest in Cornwall, England. Insert: Autumn leaf.

For additional teaching information see pp. T28A–T28D.

28.1 Plant Hormones

Plant growth and development are controlled by both external and internal factors. The external factors include light intensity, gravity, temperature, moisture, and length of day. Each species differs in its response to these factors, but in all species external factors trigger changes in internal plant physiology.

The internal factors that control plant growth are chemicals called **hormones.** Because hormones stimulate or inhibit plant growth, many scientists also refer to them as **plant growth regulators.** Hormones affect plants in many ways, from causing them to grow taller to causing leaves to change color in autumn.

Plant hormones are organic compounds that are effective in small concentrations. They are synthesized in one part of the plant and transported to **target tissues** elsewhere in the plant, where they trigger a physiological response such as growth. Botanists have identified six kinds of plant hormones. Though we will discuss these hormones one at a time, keep in mind that plant responses typically result from the interaction of several hormones.

Auxin

Auxins *(AWK-sunz)* are a class of hormones that regulate the growth of plant cells. Chemical structures of three auxins are shown in Figure 28-1. The only naturally occurring auxin is **indoleacetic** *(IN-DOLE-uh-SEED-ick)* **acid,** or **IAA.** IAA is produced in actively growing plant regions, such as shoot tips, young leaves, and developing fruits. It is then transported to other parts of the plant where it has many physiological effects.

One of the most important effects of IAA is the altering of plasticity in primary cell walls. In altering plasticity IAA regulates cell elongation. Effects on the target tissue vary with concentration. In a target tissue one concentration of IAA may stimulate cell elongation, while a greater one may inhibit elongation. The effect of an auxin also depends on the target tissue.

Synthetic Auxins

Chemists have synthesized several auxinlike compounds. These are used agriculturally. Naphthaleneacetic *(NAFF-thuh-LEEN-uh-SEED-ick)* acid (NAA) is a synthetic auxin used by farmers to keep fruit from dropping off trees. The use of NAA allows growers to harvest all their fruit in a short period of time. Naphthaleneacetic acid also stimulates the formation of roots.

Farmers use the synthetic auxin 2,4-D to kill weeds. Many broadleaf plants, such as dandelions, are sensitive to synthetic auxins, while many monocots, such as corn, are not. The 2,4-D disturbs the metabolism of the dicot cells, causing the plant to die. As a result applying the proper concentration of 2,4-D to a corn-

Indole ring — Acetic acid

Indoleacetic acid (IAA)

2,4-Dichlorophenoxyacetic acid (2,4-D)

α-Naphthalenacetic acid (NAA)

Figure 28-1 The structure of indoleacetic acid, the only naturally occurring auxin, differs in detail from two synthetic counterparts.

Figure 28-2 Lateral bud growth results when shoot tips are removed (top). Lateral bud growth is inhibited when artificially applied auxin replaces shoot tips (bottom).

Tip: Tell students that, unlike auxins, cytokinins have few commercial uses. However, current scientific research has produced new information about the role of cytokinins and their potential commercial value.

field will kill the dicotyledons but will not harm the corn. Synthetic auxins are also used to stimulate fruit development in the absence of pollination and fertilization. Spraying flowers of plants such as watermelons and tomatoes with synthetic auxins can produce seedless fruits.

Apical Dominance

In many plants the branches near the shoot tip are much shorter than those near the base. This pattern of growth results from **apical dominance.** Biologists have determined that in apical dominance the apical meristem produces auxin in a concentration that inhibits the growth of buds near the shoot tip. This limits the length of branches near the shoot apex.

Compare the plants in Figure 28-2. The shoot tips of the plant at the top have been removed, thereby eliminating the source of the auxin. As a result buds near the shoot tip have begun growing. The shoot tips of the plant at the bottom have also been removed. In this case, however, the shoot tip has been replaced with lanolin that contains an auxin. Replacing the shoot tip with auxin inhibited growth of the nearby buds and resulted in apical dominance.

Biology in Process

The Discovery of Auxin

The discovery of auxin involved the work of scientists around the world. Charles Darwin and his son Francis began studying phototropism in oat coleoptiles in the late 1870s. They observed that curvature toward light occurred in the elongating region of the coleoptile and hypothesized that this region sensed light.

The Darwins designed several experiments to test this hypothesis. They first covered the region of elongation with a cylinder of metal foil that prevented light from reaching the tissue. To their surprise this had no effect on curvature. Curvature was stopped only when the foil was placed over the tip of the coleoptile. The Darwins concluded that "when seedlings are freely exposed to lateral light some influence is transmitted from the upper to the lower part, causing the material to bend." They stated that some "influence" had to move from the tip of the coleoptile to the elongating region, causing unequal growth.

In 1926 Dutch plant physiologist Frits Went explained the Darwins' observations. Went asked: (1) Is the nature of the influence electrical or chemical? (2) Can it be isolated?

Cytokinins

Hormones called **cytokinins** *(SIE-tuh-KIE-nunz)* promote cell division. They are produced in rapidly growing regions, including roots, fruits, and seeds. Cytokinins influence many aspects of plant development, including the formation of roots and stems and the differentiation of xylem and phloem. Cytokinins seem to act by binding with RNA and thus affecting the synthesis of proteins, including those necessary for cell division.

Gibberellins

In the 1920s Japanese scientists discovered that a substance caused rice plants that were infected with a fungus of the genus *Gibberella* to grow abnormally tall. Biologists later isolated the substance responsible for this growth. They eventually found that this substance, which they called **gibberellin** *(JIB-uh-RELL-un)*, was a hormone that was also synthesized by plants themselves. Biologists have now isolated over 65 gibberellins.

Gibberellins are found in varying amounts in all parts of a plant. The highest concentration of gibberellins is found in im-

Cytokinin

Figure 28-3 The structure of cytokinin is related to its function.

The Darwins' experiment

Went's experiment

Went designed an elegant experiment. Keeping his seedlings in the dark, he first cut off the tips of several coleoptiles and placed their cut surfaces in contact with agar. After about an hour, he discarded the tips. He then placed small pieces of the agar on one side of the stumps of the plants. The coleoptile grew away from the side on which the agar block was placed. Furthermore, twice as much curvature occurred when two blocks were placed on one side of the coleoptile.

Went performed another experiment to make sure the curvature resulted from a substance that was produced by the coleoptile and transferred to the agar. On one side of a coleoptile he placed a block of agar that had not been in contact with a tip. When no curvature occurred, Went concluded that the influence for phototropism was a diffusable chemical. He also concluded that the influence could be isolated.

Went inferred that the substance in the agar stimulated growth. He named this chemical auxin, from the Greek word *auxein,* meaning "to increase."

■ What evidence allowed Went to conclude that auxin could be isolated from the plant?

Figure 28-4 Gibberellic acid, abscisic acid, and ethylene, shown in structural form, regulate various aspects of plant growth. Ethylene promotes the ripening of fruit.

Abscisic acid

Ethylene

Gibberellic acid (GA₃)

mature seeds. Gibberellins have numerous effects on target tissues, the major effect being to promote growth. Before some plants flower, the regions of the stem between the nodes, called internodes, will lengthen. This lengthening, which is called **bolting,** can also be induced by treating a plant with gibberellin. Gibberellins also stimulate seed germination and promote the formation of seedless fruits.

Other Hormones

During the 1960s botanists isolated a hormone called **abscisic** *(ab-SIZE-ick)* **acid** that inhibits the growth of buds and the germination of seeds. Abscisic acid is found in dormant buds, fruits, and roots. During a drought leaves synthesize large amounts of abscisic acid, causing stomata to close and reducing the loss of water.

Unlike other plant hormones, **ethylene** *(ETH-uh-LEEN)* is a gas produced by fruits, flowers, leaves, and roots and released into the air. Ethylene stimulates the ripening of fruit by softening cell walls and promoting the conversion of starches into sugars. Ripening fruits that give off ethylene cause nearby fruits to ripen. Growers often harvest immature tomatoes, store them, and later apply ethylene to ripen them for market.

In the 1980s botanists discovered a new class of plant hormones called **oligosaccharins** *(AHL-ih-goe-SACK-uh-rinz)*. These hormones act on primary cell walls. They regulate growth, development, and may contribute to defense against disease.

Section Review

1. How does the origin of the word *hormone* reflect their function?
2. What is the function of auxins?
3. Which hormones can be used to produce seedless fruits?
4. Explain how ethylene differs from other plant hormones.
5. List three possible adaptive advantages of apical dominance.

Answers to Section Review
1. The Greek root of the word *hormone* means to "stir up" or "set in motion" which is one function of hormones.
2. Auxins control the growth of plant cells.
3. synthetic auxin and gibberellin
4. It is a gas and can move from plant to plant.
5. Apical dominance allows plants to grow upward toward light rather than wider; it prevents upper branches from shading lower ones, maximizing light-absorbing surfaces; smaller branches at the top of the tree will keep the tree from becoming top heavy; predators will be unable to reach to shoot tips.

28.2 Plant Movements

Plants respond to environmental stimuli such as light, moisture, chemicals, gravity, and mechanical disturbances. These stimuli cause adaptive movements by influencing growth. Plant growth involves the elongation of cells, and most plant movements result from elongation triggered by environmental factors.

Tropisms

A **tropism** *(TROE-PIZZ-'m)* is a plant movement toward or away from an environmental stimulus. Each kind of tropism is named for its stimulus. For example, movements in response to light are called phototropisms. Movement toward an environmental stimulus is called a positive tropism. Thus in Figure 28-5 the shoot tips growing toward light are positively phototropic. Conversely, movement away from a stimulus is a negative tropism.

Phototropism

Phototropism is a plant growth response to light coming from one direction. Scientific analysis has shown that light coming from one direction causes auxin to move to the shaded side of the stem. In most cases the auxin causes cells on the shaded side to elongate more rapidly than cells on the lighted side of the stem. As a result the stem curves toward the light. By orienting the plant toward its light source, phototropism maximizes the amount of light that reaches the photosynthetic cells of the plant.

Thigmotropism

Thigmotropism is a growth response to contact with a solid object. Auxin and ethylene control coiling, which occurs when plants such as morning glories come into contact with an object. Thigmotropism allows a plant to "climb" over and cling to objects, increasing its chances of intercepting light for photosynthesis.

Gravitropism

Gravitropism is a plant growth response to gravity. Usually a root grows down and a stem grows up—that is, roots are positively gravitropic and stems are negatively gravitropic. Gravitropism is also known as geotropism.

Like phototropism, gravitropism appears to be regulated primarily by auxin. If a seedling is placed horizontally, auxin accumulates along the lower sides of both the root and stem. This concentration of auxin stimulates cellular elongation along the lower side of the stem, and the stem grows upward. A similar concentration of auxin inhibits cellular elongation in the lower side of the root, and as a result the root grows downward.

To learn how plant cells sense gravity, scientists have studied the position of inclusions in root cap cells called amyloplasts

Section Objectives

- *Explain the difference between tropisms and nastic movements.*

- *List the environmental stimuli to which plants respond and the tropism named for each.*

- *Describe common nastic movements, and explain their survival values.*

Tip: Explain to students that in thigmotropism cells touching the solid object shorten while cells on the opposite side elongate. The result is a pattern of coiling.

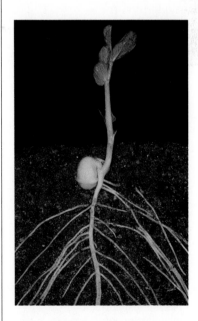

Figure 28-5 As this garden pea seedling demonstrates, roots are positively gravitropic, while shoots are positively phototropic.

Plant Responses   **425**

Figure 28-6 The dark bodies at the bottom of the cells are amyloplasts.

Answer to Think
They are nastic movements because the plant is responding to an environmental stimulus independently of the direction of that stimulus.

Think

Are the leaf responses of the Venus's-flytrap tropisms or nastic movements?

Answers to Section Review
1. Tropism refers to a movement toward or away from a stimulus. A nastic movement is independent of the direction from which the stimulus comes.
2. Amyloplasts may influence the transport of hormones that affect geotropic growth.
3. Phototropism is a response to light; shoots growing upward are phototropic. Gravitropism is a response to gravity; roots are geotropic. Chemotropism is a response to chemicals within the plant; the growth of the pollen tube into an ovule is a chemotropism. Hydrotropism is a response to water; most roots are hydrotropic. Thigmotropism is a response to touch. Coiling vines are thigmotropic.
4. negative gravitropism
5. Hormones might influence the biochemical content of cells, thus influencing changes in tissues that result in nastic movements.

(*AM-uh-loe-PLASTS*). Gravity causes amyloplasts, which contain grains of starch, to settle on the lower surface of a cell. When a plant is turned sideways, amyloplasts fall to the lower cell wall. Amyloplasts are thought to cause a movement of calcium ions, which in turn influences the transport of auxins that cause the root to curve in response to gravity.

Chemotropism and Hydrotropism

Plant growth in response to a chemical is called a **chemotropism.** For fertilization to occur, for example, the pollen tube must grow into the ovule through the micropyle. The pollen tube grows in response to chemicals produced by the pistil of a compatible flower. The growth of roots in response to water is called **hydrotropism.** One example occurs when roots encounter water-filled pipes and grow into rather than around them. Negative hydrotropism is an adaptation for avoiding flooded soils.

Nastic Movements

Plant movements that occur in response to environmental stimuli but that are independent of the direction of the stimuli are called **nastic movements.** For instance, when *Mimosa pudica*, the sensitive plant, is touched, its leaflets fold and its petioles lower within a few seconds. This response is triggered by rapid movements of potassium ions between parenchyma cells at the base of the leaflets and petiole. Stimulation causes ions to be pumped out of cells along the lower side of the petiole. Water then rapidly moves out of these cells by osmosis, and the cells shrink. Many nastic movements, such as the closing up of morning glory petals in the evening, have adaptive significance. Nastic movements allow plants to conserve water and food energy when conditions are not favorable for photosynthesis and growth.

Section Review

1. How do a tropism and a nastic movement differ?
2. How might amyloplasts function in gravitropism?
3. Define five tropisms, and give an example of each.
4. Name a kind of negative tropism shown by stems.
5. What role might hormones play in nastic movements?

Info: Nastic movements occur in stamens of some flowers. At an insect's touch the stamens droop, depositing pollen on the insect.

28.3 Photoperiodism and Vernalization

Plant responses are strongly influenced by seasonal environmental changes. For example, many trees shed their leaves in autumn and most plants flower only during certain times of the year. How do plants sense seasonal changes? Contrary to what you might suspect, most plants do not mark the seasons by changes in temperature. Instead they respond to changes in day length. This response is called **photoperiodism.**

Biologists don't know exactly how plants monitor changes in day length, but evidence suggests that this response involves a pigment called **phytochrome** *(FITE-uh-KROME)*. Phytochrome exists in two forms: red light–sensitive phytochrome, P_r, and far-red light–sensitive phytochrome, P_{fr}. The P_r absorbs red light of wavelengths of about 660 nm and in doing so is converted to P_{fr}. When P_{fr} absorbs far-red light of wavelengths of about 730 nm, it is converted to P_r. Since sunlight contains proportionally more red light than far-red light, P_r is converted to P_{fr} during the day. At night P_{fr} is slowly converted to P_r. Accordingly, the ratio of P_r to P_{fr} may be a chemical measure of the relative length of day and night. Biologists believe that the changing proportions of the two phytochromes play a role in starting the hormonal changes that cause flowering, though other factors may be involved.

Flowering

Flowering is influenced by several factors, including temperature and the availability of moisture. However, in some plants the most important factor controlling flowering is daylength. Plants vary in their response to duration of light. Many species have a **critical length,** which is the length of daylight above or below which these species of plants will flower. The adaptive value of photoperiodism is that members of a species bloom at the same time each year, when conditions for pollination are most favorable.

Long-day plants flower only when exposed to day lengths longer than the critical length for the plant. For example, wheat plants flower only when days are longer than 10 hours—during the lengthening days of late spring and early summer. Radishes, clover, irises, and beets are other long-day plants.

Short-day plants flower only when exposed to day lengths shorter than the critical length for the plant. For example, ragweed flowers only when days are shorter than 14.5 hours. Most short-day plants flower during the shortening days of the fall, though some bloom in early spring. Chrysanthemums, goldenrods, soybeans, and poinsettias are short-day plants.

Day-neutral plants are not affected by the length of days and nights. Tomatoes, dandelions, and roses are day-neutral.

Info: Nurseries can produce many flowering plants, regardless of the plant's normal blooming season, by controlling exposure to light.

Section Objectives

- *Define* critical length, *and explain the role it plays in flowering responses.*

- *Describe vernalization, and give examples of human applications of this phenomenon.*

- *Relate abscission zones to dormancy.*

- *Describe the role of phytochrome in plant responses.*

Figure 28-7 The type of phytochrome known as P_r, shown in structural form, converts to P_{fr}, giving plants a means of measuring daylight.

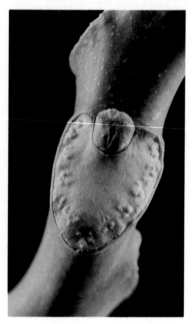

Figure 28-8 An abscission zone forms where the gingko petiole attaches to the stem (left). After the leaf falls, a leaf scar remains (right).

Vernalization and Dormancy

Many plants such as turnips, beets, and carrots must be exposed to cold before they will flower. Some seeds require extended periods of cold before they germinate. This requirement, called **vernalization,** is important for crops such as winter wheat, barley, and rye. For example, seeds of winter wheat that are planted in the short days of autumn germinate before winter. Exposure to cold winter temperatures causes the plant to flower early in the spring, and an "early" crop is produced. Farmers in areas with dry summers exploit this phenomenon to grow and harvest crops in the spring before the summer drought.

Most plants do not grow at the same rate year-round. During periods of cold or drought, plant growth ceases or is limited, and plants enter dormancy. Shorter days stimulate the formation of winter-resistant buds and **abscission** *(ab-SIZH-un)* **zones** at the base of leaf petioles. Leaves break off the plant at abscission zones. Before leaves fall, nutrients are transported back into the stem and are used again in the spring. The leaf would die in the cold anyway, so losing leaves allows the plant to retain nutrients.

Section Review

1. Distinguish between *short-day plants* and *long-day plants.*
2. How do farmers use vernalization to their benefit?
3. How does phytochrome enable plants to detect changes in seasons?
4. What is the adaptive advantage of dormancy?
5. What factors might influence flowering times in day-neutral plants?

Answers to Section Review
1. Short-day plants flower when days are shorter than their critical length. Long-day plants flower when days are longer than their critical length.
2. Farmers can plant wheat that will flower early in the spring, resulting in a crop that can be harvested before the environment becomes too hot and dry for plants to grow.
3. The ratio of the two forms of phytochrome may serve as a chemical indicator of day length.
4. Dormancy allows the seed or plant to survive conditions unfavorable for growth.
5. Vernalization and hormonal activity. Other answers may be acceptable.

Laboratory

Plant Growth and Response

Objective
To observe the effects of plant hormones on bean seedlings

Process Skills
Experimenting, collecting and analyzing data, hypothesizing

Materials
2 flower pots, 8 inches or larger; potting mixture; 10 3-day-old bean seedlings; glass-marking pencil; ruler; cotton swabs; gibberellic acid solution; water with wetting agent

Method
1 Fill two flower pots with potting mixture. Plant five seedlings in each pot. Space the seedlings evenly around the edge of the pot. The roots of each seedling should be at least 1 cm below the soil surface.
2 Use a glass-marking pencil to label each pot with your name. Water the seedlings. Label one pot *Gibberellic Acid Treated* and the other pot *Control*. Put the pots in a warm place out of direct sunlight.
3 Continue to moisten the plants periodically until they are at least 3 cm tall. This will take between three and seven days.
4 When the plants are at least 3 cm tall, make a chart to record the data you will collect on the height of the plants. Allow enough space on your chart to record five measurements for each plant.
5 Measure the height of each plant to the nearest millimeter. Measure from the soil surface to the shoot tip. Record your data.
6 Use a clean cotton swab to apply gibberellic acid solution to the leaves and shoot tips of the seedlings in the flower pot labeled *Gibberellic Acid Treated*.
7 Use a fresh swab to apply water with wetting agent to the leaves and shoot tips of the

Gibberellic acid

plants in the control group. What is the purpose of applying this solution?
8 Make sure each pot receives the same amount of light and water during the rest of the investigation.
9 Measure the height of each plant the next four times your class meets. Record the measurements on your chart.
10 After the fifth measurement, calculate the average height of the seedlings in each pot. Using the measurements from your chart, calculate the average height for each of the preceding four days. Would your results for this investigation be valid if for some reason you applied gibberellic acid solution to the seedlings in the control group?

Conclusions
1 Calculate the percentage increase in the average height of the two groups of plants from the first measurement to the last. Did the *Gibberellic Acid Treated* group or the *Control* group have the greater percentage of increase?
2 Propose a hypothesis that would explain the plant growth patterns you observed in this investigation.

Inquiry
What are some possible sources of error in this experiment?

This investigation must be done in five sessions.

Chapter 28 Review

Vocabulary

abscisic acid (424)	cytokinins (423)	indoleacetic acid (IAA) (421)	phytochrome (427)
abscission zone (428)	day-neutral plant (427)	long-day plant (427)	plant growth regulator (421)
apical dominance (422)	ethylene (424)	nastic movement (426)	short-day plant (427)
auxin (421)	gibberellin (423)	oligosaccharins (424)	target tissue (421)
bolting (424)	gravitropism (425)	photoperiodism (427)	thigmotropism (425)
chemotropism (426)	hormone (421)	phototropism (425)	tropism (425)
critical length (427)	hydrotropism (426)		vernalization (428)

Explain the differences between the terms in the following sets of related words.

1. thigmotropism, gravitropism, chemotropism
2. hormone, plant growth regulator, target tissue
3. vernalization, dormancy, photoperiodism
4. negative tropism, positive tropism, nastic movement
5. critical length, long-day plant, short-day plant

Review

1. Some plants must be explosed to cold before they will flower, a process called (a) photoperiodism (b) vernalization (c) dormancy (d) thermotropism.

2. The growth of roots above the soil surface, where oxygen is absorbed, is a (a) negative gravitropism (b) negative hydrotropism (c) positive phototropism (d) positive chemotropism.

3. Nastic movements occur (a) without a stimulus (c) toward a stimulus (b) away from a stimulus (d) independent of the direction of the stimulus.

4. Cytokinins promote cell elongation and (a) cell aging (b) cell division (c) cell storage (d) cell transport.

5. P_{fr} converts to P_r during (a) long days (b) days (c) short days (d) nights.

6. Plant growth responses specifically involve (a) cell differentiation (b) cell maturation (c) cell elongation (d) cell secretions.

7. Seedless fruits can be produced by the application of auxin or (a) cytokinins (b) gibberellins (c) ethylene (d) oligosaccarin.

8. A ripe avocado will cause other avocados to ripen through the release of (a) auxin (b) ethylene (c) abscisic acid (d) gibberellin.

9. Charles and Francis Darwin observed that phototropism caused stem curvature in (a) the apical meristem (b) the region of cellular elongation (c) the tip of the coleoptile (d) the region of cellular differentiation.

10. Abscission zones are areas where (a) leaves fall off stems (b) plants sense day length (c) abscisic acid is manufactured (d) cuttings are produced.

11. Explain the mechanism for apical dominance.

12. Describe bolting and identify what substance causes it.

13. List three effects of oligosaccharins.

14. Define *critical length,* and explain how it functions in photoperiodism.

15. How might a fruit grower use ethylene on crops?

16. What is an adaptive advantage of the leaf fall, or abscission, process?

17. Explain how a coiling thigmotropism

could result in greater absorption of sunlight for photosynthesis.

18. Name two factors that affect the influence a hormone will have.

19. Why is 2,4-D effective as a weed killer in cornfields?

20. How are hydrotropism and chemotropism similar? How are they different?

Critical Thinking

1. You are walking along a highway that is lined with broadleaf weeds. The tips of these weeds are long, pale, and spindly, with twisted stems and long internodes between leaves. What is the most probable cause of this unusual appearance?

2. How does a gaseous hormone like ethylene give a plant an adaptive advantage?

3. Why is it important that weed killers such as 2,4-D not get into the earth's groundwater supplies, lakes, and rivers?

4. Some plants track the sun to get maximize photosynthesis. Look at the picture of the sunflowers. The stems of these plants are not phototropic. What part of the plant is?

5. The seasonal loss of leaves by trees and shrubs serves the adaptive advantage of conserving nutrients. Can you think of another adaptive advantage that loss of leaves would provide?

6. The growth of most deciduous trees in the northern United States and Canada, where winters are severe, is strictly regulated by day length. That is, temperature plays no part in the regulation of their yearly growing cycle. Give a reason why it is important that these plants not have their growing cycle determined by temperature.

Extension

1. Read the article by J. Raloff entitled "Plants See Hormone as Toxic Pollutant," in *Science News,* May 18, 1985, p. 309. Give an oral report in which you point out some of the economic consequences of air pollution by ethylene gas. What group of people would be most affected by these economic consequences?

2. Plant five or six seeds of different species of plants in pots of soil. Allow them to sprout and grow until the plants are about 13 cm tall. Tip the pots on their sides and observe how long it takes the plants to point upward again. Report your findings.

3. Plant five or six beans in each of four pots of soil, and place the pots as follows: one in total darkness, one in natural daylight near a window, one in only the light of a 100-watt incandescent light bulb, and one in the light of a fluorescent lamp, preferably one with a warm light. Make a lab report in which you describe the number and type of plants used, the lights used, and the results. Incandescent bulbs are poor in blue light, whereas the light of a warm fluorescent lamp is more like daylight, since it has both red and blue light. What do your results tell you about the ability of plants to distinguish red from blue light and about the effect of light on their growth?

Intra-Science: *How the Sciences Work Together*

The Process in Biology: Feeding China's Billion People

Over one billion people—roughly one-fourth of the world's population—live in the People's Republic of China. China and the United States are about the same size geographically, but the U.S. population is approximately 245.5 million—one quarter that of China. Furthermore, only 11 percent of China's land is suitable for cultivation, compared to 23 percent in the U.S. Yet nearly 70 percent of all China's workers are farmers.

In 1983, China produced 2.5 million metric tons of beef and 17 million metric tons of pork. In comparison, the U.S. produced 10 million metric tons of beef and 6.6 million metric tons of pork. China's fishing industry produced 4.9 million metric tons, whereas the U.S. produced 2.9 million metric tons of fish. These differences illustrate one aspect of China's efforts to feed its billion people. Valuable acres are used to grow food for people instead of feed for cattle. Pork and fish are China's main sources of protein. Hog manure is also recycled as fertilizer for fields throughout the country.

The current annual rate of population increase in China is approximately 1.1 percent. This means that over 12 million people are added to the population each year. For purposes of comparison, the New York City metropolitan area contains just over 7 million people. Since 1950, the total number of crops produced in China has more than tripled. The task of feeding a billion people is being accomplished. How?

The Connection to Earth Science

The population of China is concentrated in the eastern half of the country, where the country's greatest rivers run. The agricultural land is used according to what grows best in a given area. Weather, water, soil, and terrain dictate what is grown in various regions. In China's north, for example, summers are short and hot, and winters long and cold. Wheat thrives in this climate. The southern parts of China are hot and damp, ideal for growing rice. Wheat and rice are China's main crops. The People's Republic produces over 12 percent of the world's wheat and almost 37 percent of the rice. In addition, the Chinese grow sorghum, millet, corn, and numerous vegetables such as sweet potatoes, cabbage, and soybeans.

A large number of Chinese farmers live in the Yangste River basin, where the rural population exceeds 5,000 people per square mile. For contrast, Nebraska has 21 people per square mile. Most of the regions that receive enough rainfall to support farming are hilly or mountainous. To grow crops here, stairlike terrraces are carved into the slopes, an ancient Chinese practice.

China receives over 80 percent of its rainfall during the monsoon season, from May to October. Water that used to flood river valleys has now been harnessed with dams and canals. Much of China's farmland is now under irrigation.

The Connection to Chemistry

Intensive farming depletes the soil of essential nutrients and minerals. In China, farmers expect to harvest two crops from autumn to spring, and often grow a summer crop such as soybeans or sweet potatos as well. Even for those who let their flooded fields fill with fish during the monsoons, the soil soon loses its ability to support major crops.

The chemical fertilizer industry has been growing in China since the mid-1950s. Today, chemical fertilizer is used on farms to supplement hog manure. Without fertilizer the Chinese would not be able to cultivate multiple crops.

The Connection to Other Sciences

The science of engineering has greatly benefitted China's agricultural communities. By damming major rivers, thousands of new acres have been made available for agricultural use. Irrigation systems have been devised that can deliver water and fertilizer as needed. China has 160,000 km of navigable waterways along which many agricultural products are transported to markets in major cities. Engineers have also contributed to the building of new highways and bridges that have improved the transport of food and increased China's ability to import and export goods.

In the 1950s, as efforts to improve agricultural yields were developing along with efforts to control the size of the population, the Chinese government instituted a policy requiring farmers to merge their small individual plots into giant communes. The government told the communes what and how much to grow, and while crop production increased, some people were unhappy with the dramatic changes in lifestyle required by the commune system. Traditionally, Chinese society has revolved around the family. Today, the commune system is gradually being replaced by a system in which individual families contract with the government to produce certain amounts of food. If the family has a surplus after contributing the required amount to the community, it can sell the food in the open market. Thus the community can be adequately fed and the families can benefit economically from producing crops for trade.

The Connection to Careers

As in the United States, the growth of agriculture in the People's Republic of China will continue to be influenced by political and economic factors as well as by scientific developments. To continue to feed its people, China will have to continue developing better ways to use its tillable land. Advisors from the U.S. and other countries may assist the People's Republic by sharing advances in land management and engineering. At present, China is a growing importer of foods and other goods, so there are growing numbers of brokers and businesspeople in the U.S. who are specializing in trade with China. For more information on related careers in the biological sciences, turn to page 842.

Unit 7 | Invertebrates

"We should venture on the study of every kind of animal without distaste; for each and all will reveal to us something natural and something beautiful."

Aristotle

If you were asked to list all the animals you could think of in one minute, your list would probably include very few invertebrates, animals without backbones. You may be surprised to learn, then, that about 97 percent of all animal species are invertebrates. Look at the illustrations on these two pages to get a feeling for the great diversity of this group. In this unit you will learn about each of the major phyla of invertebrates and see the myriad ways in which these animals have adapted to almost every environment on earth.

Shrimp on sea anemone

Golden beetle, *Plusiotis resplend-ens,* from Costa Rica

434

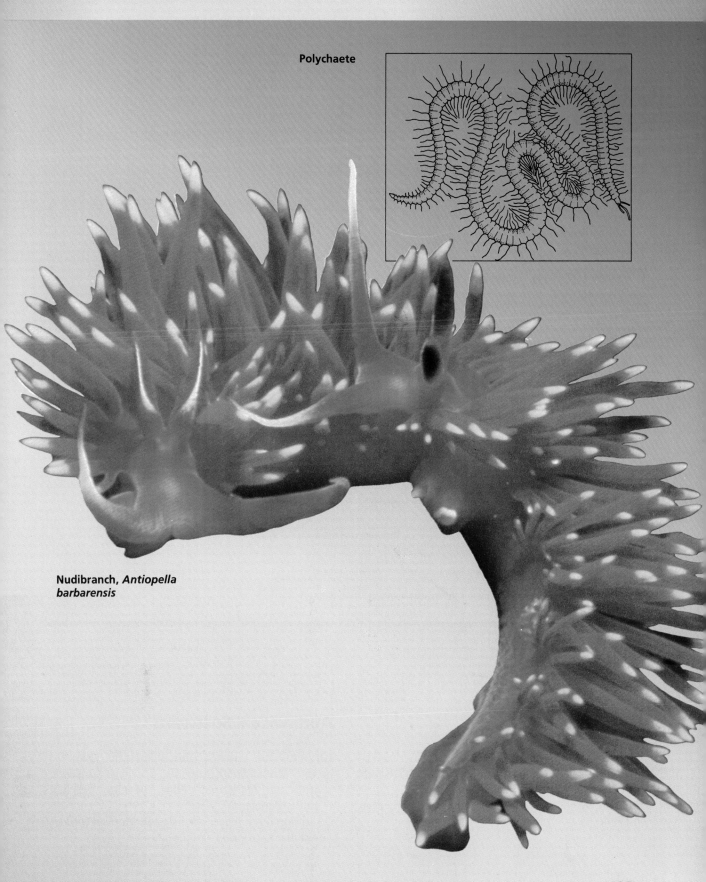

Polychaete

Nudibranch, *Antiopella barbarensis*

435

Chapter 29 | *Sponges and Cnidarians*

This chapter familiarizes students with some of the basic characteristics of invertebrates. Two phyla, sponges and cnidarians, are discussed in great detail, with an emphasis on how body structure is related to body function.

Correlation Chart

Section Number	Section Objectives	Pages	Features	Section Review
29.1	1. Define *invertebrates* and explain why they are such a diverse group.	437–439		1
	2. List three kinds of symmetry, and give examples of each.			3, 4
	3. State an advantage and a disadvantage of specialization in cells.			2
	4. Summarize the process of embryonic development.			5
29.2	1. Describe the basic body plan of a sponge.	441–443		1
	2. Describe the process of filter feeding in sponges.			2
	3. Contrast the processes of sexual and asexual reproduction in sponges.			3, 4, 5
29.3	1. List three characteristics that all cnidarians have in common.	444–447	Biology in Process, pp. 444–449	1, 4
	2. Name and describe the two body forms of cnidarians.			2
	3. Identify the three classes of cnidarians and give an example of each.		Think, p. 448	3
	4. Describe the way a hydra moves and the way it feeds.		Biology Lab, p. 449	4
	5. Summarize Darwin's theory explaining the formation of coral reefs.			5

Resources

Print

Gilbreath, Alice. *The Great Barrier Reef: A Treasure in the Sea.* Minneapolis: Dillon, 1986.

Johnson, Sylvia A. *Coral Reefs.* Minneapolis: Lerner, 1984.

Audiovisual/Software

City of Coral. Video, 57 min. Time-Life, 1986.

Simple Multicellular Animals: Sponges, Coelenterates, and Flatworms. Film, 20 min. Teacher's guide. Benchmark, 1985.

*Additional references are listed in *Modern Biology, Teacher's Program Guide.*

Section Teaching Guides

29.1 Invertebrates

Approach

The main objective of this section is to define an invertebrate and to explain its structure and development. As such, the section serves as an introduction to the more intensive phyla-based study of invertebrates in Sections 29.2 and 29.3 and in Chapters 30–35. In presenting the text, guide students toward an appreciation of the great diversity of animal life. As they read about unfamiliar life forms, encourage students to ask what possible advantage a particular body structure might have, given the animal's environment and life style.

Concept Framework

Even if students have never heard the term *invertebrate,* they will all be familiar with common invertebrates such as earthworms and bugs. You might wish to ask the students' help in writing on the chalkboard a list of familiar invertebrates. This will give you an opportunity to correct any misunderstandings. Emphasize that the animals listed on the board are related more by the absence of a structural feature than by the presence of one (except for those that belong to the same phylum). You may wish to end the discussion by asking why so many of the animals listed live in water.

Emphasis

Make certain that students understand that all animals—invertebrates and vertebrates alike—have the following characteristics in common: they are all heterotrophic, are multicellular, and go through a similar embryonic development.

Counting Invertebrates

Divide the students into two teams and challenge them to see which team can record the greatest number of invertebrate species encountered in a single 24-hour period. Define the rules of the game to include invertebrate products, such as corals, sponges, or invertebrates used as food.

Section Review Answers

(See p. 439.)

29.2 Porifera

Approach

The main objective of this section is to present the least complex phylum of invertebrates: Porifera. The text builds upon the concept of structure and function first encountered in Unit 2 on the cell. The section follows an order of presentation that will be repeated in increasing complexity in all subsequent sections on invertebrate phyla: first a description of the structure of the organism and the relationship of structure to function; then a discussion of feeding, respiration, excretion, and reproduction. In Porifera the discussion of classification is presented last, as it requires some understanding of the sponge's supporting structure. In subsequent sections, the text on classification generally precedes the text on body systems. You may wish to use the chart suggested as an Extension activity to organize the information presented in the text.

Concept Framework

Many students will know of sponges only through the plastic imitations that have now replaced the natural ones for most household uses. If possible, show the students a natural sponge, and ask them what generalizations about the organism they could make from seeing and touching the skeleton of the dead animal. Ask how many students would even guess that it was an animal. Have the students compare the natural sponge to an artificial cleaning sponge; elicit the fact that both sponges are very porous and capable of holding large amounts of water. You may wish to end the discussion by asking students to consider how this characteristic might be functionally advantageous for the live animal.

Emphasis

Make certain that students understand that water circulating through the sponge provides the animal with food, oxygen, and a means of excreting wastes. Sperm of sexually reproducing sponges also travels through the water.

Student Speaker

Invite one or more interested students to prepare a report on the commercial harvest of sponges. Ask them to make an oral presentation of their report to the class.

Section Review Answers

(See p. 443.)

29.3 Cnidaria

Approach

The main objective of this section is to present the two quite distinct body forms of cnidarians and to relate the body forms to particular stages in the life cycles of these organisms. Students will already be familiar with the concept of a life cycle from Chapters 22 and 27. The discussion of body forms builds on the discussion in Section 29.1. In presenting the text, be certain students understand that not all cnidarians alternate between polyp and medusae forms, and some spend all of their lives as one or the other. You may wish to have students continue to utilize the chart suggested in the Extension activity to organize the data presented in the text.

Concept Framework

Only students who live close to the sea may have first-hand knowledge of the Portuguese man-of-war and of stinging jellyfish. But all students will know something about stingers and tentacles, two of the characteristics common to all cnidarians. Build on students' collective knowledge by encouraging individuals to share what they know about cnidarians with the rest of the class. If possible show some examples of coral or jewelry made from coral. You may wish to end the discussion by asking students to consider the process by which scientists came to classify coral as an invertebrate animal.

Emphasis

Make certain students understand that although the three classes of cnidarians have different body forms, all have stinging cells, hollow guts, and tentacles.

Field Trip

Arrange a field trip to a freshwater pond to collect hydra. Or, alternatively, visit a marine aquarium or local pet store that stocks cnidarians. Ask students to take notes on the cnidarians they observe.

Using the Features/Darwin and Coral Reefs

The main process skills used in "Darwin and Coral Reefs" are observing and inferring. Darwin observed that coral reefs existed at far greater depths than coral could live. He inferred that the coral when alive existed nearer the surface and that the accumulated coral gradually subsided to a greater depth. Point out to students that inference always depends on observation, along with reasoning and past experience. Provide a common example of an inference based on observation, such as inferring that a mouse has taken up residence in the kitchen because cheese disappears from the countertop each night.

Section Review Answers

(Continued from p. 448.)
5. Tropical areas have the warm temperatures and abundant sunlight needed for algae inside coral cells to perform photosynthesis.

Laboratory Preparation

Background

In this lab exercise, students observe a sit-and-wait predator at work. The hydra's sessile existence is related to its radically symmetrical body plan. This animal must be prepared for prey that may approach from any direction. Students will see that tentacles do not respond to all stimuli in the same way. They can also see the nematocysts, a cnidarian trademark, and learn that along with the tentacles, the organelles also function in prey capture.

Preparation

Hydras are easily grown in a culture dish containing a 5cm sprig of *Elodea* and pasteurized spring water. (To pasteurize spring water, heat 12–16 L spring water to 65 C for 15 minutes.) Hydra may be placed in tap water if the water is set out in open containers the day before it is to be used. To develop a brine shrimp culture, spinkle brine shrimp eggs (available from pet stores) on the surface of a 4% solution of sodium chloride. Larvae will hatch in 18 to 24 hours. Remove larvae with a kitchen bulb baster and rinse in distilled water to remove any trace of salt before feeding to the hydras. Methylene blue solution can be purchased from a science materials supplier.

Safety

Impress upon your students the need for humane hydra care.

Conclusions

1. Nematocyst firing is extremely rapid. However, students should be able to see the coiled nematocyst inside the cell before firing and the uncoiled thread with a barb at the end after the nematocyst has fired.
2. The brine shrimp and the glutathione stimulated the same feeding response. A chemical in glutathione may be similar to a chemical in the brine shrimp. The hydra may be able to sense the chemical from a distance.

Inquiry

Students will devise tests of hydra responses by covering part of the culture dish, sounding a loud noise near the dish, and placing brightly colored objects in the dish.

Chapter Review

Vocabulary

1. sessile; others are parts of bilateral animals
2. spicule; others relate to embryonic development
3. nematocyst; others are part of sponges
4. gemmule; others are part of cnidarians
5. mesoglea; others are involved in reproduction

Review

11. Specialization is the adaptation of a cell for a particular function; it allows division of labor and adaptation to diverse environments.
12. concentration of nerve tissue and sensory organs in the anterior end of an organism
13. Zygote divides and forms embryo, which undergoes repeated divisions that form blastula; infolding of blastula forms gastrula, giving rise to endoderm and ectoderm.
14. They have no gastrula stage, little specialization of cells, and no true tissues or organs.
15. A sponge cannot chase down prey and so requires another method of feeding; filter feeding allows it to trap food in water that it circulates through its body.
16. Sperm are shed into water and enter another sponge through incurrent pores; collar cells trap sperm, which are then carried by amebocytes to the egg. Fertilized egg develops into larva that leaves sponge, swims about, then settles and develops into sessile adult.

17. The hydra fires nematocysts that paralyze prey and entangle it in threads attached to nematocyst barbs.
18. Hydras live in fresh water, live independently, and exist only as polyps; most other hydrozoans live in the sea, are colonial, and have a medusa stage in their life cycle.
19. Some polyps are specialized for feeding, others for digestion, and still others for reproduction.
20. Coral depend on algae within their cells to perform photosynthesis, and so they must live relatively close to the surface to allow sufficient sunlight to reach the algae.

Critical Thinking

1. Segment B; because scyphozoans are represented by the smallest number of different species among cnidarians.
2. Answers should indicate that students understand that radial symmetry is beneficial for sessile animals but that animals that live on land must move about in search of food. A bilateral body plan is much better suited to movement on land.
3. This pattern prevents self-fertilization, which could be harmful to species by increasing the frequency of harmful recessive traits.
4. The two life cycles are similar only in that they both include two distinct body forms. The major difference is that in moss a diploid form alternates with a haploid form, while in jellyfish both the medusa and the polyp are diploid, and only the gametes are haploid.
5. Sexual reproduction produces an embryo encased in a covering that protects it through the winter, while hydras formed by budding are just as vulnerable to cold as the parent hydra that formed them.
6. Sponge larvae use their flagella to swim through the water and so disperse the population, while adult sponges use their flagella to set up currents of water through the body and filter food from the water.

Chapter 29 | Sponges and Cnidarians

Introduction

The structure of any organism—from the simplest moneran to the most complex animal—is related to its way of life and its function in its environment. In this chapter, you will first learn about some basic aspects of invertebrates. Then you will study the relationships between structure and function in two of the less complex phyla of animals.

Chapter Outline

29.1 Invertebrates
- *Patterns of Symmetry*
- *Multicellular Organization*
- *Patterns of Development*
- *Classification and Evolution*

29.2 Porifera
- *Structure and Function*
- *Classification*

29.3 Cnidaria
- *Structure, Function, and Classification*
- *Hydrozoa*
- *Scyphozoa*
- *Anthozoa*

Chapter Concept

As you read, look for examples of how body structure is related to feeding, reproduction, and other functions.

Tube coral and encrusting sponge. Inset: Sea nettle jellyfish.

For additional teaching information see pages T29A–T29D.

29.1 Invertebrates

What do earthworms, butterflies, jellyfish, and giant squids have in common? About the only similarity you may note is that they are all animals. Nevertheless, these animals and about 97 percent of all other animal species are **invertebrates,** or animals without a backbone. Rather than sharing common characteristics, species of invertebrates share the common lack of a trait. Thus, it is not surprising that they are an extremely diverse group. They represent more than a dozen phyla and over a million species.

Patterns of Symmetry

Invertebrates exhibit a wide variety of shapes, which can provide clues to each one's particular way of life. One common way to identify animal shapes is by **symmetry,** or the arrangement of body parts around a point or central axis.

Some organisms, including amebas and many sponges, have no definite shape and so are called asymmetrical. Other simple organisms have **spherical symmetry**—that is, they are shaped like globes or spheres. The earliest ancestors of multicellular animals may have been hollow spheres of cells much like a colonial protozoan. This hollow body plan allows the cells to take in food and oxygen and give off wastes much more easily than if the organism were a solid ball of cells. This ability to exchange materials with the environment at any point along the organism's surface is beneficial to species that float in water.

Figure 29-1 illustrates the differences between radial symmetry and bilateral symmetry—the two types of symmetry most common among invertebrates. An organism that shows either of these types of symmetry has distinct ends. An imaginary line drawn from end to end through the center of the organism forms the longitudinal axis. An organism that shows **radial symmetry** can be divided into similar halves by any plane that passes through its longitudinal axis. In contrast, an animal with **bilateral symmetry** can be divided into similar halves by only one specific plane passing through the longitudinal axis. Thus, its left and right sides are mirror images of each other.

A radially symmetrical animal, such as a hydra, is basically cylindrical in shape. Such an animal has two different ends but no left or right side. Radial symmetry is most common in animals that move little if at all. They attach one end of their body to something stationary and feed with the unattached end.

Bilaterally symmetrical animals include worms, insects, spiders, and many other invertebrates. These animals have specialized front and rear ends and upper and lower surfaces. Biologists use the term **anterior** for the front end and the term **posterior** for the hind end. They refer to the top as the **dorsal** surface and the bottom as the **ventral** surface. Most bilaterally symmetrical ani-

Section Objectives

- *Define* invertebrates, *and explain why they are such a diverse group.*

- *List three kinds of symmetry and give examples of each.*

- *State an advantage and a disadvantage of specialization in cells.*

- *Summarize the process of embryonic development.*

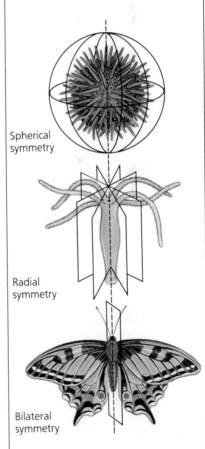

Spherical symmetry

Radial symmetry

Bilateral symmetry

Figure 29-1 Top to bottom: A sea urchin, hydra, and butterfly show spherical, radial, and bilateral symmetry, respectively.

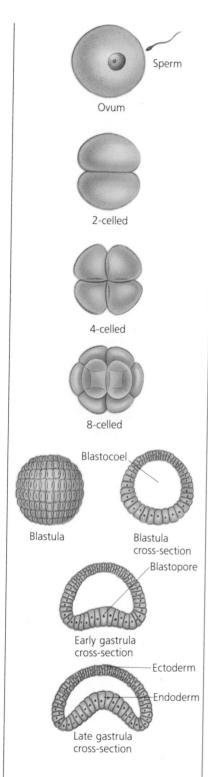

Sperm

Ovum

2-celled

4-celled

8-celled

Blastocoel

Blastula

Blastula
cross-section

Blastopore

Early gastrula
cross-section

Ectoderm

Endoderm

Late gastrula
cross-section

Figure 29-2 This diagram shows early cell division leading to gastrulation.

mals have nerve tissue and sensory organs concentrated in the anterior end. This adaptation, called **cephalization** (*SEFF-uh-luh-ZAY-shun*), allows their sensory organs to enter a new environment first and provide information about it.

Multicellular Organization

The structural characteristics of invertebrates suggest that they were probably the first multicellular animals, or **metazoans,** to evolve from the single-celled protozoans. Without this essential first step, the enormous variety of shapes and sizes in the animal kingdom may not have evolved as we see them today. Single-celled organisms can grow only so large and so complex. You may recall from Chapter 5 that the size of a cell is limited by the ratio of its surface area to its volume. This ratio must be high enough to allow food and oxygen to enter the cell and wastes to get out. As a cell grows, its volume increases faster than its surface area. Eventually it grows so large that food cannot enter fast enough to support its activities, and wastes begin to accumulate. For this reason, the cell must either divide or die.

One of the advantages of multicellular life is **specialization,** the adaptation of a cell for a particular function. Just as a contractor makes use of carpenters, electricians, and plumbers to build a house, a multicellular organism makes use of specialized cells that perform particular functions, such as digesting food or reproducing. Specialization has enabled metazoans to evolve and adapt to many environments. However, organisms pay a price for the advantages of specialization. Unlike single-celled organisms, specialized cells cannot survive on their own. They are no longer independent but rather interdependent.

Patterns of Development

The body plan and specialized structures of an animal begin to arise early in its development. In sexual reproduction, development begins with a fertilized egg, or **zygote.** The zygote divides and forms an **embryo,** the name given to an organism in its earliest stages of development. In the embryo, successive divisions produce 2 cells, then 4, 8, 16, 32, and so on. Figure 29-2 shows the development of a sea urchin as an example of successive divisions. During early division the embryo does not grow much in size but simply splits into smaller and smaller cells. Soon a hollow sphere of cells called the **blastula** (*BLAST-choo-la*) forms. This process is called blastulation.

As the blastula continues to develop, it changes shape. The cells at one point on the blastula move inward, forming a depression called the **blastopore.** Further infolding forms a cup-shaped embryo called a **gastrula.** You may think of the shape of the gastrula as similar to that of an underinflated basketball that you have pushed in on one side with your fist.

Tip: You may want to discuss highlights of invertebrate development beyond late gastrulation.

Info: **A planula is a ciliated larva.**

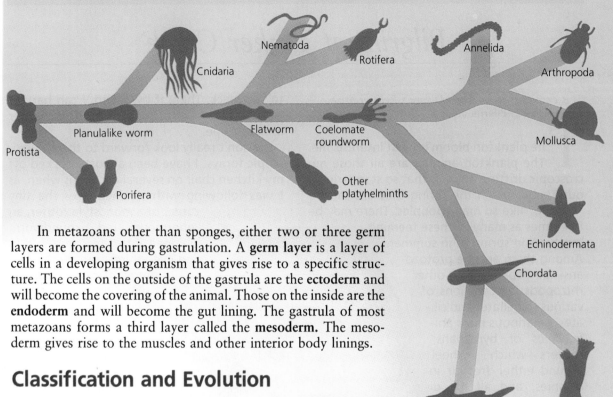

In metazoans other than sponges, either two or three germ layers are formed during gastrulation. A **germ layer** is a layer of cells in a developing organism that gives rise to a specific structure. The cells on the outside of the gastrula are the **ectoderm** and will become the covering of the animal. Those on the inside are the **endoderm** and will become the gut lining. The gastrula of most metazoans forms a third layer called the **mesoderm**. The mesoderm gives rise to the muscles and other interior body linings.

Classification and Evolution

Similarities in body plans and patterns of development help biologists classify invertebrates and hypothesize about their evolutionary history. The early fossil record of invertebrate evolution is fragmentary, and so biologists rely mainly on studies of modern species to develop family tree diagrams like the one in Figure 29-3.

In this chapter you will study two of the less complex invertebrate phyla—Porifera and Cnidaria. Porifera, or sponges, have long fascinated scientists because they so clearly represent the transition from unicellular to multicellular life. They have no gastrula stage, exhibit less specialization of cells than most invertebrates, and have no true organs. Cnidaria, which includes jellyfish, have an endoderm and an ectoderm and have evolved a diversity of body forms and specialized cells. All live in water, mostly in the oceans where their ancestors originated.

Section Review

1. Define *invertebrate*. Why is it not surprising that a group defined in this way includes a wide diversity of species?
2. Why is cell specialization important for invertebrates?
3. How do radial symmetry and bilateral symmetry differ?
4. What is the main advantage of cephalization?
5. Based on the fact that the development of sponges does not include a gastrula stage, would you expect to find endoderm and ectoderm in a sponge? Explain your answer.

Figure 29-3   This diagram shows probable evolutionary relationships among the various invertebrate phyla.

Answers to Section Review
1. An invertebrate is an animal without a backbone. It is defined by the lack of a trait.
2. Specialized structures permit adaptations to diverse environments.
3. A radially symmetrical organism can be divided into similar halves by any plane that passes through its longitudinal axis. A bilaterally symmetrical animal can be divided into similar halves by only one plane that passes through its longitudinal axis.
4. Cephalization allows an animal's sense organs to enter a new environment first and provide information about it.
5. No. Endoderm and ectoderm arise through the process of gastrulation.

Writing About Biology

Pilgrim at Tinker Creek

This excerpt is from Annie Dillard's Pulitzer-prize-winning book, Pilgrim at Tinker Creek.

The plankton bloom is what interests me. The plankton animals are all those microscopic drifting animals that so staggeringly outnumber us. In the spring they are said to "bloom," like so many poppies. There may be five times as many of these teeming creatures in spring as in summer. Among them are the protozoans—amoebae and other rhizopods, and millions of various flagellates and ciliates, gelatinous moss animalcules or byrozoans; rotifers—which wheel around either free or in colonies; and all the diverse crustacean minutiae—copepods, ostracods, and cladocerans like the abundant daphnias. All these drifting animals multiply in sundry bizarre fashions, eat tiny plants or each other, die, and drop to the pond's bottom. Many of them have quite refined means of locomotion—they whirl, paddle, swim, slog, whip, and sinuate—but since they are so small, they are no match against even the least current in the water. Even such a sober limnologist as Robert E. Coker characterizes the movement of plankton as "milling around."

A cup of duck-pond water looks like a seething broth. If I carry the cup home and let the sludge settle, the animalcules sort themselves out, and I can concentrate them further by dividing them into two clear glass bowls. One bowl I paint all black except for a single circle where the light shines through; I leave the other bowl clear except for a single black circle against the light. Given a few hours, the light-loving creatures make their feeble way to the clear circle, and the shade-loving creatures to the black. Then, if I want to, I can harvest them with a pipette and examine them under a microscope. . . .

I don't really look forward to these microscopic forays: I have been almost knocked off my kitchen chair on several occasions when, as I was following with strained eyes the tiny career of a monostyla rotifer, an enormous red roundworm whipped into the scene, blocking everything, and writhing in huge flapping convulsions that seemed to sweep my face and fill the kitchen. I do it as a moral exercise; the microscope at my forehead is . . . a constant reminder of the facts of creation that I would just as soon forget. . . .

In the puddle or pond, in the city reservoir, ditch, or Atlantic Ocean, the rotifers . . . spin and munch, the daphnia . . . filter and are filtered, and the copepods . . . swarm hanging with clusters of eggs. These are real creatures with real organs leading real lives one by one. I can't pretend they're not there. If I have life, sense, energy, will, so does a rotifer. The monostyla goes to the dark spot on the bowl: To which circle am I heading? I can move around right smartly in a calm; but in a real wind, in a change of weather, in a riptide, am I really moving, or am I "milling around"?

Write

■ Some people think that all animals have an equal right to life and that humans are not necessarily more important than other animals. What do you think? Write an essay in which you take a stand. Consider the ways that humans use other animals—as pets, as food, and as research subjects.

Excerpt from Pilgrim at Tinker Creek *by Annie Dillard. Copyright © 1974 by Annie Dillard. Reprinted by permission of Harper & Row, Publishers, Inc.*

29.2 Porifera

The sponges you use today are probably artificial, but in a limited way they imitate one of the important functions of a real sponge. They can soak up and release water because they are extremely porous—much like a real sponge. In fact the phylum name Porifera comes from a Latin word meaning "pore-bearer." Since ancient times, divers have harvested sponges in the Caribbean, Mediterranean, Aegean, and Red seas.

Early biologists thought sponges were plants, and most sponges do indeed resemble plants in some ways. They do not chase and capture food as most animals do. Instead, they are **sessile,** which means that they attach themselves firmly to surfaces and do not move. Sponges grow in many shapes and sizes. They may look like mossy mats, cactuses, or blobs of fungus. They may be only a single centimeter in length or grow to 2 m in diameter. Sponges exist in almost every color of the rainbow.

Structure and Function

The basic body plan of sponges, shown in Figure 29-4, suggests many relationships between structure and function. Simple sponges are hollow cylinders that are closed at the bottom and have an opening at the top called the **osculum** (AHS-kyoo-lum). The body wall consists of two layers of cells separated by a jelly-like substance. The covering layer is a pinnacoderm, which some biologists consider true tissue. The cylinder is lined with special **collar cells,** or choanocytes. By beating their flagella, collar cells

Section Objectives

- *Describe the basic body plan of a sponge.*

- *Describe the process of filter feeding in sponges.*

- *Contrast the processes of sexual and asexual reproduction in sponges.*

Figure 29-4 The portion of the sponge at left outlined by a rectangle is shown in internal and external detail at right. An enlarged collar cell is shown in the circle.

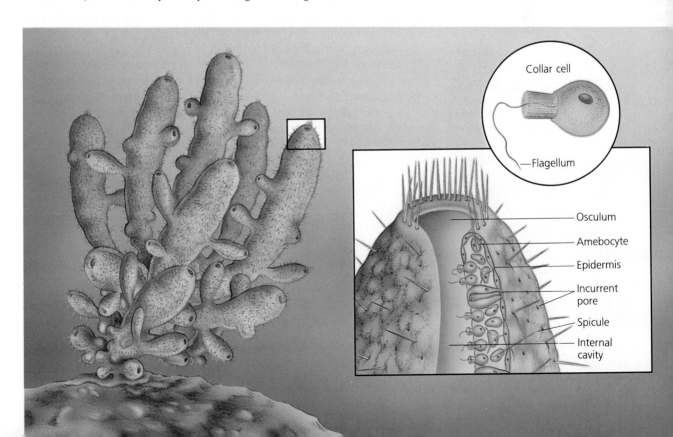

Collar cell

Flagellum

Osculum

Amebocyte

Epidermis

Incurrent pore

Spicule

Internal cavity

Tip: Ask students why a giant sponge—one standing taller than a human adult—can consume nothing larger than microscopic particles. (It is because the digestive processes of all sponges are carried out within single cells.)

Info: Amebocytes also secrete skeletal material.

draw water into the cavity through numerous pores and canals that penetrate the body. The water circulates through the interior and leaves through the osculum. This circulation helps all the body cells gather food and oxygen and excrete wastes.

The body of the sponge would collapse without some type of supporting structure. In some sponges, support is provided by a simple skeleton made up of a network of protein fibers called **spongin.** Others have skeletons made up of tiny, hard particles often shaped like spikes and stars called **spicules** *(SPICK-yools).* Spicules are composed of silicon dioxide or calcium carbonate.

Feeding

Because they are sessile, sponges cannot pursue their food. Their structure of pores and canals is thus essential to survival. The same collar cells that pump water through the body also screen food out of that water. This feeding method is called **filter feeding.** The food of sponges includes bacteria, unicellular algae and protozoans, and bits of organic matter. Collar cells engulf food and take it inside the cells to be digested.

The distribution of digested food illustrates how sponges are related to protozoan ancestors. Nutrients travel from collar cells to other specialized cells that roam about and supply the rest of the body with nutrients. Scientists call these cells **amebocytes** *(uh-MEE-buh-SITES)* because they resemble amebas. Amebocytes also carry away carbon dioxide and other wastes and release the wastes into the water passing through the sponge. The current of water carries these wastes away as it leaves through the osculum.

Reproduction

Sponges can reproduce asexually by forming small buds that break off and live separately. During periods of cold or dry weather, some freshwater sponges form specialized buds called **gemmules** *(JEM-yools).* Gemmules are food-filled balls of amebocytes surrounded by protective coats made up of organic material and spicules. The gemmules can survive harsh conditions that may kill the adult sponge that formed them. When conditions improve, the sponge cells emerge and grow into new sponges.

Figure 29-5 Specialized buds called gemmules are shown on the osculum at left. At right is a budding trumpet sponge.

Sponges also have remarkable powers of **regeneration**, the ability of an organism to regrow missing parts. In fact, a small piece of a sponge can regenerate a complete new organism. In some species even particles small enough to pass through a cloth strainer will regenerate. Sponge divers have long grown new sponges by cutting sponges into pieces, attaching them to weights, and throwing them into the sea to regenerate.

Sponges can also reproduce sexually, as you can see in Figure 29-6. The sperm are shed into the water and enter another sponge through the pores. The sperm are taken in by the collar cells and then carried to an egg by amebocytes. A fertilized egg develops into an immature form called a **larva**. The larva has flagella that enable it to escape from the sponge and swim around. Eventually it settles and attaches to an object. The cells of the larva then undergo a reorganization that forms an adult sponge.

Some species of sponges have separate sexes, but in most species each individual produces both eggs and sperm. Any organism that produces both eggs and sperm is called a **hermaphrodite** *(her-MAFF-roe-DITE)*. Self-fertilization rarely if ever occurs in hermaphroditic species, but the sperm of any individual can fertilize the egg of any other individual. Thus the probability of successful fertilization is greater than it would be if only half of the population (the females) produced eggs. Hermaphroditism is common in many invertebrates that are sessile, move slowly, or live in low-density populations.

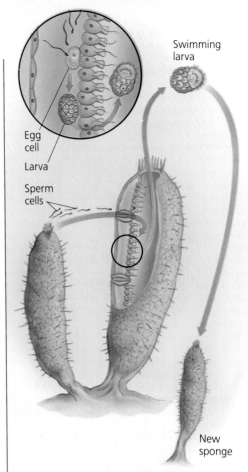

Egg cell

Larva

Sperm cells

Swimming larva

New sponge

Figure 29-6 The union of sponge egg and sperm results in a zygote that develops into a larva. The larva escapes the parent and grows into a new individual.

Classification

Differences in the composition of the skeleton provide the primary basis for classifying sponges. Biologists have identified four distinct classes of sponges:

- Calcarea have spicules of calcium carbonate.
- Hexactinella have spicules of silicon dioxide.
- Demospongiae have spongin or a combination of spongin and spicules of silicon dioxide.
- Sclerospongiae have spongin plus spicules of silicon dioxide and calcium carbonate.

Section Review

1. What is the difference between *spongin* and *spicules?* What functions do they both perform?
2. Describe the way in which sponges feed.
3. How do gemmules help some freshwater sponges survive unfavorable conditions?
4. What role do amebocytes play in the sexual reproduction of sponges?
5. Since sponges are sessile animals, how is it possible that a population of sponges can spread out into a larger area?

Answers to Section Review
1. Spongin consists of protein fibers; spicules are particles of silicon dioxide or calcium carbonate. Both provide support for sponges.
2. Flagella on collar cells help to collect food from water flowing through the sponge.
3. Their protective coats can survive harsh conditions that kill the sponge that forms them.
4. They transport sperm to the egg.
5. Larvae have flagella that enable them to move away from the parent before settling.

Section Objectives

- *List three characteristics that all cnidarians have in common.*

- *Name and describe the two body forms of cnidarians.*

- *Identify the three classes of cnidarians and give an example of each.*

- *Describe the way a hydra moves and the way it feeds.*

- *Summarize Darwin's theory explaining the formation of coral reefs.*

29.3 Cnidaria

Tiny freshwater hydra, stinging jellyfish, and flowerlike coral all belong to the Phylum Cnidaria. The name of the phylum comes from **cnidocytes** *(NIDE-uh-SITES)*, the stinging cells that characterize these organisms. The members of the phylum are called **cnidarians.** They are also known as **coelenterates** *(sih-LENT-uh-RATES)*, a name that comes from their coelenteron, or "hollow gut." In addition to their stinging cells and hollow guts, cnidarians also have flexible structures called **tentacles.**

Structure, Function, and Classification

As you can see in Figure 29-7, the body of a cnidarian may be either vase shaped or bell shaped. Scientists refer to the vase-shaped form as the **polyp** *(PAHL-ip)*, and to the bell-shaped form as the **medusa** *(muh-DOO-suh)*. The polyp is specialized for a sessile existence, the medusa for swimming. During their life cycles some cnidarians go through both polyp and medusa stages. Jellyfish, for example, go through both stages but spend most of

Biology in Process

Darwin and Coral Reefs

C harles Darwin had had little formal scientific training when he began his scientific voyage on the HMS *Beagle* in 1831. He was, however, a keen observer of natural phenomena and made careful notes on everything that attracted his attention. One source of curiosity for Darwin during the voyage was his observation that coral reefs existed in extremely deep water far out at sea. At the time no one had succeeded in explaining how this phenomenon was possible.

Darwin knew from samplings and depth measurements that corals could not live more than about 55 m below the surface of the water. He also observed that the corals had to attach themselves to something in order to grow. But what sort of foundation existed in waters that were otherwise thousands of meters deep? Darwin reasoned that the foundations were either great banks of sediment or undersea mountains.

Darwin reasoned that accumulations of sediment could not be the answer. The atolls—coral reef islands—were much too far from land; what's more, they were in very clear water. Therefore he inferred that the corals must have attached their lives as medusae. Other cnidarians live only as polyps, and still others live as mixed colonies of polyps and medusae.

As either polyp or medusa, the organism consists of two cell layers—the endoderm and the ectoderm. Between these two layers is a jellylike material called **mesoglea** (MEZZ-uh-GLEE-uh).

The relationship between the structure and function of the cnidarians' body parts is clearly seen in the way they feed. The tentacles capture the prey and paralyze it with coiled stingers called **nematocysts** (nuh-MAT-uh-SISTS), located inside the cnidocytes. The tentacles then draw the prey to the mouth. Enzymes inside the **gastrovascular cavity** (the hollow gut) break up the prey, and the specialized cells lining the cavity absorb the food. Undigested food and waste are expelled through the mouth.

Scientists classify cnidarians into three classes:

- Hydrozoa (HY-druh-ZOE-uh) include polyps, medusae, and species that alternate between the two.
- Scyphozoa (SIE-fuh-ZOE-uh), the true jellyfishes, spend most of the life cycle as medusae.
- Anthozoa (AN-thuh-ZOE-uh) live only as polyps and include anemones and corals.

Medusa

Polyp

Figure 29-7 The contrasting forms of polyps and medusae result from different arrangements of the same body parts.

Atoll

Coral buildup

Undersea mountain

themselves to undersea mountains. But if coral reefs could not grow more than 55 m below the surface, then the tops of the undersea mountains to which the corals were attached must all be at about that level. Was it possible that thousands of mountains scattered over thousands of square kilometers of ocean had all attained such uniform height? Darwin's knowledge of terrestrial mountains told him that this was impossible. Even mountains in the same chain did not exhibit such uniformity.

Through a process of inferring, Darwin arrived at an answer. In his own words:

"If, then, the foundations of the many atolls were not uplifted into the requisite position, they must of necessity have subsided [sunk] into it; and this at once solves every difficulty, for we may safely infer . . . that during a gradual subsidence the corals would be favourably circumstanced for building up their solid frameworks and reaching the surface."

Darwin's inference was challenged at the time and for many years thereafter. Then in 1952, more than 100 years after Darwin's voyage, scientists collected data that supported Darwin's inference. Scientists working on the Pacific atoll of Eniwetok drilled down through nearly a mile of solid coral. At the bottom of the bore, nearly 1,500 m below sea level, they found corals that had lived 65 million years ago. They were attached to the top of an undersea mountain that formerly was at sea level.

- Why were scientists unable to find data supporting Darwin's inference for so long?

Hydrozoa

The Class Hydrozoa includes about 3,700 species, most of which live in the oceans. However, the freshwater species that make up the genus *Hydra* have been extensively studied. Hydras range in size from only 1 to 4 cm. Most hydras are white or brown, but some appear green because of algae living beneath their outer cells. Hydras live in quiet ponds, lakes, and streams. They attach themselves to rocks or water plants by means of a sticky secretion produced by a group of cells called the basal disk.

Sometimes hydras leave one place of attachment and move to another. The basal disk can secrete bubbles of gas, which cause the hydra to float upside down on the surface. The hydra can also move by somersaulting. This peculiar movement occurs when the tentacles and the mouth end bend over and touch the bottom of the pond while the basal disk pulls free. This and other kinds of movement are made possible by the nerve net, interconnecting nerve cells that are located in the mesoglea.

Feeding in Hydras

Hydras capture prey with nematocysts. They wait for small floating animals to drift past their tentacles. Just a touch triggers a barrage of paralyzing nematocyst barbs, each on a thin thread that entangles the prey. The nerve net signals the tentacles to push the prey through the mouth and into the gut. This space is lined with specialized gastrodermal cells. Some of these cells secrete enzymes that begin the digestion of the food. Others take in the partially digested food and complete digestion.

Reproduction in Hydras

Hydras generally reproduce asexually during warm weather. In asexual reproduction small buds form on the outside of the hydra's body. These buds grow their own tentacles and then separate from the body and begin living independently.

Sexual reproduction usually occurs in the fall, when low temperatures trigger the development of eggs and sperm. The eggs are produced by meiosis along the walls of the body in swellings called **ovaries**. Motile sperm are formed by meiosis in similar swellings called **testes**. In some species eggs and sperm are produced in the same hermaphroditic individual. In others the individuals are either male or female. In either case sperm are released into the water. Some will reach the ovary of a nearby hydra, and one sperm will fertilize an egg cell. The fertilized egg, or zygote, then divides and grows into a ball of cells with a hard cover. The cover is an adaptation that protects the embryo through the winter until it develops into a new hydra in the spring.

Other Hydrozoa

Hydras are definitely not typical of the class. Hydras live independently; most other hydrozoans live in colonies. Although hydras

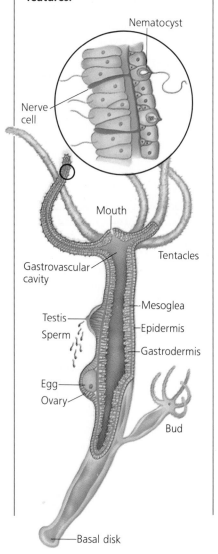

Figure 29-8   The hydra (top) has been diagrammed below to show major external and internal features.

Nematocyst

Nerve cell

Mouth

Tentacles

Gastrovascular cavity

Mesoglea

Testis

Epidermis

Sperm

Gastrodermis

Egg

Ovary

Bud

Basal disk

exist only in polyp form, most other hydrozoans go through a medusa stage. The genus *Obelia* is a colonial hydrozoan that has many polyps attached to branched stalks. Some branches are specialized for feeding, while others are specialized for reproduction.

Perhaps the most remarkable hydrozoan is the Portuguese man-of-war, a member of the genus *Physalia (fie-SAY-lee-uh)*, which exists as a colony of modified polyps and medusae. Tentacles up to several meters long dangle from a large medusa that is specialized as a gas-filled float. The float can measure as much as 30 cm across. The tentacles bear polyps specialized for feeding, digestion, and sexual reproduction. The Portuguese man-of-war preys mostly on small fish, but its nematocysts contain a poison that can be harmful and even fatal to humans.

Scyphozoa

The name Scyphozoa means "cup animals," the dominant form of the life cycle of this class, the medusa. Scyphozoans are known commonly as jellyfish, which are classified into more than 200 species. Some have cups or bells measuring as little as 2 cm across or as much as 4 m. Several species of jellyfish trail tentacles many meters long below them in the water. The nematocysts of some jellyfish carry poisons that can cause severe pain and even death. Sea wasp jellyfish have killed several dozen people in the seawaters along the northern coast of Australia.

The common jellyfish *Aurelia* is a good example of a scyphozoan whose life cycle includes both medusa and polyp forms. As you can see in Figure 29-9, adult medusae release sperm and eggs into the water, where fertilization occurs. The resulting zygote develops into a blastula. The blastula then develops into a ciliated larva called a **planula,** which settles and attaches to the ocean bottom. The unattached end develops a mouth and tentacles,

Figure 29-9 The life cycle of the common jellyfish *Aurelia* includes both medusa and polyp forms.

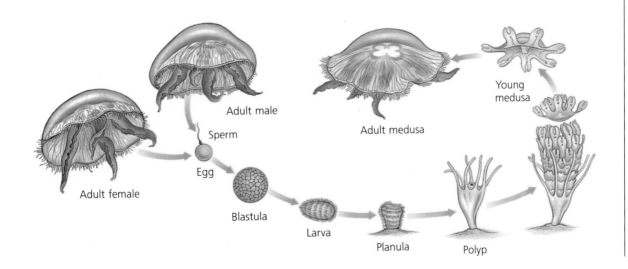

Adult male

Sperm

Egg

Adult female

Blastula

Larva

Planula

Polyp

Adult medusa

Young medusa

The word *anthozoa* means "flower animal," a description that fits sea anemones like this one very well. Why then do you think that early biologists did not mistake sea anemones for plants as they did sponges?

Answer to Think
Answers will vary but may mention that sea anemones had tentacles with which they were observed to capture prey.

Answers to Section Review
1. The name comes from *cnido-cytes*, the stinging cells characteristic of the phylum.
2. polyp and medusa
3. Hydrozoa (hydra); Scyphozoa (jellyfish); Anthozoa (sea anemone or coral)
4. Sea anemones and clown fish share food and protect each other from various predators. Corals depend on algae inside their cells for oxygen and accumulation of calcium.
(See pp. T29A–T29D.)

becoming a polyp. As the polyp grows, it forms stacks of medusae, each of which buds off and develops into a jellyfish.

Anthozoa

The word *anthozoa* means "flower animals," a fitting description for the approximately 6,100 species of this class. These brightly colored marine cnidarians include sea anemones and corals.

Sea anemones are polyps commonly found in coastal areas. They attach themselves to rocks and other submerged objects by means of a basal disk. Anemones feed on fishes and other kinds of marine life that swim within reach of their tentacles. Some sea anemones have a symbiotic relationship with the clown fish. The fish hides in the anemone's tentacles, protected by a special chemical that prevents the nematocysts from firing. The two animals share food and protect each other from various predators. The movements of the fish also help prevent sediments from harming the anemone.

Corals are small polyps, most of which live in colonies. They build rocklike reefs by cementing their calcium-based skeletons to those of adjoining polyps. When the polyps die, their hardened skeletons remain, serving as the foundation for new coral polyps. Thousands of coral polyps live together in this manner, building up large and colorful reef communities. The reef provides food and shelter for an enormous variety of fishes and invertebrates.

Corals build three kinds of reefs. Fringing reefs form close to a beach, either on the mainland or on an island. Barrier reefs form farther out. One theory holds that barrier reefs were once fringing reefs that became further separated from the mainland or island when coastal lands sank. Atolls, reefs with open lagoons in the middle, form far out at sea. Many are attached to the tops of undersea mountains or extinct volcanoes.

Coral reefs form only in the warm, clear waters at shallow depths. These conditions are required to allow photosynthesis to be carried out by certain algae that live symbiotically inside coral cells. The corals depend upon the algae as a source of oxygen and also as a means of speeding up the accumulation of calcium. The algae in turn depend upon the corals to supply vital nutrients. This symbiotic relationship is yet another example of how invertebrates have adapted to a sessile existence in the oceans.

Section Review

1. What is the origin of the word *cnidarian*?
2. Identify the two stages of life of many cnidarians.
3. Name the three classes of cnidarians, and give an example of each.
4. Describe two examples of symbiosis found among cnidarians.
5. Suggest reasons why coral reefs are most commonly found in tropical waters.

Laboratory

Feeding Behavior of *Hydra*

Objective
To observe the reactions and feeding behavior of *Hydra*

Process Skills
Experimenting, hypothesizing, observing

Materials
Culture dish, medicine dropper, *Hydra* culture, dissecting microscope, dissecting needle, depression slide, compound microscope, methylene blue, vinegar, 50-mL beaker, brine shrimp culture, glass rod, glutathione

Method
1 Fill a culture dish one-half full with water.
2 Use a medicine dropper to transfer a hydra from its container to the culture dish. The hydra must be transferred quickly. Otherwise the hydra will attach itself to the medicine dropper. Allow the hydra a few moments to settle down.
3 Place the culture dish on the stage of a dissecting microscope. Identify the animal's tentacles, mouth, and basal disk.
4 Lightly tap the dish and note any response. Gently touch the base of the hydra with a dissecting needle. Then gently touch the animal's tentacles. Are there differences in the responses at the two locations?
5 Move the hydra to a depression slide. Allow the hydra a few moments to settle down. Set a compound microscope on low power. Place the depression slide on the microscope stage. Examine the surface of the tentacles. Note the bumps on the tentacles. These are where the nematocysts are located.
6 Add a drop of methylene blue to the depression containing the hydra. Observe what happens. Then add a drop of vinegar to the depression. Observe the hydra's reaction.
7 Put the hydra back in the culture dish.

8 Fill a 50-mL beaker with water.
9 Use the dropper to pick up several brine shrimp. Rinse the brine shrimp in the beaker of water by releasing them from the dropper and immediately picking them up again.
10 Put the brine shrimp in the culture dish with the hydra. Observe what happens.
11 Dip a glass rod into glutathione and place it near a hydra that has not been fed. Do not let the glass rod touch the hydra. Observe what happens.
12 Place all hydra back into the culture.

Conclusions
1 Describe, as much as you can, the sequence of events that occur when the nematocysts discharge.
2 Compare the hydra's response to brine shrimp and to glutathione. What may have been the reason for the hydra's response to the glutathione?

Inquiry
Describe methods by which you could learn about the responses of *Hydra* to light, sound, and color.

Chapter 29 Review

Vocabulary

amebocyte (442)
anterior (437)
bilateral symmetry (437)
blastopore (438)
blastula (438)
cephalization (438)
cnidocyte (444)
cnidarian (444)
coelenterate (444)
collar cell (441)
dorsal (437)

ectoderm (439)
embryo (438)
endoderm (439)
filter feeding (442)
gastrovascular cavity (445)
gastrula (438)
gemmule (442)
germ layer (439)
hermaphrodite (443)
invertebrate (437)
larva (443)

medusa (444)
mesoderm (439)
mesoglea (445)
metazoan (438)
nematocyst (445)
osculum (441)
ovary (446)
planula (447)
polyp (444)
posterior (437)
radial symmetry (437)

regeneration (443)
sessile (441)
specialization (438)
spherical symmetry (437)
spicule (442)
spongin (442)
symmetry (437)
tentacle (444)
testes (446)
ventral (437)
zygote (438)

For each set of terms below, choose the one that does not belong, and explain why it does not belong.

1. anterior, posterior, dorsal, sessile
2. spicule, blastula, gastrula, embryo
3. amebocyte, collar cell, nematocyst, spicule
4. medusa, gemmule, polyp, cnidocyte
5. ovary, testis, zygote, mesoglea

Review

1. All invertebrates (a) live in water (b) have tentacles (c) have no backbone (d) are hermaphrodites.
2. A hydra shows (a) asymmetry (b) spherical symmetry (c) radial symmetry (d) bilateral symmetry.
3. The dorsal surface of an animal is its (a) top (b) bottom (c) side (d) front.
4. Spongin and spicules are important to a sponge because they (a) trap food (b) provide support (c) remove wastes (d) produce offspring.
5. Both feeding and sexual reproduction in sponges involve (a) gemmules and collar cells (b) spicules and gemmules (c) amebocytes and spongin (d) collar cells and amebocytes.
6. The primary basis for classifying sponges is (a) their shape (b) their color (c) the composition of their skeleton (d) the method by which they reproduce.
7. The characteristics common to all cnidarians include all of the following except (a) stinging cells (b) polyps (c) a hollow gut (d) tentacles.
8. The cnidarian body form specialized for sessile existence is the (a) polyp (b) medusa (c) planula (d) gastrula.
9. The class of cnidarians that exists only as polyps is (a) Hydrozoa (b) Scyphozoa (c) Anthozoa (d) Calcarea.
10. The movements of hydras are coordinated by their (a) mesoglea (b) nematocysts (c) nerve net (d) endoderm.
11. What is cell specialization, and how does it benefit invertebrates?
12. What is cephalization?
13. Summarize the process of embryonic development in invertebrates.
14. In what ways do sponges represent the transition from unicellular to multicellular life?
15. How is a sponge's method of feeding related to the fact that it is sessile?
16. Describe the process of sexual reproduction in sponges.

17. How does a hydra capture its prey?
18. In what ways are hydras not typical of the Class Hydrozoa?
19. What type of specialization is found among the polyps that make up a Portuguese man-of-war?
20. Why do all coral reefs form at about the same depth?

Critical Thinking

1. The pie graph shows the relative numbers of species in each class of Cnidaria. Which segment of the graph represents scyphozoans, A, B, or C? How do you know?
2. Radial symmetry is found only among species that live in water. In what way is radial symmetry not well suited to animals that live on land?
3. A hermaphroditic sponge generally produces eggs and sperm at different times. Based on your knowledge of sexual reproduction, why do you think it is evolutionarily advantageous for the sponge not to produce eggs and sperm simultaneously?
4. Compare and contrast the life cycle of *Aurelia* with alternation of generations in mosses. Be sure to list both similarities and differences.
5. Hydras generally reproduce asexually during warm weather and reproduce sexually

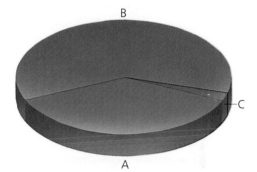

in the fall. Based on what you have learned about the hydra embryo, what do you think is the advantage to the hydra of reproducing sexually when the weather turns cool?
6. Sponge larvae have flagella on the outside, whereas adult sponges have flagella in their internal cavity. How is this structural difference related to functional differences of the larval and adult stages in the life cycle of the sponge?

Extension

1. Read "Gardens That Wriggle and Eat," *International Wildlife* (September/October 1985), pp. 52–59. Write a brief report summarizing the diversity of sea anemones shown in the article, including differences in appearance, feeding, and means of defense.
2. Create a chart that you can use to summarize information about all the phyla you will study in this unit. Use headings such as Symmetry, Feeding and Digestion, Reproduction, and Habitats. Begin your chart by filling in the information for Porifera and Cnidaria. Continue your chart as you work through the unit.
3. Natural sponges are preferred to synthetic ones in certain professions. In surgery, for example, natural sponges are desirable. Find out what methods are used to gather and prepare sponges to be used by surgeons and others. Write a report and present it to your class.
4. Use the library to research barrier reefs. Present an oral report to the class explaining how these reefs form and describing the communities of living things found on the reefs. If possible, bring along photographs of reef communities to share with the class.

Chapter 30 | *Flatworms, Roundworms, and Rotifers*

This chapter discusses the classification of body structures in flatworms, roundworms, and rotifers. Students learn how the morphology of these important organisms is modified as an adaptation to either the free-living or parasitic way of life.

Correlation Chart

Section Number	Section Objectives	Pages	Features	Section Review
30.1	1. State the distinguishing characteristics of flatworms. 2. Describe the anatomy of a planarian. 3. Compare and contrast free-living and parasitic life forms. 4. Diagram the life cycle of a fluke. 5. Describe the life cycle of a tapeworm.	453–457	Think, p. 455	1, 2 1, 2 3, 4, 5 3 4
30.2	1. Compare and contrast the body plans of flatworms and roundworms. 2. Summarize the parasitic relationships of common roundworms and humans. 3. Describe the anatomy of a rotifer. 4. Describe a classical study of nematode development.	458–461	Biology in Process, pp. 458–459 Biology Lab, p. 461	2, 5 1, 3 2, 4

Resources*

Print

Crompton, D. W. T., and S. M. Joyner. *Parasitic Worms*. London: Taylor and Francis, 1980.
Kleine, Aaron E. *The Parasites We Humans Harbor*. New York: Elsevier-Nelson, 1982.

Audiovisual/Software

Flatworms (Platyhelminthes). Film, 16 min. Britannica.
Nematode. Film, 11 min. Britannica.
Parasitism (Parasitic Flatworms). Film, 17 min. Britannica.
Worms: Flat, Round, and Segmented. Film, 15 min. Coronet.
*Additional references are listed in *Modern Biology, Teacher's Program Guide*.

Section Teaching Guides

30.1 Platyhelminthes

Approach

The main objective of this section is to demonstrate that the body plans of flatworms, though more complex than sponges and cnidarians, are still among the simplest of all animals. The text builds on the discussion of body plans in Chapter 29, and provides additional information on body types that students will use as they read about more complex animals in Chapters 31–47. The explanation of a parasite sets the stage for the discussion of symbiosis in Chapter 51. Invertebrates as agents of disease is discussed further in Chapter 44. As you present the text, encourage students to appreciate the complex life cycles and the variety of reproduction means demonstrated by these simple organisms.

Concept Framework

All students will think they know what a worm is, but if you question them you will probably find that they use the word indiscriminately to mean earthworm, roundworm, and flatworm. Use what students know about worms as a starting point to introduce the three groups of worms and to explain how those three groups differ from one another. You may wish to have students consider the advantages of cephalization for a motile animal as they read the section.

Emphasis

Make certain that students understand that in flatworms, all the cells lie close to the outer surface or the intestinal wall, allowing for the direct absorption of nutrients and oxygen and for the disposal of waste products.

Diagraming Flatworm Life Cycles

Divide students into similar sized groups and have each group diagram and label the life cycle of a different flatworm not illustrated in the text. When the diagrams are finished display them in an area of the classroom.

Section Review Answers

(Continued from p. 457.)
3. The adults and eggs lodge in the blood vessels, causing internal bleeding. The blocking of the vessels also causes tissue decay.
4. Tapeworms live on nutrients that have already been digested by their hosts. The cells of the tapeworms absorb the nutrients directly through their surfaces.
5. If the parasite killed its host, it would also die. The parasite depends on the host for nutrition and for protection against adverse conditions and predators.

30.2 Nematoda and Rotifera

Approach

The main objective of this section is to describe members of two phyla with body plans more complex than that of flatworms but still lacking the coelom of most invertebrates and vertebrates. The section builds on Section 30.1 both in the presentation of body plans and in the discussion of parasitic organisms. In presenting the text make sure that students recognize that except for the pseudocoelomate body plan, roundworms and rotifers have little in common. You may wish to refer back to Figure 30.1 in introducing the section.

Concept Framework

Most students who own dogs will be familiar with the necessity of giving the dogs pills during mosquito season to prevent heartworm infestation. Use this knowledge as a starting point for discussing the fact that parasitic roundworms inhabit virtually all vertebrates, including about 50 species that infest humans. Explain that only about 10 of these species are dangerous to humans, but these can cause serious, even fatal, diseases. You may wish to end the discussion by asking students to consider how knowledge of the life cycle of the organism can help prevent disease.

Emphasis

Make certain that students understand that some, but not all, roundworms are parasitic; other roundworms and all rotifers are free-living.

Guest Speaker

Ask a veterinarian to visit the classroom or arrange a field trip to a local veterinarian's clinic to discuss how parasitic diseases are prevented or treated in dogs and cats.

Using the Features/Roundworm Development

The main process skills used in this feature are observing and experimenting. Scientists observed the development of a single roundworm egg until they were able to trace each cell of the fully developed worm to a specific division of the parent cell. Then scientists framed hypotheses based on their observations, which were tested in experiments. To help students understand the concentration the scientists required in order to complete this study, write on the chalkboard as many successive divisions of a parent cell as the chalkboard will hold. Then ask students to help in coding the end cells, using *D, a,* and *p*.

Section Review Answers

(Continued from p. 460.)
3. by wearing shoes when walking in areas that are likely to be infested with larvae
4. to move food into the rotifer's mouth
5. It is long, flexible, and slender, with pointed tips at either end. Thus it moves easily through decomposing soil or inside an animal's organ.

Laboratory Preparation

Background

In this lab students examine the structure of a representative flatworm, *Dugesia*. The students observe the flatworm's response to simple stimuli, such as touch, gravity, and food. Through these observations, students make inferences about the anatomy and behavior of flatworms.

Preparation

Dugesia are available from a biological supply house. If pond water is not available, set out tap water in open containers the night before the investigation.

Safety

Remind students to handle the flatworms in a humane manner.

Conclusions

1. Flatworms have bilateral symmetry. The paired anterior projections and eyespots illustrate this.

2. The flatworm is propelled by the oarlike motions of its cilia.
3. Sense organs are concentrated in the anterior end of a flatworm. Students will reach this conclusion based on the observation that the animal moves more quickly away from stimuli applied to the anterior end than from stimuli applied to the posterior end.
4. It will right itself after being turned over.
5. A flatworm feeds by extending a muscular organ out of its mouth. Food is sucked into the digestive cavity through this organ.

Inquiry

1. Generalizations cannot be made on the basis of observing one specimen. The work of other class members can be used to provide an experimental sample large enough to substantiate the observations of one individual.
2. Answers will vary. Students should devise an experiment that tests for response to light and dark conditions, as well as response to light applied to the anterior and posterior ends of the animal.

Chapter Review

Vocabulary

1. parasite; the other terms relate to the body plans of organisms
2. anus; the others are body parts of the planarian
3. mastax; the others are body parts of the tapeworm
4. ganglia; the other terms relate to parasites
5. pharynx; the others are protective coverings

Review

11. The major characteristics of flatworms are that they are acoelomate, are bilaterally symmetrical, show cephalization, have no respiratory or circulatory systems, and have a digestive system with only one opening.
12. A planarian coordinates its responses to stimuli with its ganglia. Impulses are sent through a ladderlike network of nerves from the ganglia throughout the body.
13. Planarians are hermaphrodites that reproduce sexually through simultaneous cross-fertilization. They also reproduce asexually by splitting

in two and regenerating the missing parts to form two individuals.

14. Both the tegument and the cuticle will protect a parasite from being digested by its host.

15. One adaptation of a fluke is the tegument, which keeps it from being digested by its host. Another adaptation is the suckers with which it attaches itself to the host. Flukes also have a powerful pharynx that is used in feeding on the host's body fluids. Fluke larvae are protected by a hard cyst.

16. Tapeworms have a tegument and suckers, but have neither a mouth nor a digestive system. Thus they must live on nutrients already digested by the host.

17. Roundworms and rotifers both have pseudocoeloms, digestive tracts with two openings, bilaterally symmetrical body plans, and some degree of cephalization. Some roundworms are parasitic, whereas all rotifers are free-living. Rotifers have some features that are lacking in roundworms, such as a crown of cilia and a mastax.

18. *Ascaris* eggs are picked up by the host in contaminated food or water. The eggs hatch in the intestine. The larvae travel through the bloodstream, are coughed up into the mouth, and are swallowed. In the intestine they mature and reproduce. The fertilized eggs leave with the feces of the host.

19. Humans can protect themselves from trichinosis by eating pork that is free of cysts containing live larvae. This can be guaranteed by cooking the pork sufficiently to kill the larvae, by freezing the pork, or by eating pork from animals that have not been fed raw meat.

20. *C. elegans* is easily grown in the laboratory and takes only 12 hours to hatch. Furthermore, the development of the cells of the *C. elegans* eggs is predictable, both in the final position of the developing cells and in the total number of cells.

Critical Thinking

1. It is advantageous when it cannot locate a mate.

2. Answers could include chemical or tactile receptors. Any answers that suggest such a mechanism should be accepted.

3. The many new irrigation canals and the lake behind the dam provided a great deal of favorable habitat for snails, which are the hosts of the schistosoma larvae.

4. Hookworm eggs hatch in warm, moist soil, such as rice paddies. Because field workers often go barefoot, they are more likely to become infested.

5. The term *eyespot* better describes the fact that this region merely senses light and does not process images.

6. The doctor could look for proglottids or eggs in the person's feces.

7. This characteristic is adaptive in pools of water that periodically dry up.

8. The spurs might be used as a means of anchoring the rotifer.

Chapter 30 | *Flatworms, Roundworms, and Rotifers*

Introduction

As invertebrate species evolved, the complexity of their body structures increased. The sponges and cnidarians discussed in Chapter 29 are among the least complex animals. In this chapter you will learn about invertebrates that show the next level of complexity: flatworms, roundworms, and rotifers.

Chapter Outline

30.1 Platyhelminthes
- *Turbellaria*
- *Trematoda*
- *Cestoda*

30.2 Nematoda and Rotifera
- *Nematoda*
- *Rotifera*

Chapter Concept

As you read, note the increased complexity of body plans as compared with those of sponges and cnidarians. Also notice how organisms are modified for a free-living or a parasitic way of life.

Pelagic worm, *Peobius meseres*. Inset: Male and female *Schistosoma*.

30.1 Platyhelminthes

Platyhelminthes, Nematoda, and Rotifera are three phyla of invertebrates whose body plans are more complex than those of sponges and cnidarians. Scientists identify four types of body plans in invertebrates. The relationship among the germ layers differs in each body plan. Germ layers, as you learned in Chapter 29, are layers of cells that originate in the developing embryo and become specific structures in the animal. Most animals have three germ layers: the ectoderm, the endoderm, and the mesoderm. The mesoderm is the embryonic layer that forms between the ectoderm and the endoderm. In animals muscles, bones, and reproductive organs develop from the mesoderm.

The four body types, shown in Figure 30-1, are defined in part by the presence or absence of a **coelom** *(SEE-lum)*, or body cavity. **Acoelomate** *(ay-SEE-luh-MATE)*, for example, means "without a coelom." The four body types, from the least complex to the most complex, are described below:

- The acoelomate body plan with two germ layers is characterized by an ectoderm and an endoderm that are not separated by a cavity. Sponges and cnidarians have this kind of body plan. It is considered the least complex body plan.
- The acoelomate body plan with three germ layers is characterized by an ectoderm, a mesoderm, and an endoderm that are not separated by a cavity. Flatworms are acoelomates with three germ layers.
- The pseudocoelomate body plan is characterized by a **pseudocoelom,** a cavity that forms between the mesoderm and the endoderm. Roundworms and rotifers are pseudocoelomates.
- The coelomate body plan is characterized by a cavity called a coelom, which develops within the mesoderm. Mollusks, annelids, arthropods, echinoderms, and all chordates, including humans, are coelomates. This is the most complex body plan.

The term *worm* does not refer to a specific taxon of animals. Though both are called worms, flatworms and roundworms differ so significantly in their fundamental structure that scientists classify them into separate phyla.

For additional teaching information see pp. T30A–T30D.

Section Objectives

- *State the distinguishing characteristics of flatworms.*
- *Describe the anatomy of a planarian.*
- *Compare and contrast free-living and parasitic life-forms.*
- *Diagram the life cycle of a fluke.*
- *Describe the life cycle of a tapeworm.*

Tip: Impress upon students that the coelom was one of the most important evolutionary developments among animals. Lead them to see that the coelom permits the gut to move independently of the body wall. This development meant that an animal did not have to move its whole body in order to move food through the gut.

Figure 30-1 The four basic body plans of animals exhibit a progression from least complex, with only two germ layers, to most complex, with three germ layers and a coelom.

Coelom

Digestive tract

Coelomate

Digestive cavity — Two-layered acoelomate

Digestive cavity — Three-layered acoelomate

Digestive tract — Pseudocoelom — **Pseudocoelomate**

☐ Ectoderm
☐ Mesoderm
☐ Endoderm

Figure 30-2 A spade-shaped anterior and tapered posterior characterize this type of flatworm. These adaptations allow the organism to move freely and to seek its own food.

Members of the Phylum Platyhelminthes (*PLAT-ee-hel-MINTH-eez*) are called flatworms. Flatworms are bilaterally symmetrical organisms that lack respiratory and circulatory systems. In flatworms, as in most animals, the sense receptors and nerves are concentrated at their anterior end, a characteristic called cephalization. The adaptive advantage of cephalization is that an organism that enters an environment with its sense organs leading the way will have an advantage over those that do not.

The more than 13,000 species of flatworms belong to three classes: Turbellaria, Trematoda, and Cestoda. Only turbellarians are free-living. Trematodes and cestodes live as parasites on or inside other animals. The **parasite** is an organism that lives on or in another organism, called the **host,** for a long period of time. The parasite derives its nutrition and protection from the host. Yet a parasite can harm the host by depriving it of nutrition and by interfering with its internal organ systems. The majority of parasites do not kill their hosts, an adaptation that helps to ensure that the hosts will not become extinct.

Scientists believe that parasites originated as free-living organisms. During evolution some organs that were advantageous to free living became modified for parasitism; some were lost entirely. Because their food is already digested, some parasites lack mouths or well-developed digestive systems. Parasites have hooks or suckers that grip their hosts. Features that prevent parasites from being digested by their hosts are the **tegument,** a thick covering of cells, and the **cuticle,** a nonliving layer secreted by the epidermis.

Turbellaria

The majority of the 3,000 species of the Class Turbellaria live in marine environments. However, the most familiar turbellarian is the freshwater planarian *Dugesia,* which is pictured in Figure 30-3. As you study, notice the adaptations for free living.

The planarian's posterior end is tapered, and its anterior end is shaped like a spade. As an adaptation to its free-living way of life, much of the planarian's body is covered with cilia. A planarian moves by swimming with an undulating motion or by laying down a layer of mucus over which it slides, propelled by the beating of its cilia.

Figure 30-3 A planarian extends its pharynx and sucks food into its mouth and intestine. Wastes are eliminated through the mouth and via ducts lined with flame cells.

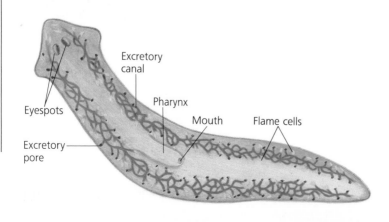

Eyespots

Excretory canal

Pharynx

Mouth

Flame cells

Excretory pore

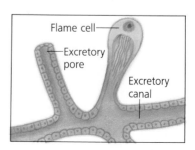

Flame cell

Excretory pore

Excretory canal

Digestion and Excretion

The planarian is a scavenger; it forages for carrion. It also preys on smaller organisms, such as protozoa. Its mouth is located in the middle of its body. A planarian feeds by extending a muscular tube, the **pharynx** *(FER-inks),* out of its mouth. Food is sucked in through the pharynx and passes into the branched intestine. Nutrients are absorbed through the intestinal wall, and undigested food is excreted through the pharynx and mouth.

Chemical wastes and excess water are eliminated through a network of ducts that run the length of the body. Each duct contains many flame cells. A **flame cell** is a cell that encloses a tuft of cilia that beats, resembling a candle flame. The beating of the cilia moves wastes into the duct to the excretory pores, where the wastes leave the body.

Nervous Control

The planarian nervous system and sense organs are also adapted for free living. It is more organized and cephalized than the nerve net of the *Hydra.* Planarians sense the intensity and direction of light with two anterior eyespots that contain photosensitive cells. Receptors for touch, taste, and smell are concentrated at the anterior end. Two anterior clusters of nerve cells, called **ganglia** *(GAN-glee-uh),* form a simple brain. The ganglia receive information from sensory cells and then send impulses to the rest of the body through a ladderlike system of two longitudinal nerves connected by tranverse nerves.

Planarians show a surprising ability to learn. Their memory, such as how to find the way through a maze, is stored chemically. One planarian can learn to pass through a maze by eating another planarian that already knows the way.

Reproduction

Planarians reproduce sexually and asexually. They are hermaphrodites—single individuals possessing both male and female sex organs. Because they are free-living, they can encounter mates. In sexual reproduction two planarians simultaneously fertilize each other. Eggs, laid in protective capsules that stick to rocks or debris, hatch in two to three weeks. Freshwater planarians reproduce asexually in the summer. They attach their posterior ends to a surface and stretch until they tear in two. Each half then regenerates its missing parts.

Trematoda

The Class Trematoda consists of about 6,000 species of parasitic flukes, most of which are shaped like leaves. Some flukes are endoparasites, meaning that they live inside their hosts. Others are ectoparasites, meaning that they live on the external surfaces of their hosts. As you read, notice the adaptations for parasitism.

Think

The planarian can reproduce itself by splitting in two and regenerating. What are the advantages and disadvantages of this form of asexual reproduction as compared with sexual reproduction?

Answer to Think
Asexual reproduction can happen quickly and does not require another member of the same species. Sexual reproduction allows for variation in the species through the recombination of genes.

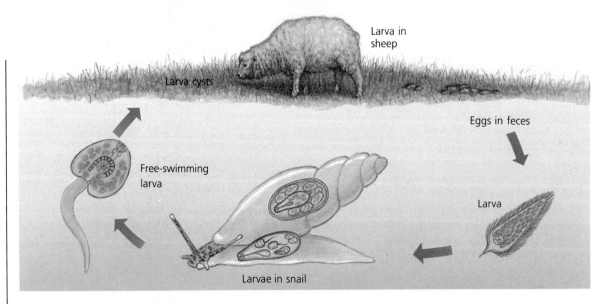

Larva in sheep

Larva cysts

Free-swimming larva

Eggs in feces

Larva

Larvae in snail

Figure 30-4 During the life cycle of the sheep liver fluke, reproduction occurs both sexually and asexually.

Structure and Reproduction of Flukes

Flukes are adapted to parasitism. They are about 1 cm long, oval shaped, and sheathed in a tough, unciliated tegument. Typically a fluke has two sucker mouths—one on its anterior and one on its ventral surface—that it uses to cling to its host. A powerful pharynx in the anterior sucker sucks in blood, cells, and fluids.

While the nervous and excretory systems of the fluke are similar to those of planarians, its life cycle and reproductive structures are more complex. Typically hermaphroditic, the fluke has a long coiled tube, or **uterus,** that stores eggs until ready for release through a genital pore. A fluke may produce tens of thousands of eggs at a time. A few survive predation and environmental destruction and hatch into larvae, or immature organisms.

The complex life cycle of the economically destructive sheep liver fluke, *Fasciola hepatica,* is illustrated in Figure 30-4. Adult flukes live in the sheep's liver and gallbladder. Flukes mate in these organs and produce eggs. The eggs enter the intestines and are eliminated with the feces of the sheep. The eggs hatch in water, and the larvae then invade snails. Inside snails, the larvae multiply asexually. The larvae leave the snail and form cysts on blades of grass. A **cyst** is a dormant larva surrounded by a hard protective covering. When sheep eat the grass, they ingest the cysts, which hatch in the sheep's digestive tract. The flukes then bore through the intestines into the blood and mature and reproduce in the liver, completing the cycle.

Schistosomiasis

One of the world's most severe public health problems is caused by three species of blood fluke in the genus *Schistosoma*. The blood fluke infests 200 to 300 million people, mostly in Asia, Africa, and South America. The resulting disease, called **schistosomiasis** *(SHISS-tuh-soe-MIE-uh-suss),* affects 20 to 30 times more people than cancer does. The life cycle of a schistosome parallels that of other flukes, with the exception that the adult lives in the human bloodstream. The spiny eggs of the schistosomes lodge in the human veins, lungs, intestines, bladder, and liver. They block blood vessels and cause internal bleeding and tissue decay that can be fatal.

Info: Schistosomiasis is epidemic in many tropical areas around lakes and rivers, and it has spread via irrigation ditches and reservoirs. In some villages near the Aswan Dam in Egypt all the children are affected.

Cestoda

The Class Cestoda consists of about 1,500 species. Most are parasitic tapeworms. All vertebrates may host tapeworms. Humans, for example, may harbor any of seven different species.

Structure

Like flukes, tapeworms are adapted for the parasitic life. They have tough outer teguments and hooks and suckers to attach themselves to their hosts. Their long, ribbonlike bodies may grow up to 12 m. The nervous system extends the length of the body.

Tapeworms lack sense organs, mouths, and digestive tracts. They absorb nutrients directly through their heavily folded teguments. The folds increase the surface area for absorption. The tapeworm grips its host with hooks and suckers on its knob-shaped head, or **scolex** *(SKOE-leks)*. It grows by producing body sections, called **proglottids** *(proe-GLAHT-udz),* immediately behind the scolex. Thus the oldest proglottids are at the posterior end. The excretory system drains the proglottid of wastes.

Life Cycle and Reproduction

Each proglottid contains both male and female reproductive organs. Cross-fertilization between two adjacent worms is typical, but self-fertilization between proglottids can also occur. After fertilization the egg-packed proglottid breaks off from the adult and is eliminated with the host's feces.

The life cycle of the beef tapeworm, *Taenia saginaton,* begins when cattle eat grass contaminated with proglottids and eggs. Larvae hatch and bore through the cow's intestine into the bloodstream. The larvae then burrow into muscle tissue and form cysts. Humans become infected when they eat beef that has not been cooked sufficiently to kill the bladder worms inside the cysts. Once the cysts are inside the human intestine, the cyst wall dissolves and releases the bladder worm, which then develops into an adult beef tapeworm.

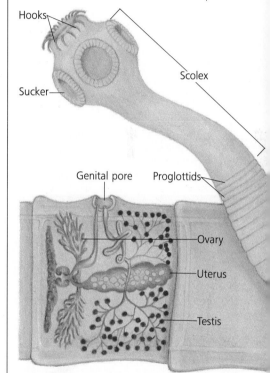

Figure 30-5 A tapeworm grows by adding proglottids behind its head. With male and female reproductive organs in each proglottid, both cross-fertilization and self-fertilization are possible.

Section Review

1. What is a *flame cell?*
2. How are stimuli sensed in flatworms?
3. How does the presence of the blood fluke *Schistosoma* affect the human body?
4. How do tapeworms obtain nutrients without a mouth or a digestive system?
5. Why is it adaptive that a parasite not kill its host?

Answers to Section Review
1. A flame cell is a planarian cell that contains cilia, which move wastes through canals to the excretory pores. The network of flame cells and canals is the excretory system.
2. through sense receptors and nerves concentrated at the anterior end of the body
(See pp. T30A–T30D.)

- *Compare and contrast the body plans of flatworms and roundworms.*

- *Summarize the parasitic relationships of common roundworms and humans.*

- *Describe the anatomy of a rotifer.*

- *Describe a classical study of nematode development.*

Tip: Discuss how improved personal hygiene and plumbing have helped reduce the incidence of human infestation by parasites.

30.2 Nematoda and Rotifera

Members of the Phyla Nematoda and Rotifera are characterized by the presence of a pseudocoelom. This body cavity is lined on the inside by endoderm and on the outside by mesoderm. The fluid-filled space contains organs, supports the body, provides the hydrostatic pressure against which muscles can contract, and serves as a storage area for wastes or for eggs and sperm.

Nematoda

Roundworms are classified in the Phylum Nematoda. The number of nematode species is estimated at 10,000 to 80,000. Roundworms have long slender bodies that taper at both ends. Most are covered with a flexible protective cuticle. They range in length from less than 1 mm to as long as the female guinea worm, which can be 120 cm. Roundworms have a digestive tract with two openings, an anterior mouth and a posterior **anus,** the opening through which wastes are eliminated. The sexes are distinct in most species.

Biology in Process

Roundworm Development

A soil-dwelling roundworm barely 1 mm long, *Caenorhabditis elegans,* is becoming one of the best-studied animals in science. Scientists are using this worm to probe a fundamental mystery: How does a single egg develop into a complete organism? The answer could help prevent human birth defects and help develop new varieties of livestock.

Observing is a fundamental scientific process and *C. elegans* is a good organism to observe. This hermaphroditic roundworm is easily raised in the laboratory and has sturdy transparent eggs that hatch in just 12 hours. Furthermore, *C. elegans* is one of the few multicellular animals that develops a fixed number of cells.

Each egg of *C. elegans* always follows the same pattern of development. The zygote is called the parent cell. The first cell division is followed by generation after generation of cell divisions until exactly 671 cells have been produced. Of these, 113 cells always die. The remaining 558 cells make up the hatchling worm. Scientists have traced the position in the adult worm of each of these cells.

Answer to Biology in Process: Observing allows scientists to keep
track of the individual cells and to know the line of succession for
each cell in the hatchling worm.

The vast majority of roundworm species are free-living on land, in salt water, and in fresh water. About 50 species are plant and animal parasites that cause enormous economic damage and physical suffering throughout the world. Humans are host to about 50 different roundworm species, three of which will be discussed here. In these roundworms notice the adaptations for parasitism.

Ascaris

Ascaris (AS-kuh-russ) is a roundworm parasite found in the intestines of pigs, horses, and humans. *Ascaris* eggs enter the body in contaminated food or water. After hatching in the intestines the larvae bore into the bloodstream and are carried to the lungs and throat. There they are coughed up, swallowed, and returned to the intestines, where they mature and mate. If the host is not treated, knots of worms can block the intestines, causing death. Larvae in the lungs cause a respiratory illness.

The adult female grows up to 30 cm. The much smaller male has a hooked posterior that holds the female during mating. A female produces up to 200,000 eggs a day. The fertilized eggs are shed through a genital pore and exit in the host's feces.

Figure 30-6   This roundworm, with its long slender body that tapers at both ends, is typical of the thousands of species in the Phylum Nematoda.

Scientists are able to locate these positions because they have observed the entire developmental process through a microscope over and over for several years. A code is assigned to each cell as it is formed. The code shows the developmental history of the cell. For example, one of the 558 cells that make up the worm's body muscle is coded *Dapppa*. The code is read backwards. It indicates that this specific cell arose from the (a) anterior half of the (p) posterior half of the (p) posterior half of the (p) posterior half of the (a) anterior half of the (D) parent cell.

Now that scientists are confident that they know the normal development process, they are beginning to perform experiments to determine the factors that control this development. One experiment is testing the effects of removing some cytoplasm from the cell. Another is designed to see how

the injection of DNA into the zygote might affect development. Gradually scientists are learning about the development process, building from the foundation established by observing roundworms.

■ How is observing important to understanding the development of C. elegans?

Info: In some Asiatic areas human feces are still used as fertilizer for mulberry trees, the leaves of which support silkworms. Because the workers do not wear shoes, the level of hookworm infestation is high.

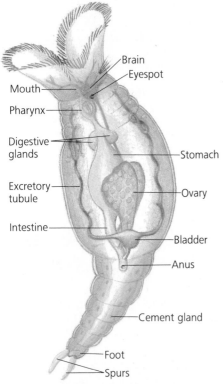

Brain
Eyespot
Mouth
Pharynx
Digestive glands
Stomach
Excretory tubule
Ovary
Intestine
Bladder
Anus
Cement gland
Foot
Spurs

Figure 30-7 Cilia surrounding the mouth of a rotifer sweep food into the mouth. Hydrostatic pressure within the pseudocoelom helps the rotifer maintain its shape.

Answers to Section Review
1. Trichinosis is a disease caused by ingestion of the *Trichinella* roundworm.
2. The presence of fluid in the pseudocoelom gives shape to these animals and provides a hydrostatic pressure for muscle contractions to work against. The cavity also contains organs and can store eggs and wastes.
(See pp. T30A–T30D.)

Ancylostoma and *Necator*

Ancylostoma and *Necator,* two genera of hookworm, seriously harm over 400 million people worldwide. *Necator* causes about 90 percent of the infections in tropical and semitropical regions. The hookworm's mouth has cutting plates that hook onto the intestinal wall. The hookworm feeds on its host's blood, causing anemia. Like *Ascaris,* the hookworm travels through the blood to the lungs and into the throat, where it is swallowed. Hookworms mate in the intestines. Their eggs are shed in the host's feces and hatch on warm moist soil. The larvae then bore through the feet of a new host. Shoes protect against hookworm infestation.

Trichinella

Roundworms of the genus *Trichinella* cause the disease **trichinosis** (TRICK-*uh*-NOE-*suss*) in humans. Humans become infected when they eat undercooked pork that is contaminated with trichina cysts. The cysts release larvae, which burrow into the wall of the small intestine, where they mature into adults. The adults then produce larvae that pass into the blood and form cysts in the muscles, causing pain and stiffness. Trichinosis is now rare in the United States. Farmers cook meat scraps before feeding them to hogs, and meat-packers freeze pork, which kills the worms.

Rotifera

Most of the 1,750 species of the Phylum Rotifera are transparent, free-swimming, microscopic animals that live in marine or fresh water. These microscopic rotifers are between 100 and 500 μm long, though the male is much smaller than the female. The crown of cilia that surrounds the rotifer's mouth, which can be seen in Figure 30-7, is an adaptation for free living. The cilia sweep food into the mouth. Under a microscope this beating of the cilia looks like rotating wheels. Rotifers feed on unicellular algae, bacteria, and protozoa. This food moves through the pharynx to the **mastax,** a muscular organ that chops up the food. The rotifer has no skeleton. It maintains its shape by the hydrostatic pressure within the pseudocoelom. The nervous system is composed of anterior ganglia and two long nerves that run the length of the body. Cephalization is indicated by the two anterior eyespots.

Section Review

1. Define *trichinosis.*
2. What is the significance of the pseudocoelom in nematodes and rotifers?
3. How can humans prevent hookworm infestation?
4. What is the function of the cilia on the rotifer's head?
5. How is the general body plan of the roundworm adapted to its way of life?

Laboratory

Observing Flatworms

Objective
To examine the morphology and behavior of *Dugesia*, a flatworm

Process Skills
Observing, inferring

Materials
Dugesia, plastic medicine dropper with the tip cut off, pond water, medicine dropper, culture dish, dissecting microscope, probe, raw liver

Method

1 Use a medicine dropper to transfer a flatworm to a culture dish. Use a dropper to cover the flatworm with pond water. Why should you use pond water here and not tap water?
2 Examine the general morphology of the flatworm under the lowest power of your microscope. How does the organism move? Write down your observations. Continue your examination under high power.
3 Observe the flatworm for several minutes. Write down your observations.
4 Make a detailed drawing of the flatworm. Compare the anterior and posterior ends of the flatworm. Compare its dorsal and ventral portions. What can you determine about the flatworm from these simple observations?
5 Use a probe to touch the posterior end of the flatworm. Touch the organism gently. If you injure the flatworm, its responses to stimuli will be altered.
6 Use the probe to touch the anterior end of the flatworm. Observe and record the animal's reaction to stimuli.
7 Use the probe to turn the flatworm onto its dorsal side. Observe and record its reaction. What are you testing for here?
8 Place a tiny piece of liver in the culture dish.

Observe the reaction of the flatworm for 2 or 3 minutes.
9 Using the probe, gently turn the flatworm over. Can you now observe directly how the flatworm feeds?

Conclusions

1 What kind of symmetry is exhibited by the flatworm? Which structural features can you use to identify this type of symmetry?
2 What external structures appear to account for the flatworm's movement?
3 Based on your observations, are the sense organs of a flatworm more heavily concentrated in the anterior or the posterior end of the animal? Explain how you can tell.
4 What evidence have you seen that a flatworm can sense gravity?
5 Based on the observations you made in step 9 describe the feeding behavior of the flatworm.

Inquiry

1 Can you generalize about the behavior of all flatworms based on your observations of one specimen? How might the observations of your classmates be used to reduce the possibility of errors in your conclusions?
2 Design an experiment to test a flatworm's sensitivity to light.

Vocabulary

acoelomate (453)	flame cell (455)	pharynx (455)	scolex (457)
anus (458)	ganglia (455)	proglottid (457)	tegument (454)
coelom (453)	host (454)	pseudocoelom (453)	trichinosis (460)
cuticle (454)	mastax (460)	schistosomiasis (456)	uterus (456)
cyst (456)	parasite (454)		

For each set of terms below, choose the one that does not belong and explain why it does not.

1. pseudocoelom, parasite, acoelomate, coelom
2. pharynx, flame cell, ganglia, anus
3. scolex, mastax, proglottid, tegument
4. schistosomiasis, trichinosis, ganglia, host
5. pharynx, tegument, cuticle, cyst

Review

1. A coelom develops within the (a) mesoderm (b) ectoderm (c) endoderm (d) acoelomate.
2. A planarian uses its pharynx to help it (a) move (b) feed (c) reproduce (d) respond to light.
3. In planarians the flame cell (a) functions to move liquid wastes through the body (b) responds to light and sound (c) is specialized for diffusion (d) controls regeneration.
4. The sheep liver fluke reproduces (a) in water (b) inside a snail (c) inside a sheep's liver (d) inside a sheep's digestive tract.
5. The function of the cyst in flukes is to (a) protect the immature worm (b) fertilize the eggs (c) facilitate movement (d) bore into the host.
6. A tapeworm uses its scolex to (a) locate food (b) communicate with other tapeworms (c) get rid of liquid wastes (d) attach itself to its host.
7. Humans ingest tapeworms by (a) drinking water containing tapeworm larvae (b) walking barefoot on contaminated soil (c) eating contaminated meat (d) wading in contaminated water.
8. A characteristic that differentiates roundworms from flatworms is (a) the presence of a mesoderm (b) a bilaterally symmetrical body plan (c) the presence of an anus (d) cephalization.
9. Trichinosis in human beings results in (a) muscle soreness (b) anemia (c) blockage of blood vessels (d) a respiratory illness.
10. The mastax in rotifers helps them to (a) move (b) sense light (c) reproduce (d) digest food.
11. What are the major characteristics of the flatworm?
12. Describe the structure and function of the nervous system of a planarian.
13. How do planarians reproduce?
14. What is the function of the tegument and cuticle in parasites?
15. What are some adaptations of flukes for a parasitic way of life?
16. How are tapeworms specialized for a parasitic way of life?
17. Compare and contrast roundworms and rotifers.
18. Describe the life cycle of the roundworm *Ascaris*.
19. How can humans protect themselves from the disease trichinosis?
20. Why is the roundworm *Caenorhabditis elegans* frequently selected for laboratory observation?

Critical Thinking

1. When would asexual reproduction be advantageous to a free-swimming flatworm?
2. Parasites are specific. This means that they must invade the correct host or they cannot survive. Suggest how members of a species find the correct host.
3. The Aswan High Dam across the Nile River in Egypt was completed in 1970. It was built to increase the supply of irrigation water, to control major flooding, and to provide a source of hydroelectric power. Since the dam was built, there has been an increase of schistosomiasis in the region. Why do you think this happened?
4. Hookworms are extremely common in China, where rice is the main crop. Rice is grown in paddies that are periodically flooded. Considering what you know about how hookworms invade the body, why might hookworm infection be so common in this part of the world?
5. Why do we use the term *eyespot* instead of *eye* to refer to the photosensitive regions at the anterior end of the planarian body?
6. A person infected with a tapeworm may show generalized symptoms, such as tiredness, loss of weight, and low blood count. These symptoms could indicate any num-

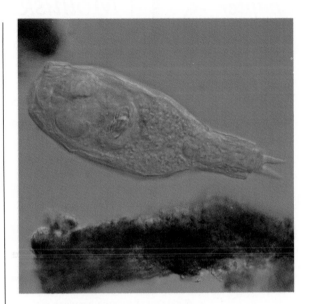

ber of diseases. How would a doctor make a positive diagnosis?

7. Some rotifers can survive being dried out for as long as four years. If their dried bodies are put into water, they revive. For what kind of environment might this characteristic be adaptive?
8. Look at the photograph above of a rotifer. Notice the two extensions, called spurs, at its posterior end. Suggest a possible function for the spurs.

Extension

1. Collect a sample of moist soil or soil and sand from the bottom of a pond. Using a dissecting microscope or a large magnifier, look for nematodes, and observe their behavior.
2. Collect a sample of pond water or make a grass infusion by placing dried grass and other plants in water and letting them stand for a few days. If these are not available, a culture of rotifers can be obtained from a biological supply house. Observe the feeding behavior of rotifers.
3. Research the types of marine species of turbellarians. Draw pictures of them,

making sure to color them accurately. List the following information under your drawings: scientific name, common name if known, habitat, area of the world where it is found, food, method of protection.

4. Read the article by Stefi Weisburd entitled "Clams and Worms Fueled by Gas?" in *Science News*, October 12, 1985, p. 231. What do the worms described in this article seem to be eating? What theory do scientists have for how these worms survive the toxic substances where they live?

Chapter 31 | *Mollusks and Annelids*

This chapter examines the similarities and differences between Phyla Mollusca and Annelida. Students learn why scientists think these phyla evolved from a common ancestor, and they consider the adaptive advantages of various body forms.

Correlation Chart

Section Number	Section Objectives	Pages	Features	Section Review
31.1	1. List three types of evidence that indicate a close evolutionary relationship between mollusks and annelids.	465–470	Think, p. 468	1
	2. Summarize the adaptive advantages of a true coelom.			2
	3. Name the major characteristics shared by most mollusks.			2
	4. Contrast respiration in land snails and sea snails.			3
	5. Identify the adaptations of clams and squids.			4, 5
31.2	1. List the distinguishing characteristics of annelids.	471–475	Biotechnology, pp. 472–473	1, 3
	2. Distinguish between the three classes of annelids.			2
	3. Identify some traits of earthworms that are adaptations for burrowing in the soil.		Biology Lab, p. 475	2, 4
	4. Compare earthworms with polychaetes.			1
	5. Name some characteristics of leeches that are leading to renewed use by physicians.			5

Resources*

Print

Hadeen, Robert A. *The Oyster: The Life and Lore of the Celebrated Bivalve*. Centreville, Md.: Tidewater, 1986.

Horsman, Paul V. *Seawatch: The Seafarer's Guide to Marine Life*. New York: Facts on File, 1985.

Audiovisual/Software

Mollusks: The Mussel Respiration and Digestion. Film, 7 min. Film guide. Benchmark, 1983.

The Portugese Man-of-War. Video, 9 min. Carolina Biological, 1986.

*Additional references are listed in *Modern Biology, Teacher's Program Guide*.

Section Teaching Guides

31.1 Mollusca

Approach

The main objective of this section is to describe the general characteristics of the phylum Mollusca and to give specific characteristics of four major classes of mollusks. The section builds upon concepts of adaptation and evolution (Chapters 15 and 16) and further develops concepts of invertebrate characteristics (Chapter 29). It also provides background information for other chapters in Unit 7. In addition, the discussion of the evolutionary relationship between mollusks and annelids draws on the material presented in Chapter 18 describing evidence used in determining phylogenetic relationships. In presenting the text, guide students to notice the adaptations that different classes of mollusks have which suit them for a particular way of life.

Concept Framework

Many students will be familiar with mollusks, such as oysters, clams, snails, and even octopuses as sources of food. Use this knowledge as the basis of a discussion of the characteristics of mollusks. Assist students in listing and describing different types of mollusks. Elicit the fact that most, but not all mollusks have shells. Tell students to note other characteristics of this phylum as they read the section.

Emphasis

Students should be able to list four characteristics common to all mollusks and to describe the distinguishing traits of members of the following classes of mollusks: Polyplacophora, Gastropoda, Bivalvia, and Cephalopoda.

Student Presentation

Students interested in cooking may wish to find out how mollusks are prepared as food in different countries. Ask these students to give a presentation discussing the availability, preparation, and nutrient content of mollusks. Culinary traditions they might research include those of France, Scandinavia, Japan, and Polynesia.

Section Review Answers

(See p. 470.)

31.2 Annelida

Approach

Presenting the general characteristics of the Phylum Annelida and specific characteristics of members of the three classes that compose the phylum is the main objective of this section. The section builds on concepts of adaptation and evolution covered in Chapters 15 and 16 and provides more detail on characteristics of invertebrates as discussed in earlier chapters in this unit. In presenting this section, guide students to realize that annelids are a very successful group, having adapted to conditions on land, in the oceans, and in freshwater.

Concept Framework

Students have probably seen earthworms burrowing in soil or crawling on sidewalks after a hard rain. Build on this experience by discussing other things students may have observed about earthworms including their appearance, where they live, what they might eat, and why they are considered important in agriculture. Tell students that though earthworms appear to be simple creatures, their internal anatomy is actually fairly complex. Suggest that students note the advanced morphology of earthworms as they read the section.

Emphasis

Students should be able to describe in detail the structure, digestion, circulation, respiration, excretion, nervous control, and reproduction of one type of annelid—the oligochaete or earthworm.

Using the Features/Medicinal Uses of Leeches

The technique described in this section is how leeches can be used for medicinal purposes. These annelids are used to help restore normal blood flow in small veins after microsurgery. Their secretions may eventually be used to inhibit blood clotting and to treat heart attack victims. You may wish to discuss the use of the leech as an example of how scientists are discovering valid medicinal uses of various species of organisms whose use was once restricted to superstition or to folk medicine. Point out that scientists are also discovering medicinal uses of various species of tropical plants. Encourage students to realize that the medicinal potential of various plants and animals is a major reason medical researchers are becoming more aware of the importance of preserving habitats rich in unstudied organisms.

Guest Speaker

Invite a speaker to discuss the importance of earthworms in agriculture. A farmer, agriculture graduate student, or a representative from the agricultural extension service would be qualified speakers. You may wish to ask the speaker to discuss how earthworms are raised commercially.

Section Review Answers

(Continued from p. 474.)

2. Adaptations of the earthworm for burrowing include a streamlined shape without appendages, presence of setae and a segmented coelom, and digestive organs adapted for extracting nutrients from the soil.
3. The annelid body is segmented; the mollusk body is not.
4. Through burrowing, earthworms loosen and aerate the soil, making water and nutrients more accessible to plants.
5. The anesthetic prevents the host from feeling the wound inflicted by the leech. This allows the leech to feed on the host's blood without being noticed.

Laboratory Preparation

Background

In this lab procedure students examine the behavior of an earthworm. Students perform a simple controlled experiment to test the animal's response to ammonium hydroxide. The students also compare the organism's degree of sensitivity at its anterior and posterior ends. Experiments also test the animal's responses to gravity and light. Based on their observations, students make inferences about the adaptive advantage of the earthworm's behavior.

Preparation

Undiluted household ammonia can be used since it is a form of ammonium hydroxide. Be sure the pans are clean. Traces of preservative can be an irritant to the earthworm. Dim the lights in the classroom so that light will not interfere with the investigation.

Safety

Remind students to handle the earthworms in a humane manner.

Conclusions

1. Answers will vary. Students may notice a greater response to stimuli applied to the anterior end of the earthworm.
2. The earthworm responds negatively to touch, ammonium hydroxide, gravity, and light.

Inquiry

Students will devise an experiment in which the reactions of earthworms to white light (the control) and their reactions to colored light are compared.

Chapter Review

Vocabulary

1. In a closed circulatory system blood circulates entirely within vessels; in an open system at least part of circulation takes place through body cavities.
2. The gizzard is the digestive organ that grinds up the earthworm's food; the radula is the structure that mollusks use to scrape food.
3. A chromatophore is a pigment-containing cell in cephalopods; a trochophore is a ciliated larval form of mollusks and annelids.
4. The mantle is the outer layer of tissue that covers a mollusk body and in most species secretes a shell; the mantle cavity is a space between the mantle and the visceral mass.
5. The gill is a structure specialized for the exchange of gases between blood and a watery environment; nephridia are excretory organs specialized to eliminate nitrogenous wastes.

Review

11. A coelom separates the muscles of the digestive system from those of the body wall, provides space where a circulatory system can function, and contains fluid that may serve as a type of skeleton against which muscles can contract.
12. head, foot, and visceral mass
13. Torsion is the twisting of a gastropod's body that occurs during larval development. Because

of torsion a gastropod's mantle cavity is located at the front of the body above the head.

14. A mollusk uses its radula to scrape up food.

15. A gill has a large surface area and a rich blood supply, both features that promote gas exchange. Cilia on the gills of some bivalves direct food into the mouth.

16. Cilia on a clam's gills direct mucus that has trapped food particles into the mouth.

17. Segmentation enables annelids to move by using muscles that alternately expand and contract different segments; it also allows it to grow larger by adding basically identical segments.

18. Microsurgeons use leeches to remove congested blood from reattached body parts in order to allow healing to occur. In the future scientists hope to use leeches to help heart attack victims.

19. Earthworms require a moist environment to maintain a moist body surface across which gases can diffuse.

20. A leech moves by alternately attaching its anterior and posterior suckers and by using muscles to effect a crude kind of walking.

Critical Thinking

1. Hermaphroditism allows a snail to mate with any other snail, rather than with just 50 percent of the snail population. This is important for the land snail since the snail moves so slowly and is so often inactive.

2. Because the egg and the sperm are sent into the water, tides, currents, and wave action may prevent their meeting, thus jeopardizing the reproductive success of the organism. Consequently, you would expect clams and other mollusks to shed many sperm and eggs.

3. The secretion will cover sand or rock fragments, which cannot be expelled by the organism, and thus prevent them from being an irritant.

4. Each section of the nerve net controls the muscles of an individual segment. This makes it possible for the two halves of the body to move in different directions.

5. Both earthworms and clams filter their environment to take out microscopic food. They differ since the earthworm's environment is soil and the clam's environment is marine, and thus their digestive anatomy is adapted for their particular environment.

6. A leech is not considered to be a parasite because it neither derives protection from its host nor lives off its host.

7. The clam's long siphons enable the clam to bury itself in sand or mud and still be able to receive oxygen and nutrients from the environment.

Chapter 31 | Mollusks and Annelids

Introduction

At first glance it's hard to see much resemblance between a feather-duster worm and an octopus. However, these animals share several important characteristics, as do the other organisms in the phyla they represent. Octopuses belong to the Phylum Mollusca, and feather-duster worms to the Phylum Annelida. In this chapter you will learn about the similarities and the differences between these two phyla and about the diversity of species in each phylum.

Chapter Outline

Chapter Concept

As you read, note how the body forms of various species of mollusks and annelids illustrate adaptions for life in particular environments.

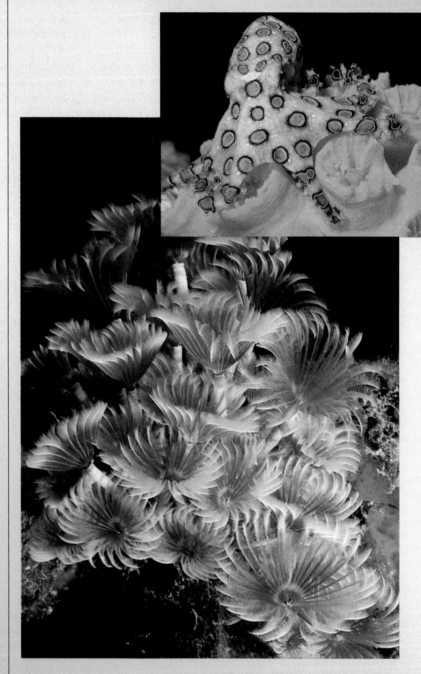

Cluster of polychaete worms known as feather dusters, *Sabella crassicornis*. Inset: Octopus, *Haptalochlaena maculosa*, Indonesia.

For additional teaching information see pp. T31A–T31D.

31.1 Mollusca

Despite their very different appearances such familiar invertebrates as chitons, clams, snails, and octopuses are all mollusks, members of the Phylum Mollusca. The word *mollusk* comes from a Latin term meaning "soft" and refers to the bodies of these animals. Although slugs and octopuses have soft bodies, most mollusks have a shell that protects and conceals their soft body.

Relationship of Mollusca and Annelida

Based on comparative studies of living species, biologists have concluded that members of the Phylum Mollusca probably share a common ancestor with the segmented worms of the Phylum Annelida. Mollusks and annelids have similar patterns of embryonic development, which you will study in Chapter 35. In addition, mollusks and annelids were probably the first major groups of organisms to have a true coelom—that is, a fluid-filled cavity within the mesoderm. The coelom provides several benefits. By separating the muscles of the gut from those of the body wall, the coelom allows food to move through the body independent of locomotion. The coelom also provides a space in which a circulatory system can function without interference from other organs. In some species the fluid in the coelom also forms a sort of internal hydrostatic skeleton against which the muscles can contract.

The strongest evidence for the common ancestry of mollusks and annelids is the characteristic larval form they share. In both groups the first stage of larval development is a pear-shaped larva called a **trochophore** *(TROCK-uh-for)*. Notice in Figure 31-1 that cilia project from both ends of the trochophore and circle the middle. These cilia propel the trochophore through the water and draw food to the mouth. Free-swimming trochophores aid in the dispersal of many marine mollusks. However, in terrestrial mollusks and in many species of marine annelids, the trochophore develops within the egg and is not free-living. The trochophore will be discussed further in Chapter 35.

Characteristics of Mollusks

Mollusks are a diverse group of more than 100,000 species. Among animal phyla only Arthropoda has more species. Some mollusks are sedentary filter feeders, such as clams and oysters. Others are predatory, such as squids and octopuses—which move about by jet propulsion and have complex nervous systems. Mollusks also have diverse connections with humans. Certain species of snails, for example, are alternate hosts of parasites, such as schistosomes, that are harbored by humans. The feeding habits of snails and slugs cause damage to crops. In contrast, humans prize many mollusks as food or for the beauty of their shells.

Section Objectives

- *List three types of evidence that indicate a close evolutionary relationship between mollusks and annelids.*

- *Summarize the adaptive advantages of a true coelom.*

- *Name the characteristics shared by most mollusks.*

- *Contrast respiration in land snails and sea snails.*

- *Identify the adaptations of clams and squids.*

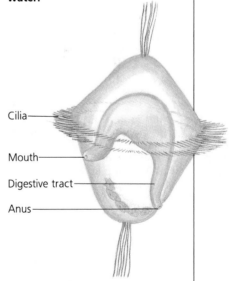

Figure 31-1   A trochophore is the first stage of development of mollusk and annelid larvae. Cilia at both ends and in the middle propel the organism through water.

Cilia

Mouth

Digestive tract

Anus

Mollusks have the following characteristics:

- All mollusks have a true coelom.
- The body has three distinct parts: the muscular foot, the head, and the visceral mass.
- Mollusks have organ systems for circulation, respiration, digestion, excretion, nerve impulse conduction, and reproduction.
- Most mollusks are bilaterally symmetrical and have one or more shells.

Body Plan

Mollusk species have numerous variations of the same basic body plan. Figure 31-2 shows one version of the anatomy of a hypothetical ancestral mollusk. Locate in the diagram the three main body parts shared by all mollusks: the muscular foot, the head, and the visceral mass. The muscular **foot** is a large organ used in locomotion. The **head** contains the mouth, sense organs, and cerebral ganglia. The **visceral mass** contains the heart and the organs of digestion, excretion, and reproduction. Covering the visceral mass is the **mantle,** an epidermal layer that in most species secretes a shell.

Organisms with a hard shell of calcium carbonate probably appeared early in the evolution of mollusks. The shell was of an adaptive value to these organisms because it protected the soft body from predators. However, it also conferred a disadvantage since the shell reduced the surface area available for gas exchange. This problem was avoided as a related structural adaptation evolved—the gills. With a large surface area and a rich supply of blood, **gills** are organs specialized for the exchange of gases with water. The delicate gills are protected within the **mantle cavity,** a space between the mantle and the visceral mass.

Biologists use structural differences to classify mollusks into four to seven classes. We will discuss four major classes: Polyplacophora (PAHL-ee-pluh-KAHF-ur-uh), Gastropoda (gass-TRAHP-uh-duh), Bivalvia (bie-VALV-ee-uh), and Cephalopoda (SEFF-uh-LAHP-uh-duh).

Figure 31-2 A hypothetical ancient mollusk (left) and a modern snail (right) are very similar. The anus on top of the snail's head results from torsion during larval development.

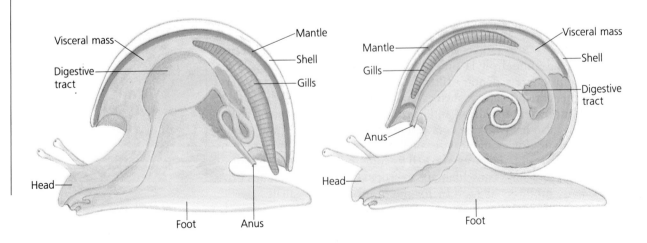

Polyplacophora

Members of the Class Polyplacophora are called chitons. These are primarily animals of the seashore that live on rocks. The word *polyplacophora* means "many plates," a reference to the main characteristic of the class: their shells are divided into eight separate, overlapping plates.

Gastropoda

The largest and most diverse class of mollusks is Gastropoda, which includes about 75,000 species of snails, slugs, abalones, nudibranchs, and conches. The name *gastropoda* means "stomach–foot." Most gastropods have a single shell, or valve, and so are called univalves. Slugs and some other gastropods have no shell.

The body plan of gastropods is based on that of ancestral mollusks. The major difference results from **torsion,** or twisting, which occurs during larval development. In torsion the visceral mass twists 180 degrees in relation to the head. Note the results of torsion by comparing the anatomy of a snail with that of the hypothetical ancestral mollusk in Figure 31-2. The twisting brings the mantle cavity to the front of the animal, thus allowing the animal to draw its head into the mantle cavity when endangered.

Snails

Snails are gastropods that live in a wide range of environments on land as well as in fresh water and oceans. Aquatic snails respire through gills in the mantle cavity. In terrestrial snails the mantle cavity acts as a modified lung that allows the animal to obtain oxygen from the air. The thin membrane that lines the cavity must be kept moist to allow gases to diffuse through it. For this reason snails are most active when the air has a higher moisture content. During dry periods a snail survives by becoming inactive. It retreats into its shell and seals the opening with a mucus plug, which keeps the snail from drying out.

Like all gastropods, snails have an **open circulatory system,** meaning that the blood does not circulate entirely within vessels. Instead blood is collected from the gills or lungs, pumped through the heart, and released directly into spaces in the tissues. From there it returns via the gills or lungs to the heart. The blood-filled space is known as a **hemocoel** (*HEE-muh-SEEL*) or blood cavity.

The main feeding adaptation of snails is the sawlike **radula,** a flexible tonguelike strip covered with chitinous teeth. Aquatic snails use the radula to scrape up algae or other food. Terrestrial snails use the radula to saw off leaves of garden plants.

Land snails are hermaphroditic, but in most aquatic species the sexes are distinct. Eggs are fertilized internally.

Snails move smoothly by the wavelike muscular contractions of the foot. Glands in the foot secrete a layer of mucus on which

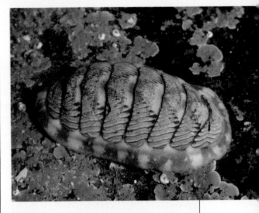

Figure 31-3 Clearly visible here are the eight separate, overlapping plates that make up the shell of this lined chiton, *Tonicella lineata.*

Info: When the tide moves out, some intertidal gastropods, such as limpets, return to their same places on rocks. There this type of gastropod uses its foot for suction and works its way into a groove it has worn down. It then pulls its shell tightly over the groove.

Figure 31-4 Inside a snail's mouth is the radula, a band of tissue covered with chitinous teeth that can scrape food from other surfaces. This SEM shows the sharp edges of these chitinous teeth.

Nudibranchs are gastropods, sea slugs that live in shallow coastal waters. They have no shell, and gas exchange takes place across the body surface. How might the structure of the nudibranch above be related to the function of respiration?

Answer to Think
Lack of a shell facilitates the diffusion of gases across the surface, and the plumed surface increases the area for such diffusion.

Tip: Distinguish between a snail's eyes, which are similar in structure to the eyespot of planaria, and the well-developed eye of a squid. If possible exhibit snails and allow students to examine their eyes with a hand lens.

the animal travels at a pace of about 3 m per hour. Two eyes at the end of delicate tentacles on the head help locate food. If danger should arise, the tentacles retract into the head.

Other Gastropods

Slugs look like snails without shells. They survive without shells because they live in moist environments, hiding in shady places by day and feeding at night. Like terrestrial snails, slugs respire through the lining of the mantle cavity.

Some gastropods show unusual modifications of the foot or radula. Instead of having a foot, pteropods, or "sea butterflies," have a winglike flap, which allows them to swim on the surface of the sea. Another adaptation is seen in gastropods with modified radulas. Oyster drills, for example, use their radulas to bore through oysters' shells and feed on the soft tissues inside.

Bivalvia

In contrast with gastropods that move about in search of food, most members of the Class Bivalvia are sessile and filter food from the water. *Bivalvia* means "two valves." Clams, oysters, scallops, and other bivalves have a shell with two valves and a muscular foot. Once this foot is in sand, blood swells the end, causing it to spread and form a hatchet-shaped anchor. The muscles of the foot then contract and pull the animal down in the sand. Bivalves do not have distinct head regions as do gastropods. As an adaptation for sessile existence, ganglia are present in the anterior region.

Each valve consists of three layers secreted by the mantle. A thin outer layer protects the shell against acidic conditions in the water. A thick middle layer of calcium carbonate crystals strengthens the shell. The smooth, irridescent inner layer protects the animal's soft body. If an irritant such as a grain of sand gets inside the shell, the mantle coats it with a secretion known as mother of pearl. Layers of this secretion form a pearl. A hinge connects the two valves of the shell. The animal can close the valves with its powerful adductor muscles. When the adductor muscles relax, the valves open.

Clams

Clams are bivalves that live buried in mud, seashore sand, or sand on the bottom of the sea. Clams have evolved adaptations for filter feeding. The mantle cavity is sealed except for a pair of hollow tubes called siphons. Cilia beating on the gills set up a current that causes water to enter the **incurrent siphon.** Cilia then propel the water over the gills and cause it to exit through the **excurrent siphon.** As the cilia move water over the gills, food—plankton from the water and organic sediments from the sea bottom—becomes trapped in a sticky mucus. Then the cilia move this mucus to the mouth. As water passes over the gills, oxygen diffuses into the blood, and carbon dioxide diffuses out.

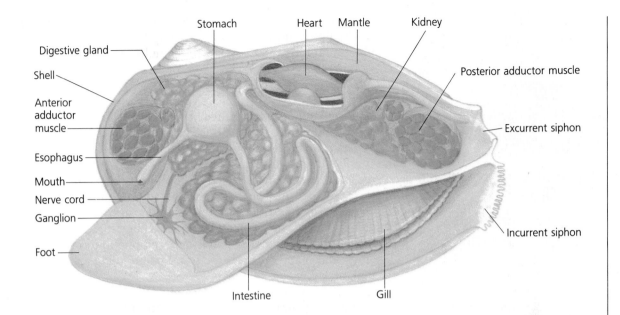

Stomach   Heart Mantle   Kidney

Digestive gland

Shell

Anterior adductor muscle

Esophagus

Mouth

Nerve cord

Ganglion

Foot

Posterior adductor muscle

Excurrent siphon

Incurrent siphon

Intestine   Gill

Clams have rudimentary sense organs. Sensory cells along the edge of the mantle respond to light and touch. Like most mollusks the ganglia that are above the mouth, in the digestive system, and in the foot are connected by two pairs of long nerve cords.

Most species of clams have separate sexes. They reproduce by shedding sperm and eggs into the water, and fertilization occurs externally. The fertilized egg becomes a trochophore larva that eventually settles to the bottom and develops into an adult.

Other Bivalves

Oysters are bivalves that become permanently attached to a hard surface early in their development. Scallops are bivalves that move through the water by opening their valves and then rapidly snapping them shut. This motion expels bursts of water, creating a form of jet propulsion. The teredo, or shipworm, is one of the few bivalves that does not filter feed. Instead, it bores into driftwood or ship timbers. Particles produced by the drilling are ingested by the teredo. The cellulose is broken down by symbiotic protozoa that live in the shipworm's intestine.

Cephalopoda

Cephalopods such as octopuses, squids, cuttlefish, and chambered nautiluses are evolutionarily advanced and complex mollusks. The name *cephalopod,* meaning "head–foot," refers to the large, well-developed head and the prominent foot divided into numerous tentacles. The head and foot are specialized for a free-swimming, predatory existence. Cephalopods capture prey with powerful tentacles equipped with strong suckers. They kill and eat their prey with the help of a radula and a sharp beak.

Figure 31-5 One valve has been stripped away here to show the anatomy of a clam. A clam is typically buried in the ocean bottom with its siphons pointing upward.

Figure 31-6 The cuttlefish, *Sepia officinalis,* has a pair of narrow fins, which border its body. It uses its eight arms and two tentacles to capture prey. The chambered nautilus, *Nautilus pompilius,* has a shell with roughly 36 chambers.

Answers to Section Review
1. a ciliated pear-shaped larval stage of mollusks and annelids
2. a coelom, all organ systems present, a three-part body (head, foot, visceral mass)
3. The mantle cavity must be moist for respiration to take place.
4. It has a well-developed brain and well-developed senses, a closed circulatory system, and tentacles that grasp prey; jet propulsion allows rapid movement.
5. Protection, water conservation, and camouflage are among the acceptable choices.

Cephalopods have a **closed circulatory system,** in which blood circulates entirely within a system of vessels. This efficient circulation rapidly transports the food and oxygen in these highly active animals.

Squids

Like all cephalopods, squids live in oceans. Most grow to about 30 cm in length, but a few species grow much longer. The giant squid may grow to 20 m long and weigh more than 3,360 kg. It is the world's largest known invertebrate.

A squid's head encloses a large, complex brain that controls a highly developed nervous system. The squid's most prominent sense organs are its pair of large eyes, similar in structure to those of humans. Squids have ten tentacles. The longest pair is used for capturing prey; the smaller pairs force the prey into the mouth. The muscular mantle propels the squid by pumping jets of water through an excurrent siphon.

Squids have defensive adaptations. They can squirt an inky substance at enemies. The substance temporarily blinds the enemy and dulls its sense of smell. Pigment cells called **chromatophores,** located on the outer layer of the mantle, enable squids to change color and blend in with their surroundings.

Like all cephalopods, squids have separate sexes. The male uses a specialized tentacle to transfer packets of sperm from its mantle cavity to the cavity of the female, where fertilization occurs. The female lays a mass of fertilized eggs encased in a gelatinous material. She then guards them until they hatch.

Other Cephalopods

Octopuses seldom grow to more than 30 cm in diameter. They have eight tentacles and resemble squids in the ways they move, feed, capture prey, and escape from enemies. Instead of using their jet propulsion to chase prey, they may crawl along the bottom with their tentacles or lie in wait in caves and rock crevices. The jet propulsion may be used to escape danger.

The chambered nautilus is the only existing cephalopod that has retained its exterior shell. The nautilus lives in an outer chamber of its shell and secretes a gas into the other chambers. By regulating the amount of gas in the chambers, the nautilus can adjust its buoyancy and so control its depth in the water.

Section Review

1. What is a *trochophore?*
2. Name three characteristics shared by all mollusks.
3. Why are terrestrial snails more active when the air around them is moist?
4. In what ways is a squid adapted for a predatory way of life?
5. Suggest three ways in which a shell is adaptive.

31.2 Annelida

Colorful feather-duster worms, common earthworms, and blood-sucking leeches are all classified in the Phylum Annelida *(uh-NELL-uh-duh)*. The name *annelid* means "little rings," which refers to the many segments that make up the body of an annelid. The diverse species of annelids share certain basic traits:

- All annelids have a true coelom.
- The body is divided into many segments. This phenomenon is called **metamerism**. Some segments fuse during development, while others remain separate segments, or metameric units.
- All organ systems are well developed in most groups.
- Most annelids have external bristles called **setae** *(SEE-tee)*.

The number of setae and the presence or absence of fleshy appendages called **parapodia** provide the basis for classifying annelids into three classes: Polychaeta *(PAHL-ee-KEET-uh)*, Oligochaeta *(AHL-uh-goe-KEET-uh)*, and Hirudinea *(HIR-yuh-DIN-ee-uh)*. Marine worms of Class Polychaeta have many setae; earthworms of Class Oligochaeta have no parapodia and few setae; and leeches of Class Hirudinea have no setae or parapodia.

Oligochaeta

We will now look closely at the earthworm's structure, feeding habits, and body systems. As you read, study the adaptations that enable the earthworm to lead a burrowing life.

Structure

The diagram in Figure 31-7 shows the segmentation that is the most distinctive feature of annelids. The more than 100 segments, or metameric units, of the earthworm's body are separated by partitions that divide the coelom. These segments are identical except where fusion has resulted in specialized sections near the

Section Objectives

- *List the distinguishing characteristics of annelids.*
- *Distinguish between the three classes of annelids.*
- *Identify some traits of earthworms that are adaptations for burrowing in the soil.*
- *Compare earthworms with polychaetes.*
- *Name some characteristics of leeches that are leading to their renewed use by physicians.*

Tip: Polychaetes are considered the most primitive of annelids. This text departs from its usual phylogenetic sequence to give greater emphasis to the more commonly studied oligochaetes.

Figure 31-7 Segmentation, as shown in this earthworm, is the most distinctive feature of annelids. The digestive tract is not segmented. Castings and dirt exit through the anus.

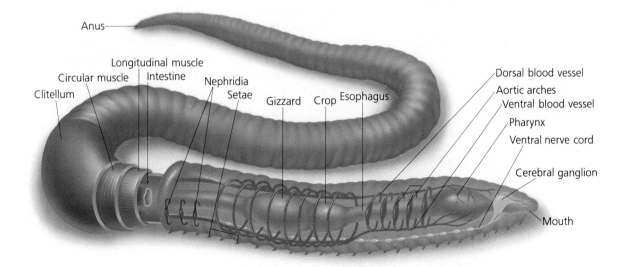

Anus — Longitudinal muscle — Circular muscle — Intestine — Clitellum — Nephridia — Setae — Gizzard — Crop — Esophagus — Dorsal blood vessel — Aortic arches — Ventral blood vessel — Pharynx — Ventral nerve cord — Cerebral ganglion — Mouth

Info: Earthworm aeration of the soil is so beneficial that their presence in hundreds of thousands per acre is a characteristic of a highly productive farm. The cultivation of these worms has grown as a business.

anterior and posterior ends. The anterior segments of the worm reflect the cephalization that is an adaptation for burrowing. The head contains the specialized sense organs.

Circular and longitudinal muscles line the interior body wall. To move, the worm anchors some segments by their setae and contracts the circular muscles in front of those segments, producing fluid pressure in the anterior coelom cavities. This pressure elongates the animal and pushes the head forward. With the anterior setae gripping the ground, the longitudinal muscles contract, pulling the posterior along. Divisions in the coelom make simultaneous contraction and expansion of segments possible.

Feeding and Digestion

Earthworms burrow and feed on soil and organic matter at the same time. They digest the organic matter and eliminate waste and undigested matter as dirt and feces called castings. The passing of dirt through the worm's body loosens and aerates the soil.

Soil is sucked into the gut by the muscular pharynx. The soil then passes through a tubelike **esophagus** to a temporary storage area called a **crop,** and from there to the **gizzard.** The thick, muscular gizzard walls contract and grind the soil, releasing and

---

Biotechnology

Medicinal Uses of Leeches

Advanced techniques in microsurgery have often allowed people who have lost a finger or a toe to have the extremity successfully reattached. These operations are so delicate that surgeons must work with tiny surgical instruments and powerful microscopes. To this sophisticated technology some microsurgeons have added an unlikely tool—the leech.

Leeches were used for medical bloodletting throughout the Middle Ages and even as late as the early 1900s. Doctors who used leeches believed that the bloodletting caused by the leeches could improve circulation and remove "bad blood" or "evil humours."

Although medical science has discredited the traditional reasons for using leeches, their use in microsurgery is just one of the new medically sound uses of leeches.

When microsurgery fails, it is often because the soft, tiny veins in the reattached parts become clotted. When that happens, the tissues die, and the part can't rejoin the body and heal. By using leeches the surgeon can sometimes improve the odds of success in difficult cases. After reattaching the part, the surgeon applies the leech. The leech starts breaking up organic matter. The gizzard is adapted for burrowing, as it grinds up soil. Food is digested and absorbed by the blood after it passes through the walls of the tubular intestines.

Circulation

Earthworms transport oxygen, nutrients, and wastes through the body via a closed circulatory system. The blood travels from the anterior to the posterior through a ventral blood vessel and then forward through a dorsal vessel. Five pairs of tubes, the **aortic arches,** link the major vessels near the anterior. Smaller vessels branch into each segment of the body. Contractions of the ventral vessel and aortic arches force blood through the body.

Respiration and Excretion

Earthworms have no gills or other respiratory organs. Oxygen and carbon dioxide diffuse directly across the skin. This exchange can take place only if the skin is moist. Thus earthworms avoid dry ground and extreme heat. Secretions of mucus and the presence of a thin cuticle also help keep the earthworm's skin moist.

Earthworms eliminate nitrogenous wastes through long tubules called **nephridia.** Coelomic fluid enters each nephridium

Tip: Discuss the phylogenetic relationship of nephridia to the kidneys of more highly developed animals. Refer to Chapter 45 for a discussion of the role of nephrons in human kidneys.

Crop
Salivary glands
Jaw

to suck out the congested blood, as shown in the diagram. A period of prolonged, localized bleeding follows. After a few days the blood in the reattached extremity is flowing normally, permitting healing to take place. Then the leech is removed.

Today some researchers are investigating potential medical applications of secretions pro-

duced by leeches. Scientists have known since the 1800s that the saliva of the European leech contains a powerful anticoagulant, a substance that inhibits blood clotting. A similar substance in the saliva of the Amazon leech dissolves clots after they have formed.

Aside from the obvious potential for treating blood clots, some researchers think that leech secretions may prove useful in the treatment of heart attack victims. The leech's anticoagulants could be used to stimulate the flow of blood in cardiac muscles that have been deprived of nourishment by a heart attack.

through a ciliated funnel opening. As fluid passes through the nephridia, some water is reabsorbed by blood vessels. The remaining fluid empties through pores on the ventral surface.

Nervous Control

Earthworms are sensitive to light, touch, moisture, chemicals, temperature, and vibrations. Light receptors concentrated in the head and tail sense direction. Most other sense organs and the nerves that control individual muscle contractions are present in each segment. The nerves in each segment form a pair of ganglia that coordinate movements with neighboring segments. A cerebral ganglion or brain in the head controls total body actions. A ventral nerve cord connects the brain with all the ganglia.

Reproduction

Earthworms are hermaphrodites, but an individual worm cannot fertilize its own eggs. Mating occurs when two earthworms join head to tail. Together they form a mucus coat around part of their intertwined bodies. Each injects sperm into the mucus. Sperm from one worm moves to the pouchlike **seminal receptacle** of the other. Simultaneously eggs in the body cavity move through oviducts to the female genital pore. After several days a mucus and chitin sheath is secreted by the **clitellum** *(klie-TELL-um)*, a swelling around the sex organs. As the worm wriggles to slip the sheath off its body, eggs and sperm are joined and fertilization occurs.

Polychaeta and Hirudinea

Polychaetes are annelids that live in virtually all marine habitats. Some are free-swimming predators that use their strong jaws to feed on small animals. Some burrowing species excavate tunnels by eating the sediment, while others dig and pump water through their bodies or scour the bottom with tentacles.

Polychaetes have numerous setae that help them move. The setae project from the parapodia. Some parapodia function in gas exchange. Polychaetes differ from other annelids by having antennae and specialized mouthparts.

Hirudinea is the smallest class of annelids, consisting of about 300 species of leeches. A leech has no setae. At each end of the body the leech has a sucker, which it uses to effect a crude walking motion. Most leeches live in calm bodies of fresh water.

Section Review

1. Distinguish between *setae* and *parapodia*.
2. In what ways are earthworms adapted for burrowing?
3. How do the annelid and mollusk body plans differ?
4. How do earthworms in the soil benefit plants?
5. Leech saliva contains an anaesthetic. What adaptive advantage do leeches derive from this anaesthetic?

Figure 31-8 Numerous setae, which project from parapodia, help polychaetes such as this move through water.

Answers to Section Review
1. Setae are bristles found on the body of most annelids; parapodia are fleshy appendages on some annelids.
(See pp. T31A–T31D.)

Behavior of Earthworms

Objective
To examine the behavior of an earthworm

Process Skills
Experimenting, inferring

Materials
Live earthworm, shallow pan, paper towel, cotton swabs, 10% aqueous ammonium hydroxide solution, blunt probe, book, black paper or piece of cardboard large enough to cover half of the pan, lamp

Method

1 Place a moist paper towel in a pan and place an earthworm on the paper towel.
2 Observe the behavior of the earthworm for a few minutes. How will these observations serve as a control for later observations?
3 Dip a cotton swab into water and hold it near the anterior end of the earthworm. Observe the response of the earthworm. Next hold the swab near the posterior end of the animal. Again observe any response.
4 Dip a cotton swab into the ammonium hydroxide solution. Hold the swab near the anterior end of the earthworm. Note any response. Next hold the swab near the posterior end of the animal. Note any response. If the earthworm responds, how can you tell if it is responding to the ammonium hydroxide and not to the cotton swab?
5 Use a blunt probe to gently touch the anterior end, then the posterior end, and then the midpoint of the earthworm. Note the animal's responses.
6 Position the earthworm with its anterior end near the middle of the pan. Prop up one end of the pan with a book so that the anterior end of the earthworm is facing uphill. Observe the earthworm for one minute, and note any response.

7 Reposition the earthworm as in step 6. This time prop the pan up with a book so that the animal's anterior end is facing downhill. What are you testing for in steps 6 and 7?
8 Position the earthworm so that its anterior end is near the middle of the pan. Place black paper or a piece of cardboard over the pan so that the posterior end of the animal is in shadow. Using a lamp, shine light on the earthworm's anterior end for five minutes. Then note the earthworm's response.
9 Reposition the earthworm as in step 8. This time cover the pan so that the anterior end of the earthworm is in shadow. Shine light on the earthworm's posterior end for five minutes. Note the earthworm's response.

Conclusions
1 Does the earthworm respond differently to stimuli applied to its anterior end than to stimuli applied to its posterior end?
2 What can you conclude about an earthworm's reaction to ammonium hydroxide, touch, gravity, and light?

Inquiry
Design an experiment to test the hypothesis that earthworms respond to various colors of light.

Chapter 31 Review

Vocabulary

aortic arch (473)	excurrent siphon (468)	mantle (466)	radula (467)
chromatophore (470)	foot (466)	mantle cavity (466)	seminal receptacle
clitellum (474)	gill (466)	metamerism (471)	(474)
closed circulatory	gizzard (472)	nephridium (473)	setae (471)
system (470)	head (466)	open circulatory	torsion (467)
crop (472)	hemocoel (467)	system (467)	trochophore (465)
esophagus (472)	incurrent siphon (468)	parapodia (471)	visceral mass (466)

Explain the difference between the terms in each of the following sets.
1. open circulatory system, closed circulatory system
2. gizzard, radula
3. chromatophore, trochophore
4. mantle, mantle cavity
5. gill, nephridia

Review

1. Mollusks and annelids share all of the following characteristics except (a) coelom (b) trochophore larva (c) segmentation (d) bilateral symmetry.
2. Gills are organs specialized for (a) gas exchange (b) movement (c) digestion (d) excretion.
3. Most species in the Class Bivalvia are (a) predators (b) parasites (c) land dwellers (d) filter feeders.
4. The only mollusks with a closed circulatory system are (a) gastropods (b) bivalves (c) cephalopods (d) snails.
5. Terrestrial snails and slugs require an environment with a high moisture content in order to carry out (a) reproduction (b) feeding (c) respiration (d) all of the above.
6. Annelids are placed into one of three classes, based on the number of their (a) setae (b) segments (c) nephridia (d) organ systems.
7. The movement of earthworms requires (a) fluid pressure in the coelom (b) muscle contractions (c) traction provided by setae (d) all of the above.
8. Earthworms respire by means of (a) gills (b) lungs (c) diffusion across the skin (d) all of the above.
9. Parapodia are a distinguishing characteristic of (a) the Class Polychaeta (b) the Class Oligochaeta (c) the Class Hirudinea (d) the Class Bivalvia.
10. Most species of leeches live (a) on land (b) in fresh water (c) in the oceans (d) inside other animals.
11. What are the functions of a true coelom in mollusks and annelids?
12. What are the three main parts of the body plan of mollusks?
13. What is torsion? What effect does it have on the location of the snail's mantle cavity?
14. What is the function of the radula in mollusks?
15. How is the structure of a gill related to its function?
16. Describe the feeding method of the clam.
17. What is the adaptive value of segmentation in annelids?
18. What use do microsurgeons make of leeches? What use might scientists make of leeches in the future?
19. Why do earthworms require a moist environment?
20. How does a leech's method of movement differ from an earthworm's?

Critical Thinking

1. Of what advantage might hermaphroditism be to the land snail?
2. Clams and other mollusks that live in water reproduce by shedding sperm and eggs into the water. How might this process affect the reproductive success of these mollusks? Would you expect the mollusks to shed many sperm and eggs or only a few?
3. In order for an oyster to make a pearl, a piece of sand or rock must get inside its shell. This causes the mantle to secrete a shiny smooth coating on it. Of what advantage might this process be to the oyster?
4. Anglers who use earthworms have observed that when an earthworm is cut in two the segments crawl in different directions. From what you know about the nervous system and muscular system of the earthworm, explain why a cut earthworm would be able to move in different directions.
5. In what way are the feeding habits of the earthworm and the clam similar? In what way do they differ?

6. Leeches suck and ingest the blood of fish and mammals, taking in two to ten times their own weight in blood. Why aren't leeches considered parasites?
7. Look at the photograph above. What is the adaptive advantage of the clam's long, extended siphons? Keep in mind the habitat of the clam.

Extension

1. As a hobby many people collect the shells of mollusks. You may be able to see some of these shells in your school or in the homes of friends or relatives. Using a book such as *The LaRousse Guide to Shells of the World,* by A. P. H. Oliver, identify five shells or pictures of shells. Draw each one, and under each drawing give the common name, the scientific name, the part of the world where it is found, and the size.
2. Read the article by David G. Gordan entitled "Devilfish of the Deep," in *Animal Kingdom,* July/August 1987, pages 39–43. Write a report in which you give details of the life cycle of the giant Pacific octopuses.
3. Research the giant squid. Make a poster in which you compare its size with that of some commonly known object or an average-sized squid. Include some interesting information for the other students such as, for example, proof that the squid is eaten by whales.
4. Annelids and mollusks show an incredible diversity of form. Write a report in which you research the structure and habitat of an annelid or mollusk species that was not covered in this chapter. How does the form of your chosen species illustrate an evolutionary adaptation to the environment in which it lives? Illustrate your report if possible.

Chapter 32 | Arthropods

This chapter discusses the characteristics of arthropods that have enabled them to adapt to most of the earth's environments. Students learn how the basic arthropod body plan evolved to perform a variety of specialized functions.

Correlation Chart

Section Number	Section Objectives	Pages	Features	Section Review
32.1	1. List three distinguishing characteristics of arthropods.	479–482	Biotechnology, p. 480	1, 2, 3
	2. Relate the structure of the exoskeleton to its function.			1, 2, 3, 5
	3. Name the four subphyla of Phylum Arthropoda, and explain the characteristics of each subphylum.			4
	4. Describe three products that researchers have developed from chitin.			
32.2	1. Give examples of crustaceans that are adapted to marine, freshwater, and land environments.	483–485		2, 3
	2. Describe the external anatomy of a crayfish, and explain the function of each major part.			1, 3, 5
	3. Summarize the processes of digestion, circulation, respiration, and excretion in crayfish.			3
	4. Describe the structure and function of the compound eye in crustaceans.			4
32.3	1. Name the three main groups of arachnids, and list their distinguishing structural differences.	486–489	Think, p. 487	1, 2, 5
	2. Explain the adaptations arachnids have for predatory life on land.		Biology Lab, p. 489	2, 3, 5
	3. List similarities and differences between millipedes and centipedes.			4

Resources*

Print

Shear, William A., ed. *Spiders: Webs, Behavior, and Evolution*. Stanford, Calif.: Stanford University Press, 1986.

Sibbald, Jean H. *Homes in the Sea: From the Shore to the Deep*. Minneapolis: Dillon, 1986.

Audiovisual/Software

Crayfish Dissection Guide. Apple II, 48K. Disk. Instruction manual. Cross.

The Joint-Legged Animal: Arthropods. Film, 19 min. Britannica.

*Additional references are listed in *Modern Biology*, *Teacher's Program Guide*.

Section Teaching Guides

32.1 Phylum Arthropoda

Approach

The main objective of this section is to focus students' attention on the characteristics that make arthropods unique and that have resulted in the evolutionary success of arthropods. This section builds on concepts of adaptation and evolution (Chapters 15 and 16) and compares arthropod structure and function to annelid adaptations (Chapter 31). This section gives students background knowledge for understanding the adaptations of insects (Chapter 33) and the evolutionary trends of invertebrates (Chapter 35). In presenting the text, guide students to realize that though all arthropods share specific characteristics, subgroups of arthropods illustrate variations in structure that indicate that arthropods might have evolved from four separate lineages.

Concept Framework

Students will be familiar with individual species in the Phylum Arthropoda, though they may not recognize that these species share common characteristics. Tap this knowledge to provide a conceptual frame for this section by showing pictures of a wide variety of arthropods. Assist students in identifying common characteristics. Encourage them to refer to these pictures as they read to visualize characteristics of members of the Phylum Arthropoda.

Emphasis

Make sure that students understand the adaptive advantages and disadvantages of the exoskeleton.

Using the Features/Protective Properties of Chitin

This feature discusses the development of various products made from chitin. Such products include varnish, a chemical that protects wheat seeds from fungal infection, a type of contact lens that protects an injured cornea and promotes healing, an artificial skin that likewise protects burned skin and promotes healing, and surgical thread that dissolves. Emphasize that because chitin does not cause an allergic reaction in most people, this material can remain in intimate contact with injured tissues, providing protection and promoting healing.

Arthropod Identification

Bring some nature magazines to class and have students find pictures of arthropods that belong to the four subphyla of arthropods. Encourage students to paste the pictures to index cards and to identify the organism and its subphylum on the back of the card. Collect and then redistribute the cards. Have students try to identify the subphylum to which the pictured arthropods belong.

Section Review Answers

(See p. 482.)

32.2 Crustacea

Approach

The main objective of this section is to describe the internal and external anatomy of the crayfish as a representative crustacean. This section builds on the concepts of structure and function (Chapter 2). In presenting this section, guide students to recognize how specific features of crayfish illustrate arthropod adaptations outlined in Section 32.1. Point out the distinguishing features of crustaceans including branched antennae and mandibles.

Concept Framework

Most students will be familiar with the crustaceans commonly known as "shellfish," including shrimp and lobsters. Use this knowledge as a starting point for introducing this section. You might explain that these "shellfish" are crustaceans and that their "shells" differ from the shells of mollusks such as clams and snails. Tell students to consider how the shells of crustacea are similar to and different from the shells of these other organisms as they read.

Emphasis

Students should be able to describe how each of the following aspects of crayfish morphology illustrates adaptation: external structure, digestive system, excretory system, circulatory system, respiratory system, nervous system, reproduction, and development.

Field Trip

Arrange for students to visit an area such as the seashore, a pond, or even a forest to observe different kinds of crustacea. You may wish to divide students into groups and assign each group a particular type of crustacean to look for. Encourage students to draw pictures of the crustaceans they observe. Direct them to take notes on each crustacean's habitat and behavior. Use a field guide to help students identify various kinds of crustaceans.

Section Review Answers

(Continued from p. 485.)
4. The functions of the compound eyes include detecting motion and sensing light.
5. They grow much faster during the first year of life as indicated by the fact that they molt seven times in this time period.

32.3 Other Arthropods

Approach

The main objective of this section is to describe the morphology of arachnids, which are representatives of the arthropod subphylum Chelicerata, and to describe centipedes and millipedes, which are representatives of the arthropod subphylum Uniramia. This section builds on the concepts presented in Sections 32.1 and 32.2. This section also gives examples of the relationship between structure and function discussed in Chapter 2. In presenting the text, guide students to identify the distinguishing features of arachnids and myriapods.

Concept Framework

Students have probably seen various kinds of arachnids and myriapods. Tap this knowledge to build a conceptual frame for this section by discussing the derivation of these words. Point out that *arachnid* comes from a Greek word meaning "spider," and that *myriapod* means "many feet." Ask students to think of examples of various kinds of arachnids and myriapods. Encourage them to think about the characteristics arachnids and myriapods share with other arthropods as they read this section.

Emphasis

Students should be able to list the characteristics of arachnids, centipedes, and millipedes. Make sure that students understand why arachnids are classified as members of the subphylum Chelicerata and why myriapods are classified as members of the subphylum Uniramia.

Making Models

Ask students to make a model of an arachnid, a centipede, or a millipede. Provide construction paper, pipe cleaners, modeling clay, and cardboard tubes that students can use for their models. Students might display their models on or against a painted backdrop to illustrate the type of environment their arthropod inhabits.

Section Review Answers

(Continued from p. 488.)
4. Centipedes are carnivorous, have flattened bodies, and have one pair of legs per segment except for the first and last two segments; millipedes are herbivorous, have round bodies, and have two pairs of legs per segment except for the last two segments.
5. Spiders feed on insects that can damage crops that humans rely on for food.

Laboratory Preparation

Background

In this lab students observe the behavior of pill bugs under different environmental conditions. The students record their observations of the pill bug's behavior under light, dark, wet, and dry conditions. Students relate their observations of pill bug behavior to the insect's natural habitat and speculate about why the pill bug's response to light differs from the response of most insects.

Preparation

Pill bugs are available from a biological supply house. Sow bugs can be substituted. When conducting the investigation, dim the classroom lights as much as practical to prevent excessive light from affecting pill bug behavior.

Safety

Remind students to handle the insects in a humane manner.

Conclusions

1. Pill bugs would tend to move away from light.
2. Pill bugs would tend to move toward moisture.
3. Answers will vary, since the choice presents a

conflict for the pill bugs. Most bugs will move toward light in order to reach moisture, but some bugs will not.

Inquiry

Students will devise various experiments to test response to such stimuli as noise and temperature. They can also investigate how light vs. dark and dry vs. moist conditions affect the pill bug's overall level of activity.

Chapter Review

Vocabulary

1. Chelicerae are fanglike appendages that inject venom; chelipeds are clawlike appendages.
2. Both have sensory receptors, but antennae are much longer and posterior to the antennules.
3. Spinnerets produce silk in spiders; swimmerets are crustacean appendages that create water currents and transfer sperm in reproduction.
4. The abdomen is the posterior section in crustaceans and arachnids; the cephalothorax is the anterior section.
5. The telson forms the tail of a crayfish; uropods are appendages that form fins on either side of the telson.

Review

11. presence or absence of antennae, chelicerae, and mandibles, as well as whether appendages are branched or unbranched
12. Muscles attached to the inside of the exoskeleton alternately contract and relax to move the joints of the appendages.
13. reduction in the number of appendages and in the number of segments
14. mandibles, maxillae, maxillipeds, and chelipeds
15. Crayfish respire through gills; spiders respire by means of tracheae or book lungs.
16. Swimmerets protect and aerate eggs during their development.
17. Spiders have chelicerae and venom to kill prey; they also use silk to build webs and entrap prey.
18. The tail of the spider abdomen contains spinnerets, which are made of hundreds of tubes. Fluid from silk glands passes through these tubes, hardening into a silk strand that can be spun.

19. Ticks and mites transmit diseases by transmitting microorganisms when they bite victims.
20. curling up into a coil and giving off noxious chemicals as defense mechanisms

Critical Thinking

1. The largest populations of algae live in lighted areas. The eyespot of the *Daphnia* senses where the light is strongest in the water and therefore where the most algae are likely to be.
2. Barnacles often attach themselves to moving objects, such as ships or whales, or they live where the water—and food—is in constant motion, such as on rocks at the tide line. They can protect themselves by drawing deep into their hard exoskeleton.
3. The fusion eliminates the vulnerable and weak joint in the place where the head and the thoracic plates join. However, the fusion also eliminates the possibility of movement at this joint.
4. This is an example of convergent evolution. Students should suggest some adaptive advantage to its looking like an ant.
5. Since the lobster's compound eyes are not able to clearly focus on images, its senses of smell and taste must compensate for its poor vision to allow the lobster to find its mates and its food.

Chapter 32 | Arthropods

Introduction

In previous chapters you have learned about many ways in which groups of invertebrates have adapted to life on land as well as in water. In this chapter you will learn about the characteristics of arthropods that have enabled them to adapt to almost all environments on earth. In the next chapter you will study the insects, the largest group of arthropods.

Chapter Outline

Chapter Concept

As you read, note the ways in which the basic arthropod body plan has become evolutionarily modified for life in specific environments.

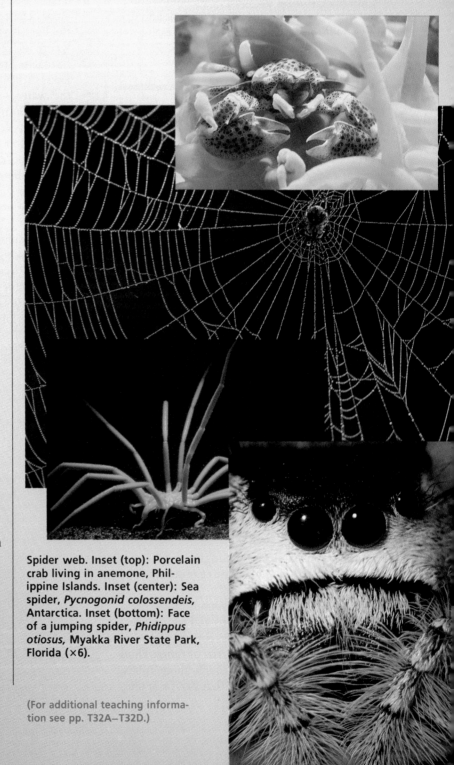

Spider web. Inset (top): Porcelain crab living in anemone, Philippine Islands. Inset (center): Sea spider, *Pycnogonid colossendeis*, Antarctica. Inset (bottom): Face of a jumping spider, *Phidippus otiosus*, Myakka River State Park, Florida (×6).

(For additional teaching information see pp. T32A–T32D.)

32.1 Phylum Arthropoda

Three-fourths of all animal species belong to the Phylum Arthropoda *(ahr-THRAHP-uh-duh)*. This diverse phylum includes insects and spiders, lobsters and crabs, as well as millipedes and centipedes. Because of the diversity of insects and their importance to people, we will examine them in more detail in Chapter 33. In this chapter we will study primarily the other groups of arthropods.

Characteristics

All members of the Phylum Arthropoda share the following characteristics:

- Arthropods have jointed appendages. **Appendages** are extensions of the body and include legs and antennae.
- The arthropod body is segmented. A pair of appendages is attached to each segment. In some species the appendages have been lost or reduced in size during the course of evolution.
- Arthropods have an exoskeleton. An **exoskeleton** is a hard external covering that provides protection and support.
- Arthropods have a ventral nervous system, an open circulatory system, a digestive system, and specialized sensory receptors.

The name *arthropod,* meaning "jointed foot," refers to the jointed appendages. Jointed appendages and an exoskeleton are the most distinctive arthropod traits. The exoskeleton provides much more structural support than the annelid cuticle and gives the internal organs better protection.

The chemical composition and the three-layered structure of the exoskeleton give it versatility and strength. The exoskeleton is composed primarily of protein and a tough carbohydrate called **chitin** *(KITE-'n)*. The waxy outer layer repels water and helps prevent desiccation in terrestrial species. The hard middle layer, which is strengthened by materials such as calcium, provides the primary protection. The inner layer, which is flexible at the joints, allows the animal to move freely. All three layers are secreted by an epidermis that lies just beneath them.

Movement and Growth

Figure 32-1 shows how muscles that are attached to the exoskeleton provide the structural basis for the wide range of movements that occurs in arthropods. Unlike the muscles of annelids, which form continuous sheets in the body wall, the muscles of arthropods occur in bundles that attach to the inside of the exoskeleton on either side of the joints. By alternately contracting and relaxing these muscles, the arthropod in essence operates a system of levers that move the body parts and appendages.

Section Objectives

- *List three distinguishing characteristics of arthropods.*
- *Relate the structure of the exoskeleton to its function.*
- *Name the four subphyla of Phylum Arthropoda, and explain the characteristics of each subphylum.*
- *Describe three products that researchers have developed from chitin.*

Tip: Have students compare the arthropod exoskeleton to the mammalian skeleton. Evaluate the versatility and efficiency of both skeleton types.

Exoskeleton

Rigid layer

Flexible layer
Flexor muscle

Extensor muscle

Figure 32-1 Muscles inside the exoskeleton enable arthropods to move their jointed legs.

Because the exoskeleton cannot enlarge as the body enlarges, the exoskeleton must be shed periodically and a new one must be formed. This process is called **molting**. Molting is also called ecdysis. The tissues of the arthropod grow until they put a great deal of pressure on the exoskeleton walls. A hormone is then produced that induces molting.

When an arthropod starts to molt, the cells of the epidermis secrete enzymes that digest the inner, flexible layer of the exoskeleton. Simultaneously the epidermis begins to synthesize a new exoskeleton, using much of the digested material. During this process the outer layer of the old exoskeleton loosens and breaks along specific lines. The old exoskeleton is then shed. The new exoskeleton, which is flexible at first, stretches to fit the now enlarged insect.

An arthropod molts many times during the course of its lifetime. Each time it molts, it becomes larger. However, during molting it is extremely vulnerable to predators because it temporarily lacks a hard shell. Terrestrial arthropods are also susceptible to desiccation during molting. For these reasons, arthropods usually go into hiding when they molt until their new exoskeleton has hardened.

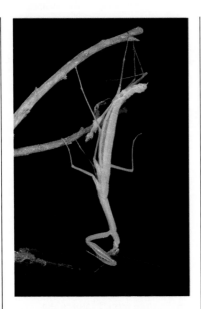

Figure 32-2 A praying mantis sheds its exoskeleton during the molting process.

Biotechnology

Protective Properties of Chitin

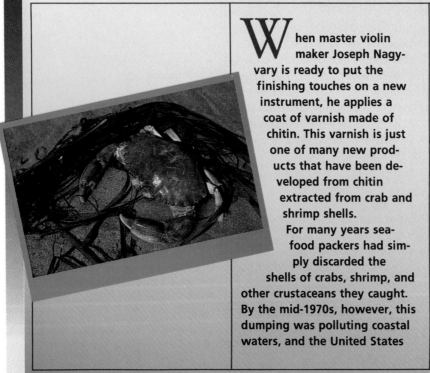

When master violin maker Joseph Nagyvary is ready to put the finishing touches on a new instrument, he applies a coat of varnish made of chitin. This varnish is just one of many new products that have been developed from chitin extracted from crab and shrimp shells.

For many years seafood packers had simply discarded the shells of crabs, shrimp, and other crustaceans they caught. By the mid-1970s, however, this dumping was polluting coastal waters, and the United States Environmental Protection Agency ordered packers to stop. The federal government then funded research to find practical uses for the shells. That research led to Professor Nagyvary's varnish and a number of other products.

Several of these products use chitin in ways that mimic its function in the exoskeleton—that is, as a protective layer. In 1986 one company began marketing a seed coating made of chitin that protects wheat seeds against fungal infections. Other researchers have developed a contact lens that could protect injured corneas while they heal.

Evolution and Classification

Animals with arthropod characteristics evolved more than 600 million years ago. Because all arthropods possess both an exoskeleton and jointed appendages, biologists have long inferred that they all evolved from a common ancestor. However, the most current studies of arthropod structure suggest that there may actually be four separate lines of arthropod evolution. On the basis of these recent studies, arthropods are now classified by scientists into four subphyla:

- Trilobita *(TRIE-luh-BITE-uh)* includes extinct organisms called trilobites.
- Crustacea *(KRUSS-TAY-shuh)* includes shrimp, lobsters, crabs, barnacles, cladocerans, ostracods, crayfish, water fleas, and copepods.
- Chelicerata *(kuh-LISS-uh-RAHT-uh)* includes spiders, scorpions, ticks, mites, sea spiders, and horseshoe crabs.
- Uniramia *(YOO-nuh-RAY-mee-uh)* is the only group that seems to have evolved on land and includes centipedes, millipedes, and all of the insects.

Info: In the course of evolution the bodies of arthropods came to have fewer segments. These segments have become fixed in number and more specialized in function.

In addition to its protective properties, chitin usually does not provoke allergic reactions. Both features have led to other medical uses of chitin. Japanese researchers have developed an artificial skin made from chitin that the body usually does not reject the way it rejects other types of artificial skin. A similar product is being tested in the United States. Physicians can use this artificial skin to protect severely burned areas of the body until new skin can form. Surgical sutures made from chitin do not have to be removed but simply dissolve.

Some filters contain chitin. Chitin molecules carry a positive electrical charge and will bind to negatively charged organic molecules. Waste water can be purified by flowing through beds of chitin-containing materials that remove organic wastes.

Research on other uses of chitin includes a chitin coating that may improve the strength and durability of paper and clothing. Some researchers are testing an ointment that may promote healing.

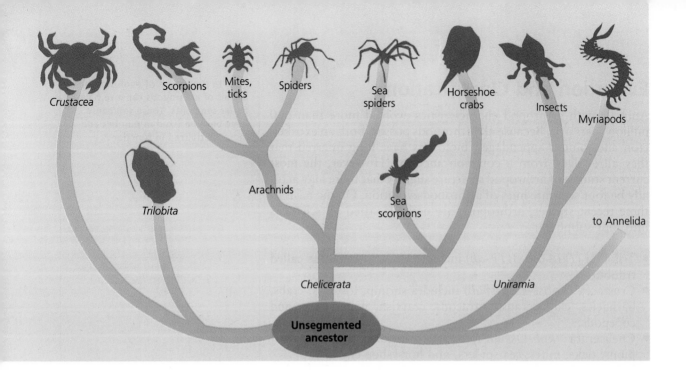

Crustacea

Scorpions

Mites, ticks

Spiders

Sea spiders

Horseshoe crabs

Insects

Myriapods

Trilobita

Arachnids

Sea scorpions

to Annelida

Chelicerata

Uniramia

Unsegmented ancestor

Figure 32-3 The four subphyla of arthropods are shown in this phylogenetic tree. Which species is extinct?

Answers to Section Review
1. a hard external covering that protects and supports the organism
2. Answers may include: jointed appendages, an exoskeleton, a segmented body, a ventral nervous system, an open circulatory system, a digestive system, and specialized sensory receptors.
3. Muscles attached to the inside of the exoskeleton on either side of the joints alternately contract and relax, thus operating the appendages like a system of levers.
4. Crustaceans have branched appendages; uniramians have unbranched appendages.
5. Swelling allows a larger exoskeleton to form with extra room for growth.

The four separate lines of arthropod evolution are illustrated in Figure 32-3. Members of the four subphyla are distinguished primarily by differences in their embryological development and differences in the morphology of structures such as appendages and mouthparts. Members of Chelicerata are distinguished from other arthropods by the absence of antennae and the presence of pincerlike mouthparts called **chelicerae.** Crustaceans are distinguished by the presence of branched antennae and chewing mouthparts called **mandibles.** The members of Uniramia are also distinguished by having antennae and mandibles, but their appendages are unbranched. *Uniramia* means "one branch."

Despite their differences, the three subphyla of living arthropods have evolved similarly. For example, ancestral arthropods had one pair of appendages per segment, but most living species have fewer appendages than this. In addition, the evolution of these groups shows a general tendency toward less segmentation of the body. For example, ancestral arthropods had many segments, but most modern adults have some segments fused together into larger structures with specialized functions.

Section Review

1. What is an *exoskeleton?*
2. List three characteristics that are shared by all arthropods.
3. How do the exoskeleton and the muscles function in arthropod movement?
4. What are the major structural differences between crustaceans and uniramians?
5. During molting an aquatic arthropod absorbs water and swells. What is the adaptive advantage of this behavior?

32.2 Crustacea

Members of the Subphylum Crustacea have hard exoskeletons that contain calcium carbonate. The approximately 25,000 species of crustaceans include crayfish, lobsters, pill bugs, sow bugs, water fleas, and barnacles. Crustacea is the only subphylum of modern arthropods that contains mostly aquatic members.

Diversity

Most crustaceans are small. Copepods no larger than a comma inhabit the surface waters of oceans, lakes, and streams. Barnacles are sessile crustaceans that attach themselves to rocks and pilings as well as to whales and sea turtles. Barnacles filter plankton from the water with 12 appendages called cirri.

Some crustaceans, such as sow bugs and pill bugs, are terrestrial. Because these animals have seven identical pairs of legs, they are called isopods, which means "same feet." Isopods usually live in damp areas, where their gills can stay moist.

The Crayfish

Crayfish are often studied as representative crustaceans because they are large and abundant. Crayfish are similar to lobsters in their internal and external structures.

External Structure

The body of the crayfish is divided into two sections, the cephalothorax and the abdomen. The **cephalothorax** consists of the fused head and the thorax; it has 13 segments and is covered by a hard carapace. The **abdomen** is divided into seven segments. The seventh segment, called the **telson,** forms a flat triangular section at the tail of the animal. Powerful abdominal muscles can jerk this tail and propel the animal rapidly backward.

Section Objectives

- *Give examples of crustaceans that are adapted to marine, freshwater, and land environments.*

- *Describe the external anatomy of a crayfish, and explain the function of each major part.*

- *Summarize the processes of digestion, circulation, respiration, and excretion in crayfish.*

- *Describe the structure and function of the compound eye in crustaceans.*

Figure 32-4 The water flea, *Daphnia* **sp. (left), the krill (center), and the sow bug (right) are members of the Subphylum Crustacea.**

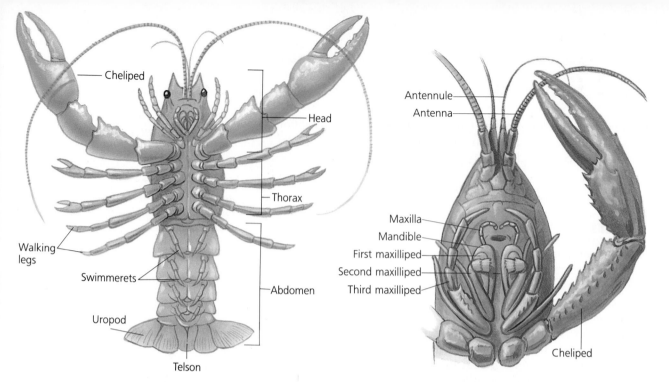

Cheliped
Head
Thorax
Abdomen
Walking legs
Swimmerets
Uropod
Telson

Antennule
Antenna
Maxilla
Mandible
First maxilliped
Second maxilliped
Third maxilliped
Cheliped

Figure 32-5 A top view (left) shows the major sections of the crayfish. Parts of its head are illustrated at the right.

Table 32-1 Crayfish Appendages

Appendages	Function
Antennules	Appendages with receptors for touch, taste, and equilibrium
Antennae	Specialized for touch and taste
Mandibles	Move up and down to crush food
Maxillae	Move side to side to tear food
Maxillipeds	"Jaws" to hold food
Chelipeds	Claws that capture food and serve as defensive weapons
Walking legs	Enable the crayfish to walk slowly
Swimmerets	Create water currents; function in reproduction
Uropods	Propel crayfish through water

A pair of appendages is attached to each segment of the crayfish. In Figure 32-5 and Table 32-1 note the location and function of the following appendages: **antennules, antennae,** mandibles, **maxillae, maxillipeds, chelipeds,** walking legs, **swimmerets,** and **uropods.** Each appendage develops from the same part of a segment, but is modified to perform a specialized function.

Digestive and Excretory Systems

Crayfish trap food with their chelipeds, tear it with the maxillae and maxillipeds, and chew it with the mandibles. The food then passes through the esophagus to the stomach, where chitinous teeth grind it into a fine paste that is mixed with digestive juices. Digestive glands absorb the mixture, and undigested particles pass through the intestines and out the anus. Excretory organs, called **green glands,** remove wastes from the blood and retain salts, which are scarce in fresh water.

Circulatory and Respiratory Systems

The crayfish has an open circulatory system. Blood flows from a dorsal sinus through small, one-way valves called ostia into the dorsal heart. The heart then pumps the blood into seven large vessels that carry it through the body. Blood leaves the vessels and fills the body cavity, where it bathes the organs and cells. Blood collects in a large ventral sinus. Other vessels then carry the blood through the gills, where it gives off carbon dioxide and takes up oxygen gas. It then returns to the dorsal sinus.

Gills are attached to each walking leg, protected in a chamber under the carapace. As the crayfish walks, water moves across the gills. Also, as the second maxillae move during feeding, the two gill bailers attached to them "bail" water over the gills.

Info: If a crayfish loses an appendage in a fight or accident, it is replaced during the next molt.

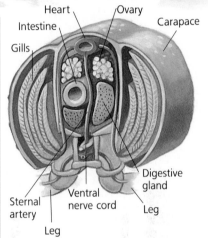

Nervous System

The crayfish nervous system includes a brain and a ventral nerve cord that runs from the brain to the tail. Nerve impulses travel to and from the nerve cord through ganglia. Nerves connect the brain with sense receptors in the antennules, antennae, and eyes.

The eyes are set on two short, movable stalks. Each eye has more than 2,000 light-sensitive lenses. Eyes with many lenses, or **compound eyes,** are highly sensitive to light and detect motion well, even though they can form only very crude images.

A crustacean senses position through the use of statocysts. Statocysts are cells that contain particles of calcium carbonate, which move when the crustacean's position changes. This movement is monitored by the nerves and interpreted by the brain.

Reproduction and Development

Crayfish usually mate in the fall. The male uses its first and second pairs of swimmerets to transfer sperm to the seminal receptacle of the female. The sperm remains there until spring, when it fertilizes the eggs as the female lays them. A sticky secretion attaches the eggs to the last three pairs of the female's swimmerets. The eggs hatch after about six weeks, having gone through several larval stages. The young look like tiny adults. They molt repeatedly—an average of seven times during the first year, then twice a year for the remaining two or three years of their lives.

Figure 32-6 The major organs within a crayfish are shown in side view (top) and in cross section (bottom).

Section Review

1. What is a *cephalothorax?*
2. What are the habitats of some crustaceans?
3. What structural adaptations of crayfish promote effective respiration in water?
4. What are the functions of compound eyes?
5. Using what you know about molting, in what year of their lives would you say crayfish grow fastest? Explain why.

Tip: If possible, bring in a live crayfish. Have the students observe how it moves, locates food, and responds to stimuli.

Answers to Section Review
1. the fused head and thorax in a crustacean
2. Answers will vary. Lobsters, crabs, and shrimp live in salt water; *Daphnia* and copepods live in fresh water; isopods live on land.
3. Gills are attached to each walking leg. When the crayfish moves or eats, water is driven over its gills.
(See pp. T32A–T32D.)

Section Objectives

- *Name the three main groups of arachnids, and list their distinguishing structural differences.*

- *Explain the adaptations arachnids have for predatory life on land.*

- *List similarities and differences between millipedes and centipedes.*

32.3 Other Arthropods

Unlike crustaceans, nearly all members of Chelicerata and Uniramia are terrestrial. In this section we will examine the major class of chelicerates, Arachnida, as well as two classes of the Subphylum Uniramia, Diplopoda and Chilopoda.

Arachnida

The Class Arachnida *(uh-RACK-nuh-duh)* is a group of more than 100,000 species, including spiders, scorpions, and mites. Most arachnids are adapted to kill prey with poison glands, stingers, or fangs.

Like crustaceans, arachnids have a body that is divided into a cephalothorax and an abdomen. Attached to the cephalothorax are four pairs of legs, a pair of chelicerae, and a pair of appendages called pedipalps. The **pedipalps** aid in chewing; in some species pedipalps are specialized to perform other functions.

Diversity

Spiders range in length from less than 0.5 mm to 9 cm in some tropical tarantula species. Spiders feed mainly on insects. Various species are adapted to capture prey in different ways. Some chase after prey, some catch prey in "trapdoors" in the ground, and some snare prey in elaborate webs.

Most scorpions live in tropical or semitropical areas; others live in dry temperate or desert regions. Scorpions differ from spiders in two ways. Scorpions have greatly enlarged pedipalps, which they hold in a forward position. They also have a large stinger over the head. Most scorpions hide during the day and hunt at night. Scorpions seize their prey with their pincerlike pedipalps. Then the stinger injects paralyzing venom, the chelicerae tear the prey, the animal is ingested, and digestion begins. Only a few species have a sting that may be fatal to humans.

Figure 32-7 The Kenya giant scorpion, *Pandinus* sp. (left, ×0.25), and the Australian black widow spider, *Latrodectus seville* (center, ×4), are deadly arachnids. The red spider mite (right, ×43) is a parasite.

Unlike other arachnids, mites and ticks have a fused cephalothorax and abdomen. Mites and ticks are the most abundant and most specialized arachnids. Ticks range in size from a few millimeters to 3 cm; most mites are less than a millimeter long.

Some mites and ticks are pests; others transmit diseases. Spider mites damage fruit trees when they suck fluid from the leaves. Many parasitic ticks pierce their host's skin to feed on blood. In the process they can transmit organisms such as those that cause Rocky Mountain spotted fever.

Structure and Function

Arachnids share many characteristics with crustaceans. For example, the nervous, digestive, and circulatory systems of both groups are structurally similar. However, arachnids are terrestrial and therefore have a unique respiratory system. Spiders respire through openings in the cuticle called **spiracles** (*SPIR-uh-kulz*). Air passes through the spiracles to the book lungs, the tracheae, or both. **Book lungs,** paired sacs in the abdomen with pagelike components, provide a large surface area for the exchange of gases. **Tracheae** (*TRAY-kee-ee*) carry air directly to the tissues.

The excretory system is also modified for life on land. The main excretory organs, called **Malpighian** (*MAL-PIG-ee-un*) **tubules,** are hollow projections of the digestive tract that collect body fluids, remove wastes, and carry wastes to the intestine. Most of the water is then reabsorbed, and the solid wastes leave the body. Some spiders also have **coxal glands,** organs that remove wastes and discharge them through an opening at the base of the leg.

Spiders have eight simple eyes rather than compound eyes. Many spiders also spin webs. The posterior tip of their abdomen contains three pairs of **spinnerets,** each made up of hundreds of microscopic tubes. Fluid from silk glands passes through the tubes and hardens into a thread that can be spun into webs. Silk is also used to build nests and egg cocoons. The young of some species use a long thread as a balloon to ride the wind to new habitats.

Think

These arachnids, called pseudoscorpions, use their pedipalps to attach themselves to beetles. Why might they do this? What trait seems to distinguish pseudoscorpions from scorpions?

Figure 32-8 Note the structures of the respiratory system that make it possible for a spider to live on land.

Internal anatomy of the spider

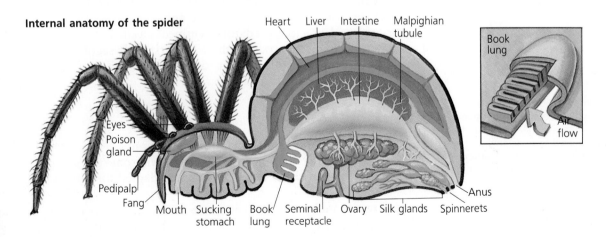

Heart Liver Intestine Malpighian tubule

Book lung

Air flow

Eyes
Poison gland
Pedipalp
Fang Mouth Sucking stomach Book lung Seminal receptacle Ovary Silk glands Spinnerets Anus

Figure 32-9 The millipede (top) of Southern Ontario is from the Class Diplopoda, while the giant centipede, *Scolopendra heros* (bottom), found in Texas, is from the Class Chilopoda.

Answers to Section Review
1. Pedipalps are a pair of appendages that aid in chewing and that can be used to seize prey.
2. Scorpions have much larger pedipalps than spiders and a pointed stinger that curls over the back.
3. It removes waste in concentrated form and so conserves water.
(See pp. T32A—T32D.)

In reproduction, a male spider gathers sperm in special sacs in the tips of his pedipalps. He places the sperm in the seminal receptacle of the female. Later the female lays eggs, which are fertilized by the stored sperm as they pass through the genital pore. The female then seals the eggs in a case of silk. The young spiders go through their first molt inside the case.

Myriapods

Centipedes and millipedes are collectively called **myriapods,** which means "many feet." Myriapods, along with insects, are members of the Subphylum Uniramia. All centipedes and millipedes are terrestrial. Centipedes and millipedes do not have a waxy cuticle. They retain moisture through behavioral adaptations, such as living in damp environments to prevent desiccation.

Millipedes

Millipedes are members of the Class Diplopoda. The term *millipede* means "thousand feet." Millipedes have two pairs of legs on each body segment except the last two. Although millipedes have many legs, they have far fewer than a thousand. The legs of a millipede are strong and well adapted for burrowing through humus and soil. However, these legs are also short, which makes them slow. When threatened, a millipede rolls its body into a coil and may also spray a noxious chemical that contains cyanide.

In cross section the bodies of millipedes are rounded. They live in soil, in logs, and under objects. Many have a strong sense of smell but poor vision. Millipedes are herbivores adapted for chewing plants and decayed matter in the soil.

Centipedes

Centipedes are members of the Class Chilopoda. The term *centipede* means "hundred legs." Centipedes differ from millipedes by having the body flattened in cross section. They also have only one pair of legs per segment, except the first one and the last two. They can also move faster because their legs are longer. Many coil up for defense. Centipedes may have anywhere from 15 to 175 pairs of legs. Voracious predators, centipedes feed on earthworms and on insects such as cockroaches. The first body segment has a pair of clawlike appendages that can inject venom into prey.

Section Review

1. Define *pedipalps*, and explain their function.
2. Name two ways in which scorpions differ from spiders.
3. How is the arachnid excretory system adapted for life on land?
4. How are millipedes and centipedes different?
5. How might the feeding habits of spiders benefit humans?

Laboratory

Behavior of Isopods

Objective
To relate the behavior of isopods under laboratory conditions to their natural behavior

Process Skills
Experimenting, predicting, inferring

Materials
2 sheets of filter paper cut to fit a petri dish, petri dish with cover, aluminum foil, 5 live pill bugs, bright lamp or flashlight, unlined paper, paper towels

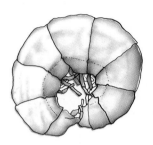

Method
1 Place a piece of filter paper in the bottom half of a petri dish.
2 Wrap half of the petri dish with aluminum foil so that half of the filter paper is shaded in darkness.
3 Place five pill bugs, which are isopods, near the light-dark boundary of the filter paper. Shine a lamp on the exposed portion of the petri dish. Can you predict the response of the pill bugs?
4 Observe the response of the pill bugs for between three and five minutes. Record your observations.
5 Place a second piece of filter paper on a paper towel. Moisten another paper towel thoroughly with water and use the second towel to dampen half of the filter paper. Do not get the filter paper so wet that the water is absorbed by the dry half. Place the filter paper in the top half of the petri dish.
6 Place the pill bugs along the boundary between the wet and dry areas. Can you predict the response of the pill bugs?
7 Observe the response of the pill bugs for between three and five minutes. Record your observations.
8 Slide the bottom half of the petri dish out from its foil cover. Remove the filter paper,

dampen half of it with water as directed in step 5, and then return the filter paper to the petri dish.
9 Place the pill bugs on the dry half of the filter paper. Return the petri dish to its foil cover so that the dry portion of the filter paper is in darkness.

10 Again shine a lamp on the exposed area of the petri dish. Can you predict the response of the pill bugs?
11 Observe the response of the pill bugs for between three and five minutes. Record your observations.

Conclusions
1 Based on your experiment, how do you think pill bugs would respond to light under natural conditions?
2 How would they respond to moisture?
3 How do you think pill bugs would behave under natural conditions when faced with the choice of moving from a dark, dry place toward moisture and light?

Inquiry
Devise an experiment that will test other pill bug responses to stimuli.

Vocabulary

abdomen (483)
antenna (484)
antennule (484)
appendage (479)
book lung (487)
cephalothorax (483)
chelicera (482)

cheliped (484)
chitin (479)
compound eye (485)
coxal gland (487)
exoskeleton (479)
green gland (484)

Malpighian tubule (487)
mandible (482)
maxilla (484)
maxilliped (484)
molting (480)
myriapod (488)

pedipalp (486)
spinneret (487)
spiracle (487)
swimmeret (484)
telson (483)
trachea (487)
uropod (484)

Explain the difference between each of the following pairs of terms.
1. chelicera, cheliped
2. antenna, antennule
3. spinneret, swimmeret
4. abdomen, cephalothorax
5. telson, uropod

Review

1. All arthropods have (a) a cephalothorax (b) spiracles (c) jointed appendages (d) antennae.
2. The exoskeleton of an arthropod (a) provides protection and support (b) plays a role in movement (c) contains chitin (d) all of the above.
3. The subphylum of modern arthropods that is primarily aquatic is (a) Trilobita (b) Chelicerata (c) Crustacea (d) Uniramia.
4. The major respiratory organs of crayfish are the (a) gills (b) lungs (c) tracheae (d) book lungs.
5. Compound eyes (a) form distinct images (b) detect motion well (c) are found in all arthropods except crayfish (d) are not very light sensitive.
6. Spiders feed mainly on (a) plants (b) decayed matter (c) insects (d) other spiders.
7. Mites and ticks differ from other arachnids by having (a) mandibles (b) a unique respiratory system (c) two pairs of antennae (d) a fused cephalothorax and abdomen.
8. Book lungs help spiders respire on land by (a) carrying oxygen directly to tissues (b) providing a large surface area for the exchange of gases (c) both a and b (d) neither a nor b.
9. Centipedes and millipedes are different in (a) the way their bodies are shaped (b) the number of legs they have on each segment (c) their feeding habits (d) all of the above.
10. Products that have been developed from chitin include all of the following except (a) artificial hair (b) artificial skin (c) varnish and filters (d) contact lenses and surgical sutures.
11. What characteristics do biologists use in order to classify arthropods into four subphyla?
12. Explain the way in which the exoskeleton and muscles produce movement in arthropods?
13. What are the evolutionary trends that are common to all three subphyla of modern arthropods?
14. Which crayfish appendages are specialized for feeding?
15. Contrast the process of respiration in crayfish and spiders.
16. What roles do the swimmerets play in crayfish reproduction?
17. In what ways are spiders adapted for a predatory way of life?

18. Describe how a spider spins a web.
19. What methods do mites and ticks use to transmit diseases?

20. What defensive behaviors are common in millipedes?

Critical Thinking

1. *Daphnia* eat algae. They also have a prominent eyespot. In what way might the eyespot be connected with the ability of the *Daphnia* to find food?
2. Barnacles are a sessile group of crustaceans. What behavioral adaptations do they have that enable them to compete for food with mobile organisms? What structural adaptations might protect them from predators?
3. The head and thoracic plates of the crayfish are fused into a single carapace. What might be the advantages and disadvantages of this fused arrangement?
4. Examine the photo at right. What type of evolution is exhibited by this spider?
5. The American lobster, *Homarus americanus*, is a nocturnal organism. Like other crustaceans the lobster has compound eyes. Marine biologists have discovered that the lobster's senses of taste and smell

are over 1,000 times more powerful than those senses in humans. The lobster uses smell and taste both to search for food and to detect mates. What adaptive advantage would these highly developed senses provide for the lobster?

Extension

1. Read the article entitled "Spotted Vaccine to Follow?" in *Science News,* January 17, 1987, p. 41. Give an oral report in which you discuss the following: the symptoms of Rocky Mountain spotted fever; the number of cases of the disease that are reported every year; where the most cases occur; the reason there may be more cases than are reported; the fatality rate for those infected with the disease; the reason a vaccine was hard to produce; the way in which scientists created a vaccine; and the animal that has been successfully protected by the vaccine.
2. Contact your local public health department and ask for information on arthropod-transmitted diseases that might be found in your area. If possible, find out how many cases of each disease have been reported in the last year.
3. Obtain a sample of pond or stream water, and examine it for *Daphnia* and ostracods. Make drawings of any arthropods you see, and identify them as accurately as you can.
4. Search in a damp, dark area around your school or home for sow bugs or pill bugs. Examine those you find with a hand lens. List characteristics you see that indicate that these organisms are arthropods.
5. Using your library, research and write a report on the life cycle of a horseshoe crab, a scorpion, or a tick.

Chapter 33 | Insects

This chapter is devoted to familiarizing students with the structural physiological, and behavior characteristics of insects. Emphasis is placed on those characteristics that have contributed to the biological success of insects.

Correlation Chart

Section Number	Section Objectives	Pages	Features	Section Review
33.1	1. List the major reasons for the success of insects. 2. Give examples of three major kinds of variations among insects. 3. Name at least ten of the major orders of insects, and give an example of each. 4. Describe the relationship of insects to human society.	493–496		2, 3, 4 4 5 1, 5
33.2	1. Summarize the function of each of the external parts of a grasshopper. 2. Summarize the function of each of the organ systems of a grasshopper. 3. Describe the processes of incomplete and complete metamorphosis, and explain how they contribute to the success of insects. 4. Give an example of three types of defense in insects.	497–502	Think, p. 502	1, 2, 3 1, 3 4 5
33.3	1. Describe the social organization and behavior of honeybees and the ways in which they behave. 2. Name three ways in which insects communicate, and give an example of each. 3. Summarize Karl von Frisch's discoveries about communication in honeybees.	503–507	Biology in Process, pp. 504–505 Biology Lab, p. 507	2, 3, 5 1, 4 4

Resources*

Print

Borror, Donald J., Dwight M. Delong, and Charles A. Triplehorn. *Introduction to the Study of Insects*. New York: CBS College Publishing, 1981.

Leahy, Christopher, *Peterson First Guides: Insects/ Simplified Field Guide to Common Insects of North America*. Boston: Houghton-Mifflin, 1987.

Audiovisual/Software

Discover Entomology. Film or video, 15 min. Entomological Society, 1987.

Insects: The Lovely and the Lethal. Film, 23 min. Britannica, 1981.

*Additional references are listed in *Modern Biology, Teacher's Program Guide*.

Section Teaching Guides

33.1 The Insect World

Approach

The main objective of this section is to familiarize students with the reasons for the great evolutionary success of insects and their major ecological importance. It builds on concepts of arthropod structure and function (Chapter 32), coevolution with flowering plants (Chapter 25), and diversity through adaptation (Chapters 15 and 16).

In presenting the text, guide students to develop an appreciation for the great positive significance of insects in ecosystems, helping them transcend stereotypical images of insects as pests. You may wish to use Table 33-1 as a tool to expand students' appreciation of insect diversity.

Concept Framework

Insects are one group of invertebrates about which all students will have some background knowledge. To tap this knowledge and provide a conceptual frame for approaching this section, engage students in a discussion of why insects are more familiar to them than the other invertebrate groups they have studied. Elicit the fact that insects are extremely abundant in terrestrial environments and interact in many ways with human populations. You may wish to end the discussion by asking students to consider what it means to say that a group of organisms is successful.

Emphasis

Make certain that students understand the significance of each of the following factors in the success of insects: small size, ability to fly, short life span.

Guest Speaker

Invite an agriculture specialist to address the class about the role of insects in human agriculture. The speaker should talk about both the beneficial and harmful effects of insects—for example, some insects pollinate crops while other insects destroy them. The discussion should also consider the advantages and disadvantages of human efforts at pest control.

Section Review Answers

(See p. 496.)

33.2 Aspects of Insect Biology

Approach

The main objective of this section is to describe how anatomy, patterns of development, and defensive strategies show adaptations that have led to the evolutionary success of insects. To help students appreciate the adaptive significance of such characteristics as wings and compound eyes, draw comparisons between insects and the arthropods discussed in Chapter 32. In discussing internal and external anatomy, stress the relationship between structure and function.

The information in this section on metamorphosis can be related back to the discussion of gene expression and morphogenesis in Chapter 12. Also identify metamorphosis as an adaptive strategy, a concept that students will also encounter when they read about echinoderms (Chapter 34) and amphibians (Chapter 37).

Concept Framework

Students will have observed many insect behaviors or characteristics that enable insects to survive in virtually any environment. Ask students to suggest examples of these characteristics, such as the housefly's speed, the grasshopper's protective coloring, and the butterfly's cocoon. To relate these examples to the text, ask students how these characteristics reflect specific adaptations of insect biology.

Emphasis

Make certain students understand the relationship between structure and function of insects' various organ systems, developmental stages, and defense strategies. Students should be able to explain how each of these aspects of biology has contributed to the success of insects.

Guest Speaker

Invite an entomology professor from a local college or other speaker who can provide additional information on the similarities and differences among insects' body parts. Suggest focusing the presentation on function in addition to form. For example, the speaker might discuss variations in mouthparts or wing structure and how these reflect adaptations to the environment.

Section Review Answers

(Continued from p. 502.)

3. the tympanum; on the first abdominal segment
4. Complete metamorphosis allows different developmental stages to become specialized for different functions, eliminates competition between larvae and adults for food and space, and increases the likelihood of survival during harsh weather.
5. The yellow jacket has warning coloration. Predators stay away from the fly because they identify it as a harmful insect. The fly displays Batesian mimicry.

33.3 Insect Behavior

Approach

The main objective of this section is to describe the social behavior of insects, particularly honeybees, and the ways in which insects communicate. It is important to convey to students that these genetically programmed behaviors are triggered by the release of hormones, which, in turn, are a result of gene expression. You might wish to review the concepts of protein production (Chapter 8) and gene expression (Chapter 12) to clarify this point. In addition, reinforce the idea that natural selection favored the evolution of social behavior and communication in insects.

Concept Framework

Anyone who has heard crickets chirping or has seen a line of ants marching toward a stale bread crust has witnessed a form of social behavior in insects. Students can apply their knowledge of social organization in humans to the study of insect behavior. For example, ask them to consider how division of labor and, consequently, communication allow human societies to operate efficiently.

Emphasis

In presenting the text stress the adaptive significance of communication and social behavior, as well as the difference between innate and learned behavior. These concepts will prepare students for the discussions of vertebrate communication and behavior in later chapters.

Using the Features/The Dance Language of the Bees

The main process skills used in the feature are observing and experimenting. Karl von Frisch observed that when a bee returned to the hive after locating a food source, it performed a series of dance-like movements. By conducting a series of experiments in which he manipulated the placement of feeding stations, von Frisch determined that the bee's dance varies according to the distance of the food source from the hive.

In discussing the feature, point out to students that because two variables are observed together does not in itself establish a relationship between the two. Von Frisch at first thought that the type of food—pollen or sweetened water—determined the type of dance performed by the bee. Only through experimentation did he determine that distance of the food from the hive was the crucial factor.

Field Trip

Arrange for the class to visit a local apiary. Touring such a facility will enable students to observe directly the social behavior of bees. In addition, it will help students appreciate the commercial significance of bees to human society. Prior to the visit, prepare a list of specific points to notice or questions to answer about the facility. Some people have life-threatening allergies to bee stings. All students who plan to participate should provide parental permission and medical clearance.

Section Review Answers

(See p. 506.)

Laboratory Preparation

Background

In this laboratory exercise, students examine the external features of the paper wasp and compare them to diagrams of the grasshopper to verify that all insects share certain basic characteristics. The genus *Polistes* is a type of social wasp, with a body plan typical of the main characteristics of the insect body, like the honeybee described in the pupils' text. Students should notice that the female ovipositor in the worker has been modified into a stinger, which is used in defending the nest from intruders. Students should know how the small sterile females become workers.

Preparation

Order the *Polistes* specimens ahead of time.

Safety

The preservative used on the specimen can be irritating to the skin. Caution students to avoid touching the specimen, and especially to avoid rubbing the eyes with contaminated hands.

Conclusions

1. A typical insect has three body segments: the head, the thorax, and the abdomen. The abdomen is divided into 11 segments. Insects also have three pairs of jointed legs that are attached to the thorax, one pair of compound eyes, and three simple eyes.
2. The ovipositor has been modified into a stinger.

Inquiry

Students should suggest that a number of representative insects be looked at to compare them morphologically to bees and wasps.

Chapter Review

Vocabulary

1. aorta; others are mouthparts
2. pheromone; others relate to metamorphosis
3. spiracle; others involved in digestion
4. behavioral variation; others are defenses involving coloration
5. nymph; others are insect body parts

Review

11. Small size may allow many species to inhabit same environment without competing.
12. different digestive enzymes or any other example involving adaptation of body systems, also different kinds of metamorphosis
13. three body parts, one pair of antennae, three pairs of jointed legs, and abdomen with 11 segments
14. by damaging crops, carrying disease, damaging wood in buildings

15. head, thorax, abdomen
16. In insects, oxygen travels directly to cells through air tubes called tracheae instead of being transported in the blood.
17. Camouflage allows insects to hide from predators by blending in with surroundings; warning coloration makes insects stand out in such a way that predators associate their appearance with their undesirability as food.
18. When levels of juvenile hormone are high, the release of molting hormone causes larva to molt. When levels of juvenile hormone drop, molting hormone causes the transition to pupa and then to adult.
19. workers, queen, drones
20. by rubbing scraper of one forewing against vein on other forewing; to attract mates and to protect territory

Critical Thinking

1. Insects with genes that confer resistance against insecticide survive and reproduce in large numbers; thus they proliferate through natural selection. This process occurs rapidly in insects because they have very short life spans.
2. Siphoning mouthparts suggest that the insect feeds on liquid food.
3. Odonata; two pairs of long, narrow membranous wings
4. A large organism probably could not transport oxygen quickly and efficiently enough through tracheae to supply all its cells; instead, the organism would require specialized organs (lungs) for gas exchange.
5. If a worker bee dies from stinging or through an attack by a predator, it has already completed most of its other functions. It would be a greater loss to the hive if death occurred early in the worker's life.
6. Answers will vary but may include: dormant stages in life cycle may have provided protection from coldest periods; ability to fly may have allowed migration to warmer areas; small size may have allowed them to inhabit protected environments; diverse food sources may not have been affected by colder climate; faster reproduction led to faster evolution.

Chapter 33 | Insects

Introduction

Insects account for about three-fourths of all animal species. They are the most successful of all the arthropods, and for this reason an entire chapter is devoted to their study. In this chapter you will learn about some of the factors that have contributed to their success.

Chapter Outline

33.1 The Insect World
- *Characteristics and Classification*
- *The Success of Insects*
- *Insects and Human Society*

33.2 Aspects of Insect Biology
- *External Structure*
- *Internal Structure*
- *Development*
- *Defense*

33.3 Insect Behavior
- *Social Behavior in Honeybees*
- *Communication*

Chapter Concept

As you read, look for the structural, physiological, and behavioral variations that have made insects such a successful group of animals.

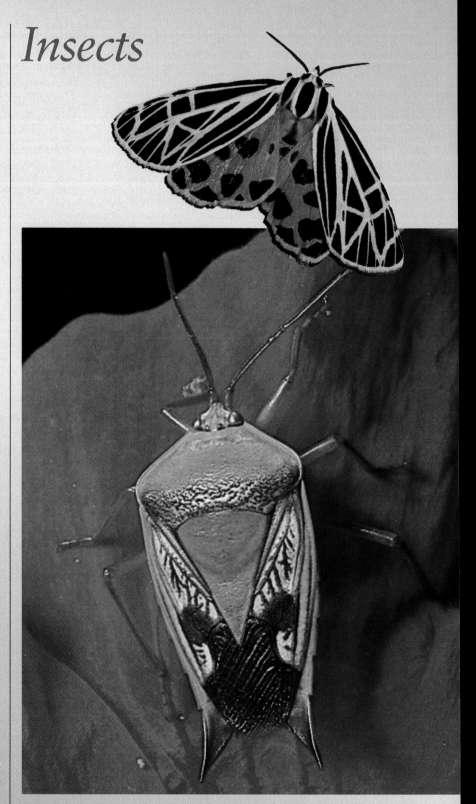

Stink bug, Papua New Guinea. Inset: Virgo tiger moth, *Apantesis virgo*, New York.

For additional teaching information see pp. T33A–T33D.

33.1 The Insect World

Above all else, the world of insects is a story of great biological success through evolution and adaptation. Insects have thrived on the earth for more than 300 million years, since long before the rise and fall of the dinosaurs. Today insects are the most successful group of animals, in terms of both diversity and numbers.

Characteristics and Classification

Many of the adaptations that have made insects successful are characteristics they share with other arthropods. For example, an exoskeleton, jointed appendages, and a segmented body are all common arthropod traits. However, insects are distinguished from other arthropods by the following characteristics:

- The body has three parts: head, thorax, and abdomen.
- The head has one pair of antennae.
- The thorax has three pairs of jointed legs and, in many species, one or two pairs of wings.
- The abdomen is divided into 11 segments. It has neither wings nor legs attached to it.

In spite of the characteristics they share, insect species show an enormous range of variations that have allowed them to succeed in diverse environments. These variations can be grouped into three categories. The first category consists of **structural variations,** such as differences in mouthparts. For example, a wasp has mouthparts adapted for chewing, while an aphid uses its mouthparts to pierce plants and suck fluids. A second group of variations, **physiological variations,** consists of differences in the way internal systems work. For instance, a female mosquito has enzymes that allow it to digest human blood; a grasshopper does not have these enzymes but has others that enable it to digest grass. Finally, **behavioral variations** are differences in the ways insects respond to their surroundings. Honeybees, for example, live in complex societies called hives. Many species of bumblebees, on the other hand, are solitary and usually live part of their lives in holes in the ground.

To make it easier to study such a vast and diverse group of organisms, taxonomists classify insects into more than 30 orders. They base their classification primarily on structural and physiological variations. Study the information about some common orders of insects in Table 33-1 on pages 494–495. How many do you recognize?

The Success of Insects

Insects live almost everywhere in the world except in the deep ocean. Water striders live on the surface of oceans and lakes, and

Info: In some primitive insect orders wings never evolved. In others wings originally present were lost.

Section Objectives

- *List the major reasons for the success of insects.*
- *Give examples of three major kinds of variations among insects.*
- *Name at least ten of the major orders of insects, and give an example of each.*
- *Describe the relationship of insects to human society.*

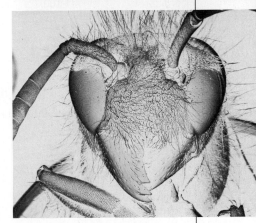

Figure 33-1 Color-enhanced SEMs show that aphid mouthparts (top, x160) are adapted for piercing and sucking while those of a wasp (bottom, x100) are adapted for chewing.

Number of species in thousands

Insects | Other Invertebrates | Vertebrates

Figure 33-2 Numbers of insect species are compared with those of other groups.

beetles live in the hottest deserts. Snowfleas survive on permanent glaciers on the world's highest mountains. Closer to home, you may find aphids on the leaves of garden plants or beetles under the bark of trees. You may even find a flea or two in the fur of your dog or cat.

As you have read, Arthropoda is the largest animal phylum, and Insecta is by far the largest class within that phylum. **Entomologists,** or scientists who study insects, have classified more than 700,000 insect species. Every year biologists describe thousands of new species. Based on current knowledge, some entomologists estimate that as many as 10 million insect species exist. Figure 33-2 shows the numbers of species in various subgroupings of the animal kingdom. Notice that the number of classified insect species is about three times that of all other classified animal species combined.

Table 33-1 Common Insect Orders

Order	Number of Species	Examples	Characteristics	Significance to Humans
Orthoptera ("straight-wing")	23,000	Grasshoppers Crickets Katydids Cockroaches	Two pairs of straight wings; chewing mouthparts	Cause damage to crops; pests in houses
Isoptera ("equal-wing")	1,800	Termites	Two pairs of membranous wings; chewing mouthparts	Destroy wood in forests and buildings; recycle resources in forests
Dermaptera ("skin-wing")	1,100	Earwigs	One or two pairs of wings; chewing mouthparts; pincer-like appendages at tip of abdomen	Damage crops and garden plants
Anoplura ("unarmed-tail")	200	Sucking lice	Wingless; piercing, sucking mouthparts	Parasitize humans and other mammals; carry disease
Hemiptera ("half-wing")	40,000	All true bugs	Two pairs of membranous wings during part of life; piercing, sucking mouthparts	Damage plants; carry disease
Homoptera ("like-wing")	20,000	Aphids Mealy bugs Cicadas	Membranous wings held like roof over body (some species wingless); piercing, sucking mouthparts	Damage crops and gardens
Ephemeroptera ("for-a-day-wing")	1,500	Mayflies	Membranous wings (forewings triangular); nonfunctioning mouthparts in adults	Serve as food for freshwater fish

Why have the insects been such a biological success? Like other arthropods, they have benefited from the evolution of an exoskeleton and jointed appendages. Additionally, they can adapt to new environments rapidly because individual insects have extremely short life spans. In many species adults live for only weeks or months. Since generations occur in rapid succession, natural selection can take place more quickly than in organisms that take longer to reproduce.

As they have evolved, insects have had a wide range of environments available to them because of their flying ability and their small size. The power of flight enables insects to disperse readily, to escape from predators, and to move into environments less accessible to other organisms. In addition, the small size of insects allows several species to inhabit different environments within a small area without competing with one another.

Tip: Discuss with students why an exoskeleton and jointed appendages have proved to be successful adaptations.

Order	Number of Species	Examples	Characteristics	Significance to Humans
Odonata ("toothed")	5,000	Dragonflies Damselflies	Two pairs of long, narrow, membranous wings; chewing mouthparts	Destroy harmful insects
Neuroptera ("nerve-wing")	4,600	Dobsonflies Lacewings	Two pairs of membranous wings; mouthparts sucking in larvae, chewing in adults	Destroy harmful insects
Coleoptera ("sheath-wing")	280,000	Weevils Ladybugs Beetles	Hard forewings, membranous hindwings; sucking or chewing mouthparts	Destroy crops; prey on other insects
Lepidoptera ("scale-wing")	110,000	Butterflies Moths	Large, scaled wings; mouthparts chewing in larvae, siphoning in adults	Pollinate flowers; produce silk; damage clothing and crops
Diptera ("two-wing")	85,000	Mosquitoes Flies Gnats	One pair of wings; sucking, piercing, or lapping mouthparts	Carry disease
Siphonaptera ("tube-wingless")	1,100	Fleas	Wingless in adults; mouthparts chewing in larvae, sucking in adults	Parasitize birds and mammals; carry disease
Hymenoptera ("membrane-wing")	100,000	Bees Wasps Ants	Two pairs of membranous wings (some species wingless); chewing, sucking, or lapping mouthparts; some species social	Pollinate flowers; make honey

Figure 33-3 The grain weevil, *Sitophilus granarius* (left, x170), shown emerging from a wheat seed in this color-enhanced SEM, competes with human beings for food. Of human benefit is the blister beetle, *Lytta magister* (right), shown pollinating a California poppy.

Answers to Section Review
1. a scientist who studies insects
2. three pairs of legs, three body parts, wings in adults of most species
3. small size, mobility provided by wings and jointed legs
4. differences in mouthparts
5. Lepidoptera are both beneficial and harmful. Many species pollinate flowers and some produce silk. A few species damage crops or clothing.

Insects and Human Society

Since insects are so abundant, it is not surprising that they affect human society in many ways. A small minority of insect species cause severe problems for people. Boll weevils, corn earworms, and other agricultural pests compete with humans for food by eating crops. Other insects spread diseases. In the tropics female mosquitoes transmit *Plasmodium*, the protozoan that causes malaria. Flies transfer the bacterium *Salmonella typhi*, which causes typhoid fever.

In spite of the problems some insects cause, it would be a serious mistake to think that the world would be better off without them. Insects play many vital roles in the environment. For example, insects pollinate more than two-thirds of the world's flowering plants; they also serve as food for a multitude of fish, birds, and mammals. We tend to think of termites as destructive pests. However, in feeding on decaying wood, they help recycle materials needed to maintain a healthy forest. Honey, silk, and shellac are some products produced by insects. Without insects to pollinate many of our crop plants, our food sources would be greatly diminished.

Section Review

1. What is an *entomologist?*
2. What features distinguish insects from other arthropods?
3. List two characteristics of insects that have helped them move into diverse environments.
4. Give an example of structural variation in insects.
5. How would you characterize the importance of the order Lepidoptera to people? Use Table 33-1 to help you answer.

33.2 Aspects of Insect Biology

The diversity in the insect world is so great that no typical insect exists. Nevertheless, in this section we will use the grasshopper as a representative example for studying details of structure and function. We will then examine patterns of development and defense that are common to most insect species.

External Structure

The African Goliath beetle grows to more than 10 cm in length, and the Atlas moth has a wingspan of more than 25 cm. However, most insects are far smaller. Among the smallest is the fairy fly, which measures 0.2 mm in length.

The diagram of an adult grasshopper in Figure 33-4 can serve as a model to help you understand the structural components of external anatomy. Like all insects the grasshopper has three body segments. The anterior segment is the head. It contains the brain and bears many of the sensory organs, such as the antennae and compound eyes. The head also has complex mouthparts that are used for gathering food. The middle segment is the thorax, to which the legs and wings are attached. The posterior segment is the abdomen, which is often specialized for reproduction and has structures for digestion, respiration, and excretion.

The thorax consists of three parts: prothorax, mesothorax, and metathorax. The prothorax and mesothorax each have a pair of walking legs that allow the grasshopper to creep along blades of grass. Attached to the metathorax is a pair of jumping legs that enables the grasshopper to leap away from danger and to launch

Section Objectives

- *Summarize the function of each of the external parts of a grasshopper.*

- *Summarize the function of each of the organ systems of a grasshopper.*

- *Describe the processes of incomplete and complete metamorphosis, and explain how they contribute to the success of insects.*

- *Give an example of three types of defense in insects.*

Figure 33-4   The grasshopper's three-part body, paired antennae, paired mandibles, and three pairs of jointed legs are typical of all insect species.

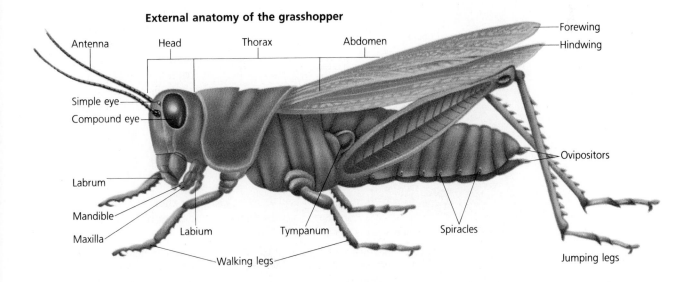

External anatomy of the grasshopper

Antenna · Head · Thorax · Abdomen · Forewing · Hindwing · Simple eye · Compound eye · Ovipositors · Labrum · Mandible · Labium · Tympanum · Spiracles · Maxilla · Walking legs · Jumping legs into flight. Spines, hooks, and pads on each foot, or **tarsus,** provide the grasshopper with grip.

The grasshopper has two pairs of wings. A pair of leathery forewings covers and protects the hindwings when the grasshopper isn't flying. Although the forewings can help the grasshopper glide during flight, it is the movement of the hindwings that actually propels the insect through the air. The grasshopper can move its wings because of muscles attached to the inside of its exoskeleton. The forewings are attached to the mesothorax; the hind wings are attached to the metathorax.

Internal Structure

Many structural and physiological adaptations can be observed by examining an insect's organ systems. The grasshopper, shown in Figure 33-6, can serve as a representative example for studying the organ systems of insects.

Digestive and Excretory Systems

Grasshoppers feed on blades of grass, and the jaws of grasshoppers are thus modified for cutting and chewing. The liplike **labium** *(LAY-bee-um)* and **labrum** *(LAY-brum)* help hold the grass in position so that the rough-edged jaws called mandibles can tear off edible bits. Behind the mandibles are the maxillae, a second set of jaws that helps hold and cut the food.

In the mouth food is moistened by saliva from the salivary glands. The moistened food then passes through the esophagus and into the crop for temporary storage. From the crop, food passes into the gizzard, where sharp chitinous plates shred it. The shredded mass then passes into the insect's stomach, called the **midgut.** There the food is bathed in enzymes secreted by pockets

Figure 33-5 The color-enhanced SEM of a grasshopper's head shows the eyes, mouthparts, and base of the antennae (x46).

Figure 33-6 The grasshopper's internal organs are representative of other insect species.

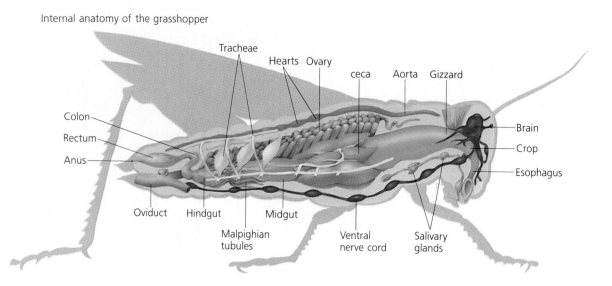

Internal anatomy of the grasshopper

Tracheae
Hearts Ovary
ceca Aorta Gizzard
Colon
Rectum
Anus
Brain
Crop
Esophagus
Oviduct Hindgut Midgut
Malpighian tubules
Ventral nerve cord
Salivary glands

of the stomach called the **gastric ceca** *(SEE-kuh)*. Digested matter then flows through the midgut wall and into the coelom, or body cavity, and is distributed by the circulatory system.

Undigested matter from the midgut travels into the **hindgut,** made up of the colon and the rectum. Meanwhile, wastes from the cells have been picked up by the blood. Malpighian tubules in the hindgut remove chemical wastes from the blood and deposit them in the rectum. All wastes then leave through the anus.

Circulatory System

Digested food and other materials reach the grasshopper's cells through an open circulatory system like that of the crayfish. Blood flows through a large vessel called the **aorta.** Muscular regions of the aorta, often called the grasshopper's hearts, are located in the posterior abdomen. They pump the blood forward through the aorta and into the part of the coelom near the head. The blood then flows back through the coelom toward the abdomen, carrying digested food to the organs. At the same time, the blood transfers cellular wastes to the Malpighian tubules and then completes the circuit back to the aorta through pores called ostia.

Respiratory System

Insects do not breathe with lungs or gills but instead take in air through tiny openings on the abdomen and thorax called spiracles. As muscles expand plates on the grasshopper's abdomen, air flows through the spiracles and enters a network of air tubes called tracheae. Note that in insects oxygen travels directly to body tissues through the tracheae and their smaller branches. In most other animals oxygen is transported by the blood. When the abdomen contracts, waste gases that have diffused out of cells collect in the tracheae. These waste gases are then expelled through the spiracles.

Nervous System

The grasshopper has a complex internal nervous system connected to the external sensory organs. Three simple eyes arranged in a row just above the base of the antennae serve simply to detect light. Two bulging compound eyes, composed of hundreds of six-sided lenses allow the insect to see in several directions at once. They can detect movement but cannot produce sharply focused images. Grasshoppers sense sounds by means of a small, nerve-rich cavity located along the first abdominal segment. This cavity is covered with a sound-sensing membrane called a **tympanum** *(TIM-puh-num)*. Sense organs for both touch and smell are located on the antennae and elsewhere on the body.

If you've ever tried to catch a grasshopper, you've seen the effect of its nervous system in action. The grasshopper's eyes detect your movement. Nerve impulses travel up nerve cords to the nerve ganglia that form the brain. Messages from the ganglia travel rapidly down other nerve cords to the muscles that control

Figure 33-7 A color-enhanced SEM of a diamondback moth shows the individual facets of the compound eyes and the ball-and-socket joint at the base of each antenna (x65).

the jumping legs. The legs flex, and in an instant the grasshopper is out of your grasp.

Reproductive System

The reproductive organs of both male and female grasshoppers are located in the abdomen. During mating the male grasshopper deposits sperm into the female's seminal receptacle, a storage pouch that holds the sperm until eggs are released. After release the eggs are fertilized internally. The female grasshopper uses two pairs of pointed organs called **ovipositors** to dig a nest hole in the soil and to deposit the eggs.

Development

All the characteristics we have observed so far are those of the adult grasshopper. However, insects go through a number of stages before they reach their adult forms. Only silverfish and a few other insects start out as smaller versions of adult insects. Most insects undergo distinct changes in both form and size as they develop. This series of developmental changes, called **metamorphosis,** is shown in Figure 33-8. Metamorphosis is the result of the processes of gene expression, described in Chapter 12.

Incomplete Metamorphosis

Grasshoppers, termites, and some other insects go through a pattern of development called incomplete metamorphosis. This process has three stages: egg, nymph, and adult. When the egg hatches, a nymph emerges. A **nymph** is an immature form that looks somewhat like the adult but is smaller, has undeveloped

Incomplete metamorphosis

Adult

Young adult

Eggs

Nymph

Figure 33-8 In incomplete metamorphosis (above) the nymph resembles the adult form. In complete metamorphosis (below) neither larva nor pupa resembles the adult.

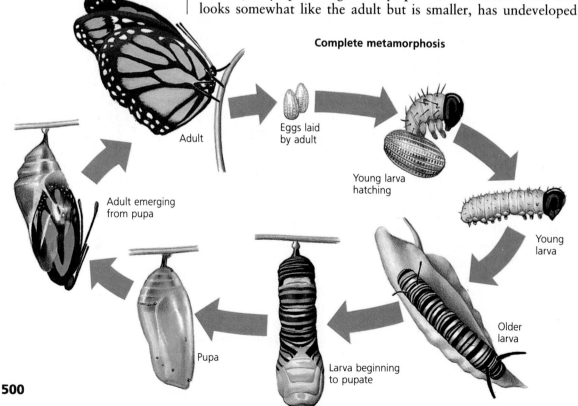

Complete metamorphosis

Adult

Eggs laid by adult

Young larva hatching

Young larva

Older larva

Larva beginning to pupate

Pupa

Adult emerging from pupa

reproductive organs, and lacks wings. The nymph molts several times until it becomes a winged adult that can reproduce.

Complete Metamorphosis

Butterflies, beetles, and most other insects go through complete metamorphosis. Complete metamorphosis includes four distinct stages: egg, larva, pupa, and adult. For example, when a butterfly egg hatches, a segmented larva emerges. The larva, commonly called a caterpillar, looks far more like a worm than an adult butterfly. Unlike a worm, however, a caterpillar has three pairs of jointed legs and several pairs of fleshy legs. A caterpillar devours leaves, grows large, and molts several times. Because caterpillars need a lot of food to grow so rapidly, they can cause much damage to the plants upon which they feed.

Typically when a caterpillar's growth is complete, it finds a sheltered spot and hangs upside down. Its body becomes shorter and thicker. Its exoskeleton splits down the back and falls off. The insect now enters an immobile stage called the **pupa.** Inside the pupa larval tissues are breaking down, and groups of cells called imaginal disks are developing into the tissues of the adult. During this process the pupa is encased in a protective covering, called a **cocoon** in moths and a **chrysalis** in butterflies. When metamorphosis is complete, a sexually mature winged adult emerges.

Hormonal Control of Metamorphosis

The process of metamorphosis is controlled by the sequential expression of genes. These genes cause the production of three hormones: brain hormone, molting hormone (called ecdysone), and juvenile hormone. Brain hormone stimulates a gland in the thorax to release molting hormone. The effect of molting hormone depends on the amount of juvenile hormone in the blood. During larval development the level of juvenile hormone is relatively high, and the release of molting hormone at this stage causes the larva to molt. However, as the insect gets older, the production of juvenile hormone decreases. When the level of juvenile hormone falls low enough, molting hormone triggers the change from larva to pupa. When juvenile hormone is no longer present, molting hormone starts the change from pupa to adult.

Importance of Metamorphosis

Metamorphosis is another adaptation that contributes to the great success of insects. In a life cycle based on complete metamorphosis, different developmental stages of the insect fulfill different functions. For example, a caterpillar is specialized for growth, while an adult butterfly is specialized for dispersal and reproduction. One of the advantages of this specialization is that it eliminates conflict between two activities that require great amounts of energy—growing and reproducing.

Metamorphosis also enhances insect survival in two other ways. First, it eliminates competition between larvae and adults

Figure 33-9 Leaf-feeding gulf fritillary caterpillars, *Agraulis vanillae* (top), do not compete with the adult form (bottom), which feeds on flower nectar.

Info: In larvae periods between molts are called instars.

Think

In any given area, monarch butterflies (top) generally emerge as adults earlier in the spring than do viceroy butterflies (bottom). Based on what you know about Batesian mimicry, why do you think it is advantageous for viceroys not to emerge first?

Answer to Think: Because monarchs emerge first, predators are more likely to encounter them and their unpleasant traits first and so learn to avoid all butterflies with a similar appearance. If viceroys emerged first, the viceroy's mimicry would be less effective.

Answers to Section Review
1. The labrum helps hold grass within the jaws; spiracles allow gases to enter and leave the body; ovipositors are used by females to lay eggs.
2. the limited ability of the thin exoskeleton to support the body
(See pp. T33A–T33D.)

for food and space. For example, a caterpillar eats leafy vegetation, but an adult butterfly feeds on flower nectar. Second, a multistage life cycle helps insects survive harsh weather. Most butterflies spend the winter encased as pupae. Likewise, the eggs of mosquitoes remain unhatched through the winter. The larvae or nymphs of other insects spend the winter under water. Some are even protected from the cold by a chemical in the blood that is similar to the antifreeze used in car radiators.

Defense

Insects have many defensive adaptations that enhance survival. Some adaptations are for aggressive defense. One of the most elaborate is that of the bombardier beetle, which sprays predators with a hot stream of a noxious chemical. The beetle can even rotate an opening on its abdomen to aim the spray at its attacker. A more common defense is the barbed stinger of bees and wasps.

Some adaptations provide a passive defense. The grasshopper's grassy green color is a good example of one form of passive defense—**camouflage,** or the ability to blend into surroundings. Camouflage enhances survival by making it difficult for predators to spot the insect. A remarkable form of camouflage is exhibited by the many species of "stick insects." These creatures look so much like twigs that they often cannot be seen unless they move.

Other insects defend themselves not by hiding but by advertising. Many poisonous or bad-tasting insects have evolved **warning coloration**—bold, bright color patterns that make them clearly recognizable and warn predators away.

In some cases evolution has resulted in several poisonous or dangerous species' having similar patterns of warning coloration. This adaptation, called **Müllerian mimicry,** encourages predators to avoid all of these species. For example, many species of stinging bees and wasps have a similar pattern of black and yellow stripes. In other cases insects that are neither poisonous nor bad-tasting have patterns that fool predators by mimicking the warning coloration of other species. For example, the viceroy butterfly looks much like the unpleasant-tasting monarch. This type of defensive adaptation is called **Batesian mimicry.**

Section Review

1. State the function of each of the following parts of a grasshopper: *labrum, tympanum, ovipositors.*
2. What factor limits the size of an insect?
3. What structure enables a grasshopper to hear? Where is it located?
4. How does complete metamorphosis benefit an insect species?
5. The yellow jacket wasp and the stingless syrphid fly have similar color patterns. How does each of these insects benefit from this coloring? What is the fly's defense called?

33.3 Insect Behavior

Just as most insects go through different developmental stages that are specialized in form and function, some insect species have different kinds of adult individuals that are specialized in form and function. For example, some species of ants, bees, and termites live in complex societies in which some individuals gather food, others protect the society, and still others only breed. This **division of labor** in insect societies creates great interdependence and a heightened need for communication. Even solitary insects such as fleas and fireflies need to communicate at certain critical times—at mating time, for example. These are all behavioral variations in insects. In this section we will look at the social behavior of honeybees and at some ways in which bees and other insects communicate.

Social Behavior in Honeybees

A honeybee hive consists of three distinct forms of individuals: the workers, the queen, and the drones. The workers are the sterile females that make up the vast majority of the hive population, which may reach more than 80,000. The workers do all of the work of the hive except reproduction. The queen is the only fertile female in the hive, and her only function is to reproduce. The drones, or male bees, are needed for mating, but otherwise they are useless to the hive. In fact, their mouthparts are too short to obtain nectar from flowers, so the workers must feed them.

As you read about the complex behavior of honeybees, keep in mind that the bees neither teach nor learn these behaviors. Instead, the behaviors are genetically determined; that is, the insects are acting out instructions stored in their genes and triggered by the substances these genes produce. This genetically controlled behavior is called **instinct** or innate behavior.

The Worker Bees

Worker bees perform many functions at different times during their brief lives, which average about six weeks in length. After making the transition from pupa to adult, workers first feed honey and pollen to the queen, the drones, and the larvae. During this stage the workers are called nurse bees. They secrete a high-protein substance called **royal jelly,** which they feed to the queen and to the youngest larvae.

After about a week as feeders, worker bees stop producing royal jelly and begin to secrete wax, which they use to build and repair the honeycomb. During this stage they may also remove dead bees and wastes from the hive, guard the hive, and fan their wings to circulate fresh air.

The workers spend the last weeks of their lives gathering nectar and pollen. A number of structural adaptations aid them in

Figure 33-10   Worker bees perform the work of the hive, while drones and the queen are involved exclusively with reproduction.

Worker bee

Drone

Queen bee

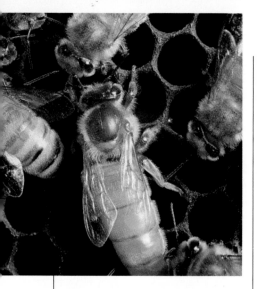

Figure 33-11 This magnified image shows worker bees feeding a queen (center).

this work. For example, the design of their mouthparts is highly efficient for lapping up nectar. Their legs have special structures that serve as pollen packers, pollen baskets, and pollen combs. The combs clean the pollen as it is packed. The sterile workers do not use their ovipositors for egg laying. Instead these structures are modified into a stinger that helps the workers protect the hive.

The Queen and the Drones

The queen bee develops from an egg identical with those that develop into the workers. The differences between the queen and the workers result from the continuous diet of royal jelly that the queen is fed throughout her larval development. In addition the queen herself secretes a substance called queen factor that prevents other female larvae from becoming sexually mature.

The queen's role is to reproduce. Although a queen bee may lay as many as a million fertilized eggs each year, she mates only once in her life. A few days after completing metamorphosis into an adult, the queen flies out of the hive and mates in the air with one or more drones. During mating millions of sperm are deposited in the queen's seminal receptacle, where they are stored for use during the five or more years of her life.

Biology in Process

The Dance Language of the Bees

Round dance

How do bees communicate so well that when one bee returns from a food source, other bees can fly directly to a fragrant flower patch several meters away? German biologist Karl von Frisch first asked himself this question more than 70 years ago.

After 25 years of observation and experimentation, he learned to translate the dance language of the bees.

To observe bees von Frisch built a glass-walled hive and placed feeding stations stocked with sweetened water near the hive. When a bee landed to feed, he dabbed paint on its thorax. This mark allowed him to identify the bee when it returned to the hive.

Again and again, von Frisch observed the painted scout enter the hive, feed several other workers, and then perform a series of dancelike movements on the comb. The scout would circle first to the right and then to the left—a behavior that von Frisch termed the "round dance." Other bees would excitedly follow the scout bee in the dance and place their antennae close to hers. Finally, the other bees would leave and locate the food.

Next von Frisch experi-

If the hive becomes overcrowded, the queen leaves with about half the workers to form a new hive. When this happens or when the queen dies, queen factor disappears from the hive. As a result, workers begin to raise new queens. When a new queen emerges, she produces new queen factor, and in response the workers destroy the other developing queens.

Drones are males that develop from unfertilized eggs. Their sole function is to deliver sperm to the queen. Through the summer the hive population includes a few hundred drones. When the honey supply begins to run low, the workers sting all the drones and clear them from the hive.

Tip: Ask students to speculate on why there are many drones in a hive even though only a few mate with the queen. (The gene pool is enhanced if only the fittest drones reach the queen for mating.)

Info: Queens are raised in special cells larger than ordinary hive cells. If a hive loses its queen, workers will begin enlarging worker cells to form emergency queen cells.

Communication

The effect of queen factor is a good example of the most common form of communication among insects—chemical communication. Queen factor is a **pheromone** *(FER-uh-MONE),* a chemical released by an animal that affects the behavior or development of other animals of the same species. Pheromones play a major role in the behavior patterns of many insects. For example, you may have noticed ants marching along in a tightly defined route. They

Answer to Biology in Process: Von Frisch had probably hypothesized that the different dances were related to differences in distance to the food and designed the experiments to test this hypothesis.

Waggle dance

mented by placing four sweetened water samples within 10 m of the hive—one north, one south, one east, and one west. He observed that although a scout might find only one sample, other bees would soon appear at all four. He concluded that the dance did not tell the bees the exact location of the food.

Von Frisch also observed a second dance, which he called the "waggle dance." Performed by pollen-laden bees returning to the hive, this dance consisted of a series of straight runs during which the bees waggled their abdomens from side to side. After each run the

bees made a semicircular turn to the right.

At first, von Frisch hypothesized that the round dance was performed by bees returning with sweetened water and the waggle dance was performed by bees returning with pollen. However, experiments with sweetened water placed

at various distances from the hive showed that the round dance was performed for food sources near the hive. The waggle dance was performed for food sources more than about 50 m away. Specific movements of the waggle dance gave information about the abundance of the source and its direction and distance from the hive. Direction was shown by the angle of the dance on the comb.

In 1973 von Frisch's work on communication won a Nobel Prize.

■ Why do you think von Frisch decided to conduct experiments with feeding stations at various distances?

Figure 33-12 Ants will follow a pheromone trail left by the ants that have preceded them.

were following a trail of pheromones left by the ants ahead of them. By secreting less than 0.01 μg of a powerful pheromone, female silkworms can attract males from several miles away.

Many insects communicate through sound. Male crickets use sound to attract females and to warn other males away from their territories. They rub a scraper on one forewing against a vein on the other forewing to produce chirping sounds. Each cricket species produces several calls that differ from those of other cricket species. In fact, because many species look similar, entomologists often use the calls to identify the species. Mosquitoes depend on sound, too. Males that are ready to mate home in on the buzzing sounds produced by females. The male senses this buzzing by means of tiny hairs on his antennae, which vibrate only to the frequency emitted by a female of the same species.

Insects may also communicate by tapping, rubbing, or signaling. Fireflies use flashes of light to find a mate. Each species of firefly has its own pattern of flashes. Males emit flashes in flight, and females flash back in response. This behavior allows male fireflies to locate a mate of the proper species. However, they must beware of female fireflies of the genus *Photuris*, which can mimic the flashes of other species. If a male of a different species responds to the flash of a *Photuris* female and attempts to mate, the female devours him. This is surely one of the more unusual behavioral adaptations in the enormously successful world of insects.

Section Review

1. What is a *pheromone?*
2. What determines whether a fertilized honeybee egg will develop into a worker or a queen?
3. How do honeybees behave when a hive is overcrowded?
4. Name some major ways in which insects communicate.
5. Current research indicates that the queen bee mates with drones from another hive rather than her own hive. In what way might this behavior benefit the honeybee society?

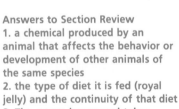

Answers to Section Review
1. a chemical produced by an animal that affects the behavior or development of other animals of the same species
2. the type of diet it is fed (royal jelly) and the continuity of that diet
3. The queen leaves and takes about half the hive with her. Then the hive raises a new queen.
4. through pheromones, sound, movement, and visual signals
5. Mating with drones from other hives provides a source of genetic diversity; drones from the queen's own hive share all of their genes with her since they develop from unfertilized eggs laid by the queen herself.

Laboratory

External Anatomy of the Wasp

Objective
To examine the external anatomy of a wasp

Process Skills
Inferring, observing

Materials
Forceps, preserved *Polistes* (paper wasp), culture dish, hand lens or dissecting microscope, blunt probe, unlined paper

Humane Alternative
A diagram of a wasp may be used as an alternative to using preserved wasps.

Method

1 The specimen you will work with is a worker wasp. Worker wasps play the same social role as worker bees. Use forceps to rinse the specimen gently but thoroughly with water to remove excess preservative. Try not to touch the wasp since the preservative can be irritating to the skin. Place the wasp in a culture dish.

2 Observe the external features of the wasp through a hand lens or dissecting microscope. Carefully move the wasp with the probe and forceps to observe the head, thorax, abdomen, legs, wings, antennae, compound eyes, and simple eyes.

3 Make a drawing that shows the major external features of the wasp. Begin at the anterior end of the body and slowly work toward the posterior end. Include as much detail in your drawing as necessary to make the features clear.

4 Label the structures listed in step 2. The wasp's most powerful muscles are located in the thorax. Why do you think this is so?

5 Observe the row of tiny openings at each side of the abdomen. These are the spiracles.

6 Observe the structure that protrudes from the last segment of the abdomen. This is the ovipositor. It indicates that the specimen is a female. Include the ovipositor in your drawing, and label it.

7 Using the tip of the probe as a guide, examine the mouth parts of the wasp. Identify the labrum, labium, mandibles, and maxillae. Make separate drawings of the structure of each mouthpart, and give it a label.

Conclusions

1 Compare your drawing of the wasp with the diagram of the grasshopper in Figure 33-3 in your textbook. Based on your observations, describe a typical insect in terms of its body segments, number of legs, number of antennae, and number and types of eyes.

2 What part of your specimen has been modified into a stinger to be used for defense?

Inquiry
What are some ways in which the morphology of bees and wasps differs from that of other insect orders? Describe a scientific method for answering this question.

Chapter 33 Review

Vocabulary

aorta (499)

Batesian mimicry (502)

behavioral variations (493)

camouflage (502)

chrysalis (501)

cocoon (501)

division of labor (503)

entomologist (494)

gastric ceca (499)

hindgut (499)

instinct (503)

labium (498)

labrum (498)

metamorphosis (500)

midgut (498)

Müllerian mimicry (502)

nymph (500)

ovipositor (500)

pheromone (505)

physiological variations (493)

pupa (501)

royal jelly (503)

structural variations (493)

tarsus (498)

tympanum (499)

warning coloration (502)

For each set of terms below, choose the one that does not belong, and explain why it does not belong.
1. labium, aorta, gastric ceca, labrum
2. pupa, cocoon, pheromone, metamorphosis
3. ovipositor, hindgut, midgut, gastric ceca
4. warning coloration, Batesian mimicry, behavioral variation, camouflage
5. aorta, nymph, tympanum, midgut

Review

1. Insect species are able to adapt to new environments rapidly because individual insects possess (a) jointed legs (b) pheromones (c) short life spans (d) exoskeletons.
2. Differences in mouthparts are an example of (a) structural variations (b) physiological variations (c) behavioral variations (d) metamorphosis.
3. Butterflies and moths belong to the order (a) Orthoptera (b) Lepidoptera (c) Coleoptera (d) Homoptera.
4. An insect's legs and wings are attached to its (a) head (b) thorax (c) abdomen (d) spiracles.
5. The function of the gastric ceca is to (a) secrete digestive enzymes (b) shred food (c) store undigested food (d) remove chemical wastes.
6. The immature form of an insect that undergoes incomplete metamorphosis is a (a) larva (b) pupa (c) caterpillar (d) nymph.
7. Metamorphosis benefits insect species by (a) allowing different developmental stages to fulfill different functions (b) eliminating competition between larvae and adults (c) promoting survival during harsh weather (d) all of the above.
8. Bees perform the round dance to inform other bees of (a) the presence of food near the hive (b) the distance to a food source (c) the direction of a food source (d) all of the above.
9. The substance called queen factor (a) causes larvae to develop into queen bees (b) attracts drones (c) prevents larvae from becoming sexually mature (d) fertilizes eggs.
10. A pheromone is a chemical that (a) regulates metamorphosis (b) digests food (c) regulates the process of cellular respiration in insects (d) affects the behavior or development of other animals of the same species.
11. How has the small size of insects contributed to their success?
12. Give an example of physiological variation in insects.
13. Name the four primary characteristics that entomologists consider when classifying insects.

14. What are three ways that insects cause harm to human society?
15. Name the three main segments of the insect's body.
16. How is the insect's system for transporting oxygen to cells different from the system that is found in most other animals?

17. What is the difference between camouflage and warning coloration?
18. Summarize the role of molting hormone in controlling metamorphosis.
19. What three types of individuals make up a honeybee society?
20. How do crickets produce sounds? What functions do these sounds serve?

Critical Thinking

1. Farmers who use insecticides to control insect pests often find that strains of insects rapidly appear that are resistant to the insecticide. How do these insecticide-resistant strains develop? Why does this happen so quickly among insects?
2. What might you infer about an insect's food source from the fact that the insect has siphoning mouthparts?
3. Study the drawing of the insect on this page. Based on the information in Table 33-1, to which order does this insect belong? What characteristic or characteristics enabled you to classify the insect?
4. In addition to the exoskeleton, another characteristic thought to limit the size of insects is their system of tracheae. Suggest a reason why this means of transporting gases might be a limit to insect size.
5. During their lives worker bees move from first feeding and caring for larvae inside the hive to later gathering nectar and de-
fending the hive. Why do you think this sequence is more advantageous for the hive than the reverse sequence might be? (Hint: Consider the fact that a worker bee has a stinger.)
6. What characteristics may have helped insects survive the major climatic changes that led to the extinction of the dinosaurs and many other species about 65 million years ago?

Extension

1. Collect a variety of insects and use a field guide to help you classify your specimens. Share your collection with the class.
2. Read the article "The Masters of Mimicry," *Discover,* June 1985, pp. 56–63. Even though the insects described in this article have effective warning coloration, some are still eaten. What might happen if their protective coloration were perfect?
3. Examine the photomicrographs of insect mouthparts in "Chewers, Lappers, Suck-
ers, and Spongers," *International Wildlife,* September/October 1986, pp. 42–47. Make a list of the types of mouthparts shown, the insects that have them, and the type of food eaten by the insect.
4. Use your school library or public library to research the topic of sociobiology. Prepare a written report on what sociobiologists mean by *altruistic behavior* and how they explain such behavior in insect societies.

Chapter 34 Echinoderms

This chapter discusses the distinguishing features of echinoderms. It emphasizes body structure and function and how echinoderms differ from other invertebrates.

Correlation Chart

Section Number	Section Objectives	Pages	Features	Section Review
34.1	1. Discuss the origin of echinoderms, and tell why scientists believe that echinoderms may be related to the chordates.	511–514		2, 5
	2. List the characteristics that distinguish echinoderms from other phyla.		Biology in Process, p. 512	1, 3
	3. Name representative species of each echinoderm class.			4, 5
	4. Describe the procedure Jacques Loeb followed to determine the cause of parthenogenesis.			
34.2	1. Describe the general body plan of the sea star.	515–517	Think, p. 516	1, 4
	2. Explain how the water-vascular system aids in movement and feeding.			2, 3
	3. Describe sexual and asexual reproduction in the sea star.		Biology Lab, p. 517	5

Resources*

Print

Barnes, Robert L. *Invertebrate Zoology*, 4th ed. New York: Holt Rinehart & Winston, 1980.

Coulombe, Deborah A. *The Seaside Naturalist: A Guide to Nature Study at the Seashore*. Englewood Cliffs, N.J.: Prentice-Hall, 1984.

Haugh, B.N., and B.M. Bell. "Fossilized Viscera in Primitive Echinoderms." *Science* 29 (August 8, 1980): 653-657.

Audiovisual/Software

Echinoderms: Sea Stars and Their Relatives. Film, 17 min. Britannica.

Life Story of the Sea Star. Film, 11 min. Britannica.

*Additional references are listed in *Modern Biology, Teacher's Program Guide.*

Section Teaching Guides

34.1 Diversity

Approach

The main objective of this section is to familiarize students with the diversity of echinoderms, focusing on those characteristics that distinguish echinoderms from other invertebrates. In discussing echinoderm evolution, draw on the concepts of evidence of evolution (Chapter 15), and classification and taxonomy (Chapter 18). You might also review the information presented in Chapter 29 on invertebrate body plan and development.

In presenting the text, compare the organization of echinoderms with that of other invertebrates such as sponges, flatworms, and mollusks. Also point out to students that the information in this section provides background for the discussion of chordate evolution in Chapter 35.

Concept Framework

Even if students do not live in an area where they can observe echinoderms in their natural habitats, students have probably seen pictures or preserved specimens of sea stars, sand dollars, and other echinoderms. Thus they know that echinoderms look quite different from any other group of animals. Use this knowledge as a starting point to introduce students to the study of echinoderms. Ask students to think of possible reasons that natural selection would have favored such characteristics as pentaradial symmetry and a water-vascular system. Tell them to look for evidence of the adaptiveness of these characteristics as they read about the evolution and classification of echinoderms.

Emphasis

In your discussion, stress the idea that echinoderms possess many characteristics that are unique to invertebrates. Students should be able to identify examples of these unique characteristics for each of the echinoderm classes.

Using the Features/Parthenogenesis in the Eggs of Sea Urchins

The main process skills used in this feature are hypothesizing and experimenting. Jacques Loeb hypothesized that the chemical process associated with parthenogenesis was similar to that involved in fertilization by sperm. He then designed a series of painstaking experiments to determine what induced parthenogenesis in sea urchin eggs.

In discussing the feature, note that Loeb did not draw hasty conclusions about the results of his research. He tested all of the variables to determine whether or not they affected the development of the sea urchin eggs. For example, after establishing that exposure to magnesium stimulated parthenogenesis, Loeb conducted further experiments that indicated that the concentration of the chemical was the key factor, not the chemical itself.

Also remind students that Loeb conducted his research during the late nineteenth century, a time when the idea that all life processes could be explained in chemical terms was a new and radical concept. Note that Loeb's work was significant, not only because it paved the way for understanding the processes of parthenogenesis and fertilization, but because it supported his belief that chemical activity formed the basis of life processes.

Guest Speaker

Invite an expert on echinoderms from a local university who can talk about the commercial importance of echinoderms to humans. For example, the speaker might discuss the fact that sea stars prey on other animals important to humans, the extensive use of sea urchin eggs in scientific research, and the use of echinoderms for food and decorative items.

Section Review Answers

(See p. 514.)

34.2 Structure and Function

Approach

This section focuses on the relationship between structure and function in the biology of the sea star, a representative echinoderm. The section builds on the concepts of cell specialization and multicellular organization (Chapter 5) and adaptation (Chapter 16). In presenting the text, encourage students to draw comparisons between the body plan, systems, development, and behavior of echinoderms with those of the other invertebrates they have read about in Chapters 29 through 33. Tell them to also look for examples of the adaptations of sea stars to marine and coastal environments.

Concept Framework

Students have most likely seen or used a water or an air pump and have an understanding of how pressure changes can create and release suction. Point out the similarities of process in these pumps and in the sea star's tube feet which function as thousands of tiny pumps. Have students refer to Figure 34-6 to see how the water-vascular system enables echinoderms to move about and obtain food.

Emphasis

Make certain that students can explain how the structure of the sea star's external and internal anatomy is related to its function. Students should also be able to identify ways in which the sea star's anatomy reflects adaptations to its environment.

Identifying Echinoderms

Bring in several samples of echinoderm skeletons to exhibit to the class. Ask students to identify the class to which each specimen belongs. Also ask them to point out key echinoderm features, such as five-part symmetry, and to identify examples of structural adaptations.

Section Review Answers

(Continued from p. 516.)

3. After opening the shell by pulling with its tube feet, the sea star inserts its stomach into the clam, partially digests the clam's soft parts, and then withdraws its stomach back into its body, where digestion is completed.
4. Fluid in the coelom distributes oxygen and nutrients. Excretion and gas exchange take place by diffusion through the tube feet and skin gills.
5. no, because each broken-off arm could regenerate into a new sea star

Laboratory Preparation

Background

In this lab students examine the similarities and differences among three echinoderms. Either living or nonliving materials can be used. However, microscopic examination and observation of living animals in sea water can be more interesting. Students compare the oral surface of the starfish with the external surface of the sea urchin to develop an understanding about echinoderm organization. Students observe the differences between brittle stars and starfish, which illustrate how different lifestyles can develop using the same symmetry but different organs. Students learn that because brittle stars use muscles for locomotion, they can move much more rapidly.

Preparation

The cost of a few demonstration specimens is consistent with other biological supply costs. The cost and problems involved in obtaining living materials for every student may be beyond the resources of most institutions; however, since the animals are not dissected, they can be reused.

Conclusions

1. Three features common to all three types of echinoderms are a spiny exterior, radial symmetry, and tube feet.
2. The sea star has no anterior and posterior ends. Its arms are arranged radially; therefore, no direction is a preferred direction.

Inquiry

Students will list advantages to radial symmetry, such as the ability to use sense organs (for example, eyespots) without a change in direction. They will also list some disadvantages, such as the inefficiency inherent in the lack of specialized anterior and posterior organs.

Chapter Review

Vocabulary

1. Evisceration is a defense mechanism used by sea cucumbers. When threatened, they eject their internal organs and regenerate lost parts later.
2. The endoskeleton of echinoids is called a test.
3. Pedicellariae are small structures resembling forceps that surround the spines and help protect and clean the sea star's body.
4. The major structures of the water-vascular system include the sieve plate, stone canal, ring canal, radial canals, tube feet, and ampullae.
5. The bipinnaria is the larval stage of an echinoderm. It is free-living and bilaterally symmetrical.

Review

11. Scientists believe that early echinoderms were bilaterally symmetrical because modern echinoderm larvae are bilaterally symmetrical.
12. Echinoderms inhabit marine environments ranging from shallow coastal waters to deep ocean trenches.
13. The short spines of the sand dollar enable the organism to burrow in sand.
14. Sea stars are of particular interest to humans because they prey on some food sources used by humans, such as clams and oysters.
15. The Class Crinoidea is most similar to the early echinoderms.
16. A sea cucumber takes food in with its tentacles, stuffs the tentacles into its mouth, and then cleans off the food.
17. The sea star wraps its stomach around the soft parts of a clam or other food and partially digests it. Then it withdraws its stomach, containing the partially digested food, into its body, where enzymes secreted by digestive glands complete the digestive process.
18. Water enters the sea star's body through the sieve plate, a small opening on the aboral surface.
19. According to Loeb's research, embryo development begins when an increase in osmotic pressure causes a loss of water in the egg.
20. During metamorphosis the sea star changes from a bilaterally symmetrical larva to a penta-radially symmetrical adult.

Critical Thinking

1. The free swimming larva allows the sea cucumber, sea lily, and other echinoderms to spread out over large areas of the ocean floor. If the larvae were as sessile as the full-grown echinoderms are, all the offspring would be clumped together in the same area. The food supply would then decrease due to overpopulation, and many members of the species would die.
2. Students might include introducing a natural enemy of the starfish into the environment. Another possibility would be poisoning the starfish with a poison, such as lime, that will not hurt the oyster.
3. A number of fossilized echinoderms exist because their hard outer shell fossilizes easily.
4. The small spines keep space between the hard coral surface and the skin of the brittle star. They also provide protection against the many predators that would see the brittle star as it climbs on the coral.

Chapter 34 | *Echinoderms*

Introduction

The spiny-skinned echinoderms are among the oldest phyla of animals on earth. They are very beautiful too, with their bright colors and five-part, or pentaradial, body plans. In this chapter you will learn about the characteristics of sea stars, sea urchins, and other echinoderm species. Scientists suggest that this small group of marine animals may share a common ancestor with the chordates.

Chapter Outline

Chapter Concept

As you read, note the unique structure of the echinoderm, and keep in mind how much echinoderms differ from the other invertebrates you've studied.

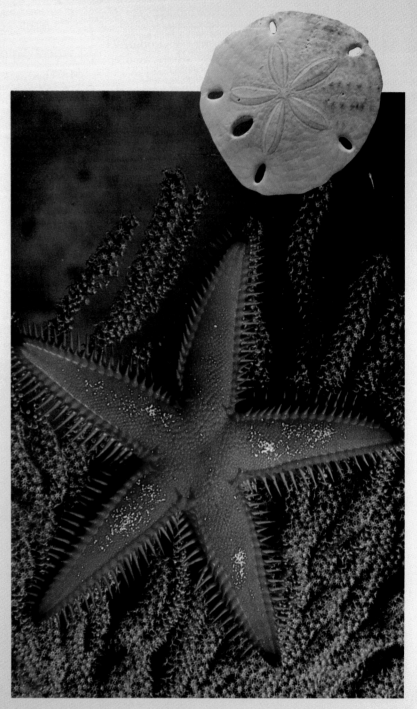

Sea star, Galápagos. Inset: Sand dollar, *Encope* sp., Baja, California.

For additional teaching information see pp. T34A–T34D.

34.1 Diversity

Members of the Phylum Echinodermata ("spiny skin") include the sea stars, brittle stars, sand dollars, and sea cucumbers. They inhabit marine environments ranging from shallow coastal waters to ocean trenches over 10,000 m deep. They range in size from 1 cm to 1 m in diameter and often have brilliant coloring.

Echinoderms bear an important evolutionary relationship to vertebrates. Evidence indicates that the echinoderms may share a common ancestor with the lower chordates and hence the vertebrates.

Characteristics

The characteristics of the echinoderms include the following, most of which are not shared by any other invertebrate phylum:

- Most undergo metamorphosis from a free-swimming, bilaterally symmetrical larva to a bottom-dwelling adult with radial symmetry. Most echinoderms have five radii or multiples thereof. Such a symmetry is called **pentaradial symmetry.**
- They have an internal skeleton called an **endoskeleton,** composed of calcium plates, which may include protruding spines.
- They have a **water-vascular system,** which is a network of water-filled canals.
- They have numerous small, movable protrusions called **tube feet** that aid in movement, feeding, respiration, and excretion.
- They have no circulatory, respiratory, or excretory systems.
- They have a nervous system but have no head or brain.
- They have two sexes, and reproduce sexually or asexually.

Evolution and Classification

The fossil record of echinoderms dates back to the Cambrian period, over 500 million years ago. Because echinoderm larvae are bilaterally symmetrical, scientists infer that they evolved from bilaterally symmetrical ancestors. Compare the larva of a hypothetical early echinoderm with the larva of a modern sea star pictured in Figure 34-1.

Section Objectives

- *Discuss the origin of echinoderms, and tell why scientists believe that echinoderms may be related to the chordates.*

- *List the characteristics that distinguish echinoderms from other phyla.*

- *Name representative species of each echinoderm class.*

- *Describe the procedure Jacques Loeb followed to determine the cause of parthenogenesis.*

Tip: The concept of symmetry can be demonstrated by placing a mirror on pictures of a person and a sea star. A person is bilaterally symmetrical and will look correct in the mirror only when it is positioned down the vertical center of the body. A sea star is pentaradially symmetrical and will look correct in the mirror whenever it is positioned so that it cuts through the radius of the sea star.

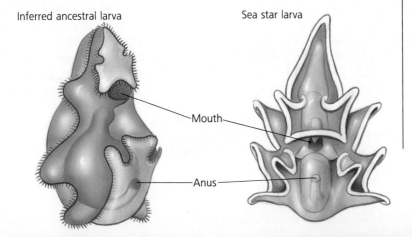

Inferred ancestral larva

Sea star larva

Mouth

Anus

Figure 34-1 Note the similarities and differences between the inferred ancestral larva and the modern larva.

Echinoderms **511**

The fossil record also indicates that the early adult forms of the early echinoderms were probably sessile, not free-living as most modern forms are. Scientists infer that radial symmetry evolved as an adaptation to this sessile existence. Conditions may have later favored free-living species, yet most modern echinoderms have retained radial symmetry. Taxonomists have divided the 6,000 species of echinoderms into five classes: Crinoidea (*kruh-NOID-ee-uh*), Asteroidea (*uh-STIR-OID-ee-uh*), Ophiuroidea (*AHF-ee-yur-OID-ee-uh*), Echinoidea (*ECK-uh-NOID-ee-uh*), and Holothuroidea (*HAHL-oe-thur-OID-ee-uh*).

Crinoidea

The Class Crinoidea ("lilylike"), which includes sea lilies and feather stars, most closely resembles the fossils of Cambrian echinoderms. Crinoids are sessile. They have a long stalk that is attached to rocks or to the sea bottom. Feather stars eventually become detached. Unlike that of other echinoderm species, the mouth of crinoids does not face the sea bottom. The five arms that extend from the body branch out to form many more arms—up to 200 in some feather star species. Sticky tube feet located at the end of each arm capture food and serve as a respiratory surface.

Figure 34-2   This adult crinoid of the genus *Comanthina*, found in the Philippines, is sessile.

Biology in Process

Parthenogenesis in the Eggs of Sea Urchins

Jacques Loeb (1859–1924), a German-born American physiologist, believed that all life processes could be explained in chemical and physical terms. Loeb was particularly interested in the process of fertilization. He wondered what chemical changes caused a fertilized egg to develop into an embryo.

Loeb knew that brief exposure to sea water could make unfertilized sea urchin eggs begin to divide. Reproduction of organisms without the fusion of gametes of opposite sexes is called parthenogenesis. Loeb hypothesized that the chemical process in parthenogenesis was similar to fertilization by sperm.

To support his hypothesis, Loeb attempted to induce parthenogenesis in the laboratory. He believed that some chemical found in sea water caused parthenogenesis to occur in nature. To test this concept, he placed unfertilized sea urchin eggs into various magnesium solutions for about two hours. He then returned them to ordinary sea water. As he predicted, one of the samples that had been exposed to magnesium began to divide.

Answer to Biology in Process: Loeb could have placed a sample of unfertilized sea urchin eggs directly into the plain sea water. If these eggs began to divide, Loeb would know sperm were present and had fertilized them.

Asteroidea

The sea stars, or starfish, belong to the Class Asteroidea ("starlike"). They live all over the world in coastal waters and along rocky shores. They are economically important because they prey on oysters, clams, and other organisms used as food by people. The biology of sea stars is discussed in detail in Section 34.2.

Ophiuroidea

The 2,000 species of basket stars and brittle stars belong to the Class Ophiuroidea ("snakelike"), the largest echinoderm class. Basket stars and brittle stars live primarily on the sea bottom, often beneath stones or in the crevices and holes of coral reefs.

Ophiuroids are distinguished by their long, narrow arms, which allow them to move very quickly. The thin, flexible arms of basket stars resemble a tangle of snakes. Brittle stars, so named because parts of their arms break off readily, can regenerate missing parts. Ophiuroids feed by raking in food with their arms; by gathering food from the ocean bottom with their tube feet; or by trapping suspended particles with their tube feet or with the aid of mucous strands found between their spines.

Figure 34-3 Brittle stars, *Ophiuroidea* sp., such as these from the Pacific Ocean, can feed by raking in food with their long, narrow arms.

In his next experiments Loeb kept the amount of magnesium in the solution constant and varied the length of time that the eggs were kept in solution. He found that exposures of 100 to 135 minutes gave the best results. However, none of the embryos exposed survived for more than two days. Next he held the exposure time constant and varied the amount of magnesium. He eventually produced virtually normal larvae.

However, was it really magnesium that caused parthenogenesis? Perhaps some other factor was involved. In other experiments Loeb found that other chemicals, such as sodium, potassium, calcium, and even sugar, induced parthenogenesis. So it wasn't the magnesium after all. Loeb determined that parthenogenesis began when the eggs lost a certain amount of water by osmosis. The concentration of the chemical in the solution was the important factor, not the chemical itself.

Loeb then conducted similar experiments with other echinoderms, as well as with chordates. He eventually succeeded in raising frogs to sexual maturity from unfertilized eggs.

Loeb's work paved the way for other scientists to study

parthenogenesis. It was later discovered that changes in the egg cell cytoplasm could be caused by many stimuli, including temperature and ultraviolet light. Such changes can induce parthenogenesis.

■ Suggest an experimental procedure Loeb might have used to ensure that cell division was not the result of fertilization by sperm present in the sea water.

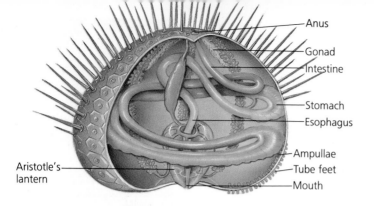

Anus
Gonad
Intestine
Stomach
Esophagus
Ampullae
Tube feet
Mouth
Aristotle's lantern

Figure 34-4 Note the spines, tube feet, and rigid endoskeleton of the sea urchin as diagrammed (left) and in the photo (right).

Figure 34-5 This sea cucumber, *Psolus chitonoides,* has ejected its intestines in self-defense.

Answers to Section Review
1. five radii or a multiple of five extending from a central body
2. Scientists believe that echinoderms share a common ancestor with chordates.
3. Crinoids are sessile and they are the only class in which the mouth does not face the sea bottom.
4. Sea urchins use their tube feet to move and the Aristotle's lantern to scrape food from hard sea bottoms. The flat, round shape and short spines of sand dollars are adaptations for burrowing in sand.
5. their great mobility and their ability to obtain food in a variety of ways

Echinoidea

About 900 species of sea urchins and sand dollars belong to the Class Echinoidea ("hedgehoglike"). As you can see in Figure 34-4, the internal organs are enclosed within a compact, rigid endoskeleton called a **test.** Most species of echinoids grind their food with a complex jawlike mechanism called **Aristotle's lantern.**

The spherical sea urchins are well adapted to life on hard sea bottoms. These echinoids use their tube feet for locomotion and feed by scraping algae from hard surfaces with teethlike structures that are part of Aristotle's lantern. They have barbs on their long spines and in some species venom that protects sea urchins from predators.

Sand dollars live along seacoasts. As their name implies, they live in sandy areas and have the flat, round shape of a silver dollar, which is an adaptation for shallow burrowing. The short spines of the sand dollar aid in locomotion and burrowing and help clean the surface of the body. Sand dollars use their tube feet to move and to scoop up food.

Holothuroidea

The armless sea cucumbers belong to the Class Holothuroidea. Most reside on the sea bottom. Because they do not have a large endoskeleton, their bodies are soft. Tube feet are present on their aboral side. A fringe of tentacles, actually modified tube feet, surrounds the mouth and sweeps up sediment and water. Sea cucumbers then stuff their tentacles into their mouths and clean the food off them.

When threatened, many sea cucumbers eject their internal organs through the anus, a defense mechanism called **evisceration.** They later regenerate their lost parts.

Section Review

1. What is *pentaradial symmetry?*
2. What is the evolutionary significance of echinoderms?
3. How do crinoids differ from other echinoderm classes?
4. Contrast the ways in which sea urchins and sand dollars are adapted to their environment.
5. What characteristics might have enhanced the evolutionary success of the ophiuroids?

Info: A dried specimen of an Aristotle's lantern taken from an echinoid resembles a small lantern.

34.2 Structure and Function

This discussion of anatomy and behavior will focus on the sea star as a representative echinoderm. The anatomy and behavior of sea stars illustrates the relationship between structure and function in many echinoderms. Also note that the structure and movement of echinoderms is unique among invertebrates.

Body Plan

The sea star's mouth is located on the underside of the body, called the **oral surface.** The top of the body is called the **aboral surface.** While some invertebrates have an external skeleton, a sea star, like all echinoderms, has an endoskeleton—a skeleton found within the body. The sharp, protective spines on the aboral surface form from calcium plates called **ossicles,** which are covered with a thin epidermal layer of cells. Groups of **pedicellariae** surround the spines. Pedicellariae, which resemble tiny forceps, help protect and clean the surface of the body.

Water-Vascular System

While most invertebrates swim, wiggle, or use appendages for movement, sea stars use a water-vascular system. This is a network of canals in which muscle contractions create hydrostatic pressure, which in turn permits movement. Water enters the **sieve plate,** a small opening on the aboral surface. It passes through the madreporite and down the **stone canal.** Use Figure 34-6 to trace the path of water from the **ring canal** encircling the mouth to the five **radial canals** that extend into each arm.

The radial canals carry water to hundreds of paired, hollow tube feet. The upper end of each foot connects to a bulblike sac called an **ampulla.** Muscles in the tube feet contract, forcing water into the ampullae and creating suction, which causes suckers at the tip of the feet to grip the surface. When the ampullae contract, water enters the feet, releasing the suction and causing the feet to

Section Objectives

- *Describe the general body plan of the sea star.*

- *Explain how the water-vascular system aids in movement and feeding.*

- *Describe sexual and asexual reproduction in the sea star.*

Info: A "skeleton" of sorts is created in the sea star when the canals fill with water, allowing muscles to contract against this temporary structure.

Tip: Students may be unable to detect the endoskeleton in a sea star specimen that has been preserved dry. Use at least a few fresh specimens for dissection or demonstration.

Figure 34-6 Trace the path of water in the water-vascular system of the sea star.

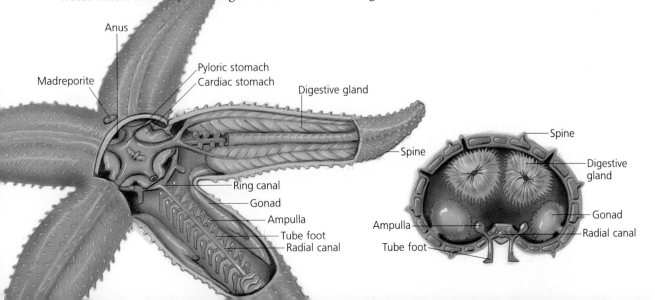

Anus

Pyloric stomach
Cardiac stomach

Digestive gland

Madreporite

Spine

Ring canal

Gonad

Ampulla

Tube foot

Radial canal

Spine

Digestive gland

Gonad

Radial canal

Ampulla

Tube foot

Figure 34-7 This ochre sea star is feeding on a California mussel along the Pacific coast.

Think

Why is it an adaptive advantage for aquatic, externally fertilizing organisms to produce millions of eggs?

Answer to Think
Even if predators eat some of the eggs or some are washed away, there is still a good chance that at least a few eggs will be fertilized.

Answers to Section Review
1. The oral surface is the side where the mouth is; the aboral surface is the opposite side. In sea stars the oral surface faces the sea bottom.
2. Water enters through the sieve plate and then goes through the stone canal, ring canal, radial canals, and the tube feet.
(See pp. T34A–T34D.)

extend. These coordinated muscle contractions enable the sea star to climb slippery rocks or to capture prey.

Feeding and Digestion

Sea stars use their tube feet to obtain food—usually mollusks, worms, and other slow-moving animals. The sea star attaches the suckers of its tube feet to both halves of a clamshell and pulls until the shell opens a crack. The sea star then turns its stomach inside out through its mouth and inserts its stomach into the clam.

Enzymes secreted by the sea star's stomach digest the clam's soft parts while they are still in the shell. The sea star then withdraws the stomach, containing the partially digested food, back into its body, where the digestive process is completed.

Other Body Systems

The sea star has no circulatory, excretory, or respiratory systems. Fluid in the coelom bathes the organs and distributes nutrients and oxygen. Gas exchange and waste excretion take place by diffusion through the thin walls of the tube feet and through the **skin gills,** hollow tubes that project from the coelom lining.

The sea star has no head or brain. A **nerve ring** surrounds the mouth and branches off into nerve cords that extend into each arm. Each arm has an eyespot that responds to light and a tentacle that responds to touch. The tube feet also respond to touch.

Reproduction

Each arm of the sea star has two gonads, which produce eggs in females and sperm in males. Fertilization occurs externally. Females produce up to 200 million eggs in one season. Each fertilized egg develops into a bilaterally symmetrical, free-swimming larva called a **bipinnaria.** After about two years the larva settles to the bottom, and metamorphosis begins. During metamorphosis the sea star develops into a pentaradially symmetrical adult.

Sea stars can also reproduce asexually by regenerating lost parts. In fact, a completely new sea star may grow from a segment of an arm as long as part of the central disk is still attached.

Section Review

1. Distinguish between the *oral surface* and the *aboral surface.*
2. Trace the path of water through the water-vascular system.
3. Describe the way a sea star grips and digests a clam.
4. How are gas exchange, distribution of nutrients, and excretion of wastes accomplished in the sea star?
5. Was the former practice of breaking the arms off of sea stars to prevent them from feeding on clams and oysters a good idea?

Laboratory

Observing Echinoderms

Objective
To examine the similarities and differences among three echinoderms

Process Skills
Observing, inferring

Materials
Preserved sea star, forceps, dissecting pan, dissecting microscope, preserved brittle star, preserved sea cucumber

Method
1 Obtain a sea star from your teacher. Using forceps to handle the sea star, rinse it thoroughly under tap water to wash away the preservative. Then place the animal in a dissecting pan. Observe its overall appearance. What type of symmetry do you observe in the sea star?
2 Locate the sieve plate—a stony disk on the upper surface. What is its function?
3 Turn the sea star over to see the groove running down the middle of each ray. What morphological feature is at the center where the grooves join? The surface of the sea star facing you is the oral side, and the opposite surface is the aboral side.
4 Find the hydraulic tube feet in the grooves of each ray. How many rows are present?
5 Observe the mouth closely. How does the animal digest food?
6 Observe the movable spines that cover the edges of the grooves. What is the function of the spines?
7 Locate the small tentacle at the tip of a ray. Find the eyespot on the tentacle. The eyespot is sensitive to light but cannot form images. How many eyespots does the animal have?
8 Place the sea star, with its aboral surface up, under a dissecting microscope. Note the small fingerlike structures among the spines.

What do you think is their function? Note the tiny pincers. What is their function?
9 Examine a preserved brittle star. How does it differ in overall appearance from the sea star? The arms of the brittle star are made of bone that is jointed like vertebrae. The arms, which are moved by muscles in the living animal, are very flexible. Examine an arm closely. In addition to their shape, what is one way in which the arms differ from the arms of a sea star? Find the tube feet.
10 Examine a preserved sea cucumber. Find the tube feet. Locate the tentacles surrounding the mouth. These tentacles collect food and place it in the mouth. Can you find the pattern of radial symmetry in this organism?

Conclusions
1 Name three features you have noted that are common to all three types of echinoderms you have examined.
2 A sea star can move equally well in any direction. Why do you think this is so?

Inquiry
Review the features and discuss the advantages and disadvantages of radial symmetry.

Chapter 34 Review

Vocabulary

aboral surface (515)	evisceration (514)	pentaradial symmetry (511)	stone canal (515)
ampulla (515)	nerve ring (516)	radial canal (515)	test (514)
Aristotle's lantern (514)	oral surface (515)	ring canal (515)	tube foot (511)
bipinnaria (516)	ossicle (515)	sieve plate (515)	water-vascular system (511)
endoskeleton (511)	pedicellaria (515)	skin gill (516)	

1. Describe the process of evisceration in the sea cucumber.
2. What term is used for the endoskeleton in echinoids?
3. What are pedicellariae?
4. List the major structures in the water-vascular system.
5. What is a bipinnaria, and what are its major characteristics?

Review

1. Scientists believe that the earliest adult echinoderms were (a) sessile and bilaterally symmetrical (b) free-living and bilaterally symmetrical (c) sessile and radially symmetrical (d) free-living and radially symmetrical.
2. Echinoderms lack all of the following organ systems except (a) a respiratory system (b) an excretory system (c) a circulatory system (d) a digestive system.
3. The echinoderm class that exhibits evisceration is (a) Class Ophiuroidea (b) Class Crinoidea (c) Class Holothuroidea (d) Class Echinoidea.
4. Crinoids superficially resemble (a) lilies (b) silver dollars (c) hedgehogs (d) cucumbers.
5. Aristotle's lantern is a jawlike structure that is characteristic of (a) ophiuroids (b) asteroids (c) echinoids (d) holothuroids.
6. The endoskeleton of the sea star is composed of calcium plates referred to as (a) tests (b) ossicles (c) ampullae (d) sieve plates.
7. The number of eggs that a female sea star can produce in any given season is (a) 20,000 (b) 200,000 (c) 2 million (d) 200 million.
8. Gas exchange and excretion of wastes take place by diffusion through the walls of the (a) ampullae (b) skin gills (c) radial canals (d) test.
9. When the ampullae of the sea star fill with water, the tube feet (a) attach to a surface (b) extend outward (c) fill with water (d) grasp food.
10. Sea stars detect light with (a) the nerve ring (b) tentacles on each arm (c) the tube feet (d) an eyespot on each arm.
11. What evidence has led scientists to believe that early echinoderms exhibited bilateral symmetry?
12. In what type of environment are echinoderms found?
13. What function do the short spines of the sand dollar serve?
14. Why are sea stars of particular interest to humans?
15. Which modern echinoderm class is inferred to be the most similar to the early echinoderms?
16. How does the sea cucumber transport food to its mouth?
17. Summarize the process of digestion in the sea star.
18. Where does water enter the sea star's body?

19. According to Jacques Loeb's research, what critical factor induces parthenogenesis in sea urchin eggs?

20. What happens during the metamorphosis of a sea star?

Critical Thinking

1. Sea cucumbers and sea lilies are relatively sessile animals. However, the larvae of these species are free-swimming. What advantage does a free-swimming larva provide these echinoderms?

2. For many years oyster farmers, who raise oysters for both food and pearls, have been plagued with starfish that invade their oyster beds and kill their oysters. Many oyster farmers have lost their entire crop and have gone out of business. Devise a way for the farmers to rid themselves of the starfish without killing their oysters.

3. Scientists have found many echinoderm fossils from the Cambrian period. However, they have found few fossils of other species that lived during the Cambrian period. What might explain the large number of fossilized starfish, sea urchins, and other echinoderms?

4. Look at this photograph of a brittle star (*Ophiothrix purpurea*) on a coral. Look carefully at the small spines sticking out from its body. Considering that these brittle stars live most of their lives on coral, name two advantages that these spines would provide.

Extension

1. Visit your nearest aquarium or a pet store that carries saltwater fish. Observe living starfish and note their special form of movement and their different colors and shapes.

2. Read the article by S. Weisburd entitled "Counting Falling Starfish in California," in *Science News,* August 17, 1985, p. 101. How could this news item be used by the conservationists on the Great Barrier Reef who are trying to stop the crown-of-thorns starfish from destroying the coral?

3. Research this question: Why do basket stars only open at night, and what do they use to protect themselves during the day?

4. Read the article by Ronald L. Shimek entitled "Sex among the Sessile," in *Natural History,* March 1987, pp. 60–63. In the local populations of sea cucumbers what are the adaptive advantages of simultaneous spawning?

5. Research and write a short paper on the different forms of evisceration among sea cucumbers. Tell how the sea cucumber uses evisceration to escape predators.

6. Research and write a short paper on a species of echinoderms that is not discussed in the preceding chapter. Focus on the particular adaptations that your chosen species shows that enable it to survive in its environment.

Chapter 35 | Evolutionary Trends in Invertebrates

This chapter explains how the various systems of invertebrates are similar even though they are widely diverse. It focuses on how invertebrates have become increasingly complex throughout evolutionary history.

Correlation Chart

Section Number	Section Objectives	Pages	Features	Section Review
35.1	1. Summarize the possible origins of invertebrates.	521–525		2
	2. Describe the embryo and larval development of invertebrates.			3, 5
	3. Summarize invertebrate symmetry, segmentation, and support.			1, 3, 4
	4. Analyze the organization of the phylogenetic tree of invertebrates.			2, 5
35.2	1. Distinguish between types of digestive systems.	526–529	Biology Lab, p. 529	1
	2. Summarize the increase in morphological complexity in respiratory systems.			
	3. Identify the different types of circulatory systems.			4
	4. Identify the increase in complexity in nervous systems.			2
	5. Identify the different types of excretory systems.			3
	6. State the adaptive advantages of the different types of reproductive systems.			

Resources*

Print

Barnes, Robert L. *Invertebrate Zoology*, 4th ed. New York: Holt, 1980.

Neisen, Thomas M. *The Marine Biology Coloring Book*. New York: Barnes and Noble, 1982.

Audiovisual/Software

Beach and Sea Animals. Film, 11 min. Britannica. *Survey of the Animal Kingdom*. Filmstrip. Teacher's guide. Educational Images, 1987.

The Vocabulary of Biology—The Invertebrates. 6 filmstrips. Nasco.

*Additional references are listed in *Modern Biology, Teacher's Program Guide*.

Section Teaching Guides

35.1 Diversity and Phylogeny

Approach

The primary objective of this section is to familiarize students with trends in invertebrate evolution. This section builds on the concepts of diversity through adaptation (Chapters 15 and 16), principles of classification and phylogeny (Chapter 18), and embryonic development of invertebrates (Chapter 29). In presenting the text, emphasize the chapter concept that invertebrate evolution has resulted in increasingly complex invertebrate groups. To help students understand how these organisms are related, you might refer to Figure 35-5 and trace the evolutionary history of invertebrates.

Concept Framework

All students are familiar with invertebrates such as insects and spiders. Tap this familiarity to provide a concrete foundation for concepts presented in this chapter by reminding students that invertebrates have no backbones. Ask students to name other familiar invertebrates. As they read this section, ask students to consider the diversity of organisms such as snails, crayfish, starfish, earthworms, and jellyfish and to note that these organisms are all classified in this group.

Emphasis

Make sure that students realize that similar characteristics among animals can indicate evolutionary relationships or convergent evolution. Although it often is difficult to determine which evolutionary process has taken place, explain to students that scientists use protein analysis as one tool to determine phylogenetic relationships.

Guest Speaker

Invite a speaker who can expand on this section's discussion of invertebrate evolutionary history. A university professor familiar with protein analysis techniques might be able to elaborate on the biochemical evidence considered in proposing phylogenetic relationships. You could also invite an expert in a particular invertebrate group, such as an entomologist, to explain the basis for the classification of particular diverse organisms.

Section Review Answers

(Continued from p. 525.)

2. Some scientists suggest that metazoans developed from flagellated protozoa that clumped together in colonies. The cells in these colonies gradually became more interdependent.

3. Examples include planula, such as the medusa stage of Hydrozoa; trochophore, such as the larval form of the snail; and bipinnaria, such as the larval form of the sea star.

4. A coelom enables an organism to move about more freely than an organism without a coelom. This improves its ability to search for food and to protect itself.

5. Indirect development insures that the larval form of the animal will not have to compete with the adult for food. Also many larval forms are free-swimming while their adult forms are sessile or slow moving. The larva is thus able to move into new environments and thereby lessen the population pressure.

35.2 Internal Anatomy

Approach

The main objective of this section is to show that invertebrate evolution has resulted in increasingly complex internal organization in invertebrates. Section 35.2 reviews the internal anatomy of the major invertebrate systems, including digestive, respiratory, nervous, circulatory, excretory, and reproductive. This section builds on the concept of diversity through adaptation (Chapters 15 and 16). In presenting the text, point out to students that the discussion of each system illustrates the fact that the less highly evolved invertebrates have simpler systems than the more highly evolved invertebrates.

Concept Framework

Students have studied the evolution and diversity of plants (Chapters 25–28). Build on that experience to provide a conceptual perspective for studying this section by asking students to apply the terms simple organism and complex organism to the various divisions of plants. Then elicit the observation that these terms describe degrees of evolutionary development and that the terms can also be applied to invertebrate evolution. Simple organisms are less highly evolved; complex organisms are more highly evolved. You might end this discussion by asking students as they read the text to arrange some familiar invertebrates, such as snails, earthworms, sponges, octopi, and beetles, in order of increasing complexity.

Emphasis

Make sure that students understand that phylogenetic trees are useful tools for outlining evolutionary pathways, but that the relationships among groups are not completely understood. Reinforce the concept that scientists use comparative studies in development and anatomy, as well as biochemical evidence, to obtain the information to construct phylogenetic trees.

Describing Invertebrate Phyla

Divide students into three groups. Make a list of major invertebrate phyla and divide the phyla among the three groups. Ask each group to make a chart for their phyla describing or drawing and labeling the internal anatomy of the digestive, respiratory, nervous, circulatory, excretory, and reproductive systems. Display the charts in a comparative wall display.

Section Review Answers

(Continued from p. 528.)

2. A closed circulatory system provides the organism with oxygen and nutrients much more quickly. This efficient system allows the organism to react to its environment more quickly.
3. Aquatic invertebrates are surrounded by water, which dilutes the ammonia gas that is released. The toxic ammonia gas could not be diffused from land animals quickly enough and these animals are not surrounded by water to absorb the gas; therefore, their systems enable them to release the ammonia in a nontoxic form and to reabsorb water that would otherwise be lost.
4. It allows the animal to receive stimuli before moving into a new environment.
5. Speed allows the animal to escape from predators and to catch numerous forms of food.

Laboratory Preparation

Background

In this lab students review the basic characteristics of invertebrates. Many features used to distinguish be-

tween major groups involve characteristics that are difficult to see in adult organisms, for example, whether an animal is a proto- or deuterostomate. Students become more aware of their environment by observing organisms that are common, but unfamiliar. Students learn about basic characteristics by comparing organisms that demonstrate similar characteristics but are very different from each other. If possible, provide some living organisms for the students to study.

Preparation

Live or preserved specimens of the organisms in this lab are available from a biological supply house.

Safety

If any live organisms are used, remind students to handle them in a humane manner.

Conclusions

1. Answers will vary according to specimens used.
2. Answers will vary according to specimens used but could include bell shape; long, flat body; jointed appendages; spine covered body; and tube feet.

Inquiry

Answers will vary, but one advantage to using preserved specimens is that it allows the students to examine the organisms closely without danger. The main disadvantage is that students are not able to observe the organism living in its native habitat, nor are they able to observe its movement.

Chapter Review

Vocabulary

1. Radial cleavage occurs when cells divide parallel with or at angles to the polar axis. Spiral cleavage occurs when cells divide in a spiral arrangement.
2. Determinate cleavage occurs when each cell of a four-celled protostome develops into a quarter of the gastrula. Indeterminate cleavage occurs when each cell of a four-celled deuterostome develops into a complete gastrula.
3. In an embryo, a protostome develops when the blastospore develops into the mouth, the embryo undergoes spiral cleavage, and the cells in the

embryo show determinate cleavage. A deuterostome develops when the blastospore develops
into the anus, the embryo undergoes radial
cleavage, and the cells in the embryo show indeterminate cleavage.

4. In direct development the young animal is born
or hatched having the same appearance as the
adult. In indirect development the organism
goes through an intermediate stage called the
larval stage.

5. In an open circulatory system blood is pumped
from vessels into the body cavity and returned to
the vessels. In a closed circulatory system the
blood remains in the vessels, and nutrients and
gases are exchanged through the membranes of
capillaries and cells.

Review

11. Scientists support that invertebrates appeared
on earth over 800 million years ago.

12. The mesoderm, which develops into the coelom,
allows the free movement of organs within the
body and enables the organism to move about
more freely.

13. Radial symmetry allows sessile organisms to receive stimuli from all directions.

14. The complete digestive system separates the organism's food from its feces, thus allowing cells
in the digestive tract to specialize for one
function.

15. The endoskeleton allows organisms to move
easily and to grow large, two characteristics
which are not possible with a hard exoskeleton.

16. Scientists base their construction of the invertebrate phylogenetic tree on all of the following:
comparative morphology, patterns of embryo
development, study of fossil records, and comparisons of larva development.

17. A closed circulatory system and a complete digestive system provide the organism with nutrients and oxygen quickly, thus allowing the
organism to increase its motility.

18. Terrestrial animals need specialized excretory
structures because they must conserve water.

19. A complex nervous system allows organisms to
monitor and respond to their environments
more readily.

20. Hermaphroditism requires an organism to
maintain both sets of gonads. This energy expenditure is only justified when the organism is
sessile or lives in underpopulated areas.

Critical Thinking

1. Invertebrates such as sponges and planarians may
be able to regenerate lost body parts because their
tissues are not highly specialized. Other invertebrates cannot regenerate lost parts because their
systems are too complex.

2. Invertebrates with hard shells were more likely to
be fossilized, since soft bodies will not be preserved. Scientists' view of invertebrate evolution
might be effected because it would be difficult to
get an accurate picture of the evolution of any
soft-bodied animals.

3. They possess cilia and a mouth that adapt them to
free-living life.

Chapter 35 | Evolutionary Trends in Invertebrates

Introduction

Invertebrates exist in all shapes and sizes, from the corals to the spiders. This diversity can make it difficult to understand how they are related. However, evolutionary and morphological studies indicate that the systems of invertebrates became increasingly complex as new invertebrate groups evolved.

Chapter Outline

Chapter Concept

As you read, notice the increasing complexity of the evolutionary adaptations in invertebrates.

Soft coral, *Dendronephthya* sp., island of Truk. Inset (top): Spiny orb weaver spider, *Gasteracantha elipsoides.* Inset (bottom): Flatworm, Palau Islands, Micronesia.

For additional teaching information see pp. T35A–T35D.

35.1 Diversity and Phylogeny

Invertebrates first appeared on earth over 800 million years ago, and by the late Precambrian era numerous invertebrates were common. The fossil record from the beginning of the Cambrian period, 500 million years ago, includes members of all the modern invertebrate phyla as well as of some phyla that are now extinct.

Origin

The first invertebrates probably either arose from or shared a common ancestor with protozoa. Biologists have drawn this inference from the fact that both invertebrates and protozoa are heterotrophic and eukaryotic. As you learned in Chapter 29, metazoans, or multicellular organisms, may have developed from colonies of protozoa that lived together. Scientists do not agree on whether the ancestors of metazoans were flagellated or ciliated protozoa. Biochemical evidence indicates that the proteins of some present-day invertebrates are more similar to those of modern flagellates than to those of modern ciliates. However, the cilia of modern invertebrates and protozoa are identical in size, structure, and movement. Some scientists suggest that at least Porifera, the phylum of sponges, shares a common ancestry with certain kinds of colonial flagellated protozoa.

Development

Although fossil evidence does not conclusively indicate the relationships among the invertebrate phyla, comparative studies in development and anatomy show that there are two types of development common to many invertebrates. One is characteristic of the mollusks, arthropods, and annelids while the other type is characteristic of echinoderms.

Early Embryo Development

Two main types of early embryo development occur in invertebrates. The embryos of the mollusks, arthropods, and annelids undergo **spiral cleavage**, in which the cells divide in a spiral arrangement. As the embryo grows, the blastopore develops into a mouth. These organisms are called **protostomes**, meaning "first mouth." The embryos of echinoderms undergo **radial cleavage**, in which the cell divisions are parallel with or at right angles to the polar axis. In echinoderms the blastopore develops into an anus, and a second opening in the embryo becomes the mouth. These organisms are called **deuterostomes**, meaning "second mouth." Compare these types of development in Figures 35-2 and 35-3 on the following pages. The members of the Phylum Chordata are also deuterostomes. Scientists use this evidence in inferring that echinoderms and chordates share a common ancestor.

Section Objectives

- *Summarize the possible origins of invertebrates.*

- *Describe the embryo and larval development of invertebrates.*

- *Summarize invertebrate symmetry, segmentation, and support.*

- *Analyze the organization of the phylogenetic tree of invertebrates.*

Tip: As the students read the Development section, have them look ahead to the phylogenetic tree on p. 525.

Info: Protozoans and metazoans have almost identical biochemical requirements of essential amino acids and vitamins.

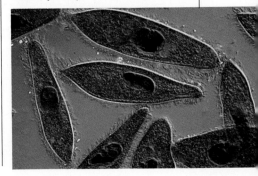

Figure 35-1 The cilia of modern invertebrates are similar to those of protozoa, such as these paramecia (×150).

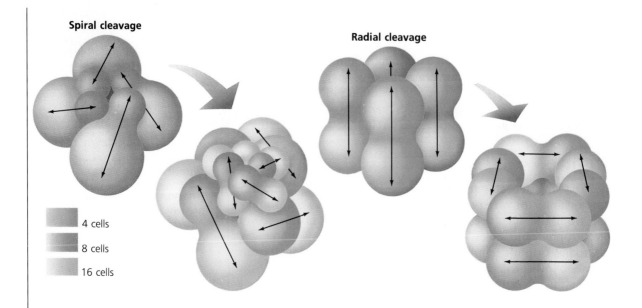

Spiral cleavage

Radial cleavage

4 cells
8 cells
16 cells

Figure 35-2 Protostomes undergo spiral cleavage during their early embryo development. Deuterostomes undergo radial cleavage during their early embryo development.

Protostomes and deuterostomes also differ in how the cells of the embryo specialize. If the cells of a four-celled protostome embryo are separated, each cell will develop into a quarter of the gastrula. Thus the fate of each cell is determined early in the development of the protostome. This type of specialization is called **determinate cleavage.** In contrast, if the cells of a four-celled deuterostome embryo are separated, each cell will develop into a complete gastrula. This type of development is called **indeterminate cleavage.**

Germ Layers and Coelom Formation

As you learned in Chapter 29, the cells in invertebrate gastrulae are organized into germ layers. Cnidarians have two germ layers, the endoderm and the ectoderm. Platyhelminths were the first invertebrates with a third germ layer, the mesoderm.

The presence of a mesoderm provides a great adaptive advantage to the organism. As well as forming the skeleton, muscles, blood, and blood vessels, the mesoderm forms the lining of the body cavity, called the coelom. The coelom allows movement of the organs within the body and enables the animal to move about more freely. All the organisms with more complex development than that of the platyhelminths have a coelom. During platyhelminth development, although the mesoderm forms various structures, it does not form a coelom.

The way in which the coelom forms in protostomes differs from the way it forms in deuterostomes. In protostomes, cells split off at the junction of the endoderm and ectoderm during gastrulation. Rapid division of these cells forms the mesoderm. This method of mesoderm formation is called **schizocoely,** or "split-body cavity." In deuterostomes, the mesoderm forms when cells

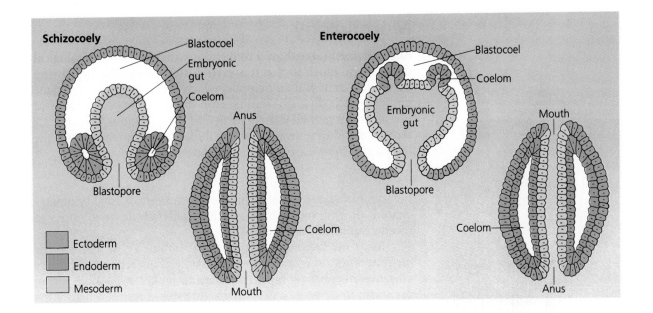

Schizocoely
- Blastocoel
- Embryonic gut
- Coelom
- Blastopore
- Anus
- Coelom
- Mouth

Enterocoely
- Blastocoel
- Coelom
- Embryonic gut
- Blastopore
- Mouth
- Coelom
- Anus

- Ectoderm
- Endoderm
- Mesoderm

at the top of the gastrula divide. This method of mesoderm formation is called **enterocoely,** meaning "gut-body cavity."

During both enterocoely and schizocoely the coelom becomes a secondary body cavity because the mesodermal cells spread out and the blastocoel disappears. During gastrulation of a roundworm, for instance, the mesoderm lines the body cavity but does not spread out to form a lining on the organs. The body cavity of the roundworm is thus called a pseudocoelom, which means "false-body cavity."

Larva Development

Invertebrates may have indirect or direct development. In **indirect development** the organism has an intermediate larval stage. You learned in Chapter 33 of the adaptive advantages of metamorphosis. In **direct development** the young animal is born or hatched having the same appearance and way of life as the adult; no larval stage occurs. Most invertebrates undergo indirect development, although a few, such as silverfish and grasshoppers, undergo direct development.

An invertebrate that has an aquatic, free-swimming larva often has one of three basic types: planula, trochophore, or bipinnaria. A planula is a free-swimming, ciliated larva with two body layers. Planulae settle to the bottom, where each develops into a colony of polyps. A trochophore is a free-swimming, bilaterally symmetrical larva. The larva of an annelid or a mollusk such as a snail is an example of a trochophore. A bipinnaria is a free-swimming larva that has a distinct pattern of development. The larva of an echinoderm such as a sea star is an example of a bipinnaria. Scientists study the kind of larva an organism has as an aid to classification.

Figure 35-3 The coelom in protostomes arises by schizocoely; the blastopore becomes the mouth. The coelom in deuterostomes arises by enterocoely; the blastopore becomes the anus.

Morphology

Adult invertebrates show a tremendous amount of morphological diversity, so much so that it is even difficult to see the similarities between a clam and an octopus, which are members of the same phylum. Such diversity results from the different adaptations of invertebrates to their environments. Many variations fall into patterns that indicate invertebrate evolution. As you read, notice the increasing complexity within the patterns.

Symmetry

Invertebrates have radial or bilateral symmetry. Radial symmetry enables drifting invertebrates such as jellyfish or brittle stars to receive stimuli from all directions.

Most invertebrates have bilateral symmetry, which is often an adaptation to a more motile life-style. This type of symmetry allows for cephalization. Cephalization makes it possible for an organism to perceive a new environment and respond before the organism moves into the environment.

Segmentation

Segmentation is the division of the body into repeating units. Segmentation is also called metamerism. Each repeating segment is called a **metameric unit.** In its simplest form, such as that seen in annelids, each unit of the body is basically the same. Within the Phylum Arthropoda there is a trend toward increasing specialization of segments, especially in the adult, and a trend toward fusing segments into larger functional units.

Support

Invertebrates have diverse means of support. The simpler invertebrates, the sponges, secrete spicules that strengthen and support their soft tissue. However, structures such as spicules do not aid in movement. The more complex invertebrates, the coelomates, are supported by the turgor pressure of their fluid-filled coelom. In roundworms, for instance, the coelom also aids in movement by providing hydrostatic pressure against which muscles can contract. Some invertebrates have true exoskeletons. The chitinous coverings of arthropods and the protective shells of some mollusks provide excellent protection, but both limit the size and impede the movement of the organism. The echinoderms are the only invertebrates with an internal skeleton. This endoskeleton allows some echinoderms to move quickly and to grow large.

Phylogeny

Look at Figure 35-5. This reconstruction of probable invertebrate evolution is based on evidence you have just read about, including comparative morphology, patterns of embryo development, study of the fossil record, and comparisons of larva development.

Figure 35-4 Compare segmentation in the polychaete worm (top) and the ghost shrimp, *Callianassa gigas* (bottom). Each unit in the body trunk of the worm is the same, but the segments in the body trunk of the ghost shrimp are fused and specialized.

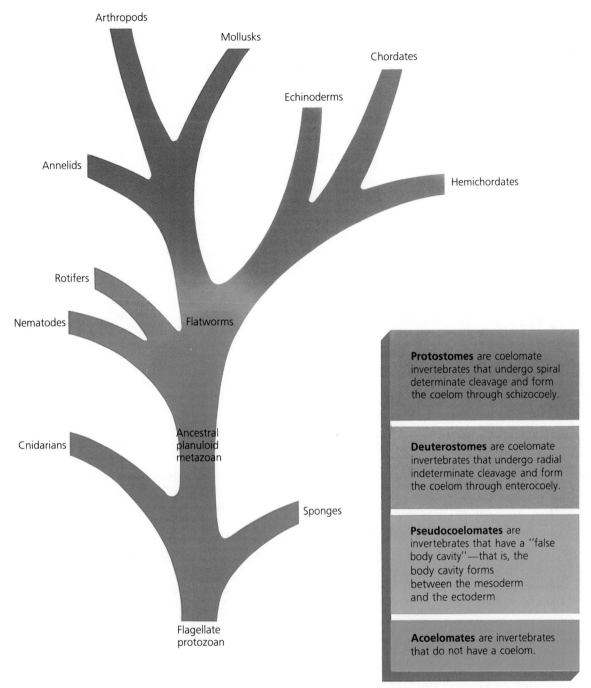

Arthropods

Mollusks

Chordates

Echinoderms

Annelids

Hemichordates

Rotifers

Flatworms

Nematodes

Cnidarians

Ancestral
planuloid
metazoan

Sponges

Flagellate
protozoan

Protostomes are coelomate invertebrates that undergo spiral determinate cleavage and form the coelom through schizocoely.

Deuterostomes are coelomate invertebrates that undergo radial indeterminate cleavage and form the coelom through enterocoely.

Pseudocoelomates are invertebrates that have a "false body cavity"—that is, the body cavity forms between the mesoderm and the ectoderm

Acoelomates are invertebrates that do not have a coelom.

Section Review

1. What is a *metameric unit?*
2. From what group of organisms do scientists suggest that metazoans originated?
3. Give an example of three invertebrate larval forms.
4. State an adaptive advantage that organisms with coeloms have over organisms without coeloms.
5. In what ways might indirect development be an adaptive advantage?

Figure 35-5 This phylogenetic tree shows the probable evolution of invertebrates. Note the evolutionary connection between echinoderms and chordates.

Answers to Section Review
1. A metameric unit is one of the repeating segments that make up organisms characterized by segmentation.
(See pp. T35A–T35D.)

- *Distinguish between types of digestive systems.*

- *Summarize the increase in morphological complexity in respiratory systems.*

- *Identify the different types of circulatory systems.*

- *Identify the increase in complexity in nervous systems.*

- *Identify the different types of excretory systems.*

- *State the adaptive advantages of the different types of reproductive systems.*

35.2 Internal Anatomy

Invertebrate phylogeny is characterized by increasingly complex internal organization. As you read this section, refer to the placement of each group on the phylogenetic tree in Figure 35-5 (p. 525), and compare the systems in Table 35-1.

Digestive Systems

In invertebrates, such as sponges, digestion occurs within individual cells. In those animals with a pseudocoelom or a coelom, the "tube within a tube" body plan makes efficient extracellular digestion possible. Food is broken down within a modified region of the body cavity, and nutrients are absorbed by specialized cells that line this region. A distinct digestive system breaks down food and thus provides energy more efficiently.

Some invertebrates, such as *Hydra*, have an **incomplete digestive system,** one opening serving as mouth and anus. Most other invertebrates have a **complete digestive system,** in which food enters through the mouth and passes through the digestive system. Undigested wastes leave through the anus. The cells along the digestive tract are specialized for the various stages of digestion.

Respiratory Systems

As the invertebrates evolved, their respiratory systems became more complex morphologically. This increase in complexity occurred as a response to the presence of protective body coverings

Table 35-1   Invertebrate Systems

	Porifera	Cnidaria	Platyhelminthes	Nematoda
Digestive	Individual cells digest food particles	Enzymes in gastrovascular cavity break down prey	Nutrients are absorbed from the environment	Digestive tract has both mouth and anus
Respiratory	Diffusion	Diffusion	Diffusion	Diffusion
Circulatory	Diffusion	Nutrients absorbed by specialized cells lining body cavity	Diffusion	Gas exchange between cells and fluid-filled cavities
Nervous	None	Nerve net	Cephalized, ladderlike system	Ganglia and nerve rings
Excretory	Wastes carried by amebocytes from cells	Wastes pass through single opening	Specialized excretory cells in free-living species	Ammonia diffuses through body wall
Reproductive	Hermaphroditic but can also regenerate or reproduce asexually	Hermaphroditic, sexual reproduction, asexual budding	Hermaphroditic and asexual	Separate sexes in most species

and, in some arthropods, as an adaptation to terrestrial life. In all invertebrates gas exchange occurs across a moist membrane. In the simplest respiratory systems the body covering acts as the moist membrane: gas diffuses from cell to cell. In the more complex systems of aquatic mollusks and aquatic arthropods, gas exchange cannot occur through the body covering because a hard protective layer of skin or the outermost layers of tissues are too thick. These animals have specialized structures called gills. In terrestrial invertebrates, which respire through the skin, body surfaces must be kept moist in order to function. The gills in terrestrial arthropods, for example, are modified to prevent desiccation. The spiracles and trachea of insects are two other adaptations for terrestrial life.

Circulatory Systems

Some invertebrates have no circulatory system. In sponges and cnidarians, for example, nutrients and gases are exchanged directly with the environment by diffusion across cell surfaces. In more complex invertebrates, such as roundworms and echinoderms, gas exchange takes place between cells and internal fluid-filled cavities. Arthropods and some mollusks have an open circulatory system, in which blood is pumped by muscularized vessels into the body cavity. The blood is then returned to the vessels from which it was pumped. Annelids have a **closed circulatory system,** in which blood remains within vessels, and nutrient and gas exchange take place between capillaries and cells. The closed circulatory system increases the efficiency of circulation by providing oxygen and nutrients to the cells more quickly.

	Mollusca	Annelida	Arthropoda	Echinodermata
Digestive	Digestive tract has both mouth and anus	Digestive tract has both mouth and anus	Digestive tract has both mouth and anus	Digestive tract has both mouth and anus
Respiratory	Gills or lungs	Diffusion in most species, gills in some	Gills in aquatic species; tracheae in terrestrial species	Diffusion
Circulatory	Open in most mollusks; closed in cephalopods	Closed	Open	Open
Nervous	Various; highly developed in cephalopods	Nerve cord connecting brain with ganglia	Specialized sensory organs	Nerve ring but no head or brain
Excretory	Wastes eliminated by nephridia	Wastes eliminated by nephridia	Wastes eliminated by nephridia	Wastes diffuse out of body
Reproductive	Most species have separate sexes; some species are hermaphroditic	Hermaphroditic	Separate sexes	Sexual reproduction and asexual regeneration

Info: Rotifers reproduce by parthenogenesis, which means that they produce eggs but the eggs develop without being fertilized.

Tip: Place some live planarians in a dish. Cover half the dish with a card and shine a light on the other half. Have students note the planarians' negative response to the light. Discuss with students the advantages of cephalization in a primitive organism.

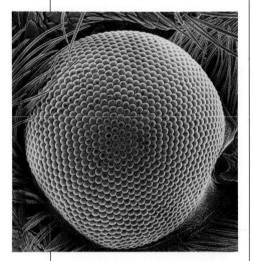

Figure 35-6 How is the highly complex compound eye of the diamondback moth, *Plutellidae lepidoptera,* an adaptation to this organism's environment?

Answers to Section Review
1. An incomplete digestive system has one opening, which serves as both the mouth and the anus. A complete digestive system has both a mouth and an anus.
(See pp. T35A–T35D.)

Nervous Systems

As invertebrates evolved, increasingly complex nervous systems enabled the animals to monitor and respond to their environments. Cnidarians, as you read in Chapter 29, have a primitive nervous system, called a nerve net. In bilateral invertebrates there is a trend toward increasing cephalization and the development of more complex sensory organs. In segmented invertebrates, such as the earthworm, large ganglia exist in the head area, though sense organs are not highly developed. In complex invertebrates, such as insects, sense organs are well developed.

Excretory Systems

In simple invertebrates nitrogenous wastes are usually excreted in the form of ammonia gas (NH_3) because it easily diffuses through the cells. In terrestrial invertebrates this poisonous gas would not leave the body quickly enough. Therefore specialized excretory structures filter NH_3 and other wastes from the coelom. These animals convert NH_3 to less toxic substances such as urea or uric acid and excrete these wastes. Another adaptation of the excretory system is the ability to reabsorb water, which is essential for terrestrial animals.

Reproductive Systems

All invertebrates are capable of some form of sexual reproduction, although many may also reproduce asexually. Some invertebrates, such as flatworms and earthworms, are hermaphroditic. Hermaphroditism allows sessile organisms or those that live in less populated areas to reproduce with any other member of their species. However, hermaphroditism requires the organism to expend large amounts of energy to maintain both male and female gonads. It is of adaptive advantage to other invertebrates to develop separate sexes and concomitantly evolve behavior patterns that bring individuals of different sexes together.

Section Review

1. What is the difference between *complete* and *incomplete digestive systems?*
2. What is the adaptive advantage of a closed circulatory system?
3. Why are simple invertebrates able to release their wastes in the form of ammonia, while terrestrial invertebrates must release their wastes in nontoxic forms?
4. What is the adaptive advantage of cephalization?
5. Many evolutionary adaptations in the various invertebrate systems provide the organism with the ability to react and move quickly. What are two advantages speed gives an animal?

Laboratory

Identifying Invertebrates

Objective
To identify representatives of the major invertebrate phyla

Process Skills
Classifying, observing, inferring

Materials
Dissecting microscope; unidentified preserved specimens, including a coelenterate, a flatworm, a roundworm, a mollusk, an annelid, an arthropod, and an echinoderm

Method
1 Perform the following series of observations for each specimen, and answer all of the following questions. Record all of your observations. Use a dissecting microscope where appropriate in order to see characteristics more clearly. If you cannot find evidence of a characteristic you are attempting to observe, write NE—for "no evidence"—after the characteristic in question. Why should you examine the same characteristics in each animal?
2 Before handling any of the specimens, make sure the specimen has been thoroughly rinsed in cool tap water in order to remove preservatives.
3 Is the animal aquatic or terrestrial? What characteristics should you examine in order to make this inference?
4 Is the animal asymmetrical, radially symmetrical, or bilaterally symmetrical?
5 Is the animal's body hard or soft, segmented or unsegmented?
6 Is the animal motile or sessile? What might you look for to help you make this inference? When possible, describe the method of movement used by the organism.
7 Does the organism have a hard external skeleton?
8 Observe the external structures of the animal's digestive tract. Does the animal have a mouth and/or an anus?
9 Use external evidence to determine the kind of respiratory system present in the organism. Does the organism have gills or lungs, or does respiration occur through the organism's external surface?
10 Note any evidence of reproductive structures. Is the organism hermaphroditic or a single sex?
11 Review your evidence to determine the phylum to which each animal belongs. Write the name of the phylum after the number of the animal.

Conclusions
1 Identify the animal specimens by genus whenever possible.
2 Name some external characteristics, besides those named in the investigation, that are helpful in classifying these animals.

Inquiry
What are some of the advantages and disadvantages of using preserved specimens rather than live animals for this investigation?

Chapter 35 Review

Vocabulary

closed circulatory
   system (527)
complete digestive
   system (526)
determinate cleavage
   (522)

deuterostomes (521)
direct development
   (523)
enterocoely (523)
incomplete digestive
   system (526)

indeterminate
   cleavage (522)
indirect development
   (523)
metameric unit (524)

protostomes (521)
radial cleavage (521)
schizocoely (522)
segmentation (524)
spiral cleavage (521)

Distinguish between the terms in each of the
following pairs.
1. radial cleavage and spiral cleavage
2. indeterminate cleavage and determinate
   cleavage
3. protostomes and deuterostomes
4. direct development and indirect development
5. open circulatory system and closed circulatory system

Review

1. Some scientists suggest that colonial flagellated protozoa were the ancestors of
metazoans because (a) present-day invertebrate proteins are similar to those of
modern ciliates (b) invertebrates are heterotrophic and eukaryotic (c) the cilia of
invertebrates and protozoa are identical in
size, structure, and movement (d) protozoans clumped together to form colonies.
2. The embryos of protostomes undergo
(a) radial cleavage (b) spiral cleavage
(c) indeterminate cleavage (d) all of the
above.
3. Scientists believe that chordates and echinoderms share a common ancestor because both are (a) protostomes (b) aquatic
organisms (c) metazoans (d) deuterostomes.
4. When the cells in a four-celled embryo are
separated and each grows into a quarter
of a gastrula, the type of cleavage is called
(a) spiral (b) radial (c) indeterminate
(d) determinate.
5. An adult that has gone through the process of changing from a larva is said to
have undergone (a) indeterminate cleavage
(b) determinate cleavage (c) indirect development (d) direct development.
6. One of the characteristics of deuterostomes is that the development of their
coelom occurs by (a) schizocoely
(b) metamerism (c) spiral cleavage
(d) enterocoely.
7. One characteristic of invertebrates such as
insects is (a) the presence of a pseudocoelom (b) radial cleavage (c) increasing specialization of body segments (d) radial
symmetry.
8. In a closed circulatory system (a) blood is
pumped through the body cavity
(b) blood is contained in blood vessels
(c) gases are exchanged with the environment (d) gases are exchanged between
cells and the body cavity.
9. Gills arose in invertebrates because (a) gas
exchange must occur across a moist membrane (b) terrestrial animals must keep
their respiratory surfaces moist in order to
breathe (c) invertebrates could not get oxygen readily by diffusion (d) a hard outer
shell prevented oxygen gas from diffusing
across the skin.
10. Hermaphroditic animals tend to (a) be
motile (b) be bilaterally symmetrical (c) be
sessile (d) live in highly populated areas.
11. When do scientists believe the first invertebrates appeared on earth?

12. What advantage does the mesoderm provide to an organism?
13. What adaptive advantage does radial symmetry provide to less complex invertebrates?
14. What adaptive advantage does a complete digestive system provide to a more complex organism?
15. How is the endoskeleton an adaptive advantage over the exoskeleton?
16. On what evidence do scientists base their construction of the invertebrate phylogenetic tree?
17. How does a closed circulatory system and a complete digestive system allow an organism to increase its motility?
18. Why do terrestrial animals need specialized excretory structures to filter wastes?
19. What adaptive advantage does a complex nervous system provide?
20. Why is hermaphroditism an inefficient use of energy for many invertebrates?

Critical Thinking

1. Many invertebrates are able to regenerate a lost body part. Suggest a reason why some invertebrates are not able to regenerate body parts.
2. Scientists have determined the phylogeny of invertebrates by studying many invertebrate features, as well as invertebrate fossils. From what you know about invertebrates, which ones, do you think, were most likely to fossilize? Why? How might this affect scientists' view of invertebrate evolution?
3. The photo on the right shows bipinnaria larvae of starfish. What adaptations for a free-living life do they show?

Extension

1. Make a table of the invertebrate phyla that have been discussed in Chapters 29–35. For each phylum list the defining feature of the phylum (e.g., soft bodies for mollusks); the evolutionary adaptations; and the type of reproductive, digestive, excretory, circulatory, and nervous systems.
2. Do research on one type of sessile invertebrate. Find out how sessility can be both an advantage and a disadvantage. Write a short paper in which you report your findings.
3. Read the article by Steven Jay Gould entitled "Glow, Big Glowworm," in *Natural History,* December 1986, pp. 10–16. Rather than seeing the larval stage as undeveloped and the adult stage as developed, how does Gould suggest that we view larvae and adults?
4. Invertebrates are incredibly diverse. Research a type of invertebrate that has not been discussed in Chapters 29–35. Present a short report to your class in which you discuss the phylogeny and evolution of your chosen invertebrate. What characteristics does this organism have that are similar to the characteristics of related invertebrates?

Intra-Science: *How the Sciences Work Together*

The Question in Biology: Can We Control Insects Safely?

Which would you rather eat—an apple riddled with worms, or an apple sprayed with poison? No doubt you would prefer an apple that had neither worms nor poison. But do you have a choice?

Of the 800,000 species of insects in the world, only about one percent are harmful pests. Left unchecked, however, these insects can cause famine by destroying food crops. This possibility becomes all too clear when you realize that pests ruin 500 million tons of grain and 130 million tons of potatoes every year in spite of pesticides and insecticides. In addition to economic damage caused by pests such as the cotton-attacking boll weevil, some insects are also carriers of disease. Mosquitoes can transmit organisms that cause malaria and yellow fever; the tsetse fly can carry protozoa that cause sleeping sickness; and the common household fly can carry organisms that cause dysentery and typhoid fever.

Obviously, insect control is necessary. A controversy exists, however, over the types of methods used. As in the example of the apple described above, it's clear that the so-called cure must not be worse than the insect problem it seeks to cure.

The war on harmful insects is fought on many fronts, among them, physical, chemical, and biological. The most successful strategy developed thus far is an integrated approach using elements of all three.

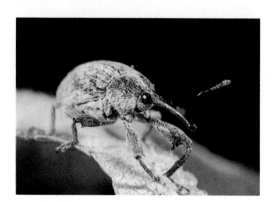

The Connection to Physical Sciences

Controlling the environment is one method of restraining insects. Draining swamps where mosquitoes breed and cleaning up rubbish and garbage heaps are effective means of combatting mosquitoes and house flies. Controlling the temperature of food storage units discourages insects that feed on harvested crops. Above 65°C most insects cannot survive; below 5°C some live, but do not feed or reproduce.

Rotating crops reduces the numbers of insects that feed on only one crop. Agronomists also advise scheduling the planting and harvesting of crops for times when insects that feed on those crops are not laying eggs or feeding. In between crops, farmers plow insect eggs underground or destroy insects and eggs by burning the field rubble.

Many countries and states enforce rigid quarantine laws in an effort to prevent insect pests from arriving by accident. Government inspectors check baggage and cargo at major points of entry.

The Connection to Chemistry

The chemical approach to pest control surged dramatically with the development of synthetic pesticides in World War II. Such pesticides were cheap, easy to apply, and killed a wide range of pests. The three main types of synthetic pesticides are phosphates, carbamates, and chlorinated hydrocarbons. The first two do not leave harmful traces on food, although the phosphates and some carbamates are poisonous to people. The chlorinated hydrocarbons include DDT and other insecticides whose use is now banned in the United States.

DDT and related insecticides remain in the soil and water for years and thus have a long-term effect on wildlife. Because DDT tends to accumulate in fatty tissues, it is passed along in the food chain. The osprey, peregrine falcon, and bald eagle were all threatened with extinction by the use of DDT. Their populations have begun to rise since the ban on DDT.

Insecticides kill by interfering with an insect's body processes. The poison enters the body through physical contact, ingestion, or inhalation. Borax, for example, wears away the exoskeleton of roaches, which die from a loss of water. Other insecticides disrupt an insect's nervous system, causing it to stop eating and to eventually die of starvation.

One limitation of insecticides is that some insects develop a resistance to their effects. The chief factor in such a resistance is, in some cases, an enzyme that detoxifies the poison. Farmers counter resistance by varying the chemicals they use.

The Connection to Biology

Biological control uses an insect's own natural enemies or knowledge about an insect's reproductive or feeding habits to control it. In the late 1880s the California citrus industry was nearly destroyed by the cottony-cushion scale. This insect was accidentally brought into the country from Australia. In 1888 growers imported the vedalia beetle, a type of ladybug, to feed on the scale. The vedalia beetle controlled the scale until about 1947, when the beetle was accidentally destroyed by the widespread spraying of DDT.

Radiation or chemicals may also be used to sterilize insects, making it impossible for them to reproduce. This technique was successful in combatting the screwworm fly that had infected U.S. livestock in the 1950s. Pheromones, the natural chemicals given out by insects that influence the behavior of other insects, are often used to bait the traps.

The Connection to Careers

The diversity of methods used to control pests provides a diversity of jobs. Pilots are needed to fly crop-dusting planes. Inspectors check passengers, cargo, and vehicles that enter the United States to prevent the entry of exotic pests. Chemists develop new insecticides that are harmless to other life forms. Agronomists and farmers experiment with crop rotation and plowing methods that reduce the number of eggs likely to hatch. Entomologists study the life cycles of insects for new methods of control, such as mass sterilization. For more information on related careers in the biological sciences, turn to page 842.

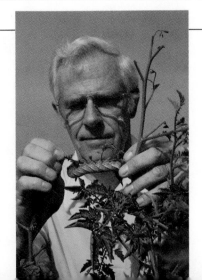

533

Unit 8 | Vertebrates

Chapters

Channel catfish, *Ictalurus punctatus*

"There is grandeur in this view of life . . . that from so simple a beginning endless forms most beautiful and most wonderful have been and are being evolved."
Charles Darwin

What do all of the animals on these two pages have in common? Perhaps the most obvious answer is that they are all vertebrates—that is, animals with backbones. However, all of these animals spend at least part of their life in water. In this unit you will learn that the major trend of vertebrate evolution is toward greater adaptation to life on land. Nevertheless, every class of vertebrates includes some species that are adapted to life in water. Can you name some mammals that live in water?

Sea turtles

King penguins, *Aptenodytes patagonica*

Chapter 36 | *Chordates and Fishes*

This chapter examines the Phylum Chordata. Students learn the features that all vertebrates share with two lesser-known chordates. They study a wide range of adaptations that have enabled fish to evolve and thrive in a water environment.

Correlation Chart

Section Number	Section Objectives	Pages	Features	Section Review
36.1	1. List the major characteristics of chordates. 2. Identify the three chordate subphyla. 3. Describe the characteristics that distinguish vertebrates from other animals. 4. Name the seven living vertebrate classes, and give an example of each one.	537–539		1, 2 3 4 5
36.2	1. List the major adaptations that enable fish to live in water. 2. Name the three fish classes, and list their distinguishing characteristics. 3. Describe the feeding behavior of lampreys and hagfish. 4. Explain how scientists measure the learning potential of sharks. 5. Compare and contrast the three types of bony fishes.	540–545	Biology in Process, pp. 542–543	1 2, 5 3 4
36.3	1. List the characteristics of body structure, fins, and integument of the yellow perch. 2. Describe the internal anatomy of the yellow perch. 3. Explain how gills function in respiration. 4. Describe the spawning behavior of fishes.	546–551	Think, p. 547 Biology Lab, p. 551	2 1, 5 3 4

Resources*

Print

Cailliet, Gregor M. *Fishes: A Field and Laboratory Manual on Their Structure, Identification, and Natural History.* Belmont, Calif.: Wadsworth, 1986.

Nilsson, Stefan, and Susanne Holmgren. *Fish Physiology: Recent Advances.* London: Croom Helm, 1986.

Audiovisual/Software

Fascinating Fishes. Film or video, 26 min. Marty Stouffer, 1984.

Fish, revised. Video or film, 14 min. Coronet, 1985.

*Additional references are listed in *Modern Biology, Teacher's Program Guide.*

Section Teaching Guides

36.1 Overview of Chordates

Approach

The main objective of this section is to introduce students to the Phylum Chordata by describing the characteristics that distinguish chordates from other animals. This section builds on the concepts of divergent and convergent evolution (Chapter 15) and phylogeny (Chapter 18). This section also draws on the discussion of echinoderm evolution in Chapter 34. In presenting the text, ask students to note the differences in embryo development between invertebrates and chordates. You might wish to review the material on invertebrate systems in Chapter 35 to reinforce the differences between the two groups of animals.

Concept Framework

Vertebrates are the group of animals most familiar to students. Ask students to name as many kinds of vertebrates as they can. To provide a starting point for studying chordates, ask students what characteristics these animals have in common. Ask students as they read this section to note that despite the diversity of vertebrates, their early developmental patterns show many similarities.

Emphasis

Make certain that students understand the evolutionary significance of the notochord, dorsal nerve cord, and pharyngeal pouches. An understanding of the evolution of these distinguishing chordate characteristics will prepare students for learning about the various adaptations of vertebrates in Chapters 37 through 40.

Section Review Answers

(See p. 539.)

36.2 Evolution and Classification of Fishes

Approach

This section focuses on the evolution of the three classes of fishes and on the adaptations that accompanied this evolution. This section builds on the concepts of natural selection (Chapter 16) and the evidence of evolution (Chapter 15). You might also review the different types of evidence used in classification (Chapter 18). In your discussion, note the different kinds of evidence, such as comparative morphology and fossil evidence, taxonomists use to classify fishes.

Concept Framework

Students know from previous chapters that evolution occurs through natural selection. To provide a framework for understanding the evolution of fishes, ask students to think of reasons that natural selection might have favored the development of a backbone, gills, and other fish characteristics. Tell them to look as they read this section for evidence that supports their ideas.

Emphasis

Students should be able to identify the distinguishing characteristics of Agnatha, Chondrichthyes, and Osteichthyes, and give examples of how each class is adapted for aquatic life.

Using the Features/Understanding Sharks

The main process skills used in this feature are observing, collecting data, and experimenting. This feature describes the work of Eugenie Clark, who applied these process skills to the study of live sharks in captivity. By analyzing the results of her experiments, Clark determined that sharks have a much greater capacity for learning than had previously been believed. Previous research on sharks had focused on anatomical studies of dead sharks and thus revealed little about shark behavior. The feature presents an opportunity to discuss the advantages and disadvantages of using live or dead subjects in scientific research. Ask students to consider what scientists can learn from the study of both live and dead subjects.

Field Trip

Plan a visit to a local aquarium so that students can observe the diversity of fishes and learn about their habitats. Prior to the trip, prepare a list of points to notice or questions to answer about the facility.

Section Review Answers

(See p. 545.)

36.3 Morphology of a Bony Fish

Approach

The main objective of this section is to describe the external and internal anatomy of a representative bony fish, the yellow perch. This section builds on the concepts of cell specialization (Chapter 5) and adaptation (Chapter 16), namely, adaptation to aquatic life. The discussions of the digestive, circulatory, respiratory, and excretory systems also relate back to Chapter 6, on homeostasis, and Chapter 7, on cell metabolism. In discussing the adaptations of bony fishes to life in water, note that the information presented in this section will provide students with a basis studying the adaptations of amphibians to both aquatic and terrestrial environments, as described in Chapter 37. Also compare the adaptations of fishes with those of other aquatic animals.

Concept Framework

Students can probably identify some of the characteristics that distinguish fishes from other vertebrates, such as fins, scales, and a streamlined body shape. Elicit this knowledge from students as a method to introduce students to the ways these characteristics show adaptations of fishes for aquatic life. Ask students as they read the text to look for examples of such adaptations.

Emphasis

Students should be able to explain how each of the following aspects of fish morphology shows adaptations for life spent in water: body structure; integument; and the digestive, circulatory, respiratory, excretory, nervous, and reproductive systems.

Guest Speaker

Invite a biologist from a local university who can talk about the adaptations of fishes to different environments, such as ponds, rivers, coastal areas, and ocean bottoms.

Section Review Answers

(Continued from p. 550.)

5. It would be difficult for bottom-dwelling fish to remain at the bottom of a body of water if a gas bladder made them so buoyant that they kept moving to a higher depth.

Laboratory Preparation

Background

This lab acquaints students with the basic characteristics of keys, as well as with the identification of fishes. Keys are artificial devices that are used to identify an organism even though it may exhibit a wide range of variation. The history of biology is filled with stories of the classification of organisms that were thought to be different species but were in reality larval forms, different sexes, or geographical or genetic variations of the same species. Because many of the characteristics that are important in separating major groups of organisms are not easily observed in preserved specimens, keys often rely on other trivial but distinctive anatomical features. Students learn that the examination of any comprehensive key will illustrate the kind of characteristics that are used for identification. Students should understand that their keys should utilize characteristics that distinguish between major groups of fishes, rather than simply identify the fish pictured.

Preparation

No advance preparation is required for this lab.

Conclusions

1. goosefish, Atlantic halibut, bay anchovy, conger eel, sea horse
2. Answers will vary but might include information about habitat, feeding habits, and behaviors.

Inquiry

1. Students might answer that they would evaluate their keys by exchanging them with other students or by using them on species other than the ones pictured.
2. Students should identify some of the problems involved in developing an unambiguous key, such as making sure that the final selection applies to only one category of objects.

Chapter Review

Vocabulary

1. Vertebral column; the others are characteristic of all chordates, while the vertebral column is found only in vertebrates.

2. Operculum; the others are parts of the brain.
3. Milt; the others have to do with the skin.
4. Pharynx; the others have to do with the circulatory system.
5. Gas bladder; the others are features of external anatomy.

Review

11. Agnatha, lamprey (accept any reasonable example); Chondrichthyes, sharks; Osteichthyes, yellow perch; Amphibia, frogs; Reptilia, snake; Aves, falcon; Mammalia, cat
12. Because the endoskeleton can grow with the body, it does not have to be shed as invertebrate exoskeletons do. The endoskeleton can support a larger body.
13. The appendicular skeleton is made up of the girdles and limbs or fins, while the axial skeleton is made up of the backbone and skull.
14. They fit them with radio transmitters and record the signals emitted.
15. Rays and skates have flattened bodies and speckled or sandy colors that help to camouflage them against the ocean bottom. The spiracles on the top of their heads allow them to take in oxygen while almost completely buried in sand. Their strong jaws and teeth allow them to crush bottom-dwelling invertebrates for food.
16. They use gills to take in oxygen when there is water in ponds and use lungs for oxygen during the dry season.
17. The growth rings indicate the age of the fish.
18. Saltwater fishes excrete excess salt through their gills and kidneys, while freshwater fishes excrete excess water.
19. The blood is pumped by the ventricle through the circulatory system to the gills, where gases are exchanged. The oxygen-rich blood is then circulated among the tissues, where gases are again exchanged, and returns to the atrium.
20. The optic lobe receives messages from the sense organs and signals muscle movement; the olfactory lobe deals with the sense of smell; the cerebrum also deals with smells; the cerebellum coordinates muscles; and the medulla oblongata regulates the internal organs.

Critical Thinking

1. Vertebrates are free moving, while the adult tunicate is sessile. On the other hand, as Figure 36-2 shows, the larval tunicate has a tail and can move about. Neoteny suggests that the evolutionary path might be traced back to an ancestral form of tunicate larva that bypassed the adult stage and began to reproduce.
2. The corkscrew shape allows for a greater surface area in the intestines for absorbing nutrients without an increase in the length of the intestines.
3. Different luminescent patterns and displays allow fish of the same species to recognize one another. Also, for some species luminescence acts as a lure, drawing prey close enough to be caught. It can also act as a repellent, frightening off prospective enemies.
4. The black dot suggests a false eye, while the real eye is camouflaged by the black bar across it. This, together with its habit of swimming backwards, might confuse a predator into mistaking the tail for the head, thus allowing the butterfly fish to escape.
5. The process of osmosis—the diffusion of water across a membrane from a less saline solution to a more saline solution—causes saltwater fish to lose water through their gills and membranes. They drink and retain water in an effort to maintain homeostasis.
6. The cod eggs far outnumber the less vulnerable bass eggs. The more protected bass eggs have a better chance of being fertilized, and fewer will be eaten by predators.

Chapter 36 | Chordates and Fishes

Introduction

Over half of the vertebrates—animals that have backbones—are fishes. This is not surprising, considering that 70 percent of the earth's surface is covered with water. In the last chapter you learned that invertebrates have a wide range of adaptations to different environments. In this chapter you will continue to study adaptations by examining fishes.

Chapter Outline

Chapter Concept

As you read, look for adaptations that have enabled fishes to evolve and thrive in a water environment.

Psychedelic dragonet, *Callionymus* sp., Coral Sea. Inset: Manta ray, *Manta birostris*, Philippines. Background: fish scales.

For additional teaching information see pp. T36A–T36D.

36.1 Overview of Chordates

Section Objectives

- *List the major characteristics of chordates.*
- *Identify the three chordate subphyla.*
- *Describe the characteristics that distinguish vertebrates from other animals.*
- *Name the seven living vertebrate classes, and give an example of each one.*

When you think of a familiar animal, you probably think of a chordate *(KORD-ate)*. The Phylum Chordata includes vertebrates—members of the Subphylum Vertebrata *(vert-uh-BRAHT-uh)*. These are animals with backbones, such as sharks, frogs, crows, dogs, and humans. Chordates also include two other small subphyla of animals without backbones, Urochordata *(yoor-uh-kor-DAHT-uh)* and Cephalochordata *(SEFF-uh-loe-kor-DAHT-uh)*. The name chordate comes from **notochord,** a firm, flexible rod of specialized cells located in the dorsal part of the chordate body. Some chordates retain the notochord all their lives. In vertebrates, however, the notochord appears only in the embryonic stage. Early in development it is replaced by the **vertebral column,** or backbone.

In Chapter 34 you learned that echinoderms are the closest invertebrate relatives to chordates. Echinoderms and chordates share several characteristics in their embryonic development. Scientists call members of these two phyla deuterostomes. **Deuterostomes** are coelomates whose embryos have radial cleavage, whose anus forms near the blastopore, and whose mesoderm arises from outpocketings of the endoderm. In most invertebrates the mesoderm arises as a separate layer between the endoderm and ectoderm. These characteristics provide strong evidence that echinoderms and chordates may share a common ancestor.

Chordates evolved in water, and their adaptations enabled them to succeed in that environment. Later the same features—an internal skeleton and highly controlled body movements and chemistry—were modified for living on land.

Characteristics

A chordate is an animal that at some stage of its development has a notochord, a dorsal nerve cord, and pharyngeal pouches.

- A notochord is a dorsal rod of specialized cells.
- A **dorsal nerve cord** is a hollow tube just above the notochord.
- **Pharyngeal pouches** are small outpockets of the anterior gut.

In vertebrates the notochord exists only in the embryo. The notochord is replaced by an **endoskeleton,** an internal skeleton that can support a larger body than can the exoskeleton of an invertebrate. The endoskeleton grows as the animal grows and does not have to be discarded by molting. In vertebrates the dorsal nerve cord develops into a spinal cord with a brain. The brain is connected to a network of complex sensory organs. In lower chordates, fishes, and amphibians, the pharyngeal pouches evolved into filter-feeding structures, which in turn evolved into **gill slits,** which later evolved into gills. In terrestrial vertebrates, pharyngeal pouches evolved into structures in the throat and ear.

Figure 36-1 The adult tunicate, *Clavelina picta,* loses its notochord during metamorphosis and so resembles a filter-feeding invertebrate more than a chordate.

Classification

Scientists divide the Phylum Chordata into three subphyla: Urochordata, Cephalochordata, and Vertebrata. Members of the first two subphyla live only in the ocean. About 2,000 species of urochordates and about 28 species of cephalochordates exist. Members of Cephalochordata retain the notochord throughout life and so are the closest living relatives of the early animals from which all chordates evolved. The third subphylum, Vertebrata, comprises about 41,000 species. More than 95 percent of all chordates are vertebrates.

Subphylum Urochordata

The hollow, barrel-shaped urochordates are commonly called tunicates because their bodies are protected by a tough covering, or tunic. Tunicates are also called "sea squirts" because they squirt out streams of water when touched. Larvae and some adults swim freely, but most adults, such as the one in Figure 36-2, attach themselves to the sea bottom. They may be solitary or colonial. Tunicates are adapted for filter feeding—straining food particles and taking oxygen gas from water drawn through their bodies by ciliary action. Being anchored and having cilia are adaptations to the filter-feeding way of life. Many such adaptations are examples of convergent evolution that are shared, as you have read, by many filter-feeding invertebrates. Despite their simple body plans tunicates have a notochord and dorsal nerve cord as larvae and gill slits both as larvae and as adults.

Subphylum Cephalochordata

Cephalochordates are marine organisms that live in shallow waters in warm southern regions. They are best represented by a blade-shaped translucent animal called *Branchiostoma* (*BRANG-kee-AHS-tuh-muh*). Some people call this organism amphioxus; its common name is lancelet. *Branchiostoma* usually stays buried in sand. Only its head protrudes, allowing it to filter feed on organisms drawn from the water and to breathe through its gill slits. *Branchiostoma's* poorly developed fins force it to swim in a spiral as it moves forward. The lancelet structure shows all three chordate characteristics: notochord, dorsal nerve cord, and gill slits. Locate these structures in Figure 36-3.

Figure 36-2 Unlike their larval forms most adult tunicates are not free swimming. Only in the larval stage do tunicates exhibit all their chordate features.

Adult tunicate

— Incurrent siphon (mouth)
— Excurrent siphon
— Pharyngeal slits
— Pharynx
— Tunic
— Intestine
— Heart
— Stomach

Larval tunicate

Notocord
Nerve cord
Gill slits

Figure 36-3 The lancelet, *Branchiostoma lanceolatum,* retains its chordate characteristics throughout its life.

Notochord
Gill slits
Nerve cord

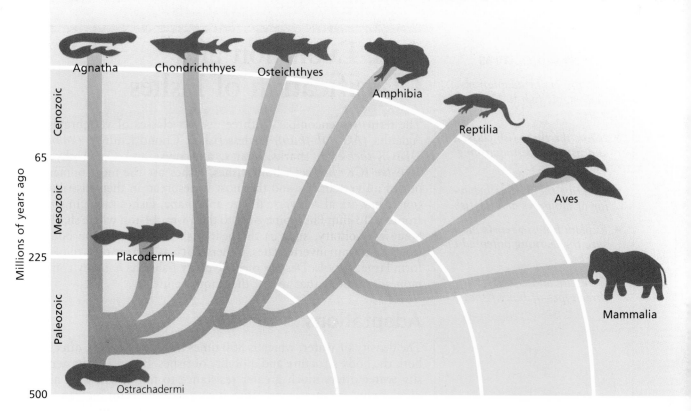

Figure 36-4 The evolution of the vertebrates has led to the seven classes that exist today.

Subphylum Vertebrata

Vertebrates are named for their vertebrae, the bones or cartilage that surround the dorsal nerve cord. Other features characterize vertebrates. Vertebrates have a brain protected by an outer **skull**, or cranium. Their backbone and skull make up the **axial** *(ACK-see-ul)* **skeleton.** Paired limbs or fins are attached to the axial skeleton by girdles of bones or cartilage. The girdles and limbs or fins make up the **appendicular** *(ap-un-DICK-yuh-lur)* **skeleton.** Muscles attached to the skeleton allow movement.

The organs of vertebrates, like those of invertebrates, are organized into ten systems: skeletal, muscular, integumentary, digestive, respiratory, circulatory, excretory, immune, nervous, and reproductive. Each performs a special function. Seven classes of vertebrates exist today. Fossil evidence indicates increasing complexity of body structure corresponding to the order in which the classes evolved. Figure 36-4 shows these relationships.

Section Review

1. How do the notochord and dorsal nerve cord differ?
2. Which invertebrate phylum is developmentally most similar to vertebrates? Explain.
3. Name and give examples of the three chordate subphyla.
4. List four distinguishing features of vertebrates.
5. Name the first living class of vertebrates found in the fossil record. Which is most recent? Refer to Figure 36-4.

Tip: Begin the discussion of vertebrates by asking students to name familiar animals. The chances are that most will be vertebrates.

Answers to Section Review
1. The notochord is a firm, flexible rod found in one or another stage of development in chordates. The dorsal nerve cord is a tube that lies just above the notochord. In vertebrates the notochord evolved into the endoskeleton and the dorsal nerve cord became the spinal cord.
2. Echinoderms. Both are deuteromes.
3. Cephalochordata, *Branchiostoma*; Urochordata, tunicate; and Vertebrata (accept any vertebrate example)
4. vertebral column, brain enclosed in a skull, axial skeleton, and appendicular skeleton (or paired limbs and fins)
5. Agnatha or jawless fish; Aves or birds

Section Objectives

- *List the major adaptations that enable fish to live in water.*

- *Name the three fish classes, and list their distinguishing characteristics.*

- *Describe the feeding behavior of lampreys and hagfish.*

- *Explain how scientists measure the learning potential of sharks.*

- *Compare and contrast the three types of bony fishes.*

36.2 Evolution and Classification of Fishes

The term *fish* encompasses three distinct classes of vertebrates: Agnatha *(AG-nuh-thuh)*, jawless fishes; Chondrichthyes *(kahn-DRICK-thee-eez)*, sharks, skates, and rays; and Osteichthyes *(AHS-tee-ICK-thee-eez)*, bony fishes. Fishes are the most numerous of all vertebrates and the most widespread in their distribution. They are also diverse in size and shape. Fishes range in size from the 13-mm Philippine goby to the 18-m, 14-ton whale shark. Some are solitary, such as the pearl fish that lives in the body cavity of certain invertebrates. Others, such as herring or sardines, form large schools. Despite their diversity fishes have many common adaptations that reflect their aquatic life.

Adaptations

The density of water, which is 800 times the density of air, affects both the body structure and mobility of fishes. Because of its density water offers much greater resistance to movement than air does. To live in water, most fishes have adaptations for buoyancy. One such adaptation consists of the trapping of gas inside their bodies. By controlling the amount of gas in their bodies, fishes regulate their vertical position in the water. Another set of adaptations enables them to swim. A streamlined shape and muscular tail enable fishes to move rapidly through the water. Paired fins allow them to maneuver easily right or left, up or down, and backward or forward. In addition, the mucus that most fishes secrete reduces friction as they swim.

Protective scales on fishes limit chemical exchanges through the skin. Instead, most exchanges between water and blood take place across the membranes of **gills**—the external respiratory organs, usually found on either side of the head. Most fishes have highly developed senses of smell and touch. A prominent adaptation present in all classes except Agnatha is the **lateral line system.** This system consists of a row of sensory structures that run the length of the fish's body on each side and that are connected by nerves to the brain. These structures detect vibrations and chemicals in the water.

Evolution

As you see in Figure 36-4, adaptations in fishes are the result of millions of years of evolution by natural selection. The first known vertebrates, small jawless fish placed in the Class Ostracodermi, were covered with heavy bony plates. They first appeared about 540 million years ago. The large bony plates remain today, modified in form, as scales. Comparative morphology shows that

Figure 36-5 Ostracoderms, the earliest vertebrates, lacked jaws and were covered with bony plates.

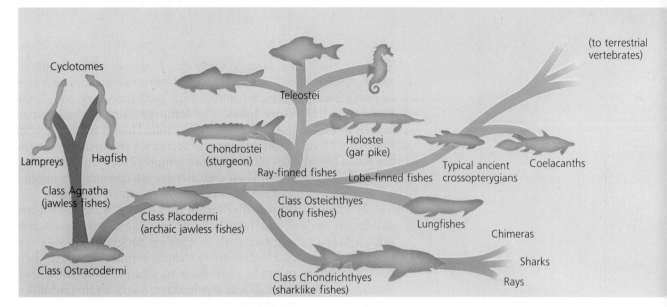

Cyclotomes

Lampreys Hagfish

Class Agnatha
(jawless fishes)

Class Ostracodermi

Class Placodermi
(archaic jawless fishes)

Chondrostei
(sturgeon)

Teleostei

Ray-finned fishes

Holostei
(gar pike)

Lobe-finned fishes

Class Osteichthyes
(bony fishes)

Lungfishes

Typical ancient
crossopterygians

Coelacanths

Chimeras

Sharks

Rays

Class Chondrichthyes
(sharklike fishes)

(to terrestrial
vertebrates)

today's jawless fishes, members of the Class Agnatha, are similar to ostracoderms. This indicates that they probably evolved from the ostracoderms. The other two classes—the Chondrichthyes and the Osteichthyes—probably also evolved from ostracoderms, but along a different phylogenetic line, as seen in Figure 36-6. Members of these two classes have jaws that presumably evolved from ostracoderm gills. Relatives of modern jawed, bony and cartilaginous fishes began to appear in the fossil record 400 million years ago.

The fossil record shows that after fishes became abundant, two other adaptations occurred in some species that were central to the evolution of other classes of vertebrates. One was the evolution of a pouch in the posterior portion of the mouth that functioned as a lung. The other was the emergence of fins supported by bony lobes projecting from the body. These adaptations occur today in the lungfish and the lobe-finned fish, two subgroups of bony fishes. The ability to breathe air and to cross land on fins enabled shallow-water species to survive droughts. Comparative morphology indicates that prototype lungs evolved into lungs in land vertebrates and that lobed fins evolved into limbs. These lobe-finned fish were the ancestors of amphibians.

Agnatha

The only existing jawless fishes are the 45 species of lampreys and hagfish that compose the Class Agnatha. *Agnatha* means "jawless." Because of their circular mouths these organisms are often called cyclostomes, meaning "round mouths." The slimy skin of lampreys and hagfish has neither plates nor scales. Both fishes have a notochord that remains throughout life. Each has an eel-like shape, a cartilaginous skeleton, and unpaired fins. Hagfish live only in the oceans. Many lampreys live in fresh water.

Some lampreys are free-living, but most are parasitic. They are adapted for sucking the blood and body fluids of other fish.

Figure 36-6 The phylogenetic tree of fish begins with jawless fishes. One line of evolution leads to terrestrial vertebrates.

Figure 36-7 The disk-shaped mouth of the lamprey, *Lampetra fluviatilis,* is adapted for attachment to prey by suction.

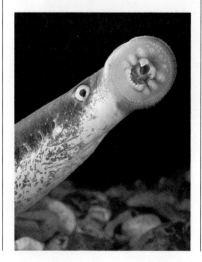

To locate their prey, lampreys rely on a highly developed sense of smell. A nasal pore on top of the head leads to olfactory sacs rich in nerve endings, which connect with specialized regions of the brain called **olfactory lobes.** The lamprey uses its disk-shaped mouth to attach itself to a fish by suction. Then it tears a hole in the fish with its toothy tongue and secretes a chemical that keeps the fish's blood from clotting. Since lampreys feed on fluids, they do not have a stomach. Their digestive system consists of a mouth, an esophagus, a straight intestine, and associated glands.

Hagfish are bottom dwellers in cold marine waters. They are scavengers of dead and dying fish on the ocean bottom. A hagfish saws a hole in these fish with its toothed tongue and eats them from the inside out. Hagfish are extremely supple and can tie their bodies in knots to evade capture or to clean off the slime they secrete in self-defense. When not feeding, hagfish remain hidden in mud burrows on the ocean floor.

Chondrichthyes

Sharks, skates, and rays belong to the Class Chondrichthyes. Like the Agnatha, Chondrichthyes have skeletons of cartilage, not

Answer to Biology in Process
Answers might include using a yellow target from the start and then changing to a white target, but not with the abruptness of Clark's original experiment.

Biology in Process

Understanding Sharks

People long feared all sharks without knowing a great deal about them. Scientists studied the anatomy of sharks by dissecting dead ones. Without observing live sharks, scientists were unable to answer some simple questions: How intelligent are sharks? How long do they live? How much food do they need?

Today scientists use a variety of sophisticated methods to collect information about the structure and the behavior of sharks. To study growth rates, researchers measure and tag individual sharks and then release them. When recaught, the sharks are again measured. To determine habitat range, scientists track sharks that have been fitted with transmitters that emit radio signals.

One American biologist, Eugenie Clark, is a world leader in shark study. She was one of the first scientists to construct a shark pen, which enabled her to observe and collect data from live sharks in captivity. In 1958 she conducted the now-classic experiments on the learning abilities of sharks. Up to that time sharks were considered somewhat stupid and poor subjects for the learning experiments that had been conducted on other animals. Clark hypothesized, however, that bone. Cartilage is a tissue made of cells surrounded by tough, flexible protein structures. Although these cartilaginous fish seem simpler in structure than bony fish, fossil evidence indicates that they probably evolved from bony fish and lost bone over time. Chondrichthyes differ from the Agnatha by having movable jaws and skeletons with paired fins.

Almost all the approximately 275 species of sharks, skates, and rays live in salt water. Most are carnivores that typically use their large olfactory organs and lateral line systems to track prey. The skin of sharks and most rays is covered with **placoid scales**—small, toothlike spines that feel like sandpaper.

Sharks

The voracious predatory habits of some shark species have given them a reputation as killers. Yet fewer than 10 percent of shark species are known to attack humans unprovoked. Whale sharks, for example, are gentle 18-m giants that graze on plankton.

Sharks swim with a side-to-side motion of their asymmetric tail fins. Just behind their heads are paired **pectoral** *(PECK-tuh-rul)* **fins,** which jut out from their bodies like the wings of a plane. These fins compensate for the downward thrust of the tail fin.

Figure 36-8 What similarities can you note between a shark's tooth (top) and placoid scales (bottom)?

the learning ability of sharks was underrated.

For her study Clark used two adult lemon sharks, which she had captured near the Cape Haze Marine Laboratory, south of Sarasota, Florida. A white target was lowered into the water at feeding time. Pushing on the target caused a bell to ring and also enabled the sharks to get at food lowered into the water on a string. Within a few days the sharks learned that ringing the bell was associated with food.

Clark varied the routine. She threw food out at faster intervals and at greater distances from the target. She altered the volume of the bell sound. Each time the sharks adjusted to the change in routine.

However, Clark learned that sharks had a limited tolerance for change. One day she altered the color of the target to a bright yellow. The male shark started as usual for the target as it was lowered into the water. Then about two feet away from the target he reacted to the change in color and did not hit the target.

■ How would you design an experiment to determine whether the shark reacted to the sudden change in color or whether the reaction was specifically to the color yellow?

Figure 36-9 The teeth of this sand tiger shark, *Odontaspis taurus,* are pointed inward in a way that keeps prey from escaping by swimming away.

Info: Skates capture prey by swimming above the prey, pushing it down, and pinning it to the ocean floor.

The shark's mouth has 6 to 20 rows of backward-pointing teeth. When a tooth breaks or wears down, a replacement moves forward. One shark may eventually use more than 20,000 teeth. The structure of each species' teeth is adapted to its feeding habits. Sharks that catch primarily large fish have big, triangular teeth with sawlike edges that hook and tear prey. Bottom-feeding species that eat mollusks and crustaceans have flattened teeth that can crush shells. Sharks that eat small fish have long, thin teeth that grasp slippery prey.

The ability of sharks to detect chemicals—that is, their sense of smell—is particularly acute. Paired nostrils on the snout have specialized nerve cells that connect with the olfactory lobes of the brain. Water entering the nostrils is continually monitored for chemicals. Sharks can detect blood from an injured animal as far as 500 m away. Sharks also have a well-developed lateral line system. In most species the sense of sight is less developed than the sense of smell.

Gas exchange requires a continuous passage of water over a shark's gills. Therefore most sharks must swim constantly, pushing water through their mouths, over their gills, and out of their gill slits.

Shark eggs are fertilized internally. During mating the male grasps the female with modified fins called claspers. Sperm runs from the male into the female through grooves in the claspers. The eggs of most species develop within the female's body, and the pups are born alive. A few species lay large yolky eggs right after fertilization.

Rays and Skates

Rays and skates have flattened bodies with winglike, paired pectoral fins and, in some species, whiplike tails. Rays have diamond- or disk-shaped bodies, while most skates have triangular bodies. Most rays and skates are less than 1 m long.

Rays and skates are primarily bottom dwellers and exhibit many adaptations to this life. Their flat shape and speckled or sandy color help to camouflage them on the ocean floor. Because their ventral mouths are often buried in sand, water enters their gills through two openings called spiracles atop their heads. Most feed on mollusks and crustaceans. Their teeth enable rays and skates to crush such prey.

Figure 36-10 The manta ray's coloring provides camouflage from above and below.

Yellow perch, *Perca flavescens*

Japanese eel, *Anguilla* sp.

Pacific herring, *Clupea pallasi*

Sockeye salmon, *Oncorhynchus nerka*

Lungfish, *Ceratodus* sp.

Latemeria, *Coelancan thus* sp.

Angler fish, *Cryptopsaras covesi*

Osteichthyes

Of the 25,000 known species of fishes, about 95 percent are in the Class Osteichthyes—the bony fishes. They make up most of the vertebrate population in fresh water and in salt water. This dominance indicates the success of their adaptations, such as their bony internal skeleton, scaly skin, sense organs, fins, and reproductive patterns. Scientists divide bony fishes into lobe-finned fishes, lungfishes, and ray-finned fishes.

Lobe-finned fishes, or coelacanths *(SEE-luh-KANTHS),* have paddlelike fins with fleshy bases. They were thought to have been extinct for more than 70 million years until South African fishermen caught one in 1938. The genus was named *Latimeria,* after the local museum curator, Mary Latimer, who recognized its biological importance. It is probably related to the ancestors of the first amphibians.

Lungfishes have lungs. **Lungs** are internal respiratory organs where gas exchange takes place between air and the blood. Lungfish also have gills, where gas exchange takes place between water and the blood. Lungfishes live in shallow tropical ponds that dry up in the summer. They burrow into the mud and cover themselves in mucus to stay moist until the pond refills.

Ray-finned fishes have fins that are supported by long bones called rays. They are diverse in appearance, behavior, and habitat. Ray-finned fishes are the most familiar fishes, and include snakelike eels, yellow perch, cave fish, herring, and lantern fish.

Section Review

1. What is the function of a fish's lateral line system?
2. Compare and contrast the three fish classes.
3. How do lampreys feed?
4. Name the three main types of bony fishes, and give a distinguishing feature of each.
5. Horn sharks, which inhabit reef areas, have large molarlike teeth at the back of their jaws and sharp teeth in front. What do you infer that these sharks eat?

Figure 36-11 The seemingly endless variety of bony fishes reflects their many successful adaptations.

Answers to Section Review
1. It detects vibrations and chemicals in water.
2. The Agnatha and the Chondrichthyes both have skeletons of cartilage while the Osteichthyes have bony skeletons. Only the Chondrichthyes and the Osteichthyes have jaws.
3. They use their sharp tongues and teeth to latch on to living fish and then suck the blood and tissues from the fish.
4. Lobe-finned fishes have fins with fleshy bases. Lungfishes have both lungs and gills. Ray-finned fishes have fins that are supported by long bones, or rays.
5. They eat both fish and shellfish. Their molars crush the shells. Their sharp teeth catch fish.

Section Objectives

- *List the characteristics of body structure, fins, and integument of the yellow perch.*
- *Describe the internal anatomy of the yellow perch.*
- *Explain how gills function in respiration.*
- *Describe the spawning behavior of fishes.*

Figure 36-12 The external features of the yellow perch, *Perca flavescens,* are representative of bony fishes. Note the growth rings on the scales shown in the inset. They indicate the fish's approximate age.

36.3 Morphology of a Bony Fish

The diversity of body shape in bony fishes reflects their varied ways of life. However, all have some characteristics in common. In this section we will examine the morphology of the yellow perch, a representative bony fish. This fish, *Perca flavescens,* grows up to 38 cm long and is common in the Great Lakes and other freshwater lakes of the eastern United States and Canada.

External Anatomy

Look at Figure 36-12 to see how the external appearance of the yellow perch shows several important adaptive features. These include the gill structures, specialized fins, and skin.

Body Structure

The yellow perch, like all bony fishes, has distinct head, trunk, and tail regions. On each side of the head is the **operculum**, a hard plate that opens at the rear and covers and protects the gills. The perch opens its mouth to take in water, and the water passes over the gills, where gases are exchanged. The water leaves through the operculum. Yellow perch have strong muscles along their dorsal backbone that thrust the tail from side to side. They can swim at speeds up to 16 km per hour.

External anatomy of a fish

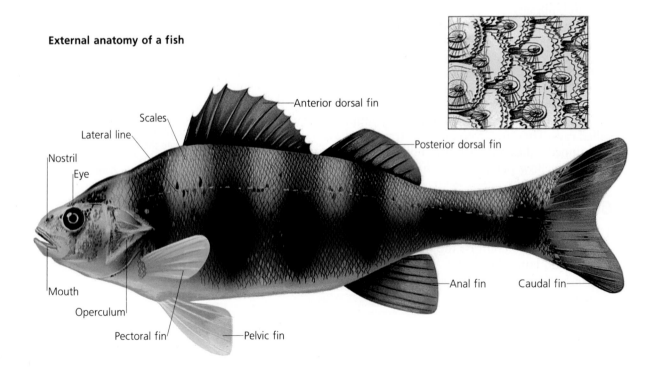

Fins

The fins of the yellow perch are adapted for swimming and guiding the fish through the water. Look again at Figure 36-12. The **caudal fin** extends from the tail. It moves from side to side and amplifies the swimming motions. Two **dorsal fins,** one anterior and one posterior, and a ventral **anal fin** help keep the fish upright and moving in a straight line. The paired **pelvic fins** and pectoral fins are more specialized than the fixed fins of sharks. The perch uses these fins to steer, brake, move up and down, and even back up. They also orient the body when at rest.

Fins are thin fan-shaped membranes richly supplied with blood. By raising and lowering its fins, the perch helps regulate its body temperature. The fins are supported by rays or spines. Rays are bony yet flexible, while spines are bony and rigid. In most fish the posterior dorsal fin has rays. The anterior dorsal fin is spiny and is often used in defense. For example, the backward-pointing spines of a perch can pierce the throat of a predator that tries to swallow it from the rear.

Integument

The **integument** *(in-TEG-yoo-munt)*, or skin, of the yellow perch is covered with scales. These are thin, round disks of highly modified bone that grow from pockets in the skin. As Figure 36-12 shows, scales overlap like roof shingles, all pointing toward the tail to minimize friction as the fish swims. Scales grow during the entire life of the fish, adjusting their growth pattern to the food supply. The scales grow quickly when food is abundant and slowly when it is scarce. The resulting growth rings give a good approximation of the yellow perch's age.

The skin of bony fishes also contains pigmented **chromatophores** *(kroe-MAT-uh-FORZ)* that create various color patterns as, for example, the bars on the sides of the yellow perch. Many deep-sea fishes are black or have phosphorescent patterns. Others blend in with the ocean bottom by changing color. Flounders, for example, can quickly match the spots of the sand found on the ocean floor.

Internal Anatomy

The yellow perch has the same major organ systems as all other bony fishes. Like other bony fishes, the yellow perch also has an organ called the gas bladder.

The bone of the yellow perch, like that of all vertebrates, is a living tissue in which cells deposit minerals, primarily calcium. Bone can grow, support many times its own weight, and heal if broken. It resists bending or breaking when stressed by muscles or blows. The major parts of the fish skeleton are the skull, spine, and ribs. The spine is made up of many bones, the vertebrae, with cartilage pads in between. This provides both strength and flexi-

Info: The spines of some fish contain chemicals that are toxic and lethal.

Answer to Think
The light-colored underbelly also serves as a camouflage. It makes the fish hard to see when viewed from below against the light-colored, sunlit water.

Think

The back of the speckled trout shown above makes the fish hard to see by both predator and prey when it lies at the bottom of a pebbly stream. What benefit might the trout derive from its light-colored undersurface?

bility. The spinal column also partly encloses and protects the spinal cord.

Digestive System

Yellow perch, like many bony fishes, are carnivores. Their jaws are armed with many sharp teeth that point inward to keep smaller fish and other prey from escaping. Strong muscles operate the jaws, which have a hinge that allows the mouth to open wide for capturing large prey. A tongue is anchored, and therefore immobile, in the bottom of the mouth. Lined with nerve cells, it helps detect chemicals in the environment.

Look at Figure 36-13. Food passes from the mouth into the **pharynx,** or throat cavity, and then moves through the tubelike **esophagus** to the stomach. Much digestion takes place in outpockets of the stomach called pyloric ceca *(SEE-kuh)*. The liver and the pancreas, located near the stomach, secrete digestive enzymes such as bile and insulin that help break down food. The short intestine has fingerlike extensions called villi that extend and increase the surface area for absorption of digested foods. Undigested material leaves through the ventral anus.

Circulatory System

Yellow perch have a highly developed circulatory system that is adapted for rapid swimming and other high-performance

Figure 36-13 The internal anatomy of a bony fish includes an organ—the gas bladder—unique to this class of vertebrates. The inset shows details of the circulatory system.

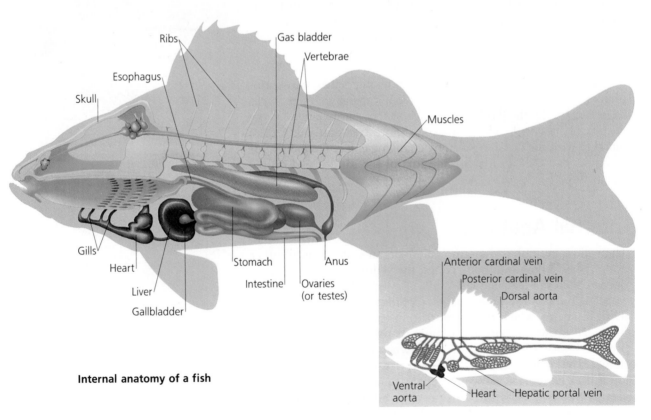

Internal anatomy of a fish

activities. Their circulatory system consists of a two-chambered heart, blood vessels, and blood containing red and white blood cells. The **atrium** is the collecting chamber of the heart, and the **ventricle** is the pumping chamber. Follow the circulatory path in Figure 36-13. The heart pumps blood through **arteries** to small, thin-walled vessels called **capillaries** in the gills. There the blood picks up oxygen gas from, and releases carbon dioxide into, the water. The blood tnen moves to the body tissues, where nutrients and wastes are exchanged. The blood returns to the heart through **veins**.

Respiratory and Excretory Systems

The fish's central respiratory organs, the gills, are adapted for gas exchange. Gills consist of four sets of curved pieces of bone on each side of the head. Each gill has a double row of thin projections called gill filaments that are richly supplied with capillaries. The large surface area allows rapid gas exchange. Swimming helps maintain a constant flow of water across the gills. Gill valves prevent particles that enter the mouth from passing over the gills.

Gills also excrete nitrogenous wastes from the body, but that task is carried out primarily by the kidneys. The kidneys filter out dissolved chemical wastes from the blood. The resulting solution, urine, is carried through a system of ducts to the urinary bladder, where it is stored and then expelled.

The gills and kidneys have the additional vital function of maintaining homeostasis between salt and water in the body. Saltwater fishes, which take in excess salt from the seawater, have special mechanisms for excreting salt. Yellow perch and other freshwater fish need more salt in their blood than is contained in the surrounding water, so they excrete excess water.

Gas Bladder

Most bony fish, including the yellow perch, have a unique organ, the **gas bladder**, or swim bladder. This thin-walled sac in the abdominal cavity contains a mixture of oxygen, carbon dioxide, and nitrogen obtained from the bloodstream. Comparative morphology suggests that the gas bladder is likely to have evolved from the lungs of lungfish-type ancestors. By regulating the amount of gas in the sac, fish adjust their overall density and thus move up or down in water or hover at a given depth.

Nervous and Sensory System

The nervous system of the yellow perch includes a brain, a spinal cord, and nerves that lead to and from all parts of the body. All fishes have more complex nervous systems than invertebrates do. This increase in complexity makes possible the more varied behavior of fishes.

The brain of the yellow perch consists of five paired lobes. The largest, at the center of the brain, are the two **optic lobes**. They receive nerve impulses from the eyes and other sense organs and signal the muscles to move. The anterior lobes of the brain are

Gill filaments

Oxygen-rich blood

Oxygen-poor blood

Water

Figure 36-14 The gill filaments provide the organism with a large surface area, thus enabling gas exchange to occur quickly.

Info: If a fish is caught at great depth and is rapidly brought to the surface, its gas bladder expands like a balloon due to the pressure drop. The internal air pressure forces the gut or the air bladder out through the fish's mouth.

Figure 36-15 Among the many fish that fertilize their eggs externally is the white sturgeon, *Acipenser transmontanus.* Note that several of these white sturgeon eggs are hatching.

Answers to Section Review
1. Both are heart chambers. The atrium collects the blood and the ventricle pushes the blood throughout the body.
2. The caudal fin helps move the fish through the water. The dorsal fin keeps the fish from rolling. The pectoral fin aids in steering.
3. Swimming and gulping of water forces water past the fish's gills, which are rich in blood vessels. Oxygen in the water and carbon dioxide in the blood are exchanged through diffusion.
4. The number is high because so few of the eggs hatch and few of those hatched live to maturity. (See pp. T36A–T36D.)

the olfactory lobes and the **cerebrum** *(suh-REE-brum)*, which respond mainly to smells. Smell is the most important sense in fish. At the posterior of the brain are the **cerebellum** *(SER-uh-BELL-um)*, which coordinates the muscles, and the **medulla oblongata** *(muh-DULL-uh ahb-lawn-GAHT-uh)*, which regulates the internal organs. The major sense organs are connected directly with the brain via **cranial nerves.**

From the medulla the spinal cord extends the length of the body and carries nerve impulses to and from the brain. **Spinal nerves** connect the spinal cord with the internal organs, muscles, and sense organs. These nerves also carry impulses to the brain from the lateral line system.

Reproduction

The reproductive systems of most bony fishes resemble that of the yellow perch. The sexes are separate. Eggs are produced by ovaries in the female, and sperm is produced by the testes in the male; both are released through an opening just to the rear of the anus. Fertilization of the eggs of yellow perch takes place externally. Young fish hatch from fertilized eggs within a few hours in warm water or after many weeks in cold water. The adaptation of laying large numbers of eggs ensures that at least a few individual yellow perch will survive to become adult fish. Most eggs are not fertilized. Many others die or are eaten. The number of eggs bony fishes lay varies considerably. A brook trout lays from 80 to 5,000 eggs, but an ocean sunfish lays about 3 million.

Some fishes bear live young. Females receive sperm from males during mating, and fertilization is internal. These "live-bearers" include fishes common in aquariums, such as guppies, platys, mollies, and swordtails. Vertebrates that bear live young are said to be viviparous.

Few generalizations can be made about the reproductive, or **spawning,** behavior of bony fish. The female yellow perch lays eggs on the lake bottom, and the male fertilizes them by releasing **milt,** a fluid containing sperm. Some species, such as sticklebacks, build crude nests from plants, sticks, and shells. Many species migrate to warm, protected shallow water to spawn. Freshwater eels from rivers in Europe migrate to the Atlantic to spawn. Adult Atlantic and Pacific salmon migrate back to fresh water to spawn. They use their sense of smell to find the stream where they were hatched. Here they spawn and die.

Section Review

1. Compare the atrium and the ventricle of a perch.
2. How do the caudal, dorsal, and pectoral fins function?
3. How do fish gills function in respiration?
4. Why do some fish lay a large number of eggs?
5. Bottom-dwelling fish often lack a gas bladder. Explain the adaptive advantage of this modification.

Tip: Tell students that fish do not have external ears. They hear by sensing vibrations picked up by a pair of inner ears.

Constructing a Dichotomous Key to Fish Species

Objective
To construct a dichotomous key to fish species using basic fish characteristics

Process Skills
Classifying, communicating

Materials
Pencil and paper

Method
1 A dichotomous key requires a choice between successive pairs of characteristics. Each subject is identified by separating out the one characteristic unique to that subject. Look at the fishes below.
2 Construct a dichotomous key that can be used to identify each species shown. Use the presence or absence of the following characteristics: spiny fins, body flattened from top to bottom, a sucking disk in place of a jaw (if the disk is not shown, assume it is absent), both eyes on the same side (surface) of the head, eyes large relative to head, long thin tail, dorsal fin that covers most of the body. Refer to p. 275 to recall how a dichotomous key should look.

Conclusions
1 Bony fishes belong to the Class Osteichthyes. All members of that class shown here have spiny fins. Name the members of the Class Osteichthyes pictured here.
2 What additional kinds of information might be helpful in classifying fishes?

Inquiry
1 How might you evaluate your key?
2 What are the problems involved in developing an unambiguous dichotomous key?

Chapter 36 Review

Vocabulary

anal fin (547)
appendicular skeleton (539)
artery (549)
atrium (549)
axial skeleton (539)
capillary (549)
caudal fin (547)
cerebellum (550)
cerebrum (550)
chromatophore (547)
cranial nerve (550)

deuterostome (537)
dorsal fin (547)
dorsal nerve cord (537)
endoskeleton (537)
esophagus (548)
gas bladder (549)
gill (540)
gill slit (537)
integument (547)
lateral line system (540)

lobe-finned fish (545)
lung (545)
lungfish (545)
medulla oblongata (550)
milt (550)
notochord (537)
olfactory lobe (542)
operculum (546)
optic lobe (549)
pectoral fin (543)
pelvic fin (547)

pharyngeal pouch (537)
pharynx (548)
placoid scale (543)
ray-finned fish (545)
skull (539)
spawning (550)
spinal nerve (550)
vein (549)
ventricle (549)
vertebral column (537)

In each set of terms below, choose the one that does not belong, and explain why.
1. notochord, vertebral column, dorsal nerve cord, gill slits
2. cerebrum, cerebellum, operculum, optic lobe
3. placoid scale, chromatophore, integument, milt
4. pharynx, atrium, ventricle, capillary
5. pectoral fin, gas bladder, operculum, lateral line system

Review

1. Chordates and echinoderms have a similar (a) body plan (b) embryonic development (c) habitat (d) feeding habit.
2. The skull functions to (a) protect the brain (b) anchor the axial skeleton (c) coordinate muscles (d) collect sensory information.
3. The density of water means that fish (a) must be heavy in order to float (b) do not need strong muscles to propel them (c) sink to the bottom when at rest (d) all of the above.
4. The lateral line system (a) keeps fish moving in a straight line (b) initiates migration (c) detects vibrations (d) acts as a camouflage.
5. Lamprey locate prey through a highly developed sense of (a) smell (b) hearing (c) sight (d) touch.
6. Sharks use claspers to (a) increase maneuverability (b) startle other fish (c) hold on to the female while mating (d) hold on to prey while feeding.
7. To move up or down, a fish uses its (a) anal fin (b) dorsal fins (c) caudal fin (d) pelvic fins.
8. The esophagus of a fish (a) creates buoyancy (b) fertilizes eggs (c) carries food from the mouth to the stomach (d) holds wastes moving from the stomach to the anus.
9. A fish's ventricle (a) pumps blood through the body (b) collects blood returning to the heart (c) facilitates gas exchange through diffusion (d) carries oxygen through capillaries.
10. Some fish species (a) build nests for their eggs (b) return to their hatching grounds to spawn (c) lay huge numbers of eggs (d) all of the above.
11. Name the seven living vertebrate classes, and give a representative animal of each.

12. Of what advantage is an endoskeleton in comparison with an exoskeleton?
13. Distinguish between the appendicular and axial skeletons.
14. How do researchers determine the habitat range of sharks?
15. Describe adaptations in rays and skates that allow a bottom-dwelling existence.

16. What is the advantage to lungfishes of having both lungs and gills?
17. How can scale pattern show a fish's age?
18. How do saltwater fish and freshwater fish maintain their salt balance?
19. Describe a fish's circulatory system.
20. Describe the function of the five lobes of a fish's brain.

Critical Thinking

1. Neotny is a phenomenon in which larvae become sexually mature while retaining larval characteristics. The term has been used to explain in part the evolution of vertebrates from ancestral chordates. Use what you know about tunicates to support this hypothesis.
2. The shark has a large corkscrew-shaped structure in its intestine called a spiral valve. How might this organ function in digestion?
3. Many species of fishes that live deep in the ocean where little or no light reaches are luminescent. What might be the advantages of such an adaptation?
4. The four-eye butterfly fish shown here often swims backward. How might the fish's behavior and coloring work to its advantage?
5. Saltwater fish drink more water and produce much less urine than freshwater fish do. How do you account for this?

6. Cod and many other ocean fish lay eggs on the surface of the water. The male largemouth bass scoops out a nest in a lake or river bottom and waits for a female to deposit her eggs. What hypothesis would you make regarding the relative number of cod and bass eggs? Why?

Extension

1. The construction of the St. Lawrence Seaway allowed sea lampreys from the Atlantic Ocean to invade the Great Lakes. There they caused extensive damage to lake trout and other commercial fish species. Use your library to find out how the lamprey population was controlled and how successful this effort was. Present your findings to the class.
2. Read the article by Kenneth E. F. Watt entitled "Deep Questions about Shallow Seas," in *Natural History*, July 1987, pp. 96–97. Write a report explaining the diversity of fish species in shallow coastal waters.
3. Study the anatomy of any fish other than a yellow perch. Identify the operculum, backbone, and each fin. Label the internal parts, and note any structural differences from the yellow perch. What conclusions can you draw about habitat and behavior on the basis of the fish's structure?

Chapter 37 | Amphibians

This chapter reviews the origins and major characteristics of Class Amphibia. Students learn about the adaptations that enable amphibians to live both on land and in water.

Correlation Chart

Section Number	Section Objectives	Pages	Features	Section Review
37.1	1. Summarize the origin of amphibians.	555–558	Biology in Process, pp. 556–557	2, 3
	2. List three conditions of life on land to which amphibians had to adjust.			3, 5
	3. List the major characteristics of amphibians.			1, 5
	4. Name the four orders of living amphibians, and give an example of each.			4
	5. Explain how the amphibian life cycle affects nutrient levels in land and water environments.			
37.2	1. List and locate the parts of a frog's external anatomy.	559–564		1, 2
	2. Summarize the digestive process in frogs.			2
	3. Trace the flow of blood through a frog's heart and lungs.			3
	4. Describe four ways in which anurans breathe.			4, 5
	5. Name the major parts of a frog's excretory system.			4
37.3	1. List the parts and functions of the frog's reproductive system.	565–567	Think, p. 566	1, 2
	2. Summarize the process of fertilization in frogs.		Biology Lab, p. 567	2, 3
	3. List the stages of metamorphosis in frogs.			4, 5

Resources*

Print

Frost, Darrel R., ed. *Amphibian Species of the World*. Lawrence, Kans.: Allen Press, 1985.

Halliday, Tim R., and Kraig Adler, eds. *The Encyclopedia of Reptiles and Amphibians*. New York: Facts on File, 1986.

Audiovisual/Software

Amphibians, revised. Film or video, 14 min. Teacher's guide. Coronet, 1985.

It's a Frog's Life. Video, 60 min. Oxford. Dist. by Carolina Biological, 1986.

Operation: Frog. Apple 64K. Commodore 64K. Disk. User's handbook. Teaching guide. Scholastic.

*Additional references are listed in *Modern Biology, Teacher's Program Guide*.

Section Teaching Guides

37.1 Overview of Amphibians

Approach

The main objective of this section is to familiarize students with the characteristics of amphibians and with the environmental conditions that favored the evolution of land-dwelling vertebrates. In presenting the text, remind students that amphibian evolution occurred because conditions favored species that could survive on land. Your discussion should draw upon the concepts of natural selection and adaptation introduced in Chapters 15 and 16. In discussing the classification of amphibians, compare the adaptations of each amphibian order to aquatic and terrestrial environments.

Concept Framework

Students know that the demands of life on land differ considerably from those of life in water. To tap this knowledge, encourage students to imagine a fish trying to survive on land. Ask them what adaptations would be necessary to meet the challenges of obtaining oxygen, water, and food, reproducing, and escaping from predators.

Emphasis

Make sure that students understand that extreme selection pressures led to the colonization of land by amphibians. Also reinforce the idea that amphibians represent an evolutionary link between water- and land-dwelling vertebrates.

Using the Features/Nutrient Cycling and the Frog's Life Cycle

The main process skills used in this feature are observing, experimenting, and collecting and interpreting data. Dianne Seale combined field observation and lab experiments to determine how the migration of frogs from water to land affects the nutrient balance in a pond ecosystem. The feature illustrates the way in which scientists use controlled experiments to interpret their observations in the field. Seale's work also demonstrates that the effect of an organism upon its environment can be measured quantitatively.

Section Review Answers

(Continued from p. 558.)

4. Anura means "without a tail." Urodela means "having a tail." Apoda means "without legs." Trachystoma means "rough mouth."

5. Land adaptations include moist skin; respiration through lungs, skin, and mouth; three-chambered heart. Water adaptations include webbed feet and clawless toes; gill breathing in larva; eggs lacking membrane or shell.

37.2 Frog Anatomy

Approach

The main objective of this section is to show how the external anatomy of frogs, as well as the major internal systems, are adapted to life spent both on land and in water. In discussing the anatomy of the frog, draw comparisons between amphibian anatomy and that of fishes, covered in Chapter 35. Also compare amphibian adaptations to terrestrial life to those of the earliest land plants, the bryophytes, as discussed in Chapter 25.

Concept Framework

Even if they do not live in an area where they can observe frog behavior, most students are familiar with frog behaviors such as jumping, insect-catching, and swimming near the surface of the water. Lead students in a discussion of frog anatomy by asking them to consider as they read the text how these behaviors reflect adaptations to life on land and to life in water.

Emphasis

Students should understand that amphibian characteristics evolved in response to the need for terrestrial organisms to conserve water and energy. Remind students that amphibians also spend part of their life in water and that amphibian anatomy reflects only a partial adaptation to life on land.

Identifying Amphibians

Divide the class into three groups. Give each group a field guide and ask each group to identify at least three amphibians that are native to the area. Then have each group list the adaptive features or behaviors that enable each amphibian to exist in the local environment.

Section Review Answers

(Continued from p. 564.)

3. Blood enters the ventricle from the right and left atria and is pumped to the lungs and to the rest of the body.
4. Carbon dioxide is eliminated through the skin and lungs. Nitrogen compounds are eliminated by the kidneys.
5. Skin would dry up and cutaneous respiration would cease. The frog would not live long because the lungs couldn't supply enough oxygen to the body.

37.3 Reproduction in Frogs

Approach

The main objective of this section is to convey the idea that amphibian adaptations to aquatic life are most apparent in reproduction and larval development. In your discussion, point out that amphibian reproduction and fertilization must take place in or near water. Students will be able to use this information as a basis for comparison with reptiles and birds (Chapters 38 and 39). Both the process and adaptive significance of amphibian metamorphosis can be compared to that of metamorphosis in insects (Chapter 33) and echinoderms (Chapter 34).

Concept Framework

Students will know that frogs undergo a dramatic transformation from tadpole to adult. Many will also know that tadpoles and frogs are most visible in the spring. Elicit this knowledge as from students to serve as a conceptual frame as they read about the frog's life cycle and the changes involved in metamorphosis.

Emphasis

Make certain that students understand that external fertilization must take place in or near water in order for the sperm to reach the eggs. This is a primary reason amphibians retain their dependence on the aquatic environment.

Guest Speaker

Invite an expert on amphibians from a local college or university to discuss the various types of amphibian reproduction and their possible evolutionary significance.

Section Review Answers

1. structures that amplify the sound of the call in males; to attract a female
2. The eggs receive a protective coating of jellylike material.
3. The male frog releases sperm over the eggs that have been released into the water.
4. Tadpoles have gills, a two-chambered heart, a tail, and can regenerate. Mature frogs have legs, no tail, a three-chambered heart, lungs, teeth, jaws, and do not regenerate.
5. It increases the likelihood that the tadpole will survive to adulthood.

Laboratory Preparation

Background

In this lab investigation, students study frog behavior that is mainly reflexive in nature. Students consider the adaptive significance of these behaviors. The frog's visual system is highly developed to preceive motion, and movement will cause the frog to jump. The anatomical differences between frog and salamander locomotion are fundamental, and students can investigate and compare these differences.

Preparation

Feed frogs crickets or other moving insects for several days before the lab. Animals that do not feed will not respond to the stimuli in the experiment. Some frogs are more nervous than others. If extras are available, use frogs that respond and feed best.

Safety

Because the frog will attempt to escape, remind students to be as quiet and inactive as possible.

Conclusions

1. The frog uses its legs and webbed feet for swimming. A fish uses its trunk muscles.
2. The frog has a nictitating membrane to protect its eyes under water, the ability to breathe both in and out of water, a dark upper body surface to camouflage it when it is out of the water and a light undersurface to make it blend in with the sky when seen from below, and legs adapted for both swimming and jumping.

Inquiry

Students should devise ways of investigating a frog's reaction to colors and movements and hypothesize regarding any relationship to feeding habits. One way to investigate this is by testing the frog's reaction to different colored paper dots attached to a length of thread.

Chapter Review

Vocabulary

1. Granular glands secrete foul-tasting or poisonous substances that protect the frog from enemies; mucus glands secrete a lubricant that keeps the skin moist.
2. Maxillary teeth line the perimeter of the upper jaw. Vomerine teeth project from the bones of the roof of the mouth.
3. The duodenum is the upper portion of the small intestine; the ileum is the middle portion.
4. The conus arteriosus is the main vessel leaving the ventricle. It branches into the right and left truncus arteriosi, which carry blood throughout the body.
5. Pulmonary respiration is breathing through the lungs. Cutaneous respiration is breathing through the skin.

Review

11. metamorphosis; thin, smooth, moist skin; webbed feet; respiration through gills, lungs, skin, and/or mouth cavity; larvae have a two-chambered heart, adults have a three-chambered heart; eggs are laid in water or in a moist environment and are fertilized externally
12. Frogs have long legs, are good jumpers, and they live in or near water. Toads have a drier skin than frogs and shorter legs.
13. Because a frog's eyes and nostrils are on the very top of its head, it can be almost totally submerged in water and still breathe and see and avoid enemies.
14. It connects the middle ear with the mouth cavity and equalizes air pressure on both sides of the eardrum.
15. Skin, lungs, and the mouth cavity are used in respiration. The methods of respiration are called cutaneous respiration (skin), pulmonary respiration (lungs), and mouth breathing.
16. When it is in water the frog expels the water that seeped into its body in a diluted urine. This process prevents the frog from bursting. When it is on land, the frog uses the diluted urine it has stored as a water source. This enables the frog to live away from water.
17. to the lungs and to the rest of the body
18. through the skin
19. carbon dioxide and nitrogen compounds
20. As tadpoles they extract nutrients from the water; then as frogs they transfer some of these nutrients back to the water. The tadpoles ensure that the phytoplankton population stayed low while the frogs ensured that the phytoplankton did not disappear.

Critical Thinking

1. They underwent adaptive radiation. All the species of amphibians evolved from a common ancestor. Speciation may have occurred as the early amphibians moved onto land to take advantage of new food sources and new habitats.
2. They would have been able to use their lobe fins and their primitive lungs to travel across land in search of water.
3. footpads, countershading, and coloration
4. To ensure, first, that the sexually mature female will find a male during the reproductive season and, second, that the frog will mate only with members of its own species.
5. a sticky tongue and two sets of teeth
6. In their environment the frogs need a keen sense of smell and sight. These areas are larger than the "thinking" area because frogs are instinctive rather than reasoning.
7. They do not have a water environment to lay their eggs in, so the eggs are not bathed in nutrients. The large yolk supplies nutrients. In addition, the environment of the hole keeps the eggs cool and moist.

Amphibians

Introduction

Between 370 million and 325 million years ago, a major biological event took place: certain species of fish successfully began to live part of their lives on land. In this chapter you will learn about the descendants of these fishes—frogs, salamanders, and other animals that belong to the group we call amphibians. This name, which comes from a Greek term meaning "double life," reflects the fact that amphibians live both on land and in water.

Chapter Outline

37.1 Overview of Amphibians
- *Evolution*
- *Characteristics*
- *Classification*

37.2 Frog Anatomy
- *External Anatomy*
- *Internal Anatomy*

37.3 Reproduction in Frogs
- *Reproductive System*
- *Fertilization*
- *Metamorphosis*

Chapter Concept

As you read, note the adaptations of amphibians that enable them to live both on land and in water.

Land eggs of the rain frog, *Eleutherodactylus* sp., Costa Rica (×40). Inset: Central newt, *Notophthalmus viridescens*, Kansas.

For additional teaching information see pages T37A–T37D.

37.1 Overview of Amphibians

The fossil record indicates that various fishes were the first vertebrates on earth. In fact, they were the only vertebrates for about 150 million years. As competition for food and space became increasingly intense, natural selection favored animals that could spend at least part of their life on land. These animals evolved into the amphibians.

Evolution

Scientists infer that amphibians evolved from lobe-finned fishes called crossopterygians *(KRAW-SAHP-tuh-RIJ-'nz)*, pictured in Figure 37-1. The fossils of these fishes show that they had short, limblike fins that they may have used to move about on land. Crossopterygians had no gills but did have internal nostrils and a primitive lung that may have enabled them to respire for short periods of time on land.

On the basis of fossil evidence, biologists conclude that amphibians appeared during the late Devonian period, about 345 million years ago. During the Permian period the evolutionary lines of amphibians diverged, one line leading to reptiles and the other leading to the modern amphibians.

The geologic record includes evidence of a number of environmental pressures and opportunities that could have led to the movement of animals onto land. The Devonian oceans, lakes, rivers, and ponds supported a tremendous number and variety of fishes. Food and space were limited, and the numerous species of fish competed intensely for them. This competition apparently was further increased by periodic droughts that dried up streams, lakes, and tidal beds. In addition primitive land plants and insects, both promising food sources, were beginning to evolve. Yet there is no evidence before this period that vertebrates lived on land, where space and oxygen were plentiful.

Characteristics

For organisms adapted to life in water, the land is a hostile environment. Gills stick together and dry out in the air, preventing the exchange of gases between air and blood. Gravity pulls more strongly on skeletal systems because air is less dense than water. In addition the temperature on land fluctuates between hot and cold much more quickly than it does in water.

Through millions of years of evolution, amphibians adapted to conditions on land. Natural selection favored legs in place of fins, stronger bones and muscles, lungs instead of gills, and skin, which in some species had keratin that reduced water loss. Species with necks could see and feed more easily, and those with oral glands could moisten the dry plants and insects they ate.

Tip: Tell students that land animals are exposed to 20 percent more available oxygen than are animals that live in water.

Section Objectives

- *Summarize the origin of amphibians.*
- *List three conditions of life on land to which amphibians had to adjust.*
- *List the major characteristics of amphibians.*
- *Name the four orders of living amphibians, and give an example of each.*
- *Explain how the amphibian life cycle affects nutrient levels in land and water environments.*

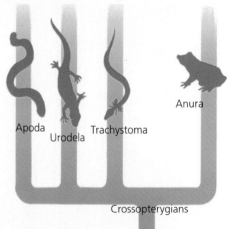

Apoda

Urodela

Trachystoma

Anura

Crossopterygians

Figure 37-1 The crossopterygian (top) is believed to be the ancestor of modern amphibians. A phylogenetic tree suggesting amphibian origins and relationships appears below it.

Amphibians are cold-blooded. Their body temperature rises and falls with that of the surrounding environment. Because of temperature extremes on land, certain protective behaviors arose in amphibians through natural selection. For example, many amphibians enter a state of dormancy, or **torpor,** when conditions are unfavorable. They often bury themselves in mud or leaves, emerging when conditions are better. Such states of inactivity are known as **hibernation** when they occur in winter and **estivation** when they occur in summer.

Amphibians have many prominent characteristics that are adaptations to a life spent both on land and in water:

- They change from an aquatic larval stage to a terrestrial adult form. This transformation is called **metamorphosis.**
- They have moist, smooth, thin skin with no scales.
- Feet, if present, are webbed. The toes lack claws.
- They use gills, lungs, skin, and mouth cavity in respiration.
- Larvae have two-chambered hearts; adults have three-chambered hearts and well-developed circulation.
- Eggs lack a membrane or shell. They are usually laid in water or in a moist environment and fertilized externally.

Answer to Biology in Process
Seale knew that nitrogen was a primary component of protein and that one-celled animals were composed largely of protein. Seale also inferred that tadpoles would remove nutrients from the pond, that frogs helped maintain nutrient balance, and that this had significance for other organisms.

Biology in Process

Nutrient Cycling and the Frog's Life Cycle

How does the life cycle of a frog—its development from egg (left) to tadpole to frog—relate to the maintenance of nutrient levels in a pond? Biologist Dianne Seale wanted to find out.

Seale combined field observations of frogs in a Missouri pond with laboratory experiments to determine how the frog's life cycle affected nutrient levels. First she studied tadpole feeding. She decided to track levels of nitrogen, because nitrogen is a component of protein, a nutrient.

Seale measured nitrogen levels in tadpoles, the pond water, and in eggs laid in the pond by adult frogs. She also monitored changes in the tadpole population. Analyzing the two sets of data, she inferred that tadpoles depleted a pond of much of its nutrients.

Tadpoles typically feed on minute particles of algae called phytoplankton, which are suspended in water. In the laboratory Seale allowed tadpoles to feed for four hours. She then dissected them and removed the contents of their gut. When she analyzed the contents of the tadpoles' intestines, she found the amount of nitrogen the tadpoles had consumed. Knowing that amount, she could measure the amount of

Classification

Biologists have identified about 2,375 living species of amphibians and have classified them into four orders. Frogs and toads make up the Order Anura ("without a tail"). Salamanders and other amphibians with legs and tails make up the Order Urodela ("visible tail"). Wormlike organisms called caecilians (*sih-SIL-yunz*) are part of the Order Apoda ("without legs"). Some aquatic amphibians belong to the Order Trachystoma ("rough mouth").

Anura

Anurans live in environments from deserts and mountains to ponds and puddles. Frogs have smooth, moist skin and short, broad bodies. Their long hind legs make them excellent jumpers. Most frogs live in or near water, though some species live in trees. Toads have dry, bumpy skin, a stocky body, and short legs. True toads belong to the Family Bufonidae, but the term *toad* is commonly used for any anuran that is particularly well adapted to dry environments. Both frogs and toads return to water to reproduce. In nearly all species eggs are fertilized externally. The fertilized eggs hatch into swimming larval forms called **tadpoles.**

Figure 37-2 *Bufo marinus* shows the bumpy skin and stocky body typical of toads.

nutrients that they had removed from the pond.

Seale also calculated the number of suspended phytoplankton particles in the pond overall. She observed that as the tadpole population developed, nutrient levels decreased. Once the tadpoles became land-dwelling frogs, however, the phytoplankton level increased dramatically.

By now Seale had a good idea of how much nutrient content the tadpole population had extracted from the pond. But why didn't the pond become permanently depleted of nutrients? Did adult frogs return nutrients to the pond

when they deposited their eggs? If so, how much nutrient content did adult frogs put back? Seale analyzed the nitrogen content of these eggs and compared it with the amount ingested by the tadpoles. The eggs, in fact, contained considerably less nitrogen.

Seale concluded that amphibians transferred a certain

amount of nitrogen from water to land. She hypothesized that other mechanisms must complete the cycling of nitrogen in order for the ecosystem of the pond to remain nutritionally balanced.

■ How did Seale infer that nutrients were cycled by studying only nitrogen? What other inferences did she make?

Figure 37-3 Two orders of amphibians are represented by the marbled salamander, *Ambystoma opacum* (left), and the caecilian, *Gymnophis multiplicata* (right).

Answers to Section Review
1. Both are types of torpor. Hibernation takes place in winter; estivation, in summer.
2. Their limblike fins permitted them to move on land. They could respire through internal nostrils and primitive lungs.
3. overcrowded water environments, drought, and the emergence of land plants and insects
(See pp. T37A–T37D.)

Urodela

Salamanders, typical members of the Order Urodela, have elongated bodies, long tails, and smooth, moist skin. They range in length from a few centimeters to 1.5 m. Compared to the anurans, salamanders are less able to remain on dry land, although some can live in dry areas by remaining inactive during the day. Many types of salamanders live either in water or under logs and stones. The aquatic species are called newts.

Like anurans, many salamanders lay their eggs in water. These eggs hatch into swimming larval forms. Other species can reproduce in damp land environments. Eggs laid on land hatch into miniature adult salamanders. Still other salamander species do not lay eggs. They have a type of internal fertilization in which females pick up sperm packets deposited by males.

Apoda and Trachystoma

Caecilians, members of the Order Apoda, compose a highly specialized group of tropical burrowing amphibians. These legless wormlike creatures average 30 cm long, but they can be up to 1.3 m long. They have very small eyes and are often blind. They eat worms and other invertebrates. The caecelian male deposits sperm directly into the female, and the female bears live young.

The Order Trachystoma contains three living species of sirens, or mud eels. Sirens live in the eastern United States and southern Europe. They have minute forelimbs and no hind limbs.

Section Review

1. Distinguish between *hibernation* and *estivation*.
2. What characteristics suggest that crossopterygians are the likely ancestors of amphibians?
3. What environmental conditions may have led to the colonization of land by amphibians?
4. What does the name of each order of amphibians mean?
5. Which amphibian traits are adaptations to land? To water?

37.2 Frog Anatomy

The amphibian ancestors that first ventured onto land faced a variety of new and harsh conditions. Those species that survived the new conditions and yet retained some aquatic characteristics gradually evolved into the modern amphibians. The anatomy of the frog reflects these adaptations.

External Anatomy

Close study of the frog's external anatomy reveals remarkable adaptations to its "double life" on land and in water. The frog's powerful hind legs are equally effective in jumping or swimming. On land frogs sit with their hind legs folded against the body, poised to jump at the first sign of danger. Most frogs can make leaps many times their body length. In water muscles working on the bones of the hind legs provide the strength for swimming.

Frog's eyes also work equally well in or out of water. Because the eyes bulge out from the head, the frog can stay submerged while literally "keeping an eye out" for predators. Eyelids that can blink protect the frog's eyes from dust and dehydration. In addition to upper and lower eyelids, a third, transparent eyelid called a **nictitating** *(NICK-tuh-TAY-ting)* **membrane** covers each eyeball and joins the lower eyelid. This membrane keeps each eyeball moist and protects the eye when it is under water. Two nostrils located near the top of the frog's head allow the frog to breathe when all but the top of its head is under water.

Frogs have eardrums, or **tympanic** *(tim-PAN-ick)* **membranes,** which are circular structures located behind each eye. The eardrums function well both in water and in air. A bone called the **columella** *(KAHL-uh-MELL-uh)* transmits sounds from the eardrum to the middle ear. A canal called a **Eustachian** *(yoo-STAY-shun)* **tube** connects each middle ear with the mouth cavity, allowing equalization of air pressure on both sides of the eardrum. Embedded in the skull is the inner ear, a minute system of sacs and canals that helps maintain balance and aids in hearing. Hearing is especially important during the mating season, when the male frog produces a distinctive mating call. The female frog's inner ear is attuned to the frequency of the male frog's call.

Section Objectives

- *List and locate the parts of a frog's external anatomy.*

- *Summarize the digestive process in frogs.*

- *Trace the flow of blood through a frog's heart and lungs.*

- *Describe four ways in which anurans respire.*

- *Name the major parts of a frog's excretory system.*

Figure 37-4 This series of exposures of a frog jumping suggests the power in the muscles of its hind legs.

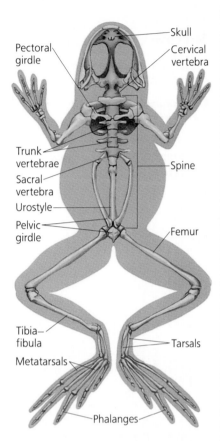

Pectoral girdle

Skull

Cervical vertebra

Trunk vertebrae

Sacral vertebra

Urostyle

Pelvic girdle

Spine

Femur

Tibia fibula

Tarsals

Metatarsals

Phalanges

Figure 37-5 This ventral view of the skeletal system of the frog illustrates a number of adaptations for living on land.

The frog's thick, moist skin serves two important functions—respiration and protection. Numerous **mucus glands** supply a lubricant that keeps the skin moist in air, a necessity for respiration through the skin. **Granular glands** secrete foul-tasting or poisonous substances that protect the frog from enemies. A frog's skin also protects against predators by providing protective coloration. Some frogs, such as the *Hyla versicolor,* can change color in order to blend with the environment.

Internal Anatomy

The frog's skeletal, digestive, circulatory, respiratory, excretory, and nervous systems reflect adaptations to land and water.

Skeletal System

The structure of early amphibians indicates that selection pressures favored a stronger skeletal system and the evolution of limbs. The frog's spine has nine vertebrae. The cervical vertebra at the anterior end of the spine allows neck movement, an adaptation that helps frogs catch prey. Posterior to this are seven trunk vertebrae, and then a single sacral vertebra that supports the hind legs. A long, slim bone called the urostyle extends from the sacral vertebra. The bones of the pectoral girdle, which form the shoulders, connect to the front legs. They also provide the primary protection to the internal organs, since the frog has no ribs. The pelvic girdle connects to the hind legs.

Digestive System

Most frogs feed on insects, and their digestive system is adapted to their diet. A frog's tongue is an excellent insect catcher. The frog simply flicks out its long sticky tongue, curls it around its prey, and pulls the insect back into its mouth. Then the frog snaps its mouth shut and swallows. Sometimes the frog pulls its eyes inward and presses them against the roof of its mouth, an action that helps push the food down its throat.

Frogs have two types of teeth that hold on to prey. A row of **maxillary teeth** line the perimeter of the upper jaw. Two **vomerine teeth** project from bones in the roof of the mouth.

Digestion in frogs takes place in the **alimentary canal,** which includes the esophagus, stomach, small intestine, large intestine,

Figure 37-6 The frog's mouth is adapted for catching, holding, and swallowing prey.

Internal nostril opening

Vomerine teeth

Maxillary teeth

Eustachian tube opening

Vocal sac openings

Tongue

Attachment of tongue

Esophagus

Glottis

and cloaca. The elastic esophagus and stomach allow the frog to swallow large amounts of small vertebrates, insects, and worms. Once food passes through the gullet to the stomach, tiny glands in the stomach walls secrete gastric juices that help break down the food. A muscle called the pyloric sphincter at the lower end of the stomach allows digested food to move into the small intestine.

The upper portion of the small intestine is called the **duodenum** *(DYOO-uh-DEE-num)*. The coiled middle portion of the small intestine is the **ileum** *(ILL-ee-um)*. A fanlike membrane called the **mesentery** holds the small intestine in place. Inside the small intestine nutrients from food broken down in the stomach pass through capillary walls into the blood, which carries them to all parts of the body.

The lower end of the small intestine leads into the large intestine. Here undigestible wastes are collected and pushed by muscle action into a cavity called the **cloaca** *(kloe-AY-kuh)*. Waste from the kidneys and urinary bladder, as well as either eggs or sperm from the sex organs, also passes into the cloaca. Waste materials exit through the **cloacal opening,** or anus.

Other glands and organs aid in the digestion process. The liver produces bile, which is stored in the gallbladder. Bile helps

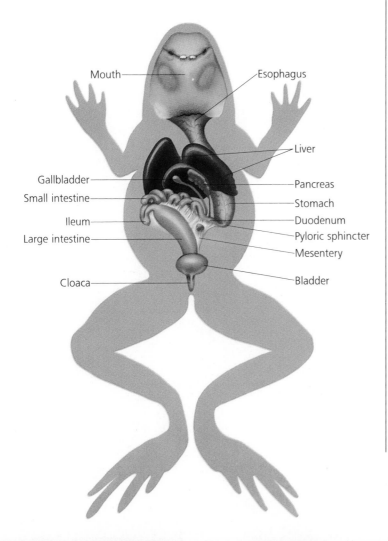

Figure 37-7 The digestive system of the frog, shown here in ventral view, is adapted for eating large amounts of small prey.

Mouth
Esophagus
Liver
Gallbladder
Pancreas
Small intestine
Stomach
Ileum
Duodenum
Large intestine
Pyloric sphincter
Mesentery
Cloaca
Bladder

break down fat into tiny globules that can be further digested and absorbed. A gland called the pancreas, located near the stomach, secretes enzymes that enter the small intestine and help break down food into products that can be absorbed by the blood.

The Circulatory System

Land-dwelling animals expend more energy than water dwellers, primarily because they must counteract the greater pull of gravity on land. An adaptation to the greater oxygen needs of land animals is a more efficient circulatory system than the fish's two-chambered heart. The amphibian's three-chambered heart, illustrated in Figure 37-8, partially separates oxygenated and deoxygenated blood which permits more oxygen-rich blood to circulate throughout the body than in fish.

The three chambers of the heart are the right atrium, the left atrium, and the ventricle. The left atrium receives oxygenated blood from the lungs, and the right atrium receives deoxygenated blood from the body. Both the atria empty into the ventricle, the main pumping chamber of the heart. In the ventricle oxygenated and deoxygenated blood mix partially and are pumped to the lungs and the rest of the body.

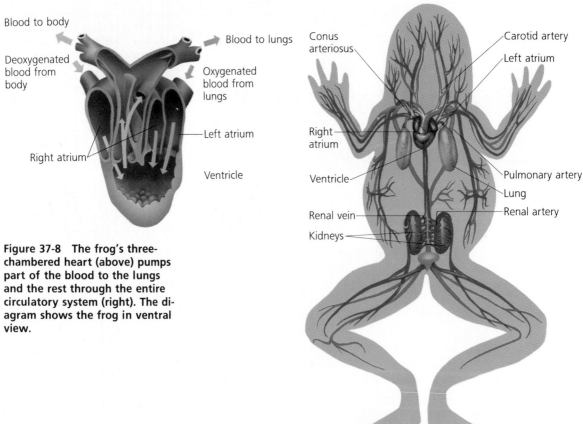

Figure 37-8 The frog's three-chambered heart (above) pumps part of the blood to the lungs and the rest through the entire circulatory system (right). The diagram shows the frog in ventral view.

Figure 37-8 traces the flow of blood through the frog's circulatory system. From the right atrium the blood enters the single ventricle. The ventricle then contracts, pumping some blood to the lungs to receive oxygen and some to the rest of the body. The blood going to the body leaves the ventricle through the **conus arteriosus,** a large vessel that lies against the front side of the heart. This vessel divides into a right and a left **truncus arteriosus,** which immediately branch again into three arches that carry blood to various parts of the body. Deoxygenated blood travels in veins back to the right atrium from the various regions of the body. Oxygenated blood returns from the lungs to the left atrium via the pulmonary veins.

The Respiratory System

Tadpoles respire, or exchange carbon dioxide and oxygen, through gills. Adult frogs lose the gills but can respire in three ways: through the lungs, through the skin, and through the mouth. No other group of animals uses all these methods of respiration.

Respiration through the lungs is called **pulmonary respiration.** A frog inhales and exhales by changing the volume and pressure of air in its mouth. Figure 37-9 shows how the frog creates a partial vacuum by lowering the floor of its closed mouth, causing air to rush into its nostrils. When a frog inhales, air is forced from the mouth into the **glottis,** the passage between the throat and the lungs. To exhale, the frog lifts the floor of its mouth, pushing air out of its nostrils.

Because the frog's lungs are small, **cutaneous respiration,** or respiration through the skin in both air and water, is very important, especially during estivation or hibernation. In the winter a frog may stay buried in the mud at the bottom of a pond for months, respiring only through its skin.

Oxygen can diffuse across the lining of the mouth and into the blood. Frogs use mouth breathing for only a relatively small amount of their respiration.

The Excretory System

Amphibians eliminate two primary types of metabolic waste products—carbon dioxide from respiration and waste compounds from the breakdown of foods. Most of the carbon dioxide is excreted through the frog's skin, although some exits through the lungs.

The kidneys, shown in Figure 37-10 on the following page, are the primary excretory organs and lie on either side of the spine against the dorsal body wall. The kidneys filter nitrogenous wastes from the blood. These wastes, flushed from the body with water, are commonly known as urine. Frog urine flows from the kidney through tiny tubes called urinary ducts to the urinary bladder and from there into the cloaca. Urine and wastes from the digestive system are eliminated through the anus.

Nostril opens

Floor of mouth drops

Nostril closes

Floor of mouth rises

Nostril closed

Floor of mouth drops

Nostril opens

Floor of mouth rises

Figure 37-9 These diagrams show the major steps in the process of pulmonary respiration.

Figure 37-10 The excretory system of the frog, shown here in ventral view, must eliminate excess water when the animal is in water and conserve water when the animal is on land.

Figure 37-10 The excretory system of the frog, shown here in ventral view, must eliminate excess water when the animal is in water and conserve water when the animal is on land.

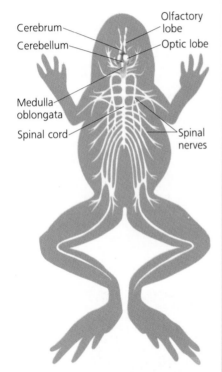

Figure 37-11 The brain of the frog is sufficiently developed to cope with both land and water environments. The diagram shows the frog's nervous system in ventral view.

Answers to Section Review
1. It keeps the eyeball moist and protects the eye when the frog is under water.
2. because they pull the eyes inward to help push the food down the throat
(See pp. T37A–T37D.)

It is vital that the frog regulate the concentration of water in its body. When the frog is in water, its permeable skin allows the water to enter its body. Frogs that live primarily in water rid themselves of excess water by excreting a large volume of very dilute urine. Frogs that live mainly on land conserve water by producing a small volume of more concentrated urine.

The Nervous System

The frog brain is more complex than the fish brain, enabling the frog to contend with a more varied environment. Use the diagram in Figure 37-11 to find the olfactory lobes, the center of the sense of smell, which lie at the anterior end of the brain. Notice that behind the olfactory lobes are the long lobes of the cerebrum, the area of the brain that controls voluntary nervous activity. The optic lobes, which control vision, lie behind the cerebrum. The cerebellum, a small band of tissue lying at right angles to the long axis of the brain, is the center of balance and coordination. The medulla oblongata lies at the back of the brain and joins the spinal cord. It controls organ functions. Ten pairs of cranial nerves extend out directly from the brain.

The spinal cord transmits signals from all parts of the body to the brain and from the brain back to the body. Encased in protective bony vertebrae, the spinal cord extends down the frog's back. As in fishes, the spinal nerves branch from the spinal cord to various parts of the body.

Section Review

1. In what way is the nictitating membrane useful to frogs?
2. Why do frogs blink when they eat?
3. Trace the flow of blood through the frog's heart.
4. How are wastes eliminated by amphibians?
5. Why would a frog die without mucus glands?

Info: Frogs that hunt at night generally have a better sense of smell than do frogs that usually hunt during the day.

37.3 Reproduction in Frogs

Section Objectives

- *List the parts and functions of the frog's reproductive system.*

- *Summarize the process of fertilization in frogs.*

- *List the stages of metamorphosis in frogs.*

The frog's life cycle, like its anatomy, reflects the amphibian's double life. Most frogs have two life stages—an aquatic tadpole stage and a partially terrestrial adult stage. Frog eggs are usually laid in water and hatch into gilled tadpoles that live in water until metamorphosis.

Reproductive System

Because both male and female frogs have internal sex organs, it is difficult to distinguish the sex of a frog during most of the year. However, as the breeding season approaches, the male frog's foreleg muscles and first fingers swell. These swellings help the male maintain his grasp on the female.

The reproductive system of the male frog includes two bean-shaped creamy white or yellowish testes located near the kidneys. During the breeding season, sperm cells develop in the testes and pass through tubes to the kidneys and urinary ducts. During mating sperm leave the body through the cloacal opening.

In female frogs a pair of large, lobed ovaries containing thousands of tiny immature eggs lie near the kidneys. During the breeding season eggs enlarge, mature, and burst through the thin ovarian walls into the body cavity. Cilia move the eggs forward into the funnellike openings of the oviducts. As the eggs pass down the oviducts, they receive protective coats of jellylike material. They remain in structures called ovisacs until ovulation is complete and then leave the body through the cloacal opening.

Fertilization

Most frogs breed once a year. In the first warm days of spring in the temperate zones, frogs emerge from hibernation. They migrate in great numbers to ponds and slow-moving streams. Males establish territories and call to females of their species.

Each species has its own mating call. Air that is driven back and forth between the mouth and the lungs vibrates the vocal cords, producing the frog's croak. Male frogs have vocal sacs that amplify their calls, so they call more loudly than females do.

When a female approaches, the male frog climbs onto her back. He grasps her firmly, just behind the forelegs, in an embrace called **amplexus**. The male will cling to her, sometimes for days, until she lays her eggs. When the female finally releases her eggs into the water, the male frog discharges his sperm over them, and direct external fertilization takes place. The frogs then separate and resume their solitary lives. Within 12 days or so of fertilization, the eggs hatch into tadpoles.

The vast majority of eggs and tadpoles are eaten by predators such as fish, birds, snakes, and turtles. Some species of frogs have

Fat bodies

Testis

Kidney

Cloaca

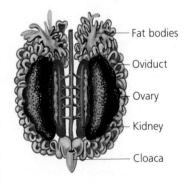

Fat bodies

Oviduct

Ovary

Kidney

Cloaca

Figure 37-12   In frogs both the male reproductive organs (top) and the female reproductive organs (bottom) are completely internal.

Eggs

Young tadpole

Tadpole with limbs

Young frog

Adult frog

Figure 37-13 External body changes during frog metamorphosis, shown here in diagram form, are paralleled by several internal adaptations to a land environment.

Think

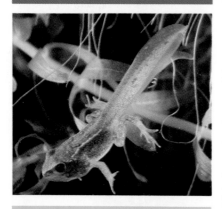

Once tadpoles begin to develop forelegs, the process of metamorphosis is completed very quickly, often in only a few hours. Why might a rapid transformation be advantageous?

Answer to Think
Tadpoles with forelegs are not able to swim well. Shortening the period when they are vulnerable to predators is advantageous.

unusual reproductive cycles that may be adaptations to severe predation. In one species, *Pipa dorsalis,* the eggs develop in pouches on the mother's back. In the genus *Rhinoderma* the male "swallows" the eggs, taking them into his vocal sacs. He later releases fully formed froglets by yawning.

Metamorphosis

Newly hatched tadpoles live off yolk stored in their bodies. They gradually grow larger and develop three pairs of gills. Like fishes, tadpoles have a two-chambered heart. Eventually, a tadpole's mouth opens, allowing it to feed. Tadpoles can also regenerate injured or lost body parts such as a leg or tail.

A tadpole grows and eventually changes from an aquatic larva into an adult. This process of change is called metamorphosis. Legs grow from the body, and the tail disappears. The mouth broadens, developing teeth and jaws. A saclike bladder in the throat divides into two sacs that become lungs. The heart develops a third chamber. The ability to regenerate disappears.

Biologists have long studied the process of metamorphosis and regeneration to learn what controls such dramatic physical changes. A hormone called thyroxine circulates throughout the bloodstream and stimulates metamorphosis. The cells of the tadpole are genetically programmed to respond to thyroxine at the appropriate stage of development. These developmental processes are important in the adaptation of all amphibians to a life spent both in water and on land.

Section Review

1. What are the *vocal sacs,* and how do they function in mating?
2. What happens to frog eggs in the oviducts?
3. How are frog eggs fertilized?
4. Compare the structures of a tadpole and a mature frog.
5. Why is metamorphosis of adaptive significance to the frog?

Answers to Section Review (See pp. T37A–T37D.)

Laboratory

Frog Behavior

Objective
To observe the behavior of a frog

Process Skills
Observing, inferring, hypothesizing

Materials
5-gallon aquarium, live frog, large jar, cardboard box, live cricket or mealworm

Method

1 Fill an aquarium half full with water. Carefully place a frog in the aquarium. How does the frog swim?

2 Frogs have a thin layer of skin called a nictitating membrane that covers their eyes when underwater. What do you think the function of the nictitating membrane is? How might you observe the nictitating membrane without harming the frog?

3 Reach into the aquarium and turn the frog onto its back. What does the frog do?

4 Note that the upper and lower surfaces of the frog's body are colored differently. How might this be advantageous?

5 Place the frog in a large jar. When the frog is comfortably sitting on the bottom of the jar, gently tilt the jar along the long axis of the frog's body. Keep in mind that receptors in the inner ear give the frog information about movement and orientation. Observe the frog's head. Next tilt the jar along the axis running between the frog's ears and observe the frog's behavior. How might this behavior be advantageous? Think particularly of how this behavior would help the frog with the activities of swimming and jumping.

6 Watch the frog breathe. As the frog inhales air into its mouth notice what happens to its throat. Is the throat expanding or contracting? What do you think is happening to the air that was in the frog's mouth?

7 Note the movement of the nostrils. Is the movement of the nostrils always coordinated with the breathing movements? Why might this be advantageous?

8 Note the location of the nostrils and eyes. The position of these features allows the frog to breathe and see while almost completely submerged in water. The frog can also breathe underwater by means of blood vessels in its skin.

9 Place the frog and cricket or mealworm in a cardboard box. Observe the feeding behavior of the frog. How are the frog's mouth and tongue adapted for food-getting?

Conclusions

1 How does the frog's style of swimming differ from that of a fish?

2 What evidence have you seen that the frog is adapted to two very different environments?

Inquiry
How might you test the reaction of a frog to different colors and movements? Formulate a hypothesis about how these reactions might relate to the feeding habits of frogs.

Vocabulary

alimentary canal (560)	cutaneous respiration (563)	maxillary teeth (560)	tadpole (557)
amplexus (565)	duodenum (561)	mesentery (561)	torpor (556)
cloaca (561)	estivation (556)	metamorphosis (556)	truncus arteriosus (563)
cloacal opening (561)	Eustachian tube (559)	mucus gland (560)	tympanic membrane (559)
columella (559)	glottis (563)	nictitating membrane (559)	vomerine teeth (560)
conus arteriosus (563)	granular gland (560)	pulmonary respiration (563)	
	hibernation (556)		
	ileum (561)		

Explain the difference between each pair of related terms.
1. granular gland, mucus gland
2. maxillary teeth, vomerine teeth
3. duodenum, ileum
4. conus arteriosus, truncus arteriosus
5. pulmonary respiration, cutaneous respiration

Review

1. Amphibians evolved from fishes that could survive (a) salt water (b) dry conditions (c) floods (d) low light.
2. Amphibians must lay eggs in water primarily because the eggs (a) need oxygen from water (b) are not laid in nests (c) do not have membranes and a shell (d) need protection from predators.
3. Metamorphosis must take place before amphibians are able to (a) swim (b) live on land (c) feed themselves (d) burrow in mud and leaves.
4. The frog's tympanic membranes are (a) eardrums (b) mouth parts (c) eyelids (d) vocal cords.
5. The urostyle is part of the frog's (a) circulatory system (b) lungs (c) spine (d) small intestine.
6. Bile is a fluid that (a) aids in circulation (b) lubricates skin (c) breaks down fats into globules (d) aids in respiration.
7. The frog's ventricle pumps (a) oxygenated blood (b) deoxygenated blood (c) waste products (d) both oxygenated and deoxygenated blood.
8. The spinal cord is (a) a part of the central nervous system (b) the strongest part of the skeletal system (c) the main part of the circulatory system (d) the source of excretion.
9. Most amphibian eggs do not survive because of (a) floods (b) heat (c) disease (d) predators.
10. In the frog brain the center of balance and coordination is the (a) cerebellum (b) medulla oblongata (c) cerebrum (d) olfactory lobes.
11. Name the common characteristics of amphibians.
12. Contrast the characteristics of a frog and a toad.
13. How does the location of the eyes and nostrils of a frog help it avoid enemies and locate prey?
14. In the frog what is the function of the Eustachian tube?
15. Name and describe the three ways in which adult anurans can breathe.
16. Keeping in mind the urine regulatory system of a frog, explain the role that urine plays in allowing the frog to live on both land and water.
17. Where does blood go after it leaves the ventricle?

18. How does a frog breathe during periods of estivation or hibernation?
19. What are the two primary types of metabolic waste products that must be eliminated by amphibians?
20. How do amphibians help maintain the nutrient balance between land and water environments? How do they help maintain the nutrient balance within a pond?

Critical Thinking

1. What pattern of evolution would you expect early amphibians to have undergone when they moved from a water environment to a land environment? Explain your answer.
2. Geologists believe that extended droughts occurred during the Devonian period. Keeping in mind the anatomy of the crossopterygians, how would these lobe-finned fishes be adapted for such dry periods?
3. Look at the picture of the tree frog on the right. What adaptations does this frog have that allow it to live in trees?
4. In each species of frog the female's inner ear is attuned to the pitch of the mating call of the male. Why is it important that the ear be so attuned?
5. Frogs feed on insects that are usually alive when the frog swallows them. What adaptation in the frog's anatomy allows the frog to capture live prey?
6. In the brains of amphibians the largest parts are the olfactory lobes and the optic lobes, the centers of smell and sight. The cerebrum, the center of reason and

thought, is small. In contrast, the cerebrum in a human brain is about 85 percent of the brain. Why does their brain configuration work well for the life-style and environment of amphibians?
7. Most amphibians lay large numbers of eggs in ponds and then abandon them. The Costa Rican tree frog, however, lays a few large eggs in a depression in the forest floor. Each egg contains all the yolk the developing organism needs for food and houses the frog from its tadpole stage until it is a fully grown frog. Explain these adaptations.

Extension

1. Observe a frog or toad eating a mealworm. Take the mealworm and place it on the end of a toothpick or probe. Holding it in front of the amphibian, watch carefully while the frog eats the worm. Describe what you observe.
2. Research the differences between frogs and toads. Make a comparison chart with headings that include habitat, food, reproduction, mouth parts, skin, feeding, and other aspects of biology and behavior. Include a drawing of a frog and a toad in which you show differences and similarities.
3. Read the article by Shannon Brownlee entitled "A Frog He Would A-Wooing Go," in *Discover,* May 1985, pp. 47–55. List the frogs discussed in the article and describe their special adaptations for reproduction.

Chapter 38 | Reptiles

This chapter explains the structural and behavioral characteristics of reptiles. It emphasizes the evolution of reptiles and the characteristics that allowed them to adapt to life on land.

Correlation Chart

Section Number	Section Objectives	Pages	Features	Section Review
38.1	1. Describe the structure of the amniote egg and explain its importance in reptile evolution.	571–575	Biology in Process, pp. 574–575	1, 2
	2. List three structural characteristics of reptiles that are adaptations to life on land.			3, 5
	3. Summarize the major developments in the evolution of reptiles.			2, 4
	4. List three types of evidence that suggest that dinosaurs may have been warm-blooded.			
38.2	1. List seven characteristics of modern reptiles.	576–579		
	2. Name the four orders of modern reptiles and give an example of each.			1, 2, 3, 4
	3. Describe the basic structure of the turtle shell and explain how it differs in land-dwelling and water-dwelling species.			2
	4. Explain how the head structure of crocodilians is related to their feeding behavior.			3
	5. List three types of defensive adaptations that are common among lizards.			4, 5
38.3	1. Summarize the evolutionary history of snakes.	580–583	Think, p. 582	2
	2. Compare the three basic ways in which snakes move.		Biology Lab, p. 583	2, 3, 5
	3. List three characteristics of snakes that are adaptations to a predatory way of life.			4
	4. Contrast the two methods by which snakes produce offspring.			1

Resources*

Print

Bakker, Robert. *The Dinosaur Heresies,* William Morrow & Co., Inc., 1986.

Ostrom, John H. and J. J. Heads. *Dinosaurs,* 2nd ed. Burlington, N.C.: Carolina Biological Supply Co., 1984.

Audiovisual/Software

The Nile Crocodile. Film or video, 29 min. Britannica, 1984.

Snakes. 2nd ed. Film or video, 15 min. Coronet, 1986.

*Additional references are listed in *Modern Biology, Teacher's Program Guide.*

Section Teaching Guides

38.1 Evolution of Reptiles

Approach

The main objective of this section is to inform students that reptiles were the first group of vertebrates to make a complete transition to terrestrial life. As you discuss the adaptations of reptiles to life on land, build on the concepts discussed in Chapter 37 on amphibians. Stress that the evolution of the amniote egg and internal fertilization were key adaptations that enabled reptiles to reproduce on land. Other reptile adaptations to land can also be contrasted with amphibian characteristics.

The evolution of reptiles reflects the theme of adaptive radiation introduced in Chapter 15. The discussion of reptile evolution also presents an opportunity to emphasize the use of scientific processes to learn about the history of life.

Concept Framework

Most students have some familiarity with dinosaurs through books, movies, and television. However, they may not understand the connection between these prehistoric creatures and the reptiles of today. Remind students that modern alligators and crocodiles are direct descendants of animals that once dominated the earth.

Emphasis

Vertebrates did not completely colonize the land until natural selection favored the ability to reproduce on land. Make sure students understand the significance of the amniote egg in the evolution of land vertebrates. Compare characteristics of amphibian and reptilian eggs to illustrate the ways in which reptilian eggs are adapted to the terrestrial environment.

Also make sure students understand that, as land animals, reptiles must have ways of conserving water and can never be too far from a water supply. Explain such features as waterproof skin and the reptilian excretory system in terms of water-conserving strategies.

Using the Features/Warm-blooded Dinosaurs?

The main process skills used in the feature on warm-blooded dinosaurs are hypothesizing, observing, gathering and interpreting data, and inferring. For example, Robert Bakker's observations of dinosaur fossils and his interpretation of data on bone development led him to hypothesize that dinosaurs were warm-blooded.

The feature provides an excellent opportunity to talk about the way accepted theories may be challenged in light of new evidence. Tell students that scientific study often involves finding new ways to test hypotheses or formulating new hypotheses as evidence is discovered or reinterpreted.

The feature also raises the question of whether modern alligators and crocodiles, which are ectothermic, evolved from endothermic dinosaurs. Ask students to formulate their own hypotheses about environmental conditions that might have favored ectothermy in reptiles.

Research Activity

Divide the students into two groups and have each group research one of the major theories of the decline of the dinosaurs—the theory that continental drift and the theory that a huge asteroid led to the mass extinction. Have the groups present their findings in class.

Field Trip

A field trip to a museum of natural history, if there is one in your area, can be an interesting and informative experience for students. They can observe fossils and other evidence used by paleontologists to trace the evolution of reptiles as well as view exhibits of what conditions are believed to have been like during the Age of Reptiles.

Section Review Answers

(See p. 575.)

38.2 Modern Reptiles

Approach

This section reviews the characteristics of the four orders of the Class Reptilia. The main objective is to show that although all reptiles share some characteristics, each order has unique adaptations to life on land. These adaptations enabled reptiles to exploit new habitats and food sources where competition was scarce. Tell students to consider these adaptations in terms of the relationship between structure and function, a concept introduced in Chapter 1.

You can also draw on the concepts of adaptation and speciation (Chapter 15) to explain why reptiles have successfully colonized all but the coldest regions of the earth.

Concept Framework

Students may have had the opportunity to observe first-hand such reptilian behaviors as camouflaging and positioning the body to regulate internal temperature. Ask students who have kept turtles or lizards as pets to share what they have learned about reptiles. Point out to students that, in the case of reptiles, "survival of the fittest" has not necessarily meant survival of the mightiest or fastest.

Emphasis

Make certain that students understand that as cold-blooded, land-dwelling animals, reptiles must be able to obtain food and defend themselves in an energy-efficient manner. Talk about such characteristics as the turtle's slow movement and retraction into a shell, the four-chambered heart in crocodiles, and the defensive strategies of lizards as energy-conserving adaptations.

Guest Speaker

Invite an ecologist or wildlife specialist to talk to the class about the ecological importance of reptiles and how the human demand for products made from reptiles may threaten their existence.

Section Review Answers

(See p. 579.)

38.3 Adaptations of Snakes

Approach

The adaptations of snakes for burrowing and escaping from predators has made them the most successful group of reptiles. The focus of this section is a description of the structural adaptations of snakes for movement, feeding and digestion, defense, and reproduction. In presenting the text, compare the movement of snakes to that of worms (Chapter 31), which have a similar body shape but no backbone. Also compare the adaptations of snakes for locating, eating, and digesting prey to those of aquatic carnivores such as echinoderms (Chapter 34) and fishes (Chapter 36).

Concept Framework

Many students may regard snakes with fear or awe as a result of exaggerated or erroneous information. Encourage students to think about such well-known but often misunderstood snake behaviors as constriction of prey and injection of venom in terms of adaptations for survival.

Emphasis

Make certain that students understand that the evolution of snakes involved modifications in internal anatomy, as well as structural and behavioral adaptations. Have students study Figure 38-10, which shows the snake's major internal organs. Point out that the organs are modified for the snake's elongated body shape.

Guest Speaker

Invite a herpetologist from a local university to speak to the class about the variety of snakes throughout the world. The discussion should focus on the adaptations of snakes to different environments, such as sidewinding across desert sands or climbing trees in a tropical rain forest.

Section Review Answers

(See p. 582.)

Laboratory Preparation

Background

This lab exercise demonstrates the change in the color of an anole in response to color brightness or color contrast in its environment. Color brightness refers to the amount of light reflected, rather than a subjective judgement. Color changes may be subtle and may be affected by adrenalin if the lizard is frightened. Students observe the structures that are important adaptations for terrestrial and arboreal living. The toe pads allow the anole to stick to surfaces in a way similar to its close relative, the gecko.

Preparation

At least one hour before the lab exercise begins, place the anoles under fluorescent light in two terraria. Place half of the anoles in a terrarium lined with black construction paper. These anoles should turn brown. Place the remaining anoles in a terrarium lined with white construction paper. These anoles should turn green. Provide water-saturated cotton balls in the terrariums for drinking. The beakers should be large enough to allow the anoles to settle comfortably on the bottom.

Safety

Anoles can run fast and are easily frightened. Remind students to handle them with care. If picked up by the tail, the tail may break off.

Conclusions

1. Although answers will vary, background color brightness is more likely to be a factor than is background color contrast in causing an anole to change color.
2. Color change is more likely to occur in patches, rather than in a solid, overall color change.

Inquiry

The sample of anoles may not have been large enough; there may not have been enough different shades of each color to make a thorough test; colors other than green or brown may have an effect on anole color change.

Chapter Review

Vocabulary

1. allantois, amnion, chorion, yolk sac
2. hemotoxin, Jacobson's organ, neurotoxin
3. *Parietal* is the name of the bone in the skull above which the parietal eye is located.
4. *Hemo* means blood; *neuro* means nerve. Hemotoxins affect the blood; neurotoxins affect the nervous system.
5. ovo = egg, vivi = living, parous = bears young; viviparous

Review

11. desiccation
12. Reptiles have double circulation that separates oxygenated from deoxygenated blood.
13. birds, mammals, and other reptiles
14. Studies of the bone structure, predator-prey ratios, and growth rates of dinosaurs suggest that they may have been warm-blooded.

15. The four orders of reptiles are Rhynchocephalia, Chelonia, Crocodilia, Squamata.
16. Crocodilians generally hunt by stealth, that is, by lurking submerged.
17. They use speed to subdue prey.
18. Like snakes, monitor lizards have forked tongues and feed on large prey.
19. rear-fanged, elapid injection, viper injection
20. Jacobson's organ detects chemicals in the air.

Critical Thinking

1. The digging of empty holes may be a protective behavior that distracts predators from the holes containing the eggs.
2. Answers may include that mammals may have preyed on tuataras or competed more effectively with tuataras for the same sources of food.
3. Students should infer that the climate of Antarctica was much warmer and at a different longitude during the Mesozoic than it is today.
4. Although reptiles as a class are remarkably adapted to land, individual species have specialized adaptations to aquatic life. In certain environments, conditions may have been more favorable for survival in water than on land because of less competition.
5. a. Tuataras and crocodilians show a much narrower temperature range. This difference may be accounted for by the fact that these groups are made up of only one or a few species that are adapted to similar environments, while snakes, lizards, and turtles each include many more species adapted to a much wider diversity of environments. b. Snakes and lizards are best able to tolerate desert temperatures.
6. Answers may include protecting eggs from predators or retaining eggs in a relatively warm and moist environment during development.
7. The venom contains neurotoxin, which interferes with the perception of pain by the nervous system. Hemotoxins would have no affect on perception of pain.

Chapter 38 | Reptiles

Introduction

Although amphibians have adapted to life on land, they remain dependent on water for reproduction. The first vertebrates to become totally at home on land were reptiles. In this chapter you will learn about the adaptations that allowed reptiles to dominate the earth for 160 million years and that continue to make them a highly successful group of animals.

Chapter Outline

38.1 Evolution of Reptiles
- *The Amniote Egg*
- *Other Adaptations to Land*
- *Origin and Evolution*

38.2 Modern Reptiles
- *Rhynchocephalia*
- *Chelonia*
- *Crocodilia*
- *Squamata*

38.3 Adaptations of Snakes
- *Movement*
- *Feeding*
- *Defense*
- *Reproduction*

Chapter Concept

As you read, look for ways in which reptiles are adapted to life on land. Pay special attention to adaptations that prevent drying out of sensitive tissues.

Vine snake, Thailand. Inset: Thorny devil lizard, *Moloch horridus*, Australia.

For additional teaching information see pages T38A–T38D.

38.1 Evolution of Reptiles

About 310 million years ago geological events were reshaping the face of the earth. As new continents emerged, a great diversity of new habitats formed. Plants, amphibians, and insects began to populate these new landmasses, but no vertebrates had yet made a permanent home on land. Onto the dry land came the reptiles—the first vertebrates to make the complete transition to life on land. Two major forces probably helped lead to this transition: (1) an increase in competition for food and space among all the life-forms in aquatic environments and (2) limited competition for the insects and plants that could be used as food on land.

After moving onto land, the reptiles evolved rapidly and dominated the land, seas, and air for 160 million years. Let us begin by taking a look at some of the adaptations that made this evolutionary success possible.

The Amniote Egg

Studies of the fossil record and comparative anatomy strongly suggest that reptiles evolved from amphibians. Although amphibians can live on land for part of their life cycle, they are forced to return to water to reproduce. With the evolution of reptiles, a new reproductive structure evolved that allowed reproduction on land. This structure is the **amniote egg**, an egg with a protective membrane and a porous shell enclosing the developing embryo. The amniote egg forms a secure, self-contained "nursery" for the developing embryo. In contrast to the naked eggs that amphibians lay in water, the amniote egg provides an internal aquatic environment that enables the embryo to survive the dry environment.

Figure 38-1 shows the complex internal structure of the amniote egg, including its four specialized membranes: amnion, yolk sac, allantois, and chorion. The egg derives its name from the **amnion** (AM-nee-AHN), the thin membrane enclosing the salty fluid in which the embryo floats. The **yolk sac** encloses the yolk, a protein-rich food supply for the developing embryo. The **allantois** (uh-LAN-tuh-wuss) stores the nitrogenous wastes produced by the embryo until the egg hatches. The **chorion** (KOR-ee-AHN) lines the outer shell and thus encloses the embryo and all the other membranes. It regulates the exchange of oxygen and carbon dioxide between the egg and the outside environment.

The entire amniote egg is surrounded by a leathery shell that may be hard in some species because of the presence of calcium carbonate. The egg is waterproof but allows gases to flow between the environment and the chorion. External fertilization of such an egg would be impossible since the sperm could not penetrate the egg. Instead the male places the sperm inside the female before the shell is formed. This strategy, called **internal fertilization,** makes water transport of sperm unnecessary.

Section Objectives

- *Describe the structure of the amniote egg and explain its importance in the evolution of reptiles.*

- *List three structural characteristics of reptiles that are adaptations to life on land.*

- *Summarize the major developments in the evolution of reptiles.*

- *List three types of evidence that suggest that dinosaurs may have been warm-blooded.*

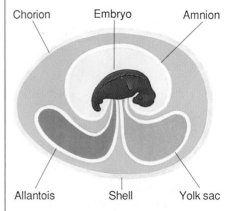

Chorion Embryo Amnion

Allantois Shell Yolk sac

Figure 38-1 The four membranes of the amniote egg surround and protect the embryo and its food supply. The shell permits gases to pass between the egg and the outside environment.

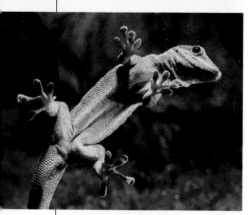

Figure 38-2 The scaly skin of a snake (top), the claws of an iguana (center), and the suction cup–equipped toes of a gecko are all adaptations for life on land.

Other Adaptations to Land

In addition to the amniote egg and internal fertilization, other adaptations helped make the transition to land possible. These include waterproof skin, modified limbs and body systems, well-developed lungs, and new methods of temperature regulation.

Waterproof Skin

A dry body covering of horny scales or plates protects reptiles. This scaly covering develops as surface cells fill with **keratin**, the same protein that forms human fingernails and bird feathers. These scales prevent water loss and protect reptiles from the wear and tear associated with living in rugged terrestrial environments. In contrast, amphibians have smooth, delicate skin. They can never be far from water or their skin will dry out.

External Structural Adaptations

Reptiles also show structural adaptations for movement on land. The limbs of some reptiles have toes with claws that permit them to climb, dig, and thus move about on a variety of terrains. Other reptiles have toes modified into suction cups that aid in climbing. In the absence of limbs, snakes rely on their scaly skin and highly developed skeletal and muscular systems for movement.

Respiration, Circulation, and Excretion

After they hatch, all reptiles breathe air through lungs, eliminating respiration through gills. Thus in reptiles the tissues involved in gas exchange are all located inside the body, where they can be kept moist in even the driest land environments.

Reptiles have a circulatory system that is more complex than that of amphibians, one that provides more oxygen to the body. Like amphibians, all reptiles have double circulation, and most species have a three-chambered heart. However, a partial division of the ventricle separates the oxygen-poor blood flowing from the body from the oxygen-rich blood returning from the lungs. In alligators and crocodiles the heart has four separate chambers, allowing total separation of oxygenated and deoxygenated blood.

The excretory system of reptiles is also well adapted to life on land. Snakes, lizards, and other land-dwelling reptiles conserve water by excreting nitrogenous wastes in dry or pasty form as crystals of uric acid.

Temperature Regulation

The rate of metabolism in the body of an organism is controlled in part by body temperature. Therefore regulating body temperature is very important. In reptiles, as in fish and amphibians, body temperature is largely determined by the temperature of their surroundings. Animals in which body temperature is determined by the environment are called **ectothermic,** or cold-blooded. Thus reptiles cannot maintain their body temperature as can birds and mammals, which are **endothermic,** or warm-blooded. Reptiles regulate their temperature by their behavior. For example, they may bask in the sun to warm up, thereby increasing their metabolic rate, or they may seek shade to prevent overheating.

Origin and Evolution

From studies of fossils and comparative anatomy, biologists infer that reptiles arose from a group of ancestral reptiles called **cotylosaurs** *(KOT'l-oe-SORZ),* which lived about 310 million years ago. Fossils indicate that these four-legged, sprawling vertebrates resembled small lizards and had teeth well suited to eating insects. In fact, the abundance of insects at the time may have been one reason the cotylosaurs flourished.

During the Permian period (280 million to 225 million years ago), the cotylosaurs began to adapt to other available environments, giving rise to new forms of reptiles. These groups included flying reptiles called pterosaurs, two groups of marine reptiles—the ichthyosaurs and plesiosaurs—and the thecodonts. The dominant land reptiles arose from thecodonts—small, lizardlike carnivores, many of which walked on their hind legs. The thecodonts were the first **archosaurs** ("ruling reptiles"), a group that later included the early crocodiles, the dinosaurs, and the reptiles that evolved into birds. Figure 38-3 shows the probable evolutionary relationships among these and other groups.

Info: By turning their bodies to get the maximum effect of the sun's rays when they need warming and the minimum effect of the rays when they need cooling, reptiles can maintain astonishingly stable body temperatures.

Figure 38-3 **The phylogenetic tree of reptiles shows animals with a wide array of adaptations diverging from common ancestors.**

Figure 38-4 An artist's rendering of a thecodont, *Euparkeria* sp., reflects scientists' view of what these animals looked like based on fossil remains.

Answer to Biology in Process: observing, hypothesizing, inferring, communicating

The Mesozoic era (225 million to 65 million years ago) is known as the Age of Reptiles. During that time reptiles—especially the dinosaurs—dominated all other forms of life. Although *dinosaur* means "terrible lizard," many were small. Yet the incredible size of some dinosaurs distinguish the group from all other forms of life. The largest known dinosaur, *Brachiosaurus*, measured 23 m long, stood 12 m tall, and weighed more than 77,000 kg. Thus *Brachiosaurus* was as long as a tennis court, as tall as a four-story building, and heavier than ten elephants.

Over 300 genera of dinosaurs have been identified around the world from fossils. These animals were adapted to a wide range of environments and different ways of life. *Brachiosaurus* and such related dinosaurs as *Diplodocus* and *Apatosaurus* (also called *Brontosaurus*) were **herbivores,** or plant eaters, and probably used their long necks to feed on leaves in treetops. *Tyrannosaurus* and other **carnivores,** or meat eaters, walked on their hind legs and used sharp teeth and huge claws to rip apart prey.

Paleontologists who study dinosaurs still have many unanswered questions. Not the least of these questions is why dinosaurs became extinct along with many other forms of animal and plant life 65 million years ago, at the end of the Cretaceous period.

Biology in Process

Warm-blooded Dinosaurs?

Were dinosaurs cold-blooded and sluggish like their modern reptilian descendants, the alligators and crocodiles? That dinosaurs were cold-blooded has long been the generally accepted theory, but today that theory is being called into question by paleontologists who are reinterpreting the fossil record.

For a long time paleontologists merely compared fossil bones of dinosaurs and classified the many different kinds. In the course of their work, paleontologists made assumptions about the lives of dinosaurs. But these were often based on what was known about alligators and crocodiles.

During his undergraduate days at Yale University in the 1960s, paleontologist Robert T. Bakker began to suspect that his peers were going astray by inferring from the habits of present-day crocodiles and alligators. He suggested that the fossil bones of dinosaurs might reveal a different biology.

In 1969 one of Bakker's professors, John H. Ostrom, set the stage for reconsidering dinosaur biology when he wrote about the clawed foot of *Deinonychus.* Fossil evidence showed that this dinosaur had

There are various schools of thought about this mass extinction. Some scientists propose that gradual climatic changes, probably brought about by the drifting of continents into new latitudes, made the earth less hospitable for many forms of Mesozoic life. Other scientists maintain that a catastrophic cosmic event was responsible. Supporters of this theory suggest that a huge asteroid hit the earth, sending so much dust into the atmosphere that the amount of sunlight reaching the earth's surface was greatly reduced. The reduced sunlight caused severe climatic changes that led to the mass extinction.

Section Review

1. Why is the reptile egg called an *amniote egg?*
2. How did the amniote egg help make reptiles completely independent of aquatic environments?
3. Name three adaptations of reptiles for life on land.
4. Name the groups of modern vertebrates that have evolved from archosaurs.
5. In what ways are the adaptations of reptiles to land like the adaptations of vascular plants to land? (See Chapter 25.)

Answers to Section Review
1. It is named for the amnion, the membrane that encloses the embryo.
2. It provided a protective watery environment in which the embryo could develop without drying up.
3. Answers may include waterproof skin, feet adapted for land movement, lungs for breathing air, an excretory system that conserves water.
4. Birds and crocodilians evolved from archosaurs.
5. Amniote eggs and internal fertilization eliminate dependence on water for reproduction, as do seeds and pollen grains. Waterproof skin prevents drying out, as does the cuticle. The circulatory system transports materials, as does the vascular system of plants.

a scythelike claw on its foot, which it may have used to kill prey. Hunters that track and kill require sustained energy to chase down their prey. They need a high metabolism of the sort present only in endothermic animals.

Based on Ostrom's inferences and on other observations, Bakker inferred that some dinosaurs may have been fleet-footed, agile, and warm-blooded. To support his theory, Bakker applied the findings of other research. For example, he used data on bone structure to distinguish between endothermic and ectothermic animals. He found that many dinosaurs had bones similar to those of warm-blooded animals.

Bakker's inference received support from the work of other paleontologists who had uncovered fossil remains of entire dinosaur herds, including eggs (above) and baby dinosaurs. These have provided a way to study the growth rate of dinosaurs. The results show a rapid rate of growth much closer to that of endothermic birds than to that of ectothermic crocodiles.

Despite all the evidence presented, Bakker's findings have by no means been entirely accepted by the paleontological community. Numerous scientists have proposed equally compelling counterarguments to his. Nevertheless, as one paleontologist said of Bakker, "He keeps you thinking," and that is one of his most important contributions to the study of dinosaurs.

■ What processes did Bakker use to test his inference that dinosaurs were warm-blooded?

- *List seven characteristics of modern reptiles.*

- *Name the four orders of modern reptiles, and give an example of each.*

- *Describe the basic structure of the turtle shell, and explain how it differs in land-dwelling and water-dwelling species.*

- *Explain how the head structure of crocodilians is related to their feeding behavior.*

- *List three types of defensive adaptations that are common among lizards.*

38.2 Modern Reptiles

Biologists have classified reptiles into 16 orders, of which 12 are extinct. The four surviving orders comprise about 6,000 species. All of these species share traits that you have learned are adaptations to land:

- The amniote egg helps protect the embryo.
- Internal fertilization replaces water transport of sperm.
- A dry body covered by scales or plates prevents water loss.
- Limbs with claws or pads aid movement on land.
- Respiration through lungs eliminates external gills.
- A three-chambered or four-chambered heart helps supply more oxygen to the body.
- Ectothermic temperature regulation controls metabolic rate.

Reptiles occur worldwide except in the coldest regions, where it is impossible for large ectotherms to survive. In recent times, human intervention has had a powerful impact on the number and distribution of reptiles. Many species of turtles, prized for food, have been hunted almost to extinction. Snakes and lizards suffer from needless killings by humans as well as from the effects of pesticides and habitat destruction.

The four living orders of the Class Reptilia are (1) Rhynchocephalia, (2) Chelonia, (3) Crocodilia, and (4) Squamata.

Rhynchocephalia

The Order Rhynchocephalia (*RING-koe-suh-FAY-lee-uh*) is an ancient one that contains only one living species, *Sphenodon punctatus*, the tuatara. Tuataras inhabit about 20 small islands off the coast of New Zealand. The native Maoris of New Zealand named the tuatara for the conspicuous spiny crest running down its back, which you can see in Figure 38-5. *S. punctatus* resembles a large lizard, growing on the average to about 60 cm long.

An unusual feature of the tuatara is the inconspicuous "third eye" on the top of its head, called a **parietal eye.** Under a thin layer of scales are a lens, retina, and nerves connecting the eye to the brain. The parietal eye seems to function as a thermostat that alerts the tuatara when it has been exposed to too much sun, thus protecting against overheating. Unlike most reptiles, tuataras are most active at low temperatures. They burrow during the day and at night feed on insects, worms, and small animals.

Chelonia

The Order Chelonia consists of about 265 species of turtles and tortoises. The term *tortoise* is generally reserved for the terrestrial Chelonia, such as the giant Galápagos tortoises. Turtles are chelonians that live in water.

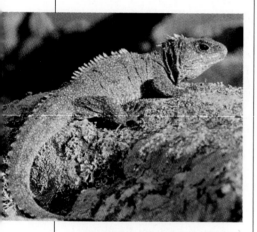

Figure 38-5 The tuatara, often called a "living fossil," has changed little from its prehistoric ancestors, which appeared on earth more than 225 million years ago. Note the crest on the head and back. The name tuatara means "spiny crest."

Chelonians have changed little in the last 200 million years. This evolutionary stability may be the result of the continuous benefit of their basic design—a body covered by a shell. In some species the shell is made of hard plates; in other species it is a covering of tough, leathery skin. In either case the shell consists of two basic parts. The **carapace** is the top or dorsal covering. In most species it is fused with the vertebrae and ribs. The **plastron** is the ventral or belly portion.

The shape of the chelonian shell is modified for a variety of ecological demands. Many tortoises have a domed carapace into which they can retract their head, legs, and tail as a means of protection from predators. Water-dwelling turtles have instead a streamlined disk-shaped shell that permits rapid turning in water. When swimming, the marine turtles are the fastest of the reptiles. The forelimbs of marine turtles have evolved into flippers, and freshwater turtles have webbed toes. These adaptations allow for maximum swimming ability both in oceans and in freshwater rivers, lakes, and ponds.

Another adaptation among chelonians is the migratory behavior of the sea and river turtles. Although these animals are adapted for life in water, they must return to land to lay their eggs. For example, Atlantic green turtles migrate from their feeding grounds off the coast of Brazil to Ascension Island in the South Atlantic—a distance of more than 2,000 km. There they lay eggs on the same beaches where they were hatched.

Crocodilia

The Order Crocodilia is composed of about 23 species of large lizard-shaped reptiles. In addition to crocodiles and alligators, the order also includes less familiar animals called caimans *(KAY-*

Figure 38-6 The carapace of the sea-dwelling Ridley turtle, *Lepidochelys* sp. (left), is streamlined for efficient swimming. The domed carapace of the Galápagos tortoise, *Geochelone elephantopus* (right), a land dweller, can serve as protection when the head, legs, and tail are withdrawn.

Tip: Ask students to hypothesize why sea turtles have flippers but freshwater turtles have feet with webbed toes. If students understand that freshwater turtles spend some time on land, they will see that webbed toes are more efficient than flippers for walking.

Crocodile

Alligator

Gavial

Caiman

Figure 38-7 The modified snouts of four species of crocodilians reflect their different feeding habits.

munz) and gavials (*GAY-vee-ulz*). Examination of the fossil record shows that crocodilians are direct descendants of the archosaurs and that many species have remained relatively unchanged for 200 million years.

All crocodilians live in or near water in tropical or subtropical regions of the world. Crocodiles are largely nocturnal animals found in Africa, Asia, and the Americas, including southern Florida. Alligators live in China and the southern United States. Caimans, which closely resemble alligators, are native to Central America but are becoming established in Florida. Gavials are a group of fish-eating crocodilians with an extremely long and slender snout adapted for snaring and eating fish. Gavials live only in India and Burma.

All crocodilians are carnivorous. They generally hunt by stealth, waiting for prey to come near and then attacking aggressively. Many of their anatomical features are adaptations that reflect this feeding behavior. Their eyes are on top of the head; the nostrils, on the top of the snout. This arrangement allows them to see and breathe while lying quietly submerged in water. A valve in the back of the mouth prevents water from entering the air passage when a crocodilian feeds under water.

Parental care is generally lacking in reptiles, though it does occur in some species. In one species of Crocodilia, the Nile crocodile, both parents carry the young in their jaws from the nest to a special nursery area, where they oversee their development. This behavior in a living reptile is interesting in light of new fossil discoveries indicating that certain dinosaurs also probably cared for their young.

Squamata

The Order Squamata consists of about 5,640 species of lizards and snakes. The distinguishing characteristics of this order are a lower jaw that is not joined directly to the skull and the presence of paired reproductive organs in the male. Squamates are more structurally diverse than members of any other order of reptiles. In this section we will look at the different kinds of lizards and see how they have radiated into many habitats. In Section 38.3 we will examine the structure and evolutionary success of the snakes.

Lizards are generally distinguished from snakes by the presence of limbs, although a few species of lizards are limbless. Common lizards include iguanas, horned toads, chameleons, skinks, and geckos. Lizards live on every continent except Antarctica. They have adapted to deserts and tropical forests. Most rely on agility, speed, and camouflage to catch prey and elude predators.

Only two species of lizards are venomous. They are the Gila monster of the southwestern United States and the beaded lizard of western Mexico. Most lizards prey on insects or on other small animals. They do not need to immobilize their prey and therefore have not developed the capacity to produce venom for survival.

Info: Although crocodilians have remained basically unchanged for millions of years, fossil evidence suggests that some were once much larger than today—up to 15 m in length.

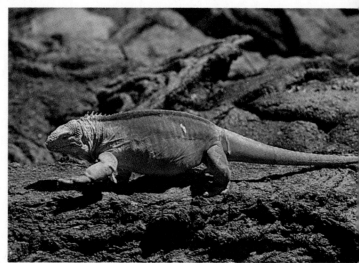

Some lizards can blend with their background by changing color. One is *Anolis carolinensis,* a common lizard of North America. Because of this ability, it is commonly called a chameleon, but true chameleons live only in Africa and Madagascar.

Blending with the background is a passive way of remaining inconspicuous. In contrast, many lizards use active displays to fend off enemies. The horned lizards of the southwestern United States, though small, are particularly impressive. They are covered with spiky armor, and when disturbed, they inflate themselves, gape their mouths, hiss, and squirt blood from their eyelids.

Many lizards, including the skinks and geckos, have the ability to lose their tails and then regenerate a new one. From observation, biologists have inferred that this trait, called **autotomy,** is an adaptation that allows the lizard to escape from predators. An unusual form of autotomy occurs in glass lizards, which get their name from the fact that their tails often shatter when grasped by a predator.

Most lizards are small, measuring less than 0.3 m in length. Iguanas can reach 1 m in length, but the largest lizards belong to the monitor family. The largest of all monitors is the Komodo dragon of Indonesia, which reaches 3 m in length and weighs up to 140 kg. Monitors are thought to be related to snakes. Like snakes, they have deeply forked tongues that serve as sense organs. They also consume large prey whole, as do snakes. A monitor lizard uses its tail as a defensive weapon.

Figure 38-8 Among the lizards, a diverse group, can be found the Gila monster, *Heloderma suspectum* (top left), the land iguana, *Conolophus subcristatus* (top right), and the skink (above).

Section Review

1. What is a *parietal eye*? Of what order of reptiles is it a distinguishing characteristic?
2. Describe the structure and function of the chelonian shell.
3. How is the position of a crocodile's eyes and nostrils related to its way of feeding?
4. What are three ways in which lizards defend themselves?
5. What factors do you think might have caused monitors to grow so much larger than other lizards?

Answers to Section Review
1. third eye on top of the head; Rhynchocephalia
2. It consists of a carapace (top) and plastron (bottom) that together protect turtles from predators. The carapace is fused with the vertebrae and ribs.
3. Eyes and nostrils on top of the head allow crocodiles to breathe and see while submerged in water.
4. camouflage, threatening display, autotomy
5. Answers might include lack of competition from mammalian predators and adaptation for feeding on larger prey.

38.3 Adaptations of Snakes

The fossil record of the evolution of snakes is far from complete. However, biologists suggest that snakes probably evolved from lizards that lived above ground during the Cretaceous period. As a way of escaping predators, some may have begun to burrow underground. Natural selection led to changes in these lizards. For example, lizards without legs and those with long and slim bodies survived. These adaptations, which made it easier to move in narrow burrows, have also made snakes highly successful reptiles.

Movement

The graceful undulations of snakes are made possible by their unique anatomy. A snake has a backbone of 100 to 400 vertebrae, each of which has a pair of ribs attached. These bones provide the framework for thousands of muscles. The muscles manipulate not only the skeleton but also the snake's skin, causing this fabric of overlapping scales to extend and contract. This interaction of bones, muscles, and skin enables a snake to move in one of three basic ways: lateral undulation, rectilinear movement, and sidewinding.

Most commonly snakes move by lateral undulation. In this method a snake moves its head to one side, which initiates a wave of muscular contractions along the vertebrae. As this lateral motion moves down its trunk, the snake uses the sides of its body to push off on pebbles, twigs, and small irregularities of the ground. The snake thus moves forward in an S-shaped path.

In rectilinear movement, a snake applies muscular force on its belly, not its sides. The trailing edges of its **scutes,** the scales on its belly, catch on bark or other rough surfaces. The body then relaxes and moves forward, inching along much like a caterpillar.

Some desert-dwelling snakes progress by sidewinding, a method of moving not forward but sideways, as shown in Figure 38-9. Because sand has few surface irregularities, neither lateral nor rectilinear motion is effective in deserts. A sidewinding snake vigorously flings its head to one side. This whiplike motion yanks along the rest of its body. Just as the hindquarters are catching up, the snake thrusts its head sideways again.

Feeding

Snakes eat animals, but they have none of the structural adaptations common to other carnivores. They don't hear or see well, and they have no limbs for subduing prey. Their small mouths and tiny teeth cannot rip and grind flesh. Their slim bodies can't even accommodate an elaborate digestive system, as you can see from Figure 38-10. Nonetheless, the evolutionary pressures that shaped them in these ways equipped them well for the predatory life.

Tip: Ask students to compare the internal anatomy of snakes with that of bony fishes.

Figure 38-9 The whiplike, side-wise movement of some desert-dwelling snakes is an adaptation to a sandy environment.

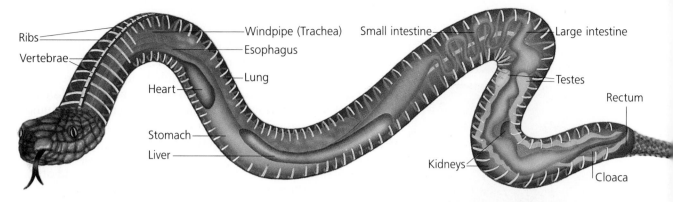

Ribs
Vertebrae
Windpipe (Trachea)
Esophagus
Heart
Lung
Stomach
Liver
Small intestine
Large intestine
Testes
Rectum
Kidneys
Cloaca

Locating Prey

Eyesight and hearing do not enhance survival in dark, quiet burrows, and these senses were not highly developed as snakes evolved. Instead, snakes evolved a sense of smell, which they use to locate prey. By flicking its forked tongue, a snake gathers chemicals from the environment. The tongue transfers these chemicals to two pits in the roof of the mouth called **Jacobson's organ**. Nerves in these pits are highly sensitive to chemicals. In this way, a snake stalks prey by following a chemical trail.

Killing Prey

Once snakes locate prey, they may, if the prey is small, just swallow it whole. Most snakes, however, employ one of two methods for killing: constriction or the injection of venom. Most snakes are constrictors. They wrap their bodies around prey to squeeze and kill it. This technique is used by boas, pythons, and anacondas—snakes that can consume small pigs and other similar-sized animals. Most North American snakes are constrictors.

Some snakes inject their prey with a toxic venom. Three methods have evolved for injecting venom. The rear-fanged snakes, such as the boomslang and twig snakes of Africa, bite the prey and use grooved back teeth to guide the venom into the puncture. Highly venomous cobras and kraits are elapids, which inject poisons through two small front fangs that act like hypodermic needles. The vipers inject venom through large front fangs. When a viper strikes, these hinged fangs swing forward from the roof of the mouth and inject venom more deeply than can the fangs of elapids. Vipers include three groups of North American snakes: rattlesnakes, water moccasins, and copperheads.

Venom is chemically complex. Its different components act in two ways. The **hemotoxins** are proteins that attack the circulatory system, destroying red blood cells and disrupting the clotting power of blood. The **neurotoxins** work on the nervous system. The disruption of nerve pathways is dangerous to respiratory and heart functions. Some venoms contain both types of toxin.

Swallowing and Digesting Prey

How do snakes consume such large and bulky prey as birds, fish, rats, and other snakes? Several structural adaptations make this

Figure 38-10 Each of a snake's several hundred vertebrae has its own pair of ribs attached. These ribs anchor the many muscles needed to manipulate the long, slender body. How are a snake's organs modified for its body shape?

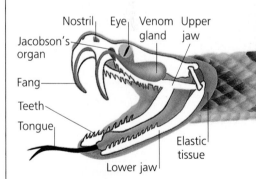

Nostril
Eye
Venom gland
Upper jaw
Jacobson's organ
Fang
Teeth
Tongue
Lower jaw
Elastic tissue

Figure 38-11 Vipers inject venom through two large fangs connected to the venom gland.

Reptiles **581**

Think

Based on what you know about the skin of reptiles, why do you think that snakes must periodically shed their skin?

Answer to Think: The skin cannot grow and must be shed as the snake increases in size.

Answers to Section Review
1. Oviparous animals lay eggs; ovoviviparous animals bear young that hatch from eggs internally.
2. The long legless form facilitated movement in underground burrows.
3. lateral undulation, rectilinear movement, sidewinding
4. injecting venom, constriction, swallowing without killing first
5. that it lives in a desert or other sandy environment

behavior possible. A snake's upper and lower jaws are hinged and move independently. When unhinged, the jaws stretch to allow the mouth to open extremely wide. While holding the prey in its small teeth, the snake places its mouth over the animal and slowly swallows it whole. While swallowing, the snake thrusts its windpipe into the throat. This adaptation lets the snake breathe while swallowing, a process that can take several hours.

The saliva begins to digest food well before it reaches the snake's stomach, which is simply an enlargement at the end of the esophagus. Digestion proceeds slowly in the stomach.

Defense

Just as natural selection resulted in structural adaptations for feeding, it also resulted in a host of modifications for defense. Camouflage, for example, is beneficial for both seeking prey and hiding from predators. Many snakes are green and blend with foliage; others are brown and hide against the bark of trees.

Rather than hiding, some snakes defend themselves by signaling their presence. Some snakes can ward off danger by rapidly changing body shape—expanding a hood, in the case of the cobras. Some snakes hiss; others make mechanical noises, such as the rattling of a rattlesnake.

Reproduction

Most male snakes rely on scent to find females of their own species. Before mating, a male and female snake may glide along side by side, with the male stroking the female with his chin and flicking his tongue over her body. Fertilization is internal.

Most snakes are **oviparous** *(oe-VIP-uh-russ)*; that is, the female lays eggs that hatch outside her body. To break out of its shell, the hatchling uses a special tooth, which is lost afterward. Other snakes are **ovoviviparous** *(OE-voe-VIE-VIP-uh-russ)*; that is, the female carries the eggs in her body throughout development and the young are born live. These eggs have a thin membrane through which water and oxygen pass from the mother. All the nourishment is provided by the yolk. The eggs hatch in the mother's body, and she gives birth to the newly hatched young. All newborns must fend for themselves, relying only on their many specialized adaptations for survival on land.

Section Review

1. Distinguish between oviparous and ovoviviparous snakes.
2. How is a snake's body structure related to its habitat?
3. What are the three main ways in which snakes move?
4. Describe two ways in which snakes subdue their prey.
5. If you observed a snake in the zoo sidewinding, what might you infer about the natural habitat of the snake?

Color Adaptation in the Anole

Objective
To analyze color change in the anole in response to environmental change

Process Skills
Observing, experimenting

Materials
Glass-marking pencil, 2 large beakers or wide-mouth jars, 2 live anoles, 6 shades each of brown and green construction paper ranging from light to dark, hand lens

Method

1 Label a beaker *Anole No. 1*. Carefully pick up an anole from one of the terrariums and place the animal in the beaker. Do not pick up the anole by its tail. Grasp it around the shoulders.

2 Place the beaker on a piece of construction paper that matches the anole's color as closely as possible. Remove the paper and label it *Original Color of Anole No. 1*.

3 Select a piece of paper the same color as the anole but in the lightest shade you can find. If you have chosen a green anole, for example, select the lightest shade of green available. The lighter the shade, the more brightness the color reflects.

4 Place the beaker on the paper selected.

5 Label a second beaker *Anole No. 2*. Select another anole of the same color as the first from the terrarium, and place the animal in the second beaker.

6 Place the beaker on a piece of construction paper that matches the anole's color as closely as possible. Remove the construction paper, label it *Original Color of Anole No. 2*, and set it aside.

7 Select paper of contrasting color as different in brightness to the anole color as possible. If you are working with brown anoles, for example, select green paper of contrasting brightness.

8 Place the beaker on the paper selected.

9 Wait about 20 minutes for both animals to respond to their backgrounds. While waiting, examine an anole's eyes, ears, scales, toes, and tail. Notice the dewlap (expanded area under the chin) in males. Make a sketch showing these features.

10 After 20 minutes, switch the paper under each beaker with the paper to which you originally matched each anole. Note the degree of color change.

11 Return each anole to its terrarium.

Conclusions

1 Which caused the greater color change, a contrasting background color or a brighter shade of the animal's own color?

2 When an anole changed color, did you notice anything unusual or unexpected about the color change?

Inquiry
What were some possible sources of error in this investigation?

Have half the class work with green anoles and the other half with brown anoles.

Vocabulary

allantois (571)	chorion (571)	internal fertilization (571)	oviparous (582)
amnion (571)	cotylosaurs (573)	Jacobson's organ (581)	ovoviviparous (582)
amniote egg (571)	ectothermic (573)		parietal eye (576)
archosaurs (573)	endothermic (573)	keratin (572)	plastron (577)
autotomy (579)	hemotoxin (581)	neurotoxin (581)	scutes (580)
carapace (577)	herbivore (574)		yolk sac (571)
carnivore (574)			

1. List the terms for the various parts of the amniote egg.
2. List the terms for the adaptations that help snakes track down and subdue prey.
3. Look up the word *parietal* in a dictionary. Why do you think the tuatara's third eye is called a parietal eye?
4. Look up the meaning of the word parts *hemo-* and *neuro-* in a dictionary. From what you know about the effects of hemotoxin and neurotoxin, explain why they are appropriately named.
5. Look up the origin of the term *ovovivipa-rous* in a dictionary. What are the meanings of the main parts of this term? What term might you use to describe animals that bear live young that don't hatch from eggs?

Review

1. The membrane that encloses the fluid around a reptile embryo is the (a) amnion (b) yolk sac (c) allantois (d) chorion.
2. The presumed ancestors of all modern reptiles were the (a) archosaurs (b) dinosaurs (c) cotylosaurs (d) tuataras.
3. The most abundant of all modern orders of reptiles is (a) Rhynchocephalia (b) Chelonia (c) Crocodilia (d) Squamata.
4. The two basic parts of a turtle shell are the (a) scutes and amnion (b) carapace and plastron (c) chorion and allantois (d) keratin and scutes.
5. A group of reptiles that does not belong to the Order Crocodilia is (a) alligators (b) tuataras (c) caimans (d) gavials.
6. The ability to lose its tail and grow a new one helps a lizard (a) reduce its need for food (b) hide from predators (c) escape from predators (d) capture prey.
7. Long legless bodies probably arose as an adaptation that helped snakes (a) live in trees (b) move in underground burrows (c) catch prey (d) swallow large animals.
8. Snakes that live in deserts generally move by (a) lateral undulation (b) rectilinear movement (c) sidewinding (d) creeping.
9. To locate prey snakes rely primarily on their sense of (a) eyesight (b) hearing (c) touch (d) smell.
10. Most snakes are oviparous, which means their young (a) hatch from eggs (b) hatch inside the mother and are born alive (c) are born alive without hatching from eggs (d) are cared for by their parents.
11. What problem of life on land was solved by the evolution of the amniote egg?
12. In what major way does circulation in reptiles differ from circulation in fish and in amphibians?
13. Which groups of modern vertebrates arose from cotylosaurs?
14. What evidence suggests that dinosaurs may have been warm-blooded?
15. List the four orders of modern reptiles.
16. What method do crocodilians generally use to hunt their prey?
17. How do biologists explain the fact that so few species of lizards are venomous?
18. What characteristics do monitor lizards

share with snakes?

19. Name three ways snakes inject venom.

20. What is the function of Jacobson's organ?

Critical Thinking

1. When a female leatherback turtle comes up on a beach to lay eggs, she first digs a deep hole, lays her eggs, and covers them with sand. Next she crawls about 100 m and digs another hole. This time she lays no eggs but just covers the hole with sand. Suggest a possible explanation for this behavior.
2. Tuataras lived on the two main islands of New Zealand until the 1800s, when British settlers brought in rats, goats, and other mammals. Why do you think the introduction of these mammals made it impossible for tuataras to survive?
3. In 1986 scientists discovered the fossil remains of a dinosaur in Antarctica. From this discovery what might you infer about the climate of Antarctica during the Mesozoic era?
4. Given that reptiles as a class have many adaptations for life on land, how do you account for the fact that some reptiles, such as sea turtles and sea snakes, live in the ocean?
5. Study the graph on this page, which shows the range of body temperatures found in active individuals of some major subgroups of reptiles.

a. Which two groups show a much narrower range of temperatures than the other groups? How might you account for this? (Hint: Consider the number of different species in each group.)
b. Which groups do you think are best able to tolerate the temperature extremes found in deserts?
6. Suggest some possible advantages that ovoviviparous reptiles may derive from carrying their eggs inside the body until after they have hatched.
7. Some pain-killing drugs are prepared from snake venom. Which type of toxin do you think is contained in the venom used? Explain your answer.

Extension

1. Read Virginia Morell's article, "Announcing the Birth of a Heresy," *Discover*, March 1987, pp. 26–50. In what ways have the fossils of the egg-laying dinosaurs discovered in Montana changed scientists' understanding of dinosaurs?
2. Use your school library or public library to research an endangered species of reptile, such as Kemp's Ridley turtle. Prepare a report describing the conditions that threaten the survival of the species and the efforts being made to save it.
3. Observe the behavior of a small lizard kept in a terrarium. Keep a log of your observations, noting differences in behavior at various times of day and before and after feeding.
4. Research the group of reptiles called amphisbaens. Write a brief report describing the similarites between amphisbaens and earthworms. Be sure to include similarities in both structure and way of life. In what ways are the two groups different?

Chapter 39 | Birds

This chapter examines the anatomical and physiological characteristics of birds that are similar to and different than other vertebrates. Students identify special characteristics that are adaptations for flight.

Correlation Chart

Section Number	Section Objectives	Pages	Features	Section Review
39.1	1. List both the avian and the reptilian characteristics of *Archaeopteryx*.	587–588		1, 2
	2. Summarize the two predominant hypotheses of bird evolution.			3, 4
	3. Identify the criteria which taxonomists use when they classify birds.			5
39.2	1. Describe the external characteristics of birds.	589–594	Think, p. 592	1
	2. Identify the major parts of a bird's skeletal system.			2
	3. List some ways in which a bird's internal organs are adapted for flight.			2, 3, 4
	4. Describe the mechanism by which birds maintain a constant body temperature.			4
	5. Explain the importance of the sense organs to avian life.			5
39.3	1. Summarize the processes of incubation and embryo development in birds.	595–599	Biotechnology, pp. 596–597	1, 2
	2. Describe the advantages of territorial and courtship behaviors.			5
	3. Explain how nest building reflects adaptations to a bird's environment.		Biology Lab, p. 599	4
	4. Identify the cues believed to aid bird migration.			3
	5. Explain how computers are used in bird conservation studies.			

Resources*

Print

Pasquier, Roger F. *Watching Birds: An Introduction to Ornithology*. Boston: Houghton-Mifflin, 1980.

Perrins, Christopher M., and Alex L.A. Middleton, eds. *The Encyclopedia of Birds*. New York: Facts on File, 1985.

Audiovisual/Software

Birds, revised. Film or video. Teacher's guide. Coronet, 1985.

Flying in Birds. Film or video, 25 min. Media Guild, 1986.

*Additional references are listed in *Modern Biology, Teacher's Program Guide*.

Section Teaching Guides

39.1 Evolution and Classification

Approach

The main objective of this section is to make students aware of the theories of how birds evolved and the methods by which taxonomists classify birds. The discussion builds on the concepts of adaptive radiation, natural selection, and varieties of fossil evidence introduced in Chapters 15 and 16. In presenting the text, compare the evolution of birds to that of reptiles (see Chapter 38), focusing on *Archaeopteryx* as an evolutionary link between birds and reptiles. You might also wish to refer students to Table 39-1 to guide them to an appreciation of bird diversity.

Concept Framework

Students will easily be able to identify the external differences between reptiles and birds. To help students understand the evolutionary relationship between reptiles and birds, talk about the similarities and differences between the two groups. You may wish to refer them to the photo of the *Archaeopteryx* fossil on p. 620. Ask students as they read the text to take note of the features of reptiles that evolved into physical features of birds.

Emphasis

Students should be able to identify the major theories of bird evolution and cite evidence in favor of each theory.

Class Discussion

Bring in pictures or slides of *Archaeopteryx,* and the early reptiles birds are believed to have evolved from. Ask the class to look for morphological evidence supporting either of the theories of avian evolution described in the text and engage the class in a discussion of which theory seems more plausible. You might wish to include a picture of *Protoavis,* the 225-million-year-old birdlike fossil that appears closely related to the dinosaurs.

Section Review Answers

(See p. 588.)

39.2 Characteristics of Birds

Approach

This section describes the external and internal characteristics of birds and particularly focuses on the ways these characteristics show adaptations for flight. In your discussion, refer to the material in Chapter 38 on reptile anatomy. Contrast reptilian characteristics with those aspects of bird anatomy that are modifications for the demands of flight. The complex anatomy of birds will also serve as a basis for comparison with mammals, the other group of endothermic vertebrates (Chapter 40). To clarify the reasons that birds require more energy than other vertebrates, you may wish to review the principles of cell metabolism, discussed in Chapter 7.

Concept Framework

Since birds inhabit virtually every region of the earth, students have most likely observed such avian characteristics as their large appetite and high level of activity. To prepare students to consider bird biology in terms of adaptations for flight, ask them to look for evidence as they read the text that birds are adapted for obtaining and metabolizing large amounts of energy as quickly as possible.

Emphasis

Make certain that students understand the relationship between the demands of flight and such avian characteristics as endothermy, feathers, air sacs, four-chambered heart, and the highly efficient digestive and excretory systems.

Guest Speaker

Invite an ornithologist from a local university to speak to the class about the ways in which birds are adapted to specific environments, such as water, deserts, rain forests, and polar regions.

Section Review Answers

(Continued from p. 594.)
1. The vanes extend from either side of the shaft and consist of many branches, or barbs. Barbules, in turn, project from the barbs.
2. The bones are hollow and are, in many cases, fused. The furculum and keeled sternum are unique to birds.
3. Food is digested and excreted quickly. The giz-

zard serves a function similar to that of the teeth and jaws of reptiles and mammals. Birds have no urinary bladder for storing liquid waste.

4. Air sacs permit oxygenated air to flow through the lungs during both inhalation and exhalation. The heart beats and circulates blood rapidly.
5. because birds generally stalk their prey from a great distance and because they hunt in the air where scents would not linger

39.3 Development and Behavior

Approach

The main objective of this section is to familiarize students with the processes of incubation and development in birds, as well as major types of bird behavior. The section builds on the concepts of natural selection and adaptation from Chapters 15 and 16. In addition, the information on territoriality, migration, and other behaviors will prepare students for the discussion in Chapter 51 of relationships and rhythms within ecosystems. In presenting the text, compare the amount of energy devoted by birds to producing and caring for offspring with that of other animal groups, especially reptiles (Chapter 38).

Concept Framework

Students will undoubtedly have observed such bird behaviors as caring for young, nest building, courtship, and migration. To tap students' knowledge of bird development and behavior, engage the class in a discussion of how different environmental characteristics might favor development of these behaviors. Ask students to look for specific examples of adaptations as they read the text.

Emphasis

Make certain that students understand how the following aspects of bird behavior have contributed to the evolutionary success of birds: incubation of eggs, care of young, territoriality, courtship, nest building, and migration.

Using the Features/Conservation of Migratory Birds

The feature on conservation of migratory birds describes the use of advanced computer technology and statistical analysis to study the migratory habits of shorebirds. These technologies are enabling conservationists to gather and interpret large amounts of data from all over the world. They hope to use this data to identify ecological trends and, as a result, prevent species from becoming endangered. In discussing the feature, encourage students to think about other ways in which computers can facilitate human efforts to protect the environment.

Field Trip

Take the class to a nearby area, such as a pond or field, where students can observe birds in their natural habitat. Depending on the time of year and the climate in your area, tell students to look for such behaviors as nesting, feeding of young, courtship, and migration.

Section Review Answers

(Continued from p. 598.)
4. Birds build nests in almost any safe place—on the ground or in trees—and use almost any available material.
5. Because birds dwell in densely vegetated areas, singing allows them to communicate without exposing themselves to predators.

Laboratory Preparation

Background

In this lab exercise, students compare the three basic types of bird feathers: quill, contour, and down. Students examine the central shaft and vane macroscopically. Using a microscope, students examine barbs, barbules, and hooks. The students compare the three feather types in relation to these structures and speculate on how these structures relate to the primary function of each type of feather.

Preparation

Prepare slides of contour feathers and slides of down feathers. Whole processed feathers are available from a biological supply house or craft supplier.

Safety

Feathers collected from fields may carry disease. If such feathers are used, remind students to wash their hands immediately after handling them.

Conclusions

1. The shaft supports the vane and makes the feather more rigid. The vanes and barbs form a broad, flat surface that increases the air resistance of the feather in flight.
2. The hooks interlock with the barbs and keep the barbs from being pushed apart by the air.
3. The quill is the largest and most rigid of the three types of feathers. The stiff shaft, broad vanes, and hooked barbules provide an air-resistant surface that is well suited for flight.
4. Down feathers have neither shafts nor hooks. As a result, down feathers spread apart easily. The space between the barbs traps air, and the trapped air insulates the bird's body.

Inquiry

Students should suggest ideas relating to the function of feathers, such as features that increase feather rigidity or increase air resistance.

Chapter Review

Vocabulary

1. *Altricial* means "one who nourishes," or "being born or hatched in an immature condition." *Precocial* means "capable of a high degree of independent activity at birth."
2. *Albumen* means "white."
3. *Furculum* means "fork."
4. *Proventriculus* means "before stomach."
5. *Incubate* means "to lie on."

Review

11. Birds are poorly represented in the fossil record because their feathers and fragile bones do not preserve well.
12. Taxonomists generally classify birds according to morphological differences in beaks, feet, plumage, bone structure, and musculature; behavior and song; and analysis of blood proteins, chromosomes, and DNA.
13. The five types of feathers are down feathers, contour feathers, quill feathers, filoplumes, and bristles.
14. The gizzard kneads and crushes food, often with the help of stones swallowed by the bird. Its function is similar to that of teeth and jaws in mammals.
15. Air sacs increase respiratory efficiency by permitting a continuous flow of air through the bird's lungs.
16. Binocular vision enables some birds to perceive depth, which is necessary for capturing prey.
17. The avian heart has a single right aortic arch, unlike the hearts of amphibians and reptiles. Also, in birds the right and left sides are completely separated.
18. The fully developed embryo uses its egg tooth to crack and scrape the shell until the opening is big enough for it to emerge.
19. Woodpeckers drill holes in trees and make their nests there. Orioles suspend their woven nests from branches. Bank swallows dig a tunnel in a sand bank. Barn swallows build a saucer of mud on the beam of a building.
20. Computers can analyze multiple variables far more quickly than a human can, and they have the capacity to store large amounts of data.

Critical Thinking

1. Reptiles are cold-blooded and cannot survive extreme temperatures. Birds are warm-blooded and can adapt to some extremes.
2. The bird must be a male, since it has a brightly colored plumage for courtship.
3. The duck is a flying animal, whereas the chicken is a domesticated animal and does not fly well. Therefore the dark meat or muscle is necessary for flight, and the chicken is bred for white meat.
4. The birds need blood with a high oxygen content because flight requires large amounts of energy and oxygen.
5. The mockingbird hatchlings will have an edge in growth over the other young. Being bigger and stronger than the younger nestlings, they will be fed first by the parent birds.
6. The hummingbirds are able to use the same nectar guides that insects do in order to find nectar in a flower.

Chapter 39 | Birds

Introduction

The ability to fly has enabled birds to inhabit virtually every area of the earth, from the equator to the poles. In this chapter you will learn about the anatomical and physiological adaptations that have made flight in these vertebrates possible.

Chapter Outline

Chapter Concept

As you read, note the characteristics of birds that are adaptations for flight.

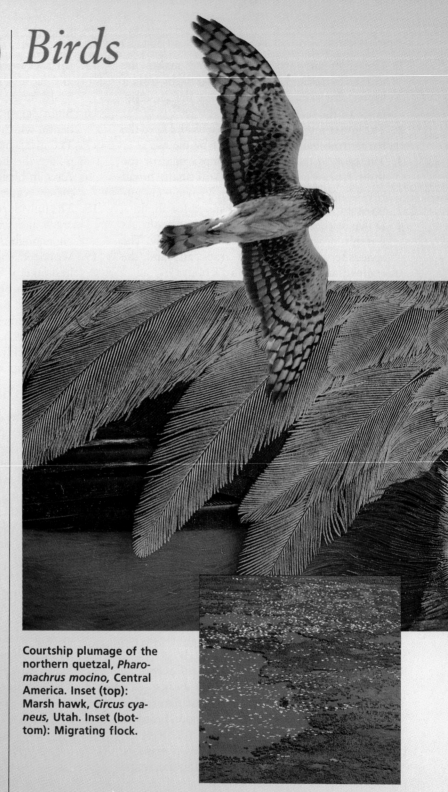

Courtship plumage of the northern quetzal, *Pharomachrus mocino*, Central America. Inset (top): Marsh hawk, *Circus cyaneus*, Utah. Inset (bottom): Migrating flock.

For additional teaching information see pp. T39A–T39D.

39.1 Evolution and Classification

Birds are vertebrates of the Class Aves. Although birds arose rather late relative to many other vertebrates, they have established themselves all over the earth. No other vertebrate group is so well adapted for life in air, on land, and in water. The evolution of flight and of endothermy, or warm-bloodedness, has enabled birds to survive in virtually every known environment.

Origin and Early Evolution

Evidence from fossils and from studies of comparative anatomy indicates that birds evolved from reptiles. Because their feathers and their fragile, hollow bones do not preserve well, birds are poorly represented in the fossil record, compared with other vertebrate groups. With so few clues scientists disagree on which particular line of reptiles gave rise to birds. However, they do concur that the fossil genus *Archaeopteryx* ("ancient wing") may represent an evolutionary link between reptiles and birds. The first *Archaeopteryx* fossil, which dates back to the Jurassic period, was discovered in 1861 in the fine limestone sediments of a quarry in southern Germany. Radioisotope dating has verified that the fossil is about 150 million years old.

Archaeopteryx possessed characteristics of both reptiles and birds, as seen in Figure 39-1. Like reptiles it had a large skull with teeth, bones that weren't hollow, claws on its forelimbs, and a long tail. Its strong legs and rounded wings indicate that it glided rather than flew. In contrast, the presence of feathers and of a **furculum** *(FUR-kyuh-lum)*, the fused collarbones commonly called a wishbone, suggests that *Archaeopteryx* was birdlike.

What group of reptiles gave rise to *Archaeopteryx*—and thus possibly to all birds? Two hypotheses prevail. One argues that birds arose about 230 million years ago from thecodonts, a group of reptiles discussed in Chapter 38. These reptiles walked on their hind legs like birds. Some, however, may have been tree dwelling. The thecodonts gave rise to a number of subgroups of reptiles. One of these, the pseudosuchians ("false crocodiles"), may have given rise to both dinosaurs and birds. As evidence for this theory scientists note that the pseudosuchian's large collarbone is homologous to the furculum of modern birds.

Another hypothesis states that birds evolved directly from a group of dinosaurs about 150 million years ago. The group, called coelurosaurian theropods, were small predators that walked on their hind legs. Their hips, forelimbs, and hind limbs resembled those of *Archaeopteryx*.

After *Archaeopteryx* the next bird fossils date from about 90 million years ago, in the Cretaceous period. *Hesperornis*, a large,

Info: Scientists disagree on whether feathers first evolved for insulation, for flight, or as a means of trapping small prey.

Section Objectives

- *List both the avian and the reptilian characteristics of* Archaeopteryx.

- *Summarize the two predominant hypotheses of bird evolution.*

- *Identify the criteria which taxonomists use when they classify birds.*

Figure 39-1 An artist's rendering, based on fossil evidence, suggests the possible appearance of three ancestors of modern birds.

Archaeopteryx

Ichthyornis

Hesperornis

Table 39-1 Representative Bird Orders

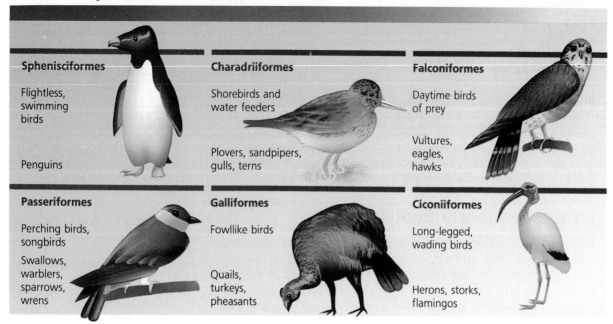

Sphenisciformes	Charadriiformes	Falconiformes
Flightless, swimming birds	Shorebirds and water feeders	Daytime birds of prey
		Vultures, eagles, hawks
Penguins	Plovers, sandpipers, gulls, terns	
Passeriformes	**Galliformes**	**Ciconiiformes**
Perching birds, songbirds	Fowllike birds	Long-legged, wading birds
Swallows, warblers, sparrows, wrens	Quails, turkeys, pheasants	Herons, storks, flamingos

flightless, diving bird, probably resembled the modern loon but had reptilian teeth. A smaller ternlike bird called *Ichthyornis* had large wings, indicating that it may have been a strong flier. The development of sustained flight may have enabled birds to colonize new areas during the Cretaceous. By about 11 million years ago in the Miocene epoch, most modern genera had evolved.

Classification

Most taxonomists classify the nearly 9,000 species of Class Aves into 27 orders, several of which are described in Table 39-1. To classify birds into orders and families, taxonomists most often use morphological evidence from beaks, feet, plumage, bone structure, and musculature. In recent years taxonomists have also begun to classify birds according to behavior and song. Technological advances in the analysis of blood proteins, chromosomes, and DNA have also contributed to scientists' understanding of avian phylogenetic relationships.

Section Review

1. What is the name for the bone that was present in *Archaeopteryx* and is unique to birds?
2. What characteristics did *Archaeopteryx* share with reptiles?
3. Summarize the two major hypotheses of bird evolution.
4. What were the main characteristics of *Hesperornis* and *Ichthyornis*?
5. Explain the fact that taxonomists commonly base their classification of birds on beaks, feet, and plumage.

Tip: Explain to students that bird wings and insect wings are analogous, not homologous, because they differ in embryonic origin.

Answers to Section Review
1. the furculum or wishbone
2. claws on forelimbs, bones that were not hollow, large skull with teeth, and a long tail
3. One hypothesis states that birds and dinosaurs both evolved from pseudosuchians, a reptile group that in turn arose from thecodonts. The other states that birds arose directly from coelurosaurian theropods, a dinosaur group.
4. *Hesperornis* was a large, toothed, flightless diving bird. *Ichthyornis* was a small ternlike bird with large wings.
5. These are the characteristics that show the greatest variation.

39.2 Characteristics of Birds

Section Objectives

- *Describe the external characteristics of birds.*

- *Identify the major parts of a bird's skeletal system.*

- *List some ways in which a bird's internal organs are adapted for flight.*

- *Describe the mechanism by which birds maintain a constant body temperature.*

- *Explain the importance of the various sense organs to avian life.*

Birds have many traits in common with other vertebrates, but the aerodynamic requirements of flight have produced a number of unique adaptations. The following characteristics distinguish birds from other vertebrates:

- The body is covered with feathers.
- The bones are thin and hollow.
- In most species the forelimbs function as wings; they are used for flight, not grasping.
- The two hind limbs, with clawed toes, support the body.
- A toothless, horny beak is present.
- Body temperature is generated and regulated internally.
- The four-chambered heart has a single right aortic arch.
- Amniote eggs are encased in hard, calcium-containing shells.
- In most species eggs are incubated in a nest.

External Characteristics

As flight enabled birds to colonize new territories across the earth, they adapted to various terrains, climates, food sources, and other conditions. The variation in the external characteristics of birds reflects these adaptations.

Feathers

Feathers are modified scales that serve two primary functions—providing lift for flight and conserving body heat. Soft, fluffy **down feathers** cover the body of nestling birds and provide an insulating undercoat in adults. **Contour feathers** give adult birds their streamlined shape and provide coloration and additional insulation. **Flight feathers** are specialized contour feathers on the wings and tail. Birds also have hairlike **filoplumes,** or pinfeathers, and dust-filtering **bristles** near the nostrils.

The structure of feathers combines maximum strength with minimum weight. Feathers develop from tiny pits in the skin called **follicles.** A **shaft** emerges from the follicle, and two **vanes,** pictured in Figure 39-2, develop from its opposite sides. At maturity each vane has many branches called **barbs.** The barbs in turn have many projections called **barbules** that are equipped with microscopic hooks. The hooks hold on to each other and give the feather its sturdy but flexible shape.

Feathers need care. In a process called preening, birds use their beaks to rub their feathers with oil secreted by a **preen gland** located at the base of the tail. Just as reptiles shed their skin, birds periodically molt their feathers. The major molt, during which the bird replaces its flight feathers, occurs in the late summer between breeding and migration. Some birds also undergo a spring molt prior to courtship.

Figure 39-2 The small rectangular area on the drawing of the complete feather is shown in detail in the drawing above.

Barb

Barbule

Hooks

Vane

Shaft

Beaks and Feet

By looking closely at a bird's beak and feet, you can infer many things about where it lives and how it feeds. Hawks and eagles have powerful beaks and clawed talons that help them capture and then rip their prey. Swifts have a tiny beak that opens wide like a catcher's mitt to snare insects in midair. Because many swifts spend most of their lives in flight, their feet are small and adapted for infrequent perching. The feet of flightless birds, on the other hand, are modified for walking and running. Figure 39-3 shows other beak and foot adaptations. In many birds, such as herons and egrets, hormonal changes cause the beak, feet, and legs to turn vivid colors. This is an adaptation that helps attract potential mates.

Internal Characteristics

Like their external features, birds' internal systems reflect many adaptations for the demands of flight. Natural selection has favored a lightweight body, yet the bird's powerful wing muscles give it unusual strength.

Skeleton and Muscles

The avian skeleton combines lightness and strength. The bones are thin and hollow. Compared with a reptile skeleton, many

Figure 39-3 Shown are (clockwise from top left) a black vulture, *Coragyps atratus*, a mountain bluebird, *Sialia currocoides*, a roseate spoonbill, *Ajaia ajaja*, two tufted puffins, *Lunda cirrhata*, and a flock of ostriches, *Struthio* sp. Suggest adaptations for the beaks and feet of each of these birds.

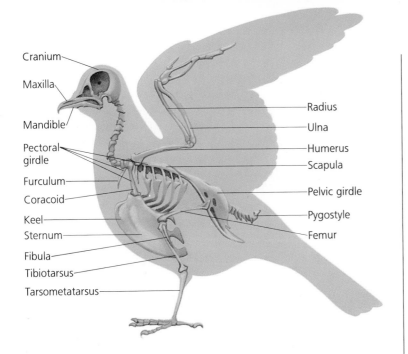

Cranium
Maxilla
Mandible
Pectoral girdle
Furculum
Coracoid
Keel
Sternum
Fibula
Tibiotarsus
Tarsometatarsus

Radius
Ulna
Humerus
Scapula
Pelvic girdle
Pygostyle
Femur

Figure 39-4   The rigid but light avian skeleton is well adapted for flight.

bones are fused, so the avian skeleton is more rigid. The rigid skeleton provides stability during flight. Note in Figure 39-4 the fused bones of the trunk and hip vertebrae and the pectoral and pelvic girdles. These fused bones form a sturdy frame that anchors the powerful breast muscles during flight and supports the leg muscles when the bird is walking or at rest. In some birds, flight muscles account for 50 percent of the body weight.

Also unique to the avian skeleton are the furculum, or wishbone, and the keel-shaped sternum. The sternum supports the large breast muscles. The humerus, ulna, and radius, along with the pectoral girdle and the sternum, support the wing. The pygostyle, the terminal vertebra of the spine, supports the tail feathers, which also play an important role in flight. The tail provides additional lift and aids in steering and braking.

Flight involves a series of complex wing movements, each one using a different set of muscles. On the downstroke each wing cuts forward and downward through the air. The upstroke moves each wing upward and backward. Air flows above and below the wing, creating lift. Muscles in the skin move the feathers, permitting precise flight maneuvers.

Endothermy

Birds are endothermic—that is, they generate and regulate body heat internally. In most birds body temperature ranges from 40° C to 42° C. Endothermy enables birds to inhabit both cold and hot climates. Rapid breathing and the digestion of large quantities of food produce the high metabolic rate necessary to generate heat. Thus birds, unlike reptiles, cannot go for long periods without eating. To help conserve body heat, birds fluff out their feathers for insulation. In aquatic birds a thin layer of fat provides additional insulation.

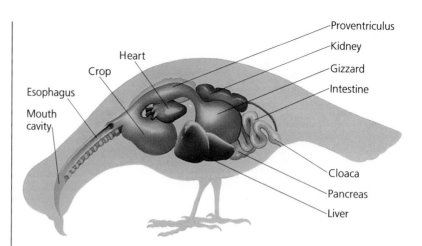

Figure 39-5 A bird's digestive and excretory systems are adapted for the rapid processing of food and metabolic wastes.

Answer to Think
Flightlessness may have arisen in areas where terrestrial predation was minimal or nonexistent, such as on islands. Flightlessness would have enabled birds to attain greater body size.

Think

Comparative anatomy studies suggest that flightless birds such as ostriches evolved from flying ancestors. What natural conditions might have favored flightlessness?

Digestive and Excretory Systems

The high energy requirements of flight and of endothermy are met by a quick and efficient digestive system. For instance, a thrush can eat blackberries, digest them, and 45 minutes later excrete the seeds. A magpie can digest a mouse in three hours. In contrast, a snake may require three days to digest a mouse of the same size.

Since birds have neither teeth, nor jaw bones, nor jaw muscles, they are not able to chew their food in their mouths. Instead food passes from the mouth cavity straight to the esophagus. An enlargement of the esophagus called the **crop** stores and moistens food. Food then passes to the two-part stomach, seen in Figure 39-5. In the first chamber, the **proventriculus** *(PRO-ven-TRICK-yuh-luss)*, gastric fluids begin breaking down the food. Food then passes to the **gizzard**, a muscular organ that kneads and crushes the food. The gizzard often contains small stones that the bird has swallowed. These aid in the grinding process. Thus the gizzard performs a function similar to that of teeth and jaws. Food then passes through the pyloric sphincter to the small intestine, also called the duodenum. Here bile and enzymes manufactured in the liver, pancreas, and intestine further break down the food. The digested food nutrients are then absorbed into the bird's bloodstream.

The avian excretory system is also efficient and lightweight. Unlike other vertebrates, most birds do not store liquid waste in a urinary bladder. Instead the two kidneys filter a nitrogenous waste called uric acid from the blood. Highly concentrated uric acid travels by ducts called ureters to the cloaca, where, along with undigested matter from the intestines, it is excreted in a semi-solid, usually white mass.

Respiratory System

The high metabolic rate of birds requires large amounts of oxygen. Yet some birds migrate thousands of miles at altitudes as high as 7,000 m, where oxygen levels are low. An elaborate and highly efficient respiratory system meets these oxygen needs.

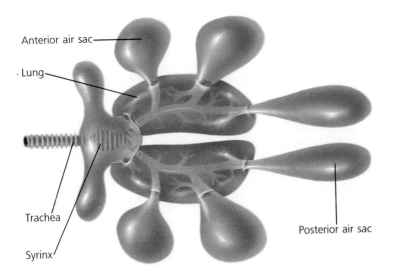

Anterior air sac

Lung

Trachea

Syrinx

Posterior air sac

Figure 39-6 Air sacs pass oxygen-rich air to the lungs as the bird exhales carbon dioxide.

Air enters the bird's body through paired nostrils located near the base of the beak. It passes down the trachea past the **syrinx** *(SIR-inks)*, or song box, and enters the two primary bronchii. From the bronchii, air moves to the lungs. About 75 percent of the air bypasses the lungs and flows directly to posterior **air sacs**, seen in Figure 39-6. Nine sacs extend out from the lungs, occupying a large portion of the bird's chest and abdominal cavity. These sacs connect with air spaces in the bones, filling the hollow bones with air. Thus the air sacs not only function in respiration but also greatly reduce the bird's density.

When the bird exhales the carbon dioxide–rich air from its lungs, oxygen-rich air is forced out of the posterior air sacs into the lungs via small tubes. The bird thus receives a supply of oxygenated air in its lungs during both inhalation and exhalation. Gas exchange then takes place through capillaries.

Circulatory System

Birds, like some reptiles, have a four-chambered heart. However, in birds the right and left sides of the heart are completely separated. The right side receives deoxygenated blood from the body and pumps it to the lungs. The left side receives oxygenated blood from the lungs and pumps it to the rest of the body. Unlike amphibians and reptiles, birds have a single right aortic arch.

In comparison with reptiles and most other vertebrates, most birds have a rapid heartbeat. The hummingbird's heart beats about 600 times a minute. The active chickadee's heart beats 1,000 times a minute. In contrast, the heart of the larger, less active ostrich averages 70 beats per minute—the same rate as a human heart.

Nervous System

Relative to their size, birds have large brains. The most highly developed areas of the bird's brain are those that control flight-related functions. The cerebellum coordinates movement. The cerebrum controls complex behavior patterns such as navigation,

Info: Because of the air sacs, a bird's respiratory system can take up as much as 20 percent of the total volume of the body.

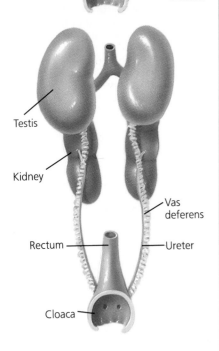

Egg

Oviduct

Kidney

Ureter

Reduced right oviduct

Shell gland

Rectum

Cloaca

Opening of oviduct

Testis

Kidney

Vas deferens

Rectum

Ureter

Cloaca

Figure 39-7 The female and male reproductive systems of a bird are shown here.

Answers to Section Review (See pp. T39A–T39D.)

mating, nest building, and care of the young. The optic lobes receive and interpret visual stimuli.

The sense organs are modified for flight and avian life. Keen vision is necessary for taking off, landing, spotting landmarks, hunting, and feeding. Most birds have good color vision that aids them in finding food. In most species the bird's large eyes are located near the sides of its head, giving the bird a wide field of vision. Birds that have eyes located near the front of the head have better binocular vision; they can perceive depth in the area where the visual fields of the two eyes overlap.

Hearing is important to songbirds and also to nocturnal species such as owls, which rely on sounds to help them locate their prey. Look at the cover of this book. Though birds lack external ears, owls possess disk-shaped feathers around their ear openings, or canals, that help direct sounds into the ear. Each ear canal leads to a tympanic membrane, or eardrum. Semicircular canals in the inner ear regulate balance and cue spatial orientation.

Compared with vision and hearing, the sense of smell has less adaptive significance for birds and is often poorly developed except in ducks and flightless birds. The sense of taste helps birds avoid bitter-tasting or toxic foods.

Reproductive System

In the male bird sperm is produced in two testes that lie beneath the kidneys. Sperm passes through small tubes called vasa deferentia into the male's cloaca. During mating the male presses his cloaca to the female's and releases sperm. As you can see from Figure 39-7, most females have a single ovary located on the left side. The ovary releases eggs into a long, funnel-shaped oviduct, where they are fertilized by sperm. Fertilized eggs move down the oviduct, where they receive protective covering and a shell.

An unfertilized egg consists of a nucleus, cytoplasm, and a yolk. When fertilized, the embryo, as shown in Figure 39-7, is suspended in **albumen,** the egg white. This liquid medium is supported by ropelike strands of material called **chalaza** *(kuh-LAY-zuh)* that are attached to the **shell membrane.** The female has a **shell gland** that secretes a protective calcium carbonate shell to surround the egg.

Section Review

1. Distinguish between *vanes, barbs,* and *barbules.*
2. In what ways does the avian skeleton differ from that of other vertebrates?
3. What features of the digestive and excretory systems are adaptations for flight?
4. How do the respiratory and circulatory systems function to meet the bird's high oxygen needs?
5. Why do you think that the sense of smell is not as important to most birds as the senses of sight and hearing?

39.3 Development and Behavior

Although both birds and reptiles lay yolky eggs, birds generally take greater care of their eggs and their young than reptiles do. Most reptiles lay their eggs and immediately abandon them, while most birds invest at least some amount of energy in the care of their young.

Incubation and Development

A female bird usually lays eggs in a nest. Either one or both parents will then **incubate,** or warm, the eggs by sitting on them and covering them with a thickened, featherless patch of skin on the abdomen called a **brood patch.** Some species show other brooding behaviors. The male emperor penguin heats the egg by placing it on his webbed feet and enfolding it with his warm abdomen. In many cuckoo species the female lays the eggs in another species' nest, leaving the incubation and feeding to other birds.

Embryo development begins when the zygote forms a plate of cells on the surface of the yolk. This plate begins to form the tissues and organs of the embryo. A yolk sac membrane grows out of the embryo's developing digestive tract and surrounds the yolk like a net. The membrane produces digestive enzymes that dissolve proteins and lipids in the yolk. Blood vessels in the yolk sac membrane then carry the nutrients to the embryo. As you can see in Figure 39-8, by the time the egg hatches, the embryo has absorbed the yolk.

The allantois, a vascular, netlike sac, also grows from the digestive tract. The allantois attaches to the chorion membrane lining the shell. Together the allantois and chorion function in respiration and help eliminate waste products. Like the reptilian embryo, the avian embryo is enclosed in an amnion, which is filled with amniotic fluid.

When hatching begins, the fully developed embryo makes a star-shaped crack in the shell with the scalelike **egg tooth** on its bill. The chick presses and scrapes the shell until the crack widens enough for the chick to emerge. The egg tooth, having served its purpose, falls off soon after the chick hatches.

Section Objectives

- *Summarize the processes of incubation and embryo development in birds.*

- *Describe the advantages of territorial and courtship behaviors.*

- *Explain how nest building reflects adaptations to a bird's environment.*

- *Identify the cues believed to aid in bird migration.*

- *Explain how computers are used in bird conservation studies.*

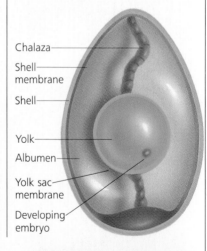

Figure 39-8 The drawing below shows internal parts of a newly fertilized egg. The photographs in the bottom row show sequential stages in embryo development.

Chalaza
Shell membrane
Shell
Yolk
Albumen
Yolk sac membrane
Developing embryo

Birds have two contrasting methods for rearing young. Those that lay many eggs and incubate them for long periods hatch **precocial** *(pree-KOE-shul)* young. These birds are active as soon as they are hatched. The mother provides some warmth and protection, but the newly hatched birds can walk, swim, and feed themselves. Ducks, quail, and other ground-nesting species produce precocial offspring. Birds that lay only a few eggs that hatch quickly produce **altricial** *(al-TRIH-shul)* young. These birds hatch blind, naked, and helpless. They depend on both parents for several weeks. The parents keep the young warm and search for food to feed them. The young of woodpeckers, hawks, pigeons, parrots, warblers, and many aquatic birds are altricial.

Behavior

The time and effort birds spend incubating and caring for their young help ensure that most of their offspring will reach maturity. Young birds obviously need protection until they develop the strength to fly and to obtain food. However, the long period of parental care may also enable birds to learn such complex behaviors as courtship, nesting, and migration.

Biotechnology

Conservation of Migratory Birds

Bird conservation efforts have generally focused on endangered species, but biologists are developing a better understanding of bird behavior, which will enable them to protect species before they become endangered. Advances in computer technology and statistical analysis have made possible large-scale studies that weren't feasible ten years ago.

Computer technology has especially benefited the study of migratory shorebirds, which include sandpipers, plovers, and curlews. Shorebird migrations often extend from the Arctic to Argentina—a round trip exceeding 25,000 km. Conservation efforts for these species obviously require international cooperation among nations throughout the Western Hemisphere.

One such effort is the International Shorebird Survey, a project established in 1974. Conservation groups in North and South America have been collecting data on dozens of shorebird species.

Volunteers record information on bird populations, habitat characteristics, weather conditions, and human activity in their areas. Researchers then use computers to identify relationships among these variables.

Territoriality and Courtship

During the breeding season many male birds establish an area that they defend against other males of their species, a behavior called **territoriality.** The male then attempts to attract a female—or in some species, several females—to share this territory. If the male is successful, no competition will exist for food sources in his territory. An example of this type of behavior is provided by the male red-winged blackbird, which marks his territory by sitting in a prominent location while trilling his song and displaying his red shoulder patches. Once a territory is established, most birds engage in a period of **courtship,** behavior that is designed to attract a mate.

A great deal of variation exists in the courtship behavior of different species of birds. Many males attract females by means of their brightly colored feathers, which they show off in various courtship displays. Some males combine song with flight displays. For example, the brilliant red vermilion flycatcher flies vertically into the air and then dives steeply. He may repeat this courtship behavior for many hours while fluttering his wings and singing a trilling song.

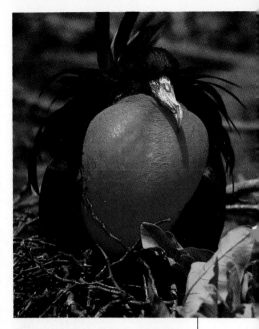

Figure 39-9 During courtship a male great frigate bird, *Fregata minor,* puffs out its red throat pouch.

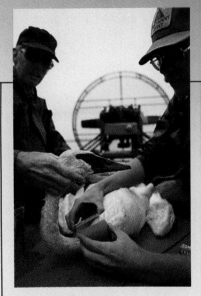

For example, determining the correlation between two variables such as population and food supply involves plotting data on a two-dimensional graph. However, far more sophisticated computations are required to factor in additional variables such as weather and human intervention.

A high-speed computer can analyze multiple variables far more quickly than a human can. The massive memory capacity of today's computer can easily store the millions of bits of information gathered over the last 15 years.

Statisticians use highly complex methods of trend analysis to interpret the data generated by the computer. As a result researchers now know a great deal about the migratory habits of shorebirds—their velocity, altitude, flight path, and gathering sites. They have also identified significant trends, such as a 70 to 80 percent drop in the population of sanderlings and short-billed dowitchers.

Have natural factors or human intervention caused this population drop? How might construction of a housing development on a prime feeding ground affect a bird population? These and many other questions still face researchers of the International Shorebird Survey. Once they determine the answers, biologists can recommend global strategies to protect shorebird species. Computers will undoubtedly play an important role in the development of these strategies.

Figure 39-10 Nest builders and the nest materials they use are (clockwise from top left) the hummingbird, mud; the osprey, *Pandion haliaetus*, branches; and the grosbeak weaver, *Amblyospiza albifrons*, grass.

Answers to Section Review
1. Precocial birds lay many eggs and have a long incubation period. The chicks are active upon hatching. Altricial birds lay few eggs and have a short incubation period. The birds hatch blind, naked, and helpless.
2. by warming the eggs with a brood patch on the abdomen
3. position of stars or the sun, topographical landmarks, magnetic field, air pressure, and low-frequency noises
(See pp. T39A–T39D.)

Nest Building

Nests hold eggs, conceal young birds from predators, provide shelter from the elements, and sometimes even serve to attract a mate. Most birds build nests in sheltered, well-hidden spots—from holes in the ground to treetops. Woodpeckers nest in tree holes that they have drilled. Orioles suspend their nests from branches, well beyond the reach of predators. Various swallow species have nesting strategies that result in minimal competition between species. Bank swallows, for example, dig a tunnel in a sand bank, whereas barn swallows build a saucer of mud on the beam of a building.

As a further adaptation to their environment, birds construct their nests of almost any material available. Twigs, grasses, feathers, and mud are the most common materials used.

Migration

Each year thousands of bird species exploit the spring and summer food resources of colder regions. Then, when temperatures drop and the food supply dwindles, these birds migrate to warmer climates. How do birds manage to navigate thousands of miles across varied terrains to the same spot year after year?

Ornithologists—biologists who study birds—know that birds rely on a variety of cues to help them navigate. Some species monitor the position of the stars or the sun. Others rely on topographical landmarks, such as mountains. Magnetic cues, changes in air pressure, and low-frequency sounds may also provide information to migrating birds. The ability of birds to "read" these cues, along with their many adaptations for flight, enables them to migrate to and inhabit virtually any environment.

Section Review

1. Distinguish between *precocial* and *altricial* birds.
2. How do most bird species incubate their eggs?
3. List at least three cues that birds may use for navigation.
4. How do a bird's nest-building habits reflect adaptations to its environment?
5. Why might wrens and cardinals rely on singing to establish territories and attract mates?

Tip: Tell students that changes in the length of daylight give birds cues to when it is time to migrate.

Laboratory

Structure and Function of Bird Feathers

Objective
To examine the structure and function of the three main types of bird feathers

Process Skills
Observing, inferring, hypothesizing

Materials
1 each of quill, contour, and down feathers; unlined paper; prepared slide of a contour feather; microscope; prepared slide of a down feather

Method
1 Examine a quill feather. Hold the base of the central shaft with one hand and gently bend the tip of the feather with your other hand. Be careful not to break the feather. Next hold the shaft and wave the feather in the air. Record your observations concerning the structure of the quill feather. Relate your observations to its possible function.
2 Examine the vane of the feather. Does the vane appear to be a solid structure? Write a description of its structure. Make a drawing of the quill feather. Label the shaft, vanes, and barbs.
3 Examine a contour feather and a down feather. Make a drawing of the contour feather and the down feather. Label your drawings as before. Do you notice a difference in the structure of the contour and the down feathers?
4 Examine a prepared slide of a contour feather under low power. Note the smaller barbs, called barbules, extending from each of the barbs.
5 How might you observe the region between the barbs? Locate the tiny hooks at the end of each barbule. Note the arrangement of the hooks on adjacent barbs. Why do you

think the hooks are so small? Make a labeled drawing of the hooks.
6 Examine the prepared slide of the down feather under low power. Locate the barbs and barbules. Switch your microscope to high power, and make a labeled drawing of the down feather.

Conclusions
1 What is the function of the shaft? What is the function of the vanes and barbs?
2 How do hooks increase the strength and air resistance of a feather?
3 How is the structure of the quill feather related to its function of aiding flight?
4 Why are down feathers more effective at keeping a bird warm than the other two feather types are?

Inquiry
Each of the feather types you have examined has a specific structure and function. Review your observations, and try to think of features that account for the efficiency of the three types of feathers.

Chapter 39 Review

Vocabulary

air sac (593)	chalaza (594)	flight feather (589)	proventriculus (592)
albumen (594)	contour feather (589)	follicle (589)	shaft (589)
altricial (596)	courtship (597)	furculum (587)	shell gland (594)
barb (589)	crop (592)	gizzard (592)	shell membrane (594)
barbule (589)	down feather (589)	incubate (595)	syrinx (593)
bristle (589)	egg tooth (595)	precocial (596)	territoriality (597)
brood patch (595)	filoplume (589)	preen gland (589)	vane (589)

Use a dictionary to find the origin of each of the following terms, and tell why the terms are appropriately named.
1. altricial, precocial
2. albumen
3. furculum
4. proventriculus
5. incubate

Review

1. The first *Archaeopteryx* fossil dates back to the (a) Miocene epoch (b) Cretaceous period (c) Jurassic period (d) Eocene period.
2. The most common basis for classifying birds is (a) morphology (b) flight patterns (c) wingspan (d) color.
3. The major molting in birds usually occurs (a) just after the end of the breeding season (b) just before the beginning of the breeding season (c) just after the fall migration ends (d) just before the first snow of winter.
4. The bone that supports the tail feathers is the (a) pelvic girdle (b) pygostyle (c) ulna (d) furculum.
5. Gas exchange in birds takes place in the (a) anterior air sacs (b) bronchii (c) syrinx (d) capillaries.
6. In order to maintain a constant body temperature, birds cannot go for long periods without (a) resting (b) molting (c) eating (d) nesting.
7. The ropelike substance that supports the embryo and surrounding materials is the (a) allantois (b) albumen (c) amnion (d) chalaza.
8. Some swallows have minimal competition because different species (a) build different types of nests (b) engage in flight display (c) breed at different times of the year (d) migrate at different times of the year.
9. Birds that produce altricial young (a) abandon their newly hatched chicks immediately (b) lay only a few eggs at a time (c) usually build nests on the ground (d) incubate their eggs for a long period of time.
10. A cue that is believed to aid birds in navigation during their annual migrations is (a) the magnetic field (b) weather (c) cloud formations (d) phases of the moon.
11. Why are birds poorly represented in the fossil record?
12. What three criteria are used to classify birds?
13. List the five types of feathers.
14. What function does the gizzard serve in the bird's digestive process?
15. How do air sacs increase respiratory efficiency in birds?
16. Why is binocular vision important to some birds?
17. What features distinguish the avian heart from the heart of amphibians and reptiles?

18. What is the function of the young bird's egg tooth?
19. Describe the nesting habits of three bird species.

20. Name two advantages to using today's high-speed computers in the study of bird populations.

Critical Thinking

1. What factors might account for the fact that birds live in the Arctic but that reptiles do not?
2. Birds are an example of animals exhibiting dimorphism: that is, the sexes differ markedly in appearance. Look at the picture of the bird on the right. From what you have learned in the chapter, is it a male or a female?
3. A large part of the meat (muscles) of the chicken is white, while all the meat of the duck is dark. What difference in the lifestyles of the two birds might account for the difference in their muscle texture and color?
4. From studying the circulatory system of birds, you know that the right and left sides of a bird's heart are completely separated. This means that the oxygenated blood is never mixed with the unoxygenated blood. What in the life-style of birds makes this complete separation in the heart necessary?
5. Mockingbirds lay their eggs in the nests of other birds. The mockingbirds hatch

slightly earlier than the other birds do, and the hatchlings are slightly larger. Why might it be advantageous to the young mockingbirds to hatch earlier and to be larger?
6. Several species of hummingbirds are able to see light that is near the ultraviolet range. Insects are also able to see ultraviolet light. Like some insects, hummingbirds are nectar eaters. Why might it be important for a hummingbird to be able to see light that is near the ultraviolet range?

Extension

1. Bird watching is a fascinating and interesting hobby. From your library or bookstore obtain a field guide to the birds in your area, and borrow a pair of binoculars. Take a long walk and try to locate as many birds as you can. Be sure to keep a notebook of the birds you find this year, including species, date, location, and habits.
2. At your grocery buy a whole chicken. Boil and bone the chicken. Lay the bones out as if you were preparing to reconstruct

the skeleton. Then identify the bones by referring to Figure 39-4.
3. Read the article by M. Philip Kahl entitled "The Royal Spoonbill," in *National Geographic*, February 1987, pp. 280–284. In the environment the spoonbill inhabits, certain selectional pressures have led to the development of its spoonlike bill. How does this unique feature benefit the spoonbill and enable it to survive in its habitat?

Chapter 40 | *Mammals*

This chapter offers an overview of the evolution and adaptation of mammals. Students become familiar with the characteristics of monotremes, marsupials, and placental mammals and speculate on the connection between characteristics and adaptation.

Correlation Chart

Section Number	Section Objectives	Pages	Features	Section Review
40.1	1. Summarize the significance of the therapsids in mammalian evolution.	603–605		1, 2
	2. Give a probable reason why the mammals survived competition with dinosaurs.			3, 4
	3. List some characteristics of mammals, and explain how these characteristics enabled mammals to adapt to different environments.			4, 5
40.2	1. List examples of both reptilian and mammalian characteristics in monotremes.	606–607		1, 2
	2. Outline the differences in reproduction between monotremes and marsupials.			2, 3, 5
	3. Describe the circumstances under which marsupials and placentals have undergone convergent evolution.			4
40.3	1. Describe the reproductive features that distinguish placental mammals from monotremes and marsupials.	608–615	Think, p. 609	1, 5
	2. Name some orders of placental mammals, and give a representative from each.		Biology in Process, pp. 612–613	2
	3. List some adaptive features of placental mammals for locomotion and feeding.		Biology Lab, p. 615	4, 5
	4. Summarize Jane Goodall's contributions to the study of wild chimpanzees.			

Resources*

Print

Savage, R. J. G. *Mammal Evolution: An Illustrated Guide*. New York: Facts on File, 1986.

Audiovisual/Software

Among the Wild Chimpanzees (Jane Goodall's research). Film or video, 59 min. Teacher's guide. National Geographic, 1984.

In the Kingdom of Dolphins. Video, 50 min. Films, Inc., 1986.

*Additional references are listed in *Modern Biology, Teacher's Program Guide*.

Section Teaching Guides

40.1 Evolution and Characteristics

Approach

The main objective of this section is to familiarize students with mammalian characteristics and to emphasize the adaptive significance of those characteristics. It builds on the concepts of diversity through adaptation (Chapters 15 and 16) and classification (Chapter 18). Remind students that mammals are the dominant land vertebrates and that they are adapted to a wide range of environments. In presenting the text, encourage students to think about how mammalian characteristics may have helped mammals become so successful.

Concept Framework

Students have learned that the dinosaurs became extinct and that mammals eventually became the dominant land animals. Use this knowledge as a conceptual starting point for this section by asking students to speculate about the traits that led mammals to survive while another group died out. You might end the discussion by encouraging students to recognize as they study this section the adaptive value of mammalian traits.

Emphasis

Make sure that students grasp the relationship between environmental factors and species survival. Discuss the climatic changes that occurred on earth at the end of the Mesozoic and at the beginning of the Cenozoic eras as continents and mountain chains arose and oceans formed. You might suggest that many reptiles could not adjust to the changing conditions.

Field Trip

Visit a museum of natural history or the zoology department of a nearby university to study vertebrate skeletons. If possible, encourage students to compare reptilian and mammalian skeletons.

Section Review Answers

(Continued from p. 605.)
4. Endothermy has allowed mammals to live in all types of environments and to be active at any time of the day. Specialized teeth allow mammals to exploit a wide variety of food sources. Viviparity allows mammals to be mobile while still protecting their unborn young within them.
5. Mammals exhibited adaptive radiation. All the species of mammals evolved from a common ancestor. Speciation occurred as the mammals moved into the myriad environments that opened once the dinosaurs had become extinct.

40.2 Monotremes and Marsupials

Approach

The main objective of this section is to investigate the possible evolutionary relationships among monotremes, marsupials, and other vertebrates. It builds on the concepts of diversity through adaptation (Chapters 15 and 16) and embryonic development (Chapter 29). Monotremes and marsupials are two groups of mammals that have unusual reproductive patterns. Point out that species in these two orders are less highly evolved than placental mammals. As a result, they account for only a small percentage of all mammalian species. In presenting the text, emphasize that these animals are good examples of the diversity among mammalian species.

Concept Framework

Students are generally familiar with common mammalian offspring such as puppies and kittens. Tap their observations to provide a conceptual perspective for studying this section by asking students to describe the characteristics of placental infants. Ask students to identify as they read this section any traits of infants, of kangaroos, or opossums. Make sure that students know that most marsupials have undergone convergent evolution in Australia with placental mammals on the other continents.

Emphasis

Students should be able to identify characteristics that distinguish monotremes and marsupials from other mammals.

Section Review Answers

(Continued from p. 607.)
3. A marsupial gives birth to immature young that crawl to a pouch on the mother's abdomen and

attach themselves to milk-secreting nipples; young remain in the pouch nursing until they can survive outside the pouch.

4. Placentals and marsupials, though only distantly related, have adapted to similar environments, use similar food sources, and are similar in body structure and behavior.

5. The backward pouch protects the developing young from being smothered with dirt when the mother burrows.

40.3 Placental Mammals

Approach

The primary objective of this section is to explore the success of placental mammals. Placental mammals have adapted to life on land, in the air, and in water. This section builds on the concept of diversity through adaptation (Chapters 15 and 16). In presenting the text, point out the reasons—dietary or locomotive adaptations, for example—for each mammalian order's successful exploitation of a particular environment.

Concept Framework

Students are familiar with a range of placental mammals. To engage students conceptually as you introduce this section, you might associate familiar representative animals with each order. Encourage students as they read to visualize a representative mammal and to identify its characteristics, rather than to memorize a list of facts.

Emphasis

Make sure that students know that placental mammals are considered to be more highly evolved than monotremes and marsupials. Emphasize the benefits of a placenta and a gestation period to unborn animals.

Field Trip

Visit a local zoo to observe representative members of the major orders of placental mammals. Ask students to chose one animal in each order and tell why it is classified in that order.

Using the Features/Observing Chimpanzees

This feature illustrates the importance of observations in scientific research. Through careful and persistent observation, Jane Goodall has learned much about wild chimpanzees, our closest living animal relative. Point out to students that observations contribute to the general knowledge from which all scientists can devise hypotheses.

Section Review Answers

(Continued from p. 613.)

5. A small litter is adaptive because the extra weight from each additional fetus would make flying more difficult for the pregnant mother. It is also easier to carry only one newborn offspring while in flight.

Laboratory Preparation

Background

In this diverse lab, students observe certain mammalian characteristics in humans, as well as other mammals. A wide variety of materials can be used to demonstrate specific features; the materials will depend on individual teaching style and availability. Students often believe that male and female animals can be distinguished by the presence of mammary glands; fetal pigs serve as good examples of the fact that sex cannot be determined on this basis. In similar fashion, a wide range of materials stimulate thought.

Preparation

Parts of the investigation can still be pursued even though the suggested materials may not be available, but they may require additional explanation.

Conclusions

1. The characteristics include mammary glands, hair, sweat glands, variable teeth, a hard palate, middle ear bones, movable facial muscles, and flexible vertebrae below the rib cage.

2. Humans seem to have less hair because their hair is relatively short, thin in diameter, and nearly colorless.

Inquiry

Answers will vary. Students may hypothesize that since other animals can move their facial muscles, the ability probably has an adaptive advantage that may well be communicative, as it is in humans.

Chapter Review

Vocabulary

1. Incisors are for biting and cutting; canines are pointed teeth for gripping, puncturing, and tearing; bicuspids are teeth with two points, for shreading and tearing; molars have flattened surfaces and are for grinding.
2. The rumen is an extra chamber in the stomach; the cecum is a blind sac at the end of the small intestine. Both harbor bacteria that aid in the digestion of plant material.
3. The echo of the sound waves helps the animal locate and identify different objects.
4. Oviparous mammals (monotremes) lay eggs, which hatch outside the body; viviparous mammals carry the young not in eggs, but inside the uterus, and they bear live young.
5. *Omni-* means "all"; *-vore* comes from a Latin word meaning "devour"; an omnivore eats everything, both plants and meat.

Review

11. Adaptations for endothermy are body hair, the diaphragm, and the four-chambered heart. These features help mammals maintain a constant body temperature.
12. The trait of having specialized teeth has enabled mammals to take advantage of diverse food sources available in a wide range of environments.
13. Most mammals are viviparous so they can protect their unborn young at all times; all newborn mammals are dependent on their mother for nourishment (milk from mammary glands). During their time of dependency the young learn survival behavior. Because they can learn, many mammals have flexible behavior patterns that allow them to adapt to changes in the environment.
14. because of the adaptive radiation of mammals that occurred after the extinction of the dinosaurs
15. All mammals have hair and produce milk to feed their young.
16. They both lay eggs, are not fully endothermic, and have a cloaca into which material from the large intestine, urinary bladder, and reproductive organs all pass.
17. Marsupials are the predominant mammals in Australia because no placental mammal inhabited the continent when it split away from the other continents about 60 million years ago. The Australian marsupials underwent adaptive radiation. On other continents placentals were able to outcompete the marsupials for resources, and the placentals underwent adaptive radiation.
18. All placental mammals carry their unborn young in the uterus, nourished by the placenta. Oxygen and nutrients pass from the mother's blood into the blood of the unborn young via the placenta. The young are born live and are able to survive outside the mother's body.
19. chimpanzee communication, eating habits, methods of hunting, and tool use
20. Lagomorphs have a double row of upper incisors, while rodents have only two incisors in each jaw.

Critical Thinking

1. Answers will vary, but students should mention one or more of the following disadvantages: adults must watch over and protect the immature young; they must gather food for the young; the reproductive rate is lower because of the necessary care for the young. Accept other answers students can explain logically.
2. Accept any of the following as reasonable expectations: students might expect kangaroos to have a chambered stomach or a blind sac off the small intestine that harbors bacteria, which aid in the digestion of cellulose; they might expect kangaroos to have large, flat molars for grinding tough grasses; they might expect kangaroos to use speed as a defense against predators. Accept other logical answers.
3. Because mice have such a short gestation period, they can reproduce rapidly; therefore, scientists can study many generations of mice in a short period of time.
4. The algae help camouflage the sloth as it hangs upside down in a tree; by colonizing the sloth's fur the algae have colonized a place that receives sunlight and is not likely to be disturbed by certain animals that feed on algae.
5. Both have a long snout, a sticky tongue, and strong claws for breaking into insect nests.

Chapter 40 | Mammals

Introduction

At the end of the Mesozoic era the ruling reptiles, the dinosaurs, became extinct. On the other hand, mammals, which had arisen at about the same time as dinosaurs, survived and adapted to the changing environment. By studying the evolution of mammals and examining their characteristics, you will learn how mammals survived great changes, diversified, and colonized varied habitats.

Chapter Outline

Chapter Concept

As you read, look for specific adaptations that have made each order of mammals a successful group.

Sperm whale, *Physeter catadon.* Below: Giraffes, *Giraffa camelopardalis.*

40.1 Evolution and Characteristics

Mammals belong to the Class Mammalia, which includes about 4,000 species. They are the dominant land vertebrates on earth. How did mammals evolve? What led to their success?

Origin

Fossil evidence and evidence from comparative anatomy indicate that mammals evolved from a group of reptiles called **therapsids** *(thuh-RAP-sidz)*. The therapsids arose about 280 million years ago, near the end of the Paleozoic era. Therapsids had both reptilian and mammalian characteristics. Like reptiles, therapsids had a jaw composed of five bones rather than of the single jawbone characteristic of mammals. However, unlike reptiles, the quickest of which ran on two legs, therapsids had mammalian skeletal adaptations that allowed them to be efficient four-footed runners. Unlike reptiles, which you have learned are ectothermic, therapsids were probably endothermic. Like mammals, therapsids had specialized teeth adapted for specialized functions.

The oldest mammalian fossil yet found is from the early Mesozoic era, about 200 million years ago. The jaw, a single bone, is the most obvious mammalian feature. Scientists infer that the mammal was only 10 cm long. Its skeleton was similar to that of a shrew. It had teeth adapted for eating insects. It was probably endothermic, had skin, and was covered with hair.

Evolution

How did early mammals survive competition with the dinosaurs during the Mesozoic era? Fossil skeletons show that early mammals had large eye sockets, evidence that these creatures may have been active at night. By hiding during the day, mammals would have avoided predation by dinosaurs. By feeding on insects, mammals would not have competed with dinosaurs for food.

Though mammals arose near the beginning of the Mesozoic, they were not abundant. Nevertheless, by the end of the Jurassic period, at least five orders of mammals had arisen. Fossils of these mammals are scarce, however, indicating that mammals probably were not as abundant then as they are today.

About 65 million years ago, at the end of the Cretaceous, the dinosaurs became extinct. This opened up new habitats that mammals could exploit without facing competition. Mammals became dominant by the beginning of the Paleocene epoch. By the end of the Eocene epoch, most modern orders of mammals had evolved. The Cenozoic era is named the Age of Mammals because of the emergence and dominance of this vertebrate class.

Section Objectives

- *Summarize the significance of the therapsids in mammalian evolution.*

- *Give a probable reason why mammals survived competition with dinosaurs.*

- *List some characteristics of mammals, and explain how these characteristics enabled mammals to adapt to different environments.*

Figure 40-1   The therapsid, which lived 280 million years ago, was a reptile that also had mammallike characteristics.

Figure 40-2   *Eozostrodon* was a mammal about 10 cm long that lived during the late Triassic period.

For additional teaching information see pp. T40A–T40D.

Mammals   **603**

Tip: Refer students to the geologic time chart on page 221 in Chapter 15.

Info: Viviparity allows mammals to be very efficient at gathering food because they don't have to stay in one place to protect their unborn young. Thus mammals have a further advantage in adapting to a wide range of environments and exploiting diverse food sources.

Figure 40-3 Teeth of various shapes enable mammals to grip, tear, shred, and grind food. Thus they can eat a wide variety of foods.

Characteristics

Mammals have certain characteristics that led to their widespread success in adapting to different environments and gave them survival advantages. Study the list of common traits:

- Mammals are endothermic.
- Mammals have hair.
- The mammalian heart has four chambers.
- Mammals have a muscle, the diaphragm, that aids breathing.
- Mammals have a single lower jawbone, and most species have four different types of teeth.
- Most species are viviparous—females carry their young in the uterus during development and then give birth to live young.
- Females secrete milk from mammary glands to feed newborn young.
- Many mammals have a highly developed brain.

Endothermy is an adaptation that has allowed mammals to be active at any time of the day or night and to colonize many severe environments, from deserts to ice fields. Several mammalian characteristics are adaptations for endothermy. Hair serves as insulation for most mammals. Variations in kinds and amounts of hair reflect adaptations to different environments.

The four-chambered heart and diaphragm are also adaptations for endothermy. Endothermic animals can regulate body temperature in any kind of environment. The **diaphragm** is a sheet of muscle below the rib cage that aids in efficient breathing. The four-chambered heart pumps blood efficiently. These adaptations result in a high metabolic rate, which helps mammals maintain a constant body temperature.

Most mammals have specialized types of teeth, as shown in Figure 40-3. Such teeth allow mammals to exploit a wide variety of food sources. Chisellike **incisors** bite and cut. Pointed **canines** grip, puncture, and tear. **Bicuspids,** teeth with two points, shear and shred. Flattened teeth called **molars** grind and crush. Variations in size and shape of teeth among different mammalian species reflect differences in diet. Different types of teeth are adaptations that result in mammals taking advantage of diverse food sources available in a wide range of environments.

Most mammal species are **viviparous**—that is, females carry and nourish the young inside their bodies for a period of time before giving birth to live young. Viviparous mammals can move around without leaving their unborn unprotected.

Newborn mammals are nourished with milk from the mother's milk-secreting glands, or **mammary glands.** Thus young mammals are dependent on the mother for food. In some species the young remain dependent for a long time, during which they mature and learn the behaviors necessary for survival. Such species have well-developed brains. Many also have flexible behavior patterns that allow them to adapt to changes in the environment.

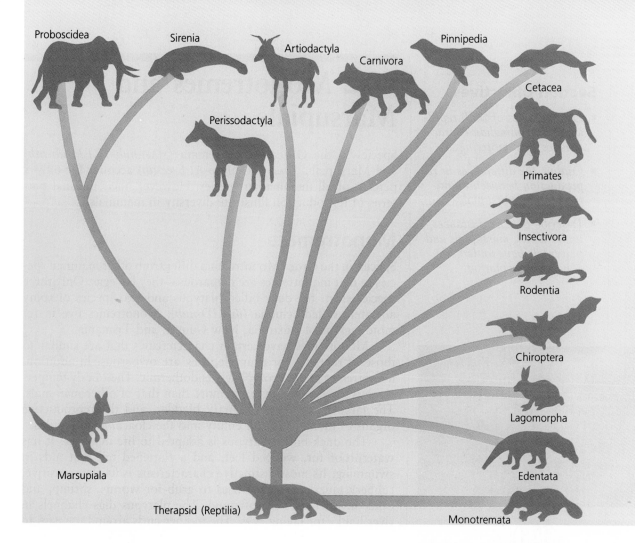

Figure 40-4 shows a phylogenetic tree with the following labels: Proboscidea, Sirenia, Artiodactyla, Carnivora, Pinnipedia, Cetacea, Perissodactyla, Primates, Insectivora, Rodentia, Chiroptera, Lagomorpha, Marsupiala, Edentata, Therapsid (Reptilia), Monotremata.

Classification

Mammals share two features that distinguish them from other vertebrates: all have hair, and all produce milk to feed their young. There are 19 orders of modern mammals in the Class Mammalia. Of these, 17 are placental mammals—mammals that nourish unborn young via the placenta. The two remaining orders are Monotremata—egg-laying mammals—and Marsupialia—pouched mammals. As you read Sections 40.2 and 40.3 study the inferred phylogenetic relationships shown in Figure 40-4.

Section Review

1. What type of animal was the therapsid?
2. Describe the oldest mammalian fossil.
3. How did mammals survive competition with the dinosaurs?
4. How have endothermy, specialized teeth, and viviparity aided mammals in adapting to varied environments?
5. What pattern of evolution did the mammals exhibit at the beginning of the Cenozoic era? Explain your answer.

Figure 40-4   The diversity of mammals is a result of evolution since the beginning of the Mesozoic. This phylogenetic tree shows the relationships of the orders discussed in this chapter.

Answers to Section Review
1. a reptile with certain mammalian characteristics
2. It was about 10 cm long, had a jawbone composed of one bone, and had teeth adapted for an insect diet.
3. Mammals avoided competition with dinosaurs by being active at night and by feeding on insects instead of plants or small dinosaurs and other animals.
(See pp. T40A–T40D.)

Section Objectives

- List examples of both reptilian and mammalian characteristics in monotremes.

- Outline the differences in reproduction between monotremes and marsupials.

- Describe the circumstances under which marsupials and placentals have undergone convergent evolution.

Figure 40-5 Both the platypus, *Ornithorhynchus anatinus* (top), and the echidna, *Tachyglossus aculeatus* (bottom), use their snouts to reach food.

Tip: Discuss with students the types of geologic evidence that support the idea of continental drift: geomagnetism, sea-floor spreading, continuities in rock strata across continents, and evidence from biogeography.

40.2 Monotremes and Marsupials

Species in the Orders Monotremata (MAH-*nuh*-TREM-*uh-tuh*) and Marsupialia (*mahr*-SOOP-*ee*-ALE-*ee-uh*) account for only 5 percent of all mammalian species. However, their unusual patterns of reproduction illustrate diversity in mammals.

Monotremata

Species in the Order Monotremata differ from other mammal species in that monotremes are oviparous—they lay eggs. Only three species exist: the duck-billed platypus and two species of spiny anteaters called echidna (*ih*-KID-*nuh*). Monotremes live in remote regions of Australia, New Guinea, and Tasmania.

Monotremes have certain characteristics that are similar to those of reptiles. For example, they are oviparous. In addition, monotremes are not completely endothermic. Their body temperature is lower and fluctuates more than that of other mammals. The large intestine, the urinary bladder, and the reproductive organs of monotremes all empty into the cloaca, as in reptiles.

The duck-billed platypus is adapted to life in water. It has waterproof fur, webbed feet, and a flattened tail that aids in swimming. Its most distinctive characteristic is its flat, sensitive, rubbery muzzle, which it uses to grub for worms, shrimp, and other invertebrates in soft mud. The platypus digs channels in river banks for laying eggs. The mother curls around the eggs to protect and incubate them. After hatching, the young lick milk from nippleless glands on the mother's abdomen.

Echidnas, commonly called "spiny anteaters," have a coat of protective spines and a long snout suited for probing into anthills. Echidnas incubate their eggs in a brood pouch on the belly.

Marsupialia

Species in the Order Marsupialia give birth to tiny, immature young that crawl to a pouch on the mother's belly immediately after they are born. There they attach themselves to milk-secreting nipples, nursing until they are mature enough to survive outside the pouch. About 250 marsupial species exist, inhabiting Australia, New Guinea, Tasmania, and the Americas. North America has only one marsupial species, the common opossum. Most marsupials live in Australia, where they are the dominant mammals.

Why are marsupials the dominant mammals in Australia, whereas placental mammals are the dominant mammals elsewhere? Geologic evidence indicates that 200 million years ago the continents were joined as a single landmass. In time the landmass split up, and the continents drifted apart. When Australia broke

Marsupials   **Placentals**

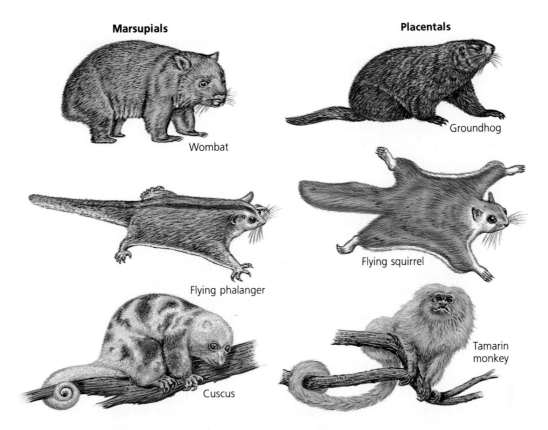

Wombat

Groundhog

Flying phalanger

Flying squirrel

Cuscus

Tamarin monkey

off from the larger landmass about 60 million years ago, no placental mammals inhabited that continent. Lacking competition, Australian marsupials underwent adaptive radiation and eventually became adapted to all the environments in Australia. On the other continents placental mammals outcompeted the marsupials for resources, and most marsupial species became extinct.

The marsupial species of Australia resemble familiar placental species. The wombat resembles a groundhog. The glider resembles a flying squirrel, and the cuscus, a monkey. These pairs of animals have adapted to similar environments, use similar food sources, and are similar in body structure and behavior. However, the marsupial and placental pairs are very distantly related. Therefore their similarities are a result of convergent evolution, adaptation to similar environments of different, isolated continents.

Section Review

1. What are some distinguishing features of the echidna?
2. What traits do monotremes share with reptiles? with mammals?
3. What makes marsupial reproduction unique?
4. Why do scientists say that placentals and Australian marsupials have undergone convergent evolution?
5. The pouch of the marsupial mole opens not toward the front of the body but toward the rear. How would such a pouch aid in the survival of this burrowing animal's young?

Figure 40-6 These paired animals demonstrate the convergent evolution of marsupials and placentals. Although very similar in many ways, they are only distantly related.

Info: Most but not all marsupials have pouches. In pouchless marsupials the young attach to nipples on the belly, where they remain for weeks or months without the protection of a pouch.

Answers to Section Review
1. The echidna is a monotreme; it lays eggs, has protective spines in its coat, and has a long snout for probing anthills.
2. Like reptiles, monotremes lay eggs, are not totally endothermic, and have a cloaca into which the large intestine, urinary bladder, and reproductive organs empty. Like mammals, monotremes have hair and feed their young with milk secreted from glands.
(See pp. T40A–T40D.)

Section Objectives

- *Describe the reproductive features that distinguish placental mammals from monotremes and marsupials.*

- *Name some orders of placental mammals, and give a representative from each.*

- *List some adaptive features of placental mammals for locomotion and feeding.*

- *Summarize Jane Goodall's contributions to the study of wild chimpanzees.*

Info: Among mammals the gestation period of larger species generally is longer than that of smaller species. For example, the gestation period for rabbits is about 1 month; for elephants, over 20 months.

Tip: Suggest to students that they notice how the teeth of placental mammals are modified to perform specialized functions.

Insectivora

40.3 Placental Mammals

Monotremes and marsupials are members of orders that are the oddities of the mammal world. Ninety-five percent of all mammal species are placental mammals. You will now study the characteristics of the major orders of placental mammals.

Characteristics

Placental mammals carry unborn young in the uterus until the young can survive in the outside world. In the uterus the young are nourished by the **placenta** *(pluh-SEN-tuh)*, an organ formed from the chorion and allantois. Oxygen and nutrients are transferred from the mother's blood, through the placenta, to the blood of the unborn mammal. The period of time during which mammals develop in the mother's uterus is the **gestation** *(jeh-STAY-shun)* **period.** Gestation periods vary in length from species to species.

Classification

Placental mammals are a diverse group, living on land, in water, and in the air. As you read, notice the adaptations that make members of each order fit their environments. Notice the adaptations for various forms of locomotion and the dietary adaptations, especially teeth, summarized in Figure 40-7. Such adaptations have contributed to the success of placental mammals.

Insectivora

The Order Insectivora *(IN-SECK-TIV-ur-uh)* consists of about 400 species, mainly shrews and moles. Insectivores are usually small animals with a high metabolic rate and are found in North America, Europe, and Asia. Most have long, pointed noses that enable them to grub for insects, worms, and other invertebrates. They have teeth adapted for picking up and piercing the prey.

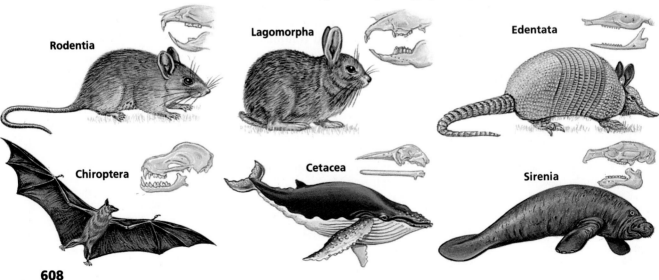

Rodentia

Lagomorpha

Edentata

Chiroptera

Cetacea

Sirenia

Insectivores are adapted to life on the ground, in trees, in water, and under ground. Shrews, for example, feed on the ground, sweeping invertebrates into the mouth with clawed paws. Moles, on the other hand, burrow under ground. They have stout limbs for digging, reduced eyes, and no external ears.

Rodentia

Rodentia *(roe-DEN-shuh)* is the largest mammalian order, having over 2,400 species. Rodents flourish on every continent except Antarctica and are adapted to a wide range of habitats. Squirrels, marmots, chipmunks, gophers, muskrats, mice, rats, and porcupines are all rodents. Guinea pigs and capybaras are South American rodents. While most mammals have four incisors in each jaw, rodents have only two. These teeth continue to grow as long as the rodent lives. Rodent incisors are sharp, an adaptation for the rodent's diet of hard seeds, twigs, roots, and bark. As it gnaws, the back surface of the tooth wears away faster than the front surface, maintaining the tooth's chisellike edge.

Lagomorpha

About 70 species make up the Order Lagomorpha *(LAG-uh-MORF-uh)*, which includes rabbits, hares, and small mountain mammals called pikas. Lagomorphs are found worldwide. They differ from rodents in that they have a double row of upper incisors, with two large front teeth backed by two smaller ones. Lagomorph teeth continue to grow throughout the animal's lifetime. Such teeth are an adaptation to the lagomorph's herbivorous diet.

Edentata

The Order Edentata *(EE-DEN-TAHT-uh)* is made up of about 30 living species, including anteaters, armadillos, and sloths. These mammals are found in southern North America, Central America, and South America. The name edentate means "without teeth." However, only anteaters are completely toothless; armadillos and sloths have teeth, though their teeth are peglike and lack enamel.

Think

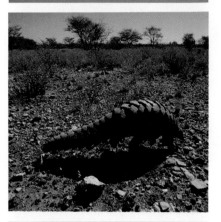

The animal in the picture is a placental mammal called the pangolin. The seven species of pangolins make up the Order Pholidota. Pangolins eat ants and termites. Like the anteater, an edentate, pangolins lack teeth. What pattern of evolution is exhibited by the pangolins and the anteaters? Explain your answer.

Figure 40-7   Examine the adaptations for locomotion and eating among the placental mammals.

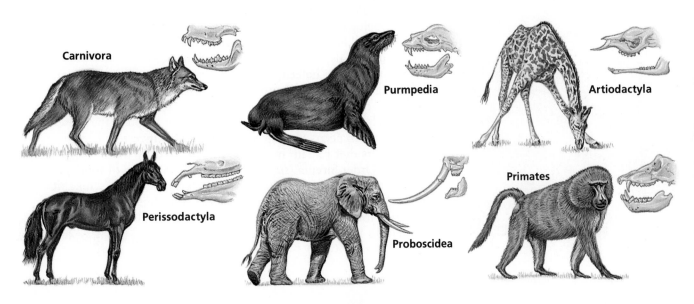

Carnivora

Purmpedia

Artiodactyla

Perissodactyla

Proboscidea

Primates

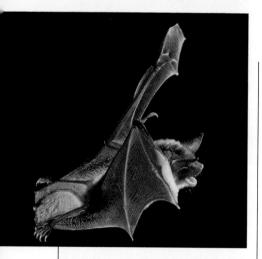

Figure 40-8 The bony wing structure of this mouse-eared bat, *Myotis myotis,* is evident as the bat swoops in on a cricket.

Figure 40-9 Instead of teeth, the upper jaw of a baleen whale contains two rows of a firm material called baleen. It holds back food as water that the whale has gulped is forced out.

Edentates have adaptations for an insectivorous diet, including a long, sticky tongue and clawed front paws. Some edentates have adaptations for other diets. Armadillos supplement their insect diet with small reptiles, frogs, mollusks, and dead animal matter. Sloths, on the other hand, are herbivores; their continuously growing teeth are an adaptation for grinding plants.

Chiroptera

The Order Chiroptera *(kie-RAHP-tuh-ruh)* is made up of over 900 species of bats. Bats are the only mammals that are true flyers. They live throughout the world except in polar environments. A bat's wing is a modified front limb in which a skin membrane stretches between extremely long finger bones to the hind limb. The bat's clawed thumb sticks out from the top edge of the wing. Bats use their thumbs for walking, climbing, or grasping.

Most bats are active at night and have a specialized way of navigating and finding food called echolocation. Bats that use **echolocation** emit high-frequency sound waves, which bounce off objects. The frequency of the returning sound waves helps the bat identify the size, distance, and rate of movement of different objects. Bats that use echolocation have small eyes and large ears. They generally feed on insects and have teeth specialized for such a diet. Some tropical bats feed on fruit or flower nectar and do not use echolocation. These bats, sometimes called "flying foxes," have large eyes and a keen sense of smell.

Cetacea and Sirenia

The Order Cetacea *(sih-TAY-shuh)* includes 90 species of whales, dolphins, and porpoises, distributed worldwide. Most cetaceans live in the oceans, but some dolphin species live in freshwater lakes and rivers. Cetaceans have fishlike bodies with forelimbs modified as flippers. They lack hind limbs, but have broad, flat tails that help propel them through the water. Cetaceans breathe through a blowhole at the top of the head.

Cetaceans are divided into two groups: toothed whales and baleen whales. Toothed whales include beaked whales, sperm whales, beluga whales, narwhals, killer whales, dolphins, and porpoises. All have from one to over a hundred teeth and prey on fish, squid, seals, and other whales. Baleen whales lack teeth. Instead they have **baleen,** thin plates of fingernaillike material, that hang down from the roof of the mouth. As the baleen whale swims, it opens its mouth and gulps huge quantities of water. Then it closes its mouth and uses its tongue to push the water through the baleen. Shrimp and other small invertebrates that make up its diet remain in the mouth, strained by the baleen.

The Order Sirenia *(sie-REEN-ee-uh)* is made up of four species of manatees and dugongs. These large herbivores inhabit tropical seas, estuaries, and rivers. Their front limbs are flippers, modified for swimming. Like cetaceans, sirenians lack hind limbs but have a flattened tail for propulsion.

Carnivora and Pinnipedia

The 250 living species of the Order Carnivora *(kahr-NIV-ur-uh)* are distributed worldwide. Dogs, cats, raccoons, bears, hyenas, and otters are some well-known carnivores. Most of the species mainly eat meat, which explains the name of the order. However, some members of this order, such as bears, feed extensively on plant material as well as meat, so they are called **omnivores** *(AHM-nih-VORZ).*

Carnivores generally have long canine teeth, strong jaws, and clawed toes—for seizing and holding prey. Most have keen senses of sight and smell, which aid in hunting. Skeletal adaptations enable terrestrial carnivores to run quickly.

Members of the Order Pinnipedia *(PIN-uh-PEED-ee-uh)* are water-dwelling carnivores with streamlined bodies for efficient swimming. About 34 species of sea lions, seals, and walruses compose this order.

Artiodactyla and Perissodactyla

Ungulates *(UN-gyuh-luts),* or hoofed mammals, are classified into one of two orders: Artiodactyla *(AHRT-ee-oe-DACK-tuh-luh)* or Perissodactyla *(puh-RISS-uh-DACK-tuh-luh).* Ungulates with an even number of toes make up the Order Artiodactyla. Deer, elk, bison, moose, cows, sheep, goats, pigs, and camels are all artiodactyls. About 150 species belong to the order. Artiodactyls live on every continent except Antarctica. Artiodactyls differ from other ungulates in that they have cloven, or split, hooves. Most are fast runners and use speed as their major defense.

Ungulates with an odd number of toes make up the Order Perissodactyla, which includes about 15 living species. Horses, zebras, rhinoceroses, and tapirs are all perissodactyls. Most species are native to Africa and Asia. However, some tapir species live in Central and South America.

Artiodactyls and perissodactyls are herbivores. Their molars tend to be large and flat, for grinding plant material. Artiodactyls have a storage chamber in the stomach called the **rumen.** Chewed and swallowed plant material is stored in the rumen, where bacteria break down cellulose. This partially digested material, called a cud, is chewed again, swallowed, and passed through the digestive system a second time. This double digestion allows artiodactyls to derive nutrients efficiently from tough grasses.

Figure 40-10 The lion, *Felis leo* (left), brown bear, *Ursus arctos* (center), and elephant seal, *Mirounga angustirostris* (right), are mainly meat eaters. Note the long canine teeth of the lion, a characteristic of most carnivores.

Artiodactyl

Cloven hoof

Perissodactyl

Three-toed foot

Figure 40-11 Split hooves with an even number of toes are characteristic of artiodactyls, including deer, *Odocoileus* sp. An odd number of toes is characteristic of perissodactyls, such as the rhinoceros, *Diceros bicornis.*

Answer to Biology in Process: Goodall made long-term observations of wild chimpanzees, carefully recorded data on the chimpanzee population, and communicated with other scientists about her research.

Perissodactyls also have adaptations that help them digest tough plant material. Members of this order have a large, convoluted **cecum** *(SEE-kum)*—a blind sac near the end of the small intestine. Like the rumen of the artiodactyl, the cecum of the perissodactyl contains bacteria that break down cellulose, releasing nutrients that the perissodactyl can digest.

Proboscidea

Members of the Order Proboscidea *(PROE-buh-SID-ee-uh)* are characterized by a boneless trunked nose or proboscis. Only two species of this order exist today—Asian and African elephants. However, species of this order were once quite common. Until 10,000 years ago woolly mammoths roamed many parts of North America.

Elephants are the largest land dwellers alive today. Weighing more than 6 tons, the elephant feeds on plants for 18 hours a day to sustain itself. The elephant uses its proboscis to gather leaves from high branches and to suck up water without lowering its heavy body. It has modified incisors, called tusks, for digging up roots and for stripping bark from branches. It has jagged molars, up to 30 cm long, for grinding plant material.

Figure 40-12 Elephants are characterized by the presence of a proboscis.

Biology in Process

Observing Chimpanzees

Evidence from comparative morphology and biochemistry reveals that chimpanzees are our closest living relatives in the animal world. Scientists have found that the DNA of humans and the DNA of chimpanzees are 99 percent equivalent. Biochemical studies have also shown that certain human and chimpanzee proteins, such as cytochrome c, are also 99 percent similar.

Despite our close phylogenetic relationship with chimpanzees, for a long time we knew very little about these primates. Until the 1960s, chimpanzees, which live in the steep mountain forests of Africa, had rarely been scientifically observed in the wild.

In 1960 a young British woman named Jane Goodall moved to East Africa to study wild chimpanzees at the Gombe Stream Reserve in Tanzania. Each day she would hike up the mountains in hopes of observing the chimpanzees. Some days the chimpanzees eluded her completely, and she returned to her camp without having made a single observation. In those early years, even when Goodall was able to sight chimpanzees, they usually ran

Primates

In Chapter 17 you studied characteristics of the Order Primates. The 200 species of living primates are classified as prosimians *(proe-SIM-ee-unz)*, including lemurs, tarsiers, and lorises; or anthropoids, including monkeys, apes, and humans. Most primates are omnivores and have teeth specialized for a varied diet. In anthropoids the ratio of brain size to body size is especially large. A complex brain has enabled anthropoids to develop complex behaviors and to live in highly organized social groups.

Section Review

1. What is a *placenta?*
2. List any five orders of placental mammals, and name a representative from each order.
3. What are some dietary adaptations of whales?
4. Distinguish between artiodactyls and perissodactyls.
5. Unlike most mammals of their size, bats usually give birth to only one young at a time. How might small litter size be an adaptive trait in these flying mammals?

Answers to Section Review
1. a membrane that transfers oxygen and nutrients from the mother's blood to the blood of the unborn mammal
2. Answers will vary. Accept any five of the orders discussed in Section 40.3, and accept any representative animal so identified.
3. Toothed whales have teeth that allow them to prey on fish, squid, seals, and other whales. Baleen whales have thin plates of fingernaillike material that strain invertebrates from the water the whale takes in.
4. Artiodactyls have an even number of toes and a rumen, an extra chamber in the stomach that aids digestion of plant material. Perissodactyls have an odd number of toes and a cecum, a blind sac at the end of the small intestine that aids digestion of plant material.
(See pp. T40A–T40D.)

away from her or made threatening gestures if she attempted to get close to them.

Despite frustration and hardship Goodall persisted with her work. In time the chimpanzees allowed her to view them from a distance. However, four years passed before the chimpanzees allowed her to make abundant, close-range observations.

For about 30 years Goodall has observed the chimpanzees of Gombe. She has learned that chimpanzees are highly intelligent and intensely social animals. Like humans, each chimpanzee has a distinct personality. Chimpanzees communicate through gestures, postures, facial expressions, and vocalizations.

Although they mainly eat fruit, chimpanzees sometimes hunt for small animals, such as infant baboons, in an organized, cooperative way. They also make and use simple tools, stripping twigs, which they insert into termite nests. They pull out the insect-covered twig and eat the insects.

With these and other observations, Goodall has given us a clearer picture of the life of the chimpanzees. By communicating about her research, she has inspired others to make detailed studies of chimpanzees and other apes in their natural habitats. These studies are helping us gain a better understanding of primates, how they live and how they might have evolved.

■ What scientific processes are involved in Jane Goodall's research of wild chimpanzees? Explain.

Writing About Biology

Gorillas in the Mist

The following account comes from Gorillas in the Mist *by Dian Fossey.*

On my first visit to Kabara in 1963 I was fortunate in meeting Joan and Alan Root, photographers from Kenya who were camped at the meadow while working on a photographic documentary of the mountain gorillas. Both Joan and Alan kindly overlooked the intrusion of a somewhat hobbly and inquisitive American tourist into their secluded mountain workshop and allowed me to accompany them on some of their extraordinary contacts with the . . . gorillas of Kabara. It was only because of their generosity, coupled with the skill of Sanwekwe, a Congolese park guard and tracker, that I was able to contact and photograph the animals during that brief visit. . . .

I shall never forget my first encounter with gorillas. Sound preceded sight. Odor preceded sound in the form of an overwhelming musky-barnyard, humanlike scent. The air was suddenly rent by a high-pitched series of screams followed by the rhythmic rondo of sharp *pok-pok* chestbeats from a great silverback male obscured behind what seemed an impenetrable wall of vegetation. Joan and Alan Root, some ten yards ahead on the forest trail, motioned me to remain still. The three of us froze until the echoes of the screams and chestbeats faded. Only then did we slowly creep forward under the cover of dense shrubbery to about fifty feet from the group. Peeking through the vegetation, we could distinguish an equally curious phalanx of black, leather-countenanced, furry-headed primates peering back at us. Their bright eyes darted nervously from under heavy brows as though trying to identify us as familiar friends or possible foes. Immediately I was struck by the physical magnificence of the huge jet-black bodies blended against the green palette wash of the thick forest foliage.

Most of the females had fled with their infants to the rear of the group, leaving the silverback leader and some younger males in the foreground, standing tense with compressed lips. Occasionally the dominant male would rise to chestbeat in an attempt to intimidate us. The sound reverberated throughout the forest and evoked similar displays, though of lesser magnitude, from gorillas clustered around him. Slowly, Alan set up his movie camera and proceeded to film. The openness of his motions and the sound of the camera piqued curiosity from other group members, who then tried to see us more clearly. As if competing for attention, some animals went through a series of actions that included yawning, symbolic-feeding, branch-breaking, or chestbeating. After each display, the gorillas would look at us quizzically as if trying to determine the effect of their show. It was their individuality combined with the shyness of their behavior that remained the most captivating impression of this first encounter with the greatest of the great apes. I left Kabara with reluctance but with never a doubt that I would, somehow, return to learn more about the gorillas of the misted mountains.

Write

- Write an informative essay based on Dian Fossey's observations of gorilla behavior.

 From Gorillas in the Mist *by Dian Fossey. Copyright © 1983 by Dian Fossey. Reprinted by permission of Houghton Mifflin Company.*

Laboratory

Mammalian Characteristics

Objective
To examine certain distinguishing characteristics of mammals

Process Skills
Observing, inferring, hypothesizing

Materials
Photographs of male and female fetal pigs, hand lens, microscope slide of mammalian skin, labeled picture of a mammalian internal ear

Method
1 Mammals get their name from their mammary glands. Observe pictures of male and female fetal pigs. Are mammary glands present in both the male and the female pigs? Can you differentiate between the sexes of the pigs on the basis of mammary glands?
2 Use a hand lens to examine several areas of your skin that appear to be hairless. Are these areas actually hairless?
3 Compare the growth patterns of a hair on your head and a hair on your arm. Are the growth patterns different? Which pattern do you think applies to most other mammals? What is the main function of hair?
4 Examine a slide of mammalian skin under a microscope set at low power. What kind of glands in mammals enable them to cool off? Locate and observe these glands on the slide.
5 The four kinds of mammalian teeth are found in the human mouth. What are the four kinds of teeth? How many of each do you have? It is customary to count only those on one side of the mouth.
6 Breathe with your mouth closed. Do you feel a flow of air into your mouth? You do not feel a flow of air because, like other mammals, you have a hard palate (the roof of the mouth) which separates the mouth from the nasal cavities.
7 Try to move your ears. Why is this ability more important for certain other mammals than it is for humans?
8 Can you move the skin anywhere else on your head? Where? What is the function of these movements in humans?
9 Examine a picture of the internal ear. Locate the three bones in the middle ear. These bones are found only in mammals.
10 Can you turn your chest without moving the rest of your body? What structures make this possible? Flexible vertebrae below the rib cage are a mammalian characteristic.

Conclusions
1 Name five characteristics that distinguish mammals from other vertebrates.
2 Can you explain why humans seem to have much less hair than either the gorilla or chimpanzee?

Inquiry
Other mammals can move their facial muscles. Do you think that other animals use this ability to communicate? Form a hypothesis regarding this question, and determine how to defend it.

Chapter 40 Review

Vocabulary

baleen (610)
bicuspid (604)
canine (604)
cecum (612)

diaphragm (604)
echolocation (610)
gestation period (608)
incisor (604)

mammary gland (604)
molar (604)
omnivore (611)
placenta (608)

rumen (611)
therapsid (603)
ungulate (611)
viviparous (604)

1. List and describe the four types of teeth.
2. Contrast the rumen of the artiodactyl with the cecum of the perissodactyl.
3. Why is echolocation a good name for the type of behavior it describes?
4. Contrast the reproductive behavior of oviparous and viviparous mammals.
5. Find the meaning of the word parts that compose the word *omnivore*. Explain why omnivores are appropriately named.

Review

1. Scientific evidence indicates that therapsids had characteristics of both mammals and (a) amphibians (b) birds (c) reptiles (d) vertebrates.
2. Evidence that early mammals were active at night includes the fossil mammal's (a) single jawbone (b) large eye sockets (c) four-chambered heart (d) mammary glands.
3. After the dinosaurs became extinct, mammal species flourished because (a) many new habitats were suddenly available to them (b) the climate became warmer and wetter (c) the reptiles had undergone adaptive radiation (d) the reptiles had become endothermic.
4. The diaphragm allows mammals to (a) carry the young inside the uterus (b) pump blood efficiently (c) provide nourishment for their young (d) breathe efficiently.
5. Two types of monotremes include the echidna and (a) the edentate (b) the opossum (c) the manatee (d) the duck-billed platypus.
6. Marsupials differ from monotremes in that marsupials (a) lay eggs (b) carry developing young in a pouch (c) nourish unborn young via the placenta (d) nourish their newborns with milk from nippleless glands.
7. Most mammals are (a) monotremes (b) marsupials (c) placentals (d) lagomorphs.
8. The Order Chiroptera is made up of (a) toothless mammals (b) marine mammals (c) flying mammals (d) arboreal mammals.
9. Most members of the Order Carnivora eat (a) meat (b) plants (c) fruits and nuts (c) insects.
10. Artiodactyls and perissodactyls are both (a) carnivores (b) herbivores (c) omnivores (d) insectivores.
11. What are some mammalian adaptations for endothermy?
12. How has the trait of specialized teeth allowed mammals to adapt to a wide range of environments?
13. In what ways do mammals reproduce and raise their young differently from reptiles?
14. Why is the Cenozoic era called the "Age of Mammals"?
15. Name two characteristics common to all mammals.
16. In what ways do monotremes resemble reptiles?
17. Why are marsupials the dominant mammals in Australia and placentals the dominant mammals elsewhere?
18. What features distinguish placental mammals from other types of mammals?

19. What are some behaviors Jane Goodall has observed in her study of the chimpanzees of Gombe Stream Reserve?

20. How do the teeth of rodents differ from those of lagomorphs?

Critical Thinking

1. Some mammal species must care for their young for many years until the young reach maturity and can survive on their own. Section 40.1 indicates how this pattern of rearing young has advantages over the pattern by which reptiles rear their young. List some ways the mammalian pattern might be disadvantageous, compared with the reptilian pattern.
2. Kangaroos are grazing animals that feed on the grasses covering the plains of Australia. Notice that kangaroos live in a similar environment and feed on the same type of plants as deer do. What morphological and behavioral similarities might you expect to find between kangaroos and deer? Consider adaptations for diet, locomotion, and defense.
3. The gestation period of mice is relatively brief, about 21 days. Why might this short gestation period make mice an ideal laboratory animal for experiments dealing with mammalian development and heredity?
4. Sloths are arboreal edentates that spend most of their lives hanging upside down from tree branches in the tropical forests

of Central and South America. Most sloths have green algae growing in tiny pits in their hair. What advantage might algae growth in the hair confer to the sloth? What advantage might the green algae gain by colonizing the sloth's fur?

5. The photograph above shows an aardvark, the single species in the mammalian Order Tubulidentata. Like many edentates, aardvarks feed on nesting insects such as termites. Study the picture, and review the characteristics of insectivorous edentates. In what ways do aardvarks and edentates illustrate convergent evolution?

Extension

1. Choose an undomesticated mammal species, and write a report on its classification; habitat; morphology; life cycle; and adaptations for feeding, defense, and reproduction. Discuss its evolution if information is available. You may wish to visit a zoo to observe the mammal. If possible include labeled drawings or photographs of the mammal to illustrate your report.
2. Read the article by George B. Schaller entitled "Secrets of the Wild Panda," in

National Geographic, March 1986, pp. 284–308. What factors contribute to the endangerment of this unique species?

3. Obtain a live mammal, such as a mouse, hamster, or gerbil, from a pet store or your school. Get instructions on the care and feeding of the mammal. Then observe the mammal for several weeks or more if possible. Take notes on its feeding, sleeping, and breeding behavior. Share what you learn with your class.

Chapter 41 | *Evolutionary Trends in Vertebrates*

This chapter explains how vertebrates have evolved into increasingly complex organisms, adapting to nearly every habitat on earth. The behavioral, social, and physical adaptations which have led to their success are discussed.

Correlation Chart

Section Number	Section Objectives	Pages	Features	Section Review
41.1	1. Summarize the probable origin of vertebrates. 2. Describe the major trends in vertebrate evolution. 3. Explain why adaptations that are simplifications do not necessarily indicate that an organism is phylogenetically more primitive.	619–620		3, 4 2 1, 5
41.2	1. Describe the major adaptations in embryo development that help vertebrates to live on land. 2. Explain the way in which adaptations in the skeletal, integumentary, circulatory, and excretory systems made vertebrates better suited to live on land. 3. List some adaptations in vertebrate reproduction that help ensure the survival of the young.	621–626	Think, p. 622	1, 2 3, 5 2, 4, 5
41.3	1. Describe the major trend in vertebrate behavior. 2. Explain the process of imprinting. 3. Describe the apparent adaptive advantage of one example of social behavior.	627–629	Biology Lab, p. 629	4, 5 2 1, 3, 4, 5

Resources*
Print

King, Gillian M., and David R. N. Custance. *Colour Atlas of the Vertebrate Anatomy: An Integrated Text and Dissection Guide.* Palo Alto, Calif.: Blackwell Scientific Publications, 1982.

Romer, Alfred S., and Thomas S. Parson. *The Vertebrate Body,* 6th ed. New York: Saunders College Publishing, 1986.

Audiovisual/Software

Animals With Backbones. Film, 10 min. Coronet.
Survey of the Animal Kingdom: the Vertebrates. 6 filmstrips and cassettes, 30-40 min. each. Teacher's guide. Educational Images, 1986.
*Additional references are listed in *Modern Biology, Teacher's Program Guide.*

Section Teaching Guides

41.1 Origin and Phylogeny

Approach

The primary objective of this section is to familiarize students with a probable pathway of vertebrate evolution. This section builds on the concepts of diversity through adaptation (Chapters 15-16), of classification (Chapter 18), of the evolutionary trend toward increasingly complex organisms. In presenting the text, point out that scientists have reconstructed multiple pathways of vertebrate evolution.

Concept Framework

At this point in the course most students are familiar with theories of evolution and taxonomy. Build on this familiarity to provide a conceptual framework for studying this section by reminding students that this section deals with the evolution of vertebrates. Engage students in a brief discussion of the traits humans share with other vertebrates (such as those discussed in Chapters 36–40). You might ask students to keep these vertebrate traits in mind as they study the material.

Emphasis

Make sure that students understand that adaptations do not necessarily lead to more complex systems.

Field Trip

Visit a museum of natural history to study evolutionary trends in vertebrates. Ask students to pay attention to ancestral forms of still-existent organisms and to keep the following question in mind: Has evolution resulted in simplification of more complex structures?

Section Review Answers

(See p. 620.)

41.2 Development and Morphology

Approach

The main objective of this section is to give students an overview of the evolution of structural and physiological adaptations in vertebrates. It builds on the concepts of diversity through adaptation (Chapters 15–16) and vertebrate diversity (Chapters 36–40). In presenting the text, emphasize the major evolutionary trends in vertebrate development and morphology.

Concept Framework

In Chapters 36–40 students studied characteristics of five classes of vertebrates. Tap this knowledge to establish a conceptual foundation for studying this section by pointing out that this section discusses similar material but with a different emphasis. Then assist students in making comparisons between aquatic and terrestrial vertebrates to emphasize the adaptations that enable vertebrates to live on land. Ask students as they read to note how one characteristic varies among the classes rather than focusing on a class or an organism.

Emphasis

Make certain that students appreciate that a highly developed nervous system allows an organism to respond to stimuli in a variety of ways increasing its chances of survival.

Vertebrate Systems Comparison Charts

Divide the class into eight groups. Assign each group one of the following: skeletal and muscular systems, integumentary systems, circulatory systems, respiratory systems, digestive systems, excretory systems, nervous systems, and reproductive systems. Ask each group to make a chart that briefly describes the features of their assigned system in each of the five vertebrate classes studied in Chapters 36–40. Hang the charts in a comparative wall display.

Section Review Answers

(See p. 626.)

41.3 Behavior

Approach

The main objective of this section is to illustrate that the evolutionary trend in vertebrate behavior is towards increasingly complex, better-integrated responses to stimuli. This section builds on the concepts of invertebrate evolution presented in Chapter 40, and the discussion of vertebrate evolution presented in this chapter. Remind students that the general evolutionary trend is from simple to complex, although some adaptations may lead to a simplification of some structures. You might ask students as they read to consider why complex behaviors are adaptive.

Concept Framework

Students know that most human infants and children recognize and imitate their parents or primary care givers. However, in some animal groups, offspring never see their parents. To use students' knowledge as conceptual starting point for studying this section, engage them in a discussion of animal behavior by asking whether parent-infant behavior is learned or genetic. As students read the text, ask them to keep in mind that learning and heredity both influence animal behavior.

Emphasis

Make certain that students recognize that insight learning is a complex behavior characteristic of anthropoids.

Field Trip

Visit a local zoo to observe different types of vertebrate social behavior. Ask students to watch for visual displays such as those of elephants and macaques, as shown in Figure 41-10.

Section Review Answers

(See p. 628.)

Laboratory Preparation

Background

In this lab students observe how vertebrates have adapted to their various environments, with the focus on adaptations in locomotion. Students learn that forward force is generated by fish when the body and tail are pushed against the water and an opposite force is generated that is strongest in a direction perpendicular to the water, i.e., in the direction from which the fish is pushing. Because the fish is flat laterally and narrow from a forward point of view, this force does not move the fish sideways, but it does move it forward. Students consider the advantages of lifting the body off the ground so as to use as little energy as possible to move. Humans are an example of this kind of efficiency.

Preparation

Suggested vertebrates for this investigation are a goldfish, a newt, an anole, and a garter snake. However, many other examples of vertebrates will be acceptable. Observe the anole in a covered terrarium to prevent its escape.

Safety

Remind students to handle animals in a humane manner.

Conclusions

1. The salamander and the snake move on land by undulation, just as the fish moves in water.
2. Legs placed directly under the body can move more efficiently.

Inquiry

Students will think of ways the investigation could be improved, such as by observing the skeletons of vertebrates in order to study the details of limb structure.

Chapter Review

Vocabulary

1. Oviparous organisms, such as reptiles and birds, have embryos that are protected by a shell and which develop outside the body, nourished only by the yolk. Ovoviviparous organisms, such as some fishes, amphibians, and reptiles, have embryos that are also nourished only by a yolk, but which remain in the female until they hatch. Viviparous organisms, such as most mammals, have embryos that are nourished by the mother through the placenta until they are born.
2. If resources are plentiful, those organisms with direct development can easily share those resources. In mammals and some birds, direct development also allows the parents to provide resources to their young.
3. behavior that is based on learning
4. social communication and hierarchies as well as basic social behaviors, such as territoriality, dominance, sexual behavior, and play
5. birds that give warning calls and porpoises that help their injured relatives

Review

6. small filter-feeding organisms that resembled some cephalochordates but had a backbone

7. Simpler body structures are often adaptations within more complex systems, which would not necessarily indicate that they were phylogenetically more primitive.

8. increase in adaptations to the terrestrial environment, such as membranes and egg shells that protect the embryo

9. Indirect development occurs in lower chordates and amphibians. Direct development occurs in reptiles, birds, and mammals. If resources are plentiful, direct development allows the adults and young to share them. If resources are scarce, indirect development prevents competition between the adults and young.

10. An increase in the size of the pectoral and pelvic girdles and their accompanying muscles gave terrestrial vertebrates better mobility and support. The legs of vertebrates are also directly under their bodies, which allows them to move faster and with a longer stride.

11. The flippers of many marine mammals and the fins of fishes may be similar in outer structure, but the bone structure of the flipper is actually a modification of the mammalian limb. Other similarities to mammals indicate that marine mammals and terrestrial mammals are closely related.

12. Integuments protect against injury and loss of water, and insulate the body.

13. Terrestrial higher vertebrates convert the toxic ammonia wastes from their cells into urea. They also have tubules in their kidneys that reabsorb most of the water that filters from the blood.

14. Control of the nervous system of higher vertebrates has become concentrated in centers in the brain, thus allowing better-coordinated responses. The highly developed cerebrum of mammals gives them the ability to learn from experience and, in humans, to have foresight and judgment.

15. Both better ensure the survival of the young. Internal fertilization makes fertilization more likely. Internal development allows the mother to nourish the embryo and protect it from predators and the environment.

16. instinctual behavior

17. the pecking order of chickens; to make group interactions stable and orderly

18. because survival on its own might be far more difficult

19. the song of a bird, which must be heard before it can be repeated

20. when there is a clear advantage to the group in that behavior

Critical Thinking

1. A study of the bone structure of the limbs of both organisms would show that the flippers of marine turtles are actually modifications of the limbs of terrestrial turtles and that therefore marine turtles probably descended from terrestrial turtles.

2. to establish its territory for birds of its own and other species, and to attract mates

3. Because the skeletons of birds are extremely fragile, few of them have been preserved in the fossil record.

4. It demonstrates the lasting effects of the imprinting process.

5. It establishes the pack hierarchy, and the avoidance of serious injury best ensures that the group as a whole survives.

Chapter 41 | *Evolutionary Trends in Vertebrates*

Introduction

Since the evolution of the first vertebrate species, vertebrates have become adapted to nearly every habitat on earth. In the previous five chapters you studied adaptations that characterize the five classes of vertebrates. In this chapter you will study evolutionary trends in vertebrates.

Chapter Outline

Chapter Concept

As you read, look for adaptations of terrestrial vertebrates to their environments.

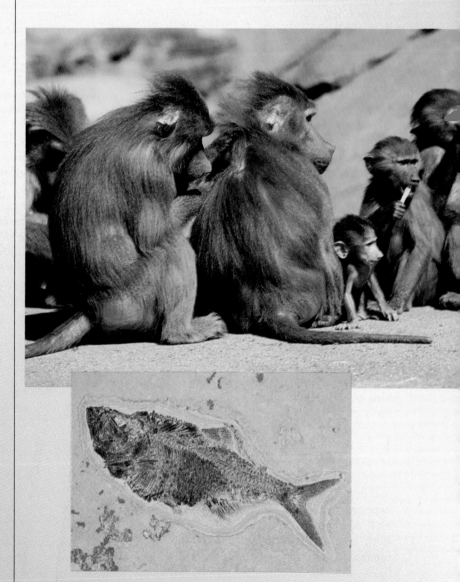

Hamadryas baboons, *Papio hamadryas*, Africa. Inset: Fossil fish of the genus *Diplomystus*, Wyoming.

For additional teaching information see pp. T41A–T41D.

41.1 Origin and Phylogeny

Life on earth began long before the first vertebrates evolved. Fossil evidence indicates that the first prokaryotic cell appeared more than 3.5 billion years ago; the first eukaryotic cell, 2 billion years later; the first primitive vertebrate, another billion years later. Vertebrates have since evolved into a diverse group that consists of the five classes you read about in Chapters 36–40.

Origin

Where did vertebrates come from? You know from Chapter 36 that vertebrates are a subphylum of the Phylum Chordata and that both chordates and echinoderms are deuterostomes. Deuterostomes are organisms in which, among other things, the embryo divides by radial cleavage. Because chordates and echinoderms share developmental characteristics, scientists infer that they evolved from a common ancestor. Figure 41-1 shows a possible evolutionary pathway leading to vertebrates. Many biologists cite the structural similarities between the organisms shown as evidence that chordates and echinoderms evolved from an echinodermlike organism.

The adult form of lancelets is morphologically similar to the larval form of tunicates. From this similarity scientists infer that the primitive cephalochordates evolved from a larval form of a tunicatelike organism. Scientists also infer that the first vertebrates evolved from primitive cephalochordates. The early vertebrates, however, were distinguished by a backbone.

Section Objectives

- *Summarize the probable origin of vertebrates.*

- *Describe the major trends in vertebrate evolution.*

- *Explain why adaptations that are simplifications do not necessarily indicate that an organism is phylogenetically more primitive.*

Tip: Give students separate pictures of a chordate, an echinoderm, an annelid, and so on and have them place the pictures in order from "primitive" to "modern" before you talk about "accepted" phylogenies.

Figure 41-1 This is one of several possible pathways of vertebrate evolution.

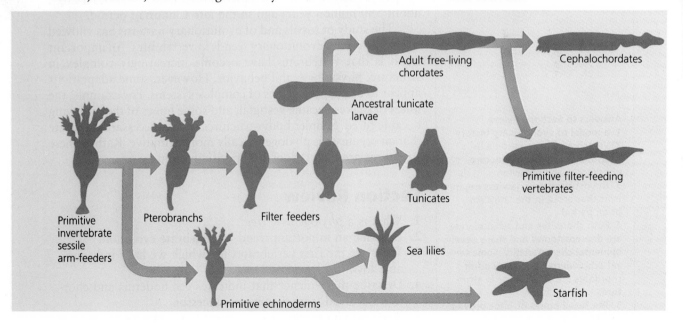

Primitive invertebrate sessile arm-feeders

Pterobranchs

Filter feeders

Ancestral tunicate larvae

Adult free-living chordates

Cephalochordates

Primitive filter-feeding vertebrates

Tunicates

Primitive echinoderms

Sea lilies

Starfish

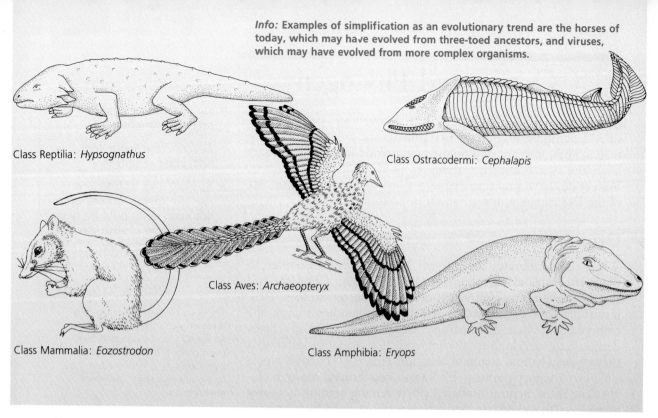

Info: Examples of simplification as an evolutionary trend are the horses of today, which may have evolved from three-toed ancestors, and viruses, which may have evolved from more complex organisms.

Class Reptilia: *Hypsognathus*

Class Ostracodermi: *Cephalapis*

Class Aves: *Archaeopteryx*

Class Mammalia: *Eozostrodon*

Class Amphibia: *Eryops*

Figure 41-2 Each of these five organisms is the first representative of its class found in the fossil record.

Answers to Section Review
1. a model of evolutionary history
2. Vertebrates have become increasingly complex in structure, physiology, and behavior.
3. the ostracoderms, jawless organisms that arose in the late Cambrian period
4. Both chordates and echinoderms are deuterostomes and share developmental characteristics. Some larval echinoderms and some adult chordates are also similar in structure.
5. The fused bones of birds provide them with a light and stiff structure for flight.

Phylogeny

Using evidence from fossils and comparative anatomy, scientists have reconstructed possible pathways of vertebrate evolution. The first vertebrates were probably filter-feeding organisms. The first vertebrates for which we have indisputable fossil evidence are the ostracoderms, jawless organisms resembling fishes that arose about 540 million years ago in the late Cambrian period.

The study of fossils and of evolutionary patterns has allowed scientists to infer evolutionary trends in vertebrates. An important trend is that vertebrates have become increasingly complex in structure, physiology, and behavior. However, some adaptations appear to be simplifications of complex systems. For example, the limbs of snakes became vestigial, and some bones of the bird wing became fused. Simpler body structures do not necessarily indicate that an organism is phylogenetically more primitive. Rather, these structures are adaptations for specific habitats.

Section Review

1. What is a *phylogenetic tree*?
2. Describe an important trend in vertebrate evolution.
3. What are the first vertebrates for which we have indisputable fossil evidence?
4. Describe the evidence that indicates echinoderms and chordates evolved from a common ancestor.
5. What do you think is the adaptive advantage of fused wing bones in birds?

41.2 Development and Morphology

As they have evolved, vertebrates have become more complex in their structure and physiology. As you read about the evolution of vertebrate adaptations, look for those that are responses to the selective pressures of a terrestrial environment. Also refer to the discussion of structures in the chapters on the five classes.

Development

The main evolutionary trend in embryo development is an increase in adaptations to the terrestrial environment. For example, the eggs of fishes and amphibians must develop in water, since they dry out quickly when exposed to air. The eggs of reptiles and birds have four embryonic membranes and a leathery or hard shell that enable the embryos to develop on land. Most mammals do not have shells around their eggs. Instead the mammal embryo is protected inside the mother's body until birth.

Figure 41-3 shows the developing embryos of a frog, a bird, and a mammal. In frogs the entire embryo divides repeatedly to give rise to a hollow blastula. The gastrula stage is characterized by more rapid cell divisions at one pole of the embryo than at the other. In birds the early divisions are restricted to one side of the blastula. The cavity in the blastula, or blastocoel, is smaller; and the embryo develops from a much smaller segment of the gastrula. In mammals the embryo develops from a much smaller mass of cells than it does in frogs or birds. The bird embryo tends to have a

Section Objectives

- *Describe the major adaptations in embryo development that help vertebrates to live on land.*

- *Explain the way in which adaptations in the skeletal, integumentary, circulatory, and excretory systems made vertebrates better suited to live on land.*

- *List some adaptations in vertebrate reproduction that help ensure the survival of the young.*

Figure 41-3   Compare the development of these three vertebrate embryos. What are the major differences?

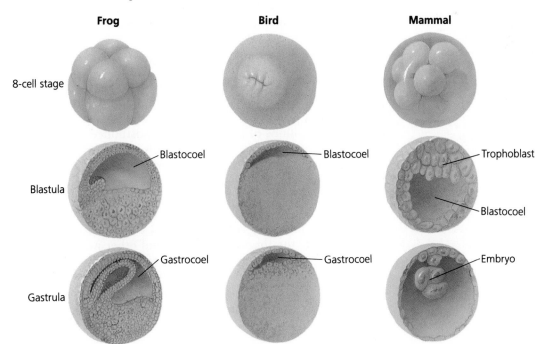

	Frog	Bird	Mammal
8-cell stage			
Blastula	Blastocoel	Blastocoel	Trophoblast / Blastocoel
Gastrula	Gastrocoel	Gastrocoel	Embryo

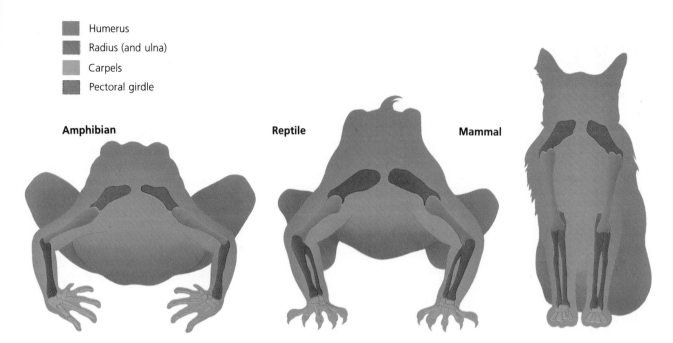

Humerus
Radius (and ulna)
Carpels
Pectoral girdle

Amphibian　　　　　**Reptile**　　　　　**Mammal**

Figure 41-4 Examine the trends in limb position illustrated by these terrestrial vertebrates.

large yolk that provides nourishment. In most mammals, where the embryo is nourished through the placenta, there is no nutritional need for a large yolk.

Another evolutionary trend in vertebrates is the shift from indirect development to direct development. Remember from Chapter 35 that in indirect development the larvae of an organism are different in structure and physiology from the adults. This pattern occurs in the lower chordates and in amphibians. In direct development the young are similar to the adults. This pattern occurs in reptiles, birds, and mammals. This similarity allows the young and the adults to share the same niche and resources—a distinct advantage if those resources are plentiful.

Skeletal and Muscular Systems

As terrestrial vertebrates evolved from aquatic vertebrates, their limbs and associated muscles were modified by natural selection in ways that gave them better support and greater mobility. For example, the limbs of amphibians are positioned away from the body. However, the legs of mammals are positioned directly under their bodies, which allows them to move faster and with a longer stride. Figure 41-4 illustrates this trend. In addition, support and mobility were improved by an increase in the size of the pectoral and pelvic girdles and their accompanying muscles.

Limbs have also become adapted to special environments. For example, the flippers of whales and other marine mammals resemble the fins of fishes. However, an examination of the bone structure of a flipper reveals that it is actually a modification of the mammalian limb. This suggests that marine mammals evolved from land-dwelling mammals that returned to the water.

Integumentary Systems

The major evolutionary trend in vertebrate integumentary systems has been an increased ability of the skin to withstand desiccation. In general, the tough outer coverings of terrestrial vertebrates protect against excessive loss of water from the body, while those of fishes and amphibians are adapted only to moist environments. Integuments also serve other purposes. For example, the scales of a reptile protect it from predators. The smooth skin of an amphibian is a major respiratory organ that allows the exchange of gases. The feathers of birds, which evolved from scales, and the hair of mammals are adaptations that efficiently insulate the body.

Circulatory Systems

As you can see in Figure 41-5, the major evolutionary trend in the vertebrate circulatory system is toward a greater number of chambers in the heart. In general, endotherms such as birds and mammals have higher energy requirements than ectotherms such as fishes and amphibians. As you read in Chapter 36, fishes have only a two-chambered heart. The four-chambered heart of endotherms, however, more efficiently separates oxygenated and deoxygenated blood, thereby helping to provide endotherms with the extra oxygen their high rate of respiration requires.

Info: The heart is particularly useful for showing evolutionary transitions. Note that the turtle heart has four chambers and shows a marked transition from the reptilian and mammalian systems.

Figure 41-5 Compare the number and arrangement of chambers in these four representative vertebrate hearts.

Fish — Atrium, Ventricle

Frog — Atrium, Ventricle

Mammal — Atrium, Septum

Bird — Atrium, Ventricle, Septum

Respiratory Systems

The major evolutionary trend in vertebrate respiratory systems is from gills to lungs. The gills of fishes and larval amphibians are efficient in moving water, but would quickly dry out in air. The lungs of amphibians, which also respire through their skin, have far less surface area than those of reptiles. In the larger or more active birds and mammals higher energy demands require a more efficient respiratory system. For example, flying requires a great deal of energy. The air sacs in birds are adaptations that maintain a constant supply of air in the lungs while adding buoyancy.

Digestive Systems

As vertebrates have evolved, the structure of the digestive system has become adapted to many types of environments. The stomach, for example, is modified depending on an organism's diet. In herbivores the digestive system is long and contains storage sacs, called ceca, that hold food while it is digested by bacteria.

Excretory Systems

As vertebrates evolved, the excretory system has become more complex. In most vertebrates two kidneys near the dorsal surface filter wastes from the blood and help regulate the concentration of salts and water in the blood. Figure 41-6 shows modifications in the excretory pathway. Notice that most mammals lack a cloaca. Wastes from their kidneys collect in a urinary bladder and are eliminated through the urethra. In some vertebrates, such as amphibian larvae and fishes, nitrogenous wastes leave the cell as ammonia gas. Terrestrial vertebrates conserve water when they convert ammonia gas to the less toxic and solid urea before it is excreted. They also have kidneys with tubules that reabsorb most of the water that filters from the blood. Birds conserve additional water by eliminating their wastes as uric acid crystals.

Figure 41-6 The excretory systems of fishes, birds, and mammals differ in morphology and physiology.

Cloaca
Kidney
Ureter
Urinary bladder

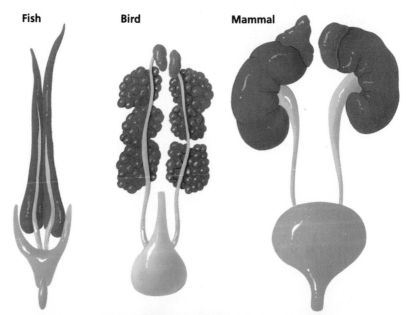

Fish   Bird   Mammal

Nervous Systems

The general evolutionary trend in vertebrate nervous systems is toward an increasingly elaborate system, which allows more varied and intricate responses to the environment. Also there is a strong trend toward the concentration of control of most functions in specific centers in the brain, thus allowing for better-coordinated responses. As Figure 41-7 shows, the cerebellum and cerebrum have become larger as vertebrates have evolved. In lower vertebrates the brain stem is the main center of the brain; trunk movements remain mostly reflex actions. In more complex vertebrates the cerebellum plays a greater role than it does in less complex vertebrates in coordinating motor activities. The degree of development of the cerebellum generally corresponds to the intricacy and precision of the movements the organism is able to perform. The highly developed cerebrum in mammals gives them the ability to learn from experience, and humans, in particular, have heightened abilities of foresight and judgment. In general, the size of the optic lobe corresponds to the importance of functioning vision in the organism.

Tip: Review the coverage of vertebrate nervous systems in Chapters 36–40 and have students point out additional examples of increasing complexity.

Figure 41-7   Specific areas of the vertebrate brain have grown more complex and have increased in size over evolutionary time.

Fish

Reptile

Bird

Mammal

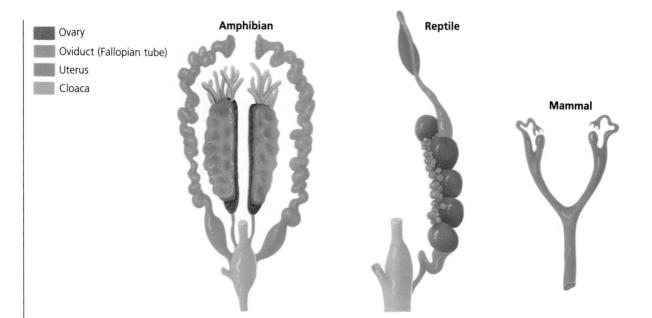

■ Ovary
■ Oviduct (Fallopian tube)
■ Uterus
■ Cloaca

Amphibian

Reptile

Mammal

Figure 41-8 Compare the female reproductive systems in these three representative vertebrates.

Reproductive Systems

In vertebrates the evolutionary trend in reproduction is toward internal fertilization and internal development, both of which better ensure the survival of the young. In most fishes and amphibians eggs and sperm are released directly into the water, where fertilization takes place. In reptiles, birds, and mammals the egg and sperm unite within the body of the female, increasing the likelihood that the egg will be fertilized.

In general, the evolutionary trend has been from ovoviviparity to oviparity to viviparity. Some fishes, amphibians, and reptiles are ovoviviparous. Their eggs are nourished only by the yolk but remain inside the female until they hatch. Many reptiles and birds are oviparous. Their embryos are protected by a shell and develop outside the body, nourished only by the yolk until they hatch. Most mammals are viviparous. Their young are nourished by the mother through the placenta until they are born, which protects the young against predators and harsh climates.

Section Review

1. Describe the difference between *indirect development* and *direct development*.
2. Why is there no large yolk in mammal embryos?
3. What is the major evolutionary trend in respiration? Describe one beneficial adaptation in a terrestrial vertebrate.
4. What adaptive advantages are gained in the transition from ovoviviparity to oviparity to viviparity?
5. Why might marine mammals, whose ancestors were adapted to land, have become adapted to aquatic conditions?

Answers to Section Review
1. In indirect development the larvae of an organism are different in structure from the adults. In direct development the young are similar.
2. Mammal embryos are nourished through the placenta and therefore do not have a large yolk.
3. The major trend is toward lungs with a larger surface area for gas exchange. One example is the air sac in birds, which maintains a constant supply of air in the lungs.
4. Nourishment in ovoviviparous and oviparous organisms is limited to that provided by the yolk, whereas viviparous mothers nourish embryos continually through the placenta.
5. Natural selection favors features that give an organism an advantage in new habitats, especially those that offer little competition. The adaptations of early marine mammals must have allowed them to successfully exploit the resources of the ocean.

41.3 Behavior

Just as the pressure of natural selection has caused the evolution of increasingly complex structures in vertebrates, that pressure has also caused the evolution of complex behavior. The general trend in vertebrate behavior has been toward more complex, better-integrated behavior that allows more flexible responses to different environmental situations. In higher vertebrates the behavior of both individuals and groups of individuals is increasingly modified through learning.

Individual Behavior

In many individuals behavior is often predictable. For example, Konrad Lorenz (1903–), an Austrian zoologist, observed that all young geese followed their mother. He hypothesized that this behavior was genetically programmed. To test his hypothesis, he removed a group of recently hatched goslings from their mother. To his surprise, the goslings followed him as if he were the mother. Lorenz also discovered that the goslings would follow a dog, a cat, or a toy train—as long as it was the first moving thing they saw within their first 36 hours of life. Lorenz called this process **imprinting.** Lorenz discovered that after the 36-hour **critical period** imprinting would not occur even if the goslings were placed with the mother.

While the individual is genetically predisposed to learn behaviors that are characteristic of its species, the development of any such behavior requires the proper stimulus. For example, to be able to sing the song of its species, most birds must hear the song first.

Insight learning is a major example of the more complex behavior that is highly developed in anthropoids. **Insight learning** allows individuals to apply what they have learned in one situation to other situations. A few species, most notably humans, are also able to transmit learned behavior symbolically from generation to generation. Changes in behavior that are based on learning are said to be culturally induced.

Social Behavior

The adaptive interactions among members of the same species are called **social behavior.** In vertebrate societies the organization is not so rigid as that in invertebrate societies. For example, in an ant society most of the individuals are sterile workers; they are chemically induced to assume certain roles and have no ability to reproduce. Nonetheless, vertebrates often form hierarchies, such as the dominance order of male wolves or the hierarchies of females in baboon societies. These hierarchies are adaptations that tend to make those groups stable and orderly.

Section Objectives

- *Describe the major trend in vertebrate behavior.*

- *Explain the process of imprinting.*

- *Describe the apparent adaptive advantage of one example of social behavior.*

Figure 41-9   These brown bears, *Ursus arctos,* **fishing in the McNeil River of Alaska are exhibiting learned behavior.**

Tip: Have students try to list "inherited" human behaviors. In discussion, they will find very few that cannot be attributed to learning.

Figure 41-10 Elephants, *Loxodonta africana* (left), wrap trunks in an effort to establish dominance. Two factions of macaques, *Macaca fasicularis* (right), vie for leadership of the troop.

In all social behavior communication is necessarily important. Communication is the means by which one individual influences the behavior of others in the group. Among vertebrates there are three basic methods of communication: sound, visual displays, and chemical signals. Again, the trend in communication is toward increasingly complex methods.

Ethologists are scientists who study social communication and hierarchies as well as other kinds of basic social behaviors, such as territoriality, dominance, sexual behavior, and play. They then form hypotheses about the way in which each kind of behavior makes a species adapted to its environment. For example, ethologists may determine the degree of altruism in a society. **Altruistic behavior** is a sacrificing of immediate gratification by an individual that results in an immediate benefit for one or more other individuals. A clear advantage exists for the larger group in such behavior. A bird that gives a warning call to others in a flock that a predator is near, or a porpoise that pushes its injured mate to the surface so it can breathe, improves the chances for survival of members of its species. The individual may be at risk in a particular situation, but in another similar situation might benefit from such behavior by others.

Section Review

1. Define *insight learning*.
2. Describe Konrad Lorenz's experiment with young geese. What does this experiment tell you about instinctual behavior?
3. What are the three basic methods of communication in vertebrates?
4. Describe altruistic behavior.
5. Why is communication so important in social behavior? Why might it be increasingly important in higher vertebrates?

Answers to Section Review
1. Insight learning allows individuals to apply what they have learned in one situation to another.
2. Lorenz observed that goslings follow the first moving object they see within their first 36 hours of life. After that period this imprinting will no longer occur. Lorenz inferred that this behavior was genetically programmed.
3. sound, visual displays, and chemical signals
4. Altruistic behavior is the sacrifice of one individual for the benefit of others. There is a clear reproductive advantage to the group in such behavior.
5. Communication is the means by which one individual influences the behavior of others in the group. In higher vertebrates, in which behavior is increasingly flexible, in which more responses are possible, and in which members of the community are more interdependent, the communication by the individual of his or her responses is very important to the group.

Laboratory

Comparing Locomotion in Vertebrates

Objective
To relate form of locomotion to body structure in vertebrates

Process Skills
Observing, inferring

Materials
Live fish, salamander, lizard, and snake; large aquarium; large terrarium

Method

1 In the following investigation, record all your observations. First observe a fish by looking down on the fish from a position above. Describe the overall motion of the fish's body. Now observe the fish from the side. Describe the manner in which the fish coordinates its muscles.

2 Place a salamander in an aquarium and observe it from above and then from the side as it swims.

3 Place the salamander in a terrarium and observe the animal as it walks, both from above and from the side. Does the salamander have difficulty walking on land? What impedes its movement?

4 Place a lizard in the terrarium. How does the lizard's movement differ from the movement of the salamander?

5 Place a snake in water and observe it from above as it swims.

6 Remove the snake from the water and place it in the terrarium. Does the snake use its muscles on land the same way it does in water? What does it push against? It is difficult to observe the function of the snake's scales in locomotion. The scales keep the snake from backsliding.

7 Observe the legs of a mammal, in this case, yourself. In addition to the fact that you walk on two legs, do you notice any structural differences of the legs compared with the legs of other vertebrates?

8 To discover more about the changed position of the limbs in relation to the body in mammals, perform the following exercise: Hold your arms straight out to the sides with your palms facing forward. With the palms still facing forward, bring your thumbs together with your fingers pointing up. Now, with your arms still extended forward, turn your palms so that they face the ground. In performing this exercise, what do you notice about your wrists? Other mammals have this feature on their forelegs.

Conclusions

1 Of the terrestrial vertebrates you examined in this investigation, which resemble the fish in their method of locomotion? In what way?

2 What is the advantage of the placement of a mammal's leg in relationship to the rest of the body?

Inquiry
Discuss ways in which this investigation could have been improved.

Chapter 41 Review

Vocabulary

1. Describe the advantages to organisms of ovoviviparity, oviparity, and viviparity. Give an example of an organism that fits each category.
2. Describe the advantages that direct development gives to organisms.
3. Describe what is meant by culturally induced behavior.
4. What does an ethologist study?
5. Give one example of altruistic behavior among vertebrates.

Review

1. Direct development (a) allows the parent and its young to survive in different environments (b) prevents the parent and its young from competing for the same resources when these are scarce (c) allows the parent and young to share the same niche (d) all of the above.
2. The tough outer skin of vertebrates (a) protects against injury and the excessive loss of water (b) insulates the body (c) allows the absorption of small amounts of water (d) all of the above.
3. Endotherms, which often have four-chambered hearts, (a) have lower energy requirements than ectotherms do (b) have higher energy requirements than ectotherms do (c) do not require extra oxygen (d) require greater numbers of erythrocytes.
4. The evolving behavior in higher vertebrates (a) allows less flexible responses than that in lower vertebrates (b) is dominated by instinctual behavior (c) is increasingly modified through learning (d) none of the above.
5. Culturally determined behavior is based on (a) changes in gene frequencies (b) changes in the environment (c) learning (d) changes in basic human capacities.
6. Describe the probable first vertebrates.
7. Some structural adaptations are actually simplifications. Do such simplifications indicate that an organism should be placed in a more primitive position in the phylogeny? Why or why not?
8. What is the major evolutionary trend in embryo development? Give one example that shows this trend.
9. Describe the vertebrate groups that are characterized by direct and indirect development. List the advantages each group derives from its form of development.
10. Describe the changes in limb structure that gave terrestrial vertebrates greater mobility and support.
11. Explain why the resemblance between the flippers of whales and other marine mammals and the fins of fishes does not indicate that the organisms are closely related.
12. List three ways in which the integuments of terrestrial vertebrates serve as adaptations to a dry environment.
13. Give two examples of adaptations in the excretory system of higher vertebrates, and explain why they are important.
14. Describe two ways in which the nervous system of higher vertebrates has become increasingly complex. Explain why these adaptations are important.
15. Describe the adaptive advantages of internal fertilization and development in higher vertebrates.
16. What kind of behavior dominates in lower vertebrates?
17. Give one example of a hierarchy that exists within a vertebrate society. What is

the function of such hierarchies within vertebrate societies?

18. In vertebrate hierarchies why might a subordinate individual, who might eat last or rarely have an opportunity to reproduce, remain with the group?

19. Give one example of a stimulus that is required if an individual is to learn the behavior characteristic of its species.

20. Explain the conditions under which natural selection might not act against altruistic behavior.

Critical Thinking

1. As you learned in Chapter 38, there are terrestrial and marine turtles. Terrestrial turtles have clawed limbs; marine turtles have paddlelike flippers. How would you go about deciding which organism is probably phylogenetically more primitive?

2. The songs of various species of birds, which are often sung repeatedly by individuals, are clearly distinguishable. What are two possible adaptive advantages of singing distinct songs?

3. There are very few known fossils of birds. Why might this be?

4. Klaus Immelman, a biologist in Germany, did experiments in which he allowed young zebra finch males to be raised by Bengalese finch foster parents. These species are quite different in color and size and do not engage in courtship in nature. Yet when they matured, if given a choice between female Bengalese and female zebra finches, the zebra finch males would

court only Bengalese finches. How might you explain this behavior?

5. Examine the photo above. When individual wolves threaten each other, one will usually back down and any fighting will stop. Only rarely will wolves injure other individuals in the pack. What is the adaptive advantage of such behavior?

Extension

1. Read the article by Richard Ellis entitled "Why Whales Strand," in *Oceans,* June 1987, pp. 24–29. Give an oral report that includes answers to the following questions: What theories have been proposed to explain why whales sometimes beach themselves? Which of these theories does Ellis find most plausible? Why? Why can't whales survive on land? What have scientists learned about living whales from studying those that are beached? What does Ellis conclude about human efforts to save whales?

2. Marine mammals such as whales and porpoises must be able to survive intense pressures when they dive in search of food. Using your library, research and write a report on the adaptations that these marine mammals have evolved that allow them to survive these conditions.

3. Visit a zoo, a farm, or a kennel where you can observe the interactions of individuals of the same species. Observe these interactions for at least a couple of hours. Make a list of the interactions you see and suggest the function of each.

Intra-Science: *How the Sciences Work Together*

The Question in Biology: Can We Save the Salmon?

In this unit you read about the diversity of behaviors exhibited by vertebrates. One example of vertebrate behavior is the annual migration of hundreds of thousands of salmon from the Atlantic and Pacific oceans to the inland riverbeds where they were hatched. For some this is a distance as long as 3200 km. In their hatching grounds, the salmon lay and fertilize their eggs.

At the time of migration, which varies by species, the estuaries, or mouths of rivers, on the Pacific and Atlantic coasts teem with salmon. The salmon remain

in these waters while their systems adjust to the change from a saltwater to a freshwater environment. Sharks, seals, and killer whales as well as commercial and sports fishers all catch great numbers of salmon in these estuaries.

The surviving fish swim upstream through the fast rivers, fighting against rushing currents and jumping over waterfalls in leaps as high as 3 m. Pacific coast salmon also try to elude the brown bears, which gather along the banks of the rivers.

About a century ago, the salmon were so plentiful that a Smithsonian Institution observer wrote: "Their numbers are beyond calculation and seem to be so great as to challenge human ingenuity to affect them in any way."

"Human ingenuity," however, has affected the salmon's numbers—in a negative way. Some observers have even predicted the imminent extinction of the Pacific coho salmon.

Many factors have contributed to this problem. While commercial overfishing has been partly responsible, activities on the land are the major contributors to the salmon's demise. Deforested land has eroded into streams and buried some of the salmon nurseries; industrial pollutants have washed into rivers and poisoned some fish; massive dams, built to produce hydroelectric power, impede the fish's journey upstream.

The salmon, which is North America's most important food fish, needs to be protected by the application of scientific methods to ensure its survival. Preserving it will require the efforts of people working in many branches of science.

The Connection to Chemistry

In the course of its migration a salmon moves from saltwater to freshwater and its body undergoes changes to adjust to the new environment. The properties of the cell membranes in the gills change; the kidneys undergo physiological changes; and the sodium-potassium pump is modified to maintain homeostasis in fresh water. These processes allow the salmon's cells to remain in biological balance.

After it has adjusted to fresh water the salmon literally follows its nose to its spawning grounds. When a salmon hatches, the chemical makeup of the water

around it is imprinted in its memory. As an adult, the salmon follows streams with a chemical makeup that most closely corresponds to this memory.

Scientists who work to replenish populations of salmon place scent chemicals in the water of fish hatcheries. Young salmon will remember this scent, and after they mature in ocean waters, they find their way back to the hatchery. There, workers retrieve them from the water and strip them of eggs or milt. The sperm and eggs are then mixed, and another generation of fish are hatched.

The Connection to Physics

Even a salmon with its great jumping ability cannot leap over a dam 100 m or more high. However, the fish will keep trying until it dies. In some locations fishways have been built to provide the fish with a route around high dams.

One type of fishway, similar to a ship lock, takes advantage of gravity. Water flows from a high lock into a lower lock where the fish have gathered. As the water level in the lower lock rises, so do the fish. In this way fish are gradually raised from one lock to the next until they reach the level of the river behind the dam.

Ladders, another type of fishway, consist of a series of pools, each one about a foot higher than the previous one. The salmon leap from pool to pool. This type of fishway requires careful control. If the water flows down the steps too quickly, the fish cannot jump against the flow. If the water flows too slowly, the fish may not try to leap to the next level because they always seek the maximum water flow.

Principles of physics have also been used to build artificial spawning channels. These channels provide the cold, oxygen-rich water that salmon need in order to spawn.

The Connection to Other Sciences

Bionic engineers solve many problems by studying natural systems. These scientists hope to learn how salmon manage to swim thousands of kilometers in the ocean, then find the mouth of their home river. As you know, the salmon relies on its sense of smell to navigate rivers. While still in the ocean, however, scientists think a salmon finds its way by using a kind of built-in magnetic compass and an ability to sense temperature differences and chemical variations.

The Connection to Careers

Salmon is a major food fish and thus provides jobs for the commercial fishers who catch it, the cannery workers who process it, the truckers who transport it, and the merchants who sell it. The fishways are designed by engineers and built by construction workers. The hatcheries employ fisheries biologists, chemists, and support staff. Ecologists and animal physiologists study the life cycle and adaptations of this unique migrating fish. For more information about related careers in the biological sciences, turn to page 842.

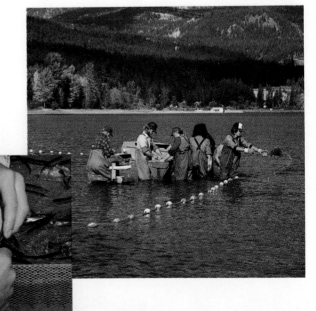

Unit 9 | Human Biology

Chapters

"I sing the body electric."
Walt Whitman

The images on the facing page present two contrasting views of parts of the human body—one medieval and artistic, the other modern and technological. However, the highly developed brain shown in the magnetic resonance image is a uniquely human characteristic that underlies all human achievement, including both art and technology. In this unit you will learn many details about the structure and function of the human body. You will also develop a deeper appreciation for such distinctively human characteristics as upright posture, opposable thumbs, stereoscopic vision—and, of course, the human brain. In what ways are these characteristics important to the construction worker and surgeon shown?

Anatomical drawings by Michelangelo (1475–1564); magnetic resonance image of human brain (inset)

Construction worker on skyscraper

Surgeon's hands

Chapter 42 | Skeletal, Muscular, and Integumentary Systems

This chapter studies the bone, muscle, and skin structures of the human body. Students discover the many ways in which these structures perform the functions of support, movement, and protection.

Correlation Chart

Section Number	Section Objectives	Pages	Features	Section Review
42.1	1. Name the main body cavities, and tell where they are located.	637–639		1
	2. Describe the four types of tissue in the human body.			2, 3, 4
	3. List the primary organ systems, and describe their functions.			5
42.2	1. Describe the functions of the skeleton.	640–644		2
	2. List the parts of the skeleton.			
	3. Identify the parts of a long bone.			
	4. Summarize bone growth.			1, 3, 5
	5. List the types of joints, and give an example of each.		Biotechnology, pp. 642–643	4
	6. Explain how arthroscopic surgery differs from other types of surgery.			
42.3	1. Distinguish between the three types of muscles.	645–648		2
	2. Describe a skeletal muscle cell.			1
	3. Describe muscle contraction and relaxation.			3, 4
	4. Explain how muscles move bones.			5
42.4	1. Describe the functions of the skin.	649–651	Think, p. 650	1
	2. Distinguish between the different layers of skin.		Biology Lab, p. 651	2, 5
	3. Identify two glands found in the skin, and describe their functions.			4

Resources*

Print

Powis, Raymond L. *The Human Body and Why It Works*. Englewood Cliffs, N.J.: Prentice-Hall, 1985.

Smith, Anthony. *The Body*. New York: Viking, 1986.

Audiovisual/Software

Human Biology: Design for Living. Film or video, 26 min. Teacher's guide. Humanities and Sciences, 1985.

Muscles and Joints: Moving Parts. Film, 26 min. Teacher's guide. Humanities and Sciences, 1985.

The Skeletal System. Disk. Apple 48K, Commodore 64. Scholastic.

*Additional references listed in *Modern Biology, Teacher's Program Guide*.

Section Teaching Guides

42.1 The Human Body Plan

Approach

The main objective of this section is to describe the way the human body is organized into tissues, organs, and organ systems. This section draws on the discussion of multicellular organization (Chapter 5), which provides a general description of cell specialization in multicellular organisms. You might wish to begin the discussion by reviewing this material from Chapter 5.

The description of the human body plan provides a basis for comparison with the body plan of invertebrates (Chapter 29) and of other vertebrates (Chapter 36). Finally, the overview of human organ systems will prepare students for the more detailed discussions of these systems in Chapters 43–47.

Concept Framework

Students know that their bodies consist of many different types of cells and tissues. Tap this knowledge in order to provide students with a framework for understanding the organization of the human body by asking them to identify as many different kinds of human tissue as they can. You may wish to have students refer to Figure 42-2, which shows the four categories of body tissue, and determine whether the tissues they have named are connective, nervous, muscle, or epithelial.

Emphasis

Make sure that students understand that, despite the high degree of cell specialization within the human body, proper functioning of the body requires close interaction among tissues, organs, and organ systems. This knowledge will enhance students' understanding of how the various systems of the human body work.

Identifying Organs of the Body

Bring to class pictures or slides of a variety of human organs to show to the class. Ask students to name the organ and the organ system to which it belongs. Also ask them to determine whether the organ consists of connective, nervous, muscle, or epithelial tissue, or some combination of these.

Section Review Answers

(See p. 639.)

42.2 Skeletal System

Approach

The main objective of this section is to describe the structure, function, and development of the human skeletal system. Since bone consists of both inorganic and organic material, the description of bone structure builds on concepts of inorganic and organic chemistry presented in Chapters 3 and 4.

In relating the structure of the skeleton to its function, you might find it useful to compare the human endoskeleton to the exoskeletons found in arthropods and mollusks, noting the ways both structures support and protect the body. You might also wish to compare the human skeletal system with that of other vertebrates, particularly birds (Chapter 39). In your discussion, point out the way the skeletal systems of each group reflect adaptations to their environments.

The discussion of ossification—the process of bone development—builds on the discussion of cell growth and specialization in Chapter 5. It also relates to the discussion of human fetal development in Chapter 47.

Concept Framework

Many students have experienced some type of injury to their skeletal system, such as a broken bone, sprain, or torn ligament or cartilage. Ask if any students who have had such an injury are willing to share their experiences with the class. Those students who are willing might explain what caused the injury, which bone, tendon, or ligament was hurt, what type of treatment they received, and what adjustments were necessary during the recovery process. If the injury was to part of the axial skeleton, ask students what additional injuries might have resulted if there were no bones to protect their internal organs.

Use this discussion as a springboard for discussing this section. Elicit from students the fact that the skeleton is essential to the support and movement of the body. As students read ask them to consider that the skeletal system withstands a great deal of stress and in doing so it prevents damage from occurring to internal organs.

Emphasis

Students should be able to identify the major functions of the skeletal system—support, protection, storage, and blood cell production. They should also be able to explain how the various components of bones and joints are related to these functions.

Using the Features/Arthroscopic Surgery

The technology described in this feature is the use of arthroscopic surgery to treat injuries to ligaments, tendons, and cartilage. Arthroscopic surgery has been made possible through the development of the arthroscope, fiber-optic technology, and special surgical instruments. The procedure involves inserting an arthroscope and a fiber-optic beam through a small incision in the skin. A system of mirrors and lenses enables the surgeon to view an image of the injured area on a screen. The surgeon then uses small cutting instruments and suction tubes to repair or remove the damaged tissue.

Arthroscopic surgery has two primary advantages to the patient. First, arthroscopic surgery is generally performed on an outpatient basis, under a local anesthetic. Second, because the incision is so small, less healing time is required than if the surgeon had to cut through large amounts of tissue. In discussing this feature, you may wish to mention other advances in surgical technology that have similar advantages. The use of lasers in eye surgery is one example.

Identifying Bones and Joints

Bring a model of a human skeleton to class. Ask students to identify which bones are primarily involved in support and movement and which primarily function to protect internal organs. Also ask students to point out examples of a hinge joint, ball-and-socket joint, pivot joint, angular joint, and gliding joint. You may wish to have interested students devise models of these types of joints.

Guest Speaker

Invite a physician or a professor from a local college or university to speak to the class about the effect of the aging process on the skeletal system. The discussion might focus on such disorders as osteoporosis and rheumatoid arthritis. Ask the speaker to address the role of nutrition and heredity in the development of such disorders.

First Aid for Bone Injuries

Bring to class a first-aid manual and a supply of cloth bandages and splints. Demonstrate to the class how to wrap injuries to different parts of the body. You might wish to divide the class into pairs and have students practice some of these techniques. Ask students to suggest items that might be usable as temporary splints in an emergency situation, such as rolled-up magazines or broom handles.

Section Review Answers

(Continued from p. 644.)

4. A ball and socket joint allows circular motion. Hinge joints allow back-and-forth motion.
5. It permits greater flexibility during birth.

42.3 Muscular System

Approach

The main objective of this section is to identify the types of muscle in the human muscular system and to explain how skeletal muscles contract and enable the body to move. Muscle contraction and relaxation occurs as a result of a series of chemical reactions. Consequently the discussion builds on the concepts of biochemistry in Chapter 4. In addition, it will probably be helpful to review with students the concepts of cell energetics and the role of ATP in cell metabolism, as discussed in Chapter 7.

To reinforce the idea of the relationship between structure and function, ask students to note the ways in which the muscular system permits movement in animals whose morphology is very different from that of humans. For example, you might compare the human muscular system with that of fish (Chapter 36), snakes (Chapter 38), and birds (Chapter 39).

Concept Framework

Students know that movement is accomplished by muscles. However, they might not realize that even the simplest movements require the coordination of several muscles in the body. To increase their understanding of the complexity of the human muscular system, ask them to perform some simple activity, such as bending over to pick up a book. Ask them to think about which muscles are involved in this motion—which ones are contracting and which ones are relaxing. Tell students to identify these muscles by referring to Figure 42-8.

Students have probably experienced the form of discomfort known as muscle fatigue. Ask them to consider when fatigue is likely to occur and what might cause it. Tell them to look for possible answers to this question as they read about the biochemical basis of muscle action.

Emphasis

Make certain that students understand that the muscular system interacts closely with other systems of the body. They should be able to give several examples of how each type of human muscle—skeletal, cardiac, and smooth—affects the functioning of other parts of the body. Students will be able to use this knowledge as they read about other organ systems in Chapters 43–47.

Field Trip

Visit a local health club or athletic training facility so that students can see the various types of equipment used in developing the muscular system. Request that the tour include an explanation of how the different machines and regimens are designed to benefit the body in different ways (for example, to strengthen the upper body or to increase endurance). Prior to the trip assist students in preparing a list of questions to answer or points to notice about the facility.

Researching Muscular System Diseases

Divide the class into small groups and have each group research one of the many diseases that affects the human muscular system. Examples include muscular dystrophy, multiple sclerosis, and ALS. Compile a chart to display in class that lists the most frequent victims of the disease, and its causes, symptoms, and treatments, if any.

Section Review Answers

(See p. 648.)

42.4 Integumentary System

Approach

The main objective of this section is to describe the functions of the integumentary system, which consists of the skin (dermis and epidermis), hair, and nails. Skin plays an important role in maintaining proper fluid levels within the body, eliminating waste products, and regulating body temperature. Thus the discussion of the integumentary system draws on the concepts of cell transport and homeostasis described in Chapter 6.

A comparison of the human integumentary system with that of other animal groups will reinforce the idea that an organism's integumentary system shows adaptations to its particular environment. Consequently you might wish to compare human skin, hair, and nails to fish and reptilian scales (Chapters 36 and 38), amphibian skin (Chapter 37), bird feathers (Chapter 39), and mammalian fur (Chapter 40).

Finally, point out to students that, as the largest organ of the body, the skin is closely involved in the functioning of several organ systems that will be discussed in later chapters. Tell students to look for ways that the skin is connected with the respiratory, excretory, nervous, and endocrine systems.

Concept Framework

All students know what it is like to perspire, to shiver, to have dry or oily skin, or to get a sunburn. To prepare students for learning about the functions of the integumentary system, ask students to think about what environmental factors cause these conditions. Ask them what they think would happen if the skin did not respond in these ways. Also encourage students to consider the importance of hair and nails by asking them what would happen if humans did not have these features.

Emphasis

Students should be able to name the major components of the integumentary system and identify the functions of each component. They should also be able to explain the relationship between the integumentary system and other organ systems in the human body.

Field Trip

Visit the burn unit of a local hospital so that students can learn about the treatment of severe burns. Request that the tour include a discussion of the degrees of burns and the treatments required for each. The discussion should also explain why damage to the integumentary system as the result of severe burns presents such a risk to the victim. Prior to the trip, prepare a list of questions to answer or points to notice about the facility.

Guest Speaker

Invite a dermatologist to speak to the class about skin care and the treatment of various skin disorders. The discussion should include information about the effects of such factors as exposure to sun, wind, and extreme temperatures, diet, and emotional stress on the skin.

Section Review Answers

(See p. 650.)

Laboratory Preparation

Background

In this lab, students examine the structure and function of the skeletal and muscular systems of the chicken wing through simple dissection. Students carefully observe the muscle, tendons, ligaments, and bones of the wing. Simple wing movements illustrate the interaction of opposing muscle pairs, tendons, ligaments, and joints. Further dissection reveals the nature of cartilage and its contribution to the smooth action of a typical joint. Based on their observations, students then describe how each of these structures works with others to produce movement.

Preparation

Use fresh, raw chicken wings that have been soaked in 70% ethanol and refrigerated overnight to reduce their fat content. Show students where to make the first cut by drawing a dotted line on the wing from the exposed end of the upper wing to the first joint of the wing, as suggested near the labeled photograph.

Safety

Remind students to use the scalpels and scissors with caution.

Ethanol is flammable. If spilled, flush affected areas immediately with water.

Conclusions

1. Tendons are thick, white, fibrous band or cords, while muscle is smooth, pinkish tissue. The tough, fibrous tissue of the tendon resists pulling and tearing during movement. Muscle tissue is softer and more flexible.
2. Ligaments hold the bones firmly together. The smooth, flexible cartilage prevents rubbing and cushions impact. The bone ends are formed so that they fit smoothly against each other.
3. Answers will vary but should include a discussion of the action of opposing pairs of muscles across a joint between a stationary and a movable bone. The discussion should also mention the function of the ligaments in binding bones together and the tendons that attach muscles to bones.

Inquiry

Students might suggest that the wings of a bird that flies will have longer bones than those in a chicken's wing for support in flight and a larger muscle mass to provide strength for flying. Students might also suggest that the bones of such a bird will be lighter, thus making flight easier.

Chapter Review

Vocabulary Section

1. The matrix surrounds the cells in connective tissue and is instrumental in determining the form of the connective tissue.
2. Compact bone is the dense part of a bone that contains rings of minerals and osteocytes. Spongy bone is a loose, airy network of connective tissue that contains red marrow.
3. A sarcomere consists of two z bands, each attached to an actin filament. A myosin filament is centered between the actin filaments.
4. Skeletal muscle is found throughout the external part of the skeleton. Cardiac muscle is found in the heart. Smooth muscle is found in the walls of organs such as the stomach, intestines, and blood vessels.
5. The epidermis is primarily dead cells and is found on the external surface of the skin. The dermis is rich with living cells and is found under the epidermis.

Review

11. The abdominal cavity contains the stomach, intestines, the reproductive organs, and the excretory organs.
12. An organ system contains various organs that work together to function.
13. The five functions of the skeleton are as follows:

it works with bones to move the body, protects delicate internal organs, supplies the body with red blood cells, supports the body, and gives the body shape.

14. The interconnected Haversian canals provide a pathway for blood vessels and nerves to reach the osteocytes.

15. The bursa and the synovial fluid prevent the bones of a joint from rubbing together.

16. Skeletal muscle consists of muscle fibers that are composed of myofibrils that are composed of bands of actin and myosin filaments.

17. Arthroscopic surgery minimizes recovery time because a local anesthesia is used and only a small incision needs to be made.

18. A motor unit initiates the contraction of a sarcomere.

19. Hair grows from specialized cells at the base of a hair follicle. Oil secreted by glands in the skin prevents hair from drying out.

20. The function of melanin is to absorb ultraviolet light.

Critical Thinking

1. The larger oval-shaped pelvis of the female gives the baby more room when passing through the birth canal. The malleable nature of the baby's skull bones allows the skull to accommodate growth of the brain.

2. It takes place in young horses and makes the bones harder and able to endure more stress.

3. Cells leave the red marrow and move into the blood vessels that are in the compact bone and periosteum and into the main circulatory system.

4. Blood vessels need to constrict in order to help push blood through the circulatory system. The stomach and small intestine can constrict in two directions and thus can break up food as it passes through the digestive system.

5. Too much bathing can remove the fatty acids and thus reduce the skin's natural defense against bacteria.

6. The overlapping helps eliminate cracks that dirt or water could pass through.

Chapter 42 | Skeletal, Muscular, and Integumentary Systems

Introduction

The human body is able to make graceful leaps and powerful jumps. The skeleton provides a framework as strong as metal yet far lighter. Muscles provide power and flexibility. The skin protects the body yet keeps it cool. Without the interaction of these systems, the human body would be as graceless as a robot.

Chapter Outline

Chapter Concept

As you read, note how structure is related to function in bones, muscles, and skin.

Striated fibers of skeletal muscle. Inset: A runner.

For additional teaching information see pp. T42A–T42F.

42.1 The Human Body Plan

The human body begins to take shape in the earliest stages of embryonic development. The cells divide and the embryo takes on the bilaterally symmetrical form that parallels that of the other vertebrates. The unique quality of the human body is the development of the structures into a form that supports an upright body position.

Body Cavities

The human body can be divided into two main cavities, the **dorsal cavity** and the **ventral cavity**. Each of these cavities, shown in Figure 42-1, contains one or more body organs. The dorsal cavity, which contains the organs of the nervous system, can be further subdivided into the cranial and spinal cavities. The **cranial cavity** surrounds and protects the brain. The **spinal cavity** surrounds the vertebral column and protects the spinal cord.

The ventral cavity is divided into two regions by a wall of muscle called the diaphragm. The upper region, or **thoracic cavity,** contains the heart, lungs, trachea, and esophagus. Notice that the ribs, breastbone, and spine form a frame that protects these organs. The lower region, or **abdominal cavity,** contains the organs of the digestive, reproductive, and excretory systems.

Each body system is made up of many different organs. The organs consist of various tissues, which in turn are made up of groups of similar cells. The organization of all of these components of the body is an evolutionary adaptation to the upright posture of humans.

Section Objectives

- *Name the main body cavities, and tell where they are located.*

- *Describe the four types of tissue in the human body.*

- *List the primary organ systems, and describe their functions.*

Figure 42-1 **The dorsal cavity includes the cranial and spinal cavities. The ventral cavity includes the thoracic and abdominal cavities. Cross sections through the chest (top right) and the abdomen (bottom right) show the position of the organs in the body.**

Dorsal cavity

Cranial cavity

Spinal cavity

Ventral cavity

Spinal cavity

Lungs

Heart

Thoracic cavity

Diaphragm

Abdominal cavity

Spinal cavity

Abdominal cavity

Intestines

Body Tissues

A **tissue** is a group of cells that are similar in structure and function. The human body has four main types of tissue: muscle, nervous, epithelial, and connective tissue, as shown in Figure 42-2.

Muscle tissue is composed of cells that contract. Skeletal muscle tissue moves bones; cardiac muscle tissue, found in the heart, pumps blood through the body; smooth muscle tissue, found in digestive organs, moves food through the body.

Nervous tissue consists of cells that transmit messages through the body. Nerve cells respond to changes occurring in all parts of the body. The brain, spinal cord, sensory organs, and nerves are composed of nervous tissue. The nervous system is discussed in Chapter 46.

Epithelial (*EP-uh-THEE-lee-ul*) **tissue** protects, secretes, and absorbs. Epithelial cells bind tightly together, forming a solid sheet that covers, protects, or lines a body part. Examples of epithelial cells include the skin and the cells that line the organs of the body cavity. The cells in each type of epithelial tissue are shaped and arranged so as to perform a particular function. For example, the thin, flat cells that line the inside of the mouth are arranged like tiles on a floor. These cells shield the cells that lie under them.

Connective tissues help support the body. Connective tissues include bone, cartilage, tendons, fat, blood, and lymph. All connective tissues are embedded in or surrounded by a substance called a matrix. The matrix can be solid, semisolid, and in some cases, liquid. Bone cells are surrounded by a hard, crystalline matrix. Cartilage and tendon cells are surrounded by a matrix that is less solid than bone. Fat cells are surrounded by a semisolid matrix; and blood and lymph cells, by a liquid matrix.

Info: Epithelial cells in the nose and mouth are specialized to receive sensory information.

Figure 42-2 These four diagrams show representative cells of the four main types of tissues. The diagram of nervous tissue (upper right) shows one nerve cell.

Muscle tissue

Nervous tissue

Epithelial tissue

Connective tissue

Table 42-1   Overview of Organ Systems

System	Major Structures	Functions
Skeletal	Bones	Structure, support, movement
Muscular	Muscles	Structure, support, movement
Integumentary	Skin, hair, nails	Protection, excretion
Circulatory	Heart, blood vessels	Transport
Respiratory	Lungs, air passages	Gas exchange
Immune	Lymph nodes, T and B cells	Defense against disease
Digestive	Stomach, intestines, mouth, esophagus	Digestion, absorption
Excretory	Kidneys, intestines, skin	Removal of waste, reabsorption
Nervous	Brain, spinal cord, nerves	Sense, coordination, control
Endocrine	Glands, hypothalamus	Control, regulation of homeostasis
Reproductive	Gonads	Production of offspring

Organs and Organ Systems

Tissues are arranged into organs. An **organ** consists of different tissues that together carry out a specific function. For example, the heart is an organ in which muscle, nerve, epithelial, and connective tissues work together to move blood through the body. The stomach is an organ in which muscle, nerve, epithelial, and connective tissues work together to digest food.

Usually, a single organ, such as the heart, does not function in isolation. Rather, groups of organs work together in an **organ system**. In the digestive system, for example, the small intestine, stomach, liver, and pancreas all contribute to the process of breaking food down into nutrients that can be used by the cells. You will find a list of the major body organs and organ systems, with their functions, in Table 42-1. As you study the table, think about the ways in which the different organ systems interact in the human body.

Section Review

1. What is the *dorsal cavity*?
2. Name the four different types of tissues found in the body, and give an example of each.
3. Describe the functional differences between nervous tissue and muscle tissue.
4. What is the relationship between an organ and a tissue?
5. When a person is running, what interactions occur among organ systems?

Tip: Explain to students that an organ system is analogous to a school. A cell is like a student, a tissue is like a biology class, and an organ is like all biology classes.

Answers to Section Review
1. the body cavity that contains the organs of the nervous system
2. connective tissue: bone, cartilage, tendons, fat, blood, and lymph; nervous tissue: nerve cells; muscle tissue: skeletal, cardiac, and smooth muscle; epithelial tissue: skin and cells lining the organs of the body cavity
3. Muscle tissue contracts; nerve tissue transmits messages.
4. An organ consists of different tissues.
5. Answers could include that the circulatory and respiratory systems supply the muscular system with oxygen and nutrients.

Section Objectives

- *Describe the functions of the skeleton.*

- *List the parts of the skeleton.*

- *Identify the parts of a long bone.*

- *Summarize bone growth.*

- *List the types of joints, and give an example of each.*

- *Explain how arthroscopic surgery differs from other types of surgery.*

42.2 Skeletal System

The human body contains 206 bones, which are organized into an internal framework called the **skeleton.** Look at Figure 42-3. Examine the variation in size and shape among the bones. Because the human skeleton is an internal structure, biologists refer to it as an **endoskeleton.**

Skeletal Function

The bones that make up the skeleton function in a variety of ways. Bones and groups of bones support the muscles and organs, give shape and structure to the body, and protect delicate internal organs. Notice, for example, that the ribs curve and form a cage that protects the heart and lungs. Similarly the skull bones, together called the cranium, form a protective case for the brain. Bones store calcium and phosphorous, important minerals used by the body in certain vital metabolic processes. In addition, the internal portion of certain bones manufactures blood cells.

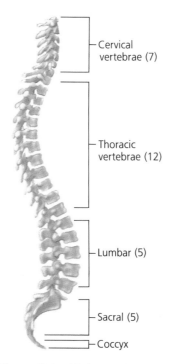

Cervical vertebrae (7)

Thoracic vertebrae (12)

Lumbar (5)

Sacral (5)

Coccyx

Figure 42-3 Of the 206 bones in the human body 64 are in the hands and arms. How does the size of the bones in the hand adapt the hand for its functions?

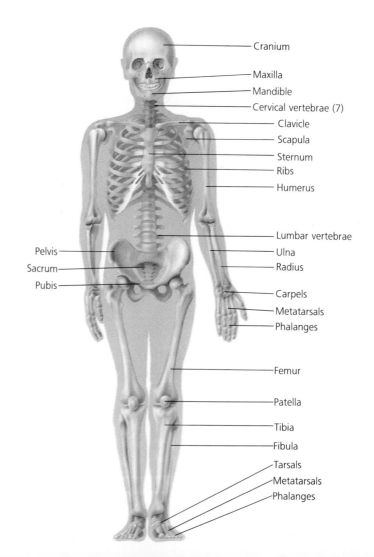

Cranium

Maxilla

Mandible

Cervical vertebrae (7)

Clavicle

Scapula

Sternum

Ribs

Humerus

Lumbar vertebrae

Ulna

Radius

Pelvis

Sacrum

Pubis

Carpels

Metatarsals

Phalanges

Femur

Patella

Tibia

Fibula

Tarsals

Metatarsals

Phalanges

Skeleton and Bone Structure

The skeleton is composed of two parts—the axial skeleton and the appendicular skeleton. The **axial skeleton,** consisting of about 80 bones, includes the spine, ribs, sacrum, sternum, and cranium. The **appendicular** *(AP-un-DICK-yuh-lur)* **skeleton,** which contains 126 bones, includes the bones of the arms, legs, pelvis, and shoulders. Study these bones in Figure 42-3.

Bones are made up of both organic and inorganic material. Refer to Figure 42-4 to help you understand the internal structure of the long bones of the arms and legs.

The tough outer membrane of the bone is called the **periosteum** *(PER-ee-AHS-tee-um)*. This membrane contains a network of blood vessels that supplies living bone cells with O_2 and nutrients and carries away CO_2.

Under the periosteum is a hard material called **compact bone.** In the cross section shown in Figure 42-4 you will notice that compact bone is composed of rings of mineral crystals and protein fibers. Living bone cells called **osteocytes** *(AHS-tee-uh-SITE)* are interspersed throughout these mineral rings. In the center of each ring is a channel called a **Haversian** *(huh-VUR-zhun)* **canal,** which contains nerves and blood vessels. The blood vessels that run through the interconnected Haversian canals nourish the osteocytes. A thick layer of compact bone enables the shaft of the long bone to endure the great amount of stress it receives upon impact with a solid object.

Interior to the compact bone is a network of connective tissue called **spongy bone.** Spongy bone also fills the interior of the knoblike ends of the long bones of the arms and legs. The lacy structure of spongy bone adds strength to the bone without adding much weight. This is an important characteristic, because bones must support the weight of the body and yet be light enough for a person to move easily.

Some bones contain **bone marrow,** which is either red or yellow. Red bone marrow—consisting of blood vessels, fibers, and cells—manufactures erythrocytes and white blood cells. It is found in spongy bone and the ends of long bones, as well as in the ribs, vertebrae, sternum, and pelvis. Yellow bone marrow—consisting mostly of fat cells—serves as an energy reserve. It fills the shafts of the long bones.

Info: The prefixes *os-, oss-,* and *osteo-* refer to bones.

Figure 42-4 This cross section shows the internal structure of a long bone. The small box in the cross section is shown in greater detail in the large box.

Haversian canal

Osteocytes

Mineral rings

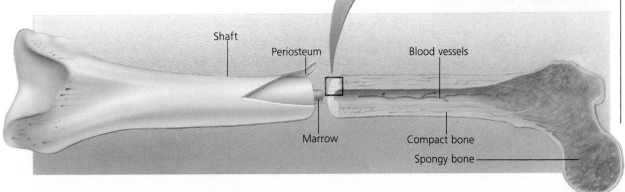

Shaft

Periosteum

Blood vessels

Marrow

Compact bone

Spongy bone

Figure 42-5 A child's fingers grow when bone replaces cartilage at the epiphyseal plates.

Bone Development

Ossification (*AHS-uh-fuh-KAY-shun*) is the process by which bones develop. There are two types of ossification—the conversion of cartilage into bone and the formation of membrane layers.

Cartilage is a tough, flexible connective tissue. During the first month of human fetal development the entire skeleton is made up of cartilage. During the second month osteocytes begin to develop in this cartilage. The osteocytes release minerals that lodge in the empty spaces between the cartilage cells. Eventually most of the cartilage is replaced by bone. However, some cartilage remains between bones, at the end of the nose, in the external ears, and on the inside of the tracheae. It is the cartilage that makes these areas flexible.

A few bones, such as those in the clavicle and in some parts of the skull, develop directly into hard, nonmembranous bone from membranous connective tissue without a distinctive cartilaginous phase. The membranes of the tissue ossify and become flat plates of bone. For example, the membranes in the head are replaced by the long plates that form the skull. The plates have suture lines between them.

Biotechnology

Arthroscopic Surgery

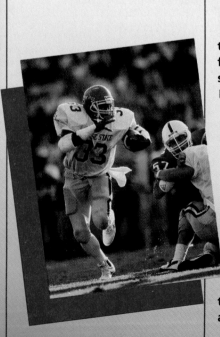

Until recently, athletes who injured ligaments, tendons, or cartilage were forced to undergo extensive surgery to treat their injuries. For the athlete this could mean weeks without training, coaching, or competition.

Advances in medical science, specifically in the areas of fiber optics and instrument development, have resulted in a procedure known as arthroscopic surgery. Unlike previous forms of surgery, arthroscopic surgery requires only a tiny incision. The surgeon can usually treat the injured area without affecting any adjacent tissues.

Knee surgery is particularly well suited for the arthroscopic technique. Many knee problems are caused by torn cartilage or by a piece of tissue that interferes with the functioning of the joint. The joint can become weak when the athlete overtrains or uses improper exercising form. Often the athlete can overcome such injuries by resting, by slowly strengthening a weak muscle, or by correcting improper form with coaching. Extreme cases may require surgery.

In performing arthroscopic surgery the surgeon makes a small incision in the skin and inserts an arthroscope to view

Bones continue to develop after birth. In Figures 42-5 and 42-6 notice that between early childhood and late adolescence bone cells replace the cartilage in the fingers and the fingers grow longer. Growth at the epiphyseal plate on each bone causes the elongation. The **epiphyseal** (*EP-uh-FIZZ-ee-ul*) **plate** is a band of cartilage located near the distal end of a long bone. Cartilage cells within the epiphyseal plate divide and form new cartilage cells. These cells are replaced by bone at the bottom of the epiphyseal plate, allowing the plate to continue to grow at the distal end. Growth continues until the bone reaches its full length, at which time bone has replaced all the cartilage, including the cartilage in the epiphyseal plate. After this, no further longitudinal growth is possible in the bone.

Joints

The place at which two bones meet is called a **joint.** Tough bands of connective tissue called **ligaments** hold the bones of a joint in place. The ligaments also stretch as the bones move. There are three kinds of joints: fixed joints, semimovable joints, and movable joints.

Figure 42-6   This joint, called a ball-and-socket joint, allows circular motion.

the injured area. The doctor also inserts a special fiber-optic light. This powerful light provides a bright, focused beam that illuminates the injured tissues. Through a precisely coordinated system of mirrors an image is reflected through the scope. This image can be reflected into a special lens and then projected onto a screen. Thus the doctor can continuously refer to the projected image while conducting the surgery.

After locating the damaged tissue the surgeon begins to remove or repair the area. If the problem is a flap of tissue that must be removed, the surgeon uses a tiny cutting instrument to do the work. A special suction tube pulls the pieces out of the working area. Once the work is completed, the surgeon removes his instruments and closes the incision with a couple of stitches.

Arthroscopic surgery has many advantages. The first is that the surgery can usually be done on an outpatient basis. That is, a patient can come to the hospital or clinic in the morning and go home the same day. The second is that most arthroscopic surgery can be done with the use of only a local anesthetic. This type of procedure eliminates the side effects that may occur from general anesthesia. Best of all, the recovery time for arthroscopic surgery is usually short; because surgeons do not have to cut through masses of tissue, there is less tissue that has to heal.

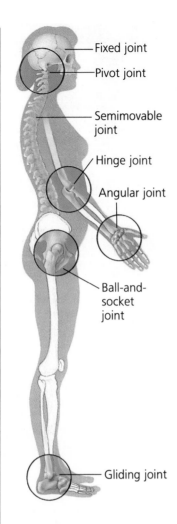

- Fixed joint
- Pivot joint
- Semimovable joint
- Hinge joint
- Angular joint
- Ball-and-socket joint
- Gliding joint

Figure 42-7 In addition to fixed joints and semimovable joints, the human body includes five types of movable joints: hinge, ball-and-socket, pivot, angular, and gliding.

Tip: List on the chalkboard the bones students have broken. Ask students to identify the scientific name of each.

Answers to Section Review
1. the process by which bones develop
2. supporting muscles and organs, giving shape and structure to the body, and protecting organs
3. Through ossification cartilage cells are generated and replaced by bone cells.
(See pp. T42A–T42F.)

Types of Joints

The bones of the skull meet at fixed joints, the sutures, as shown in Figure 42-7. A small amount of connective tissue remains in the suture and helps absorb any impact on the head and prevent broken bones. The skull bones move very little.

Semimovable joints hold the bones of the vertebral column in place and allow the body to bend and twist. In Figure 42-7 you can see that the bones of the spine are separated by cartilaginous disks. Some ribs are attached to the sternum by long strands of cartilage, allowing the chest to expand during breathing.

The remaining joints shown in Figure 42-7 are examples of movable joints. A hinge joint, such as the elbow, allows bones to move forward and backward. A ball-and-socket joint, such as a hip joint, allows circular motion. A pivot joint, such as the top bones of the vertebral column, allows the head to move from side to side and up and down. Angular joints found in the wrists and ankles allow the hands and feet to make twisting and snapping movements. Finally, gliding joints such as the small bones of the wrist and ankle allow the bones to slide against each other.

Functioning of Joints

Joints are often subjected to a great deal of pressure and stress. Connective tissue near the joints secretes synovial fluid, which cushions the bones. This cushioning helps prevent the ends of the bones from wearing away. A fluid-filled sac called a **bursa** is found in the knee and shoulder joints. This sac adds an additional cushion between the bones.

Sometimes all of these protective structures are not enough to prevent injury to a joint. One type of injury, called bursitis, occurs when the bursa becomes inflamed as a result of the overuse of a joint. Bursitis is especially common among athletes. Another type of injury, called a sprain, occurs when the ligaments surrounding the joint are torn or overstretched. Rest, compression, and the safe use of ice are usually recommended to treat a sprain.

One of the most mysterious causes of painful and damaged joints is rheumatoid arthritis. Scientists are not sure what causes this type of arthritis. However, they know that in arthritis sufferers the synovial membranes become inflamed and grow thicker. Eventually, fibrous tissues in the joints may become ossified, thus immobilizing the joint and fusing the bones together.

Section Review

1. What is *ossification?*
2. What are three functions of bones?
3. How do bones elongate?
4. How is a ball-and-socket joint different from a hinge joint?
5. What is the advantage of a cartilaginous skeleton during prenatal development?

42.3 Muscular System

The bones that make up the supporting framework cannot move alone. It is the muscles that are attached to the bones that cause the body to move. When the body moves, the integration of the various body systems is clearly evident.

Muscle Types

A **muscle** is a contractile organ consisting of many cells. The human body contains three types of muscle tissue: skeletal muscle, cardiac muscle, and smooth muscle.

Skeletal **muscle** is attached to the periosteum of bone, either directly or by a tough connective tissue called a **tendon.** Figure 42-8 shows the different sizes and shapes of human skeletal muscles. Skeletal muscles have **striations,** dark bands located at right angles to the long axis of the muscle. Because their contractions can usually be controlled, skeletal muscles are referred to as voluntary muscles.

Section Objectives

- *Distinguish between the three types of muscles.*

- *Describe a skeletal muscle cell.*

- *Describe muscle contraction and relaxation.*

- *Explain how muscles move bones.*

Figure 42-8 Skeletal muscle tissue is shown in these diagrams of the front (left) and back (right) of the human body.

Frontalis

Trapezius

Deltoid

Pectoralis major

Biceps brachii

Triceps brachii

Abdominal muscles

Sartorius

Rectus femoris

Gastrocnemius

Trapezius

Deltoid

Triceps brachii

Gluteus maximus

Biceps femoris

Gastrocnemius

Achilles tendon

645

Figure 42-9 In this photograph of a cardiac muscle (×100) the small vertical lines in each muscle fiber are striations.

Figure 42-10 For a muscle to contract, chemical reactions must take place that cause the sarcomere to shorten.

Cardiac muscle, which makes up the walls of the heart, is also striated. Cardiac muscle is controlled by a special nerve center called the sinoatrial node, which sends electrical signals through the cells. Because its movements cannot be controlled, cardiac muscle is called involuntary muscle.

Smooth muscle, which is found in the walls of the stomach, intestines, and blood vessels, does not have striations. It is made up of long spindle-shaped cells that contain a single nucleus. Smooth muscles are involuntary muscles.

The Muscle Cell

Each skeletal muscle cell is called a **muscle fiber.** Each fiber is multinucleate and contains from 1,000 to 2,000 threads of protein called **myofibrils** *(MIE-oe-FIBE-rulz)*. Each myofibril appears banded. Look at Figure 42-10 and notice the small Z bands in the myofibril. The area between Z bands is called a sarcomere. The **sarcomere** *(SAHR-kuh-mir)* is the functional unit of contraction. The striated appearance of the sarcomere results from the presence of two types of protein filaments, myosin and actin. The

Muscle

Muscle fibers

thicker **myosin** *(MIE-uh-sun)* filaments have lateral extensions called cross-bridges. The cross-bridges are attached to the thinner **actin** filaments that lie beside the myosin filaments. Study the relationship between actin and myosin in Figure 42-10. Notice that the ends of the actin molecule are attached to the Z bands.

How Muscles Contract

As you read about muscle contractions, refer to Figure 42-10. A contraction begins when a nerve impulse reaches a muscle. The junction of a nerve branch and a muscle is called a motor end plate. A number of closely arranged motor end plates form a motor unit. A stimulus reaching the motor end plate causes the cell membrane of the muscle fiber to release calcium into the muscle cytoplasm. The calcium breaks a chemical bond between the actin and the heads of the myosin cross-bridges and triggers the breakdown of ATP in the cross-bridges, thus releasing energy.

Using the energy of ATP, the heads of the cross-bridges attach to a new position on the actin filament. This pulls the actin filament toward the center of the sarcomere. The heads then

Figure 42-11 Smooth muscles (×100) are involuntary and not striated.

Info: Most muscle enlargement occurs when muscle cells increase in size rather than through production of new cells.

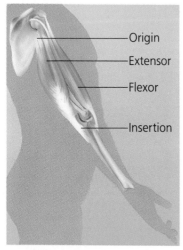

Figure 42-12 When the biceps contract (top), the arm bends. When the triceps contract (bottom), the arm extends.

Tip: Point out to students that when they lift an object that is much lighter than expected, the overreaction is the result of too many muscle fibers contracting.

Answers to Section Review
1. the unit of contraction in a skeletal muscle
2. Skeletal muscle moves bones and is voluntary; cardiac muscle pumps blood and is involuntary.
3. Sarcomeres shorten in synchrony; muscle fibers therefore contract, as does the muscle.
(See pp. T42A–T42F.)

detach themselves, move back down the actin filament, and reattach to a new position. The process is then repeated. Because actin filaments are attached to the Z bands, the Z bands move closer to each other and the entire sarcomere contracts. The synchronized shortening of sarcomeres along the full length of a muscle fiber causes the whole fiber, and hence the muscle, to contract. The contraction process continues until calcium and ATP supplies are depleted or nerve stimulation stops.

Experiments have revealed that the relaxing of muscle fibers requires energy also. The energy is used to enable the heads of the cross-bridges to detach themselves and move back to their original positions.

When a single muscle fiber is stimulated, an all-or-none response occurs. This means that the fiber either contracts or it doesn't. No in-between state exists. How, then, can you contract your muscles tightly enough to lift a dumbbell or gently enough to lift a penny? The force of a contraction is determined by the number of fibers stimulated. As more fibers are activated, the force of the contraction increases.

How Muscles Move Bones

When skeletal muscles contract, they pull at the base of a joint. For example, the biceps, the large muscle at the front of the upper arm, is connected by tendons to the forearm at one end and to the shoulder bone at the other. When the biceps contract, the forearm moves while the shoulder remains still. The point of attachment on the moving bone—in this case the forearm—is called the **insertion**. The point of attachment on the stationary bone—in this case the bones of the shoulder—is the **origin**.

Most skeletal muscles function in opposing pairs. One muscle in a pair flexes a joint, and the other muscle extends it. The opposing muscles can be organized into four broad categories: flexors, extensors, abductors, and adductors. Flexors, such as the biceps, bend joints. Extensors, such as the triceps, straighten joints. Abductors, such as the deltoids, move a limb away from the body, as when you move your arm out to the side. Adductors, such as the pectorals, move a limb toward the body, as when you bring your arms together.

Section Review

1. What is a *sarcomere?*
2. What are some functional differences between skeletal and cardiac muscles?
3. What happens within the muscle fibers to cause a muscle to contract?
4. What factor affects the strength of a muscle contraction?
5. How might injury to one of the muscles in a pair affect movement?

42.4 Integumentary System

The integumentary system, consisting of the skin, hair, and nails, acts as a protective barrier between the body and the outside world. It also functions to retain body fluids, protect against disease, eliminate waste products, and regulate body temperature. As you read the following sections, refer to Figure 42-13.

Epidermis

The skin, which is the largest organ of the human body, is made up of the epidermis and dermis. The **epidermis,** or outer layer of the skin, is composed of about 20 layers of dead cells. These cells are constantly being shed and replaced by new cells. The cells of the epidermis are filled with a protein, called **keratin,** that gives the skin its rough, leathery texture. Keratin also gives the skin its waterproof quality.

Dermis

Dermis, the inner layer of the skin, is composed of living cells and contains nerves, blood, and lymph vessels. The numerous nerves make it possible for the skin to receive many kinds of environmental signals. The blood vessels release heat and help the body maintain a comfortable temperature. The lymph helps the skin fight against infection.

A layer of fat cells lies below the dermis. These cells enable the skin to store food for energy. They also provide protection to the body and insulate it against heat loss.

Section Objectives

- *Describe the functions of the skin.*
- *Distinguish between the different layers of skin.*
- *Identify two glands found in the skin, and describe their functions.*

Info: Keratin is also found in the feathers and the beaks of birds.

Figure 42-13 Cells in the dermis, the layer of skin beneath the epidermis, are adapted to perform a variety of functions.

Dermis —
Oil gland —
Muscle fibers —
Hair follicle —
Subcutaneous tissue —
Hair
Pore
Epidermis
Nerves
Sweat gland
Blood vessels
Fat cells

Think

The person in the picture has just completed a strenuous workout. Even though the weather is warm, he is wearing a jacket. What function might the jacket serve in helping the man regulate his body temperature?

Answer to Think
Putting on a jacket helps the woman retain heat and prevents her from losing too much water. This will keep her from becoming chilled and will help her body return to its normal temperature.

Answers to Section Review
1. a pigment that absorbs harmful ultraviolet rays
2. The epidermis, the outer layer of the skin, is made up of dead cells that contain keratin. The dermis, the inner layer of the skin, is composed of living cells and contains nerves, blood, and lymph vessels.
3. Hair grows from cells at the base of the hair follicles.
4. Sweat glands help rid the body of wastes and maintain body temperature; oil glands help keep skin soft and repel water.
5. Shedding dead skin cells carries environmental toxins away from the body and allows for cells to move from the dermis to the epidermis as they age.

Hair and Nails

Hair is produced by cells at the base of the **hair follicles,** deep pockets that extend into the dermis. Blood vessels surrounding the hair follicles nourish the hair root. The shaft of hair that extends beyond the skin consists mostly of keratin and requires no nourishment. Oil secreted by glands in the skin prevents the shaft from drying out and breaking off. Individual hairs usually grow for several years and then fall out. However hair on the scalp can grow continuously for many years.

Nails, which protect the ends of the fingers and toes, form from cells at the end of a deep fold of epidermis. As new cells form at the base of the nail, the nail grows longer. Like hair, nails are composed primarily of keratin.

Glands

The skin contains exocrine glands, glands that release secretions through ducts. The main exocrine glands of the skin are the **sweat glands** and the **oil glands.**

The skin functions as an excretory organ by releasing excess water, salts, and urea through the sweat glands. By releasing excess water the skin also helps regulate body temperature. When the body temperature rises, the circulation increases, and the skin becomes warm and flushed. The sweat glands then release sweat. As the water in the sweat evaporates the skin is cooled.

The oil glands secrete a substance known as sebum. This substance helps soften the skin and keeps it waterproof. These glands are usually associated with hair follicles. The production of sebum is controlled by hormones.

Among humans there is a great variety in skin color. A pigment called **melanin,** which is produced in the lower layers of the epidermis and which absorbs harmful ultraviolet light, helps determine the color of the skin. If a large amount of melanin is produced, the skin will be dark; if a small amount is produced, the skin will be light. In some individuals exposure to sunlight temporarily increases the production of melanin, thus causing the skin to become darker, called a tan. People with fair skin, which cannot produce much melanin, need to minimize their exposure to the sun to protect against sunburn.

Section Review

1. What is *melanin?*
2. Differentiate between the dermis and the epidermis.
3. How does hair grow?
4. Explain the function of the two types of exocrine glands of the skin.
5. How is the shedding of cells by the skin an adaptive advantage?

Tip: Using a microscope, demonstrate to students how tiny drops of sweat are visible at the top of each pore of a warm hand.

Laboratory

Observing Muscles and Bones

Objective
To examine the structure and the function of the skeletal and muscular systems of chicken wing

Process Skills
Observing, inferring

Materials
Chicken wing, dissecting pan, scissors, forceps, scalpel, paper towel, unlined paper

Method

1. Place a chicken wing in a dissecting pan. Insert the tip of a scissors under the skin at the exposed end of the upper wing. Keeping the scissors away from the muscle, cut the skin to the first joint of the wing.

2. Use your fingers and a forceps to gently pull the skin away from the muscle. As you pull the skin away you will see a thin, clear layer of tissue, called the mesentery. What is the function of the mesentery?

3. Use scissors, forceps, and a scalpel to remove the skin completely from the muscles and joints of the wing. Then carefully cut away the mesentery. Rinse the wing with water. Blot the muscles dry with a paper towel wherever oil globules remain.

4. Observe the muscles and the tendons of the wing. How does the structure of a tendon relate to its function?

5. Hold the upper portion of the wing while moving the lower portion. Which pair of opposing muscles controls this movement?

6. While holding the lower wing, move the wing tip. Note the movement of the muscles and tendons. Make a detailed drawing that shows all of the muscles and tendons in the wing. Identify on your drawing the pairs of opposing muscles that control the movement of the lower wing and the wing tip.

7. Carefully remove the muscles from the bones. How does the structure of muscle relate to its function?

8. Examine the joints of the wing. Note the ligaments, the tough white tissue that connects two bones. How does the structure of ligament relate to its function? Make a drawing of all the bones and ligaments in the wing.

9. Separate the joints. Examine the shapes of the bone ends and the cartilage that covers the bone ends. How does the structure of cartilage relate to its function?

Conclusions

1. How does tendon tissue differ from muscle tissue? What are the structural or organizational characteristics of these tissues that may account for these differences?

2. How does the structure of a joint enable it to move smoothly?

3. Explain how muscles, tendons, bones, ligaments, and joints combine to allow the back and forth movement of the lower wing.

Inquiry
The chicken is not capable of sustained flight. Formulate a hypothesis about how the wing structure of birds that fly differs from the wing structure of flightless birds.

Vocabulary

abdominal cavity (637)

actin (647)

appendicular skeleton (641)

axial skeleton (641)

bone marrow (641)

bursa (644)

cardiac muscle (646)

cartilage (642)

compact bone (641)`

connective tissue (638)

cranial cavity (637)

dermis (649)

dorsal cavity (637)

endoskeleton (640)

epidermis (649)

epiphyseal plate (643)

epithelial tissue (638)

hair follicle (650)

Haversian canal (641)

insertion (648)

joint (643)

keratin (649)

ligament (643)

melanin (650)

muscle (645)

muscle fiber (646)

muscle tissue (638)

myofibril (646)

myosin (647)

nervous tissue (638)

oil gland (650)

organ (639)

organ system (639)

origin (648)

ossification (642)

osteocyte (641)

periosteum (641)

sarcomere (646)

skeletal muscle (645)

skeleton (640)

smooth muscle (646)

spinal cavity (637)

spongy bone (641)

striation (645)

sweat gland (650)

tendon (645)

thoracic cavity (637)

tissue (638)

ventral cavity (637)

1. What is the matrix of connective tissue?
2. Distinguish between compact bone and spongy bone.
3. List the components of a sarcomere.
4. Distinguish between skeletal, cardiac, and smooth muscle.
5. What is the difference between the dermis and the epidermis?

Review

1. The thoracic cavity contains the (a) brain (b) intestines (c) respiratory organs (d) spine.
2. The connective tissue is embedded in a substance called its (a) matrix (b) organ (c) cell (d) bone.
3. The periosteum (a) produces erythrocytes (b) covers the bone (c) contains marrow (d) contains Haversian canals.
4. During ossification (a) cartilage replaces bone cells (b) bones become stronger (c) bone replaces cartilage (d) marrow is produced.
5. Hinge joints allow (a) back-and-forth movement (b) circular movement (c) no movement (d) gliding movement.
6. Cardiac muscle is (a) voluntary muscle (b) involuntary muscle (c) a type of smooth muscle (d) found in the lungs.
7. Actin and myosin are (a) found in joints (b) types of cartilage (c) types of muscle cells (d) parts of the sarcomere.
8. Skeletal muscles (a) push and pull bones (b) push bones (c) pull bones (d) cannot be controlled.
9. The epidermis (a) is primarily made up of dead cells (b) contains blood vessels (c) contains nerve cells (d) contains sweat glands.
10. The dermis (a) is primarily made up of dead cells (b) contains keratin (c) is the top layer of the skin (d) contains oil glands.
11. What organs are found in the abdominal cavity?
12. What is the relationship between an organ and its organ system?
13. What are the five functions of the skeletal system?
14. Explain the role of the Haversian canals in the bones.
15. What prevents the bones from rubbing together at a joint?
16. Describe the parts of a skeletal muscle.

17. What are some advantages of arthroscopic surgery?
18. What happens when a motor unit is stimulated?

19. How does hair grow? What substance keeps it from drying out? What produces this substance?
20. What is the function of melanin?

Critical Thinking

1. During a normal birth, the baby must pass through the mother's pelvis. The pelvis of the female is much larger in diameter and more oval in shape than that of the male. In addition, the skull bones of a newborn are not completely ossified but are somewhat plastic. Give two reasons why these skeletal properties make birth easier.
2. If young thoroughbred horses are raced too hard and too fast early in life, their leg bones may break. What can you infer about the process of ossification in horses?
3. As you have learned, red bone marrow produces erythrocytes. How are these solids transported to other parts of the body? Use the picture to help you explain your answer.
4. Blood vessels have one layer of smooth muscle that encircles the vessel wall. The stomach and small intestine have one layer of circular smooth muscle and one layer of longitudinal smooth muscle. How

does this muscle arrangement reflect the function of each structure?
5. In addition to oil, the oil glands of human skin secrete fatty acids that help kill bacteria. Keeping this in mind, why can it actually be unhealthy to wash your skin too much?
6. Both the epithelial cells of the skin and the shingles of a roof are flat and overlapping. How does overlapping help both cells and shingles perform their respective protective functions?

Extension

1. Read the article by Lawrie Mifflin entitled "Gymnastics May Be Hazardous to Your Health," in *Seventeen,* January 1985, p. 81. Give an oral report to the class in which you explain why overuse injuries are more common in young people than in adults.
2. Read the article by Claudia Wallis entitled "Treating Reagan's 'Pimple,'" in *Time,* August 19, 1985, pp. 56–57. Give an oral report to the class in which you explain skin cancer. Be sure to mention the possible causes of skin cancer.
3. Research and write a report on the effects of a particular exercise on the human body. Choose one exercise, such as running, biking, gymnastics, tennis, or wrestling. What type of skills are necessary for your chosen sport: endurance, strength, speed, fine motor coordination? Do champion athletes in your chosen sport show particular physiological adaptations that enable them to excel? For instance, is a high percentage of their muscle mass in their upper body? Do they have a very low percentage of body fat?

Chapter 43 | *Circulatory and Respiratory Systems*

This chapter describes how the human circulatory and respiratory systems function, and how the structures of these systems are complex adaptations for transporting nutrients to and wastes from cells throughout the body.

Correlation Chart

Section Number	Section Objectives	Pages	Features	Section Review
43.1	1. List the different types of vessels found in the circulatory system.	655–661	Biology in Process, pp. 658–659	1, 4
	2. Describe the movement of blood through the heart.			2
	3. Explain how the sinoatrial and atrioventricular nodes control each heartbeat.		Think, p. 660	2
	4. Distinguish between the two primary subsystems of circulation.			3, 5
	5. Explain how William Harvey discovered the circular flow of blood.			
43.2	1. List the components of blood.	662–664		1
	2. Describe the physical characteristics of the three types of blood solids.			2
	3. Summarize the process of blood clotting.			3
	4. Explain what determines the compatibility of blood types for transfusion.			4, 5
43.3	1. Summarize the skeletal and muscular changes that occur during breathing.	665–669		
	2. Name and describe the organs of the respiratory system.		Biology Lab, p. 669	1, 2
	3. Explain how the structure of alveoli is related to the function of diffusion of gases.			3, 4
	4. Describe the physical effects of exercise on the breathing rate.			5

Resources*

Print

Hole, John W. 4th ed. *Human Anatomy and Physiology*. DuBuque: Wm. C. Brown, 1987.

Tiger, Steven. *Heart Disease*. (From the *Understanding Disease* Series.) New York: Messner, 1986.

Audiovisual/Software

Artificial Heart. Video, 57 min. Time-Life (Part of the *Nova* Series.), 1984.

The Heart: A Mighty Pump. Apple 48K. Disk. Scholastic.

*Additional references are listed in *Modern Biology, Teacher's Program Guide*.

Section Teaching Guides

43.1 Circulatory System

Approach

The main objective of this section is to describe the structure and function of the human circulatory system including the heart and the blood vessels. The section builds on concepts of tissue, organs, and organ systems presented in Chapter 5 and on concepts of muscle types presented in Chapter 42. The section will help students to better understand subsequent chapters in Unit 9. In presenting this section, guide students to recognize the general function of the circulatory system. Help students understand how the structures of the circulatory system work together in performing their functions.

Concept Framework

Students know that the heart is the organ that pumps blood. If possible, bring in a stethoscope and allow students to listen to their heartbeats. Encourage students to listen for a "lubb dup" sound. Tell students to note the explanation for what causes this sound as they read the section.

Emphasis

Make certain that students are able to describe the morphology of the human circulatory system and lymphatic system. They should also be able to relate the structure of the organs of these systems to their functions.

Using the Features/William Harvey and Circulation

The main process skills highlighted in this feature are observing, experimenting, and inferring. The feature describes the state of understanding of the human circulatory system before and after Harvey's observations and experiments. The feature outlines how Harvey inferred the circular motion of blood flow through the closed system. In presenting the text, stress Harvey's observation of blood flow in live animals. Make sure students can summarize Harvey's reasoning.

Demonstration

Ask a volunteer to let you or another student measure his or her pulse rate for fifteen seconds. Let the class calculate the resting pulse rate in beats per minute. Direct the volunteer to walk around the room for two minutes. Immediately measure the pulse rate for fifteen seconds and ask the class to calculate the walking pulse rate in beats per minute. Finally, ask the volunteer to run in place for two minutes. Immediately measure the pulse rate for fifteen seconds and ask the class to calculate the running heart rate in beats per minute. Discuss what the demonstration shows about relationship between levels of exercise and pulse rate. Talk about the cause of this correlation.

Section Review Answers

(Continued from p. 661.)
4. The circulatory system distributes oxygen, hormones, and nutrients throughout the body and carries away CO_2 and other wastes.
5. The walls of a capillary are only a single cell thick, thus allowing easy diffusion of gases in and out.

43.2 Blood

Approach

The main objective of this section is to make students aware of the characteristics of blood and the functions of the components that make up blood. This section builds on the concept of diffusion presented in Chapter 6 and helps students better understand the general functions of the circulatory system presented in Section 43.1 of this chapter. The section also prepares students to understand the functions of the immune system, covered in Chapter 44. In presenting the text, guide students to realize that blood is a complex tissue composed of solid particles suspended in a solution.

Concept Framework

Students are aware that blood flows as a liquid from a cut. Point out that blood is really a suspension of blood cells and other particles floating in a watery solution. Encourage students to compile a chart of the components of the blood and all their functions as they read.

Emphasis

Make certain that students know the components of the blood and understand that blood types are determined by chemicals on the surfaces of erythrocytes.

Exhibit

Obtain slides containing erythrocytes and different types of leukocytes and set up an exhibit of these slides for students to look at. Encourage students to note the shape of the erythrocytes and the presence of nuclei, which show up as stained portions, in leukocytes. You may wish to have students make drawings of each type of blood solid.

Section Review Answers

(See p. 664.)

43.3 Respiratory System

Approach

The main objective of this section is to describe the structure and function of the respiratory system. This section further develops concepts presented in Sections 43.1 and 43.2 of this chapter and presents important background information students will need for understanding the harmful effects of tobacco discussed in Chapter 48. In presenting the text, guide students to realize that exchange of gases in the body depends on the interaction between the respiratory and circulatory systems.

Concept Framework

Students are aware that the lungs function to bring oxygen into the body. Bring out this knowledge letting them outline the path air takes into the lungs and discuss how oxygen is transferred to the blood. You may wish to refer them to a poster of the respiratory system or to Figure 43-10 on page 665. Tell students to note how the process of breathing occurs and how gases are exchanged in the lungs as they read the text.

Emphasis

Students should understand how the respiratory system and the circulatory system work together to transport oxygen to the cells and to remove waste products.

Field Trip

Take students on a field trip to a hospital or medical school and arrange for a doctor, nurse, or medical technician to explain how a heart-lung machine and an iron lung (or negative-pressure respirator) work.

Ask them to discuss the conditions under which these machines are used. If possible discuss the similarities and differences between the ways these machines operate and the ways the heart and lungs work.

Section Review Answers

(See p. 668.)

Laboratory Preparation

Background

This lab procedure allows students to measure and learn about lung capacity and the process of breathing. Students inflate round balloons according to instructions and measure their diameters. Students then convert the diameters to volumetric units. Students make direct inferences from these volume calculations to lung capacity, reserve volume, expiratory reserve volume, and vital capacity. Students need to be familiar with the concepts of tidal volume, expiratory reserve volume, and vital capacity. They should also be familiar with the mathematical relationship between the diameter and the volume of a sphere.

Safety

Caution students to breathe normally between steps to avoid hyperventilation.

Conclusions

1. The procedure with the balloon can measure only exhaled air. Inspiratory reserve volume refers to inhaled air.
2. To calculate inspiratory reserve volume, add the average expiratory reserve volume to the average tidal volume. Subtract this sum from the vital capacity.
3. A variation from the norm in lung capacity can suggest the presence of respiratory problems.

Inquiry

Answers will vary. Some factors affecting lung capacity include a person's size, health, sex, and level of physical fitness. An additional factor is the altitude to which a person is acclimatized. For example, Andean Indians who live at high altitudes generally have much larger lung capacities per body volume than people living at sea level.

Chapter Review

Vocabulary

1. Systole is the first phase of a heartbeat, in which the ventricles contract; diastole is the second phase, in which the ventricles relax.
2. Arteries, capillaries, and veins are all vessels that transport blood.
3. Total lung capacity is the total amount of air a person's lungs can hold. Vital lung capacity is the maximum amount of air a person can inhale and exhale.
4. These are all contributing factors to the formation of blood clots.
5. sinoatrial node

Review

11. Blood enters the right atrium, travels to the right ventricle, goes to the lungs, enters the heart again through the left atrium, goes to the left ventricle, and is then pumped out into the aorta.
12. Oxygenated blood leaves the heart through the aorta and travels through the arteries to the arterioles and then to the tiny capillaries. Oxygen and nutrients pass through the thin capillary walls into the cells.
13. The pulmonary artery is the only artery that carries deoxygenated blood.
14. The lymphatic system helps remove excess intercellular fluid and is also part of the infection-fighting system.
15. They do not have nuclei.
16. The blood will agglutinate and block the recipient's blood vessels.
17. Injections of anti-Rh antibodies help prevent the damage caused by Rh incompatibility.
18. During inspiration the diaphragm flattens and the intercostal muscles pull the ribs up and out. During expiration the diaphragm relaxes and the intercostal muscles relax, lowering the ribs.
19. Human vocal sounds are produced by air passing through the vocal cords in the larynx.
20. High level of CO_2 in the blood stimulates the brain to increasing the breathing rate.

Critical Thinking

1. A shortage of erythrocytes would cause a lack of energy because the body would not have enough oxygen. This lack of oxygen is caused by a shortage of hemoglobin, the part of the erythrocyte that carries oxygen.
2. Polio affects the muscles that control breathing. Since these muscles cannot move, the polio victim cannot breathe.
3. The blood on the right side of the heart would mix with blood on the left side of the heart. This means that blood high in oxygen would mix with blood high in carbon dioxide and not enough oxygen would get to the organs.

Chapter 43 Circulatory and Respiratory Systems

Introduction

A single-celled organism is often bathed in a nutrient- and oxygen-rich fluid. Its survival depends on the exchange of substances through the cell membrane. For humans the process of nutrient and gas exchange is not so simple. The human circulatory and respiratory systems transport substances to and from cells embedded deep within the body.

Chapter Outline

Chapter Concept

As you read, identify ways in which the structures of the circulatory and respiratory systems are related to their functions of transport and exchange of materials.

Human tricuspid valve. Inset: Color-enhanced SEM of bronchial mucus gland (×1,172).

43.1 Circulatory System

Human life depends on the distribution of oxygen, hormones, and nutrients to cells in all parts of the body and on the removal of carbon dioxide and other wastes. These tasks are partially carried out by the **circulatory system,** which consists of the heart, an intricate network of blood vessels, and blood. The blood moving through the vessels serves as the transport medium for oxygen, nutrients, and other substances.

The Heart

The **heart** is a muscular organ that pumps blood through the body. About four weeks following fertilization, the fetal heartbeat can be heard as it pumps blood through the developing fetus. The heart continues to beat without interruption more than 2.5 billion times in the average life span. The typical adult heart beats 72 times and pumps about 5.5 L of blood per minute.

The heart rate can change, however, depending on a person's activity level. During strenuous exercise tissues need more oxygen. The heart responds by beating faster and pumping more vigorously. At its maximum rate the heart can reach between 180 and 195 beats per minute and force 17.5 L of blood through its chambers. That's about three times the normal load.

Structure

The heart consists of cardiac muscle, nervous tissue, and connective tissues. The adult heart is about the size of a fist. It lies within the thoracic cavity, behind the sternum and between the two lungs. A tough saclike membrane called the **pericardium** *(PER-uh-KAHR-dee-um)* surrounds the heart and secretes a fluid that reduces friction as the heart beats.

The **septum** divides the heart lengthwise into two sides. The right side pumps blood to the lungs, and the left side pumps blood to the other parts of the body. Each side of the heart is divided into an upper and lower chamber. Each upper chamber is called an **atrium** *(AY-tree-uhm)*, and each lower chamber is called a **ventricle** *(VEN-trih-kul)*.

In each side of the heart blood flows from the atrium to the ventricle. One-way valves separate these chambers and prevent blood from moving in the wrong direction. The atrioventricular (a-v) valves separate the atria from the ventricles. The a-v valve on the right side is the tricuspid valve, and the a-v valve on the left is the bicuspid valve. Blood also flows out of the ventricles. Two semilunar (s-v) valves separate the ventricles from the large vessels through which blood flows out of the heart. Each of the valves consists of flaps of tissue that open as blood is pumped out of the ventricle. Locate the tricuspid, bicuspid, and semilunar valves in Figure 43-1.

For additional teaching information see pp. T43A–T43D.

Section Objectives

- *List the different types of vessels found in the circulatory system.*

- *Describe the movement of blood through the heart.*

- *Explain how the sinoatrial and atrioventricular nodes control each heartbeat.*

- *Distinguish between the two primary subsystems of circulation.*

- *Explain how William Harvey discovered the circular flow of blood.*

Tip: Point out that, compared with the two-chambered fish heart, the human heart may be thought of as a two-heart system.

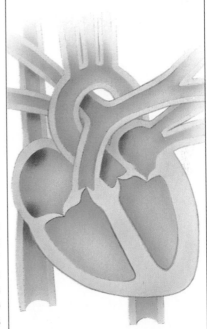

Figure 43-1 Valves control the flow of blood between sections of the heart and from the heart to the rest of the body.

Figure 43-2 Notice the pathway
of blood through the heart and
the major blood vessels leading
to and from the heart.

Arteries leading to the head
and upper body

Pulmonary
artery

Aorta

Pulmonary
artery

Superior
vena cava

Pulmonary
veins

Pulmonary
veins

Inferior
vena cava

Figure 43-3 Each heartbeat
starts with a nerve impulse from
the sinoatrial node. Within a
tenth of a second the impulse
reaches the atrioventricular node.

Circulation in the Heart

As shown in Figure 43-2, blood returning to the heart, which has a
high concentration of carbon dioxide and a low concentration of
oxygen, enters the right atrium. The atrium pumps it through the
tricuspid valve into the right ventricle. The muscles of the right
ventricle contract and force the blood through the right semilunar
valve and into vessels leading to the lungs. In the lungs, carbon
dioxide leaves the circulating blood and oxygen enters it. The
oxygenated blood returns to the left atrium of the heart. The oxygenated blood is then pumped through the bicuspid valve into the
left ventricle. Strong contractions of the muscles of the left ventricle force the blood through the semilunar valve, into a large blood
vessel, and then throughout the body.

Control of Heartbeat

The heart consists mostly of cardiac muscle, which functions differently from skeletal muscle. Unlike skeletal muscle, in which
each cell contracts individually, cardiac muscle cells contract in
waves. When the first cell is stimulated, it excites neighboring
cells. Those cells in turn stimulate surrounding cells. This chain
reaction proceeds until all the cells contract. As the activity
spreads, the atria and ventricles contract in synchronous rhythm.

The initial stimulus of this wavelike contraction begins in the
sinoatrial *(sie-noe-AY-tree-ul)* **node** (s-a node), a specialized re-

gion of muscle and nerve tissue located in the right atrium, as shown in Figure 43-3. The s-a node is often called the pacemaker because it regulates the rate of contraction of the entire heart. The s-a node generates an electrical charge that stimulates the atria to contract. About a tenth of a second later the impulse reaches the **atrioventricular** *(AY-tree-oe-ven-TRIK-yuh-lur)* **node** (a-v node), located in the septum between the atria. The a-v node delays the impulse for about a millisecond and then sends it into the ventricles. This electrical charge causes the ventricles to contract, completing the full beat.

The beat just described can also be described as having two phases. In one phase, called **systole**, the ventricles contract and force blood into the arteries. In another phase, called **diastole**, the ventricles relax, and the blood flows in from the atria. The result is the characteristic "lubb dup" we call a heartbeat.

Blood Vessels

Each beat of the heart forces blood through blood vessels. The large, muscular vessels that carry blood away from the heart are called **arteries**. The thick walls of the arteries have three layers, as shown in Figure 43-4: an inner endothelium, a middle layer of smooth muscle, and an outer layer of connective tissue. This structure gives arteries a combination of strength and elasticity that allows them to stretch as blood under high pressure enters them from the heart. You can feel this stretching of arteries as the pulse in your wrist.

When the left ventricle contracts, it forces blood through the left semilunar valve into the body's largest artery, the **aorta**. From the aorta blood travels through a network of smaller arteries, which in turn divide and form even smaller vessels called **arterioles**. The arterioles branch into a fan of tiny vessels called **capillaries**, shown in Figure 43-5. Capillary walls are only one cell thick. Diffusion of gases and nutrients takes place across the thin capillary walls from areas of greater concentration to areas of lesser concentration. If the concentrations of oxygen and nutrients are higher in the blood than in the surrounding cells, these substances will diffuse from the blood into the cells. In this way cells take in oxygen and nutrients and in a similar manner discharge wastes into the blood.

Deoxygenated blood flows through capillaries that merge and form larger vessels called **venules**. Several venules in turn unite to form a **vein**, a large blood vessel that carries blood to the heart. From the lower part of the body the venules transport blood into veins that merge to form the **inferior vena cava** *(VEE-nuh KAY-vuh)*. Venules from the upper part of the body merge to form veins, which in turn unite to form the **superior vena cava**. These large veins then carry blood back to the heart.

The walls of veins, which are also composed of three layers, are thinner and less muscular than those of arteries. By the time

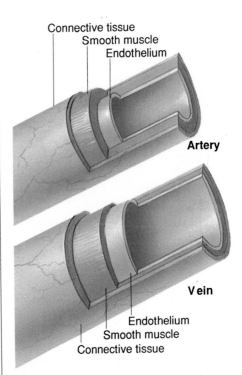

Connective tissue
Smooth muscle
Endothelium

Artery

Vein

Endothelium
Smooth muscle
Connective tissue

Figure 43-4 The muscle layer of an artery is thicker and stronger than the muscle layer of a vein. Why is this so?

Figure 43-5 Erythrocytes move in single file through the capillaries, as shown in this photo (×1,200).

blood reaches the veins, it is under much less pressure than it is in the arteries. Thus the veins are not as strong or as flexible as the arteries. Many veins pass through skeletal muscles; during movement these muscles contract, squeezing blood through the veins. Valves prevent the blood from moving backward or downward. These valves, which are usually located above the point at which two veins meet, allow blood to flow in only one direction through the veins. However, when muscle tone is poor, a person's valves may not fully come together, which will then allow some blood to seep backward through the valves.

Patterns of Circulation

The English scientist William Harvey (1578–1657) first showed that the heart and blood vessels formed one continuous, closed system of circulation. To communicate effectively, scientists often say this system is divided into various subsystems. The two primary subsystems of circulation are pulmonary circulation, in which the blood travels between the heart and lungs, and systemic circulation, in which the blood travels between the heart and all other body tissues.

Biology in Process

William Harvey and Circulation

Today it may seem obvious, but until the seventeenth century scientists did not know that blood flowed in a closed pattern through the body or that the blood was continuously recycled. Changes in scientific methods of experimentation during the fifteenth and sixteenth centuries led to these discoveries. During this period, in order to better understand the anatomy of the body, many scientists began to dissect and observe human cadavers.

By the beginning of the sixteenth century scientists had discovered that the heart was separated into left and right sides. Before that time scientists had believed that blood seeped through pores in the septum. They also discovered that there are valves in the veins, although they did not understand the function of these valves.

William Harvey, an English physician, began to research the movement of the blood. He also determined that there were no pores in the septum. If the septum was closed and the veins contained valves, how could these facts be used to explain how blood traveled through the body? Harvey did his research during the first half of the seventeenth century.

Pulmonary Circulation

Pulmonary circulation is the movement of blood from the heart to the lungs and back to the heart. Blood from the right ventricle flows through the right semilunar valve into the **pulmonary artery** and then to the lungs. The pulmonary artery is the only artery that carries deoxygenated blood. At the lungs the pulmonary artery divides into two smaller arteries, one leading to each lung. These arteries branch into arterioles and then into capillaries. Blood in the capillaries releases carbon dioxide and absorbs oxygen. Oxygenated blood then flows into venules, which merge into the **pulmonary veins** that lead to the left atrium of the heart. The pulmonary veins are the only veins that carry oxygenated blood.

Systemic Circulation

Systemic circulation is the movement of blood to all parts of the body except the lungs and then back to the heart. Blood from the left ventricle enters systemic circulation through the aorta. Systemic circulation is further divided into several other subsystems. These subsystems are coronary circulation, renal circulation, and hepatic portal circulation.

Answer to Biology in Process
Harvey used research, observation of cadavers and living organisms, collection of data, experimenting, and drawing conclusions.

dissecting and observing human cadavers as did earlier scientists. However, he carried his experiments one step further. In addition to observing the anatomy of a cadaver, Harvey studied the anatomy of living organisms. By collecting data on the circulation of live animals, he reasoned, he would be able to draw inferences about the way blood moved through the human body.

In his research Harvey obstructed the blood flow in live animals. When he pinched the vena cava of a live snake, for instance, he observed that the heart would empty its contents. When he blocked the flow of blood into the aorta, the snake's heart would fill to capacity.

Convinced that blood from the veins traveled to the arteries, Harvey began to wonder why the veins didn't empty immediately and the arteries overflow. Using his knowledge that blood did not flow through the septum and that valves prevented blood from flowing backward, Harvey hypothesized that blood flowed in a circular motion. In his work Harvey wrote that blood could not flow from veins to arteries unless it returned first from the arteries back into the veins and then flowed to the right ventricle of the heart. "I began," Harvey wrote, "to consider whether [the blood] might have a kind of motion, as it were, in a circle, and this I afterward found to be true."

■ What processes did Harvey use to discover the circulation of the blood?

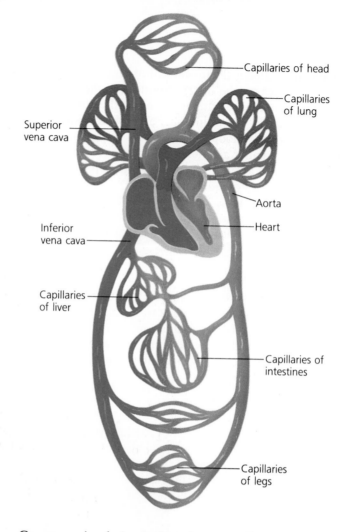

Figure 43-6 Pulmonary circulation carries blood from the heart to the lungs and back to the heart. Blood circulates through the rest of the body via the systemic system.

Capillaries of head

Capillaries of lung

Superior vena cava

Aorta

Heart

Inferior vena cava

Capillaries of liver

Capillaries of intestines

Capillaries of legs

Think

Endurance athletes often have larger hearts, slower resting heart rates, and an increased coronary circulation as a result of endurance training. How would these modifications help the heart deal with the demands placed on it by endurance activities?

Answer to Think
The larger heart means stronger contractions and thus greater blood flow from the heart. The slower resting rate indicates greater efficiency. The increased circulation means more blood—thus more endurance—for the heart.

Coronary circulation is the subsystem of systemic circulation that supplies blood to the heart itself. Two **coronary arteries** branch off from the aorta and transport blood to arterioles that penetrate the tissue of the heart. The blood then returns to the right atrium through a large vein called the coronary sinus.

With all the blood traveling through the heart, you might think that the heart would have no need for its own arteries. However, heart muscle is thick, and every cell needs oxygen. If, for example, an artery becomes clogged and the blood supply to part of the heart is reduced, muscle cells will die. An artery may become blocked by a blood clot or by atherosclerosis, a disease characterized by the buildup of fatty materials on the walls of the coronary artery. If either blockage totally shuts off the flow of blood to the heart, a heart attack will result.

Renal circulation is the subsystem of systemic circulation that moves blood through the kidneys and back to the heart. Two renal arteries branch from the aorta, one entering each kidney. The kidneys filter nitrogenous wastes from the blood, which then leaves through the renal veins. The renal veins join the inferior vena cava.

Info: During systole blood is first distributed to the tissue of the heart itself via the coronary arteries.

Hepatic portal circulation is the subsystem of systemic circulation that moves the blood through the digestive tract and liver. Hepatic portal circulation involves two vessels. One, the hepatic artery, is the branch of the aorta that delivers oxygenated blood. The other, the hepatic portal vein, brings blood with nutrients from the digestive tract to the liver. Unlike other veins, the hepatic portal vein branches into capillaries that distribute nutrients to liver tissue. Ultimately blood from the hepatic artery and blood from the hepatic portal vein mix and leave the liver through the hepatic vein, which joins the inferior vena cava.

Lymphatic System

Another part of the body's circulatory system is the **lymphatic system,** which is a series of one-way vessels that carry intercellular fluid from the tissues to the bloodstream. The lymphatic and circulatory systems are together often referred to as the body's vascular system.

Some fluid from the blood passes through the capillary walls along with oxygen and nutrients. Some, but not all, of the fluid returns to the blood along with wastes that diffuse through the capillary walls. Some fluid remains in the intercellular spaces. A buildup of this fluid would lead to bloating of the tissues, and so it must be removed. This fluid collects in open-ended vessels of the lymphatic system. Fluid from the blood that has entered the lymph vessels is called **lymph.**

Lymph capillaries transport lymph to lymph vessels and on to two large ducts, the right lymphatic duct and the thoracic duct. The two ducts empty into large veins in the upper chest.

The lymphatic system also plays a role in immunity. Small clumps of cells called nodes are located on the lymphatic vessels like beads on a string. Notice in Figure 43-7 that these lymph nodes are concentrated in the neck, armpits, inner elbow, and groin. Nodes filter the lymph as it passes, catching foreign particles, microorganisms, and other tissue debris. They also produce cells called lymphocytes that help fight disease. When a person has an infection, the nodes enlarge because they are manufacturing more lymphocytes.

Section Review

1. Distinguish between an *artery* and a *vein*.
2. What prevents blood from moving in the wrong direction in veins?
3. Trace the path that the blood follows in systemic circulation.
4. How would you define the purpose of the entire circulatory system?
5. How does the structure of a capillary enable it to transport substances to cells?

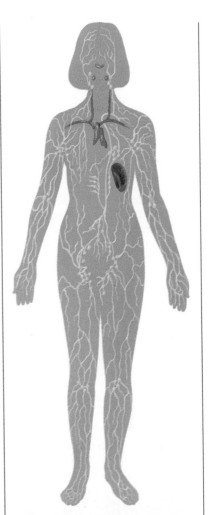

Figure 43-7 The lymphatic system carries fluid that collects in intercellular spaces back into the blood system.

Answers to Section Review
1. An artery transports blood away from the heart; a vein transports blood to the heart.
2. One-way valves close after blood flows through them, preventing it from flowing the wrong way.
3. Blood leaves the heart, enters the aorta, and travels through decreasingly smaller arteries until it enters capillaries. The capillaries fuse and form venules which fuse and form increasingly larger veins that lead to the vena cava, which enters the heart.
(See pp. T43A–T43D.)

Circulatory and Respiratory Systems   **661**

43.2 Blood

Blood is a liquid connective tissue that forms the transport medium of the circulatory system. The major function of the blood is to transport nutrients and oxygen to the cells and to carry away carbon dioxide and nitrogenous wastes from the cells. Blood also transfers body heat to the body surface and plays a major role in defending the body against disease.

Composition of Blood

Blood is composed of a liquid called plasma and blood solids. Blood solids consist of erythrocytes, also called red blood corpuscles, white blood cells, and platelets. Plasma makes up 55 percent of the blood, and blood solids make up the remaining 45 percent.

Plasma

Plasma is a sticky, straw-colored fluid that is about 90 percent water. Cells receive nourishment from dissolved substances carried in the plasma. These substances—including vitamins, minerals, amino acids, and glucose—are absorbed from the digestive system and transported to the cells. Distribution of these nutrients is essential for the cells to function properly. Plasma also carries hormones and brings wastes from the cells to the kidneys or lungs to be removed from the body.

The various proteins found in plasma have important functions. One protein, called **fibrinogen,** is essential for the formation of blood clots. Serum albumin plays an integral role in the regulation of osmotic pressure between plasma and blood cells and between plasma and tissues. Other proteins, called serum globulins, include antibodies that help the body fight disease.

Erythrocytes

Red blood corpuscles, or **erythrocytes** *(uh-RITH-ruh-SITES),* illustrated in Figure 43-8, are vital to the survival of cells in all parts of the body because they transport oxygen. When erythrocytes are formed in the red marrow of bones, an iron-containing protein called **hemoglobin** *(HEE-muh-GLOE-bun)* also forms. Hemoglobin is the molecule that actually transports oxygen and carbon dioxide. During the formation of the corpuscle, its cell nucleus and organelles disintegrate. The mature erythrocyte is therefore merely a membrane sac containing hemoglobin.

Because they lack nuclei, erythrocytes cannot repair themselves and have a limited survival period, usually 120 to 130 days. Of the more than 30 trillion red blood corpuscles circulating throughout the body at one time, 2 million disintegrate every second. The remains of the membranes are removed by the liver and spleen. The iron portion of the hemoglobin molecule is carried in the blood to the marrow, where it is reused for new corpuscles.

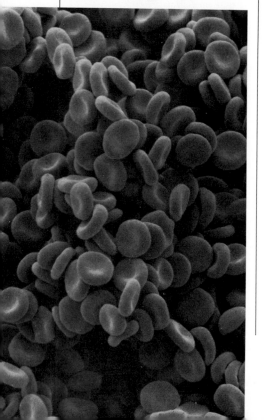

Figure 43-8   Each disk-shaped erythrocyte in this SEM is filled with hemoglobin (×2,500).

Leukocytes

White blood cells, or **leukocytes** (*LOO-kuh-SITES*), are cells that help defend the body against disease. They are formed in red marrow, lymph nodes, and the spleen. Leukocytes are larger than erythrocytes and significantly less plentiful. A drop of blood, about 0.05 mm^3, usually contains about 5 million erythrocytes but only 7,000 leukocytes. Notice in Figure 43-9 that a leukocyte may be irregularly shaped and may have a rough outer surface. In contrast with the short-lived erythrocytes, leukocytes may function for years.

Some types of white blood cells are **phagocytes**—that is, cells that engulf invading microorganisms. Other white blood cells produce **antibodies,** proteins that destroy foreign substances. When a person has an infection, the number of white blood cells can increase tenfold.

Platelets

Look again at Figure 43-8. Locate the **platelets** that are about 2 μm in diameter. Platelets, also known as thrombocytes, are not whole cells but parts of cytoplasm that have broken off from larger cells formed in the marrow. Platelets lack a nucleus and have a life span of between 7 and 11 days.

Platelets are essential to the formation of a **blood clot,** a mass of interwoven fibers and blood cells that prevents excess loss of blood from a cut. When a blood vessel tears or rips, platelets congregate at the damaged area, stick together, and form a small plug. The vessel constricts, slowing blood flow to the area. Then special clotting factors are released from the platelets. These factors interact with prothrombin, a protein in plasma, to form the enzyme thrombin. This enzyme immediately reacts with a blood protein called fibrinogen. Together thrombin and fibrinogen form long sticky threads called **fibrin,** which forms a net that traps erythrocytes. This structure of fibrin and erythrocytes hardens into a clot or scab.

Blood Types

The blood carries many chemicals, which have various functions. One group of chemicals, called **antigens,** can stimulate the production of antibodies. Three of the most important human antigens are called A, B, and Rh. The A–B–O system of blood typing is based on the A and B antigens.

A–B–O System

The A–B–O system is a means of classifying blood according to the antigens located on the surface of the erythrocytes. An individual's erythrocytes may carry an A antigen, a B antigen, both A and B antigens, or no antigen at all. These antigen patterns are called A, B, AB, and O, respectively. When A blood is added to B

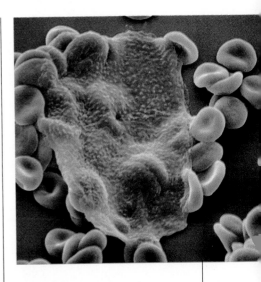

Figure 43-9   One type of phagocyte, called a macrophage, is shown destroying old erythrocytes in this color-enhanced SEM (×2,500).

Tip: Tell students that when the clotting factor in the blood is deficient, a person suffers from a condition called hemophilia. Hemophilia basically is a male disease, although a mild form may also affect females.

Table 43-1 Blood Types

Type	Genotype	Antigen on the RBC	Antibodies in Plasma	Can Get Blood from	Can Give Blood to
A	$I^A i^O$ or $I^A I^A$	A	Anti-B	O, A	A, AB
B	$I^B I^B$ or $I^B i^O$	B	Anti-A	O, B	B, AB
AB	$I^A I^B$	AB	None	A, B, AB, O	AB
O	$i^O i^O$	None	Anti-A, Anti-B	O	A, B, AB, O

blood, the antibodies in the B blood attack the foreign antigens in A blood, and the blood clumps, or agglutinates. In a body the agglutinated blood will block the vessels. For this reason transfusion recipients must receive blood that is compatible with their own. Table 43-1 explains which blood types are compatible.

Rh Factor

The **Rh factor** is another type of blood antigen. Eighty-five percent of the U.S. population is Rh-positive (Rh$^+$), meaning that they have Rh antigens on their blood cells. A person without Rh antigens in their blood is termed Rh-negative (Rh$^-$). If Rh$^+$ blood is put into the body of an Rh$^-$ person, antibodies to the Rh antigen will react with the antigen and agglutination will occur.

Complications arising from Rh incompatibility may occur during pregnancy. When the mother is Rh$^-$ and the father is Rh$^+$, the child may inherit the dominant Rh$^+$ allele from the father. Because the blood supply of mother and fetus is separate, only a small amount of the fetus's Rh$^+$ blood reaches the mother's bloodstream, usually during delivery. However, the mother then develops antibodies to the Rh factor. If a second Rh$^+$ child is later conceived, serious damage may result when the mother's antibodies attack the child's blood. This condition is called erythroblastosis fetalis. The fetus may die, or if the child is born alive, it may need an immediate transfusion of Rh$^+$ blood.

In the past several decades these concerns have largely been overcome. Now the Rh$^-$ mother is given anti-Rh antibodies within three days after the birth of her first Rh$^+$ child. These antibodies destroy the Rh antigens in her blood before her body makes antibodies, eliminating any danger to a second child.

Section Review

1. What is *plasma?*
2. What are the three types of blood solids?
3. What is the final result of the clotting process?
4. Which blood type can be donated to all others? Why?
5. Why should a pregnant woman know her blood type?

Info: A transfusion may consist of whole blood (fluid plus cells), packed cells (cells only), serum (plasma only), or platelets.

Answers to Section Review
1. the liquid part of the blood
2. erythrocytes, leukocytes, and platelets
3. long, sticky threads called fibrin, which weave a net and trap erythrocytes and then harden into a scab
4. type O, because it does not have any antigens
5. in order to prevent Rh incompatibility problems

43.3 Respiratory System

You have read how the blood transports oxygen from the lungs to cells and carries carbon dioxide from the cells to the lungs. It is the function of the **respiratory system** to transport gases to and from the circulatory system.

External respiration is the exchange of gases between the atmosphere and the blood. **Internal respiration** is the exchange of gases between the blood and the cells of the body. Cellular respiration is the breakdown of glucose in the cell. In this section we will examine the structures and mechanisms involved in the exchange of gases between cells and the external environment.

The Lungs

The two triangular **lungs** are the sites of gas exchange between the atmosphere and the blood. The right lung has three divisions, or lobes, and is slightly larger than the two-lobed left lung.

The lungs are located inside the thoracic cavity, bounded by the rib cage and the **diaphragm,** a sheet of muscle at the bottom of the thoracic cavity. Lining the entire cavity and encasing the lungs are pleura, double membranes that secrete a mucus that decreases friction from the movement of the lungs during breathing. Figure 43-10 illustrates the organs of the respiratory system.

Section Objectives

- *Name and describe the organs of the respiratory system.*

- *Summarize the skeletal and muscular changes that occur during breathing.*

- *Explain how the structure of alveoli is related to the function of diffusion of gases.*

- *Describe the physical effects of exercise on the breathing rate.*

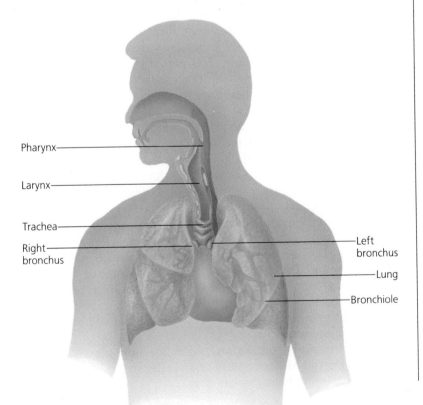

Figure 43-10 Air passes through these main parts of the respiratory system each time a person breathes.

Pharynx

Larynx

Trachea

Right bronchus

Left bronchus

Lung

Bronchiole

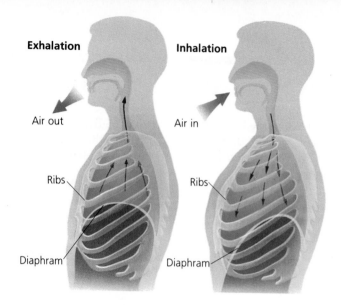

Exhalation

Air out

Ribs

Diaphram

Inhalation

Air in

Ribs

Diaphram

Figure 43-11 Muscles of the diaphragm and the intercostal muscles between the ribs control the movement of the rib cage during breathing.

Mechanism of Breathing

Breathing is the process of inspiring and expiring air. When you breathe, air enters and leaves the lungs. Each phase of breathing involves interactions of the diaphragm, abdomen, and ribs. **Inspiration** is the process of taking air into the lungs. When you take a deep breath, you can feel your chest expand. Your diaphragm flattens as the muscles contract. This pushes down on the abdomen, as shown in Figure 43-11. Muscles in the abdominal wall in turn relax. This action provides room for the flattened diaphragm. At the same time, the intercostal muscles contract, moving the ribs up and outward. This enlarges the chest.

When the diaphragm flattens and the ribs are pushed out, the volume of the thoracic cavity increases. This causes air pressure within the cavity to decrease. The pressure inside the thoracic cavity thus becomes lower than the air outside of the body. As a result air from the atmosphere moves into the lungs and equalizes the pressure.

During **expiration** the reverse process takes place. The diaphragm and rib muscles relax while the abdominal muscles contract. The size of the thoracic cavity then decreases. The air pressure inside the cavity is then greater than the air pressure outside the body. This pressure difference forces air out of the lungs until the pressure is again equalized.

Each of the lungs weighs about 600 g. Together they can hold between 5 and 6 L of gas, the **total lung capacity**. When you breathe, you normally inhale only about 0.5 L instead of the entire 6 L of total lung capacity. Even during deep breathing you can't completely empty your lungs. Through strenuous exercise it is possible to increase the capacity of a single breath to about 4.5 L. The maximum amount of air a person can inhale and exhale is called **vital lung capacity**.

One completed inhalation and exhalation makes up one respiratory cycle. Most people at rest breathe at a rate of 16 to 24 cycles per minute, but rates can vary.

The Passage of Air

External respiration begins at the mouth and nose. Air filters through the small hairs of the nose and passes into the nasal cavity located above the roof of the mouth. In the cavity mucous membranes warm and moisten the air, a process that prevents damage to the delicate lung tissue. The walls of the nasal cavity are lined with cilia. These catch inhaled particles and sweep them into the throat, where they are swallowed.

The moistened, filtered air then moves into the **pharynx,** a tube at the back of the nasal cavities and mouth. The pharynx contains passageways for both food and air. When food is swallowed, a flap of cartilage, called the epiglottis, presses down and covers the opening to the air passage. During inhalation the epiglottis is in an upright position, and air passes into a cartilaginous tube called the **trachea,** or windpipe. The trachea is about 10 to 12 cm long and has walls lined with ciliated cells.

The **larynx,** or voice box, is located at the upper end of the trachea. Sounds are produced when air is forced past two ligaments—the vocal cords—that stretch across the larynx. The pitch and volume of sounds varies with the amount of tension on the vocal cords and the amount of air being forced past them.

The trachea branches into two **bronchi** *(BRAHNG-kie),* each of which leads to a lung. Within the lungs, the bronchi branch into smaller and smaller tubes. The walls of the bronchi consist of smooth muscle and cartilage and are lined with cilia and mucus. The smallest tubes, the **bronchioles** *(BRAHNG-kee-OELZ),* are lined with mucus and cilia. Eventually the bronchioles end in clusters of tiny air sacs called **alveoli** *(al-VEE-oe-LIE).* A network of capillaries surrounds each alveolus. It is here that the exchange of gases takes place. Each lung contains nearly 300 million alveoli, each measuring from 0.1 to 0.2 mm in diameter. This results in a total surface area of about 70 m^2 per lung.

Gas Exchange and Transport

Within the lungs gases are exchanged between the alveoli and the blood in the capillaries. Oxygen moves into the bloodstream, and carbon dioxide moves into the air passages. How does this exchange occur?

Gas Exchange in the Lungs

When air moves into the lungs, oxygen from the air dissolves in the mucus secreted by the lining of the alveoli. The dissolved oxygen then crosses both the thin alveolar membranes and the capillary walls and enters the blood.

What determines the direction in which oxygen and carbon dioxide move? Air moving into the alveoli is rich in oxygen and contains little carbon dioxide. Blood in the capillaries surrounding the alveoli is low in oxygen and contains high levels of carbon

Tip: Use a pad of paper to demonstrate how the alveoli provide a great surface area. Arrange the individual sheets of paper side by side over a large area. You may wish to have students calculate the area covered by a single pad.

Figure 43-12 Blood vessels surround the millions of air sacs, or alveoli, in the lungs.

Figure 43-13 Oxygen and CO_2 diffuse through the alveolar membranes and capillary walls.

dioxide. Gases move by diffusion from areas of higher concentration to areas of lower concentration. The differences in concentration cause CO_2 to diffuse from the blood into the alveoli and the oxygen to diffuse from the alveoli into the blood.

Transport and Exchange of Gases

When oxygen diffuses into the blood, only a small amount dissolves in the blood plasma, not enough to nourish the tissues. Hemoglobin makes up the difference. Each hemoglobin molecule can carry up to four molecules of oxygen, which bind with its central iron atom. Hemoglobin transports about 97 percent of the oxygen in the blood. When oxygenated blood reaches tissues, hemoglobin releases O_2. The O_2 diffuses out of the capillaries and into the surrounding cells. Similarly, CO_2 diffuses out of a cell and into the bloodstream. When CO_2 leaves cells and diffuses into the blood, only about 8 percent of the total CO_2 actually dissolves in the blood plasma. About 25 percent binds to hemoglobin. The remaining 67 percent is converted to bicarbonate ions:

$$CO_2 + H_2O \rightarrow H_2CO_3 \rightarrow HCO_3^- + H^+$$

Bicarbonate ions (HCO_3^-) dissolve in plasma. When the blood reaches the lungs, a reverse reaction takes place. CO_2 leaves the capillaries by diffusion and is exhaled into the atmosphere.

Regulation of Breathing

What controls the rate of breathing? Because cells use oxygen for respiration, the rate of oxygen use depends on the activity of the cells, which in turn depends on the activity of the body. For example, cells consume less oxygen when a person is sleeping than when exercising, since a sleeping individual needs less energy to function. **Basal metabolic rate** (BMR) is a measure of the energy required to maintain the body's functions.

As the cells use oxygen in aerobic respiration, they release carbon dioxide into the bloodstream. As metabolism increases, high levels of CO_2 in the bloodstream stimulate nerve cells to send a message to the brain, which in turn sends messages to respiratory muscles to increase the breathing rate. When the CO_2 concentration returns to its normal level, the nerve cells signal the brain to return to a normal breathing rate.

Section Review

1. How do *external* and *internal respiration* differ?
2. Where is air warmed and moistened before it enters the lungs?
3. How is oxygen transported in blood to the cells?
4. How does CO_2 leave the cells?
5. What is the adaptive value of having the organ of gas exchange inside the body?

Info: One of the serious effects of alcoholism is disruption of the breathing control mechanism in the brain, occasionally even causing death.

Answers to Section Review
1. External respiration is the exchange of gases between the atmosphere and the blood. Internal respiration is the exchange of gases between the blood and the cells.
2. in the mouth and nasal cavity
3. by binding to the hemoglobin in erythrocytes
4. CO_2 diffuses from the cells into the blood.
5. The lungs are kept moist, thus allowing the organism to live in dry environments.

Laboratory

Measuring Lung Capacity

Objective
To measure the tidal volume, expiratory reserve volume, and vital capacity of your lungs

Process Skills
Measuring, collecting and interpreting data

Materials
Round balloon, ruler

Method
1 Prepare a chart for your data. Head three columns *Tidal Volume*, *Expiratory Reserve Volume*, and *Vital Capacity*. Allow space for five entries under each heading.
2 Blow up and deflate a balloon a few times to stretch it out.
3 Inhale normally, and then exhale a normal breath into the balloon.
4 Pinch the balloon shut. Hold it in place on the counter while your partner measures its diameter at the widest point. Record this figure under *Tidal Volume*. Tidal volume is the volume of air that enters and exits your lungs during a single, normal breath.
5 Repeat steps 3 and 4 two more times. Record your data. Calculate the average of the three measurements. Enter this figure on your chart. Does the figure actually indicate your average tidal volume? Why or why not?
6 Convert the average balloon diameter into cubic centimeters of lung volume using the conversion graph provided. Record your results under the average volume at the bottom of the *Tidal Volume* column.
7 Inhale and exhale normally. Then blow the remaining air in your lungs into the deflated balloon. Pinch the balloon shut. The volume of air that can be exhaled by force after exhaling normally is called the expiratory reserve volume. Measure the diameter of the balloon and record your results.

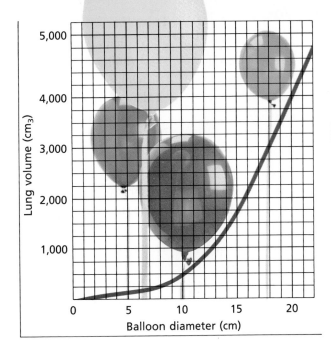

8 Repeat step 7 two more times and calculate the average of the three measurements. Convert the result into average volume.
9 The maximum amount of air you can inhale and exhale is your vital capacity. Inhale as deeply as you can. Then exhale as much air as you can into the deflated balloon. Pinch the balloon shut. Measure the diameter of the balloon and record your results.
10 Repeat step 9 two more times. Calculate the average of the measurements. Convert the result into average volume and record.

Conclusions
1 The amount of air your lungs can take in beyond your normal tidal volume is your inspiratory reserve volume. Why can you not measure this volume directly?
2 How might you calculate this volume?
3 Why is it sometimes important to know a person's lung capacity?

Inquiry
What factors might affect lung capacity?

Chapter 43 Review

Vocabulary

alveoli (667)
antibody (663)
antigen (663)
aorta (657)
arteriole (657)
artery (657)
atrioventricular node (657)
atrium (655)
basal metabolic rate (668)
blood clot (663)
bronchi (667)
bronchiole (667)
capillary (657)
circulatory system (655)
coronary artery (660)

coronary circulation (660)
diaphragm (665)
diastole (657)
erythrocyte (662)
expiration (666)
external respiration (665)
fibrin (663)
fibrinogen (662)
heart (655)
hemoglobin (662)
hepatic portal circulation (661)
inferior vena cava (657)
inspiration (666)
internal respiration (665)

larynx (667)
leukocyte (663)
lungs (665)
lymph (661)
lymphatic system (661)
pericardium (655)
phagocyte (663)
pharynx (667)
plasma (662)
platelet (663)
pulmonary artery (659)
pulmonary circulation (659)
pulmonary vein (659)
renal circulation (660)

respiratory system (665)
Rh factor (664)
septum (655)
sinoatrial node (656)
superior vena cava (657)
systemic circulation (659)
systole (657)
total lung capacity (666)
trachea (667)
vein (657)
ventricle (655)
venule (657)
vital lung capacity (666)

1. Distinguish between systole and diastole.
2. What do arteries, capillaries, and veins have in common?
3. What is the difference between total lung capacity and vital lung capacity?
4. Explain the relationship between pro-thrombin, thrombin, fibrin, and fibrinogen.
5. Identify the structure that initially stimulates cardiac muscles to contract.

Review

1. The wall that divides the heart lengthwise is the (a) ventricle (b) pericardium (c) septum (d) atrium.
2. During systole blood moves from the (a) ventricles to the atria (b) atria to the arteries (c) atria to the ventricles (d) ventricles to the arteries.
3. Pulmonary circulation involves movement of the blood to and from the (a) lungs (b) brain (c) liver (d) kidneys.
4. The lymphatic system (a) interacts with the respiratory system (b) helps the body fight infection (c) consists of a series of two-way vessels (d) transports blood from the heart to the lungs.
5. One function of plasma is (a) to carry substances that nourish cells (b) to aid in the formation of blood clots (c) to carry the majority of the oxygen supply of the blood (d) to defend against diseases.
6. The function of fibrin is (a) to transport oxygen (b) to destroy foreign substances (c) to stimulate the production of antibodies (d) to form a blood clot.
7. In the A–B–O system blood is classified according to (a) Rh factor (b) antigens located on the erythrocytes (c) number of erythrocytes (d) ability to clot.

8. During internal respiration gases are (a) exchanged between the atmosphere and blood (b) exchanged between blood and the cells (c) produced by the cells (d) warmed and moistened.
9. Gases diffuse (a) from areas of high concentration to areas of low concentration (b) from areas of low concentration to areas of high concentration (c) directly from the cells to the air passages (d) from the alveoli to the cells.
10. Cilia (a) move air molecules (b) moisten the air passages (c) sweep foreign particles into the stomach (d) sweep particles out of the air passages.
11. Describe the movement of blood through the heart.
12. Explain how the blood vessels that carry oxygenated blood to the cells are organized.
13. How is the pulmonary artery unique?
14. What are two major roles of the lymphatic system?
15. Why do erythrocytes have a limited life span?
16. What can happen to the blood of a recipient who receives a type of blood that is incompatible with his or her own?
17. What recent development has reduced the hazards of Rh incompatibility?
18. Describe the movement of the intercostal muscles of the rib cage and the diaphragm during inspiration and expiration.
19. How are human vocal sounds produced?
20. What is one factor that stimulates the brain to increase the breathing rate?

Critical Thinking

1. Anemia is a condition in which the patient has too few erythrocytes. The most common symptom is a lack of energy. Why would anemia have this symptom?
2. Polio is a disease that paralyzes muscles by affecting the nerves that make them move. Before polio vaccine was developed, many people who were stricken by polio died because they could not breathe. From what you know about the respiratory system, why might this happen?
3. Some babies are born with a hole in the septum between the two atria. Look at the drawing of the heart. Keeping in mind

what you know about blood flow through the heart, why would this condition be harmful to the baby?

Extension

1. Read the article by Ivars Peterson entitled "Heart Valves by Computer" in *Science News*, September 27, 1986, pp. 204–206. What is the main reason for valve replacement?
2. Using the library, research the Jarvik-7 artificial heart. Write a paper about the operation, the likely candidates for the operation, the number of people who have undergone the operation, and the number of operations that have been successful.
3. Construct a model of the lungs, using any materials available. Some suggestions are clay, sponge, paper, cardboard, or other art supplies. Make sure that all of the parts show: lungs, trachea, and bronchi.

Chapter 44 | Infectious Diseases and the Immune System

This chapter focuses on how the immune system fights infectious diseases. Students learn how infectious disease is transmitted and how research has enabled scientists to develop drugs to prevent, treat, and cure disease.

Correlation Chart

Section Number	Section Objectives	Pages	Features	Section Review
44.1	1. Summarize the four steps of Koch's postulates. 2. Identify five types of pathogens. 3. Describe four mechanisms of disease transmission.	673–674		1, 2 3 4, 5
44.2	1. Explain how the skin functions as a defense against disease. 2. Distinguish between a specific and a nonspecific response. 3. Describe the actions of B cells and T cells in a primary immune response. 4. Describe the relationship between vaccination and immunity. 5. Describe the methods Jonas Salk used to develop a vaccine for the polio virus.	675–679	Think, p. 678 Biology in Process, pp. 680–681	2 2, 4 3, 4, 5 1
44.3	1. Name the two chief methods of transmitting AIDS. 2. Describe how HIV affects the immune system.	680–683	Biology Lab, p. 683	4, 5 1, 2, 3

Resources*

Print

Elting, Mary. *The Macmillan Book of the Human Body.* New York: Aladdin (Macmillan), 1986.

Hancock, Graham, and Enver Carim. *AIDS: The Deadly Epidemic.* North Pomfret, Vt.: Gollanez (dist. by David & Charles), 1986.

Scott, Andrew. *Pirates of the Cell: The Story of Viruses from Molecule to Microbe.* New York: Basil Blackwell, 1985.

Audiovisual/Software

AIDS: What Everyone Needs to Know. Film or video, 18 min. Teacher's guide. Churchill, 1986.

Infection and Immunity, Film or video, 13 min. Teacher's guide. I.F.B., 1984.

The Mechanisms of Defense: Internal Defenses. Film, 26 min. Humanities and Sciences, 1985.

*Additional references are listed in *Modern Biology, Teacher's Program Guide.*

Section Planning Guides

44.1 Infectious Disease

Approach

The main objective of this section is to present the ways in which pathogens enter the body and cause disease. This section builds on the concepts of scientific processes (Chapter 2) and on students' knowledge of viruses (Chapter 19), bacteria (Chapter 20), and arthropods (Chapters 32 and 33). In presenting this section, guide students to realize that many types of organisms are known to cause infectious diseases and that these diseases are transmitted in a number of different ways.

Concept Framework

Students will be familiar with the symptoms and the mode of transmission of the common cold. Tap this knowledge to springboard a discussion of the ways diseases are transmitted. As students read the section, encourage them to consider the methods scientists could use to identify pathogens that cause colds.

Emphasis

Make sure that students understand how different pathogens harm the body and how pathogens are transmitted through air, water, human contact, and arthropods.

Library Research

Ask students to use library resources to find out about one of the diseases listed in Table 44-1. Encourage them to find out about the symptoms of the disease, where the disease is common, who gets it, and how it can be cured or how the symptoms can be alleviated. Ask students to present this information in an oral report to the class.

Section Review Answers

(See p. 674.)

44.2 Immune System

Approach

The main objective of this section is to present ways in which the body defends itself against pathogens. This section builds upon the concepts of the structure and function of skin, discussed in Chapter 42, and the structure and function of blood cells and of the respiratory system discussed in Chapter 43. In presenting the text, guide students to realize that without the immune system, the pathogens to which a person is exposed would eventually overwhelm the body.

Concept Framework

Students may have a basic understanding of how the electronic security system of a building works. Tap this knowledge to provide a conceptual framework for this section by making an analogy between a building's security system and the body's immune system. Guide students to realize that both systems have the same basic function. Tell students to note how the human "security system" defends against "invaders" as they read.

Emphasis

Students should understand how specific and nonspecific defenses work together to protect the body from disease.

Using the Features/Jonas Salk and the Polio Vaccine

The main scientific process skills discussed in this feature are observing and experimenting. This feature describes how Salk researched the viruses that cause polio and developed a vaccine against the disease. Encourage students to note that Koch and Salk used similar methods to investigate causes and cures of diseases.

Making Posters

Divide the class into two groups. Ask one group to make a poster illustrating nonspecific defenses. Ask the other group to make a poster illustrating specific defenses. Let students display their posters.

Section Review Answers

(Continued from p. 679.)
3. Suppressor cells prevent antibodies from being produced after the pathogen has been eliminated from the blood.
4. Antibodies are released in large quantities, binding to the allergen and causing the release of histamines.
5. Doctors treat strep throat and other bacterial infections with an antibiotic.

44.3 AIDS

Approach

The main objective of this section is to describe how the human immunosuppressive virus (HIV) disrupts the immune system, leading to acquired immune deficiency syndrome (AIDS). This section builds on the concepts of the structure and the function of viruses (Chapter 19) and on the concepts of disease transmission and immune function discussed in 44.1 and 44.2. In presenting the text, guide students to realize that HIV attacks the immune system, rendering a person defenseless against most other pathogens.

Concept Framework

From studying Chapter 19 students will be familiar with the retrovirus, the virus category virus to which HIV belongs. Form a framework for discussing this section by reviewing how this type of virus reproduces. As students read this section, encourage them to note how reproduction of HIV eventually destroys the immune system.

Emphasis

Students should understand how HIV disrupts the immune system by lodging in the T4 cells and causing them to become ineffective.

Invite a physician or other health-care professional to class to discuss the latest information on the affects of the HIV virus in causing AIDS. Ask the speaker to emphasize ways to minimize the risk of contracting AIDS.

Section Review Answers

(Continued from p. 681.)

3. The virus is engulfed by a phagocyte, which cannot destroy it. T4 cells attack but do not destroy the virus. The virus lodges in T4 cells. In time the virus reproduces inside the cells, destroying the cells' capacity to fight disease.
4. They might use contaminated syringes and become infected.
5. Abstaining from intimate sexual contact and from unsupervised drug injection or exercising extreme caution with these activities can decrease the risk of contracting AIDS.

Laboratory Preparation

Background

This lab procedure familiarizes students with some common pathogens, as well as diseases and symptoms associated with these pathogens. Students examine pictures of viruses and slides of protozoans. Students should be familiar with the life cycle of the sporozoan *Plasmodium* before they study the slides. Also, since they examine living nonpathogenic bacteria and classify them according to shape, students should know the three fundamental shapes of bacteria: baccilli, spirilla, and cocci.

Preparation

The bacterial cultures and prepared slides are available from a biological supply house.

Conclusions

1. Viruses are too small to be seen with a compound microscope.
2. *Bacillus megaterium* is rod-shaped; *Spirillum volutans* looks like a coiled rope; *Streptococcus lactis* is spherical. Studying nonpathogenic bacteria allows students to observe these bacterial shapes, as well as bacterial structures which are similar to those of pathogens.
3. The organisms belong to the Kingdom Protista.

Inquiry

Students may disagree on the safety of studying live pathogens, even when precautions are taken. However, the discussion will give them an opportunity to review laboratory safety practices and sterile techniques.

Chapter Review

Vocabulary

1. Specialized T cells, called helper cells, recognize a foreign antigen and rapidly divide to form many cells. Some of these cells destroy the offending antigen. Others called memory cells stay in the blood to recognize and destroy the antigen should it appear again.
2. interferon
3. Neutrophils are small white blood cells with

irregular nuclei that can consume and digest small number of bacteria, and macrophages are large white blood cells that can eat and eliminate up to 100 bacteria.

4. *Vacca-* means "cow," and the original vaccinations were done by infecting people with cowpox, a disease contracted by cows. *Patho-* means disease and a pathogen is anything that causes disease.

5. T cells are lymphocytes that attack antigens directly, and B cells are lymphocytes that produce chemicals that bind with antigens and render them harmless.

Review

11. Koch's postulates are: isolate the organism, grow the organism, inoculate a new host, and compare the two organisms.

12. The five types of pathogens are bacteria, viruses, invertebrates, fungi, and protozoa.

13. The four means of pathogen transmission are air, water, direct contact, and arthropod vectors.

14. An inflammatory response includes both the action of neutrophils and macrophages swallowing the pathogens and an increase in body temperature.

15. B cells and T cells attack an antigen.

16. Antibodies destroy antigens.

17. Memory cells are clones of the B cells that originally responded to a particular antigen; they remain in the body and attack the same antigen if it enters the body again.

18. Jonas Salk developed a vaccine to prevent polio.

19. The immune system is unable to fight disease and the person becomes susceptible to pathogens that would normally be fought off.

20. The dormant period is anywhere from a few months to about 10 years.

Critical Thinking

1. Because the heat from a fever can often kill the pathogen causing the infection, reducing the fever makes this part of the natural defense system of the body ineffective. Also, because a fever is a sign of an infection, if this symptom is masked, the underlying problem may never be discovered.

2. AIDS and some forms of cancer are caused by viruses. Interferon triggers the production of an enzyme that allows a cell to recognize these viruses as foreign invaders. This enzyme then prevents the invading viruses from reproducing.

3. This treatment works because the person is gradually desensitized to the allergen. Injecting slightly greater amounts of the allergen into the blood with each treatment allows an allergic individual to develop a tolerance to it.

4. T cells recognize body cells that have been invaded by cancer because the cancerous cells register as "nonself." This allows them to attack the cancer by launching a cell-mediated response.

5. Because the skin is destroyed and fluid is lost, many of the white blood cells and other necessary parts of the immune system are not present.

6. Normally when an antigen is introduced into the body, T cells rush to the site and destroy the intruder. Memory cells are left behind to recognize and eliminate that antigen if it should appear again. When the immune system of a patient is barraged with antibiotics, it does not develop its own antibodies to fight infections. Since no memory cells are formed, the immune system has no antibodies against that antigen should it reappear.

Chapter 44

Infectious Diseases and the Immune System

Introduction

The human body is exposed to a number of vectors that cause disease. The immune system adapts and changes to fight off these diseases. In addition, advancements in research techniques have helped scientists understand the normal workings of the immune system, and its disruption by organisms such as the virus that causes AIDS.

Chapter Outline

44.1 Infectious Disease
- *Koch's Postulates*
- *Pathogens*
- *Disease Transmission*

44.2 Immune System
- *Nonspecific Defenses*
- *Specific Defense*
- *Immunity and Immune Disorders*

44.3 AIDS
- *Cause*
- *Transmission*
- *Viral Action*
- *Effects of Immune Disruption*
- *Risk and Prevention*

Chapter Concept

As you read, identify ways to diagnose, prevent, treat, and cure diseases.

Pseudopodia of a macrophage ensnaring a bacterium that is tinted green (×14,740). Inset: Adenovirus, the virus that causes the common cold.

For additional teaching information see pp. T44A–T44D.

44.1 Infectious Disease

The human body is continuously under attack by disease-causing organisms. The body's immune system consists of a number of mechanisms that fight disease. This chapter will focus on infectious diseases and how the immune system combats them. An **infectious disease** is one caused by the entrance into the body of an organism and its multiplication there. An infectious disease may spread from host to host. Any agent that causes an infectious disease is called a **pathogen.**

Koch's Postulates

Robert Koch (1843–1910), a German physician, was the first scientist to establish a step-by-step procedure for identifying the pathogen responsible for a given disease. Koch focused his investigation on a cattle disease called anthrax. Observing that infected cows had swarms of bacteria in their blood, he hypothesized that these bacteria were the cause of anthrax.

To test his hypothesis, Koch isolated these bacteria from the infected host, grew them in a laboratory culture, and then inoculated a healthy cow with the bacteria. The new host then developed anthrax. Koch compared the bacteria from the original host with the bacteria in the new host and found them the same. He concluded that the bacteria caused the disease. The steps that Koch formulated for determining the cause of a given disease, known as **Koch's postulates,** are described in Figure 44-1.

Pathogens

Scientists have used Koch's postulates to identify thousands of pathogens. The most common are bacteria, viruses, protozoa, fungi, and invertebrates. Each type affects the body differently. Table 44-1 on the next page lists common pathogens and diseases.

Many pathogens produce **toxins,** chemical substances harmful to the normal functioning of cells. Among the most powerful toxins are the **neurotoxins,** substances that disrupt the transmission of nerve impulses. For example, the bacterium associated with botulism produces a neurotoxin that causes double vision

Section Objectives

- *Summarize the four steps of Koch's postulates.*

- *Identify five types of pathogens.*

- *Describe four mechanisms of disease transmission.*

Tip: Which diseases are not caused by a virus or other invading organism? (conditions caused by genetic imbalances, which are inherited and not acquired)

Figure 44-1 The bacterium that causes anthrax, *Bacillus anthracis* (left, × 42,268), was identified as a pathogen by means of Koch's postulates (right).

Koch's Postulates

1. Isolate the organism suspected of causing the disease.

2. Grow the organism in laboratory cultures.

3. Inoculate a healthy animal with the cultured organisms. Observe whether the animal contracts the disease.

4. If the animal contracts the disease, isolate the organism that caused it. Compare this organism with the one isolated in step 1.

Table 44-1   Common Diseases and Their Causes

Disease	Pathogen	Method of Transmission
Athlete's foot	Fungus	Contact with contaminated objects such as shower floors
Common cold	Any of 100 viruses	Close contact with infected person; airborne spread of virus
Hepatitis A	Virus	Person-to-person contact; eating contaminated food and water
Influenza	Any of 100 viruses	Person-to-person contact; contact with contaminated objects
Malaria	Protozoa	Bites from mosquitoes carrying the protozoa; transfusion of infected blood
Measles (rubeola)	Virus	Contact with infected saliva or mucus; airborne spread of virus
Pneumonia	Virus or bacteria	Contact with infected saliva or mucus
Syphylis	Bacterium	Sexual contact

Info: Water treatment plants are constantly being checked for coliform bacteria. The bacteria themselves are not harmful, but if found in water, they indicate contamination by sewage.

Answers to Section Review
1. a condition caused by viruses, bacteria, fungi, or protozoans entering the body and multiplying
2. by isolating the bacteria, growing it in a culture, inoculating a new host, and comparing the bacteria with the bacteria in the original host to see if they were the same
3. Neurotoxins disrupt the transmission of nerve impulses.
4. If waste containing pathogens is released into water used for normal household purposes, the water supply is contaminated.
5. The insect lands on the water, either ingests or picks up the pathogen, carries the pathogen to a human, and injects the pathogen into the human host.

and severe pain. Viruses and certain bacteria also cause diseases by destroying cells. Invertebrates cause diseases by burrowing into tissues, such as muscle or intestinal tissue. They cause pain while interfering with the normal function of the tissues.

Disease Transmission

Pathogenic viruses and microorganisms may spread from host to host. Scientists have identified four main mechanisms of transmission: air, water, human contact, and arthropods.

An organism may be transmitted through air in spores or cysts or in droplets. For example, when a person with a cold sneezes or coughs, droplets, which contain pathogens, are released into the air.

Pathogens may also be transmitted through water. This can occur when wastes containing pathogens are released into water that is then used for other purposes. If a leaky septic tank contaminates a water supply, an entire community can be infected.

A contaminated person can transmit pathogens directly by touching another person or indirectly by touching some object that is in turn touched by another person.

Pathogens can be transmitted through an arthropod vector. Arthropods such as lice, mosquitoes, ticks, and fleas carry pathogens and transmit them to hosts when penetrating the skin.

Section Review

1. What is an *infectious disease?*
2. How did Koch test his hypothesis about the cause of anthrax?
3. Describe the general effect of a neurotoxin.
4. How are pathogens transmitted by water?
5. How might insects transmit pathogens from water to humans?

44.2 Immune System

The body is able to defend itself against pathogens by means of a complex set of defenses. These defenses make up the body's immune system. The reaction of the body against a foreign substance is called an **immune response.** Both nonspecific and specific defenses exist.

Nonspecific Defenses

Nonspecific defenses are those that act against all disease-causing organisms in the same way regardless of the biological nature of the organism. These defenses include the skin and the mucous membranes of the nose and mouth, which act as mechanical and chemical barriers to pathogens. Other nonspecific defenses are phagocytes, fever, and interferon, all of which act against pathogens that do enter the body.

Skin and Mucous Membranes

The layer of overlapping dead cells on the surface of the skin prevents pathogens from entering the body. A break in the skin may permit pathogens to enter body tissues. Blood clots then form a temporary barrier until new skin seals the wound.

The sweat, oils, and waxes produced by the skin contain chemical substances, such as lactic acid, that are poisonous to many bacteria and fungi. However, not all bacteria are harmful. Large populations of harmless bacteria living on the skin actually aid in protecting the body by inhibiting the growth of pathogens.

Mucous membranes defend the body against pathogens that enter through the mouth and nose. Mucus and tiny hairs in the mouth and nose trap microorganisms. Enzymes in the mucus destroy pathogens. Pathogens that do reach body organs are met with other nonspecific defenses. For example, pathogens that enter the stomach are usually destroyed by digestive acids.

Phagocytes and Fever

An **inflammatory response** is one in which white blood cells gather and engulf the foreign substances and the body temperature rises. These white blood cells are phagocytes. Humans have two types of phagocytes—neutrophils and macrophages. **Neutrophils** (NYOO-truh-FILLZ) are small white blood cells with irregular nuclei. Neutrophils ingest small numbers of bacteria. **Macrophages** (MACK-ruh-FAJ-uz) are large white blood cells that can engulf hundreds of bacteria. The action of the neutrophils and macrophages is usually accompanied with an increase in body temperature, or **fever.** If the fever is high enough, it can kill pathogens. A fever is not a disease but a symptom. It is a sign that the body is responding to an infection. However, if the infection is severe, fever itself can cause death.

Section Objectives

- *Explain how the skin functions as a defense against disease.*

- *Distinguish between a specific and a nonspecific response.*

- *Describe the actions of B cells and T cells in a primary immune response.*

- *Describe the relationship between vaccination and immunity.*

- *Describe the methods Jonas Salk used to develop a vaccine for the polio virus.*

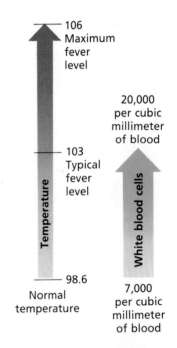

Figure 44-2 The immune response involves both an elevation in temperature and an increase in the number of white blood cells.

Interferon

Cells respond to a viral infection by producing a protein called **interferon** *(INT-ur-FIR-ahn)*, which inhibits the reproduction of viruses. Interferon triggers the production of an enzyme that enables a cell to recognize a virus as a foreign substance. Interferon thus prevents the virus from reproducing. Interferon affects any type of virus. Through DNA-splicing techniques genetic engineers have made large amounts of human interferon.

Specific Defense

In a specific defense one or more components of the immune system attack a specific pathogen. The immune system identifies specific chemicals on the surface of a pathogen and then targets that pathogen for destruction. Any substance that causes a specific immune response is called an **antigen**.

Specific defenses are of two types, one using cells, and the other using proteins. Both involve **lymphocytes** *(LIM-fuh-SITES)*, which are special white blood cells. They exist in two primary forms—T cells and B cells. Their responses to antigens may be primary or secondary.

Primary Response

A **primary immune response** is a specific defense involving both B cells and T cells. **T cells** attack antigens directly. **B cells** produce chemicals that render antigens harmless. Both types of lymphocytes develop in the bone marrow but are named for the place where they differentiate. B cells differentiate in the bone marrow; T cells migrate to the thymus before differentiating.

Figure 44-3 In the primary immune response B cells and T cells act together to destroy the antigen.

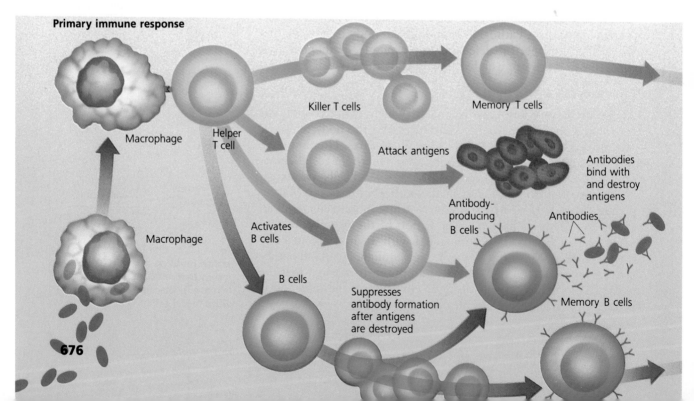

Primary immune response

Macrophage

Helper T cell

Killer T cells

Memory T cells

Attack antigens

Macrophage

Activates B cells

Antibody-producing B cells

Antibodies bind with and destroy antigens

Antibodies

B cells

Suppresses antibody formation after antigens are destroyed

Memory B cells

The response of T cells is a **cell-mediated response.** Figure 44-3 shows this response. Specialized T cells called **helper cells** initiate a cell-mediated response. Each T cell can distinguish the "self" (other lymphocytes) from the "nonself" (foreign substances). When a T cell recognizes a foreign antigen, it rapidly divides to form many cells. Some destroy the antigen. Others, called **memory cells,** stay in the blood. If the same antigen appears again, memory cells recognize and destroy it.

B cells respond by producing antibodies—Y-shaped proteins that bind to and destroy specific antigens. Each B cell synthesizes an antibody that acts only against a particular antigen. When a B cell binds to an invading antigen, the cell is stimulated to divide. Each resulting cell, called a clone, produces the same antibody and thus destroys the same type of pathogen.

Secondary Response

Initially scientists believed that the functions of T cells and B cells were separate. Now they know that the functions of the two types of cells are coordinated. They have discovered that T cells regulate the actions of B cells in the secondary immune response. A **secondary immune response** is the response to a subsequent infection by the same pathogen.

After the pathogen has disappeared from the blood, modified T cells called **suppressor cells** stop the B cells from producing antibodies. Though B cells stop synthesizing antibodies, several clones called memory B cells remain in circulation. As you know, the T cells have also formed memory T cells. Therefore, when the body encounters the same pathogen in the future, the memory T and B cells proliferate quickly and fight the infection before it overwhelms the system.

Tip: Ask students how many of them have had chicken pox. Then ask how many have had it twice. If anyone has had the disease more than once, two strains are indicated.

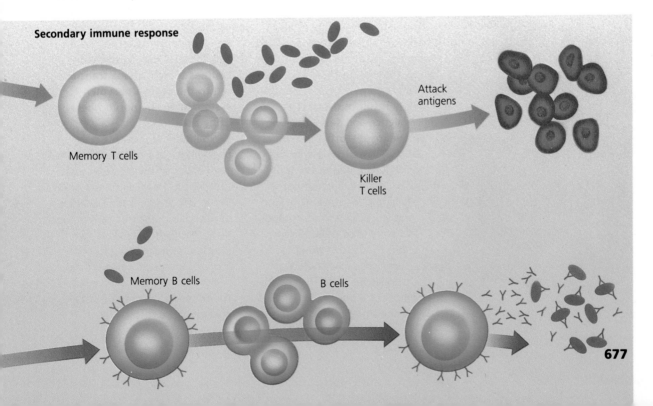

Secondary immune response

Memory T cells

Killer T cells

Attack antigens

Memory B cells

B cells

677

Think

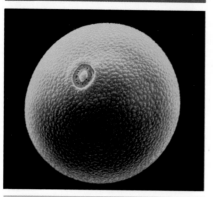

Pictured above is an airborne allergen called a pollen grain. How might such an allergen enter the body? How could the body rid itself of this allergen?

Immunity and Immune Disorders

A person who has a resistance to a specific pathogen is said to have an **immunity** to it. **Natural immunity** is immunity that is present at birth. **Acquired immunity** is developed after birth.

One way of acquiring immunity is through vaccination. In **vaccination** (VACK-suh-NAY-shun) an individual is exposed, usually by means of injection, to a killed or weakened form of a pathogen in order to produce immunity to the disease caused by the pathogen.

Antibiotics (ANT-ih-bie-AHT-icks), drugs that fight bacterial infection, have made it possible for scientists and physicians to prevent and treat a large number of diseases. Penicillin, one of the best-known antibiotics, was discovered in 1929 by Sir Alexander Fleming (1881–1955), a British bacteriologist.

The immune system is subject to various disorders. Normally suppressor cells prevent B cells from producing antibodies against nonpathogenic substances. In some individuals, however, the lymphocytes react to harmless antigens. These antigens, or **allergens**, include foods, pollen, and substances on animal fur. A reaction to an allergen is called an **allergy**.

Biology in Process

Jonas Salk and the Polio Vaccine

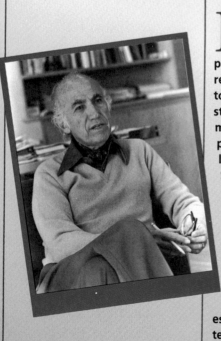

For centuries poliomyelitis affected people in all parts of the world. Polio, which remains incurable, causes symptoms that range from pain and stiffness to paralysis. It can ultimately result in death. The prevalence and severity of poliomyelitis disturbed Dr. Jonas Salk, an American physician. In 1942 Salk began to study polio at the University of Michigan. He concentrated on the cause and physiological effects of polio. Salk identified three polio viruses that all acted in a similar manner. He hypothesized that a polio virus entered the body through the nose or mouth and was carried to the intestines. Through observation and experimentation Salk learned that from the intestines the virus was transmitted to the central nervous system either by nerve fibers or by the blood.

Salk knew from his own previous experiments and from the results of his predecessors that when the virus enters a nerve cell, it alters the function of the cell, causing it to reproduce the virus. The virus can multiply rapidly enough to destroy or kill most adjacent cells. If enough cells are destroyed, the nervous system is affected, and paralysis results.

Info: Other autoimmune diseases include rheumatoid arthritis, pernicious anemia, and systemic lupus erythematosus. Some forms of diabetes and Addison's disease are also suspected autoimmune diseases.

At the first exposure to an allergen an individual may not have a noticeable reaction. Nevertheless, B cells may synthesize antibodies and develop memory cells. At the second exposure the individual produces a secondary immune response. Antibodies and allergens bind together and cause the release of **histamine** (*HISS-tuh-MEEN*), a chemical associated with the sneezing, watery and itchy eyes, and wheezing of allergies. Drugs called **antihistamines** help counteract the effect of histamines.

In another type of disorder, **autoimmune disease,** the immune system produces antibodies against cells in the body because it fails to recognize them as self. In some cases, an antigen resembles one of the body's own substances. Antibodies from the immune system may destroy this substance as well as the foreign antigen.

Section Review

1. Define *natural immunity* and *acquired immunity*.
2. Describe three nonspecific responses of the immune system.
3. What is the role of suppressor cells?
4. Describe what happens in an allergic response.
5. How might a doctor treat strep throat, a bacterial infection?

Answer to Think
Pollen grains most commonly enter the body through the nose and mouth. As a first defense the cilia and the mucus lining the air passages would stop the pollen grain. If the pollen grain survived this attack, the immune system would attempt to render the allergen harmless.

Answers to Section Review
1. Natural immunity is immunity that is present at birth. Acquired immunity is developed after birth.
2. The skin acts as a physical and chemical barrier to pathogens; the cilia and the mucus that line body passages trap pathogens before they can enter the body; phagocytes, together with fever, kill off pathogens.
(See pp. T44A–T44D.)

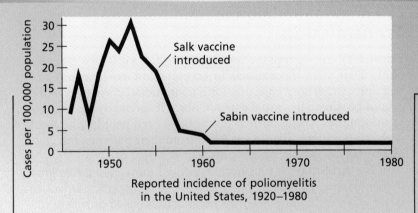

Reported incidence of poliomyelitis
in the United States, 1920–1980

After completing his initial investigations Salk had enough information about the virus to begin to develop a vaccine. He researched the work of earlier epidemiologists and incorporated their findings into his plan. Most vaccines are made from actual viruses that have been weakened or killed but can still produce the immune response in a new host. Salk discovered he could kill the three polio viruses by using a formaldehyde solution. Through carefully controlled experimentation Salk produced a vaccine from the treated viruses. The next step was to determine whether the vaccine would work in the human body. He injected himself, his wife, and his three sons. The results were successful.

True success of the vaccines could not be determined by this limited test. Therefore in 1954, with money and personnel from the National Foundation for Infantile Paralysis (later called the March of Dimes Foundation), the Salk vaccine was administered to 1,830,000 American school children. Follow-up studies proved that the vaccine was not only safe but effective. In April 1955 the vaccine was made available to the general population.

Shortly after Salk's initial test, another American physician, Albert Sabin, used further research to develop an oral vaccine. This made the administration of the vaccine easier and less painful.

■ Why was it important for Salk to study prior research?

Infectious Diseases and the Immune System   **679**

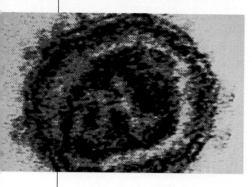

Figure 44-4 These three photos show increasingly greater magnifications of the AIDS virus. The bottom magnification is ×120,000.

Info: A hormone called GM-CSF is being used experimentally in AIDS patients to enhance the production of white blood cells, fortifying the immune system.

44.3 AIDS

The responses of the cell-mediated system and the antibody system keep most diseases from being expressed in the human body. However, disorders of the immune system do occur. One of the most serious disorders is called **AIDS,** which stands for **acquired immune deficiency syndrome.**

Cause

AIDS is a viral disruption of the proper workings of the immune system. The virus that causes the disruption is shown in Figure 44-4. It has been named HIV, which stands for human immuno-deficiency virus. It is also called HTLV-III, which stands for human T-cell lymphotrophic virus, and LAV, which stands for lymphadenopathy-associated virus.

The virus is a retrovirus, which means that it contains RNA as its genetic complement. HIV has two strands of RNA that form the template for DNA when acted on by the enzyme reverse transcriptase. The resulting DNA has seven genes that code for the formation of new viruses after infection. As you will see, the virus acts on cells of the immune system.

Transmission

Before HIV can disrupt the immune system, it must first enter the body. Two main mechanisms of transmission exist: sexual contact and intravenous injection with contaminated syringes. AIDS cannot be spread by casual contact, nor is it spread by insects such as mosquitoes. Using normal precautions even people who work with AIDS patients do not get AIDS from their patients. Infection accompanying blood transfusion is now rare.

The virus may be transmitted during intimate sexual contact involving an exchange of body fluids. Transmission is especially likely when cells along body openings are damaged during sexual acts, thereby allowing the virus to enter the bloodstream directly.

Because the virus is also readily transmitted during intravenous injection with contaminated needles, AIDS is prevalent among users of illegal drugs. Injection of substances under a physician's supervision poses no threat of transmission.

Viral Action

As you read about the mechanisms by which HIV affects the immune system, refer to Figure 44-5. When the virus enters the bloodstream, the immune system recognizes it as an antigen. First the phagocyte engulfs but is unable to destroy the virus. A type of T cell called a T4 cell then attacks the virus, but the virus enters the T4 cells unharmed. At the same time antibodies made by the B

Tip: Suggest that students review the discussion of phagocytes in Chapter 43.

cells also attack the viruses in the blood, but the viruses are unharmed by antibody activity.

The virus may remain dormant in the T4 cells for up to ten years. At some point, however, the virus undergoes a lytic cycle of reproduction and creates many new viruses inside the T4 cells. As these viruses are released, they infect additional T4 cells and eventually destroy the capacity of the T4 cells to fight disease.

Effects of Immune Disruption

Once the T4 cells are destroyed, the immune system cannot fight diseases. People infected with the HIV virus then become susceptible to pathogens to which they normally would be immune. Many people with AIDS have secondary, and usually fatal, infections from *Pneumocystis carinii*, a bacterium that causes pneumonia. Another disease that occurs in an immune-deficient body is a form of cancer called Kaposi's sarcoma. Many other diseases also develop in HIV-infected individuals, diseases that cannot be suppressed and that ultimately lead to death.

Sometimes a full-blown AIDS-related infection is preceded by similar though less virulent diseases. Patients with such diseases are said to have **ARC**, or an **AIDS-related complex.**

Risk and Prevention

AIDS attacks both males and females. At particular risk, as indicated by disease statistics, are intravenous drug users and those homosexual men who engage in unsafe sexual activity. Women also are susceptible to infection via some forms of intimate sexual contact. At present, far fewer U.S. women than men have AIDS.

Since AIDS is not spread by casual contact, transmission can be prevented by abstaining from intravenous injection unless under strict supervision of a physician and by abstaining from sexual activity. Any user of injected illegal drugs runs a severe risk of contracting AIDS. Sexual activity carries additional risks for AIDS. According to the Surgeon General of the United States, if sexual activity occurs, the use of condoms will reduce but will not eliminate the threat of transmission.

A blood test that detects antibodies to HIV now exists. The test reveals whether a person has been exposed to the virus.

Section Review

1. What does *AIDS* stand for?
2. Why does AIDS lead to other diseases?
3. Describe the action of the HIV virus after it enters the bloodstream.
4. Why are intravenous drug users at risk for AIDS?
5. How can behavioral changes modify an individual's chance of getting AIDS?

Figure 44-5 After the HIV virus enters the bloodstream, its RNA is translated to DNA, which fuses with T4 cell DNA. For simplicity only one DNA strand has been shown.

Virus enters bloodstream

Virus attaches to receptor on T4 cell

Viral RNA enters T4 cell

Enzyme translates RNA into DNA, which is incorporated into the chromosomes of T4 cell

After a period of dormancy viral DNA is expressed, causing cell to produce new viruses, which lyse the T4 cell

Answers to Section Review
1. acquired immune deficiency syndrome
2. because the HIV virus acts to disrupt the workings of the immune system
(See pp. T44A–T44D.)

The AIDS Virus

The following is the introduction to Robert C. Gallo's article from the January 1987 issue of Scientific American.

I t is a modern plague: the first great pandemic of the second half of the 20th century. The flat, clinical-sounding name given to the disease by epidemiologists—acquired immune deficiency syndrome—has been shortened to the chilling acronym AIDS. First described in 1981, AIDS is probably the result of a new infection of human beings that began in central Africa, perhaps as recently as the 1950's. From there it probably spread to the Caribbean and then to the U.S. and Europe. By now as many as two million people in the U.S. may be infected. In the endemic areas of Africa and the Caribbean the situation is much worse. Indeed, in some areas it may be too late to prevent a disturbingly high number of people from dying.

In sharp contrast to the bleak epidemiological picture of AIDS, the accumulation of knowledge about its cause has been remarkably quick. Only three years after the disease was described its cause was conclusively shown to be the third human retrovirus: human T-lymphotropic virus III (HTLV-III), which is also called human immunodeficiency virus (HIV). Like other retroviruses, HTLV-III has RNA as its genetic material. When the virus enters its host cell, a viral enzyme called reverse transcriptase exploits the viral RNA as a template to assemble a corresponding molecule of DNA. The DNA travels to the cell nucleus and inserts itself among the host's chromosomes, where it provides the basis for viral replication.

In the case of HTLV-III the host cell is often a T4 lymphocyte, a white blood cell that has a central role in regulating the immune system. Once it is inside a T4 cell, the virus may remain latent until the lymphocyte is immunologically stimulated by a secondary infection. Then the virus bursts into action, reproducing itself so furiously that the new virus particles escaping from the cell riddle the cellular membrane with holes and the lymphocyte dies. The resulting depletion of T4 cells—the hallmark of AIDS—leaves the patient vulnerable to "opportunistic" infections by agents that would not harm a healthy person.

How HTLV-III manages to replicate in a single burst after lying low, sometimes for years, is one of the most fundamental questions confronting AIDS researchers. Another important question is the full spectrum of diseases with which the virus is associated. Although most of the attention given to the virus has gone to AIDS, HTLV-III is also associated with brain disease and several types of cancer.

In spite of such lingering questions, more is known about the AIDS virus than is known about any other retrovirus. The rapidity of that scientific advance was made possible partly by the discovery in 1978 of the first human retrovirus, HTLV-I, which causes leukemia. In its turn the new knowledge is making possible the measures that are desperately needed to treat AIDS and prevent its spread.

Write
- **AIDS has been described by some experts as a disease that could change the world. Write an essay in which you discuss some of the ways that this could happen.**

From "The AIDS Virus" by Robert C. Gallo, in Scientific American, *January 1987.*

Laboratory

Observing Pathogens

Objective
To study agents that cause human diseases

Process Skills
Observing, inferring, classifying

Materials
Glass-marking pencil; 3 microscope slides; live cultures of *Bacillus megaterium, Spirillum volutans*, and *Streptococcus lactis*; 3 medicine droppers; 3 coverslips; compound microscope; prepared slides of *Staphylococcus aureus, Plasmodium vivax*, and *Trypanosoma gambiense*

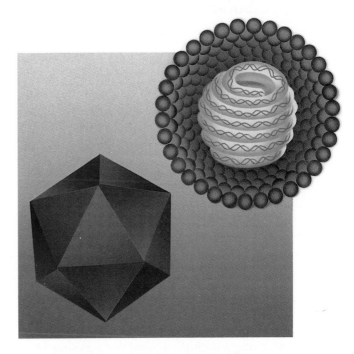

Method
1. Examine the illustrations of the influenza (top) and polio (bottom) viruses at right. Contrast their shapes.
2. Use a glass-marking pencil to label three microscope slides *Bm, Sv,* and *Sl.* These labels stand for *Bacillus megaterium, Spirillum volutans,* and *Streptococcus lactis,* which are all harmless bacteria. These organisms exhibit the basic bacterial shapes.
3. Obtain a culture of *Bacillus megaterium* from your teacher. Use a clean dropper to place a drop of culture on the appropriate slide. Using correct wet-mount technique, place a coverslip over the slide.
4. Examine the culture under a microscope set on low power. Observe the shape of this organism. Locate the nucleus, cytoplasm, and cell wall. Switch the microscope to high power to observe these features in detail.
5. Repeat steps 3 and 4 using a clean dropper and a culture of *Spirillum volutans.*
6. Repeat steps 3 and 4 using a clean dropper and a culture of *Streptococcus lactis.*
7. Examine a prepared slide of *Staphylococcus aureus* under a microscope. Is this organism a virus or a bacteria? How can you tell?
8. Obtain a prepared slide of *Plasmodium vi-*

vax from your teacher. *Plasmodium,* which is transmitted by mosquitoes, causes malaria. Set the microscope to low power and carefully scan the slide to observe different stages in the life cycle of *Plasmodium.* How many stages can you find?
9. Examine a prepared slide of *Trypanosoma gambiense. Trypanosoma* causes African sleeping sickness.

Conclusions
1. Why did you not examine prepared slides of the polio and influenza viruses?
2. What are the shapes of the three living organisms you observed? What are you able to learn about pathogens using information from the study of nonpathogenic bacteria?
3. From your observations, to what kingdom do you think *Plasmodium vivax* and *Trypanosoma gambiense* belong?

Inquiry
Are there techniques with which you could safely examine live cultures of pathogens?

Vocabulary

acquired immune deficiency syndrome (AIDS) (680)

acquired immunity (678)

AIDS-related complex (ARC) (681)

allergen (678)

allergy (678)

antibiotic (678)

antigen (676)

antihistamine (679)

autoimmune disease (679)

B cell (676)

cell-mediated response (677)

fever (675)

helper cell (677)

histamine (679)

immune response (675)

immunity (678)

infectious disease (673)

inflammatory response (675)

interferon (676)

Koch's postulates (673)

lymphocyte (676)

macrophage (675)

memory cell (677)

natural immunity (678)

neurotoxin (673)

neutrophil (675)

pathogen (673)

primary immune response (676)

secondary immune response (677)

suppressor cell (677)

T cell (676)

toxin (673)

vaccination (678)

1. Define *helper cell* and *memory cell*.
2. Identify the protein that interferes with viral replication.
3. Compare a neutrophil and a macrophage.
4. Using a dictionary, find the meaning of the word part *vacca-* in *vaccination* and *patho-* in *pathogen*. Explain why those terms are appropriately named.
5. Explain the relationship between T cells and B cells.

Review

1. Robert Koch (a) established procedures for identifying the cause of disease (b) identified specific neurotoxins (c) discovered interferon (d) pioneered the process of vaccination.
2. Pathogens can be (a) bacteria (b) viruses (c) invertebrates (d) any of the above.
3. Toxins (a) always affect the brain (b) affect the normal functioning of the cell (c) burrow into tissues (d) destroy muscle tissue.
4. Nonspecific defenses include the (a) cell-mediated response (b) primary immune response (c) inflammatory response (d) secondary immune response.
5. Skin acts as a defense against infection by (a) forming a physical barrier to pathogens (b) forming a chemical barrier to pathogens (c) forming blood clots (d) all of the above.
6. A cell-mediated response involves the (a) recognition of "self" and "nonself" (b) production of interferon (c) general reaction to all pathogens (d) action of cilia lining the air passages.
7. Antibodies are (a) lipids (b) B cells (c) Y-shaped proteins (d) T cells.
8. AIDS is transmitted mainly through (a) blood transfusions with contaminated blood (b) intimate sexual contact that involves the exchange of body fluids (c) intravenous injection with contaminated syringes (d) both b and c
9. Vaccinations (a) prevent diseases (b) cure diseases (c) treat diseases (d) cause diseases.
10. Penicillin (a) causes infections (b) prevents infections (c) cures infections (d) kills viruses.
11. Name the four Koch's postulates.
12. What are five types of pathogens?
13. Name four ways in which pathogens are transmitted.
14. Describe an inflammatory response.

15. Describe the primary immune response.
16. What function do antibodies serve?
17. What are memory cells, and what is their role in the prevention of disease?
18. What was the result of Jonas Salk's research on poliomyelitis?

19. Explain what occurs in the immune system of a person infected with AIDS once all the T4 cells have been destroyed by the virus.
20. How long can the AIDS virus remain dormant in a person who is carrying it?

Critical Thinking

1. Many people take fever-reducing drugs as soon as their temperature exceeds 98.6° F. Why might it not be a good idea to immediately reduce a fever with drugs?
2. Interferon has been used to treat both cancer and AIDS. Why is interferon used in the treatment of each of these two diseases?
3. Doctors often treat allergies by injecting small amounts of allergen into the blood of the affected person. Why do you think this is an effective treatment?
4. Look at the photo at right of a T cell. T cells attack cancer cells. What causes T cells to launch a primary immune response against cancer?
5. When a person is burned over large areas of the body, the immune system often fails to fight off infections. Why might the immune system fail in these circumstances?

6. Patients are often treated with antibiotics to fight bacterial infections. Repeated exposure to penicillin and other antibiotics has been shown to cause weakness to the immune system. By what mechanism might this weakness occur?

Extension

1. Read the article by Gina Kolata entitled "Can Immune Boosters Beat Cancer and AIDS?" in *Discover*, September 1987, pp. 68–74. What are the possible uses for GM-CSF, and what are some disadvantages of its use?
2. Use the *Physicians' Desk Reference* (PDR) at your local library to research the large numbers of specialized antibiotics made by drug companies. Make a large chart on poster board using the following headings: brand name of the drug, generic name of the drug, uses of the drug, company that manufactures the drug, and disease the drug controls.

3. Obtain two cultures of the common bacteria *Escherichia coli*. Place each culture in a petri dish containing nutrient agar. In one dish place a paper disk that has been soaked with ampicillin, a variation of the antibiotic penicillin. Cover both dishes. Label the culture with the ampicillin *Experimental*, and label the one without the ampicillin *Control*. Keep both dishes out of direct sunlight and at room temperature. After two days, open both of the dishes and observe the cultures to see if the bacteria has grown. Record your observations.

Chapter 45 | Digestive and Excretory Systems

This chapter teaches students how the human body converts food into the nutrients body cells need, and how it excretes useless and toxic by-products. It emphasizes how the structure of digestive and excretory organs relates to their function.

Correlation Chart

Section Number	Section Objectives	Pages	Features	Section Review
45.1	1. Define *nutrient,* and distinguish organic from inorganic nutrients.	687–693	Think, p. 688	1
	2. Describe the molecular structure of carbohydrates, proteins, and fats.		Biology in Process, pp. 692–693	2, 3
	3. Distinguish between vitamins and minerals, and summarize the functions they perform in the body.			5
	4. List some of the important functions of water in the body.			4, 5
	5. Name the events that led to the discovery of vitamin B_1.			
45.2	1. Define *alimentary canal,* and name the major organs of the digestive system.	694–699		1, 2
	2. Relate the structure of each digestive organ to its function in mechanical digestion.			2, 3, 4
	3. Identify the source of each major digestive enzyme, and describe the function of the enzyme.			
	4. Summarize the process of absorption in both the small and large intestine.			4, 5
45.3	1. Define the term *excretion,* and list the functions of each of the major excretory organs.	700–703	Biology Lab, p. 703	1, 2, 5
	2. Identify the major parts of the kidney.			3
	3. Relate the structure of a nephron to the functions it performs.			3
	4. Explain how the processes of filtration, reabsorption, and tubular secretion help maintain homeostasis.			4
	5. Name the main parts of the urinary system.			2

Resources*

Print

Guyton, Arthur. *Anatomy and Physiology.* Philadelphia: Saunders, 1985.

Audiovisual/Software

Digestive System. Film or video, 16 min. Coronet.

*Additional references are listed in *Modern Biology, Teacher's Program Guide.*

Section Teaching Guides

45.1 Nutrients

Approach

To understand how the process of digestion converts food into energy, students need to know what nutrients are and how they enable the body to function properly. Thus the main objective of this section is to describe the six major types of nutrients, where they come from, and how the human body utilizes them.

The discussion relates back to Chapter 4, on biochemistry, which describes the structure and function of proteins, saccharides, and other molecules basic to life. In discussing the chemistry of food metabolism, you can also draw on the concepts of cell energetics presented in Chapter 7.

Concept Framework

Most students have some understanding of the basic food groups and are aware that nutrition directly affects a person's overall health and appearance. To tap this knowledge and provide background for understanding the biochemical processes involved in utilizing nutrients, engage students in a discussion of specific foods and their effect on body organs.

Emphasis

Make sure students can identify the six types of nutrients, their sources, their functions, and what happens if a nutrient is missing from a person's diet. Refer students to Tables 45-1 and 45-2 to get an idea of the many functions of vitamins and minerals.

Research Activity

Divide the class into small groups and have each group research the typical diet of a specific nation or region of the world. Have students prepare a chart that compares the nutritional strengths and weaknesses of each of these diets.

Using the Features/The Discovery and Synthesis of Vitamin B₁

The main process skills used in the feature on the vitamin B_1 are observing, hypothesizing, experimenting, and inferring. Christiaan Eijkman's hypothesis that the cause of beriberi was associated with polished rice resulted from his observation of chickens who had the disease and then recovered.

This feature illustrates the process by which scientists devise experiments to test a hypothesis. Ask students to identify the dependent and independent variables in Eijkman's and Gerrit Grijns's experiments. Also ask students how Robert Williams used experimentation to isolate and synthesize vitamin B_1—the cure for beriberi.

Section Review Answers

(See p. 693.)

45.2 Digestive System

Approach

The focus of this section is on the components of the digestive system and how each organ contributes to the process of breaking down food into molecules that the body can use. The digestive process is essentially a series of mechanical breakdown and enzymatic reactions. Hence you might wish to review the discussion of enzymes in Chapter 4. You can also relate this section back to the discussion of circulation (Chapter 43), as the circulatory system distributes nutrients throughout the body.

Concept Framework

All students know that food is digested, although they are probably more aware of the process when indigestion takes place. To prepare students for a discussion of the biochemical aspects of digestion, ask them which foods they think are most easily digested. Also ask students to consider the characteristics of easily digestible foods.

Emphasis

Students should be able to identify and describe the structure of the major components of the digestive system and tell how each is involved in the breakdown of proteins, carbohydrates, and fats.

Guest Speaker

Invite a nutritionist from a local hospital to speak to the class about diet and health care. The speaker should give examples of how modifications in diet can help control ulcers, diabetes, and other diseases related to the digestive system.

Section Review Answers

(See p. 699.)

45.3 Excretory System

Approach

The main objective of this section is to describe the process by which metabolic wastes are removed from the human body. Explain to students that the human excretory system is a highly complex and efficient system that removes waste products from cellular activity from the bloodstream and eliminates them from the body. In presenting the text, refer back to the discussions of homeostasis (Chapter 6) and cellular respiration (Chapter 7). Also relate the excretory process to the processes of digestion (Section 45.2) and to the processes of respiration and circulation (Chapter 43).

Concept Framework

Students will know that urinalysis is one of the basic procedures used to identify health problems. In addition, the issue of having urine tested to detect the presence of alcohol or other drugs has been a newsworthy story in recent years and one that is likely to be of interest to students. You might use these topics to lead into your discussion of the function of the excretory system—to remove wastes and impurities from the bloodstream. Ask students to consider the consequences if the excretory system does not function properly.

Emphasis

Students should be able to explain how the structure of each component of the kidney is related to the functions of filtration, reabsorption, and secretion. They should also understand the significance of the excretory system in terms of maintaining homeostasis within the body.

Guest Speaker

Arrange for someone who works at a local kidney treatment center to speak to the class about the use of dialysis machines in treating kidney patients. The speaker should explain how the dialysis machine performs the function of the kidneys.

Section Review Answers

(Continued from p. 702.)

4. During filtration the blood gets rid of materials, whereas during reabsorption the blood regains materials.

5. Food waste that leaves the body via the large intestine is, for the most part, not a product of cellular activity, and technically excretion is the process through which only the waste products of cellular activity are expelled.

Laboratory Preparation

Background

This lab procedure demonstrates the digestion of protein in the stomach. Students learn what components must be present in the digestive fluid to cause the breakdown of protein. Students observe that each component alone cannot promote digestion. Students should understand that water is an important part of the digestive mixture.

Preparation

To prepare a 1% pepsin solution, mix 1 g of pepsin (available in powder form) into 99 mL of distilled water. To prepare a 0.2% solution of hydrochloric acid, pour 20 mL of concentrated hydrochloric acid into 1L of distilled water. To prepare a 1% sodium bicarbonate solution, mix 1 g of baking soda into 99 mL of distilled water.

Safety

Hydrochloric acid can cause severe burns. Caution students to handle it carefully and to avoid breathing its fumes. If spills occur, flush the affected area immediately with water.

Conclusions

1. Test tube 5 most closely simulates stomach conditions. The test tube contains hydrochloric acid, pepsin, and water, all of which are present in the human stomach and necessary for digestion.
2. They are controls. They show that neither water alone, pepsin alone, not hydrochloric acid alone is effective in promoting digestion.
3. It shows that pepsin does not digest proteins in a basic environment.

Inquiry

1. Students should devise a version of the experiment that will test the effectiveness of pepsin, hydrochloric acid, and sodium bicarbonate in the digestion of fat.

Chapter Review

Vocabulary

1. *Mineral* names an inorganic nutrient, which does not belong in a group of organic nutrients.
2. *Bolus* names the matter that passes through the alimentary canal, which does not belong with terms naming structures of this canal.
3. *Triglyceride* names a fat molecule, which does not belong in a group of carbohydrate molecules.
4. *Absorption* names a function performed by digestive organs, which does not belong with functions performed by excretory organs.
5. *Ileum* names a structure of the small intestine, which does not belong in a listing of the structures of the kidney.

Review

11. Scientific studies have indicated that consuming a high level of saturated fats seems to trigger an increase in the production of cholesterol, and elevated cholesterol levels may lead to clogging of the arteries, a condition that may result in heart disease or strokes.
12. The synthesis of vitamin D begins with the conversion of cholesterol to provitamin D by intestinal enzymes. Provitamin D is stored in the skin, where exposure to the ultraviolet light in sunlight converts it into vitamin D.
13. Eijkman's experiments showed that chickens fed only polished rice began to suffer symptoms like those of beriberi, whereas chickens fed only unpolished rice remained unaffected. Grijns inferred from this that the rice hull contained an essential nutrient.
14. The pancreas and liver are the only two digestive organs that are composed of lobes and that are not located along the alimentary canal but instead release secretions into it.
15. The mechanical digestive work of the stomach is the churning action it performs that mixes food particles with gastric fluid.
16. Ammonia is so toxic that it could not remain long in the body without harming cells. Therefore the body protects itself against ammonia poisoning by converting ammonia to the less harmful substance urea.
17. The three main parts of the kidney are the cortex, medulla, and the renal pelvis.
18. The excretory processes that maintain homeostasis are filtration, reabsorption, and secretion.
19. Each nephron is made up of a glomerulus, a renal tubule, and a Bowman's capsule.
20. The urinary system consists of two ureters, a urinary bladder, and the urethra.

Critical Thinking

1. There was probably a lack of something in the diet of rice and water. The correct answer is that protein was missing, but the students may use other food supplements, as long as they figure that an important part of the diet was missing.
2. Probably in the small intestine, where lactase breaks up lactose. But students could also say that it might be a problem with absorption of this specific sugar.
3. A person with serious diarrhea could die of dehydration.
4. The sharper teeth of carnivores are specialized for puncturing and tearing. The flattened teeth of herbivores are specialized for grinding. The teeth of humans are adapted for both functions.
5. The design should include two hookups to the body, one to take the unfiltered blood away from the body, the other to return the filtered blood to the body. The machine itself should have some mechanism to filter the blood and another to provide fluid balance in the body.

Chapter 45 | Digestive and Excretory Systems

Introduction

Carrots, fish eggs, hamburgers, blackberries, cow's milk—the human body is able to convert each of these foods into nutrients body cells need to function, grow, and replicate. In this chapter you will learn just how the body does that, as well as how it rids itself of the useless or toxic by-products of cellular activity.

Chapter Outline

45.1 Nutrients
- *Carbohydrates*
- *Proteins*
- *Fats*
- *Vitamins*
- *Minerals*
- *Water*

45.2 Digestive System
- *Mouth*
- *Esophagus*
- *Stomach*
- *Liver*
- *Pancreas*
- *Small Intestine*
- *Large Intestine*

45.3 Excretory System
- *Kidneys*
- *Urinary System*

Chapter Concept

As you read, notice how the structure of each digestive and excretory organ relates to its function.

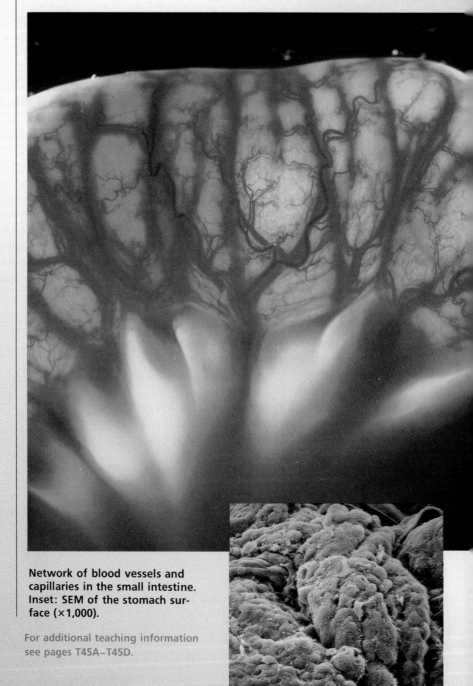

Network of blood vessels and capillaries in the small intestine. Inset: SEM of the stomach surface (×1,000).

For additional teaching information see pages T45A–T45D.

45.1 Nutrients

All the different kinds of food in the world are composed of the same six basic ingredients: carbohydrates, proteins, fats, vitamins, minerals, and water. These ingredients, called **nutrients,** are the chemical substances that organisms need in order to grow and function properly. Carbohydrates, proteins, fats, and vitamins are called organic nutrients because they are synthesized within living organisms. Minerals and water are inorganic nutrients and are not synthesized within living organisms.

Few foods contain all the nutrients. In fact most foods contain a concentration of just one or two of them. Nutritionists classify foods with similar nutrients into a single group within a classification system based on four food groups: (1) meat, (2) milk, (3) fruits and vegetables, and (4) bread and cereals.

Carbohydrates, proteins, and fats are involved in specialized functions. However, all of these nutrients can also provide the body with the energy that it needs to maintain itself, to function, and to grow. A **calorie** is the amount of heat required at standard atmospheric pressure to raise the temperature of 1 g of water from 14.5° C to 15.5° C. A kilocalorie is 1,000 calories. Scientists use the term **Calorie,** with a capital C, to express the amount of energy that a particular food provides. A Calorie is actually the same as a kilocalorie. When you read that a food has 150 Calories, it actually has 150,000 calories. Scientists determine the number of Calories a food contains by measuring the amount of heat a specific quantity of food releases when burned in an instrument called a calorimeter.

Carbohydrates

Carbohydrates are compounds of carbon, hydrogen, and oxygen that provide most of the body's energy when they are broken down in cellular respiration. Although proteins and fats can also supply energy, the body can most easily use the energy provided by carbohydrates. Carbohydrates contain sugars that are converted into the usable energy of ATP. Proteins and fats must be subjected to many more chemical processes before the body can obtain energy from them.

What sugars provide energy? The fructose and the glucose (also known as dextrose) in fruit and honey are simple sugars, **monosaccharides** (*MAHN-uh-SACK-uh-RIDEZ*), that are absorbed directly into the bloodstream. Sucrose (cane sugar), maltose, and lactose (milk sugar) are **disaccharides** (*die-SACK-uh-RIDEZ*), sugars that consist of two monosaccharides that are chemically linked. Disaccharides split into two monosaccharides when water breaks the chemical bond that joins them. This splitting of a molecule through reaction with water is called **hydrolysis** (*hie-DRAHL-uh-suss*).

Section Objectives

- *Define* nutrient, *and distinguish organic from inorganic nutrients.*

- *Describe the molecular structure of carbohydrates, proteins, and fats.*

- *Distinguish between vitamins and minerals, and summarize the functions they perform in the body.*

- *List some of the important functions of water in the body.*

- *Summarize the events that led to the discovery of vitamin B_1.*

Figure 45-1 This diagram shows the hydrolysis of sucrose, which occurs in the presence of enzymes.

Sucrose
+
Water (H_2O)

Glucose   Fructose

Answer to Think
The dieter would get insufficient carbohydrates. The body would then need to convert much of the body's proteins and fats into glucose. The individual's fat stores would diminish, eventually making the person more susceptible to cold and to injury to vulnerable internal organs. Lack of sufficient cellulose could impair digestive function. Insufficient intake of vegetables could interfere with normal liver function.

Think

Many people go on fad diets that consist primarily of selections from one food group. Anyone who follows such a limited diet soon develops a nutrient deficiency. In a diet consisting only of the foods shown above, what nutrients would be lacking? What effects might these deficiencies produce?

The carbohydrates that require the longest digestion time are polysaccharides. **Polysaccharides** (*PAHL-ih-SACK-uh-RIDEZ*) consist of many simple sugars bonded together. The starch found in rice, corn, and many other grains and vegetables is a polysaccharide made up of long chains of glucose molecules. The body chemically dismantles these long chains into individual glucose units, which can be absorbed by the blood and carried to the tissues.

Many plant foods contain cellulose, a polysaccharide that the body cannot dismantle into its individual component sugars. Cellulose, the substance that forms the walls of plant cells, is nevertheless an extremely important part of the human diet. Although indigestible itself, cellulose provides fiber that aids in digestion.

Proteins

Proteins are the major structural and functional material of body cells. Proteins from food help the body to grow and to repair tissues. Also, proteins called **enzymes** are organic catalysts in the myriad chemical reactions that take place in the body. You will recall from Chapter 4 that a catalyst is a substance that speeds up a chemical reaction but does not change the end products. Thus enzymes speed up the thousands of different cell reactions.

As you remember from Chapter 4, proteins consist of long chains of amino acids. The human body uses about 20 kinds of amino acids to construct the proteins it needs. The body manufactures many of these amino acids, but it cannot produce all of them in the quantities that it needs. Amino acids that the body produces are termed nonessential amino acids because you do not have to get them from your food. Amino acids that must be obtained from food are called essential amino acids. Ten amino acids are essential to children and teenagers, while only eight are essential to adults. The two additional amino acids children require function in reactions involving growth.

Some foods contain all of the essential amino acids. Most animal proteins—eggs, milk, fish, poultry, beef—contain all the essential amino acids. Most plants, on the other hand, do not. However, eating certain combinations of two or more plant foods can supply all the essential amino acids. Many such combinations exist. Vegetarians who eat no animal products get their proteins by eating two or more plant foods or by combining seeds or grains with legumes. Examine Figure 45-2 to discover just a few of these combinations.

Fats

Fats are organic molecules that the body utilizes to obtain energy and to build cell membranes and other cell parts. The body stores excess fat from the diet in special tissues under the skin and around the kidneys and liver. Excess carbohydrates may be converted into fat for storage.

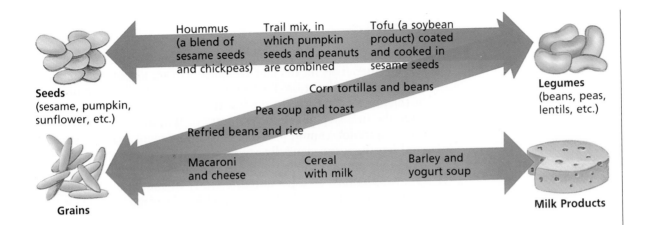

Seeds (sesame, pumpkin, sunflower, etc.)

Hoummus (a blend of sesame seeds and chickpeas)

Trail mix, in which pumpkin seeds and peanuts are combined

Tofu (a soybean product) coated and cooked in sesame seeds

Corn tortillas and beans

Pea soup and toast

Refried beans and rice

Legumes (beans, peas, lentils, etc.)

Macaroni and cheese

Cereal with milk

Barley and yogurt soup

Grains

Milk Products

Why does the human body store fat? After all, excessive fat can strain the heart. The answer is that stored fats are actually beneficial unless they are excessive. A light layer of body fat beneath the skin provides insulation in cold weather. Fat surrounding vulnerable organs, such as the kidneys and liver, acts as protective padding. Most important, fat reserves form a backup fuel supply. While you cannot live just off your body fat, the body can use this fat for energy when carbohydrates are unavailable.

Figure 45-2 Combining legumes with either seeds or grains or combining grains with milk products will furnish all the essential amino acids.

Structure of Fats

To use fats, the body must first break down each fat molecule into its parts—glycerol and fatty acids. About 98 percent of the fats in food are **triglycerides** *(trie-GLISS-uh-RIDEZ)*, molecules formed from three molecules of fatty acids and one molecule of glycerol. The glycerol molecule is the same in all fats, but the fatty acids differ in both structure and composition. The body converts some fatty acids into other fatty acids. Which fatty acids the body produces depends upon its needs at the time.

Types of Fats

Scientists classify fats as saturated or unsaturated on the basis of structural differences in their fatty acids. A saturated fatty acid has all its carbon atoms connected by single bonds and thus contains as many hydrogen atoms as possible. An unsaturated fatty acid has at least one double bond between carbon atoms. If there are two or more double bonds, the fatty acid is called polyunsaturated. In general, animal fats are saturated and plant oils are unsaturated. However, some vegetable oils, such as palm oil and coconut oil, are composed primarily of saturated fats.

Scientific studies have indicated that consuming a high level of saturated fats seems to trigger an increase in the production of **cholesterol,** a fatlike molecule in the body. In turn, an elevated blood cholesterol level may lead to the formation of fatty deposits that can clog arteries, hardening them and restricting blood flow. The result of this condition may be heart disease or a stroke.

Tip: Review with students the basic steps in respiration, as explained in Chapter 7, so that they will understand the ultimate use of nutrients.

Vitamins

Vitamins are complex organic molecules that serve as cofactors for enzymes. This means that vitamins activate the enzymes and assist with their function. For example, the presence of vitamin B_1 or thiamine, is essential to the function of two enzymes that catalyze the first step in aerobic respiration. Without thiamine, aerobic respiration cannot proceed. Because vitamins perform such vital functions and generally cannot be synthesized by the body, a daily diet should include the proper amounts of all vitamins.

All vitamins dissolve in either water or fat. The fat-soluble vitamins include vitamins A, D, E, and K. The water-soluble vitamins are vitamins B and C. Fat-soluble vitamins are absorbed and stored like fats. As with fats, your body can amass a reserve of fat-soluble vitamins. However, unpleasant physical symptoms

Table 45-1 Vitamins

Vitamins	Best sources	Essential for	Deficiency diseases and symptoms
Vitamin A (fat soluble)	Fish-liver oils, liver and kidney, green and yellow vegetables, yellow fruit, tomatoes, butter, egg yolk	Growth, health of the eyes, structure and functions of the cells of the skin and mucous membranes	Retarded growth, night blindness, susceptibility to infections, changes in skin, defective tooth formation
Vitamin B_1 (thiamin; water soluble)	Meat, soybeans, milk, whole grain, legumes	Growth; carbohydrate metabolism; functioning of the heart, nerves, muscles	Beriberi: loss of appetite and weight, nerve disorders, faulty digestion
Vitamin B_2 or G (riboflavin; water soluble)	Meat, fowl, soybeans, milk, green vegetables, eggs, yeast	Growth, health of the skin and mouth, carbohydrate metabolism, functioning of the eyes	Retarded growth, dimness of vision, inflammation of the tongue, premature aging, intolerance to light
Vitamin B_3 (niacin; water soluble)	Meat, fowl, fish, peanut butter, potatoes, whole grain, tomatoes, leafy vegetables	Growth; carbohydrate metabolism; functioning of the stomach, intestines, and the nervous system	Pellagra: smoothness of the tongue, skin eruptions, digestive disturbances, mental disorders
Vitamin B_{12} (water soluble)	Green vegetables, liver	Preventing pernicious anemia	A reduction in number of red blood cells
Vitamin C (ascorbic acid; water soluble)	Fruit (especially citrus), tomatoes, leafy vegetables	Growth, strength of the blood vessels, development of teeth, health of gums	Scurvy: sore gums, hemorrhages around the bones, tendency to bruise easily
Vitamin D (fat soluble)	Fish-liver oil, liver, fortified milk, eggs, irradiated foods	Growth, calcium and phosphorus metabolism, bones and teeth	Rickets: soft bones, poor development of teeth, dental decay
Vitamin E (tocopheral; fat soluble)	Wheat germ oil, leafy vegetables, milk, butter	Normal reproduction	(Undetermined)
Vitamin K (fat soluble)	Green vegetables, soybean oil, tomatoes	Normal clotting of the blood, liver functions	Hemorrhages

and even death can result from storing too much of a particular vitamin as well as from having too little. Continual megadoses of vitamin A, for instance, can result in severe nausea and a yellow skin color. The body cannot store the water-soluble vitamins and therefore excretes any surplus amounts.

The only vitamin that the body can synthesize in large quantities is vitamin D. This synthesis begins with the conversion of cholesterol to provitamin D by intestinal enzymes. Provitamin D is stored in the skin, where exposure to the ultraviolet light in sunlight converts it into vitamin D. People who don't spend a lot of time in the sun get their vitamin D from food.

With the exception of vitamin D, human beings must obtain the vitamins they need from food sources and vitamin supplements. For this reason a knowledge of which foods contain which vitamins is important. Study Table 45-1 to learn some of the food sources of the vitamins listed.

Info: The vitamin needs of various animal species differs because species vary in the enzyme cofactors that they can synthesize for themselves. Many animals, for example, synthesize their own vitamin C.

Minerals

In the context of nutrition, **minerals** are inorganic substances required for the normal function of the body. Minerals such as calcium, magnesium, and iron are drawn from the soil and become part of plants. Animals that feed on plants, in turn, extract the minerals and incorporate them into their bodies. Table 45-2 lists the primary sources and functions of a few of the minerals most essential to human beings. Iron, for example, is nec-

Table 45-2   Minerals

Minerals	Source	Essential for
Sodium salts	Table salt, vegetables	Blood and other body tissues
Calcium salts	Milk, whole-grain cereals, vegetables, meats	Deposition in bones and teeth; heart and nerves
Phosphorus salts	Milk, whole-grain cereals, vegetables, meats	Deposition in bones and teeth; formation of ATP, nucleic acid
Magnesium salts	Vegetables	Muscle and nerve action
Iron salts	Leafy vegetables, liver, meats, raisins, prunes	Formation of red blood corpuscles
Iodine	Seafoods, water, iodized salt	Production of thyroid gland secretion
Potassium salts	Vegetables, citrus fruits, bananas, apricots	Maintain acid–base balance; growth

essary for the formation of red blood corpuscles, while potassium maintains the body's acid–base balance and aids in growth. Both are found in certain fruits and vegetables; iron is also found in meats. Iodine—found in seafoods, water, and iodized salt—is processed and secreted by the thyroid gland. Minerals are excreted through the skin in perspiration as well as through the kidneys in urine.

Water

The importance of water to human life may be inferred from the fact that water accounts for over half of your body weight. Most of the reactions that maintain life can take place only in water. Water makes up over 90 percent of the fluid part of the blood, which carries essential nutrients to all the parts of the body. It is also the medium in which waste products are dissolved and carried away from body tissues.

Water also helps regulate body temperature. It absorbs the heat released in cellular reactions and distributes the heat throughout the body. When the body needs cooling, the problem is solved through perspiration. When this perspiration—a water-

Biology in Process

The Discovery and Synthesis of Vitamin B$_1$

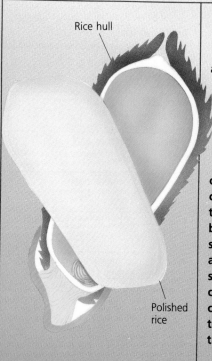

Rice hull

Polished rice

In the early 1600s, a terrible disease began to spread among people in parts of Asia. The disease was called beriberi. Its victims became so weak that they were unable to do anything. Many of them died.

In the 1890s Dutch physician Christiaan Eijkman began conducting experiments in Java to determine the cause of beriberi. By chance Eijkman observed a group of chickens that appeared to have a disease similar to beriberi. Intending to continue his observation of the chickens, he had them moved to another location, but then the disease disappeared. What

could be the reason for this? Eijkman observed that in their original location the chickens had been fed polished rice; after they were moved, they received unpolished rice. Eijkman hypothesized that the cause of beriberi was associated with the polished rice. Testing his hypothesis, Eijkman found that chickens fed polished rice developed symptoms like those of beriberi, while chickens fed unpolished rice did not.

Dutch physician Gerrit Grijns continued Eijkman's experiments. Grijns concluded that the rice hull, which was ground away by polishing, contained an essential nutrient. In

Answer to Biology in Process: Eijkman had to test his hypothesis in order to be sure that it was the polished rice that was causing beriberi.

based substance—evaporates from the skin, heat is drawn away from the body in the process.

Usually the water lost through your skin and kidneys is easily replaced by drinking water, by consuming moist foods, or by inhaling air, since air contains water. If its water supply is not replenished, the body begins to take water from intercellular spaces to perform vital functions within the cells. Eventually the body has to draw water from the cells themselves. As a cell loses water, the cytoplasm becomes more concentrated until finally the cell can no longer function. This condition is referred to as **dehydration** (DEE-hie-DRAY-shun).

Section Review

1. Define *nutrient*.
2. What are two functions performed by proteins?
3. Name three reasons why a certain amount of body fat is beneficial.
4. Why is water a vital nutrient?
5. What do you think would happen to a vegetarian who consumed only water, unpolished rice, and fruits?

Answers to Section Review
1. a chemical substance that the body needs in order to grow and function properly
2. Possible answers are: they help the body grow, repair tissues, and act as enzymes.
3. It provides insulation in cold weather, protects organs, and serves as a backup fuel supply.
4. Most reactions that maintain life take place only in water. Water is also the medium in which waste is dissolved and carried away from body tissues and in which nutrients are carried throughout the body.
5. He or she would suffer from the symptoms of a protein and mineral deficiency and would be less able to function properly.

other words, he inferred that beriberi was the result of a nutrient deficiency.

In 1910 a U.S. Army doctor, Captain Edward Vedder, wanted to isolate the curative ingredient for beriberi so that it could then be synthesized. Once a substance can be synthesized, it can usually be produced in large quantities at low cost. Vedder introduced his idea to a young chemist named Robert R. Williams.

During the 1920s and 1930s, Williams worked on this project primarily in his spare time in a garage laboratory. To isolate the nutrient, he chemically separated compounds in a rice hull extract and then tested the curative power of each compound on animals with beriberi. In 1933 Williams and his associates isolated the single curative nutrient, a vitamin they identified as B_1.

The Williams team spent the next three years trying to synthesize this vitamin. However, to build a molecule of it themselves, they needed to learn a great deal about its composition and structure. First, they identified the elements it contained and the amount of each. Then they split a molecule, studied its fragments, and inferred how each atom was joined. Finally in 1936 the team synthesized the vitamin, and a proud Williams gave it the name by which it is commonly known today, *thiamine*. Thiamine is shown here in crystalline form.

■ Why was it important for Eijkman to test his hypothesis about polished rice after his initial chance observation of the chickens?

Section Objectives

- *Define* alimentary canal, *and name the major organs of the digestive system.*

- *Relate the structure of each digestive organ to its function in mechanical digestion.*

- *Identify the source of each major digestive enzyme, and describe the function of the enzyme.*

- *Summarize the process of absorption in both the small and large intestine.*

45.2 Digestive System

Before your body can use the nutrients in the food you consume, the cells must break down the nutrients both physically and chemically. This process of breaking down food into molecules the body can use is called **digestion**. Digestion occurs in the **alimentary canal,** or digestive tract, which begins at the mouth and winds through the body to the anus. Located along the alimentary canal are the stomach and the other organs that aid in digestion. Other digestive organs, such as the liver and pancreas, are not part of the alimentary canal but deliver secretions into the canal through ducts. As you read, locate each of the structures of the digestive system in Figure 45-3.

Mouth

When you take a bite of food, you begin the mechanical phase of digestion. In this phase the body physically breaks down chunks of food into small particles. Mechanical digestion increases the surface area of food on which digestive enzymes can act. Inci-

Figure 45-3 Digestion in the alimentary canal is aided by secretions from other organs.

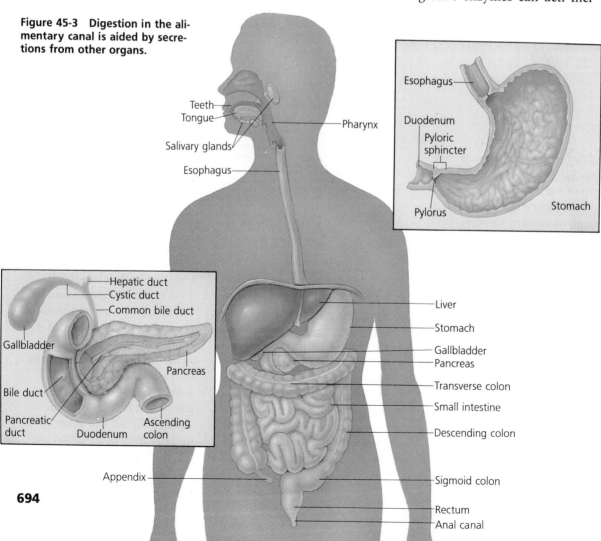

694

sors—sharp, flat front teeth—cut the food. Then the broad, flat surfaces of molars, or back teeth, grind it up. The tongue helps keep this food between the chewing surfaces of the upper and lower teeth by manipulating it against the **hard palate,** the bony, membrane-covered roof of the mouth. This is a distinctly different structure from the **soft palate,** the area located just behind the hard palate, made of folded membranes and separating the mouth cavity from the nasal cavity.

While the mechanical phase of digestion is occurring, the chemical phase of digestion is also taking place. Preparations for this phase begin even before the first bite of food is taken. The mouth starts to water—that is, the salivary glands increase their production of **saliva,** a mixture of water, mucus, and a digestive enzyme called salivary amylase. In addition to the many tiny salivary glands located in the lining of the mouth, there are three pairs of larger salivary glands. Locate the three sets of salivary glands in Figure 45-3.

The mucus in the saliva softens and lubricates food and helps hold it together. In addition, the salivary amylase begins the chemical digestion of carbohydrates by breaking down some starch into the disaccharide maltose.

Esophagus

Once food has been thoroughly chewed, moistened, and rolled into a **bolus,** or ball, it is forced by the swallowing action of the tongue into the pharynx. The pharynx *(FER-inks)* is the open area that begins at the back of the mouth and serves as a passageway for both air and food. As Figure 45-4 shows, during swallowing a flap of tissue called the **epiglottis** prevents food from entering the trachea, or windpipe. Instead the bolus passes into the **esophagus,** the approximately 25-cm-long muscular tube that connects the pharynx with the stomach. The esophagus has two muscle layers—a circular layer that wraps around the esophagus and a longitudinal layer that runs the length of the tube. Alternating contractions of these muscle layers push the bolus through the esophagus into the stomach. This series of rhythmic muscular contractions and relaxations is called **peristalsis.**

Stomach

The **stomach,** an organ of both mechanical and chemical digestion, is located in the upper left side of the abdominal cavity, just below the diaphragm. It is an elastic bag that is J-shaped when full and that lies in folds when empty. In addition to circular and longitudinal muscles, the walls of the stomach have a third, diagonal layer of muscles. Figure 45-6, shown on the following page, illustrates the arrangement of these muscle layers. Together these muscles can twist the stomach and churn its contents. This churning helps the stomach perform mechanical digestion.

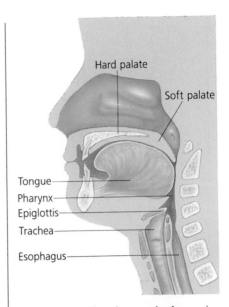

Figure 45-4 The pharynx is the only passage shared by the digestive and respiratory systems. The soft palate and the epiglottis keep the processes of eating and breathing separate in critical areas.

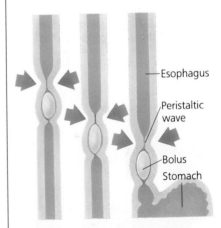

Figure 45-5 A peristaltic wave moves food downward by means of two-way muscular action in the esophagus.

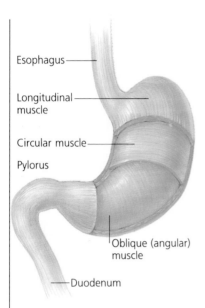

Esophagus

Longitudinal muscle

Circular muscle

Pylorus

Oblique (angular) muscle

Duodenum

Figure 45-6   Three types of stomach muscles—each oriented in a different direction—permit the stomach to twist and churn, movements that mechanically break up food.

Figure 45-7   This photo shows two of the gastric glands that produce stomach secretions through gastric pits (×400).

The inner lining of the stomach is a thick, wrinkled mucous membrane dotted with small openings called gastric pits. **Gastric pits** are the open ends of gastric glands that produce the cells that release secretions into the stomach. Some of these glands secrete mucus, others secrete digestive enzymes, and still others secrete hydrochloric acid. The mixture of these secretions forms the acidic **gastric fluid.**

Chemical Digestion

Gastric fluid carries out chemical digestion in the stomach. The chemical breakdown of carbohydrates that started in the mouth continues in the stomach. The stomach also begins the chemical digestion of proteins. This process starts as the digestive enzyme pepsin splits complex protein molecules into shorter chains of amino acids called peptides. Pepsin is secreted as an inactive fluid called pepsinogen, which is converted into pepsin at a low pH. The presence of hydrochloric acid within the stomach ensures a low pH. In addition to helping to transform pepsinogen into pepsin, hydrochloric acid dissolves minerals and kills bacteria that enter the stomach along with food.

The mucus secreted in the stomach is vital to the survival of this organ. It forms a coating that protects the lining from hydrochloric acid and prevents pepsin from digesting the proteins that make up the stomach tissue. In some people the mucus coating of the stomach tissue breaks down, allowing digestive enzymes to eat through part of the stomach lining. The sore that results is called an ulcer.

Formation of Chyme

Food enters the stomach when the cardiac sphincter opens. The **cardiac sphincter** is a circular muscle located between the esophagus and the stomach. Once the food is in the stomach, the cardiac spincter closes to prevent the food from regurgitating back into the esophagus. Food usually remains in the stomach for two to three hours. During this time muscle contractions in the stomach walls churn its contents, breaking up food particles and mixing them with gastric fluid. This process forms a mixture called **chyme,** a name that comes from the Greek word meaning "juice." Chyme usually contains fats, sugars, maltose, starches, vitamins, minerals, coagulated milk protein, peptides, and proteins that were not changed by pepsin.

Peristalsis forces chyme out of the stomach and into the small intestine. The **pyloric sphincter,** a circular muscle between the stomach and the small intestine, regulates the flow of chyme. Each time the pyloric sphincter opens, about 5 to 15 mL of chyme moves into the small intestine until finally the stomach is empty.

When chyme enters the small intestine, it mixes with secretions from the liver and pancreas. Before you learn about these secretions, you should know something more about the location, structure, and function of the liver and pancreas.

Liver

The **liver** is a large organ located to the right of the stomach in the upper right area of the abdominal cavity just below the diaphragm. The liver plays a vital role in the digestion of fats by secreting bile. **Bile** is not a digestive enzyme but an emulsifying agent that breaks fat globules into small droplets, forming a milky emulsion. This process exposes a greater surface area of fats to the action of digestive enzymes and prevents small fat droplets from rejoining into large globules.

The bile secreted by the liver passes from the liver through a Y-shaped duct shown in Figure 45-3. The bile travels down one branch of the Y, the hepatic duct, and then up the other branch, the cystic duct, to the **gallbladder,** a saclike organ that stores and concentrates bile. When chyme is present in the small intestine, the gallbladder releases bile through the common bile duct into the small intestine.

Pancreas

The **pancreas** is an organ that lies behind the stomach, against the back wall of the abdominal cavity. It performs two highly different functions. Chapter 47 discusses its function in regulating the level of glucose in the blood. As part of the digestive system the pancreas secretes pancreatic fluid, which contains digestive enzymes that help complete the breakdown of nutrients in the chyme. This pancreatic fluid enters the small intestine through the pancreatic duct, which joins the common bile duct just before it enters the intestine.

Pancreatic fluid contains sodium bicarbonate, which changes the pH of the chyme from acid to base. Once the pH has changed, the many enzymes in the pancreatic fluid are activated. Pancreatic amylase splits into disaccharides the molecules of starch or glycogen that were not acted upon by salivary amylase. Pancreatic lipase breaks fat into fatty acids and glycerol. Trypsin, chymotrypsin, and carboxypeptidase split proteins into peptides.

Figure 45-8 Chyme entering the small intestine is acted on by pancreatic digestive secretions. At the same time bile from the liver enters by way of the gallbladder to emulsify fats.

Pancreas: Pancreatic fluid
Pancreatic amylase, pancreatic lipase, trypsin, chymotrypsin, carboxpeptidase

Liver and gallbladder: Bile
Bile, some waste materials

Stomach: Chyme
Fats, sugars, maltose, starches, coagulated milk protein, peptides, proteins

Duodenum

Small Intestine

If you could stretch the small intestine to its full length, you would find it is nearly 7 m long. The **duodenum** *(DYOO-uh-DEE-num)*, the first section of this coiled tube, makes up only the first 25 cm of that length. The **jejunum** *(juh-JOO-num)*, the middle section, is about 2.5 m long. The **ileum** *(ILL-ee-um),* which makes up the remaining portion of the small intestine, is approximately 4 m in length.

The secretions from the liver and pancreas enter the duodenum, where they continue the chemical digestion of the chyme. When the secretions from the liver and pancreas along with the chyme enter the duodenum, they trigger intestinal mucous glands to release large quantities of mucus. This mucus protects the intestinal wall from protein-digesting enzymes and the acidic chyme. Glands in the mucous lining of the small intestine release the following enzymes:

- Peptidase completes protein digestion by breaking down peptides into amino acids.
- Maltase, lactase, and sucrase split the disaccharides maltose, lactose, and sucrose into monosaccharides.
- Intestinal lipase splits fats into glycerol and fatty acids.

The end products of digestion—amino acids, monosaccharides, glycerol, and fatty acids—are then absorbed into the circulatory system through blood and lymph vessels in the lining of the small intestine. The structure of this lining provides a huge surface area through which absorption takes place. **Absorption** is the process by which the end products of digestion are transferred into the circulatory system. The highly folded lining is covered with millions of fingerlike projections called **villi.** The cells covering the villi, in turn, have extensions on their cell membranes called **microvilli.** The folds, villi, and microvilli give the small intestine a

Figure 45-9 Villi, as shown in the SEM (left, ×1,000) and the diagram (right), expand the absorption area of the small intestine. What other processes could be added to the diagram to complete the picture?

Villus

Capillaries

Intestinal gland

Lacteal

surface area of about 250 m², roughly the size of a tennis court. Nutrients are absorbed through this surface by means of diffusion and active transport.

Inside each villus are capillaries and tiny lymph vessels called **lacteals.** Glycerol and fatty acids enter the lacteals, are carried through the lymph vessels, and eventually reach the bloodstream through lymphatic ducts near the heart. Amino acids and monosaccharides enter the capillaries and are carried to the liver. The liver detoxifies the blood and removes any excess glucose, converting it to glycogen for storage. The filtered blood then carries the nutrients to all parts of the body.

Large Intestine

Once absorption in the small intestine is complete, peristalsis moves the remaining material on to the large intestine. The **large intestine,** or **colon,** is the final organ of digestion. Study Figure 45-3 to identify the four major parts of the colon: ascending colon, transverse colon, descending colon, and sigmoid colon. The rectum and the anal canal that it leads into are the names given to the very short final portions of the large intestine. Find these portions of the large intestine in Figure 45-3.

The material that enters the large intestine contains an abundance of water and minerals. Cellulose present in this material draws even more water into the intestine. This influx of water increases the mass of the material in the small intestine and expands the large intestine walls, triggering muscle contractions. These contractions form the rhythmic peristaltic wave that moves the watery mass through the length of the intestine and eventually out of the body. As this matter moves through the intestine, the body then absorbs water and minerals, solidifying the mass in the process. The solidified material is called **feces.**

As the fecal matter solidifies, mucous secretions of the intestinal wall lubricate the wall to make the passing of the feces less abrasive. Mucus also binds the fecal matter together and neutralizes any acids that are released when bacteria act upon the feces. Such bacteria also aid in the manufacture of vitamin K. The feces are eliminated through the anus.

Section Review

1. What is the *alimentary canal?*
2. Name one structure involved in the mechanical phase of digestion and explain its function in mechanical digestion.
3. What is the function of bile?
4. How is the structure of the small intestine related to the function of absorption?
5. Describe how cellulose aids in digestion. What do you imagine would be the consequence of not eating enough cellulose?

Figure 45-10 This X ray shows the large intestine, or colon. Try to identify the four major parts of the colon in this X ray.

Answers to Section Review
1. The digestive tract along which lie most of the organs that aid in digestion. It begins at the mouth and ends at the anus.
2. Any of the following is acceptable: incisors, which cut food; molars, which grind food; tongue, which keeps food between the teeth; stomach, which churns food.
3. It acts as an emulsifying agent by breaking fat globules into small droplets. This exposes a greater surface area of the fats to the action of digestive enzymes.
4. By means of villi and microvilli, the small intestine provides a huge surface area on which absorption can take place.
5. Cellulose aids digestion by drawing water into the large intestine, expanding the intestine walls and triggering muscle contractions that move the feces through the intestine and out of the body. Not eating enough cellulose would retard the movement of food through the alimentary canal.

Section Objectives

- *Define the term* excretion, *and list the functions of each of the major excretory organs.*

- *Identify the major parts of the kidney.*

- *Relate the structure of a nephron to the functions it performs.*

- *Explain how the processes of filtration, reabsorption, and tubular secretion help maintain homeostasis.*

- *Name the main parts of the urinary system.*

45.3 Excretory System

The body must rid itself of the waste products of cellular activity. This process of removing metabolic wastes, called **excretion,** is just as vital as digestion to maintaining the body's internal environment. Thus the excretory system not only eliminates wastes but also plays a crucial role in maintaining homeostasis by regulating the balance of water and other substances in the blood.

The main waste products that the body must eliminate are, first, carbon dioxide and water from cellular respiration and, second, nitrogenous compounds from the breakdown of proteins. The lungs excrete most of the carbon dioxide and some of the water through respiration. Some water and excess minerals are lost in the form of perspiration. Most water and nitrogenous wastes are eliminated by the kidneys, the main excretory organs of the body.

The most common nitrogenous waste is ammonia, a substance so toxic that it could not remain long in the body without harming cells. The body is protected from ammonia poisoning by the liver, which removes ammonia from the blood and converts it into a less harmful substance called urea. The urea enters the bloodstream and is then removed by the kidneys.

Kidneys

Humans have two **kidneys,** bean-shaped excretory organs each about the size of a clenched fist. The kidneys are located in the small of the back, one behind the stomach and the other behind the liver. The longitudinal section in Figure 45-11 shows the three main parts of the kidney. The **cortex** is the outermost portion of the kidney, which makes up about a third of its tissue mass. The **medulla** is the inner two-thirds of the kidney. The **renal pelvis** is a funnel-shaped structure in the center of the kidney. Also notice in

Figure 45-11 The excretory system (left) involves the kidneys. Note the kidney and nephron diagramed at right.

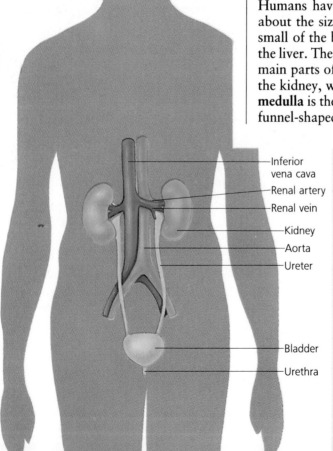

Inferior vena cava
Renal artery
Renal vein
Kidney
Aorta
Ureter
Bladder
Urethra

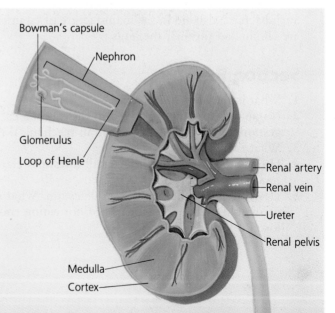

Bowman's capsule
Nephron
Glomerulus
Loop of Henle
Renal artery
Renal vein
Ureter
Renal pelvis
Medulla
Cortex

the figure that blood enters the kidney through a renal artery and leaves through a renal vein.

A complex series of processes in the kidney removes waste from the blood and adjusts its chemical makeup. The substances—toxins, urea, water, and mineral salts—removed from the blood by these processes form an amber-colored liquid called **urine.** The urine flows from the kidneys into the urinary system for storage and eventual elimination from the body.

Nephrons

Take a close look at the structure of the functional unit of the kidney, the **nephron,** shown in Figure 45-12. Each kidney consists of more than a million nephrons. Each nephron has a cup-shaped structure called **Bowman's capsule** that encloses a bed of capillaries. This capillary bed, called a **glomerulus** *(gluh-MER-yuh-luss),* receives blood from the renal artery. High pressure forces fluids from the blood through the capillary walls into Bowman's capsule. The material filtered from the blood then flows through the **renal tubule,** a long tube with permeable walls. The remaining blood flows through a network of capillaries that wraps around the renal tubule.

As the blood and fluid flow through a nephron, the composition of both is modified by the exchange of materials among the renal tubule, the capillaries, and the extracellular fluid. Various types of exchange take place in the four different parts of the renal tubule: the proximal convoluted tubule, the loop of Henle, the distal convoluted tubule, and the collecting duct. To understand how the structure of each part of the nephron is related to its function, let's look at the three major processes that take place in the nephron: filtration, reabsorption, and secretion.

Filtration

The process through which materials from the blood are forced out of the glomerulus and into Bowman's capsule is called **filtration.** Pressure forces water, nitrogenous wastes, glucose, and mineral salts through the thin capillary walls. About one-fifth of the fluid portion of the blood filters into the capsule. The rest remains in the capillaries along with proteins and cells, which are too large to pass through the capillary walls. Thus the filtrate—the fluid that enters the nephron—resembles blood plasma without the large protein molecules found in plasma.

Reabsorption and Secretion

The body needs to retain many of the substances that have been removed from the blood by filtration. Thus as the filtrate flows through the renal tubule, these materials return to the blood by passing through the walls of the tubule and entering the surrounding capillaries. This process is called **reabsorption.** Most reabsorption occurs in the proximal convoluted tubule. In this region about 75 percent of the water in the filtrate returns to the capil-

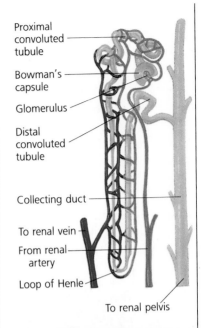

Proximal convoluted tubule

Bowman's capsule

Glomerulus

Distal convoluted tubule

Collecting duct

To renal vein

From renal artery

Loop of Henle

To renal pelvis

Figure 45-12 A nephron is shown in the diagram (top). A glomerulus is shown in the center of the SEM (below). Surrounding the glomerulus are renal tubules and red blood corpuscles (×250).

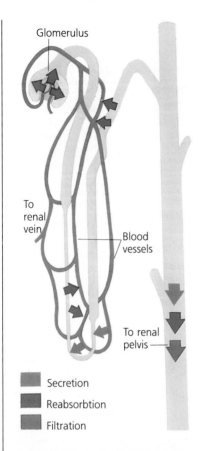

Glomerulus

To renal vein

Blood vessels

To renal pelvis

Secretion

Reasorbtion

Filtration

Figure 45-13 Color-coded arrows indicate where in the nephron secretion, reabsorption, and filtration occur.

Info: Mammals and birds are the only organisms able to produce urine more concentrated than their interstitial fluid. They are also the only two groups whose nephrons include a loop of Henle, the structure which permits this concentration to take place.

Answers to Section Review
1. the process of removing metabolic wastes
2. Possible answers are: the lungs excrete carbon dioxide and some water through respiration; and the skin excretes water and excess minerals through perspiration.
3. the proximal convoluted tubule, the loop of Henle, the distal convoluted tubule, and the collecting duct
(See pp. T45A–T45D.)

laries by osmosis. Glucose and minerals such as sodium, potassium, and calcium are returned to the blood by active transport.

Some additional reabsorption occurs in the distal convoluted tubule. Also in this region of the tubule some substances pass from the blood into the filtrate through a process called secretion. The pH of the blood is also adjusted by secreting hydrogen ions into the filtrate. The fluid and wastes that remain in the renal tubule form urine. The urine from several renal tubules flows into a collecting duct.

In the collecting duct urine is further concentrated by the osmosis of water through the wall of the duct into the extracellular fluid. This process allows the body to conserve water. In fact, osmosis in the collecting duct, together with reabsorption in other parts of the tubule, returns to the blood about 99 of every 100 mL of water in the filtrate.

You may have noticed that in tracing the flow of filtrate through the nephron, we skipped over the loop of Henle. The function of the loop of Henle is closely related to that of the collecting duct. Water moves out of the collecting duct because the concentration of sodium chloride (NaCl) is higher in the extracellular fluid of the medulla than it is in the fluid inside the collecting duct. This high concentration of sodium chloride is maintained by the loop of Henle. Cells in the wall of the loop actively transport negatively charged chloride ions from the filtrate into the extracellular fluid. Positively charged sodium ions follow the chloride ions into the fluid through electrical attraction. This ensures that the NaCl concentration of the extracellular fluid remains high and thus promotes the reabsorption of water from the collecting duct. Understanding this relationship between the loop of Henle and the collecting duct will help you understand how these two structures function together to concentrate wastes.

Urinary System

Urine from the collecting ducts flows through the renal pelvis and into a narrow tube called a **ureter.** A ureter leads from each kidney to the **urinary bladder,** a muscular sac that stores urine. Muscular contractions of the bladder force urine out of the body through a tube called the **urethra.**

Section Review

1. What is *excretion?*
2. Name one organ other than the kidney that is involved in excretion, and describe what and how it excretes.
3. What are the structures of the renal tubule?
4. What is the most basic difference between filtration and reabsorption?
5. Given the definition of excretion, why do you think the large intestine is not classified as a major excretory organ?

Laboratory

A Model of Human Digestion

Objective
To construct a model of chemical digestion in the human stomach

Process Skills
Modeling, observing, predicting, inferring

Materials
Glass-marking pencil, 5 test tubes with stoppers, test tube rack, scalpel, cooked egg white, 10-mL graduated cylinder, 1% pepsin solution, 0.2% hydrochloric acid, 1% sodium bicarbonate solution, red and blue litmus paper

Method
1 With a glass-marking pencil, number five test tubes *1*, *2*, *3*, *4*, and *5*. Place the test tubes in a test tube rack.
2 Use a scalpel to chop about 3/4 of a teaspoon of cooked egg white into fine pieces.
3 Place equal amounts of egg white in each test tube. Egg white is mainly protein.
4 Measuring with a graduated cylinder, add 10 mL of water to test tube 1, 10 mL of 1% pepsin solution to test tube 2, and 10 mL of 0.2% hydrochloric acid to test tube 3.
5 Use the graduated cylinder again to add 5 mL of 1% pepsin solution plus 5 mL of 1% sodium bicarbonate solution to test tube 4. Add 5 mL of 1% pepsin solution plus 5 mL of 0.2% hydrochloric acid to test tube 5.
6 Place stoppers in the test tubes. Gently shake each test tube and set it back in the rack. Why do you shake the tubes?
7 On a sheet of paper, make a table four columns wide. Head the columns *Test Tube Number*, *Contents*, *pH*, and *Degree of Digestion*. Fill in the first two columns and save the table for later use. Which test tube would you predict will show the most digested protein in 48 hours?
8 Store the test tube rack for 48 hours in a safe

place where its contents will be kept at room temperature.
9 After 48 hours, remove the stoppers from the test tubes and test each solution for pH. To do this, add a strip of red litmus paper to each test tube. If there is no color change, add a strip of blue litmus paper. Remember that an acid turns blue litmus paper red, a base turns red litmus paper blue, and a neutral solution does not change the color of either kind of paper. Record your results.
10 Examine the degree of digestion, if any, of the cooked egg white in each test tube. Record your observations in the table.

Conclusions
1 Which test tube most closely simulates the chemical conditions in the human stomach? In what ways?
2 What information do test tubes 1, 2, and 3 contribute to the investigation?
3 What information does test tube 4 contribute to the investigation?

Inquiry
How do you think the results of this experiment would be different if you digested a fat, such as vegetable oil? Design your own experiment to test fat digestion.

This investigation must be done in two sessions.

Chapter 45 Review

Vocabulary

absorption (698)	dehydration (693)	hydrolysis (687)	protein (688)
alimentary canal (694)	digestion (694)	ileum (698)	pyloric sphincter (696)
bile (697)	disaccharide (687)	jejunum (698)	reabsorption (701)
bolus (695)	duodenum (697)	kidney (700)	renal pelvis (700)
Bowman's capsule (701)	enzyme (688)	lacteal (698)	renal tubule (701)
calorie (687)	epiglottis (695)	large intestine (699)	saliva (695)
Calorie (687)	esophagus (695)	liver (697)	soft palate (695)
carbohydrate (687)	excretion (700)	medulla (700)	stomach (695)
cardiac sphincter (696)	fat (688)	microvilli (698)	triglyceride (689)
cholesterol (689)	feces (699)	mineral (691)	ureter (702)
chyme (696)	filtration (701)	monosaccharide (687)	urethra (702)
colon (699)	gallbladder (697)	nephron (701)	urinary bladder (702)
cortex (700)	gastric fluid (696)	nutrient (687)	urine (701)
	gastric pit (696)	pancreas (697)	villi (698)
	glomerulus (701)	peristalsis (695)	vitamin (690)
	hard palate (695)	polysaccharide (688)	

For each set of terms below, choose the one that does not belong, and explain why it does not belong.
1. carbohydrate, protein, fat, mineral
2. pharynx, epiglottis, bolus, esophagus
3. monosaccharide, glucose, polysaccharide, triglyceride
4. absorption, filtration, secretion, reabsorption
5. nephron, ileum, glomerulus, renal tubule

Review

1. The primary function of carbohydrates is to (a) break down molecules (b) aid in digestion (c) supply the body with energy (d) regulate the flow of chyme.
2. Cellulose (a) builds body tissue (b) is a monosaccharide (c) is used for energy (d) aids in digestion.
3. Proteins consist of (a) catalysts (b) enzymes (c) amino acids (d) carbon, hydrogen, and oxygen.
4. The body needs vitamins because they (a) supply it with energy (b) serve as cofactors for enzymes (c) function as enzymes (d) act as hormones.
5. The best way to prevent dehydration is by (a) inhaling air (b) drinking water (c) not drinking water (d) perspiring.
6. The epiglottis is important because it

(a) prevents food from going down the trachea (b) separates the pharynx from the nasal cavity (c) is the passage through which food travels to the stomach (d) regulates the flow of chyme.
7. The gallbladder (a) creates bile (b) stores urine (c) stores, concentrates, and secretes bile (d) is made up of nephrons.
8. During absorption the lacteals absorb (a) glycogen (b) glycerol and fatty acids (c) amino acids and monosaccharides (d) lactose.
9. The organs primarily involved in the excretion process are the (a) kidneys and stomach (b) liver and pancreas (c) nephron and glomerulus (d) kidneys and liver.
10. During secretion in the kidney, substances move from (a) filtrate to blood (b) blood

to blood (c) blood to filtrate (d) filtrate to filtrate.

11. Why can consuming a great deal of saturated fat be harmful?
12. Briefly explain the synthesis of vitamin D in the body.
13. What data did Eijkman's experiments yield, and what did Grijns infer from these data?
14. In what two ways do the liver and the pancreas differ from other digestive organs?
15. What is the mechanical digestive work performed by the stomach?
16. Why does the body convert ammonia into urea?
17. What are the three main parts of the kidney?
18. What are the excretory processes that maintain homeostasis?
19. What are the structures of a nephron?
20. What are the main structures of the urinary system?

Critical Thinking

1. In some countries many children suffer from a type of malnutrition called kwashiorkor. They have swollen stomachs and become increasingly thin until they die. Even when given rice and water, they still die. What type of nutritional deficiency might they be suffering from?
2. Some people cannot drink milk because they are unable to digest lactose, the sugar in milk. Doctors think this inability involves specific areas of the digestive system. What areas do you think these are?
3. What can happen to a person who gets serious diarrhea? (Hint: Why is it important that the large intestine reabsorb water and not excrete it?)
4. Look at the pictures of the teeth of different animals. What can you tell about the human diet by comparing the teeth of humans with those of the other animals shown here?

5. When the kidneys stop functioning, urea builds up in the blood and poisons the body. A person whose kidney has failed will eventually die if the urea is not somehow removed. For it to be removed, the patient must be attached to an artificial kidney called a dialysis machine. Suggest a design for the major components of a dialysis machine based on the way a normal kidney works.

Extension

1. Make a chart in which you keep the Calorie counts of what you eat for breakfast, lunch, and dinner for an entire week. Compare your daily caloric intake with that of the other students in your class.
2. Look in the *Readers' Guide to Periodical Literature* for titles of recent articles in which the authors claim that certain diet changes can help prevent heart attacks and cancer. Write a brief report stating the findings of a few of these studies.
3. Read the article by B. Bower entitled "The 2-Million-Year-Old Meat and Marrow Diet Resurfaces" in *Science News*, Vol. 131, January 3, 1987, p. 7. How does the evidence found by Henry Bunn and Ellen Kroll seem to conflict with the teaching of dieticians today?

Chapter 46 | Nervous System and Sense Organs

This chapter describes the structure of the nervous system and the transmission of nerve impulses. Students learn how the nervous system controls human adaptation to the environment.

Correlation Chart

Section Number	Section Objectives	Pages	Features	Section Review
46.1	1. Identify the primary divisions of the nervous system.	707–712	Biotechnology, pp. 710–711	2, 5
	2. Describe the function of receptors, conductors, and effectors.			1
	3. Summarize the functions of the three regions of the brain.		Think, p. 712	3, 4, 5
	4. Explain how advances in photographic techniques have helped scientists learn more about the brain.			
46.2	1. Relate the structure of a neuron to its function.	713–717		2
	2. Summarize the electrical and chemical conditions of a neuron in the resting state.			1
	3. Describe a nerve impulse.			2, 3
	4. Describe the role of neurotransmitters in the transmission of impulses across synapses.			3, 4, 5
46.3	1. Describe a sense receptor, and list the five primary types.	718–723		1, 5
	2. Describe the roles of rods and cones in vision.			2
	3. Identify the parts of the ear responsible for hearing and for maintaining balance.			3
	4. Explain the chemical relationship between taste and smell.		Biology Lab, p. 723	4

Resources*

Print

Kuffler, Stephen W., John G. Nicholls, and A. Robert Martin. *From Neuron to Brain: A Cellular Approach to the Function of the Nervous System,* 2nd ed. Sunderland, Mass.: Sinauer, 1984.

Restak, Richard. *The Brain.* New York: Bantam, 1984.

Audiovisual/Software

The Human Brain. Film or video, 24 min. Teacher's Guide. Britannica, 1983.

The Senses: Eyes and Ears. Film or video, 26 min. Humanities and Sciences. 1985.

*Additional references are listed in *Modern Biology, Teacher's Program Guide.*

Section Teaching Guides

46.1 Structure

Approach

The main objective of this section is to describe the structure and functions of the main components of the human nervous system. The nervous system is quite complex, and this section stresses how structures of the nervous system reflect their functions in transmitting, integrating, interpreting, and relaying messages.

This section builds on many concepts presented in previous chapters. For example, it builds on the concept of homeostasis, presented in Chapter 6, in discussing the function of the hypothalamus. This section also extends the concepts presented in Chapter 42 on how muscles function. Finally, the section builds on the main concepts of Chapters 15 and 16 by giving a concrete example of the adaptive significance of reflexes.

This section prepares students to understand how the endocrine system and the nervous system work together in regulating other body systems. In presenting the material, help students to retain a perspective on the overall function of the nervous system in allowing a person to respond to changes in the internal and external environment. Assist students in relating the structures of the nervous system to their functions. Refer students to Figure 46-1 to help them get an overview of the different parts of the nervous system. Make sure that students learn the specific functions of each part of the nervous system listed in the diagram.

Concept Framework

Students will be familiar with the brain as the organ that allows them to think. They will also be aware of nerves as structures that allow them to process information and to respond or not to respond to the environment.

Tap this knowledge by writing the words *brain* and *nerves* on the chalkboard. Elicit students' associations with these words and write them down on the chalkboard. Point out to students that their brains do far more than allow them to think and that their nerves allow them to respond to stimuli through intricate interconnections with muscles.

You may wish to end the discussion by asking students as they read this section to look for comparisons between the nervous system and a computer network. Stress that the brain and nervous system are infinitely more complex, integrated, and sensitive than even the largest, most powerful computer.

Emphasis

Make sure that students understand how the central nervous system and the peripheral nervous system work together to enable smooth, efficient communication among all parts of the body. You may wish to ask students to list the parts of each of these systems and define the function of each part in terms of detecting, transmitting, integrating, interpreting, and relaying nerve impulses.

Using the Features/New Imaging Techniques

This biotechnology feature presents improved imaging techniques such as computer tomography (CT) scanners, magnetic resonance imaging (MRI), and the scanning electron microscope (SEM). These techniques allow scientists to study the body, including the nervous system, in greater detail than they could before.

The CT scanner uses an X-ray beam and a computer to produce three-dimensional images of the body. The MRI also produces a three-dimensional image utilizing the magnetic properties of hydrogen-containing molecules and a computer. The SEM uses electrons to produce an image of a specimen's surface that is relayed to a viewing screen. Point out to students that all of these techniques provide scientists and physicians with improved images, and thus improve the ability to understand, diagnose, and treat diseases and malfunctions.

You may wish to discuss how the various imaging techniques presented in this feature can be used specifically to study the nervous system. Point out to students that CT and MRI can be used to study the structure of the brain and spinal cord. The SEM can help reveal the structure of individual nerve cells.

Charts of the Brain

Divide students into three groups. Ask each group to prepare a diagram of one of the regions of the brain and to make an accompanying chart describing the structure and function of that region of the brain. Let students present their diagrams and charts to the rest of the class. Use these charts as a basis for a class discussion of the structure and function of the various parts of the brain. Ask students how these different parts of the brain help coordinate various activities of everyday life.

Reflex Demonstrations

Ask volunteers to research and demonstrate reflexes such as the knee jerk reflex, the lifting of the foot in response to a sharp object, or the pulling away of the hand from a heat source. Direct volunteers to be prepared to show the path nerve impulses follow by tracing their path on a large diagram of the nervous system or the body. Ask them to try to find out how each reflex is adaptive. Give students time to demonstrate the reflexes and to share what they have learned with other students.

Library Research

Encourage interested students to use recent periodicals to find out the state of research into artificial intelligence. Students might look into ways in which computers have helped scientists learn more about how the brain works. Suggest that students discuss this information with the class.

Section Review Answers

(Continued from p. 712.)

1. Sensory neurons carry impulses from receptors to the spinal cord; motor neurons carry impulses from the spinal cord to the effectors; association neurons maintain neural connections inside the spinal cord.

2. The central nervous system consists of the brain and spinal cord; the peripheral nervous system has 12 pairs of cranial nerves and 31 pairs of spinal nerves that radiate out from the central nervous system.

3. The cranium encloses the brain in a hard case of bone; the spinal column is likewise encased in the hard, protective bones that form the vertebral column. Surrounding both the brain and the spinal cord are three layers of meninges. The meninges cushion the brain and spinal cord. They include a clear, watery substance called cerebrospinal fluid, which helps absorb energy from impact, protecting from shock.

4. The cerebrum controls voluntary movement of the skeletal muscles and conscious responses to sensory stimuli. The cerebellum receives stimuli from tendons, joints, eyes, and ears, and uses these stimuli to regulate balance and coordinate muscle contractions of motor responses. The brain stem connects the cerebrum with the spinal cord, relaying impulses and helping to control many different types of body actions.

5. The various subdivisions of the human nervous system allow each part to become increasingly more specific and refined in its responsibilities and make it possible for a smoother transition to take place between systems.

46.2 Transmission of Nerve Impulses

Approach

The main objective of this section is to present students with a description of neurons and the details of how nerve impulses are transmitted. The mode of impulse transmission is electrochemical. Therefore the section requires an understanding both of electrical potential and also of various cell processes that allow nerve cells to regulate the movement of charged particles across the cell membrane. Thus, the section builds on a number of concepts presented in earlier chapters.

For example, in describing the structure of a neuron, this section builds on the concepts of structure and function (Chapter 1) and on cell structure and tissue specialization (Chapter 5). In explaining how sodium and potassium ions create an electrical potential across the membrane of a neuron, this section expands upon the concepts of basic chemistry (Chapter 3). In discussing the polarization of neurons, the section reviews the concepts of movement of substances across cell membranes including the sodium-potassium pump (Chapter 6).

In presenting this section, guide students to understand how the laws of chemistry and physics relate to biological functions in the case of nerve impulse transmission. You may wish to refer students to Figure 46-6, to help them visualize the process of impulse transmission in neurons. Refer students to Figure 46-7 to help them understand how neurotransmitters are released at a synapse.

Concept Framework

Students will be familiar with various sources of electrochemical energy, such as batteries. Tap this knowledge to assist students in listing sources of electrochemical energy. Discuss how these energy sources are used.

Prepare students for understanding electrochemical energy by telling them that the separation of charged particles, which are ions of various chemicals, constitutes the source of potential energy in a battery. Review the concepts of potential and kinetic energy to help students understand how energy is stored and released in such a system. Tell students that a battery and a nerve cell store and release potential energy in similar ways. As students read this section, suggest that they pay close attention to the effects of ion movement across nerve cell membranes in the transmission of nerve impulses.

Emphasis

Students should understand how the structure of the neuron is directly related to its function in impulse transmission. Make sure that students can outline the role of each of the following components in the transmission of nerve impulses: the axon, the dendrite, the myelin sheath, and the synapse. Students should also realize that stimulation of a neuron results in an all-or-none response.

Guest Speaker

Invite a biologist or physician to class to discuss neurological disorders such as depression and Parkinson's disease caused by imbalances of neurotransmitters.

Neuron Models

Encourage interested students to make a paper, fabric, or mechanical model of a neuron and use it to demonstrate how nerve impulses are transmitted in neurons. Students might use circles of one color to represent sodium ions and circles of a different color to represent potassium ions. Larger circles might represent neurotransmitters released at a synapse. Give students time to present and explain their models to the class.

Section Review Answers

(See p. 717.)

46.3 The Senses

Approach

The main objective of this section is to familiarize students with the sense organs and to explain how each sense organ receives certain types of stimuli and converts the stimuli to nerve impulses to be interpreted by the brain. This section builds on the concepts presented in 46.1 and 46.2. In presenting this section, guide students to realize how the sense organs and the brain work together to allow a person to detect stimuli and to integrate and interpret resulting nerve impulses. You may wish to refer students to Figure 46-10 as they read the section on vision. Encourage students to use Figure 46-12 to trace the path sound energy follows through the ear.

Concept Framework

All students are familiar with the primary functions of the various sense organs. Tap this knowledge to help build a conceptual framework for the material in this section by assisting students in listing the sense organs. Then lead students in a discussion of the importance of the eyes, ears, taste buds, nose, and skin. Tell students that as they read this section they should make sure to note how the various types of receptors in these sense organs detect and transmit stimuli to the brain.

Emphasis

Make sure that students can explain how each of the following sense organs works to detect different kinds of stimuli and transmit nerve impulses to the brain: the eyes, the ears, the taste buds, the nose, and the skin. Students should also understand the adaptive significance of the development of each of the different kinds of sense organs.

Sense Organs Group Reports

Divide students into five groups. Direct each group to use library resources to research one of the major sense organs. Let the group divide up the topic and compile a written report on the sense organ. The report might include diagrams of the sense organ and the receptors that detect stimuli and an explanation of the role of the brain in interpreting the stimuli. Make sure that each group divides up the topic so that each student takes part in researching and writing the report. Display the reports in the class and give students time to read the reports written by other groups.

Smell Survey

Encourage interested students to design a survey to test the sense of smell. Tell students to collect various substances that have different odors, such as flowers, garlic, musk, and spices. Some substances should have a strong odor and other substances should have a mild odor. Students should place the substances in containers so that the substances can be smelled without being viewed. For example, students might cut small holes in the lid of a jar, place the substance in the jar, and cover the jar with black paper.

After labeling the jars, students can write up a multiple-choice survey sheet and distribute it to the class. Give other students time to fill out the survey in class. You may wish to invite students from other classes to participate in the survey. Let students who prepared the survey tally and present their results. Direct them to focus on the substances people had the easiest time identifying and those that were more difficult to identify.

Optical Illusions

Ask students to collect examples of optical illusions. Give them time to present their examples to the class and to discuss what makes each example interesting. Talk about the factors that make it difficult for the brain to interpret what the eye sees. You may wish to set up a bulletin-board display of the optical illusions either in the classroom or in the hallway for students to examine in more detail.

Section Review Answers

(See p. 722.)

Laboratory Preparation

Background

In this lab students map the areas of the tongue that contain specific taste buds for salty, sweet, sour, and bitter tastes. The students conduct the experiment in a "double blind" design so that neither the experimenter nor the taster know the chemical until mapping begins. This technique contributes to the "honesty" of the experiment as an investigation, rather than as a demonstration of the known areas on the tongue. The same experimental design is often used to demonstrate the true efficacy of a drug.

Preparation

To prepare a salty solution, add 10 g sodium chloride to 90 mL water. Label the solution A. To prepare a

sweet solution, add 5 g sucrose to 95 mL water. Label the solution B. To prepare a sour solution, add 1 part white vinegar to 2 parts water. Label the solution C. To prepare a bitter solution, add 0.1 g quinine sulfate to 99.9 mL water. Label the solution D. (Quinine sulfate is available from a biological supply house.) For the tasteless solution, use plain tap water and label the solution E.

Conclusions

1. The tastes in order are salty, sweet, sour, bitter, and tasteless.
2. Salt is tasted in area 1 and along the front sides of the tongue. Sweet is tasted in these same areas. Sour is tasted only in areas 2 and 4. Bitter is tasted only in area 5.
3. Yes. There is some overlap between salty and sweet and between salty and sour.

Inquiry

1. Answers will vary but may include forgetting to rinse between tests, touching the wrong area of the tongue, and tasting residual tastes in the mouth from food or chewing gum.
2. Students devise an experiment similar to the one just performed, but the solutions are administered while holding the nose closed.

Chapter Review

Vocabulary

1. central nervous system; all other terms relate to the peripheral nervous system
2. effector; all others are parts of a neuron
3. spinal cord; all other terms are specific parts of the brain
4. polarization; all other terms refer to the sequence of impulse transmission
5. retina; all others are parts of the ear

Review

11. The cerebrospinal fluid forms a fluid cushion to protect against shock.
12. Receptors respond to stimuli, conductors carry the impulse to and from the central nervous system, and effectors respond to the impulse.
13. The right hemisphere controls the left side of the body and is thought to control artistic abilities;

the left hemisphere controls the right side of the body and is thought to control reading, writing, and analytical abilities.

14. The midbrain controls reflex movements of the head; the pons is primarily a relay station between the cerebrum and the cerebellum; the medulla oblongata relays impulses from the spinal cord and controls homeostatic activities.

15. If a stimulus is strong enough, an impulse will be generated. If not, there will be no impulse.

16. Two types of macroscopic imaging devices are the MRI and CT scanner.

17. When a receptor cell responds to a stimulus, the impulse travels along the primary nerve for that sense to the corresponding region in the brain, where it is translated.

18. Movement of the fluid inside the semicircular canals stimulates mechanoreceptors, which transmit a message to the brain about the position of the body.

19. Tastes are distinguished by different types of taste buds.

20. The movement of body hair stimulates mechanoreceptors at the base of the follicle to send information to the brain.

Critical Thinking

1. Some of the repeated rapid impulses from the brain reach the muscles and cause intense contractions. Since the impulses are so rapid, the supply of neurotransmitters at the synapses is quickly used up and the impulses cease.

2. Potassium is necessary for the nerves to conduct impulses. If the potassium level is low, the impulses are affected, and they in turn affect the functioning of the brain, muscles, heart, and other organs.

3. Answers will vary as to the functions of the areas of the brain. The size of the cerebellum in the goose indicates that the movements of this bird are intricate. The cerebrum is moderately well developed as memory and learning ability are relatively important in birds. The optic lobe of the goose is very large, indicating that eyesight is important. The olfactory bulb is small, indicating that the sense of smell is not.

Chapter 46 | Nervous System and Sense Organs

Introduction

Human mental and physical capabilities are controlled by a highly evolved system of cellular communication called the nervous system, within which a carefully organized division of labor allows each part to operate effectively.

Chapter Outline

46.1 Structure
- *Organization*
- *Brain*
- *Spinal Cord*
- *Peripheral Nervous System*

46.2 Transmission of Nerve Impulses
- *Neuron Structure*
- *Nerve Impulses*
- *Transmission across Synapses*
- *Effects of Neurotransmitters*

46.3 The Senses
- *Receptors and Sense Organs*
- *Vision*
- *Hearing and Balance*
- *Taste and Smell*
- *Other Senses*

Chapter Concept

As you read, notice that the nervous system is composed of cells, tissues, and organs.

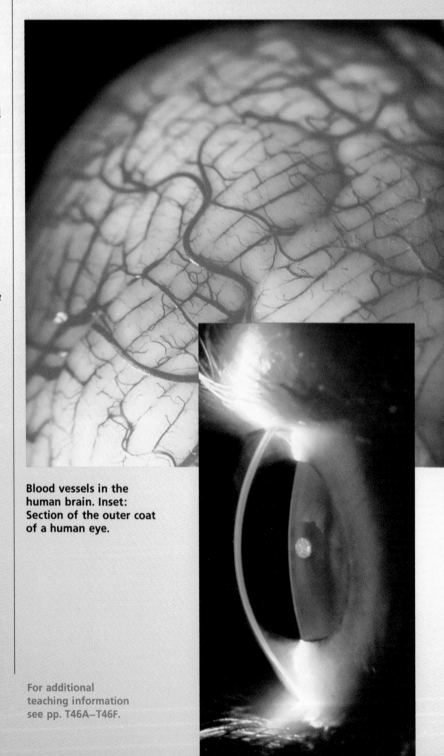

Blood vessels in the human brain. Inset: Section of the outer coat of a human eye.

For additional teaching information see pp. T46A–T46F.

46.1 Structure

The human nervous system enables smooth, efficient communication among all parts of the body. Just as a computer network has interconnected terminals that generate and receive messages in a coordinated effort, so the human nervous system has component parts that process information efficiently.

Organization

The nervous system is composed of cells called **neurons.** Groups of related neurons are organized into **nerves.** Nerves are organized into two major subsystems—the central nervous system and the peripheral nervous system. The brain and the spinal cord make up the **central nervous system.** The spinal cord receives messages from other parts of the body and carries them to the brain, where they are integrated and interpreted. Responses are then transmitted back to the body. The **peripheral nervous system** consists of nerves that radiate out from the brain and spinal cord to the rest of the body. These nerves connect the central nervous system with all parts of the body. A mass of neurons outside the central nervous system is called a **ganglion.**

There are three general categories of peripheral nerve cells: receptors, conductors, and effectors. **Receptors** are cells, or groups of cells, that receive information from both internal and external stimuli. **Conductors** are specialized nerve cells that transmit messages from receptors to the central nervous system. **Effectors** are the cells that receive messages from the central nervous system and carry them to an area of the body. Muscles are activated by effectors and contract in response to the signal of an effector. Glands are activated by effectors and secrete hormones in response to the signal of an effector.

Section Objectives

- *Identify the primary divisions of the nervous system.*

- *Describe the function of receptors, conductors, and effectors.*

- *Summarize the functions of the three regions of the brain.*

- *Explain how advances in photographic techniques have helped scientists learn more about the brain.*

Info: Sometimes the term *receptor* is used to refer to the entire sense organ rather than to the cells.

Tip: Have students trace the path of an impulse through the receptors, conductors, and effectors. Use examples such as how the body responds when a person smells smoke or how the body reacts when a driver sees another car coming head on.

Figure 46-1   The nervous system consists of the central nervous system and the peripheral nervous system.

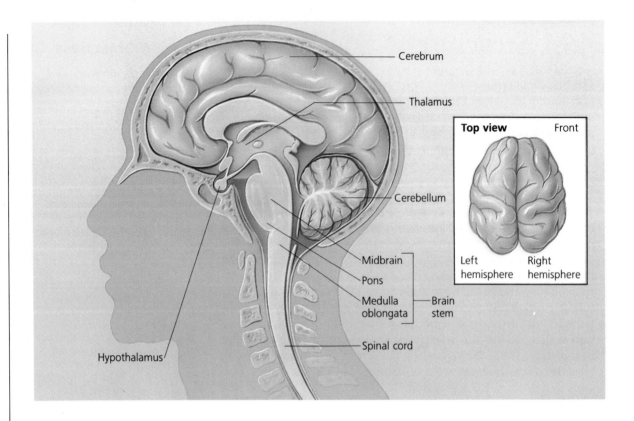

Figure 46-2 labels: Cerebrum, Thalamus, Cerebellum, Midbrain, Pons, Medulla oblongata, Brain stem, Spinal cord, Hypothalamus

Top view — Front, Left hemisphere, Right hemisphere

Figure 46-2 A side view of the brain shows its main parts. The top view shows the main sections of the brain, the left and right hemispheres.

Brain

The human **brain** weighs an average of 1.4 kg, or about 2 percent of the total body weight. Nevertheless, the brain contains approximately 100 billion neurons. The network of blood vessels that nourishes the neurons is so dense that 20 percent of the oxygen used by the body at rest is used in the brain. As Figure 46-2 shows, the brain consists of three regions: the cerebrum, the cerebellum, and the brain stem.

Cerebrum

The **cerebrum** *(suh-REE-brum)* is the largest portion of the brain. It is made up of two **cerebral hemispheres.** These are connected by nerves that form a structure called the **corpus callosum.** Deep grooves separate each hemisphere into four lobes: the frontal, parietal, temporal, and occipital lobes.

The highly folded outer layer of the cerebrum is called the **cerebral cortex.** The folds, or convolutions, maximize the surface area of the brain within the limits imposed by the cranium, or skull. Scientists estimate that the cerebral cortex contains nearly 75 percent of the body's total number of neuron cell bodies. The neuron cell bodies in the outer portion of the cerebral cortex are often referred to as gray matter. Beneath the cerebral cortex are large masses of white matter, bundles of nerves that connect the cortex with other parts of the nervous system.

Through experimentation scientists have determined that separate areas of the cerebral cortex control motor and sensory activities. Both hemispheres of the cerebrum control voluntary movement of skeletal muscles, and both intepret sensory stimulation. In general, impulses from and to the right half of the body are processed in the left half of the brain, and vice versa. Some evidence suggests that the right hemisphere controls both verbal and nonverbal artistic abilities and that the left hemisphere controls such skills as reading, writing, and the ability to analyze.

Cerebellum

Just below the occipital lobes of the cerebrum lies the **cerebellum** *(SER-uh-BELL-um)*, a region of the brain that plays a vital role in the coordination of muscle action. The cerebellum receives sensory impulses from muscles, tendons, joints, eyes, and ears. After receiving these impulses the cerebellum sends motor impulses down through the spinal cord to stimulate or inhibit skeletal muscles. The cerebellum also controls balance, an involuntary action, by communicating with sense receptors in the eyes and ears. It relays information regarding body position to appropriate muscles and controls posture by keeping skeletal muscles in a constant state of partial contraction.

Brain Stem

The third major region of the brain, the **brain stem,** connects the cerebrum with the spinal cord. The brain stem relays information to and from the cerebrum and thus plays a role in controlling body actions.

The brain stem has three main sections: the midbrain, the medulla oblongata, and the pons. The **midbrain** controls reflexes such as changing the size of the pupils.

The **medulla oblongata** *(muh-DULL-uh AHB-lawn-GAHT-uh)* is a group of neurons that serves as both a nerve relay station and a control center. Nerve fibers in the medulla oblongata cross over in such a way that impulses from the right side of the body are relayed to the left cerebral hemisphere, and vice versa. Also within the medulla oblongata are centers that control various homeostatic activities, including heart rate, breathing rate, and the flow of blood through the blood vessels.

The **pons** serves as a relay station between the nerves of the cerebrum and those of the cerebellum. It also relays messages between the midbrain and the medulla oblongata.

Two structures near the upper brain stem work with the central nervous system. The **thalamus** directs all incoming sensory impulses except odors to the proper region of the cerebral cortex. The hypothalamus, located just below the thalamus, is connected by nerve fibers with the thalamus, cerebral cortex, and parts of the brain stem. The interaction of the hypothalamus and the pituitary gland of the endocrine system help maintain chemical homeostasis in the body.

Figure 46-3 Each hemisphere of the brain has four lobes. Note that control of various functions is located in different areas of the brain.

The thalamus, hypothalamus, and cells deep within the gray matter are connected by a group of neurons that are located near the corpus callosum. Together these make up the **limbic system,** a system that helps regulate the emotions. Research indicates that the limbic system detects changes in the physiological state of the individual and responds by producing feelings, such as fear or anxiety, that may ensure the person's safety.

A network of nerve fibers called the **reticular formation** runs through the brain stem and the thalamus. It acts as a filtering system for incoming sensory impulses. By controlling the numbers of impulses that are sent, the reticular formation enables a person to sleep through the noise of a radio or television but to be awakened by a knock at the door.

Protection

The brain and spinal cord are surrounded by three protective layers called the **meninges** *(muh-NIN-jeez)*. The outer layer, the **dura mater** *(DYOOR-uh MATE-ur)*, consists of connective tissues, blood vessels, and nerves. The dura mater lines the inside of the skull and also forms a tube that surrounds the spinal cord. The middle layer, the **arachnoid layer,** is elastic and weblike. The thin

Info: Spinal meningitis is a disease caused by bacteria or viruses that cause the meninges to become inflamed. Meningitis can lead to deafness, mental retardation, or death.

Biotechnology

New Imaging Techniques

In studying the human body medical scientists for many years relied on light microscopes, which revealed a world invisible to the unaided eye. The invention and refinement of various imaging techniques, such as computed tomography (CT) scanners and magnetic resonance imaging (MRI) for macroscopic work and the scanning electron microscope for microscopic work, now enable medical scientists to study the body in more detail than was imaginable before.

CT scanners use an X-ray beam that revolves around the patient to transmit cross-sectional images to a computer terminal. Numerous images are then combined to form a three-dimensional picture (left).

MRI utilizes the magnetic properties of molecules containing hydrogen, a magnetic scanning device, and a computer. The scanner excites the protons of hydrogen molecules in the body. The movement of the protons generates a radio wave, which is translated into a three-dimensional image by the computer (right).

Lennart Nilsson, a medical and biological photographer, has perfected a technique for viewing human cells or tissues inner layer, the **pia mater** *(PIE-uh* MATE-*ur)*, contains many nerves and blood vessels. It adheres to the brain and spinal cord. A clear, watery substance called **cerebrospinal fluid** separates the arachnoid layer and the pia mater and provides a fluid cushion that protects the brain and spinal cord from shock. Parts of the skeletal system also protect the central nervous system. The cranium encloses the brain, and the vertebrae surround the spinal cord.

Spinal Cord

The **spinal cord** is a column of nerve tissue that starts in the medulla oblongata and runs down through the vertebral column. It contains an outer sheath of white matter and a rigid inner core of gray matter. Thirty-one pairs of spinal nerves originate in the spinal cord and branch out into both sides of the body. Each spinal nerve consists of a dorsal root and a ventral root. The dorsal root contains **sensory neurons,** nerve cells that carry impulses from receptors to the spinal cord. The ventral root contains **motor neurons,** nerve cells that carry impulses from the spinal cord to the effectors. **Association neurons,** also called interneurons, maintain neural connections within the spinal cord.

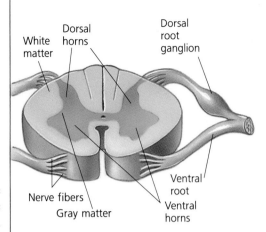

Figure 46-4 **An H-shaped section of gray matter lies in the center of the spinal cord. Dorsal and ventral roots connect spinal nerves to the cord.**

with the scanning electron microscope. He removes cells or tissues from biopsy or autopsy specimens and places them in a nutrient-rich solution, where they can continue to live for a short time in a controlled laboratory setting. Nilsson carefully preserves and dehydrates a sample of these cells. He also

applies a thin, protective metal coating to the sample and places it inside a vacuum chamber that is near the bottom of his SEM.

As its name indicates, the scanning electron microscope uses electrons rather than light to produce an image (above). A filament near the top of the

microscope generates the electrons that are directed toward the sample. These electrons bounce off the metallic coating on the sample. Special sensors in the microscope respond to the reflected electrons. The electrons are converted into light particles and then into an electrical signal. This electrical signal is amplified and relayed to a viewing screen for Nilsson to study.

The images produced by the scanning electron microscope are usually gray or blue. Color may be added to the images for better contrast or to highlight specific parts of a cell or tissue.

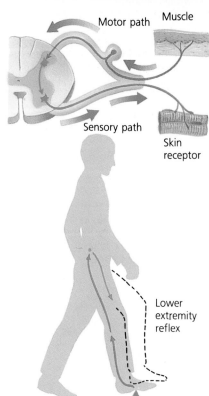

Motor path   Muscle

Sensory path

Skin receptor

Lower extremity reflex

Use the picture above to describe a reflex action. Identify the stimuli, and explain how the nervous system will communicate information about the situation.

Answer to Think
The object is sharp. Receptors in the foot will detect the object and relay the information to the spinal cord. Then an impulse will immediately signal certain muscles to contract and pull the foot away. The message will travel to the brain, and the person will comprehend the sensation.

Answers to Section Review
(See pp. T46A–T46F.)

Peripheral Nervous System

Communication between the central nervous system and the rest of the body takes place via the branches of the peripheral nervous system. The basic organization of the peripheral nervous system is shown in Figure 46-1. In humans 12 pairs of cranial nerves connect the brain with various parts of the head and upper neck. The 31 pairs of spinal nerves connect the spinal cord with all the remaining parts of the body.

One important function of the peripheral nervous system is the relaying of impulses in **reflex actions**—involuntary and usually self-protective movements. For example, if you touch a hot stove, it is to your distinct adaptive advantage to be able to draw away from the heat automatically without first having to think about the situation. The pathway of such impulses, called a **reflex arc,** involves a minimum of one sensory, one association, and one motor neuron. In a reflex arc the spinal cord is responsible for transmitting the response that pulls your hand away. Later, sensory impulses will reach your brain, and you will feel pain or irritation and become aware of what has happened.

The peripheral nervous system is divided into two subsystems—the somatic and the autonomic nervous systems. The **somatic nervous system** consists of the cranial and spinal nerves that control the voluntary movement of skeletal muscles. The **autonomic nervous system** consists of the nerves that control involuntary actions, such as respiration and heart rate.

The autonomic nervous system can be further divided into the sympathetic and parasympathetic nervous systems. The **sympathetic nervous system** controls the stimulation of internal organs during conditions of high stress or increased activity. The **parasympathetic nervous system** controls internal organs during routine conditions. When a stressful condition arises, the sympathetic nervous system may generate impulses that cause the heart rate and breathing rate to increase. When the situation returns to normal, the parasympathetic nerves signal the organs to revert to routine levels of activity. The alternating actions of these two systems help the body maintain homeostasis.

Section Review

1. Distinguish between *sensory, motor,* and *association neurons.*
2. Distinguish between the structure of the central and peripheral nervous system.
3. How are the brain and spinal cord protected from injury?
4. How do the functions of the cerebrum, cerebellum, and brain stem differ?
5. How might the various parts of the nervous system be related to the increased motor control and higher-level thinking abilities humans possess that less complex animals do not possess?

46.2 Transmission of Nerve Impulses

As you learned in Section 46.1, the structure of the human nervous system is extremely complex. The ability of the nervous system to monitor and respond to the environment depends on the transmission of impulses between cells. This intricate process is effected by the flow of electrochemical energy.

Neuron Structure

Neurons carry electrical signals, called **impulses,** throughout the nervous system. As shown in Figure 46-5, a neuron consists of a nucleated cell body and two kinds of threadlike extensions, dendrites and axons. Most neurons have many dendrites but only one axon. **Dendrites** are branched cell components that receive impulses from other neurons or receptors and carry the impulses to the cell body. The **axon** is an elongated extension that carries impulses away from the cell body. Notice in Figure 46-5 that the end of the axon is divided into fingerlike projections called axon terminals. An electrical impulse travels from the end of a dendrite, through the cell body, down the axon, and into the axon terminals. From there it travels by chemical messenger across the small space between two neurons, called a **synapse,** to an organ, a gland, or the dendrite of another neuron.

In many neurons the axon is surrounded by a layer of cells called **Schwann cells.** These cells form a structure known as the **myelin sheath,** which insulates and protects the axon. The axon is exposed at gaps in the myelin sheath called **nodes of Ranvier** *(RAHN-vee-ay).*

Section Objectives

■ *Relate the structure of a neuron to its function.*

■ *Summarize the electrical and chemical conditions of a neuron in the resting state.*

■ *Describe a nerve impulse.*

■ *Describe the role of neurotransmitters in the transmission of impulses across synapses.*

Info: Nerve impulses move much more slowly than electricity although an impulse is an electrical signal. Impulses move at about 90 m per second, and electricity moves at approximately 285,000 km per second.

Figure 46-5 Many of the individual parts of a neuron shown in the illustration are visible in the SEM image of a neuron from a human cortex (×890).

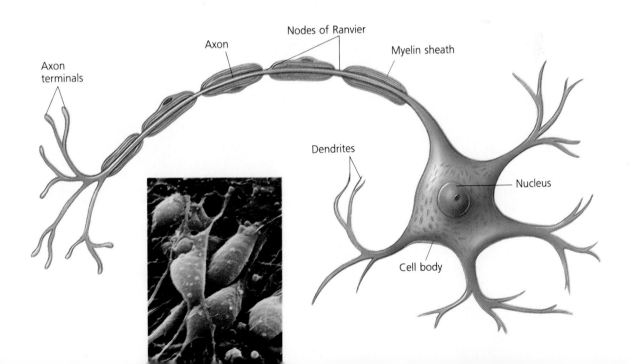

Axon terminals

Axon

Nodes of Ranvier

Myelin sheath

Dendrites

Nucleus

Cell body

Nerve Impulses

A nerve impulse is actually a flow of electrical charge along the cell membrane of a neuron. This flow results from changes in the concentration of certain ions on either side of the cell membrane. Transmission usually begins at a receptor cell and flows from neuron to neuron and eventually reaches an effector that stimulates a gland or causes a muscle to contract.

The Resting State

A neuron that is not conducting an impulse is in a **resting state.** In the resting state the area inside the nerve cell membrane has a much higher concentration of positively charged potassium (K^+) ions and negatively charged organic ions than the fluid outside the neuron does. At the same time the concentration of positively charged sodium (Na^+) ions is much higher outside the cell than inside. These differences are maintained by the selectively permeable cell membrane, which is highly impermeable to Na^+ and organic ions but allows K^+ ions to diffuse freely through it. The movement of ions takes place through a channel in the cell membrane. As shown in Figure 46-6, the differences in ion concentration give the inside of the membrane a negative charge relative to the outside of the membrane. This electrochemical difference, called a potential, is about -70 millivolts (mV). A volt is a measure of the difference in electrical potential. During the resting state the neuron is said to be polarized.

As mentioned, the polarization of the neuron results from the uneven distribution of potassium and sodium ions on the two sides of the membrane. This distribution is achieved both by the selective permeability of the cell membrane and by a mechanism

Figure 46-6 The transmission of a nerve impulse involves a change in electrochemical potential across a nerve cell membrane.

called the **sodium–potassium pump.** The sodium–potassium pump, described in Chapter 6, moves sodium ions to the outside of the membrane and potassium ions to the inside. The pump is an active transport mechanism powered by the chemical energy of ATP. It moves three sodium ions to the outside of the membrane for every two it pumps to the inside. Because potassium can diffuse passively across the membrane, there are more potassium ions on the inside of the membrane than sodium ions on the outside. The result is a resting neuron that has an electrical potential across its membrane.

Impulse Transmission

When a neuron is stimulated, an all-or-none response occurs. An impulse is either generated or it is not. If the impulse is generated, a sudden change occurs in the permeability of the membrane at the point of stimulation. The membrane becomes permeable to Na^+ ions, which then diffuse rapidly from the outside to the inside of the membrane. As a result the inner surface of the membrane becomes more positively charged than the outer surface. This reversal of the resting-state electrical potential is called depolarization. At first the depolarization occurs in only a short segment of the membrane, but it immediately disrupts the adjacent part of the membrane, which then also becomes depolarized. As Figure 46-6 shows, a wave of depolarization, called an **action potential,** moves along the membrane. This action potential is the impulse.

The presence of the myelin sheath, a structure found primarily in vertebrates, affects the speed of an impulse. Electrical stimulation cannot take place through the myelin sheath. Therefore the impulse moves down the axon until it reaches one of the nodes of Ranvier. Here depolarization occurs very quickly and creates a

Tip: You might compare the wave of depolarization that occurs during the transmission of an impulse, called action potential, to a falling line of dominoes. Unlike toppled dominoes, once the impulse has traveled across the axons, the axons almost immediately reset themselves and are ready to receive another impulse.

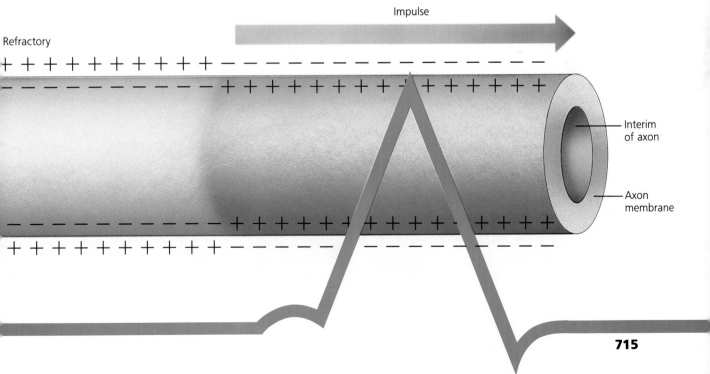

new action potential, which jumps across the gap of the myelin sheath to the next section with little loss of energy. At each gap the energy of the original stimulus is regenerated, providing a constant, strong stimulus along the length of the axon.

The frequency at which impulses travel along the axon is limited in part by the **refractory period,** the time that follows the passage of a nerve impulse. During the refractory period the neuron is returning to its resting potential and cannot be stimulated. The period of time in which a neuron will not accept an impulse is about 0.0004 of a second.

Transmission Across Synapses

When a nerve impulse travels along the length of an axon, it eventually reaches the axon terminals. The axon terminals lie near an organ, a gland, or the dendrite of another neuron, but do not touch it. The small space between the axon and the next neuron is called the synapse.

Impulses are carried across a synapse by chemical messengers called **neurotransmitters,** as shown in Figure 46-7. These chemicals are stored in synaptic vesicles embedded in a **bouton** *(boo-TAHN),* a bulblike structure at the tip of an axon terminal. The bouton is enclosed by a presynaptic membrane, which lies almost adjacent to the postsynaptic membrane of, for example, the dendrite of another neuron.

When an impulse reaches the axon terminal, the synaptic vesicles fuse with the presynaptic membrane and release neurotransmitters into the gap. After their contents are released, the vesicles usually pull away from the membrane and return to the cytoplasm, where they once again are filled with neurotransmitters.

Neurotransmitters travel across the gap and bind to receptor molecules in the postsynaptic membrane. The interaction of the neurotransmitter and the receptor creates an action potential that then travels down the receiving dendrite to the cell body.

Figure 46-7 Neurotransmitters released into a synapse carry a nerve impulse to the next neuron.

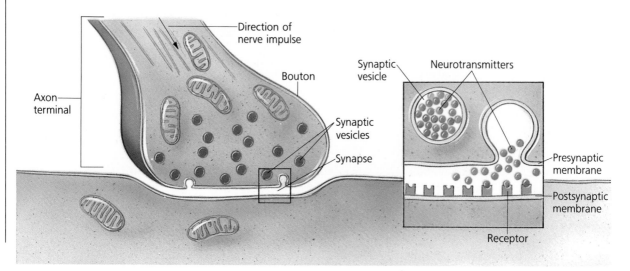

Direction of nerve impulse

Bouton

Axon terminal

Synaptic vesicles

Synapse

Synaptic vesicle

Neurotransmitters

Presynaptic membrane

Postsynaptic membrane

Receptor

Figure 46-8 Two end plates of a nerve fiber are shown adjacent to a motor nerve in this SEM image (×3,000).

In the synapse neurotransmitters are destroyed by the action of enzymes. It is important for the neurotransmitters to be destroyed so that they will not continue to stimulate the receiving neuron.

Impulses are transmitted along neurons until they reach a muscle or a gland. An example is shown in Figure 46-8. In the muscle or gland the impulse causes the muscle to contract or the gland to release a hormone.

Effects of Neurotransmitters

The approximately 30 different neurotransmitters that have been identified by scientists cause one of two responses. They either stimulate the action potential in a second cell, as you just learned, or they inhibit it.

Research findings on the effects of neurotransmitters have led investigators to a greater understanding of certain mental and physical disorders. Neurotransmitters have been used in the treatment of such nervous disorders as Parkinson disease. Biochemists have found that low levels of serotonin and norepinephrine, two neurotransmitters found in the brain, may be responsible for serious cases of depression. Doctors have successfully treated such imbalances with a substance that prevents the reabsorption of serotonin and norepinephrine.

Section Review

1. What is a *resting state?*
2. How do impulses travel along myelinated axons?
3. How are impulses transmitted between adjacent neurons?
4. Describe the two possible effects that neurotransmitters may have at a synapse.
5. How might research on the effects of neurotransmitters be useful in the treatment of mental disorders?

Answers to Section Review
1. a state in which a neuron is not conducting an impulse
2. Impulses travel along the membrane until they reach a node of Ranvier, where depolarization occurs and a new action potential is created that jumps across the gap of the myelin sheath.
3. Impulses are carried across the synapse that separates two neurons by neurotransmitters.
4. Neurotransmitters can either inhibit or stimulate a response in the second cell.
5. Psychiatrists might be able to treat many mental disorders through careful administration and regulation of neurotransmitters.

- *Describe a sense receptor, and list the five primary types.*

- *Describe the roles of rods and cones in vision.*

- *Identify the parts of the ear responsible for hearing and for maintaining balance.*

- *Explain the chemical relationship between taste and smell.*

Figure 46-9 The bright red areas in this human brain scan indicate increased activity in those sections of the brain.

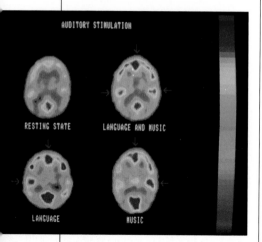

46.3 The Senses

The human environment is filled with both internal and external stimuli. Humans are able to distinguish among the many different types of stimuli by means of a highly developed system of nerves, specialized cells, and organs. The human nervous system is adapted to monitor and respond to a vast range of environmental phenomena.

Receptors and Sense Organs

Receptors are cells that receive information from the environment and then send impulses via conductors to the central nervous system. Receptors may be categorized on the basis of the types of stimuli they respond to:

- Mechanoreceptors detect movement, pressure, or tension.
- Photoreceptors detect variations in light.
- Chemoreceptors detect chemicals.
- Thermoreceptors respond to both internal and external heat and cold.
- Pain receptors respond to tissue damage.

The structure of the receptor cell enables it to interpret the type of energy being generated by a stimulus. The receptor cell then converts that energy into an electrical impulse that is transmitted through the nervous system and is finally interpreted by the brain.

Receptors are the neural components of the **sense organs,** the eyes, ears, nose, mouth, and skin. When a particular sense organ receives stimulation, the receptor converts the stimulus into an electrical impulse that is sent to a specific region of the brain. The impulses generated by the different sense organs are electrically similar, but the region of the brain to which they are directed will vary.

As you learned in Section 46.1, the brain has a specific region for each sense. Thus all impulses received by the auditory region of the temporal lobe are interpreted as sounds whether or not a sound wave was the original source of energy. For example, some people suffer from tinnitus, a sensation of noise such as a ringing or a roaring sound, caused by conditions such as a damaged tympanic membrane.

Vision

The eyes are specialized organs that function by receiving light and transmitting impulses to the visual cortex of the brain. All the structures of the eye act together to focus light on the **retina,** the light-sensitive inner layer of the eye. The **pupil,** an opening in the **iris,** controls the amount of light that enters the eye. The pupil

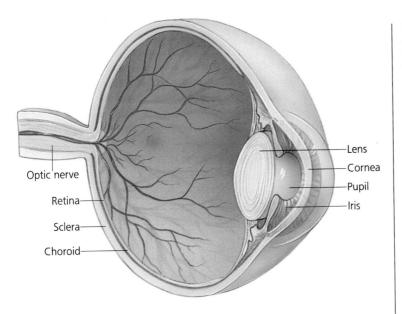

Optic nerve

Retina

Sclera

Choroid

Lens

Cornea

Pupil

Iris

Figure 46-10 Light entering the eye travels through the iris and lens to the retina, which activates impulses to the optic center of the brain.

becomes larger when light is dim and smaller when it is bright. These involuntary responses are controlled by muscles in the pigmented iris.

After light passes through the pupil, it travels to a convex crystalline structure called the **lens.** Attached to the lens are muscles that adjust its shape to bend the rays of the incoming light. This bending focuses onto the retina the complete image formed by the light.

Lying deep within the retina are rods and cones, photoreceptors that are essential in creating the vast array of visual images you perceive. There are nearly 125 million **rods** in a single retina. These are stimulated by weak light. Rods contain rhodopsin, which is a light-sensitive pigment that allows the rods to respond to weak light. Rods cannot respond to the strong light associated with color, which is why we cannot distinguish colors in dim light. The 7 million **cones** in the retina are stimulated by bright light. The cones initiate the production of sharp images and respond differently to different colors. Humans have three kinds of cones. Each type of cone contains a pigment that absorbs different wavelengths of light. When the impulses from these three kinds of cones are integrated, a person is able to see all the colors in the visible spectrum. Color blindness, which is the inability to distinguish certain colors, is caused by a chemical disorder in the cones.

Each photoreceptor responds to light from a single point in the image and then sends out an electrical signal. Impulses from all of the stimulated photoreceptors travel through nerves to ganglia on the surface of the retina. From the ganglia more than a million nerve fibers carry impulses to the **optic nerve** and then on to the occipital lobe of the cerebrum, where the impulses are interpreted as vision.

Tip: Students are often surprised to discover that the pupil is not made out of black protein but is actually a hole. You might also point out that the pupil becomes larger only because the size of the iris decreases.

Figure 46-11 Rods and cones in the retina enable humans to see shades of gray and colors. The cells shown in this SEM are magnified 2,500 times.

Hearing and Balance

The ear is specialized for two functions: detecting sound and maintaining balance. Sound, a form of energy consisting of air vibrations, is directed by the outer structures of the ear into the ear, where the vibrations stimulate mechanoreceptors.

Air enters the ear through both the **auditory canal** and the **Eustachian** *(yoo-STAY-shun)* **tube,** balancing the pressure on both sides of the **tympanic** *(tim-PAN-ick)* **membrane,** or eardrum. Air vibrations passing through the auditory canal cause the tympanic membrane to vibrate. These vibrations are transmitted to the three small bones of the middle ear: the hammer, the anvil, and the stirrup. The stirrup transfers them to a membrane called the **oval window** separating the middle ear from the inner ear.

The inner ear contains the **cochlea** *(KOE-klee-uh),* a coiled tube filled with fluid and lined with hair cells. Vibrations of the oval window set up vibrations in the fluid of the cochlea. The movement of this fluid stimulates the hair cells to produce impulses that travel along the **auditory nerve** to the brain. The hair cells respond to variations in pitch. The intensity of a sound is determined by the number of cells stimulated: the more cells that are stimulated, the louder the sound. Impulses from the sensory neurons of the hair cells travel through the auditory nerve to the auditory region of the brain, where they are interpreted as sound.

Balance is maintained with the help of mechanoreceptors in the three **semicircular canals** of the inner ear. These canals are perpendicular to one another and, like the cochlea, are filled with fluid and lined with hair cells. These hair cells are embedded in a gelatinous material that has particles of calcium carbonate on its surface. When the head moves at a different speed or in a different direction these particles move. This movement stimulates the hair cells. Neurons detect the signal and relay impulses to the cerebellum, where motion and position in space are interpreted. The cerebrum then sends responses to maintain balance.

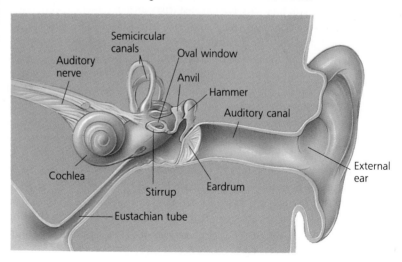

Figure 46-12 Air vibrations entering the ear are transformed into impulses that travel through the ear to the hearing center of the brain. Balance is maintained through the semicircular canals.

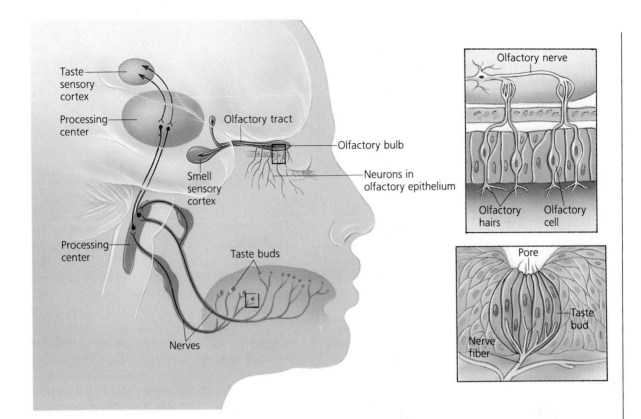

Figure 46-13 labels:
- Taste sensory cortex
- Processing center
- Olfactory tract
- Olfactory bulb
- Smell sensory cortex
- Neurons in olfactory epithelium
- Processing center
- Taste buds
- Nerves
- Olfactory nerve
- Olfactory hairs
- Olfactory cell
- Pore
- Taste bud
- Nerve fiber

Taste and Smell

Specialized chemoreceptors allow humans to perceive variations in tastes and odors. The chemoreceptors for taste are clustered in the **taste buds.** Most of the 10,000 taste buds are embedded between bumps called **papillae** *(puh-PILL-ee)* on the tongue; additional taste buds are found on the roof of the mouth and in the throat.

Chemicals that dissolve in saliva stimulate different kinds of taste. This chemical–saliva mixture enters the taste bud through a small opening and stimulates the nerve endings that line the inner surface. The nerve endings begin an impulse that travels to the cerebrum, where it is interpreted as taste. Humans can taste only salty, sweet, sour, and bitter flavors or combinations of these flavors. Other flavors are perceived by receptors in the nasal passages.

Specialized chemoreceptors called olfactory receptors are located in the olfactory epithelium of the nasal passage, as shown in Figure 46-13. These cells, dendrites of sensory neurons, lie within the mucus lining of the epithelium. Specific olfactory receptors respond to different chemicals. Chemicals as gas or particles enter the nose and dissolve in watery mucus before stimulating the olfactory receptors. These receptors generate impulses that travel through the **olfactory nerve** to the olfactory region of the cerebral cortex, where they are interpreted as odors.

Figure 46-13 Special sensors in the mouth and nasal passages make taste and smell possible. Nerve impulses travel to the appropriate centers of the brain.

Figure 46-14 Hairs emerge from hair follicles through the dead layers of epithelial cells on the skin. Dendrites on mechanoreceptors at the base of a hair follicle, shown in this SEM, detect movement of the hair and set up nerve impulses (×265).

Info: Thermoreceptors can be fatigued by overstimulation. When a person is in a hot bath for a long time, the thermoreceptors that sense heat become fatigued, thus sending the brain a message that the water is not as hot as it was earlier.

Answers to Section Review
1. Chemoreceptors detect chemicals; mechanoreceptors detect movement, pressure, and tension.
2. Rods enable a person to see light as shades of gray; cones are sensitive to bright light, produce sharp images, and help a person to distinguish colors.
3. Vibrations travel through the outer ear and are channeled into the middle ear and through the various bones and membranes of the middle and inner ear.
4. Both have chemoreceptors.
5. Humans rely heavily on their hands to gain information and to do tasks.

Other Senses

Mechanoreceptors located throughout the skin make it possible to sense touch, pressure, and tension. In humans, the receptors for touch are concentrated in the face and fingertips. Body hair also plays an important role in the ability to sense touch. Large numbers of mechanoreceptors are found at the base of the hair follicles. The dendrites of these mechanoreceptors detect movement of the hair and relay impulses to the central nervous system. When a stimulus is constant, many receptors stop sending impulses to the brain. For example, when you put on your watch or ring in the morning, you will be aware of its pressure. Soon, however, you "forget" that it's there. Only when the pressure changes in some way will the receptors start sending impulses again.

Two types of specialized thermoreceptors monitor temperature. Cold receptors respond to temperatures below 45° C. Heat receptors respond to temperatures above 10° C.

Pain receptors are the dendrites of the sensory neurons that are located in the base of the epidermis. Pain receptors are stimulated by mechanical, thermal, electrical, or chemical energy. The type and number of pain receptors vary throughout the body. For example, the hands have a high concentration of pain receptors.

Section Review

1. Distinguish between *chemoreceptors* and *mechanoreceptors*.
2. Explain the role of rods and cones in the perception of images by humans.
3. How are sound vibrations transmitted through the ear?
4. What mechanisms do the sense of taste and the sense of smell have in common?
5. What is the importance of the high concentration of pain receptors in the hands?

Mapping Tongue Taste Receptors

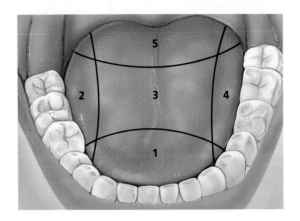

Objective
To discover the location of various taste receptors on the human tongue

Process Skills
Collecting and interpreting data, hypothesizing, experimenting

Materials
6 paper cups, marking pencil, 5 bottles of prepared solutions numbered 1 through 5, container of tap water, paper towels, 5 sterile cotton swabs

Method
1 Prepare a table to record your data. Head columns *Solution Number, Tongue Area, Sour, Sweet, Bitter, Salt,* and *No Taste.* Under *Tongue Area* number rows 1 through 5. Repeat this numbering system four more times. You should have 25 rows, five groups of five. These numbers correspond to the five areas of the tongue shown in the figure. Under *Solution Number* opposite the first group write the number 1. Opposite the second group write the number 2 and so on.
2 Use a marking pencil to number five paper cups 1 through 5. Put the sixth cup aside. Fill each of the five numbered paper cups with liquid from the solution bottle having the corresponding number.
3 Place a container of tap water at your work station.
4 Using the reserve paper cup, rinse your mouth with tap water. Blot your tongue with a clean paper towel.
5 You will work with a partner during this investigation. Your partner will apply the five solutions to your tongue and you will record your own reactions. Have your partner dip a sterile cotton swab into solution 1 and apply the swab to each of the five tongue areas shown in the figure. Your partner should use a fresh swab for each solution. Why is this procedure important? After each application mark the type of taste sensation experienced in the appropriate area of the chart. Do not let your partner see what you have recorded.
6 Repeat steps 4 and 5 for each remaining solution. Be sure to rinse your mouth and blot your tongue before testing a new solution. Why are these procedures important?
7 Change places with your partner and administer the five solutions to your partner as directed in steps 4 and 5.
8 Share your information with the class.

Conclusions
1 What are the tastes of the five liquids?
2 Examine your table. What area or areas of the tongue are receptive to each taste?
3 Based on your observations, is there evidence that the taste areas overlap? Explain your answer.

Inquiry
1 What are possible sources of error in this experiment?
2 Devise an experiment that tests the relationship between taste and smell.

Note that local health regulations may preclude the use of quinine sulfate.

Chapter 46 Review

Vocabulary

action potential (715)
arachnoid layer (710)
association neuron (711)
auditory canal (720)
auditory nerve (720)
autonomic nervous system (712)
axon (713)
bouton (716)
brain (708)
brain stem (709)
central nervous system (707)
cerebellum (709)
cerebral cortex (708)
cerebral hemisphere (708)
cerebrospinal fluid (711)
cerebrum (708)

cochlea (720)
conductor (707)
cone (719)
corpus callosum (708)
dendrite (713)
dura mater (710)
effector (707)
Eustachian tube (720)
ganglion (707)
impulse (713)
iris (718)
lens (719)
limbic system (710)
medulla oblongata (709)
meninges (710)
midbrain (709)
motor neuron (711)
myelin sheath (713)
nerve (707)
neuron (707)

neurotransmitter (716)
nodes of Ranvier (713)
olfactory nerve (721)
optic nerve (719)
oval window (720)
papilla (721)
parasympathetic nervous system (712)
peripheral nervous system (707)
pia mater (711)
pons (709)
pupil (718)
receptor (707)
reflex actions (712)
reflex arc (712)
refractory period (716)
resting state (714)

reticular formation (710)
retina (718)
rod (719)
Schwann cell (713)
semicircular canal (720)
sense organ (718)
sensory neuron (711)
sodium–potassium pump (715)
somatic nervous system (712)
spinal cord (711)
sympathetic nervous system (712)
synapse (713)
taste bud (721)
thalamus (709)
tympanic membrane (720)

In each set of terms below choose the term that does not belong, and explain why.
1. peripheral nervous system, somatic nervous system, central nervous system, parasympathetic nervous system
2. axon, effector, dendrite, node of Ranvier
3. cerebrum, spinal cord, brain stem, cerebellum
4. polarization, resting potential, refractory period, action potential
5. retina, auditory nerve, semicircular canal, Eustachian tube

Review

1. The nervous system consists of (a) the brain and spinal cord (b) 31 pairs of spinal nerves (c) many cranial nerves (d) the central and peripheral nervous systems.
2. The cerebrum (a) controls thinking (b) coordinates muscle action (c) controls balance (d) maintains heart rate.
3. The spinal cord (a) only receives messages (b) only sends messages (c) relays messages to and from the brain (d) interprets messages.
4. In a reflex arc the impulse travels (a) immediately to the brain (b) to the spinal cord and the brain (c) to the spinal cord and out to an effector (d) directly to a sense organ.
5. During the resting state of a neuron (a) the outside is negatively charged (b) the inside is negatively charged (c) both sides are equally charged (d) the inside is positively charged.

6. The speed of an impulse is determined by (a) the length of the axon (b) the diameter of the axon (c) the presence of a myelin sheath (d) both *b* and *c*.
7. Neurotransmitters (a) carry impulses across the synapse (b) inhibit a second cell (c) stimulate a second cell (d) all of the above.
8. Receptor cells (a) send impulses (b) generate responses (c) respond to stimuli (d) stimulate specific regions of the brain.
9. Photoreceptors are stimulated by (a) air (b) pressure (c) chemicals (d) light.
10. Mechanoreceptors respond to (a) pain (b) pressure (c) temperature (d) all of the above.
11. How does cerebrospinal fluid protect the brain?
12. What is the relationship among receptors, conductors, and effectors?
13. How do the activities of the right cerebral hemisphere differ from the activities of the left cerebral hemisphere?
14. What are the roles of the three sections of the brain stem?
15. How does the term *all-or-none response* relate to impulse transmission?
16. What are two types of macroscopic imaging devices?
17. Describe the basic pathway from a sense receptor to the brain.
18. What is the role of the ear in maintaining balance?
19. How are different tastes distinguished?
20. Why is body hair important to the sense of touch?

Critical Thinking

1. Epilepsy affects one of every 200 Americans. Brain cells normally produce small bursts of impulses in varying patterns. In an epileptic person large numbers of brain cells occasionally send rapid bursts of impulses simultaneously. The individual then may jerk and convulse. From what you know about nerves, how might you explain these symptoms? How might you explain the cessation of the symptoms after a short period of time?
2. For some people too little potassium in the bloodstream is a serious medical problem. Symptoms of this condition include an increased heart rate and headaches. How might you explain these symptoms?

3. Look at the diagram above of the brain of a goose. Using what you know about the human brain, what would you assume the size of the portions of this brain indicate about their relative importance?

Extension

1. Read the article by Avery N. Gilbert and Charles J. Wysocki entitled "The Smell Survey—Results," in *National Geographic*, October 1987, pp. 514–525. How do the findings support claims that the sense of smell triggers more reactions and memories than any other sense does?
2. Research the way anesthetics work, and write a report in which you discuss the chemicals used, the effect of the chemicals on the brain, and the effect of body weight and the passage of time on the action of an anesthetic.

Chapter 47 | *Endocrine and Reproductive Systems*

This chapter investigates the development of a single cell into a mature human being. Students learn how the endocrine system controls the continued growth of the organism and how it helps maintain homeostasis.

Correlation Chart

Section Number	Section Objectives	Pages	Features	Section Review
47.1	1. Describe a negative feedback mechanism.	727–734		2, 4
	2. Explain the difference between an exocrine gland and an endocrine gland.			
	3. List the major endocrine glands, and describe the function of each.			2, 3, 5
	4. Describe the chemical action of two types of hormones.			1, 4
47.2	1. Identify the major parts of the male and female reproductive systems and state the function of each.	735–738		1, 2, 3, 5
	2. List similarities and differences between eggs and sperm.			1
	3. Summarize the role of hormones in regulating the events of the menstrual cycle.			4, 5
47.3	1. Describe the process of fertilization.	739–745	Biotechnology, pp. 740–741	
	2. Summarize the changes that take place during embryo development.		Think, p. 744	1, 2, 3, 5
	3. List the three stages of pregnancy and describe characteristics of each stage.			1, 3, 4, 5
	4. Describe the process of in vitro fertilization.		Biology Lab, p. 745	

Resources*

Print

Bodanis, David. *The Body Book: A Fantastic Voyage to the World Within*. Boston: Little, Brown, 1984.

Miller, Jonathan, and David Pelham. *The Facts of Life*. New York: Viking, 1984.

Audiovisual/Software

Diabetes: Beating the Needle. Film or video, 28 min. Filmakers (Part of *The Nature of Things* Series), 1984.

The Body in Focus. Apple 48K. Disk. Scholastic.

The Miracle of Life. Film or video, 57 min. Time-Life (Part of the *Nova* Series), 1983.

*Additional references are listed in *Modern Biology, Teacher's Program Guide*.

Section Teaching Guides

47.1 Endocrine System

Approach

The main objective of this section is to explain how hormones regulate chemical activity within the human body. The endocrine system is a highly complex system, and the material discussed in this section builds on several concepts presented in earlier chapters.

For example, in explaining how the parathyroid, pituitary, and other glands operate through a negative feedback mechanism, you might relate the discussion to the idea of homeostasis (Chapter 6), pointing out to students that hormones help maintain proper chemical balances within the body. Reviewing the principles of biochemistry (Chapter 4) and cell specialization (Chapter 5) will provide students with a framework for understanding how hormones work upon the body and why coordination of the activities of the different organ systems is essential to keep the body healthy and functioning properly.

All of the systems students have read about in Chapters 42–46 are regulated to some degree by the endocrine system, as is the reproductive system, discussed in Section 47.2. To give students an idea of the extent of hormonal influence on the human body, have them refer to Table 47-1. In presenting the text, note that several endocrine glands, such as the pituitary, secrete hormones that act on other endocrine glands. Ask students to consider why this extra regulatory step, which requires the body to expend additional energy, would have adaptive significance.

Concept Framework

Students can observe the effects of one type of hormonal regulation simply by looking around the classroom and noting the variations in height and weight among members of the class. In addition, students have certainly experienced the physiological and emotional effects of hormonal activity, even if they are unaware of the actual chemistry involved. For example, students most likely are aware that hormones control sexual functioning. They may also have said at some time that their "adrenaline is racing" if they were participating in athletic activities or if they were in a state of agitation or excitement.

To build a conceptual framework in order to introduce students to this section on the endocrine system, and especially negative feedback mechanisms, ask students what physical changes take place in their bodies when they encounter a stressful situation. Also ask them what changes take place when the source of stress is removed. Tell students that as they read the text they should look for examples of the way the endocrine system "instructs" the other organs of the body to meet the physiological demands of a stressful situation. Also guide them to an appreciation of the way the endocrine system functions to maintain normal levels of chemical activity, even when there are not special demands placed on the body.

Emphasis

Make certain that students can identify the location of the endocrine glands in the body, the major hormones produced by each gland, and the primary functions of each hormone. Students should also be able to explain how feedback mechanisms, particularly negative feedback mechanisms, regulate the activity of the endocrine glands. Finally, students should be able to predict the possible consequences to the body should one of these feedback mechanisms malfunction.

Endocrine Gland Diagrams

Divide the class into small groups. Have each group prepare a diagram showing the location of one of the major endocrine glands in the body, the hormones it produces, and its target tissues and organs. Set up a classroom display showing all of these diagrams.

Guest Speaker

Invite an athletic coach from a local university or a physician from a sports medicine facility to talk to students about the use of steroids in athletics. The discussion should include information on how steroids affect the body and why scientists warn against their use as a means to enhance athletic ability.

Checking the Effects of Stress

At the beginning of class, expose students to some mild form of stress. For example, you might announce a pop quiz that tests their knowledge of the human endocrine system. Administer the quiz to students and then immediately following the quiz, have students check and record their pulse rates and examine the pupils of each other's eyes to see if they are dilated. Then tell students that the quiz doesn't count. Later in the class period, have students again check and record their pulse rates and look at the pupils of their eyes. Have students compare their two sets of results that depict their reactions immediately following exposure to stress and after the source of stress has been removed. Ask students to account for any differences they observe in the two sets of data.

Section Review Answers

(See p. 734.)

47.2 Reproductive System

Approach

The main objective of this section is to describe the structure and function of the major parts of the male and female reproductive systems. The section builds on the concepts of cell structure and function in Chapter 5 and of hormonal regulation in Section 47.1. In addition, you might wish to review the process of meiosis, as discussed in Chapter 9, to remind students how gametes are produced. In discussing the menstrual cycle, focus on the role of the endocrine system in regulating the cycle and point out examples of negative feedback mechanisms. For example, the presence of steroid hormones in the bloodstream causes the pituitary to stop producing follicle-stimulating hormone (FSH).

The recombination of genetic material provides for the variation in organisms that leads to natural selection, which in turn helps ensure survival of a species. This theme has been brought out in previous chapters on inheritance (Chapters 10 and 11) and evolution (Chapters 15 and 16). In discussing the human reproductive system, reinforce this concept by pointing out to students that the large number of gametes produced by humans, particularly males, makes possible a virtually unlimited number of genetic combinations with each new generation. A review of Chapter 11, which describes human inheritance patterns, will also enhance students' understanding of how genetic variation is achieved through human reproduction.

Concept Framework

Students know the anatomical differences between males and females. They also know that a new human being is formed from the union of an egg from a female and a sperm cell from a male. Elicit students' background knowledge about reproduction to use as a foundation for introducing them to the study of the human reproductive system. Ask students to consider the relationship between structure and function in male and female gametes and reproductive organs.

For example, you might elicit the fact that the structure of a sperm cell—a packet of DNA equipped with energy-providing mitochondria and a flagellum—is adapted for motility, thus increasing the likelihood that it will reach an unfertilized ovum. Students might also mention that the structure of the uterus permits the nourishment and protection of a developing fetus. Tell students to look for evidence as they read this section that in terms of both structure and function, the human reproductive system is adapted for passing genetic information on to the next generation—in other words, for ensuring evolutionary success.

Emphasis

Students should be able to identify the major parts of the male and female reproductive systems and to explain how they function in producing, storing, and releasing eggs and sperm. Students should also be able to explain role of the endocrine system in relation to the reproductive system, particularly in how it regulates the menstrual cycle.

Comparing Uteri

Obtain a preserved pig uterus from a biological supply house and display it to the class. Have students refer to Figure 47-11 in the textbook, which shows the structure of a human uterus. Assist students in making structural comparisons between the uteri from the two different species. Explain to students that swine uteri are of the bicornuate type—that is, the horns of the uterus are folded and are quite long in relation to the body of the uterus which is very short. You may wish to point out to students that the length of the uterine horns is an anatomic adaptation for producing offspring in litters. If the specimen was obtained from a pregnant pig, you might have students count the number of embryos present in the uterus.

Section Review Answers

(See p. 738.)

47.3 Fertilization and Development

Approach

The main objective of this section is to summarize the changes that take place from fertilization to embryo development to birth. The discussion of embryo development relates back to Chapter 5, on cell growth and cell specialization, and Chapter 12, which describes development as the result of sequential gene expression. Remind the class that taxonomists carefully study embryological development in different animal species and look for similarities in developmental patterns as evidence of phylogenetic relationships (see Chapter 18).

In presenting the section on human embryo development, encourage students to study the photographs in Figure 47-17 so that they can observe how development progresses throughout pregnancy. Also note that the discussion of placental mammals in Chapter 40 can serve as a basis of comparison between the embryo development of humans and other mammals.

Concept Framework

Students know that it takes nine months—a long period of time in comparison to most animals—for a single fertilized egg to develop into a fully formed human being. They also know that a tremendous number of changes take place in the developing fetus during this time.

To provide a conceptual framework for understanding the processes of fertilization and development, encourage students to consider these processes in terms of their adaptive significance. Ask students to think about the adaptive significance of such aspects of human development as the chemical activity following fertilization, the membranes surrounding the developing fetus, and the long gestation period. Tell them to look for other such adaptations as they read the text.

Emphasis

Students should be able to trace the development of a human being through the stages of fertilization, cleavage, implantation, and development of the embryo and fetus. They should know what initiates each of these stages and be able to identify the important changes that take place at each stage.

Using the Features/In Vitro Fertilization

The technology described in this feature is the use of in vitro fertilization to enable infertile couples to conceive children of their own. In vitro fertilization is the procedure carried out in a laboratory in which an egg from the female is fertilized in a glass Petri dish by sperm from the male. The fertilized egg is then implanted in the prospective mother's uterus.

In vitro fertilization has helped couples in which either partner has some physical condition that either prevents or greatly reduces the likelihood of in utero fertilization. In discussing this feature ask students to note the high percentage of attempts at in vitro fertilization that fail, either because implantation does not occur or because of miscarriage. Ask students to consider the possible difficulties with such a procedure that might explain the low success rate.

A discussion of in vitro fertilization also provides an opportunity to talk about other procedures that enable infertile couples to have children. Thus you might wish to explain how such practices as artificial insemination and surrogate motherhood work, and consider the advantages and disadvantages of using these procedures.

Prenatal Care: Past and Present

Obtain copies of old magazines or books from the library that contain articles on proper prenatal care and delivery methods, and share your findings with the class. Compare the beliefs of past decades with those commonly held today. Engage students in a discussion of how our expanding scientific knowledge has changed thinking in the medical community about the best ways to ensure a healthy pregnancy. The older the material you can find, the better, but even articles from the 1950s and 1960s will demonstrate how the vast amounts of new evidence scientists have obtained in recent years have led to better and more comprehensive prenatal care.

Class Discussion

A current subject of controversy is the increase in number of deliveries by Caesarian section in recent years. While some obstetricians point out that Caesarian deliveries have saved many infants from life-threatening complications, others question the wisdom of subjecting women to a major surgical procedure that might not always be necessary. Present evidence supporting both sides to the class and have students discuss and evaluate the advantages and disadvantages of Caesarian deliveries.

Field Trip

If possible, visit the obstetric ward of a local hospital so that students can view such facilities as a birthing room and the nursery, and equipment such as ultrasound machines and fetal monitors. Prior to the trip, give students a list of points to notice or questions to answer about the facility.

Guest Speaker

Invite a speaker who can describe some of the popular methods of childbirth, such as Lamaze, Bradley, and others. The discussion should include information on the type of preparation involved and what happens during the actual delivery. The speaker should also note the advantages and disadvantages associated with each method.

Section Review Answers

(See p. 744.)

Laboratory Preparation

Background

In this lab procedure, students measure the effect of thyroxine on frog metamorphosis. Thyroxine is a hormone that affects many facets of growth, including the transition from gills to lungs, the decrease in tail size, and the growth of limbs. Iodine is a part of the thyroxine molecule, and the presence of more or less iodine will influence the speed of the transformation from tadpole to adult frog. Students will be able to draw inferences about the effect of thyroxine on frog development but will not be able to "prove" that it is one of the reasons for frog metamorphosis. This procedure demonstrates the effect of two doses versus a control but does not provide proof that thyroxine causes metamorphosis.

Preparation

Check with a biological supply house well in advance of the investigation since developing tadpoles are not always available. To prepare thyroxine solution, dissolve 10 mg crystalline thyroxine in 5 mL of 1% sodium hydroxide solution. Dilute this solution by adding distilled water to make a total volume of 1 L.

Keep the solution refrigerated. If pond water is not available, set out open containers of water the night before the lab session. The spinach should be canned baby food. This procedure works well when students are divided into groups of three. Results may vary due to a variety of natural factors in addition to measuring errors.

Safety

Remind students to handle tadpoles humanely.

Conclusions

1. Thyroxine increases the rate of metamorphosis.
2. The stronger concentration causes the greatest change.
3. The body length gets longer, and the tail length gets shorter.

Inquiry

Students will design and conduct an experiment similar to the one completed but using iodine in place of thyroxine. They will then offer hypotheses concerning the results.

Chapter Review

Vocabulary

1. uterus; others are glands
2. semen; others are hormones
3. hypothalamus; others are organs in the male reproductive system
4. prostate; others are organs in the female reproductive system
5. placenta; others are prenatal development stages

Review

11. Exocrine glands release their secretions through ducts, and endocrine glands release their secretions directly into the blood.
12. Steroid hormones go through the cell membrane and bind with a special receptor. The receptor-steroid complex enters the nucleus, where it stimulates genes to synthesize messenger RNA.
13. The thyroid gland helps regulate a person's basal metabolic rate.
14. the pancreas secreting insulin
15. Sperm have flat round heads that contain a

nucleus, narrow middles that contain mitochondria, and tails or flagella that enable the sperm to move.

16. Ova are produced in the ovaries.

17. Unfertilized eggs are discharged from the uterus during menstruation.

18. In vitro fertilization is fertilization that takes place in the laboratory in a glass dish and it is a technique for helping couples who have fertility problems.

19. The developing fetus receives nourishment through the placenta, via the chorionic villi.

20. The cervix softens and enlarges.

Critical Thinking

1. The thyroid gland is in the area of the tonsils. After many years, the X ray exposure was suspected of causing thyroid cancer in those people treated for tonsilitis.

2. If a gland does not secrete enough, or too much, of a particular hormone, then the concentration in the blood of the substance controlled by that hormone would become excessive or insufficient. Supplemental hormones could help control such a problem.

3. The pituitary regulates the activity of other glands. It forms the main link between the endocrine and nervous systems. It also secretes hormones that control growth, which are critical in children.

4. The development of the embryo depends on a diploid set of chromosomes—one haploid set from a sperm and one set from the egg. A third set could prevent normal development.

5. A fetus needs calcium to develop bones.

6. The fetus is sucking its thumb; this is needed to develop the sucking response needed after birth.

7. Arms and legs were developing early in pregnancy.

Chapter 47

Endocrine and Reproductive Systems

Introduction

The fetus emerges from the womb and with a gulp of air becomes a breathing human being. During the nine months prior to birth, a single cell grew and changed into an eight-pound baby. In this chapter you will learn how these changes took place. You will also learn how the endocrine system controls the changes that take place as the baby grows.

Chapter Outline

47.1 Endocrine System
- *Feedback Mechanisms*
- *Endocrine Glands*
- *Hormones*

47.2 Reproductive System
- *Male Reproductive System*
- *Female Reproductive System*

47.3 Fertilization and Development
- *Fertilization*
- *Formation of Germ Layers*
- *Development*
- *Birth*

Chapter Concept

As you read, identify the ways that the endocrine system helps maintain homeostasis and ways that it regulates changes in the reproductive systems of males and females.

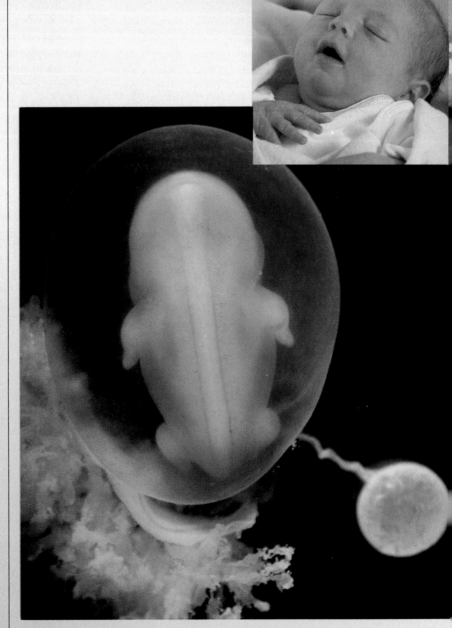

A human embryo at six weeks. Inset: A newborn baby.

For additional teaching information see pp. T47A–T47F.

47.1 Endocrine System

The endocrine system consists of glands that transmit chemical messages throughout the body. These chemical messages are called **hormones.** Hormones are substances that are produced in one part of the body and specifically influence the activity of cells in another part of the body. The glands secrete tiny amounts of these hormones into the blood. The blood then carries them to every part of the body. However, each hormone affects only specific cells, called **target cells,** that are supplied with receptors for that hormone.

Feedback Mechanisms

Most endocrine glands have a self-regulating system of checks and balances. This system depends on a **feedback mechanism.** This is a mechanism in which the end product of a series of steps controls the first step in the series. If the end product promotes the first step, the mechanism is called **positive feedback.** If the end product inhibits the first step, the mechanism is called **negative feedback.**

Most endocrine glands have negative feedback mechanisms. To understand how negative feedback works, think of the thermostat that controls the temperature in your home. When the temperature drops below a set level, the thermostat turns on the furnace. The furnace then produces heat, which warms the air. The end product––higher air temperature—is detected by the thermostat, which then turns off the furnace.

The first step in an endocrine feedback mechanism is the secretion of a hormone by a gland. The end product is a change in the concentration in the blood of the products controlled by that hormone. Follow the example of a negative feedback mechanism in Figure 47-1. When the blood glucose level in the body goes up, cells in the pancreas are stimulated and produce insulin. This insulin causes muscles and the liver to absorb glucose and store it in the form of glycogen. The reduced glucose level in the blood causes the cells to stop secreting insulin. As you read about the various endocrine glands, look for examples of how a negative feedback mechanism helps maintain homeostasis in the body.

Section Objectives

- *Describe a negative feedback mechanism.*
- *Explain the difference between an exocrine gland and an endocrine gland.*
- *List the major endocrine glands, and describe the function of each.*
- *Describe the chemical action of two types of hormones.*

Figure 47-1 Cells in the pancreas are stimulated to secrete insulin as part of a negative feedback mechanism.

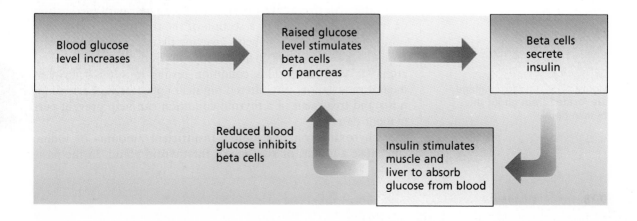

Blood glucose level increases → Raised glucose level stimulates beta cells of pancreas → Beta cells secrete insulin → Insulin stimulates muscle and liver to absorb glucose from blood → Reduced blood glucose inhibits beta cells

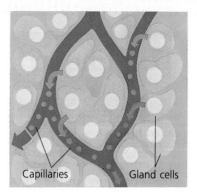

Duct

Gland

Capillaries   Gland cells

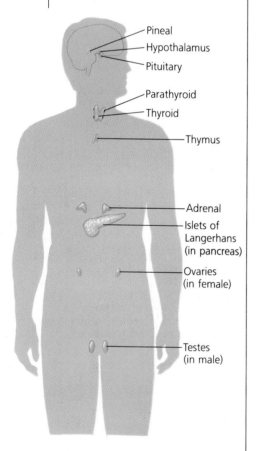

Figure 47-2 Exocrine glands (left) release their secretions through ducts while endocrine glands (right) secrete directly into the blood.

Pineal
Hypothalamus
Pituitary
Parathyroid
Thyroid
Thymus
Adrenal
Islets of Langerhans (in pancreas)
Ovaries (in female)
Testes (in male)

Figure 47-3 Endocrine glands are located throughout the human body.

Endocrine Glands

The body has two types of glands. **Exocrine** *(ECK-suh-KRIN)* **glands,** such as sweat glands, are glands with ducts. **Endocrine** *(EN-duh-KRIN)* **glands** have no ducts and secrete directly into the blood. Figure 47-2 shows a comparison of these glands.

Endocrine glands are located throughout the body and include the thyroid, the parathyroids, the adrenals, the pineal, the gonads, and the thymus. In addition, specialized cells also function as endocrine glands. These include the hypothalamus in the brain, the islets of Langerhans in the pancreas, and the digestive glands of the stomach and small intestine. Look for these glands in Figure 47-3 as you read about the hormones they produce.

Thyroid Gland

Notice in Figure 47-4 that the two-lobed **thyroid gland** lies in the neck near the lower part of the larynx. The thyroid gland secretes the hormone thyroxine. Thyroxine plays a vital role in regulating such activities as protein synthesis and ATP production.

If the thyroid gland becomes overactive and produces too much thyroxine, a condition called **hyperthyroidism** can result. People with this condition are physically overactive, their blood pressures are high, their hearts beat faster than normal, and their body temperatures are higher than normal. The secretion of too little thyroxine results in **hypothyroidism,** a condition in which the heart rate and other bodily functions are slower than normal. Victims of hypothyroidism tend to be overweight and often tire easily. Hypothyroidism and hyperthyroidism can be detected by measuring either basal metabolic rate (BMR) or blood thyroxine level. An underactive thyroid can be treated with supplementary thyroxine. An overactive thyroid can be treated with medication or by surgical removal of a section of the thyroid.

In infants, thyroxine helps growing tissues develop properly. If an infant suffers from an underactive thyroid, a condition called cretinism can occur. This condition results in stunted growth, mental retardation, and altered physical appearance. Early diagnosis and treatment of a thyroid condition can help prevent permanent damage.

Thyroxine contains iodine. Insufficient amounts of iodine can cause a goiter, or swelling of the thyroid gland. In the past,

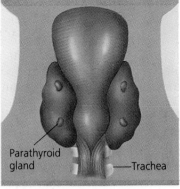

Trachea — Thyroid gland Parathyroid gland — Trachea

Figure 47-4 The thyroid gland (left) is located in the neck. A dorsal view of the same gland (right) reveals that four parathyroid glands are embedded in the back of the thyroid.

goiters frequently occurred in people living in areas where the soil contained low levels of iodine. Today, the occurrence of goiter is lower as a result of the addition of iodine to water supplies and table salt.

Tip: Ask students which hormones would probably most affect a person undergoing a lie detector test. (adrenaline and noradrenaline)

Parathyroid Glands

Four **parathyroid glands** are embedded in the back of the thyroid gland, two in each lobe. These glands secrete parathyroid hormone, which regulates the levels of calcium and phosphate ions in the blood. A proper balance of these two ions is essential for normal bone growth, muscle tone, and nerve activity.

The parathyroid glands regulate calcium and phosphate ion levels through a negative feedback mechanism. Refer back to the diagram of this kind of mechanism in Figure 47-1. When the calcium level in the blood drops, the parathyroids increase their production of parathyroid hormone. This hormone stimulates target cells in the bones to release calcium and phosphate ions into the blood. As the blood level of calcium and phosphate ions rises, the parathyroid glands decrease their hormone production. In this way, the levels of these important ions are kept in balance.

Adrenal Glands

Two **adrenal glands** are located one on top of each kidney. Each adrenal gland has an inner part, or medulla *(muh-DULL-uh)*, and an outer layer, or cortex. The medulla and cortex function as separate endocrine glands.

The medulla of the adrenal gland produces two hormones, adrenaline and noradrenaline, which play an active role in the body's reaction to stress. When a person is in a physically or emotionally stressful situation, the medulla secretes adrenaline and noradrenaline into the bloodstream. These hormones cause the heart to beat faster, the bronchial tubes to enlarge, and the pupils of the eyes to dilate. In turn, blood pressure rises, and surface blood vessels constrict. At the same time, more blood is pumped to the muscles, brain, and heart. Intestinal processes slow down, and the liver releases stored sugar, which is oxidized for extra energy. These reactions help provide the additional energy needed during a stressful situation.

The cortex of the adrenal gland produces the hormone hydrocortisone, which regulates certain phases of carbohydrate,

Figure 47-5 This light micrograph (×43) shows that the cells of the cortex and the cells of the medulla are morphologically distinct.

fat, and protein metabolism. Another hormone, aldosterone, helps maintain the water and salt balance in the body by stimulating the reabsorption of sodium ions by the kidneys.

Gonads

Gonads are gamete-producing organs. The **ovaries** of the female, which produce eggs, and the **testes** of the male, which produce sperm, are gonads. Gonads also produce sex hormones. The ovaries secrete hormones called estrogens *(ESS-truh-jenz)* and progesterone *(proe-JESS-tuh-RONE)*. The testes secrete a group of hormones called **androgens,** the main one being testosterone. Sex hormones are responsible for changes during puberty that result in secondary sex characteristics. In males the voice deepens, the chest broadens, and more hair grows on the body and face. In females the breasts grow and the hips widen. Menstruation also begins. In Section 47.2 you will learn about the hormones and feedback mechanisms that regulate menstruation.

Pituitary Gland and Hypothalamus

The **pituitary gland** is located at the base of the brain. It secretes hormones that control body growth and regulate the activity of the other endocrine glands. The pituitary has an anterior lobe and a posterior lobe. As you can see in Figure 47-6, both lobes are adjacent to the **hypothalamus,** a part of the brain. Nerve cells in the hypothalamus extend into the posterior lobe of the pituitary. The anterior lobe is connected with the hypothalamus by blood vessels. The pituitary and hypothalamus form the main link between the endocrine system and the nervous system.

The hypothalamus secretes hormones called releasing factors into the anterior lobe via the blood vessels. There are six releasing

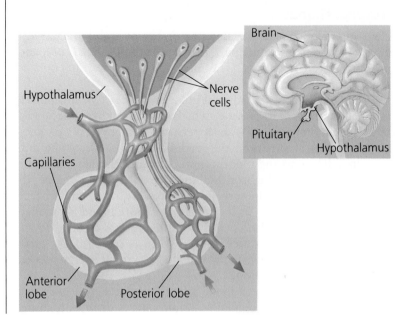

Figure 47-6 The hypothalamus communicates with the pituitary gland by means of both nerve cells and capillaries. The inset shows the location of the pituitary in relation to the brain.

factors. Each factor controls the release and therefore the level of one hormone produced in the anterior lobe of the pituitary gland.

One of these hormones is growth hormone, which controls the growth of the body's skeletal and muscular systems. Another is prolactin, which stimulates milk production following the birth of a child. The other four hormones regulate the secretions of other endocrine glands. Refer to Table 47-1 to learn the functions of these hormones.

Table 47-1
Hormones Released by Glands of the Endocrine System

Gland	Hormone	Function of Hormone
Thyroid	Thyroxine	Alter basal metabolic rate
Parathyroids	Parathyroid hormone	Maintain Ca and P levels
Adrenal:		
▪ **Cortex**	Corticoids	Maintain salt-water balance
▪ **Medulla**	Adrenaline, noradrenaline	Initiate responses to stress
Pituitary:		
▪ **Anterior lobe**	Growth hormone	Regulate growth
	LH	Regulate development of sex organs and hormones, control ovulation
	LTH	Control growth of mammary glands, control milk production, maintain corpus luteum
	FSH	Control gamete production
	ACTH	Stimulate adrenal cortex
	TSH	Stimulate production of thryoxine
▪ **Posterior lobe**	Oxytocin	Regulate blood pressure, stimulate smooth muscle contraction, initiate uterine contractions during childbirth, regulate milk production
	Vasopressin	Control reabsorption by kidneys
Islets of Langerhans	Glucagon, insulin	Maintain blood glucose levels
Ovaries	Estrogens	Produce female secondary sex characteristics, affect mature female body functions
	Progesterone	Maintain growth of uterine lining
Testes	Androgens (testosterone)	Produce male secondary sex characteristics
Thymus	Thymosins	Bolster children's immune systems
Digestive glands	Gastrin	Stimulate acid production
	Secretin	Influences digestive activity
Pineal gland	Melatonin	Influences biorhythms

The posterior lobe of the pituitary gland secretes two hormones—oxytocin and vasopressin. These secretions are also under the control of the hypothalamus. For example, when nerve cells in the hypothalamus detect that the concentration of water in the blood is low, they send nerve signals to the posterior lobe to secrete more of the hormone vasopressin. The increased concentration of vasopressin causes the blood vessels in the kidney to reabsorb more water. As a result the concentration of water in the blood rises. The hypothalamus detects this change and signals the posterior lobe to stop producing vasopressin. Oxytocin causes the muscles of the uterus to contract during childbirth.

Thymus

The thymus is located under the breastbone, between the lungs. In children the thymus is large, but it diminishes in size as a child grows older. The thymus produces hormones called thymosins that stimulate the development of infection-fighting antibodies and bolster the child's immune system.

Islets of Langerhans

The pancreas primarily is an exocrine gland that produces enzymes and secretes them into the small intestine via ducts. However, special cells in the pancreas called the **islets of Langerhans** function as endocrine glands. These cells, pictured in Figure 47-7, secrete two hormones, insulin and glucagon.

Insulin is a hormone that lowers the blood sugar level by stimulating cells to absorb glucose and by causing the liver and muscles to convert glucose to glycogen. In contrast, glucagon raises the blood sugar level by stimulating the breakdown of glycogen into glucose. In a healthy person the balance between insulin and glucagon maintains normal carbohydrate metabolism.

Without sufficient insulin too much glucose remains in the blood. This condition is called **diabetes mellitus** *(MELL-uh-tuss)*. In the United States more than one million people have one of the two forms of diabetes. Type 1, juvenile-onset diabetes, occurs when cells in the islets of Langerhans secrete little or no insulin. Regular injections of insulin help regulate the imbalance of glucose in the blood. Type 2, maturity-onset diabetes, is milder than Type 1. Victims of Type 2 diabetes may have normal levels of insulin in the blood but cannot use it, probably because their cells have a low number of receptors for insulin molecules. Case studies have revealed that heredity and obesity are contributing factors to occurrence of Type 2 diabetes, which can be treated by a carefully controlled diet. If diabetes is not treated, the victim will starve since his or her cells will not be able to absorb glucose.

Another disorder, called **hypoglycemia**, is the presence of too little glucose in the blood. Excess insulin causes the liver to store sugar that should be delivered to cells and results in an imbalance of blood glucose levels. This imbalance can cause a person to feel tired and dizzy at one time and then nervous and active at another.

Islet of Langerhans

Glucagon-secreting cell   Insulin-secreting cell

Figure 47-7 The islets of Langerhans play a crucial role in the regulation of blood glucose. The mechanism also involves other hormonal systems.

Other Glands

Endocrine glands in the stomach and lining of the small intestine produce hormones that aid digestion. After a person has eaten, cells in the stomach lining secrete gastrin, a hormone that stimulates other stomach cells to produce hydrochloric acid. Cells in the lining of the small intestine produce a hormone called secretin. This hormone affects the pancreas, the stomach, and the liver. Another gland, the pineal, is located in the forebrain. It produces melatonin, a hormone thought to be involved in biorhythms. Melatonin is also believed to influence maturation, perhaps by inhibiting the release of certain sex hormones until puberty.

Hormones

The maintenance of homeostasis and other vital functions performed by endocrine glands involve complex chemical interactions between hormones and other molecules in the body. Hormones differ in their chemical structure but are of two main types, protein hormones and lipid hormones.

Protein Hormones

Most hormones are protein hormones. They are either complete proteins, or polypeptides, amino acids, or amines. The chemical structure of one protein hormone is shown in Figure 47-8. A protein hormone acts by first attaching itself to a receptor on the membrane of a target cell. As you can see in Figure 47-8, the hormone then activates an enzyme in the cell membrane. This enzyme converts ATP molecules inside the cell to cyclic AMP. The cyclic AMP initiates changes within the cell by activating specific enzymes. The hormone adrenaline, for example, causes cyclic AMP to form in liver cells. As a result the cells produce enzymes that transform glycogen into glucose. The protein hormone is therefore called a first messenger; the cyclic AMP is called a second messenger.

Figure 47-8 Protein hormones, such as oxytocin, shown here in structural form, remain outside the cell but cause changes within by initiating a series of reactions involving enzymes and ATP.

Oxytocin

cys-tyr-ile-glu-asn-cys-pro-leu-gly

└────S────S──┘

Cortisol

Figure 47-9 Lipid hormones, such as cortisol, shown here in structural form, penetrate the cell wall before initiating cell changes.

Answers to Section Review
1. Hormones are substances produced in one part of the body that specifically influence the activity of cells in another part of the body.
2. the thyroid gland
3. The heart rate increases, breathing quickens, the pupils dilate, and the bronchial tubes enlarge.
4. Nerves connect the hypothalamus with the posterior pituitary; secretions from the hypothalamus travel through the blood to the anterior pituitary.
5. The hypothalamus has special cells that function like a gland, but it is primarily nerve tissue.

Lipid Hormones

A second type of hormone is a lipid hormone or **steroid.** You may have read about steroid abuse among athletes. Certain steroids, specifically ones derived from testosterone, were thought to increase muscle size. Scientists have not found evidence to support this claim. However, they have found that steroid abuse may contribute to liver disease and liver cancer, intestinal bleeding, and sexual dysfunction.

Notice in Figure 47-9 that, unlike protein hormones, steroids diffuse through the cell membrane and enter the cytoplasm. The steroid then binds to a receptor molecule to form a receptor-steroid complex that enters the cell's nucleus. Biologists hypothesize that this action stimulates genes to synthesize messenger RNA. The messenger RNA then moves into the cytoplasm, where it forms proteins that function as enzymes, promoting reactions that bring about the changes associated with the hormone.

Prostaglandins

A special group of lipids, called **prostaglandins,** also function as cell regulators. Unlike lipid and protein hormones, prostaglandins are not produced by specific glands. These powerful substances are produced in small quantities by specialized cells in many parts of the body. They act locally rather than through blood transport. The effects of prostaglandins include relaxation of the smooth muscles of air passages and blood vessels, contraction of the uterine and intestinal walls, regulation of blood pressure, and stimulation of the inflammatory response to infection.

Section Review

1. What are *hormones,* and what do they do?
2. Which gland controls a person's basal metabolic rate?
3. What physical changes occur during stress?
4. How does the hypothalamus control the pituitary gland?
5. How is the hypothalamus similar to an endocrine gland?

Info: Since aspirin reduces joint inflammation in patients with rheumatoid arthritis, it is considered a prostaglandin inhibitor.

47.2 Reproductive System

Although the gonads play important regulatory roles as part of the endocrine system, their primary function is to produce gametes. The reproductive system in both males and females is composed of a complex set of specialized organs that not only produce gametes but store and release these sex cells in ways that make reproduction possible.

Male Reproductive System

The male reproductive system produces, stores, and releases the male gametes, or **sperm.** Males begin to produce sperm during puberty and continue to do so throughout most of their lives.

Sperm

Notice the shape of the sperm cell in Figure 47-10. This shape reflects the function of sperm—to move and carry genetic information to the egg. The large head region of the sperm contains DNA. The narrow middle part of the cell contains mitochondria that provide the chemical energy that moves the flagellum. The wavelike motion of the flagellum propels the sperm forward.

Male Reproductive Structures

Sperm begin life in two oval-shaped organs called the **testes.** These organs are contained within the **scrotum,** a pouch of skin formed from the lower part of the abdominal wall. The temperature within the scrotum is slightly lower than that inside the abdomen, a condition that promotes the formation and survival of sperm. Tightly coiled tubes called seminiferous (SEM-*uh*-NIF-*uh-russ*) tubules are located inside the testes. Here special cells divide by meiosis and form sperm. The sperm are released into an elongated sac called the epididymis (EP-*uh*-DID-*uh-muss*), where they mature and are stored. Maturation takes place within 18 hours.

Sperm leave the epididymis and travel in a duct called the vas deferens *(VASS-DEFF-uh-runz)* into the urethra, a tube within the penis. In the urethra the sperm mix with secretions from the seminal vesicles, the prostate gland, and the small Cowper's gland to form a fluid called **semen.** These secretions nourish and protect the sperm. Semen contains a high concentration of the simple sugar fructose, which sperm metabolize to provide energy for movement. Urine, which is highly acidic, also passes through the urethra and can leave residues that might kill an unprotected sperm cell. However, semen neutralizes the acidity and thus protects the sperm cells. During orgasm the semen is forcefully expelled from the body by strong muscular contractions of the sperm ducts. This process is called ejaculation. The external part of the male reproductive system is the penis. The penis contains the external portions of the urethra. The penis is the organ by

Figure 47-10 A sperm cell is essentially a packet of DNA equipped with a flagellum that provides motility and mitochondria that provide the energy for movement.

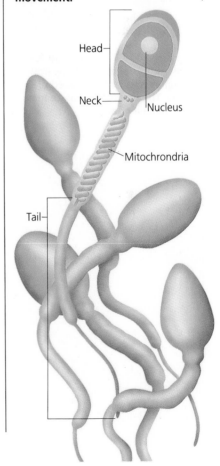

Head

Neck

Nucleus

Mitochrondria

Tail

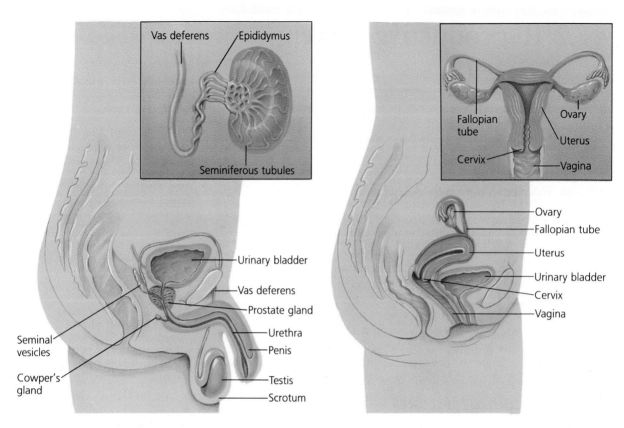

Figure 47-11 The male and female reproductive systems are complex sets of specialized organs. The insets emphasize male and female gonads.

which sperm is introduced into the female. It contains spongy tissue that becomes turgid and erect when filled with blood. The male reproductive system can be seen in Figure 47-11.

Female Reproductive System

The ovaries produce and store eggs, the female gametes. An egg, or **ovum,** is barely visible to the unaided eye, and yet one egg is about 75,000 times larger than a single sperm cell. Each female is born with over 400,000 immature eggs in her ovaries and will not produce any new eggs during her lifetime. In most women only about 400 eggs actually mature. Other structures of the female reproductive system include the Fallopian tubes, the uterus, the cervix, the vagina, and the external genitalia. Females have two **Fallopian tubes,** one located next to each ovary. Each functions to carry eggs from the ovary to the uterus, and each can be the site of fertilization. The **uterus** is a muscular structure that functions to house the developing fetus if fertilization occurs. The lower entrance to the uterus is called the cervix. The vagina is a tube leading from the cervix to the outside of the body. It is the canal that accepts the penis during intercourse and through which the fetus passes during childbirth. The external genitalia include the inner and outer labia and the clitoris. The female reproductive system can be seen in Figure 47-11.

Menstrual Cycle

Each month the female reproductive system goes through a series of changes called the **menstrual cycle.** The word *menstrual* comes from the Latin word *mensis,* which means "month." For most women, the menstrual cycle occurs over the course of about 28 days. During this time, an egg matures and is positioned to meet with a sperm cell in the Fallopian tube. If the egg is not fertilized, it is then discharged.

The menstrual cycle has four phases: (1) the follicular phase, (2) ovulation, (3) the luteal phase, and (4) menstruation. As you read about the menstrual cycle, refer to Figure 47-12.

Follicular Phase

The follicular phase begins when the hypothalamus produces a releasing factor. The releasing factor stimulates the anterior lobe of the pituitary gland to release follicle-stimulating hormone, or FSH. FSH is transported via the bloodstream and causes a group of ovarian cells, called a follicle, to form around the egg. Simultaneously, FSH promotes the production of the hormone estrogen, which is released into the bloodstream. Estrogen in turn stimulates the lining of the uterus to thicken. In addition, estrogen causes the pituitary gland to produce luteinizing hormone, or LH. This hormone then causes the maturation of the egg.

Tip: Point out that follicle-stimulating hormone and luteinizing hormone are also produced in males. The target organs in males are different from those in females since males do not produce follicles or experience luteal phases.

Figure 47-12   The four phases of the menstrual cycle occur over about a 28-day span.

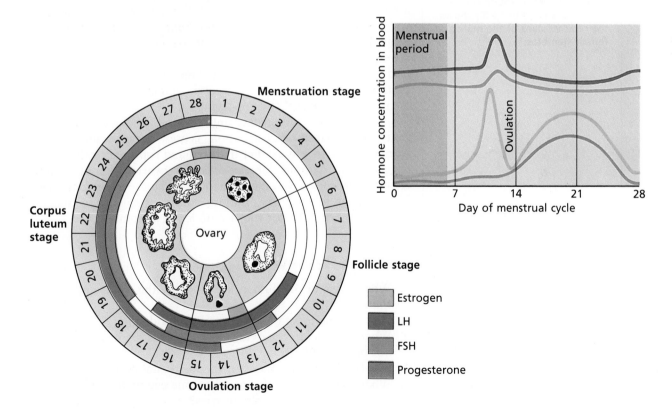

Endocrine and Reproductive Systems   **737**

Figure 47-13 This color-enhanced SEM of sperm surrounding an egg (×3,400) shows the relative sizes of the male and female gametes.

Answers to Section Review
1. the female and male gametes
2. Semen neutralizes acidic urine residues, protecting the sperm. Its high concentration of fructose nourishes the sperm.
3. The ovum goes from the ovary to the Fallopian tube to the uterus and is then discharged from the body.
4. the follicular stage, ovulation, the luteal phase, and menstruation
5. More than one egg might be fertilized and more than one embryo might develop.

Info: **Many women suffer unpleasant physical symptoms at menopause. These can often be relieved with estrogen replacement therapy.**

Ovulation

When the egg has matured, the follicle moves to the wall of the ovary, ruptures, and releases the ripe egg. This event, called **ovulation,** is the second stage of the menstrual cycle. During ovulation tiny fingerlike projections draw the egg from the ovary into the Fallopian tube. The egg remains in the Fallopian tube for about four days and can be fertilized during this time. Figure 47-13 shows an egg surrounded by sperm. The egg, or zygote if the egg has been fertilized, then moves through the Fallopian tube to the uterus, which is now thick and lined with mucus.

Luteal Phase

During the luteal phase, the ruptured follicle left on the surface of the ovary develops into a new structure called the **corpus luteum.** Another hormone produced by the pituitary, luteotropic hormone (LTH), stimulates the corpus luteum to send out steroid hormones, including estradiol and progesterone, which establish an even thicker lining on the inside wall of the uterus. In addition, the increased levels of these hormones in the blood cause the pituitary gland to stop producing FSH. The uterine lining will enable the uterus to nourish and protect the fertilized egg.

Menstruation

If the egg was fertilized in the Fallopian tube, the resulting embryo attaches to the lining of the uterus. If the egg was not fertilized, it will not attach to the lining. In this case, the corpus luteum stops producing progesterone and the soft tissues of the uterine lining are sloughed off. These tissues along with blood and the unfertilized egg are discharged in the last stage of the cycle, called **menstruation.** They exit through the cervix and the vagina. Following menstruation the cycle begins again.

Most women menstruate until around age 45. At this time, called menopause, menstruation ceases because most of a woman's follicles have either matured and ruptured or they have degenerated. Without follicle cells the ovaries cannot secrete enough estrogen or progesterone to maintain the menstrual cycle. The anterior lobe of the pituitary continues to secrete follicle-stimulating hormone for the remainder of a woman's life.

Section Review

1. What phase of the human life cycle are the *ovum* and *sperm?*
2. Describe the function of semen in reproduction.
3. What happens to the ovum after it leaves the ovary if it has not been fertilized?
4. What are the four phases of the menstrual cycle?
5. What might happen if more than one egg were simultaneously released from the ovaries?

47.3 Fertilization and Development

The primary function of human reproductive systems is to create new individuals. This occurs when the male gamete is successful in reaching and penetrating the female gamete, resulting in the formation of a zygote that develops into a human being.

Fertilization

Fertilization occurs when a sperm combines with an ovum in a Fallopian tube. Sperm enter the female during the act of sexual intercourse, in which the male places his penis in the female's vagina. Muscular contractions during male orgasm eject semen from the body, through the urethra in the penis, and into the female. Hundreds of millions of sperm cells are released at once.

The egg in the Fallopian tube releases a chemical into the surrounding fluid that attracts the sperm. When sperm reach the egg, they first stick to the jellylike substance on the outer surface of the egg. Many sperm may attach themselves but only one sperm will actually penetrate the surface.

The egg membrane engulfs the head of a single sperm and the sperm nucleus breaks out of the head. Once this has occurred, a membrane forms around the egg and prevents any other sperm from entering. The sperm nucleus then fuses with the egg nucleus.

Each gamete contains 23 chromosomes, the haploid (1N) number. Thus the union of sperm and egg causes the zygote to have 46 chromosomes, the diploid (2N) number. The presence of

Section Objectives

- *Describe the process of fertilization.*

- *Summarize the changes that take place during embryo development.*

- *List the three stages of pregnancy and describe characteristics of each stage.*

- *Describe the process of in vitro fertilization.*

Tip: Tell students that although the 300 million to 500 million sperm produced in a single ejaculation seem unnecessarily high, males producing lower counts have high failure rates in fertilization.

Figure 47-14 Fertilization of the egg takes place in the Fallopian tube, where the egg undergoes its first series of divisions before implantation in the uterus.

Figure 47-15 The blastocyst, composed of the embryo and the developing placenta, is implanted in the uterine wall.

the diploid set of chromosomes initiates embryo development. You may remember from Chapter 12 that development is the result of sequential gene expression.

Cleavage and Implantation

Immediately following fertilization, a phase known as cleavage occurs in which the zygote goes through many mitotic cell divisions while still in the Fallopian tube. These divisions produce a ball of cells called a morula. As the cells of the morula divide, they release fluid into the center of the sphere. The structure is now called a **blastocyst,** a sphere of cells with a large, fluid-filled cavity at its center. The outer layer of cells of the blastocyst, called the trophoblast, releases an enzyme that breaks down the epithelial tissue of the uterus and enables the blastocyst to embed itself in the thick lining. This process is called implantation.

The embryo will form from a mass of cells on the inner surface of the blastocyst, as shown in Figure 47-15. This inner mass of cells has three **primary germ layers:** the ectoderm, the mesoderm, and the endoderm. These three layers will eventually form all the body organs. The ectoderm forms the skin, the skin glands,

Biotechnology

In Vitro Fertilization

One out of seven married couples in the United States is infertile. These couples have not conceived a child even though they have tried for more than a year. For some of these couples, in vitro fertilization can help.

Infertility is the result of some condition in the male, the female, or both partners. For example, the man may produce low numbers of sperm or the woman may have blocked Fallopian tubes. In the past little could be done to help infertile couples who wanted a child of their own. Today, however, doctors have successfully induced pregnancy by using a method of in vitro fertilization. The babies that are the result of this process are often called "test-tube babies."

The woman who undergoes in vitro fertilization is first given a fertility drug or hormone. This stimulates her ovaries to release more than a single egg at one time. The doctor then uses an instrument to observe the woman's ovaries. When the doctor sees the mature egg follicles, the doctor inserts a hollow needle and gently removes the follicles.

The doctor then places these follicles in a glass Petri

most cartilage, the nervous system, the pituitary gland, the lining from the mouth to the pharynx, a portion of the lining of the rectum, and the adrenal medulla. The mesoderm forms the connective tissue, the bone, most muscles, the kidneys and their ducts, the gonads and their ducts, the blood and blood vessels, the heart, and the lymph system. The endoderm forms the lining of the alimentary canal from the pharynx to the rectum, the thyroid, the parathyroids, the trachea, the lungs, the bladder, and the liver.

Development

The fertilized egg develops inside the uterus during a nine-month period called **gestation** or **pregnancy.** In the first eight weeks of its development, the organism is referred to as an **embryo.** From eight weeks until birth, it is called a **fetus.** Physicians commonly divide a pregnancy into three equal periods called **trimesters.**

Membranes

Throughout pregnancy the developing child is surrounded and protected by four membranes that develop from the trophoblast. The first of these membranes is called the **chorion** *(KOR-ee-ahn).*

dish. *In vitro* means "in glass." Next the doctor applies a drop of the husband's sperm to the egg follicles. The eggs are then incubated until fertilization and mitotic cell divisions have taken place. This process requires 48 to 72 hours.

After a number of cell divisions have taken place, the doctor inserts the tiny embryo into the woman's uterus, where it may attach to the thickened uterine lining and develop normally.

Yet even if fertilization does occur, only a third of all women whose eggs are fertilized in vitro will become pregnant. Of those that do become pregnant, about one-third will suffer miscarriages during the first three months of their pregnancy.

The first child born through the in vitro fertilization process was Louise Brown, who was born in England in 1978. Since her birth the number of clinics using the in vitro fertilization technique has multiplied. In the United States alone there are about 136 medical clinics using this procedure. In vitro fertilization clinics also exist in many other countries around the world.

Blood to mother
Villi
Blood from mother
Umbilical cord To Fetus
Fetal vessels
Maternal vessels
Placenta

Figure 47-16 Food and oxygen from the mother and waste products from the fetus pass through the placenta, the lifeline between mother and fetus.

This membrane possesses a number of small, fingerlike projections called chorionic villi. Food and oxygen from the mother and waste products from the embryo are exchanged through the membranes of the chorionic villi. The area where the chorionic villi meet the maternal blood supply is called the **placenta.**

A second membrane, called the **amnion,** is a fluid-filled sac that surrounds the developing embryo. This sac is filled with amniotic fluid, which cushions the embryo and keeps it moist.

A third membrane is the **yolk sac.** The yolk sac provides some nourishment during early embryo development.

The fourth membrane surrounding the embryo is the **allantois** *(uh-LAN-tuh-wus).* Along with the chorion, the allantois lengthens to become the **umbilical cord,** a cord that contains arteries and veins that carry blood between the embryo and the placenta.

Cells within the placenta secrete chorionic gonadotropic hormone, which keeps the corpus luteum functioning. The corpus luteum produces estradiol and progesterone, two hormones that maintain the thick lining of the uterus. The high estrogen levels also stop follicle-stimulating hormone and luteinizing hormone production by the pituitary. As a result ovulation and menstruation do not occur during pregnancy.

The placenta, shown in Figure 47-16, is the life-supporting link between mother and fetus. The placenta is also shown in the bottom photograph in the left-hand column of Figure 47-17. Most of what the mother ingests passes into the fluid that surrounds the capillaries in the placenta. These substances can then pass into the capillaries and into the fetal blood supply.

First Trimester

The most dramatic changes in the development of a child take place in its first six to eight weeks. In the first two to three weeks, the human embryo resembles the embryos of other animals. But by the fifth week human features exist.

The brain, spinal cord, and nervous system begin forming in the third week. The heart begins to beat at 21 days and develops a smooth rhythm at 28 days. You can locate the heart as the dark mass in the chest area of the five-week-old embryo in Figure 47-17. By the fifth week eyes, ears, nasal organs, arms, legs, and the digestive system begin to develop. At six weeks the fingers, toes, and the external ears form. Brain waves occur at this time. The embryo also begins to move, although it is so small the mother cannot feel it turning. Figure 47-17 also shows the structures that have developed at six weeks. Compare this photograph with the photograph of the five-week-old embryo to see how quickly the embryo develops.

When the first trimester ends the embryo is only about 5 cm long, but most of its organ systems are established. Throughout its development the embryo, and later the fetus, is sensitive to viruses or toxins that may be passed to it from the mother.

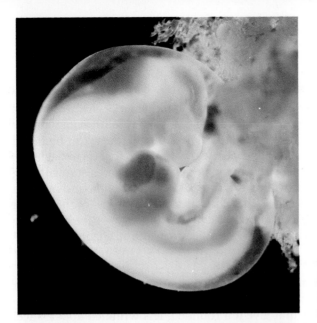

Figure 47-17 These photographs show (beginning at the left and proceeding counterclockwise) a 5-week embryo with a visible heart; a 6-week embryo, an 11-week fetus with the surrounding placenta clearly shown; a 12-week fetus: and a 22-week fetus.

Answer to Think
The fetus receives oxygen from the maternal blood supply. The surrounding tissues are filled with fluid, so the fetus would not be able to breathe in utero. Thus the lungs are not needed until birth.

Answers to Section Review
1. An embryo is an organism prior to eight weeks of development. A fetus is an organism after all the organs have begun to develop, that is, from eight weeks until birth.
2. A blastocyst is a ball of cells with a fluid-filled center formed at an early stage of development. It contains a mass of cells on the inside that will develop into an embryo.
3. Nourishment from the maternal blood supply travels to the embryo through the chorionic villi of the placenta. The amnion moistens and cushions the embryo.
4. The organs of the embryo develop during the first trimester. The fetal heart beat can be heard during the second trimester. The fetus grows in size and weight during the third trimester.
5. Any substance entering the mother's blood can travel through the placenta to the fetus. Therefore the mother should eat well during pregnancy and avoid ingesting any harmful substances.

Second Trimester

In the second trimester the mother's abdomen begins to swell as her uterus enlarges. The fetus' skeleton begins to form and a layer of soft hair grows over its skin. The fetal heartbeat can be heard with the aid of a stethoscope. At this time the fetus also begins to wake and sleep. The mother may now feel the fetus start moving about. The fetus swallows, hiccups, sucks its thumb, and makes a fist. It also kicks its feet and curls its toes. By the end of the second trimester the fetus is about 32 cm long and its eyes are open. The fetus at 12 weeks and 22 weeks can be seen in Figure 47-17.

Third Trimester

In the third trimester the fetus is becoming modified to survive in the outside world. It grows quickly in size and weight, and develops fat deposits that give it a rounded, less wrinkled appearance. The fetus can see light and darkness through its mother's abdominal wall and reacts to music and loud sounds. Ultrasound studies have led some researchers to believe that learning may take place before birth occurs.

Birth

Birth begins about 270 days after the egg has been fertilized by the sperm cell. The pituitary gland of the fetus, prostaglandins in the fetal membranes, and glands within the mother's body all release hormones that initiate childbirth. Oxytocin is secreted in large amounts, causing the smooth muscles of the uterus to contract. The amniotic sac breaks, and the fluid it contains flows out through the vagina. This is called breaking water. The tissues of the normally narrow cervix relax and enlarge, making it easier for the fetus to pass through. During the birth process, strong uterine contractions help push the baby through the cervix, out of the uterus, down the greatly expanded vagina, and out through the external genitalia.

Following birth the umbilical cord is tied and then cut. At this time, the baby's lungs expand for the first time and the newborn begins to breathe on its own. The remains of the placenta and amnion, the **afterbirth**, are then expelled from the mother's body about ten minutes after the birth of the baby.

Section Review

1. Describe the difference between an *embryo* and a *fetus*.
2. Describe a blastocyst.
3. How is the embryo nourished and protected during its development?
4. What are some characteristics of the embryo and fetus in the different stages of pregnancy?
5. Why is it important for a pregnant woman to follow good health practices?

Laboratory

Effects of Thyroxine on Frog Metamorphosis

Objective
To determine the effect of thyroxine on the development of tadpoles

Process Skills
Measuring, inferring, experimenting

Materials
Glass-marking pencil, 3 600-mL beakers, pond water, 10-mL graduated cylinder, 0.01% thyroxine stock solution, strained spinach, graph paper marked in 1-mm squares, petri dish, small fish net, 9 tadpoles with hind legs just beginning to show, 3 pencils in different colors

Method

1 Use a glass-marking pencil to mark three beakers *A, B,* and *C.* Label each beaker with your initials.
2 Add 500 mL of pond water to each beaker.
3 Use a graduated cylinder to measure 10 mL of thyroxine solution, and add the solution to beaker A. Add 5 mL of thyroxine solution to beaker B. Add nothing to beaker C.
4 Add about 1 cm^3 of spinach to each beaker.
5 Place a sheet of graph paper, ruled side up, underneath a petri dish.
6 Catch a tadpole with a fish net and place the tadpole in the petri dish. Measure the tadpole's total length, tail length, and body length in millimeters by counting the number of squares it covers on the graph paper. Record your measurements. Place the tadpole in beaker A.
7 Repeat step 6 with two more tadpoles.
8 Repeat step 6 with six more tadpoles, this time placing three tadpoles in beaker B and three in beaker C. What is the purpose of putting tadpoles in water without the thyroxine solution?
9 Feed the tadpoles about 1 cm^3 of spinach every other day. Be careful to avoid overfeeding. Change the water every four days, adding thyroxine solution to beakers A and B at the original concentrations.
10 Measure the tadpoles twice a week for three weeks. Record your data. Calculate the average growth per week for each group of tadpoles.
11 Graph your data using different colored pencils for group A, group B, and group C tadpoles. How should you label the axes of your graph?

Conclusions
1 What is the effect of thyroxine on tadpole metamorphosis?
2 Which concentration of thyroxine caused the greatest visible change in the tadpoles?
3 How do average body length and tail length change during metamorphosis?

Inquiry
Design and conduct an experiment that shows the effect, if any, of adding iodine to tadpole water. Speculate on the reason for the results.

Vocabulary

adrenal gland (729)
afterbirth (744)
allantois (742)
amnion (742)
androgen (730)
blastocyst (740)
chorion (741)
corpus luteum (738)
diabetes mellitus (732)
embryo (741)
endocrine gland (728)
exocrine gland (728)
Fallopian tube (736)

feedback mechanism (727)
fertilization (739)
fetus (741)
gestation (741)
gonads (730)
hormone (727)
hyperthyroidism (728)
hypoglycemia (732)
hypothalamus (730)
hypothyroidism (728)
insulin (732)
islets of Langerhans (732)

menstrual cycle (737)
menstruation (738)
negative feedback (727)
ovary (730)
ovulation (738)
ovum (736)
parathyroid gland (729)
pituitary gland (730)
placenta (742)
positive feedback (727)
pregnancy (741)

primary germ layer (740)
prostaglandins (734)
scrotum (735)
semen (735)
sperm (735)
steroid (734)
target cells (727)
testis (730)
thyroid gland (728)
trimester (741)
umbilical cord (742)
uterus (736)
yolk sac (742)

For each set of terms below, choose the one that does not belong and explain why it does not belong.
1. adrenal, pituitary, uterus, parathyroid
2. estrogen, insulin, corticoid, semen
3. vas deferens, hypothalamus, scrotum, urethra
4. prostate, vagina, uterus, cervix
5. embryo, placenta, blastocyst, fetus

Review

1. The endocrine system (a) affects only the nervous system (b) helps maintain homeostasis (c) affects only the reproductive system (d) primarily utilizes a positive feedback mechanism.
2. In a negative feedback system the (a) end product inhibits the first step (b) end product inhibits the last step (c) end product stimulates the first step (d) end product stimulates the last step.
3. Steroid hormones (a) attach to a receptor on the cell membrane (b) convert ATP to cyclic AMP (c) diffuse through the cell membrane (d) increase muscle size.
4. The correct pathway of sperm is (a) testes to vas deferens to epididymis (b) epididymis to urethra to vas deferens (c) urethra to vas deferens to testes (d) testes to epididymis to vas deferens.

5. The correct pathway of an ovum is (a) ovary to Fallopian tube to uterus (b) Fallopian tube to ovary to urethra (c) uterus to ovary to Fallopian tube (d) ovary to Fallopian tube to urethra.
6. Fertilization takes place in the (a) vagina (b) cervix (c) uterus (d) Fallopian tubes.
7. During menstruation (a) the egg moves into the Fallopian tube (b) the corpus luteum develops (c) the unfertilized egg is discharged (d) the egg matures.
8. For fertilization to occur (a) many sperm must penetrate the egg (b) only one sperm must penetrate the egg (c) many eggs must be released (d) the uterus must be enlarged.
9. By the end of the first trimester (a) the fetus can suck its thumb (b) all of the organs have begun to form (c) the fetus uses

its lungs to breathe (d) the brain is fully developed.

10. The placenta is (a) the site of yolk development (b) the site of nutrient and gas exchange between the mother and the fetus (c) a structure that protects against injury (d) the tip of the uterus.

11. Explain the difference between an exocrine gland and an endocrine gland.

12. Describe the interaction between a steroid and its receptor cell.

13. Which gland regulates basal metabolism?

14. Give an example of a biological negative feedback mechanism.

15. Describe the structure of a sperm cell.

16. Where are ova produced?

17. What happens to an egg if it is released but is not fertilized?

18. What is in vitro fertilization?

19. How does the developing fetus receive nourishment?

20. What changes does the cervix undergo during childbirth?

Critical Thinking

1. A number of years ago a widely used treatment for tonsillitis, an inflammation of the tonsils, was to expose the neck area to X rays to shrink the tonsils. Explain why this procedure was later suspected of being dangerous and was therefore discontinued.

2. Using the concept of a negative feedback mechanism, explain how a hormone imbalance might occur and how the imbalance could be treated.

3. Why might damage to the pituitary gland be considered far more serious than damage to one of the other endocrine glands?

4. What do you think might happen if more than one sperm was able to penetrate the egg membrane?

5. Why do you think that doctors recommend that pregnant women drink milk?

6. What is the fetus doing in the photograph at right? Suggest an adaptive advantage for this activity.

7. About 20 years ago a large number of babies were born with serious limb birth defects after their mothers had taken a tranquilizer called thalidomide. Other mothers who took the drug later on in pregnancy had normal children. What does this tell you about the sequence of development of the fetus?

Extension

1. Research some of the physical and hormonal causes of infertility. Write a report describing the conditions, and explain how medical science is helping overcome infertility.

2. Read the article by Shannon Brown Lee entitled "Lord of the Flies" in *Discover*, April 1987, pp. 26–40. What does the discovery of the homeobox have to do with human embryology?

3. Steroids can be used to treat certain illnesses. Using the Physician's Desk Reference (PDR), list the names and uses of the specific steroids.

Chapter 48 | *Tobacco, Alcohol, and Other Drugs*

This chapter teaches students about the physical and behavioral effects of tobacco, alcohol, and other drugs. They learn about the major types of drugs and addictions.

Correlation Chart

Section Number	Section Objectives	Pages	Features	Section Review
48.1	1. Identify two harmful chemicals produced when tobacco is burned. 2. Describe the physical effects of inhaling tobacco smoke. 3. List three diseases that can result from smoking or chewing tobacco. 4. Explain what can happen to the fetus if a pregnant woman smokes.	749–750		2 2, 3 1, 3, 4 5
48.2	1. Define *alcohol.* 2. Summarize how alcohol affects the body. 3. Describe long-term results of excessive alcohol consumption.	751–752		1, 2, 3, 5 4, 2
48.3	1. Distinguish among the four major types of drugs. 2. List one example of each type of drug, and describe how it affects the body. 3. Explain the difference between physical and psychological addiction. 4. Explain a possible source of error in a study using the single-blind method.	753–757	Think, p. 754 Biotechnology, pp. 754–755 Biology Lab, p. 757	1, 3 3, 5 2

Resources*

Print

Gussin, Gilda, and Ann Buxbaum. *Self-Discovery: Alcohol and Other Drugs—Using Skills to Make Tough Choices.* Boston: Management Sciences for Health, 1984.

Woods, Geraldine. *Drug Use and Drug Abuse,* revised ed. New York: Watts, 1986.

Audiovisual/Software

Choices: Alcohol, Drugs, or You. Film or video, 24 min. Teacher's guide. Barr, 1983.

Drinking and Driving: What Teens Should Know. 2 filmstrips and cassettes. 12–14 min. each. Teacher's guide. Pleasantville.

One for My Baby. Film or video, 28 min. AIMS, 1984.

Smoking: A Research Update. 2 filmstrips and cassettes, 12–15 min. ea. Teacher's guide. Pleasantville.

*Additional references are listed in *Modern Biology, Teacher's Program Guide.*

Section Teaching Guides

48.1 Tobacco

Approach

The main objective of this section is to familiarize students with the effects of tobacco on the human body. It builds on the concepts of the respiratory and the circulatory systems (Chapter 43). Make sure that students understand that tobacco contains a powerful stimulant, nicotine. In presenting the text, emphasize that despite tobacco's wide use and easy availability, tobacco is a dangerous substance containing a powerful drug.

Concept Framework

In studying previous chapters, students have learned how a healthy body functions. Build on that knowledge to establish a conceptual framework for studying this section by asking students to trace the path of air through the respiratory system. Point out that cilia lining the air passages move foreign particles out of the body. Ask what might happen if the cilia did not perform their function. Then tell students that tobacco smoke paralyzes the cilia. You might end the discussion by asking students as they read to consider why smokers are more susceptible to diseases such as chronic bronchitis and emphysema.

Emphasis

Make certain that students understand that inhaling sidestream smoke is as dangerous as smoking cigarettes. Depending on your region, you might compare the air pollution from smoking to the air pollution from other sources.

Guest Speakers

Invite several tobacco users to speak with the class about their tobacco habits, the addiction they feel, and the difficulty they experience in trying to quit smoking cigarettes or other tobacco products. You might end the discussion with a question and answer session.

Section Review Answers

(Continued from p. 750.)
3. Tar paralyzes the cilia that move disease-causing microorganisms out of the body.

4. lung cancer, heart disease, chronic bronchitis, emphysema
5. Pregnant women who smoke risk miscarriages more than do other pregnant women. Their babies are born smaller and less robust and may eventually develop heart disease.

48.2 Alcohol

Approach

The main objective of this section is to illustrate the effects of alcohol on the human body. This section builds on concepts of human physiology (Chapters 42–47). In presenting the text, remind students that ethanol, a depressant, is the drug found in alcoholic beverages. You may wish to outline the effects of a depressant on the human body. You might also ask students to speculate about the combined effects of smoking and drinking alcohol.

Concept Framework

Students have heard that drinking and driving do not mix. To use this awareness as a conceptual starting point for this section, point out that despite their widespread use and availability, alcoholic beverages contain a powerful depressant, ethanol. Engage students in a discussion of alcohol's effects by asking why it is dangerous to drink and then drive. Elicit the fact that a depressant inhibits the central nervous system and therefore slows a person's responses. Ask students as they read to consider how alcohol effects other parts of the body.

Emphasis

Make sure that students understand that a variety of factors affect a person's BAC. Two people who have drunk the same amount of alcohol will not necessarily have the same BAC.

Guest Speaker

Invite a speaker from an organization such as Alcoholics Anonymous to discuss alcoholism with students. Ask the speaker to emphasize that alcoholism is a disease and that physicians are unsure as to why some people are predisposed to becoming alcoholics.

Section Review Answers

(Continued from p. 752.)
5. Such a person would suffer some lack of coordination and should not drive a car.

48.3 Other Drugs

Approach

The primary objective of this section is to familiarizes students with the four major types of drugs. It also distinguishes prescription drugs from over-the-counter drugs. The section builds on concepts of nervous system physiology (Chapter 46). In presenting the text, emphasize that many drugs are necessary for medical purposes and are safe when taken as directed. Point out, however, that many medical drugs as well as most illegal drugs are addictive.

Concept Framework

Most students are familiar with over-the-counter drugs such as aspirin and cough syrup. In fact, most students have taken over-the-counter medication at some point. Use this knowledge as a starting point for studying this section and then ask students to name some over-the-counter drugs and their effects. Encourage discussion of the benefits of these drugs when they are taken as directed and the dangers of these drugs when abused. You might wish to end the discussion by asking students to name some illegal drugs and tell why these drugs are illegal.

Emphasis

Make certain that students understand the differences between psychological and physical addictions. Point out that both are equally dangerous.

Guest Speaker

Invide a pharmacist to speak to the class about the medicinal uses and possible side effects of some common prescription drugs and some over-the-counter drugs. You might ask the speaker to explain the steps involved in developing a drug for human use.

Using the Features/The Double-Blind Test

The methodology presented in this feature is the double-blind technique for testing a drug's effectiveness. The double-blind technique is more accurate than the older single-blind method because neither patients nor those administering the drugs know who is receiving a placebo. In presenting this feature, make sure that students understand the importance of thoroughly testing new drugs before approving them for sale.

Section Review Answers

(See p. 756.)

Laboratory Preparation

Background

In this lab, students observe the effects of common chemicals and drugs on the behavior of both cell contents and small organisms. By varying concentrations, students learn that quantity is an important variable. Different organisms may respond differently, but cessation of cytoplasmic activity at the cellular level is a basic effect. Plant defensive compounds, such as nicotine, probably evolved as harmful agents for herbivores.

Preparation

To prepare 1% ethanol, add 3 mL 70% ethanol to 197 mL distilled water. To prepare 2% ethanol, add 6 mL 70% ethanol to 194 mL distilled water. To prepare 5% ethanol, add 14 mL 70% ethanol to 186 mL distilled water. To prepare 20% ethanol, add 57 mL 70% ethanol to 143 mL distilled water. To prepare caffeine solution, add 1 No-Doz tablet per 20-30 mL distilled water. To prepare nicotine solution, soak 3 Camel cigarettes (remove paper) overnight in 30 mL distilled water. If pond water is not available, place tap water in an open container and let sit overnight. *Elodea* is available from pet stores or can be collected from ponds or streams. *Vorticella* cultures can be grown in pond water.

Conclusions

1. Observed responses will vary. In general, cytoplasmic streaming will cease. At higher concentrations death will occur. Variations may occur based on the amount of alcohol actually wicked across or the location of observed cells in the plant.
2. The *Vorticella* exhibit avoidance behavior.
3. In most cases death occurs.

Inquiry

Sources of error may include inaccurate measurements used to prepare solutions, variations in wicking action, differing conditions of the experimental plants and animals, and differences in elapsed time before observations are recorded.

Chapter Review

Vocabulary

1. They are the two drugs found in tobacco.
2. emphysema and chronic bronchitis
3. psychoactive drugs, narcotics, stimulants, and depressants
4. Blood alcohol concentration refers to the level of alcohol in the bloodstream; rate of consumption, ability to metabolize alcohol, body weight, and amount of food in the stomach are factors involved in BAC.
5. A person who has become addicted to a drug may develop a tolerance. If the person tries to quit using the drug or if the drug is unavailable, the person may suffer withdrawal symptoms.

Review

11. Nicotine is absorbed through the alveoli and through the tissues lining the mouth.
12. The harmful effects of sidestream smoke are the same as for the person who smokes directly.
13. low birthweight, miscarriage, early death of infant, and high risk of the child suffering from congenital heart disease
14. Excessive use of alcohol forces the cells of the liver to use alcohol as an energy source, and eventually they are unable to function properly. The result is accumulation of fat deposits that may lead to fatty liver and, if untreated, cirrhosis.
15. As BAC increases, a person is less able to think clearly and make appropriate decisions. The person becomes more and more disoriented and eventually may lose consciousness.
16. high blood pressure, heart disease, gastritis, metabolic disorders, fatty liver, cirrhosis, and alcoholism
17. If fatty liver remains untreated it can eventually result in cirrhosis and permanent damage to the liver cells.
18. Psychoactive drugs are drugs that alter sensory perceptions and thought processes.
19. The placebo is administered without the knowledge of either the patient or the doctors and nurses to evaluate the effectiveness of the drug being tested without biasing the results.
20. Cocaine can cause heart disorders, hallucinations, and death.

Critical Thinking

1. Answers may vary, but students should show evidence that their estimates are real.
2. Among the early effects of alcohol are increased feelings of relaxation and lightheadedness. Persons at this stage of intoxication may appear carefree or happy.
3. Methadone probably mimics heroin's interference with normal nerve transmission.
4. Answers will vary, but students might include the fact that aspirin in some cases causes intestinal bleeding and that an overdose of aspirin can be fatal. Students might also say that aspirin is sometimes very helpful in the treatment of diseases such as heart conditions and arthritis.
5. The nicotine and the drugs enter into the fetus's bloodstream through the placenta.
6. Answers may vary, but students should see that the testing is very thorough in the United States. However, it probably makes it difficult for drugs developed in other countries to be used here by people who need them.

Chapter 48

Tobacco, Alcohol, and Other Drugs

Introduction

The human body is the product of many organ systems working together as a functional organism. Its health depends on the performance level of each organ system. Medical science has developed many different kinds of drugs that greatly improve the body's ability to overcome illness and maintain health. However, the use of tobacco and alcohol and the misuse of other drugs can interfere with normal body functions and cause irreversible damage. In this chapter you will learn how drugs interfere with normal body functions.

Chapter Outline

Chapter Concept

As you read, try to identify ways each substance interferes with the body's ability to function properly.

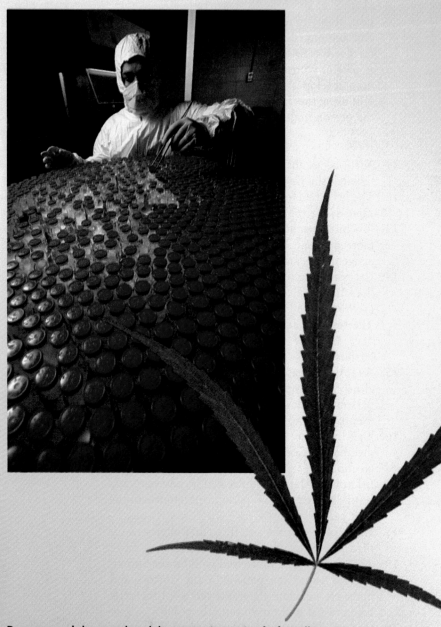

Drug research in a modern laboratory. Inset: Leaf of marijuana, *Cannabis sativa*.

For additional teaching information see pp. T48A–T48D.

48.1 Tobacco

A **drug** is any chemical taken into the body that alters the normal processes of either the mind or the body. Tobacco and alcohol are drugs, as are cocaine, marijuana, and nicotine.

Tobacco comes from *Nicotiana tabacum,* a tall, leafy plant of the Solanaceae, the nightshade family. After being dried and crushed, tobacco leaves are smoked in cigarettes, cigars, and pipes. Tobacco is also inhaled or chewed. Scientific experiments have repeatedly shown that tobacco can have severe effects on the body. Each year an estimated 350,000 Americans die from tobacco-related diseases. These deaths are mostly the result of the more than 2,000 potentially toxic chemical compounds produced when tobacco is burned. The most potent of these compounds are nicotine and tar.

Nicotine, the major drug found in tobacco, is a **stimulant,** a drug that increases the activity of the central nervous system. Nicotine can cause **addiction**—the user can become dependent on the presence of the drug and thus find it difficult to do without. **Tar,** a complex mixture of chemicals and smoke particles, settles into the lungs. The result is a reduction in breathing capacity and increased susceptibility to infections.

Effects of Tobacco

When tobacco smoke is inhaled, nicotine is absorbed into the bloodstream through the lining of the mouth and through the lungs. It is transported to the brain in 7.5 seconds. Effects on the circulatory system include an increase in blood pressure, an increase in heart rate by up to 33 beats a minute, a decrease in oxygen supply to body tissues, and a decrease in circulation to the hands and feet.

Tar and other particles in smoke paralyze the cilia that line the air passages. Normally cilia move particles out of the air passages and protect them from disease-causing microorganisms. Tar irritates the nose, throat, trachea, and bronchial tubes, causing sore throats and coughing. Figure 48-2 on the following page illustrates other effects of smoking.

Tobacco and Disease

Scientific studies have shown that smokers are more susceptible than nonsmokers to many kinds of diseases and may also have a shorter life expectancy. Nearly 90 percent of all lung cancer deaths can be attributed to smoking, and approximately 25 percent of all heart attacks are associated with the use of tobacco.

Many smokers are highly susceptible to **chronic bronchitis,** an inflammation of the bronchi and bronchioles, and **emphysema,** a degenerative lung disease. Emphysema results from the rupture

Tip: Tell students that it is the carbon monoxide in cigarette smoke that interferes with the supply of oxygen to the blood.

Section Objectives

- *Identify two harmful chemicals produced when tobacco is burned.*
- *Describe the physical effects of inhaling tobacco smoke.*
- *List three diseases that can result from smoking or chewing tobacco.*
- *Explain what can happen to the fetus if a pregnant woman smokes.*

Figure 48-1   Cancer can transform a healthy lung (top) into a diseased lung (bottom).

Mouth
- Causes offensive breath odor
- Dulls taste buds
- Causes cancer of the mouth and larynx
- Stains teeth

Air passages
- Paralyzes cilia and contributes to upper respiratory disease
- Causes lung cancer

Circulatory system
- Causes blood vessels to constrict, forcing an increase in heart rate and blood pressure
- Causes vascular disease, affecting circulation to the arms, hands, feet, and legs

Digestive system
- Increases risk of peptic ulcers, decreases effectiveness of treatment, and slows rate of healing

Excretory system
- Increases risk of bladder cancer

Figure 48-2 The health risks brought on by smoking extend to many parts of the body.

**Table 48-1
Tobacco Use Statistics**

- The average smoker will spend $10,000–20,000 on cigarettes during a lifetime
- Average life span decreases by 5½ years if one pack a day is smoked, 7 years if two packs a day are smoked
- 25,000 injuries and $313,000 in property damages per year result from fires caused by smoking

Answers to Section Review
1. a degenerative lung disease that results from ruptured alveoli
2. Nicotine causes an increase in heart rate and blood pressure and a decrease in circulation to the hands and feet.
(See pp. T48A–T48D.)

of the alveoli, the air sacs where gas exchange occurs. Users of smokeless tobacco have a higher rate of lip, gum, and mouth cancer than those who do not use smokeless tobacco.

Nonsmokers who are around smokers also have increased health risks. Tobacco smoke flowing from cigarettes or exhaled by a smoker is called **sidestream smoke.** Laboratory experiments have shown that sidestream smoke puts anyone who inhales it at risk for the same diseases as those who smoke.

The consequences of smoking during pregnancy are profound. Pregnant women who smoke are twice as likely as nonsmoking mothers to suffer miscarriages. Their babies are often much smaller and less robust than other babies and are twice as likely to die in the first few months of life. The long-term risks for children whose mothers smoke during pregnancy include heart disease.

Section Review

1. What is *emphysema?*
2. In what ways does nicotine affect the body?
3. How does tar contribute to a smoker's increased susceptibility to colds and infection?
4. What are some diseases caused by smoking tobacco?
5. Why might it be harmful for a pregnant woman to work in an office that does not have smoking restrictions?

Info: Because emphysema interferes with gas exchange, the smoker cannot eliminate sufficient carbon dioxide or receive enough oxygen.

48.2 Alcohol

Ethanol (C_2H_5OH) is the drug found in beer, wine, liquor, and certain other beverages. Ethanol is produced by **alcoholic fermentation,** the anaerobic action of yeast on the sugars found in fruits or grains. Ethanol is a **depressant,** a drug that decreases the activity of the central nervous system and lowers the level of activity of many body functions. **Alcoholism** is the disease of being addicted to the depressing effects of ethanol.

Effects of Alcohol

Following consumption, alcohol is immediately absorbed from the stomach and intestines into the bloodstream and then transported to the brain and other body organs. When alcohol diffuses into the tissues of the liver, it is oxidized by the cells, resulting in a series of physiological changes. First, oxidation produces heat, which increases blood temperature. This stimulates the brain to increase circulation to the skin, resulting in a decrease in blood flow to the internal organs.

When alcohol enters the control centers of the brain, coordination is impaired, speech may become slurred, and reaction time is greatly slowed. Changes in other body organs include increased perspiration as a response to the rise in blood temperature at the skin surface. Water in the kidneys is reabsorbed to compensate for the loss of water through the skin. As a result nitrogenous wastes collect in the kidneys, interfering with the normal filtration and secretion processes. Drinking alcohol removes water from the body, resulting in general dehydration.

The severity of these effects depends largely on **blood alcohol concentration** (BAC), a measurement of the amount of alcohol in the blood. Table 48-2 describes some of the physical and mental

Section Objectives

- *Define* alcohol.
- *Summarize how alcohol affects the body.*
- *Describe long-term results of excessive alcohol consumption.*

Table 48-2 Blood Alcohol Level and Its Effects

Drinks per Hour	BAC (percent)	Effects
1*	0.02–0.03	Slight feeling of relaxation and warmth
2	0.05–0.06	Increased feeling of relaxation; slight lack of coordination
3	0.08–0.09	Loss of coordination; lightheadedness; slight trouble talking, thinking
4	0.11–0.12	Decreased reaction time; lack of judgment and other mental abilities
5	0.14–0.15	Slurred speech; blurred vision; lack of motor control; impaired judgment, thinking
7	0.20	Mental confusion; loss of motor control
14	0.40	Unconsciousness
17	0.50	Deep coma likely
20	0.60	Breathing ceases; death likely

*1 drink = ¾ ounce (22.1 mL) of alcohol

Info: Although alcoholics can often be cured, physicians do not know exactly why people become alcoholics.

Table 48-3
Alcohol Use Statistics

Alcohol contributes to:
- Over 50% of all suicides
- 70% of all murders
- 50% of all arrests
- Nearly 50% of all fatal crashes involving young people
- Over 50% of all fire-related deaths
- Over 50% of all drownings

750,000 teens are alcoholics.

Figure 48-3 Cirrhosis causes scarring in liver cells.

Answers to Section Review
1. a measure of the amount of alcohol in the blood
2. Heat is produced, increasing the blood temperature and stimulating the brain to increase circulation to the skin, which decreases blood flow to other body organs.
3. slurred speech, blurred vision, and lack of motor control
4. low birthweight, cleft palate, general weakness, heart defects, lack of coordination, and learning difficulties
(See pp. T48A–T48D.)

changes that occur as BAC levels increase. The BAC can be estimated from a breath test, which measures alcoholic vapors given off by the lungs. The amount of alcohol consumed is not the only factor determining BAC. Rate of consumption, ability to metabolize alcohol, body weight, and amount of food in the stomach are other factors that affect BAC.

The effects of alcohol are compounded when it is used with other drugs. Aspirin taken while drinking can cause the lining of the stomach to bleed. Alcohol combined with another depressant can slow body functions to a dangerous level and result in death.

Alcohol and Disease

Long-term use of alcohol damages the body. People who consume alcohol in large amounts and over long periods of time may be more likely than nonusers to suffer from high blood pressure and some forms of heart disease. Excessive alcohol use can irritate the lining of the stomach, a condition called **gastritis,** and eventually can lead to stomach ulcers. In some individuals, alcohol takes the place of nutritious foods, causing metabolic disorders.

Chronic alcohol use forces the liver to use alcohol as an energy source instead of relying on fatty acids. Eventually liver cells are unable to function properly and they accumulate fat deposits. As a result the liver begins to swell, causing **fatty liver,** a condition that occurs in 75 percent of all alcoholics. If the drinker abstains from alcohol, the liver can usually return to normal functioning. If drinking continues, the drinker may develop alcoholic hepatitis, an inflammation of the liver, or even **cirrhosis** *(suh-ROE-suss),* a condition in which normal liver tissues are replaced by scar tissue, as shown in Figure 48-3. Cirrhosis results in severe impairment of liver function and can be fatal.

Children born to women who drink during pregnancy may suffer from **fetal alcohol syndrome** (FAS), a condition that can cause physical and mental disabilities in newborn babies. Immediate effects of FAS are low birthweight, below-average length, cleft palate, general weakness, and heart defects. The long-term effects include lack of coordination, slow growth, and learning difficulties due to mental retardation, weak attention span, and hyperactivity.

Section Review

1. What is *blood alcohol concentration?*
2. What happens when alcohol is oxidized by the liver cells?
3. What are some typical behaviors of a person who has a BAC of 0.15 percent?
4. What are some of the effects of fetal alcohol syndrome?
5. Would a person be able to safely drive a car after two drinks in one hour? Explain your answer.

Tip: Tell students that over 200,000 people die of alcohol-related diseases each year.

48.3 Other Drugs

In Sections 48.1 and 48.2 you learned that nicotine and alcohol act on the central nervous system. Nicotine is a stimulant, and alcohol is a depressant. There are two other major types of drugs. **Psychoactive drugs** are chemical compounds that change sensory perception and thought processes. **Narcotics** are those drugs made from opium that suppress the cerebral cortex. Because of their highly addictive nature, narcotics can be obtained legally only with a prescription from a doctor. Table 48-4 describes the effects of various drugs.

Medical Uses of Drugs

Some drugs are administered for medical purposes. These drugs are used to improve health. **Prescription drugs** are those drugs that can be obtained only on written orders from a doctor. The doctor specifies, or prescribes, the dosage and the length of time the medication must be taken and then monitors the effects of the drug on the patient.

Section Objectives

- *Distinguish among the four major types of drugs.*

- *List one example of each type of drug, and describe how it affects the body.*

- *Explain the difference between physical and psychological addiction.*

- *Explain a possible source of error in a study using the single-blind method.*

Table 48-4   Short-Term and Long-Term Effects of Drugs

Drug	Dependence Potential	Short-term Effects	Long-term Effects
Stimulants			
Caffeine	Probable	Increased heart rate, increased blood pressure	Irregular heartbeat, high blood pressure, headaches, stomach disorders, nervousness; for cocaine, paranoia and death
Amphetamines	High		
Cocaine	High		
Depressants			
Barbiturates	High	Drowsiness, decreased coordination	Depression, emotional instability, hallucinations, death
Tranquilizers	High		
Methaqualone	High		
Phencyclidine hydrochloride (PCP)	Probable	Illusions, poor time–space perception, violent behavior	
Psychoactive drugs			
Lysergic acid diethylamide (LSD)	Probable	Hallucinations	Psychosis
Cannabis sativa (marijuana)	Probable	Short-term memory loss, disorientation	Lung damage, loss of motivation
Narcotics			
Heroin	High	Drowsiness, respiratory depression, nausea, constricted pupils	Convulsions, coma, death
Codeine	High		
Morphine	High		

One method of administering drugs is transdermally, by a patch placed on the skin that releases a continual supply of the medication. For example, antinausea medication, often used by persons susceptible to motion sickness, can be administered in this way. What do you think are the advantages of such a method?

Some drugs can be purchased without a prescription. These are called **over-the-counter drugs.** The labels on these drugs contain information regarding correct usage. Over-the-counter drugs are safe when taken in the recommended dosage but can be dangerous if too much is taken. As an example, more than 100 children in the United States die each year from accidental overdoses of aspirin.

Both prescription and over-the-counter drugs can be administered in a variety of ways. The most common way is orally, in pill or liquid form. Oral drugs are absorbed by the small intestine and then enter the bloodstream. Drugs can also be injected directly into the bloodstream or into muscle tissue. Other methods of administration effective in certain circumstances include placing the drug under the tongue, inhaling the drug, or placing the drug into the rectum in the form of a suppository.

Drug Addiction

When taken correctly, both prescription and over-the-counter drugs can be beneficial. However, if a drug is taken for longer than the recommended time, a patient may develop an addiction

Biotechnology

The Double-Blind Test

Before a medical drug appears on store shelves or is prescribed by a doctor and dispensed by a pharmacist, it has to be tested. The drug testing process is important for a variety of reasons, including safety, effectiveness, and special individual considerations. Just as each of us differs in appearance so do we differ in our responses to specific drugs. The range of effects of a given drug on a variety of types of recipients must be thoroughly understood.

Researchers must test a new drug extensively to make sure it will be safe and effec-

tive. The procedures for testing drugs must be followed before the U.S. Food and Drug Administration (FDA) will approve the drug for sale.

To test drugs' effectiveness scientists first developed the placebo method. They would divide test subjects into two groups. One group, the treatment group, would receive the experimental drug. The second group, the control group, would, without knowing it, be given a pill containing sugar or another ingredient considered ineffective. This ineffective pill is called a placebo.

When testing was finished, researchers would compare the

to it. An emotional dependence on a drug is called a **psychological addiction.** Changes in body chemistry that result in a demand for a steady supply of a drug create a **physical addiction.** These two types of addictions often occur together.

When physical addiction causes the body to become less responsive to the drug, the user is said to have developed a **tolerance.** The user's body needs increasingly large amounts of the drug to maintain the feelings previously achieved with a smaller dose. If the drug supply is cut off, the addict will go through **withdrawal,** the physical and mental reactions to the lack of the drug. The symptoms of withdrawal vary, depending on the drug being used and duration of use. Symptoms may include nausea, headache, insomnia, breathing difficulties, depression, mental instability, and seizure. Addicts undergoing withdrawal are often hospitalized so that their responses can be monitored.

Illegal Drugs

Some drugs listed in Table 48-4 are illegal. Any use of these illegal drugs is considered drug abuse. Marijuana, which comes from the flowers, leaves, and seeds of *Cannabis sativa*—the Indian hemp

Answer to Think
Transdermal administration allows the drug to be absorbed into the system more quickly than oral administration does and is less painful than injection. Also, since motion sickness often results in vomiting, antinausea medication administered orally may never have a chance to take effect.

Single-blind test results

Double-blind test results

results of the two groups. A few members of the group receiving the placebo got better simply because they believed they had been given the real drug. However, if there were many more recoveries in the treatment group than in the group that had received the placebo, the new drug would be considered effective.

The trouble with this method of testing—called the single-blind method—is that control group patients were often able to guess from the behavior of the person administering the test whether they were receiving the placebo. As shown above, this artificially lowered the number of recoveries in the control group relative to the treatment group and interfered with the results.

To avoid this problem, today's researchers use a double-blind test. In the double-blind testing technique, neither the patient nor the doctor or nurse, knows which pill—the real drug or the placebo—is being administered. Both preparations are identified by code numbers and these can be interpreted only by the people who make up the drugs. At the end of the test, the secret code is revealed and the patients of each group are identified and their responses to the drug are compared. In this way, there is much less chance that the results of the drug test will be altered by the knowing look of a doctor or nurse who's aware of the secret.

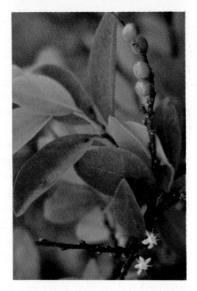

Figure 48-4   Three plants that are the source of dangerous drugs are (left to right) the Indian hemp plant, *Cannabis sativa*, the opium poppy, *Papaver somniferum*, and the coca plant, *Erythroxylon coca*.

Tip: Tell students that marijuana is unpredictable. It can produce a sense of well-being but it can also intensify depression.

Answer to Section Review
1. drugs that can be bought only on a doctor's orders
2. The person can become addicted to the drug.
3. Drugs differ in the way they affect a person physically and mentally and in their potential for dependence.
4. Answers may vary but might include oral administration, injection, inhalation, or suppository insertion.
5. A person might use one drug to counteract or to enhance the effects of another drug.

plant—is a psychoactive drug. The active ingredient in marijuana is Δ9-tetrahydrocannabinol, or THC. When smoked, or when heated and eaten, marijuana acts on the nervous system, producing feelings of disorientation in time and space. Continuous use of marijuana can lead to psychological addiction and to a suppressed activity level, often referred to as **amotivational syndrome.**

Opium comes from the fruit of the poppy, *Papaver somniferum*. Morphine and codeine, two types of prescription painkillers, are both made from opium. Heroin, considered one of the most highly addictive drugs, is a compound made from morphine.

Cocaine is a white powder extracted from the leaves of the coca plant, *Erythroxylon coca*. It is a powerful stimulant that raises heart rate, blood pressure, and body temperature. Cocaine acts by changing the chemistry of nerve synapses in the brain. It is extremely addictive, both psychologically and physically. Prolonged use can cause hallucinations, psychosis, and death. Inhaling cocaine also causes damage to the nasal septum. Recently new and highly addictive forms of cocaine such as crack have emerged.

One of the most dangerous depressants is phencyclidine hydrochloride, or PCP. PCP is sometimes added to marijuana, often without the marijuana user being aware of its presence. Use of PCP can cause paranoid and violent behavior. PCP has been banned for human use.

Alcohol, tobacco, and other drugs have great potential for abuse. While drugs are often necessary for proper medical care, their abuse can cause serious physical and mental problems.

Section Review

1. What are *prescription drugs?*
2. What can happen if a person takes a prescription drug for longer than the recommended time?
3. What are some differences among various types of drugs?
4. Name three methods of administering drugs.
5. Why might a person use more than one drug at a time?

Info: Many derivatives of opium are used as pain relievers. However, heroin, the strongest opiate of all, is not used in medicine.

Laboratory

How Drugs Affect Living Organisms

Objective
To observe the influence of drugs on living organisms

Process Skills
Experimenting, communicating

Materials
Elodea; forceps; 11 microscope slides; pond water; pipettes; coverslips; microscope; ethanol in 1%, 2%, 5%, and 20% solutions; paper towels; *Vorticella;* caffeine solution; nicotine solution

Method
1 Find the tiniest leaves at the tip of an *Elodea* shoot. Using forceps, place one leaf on a slide. With a pipette, add a drop of pond water. Cover the leaf with a coverslip.
2 Observe the *Elodea* cells under high power of a microscope. A microphoto of *Elodea* leaf cells is shown at the right. Look for evidence of cytoplasmic streaming. What is the function of this slide?
3 Repeat step 1 with a new slide. Use a pipette to add two or three drops of 1% ethanol at the edge of the coverslip. Gently blot up water from the other side of the coverslip using a piece of paper towel. This creates a wick effect that draws the ethanol under the coverslip. Do you still see evidence of cytoplasmic streaming? Record your observations.
4 Using a fresh *Elodea* leaf and a clean slide each time, repeat step 3 using the 2%, 5%, and 20% ethanol solutions in turn. What should you look for? Continue to record your observations.
5 With a clean pipette, transfer some of the culture of *Vorticella* to a clean slide.
6 Add a drop of pond water, and cover the preparation with a coverslip. Now use low power of the microscope to observe the organism's behavior. Note the movement of the cilia. Can you see evidence of cytoplasmic streaming? Record your observations.
7 Using a new *Vorticella* and a clean slide each time, repeat step 6 three times. Continue to note and record the organisms' reactions as you perform each step.
8 Using a clean slide and new *Vorticella,* repeat step 6 using first the caffeine and then the nicotine solution.
9 Share your results with your classmates. Why could this be important?

Conclusions
1 Did the entire class observe the same response of *Elodea* cytoplasmic streaming to alcohol? Why might different class members have observed different responses?
2 What was the first response of *Vorticella* to the caffeine and nicotine solutions?
3 What was the final response of *Vorticella* to the caffeine and nicotine solutions?

Inquiry
What are some possible sources of error in this series of experiments?

Chapter 48 Review

Vocabulary

addiction (749)
alcoholism (751)
alcoholic
   fermentation (751)
amotivational
   syndrome (756)
blood alcohol
   concentration (751)
chronic bronchitis
   (749)

cirrhosis (752)
depressant (751)
drug (749)
ethanol (751)
emphysema (749)
fatty liver (752)
fetal alcohol
   syndrome (752)
gastritis (752)
narcotic (753)

nicotine (749)
over-the-counter drug
   (754)
physical addiction
   (755)
prescription drug
   (753)
psychoactive drug
   (753)

psychological
   addiction (755)
sidestream smoke
   (750)
stimulant (749)
tar (749)
tolerance (755)
withdrawal (755)

1. How are nicotine and tar related?
2. Identify two respiratory diseases caused by smoking.
3. List the four major types of drugs.
4. What is *blood alcohol concentration,* and what are some factors affecting it?
5. Describe the relationship between addiction, tolerance, and withdrawal.

Review

1. Nicotine is a (a) psychoactive drug (b) stimulant (c) narcotic (d) depressant.
2. Tar (a) causes an increase in heart rate (b) is a stimulant (c) paralyzes cilia (d) is transported by the brain.
3. Emphysema is (a) a degenerative lung disease (b) a form of cancer (c) an inflammation of the kidneys (d) a result of chewing tobacco.
4. Alcohol is a (a) stimulant (b) depressant (c) narcotic (d) psychoactive drug.
5. BAC specifically measures (a) the rate at which a person has been drinking (b) a person's genetic ability to metabolize alcohol (c) the amount of alcohol in a person's blood (d) the time at which a person was drinking.
6. Fetal alcohol syndrome can cause (a) low birthweight (b) heart defects (c) learning difficulties (d) all of the above.
7. Aspirin is one kind of (a) over-the-counter drug (b) illegal drug (c) depressant (d) psychoactive drug.
8. Psychological addiction (a) occurs when the body becomes dependent on a substance (b) is an emotional dependence on

a drug (c) occurs after tolerance to a drug has developed (d) is caused only by illegal drugs.
9. Depressants include (a) cocaine (b) tar (c) nicotine (d) alcohol.
10. Narcotics cause (a) increased heart rate (b) increased alertness (c) decreased alertness (d) lung damage.
11. How is nicotine absorbed into the bloodstream?
12. Describe the harmful effects of sidestream smoke.
13. What are some of the harmful effects of smoking during pregnancy?
14. Describe the specific effects of alcohol on the cells of the liver.
15. In what ways are mental capabilities diminished as BAC increases?
16. What are some potential effects of long-term, excessive alcohol use?
17. What is the relationship between fatty liver and cirrhosis?
18. What are psychoactive drugs?
19. Explain the use of a placebo in a double-blind study.
20. Describe the harmful effects of cocaine.

Critical Thinking

1. Caffeine is a stimulant commonly found in coffee, tea, cola drinks, and cocoa. Authorities say that people who consume 500 mg or more of caffeine per day may suffer harmful effects. Look at the chart on the right, which shows the milligrams of caffeine contained in given amounts of various beverages and chocolates. Estimate your daily caffeine intake, listing sources and the times of day you ingest it. Is your caffeine intake too high?

2. Alcohol is known to be a depressant. How do you explain the fact that consumption of alcohol sometimes results in a carefree attitude and a feeling of elation?

3. Methadone, a synthetic narcotic drug, is often used in the treatment of heroin addiction. Methadone blocks the body's craving for heroin. Based on your knowledge of the nervous system, suggest a mechanism by which this is possible.

4. Aspirin is one of the most commonly used drugs. Look into some of the positive and negative effects of aspirin and write a short paragraph about your findings.

5. If a pregnant woman smokes cigarettes or takes drugs, the fetus can be affected. Explain how nicotine and drugs get into the bloodstream of the unborn child.

6. In Europe in the 1960s many pregnant women took the drug thalidomide. Although tests were conducted using labora-

Amounts of Caffeine in Some Beverages and Foods (in milligrams)

Coffee (5 oz.)	
Brewed, drip method	60–180
Brewed, percolator	40–170
Instant	30–120
Decaffeinated, brewed	2–5
Decaffeinated, instant	1–5
Tea (5 oz.)	
Brewed, major U.S. brands	20–90
Brewed, imported brands	25–110
Iced (12 oz.)	67–76
Instant	25–50
Chocolate and cocoa	
Cocoa (5 oz.)	2–20
Chocolate milk (8 oz.)	2–7
Milk chocolate (1 oz.)	1–15
Semisweet chocolate (1 oz.)	5–35
Chocolate flavored syrup (1 oz.)	4
Soft drinks (12 oz.)	
Cola drinks and cola–fruit drinks	30–46
Other soft drinks	52–58

tory animals, many children were born with severe deformities to women who had taken thalidomide. In the United States this drug was not available, because it had not undergone the lengthy testing required for approval. Discuss the advantages and disadvantages of requiring a long testing period.

Extension

1. Read the article by Denise Grady entitled "Look Doctor, I'm Dying. Give Me the Drug," in *Discover*, August 1986, pp. 78–86. This article discusses some of the problems involved in testing drugs. What specific problem is cited in the article?

2. In the local library or a doctor's office ask to see the *Physicians' Desk Reference*. This book shows all the prescription drugs made by drug companies in the United States. Make a chart of ten different drugs. Headings should include name of the drug, generic name, drug manufacturer, uses, dosage, and side effects.

3. Both private groups and local and federal government agencies can give you information about preventing addiction. Send for information about the Say No to Drugs campaign and about Students Against Drunk Driving (SADD).

Intra-Science: *How the Sciences Work Together*

The Process in Biology: *The Function of the Heart*

In this unit you studied the circulatory system. As you know the principle function of the circulatory system is to maintain homeostasis by providing oxygen and nutrients to all parts of the body and to transport metabolic wastes. Many invertebrates have open circulatory systems: blood does not remain in veins at all times. Instead blood filters through tissues, and body movements keep the blood circulating. Vertebrates have closed circulatory systems in which blood flows through a system of vessels. In vertebrates, the process of circulation is controlled by the heart.

The structure of the heart varies among vertebrates. Fishes have a heart with only two chambers. Amphibians and most reptiles have a heart with three chambers. Crocodilians, birds, and mammals have four-chambered hearts. More advanced vertebrates have double circulation in which blood is pumped from the heart to the lungs and back to the heart before leaving the heart for the body.

The heart of the blue whale, the largest creature ever to live on earth, is about the size of a small car. It pumps about 2,000 gallons of blood through arteries large enough for a child to swim through.

In some organisms, the amount of blood purnped to various parts of the body changes as environmental factors change. For example whales and other marine mammals

Mammal

Fish   **Reptile**   **Bird**

have circulatory systems adapted to pressure changes that happen during long dives. In these marine mammals, as well as warm-blooded fishes, a modified network of blood vessels surrounds the muscle tissue. This network sends extra oxygen to the swimming muscles.

The Connection to *Physics*

The sinoatrial (s-a) node and the atrioventricular (a-v) node stimulate cardiac muscle cells by emitting electrical charges. The muscles respond in a wave of sequential contractions. The science of physics allows researchers to observe and describe these waves.

Imagine a wavy line running across this page. This line can represent a periodic wave. A periodic wave, such as that generated by the nodes in the heart, consists of a series of pulses and has two main features, amplitude and frequency. The amplitude of a wave has to do with its volume or capacity. The distance your wavy line travels above or below a straight center line is its amplitude. The amplitude of the wave of contractions in the heart reflects

the amount of blood passing through the atria and ventricles of the heart.

In physics, an increase in amplitude accompanies an increase in the energy being transferred by a wave. The greater the amplitude of the wave, the more blood is pumped with each contraction. Thus the harder the heart works, the more blood it pumps.

A wave also has a frequency, which is related to time. In this case, the frequency of the periodic wave is represented by the number of waves running across the page or the number of pulses per minute. The pulse you can feel in your wrist or neck results from a surge of blood due to the contraction of the heart's ventricles.

The Connection to *Chemistry*

The most important chemical process in the blood is the binding of oxygen and carbon dioxide to hemoglobin, which transports these molecules throughout the body. The hemoglobin molecule is composed of four polypeptide chains. Each chain contains a pigment that has an iron molecule in the center. Each of the iron molecules can combine with one molecule of oxygen, so the hemoglobin molecule can carry four oxygen molecules. Chemists have noted that after the first oxygen molecule attaches, the hemoglobin molecule changes shape in a way that facilitates the attachment of the other oxygen molecules.

In addition to having hemoglobin in their red blood corpuscles, fish and marine mammals have a pigment called myoglobin in their muscles that binds to oxygen. Myoglobin has a greater tendency than hemoglobin to combine with oxygen, and its concentration in swimming muscles permits deep dives and the perpetual swimming behavior of some fish. Further studies of myoglobin chemistry can lead to a greater understanding of the structure of globular proteins such as hemoglobin.

The Connection to *Other Sciences*

The circulatory system is of great significance to the medical sciences. Eating fatty or starchy foods, smoking, drinking excessive amounts of alcohol, taking harmful drugs, and getting too little exercise can all contribute to heart disease. Lipid deposits can clog arteries, which reduces the diameter of the vessels and thereby increases blood pressure.

Physicians are trained to hear what may seem like subtle differences in heartbeats. An irregular heartbeat, or arrhythmia, may be caused by damage to the heart muscle or to disorders of the electrical charging system that controls pumping in heart chambers. Specific arrhythmias can be identified in electrocardiograms because they create particular changes in the periodic wave of contractions in the heart.

The circulatory system interacts with other systems in the body in maintaining homeostasis. Scientists studying respiration, nutrition, ecology, evolutionary biology, and the nervous system must all be aware of the mechanisms by which oxygen is delivered to tissues and carbon dioxide carried away. A reduction in functioning in the circulatory system has immediate effects on other systems.

The Connection to *Careers*

An understanding of the circulatory system is critical for all health professionals. Doctors and nurses test blood pressure and pulse, listen to heartbeats, and measure blood components in efforts to determine a patient's condition. Dieticians and rehabilitation therapists use knowledge of the circulatory system to design nutrition and exercise programs for people who suffer from heart disease or are at a high risk for developing such disease.

Heart specialists, or cardiologists, are experts on the circulatory system. They may work in rehabilitation, research or surgery. In recent years, some have helped engineer the artificial heart that has been successfully implanted in several individuals.

Athletes, trainers, and physical education teachers must have a knowledge of the workings of the circulatory system in order to exercise and coach others safely.

Designers of equipment and clothing must be aware of circulatory needs so that their products do not impede blood flow. For more information on related careers in the biological sciences, turn to page 842.

Unit 10 | Ecology

Chapters

"No part of life can be considered apart from any other; there is no such thing as an organism without an environment."

Lewis Mumford

You've probably never thought of a layer of moisture on moss as an environment, but it is in fact an environment in which tiny invertebrates called tardigrades make their home. The relationships between organisms and their environments make up the subject matter of ecology. Ecologists may choose to study tiny microenvironments like the surface of a moss leaf or larger environments like the volcanic craters where the silversword plant grows. The largest environment of all is the biosphere, the thin layer of the earth's surface where all organisms live.

Tardigrade, *Echiniscus spiniger*

Dewdrops on moss, *Polytrichum hirsutum*

762

Silversword plant, *Argyroxiphium sandwicense,* in Hawaii

The biosphere from space

Chapter 49 | *The Biosphere and Biomes*

This chapter discusses the diversity and distribution of life forms and examines their relationships with one another and with the environment. Students learn about major terrestrial and aquatic biomes and how life has adapted to them.

Correlation Chart

Section Number*	Section Objectives	Pages	Features	Section Review
49.1	1. Define the term *biosphere*. 2. Distinguish between populations, communities, and ecosystems. 3. Name the types of abiotic factors in an ecosystem. 4. Name three types of interactions among biotic factors in an environment. 5. List some ways biotic and abiotic factors in an environment interact.	765–766		1 2 2 3, 4 4, 5
49.2	1. Describe the major terrestrial biomes. 2. Give examples of the adaptation of plants in a biome. 3. Give examples of the adaptation of animals in a biome. 4. Describe ways in which ecologists use the LANDSAT satellite.	767–773	Think, p. 768 Biotechnology, pp. 770–771	1 2, 3 4, 5
49.3	1. Name three areas of the marine biome, and list their characteristics. 2. Give examples of adaptations of organisms to the marine biome. 3. Name two freshwater biomes, and describe how they differ.	774–777	Biology Lab, p. 777	1, 2 3 4, 5

Resources*

Print

Barnes, R.S.D., and R.N. Hughes. *Introduction to Marine Exology*. Oxford, England: Blackwell Scientific, 1982.

Sibbold, Jean H. *Homes in the Sea; From the Shore to the Deep*. Minneapolis: Dillon, 1986.

Audiovisual/Software

Succession: From Sand Dune to Forest. Film, 16 min. Britannica.

The Fire's Inch. Film or video, 28 min. Filmakers, 1983.

*Additional references are listed in *Modern Biology, Teacher's Program Guide*.

Section Teaching Guides

49.1 Ecological Units

Approach

The main objective of this section is to define the ecological units that make up the biosphere and to describe the ways abiotic and biotic factors within the biosphere interact. This section provides the background for Chapters 50–53, in which students will learn about relationships among living and nonliving things in ecosystems, communities, and populations. In presenting the text, reinforce the idea that all organisms interact with and adapt to their environment.

Concept Framework

Students are familiar with the fact that complex organisms are composed of many individual cells that function as a unit. This knowledge can be used as a point of comparison in discussing the organization of the biosphere. Aid students in identifying similarities between the organization of cells, the smallest units, into tissues and organs, and the organization of interacting species into communities and ecosystems.

Emphasis

Make certain that students understand the significance of both abiotic and biotic factors in the biosphere. They should be able to name several abiotic and biotic factors and give examples of how each type of factor affects the other.

Section Review Answers

(See p. 766.)

49.2 Terrestrial Biomes

Approach

This section describes the major terrestrial biomes, focusing on the adaptations of plants and animals to their particular environment. The section builds on the concepts of adaptation in Chapter 16 and of the conditions necessary for life on earth in Chapter 14. In addition, because the animal life in a biome depends largely on the available plant life, you might wish to review the biological importance of plants, as discussed in Chapter 24.

Concept Framework

Students will be familiar with the climate and geography of their own area. Encourage them to think about how the predominant plant and animal life in their environment shows adaptations to these conditions. Ask them to consider as they read the text how an extreme change in some abiotic factor, such as temperature or water supply, would affect plants and animals.

Emphasis

Students should be able to describe the climatic and geographic characteristics of each major biome and identify the ways in which plants and animals are adapted to these conditions.

Using the Features/Remote Sensing Satellites

The main technology described in the feature on remote sensing is the use of LANDSAT satellite images to observe ecological characteristics of large areas of the earth. The feature describes the use of these images for such purpose as measuring the amount of vegetation in the Arabian desert or locating suitable mountain habitats for grizzly bears. The feature also explains how scientists can compare images of the same area at different times to spot changes in an environment. Students can see another application of this technology in Figure 53-4, where satellite images illustrate the diminishing ozone layer over the southern hemisphere. Ask students to think of other possible applications of remote sensing technology.

Research Activity

Divide the class into seven groups and assign one of the major terrestrial biomes to each group. Have students collect data on climate and geography, along with pictures that show the representative adaptations of plants and animals to each biome. Arrange the information in a classroom display.

Section Review Answers

(Continued from p. 773.)

2. The Arctic is warmer because the ice, which is heated by underlying water, is thinner and melts in places in the summer.
3. Dominant coniferous trees include spruces, pines, firs, and cedars. Dominant deciduous trees in-

clude birches, beeches, maples, oaks, hickories, sycamores, elms, willows, and cottonwoods.

4. Answers will vary but can include the following: desert—kangaroo rat, camel, squirrel, snake, rabbit; grassland—grasshopper, antelope, elephant, giraffe, cheetah, lion, kangaroo; tundra—mosquito, duck, goose, caribou, owl.

5. More food sources, such as leaves and insects, are available in sunlit areas of the canopy.

49.3 Aquatic Biomes

Approach

The main objective of this section is to describe how organisms are adapted to marine and freshwater biomes. Since the chemical balance of the water and availability of light are key abiotic factors in aquatic biomes, it may be useful to review the concepts of homeostasis in Chapter 6 and photosynthesis in Chapter 7. In presenting the text, recall earlier discussions of the relationship between structure and function in aquatic organisms such as mollusks (Chapter 31), echinoderms (Chapter 34), and fishes (Chapter 36), and consider these as adaptations to the environment.

Concept Framework

Students know that oceans are large, salty bodies of water and that freshwater biomes generally consist of smaller bodies of water that contain little salt. If students live near a body of water, they are also familiar with the plants and animals that live there. Encourage students to apply this knowledge by considering how such abiotic factors determine the types of organisms that can survive in marine and freshwater environments.

Emphasis

Make certain that students know how abiotic factors differ in marine and freshwater biomes. They should also be able to describe the adaptations of aquatic organisms to each type of biome.

Guest Speaker

Invite an oceanographer from a local university to talk about the different life forms that exist in different oceanic zones, including coastal areas. The discussion should focus on adaptations to various environmental conditions.

Section Review Answers

(See p. 776.)

Laboratory Preparation

Background

In this lab students relate the activity of *Elodea* to changes in its environment and observe how such environmental changes may affect other members of the community. Students use a simple chemical test to obtain a quantitative indication of the oxygen content of the environment. The students use this data to draw conclusions of the effect of *Elodea* on the environment, and they speculate on how the organisms can produce effects which are both beneficial and harmful to other community members.

Preparation

24 hours before this procedure, prepare two bottles of *Elodea* for each team of students. For each team place 2 sprigs of *Elodea* in 100 mL water in a 180 mL wide-mouth bottle, seal the bottle, label it A, and place under a 200 to 300 watt bulb. Repeat the process, label the second bottle B, and place it in total darkness. To prepare 25% manganous sulfate solution, add 350 g manganous sulfate to 1050 mL distilled water. To prepare 40% potassium hydroxide-potassium iodide solution, add 500 g potassium hydroxide to 1 L distilled water, then add 150 g potassium iodide. To prepare .003% sodium thiosulfate solution, add 3.1 g sodium thiosulfate to 1 L distilled water.

Safety

Sulfuric acid is a strong corrosive. If spilled on clothing, remove clothing at once. If spilled on skin, immediately flush the area with water. Avoid breathing the fumes.

Conclusions

1. The sample placed under bright light contains more oxygen than the sample placed in the dark. The sample placed under bright light affects its environment by providing more oxygen to support other organisms than does the other sample.
2. The sample placed under bright light added oxygen to the water through photosynthesis.

3. Each bottle acts as a control for the other, allowing comparisons to be made between the light and dark environments.

Inquiry

Students should suggest a water pollutant, such as sulfuric acid and hypothesize about its effect on photosynthesis. The hypothesis can be tested by adding sulfuric acid to the water in bottles A and B, then repeating the experiment using only bottle A. The new results can be compared to the original results.

Chapter Review

Vocabulary

1. A population is all the members of a species living in the same area. A community made up of all the populations in one area.
2. Abiotic factors are the nonliving parts of an environment such as sunlight and rainfall.
3. Symbiosis means the interdependence of one species upon another; predation involves one species eating another.
4. The pelagic zone is the open ocean, and the benthic zone is the ocean bottom.
5. *Eu-* means "good," *oligo-* means "few," and *-troph* means "nourishment." Thus eutrophic lakes are rich in nutrients; oligotrophic lakes have few nutrients.

Review

11. Answers will vary but may include the following: an island or a mountain that is created by a volcano, or river that changes course.
12. Antarctic ice is over land; Arctic ice is over water. The latent heat of water melts the ice from below, making the Arctic ice thinner.
13. Grasses are highly efficient at converting solar energy to food energy.
14. They can quickly utilize moisture, which is available only for a short time.
15. Temperatures in the temperate rain forest are more moderate than those in the tropical rain forest. Most trees of the temperate rain forest are conifers, and most animals live on the ground.
16. Some adaptations of animals to tree-dwelling life are the ability to grasp vines and branches with hands or tail, stereoscopic vision, and the ability to fly or glide.
17. The three areas of the pelagic zone are the neritic zone, oceanic zone, and abyssal zone.
18. Plankton are the organisms that drift freely in open waters.
19. Sulfur is the source of energy in thermal vent communities.
20. Answers will vary but can include carp, catfish, largemouth bass, and pike.

Critical Thinking

1. Tropical rain forest biome, since the amount of rainfall is high and the temperature is warm.
2. Many of the animals died off as their food supply dwindled. Those that survived would have experienced a decreased reproduction rate until a new balance was reached between plant and animal populations.
3. Both conifers and succulents have reduced leaves. Conifers have needles with sunken stomata and with a waxy coating that prevents water loss. Succulents have spines rather than true leaves, and photosynthesis takes place in the large stem. The stem of succulents, like the needles of conifers, have a waxy protective covering.
4. Abiotic factors include the heat generated from the earth's crust, the lack of light, and the abundant supply of minerals. These allow bacteria to thrive without being able to photosynthesize. A variety of marine animals can then live off of these bacteria.
5. Animals that graze on grassland would eat this type of plant.
6. Nutrient levels are much higher in estuaries than in the open ocean. On the other hand, estuaries are more unstable environments and many predators inhabit these areas.

Chapter 49 | The Biosphere and Biomes

Introduction

From the highest mountain peak to the deepest ocean trench, nearly every part of the earth supports some form of life. Much of the diversity of living things reflects adaptations to the earth's varying climate and geography. In this chapter you will learn about the major distribution patterns of plant and animal life found throughout the biosphere—that area of the earth where life exists.

Chapter Outline

Chapter Concept

As you read, note how plants and animals are adapted to the various climatic and geographic conditions found on the earth.

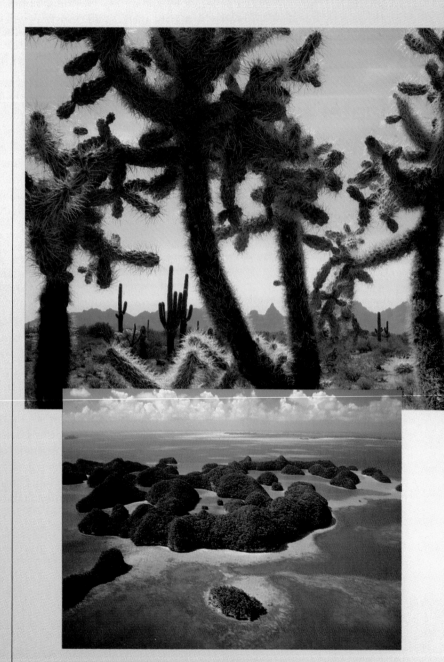

Organ pipe cactus, *Lamaireocereus* sp., Arizona. Inset: The marine biome.

For additional teaching information see pp. T49A–T49D.

49.1 Ecological Units

Section Objectives

- *Define the term* biosphere.

- *Distinguish between populations, communities, and ecosystems.*

- *Name the types of abiotic factors in an ecosystem.*

- *Name three types of interactions among biotic factors in an environment.*

- *List some ways biotic and abiotic factors in an environment interact.*

No organism—no rhinoceros, duck, oak tree, or bacterium—exists alone. Each is part of an intricately linked system of living and nonliving elements. The study of the relationship between organisms and their environment is called **ecology.** Ecologists study not only how organisms act together but also how they are adapted to their environments. Adaptation, which results from evolution by natural selection, is a factor that unifies the structure, physiology, and behavior of all organisms throughout the biosphere.

The Biosphere

The **biosphere** is that area of the earth where life exists. Insects and spores of microorganisms and plants have been found 8 km high in the atmosphere. Other organisms such as sea urchins and brittle stars live 8 km below the surface of the ocean. Almost all life, however, exists on or within a few meters of the earth's surface. The biosphere is so thin that if the earth were the size of a basketball, the biosphere would be much thinner than this page.

Ecologists learn about the biosphere by studying smaller and simpler ecological units within it. These simpler levels serve as models for the larger, more complex biosphere. In Sections 49.2 and 49.3 you will learn about large ecological units of the biosphere called biomes. A smaller ecological unit is a **population,** all the members of a species that live in the same area and make up a breeding group. For example, the moose on Isle Royale, a national park in Lake Superior, constitute a population. They interact with each other and with other species on the island, including the timber wolves that kill and eat moose. The moose population, the wolf population, and all the other living things on the island form a community. A **community** includes all the populations in an area.

Ecosystems

Ecologists also study ecosystems. An **ecosystem** is an ecological unit that includes all the interacting parts of an environment in an area. A lake, a prairie, and a cave are all examples of ecosystems. The nonliving components of an ecosystem are called **abiotic** *(ay-bie-AHT-ick)* **factors,** and the living components are called **biotic** *(bie-AHT-ick)* **factors.**

Abiotic Factors

Abiotic factors in an ecosystem include sunlight, precipitation, temperature, the slope and drainage of the land, and the chemistry of soils and of the atmosphere. These physical factors interact with one another. For example, a season of heavy rains can cause

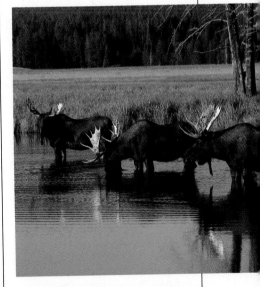

Figure 49-1   Moose are just one of the many populations that interact to make up the Isle Royale community.

a river to flood its banks, depositing sediments on nearby land. When the flood recedes, the sediments will have changed the soil composition. Other interactions of abiotic factors can produce more dramatic results. A river can completely change course, or, as Figure 49-2 illustrates, volcanic activity can suddenly create a mountain or an island.

Biotic Factors

Biotic factors in an ecosystem include plants, animals, protists, and all other living organisms. Like abiotic factors, biotic factors also interact in many ways. All organisms compete for food, water, space, and other resources. One type of competition occurs between members of the same species, and another type of competition occurs between different species. For example, two different species in the same community may eat the same plants.

Two other forms of interaction of biotic factors are predation and symbiosis. In **predation** one animal, the predator, kills and eats another animal, the prey. In the Isle Royale community wolves are the predators and moose are the prey. **Symbiosis** *(SIM-BIE-OE-suss)* is a close relationship between two organisms of different species. Usually at least one of the species benefits from the relationship. Chapter 51 will provide a more detailed discussion of predation and of the various types of symbiosis.

Abiotic and Biotic Interactions

Abiotic and biotic factors in ecosystems also interact with each other. For example, climate and soil conditions determine which plants will live in a certain area. Plant life in turn affects the distribution of animals. For example, the grass that grows in Siberia must withstand that region's cold, dry climate. The teeth and digestive system of the saiga antelope are adapted for eating this grass. In contrast, deer that reside in the wetter forests of the eastern United States are adapted for browsing on the leaves of the trees that are found there.

Living organisms also affect abiotic factors. For example, plants replenish oxygen in the atmosphere by the process of photosynthesis. The interaction of biotic and abiotic factors results in the great diversity of ecosystems in the biosphere.

Section Review

1. What is the *biosphere?*
2. What is an ecosystem? Name two abiotic and two biotic factors in an ecosystem.
3. Name and describe two types of interactions among biotic factors in an environment.
4. What major factor determines the kinds of animals that live in an ecosystem?
5. What do you think would happen to other animals on Isle Royale if the moose disappeared?

Figure 49-2 Surtsey, an island off the coast of Iceland, emerged from the ocean in 1963. Built of lava from volcanic eruptions, it continued to grow until the eruptions ceased four years later.

Info: In 1967 Surtsey occupied about 3 km². Since then, wind and water have eroded about a third of the island.

Answers to Section Review
1. the area of the earth where life exists
2. All the biotic and abiotic factors in an area. Abiotic factors include rainfall, temperature, sunlight, and the chemistry of the environment. Biotic factors include all organisms.
3. Answers may include competition, which is a contest among organisms for food or water; predation, which involves the feeding of one species on another; and symbiosis, which is a close, usually beneficial, relationship between two organisms of different species.
4. the kinds of plants that grow there
5. Either the wolves would die out from lack of food or they would need to find other species to prey on.

49.2 Terrestrial Biomes

Abiotic factors such as temperature, rainfall, and soil determine which organisms will thrive in a certain area. A **biome** is a large area identified by the presence of characteristic plants and animals. Biomes are commonly identified by their dominant plant life—for example, hardwood trees are the dominant form of plant life in the deciduous forest biome.

Because abiotic factors change gradually, biomes seldom have distinct boundaries, but Figure 49-3 shows an approximate distribution of the world's biomes. As climate varies over the earth's surface, deserts tend to grade into grasslands, and cold-climate forests grade into warmer-climate forests. Ecologists separate biomes into two very broad classifications: terrestrial biomes, or those on land, and aquatic biomes, or those of water. Table 49-1 on the following page lists the major land biomes and their average temperature and rainfall. Organisms that live in a biome are adapted to the environment of the biome. As you read, pay attention to these adaptations.

Polar Biomes

The biomes found at the earth's coldest extremes are the **polar biomes.** They surround the North and South poles. Similar conditions exist on the tops of the highest mountains. The poles receive almost no precipitation, so although ice is abundant, fresh water is scarce. The sun barely rises in the winter months, and fierce winds sweep over the land. Despite the harsh conditions, however, life does exist in the polar biomes.

Section Objectives

- *Describe the major terrestrial biomes.*

- *Give examples of the adaptation of plants in a biome.*

- *Give examples of the adaptation of animals in a biome.*

- *Describe ways in which ecologists use the LANDSAT satellite.*

Figure 49-3 The biomes of the world occur in approximately the areas shown. Note the inclusion of the broadleaf evergreen forest.

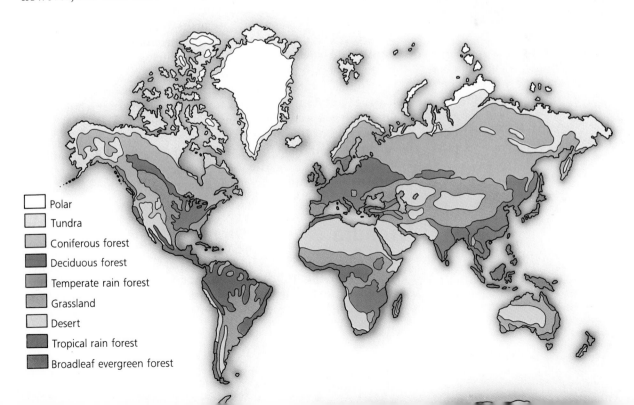

- Polar
- Tundra
- Coniferous forest
- Deciduous forest
- Temperate rain forest
- Grassland
- Desert
- Tropical rain forest
- Broadleaf evergreen forest

Table 49-1 Terrestrial Biome Characteristics

Biome	Average Yearly Temperature Range	Average Yearly Precipitation	Soil	Vegetation
Polar	−40°C–4°C	< 10 cm	Sparse, very low in nutrients, frozen much of year	Mosses, lichens, small flowering plants along coast
Tundra	−26°C–4°C	< 25 cm	Thin, moist topsoil over permafrost; nutrient-poor; slightly acidic	Mosses, lichens, dwarf woody plants
Coniferous forest	−10°C–14°C	35–75 cm	Low in nutrients, highly acidic	Needle-leafed evergreen trees
Deciduous forest	6°C–28°C	75–125 cm	Moist, moderate nutrient levels	Broad-leafed deciduous trees and shrubs
Grassland	0°C–25°C	25–75 cm	Very rich in nutrients, deep layer of topsoil	Dense, tall grasses in moist areas; short clumped grasses in drier areas
Desert	24°C–34°C	< 25 cm	Dry, sandy, nutrient-poor	Succulent plants; scattered grasses and sagebrush
Tropical rain forest	25°C–27°C	200–400 cm	Thin, moist, low in nutrients	Broadleaf evergreen trees and shrubs
Temperate rain forest	10°C–20°C	200–400 cm	Moist, nutrient-rich, highly acidic	Giant needle-leafed evergreen trees

Think

Could there be reasons, other than the warmer temperatures, that explain why more organisms have adapted to life in the Arctic than to life in the Antarctic? Consider the effect of geographic position on each of the two areas.

Answer to Think
More animals migrate to the Arctic and thus take with them more seeds from other biomes. Also, the closer proximity of the Arctic to land means that seeds can be carried to the Arctic by winds and that animals can more easily enter and leave the Arctic environment.

In the Arctic polar biome an ice cap lies over the Arctic Ocean, where the latent heat of water keeps the ice no thicker than 5 to 30 m. During the summer months, when temperatures can rise to 25° C, some of the ice melts, revealing scattered patches of soil. More than 100 species of flowering plants, along with mosses and lichens, live here. Animals such as gulls, walruses, and polar bears also inhabit the Arctic coast.

In the Antarctic polar biome an ice cap lies over a large landmass; therefore the ice cap is not warmed by water's latent heat. Only a few species of lichens and mosses live along the coast. Three types of flowering plants also grow on the rocky tip of the Antarctic peninsula. Only bacteria and some small insects inhabit the vast Antarctic interior. Seals, whales, and penguins are found on or near the Antarctic coast. Penguins breed in winter so that their young will hatch in the spring.

Tundra

The **tundra,** a treeless biome blanketed by snow in the winter, forms a continuous belt across northern North America, Europe, and Asia. Tundralike areas also exist on mountains above the tree line, the highest point at which trees can grow. **Permafrost,** a permanently frozen layer of ground over 500 m thick, characterizes the northern tundra. Even the surface soil above the permafrost remains frozen for all but eight weeks of the year.

In the brief summer the tundra landscape explodes with life, becoming a patchwork of ponds, bogs, and soggy soil. Tundra plants, which are short and often woody, appear, as do swarms of mosquitos and blackflies. Ducks, geese, and predatory birds arrive in great numbers. By mid-September the cold returns.

Figure 49-4 shows some adaptations of plants and animals to the cold, snowy tundra environment. Dwarf willows are only about 0.3 m tall, an adaptation to the cold winds that sweep across the tundra. The large size of caribou creates a low surface to volume ratio that helps prevent heat loss. Caribou are also swift runners with great stamina, adaptations that allow them to avoid predators and migrate long distances.

Coniferous Forest

The **coniferous forest** is a biome that is dominated by conifers, cone-bearing evergreen trees such as pines, firs, spruces, and cedars. In the northern hemisphere this biome lies south of the tundra. Plants are adapted for long, cold winters; short summers; nutrient-poor soil; and frequent droughts. Their waxy needles remain on the trees all winter. The shape of their needles is an adaptation that reduces water loss. In addition, their stomata are recessed in the body of the leaf, which helps conserve water.

Typical mammals of this biome include moose, bears, and lynx. Many of these animals rely on stored body fat during the cold months. Although some animals reside in the forest year round, others migrate to warmer climates in the fall and return in the spring. Most invertebrate species hibernate six to eight months of the year, taking advantage of the insulating snow cover.

Large tracts of coniferous forest cover northern parts of Europe, Asia, and North America. Ecologists subdivide the coniferous forest in North America into three areas:

- The taiga *(TIE-guh)*, the northernmost band, occupies northern Canada and Alaska. The widely spaced trees include spruce and balsam fir, whose cone-shaped crowns shed heavy snows.
- The coniferous belt is found south of the taiga and extends across much of Canada as well as northern parts of the United States. Ecologists call this the "spruce-moose" belt, named for two prominent resident species.
- The southern pine forests cover much of the southeastern United States. In addition to conifers, oaks and other broadleaf trees are found in this more temperate zone. In areas with frequent forest fires, southern pines dominate. The pitchy bark of these trees scorches but does not burn, protecting the cells in the sapwood beneath it. Fires clear the land. Pine seedlings that germinate after a fire grow well on the unshaded land.

Figure 49-4 Dwarf willows, *Salix* sp., and blueberries, *Vaccinium* sp., appear during the brief summer on the tundra of Alaska (top). Caribou, *Rangifer tarandus*, (bottom), are adapted to tundra life.

Deciduous Forest

The **deciduous forest** is a biome characterized by the presence of trees that lose their leaves in the fall. Deciduous forests stretch across the eastern United States, much of Europe, and parts of Asia, South America, Africa, and Australia. These regions have pronounced seasons, with precipitation evenly distributed throughout the year. The moderate rainfall and rich soil in this biome support vast numbers of plant and animal species.

Deciduous trees have broad, thin leaves with large surface area that permits maximum light absorption. Dominant species include birches, beeches, maples, oaks, hickories, sycamores, elms, willows, and cottonwoods. White-tailed deer, foxes, raccoons, and squirrels are typical mammals. Black bears live in some less populated regions. Some animals of the deciduous forest biome hibernate during the cold winters. Others, like hundreds of bird species such as warblers, spend the warm months in the deciduous forest and migrate to warmer climates for the winter. About one-fourth of the bird species that live in the deciduous forest, including the cardinal and bluejay, remain in the forest throughout the winter.

Biotechnology

Remote-Sensing Satellites

Sometimes ecologists want to study huge parts of biomes. For example, an ecologist might want to determine the amount of vegetation in the Arabian desert or the distribution of photosynthetic plankton along a large coastal area. Until recently such studies were difficult because of the sheer size of territory to be covered.

Today ecologists are studying large ecosystems in great detail using the technique of remote sensing. With the help of sophisticated remote-sensing technology aboard earth-orbiting LANDSAT satellites, such as the one at left, that can travel about 700 km above the earth's surface, scientists can record detailed ecological information that they could not otherwise obtain.

Although remote-sensing devices help produce images that look like high-altitude photographs, they do not usually produce traditional photographs of the earth's surface. They instead detect various kinds of electromagnetic radiation (EMR) from visible light, ultraviolet light, infrared light, and microwave radiation.

Plants, water, soil, rock, and other materials absorb,

Grassland

The **grassland** is a biome dominated by grasses. This biome covers about one-fourth of the land surface of the earth. Grassland areas vary greatly in their temperature range. They occur at about the same latitude as deciduous forests but do not receive enough rainfall to support trees. Grasslands are known by various names in different regions of the world—prairie in North America, steppes in Asia, pampas in South America, and veldt in South Africa. The savanna is a grassland with scattered trees found in tropical and subtropical areas.

Some grasslands produce much food and support large numbers of animals. Dominant grazing mammals include bison and antelope in North America; antelope, elephants, and giraffes in Africa; and kangaroos in Australia. In South America the rhea, a huge flightless bird similar to the ostrich, is a major grazing animal. Billions of grass-eating insects also inhabit this biome, including numerous species of grasshoppers. Grasses can survive not only the grazing of animals but also the occasional fires that sweep across the plains. Note in Figure 49-5 that the apical meristem of a grass plant lies underground, safe from damage.

Figure 49-5 Because the growing part of a grass plant is underground, the plant can survive prairie fires.

emit, or reflect characteristic kinds of EMR. The remote-sensing technology measures these variations in EMR. After analyzing differences in EMR, scientists can then interpret their findings as biological and geographical differences.

The EMR picked up by LANDSAT sensors are converted to digital form and beamed back to a receiving station on earth to be stored on magnetic tape. Computers can then process this information to produce photographs like the ones on these pages. Usually, scientists will instruct the computer to highlight the variations in EMR with bright colors.

For example, bright red might represent forested areas and bright blue might represent bare areas. Although they do not represent colors our eyes would see, the images make comparisons easier.

An ecologist also can overlap images taken at different times to study changes in an area over time. For example, an ecologist might overlap an image of a portion of rain forest taken January 15 with an image from the following April 15. The ecologist can then determine if areas of forest were cleared during that period.

In recent years scientists have used remote sensing to survey the amounts of photosynthetic plankton in portions of the Pacific Ocean. Others have used remote sensing to assess suspected air pollution damage to large areas of forest. Still others have used the satellites to find ecosystems suitable for grizzly bears in the Rocky Mountains.

Figure 49-6 Organisms well adapted to desert life are (left to right) the black-tailed jackrabbit, *Lepus insularis,* the saguaro cactus, *Cereus gigantea,* and the Kalahari ground squirrel, *Paraxerus* sp. Jackrabbits spend the hottest hours of the day in cool, underground burrows.

Tip: The higher one climbs on a mountain, the more the environment resembles more northerly biomes. For example, a climb can reveal tropical forest, deciduous forest, coniferous forest, and tundra biomes by turn.

Figure 49-7 What are the conditions to which trees that form a rain forest canopy must adapt?

Canopy

Understory

Forest floor

Desert

Deserts are dry areas where rainfall averages less than 25 cm per year. Rain may fall on Africa's Sahara Desert only once every few years. Yet most deserts are not barren, and most harbor many species of plants and animals. Many deserts are not hot year round. In the desert of Idaho and northern Nevada, winter temperatures often plunge below freezing.

Organisms are adapted to the desert climates. Some desert plants absorb water through deep root systems, while cactuses and other succulents have shallow, diffuse roots. The saguaro cactus of Arizona and Mexico stores water and carries out photosynthesis in thickened stems and branches. A single saguaro can take up and store as much as a ton of water. A thick, waxy cuticle covering the stem helps prevent water loss. The leaves of the saguaro have evolved into sharp protective spines, seen in Figure 49-6, an adaptation that protects against thirsty predators.

Within days or even hours of the infrequent rains, a desert becomes awash with color. Plants have little time to exploit the sudden moisture for photosynthesis and cell growth. Thus they are adapted to quick growth, flowering, and seed production.

Like plants, desert animals are also adapted to their environment. The camel absorbs water from food and produces dung so dry that it can be used to fuel a fire hours after it is excreted. Many desert animals also show behavioral or physiological adaptations. For instance, the Kalahari ground squirrel uses its bushy tail as an umbrella, holding it over its body for shade.

Rain Forests

There are two types of rain forest biomes—the tropical rain forest and the temperate rain forest. The **tropical rain forest** biome is found near the equator where rainfall and sunlight abound. These rain forests may get as much rain in a month as some grasslands do in a year. One small area may support 100 species of plants, all of which compete for sunlight. Thus evolution has resulted in a distinct stratification of vegetation in rain forests, as shown in Figure 49-7. The tallest trees, which reach heights of 30 to 45 m,

Info: Deserts are found on every continent except Europe.

form the canopy. Lower levels of growth make up the understory. The bottom vegetative layer is the forest floor, also called the ground layer. Though you may think of the tropical rain forest as an impenetrable jungle, much of the forest floor is surprisingly cool and free of vegetation since little sunlight reaches it. The very dense growth known as jungle is actually a specific kind of community found only along riverbanks where sunlight breaks through to the forest floor.

Animal life is also rich and diverse, with hundreds of thousands of insect species; colorful birds such as parrots and toucans; apes, monkeys, and predatory cats in South America; and the clouded leopard in Asia. Animals of the tropical rain forest inhabit various levels of the forest. Some are adapted to tree-dwelling life. Primates, for example, use their hands and tail to grasp branches and vines. Stereoscopic vision improves their ability to judge distances. Some animals, such as the flying lizard, glide from tree to tree. Brilliant plumage helps distinguish a tropical bird from its thousands of competitors.

Moderate temperatures and high humidity characterize the **temperate rain forest.** This biome extends along the west coast of North America from central California to southern Alaska. Plant life includes conifers such as the redwood and the Sitka spruce. Unlike mammals of the tropical rain forest, deer, elk, rodents, and most other temperate rain forest species live on the ground.

Figure 49-8 Three organisms found in tropical rain forests are (clockwise from top left) a tropical butterfly, *Marpesia iole,* of Costa Rica; the gibbon, *Hylobates* sp., of southeast Asia; and the scarab beetle, *Scarabaeidae* sp., of Papua New Guinea.

Answers to Section Review
1. A terrestrial biome is a large land area of the biosphere that has a distinct pattern of dominant plants and animals. Major terrestrial biomes are: polar biome, tundra, coniferous forest, deciduous forest, grassland, desert, tropical rain forest, and temperate rain forest.
(See pp. T49A–T49D.)

Section Review

1. Define *terrestrial biome*, and name the major ones.
2. Why is life in the Arctic more abundant than life in the Antarctic?
3. Name three dominant tree species in the coniferous forest biome and three in the deciduous forest biome.
4. List at least two animals that live in each of these biomes: desert, grassland, and tundra.
5. Why do most tropical rain forest animals live in the trees?

49.3 Aquatic Biomes

Aquatic biomes occupy the majority of the earth's surface. As with land biomes, organisms are evolutionarily adapted to the various abiotic and biotic factors. For example, algae are adapted to the intensity of sunlight that penetrates the water and determines the rate of photosynthesis. As you read look for the adaptations that enable organisms to live in aquatic biomes.

Marine Biome

The **marine biome** is the earth's ocean and its associated areas. Ecologists have subdivided marine biomes into three areas: ocean, intertidal zones, and estuaries.

Ocean

The ocean covers about 70 percent of the earth and has an average depth of 3.7 km. In some places the ocean is 11.5 km, about 7 miles, deep. The water contains about 3.5 percent salt, mostly NaCl, a fact that profoundly affects the biology of organisms living there. Ecologists divide the ocean into two zones: the open ocean, or **pelagic zone,** and the ocean bottom, or **benthic zone.** As shown in Figure 49-9, the ocean is further divided into the **photic zone,** where light penetrates, and the **aphotic zone,** where no light penetrates. The photic zone extends to a depth of about 200 m; photosynthesis occurs most efficiently above this depth.

The pelagic zone is divided into two subareas. The first subarea, the **neritic zone,** extends over the continental shelf. Since light penetrates these waters and strong currents, called upwellings, carry minerals from the ocean bottom, the neritic zone supports the greatest amount of marine life. These waters are rich in plankton—the protozoa, algae, and invertebrates on which larger animals feed. Numerous fish, squid, sea turtles and other animals live in these waters.

Figure 49-9 The various zones of the ocean support different forms of life.

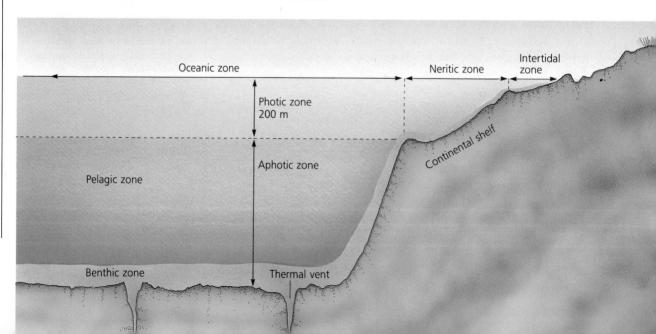

The second subarea, the **oceanic zone,** the deep water of the open sea, is less populated. Even in photic areas, mineral levels are too low to support much life. Animals of the oceanic zone feed primarily on sinking plankton and dead organisms.

Organisms living below 2,000 m are called abyssal organisms and are some of the strangest on earth. Deep-sea fish have slow metabolic rates and reduced skeletal systems that are adaptations to near-freezing temperatures and tremendous pressure. Their large jaws and teeth and expandable stomach can accommodate the rare prey that they can catch at these depths. Even more unusual organisms inhabit thermal vent communities. Near these areas seawater and sulfur gases seep through vents in the earth's crust. These hot springs, rich in minerals and often exceeding 700° C, support about 200 species of bacteria in the surrounding area. The bacteria are chemosynthetic—that is, they manufacture food with chemical energy, rather than light energy. Most convert sulfur compounds to useable energy. An assortment of unique clams, crabs, and worms feed on these bacteria.

Intertidal Zones

Along ocean shores, the tides produce a rhythmic rise and fall of water in an area called the **intertidal zone.** Organisms in this zone are adapted to periodic exposure to air. Crabs prevent dehydration by burrowing into sand or mud. Clams, mussels, and oysters filter plankton from the water during high tide, then retreat into their shells at low tide. At low tide, shorebirds descend on the beaches to snatch this abundant food. One bird, the oystercatcher, uses its long beak to pry open clam shells.

These animals are also adapted to the force of waves crashing onto the shore. Sea anemones cling to rocks with a muscular disk. Sea stars gain footing with the aid of tube feet. The streamlined shape of most chitons and bivalves helps protect these mollusks from the pounding waves.

Estuaries

An **estuary** is a biome found throughout the world where freshwater rivers and streams flow into the sea. Examples of estuarine communities include bays, mud flats, and salt marshes. The shallow waters ensure plenty of light, and rivers deposit large amounts of mineral nutrients. However, the interplay between fresh and salt water causes great variation in temperature and salinity. In addition, like the intertidal zone, much of the estuary is exposed to the air during low tide.

Estuarine life is adapted to frequent change. For example, some mangrove trees have special glands on their leaves that eliminate excess salt water taken up by the roots. Soft-shell clams lie buried in mud, with only their long siphons protruding above the surface. The siphon filters plankton from the water at high tide and also detects predators at low tide, contracting whenever it senses danger.

Figure 49-10 These Galápagos rift tube worms are part of a thermal vent community. The photo appears washed out because, at a depth of 2,000 m, it had to be taken with artificial light.

Figure 49-11 Sea stars have adaptations for life in the intertidal zone.

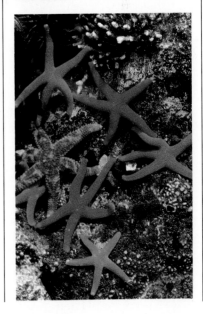

The Biosphere and Biomes   **775**

Freshwater Biome

Low levels of dissolved salts characterize the **freshwater biome.** The salt content of fresh water is about 0.005 percent, compared with 3.5 percent in oceans. The freshwater biome includes bodies of water as huge as Lake Superior and as small as a half-acre pond. They include clear mountain streams in the Himalayas or the Rockies and slow, turbid rivers like the lower Mississippi or the Mekong River of southeast Asia.

Lakes and Ponds

Ecologists divide lakes and ponds into two categories. **Eutrophic lakes** are rich in organic matter and vegetation, making the waters relatively murky. The giant water lily pictured in Figure 49-12 is growing in a shallow, eutrophic pond. **Oligotrophic lakes** contain little organic matter. The water is much clearer, and the bottom is usually sandy or rocky.

Fishes inhabit both eutrophic and oligotrophic lakes. Carp and catfish are typical of murky eutrophic waters. Largemouth bass and pike survive in more moderately eutrophic waters. Lake trout are found in oligotrophic lakes. Freshwater lakes and ponds also support mammals such as the otter and muskrat, and birds such as ducks and loons.

Rivers and Streams

A river is a body of water that flows down a gradient, or slope, toward its mouth. A river can be as short as a few kilometers or as long as the Nile, which runs from east Africa to the Mediterranean, a distance of 6,738 km.

Gradient is a key abiotic factor in a river. Water flows swiftly down steep gradients, and organisms here are adapted to withstand powerful currents. These include the larvae of caddis flies and the nymphs of mayflies, stoneflies, and dobson flies, which cling to the rocky bottom. Brook trout and other fishes are strong enough to face upstream while feeding on drifting invertebrates. Slow-moving rivers and their backwaters are richer in nutrients than fast waters are and therefore support a greater diversity of life. Rooted plants and the fishes that feed on them are adapted to the weak currents.

Section Review

1. Distinguish between the *photic* and *aphotic zones*.
2. Identify and describe the three parts of the marine biome.
3. How do organisms survive the breaking of waves in an intertidal zone?
4. List two ways that the gradient of a river limits the types of organisms that live in the river.
5. What might happen to organisms living in a fast-moving river if a dam is built on the river?

Figure 49-12 The Amazon water lily, *Victoria amazonica*, is adapted to life in shallow, eutrophic ponds.

Answers to Section Review
1. The photic zone is the ocean region where light penetrates. The aphotic zone is the region where no light penetrates.
2. Oceans are large, deep bodies of salt water. Intertidal zones are coastal areas that are periodically covered and uncovered by tidal waters. Estuaries are areas where fresh water and salt water mix as streams and rivers meet the sea.
3. by attaching themselves to rocks and shells
4. The gradient affects current strength and nutrient availability.
5. A dam will create a slower river and make survival difficult for organisms adapted for life in fast-flowing streams.

Laboratory

Organisms and the Environment

Objective
To examine how organisms alter the physical composition of their environment

Process Skills
Experimenting, hypothesizing

Materials
Two bottles of *Elodea* in water, labeled *A* and *B*; 1-mL pipette; 2-mL pipette; 25% aqueous manganous sulfate solution; 40% aqueous potassium hydroxide–potassium iodine solution; concentrated sulfuric acid solution; 2 medicine droppers; 0.003% aqueous sodium thiosulfate solution

Method

1 In this investigation you will use several chemical tests to determine the concentration of oxygen in two samples of water. Dissolved oxygen in the water samples will react with the chemicals and produce iodine. The amount of iodine is proportional to the amount of oxygen in the samples.

2 Bottle A has been placed under bright light for 24 hours. Bottle B has been placed in the dark for 24 hours.

3 Uncap each bottle and remove the *Elodea* carefully, trying not to agitate the water. Gently shake off each sprig over the bottle to keep as much water in the bottle as possible. Why are you trying not to agitate the water?

4 Use a 1-mL pipette to add 0.5 mL of manganous sulfate solution to each bottle. Rinse the pipette thoroughly with water. Use a 2-mL pipette to add 1.5 mL of potassium hydroxide–potassium iodide solution to each bottle. Rinse the pipette. Cap the bottles, and shake them to mix the solutions.

5 Uncap each bottle and use the 1-mL pipette to add 0.5 mL of sulfuric acid to each bottle.

Your samples should now appear brown in color. This indicates the presence of iodine in the solution.

6 An indication of the amount of iodine in the water in each bottle can be made by using sodium thiosulfate solution. The sodium thiosulfate reacts with iodine and removes it from the solution. The more thiosulfate solution you add before the sample becomes colorless, the more iodine is present in the sample. Use a clean dropper to add thiosulfate solution to each bottle. Count the number of drops required to make each sample colorless. Record your observations. What does this test tell you about the amount of dissolved oxygen in each sample?

Conclusions

1 Why does the oxygen content of the two samples differ? Explain your answer.

2 How do the two samples of *Elodea* differ in the way they affect their environment?

3 Why is it necessary to use two bottles, one placed in the dark and one placed in the light, to draw valid conclusions from this investigation?

Inquiry
Think of a pollutant you might add to the water in bottle A. Form a hypothesis regarding its effect on the environment in the bottle and devise a means of testing the hypothesis.

Chapter 49 Review

Vocabulary

abiotic factor (765)
aphotic zone (774)
benthic zone (774)
biome (767)
biosphere (765)
biotic factor (765)
community (765)
coniferous forest (769)

deciduous forest (770)
desert (772)
ecology (765)
ecosystem (765)
estuary (775)
eutrophic lake (776)
freshwater biome (776)
grassland (771)

intertidal zone (775)
marine biome (774)
neritic zone (774)
oceanic zone (775)
oligotrophic lake (776)
pelagic zone (774)
permafrost (768)
photic zone (774)

polar biome (767)
population (765)
predation (766)
symbiosis (766)
temperate rain forest (773)
tropical rain forest (772)
tundra (768)

1. What is a population, and how does the concept of population relate to that of community?
2. What are some abiotic factors that influence an ecosystem?
3. What is the difference between symbiosis and predation?
4. Distinguish between the pelagic and benthic zones of the ocean.
5. Use a dictionary to find the meaning of the word parts *eu-*, *oligo*, and *-troph*. Explain why eutrophic and oligotrophic lakes are appropriately named.

Review

1. An ecosystem includes (a) all the members of one species (b) the living and nonliving factors in an environment (c) all parts of the earth where life exists (d) all members of a species in the same area.
2. Abiotic factors in an ecosystem can include (a) plants (b) animals (c) sunlight (d) microorganisms.
3. Permafrost is (a) a glacier (b) Arctic ice (c) permanently frozen ground (d) ground that thaws in the spring.
4. The neritic zone is in the (a) tundra (b) tropical rain forest (c) intertidal zone (d) ocean.
5. Organisms in the deserts must be adapted to low levels of (a) water (b) food (c) heat (d) salt.
6. An estuary is an area where (a) deserts become grasslands (b) coniferous forests meet deciduous forests (c) salt water meets fresh water (d) rain forests become grasslands.
7. Eutrophic lakes are (a) clear (b) murky (c) swift (d) small.
8. Adaptations of life in a river are primarily determined by (a) gradient (b) salinity (c) plankton (d) predation.
9. Plants and animals become adapted to life in particular biomes through a process of (a) symbiosis (b) competition (c) growth (d) evolution by natural selection.
10. Ecologists use remote-sensing technology to measure (a) heat waves (b) variations in EMR waves (c) variations in climate (d) changes in rainfall.
11. Give an example of a relatively permanent physical change that is caused by abiotic factors.
12. Why is the ice in the Antarctic thicker than the ice in the Arctic?
13. Why can some grasslands support a large amount of animal life?
14. What is the advantage of a short life cycle for desert plants?
15. How does the temperate rain forest differ from the tropical rain forest?
16. What are some adaptations of animals in the tropical rain forest to tree dwelling?

17. What are the three subareas of the pelagic zone?
18. What are the organisms that drift freely in open waters?
19. What is the source of energy in the thermal vent communities?
20. Name two types of fish that can live in eutrophic water.

Critical Thinking

1. Examine the graph at the right. From what you know about the biomes of the world, which biome do you think this graph represents? List the reasons for your answer.
2. Deciduous forests once covered a large portion of eastern North America. However, humans have cleared much of these forests to make way for cities and towns. How do you think this shrinking area has affected plants and animals in the remaining forests?
3. Although plants of the coniferous forest and desert biomes seem to have little in common, both contend with dry conditions and a brief growing season. Compare the ways in which conifers and succulent plants are adapted to these conditions.
4. What are the significant abiotic factors in the thermal vent communities? How do they determine the types of life found there?
5. The plant *Artemisia gerardi* has diffuse roots and will continue to grow even if

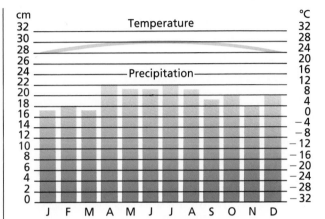

subjected to periodic fire and drought. What kind of animal do you think would feed on this plant?
6. Estuaries serve as breeding sites for thousands of species of marine animals. What characteristics of this biome would make estuaries advantageous areas for fishes, mollusks, crustaceans, and other aquatic animals to reproduce? What might be some disadvantages of reproducing in estuarine areas?

Extension

1. Read the article by Fred Bruemmer entitled "Life on the Permafrost" in *Natural History,* April 1987, pp. 30–38. Make a chart with the following information about the tundra biome: altitude, rainfall, temperature, plant life, and animal life.
2. Research the changes in the environment surrounding Mount St. Helens after the eruptions in 1980. Write a report on the efforts of ecologists to restore the area.
3. Visit a biological community in your area. Choose a lake, pond, seashore, marsh,

forest, park, or even an abandoned railroad track. Draw a selective sample of plants you see there and try to identify them from books in the library.
4. Pick a small community—a tree, lawn, window sill, or part of your back yard. With string, mark off a 1-foot-square area and count carefully the plants and animals found within this area. Note the number of species found, and identify as many as you can. Compare your findings with those of others in the class.

Chapter 50 | Ecosystems: Structure

This chapter explains the structure of the ecosystem by analyzing the flow of energy and the cycling of nutrients. Models are used to illustrate the ways that biotic and abiotic factors interact.

Correlation Chart

Section Number	Section Objectives	Pages	Features	Section Review
50.1	1. Distinguish between habitat and niche. 2. Describe the function of producers in an ecosystem. 3. Name the types of consumers, and describe the feeding habits of each. 4. Explain how decomposers make nutrients available to the ecosystem.	781–784	Think, p. 783	1 2, 5 3 4
50.2	1. Describe the flow of energy through a food chain. 2. Explain how food chains interact to form a food web. 3. Identify the three types of ecological pyramids and explain what they measure.	785–787		2 1 3, 4, 5
50.3	1. Trace the cycle of water between the ear h and the atmosphere. 2. Summarize the major steps in the nitrogen cycle. 3. Explain how oxygen and carbon are cycled through an ecosystem. 4. Describe the role of vegetation in nutrient cycling in a forest.	788–793	Biology in Process, pp. 788–789 Biology Lab, p. 793	2 5 1, 3, 4

Resources*
Print

Brown, Lester R., et al. *State of the World, 1986: A Worldwatch Institute Report on Progress toward a Sustainable Society.* New York: Norton, 1986.

Perry, Donald. *Life Above the Jungle Floor: A Biologist Explores a Strange and Hidden Treetop World.* New York: Simon & Schuster, 1986.

Sibbald, Jean H. *Homes in the Sea: From the Shore to the Deep.* Minneapolis: Dillon, 1986.

Audiovisual/Software

A Small Body of Still Water. Film or video, 16 min. New Dimensions, 1984.

The Ecology of a Stream. Filmstrip or slides with cassette, 12 min. Teacher's guide. Carolina Biological, 1983.

Habitat Tidal Flat. Film, 17 min. Britannica.

*Additional references are listed in *Modern Biology, Teacher's Program Guide.*

Section Teaching Guides

50.1 The Structure of Ecosystems

Approach

The main objective of this section is to describe the various roles of individual organisms in an ecosystem. Chapter 49 introduced the concept of ecological units within the biosphere, and this section describes in greater detail the largest ecological unit—the ecosystem. The information presented here will prepare students for discussions of the dynamics of communities and populations in Chapters 51 and 52. In your discussion, draw on the examples given in the text, such as the clearing of tropical forests in Brazil, to bring out the idea that each organism affects all other components of an ecosystem.

Concept Framework

Students will be familiar with a variety of organisms in their own environment. Build on this knowledge to develop a conceptual framework from which to study ecosystem structure. Elicit from students how these organisms might be interacting in their own environment. Select one or two prominent species in your area, such as a type of grass or a bird, and ask students to describe their niche in as much detail as possible. Answers should include information about the organisms' habitat and trophic level.

Emphasis

Make sure that students understand how an organism's habitat, niche, and trophic level each figure in defining the organism's role within the ecosystem.

Observing Feeding Adaptations

Bring in photographs or slides of a variety of animals that show feeding adaptations such as mouth or foot structure. Ask students to determine whether the animals are primary consumers, secondary consumers, scavengers, or some combination of the above.

Section Review Answers

(Continued from p. 784.)
4. to break down dead tissue and wastes, thus returning nutrients to the soil
5. water, oxygen, carbon dioxide levels, and light

50.2 Energy Flow

Approach

The main objective of this section is to explain how energy flows through an ecosystem, and to describe methods of measuring this energy flow quantitatively. Since energy flow occurs through the eating of one organism by another, you might wish to review the concepts of forms of energy (Chapter 3) and cell metabolism (Chapter 7). The material presented in this section provides background for the discussions of predation and competition in Chapter 51 and of limits to population growth in Chapter 52.

Concept Framework

Students know that all organisms, including humans, require energy to carry on metabolic processes. Build upon this knowledge to reinforce the idea that organisms store only part of the energy they consume, ask students to think of the various ways the human body obtains, uses, and stores energy. Tell them to apply their knowledge of energy consumption as they read about the flow of energy through ecosystems.

Emphasis

In order to understand the dynamics of ecological relationships, students need to remember that energy is always lost from one trophic level to the next. To reinforce this concept, refer students to the diagrams of ecological pyramids on pages 786 and 787 (Figures 50-8, 50-9, and 50-10).

Diagraming Food Webs

Assist students in diagraming a food web for organisms in your area on the chalkboard. Generate a discussion about the importance of food chains and food webs in your area. The discussion should focus on the possible consequences of disrupting a food web, for example, through human activity or a natural catastrophe.

Section Review Answers

(Continued from p. 787.)
4. because energy is always lost from one trophic level to the next
5. It is probably a rain forest because of the very high biomass and productivity; a grassland would not have as great a biomass.

50.3 Biogeochemical Cycles

Approach

This section focuses on the cycling of inorganic materials—namely water, nitrogen, oxygen, and carbon—through the biosphere. The discussions of the water and oxygen-carbon cycles are based on previous discussions of biochemistry (Chapter 4) and photosynthesis and respiration (Chapter 7). The discussion of the nitrogen cycle builds on the concepts of biochemistry and protein synthesis (Chapter 8). In describing these biogeochemical cycles, note that organisms at each trophic level, along with abiotic components of the biosphere, are involved in the cycling of inorganic materials.

Concept Framework

Students know that water, oxygen, and other inorganic materials are essential to life. In light of this knowledge ask students to consider what would happen if these materials did not constantly recycle through the biosphere. Elicit the fact that biogeochemical cycles exist because organisms at one trophic level depend on the metabolic wastes of organisms at other trophic levels.

Emphasis

In presenting the text, stress the idea that, whereas energy flows in one direction through an ecosystem, nutrients and other inorganic materials are constantly recycled throughout the biosphere.

Using the Features/Nutrient Cycling and Deforestation

The main process skills used in this feature are collecting and interpreting data, experimenting, and communicating. The collection of massive amounts of data on precipitation, runoff, and nutrient levels provided information not only about the role of vegetation in an ecosystem, but on the possible effects of human intervention. In discussing the feature, note that Gene Likens and his colleagues used experiments to devise solutions to the problem of nutrient runoff, not merely to identify the problem. Also point out the significance of communicating such information to the public. You might mention that Likens is currently researching the effects of acid rain on forests, another ecological problem of great concern to humans.

Discussing the Greenhouse Effect

Conduct a class discussion in which students discuss the possible effects of the greenhouse effect. Tell students that some ecologists believe that temperature increases resulting from increased levels of carbon dioxide could have a catastrophic effect on the earth. Others believe that the increased carbon dioxide levels will stimulate the rate of photosynthesis, thus achieving equilibrium in the biosphere. Ask students to think of evidence supporting both theories.

Section Review Answers

(See p. 791.)

Laboratory Preparation

Background

In this lab exercise students discover the diversity of life in an aquatic ecosystem and identify the ecosystem's producers and consumers. Since students examine water samples and identify microorganisms, they should be familiar with common phytoplankton, zooplankton, and invertebrates. Students should be aware that producers contain chlorophyll and usually are sessile, while consumers do not contain chlorophyll and usually are motile.

Preparation

If pond water is available, the organisms will be gathered with the pond water. If pond water is not available, plankton and microscopic invertebrates can be ordered from a biological supply house. Make additional references available so that students can identify organisms other than the ones pictured here.

Conclusions

1. Producers should include phytoplankton and plants. Consumers should include zooplankton and microscopic and macroscopic animals. Producers are green because they contain chlorophyll.
2. There were probably no decomposers observed in the water sample. Such organisms tend to be bacteria, which are too small to be seen with a compound microscope.
3. There are more producers. Producers manufacture all of the organic nutrients needed to support the aquatic ecosystem.

Inquiry

Students will form a hypothesis and devise a means of testing it, such as setting up an aquarium with the necessary producers without introducing consumers to the system.

Chapter Review

Vocabulary

1. Gross primary productivity is the total energy production for an ecosystem; net primary productivity is gross productivity minus the energy used for respiration.
2. Carnivores feed on living animals; scavengers feed on dead animals.
3. Primary consumers eat producers; secondary consumers eat primary consumers.
4. Primary producers manufacture energy for the ecosystem; decomposers return nutrients to the ecosystem by breaking down dead tissues.
5. In nitrification, nitrites and nitrates are formed from the oxidation of ammonia. In denitrification, nitrites and nitrates are broken down into atmospheric nitrogen.

Review

11. The major producers in a prairie, a tropical rain forest, and a lake are grasses, leaves in the canopy, and phytoplankton, respectively.
12. An herbivore feeds on primary producers; a carnivore feeds on consumers; and an omnivore feeds on producers and consumers.
13. The vulture's adaptations for scavenging include a powerful beak for tearing flesh and a lack of head feathers to prevent bacterial infections.
14. The energy not stored by primary producers is either reflected or used for the life processes of the plants.
15. Changes in an environment can have far-reaching effects because of the interrelatedness of habitat, niche, and food webs.
16. A numbers pyramid might appear inverted for ecosystems such as a forest, in which a few trees support thousands of consumers.
17. A biomass pyramid measures the biomass or dry weight of organisms in an ecosystem.
18. The two major processes involved in the oxygen–carbon cycle are photosynthesis and respiration.
19. Energy flows through an ecosystem in one direction, with some loss at each trophic level. Nutrients are constantly recycled through an ecosystem.
20. Nitrates can become available to an ecosystem through nitrification, lightning energy, or the erosion of nitrate-rich rocks.

Critical Thinking

1. An omnivore can choose from many types of food. If its environment changes, it may choose another type.
2. Sharks are important because they cull weak and dying fish. The strongest fish are then left to reproduce and carry on the species. If large numbers of sharks disappeared, the weak fish would also be reproducing, decreasing the vitality of the species.
3. When insecticides are sprayed on plants, they may be eaten by a primary consumer. The insecticide residues are often bound up inside the organism's cells where they accumulate. These residues are then concentrated as they bind to the cells of successive consumers.
4. An individual might move to a new ecosystem or might change the type of food it ate. Those members of the species who could adapt to the new conditions would carry on the species.
5. Farmers grow alfalfa or soybeans because they fix nitrogen, thus rebuilding the soil the corn has depleted.
6. The consumers in the picture are the birds and the seals. They both consume fish, which consume other fish or algae and plankton.

Chapter 50 | Ecosystems: Structure

Introduction

The biotic and abiotic factors of an ecosystem interact in highly complex ways. The transfer of energy and nutrients is the basis for many of these interactions. By analyzing the flow of energy and the cycling of nutrients through ecosystems, ecologists can gain a better understanding of ecosystem structure.

Chapter Outline

Chapter Concept

As you read, note the way energy relationships determine the structure of ecosystems.

Comet Geyser in Yellowstone National Park, Wyoming. Inset: Bald eagle, *Haliaeetus leucocephalus.*

For additional teaching information see pp. T50A–T50D.

50.1 The Structure of Ecosystems

You learned in Chapter 49 that an ecosystem is a unit of the biosphere consisting of interacting biotic and abiotic factors. A single forest—its climate, soil, trees, shrubs, bacteria, invertebrates, and vertebrates—makes up one ecosystem. A lake beside the forest and a meadow beyond the lake are also distinct ecosystems. Each organism in an ecosystem has a specific role. Three important aspects of the ecology of an organism are its habitat, its niche, and its trophic level.

Habitat

To understand an organism's role in the structure of an ecosystem, ecologists often begin by studying the organism's **habitat**—the physical area in which an organism lives. They will study the climate, topography, soil and water chemistry, and plant and animal life of the habitat. An organism's habitat may include many different kinds of areas. The habitat of the bald eagle, for example, may include both the deciduous forests and the prairies of central Wisconsin. The mouse's habitat may include a forest, a garden, and the walls of a house. The monarch butterfly's habitat may be the forests of the Great Lakes region during the warm months and the forests of Mexico during the winter.

A drastic change in the abiotic or biotic factors of a habitat that affects one organism can have an effect on the whole ecosystem. For instance, as Figure 50-1 shows, dumping chemicals into a pond changes the water chemistry, which may disturb the biotic and abiotic interactions of the plants and animals in the pond. Cutting down a patch of forest may result in the death of the animals that live there. In Brazil, for example, the tropical forest along the Atlantic coast once covered an area as large as Texas and New Mexico combined. Today only 2 percent of the forest habitat remains, the rest having been cleared to make way for cities, towns, and farms. As a result of the destruction of their habitat, 14 of the 21 species and subspecies of monkeys that live there are now considered endangered species.

Niche

A **niche** is the way of life of a species. An organism's niche includes its habitat, feeding habits, reproductive behavior, and all other aspects of its biology.

Consider the niche of the North American beaver. Beavers live in coniferous and deciduous forest habitats in the Northern Hemisphere. A beaver cuts down trees and builds dams that create temporary ponds. It eats plants. It reproduces. It adds nutrients to

Section Objectives

- *Distinguish between habitat and niche.*
- *Describe the function of producers in an ecosystem.*
- *Name the types of consumers, and describe the feeding habits of each.*
- *Explain how decomposers make nutrients available to the ecosystem.*

Tip: Help students understand the difference between the habitat and the niche of an organism by using a relatively common organism such as a spider. List features of the organism's habitat. Then point out that the habitat is an important part of the organism's niche.

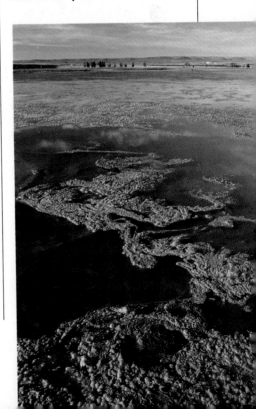

Figure 50-1 Dumping phosphorus into this pond has resulted in a bloom of algae that seriously affects the lives of other organisms in the pond.

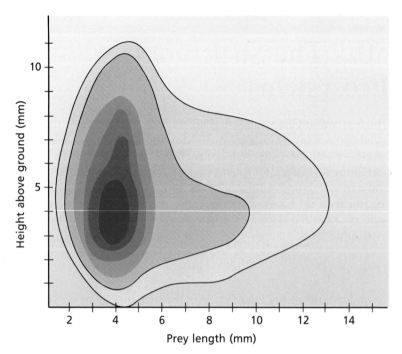

Figure 50-2 The contour lines on the graph show the frequency of capture of prey in trees by the blue-gray gnatcatcher. Prey— mostly insects—are most often captured between 3 and 5 m above ground. Prey averaging about 4 mm in length are the most frequently captured.

Info: Productivity is concerned with rate. It does not refer to yield, which is the measurable weight of a crop.

the environment not only by pulling plants into the water but by excreting waste products. It is also host to a number of tick and mite parasites. The male beaver establishes a territory and defends it against other males. These and many more features of the biology of a beaver make up its niche.

An organism's niche can be quantified. For instance, the graph in Figure 50-2 illustrates just two characteristics of the niche of the blue-gray gnatcatcher, a type of warbler that lives on insects. Hundreds of similar graphs would be needed to fully quantify its niche.

Trophic Levels

To understand the role of an organism in an ecosystem, ecologists must study the energy flow among the organisms of the ecosystem. Energy enters an ecosystem primarily from the sun. Energy flows between organisms when one organism eats another. The relationship between what an organism eats and what it is eaten by defines its **trophic level,** its feeding level, in an ecosystem.

Producers

Autotrophic organisms, which include plants, algae, and photosynthetic and chemosynthetic bacteria, manufacture their own food, using solar or chemical energy. Because autotrophs manufacture all the organic nutrients for an ecosystem, they are called **primary producers.** In a tropical rain forest, for example, tall trees produce most of the nutrients. In a midwestern prairie or an African veldt grasses produce most of the nutrients. In aquatic ecosystems species of phytoplankton, such as diatoms and dinoflagellates, are the major producers.

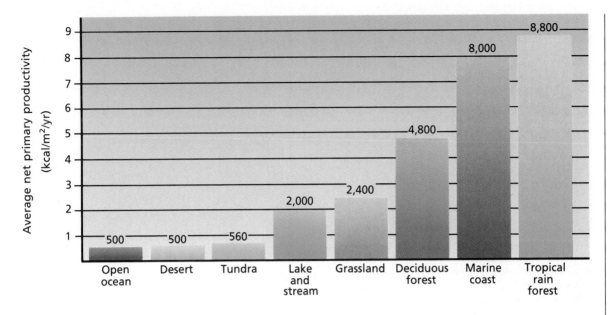

Ecosystem type

The rate at which solar energy is converted into organic compounds is called the **primary productivity** of the ecosystem. Primary productivity is usually expressed as the rate at which energy is stored as organic matter. The units are kilocalories per square meter per year (kcal/m²/yr). **Gross primary productivity** is the total amount of energy produced, including the energy used by the plants for their own respiration. **Net primary productivity** is the rate at which plants store energy that is not used in plant respiration. Over time net primary productivity would show up as plant material that can be used as food by animals.

Figure 50-3 shows that the primary productivity of different ecosystems can vary greatly. For example, productivity in a tropical rain forest exceeds that of the same area of tundra by 16 times. Although rain forests occupy only 5 percent of the earth's surface, they account for almost 30 percent of its productivity.

Consumers

Animals and protozoa are heterotrophs. Unlike autotrophs, heterotrophs cannot manufacture their own food. Instead they obtain energy from the chemical bonds in the nutrients they eat. Ecologists call heterotrophs **consumers.** They often call any organism that eats a producer a **primary consumer.** A **secondary consumer** is an organism that eats a primary consumer.

Consumers are also classified according to the type of food they eat. **Herbivores** obtain their nutrients by eating primary producers. The grass-eating antelope is an herbivore. So are the minute zooplankton that feed on phytoplankton floating in ocean or lake waters. **Carnivores** obtain nutrients by eating other consumers. Lions, eagles, cobras, and preying mantises are all carni-

Figure 50-3 The amount of energy stored by plants after respiration has taken place is greatest in tropical rain forests and least in deserts and open oceans.

Think

What adaptations does this sparrow hawk, *Falco sparverius,* have for carnivory?

Answer to Think: The sparrow hawk's adaptations include grasping talons, acute vision, and a bill modified for tearing flesh.

Figure 50-4 A white-backed vulture, *Gyps africanus*, a scavenger, feeds on a dead giraffe at Masai Mara, Kenya.

Figure 50-5 The bracket fungus, *Polyporus versicolor*, is a decomposer that breaks down dead tissue in tree trunks and wooden structures.

Answers to Section Review
1. A habitat is the physical area in which an organism lives; a niche is the way of life of a species in an ecosystem.
2. They produce all the organic nutrients for an ecosystem.
3. Zooplankton, caterpillars, antelopes, and other herbivores are primary consumers. Carnivores, such as lions and cobras, and scavengers, such as vultures, are secondary consumers. (Accept other correct examples.)
(See pages T50A–T50D.)

vores. **Omnivores** obtain nutrients by eating both primary producers and consumers. Grizzly bears, whose diet ranges from berries to salmon, are omnivores. Biologically humans are also omnivores. Whenever one organism eats another, energy in the form of the chemical bonds of nutrients is transferred from one organism to another.

Scavengers are consumers that feed on organisms that have recently died. Vultures and crows, which eat animal remains, are scavengers. Notice the vulture's head in Figure 50-4. It has a heavy, powerful beak that is adapted for tearing flesh. Its lack of head feathers is an adaptation that reduces the possibility that its feathers might be infected by bacteria as it sticks its face into decaying meat.

Decomposers

When leaves fall off trees or when animals die, they are acted upon by decomposers. **Decomposers,** such as bracket fungi shown in Figure 50-5, are organisms that break down dead tissues and wastes into nitrogen, carbon, phosphorus, and other elements and compounds. Autotrophs then extract these materials from the soil and use them to manufacture food. Thus the process of decomposition recycles chemical nutrients, making them available to the producers once again. This cycle is repeated over and over in the same ecosystem.

Section Review

1. Distinguish between a habitat and a niche.
2. Why are primary producers an essential component of an ecosystem?
3. Give three examples of a primary consumer and three examples of a secondary consumer.
4. What is the function of a decomposer?
5. What abiotic factors might account for differences in net primary productivity in ecosystems?

50.2 Energy Flow

Although energy flows in various patterns through an ecosystem, the total energy of one trophic level of an ecosystem never flows to the next level. In this section, you will see that the lower trophic levels of an ecosystem always contain more total energy than do the higher trophic levels.

Food Chains

A **food chain** is the specific sequence in which organisms obtain energy in an ecosystem. In a freshwater ecosystem, for example, a food chain may begin with a primary producer, such as an autotrophic alga, which is eaten by a primary consumer, such as a tiny insect nymph. A fish then eats the insect nymph. An otter eats the fish. When the otter dies, bacteria feed on the carcass, releasing nutrients back into the ecosystem. Then other algae use the nutrients and energy from the sun to make food. As a result of this series of interactions a specific sequence of production and consumption exists in the ecosystem.

Figure 50-6 shows a food chain in a grassland ecosystem. The food chain begins with the autotrophic grass, which is the primary producer. The chain continues with the primary consumer of the grass—an herbivorous zebra. Then the chain continues with a carnivorous lion and a scavenging vulture. Again bacteria perform the final job of decomposing carcasses, dead leaves, and wastes into basic nutrients.

Food chains usually have only four or five levels. In the example above, the vulture may seem to be the highest-level consumer. However, most large animals are hosts for thousands of parasites of many different species. These parasites, such as ticks, insects, mites, worms, and lice, can be fifth- or sixth-level consumers. If they have their own parasites, and those parasites also have parasites, a food chain can consist of as many as eight or nine trophic levels.

As shown in Figure 50-6, energy flows through a food chain. However, only a portion of the energy available to organisms at each level of the food chain is stored. Only this stored energy is available to organisms at the next trophic level. For example, a plant converts only about 1 to 5 percent of the solar energy that falls on it. The rest is reflected off the plant, used in the life processes of the plant, or lost in the form of heat. In addition, a zebra that then feeds on this plant stores only about 10 to 20 percent of the energy it receives from the plant. In turn, a lion that eats the zebra stores only a small percentage of the energy it gets. On the average an organism stores only about 15 percent of the energy available to it from the trophic level beneath it. Thus the amount of energy available to the higher trophic levels is much smaller than the amount available to lower levels.

Section Objectives

- *Describe the flow of energy through a food chain.*
- *Explain how food chains interact to form a food web.*
- *Identify the three types of ecological pyramids and explain what they measure.*

Producer

Primary consumer

Secondary consumer

Scavenger

Figure 50-6 The food chain in a grassland ecosystem begins with grass (a producer) and ends with scavengers.

Decomposers

Figure 50-7 The illustration shows the relationships between a variety of producers, consumers, and decomposers in a grassland food web.

Food Webs

Feeding relationships and energy flow among species in an ecosystem are actually too complex to be represented by a single food chain. Many consumers eat more than one type of food. In addition, more than one species of consumer may feed on the same organism. Thus in nature many food chains interlink and overlap within a single ecosystem. Both otters and eagles, for example, may eat the same species of fish. A diagram of the feeding relationships among all the organisms in a single ecosystem would look much like a web. For this reason, the interrelated food chains in an ecosystem are called a **food web**. Figure 50-7 shows a simplified food web.

Figure 50-8 A numbers pyramid for an acre of bluegrass shows that 1.5 million primary producers are needed to support a single tertiary consumer. What might that consumer be?

Grasslands

1
90,000
200,000
1,500,000

Tertiary consumers

Secondary consumers

Primary consumers

Primary producers

Ecological Pyramids

To simplify the complexities of an ecosystem, ecologists often graph the relationships in an ecosystem as a pyramid, with the primary-producer trophic level at the base and the consumer levels above it. Common pyramid models include numbers pyramids, biomass pyramids, and energy pyramids.

Numbers Pyramids

Numbers pyramids compare the number of individuals at each trophic level. Figure 50-8 shows a numbers pyramid for an acre of bluegrass. Ecologists determined that 5,842,424 primary producers, 707,624 primary consumers, and 354,904 secondary consumers, such as ants, spiders, and predatory beetles, existed in the acre of bluegrass. At the top of the pyramid there were three birds and three moles.

Info: An ecosystem is generally limited to four or five trophic levels for several reasons: (1) Animals at a higher level do not eat all the available food at the lower level. (2) There is a great amount of "waste" at every level. (3) Energy is expended during metabolism.

Grasslands biomass (g/m²)

Carnivore (0.1)

Herbivore (0.6)

Producer (470.0)

Freshwater biomass (g/m²)

Zooplankton (21)

Phytoplankton (4)

Numbers pyramids often reveal that organisms at the base of the pyramid are the most numerous. In some cases, however, a numbers pyramid can seem inverted. An inversion seems to exist when an organism that occurs in the greatest number does not fall at the bottom of the pyramid. For example, in a forest ecosystem a single large producer, such as an oak tree, may support thousands of caterpillars and hundreds of secondary consumers, such as spiders and birds.

Biomass Pyramids

Biomass pyramids compare the amount of biomass at each trophic level. **Biomass** is the total dry weight of the organisms in the ecosystem. Figure 50-9 shows the biomass pyramid for a grassland area. Note that in the grassland ecosystem the biomass of the primary producers is over 783 times greater than that of the primary consumers. In contrast, phytoplankton in a marine ecosystem have less than one-fifth the biomass of the zooplankton they support.

Energy Pyramids

Energy pyramids compare the total amount of energy, measured in kilocalories, in each trophic level. Unlike numbers pyramids and biomass pyramids, energy pyramids are never inverted, because energy is always lost from one trophic level to the next higher. Figure 50-10 shows an energy pyramid for a river in Florida. Note that as energy flows through the food chain, there is a dramatic drop in the amount of energy each trophic level retains.

Section Review

1. How does a *food chain* differ from a *food web?*
2. What percentage of the available solar energy is stored by producers and consumers?
3. Distinguish between numbers, biomass, and energy pyramids.
4. Why are energy pyramids never inverted?
5. Suppose that a set of ecological pyramids for an ecosystem reveals that producers have extremely high biomass and productivity. What type of ecosystem might these pyramids represent?

Tip: Explain to students that the term *biomass* is a biological convention. The use of this term does not imply that mass and weight are synonymous.

Figure 50-9 The biomass pyramids for a grassland and a freshwater biome are different. Why might this be so?

Figure 50-10 The illustration shows an energy pyramid for a river ecosystem. Note that energy levels are expressed in thousands of kilocalories per square centimeter per year.

Net production

Respiration

Secondary carnivore

21

Primary carnivore

383

Herbivore

3,368

Producer

20,831

Answers to Section Review
1. A food chain is a specific sequence of energy transfer; a food web is an interlinked network of food chains.
2. Producers store 1 to 5 percent of available solar energy; consumers store 10 to 20 percent.
3. Numbers pyramids show numbers of organisms; biomass pyramids show the dry weight of living organisms; energy pyramids show energy production for each trophic level.
(See pp. T50A–T50D.)

Section Objectives

- *Trace the cycle of water between the earth and the atmosphere.*

- *Summarize the major steps in the nitrogen cycle.*

- *Explain how oxygen and carbon are cycled through an ecosystem.*

- *Describe the role of vegetation in nutrient cycling in a forest.*

Tip: Review the aerobic and anaerobic respiration processes.

50.3 Biogeochemical Cycles

The process by which inorganic materials move from the atmosphere or soil into living organisms and back again is called a **biogeochemical cycle.** Some materials, including nitrogen, water, oxygen, and carbon, reach ecosystems via the atmosphere. Rocks and soil particles contain other important elements, such as calcium and phosphorus, which are usually chemically bound as mineral salts. The weathering of rocks and soil by precipitation releases these minerals. In addition, every time you breathe, there is a good chance that you are taking in oxygen once breathed by a dinosaur. Your teeth probably contain calcium that was once part of a long-extinct invertebrate.

Water Cycle

Chemical reactions and mineral transport within most organisms cannot occur without water. Plants absorb water through their roots and eliminate water through their stomata. Animals drink water or obtain it from their food. They return water to the bio-

---

Biology in Process

Nutrient Cycling and Deforestation

What effect do trees have on nutrient cycling in a forest? How does their removal affect the nutrient balance in the ecosystem? Through long-term studies Cornell University ecologist Gene Likens and his colleagues have attempted to answer these questions.

Since 1963 Likens' team has studied nutrient cycling and the effects of deforestation in the Hubbard Brook Experimental Forest in New Hampshire. Because solid granite bedrock lies under the thin soil of the area, water cannot seep into the ground. Water leaves this ecosystem only through evapora-

tion and transpiration or by flowing out through one of six small creeks in the forest.

Likens and his colleagues obtained runoff samples by building metal V-shaped weirs across the creeks. All water flowing out of the valleys had to flow over one of the weirs. To measure the water entering the ecosystem, they set up rain and snow collectors.

Next they measured the nutrient levels of the precipitation; the creek water; and the plants, animals, and soil in the forest. Interpreting the massive amount of data provided a picture of nutrient uptake, recycling, and retention.

sphere when they exhale, perspire, or excrete. Of the nearly 1.5 billion km³ of water on the earth, less than 1 percent is available to land organisms.

Water that sustains life on land cycles constantly between the air and the earth. Most of this cycling takes place through evaporation, transpiration, and precipitation. Evaporation occurs when solar energy causes water to move from the earth into the atmosphere. Plants also release water into the atmosphere during transpiration. The amount of water the atmosphere can hold depends on such abiotic factors as temperature and air pressure. Once the atmosphere becomes saturated with water, precipitation occurs in the form of rain, snow, sleet, hail, or fog.

Nitrogen Cycle

Plants and animals use nitrogen to build structural proteins and enzymes. Nitrogen gas makes up about 78 percent of the atmosphere. However, most living things cannot use atmospheric nitrogen (N_2). They must rely instead on nitrogen compounds found in soils, such as the nitrites (NO_2^-) and nitrates (NO_3^-). The process that converts nitrogen gas into compounds that plants and

Answer to Biology in Process
They first studied the nutrient levels in all parts of the ecosystem, and then removed trees from one part of the forest, and studied the impact of deforestation on nutrient levels.

The ecologists discovered that the ecosystem retained most of its own nutrients. In other words, the amount of nutrients held in the ecosystem far exceeded the amount that entered or exited. For example, plants recycled almost all of the inorganic nutrients from decomposed leaves and twigs. In addition, they found that the amount of nutrients flowing out of the ecosystem nearly equaled the amount deposited from the atmosphere.

Once Likens and his colleagues determined the efficiency of nutrient cycling in the forest, they designed experiments to test the effects of deforestation. They cut down all the trees and shrubs in one of the six valleys and then collected data on nutrient runoff.

The flow of water from the creek immediately increased by 40 percent. More important, nutrients began to flow out of the ecosystem at a dramatically increased rate. In fact, loss of

nitrates increased by more than 60 times after deforestation. Clearly vegetation plays a key role in maintaining nutrient levels in the ecosystem.

In further experiments the scientists tried to find ways to prevent nutrient runoff. They proved that by leaving intermittent strips of trees, the nutrient loss was greatly reduced. By writing their findings in books and journals, the scientists communicated to foresters the advantages of reducing the rate of deforestation.

■ What process did Likens and his colleagues use to determine how trees affect nutrient cycling in a forest ecosystem?

Figure 50-11 Since only a few microorganisms can fix atmospheric nitrogen, the movement of nitrogen through an ecosystem is more like that of a mineral than a gas.

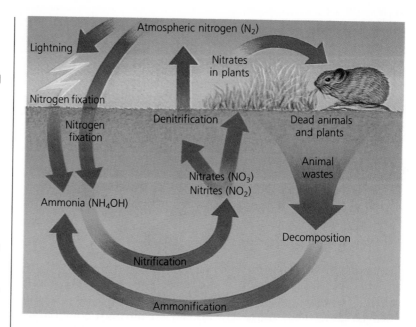

Info: Organisms need these elements in large amounts: carbon, hydrogen, oxygen, nitrogen, phosphorus, and sulfur. They need these in smaller amounts: sodium, potassium, manganese, calcium, iron, magnesium, chlorine, iodine, cobalt, and boron. All of these elements are subject to geochemical cycling.

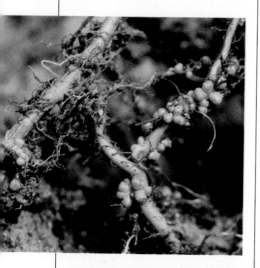

Figure 50-12 Nodules containing *Rhizobium japonicum,* nitrogen-fixing bacteria, adhere to a red alder plant. Bacteria like these convert nitrogen into ammonium compounds.

animals can use is called the nitrogen cycle. The cycle includes four major processes: nitrogen fixation, ammonification, nitrification, and denitrification.

During **nitrogen fixation** bacteria such as *Rhizobium japonicum* convert nitrogen gas (N_2) into ammonium compounds (NH_4OH). These bacteria live on the roots of legumes, which include alfalfa, clover, peas, and beans. Some blue-green bacteria can also fix nitrogen.

In the second process, **ammonification,** bacteria break down nitrogen-containing amino acids from animal wastes and dead organisms, forming ammonia compounds.

During **nitrification,** chemosynthetic bacteria oxidize ammonia compounds to produce nitrites and nitrates. Nitrates can also enter the soil from other sources. The enormous energy in a bolt of lightning can cause nitrate ions to form from atmospheric nitrogen. The erosion of nitrate-rich rocks also creates nitrates. Plants use the nitrites and nitrates to form amino acids.

In the final process, **denitrification,** anaerobic bacteria break down nitrates, releasing nitrogen gas back into the atmosphere. The process of converting atmospheric nitrogen into nitrites and nitrates requires much energy and so occurs slowly. As a result nitrogen availability often limits the productivity of an ecosystem. The nitrogen cycle is illustrated in Figure 50-11.

Oxygen–Carbon Cycle

During photosynthesis plants use carbon dioxide (CO_2) along with water and solar energy to make glucose and release oxygen. Both plants and animals use oxygen to break down glucose during cellular respiration. The by-products of respiration are carbon

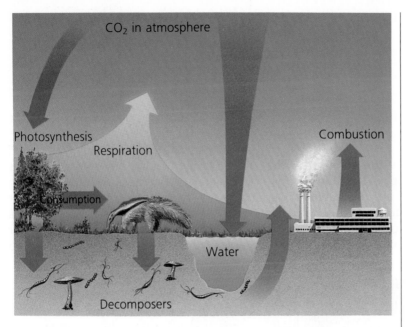

Figure 50-13 The processes of photosynthesis and respiration cause the balanced cycling of oxygen and carbon through an ecosystem.

dioxide and water. Together photosynthesis and cellular respiration form the basis of the oxygen–carbon cycle. Decomposers also contribute to the cycle by releasing carbon dioxide back into the atmosphere when they break down organic compounds. Figure 50-13 traces the path of oxygen and carbon through an ecosystem.

The processes of photosynthesis and respiration generally balance each other. In other words, the levels of oxygen and carbon dioxide remain fairly constant. However, since about 1850 humans have increased their consumption of fossil fuels. These fuels, principally petroleum products and coal, come from the stored carbon compounds found in the transformed remains of prehistoric plants and animals. The combustion of fossil fuels has created a slow but steady increase in atmospheric carbon dioxide levels. The increase in levels of CO_2 in the atmosphere may have adverse environmental effects, as this causes an increase in the ability of the atmosphere to trap heat. This trapped heat may cause temperatures around the world to gradually rise, a phenomenon called the **greenhouse effect.** This kind of long-term disruption of the natural cycling of nutrients can change the habitat of many species.

Section Review

1. What is the *greenhouse effect?*
2. Trace the cycling of water between air and earth.
3. List the two main processes of the oxygen–carbon cycle.
4. How has the increased consumption of fossil fuels affected the oxygen–carbon cycle?
5. What abiotic factors might affect the rate of nitrogen cycling?

Answers to Section Review
1. the trapping of heat within the atmosphere due to an increase in levels of CO_2
2. Water enters the atmosphere through evaporation and transpiration. It returns to the earth as precipitation.
3. photosynthesis and respiration
4. It has increased the level of carbon dioxide.
5. Water and oxygen levels, temperature, and soil chemistry can affect the activity of bacteria and therefore the rate of nitrogen cycling.

Writing About Biology

Arctic Dreams

This excerpt is from the first chapter of Barry Lopez's prize-winning book, Arctic Dreams.

The far northern trees, like the animals, constitute a very few species—willows growing in valleys where they are protected from the wind and a dwarf form of birch. Along the tree line itself, the only successful strategists are species in the pine and birch families. Their numbers thin out over a span of several miles, with trees persisting farther north in isolated patches where there is a fortuitous conjunction of perennially calm air, moisture, and soil nutrients. Islands of trees in the tundra ocean.

The growth of trees in the Arctic is constrained by several factors. Lack of light for photosynthesis of course is one; but warmth is another. A tree, like an animal, needs heat to carry on its life processes. Solar radiation provides this warmth, but in the Arctic there is a strong correlation between this warmth and closeness to the ground. In summer there may be a difference of as much as 15° F in the first foot or so of air, because of the cooling effect of the wind above and the ability of dark soils to intensify solar radiation. To balance their heat budgets for growth and survival, trees must hug the ground—so they are short. Willows, a resourceful family to begin with, sometimes grow tall, but it is only where some feature of the land stills the drying and cooling wind.

Lack of water is another factor constraining the development of trees. No more moisture falls on the arctic tundra in a year than falls on the Mojave Desert; and it is available to arctic plants in the single form in which they can use it—liquid water—only during the summer.

Permafrost, the permanently frozen soil that underlies the tundra, presents arctic trees with still other difficulties. Though they can penetrate this rocklike substance with their roots, deep roots, which let trees stand tall in a windy landscape, and which can draw water from deep aquifers, serve no purpose in the Arctic. It's too cold to stand tall, and liquid water is to be found only in the first few inches of soil, for only this upper layer of the ground melts in the summer. (Ironically, since the permafrost beneath remains impervious, in those few weeks when water *is* available to them, arctic trees must sometimes cope with boglike conditions.)

Trees in the Arctic have an aura of implacable endurance about them. A cross-section of the bole of a Richardson willow no thicker than your finger may reveal 200 annual growth rings beneath the magnifying glass. Much of the tundra, of course, appears to be treeless when, in many places, it is actually covered with trees—a thick matting of short, ancient willows and birches. You realize suddenly that you are wandering around on *top* of a forest.

Write

- The trees of the Arctic tundra are so tiny that in relation to them a human seems like a giant. In the tropics, however, enormous trees, ferns, and other plants dwarf any person. Using Barry Lopez's observations as a basis for reasoning, write an explanation for the tremendous size—and variety—of plants in the tropics.

 Barry Lopez, excerpted from Arctic Dreams. *Copyright © 1986 Barry Holstun Lopez. Reprinted with the permission of Charles Scribner's Sons, an imprint of Macmillan Publishing Company.*

Laboratory

Producers and Consumers

Objective
To identify the producers and consumers in an aquatic ecosystem

Process Skills
Classifying, inferring, hypothesizing

Materials
10-gallon (or larger) aquarium filled with pond water, 1-mL pipette, 2 microscope slides, 2 coverslips, microscope, unlined paper

Method

1 Observe the pond water in the aquarium. Does the appearance of the water differ from the top of the aquarium to the bottom? Note the amount of light that reaches the water at the bottom. Would light availability effect the depth at which particular organisms are found? Why or why not? Use a pipette to take a sample of water from the surface.

2 Place a drop of the surface water on a microscope slide. Cover the slide with a coverslip. Study the sample with a microscope under low power. When you locate an organism, switch to high power, and examine the organism closely. Make a drawing of the organism. Return to low power, and continue to search for organisms.

3 The microcommunity of an aquatic ecosystem consists of plankton (microscopic organisms, primarily protists and bacteria, that form the base of the food chain) and small invertebrates. Phytoplankton make their own food and zooplankton use other organisms for food. Use the figure to identify microorganisms such as phytoplankton, zooplankton, and invertebrates. Which organisms are photosynthetic?

4 Determine whether each microorganism is motile or sessile. Add this information beneath your drawing. Is there any relation-

ship between motility and photosynthetic activity?

5 Use the pipette to obtain a sample of water from near the bottom of the aquarium.

6 Repeat steps 2, 3, and 4 with the second drop of water.

7 Note and record any differences you observe in the numbers and kinds of organisms that inhabit each area of the aquarium.

Conclusions

1 Which organisms are producers and which are consumers? How can you tell?

2 Ecosystems depend on decomposers to break down dead tissues and waste materials. Did you see any decomposers in your water samples? Why or why not?

3 Are there more producers or consumers in an aquatic ecosystem? Why do you think this is so?

Inquiry
What results would you expect to observe if a group of primary consumers were removed from an aquatic ecosystem? How could you test your hypothesis?

Chapter 50 Review

Vocabulary

ammonification (790)	food chain (785)	net primary productivity (783)	primary producer (782)
biogeochemical cycle (788)	food web (786)	niche (781)	primary productivity (783)
biomass (787)	greenhouse effect (791)	nitrification (790)	scavenger (784)
carnivore (783)	gross primary productivity (783)	nitrogen fixation (790)	secondary consumer (783)
consumer (783)	habitat (781)	omnivore (784)	trophic level (782)
decomposer (784)	herbivore (783)	primary consumer (783)	
denitrification (790)			

Explain the difference between the terms in each of the following sets.

1. gross primary productivity, net primary productivity
2. carnivore, scavenger
3. primary consumer, secondary consumer
4. primary producer, decomposer
5. nitrification, denitrification

Review

1. The physical area in which an organism lives is its (a) trophic level (b) niche (c) ecosystem (d) habitat.
2. An organism's niche in an ecosystem does not include its (a) feeding habits (b) reproductive behavior (c) habitat (d) biomass.
3. The major producers found in aquatic ecosystems are (a) phytoplankton (b) chemosynthetic bacteria (c) aquatic plants (d) zooplankton.
4. Decomposers benefit an ecosystem by (a) manufacturing energy (b) returning nutrients to the soil (c) controlling the population (d) removing toxic substances.
5. The greatest amount of energy in an ecosystem is available to (a) producers (b) primary consumers (c) secondary consumers (d) decomposers.
6. If you measure the biomass of an ecosystem, you are measuring the (a) amount of energy produced (b) amount of nutrients (c) dry weight of organisms (d) dry weight of all organic material.
7. The percentage of the earth's water supply that is available to land organisms is (a) more than 50 percent (b) less than 1 percent (c) 10 percent (d) 5 percent.
8. The combustion of fossil fuels has increased atmospheric levels of (a) nitrates (b) carbon monoxide (c) ammonium compounds (d) carbon dioxide.
9. Most plants obtain their nitrogen from (a) ammonium compounds (b) nitrogen gas (c) nitrites and nitrates (d) ammonia.
10. The removal of trees from the Hubbard Brook Experimental Forest resulted in (a) several resident animal species' becoming endangered (b) severe flooding in the valleys (c) increased nutrient runoff (d) higher nutrient levels in the soil.
11. Name the major primary producers in a prairie, a tropical rain forest, and a lake.
12. Distinguish between an herbivore, a carnivore, and an omnivore.
13. How is a vulture adapted to scavenging?
14. What happens to the solar energy not stored by primary producers?
15. Why do changes in an environment often have far-reaching effects?
16. Why might a numbers pyramid appear inverted?

17. What do biomass pyramids measure?
18. Name the two major processes involved in the oxygen–carbon cycle.
19. How does the transfer of energy in an ecosystem differ from the transfer of nutrients?
20. Describe three ways in which nitrates become available to an ecosystem.

Critical Thinking

1. Human beings, raccoons, and bears are omnivores. What adaptive advantage might this give them?
2. During the 1970s, shark-killing expeditions became a fad. Why are sharks important in the food chain of the ocean, and what do you suppose might happen to other fishes if large numbers of sharks disappeared?
3. When insecticides are sprayed on fields or forests, residues of the insecticide show up in the bodies of the organisms living in these habitats. The most concentrated residues usually show up in the bodies of organisms on the highest trophic levels of the food chain. Why do you think this is so?
4. If an organism's habitat is destroyed, how do you suppose this would affect an individual within the species? How might this

affect the species as a whole if the damage is widespread?
5. Explain why farmers often grow alfalfa, soybeans, or other legume crops in a field after they have grown corn.
6. Which organisms in the picture above are consumers? What other organisms would you guess are part of their food chain?

Extension

1. Build your own ecosystem in an aquarium or a terrarium. Be sure to include a representative consumer, producer, scavenger, and decomposer.
2. Study an ecosystem near your home or school. Make a list of all the consumers, producers, scavengers, and decomposers you observe. Draw an ecological pyramid that includes them all.
3. Read the article by Shana Alexander entitled "The Serengeti," which appeared in the *National Geographic*, May 1986, p. 585. List at least three ways the poachers described in this article have affected, either directly or indirectly, the food chain of the Serengeti.
4. Use your library to research the ecosystem that exists around the floating and free-living plankton in the ocean. Draw the plankton types you learn about, and make a diagram of the food chain that begins with them.
5. Some scientists believe that the growing amount of carbon dioxide in the atmosphere will cause the greenhouse effect, resulting in rising world-wide temperatures. Other scientists argue that other kinds of pollution are creating large amounts of dust and other particles in the atmosphere. This pollution, they argue, will counteract the greenhouse effect. Use your library to research both of these theories, and make a list of the arguments on both sides.

Chapter 51 | Ecosystems: Relationships

This chapter familiarizes students with the relationships and predictable patterns of living things. Environmental change is stressed as both a cause and a result of the interaction among organisms.

Correlation Chart

Section Number	Section Objectives	Pages	Features	Section Review
51.1	1. Define *competition,* and distinguish between intraspecific and interspecific competition.	797–799		1
	2. State the competitive exclusion principle.			2
	3. Describe the importance of predator–prey relationships in an ecosystem.			5
	4. Identify the three types of symbiosis, and give an example of each.			3, 4
51.2	1. Define *circadian rhythms, annual rhythms, migration,* and *tidal rhythms.*	800–802		1, 4
	2. Explain the adaptive advantages of rhythmic patterns.			2, 5
	3. Distinguish hibernation from estivation.		Think, p. 801	3
51.3	1. Define *primary* and *secondary* succession, and give an example of each.	803–807	Biology in Process, pp. 804–805	1, 3, 5
	2. Distinguish between seral and climax communities.			2
	3. Describe the succession process in lakes.		Biology Lab, p. 807	4
	4. Summarize the evidence that Cowles used to develop his model of succession in sand dunes.			

Resources*
Print

Dethier, Vincent G. *The Ecology of a Summer House.* Amherst, Mass.: University of Massachusetts Press, 1984.

Zann, Leon P. *Living Together in the Sea.* Neptune City, N.J.: T.F.H. Publications, 1980.

Audiovisual/Software

Biological Clocks. 2 filmstrips with cassettes, 17-19 min. each. Teacher's guide. National Geographic, 1984.

The Islands: Watery Outposts of Australia. Video, 25 min. Centre, 1986.

*Additional references are listed in *Modern Biology, Teacher's Program Guide.*

Section Teaching Guides

51.1 Biotic Relationships

Approach

The main objective of this section is to define and give examples of three major types of biotic relationships. The section builds upon concepts from Chapter 50. In presenting the text, guide students to realize that biotic relationships center around access to resources such as food, water, sunlight, and space or territory.

Concept Framework

All students have experienced competition at different times in their lives. To build a conceptual framework for understanding competition in biotic relationships, engage students in a discussion of competition. Elicit from students examples of competition they might have experienced. Tell students that competitive relationships are a part of the three types of biotic relationships they will be reading about in this section. Encourage them to note examples of competition among organisms and to figure out the kinds of things organisms compete for as they read the section.

Emphasis

Make sure students can give examples of the following biotic relationships: competition, including intraspecific and interspecific competition; predation; and symbiosis, including parasitism, commensalism, and mutualism.

Bulletin-Board Display

Divide the class into three groups. Direct each group to design a bulletin-board display of one of the three biotic relationships. Provide magazines for students to cut up. Tell them to write explanatory captions for the pictures.

Section Review Answers

1. Intraspecific competition takes place between organisms of the same species; interspecific competition takes place between organisms of different species.
2. The competitive exclusion principle holds that complete interspecific competitors cannot coexist. One example might be the paramecia in Gause's experiments, which demonstrated that the two species could not survive in the same culture.
3. Parasites such as roundworms may survive in the digestive tract of a host such as a deer. The parasites consume the host's food, depriving it of nutrients, causing the host to weaken. The parasite benefits from the relationship, but the host does not.
4. In commensalism only one organism benefits; in mutualism both benefit.
5. Predation establishes a long-term balance in numbers between predator and prey.

51.2 Rhythmic Patterns

Approach

The main objective of this section is to discuss the adaptive advantages of rhythmic patterns of organisms. The section builds upon knowledge of biotic relationships presented in Section 51.1. In presenting the chapter, guide students to notice the period or time involved in circadian rhythms, annual rhythms, migration, and tidal rhythms.

Concept Framework

Students will be aware of patterns in their lives that follow periodic rhythms. To draw on this knowledge, ask students to describe some of these rhythmic patterns. Build on their knowledge by explaining that some of these rhythms correlate with changes in the environment and others do not. Encourage students to note the different rhythmic patterns organisms experience as they read.

Emphasis

Students should be able to give examples of the following rhythms: circadian rhythms, annual rhythms, migratory rhythms, and tidal rhythms.

Library Research

Ask each student to choose a migratory species, such as certain whales, reindeer, or birds. Ask them to do library research on the migratory patterns of these animals and to prepare a presentation for the class. Encourage them to discuss why the animals migrate, what routes they follow, and how the animals might navigate. Have students include a map of the migration pattern in their presentations.

Section Review Answers

(See p. 802.)

51.3 Succession

Approach

Describing the steps of succession in different ecosystems is the main objective of this section. It builds upon concepts of adaptation and evolution from Chapters 15 and 16 and on concepts of ecosystems from Chapters 49 and 50. In presenting the text, note that succession continues until a climax community is established. Point out that such a community can remain stable for thousands of years unless the environment is significantly disturbed.

Concept Framework

Most students have probably observed examples of succession in their neighborhood or in places they have visited. To tap into this knowledge, discuss how areas such as fields, empty lots, or even lawns or gardens change over time if left undisturbed. If possible, bring in photographs of such areas taken at intervals of ten or more years. Ask students to note the patterns of changes in ecosystems as they read.

Emphasis

Make sure that students can define *succession* and describe the process of primary succession, secondary succession, and succession in lakes.

Using the Features/Developing a Model of Succession

The main process skills illustrated in this feature are observing, inferring, and modeling. The feature describes how Dr. Henry Cowles developed a model of succession in a dune ecosystem. He inferred that as the level of the lake dropped over the last 12,000 years, new levels of land were exposed and were colonized. Thus communities nearest the lake's edge would exhibit the early stages of succession and those furthest from the lake's edge would exhibit the latest. Refer students to the illustration on pp. 804–805 to understand the order of succession discussed in the feature.

Field Trip

Take students on a field trip to a lake shore, dune, field, pine forest or other seral community. Ask them to identify different species of plants and to look for evidence of succession. You may wish to suggest that they take photographs or make sketches of such evidence to display in class.

Section Review Answers

(See p. 806.)

Laboratory Preparation

Background

This lab introduces students to two biological issues. One is nitrogen fixation; nitrogen is the major limiting nutrient in most soils. Without nitrogen-fixing bacteria, the lack of available nitrogen for plant growth would noticeably change vegetation worldwide. The other issue is symbiosis; symbiosis is not only important for legumes and nitrogen fixation but is also responsible for the formation of coral reefs, for the presence of chloroplasts and mitochondria in plant cells, and for the presence of mitochondria in animal cells.

Preparation

Rhizobium and legume seeds are available from a biological supply house. Noninfected plants, used as a control, must be grown from sterile seeds in sterile soil. Label the plants *Infected* and *Noninfected* so that students can select one of each.

Safety

Remind students to handle the scalpels with care.

Conclusions

1. Nonsterile soil might contain *Rhizobium* that could invade the young plants.
2. When the legumes were harvested, the bacteria in their roots would be left in the soil to nourish the nonlegume crop the following year. Some usable nitrogen would also be left in the soil.

Inquiry

Students can infect two containers of soil with *Rhizobium*, one container without plants and one container with legumes specific for that type of *Rhizobium*. After a certain amount of growing, students can begin testing the container without plants to see whether the *Rhizobium* can thrive alone.

Chapter Review

Vocabulary

1. *Dian* comes from *dias,* which means "day." Combined with *circa* it forms a word meaning "about one day."
2. *Mensal* comes from *mensa,* the Latin word for table. *Commensalism* refers to a relationship between organisms in which each obtains food from the same place, or "at the same table."
3. *Eutroph* comes from *eutrophos,* which means "well nourished"; eutrophication is the process by which a lake becomes nutrient rich.
4. *Necro* originally meant "dead body." *Necromass* is material from dead organisms.
5. *Hiber* comes from a word for winter; *hibernation* is a state of dormancy during winter.

Review

11. Gause used two species of paramecium that competed for the same food (*Paramecium caudatum* and *P. aurelia*). He placed them in separate cultures and found that each grew well. Then he combined them in one culture and found that only *P. caudatum* survived.
12. These three do not feed in the same place on a tree. Blackburnian warblers feed in the canopy; bay-breasted warblers in the middle layers of the tree; and myrtle warblers, near the ground.
13. In both relationships one organism feeds on another organism. However, a predator catches and kills its prey; a parasite lives within the organism it feeds on and can only feed if the host organism remains alive.
14. A diurnal animal is most active during the day, and a nocturnal animal is most active at night.
15. To discover whether physical factors, such as daylight, control human circadian rhythms or whether the rhythms are controlled by something else, scientists placed human subjects in darkened rooms and noted their circadian rhythms.
16. It refers to an undefined internal timing mechanism in organisms.
17. Migration gives organisms an advantage in exploiting nutrients, climatic conditions, and habitats available only seasonally.
18. The interactions of pioneer species with the environment change the microenvironment to one that is ultimately unfavorable to the growth of the seedlings of the pioneer species but favorable to the growth of other species.
19. Necromass becomes soil, which can fill in lakes and support terrestrial plants.
20. Lichens were the pioneer species.

Critical Thinking

1. Answers may vary. The delay may allow the individual to gain the strength and status to compete with older adults of its own species. It may give the individual time to become expert at feeding itself and therefore its future young. It may also allow the individual to become expert at protecting its territory and therefore its future mate against competitors of its own species and its young against competitors of other species. The more adept the individual is at such things when it reproduces, the more likely that the species will survive.
2. Disease kills animals and plants that have immunological or genetic weaknesses. This reduces the species population and lessens intraspecific competition. In the long term the culling of the weak strengthens the species in interspecific competition.
3. The owls must move south when the lemmings are scarce—every four years on average.
4. Because people have different circadian rhythms, some would adjust better to night work, others to day work.
5. Herbaceous plants, which can seed themselves and which grow quickly in the fertile volcanic dust, would be the first to appear. They would most likely be followed by shrubs and much later, pines. (Scientists estimate that it may take 40 or more years to rebuild the food chain even though the U.S. Forest Service is fertilizing and seeding the area.)

Chapter 51 | Ecosystems: Relationships

Introduction

You have seen that energy and nutrients flow through an ecosystem in predictable patterns. Similarly, individuals and populations in an ecosystem interact in predictable ways. In this chapter you will learn about the major types of interactions among organisms and about some of the ways in which organisms respond to and alter their environment.

Chapter Outline

51.1 Biotic Relationships
- *Competition*
- *Predation*
- *Symbiosis*

51.2 Rhythmic Patterns
- *Circadian Rhythms*
- *Annual Rhythms*
- *Migratory Rhythms*
- *Tidal Rhythms*

51.3 Succession
- *Primary Succession*
- *Secondary Succession*
- *Succession in Lakes*

Chapter Concept

As you read, look for predictable patterns in the ways organisms live together, respond to rhythmic changes in the physical environment, and alter the environmental conditions in an ecosystem.

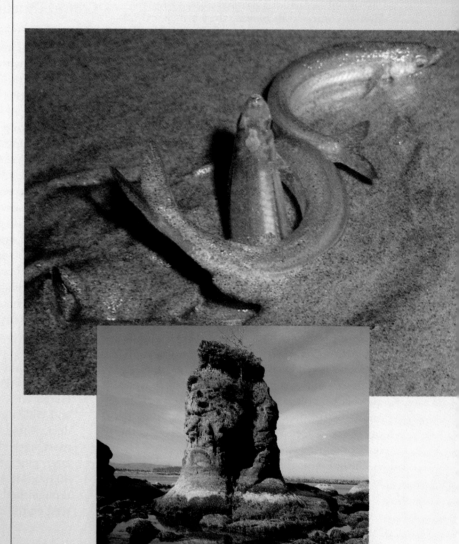

Grunion, *Leuresthes tenuis*, depositing eggs, southern California. Inset: Sea stacks at low tide, Tillamook Bay, northern coast of Oregon.

For additional teaching information see pp. T51A–T51D.

51.1 Biotic Relationships

All relationships among organisms are called **biotic relationships.** In studying food webs in Chapter 50, you learned about one type of biotic relationship. In this section we will take a look at other major biotic relationships, including competition, predation, and symbiosis.

Competition

One of the primary ways that organisms interact is through competition. **Competition** is the use or defense of a resource by one individual that reduces the availability of that resource to other individuals. For example, a brook trout may compete for habitat by defending an area directly around a fallen log. When other fish attempt to move into this territory, the brook trout fights to hold it. If the trout is successful, it retains access to food and space. Fish that do not gain this prime habitat may be less successful in the struggle to survive and reproduce.

Biologists recognize two main types of competition—intraspecific and interspecific competition. **Intraspecific competition** occurs between organisms of the same species. **Interspecific competition** occurs between organisms of different species.

Intraspecific Competition

Intraspecific competition is very keen because members of the same species require the same food, space, and mates. For example, in many mammalian species, such as the musk-ox and baboon, the males fight for the social status that will allow them to mate. Beech trees likewise compete for access to nutrients and water in the soil and for the sunlight needed for photosynthesis.

In a population the individuals best adapted to survive will pass their genetic material on to offspring at a greater rate than less fit competitors. Thus intraspecific competition, a constant process in nature, is one of the driving forces of evolution.

Interspecific Competition

Interspecific competition is often less intense than intraspecific competition because individuals of different species do not compete for exactly the same kinds of food, space, or mates. However, when two species do compete for a certain natural resource, one will succeed in obtaining it, and the other will not. This was demonstrated in an experiment conducted in the 1930s by Soviet ecologist G. F. Gause. Gause used two species of paramecia that compete intensely for one kind of food. Figure 51-1 shows the results of Gause's experiments. Note that each species grew well in a separate culture, but when Gause placed the two species in the same culture, the population of *Paramecium caudatum* always died, and the population of *P. aurelia* always survived.

Section Objectives

- *Define* competition, *and distinguish between intraspecific and interspecific competition.*

- *State the competitive exclusion principle.*

- *Describe the importance of predator–prey relationships in an ecosystem.*

- *Identify the three types of symbiosis, and give an example of each.*

Info: In other experiments Gause introduced a new variable, the cleanliness of water. When he grew the species in culture tubes in which wastes had been allowed to accumulate, *P. caudatum* survived and *P. aurelia* died.

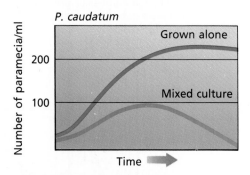

Figure 51-1 Note that both species of paramecia thrived when they were grown separately but that *P. caudatum* always died when the two species were mixed.

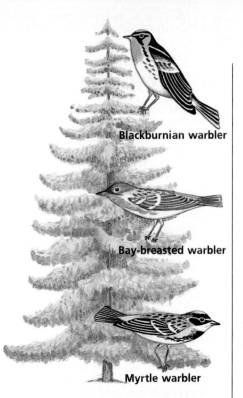

Figure 51-2 These species of warblers have adapted so that each species eats primarily at a different level of the tree. This adaptation has reduced the inter-specific competition among them.

Blackburnian warbler

Bay-breasted warbler

Myrtle warbler

Figure 51-3 Notice the cycles of predation: As predators eat the prey, the number of predators increases and the number of prey decreases. As the food supply drops, the number of predators decreases and the number of prey increases.

Gause's work and similar experiments by other scientists established that when two or more species compete for the very same resource, all but one will eventually fail as a competitor. In other words, two competitors cannot coexist on the same limiting resource. Biologists call this idea the **competitive exclusion principle.** Figure 51-2 illustrates one instance of adaptation of habits that reduces competition. The three species of warblers shown eat similar types of food and can occupy the same forest, but they differ in their feeding patterns. Each species feeds at a different height in the trees. Such behavioral adaptations greatly reduce—but do not entirely eliminate—the competition for food.

Predation

Another kind of biotic relationship is predation, in which one organism feeds upon another. In any ecosystem the relative numbers of predators and prey vary from year to year. Over time, however, a biological balance is established. This pattern is shown in Figure 51-3, which charts the numbers of predator mites, *Typhlodromus,* and prey mites, *Eotetranychus,* during an experiment over an eight-month period.

Predators and prey are part of the food web of an ecosystem. Therefore a change in the number of either affects the entire ecosystem. In some areas, for example, sheep are preyed upon by coyotes. However, coyotes also prey on other animals, such as jackrabbits. Therefore if ranchers kill coyotes to protect their sheep, a few more sheep may survive, but the number of jackrabbits in the area may rise, allowing them to compete with the sheep for food and other necessary resources.

Symbiosis

Symbiosis is a biological relationship in which two dissimilar organisms live together in a close association. Biologists distinguish three main types of symbiosis: parasitism, commensalism, and mutualism.

Tip: Ask students to describe possible sources of error in the collection of data on the predator and prey mites in Figure 51-3.

Parasitism

Parasitism is a close, long-term symbiotic relationship in which one organism obtains its nutrition from another organism. The organism receiving the benefits is a **parasite**. The organism providing them is a **host**. For example, the parasitic worms in the digestive tract of a white-tailed deer survive by consuming some of the deer host's own partially digested food. In time the parasite's food consumption may be enough to deprive the deer of nutrition. This deprivation weakens the host, making it less fit to survive. In Figure 51-4 you can see the effects of mistletoe parasitizing a mesquite tree.

Commensalism

Commensalism is a form of symbiosis in which one organism benefits and the other neither benefits nor suffers harm. **Epiphytes** *(EP-uh-FITES)*, plants that grow on other plants, are examples. Among these are the orchid species that grow high in the trees of tropical forests. The trees provide the orchids with the support to grow and allow them to capture more sunlight than they would on the forest floor. The trees are not harmed, since the orchids neither feed on their tissues nor prevent significant amounts of sunlight from reaching their leaves. Figure 51-4 shows another commensal relationship, between shrimp and a sea cucumber.

Mutualism

Mutualism is a form of symbiosis in which both organisms benefit from living together. For example, the enzymatic action of bacteria that live in the digestive tracts of cattle enables the cattle to digest grasses. By breaking down the cellulose in the grasses, the bacteria make it possible for the cattle to use the nutrients the grasses contain. At the same time the bacteria take in nutrients for themselves—but not enough to harm the host. Both organisms therefore benefit from this close association. The clam and the algae shown in Figure 51-4 have a similar relationship. While the clam provides the algae with a place to anchor, the algae camouflage the clam, thus providing protection.

Section Review

1. Define *intraspecific competition*, and explain how it differs from *interspecific competition*.
2. State the competitive exclusion principle, and give one example of how it works.
3. Briefly describe one example of a parasitic relationship. Be sure to identify and define both the parasite and the host.
4. How does commensalism differ from mutualism?
5. In what way does predation result in homeostasis in the ecosystem?

Answers to Section Review (See pp. T51A–T51D.)

Figure 51-4 Various symbiotic relationships are shown here: parasitism (top), a mistletoe, *Viscum album*, in a mesquite tree, *Phoradendron flavescens*; commensalism (center), a shrimp and sea cucumber, *Periclimenes imperator*; and mutualism (bottom), a clam and algae.

- *Define* circadian rhythms, annual rhythms, migration, *and* tidal rhythms.

- *Explain the adaptive advantages of rhythmic patterns.*

- *Distinguish hibernation from estivation.*

Info: One theory on why animal circadian rhythms evolved holds that the rhythms enable animals to anticipate and prepare for daily events. The ability to do so increases animals' efficiency and conserves their energy.

51.2 Rhythmic Patterns

Living things interact not only with other organisms but also with their physical environment. In this section we will examine biological rhythms, the periodic responses of organisms to periodic changes in their physical environment. For example, many organisms, such as dogs and humans, are **diurnal** (*die-URN-'l*)—that is, they are active mainly during the day. Other organisms, such as raccoons and owls, are **nocturnal**—that is, they are active mainly at night. This daily alternation between periods of activity and inactivity is just one of many rhythmic patterns.

Rhythmic patterns are adaptations that often limit competition, facilitate food gathering, or help organisms respond to changes in climate. Rhythmic patterns also allow different species to occupy the same habitat but to utilize its resources at different times. For example, when the owl hunts by night and the hawk by day, competition for the same food, mice, is limited.

Circadian Rhythms

When the physiological or behavioral pattern of an organism occurs in cycles of about 24 hours, the pattern is called a **circadian rhythm.** If, for example, an animal sleeps during the day and is active during the night it follows a circadian rhythm. Likewise, a plant whose flower petals open during the day and close at night exhibits a circadian rhythm.

Scientists have investigated whether human circadian rhythms are controlled by sunlight, by other physical factors, or by some internal physiological mechanism. Human subjects placed in darkened rooms maintain regular circadian rhythms of between 22 and 26 hours. Scientists therefore suggest that while these rhythms can be affected by social needs or by cycles of daylight, they are controlled by some internal timing mechanism, referred to as a **biological clock.** Many organisms seem to respond to a biological clock. The exact nature of this internal control is unknown, but it may involve fluctuating levels of hormones.

Figure 51-5 In a test of circadian rhythms in humans assigned to eight-hour shifts, highest efficiency occurred between 7 A.M. and 11 A.M. and lowest efficiency at 3 A.M.

Tip: How is endospore formation similar to hibernation? (Both provide an organism with a greater chance of survival during adverse climatic conditions and low food supply.)

Annual Rhythms

A physiological or behavioral pattern that recurs in yearly cycles is called an **annual rhythm.** The reproductive cycles of many plants and animals, for example, are annual rhythms. Female bears give birth during the winter in the protection of the den. Ducks nest and lay eggs in the spring and summer, which allows the young to grow strong enough to survive migration in autumn. Many plants also reproduce in the spring, allowing the next generation to grow and set seeds before the next cycle of cold weather.

Hibernation, a state of severely reduced physiological activity occurring during the winter, is another kind of annual rhythm. This adaptation allows an otherwise active creature to survive during the months when temperatures are extremely cold and food is scarce. When the woodchuck nestles into its protected den as winter approaches, its metabolism, breathing rate, and heartbeat slow. This sleeplike state can be maintained with the use of very little energy.

Estivation (*ES-tuh-VAY-shun*) is an annual rhythm characterized by severely reduced physiological activity during the summer. Animals such as toads and box turtles bury themselves in cool mud and estivate during the hot summer when ponds dry up. Estivation can last for days, weeks, or even months.

Migratory Rhythms

Many species of animals move from one community or biome to another with the change of seasons. Such seasonal movement is a kind of rhythm called **migration.** Migration is a behavioral adaptation that allows animals to exploit the nutrients, climatic conditions, and habitats that are available only seasonally. In summer the northern tundra biome becomes a complex of ponds and wetlands. Dozens of species of waterfowl migrate north to the tundra, where they find a rich supply of vegetation and invertebrates to eat and space in which to hatch and rear their young. The birds then migrate south in the fall to avoid the Arctic winter, which is too cold and dry to support much life. The arctic tern may make the longest journey of any migratory species. Every year it flies 36,000 miles between the North Atlantic, where it breeds, and the Antarctic.

Birds are not the only animals that migrate. Reindeer migrate between the northern taiga and the tundra. Monarch butterflies migrate between the mountains of Mexico and the northern United States and parts of Canada. Many kinds of whales migrate between polar seas, where they feed during the summer, and tropical waters, where they mate and raise young during the winter.

Mountain-dwelling animals may migrate vertically. For example, the bighorn sheep and elk that live in the Rocky Mountains move up the slopes with the onset of summer and descend into the valleys again in winter.

Think

How might you determine whether the degree of cold and the availability of food would cause an organism to hibernate during the winter?

Answer to Think
A study of the habitat of an organism—the temperature range and availability of food—and the dates of hibernation could help you determine the conditions that would cause an organism to hibernate. Experiments involving the variation of the temperature in zoos might also yield clues.

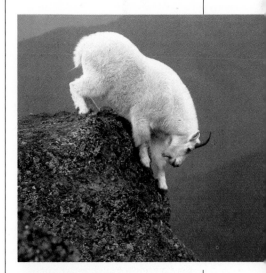

Figure 51-6   A mountain goat, *Oreamnos americanus,* migrates vertically with the change in seasons.

Figure 51-7 Starfish and sea urchins are two types of organisms that can live in deep water at high tide and shallow water at low tide.

Tidal Rhythms

The position of the sun and moon causes tides. When cyclical changes in the physiology or behavior of an organism correspond to the rise and fall of the sea, the cycles are **tidal rhythms.** As you read in Chapter 49, organisms that live in ecosystems along coastlines have adapted to the rhythms of the tides. During low tide small fishes and crustaceans may find protection in tidal pools, and clams seal in moisture by clamping their shells shut.

Barnacles are examples of organisms with a tidal rhythm. Barnacles, mollusks that are sessile, are adapted to tidal conditions. They secrete a substance that glues their shells to rocks or to the shells of other animals. Opposite the end where it is attached, the barnacle shell is open. When the tide is low, the barnacle is exposed to air. The barnacle then pulls deep into its shell to prevent its body from becoming desiccated. When the tide is high, the barnacle extends itself out of its shell and feeds by filtering microorganisms and small invertebrates from the water.

Regular tides also affect organisms that live in the open ocean. Many fish feed in tidal pools at high tide and leave the pools as the tide recedes.

Section Review

1. Define *circadian rhythms.*
2. What are three adaptive advantages of rhythmic patterns or cycles?
3. Compare and contrast hibernation and estivation.
4. Briefly describe the adaptation of a barnacle to the movement of tides.
5. The lack of a regular human reproductive pattern is an exception to the usual seasonal cycle of many animals. What advantage might humans gain by not having an annual reproductive rhythm?

Answers to Section Review
1. A circadian rhythm is a daily change in an organism's physiology or behavior.
2. They eliminate competition, facilitate food gathering, and help an organism adjust to climatic changes.
3. Both are states of severely reduced physiological activity, but hibernation occurs during winter and estivation during summer.
4. During low tide barnacles retreat into their shells and close them to prevent desiccation. During high tide they emerge from the shell and feed.
5. Answers will vary but should reflect an understanding that there may be a selective advantage in being capable of reproducing at any time of year over organisms tied to an annual system.

51.3 Succession

Just as biologists can often predict the periodic responses of organisms, they can also often predict specific patterns of change in plant and animal populations. The gradual, sequential replacement of populations in an area is called **succession.**

Figure 51-8 illustrates the process of succession. The grasses shown are the **pioneer species,** or the first species to colonize a new habitat. All the pioneer species in an area make up a pioneer community. Each set of species in the community changes the environment in ways that ultimately make it unfavorable for the survival and reproduction of those species. Yet these changes allow other species to survive and reproduce. As a result the pioneer community will be replaced by a new community, which will later be replaced by another. Each intermediate community that arises through this process is called a **seral** *(SIR-ul)* **community.** The repeated replacement of seral communities eventually leads to the establishment of a **climax community,** a community that will remain stable as long as the area is undisturbed. What plants dominate the climax community in Figure 51-8?

The soil, climate, and other abiotic factors in a region determine the organisms that will make up a climax community. The forest in Figure 51-8 is a typical climax community in temperate areas of the eastern United States.

Section Objectives

- *Define* primary *and* secondary succession, *and give an example of each.*

- *Distinguish between seral and climax communities.*

- *Describe the succession process in lakes.*

- *Summarize the evidence that Cowles used to develop his model of succession in sand dunes.*

Figure 51-8 From a pioneer community of grasses, this area has progressed to a climax community. Two seral communities preceded the climax community.

		Characteristics	Effects on Environment
Climax community	Maples and beeches	Shade-tolerant trees whose saplings thus survive at higher rate than aspens and birches	Establishes stable environment favorable for its own survival
Seral communities	Aspens and birches	Rapidly growing trees that require more nutrients and water than shrubs	Shade and high nutrient uptake cause shrubs to die off: shade eventually inhibits growth of saplings
	Shrubs	Species that thrive in sun; longer life cycle and more extensive root systems than pioneers	Crowd out pioneer species; further stabilize and enrich soil
Pioneer species	Grasses	Hardy species with well-dispersed seeds that remain viable, grow rapidly in full sun	Stabilize and enrich soil
	Bare land		

Increasing species diversity

Increasing biomass

Tip: Emphasize to students that dominant plants eventually establish themselves and their dependent animals at the expense of many other species that do not survive.

Answer to Biology in Process From geology Cowles learned about sand dunes and the water level of Lake Michigan. Thus he was able to infer that the series of dunes were a record of the process of succession in the dune ecosystem.

Primary Succession

Primary succession is the sequential replacement of populations in an area that has not previously supported life, such as bare rock or a sand dune. The transformation of a barren environment into a climax community may require a thousand years or more.

When glaciers last retreated from eastern Canada about 12,000 years ago, they left a huge stretch of barren bedrock from which all the soil had been scraped. This geologic formation, called the Canadian Shield, was inhospitable to plants and to most animals.

Repeated freezing and thawing broke this rock into smaller pieces. In time lichens colonized the barren rock. Acids in the lichens and mildly acidic rain leached out nutrient minerals from the rock. Eventually the **necromass,** or dead organic matter, from the lichens and minerals from the rock began to form a thin layer of soil in which a few herbaceous plants could grow. These plants then died, building the thickness of the soil. Soon shrubs could grow and then trees. Today much of the Canadian Shield is dense with pine, balsam, and spruce trees whose roots cling to soil that in some areas is still only a few inches deep.

Biology in Process

Developing a Model of Succession

Because succession in an environment may take thousands of years, no biologist can observe the complete process. However, as ecologist Henry Cowles tramped the sand dunes at the southern end of Lake Michigan in the 1890s, he made an observation that allowed him to develop a model of primary succession.

Cowles was a graduate student in ecology, but his undergraduate training had been in geology. He knew that the series of sand dunes at the tip of Lake Michigan had formed when the glaciers retreated about 12,000 years ago, leaving the lake level much higher than it is today. As the lake level fell over time, a series of "raised beaches" was left.

Cowles observed the earliest stages of succession near the lake, and by walking up

Oak   Pine   Cottonwood

In the Galápagos Islands ecologists have gained further insights into primary succession. These relatively new islands were created by volcanic eruptions about 2.5 million years ago. Rainfall gradually weathered the volcanic rock into tiny mineral particles. Ecologists assume that marine organisms and sea birds were among the first life-forms to reach the islands. Bird droppings, called guano, provided a habitat for bacterial spores deposited by the wind. In time other organisms arrived, blown by winds or carried on ocean currents. Some of these organisms thrived and established new communities. As more new species arrived, new seral communities formed, and the process of primary succession continued.

Info: The eruption of Mount St. Helens defoliated practically the entire mountain and exterminated all the local animal life. Recent inspection indicates that plants are beginning to grow once again in the residual volcanic ash.

Secondary Succession

Secondary succession is the sequential replacement of populations in disrupted habitats that have not been totally stripped of soil and vegetation. The disruption may stem from a natural disaster, such as a forest fire or volcanic eruption, or from human activity, such as farming, logging, or mining. After a forest fire, for example, succession might follow the pattern shown in Figure 51-8.

through the dunes he "went backward" in successional time. Since the soil was the same and the climate had changed little, he had a natural laboratory.

As Cowles moved away from the bare sand of the shore, he found the following sequence of plant communities: marram grass; other grasses; sand cherry, willow, and cottonwood; jack pine and white pine; and black oak. Cowles hypothesized that this sequence formed the pattern of succession in the dune ecosystem.

Further studies by Cowles and others helped build this model of succession. Marram grass is adapted as a pioneer plant on dunes; its long, horizontal stems spread quickly and form a network that anchors the sand. Then the marram dies out as later colonists move in. The extensive roots of all of these plants make the dune stable.

Carbon dating of these dunes revealed that about 100 years after the first deciduous trees appeared, pines began to grow. Another 100 to 150 years later the first black oaks, the

climax tree species, arrived and shaded out the pines.

In 1899 Cowles speculated that the succession process on the dunes might continue until beech and maple forests overtook the black oaks. Subsequent studies by other scientists suggest that the acid dune soil will probably not allow the growth of beech–maple forests. This hypothesis is supported by the fact that even the dunes farthest from the lake are still dominated by black oaks.

■ How did Cowles's knowledge of geology help him interpret his observations?

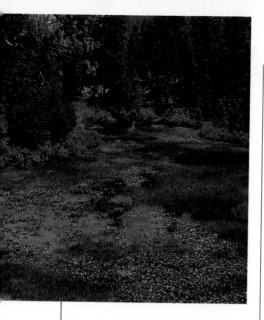

Figure 51-9 This lake is gradually being filled by vegetation.

Answers to Section Review
1. Succession is the gradual, sequential replacement of populations in a habitat.
2. A seral community is part of the evolution of a habitat; a climax community will eventually remain stable if the area is undisturbed.
3. Answers may vary. A forest fire is followed by weed pioneer species; these are succeeded by shrubs that like sun, such as the blueberry; these are succeeded by aspen, birches, cottonwoods; these are succeeded by shade-tolerant trees, such as maple and beeches.
4. Nutrients from the land and the atmosphere flow into the lake; the lake grows murky; sediment accumulates around plants growing at the shore; aquatic arthropods and fish populate the lake; soil and necromass begin filling the lake.
5. Interspecific competition is a driving mechanism of succession. As the environment changes, different species become more effective in competing for the same resources and so displace the species that made up the previous seral community.

Old field succession is the replacement of populations in abandoned farm fields. It commonly takes about 100 years to produce a stable community of trees, such as beech and maple, in an old field. Secondary succession in an abandoned field in the eastern United States begins with annual grasses, mustards, and dandelions. The succession proceeds with perennial grasses and shrubs, continues with trees such as aspens and dogwoods, and ends with a climax deciduous forest of beech and maples.

In many areas environmental conditions are such that a true climax community never forms. For example, the natural climax community in many grassland biomes would be a forest, but periodic fires prevent forests from developing. These grasslands are called fire-controlled climax communities.

Succession in Lakes

Lakes undergo succession that slowly transforms them from crystal clear bodies of water into dry land. A tiny prairie pond may be transformed into dry land in a few dozen years. Changes in the largest lakes take place over millions of years.

The process of lake succession, shown in Figure 51-9, involves eutrophication. **Eutrophication** *(yoo-TROE-fuh-KAY-shun)* is the increase in nutrients in an environment. Early in its existence a lake may be so low in nutrients that relatively few organisms can survive in it. Such a lake is termed **oligotrophic.** As nutrients such as nitrogen and phosphorus flow into the lake from surrounding land and from the atmosphere, the biomass in the lake gradually increases and the lake becomes murkier. Meanwhile, sediment accumulates around the roots of the cattails and rushes that have begun growing along the shore of the lake. Also, an increasing number of aquatic arthropods and fish begin to populate the lake.

In time the accumulated soil from the surrounding watershed along with the necromass begins to fill in the lake. The lake has then gone from an oligotrophic condition to a **eutrophic** condition, one in which there are many nutrients. Eventually the lake may become so full of rich sediment that it becomes a marsh, then a swamp, and finally dry land—which may then proceed through successional stages to become a dense forest.

Section Review

1. Define *succession.*
2. How do seral and climax communities differ?
3. Give one example of secondary succession. Be sure to indicate the disturbance that preceded it, and identify the pioneer species.
4. Briefly describe the eutrophication of a lake.
5. How is interspecific competition related to the process of succession?

Laboratory

Nitrogen Fixation in Root Nodules

Objective
To examine the structure and function of root nodules in legumes

Process Skills
Observing, inferring

Materials
Living alfalfa, bean, or other leguminous plant infected with symbiotic *Rhizobium;* the same type of plant without *Rhizobium* infection; scalpel; microscope; prepared slide of cross-section of legume nodules

Method

1 Carefully dig a plant from the pot labeled *Noninfected.* Shake free any soil that clings to the roots. Examine the root system closely. Set the plant aside.

2 Dig a plant from the pot labeled *Infected,* and repeat step 1. Locate a root nodule on the root system as shown in the photo at the right ($\times 20$). Inside the nodules are cells filled with *Rhizobium* bacteria. The bacteria fix atmospheric nitrogen (N_2) by converting it into ammonia compounds (NH_4OH). Excess ammonia is expelled into the soil and plant tissue surrounding the nodules. Given the need of all green plants for nitrogen, how would you describe the relationship between the bacteria and the legume?

3 Using a scalpel, carefully cut open one of the root nodules.

4 Examine the inside of the nodule carefully. What color is it? If the infecting strain of *Rhizobium* is able to fix nitrogen in this species of legume, the inside of the nodule will be pink. If the strain cannot fix nitrogen, the inside will be green. The pink color indicates the presence of a plant protein necessary for the fixation of nitrogen.

5 Use the scalpel to cut through several other nodules. Are you able to find both types of nodules on the roots of the same plant? Does the fact that some of the nodules are not effective at fixing nitrogen mean that they do not function in the relationship between the bacteria and the legume? What role might the uninfected nodules play?

6 Use a microscope to examine prepared slides of sections of root nodules. Use low power to locate an area you wish to view, and then switch to high power. Determine whether all the cells are infected with bacteria.

Conclusions

1 If you were to grow legumes without root nodules to use as experimental controls, why should you plant the seeds in sterile soil?

2 Why would a farmer want to plant legumes in a particular field one year and a nonlegume crop the next?

Inquiry
In the course of this investigation you have discovered how legumes benefit from this relationship. Devise an experiment that shows whether *Rhizobium* also benefits.

Label the plants *Infected* and *Noninfected* so that students can select one of each.

Chapter 51 Review

Vocabulary

annual rhythm (801)
biological clock (800)
biotic relationships (797)
circadian rhythm (800)
climax community (803)
commensalism (799)
competition (797)

competitive exclusion principle (798)
diurnal (800)
epiphyte (799)
estivation (801)
eutrophic (806)
eutrophication (806)
hibernation (801)
host (799)
interspecific competition (797)

intraspecific competition (797)
migration (801)
mutualism (799)
necromass (804)
nocturnal (800)
old field succession (806)
oligotrophic (806)
parasite (799)

parasitism (799)
pioneer species (803)
primary succession (804)
secondary succession (805)
seral community (803)
succession (803)
tidal rhythm (802)

Use a dictionary to find the origin of each of the terms below. Explain how the root meaning of each underlined part relates to the meaning of the term.
1. circa<u>dian</u>
2. <u>com</u>mensalism
3. <u>eutro</u>phication
4. <u>necro</u>mass
5. <u>hiber</u>nation

Review

1. Intraspecific competition may occur between (a) musk-oxen and baboons (b) red-winged blackbirds and beech trees (c) two baboons (d) starlings and blue-jays.
2. Competition, predation, and symbiosis are all types of (a) parasitic relationships (b) mutualistic relationships (c) biotic relationships (d) annual rhythms.
3. Epiphytes are (a) a type of orchid (b) plants that grow on other plants (c) insects that live high up on trees (d) bacteria that live in the digestive tract of grazing animals.
4. Estivation (a) occurs during months when conditions are favorable for the organism's physiology (b) occurs during the winter (c) always lasts for months (d) occurs during the summer.
5. Seasonal movement is known as (a) hibernation (b) migration (c) estivation (d) succession.
6. Regular tidal rhythms (a) only affect organisms in tidal pools (b) do not affect organisms that do not live in the intertidal zone (c) may affect organisms that don't live in the intertidal zone (d) affect all organisms.
7. Primary succession is the sequential replacement of populations starting (a) on bare rock (b) in a disrupted habitat (c) after a forest fire (d) only on dry land.
8. The biological community on the Galápagos Islands is the result of (a) primary succession (b) secondary succession (c) eutrophication (d) old field succession.
9. Henry Cowles developed his model of succession by (a) experimenting with paramecia (b) studying the fossil record (c) observing sand dune communities (d) using computers.
10. Old field succession (a) is a type of primary succession (b) results in a fire-controlled climax community (c) begins with pioneer tree species (d) is a type of secondary succession.

11. Briefly describe the experiment that led G. F. Gause to establish the competitive exclusion principle.
12. Explain how three species of warblers that consume similar types of food can occupy the same forest without violating Gause's principle.
13. Compare and contrast predation and parasitism.
14. What is the difference between a diurnal animal and a nocturnal one?
15. Describe one study in which scientists studied circadian rhythms in humans. What was the purpose of the study?
16. What does the term *biological clock* mean?
17. Name three adaptive advantages of migratory behavior.
18. Explain what occurs so that an undisturbed community of pioneer species does not remain stable.
19. How does necromass aid in succession?
20. What was the pioneer species in the area known as the Canadian Shield?

Critical Thinking

1. A number of birds and animals reach adult size long before they begin mating and reproducing. What competitive advantages might the delay in reproduction give individuals?
2. In any community of animals, including humans, disease will claim a certain number of the population. How might disease affect competition? In your answer be sure to consider both intraspecific and interspecific competition.
3. Normally, snowy owls hunt lemmings in the Arctic. However, about every four years the owls move south to hunt in southern Canada and the northern United States. How would you explain the migratory rhythm of the owls?
4. Men and women who work on a police force or in a fire department are often required to work nights. Some of them find it very difficult to adjust to this irregular

schedule. How might you explain their difficulty?
5. Look at the photo above. Ferns, asters, and fireweeds were some of the first species to grow out of the ash from the eruption of Mt. St. Helens in Washington. Why might this type of plant have emerged first? What general kinds of plants might follow?

Extension

1. Keep a log of your waking and sleeping hours for one week. Compare your data with those of other students. Notice any variations, and consider what they might say about human circadian rhythms.
2. Read the article by Merlin D. Tuttle entitled "Africa's Flying Foxes," in *National Geographic,* April 1986, pp. 540–558. What circadian rhythms, annual rhythms, and migrations are discussed?
3. Write a report on the migration of the monarch butterfly. Draw a map of the migration route and destinations, and include the time required for the butterfly to make its trip.

Chapter 52 | *Populations*

This chapter studies factors that cause variations in the growth of populations. Students learn how to analyze a growth curve, what factors limit growth of a population, and how to predict population growth patterns.

Correlation Chart

Section Number	Section Objectives	Pages	Features	Section Review
52.1	1. Define *biotic potential*.	811–814	Biology in Process, pp. 812–813	1
	2. Differentiate between J-shaped and S-shaped growth curves, and explain the phases of population growth they illustrate.			2, 3
	3. Distinguish between density-dependent and density-independent factors, and explain how each may limit population growth.			4, 5
	4. State Edward Wilson and R. H. MacArthur's hypothesis pertaining to island biogeography.		Think, p. 814	
52.2	1. Explain how the domestication of plants and animals has affected human population growth.	815–817		1
	2. Summarize the growth that has taken place in the human population since people learned to domesticate plants and animals.			5
	3. Compare and contrast three different models of human populations.		Biology Lab, p. 817	2, 3, 4

Resources*

Print

Attenborough, David. *The Living Planet*. Boston: Little, Brown, and Co., 1985.

Begon, Michael, and Martin Mortimer. *Population Ecology: A Unified Study of Animals and Plants*, 2nd ed. Boston: Blackwell Scientific (dist. by Sinauer), 1986.

Crawley, M. J. *Herbivory: The Dynamics of Animal-Plant Interactions*. Berkeley, Calif.: University of California Press, 1983.

Audiovisual/Software

Animal Populations: Nature's Checks and Balances. Film or video, 22 min. Teacher's guide. Britannica, 1983.

Population Ecology. Filmstrip or transparencies. Teacher's guide. Biology Media, 1986.

Population Growth. Apple II, 48K. Includes diskette with backup and 150-page text. COMPress. Carolina Biological.

*Additional references are listed in *Modern Biology*, *Teacher's Program Guide*.

Section Teaching Guides

52.1 Population Growth

Approach

The main objective of this section is to familiarize students with the factors that govern population growth. The text draws on Darwin's theory of evolution by natural selection (Chapter 15). In addition, the information presented in Chapters 50 and 51 on ecosystem structure and relationships will provide useful background for understanding the concepts of carrying capacity and limits to growth.

Concept Framework

Students have probably observed changes in the population of some organism familiar to their area, such as mosquitoes or certain weeds. To provide a framework for understanding this section, ask students to suggest reasons that the population of a local organism might change. Then as students read this section ask them to consider the factors that limit a population from reaching its biotic potential.

Emphasis

Make certain that students know that many populations of organisms may inhabit the same ecosystem and that these populations interact with one another. Consequently, changes in one population may cause changes in another.

Using the Features/Island Biogeography

The main process skills used in this feature are hypothesizing, experimenting, collecting and analyzing data, and observing. To test the hypothesis that size of an island affects the number of species on the island, Wilson and Simberloff collected data on arthropod populations on a group of islands, removed the populations, and observed recolonization. The results of their experiments supported the hypothesis and led to further predictions that the "island effect" would apply to areas isolated by barriers other than water.

Field Trip

Visit a nearby site such as a pond, field, or forest, and ask students to make a list of all the populations they observe living there. Have students identify the various factors that might affect the size of these populations and classify the factors as either density-dependent or density-independent.

Section Review Answers

(Continued from p. 814.)

2. J-shaped curves illustrate only lag and exponential phases of population growth; S-shaped curves track these and the period of relative stability in numbers of organisms that follows the exponential phase, thus indicating the carrying capacity of the ecosystem in which the population exists.
3. the maximum number of individuals of a species that the ecosystem can support
4. Density-dependent factors are related to the density of the population, whereas density-independent factors are not. Both may affect population size.
5. In dense populations waste materials—the breeding ground of parasites and disease—tend to accumulate increasing the likelihood that organisms will come into contact with parasites and disease. Since parasites and disease are usually transmitted by physical contact, in a dense population these are more easily passed between individuals.

52.2 Human Populations

Approach

The main objective of this section is to describe factors that govern human population growth. The discussion of domestication of plants and animals builds on the concepts of controlled breeding (Chapter 13) and plants as food (Chapter 24). You might wish to review the discussions of genetic disorders (Chapter 11) and infectious diseases (Chapter 44), focusing on disease as a factor in population control.

Concept Framework

Students are most likely aware that the human population is growing rapidly. Refer students to Figure 52-3, which shows the growth curve for the human population, and ask them to identify the lag and exponential phases. Also ask students as they read to consider reasons for, and possible consequences of, the current exponential growth of the human population.

Emphasis

Make certain that students understand that the principles governing population growth, described in

Section 52-1, apply to human as well as nonhuman populations.

Population Growth Around the World

Assign a different country to each student, choosing a variety of developed and undeveloped nations from around the world. Have students research the population growth of their assigned country throughout this century, including projections of future growth. Have students construct a population growth curve based on their findings. Put together a classroom display that compares the findings of each student.

Section Review Answers

(See p. 816.)

Laboratory Preparation

Background

In this lab students study the effect of temperature on population growth in a *Tetrahymena* community, which grows rapidly and uniformly as it moves through the medium. Students should understand how to calculate actual numbers from diluted samples. Natural agents that may affect population growth, for example, temperature, pH, or salinity, produce better data than agents that may kill all the organisms.

Preparation

Tetrahymena can be ordered from a biological supply house. Both the culture and culture medium must be sterile. To prepare *Tetrahymena* medium, add 5 g proteose peptone, 5 g tryptone, and 0.2 g K2 HPO4 to 1 L of distilled water. Adjust the pH to 7.2 before autoclaving the mixture. For sterilization in steps 1 and 5, students should flame the top edge of the culture container—pass it through the flame of a Bunsen burner—after removing the stopper or top. They should also flame the pipette before using it, and should flame the top edge of the culture tube before replacing the stopper. In step 6, they should repeat these procedures. In addition, they should flame the edge of the open culture medium tube after removing the cap. They should flame the edge of the tube and culture medium tube cap before replacing it. Assign students three temperature groups: one group cultivating their cultures at approximately 20° C (room

temperature), one group about 10° lower, and one group 10° higher. If an incubator is not available, find three locations that fit these temperature conditions approximately. Do not set any cultures in sunlight. Students should graph their results on logarithmic graph paper. This investigation requires five sessions.

Safety

Remind students to use caution with the Bunsen burner. Freshly-sterilized equipment is very hot and can cause burns if touched. Have first aid cream available.

Conclusions

1. The higher the temperature, the greater the population growth.
2. Temperature is a density-independent factor. Its effect on population growth does not change as population density changes.

Inquiry

Students will design an experiment similar to the one performed but with two cultures introduced simultaneously into the medium. If they use one of the temperatures in the investigation already completed, they can compare results.

Chapter Review

Vocabulary

1. Population growth is an increase in the number of individuals in a given species in a given area over time. The increase over a specified time is the population growth rate.
2. Population growth rate is a general rate term. Biotic potential is the maximum rate at which a population could possibly grow.
3. The lag phase of population growth may precede the exponential phase and differs from it in that it is a slow period of growth. In the exponential phase the population size doubles in an increasingly shorter time.
4. The carrying capacity of an ecosystem is the approximate number of individuals in a given population which that ecosystem can support. This capacity can be inferred by studying an S-shaped curve of that population's growth.

5. Population density is the number of individuals in a given population per the amount of space. Density-dependent factors affect the growth of a population only when that population reaches a critical density.

Review

11. $\dfrac{\text{change in the number of individuals}}{\text{time}}$

12. Population growth can occur at rates so great that after a while not enough resources exist to support all the members of that population.

13. a shortage of food, a shortage of space, and an accumulation of waste

14. A lag phase is the initial phase in population growth for any population that begins with so few individuals that an increase in the numbers of individuals can occur only very slowly.

15. By charting the number of individuals in a population over time you could determine when that population reached a period of relative stability. The number of individuals in that population during this period is the number of individuals that ecosystem can support, or its carrying capacity.

16. The butterfly population is subject to density-independent factors like sudden snowstorms that devastate it. However, since the carrying capacity of the ecosystem is not exceeded when only a few butterflies remain, the population is able to rebound quickly until it levels out once again at the carrying capacity—until the next snowstorm hits.

17. $\dfrac{\text{number of individuals}}{\text{area}}$

18. availability of oxygen; the food supply of a population of sunfish becomes so plentiful that the population of sunfish increased in number until a shortage of oxygen resulted

19. It was an ideal site because all of the islands were formed at the same time and were populated at the time of formation by a different mix of the same species. Therefore, the only notable differences between the islands would be their size and their distance from the mainland.

20. You would need to know the structure of the population—the age and sex of each of its members, or a rough approximation of this; information on density-dependent factors, such as availability of food and of space.

Critical Thinking

1. DDT became ineffective as the insect population gradually developed a resistance to the insecticide. Although initially most insects were killed off by DDT, a small number were resistant to it and survived. Those survivors reproduced and passed this trait to some of their offspring. This cycle continued until eventually the entire insect population was unaffected by DDT.

2. From the map you can infer that pinnipeds are affected by availability of food and by predators. Since they feed on fish along the coastlines, if the pinniped population gets too large there will not be enough fish to feed on. Since they live in the water, pinnipeds are prey to killer whales and other predators, which reduce the size of the pinniped population.

3. Humans kill off animals directly by hunting them and indirectly by destroying their habitats through building cities and farming, as well as by introducing pesticides into the ecosystem. Humans can have a positive effect on their environment by land conservation efforts and limiting hunting, which helps animal populations to remain stable or increase.

Chapter 52 | Populations

Introduction

Cheetahs and lupines are members of populations. Although lupines are thriving, cheetahs are endangered. Environmental conditions and human interference have reduced the number of cheetahs to a dangerously low level. The methods used to save such organisms from extinction depend on knowledge of the biology of populations, the subject of this chapter.

Chapter Outline

52.1 Population Growth
- *Biotic Potential*
- *Growth Curves*
- *Limits to Growth*

52.2 Human Populations
- *Human Population Growth*
- *Human Population Structures*

Chapter Concept

As you read, notice how mathematical models are used to examine variations in the growth of populations.

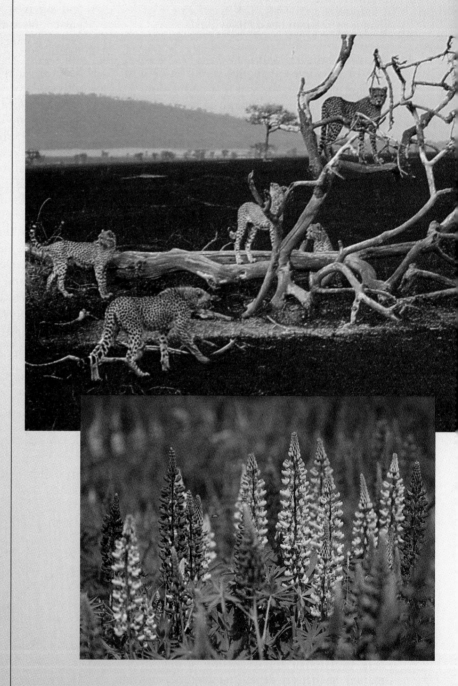

Cheetahs, *Acinonyx jubatus*, central Africa. Inset: Lupines, *Lupinus* sp.

For additional teaching information see pp. T52A–T52D.

52.1 Population Growth

A population is a group of individuals of the same species that live in the same area. An increase in a population is called **population growth**. The **population growth rate** is the change in the number of individuals in a population over time. This rate is determined by using the following formula:

$$\text{Population growth rate} = \frac{\text{change in number of individuals}}{\text{time}}$$

A population may grow at such a great rate that available resources become insufficient to support all its members. It was Charles Darwin's observation of this and related phenomena that led him to form his theory of evolution by natural selection. Darwin's theory was that when natural resources were limited, the fittest members of a given population would survive and the least fit would perish. Keeping this theory in mind as you read about population growth will help you understand the factors that may affect it.

Biotic Potential

The **biotic potential** of a population is the rate at which a population will grow if all individuals survive and reproduce at maximum capacity. Biotic potential can be reached when organisms capable of reproducing are put into an ideal environment, one with unlimited resources and space and without such hazards as disease and predators.

The high reproductive capability of the common housefly, *Musca domestica*, can be used as an example of how biotic potential can be achieved. One housefly produces 120 eggs in one year. How many houseflies can be produced in seven generations? If each fly lives only one generation, if half the eggs of each generation are female, and if each female produces 120 eggs a total of 5,598,720,000,000 flies will result. Table 52-1 shows the number of flies that will be produced in each of the seven generations if biotic potential is achieved.

In nature populations rarely achieve their biotic potential for any sustained period. Eventually the number of individuals declines because of shortage of food, shortage of space, predation, or accumulation of waste. For example, forest managers often introduce trout into lakes that contain no large fish. Because the trout have no competition for food or space, their population grows rapidly—but only for a while. The number of fish soon becomes so great that intraspecific competition for food and space ensues. This competition results in a decline in the number of trout. Once a balance is reached between the number of individuals and the amount of resources available, this decline ceases, and the population becomes stable.

Section Objectives

- *Define* biotic potential.

- *Differentiate between* J-shaped and S-shaped growth curves, and explain the phases of population growth they illustrate.

- *Distinguish between* density-dependent and density-independent factors, and explain how each may limit population growth.

- *State* Edward Wilson and R. H. MacArthur's hypothesis pertaining to island biogeography.

Info: If a pair of houseflies reached their full biotic potential, they would produce 25 million descendants in only two months. This number would fill a room the size of a large closet.

Table 52-1 Biotic Potential of the Housefly

Generation	Population
1	120
2	7,200
3	432,000
4	25,920,000
5	1,555,200,000
6	93,312,000,000
7	5,598,720,000,000

Figure 52-1 Population growth can be charted on a J-shaped curve. It shows that population grows slowly during the early lag phase and then very rapidly during the exponential phase.

Tip: Have students plot a growth curve based on the population of their town or city.

Growth Curves

A useful tool for studying the patterns of population growth is the growth curve. A **growth curve** is a graph showing the number of individuals in a population over time. The **J-shaped curve** is a growth curve that tracks two phases of population growth—the lag phase and the exponential phase.

During the **lag phase,** little or no increase occurs in a population. For example, when only one male and female sunfish are introduced into a fertile pond, the sunfish population increases slowly because only one female is producing young. However, as more females are born and in turn reproduce, the rate of population growth increases. If there is enough food and space for the sunfish to thrive, the population will eventually increase at an exponential rate.

During the **exponential phase** a population increases so rapidly that the number of individuals doubles in a specific time interval and keeps doubling in increasingly shorter periods of time. Carefully study the J-shaped curve in Figure 52-1. Note the difference in the rate of growth of the sunfish population during the lag phase and the exponential phase.

Biology in Process

Island Biogeography

Can the number of plant and animal populations on an island be predicted from the size of the island or its distance from the mainland? In the late 1960s ecologists collected and compared data on the number of species that had survived on each of the many islands of the New Guinea shelf. These islands at one time had each been populated by the same mix of species. Over time they had been subject to virtually the same density-independent factors. But the islands were of different size and distance from the mainland. The ecolo-

gists discovered a direct relationship between the size of an island and the number of species surviving on that island. They found that the smaller the island was, the fewer species it had.

This result led ecologists Edward Wilson and R. H. Mac-Arthur to hypothesize that the number of species that could be supported on an island reached an equilibrium and that this number was affected by both the size of the island and its distance from the mainland. Wilson and Daniel Simberloff then devised an experiment to test the validity of the hypothesis.

An **S-shaped curve** is a growth curve that depicts the period of relative stability in a population that occurs after its lag and exponential phases. This curve therefore gives an indication of the carrying capacity of the ecosystem. The **carrying capacity** is the maximum number of individuals that the ecosystem is capable of supporting. The S-shaped curve in Figure 52-2 reveals that the number of sunfish in the pond usually did not exceed 100. Thus for the sunfish population in the example the carrying capacity of the pond is about 100. Notice that the J-shaped curve is the lower part of the S-shaped curve.

When the environment is stable, the maximum number of individuals in a population fluctuates near the carrying capacity of the environment. If the environment becomes unstable, the fluctuations become more radical. For example, high in the Rocky Mountains the size of the butterfly population is usually stable, although it may vary somewhat with available food and space. However, freak snowstorms sometimes cause a sudden drop in the number of individuals. Nevertheless, because the number of butterflies is then well under the carrying capacity of the ecosystem, the population usually increases quickly until it again approaches the carrying capacity.

Figure 52-2 An S-shaped curve shows that population growth levels off after the exponential phase. The top of this curve indicates the carrying capacity of an ecosystem.

They examined arthropod communities on six tiny mangrove islands off the coast of Florida. First, they collected data on the kind and number of arthropod populations on each island. Then, using scaffolding and plastic sheets, Wilson and Simberloff erected a tent over each island. Next they sprayed the islands with pesticides that killed off all the invertebrates. Finally they removed the tents and observed the recolonization. After a period of six months the same number of species—although not always the same species—had repopulated the islands. Continued observations re- vealed that after reaching an equilibrium the diversity in species remained constant, thus supporting the hypothesis.

More recently biologist William Newmark observed that in wilderness areas and in continental national parks, similar patterns existed. Although these habitats were not islands in the literal sense—they were not enclosed by water—they were surrounded by cities, highways, and ranches. Even the scent of humans could create an islandlike environment. Newmark therefore predicted that if these areas were truly islandlike, then Wilson and MacArthur's "island effect" would hold true for them too. Newmark discovered that the number of mammal species in wilderness preserves did in fact decline as barriers encroached on them. His findings suggest that the island effect holds true for these areas.

■ What aspect of Wilson and MacArthur's hypothesis was not tested in the experiment designed by Wilson and Simberloff?

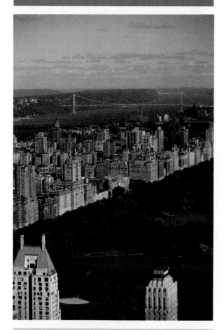

Answer to Think
Since this park is, in effect, an island, the number of species it can support will decline until an equilibrium is reached. This number would be affected by the size of the park and its distance from other areas in which wildlife exists. Density-dependent factors would also have an effect.

Answers to Section Review
1. Biotic potential is the rate at which a population would grow if all individuals survived and reproduced at maximum capacity. (See pages T52A–T52D.)

Limits to Growth

Ecologists label factors that can limit population growth as density-dependent or density-independent. Here the word *density* refers to **population density,** the number of individuals in a population in a given area at a given time. For example, if a sunfish pond holds 1,000 m³ of water, 100 sunfish represent a population density of one fish per 10 m³, as shown in the following formula:

$$\text{Population density} = \frac{100 \text{ sunfish}}{1{,}000 \text{ m}^3 \text{ of water}} = \frac{1 \text{ sunfish}}{10 \text{ m}^3 \text{ of water}}$$

Density-dependent Factors

Density-dependent factors are factors that affect populations in different ways depending on population density. For example, when 20 sunfish are put in a pond with 1,000 m³ of water, population density is one fish per 50 m³ of water, and food is abundant. As the population density increases, less food is available to each individual. This results in intense competition for food and the elimination of some sunfish. Thus the availability of food is a density-dependent factor that controls population size. Other common density-dependent factors may include the availability of space and light, parasitic infections, and disease.

The number of predators in an ecosystem may also be a density-dependent factor. Usually a predator population increases after the prey population has increased. So the increased number of predators prevents another increase in the number of prey.

Abiotic factors may be density-dependent. For example, if the food in our sunfish pond were unlimited, the population might increase until a shortage of oxygen resulted. The availability of oxygen would then be a density-dependent factor. However, it is one that will affect population size only when density is high.

Density-independent Factors

Density-independent factors are factors that affect populations regardless of population density. These factors, which are usually abiotic, include changes in weather, temperature, and humidity; variations in the amount of sunlight; and the amount of available energy. For example, a population may be entirely wiped out by a forest fire or decimated by an earthquake or a flood.

Section Review

1. Define *biotic potential.*
2. In what ways do J-shaped and S-shaped growth curves differ?
3. What is the carrying capacity of an ecosystem?
4. What is the difference between density-dependent and density-independent factors? How are they similar?
5. How might parasitic infections and diseases be density-dependent factors?

52.2 Human Populations

Early humans foraged for the plants they ate and hunted for the animals they used for food and clothing. Later they learned to cultivate crops and to tame wild animals. Plants and animals that are adapted or trained to live in a human environment are referred to as **domesticated.** By domesticating plants and animals people were often able to increase their food supply, even to create surpluses that could be stored. Domestication gave people a more reliable and constant supply of food, which was one important factor in determining human population size.

Human Population Growth

From about 6000 B.C. until about A.D. 1800, the number of people in the world increased steadily. This increase, however, was relatively slow. Then in the middle of the eighteenth century the population began to increase dramatically. This exponential growth rate can be largely explained by the increase in food production, the rise of industry, and advances in medicine. In the field of medicine alone, the development of immunology, the discovery of anesthetics, and the introduction of antiseptic surgery increased human life expectancy in many parts of the world. Look at Figure 52-3 to see how the world population has grown since the 1600s and how the population is projected to grow from the present to the year 2000.

In the year 1650 there were about 500 million people in the world; today, according to United Nations estimates, there are over 5 billion people. The human population is increasing by approximately 70 million per year. This growth rate may eventually be limited by the same density-dependent factors that limit the growth of other populations—the availability of natural resources.

Human Population Structures

The growth rate for a human population in a given area is dependent on a number of factors. To make predictions about population growth, population biologists must consider the composition of a population. The following hypothetical example, although oversimplified, will illustrate the point: Suppose that you want to predict the growth patterns for three small villages, all with the same number of people. One is made up of families with children, parents, and grandparents—a population of mixed age and sex. The second is made up solely of children. The third is made up only of elderly people. If you considered only external limits to growth, such as the availability of food and the potential for natural disasters, you would conclude that each of the three populations had the same potential for growth.

Section Objectives

- *Explain how the domestication of plants and animals has affected human population growth.*

- *Summarize the growth that has taken place in the human population since people learned to domesticate plants and animals.*

- *Compare and contrast three different models of human populations.*

Tip: Have students list probable causes of death of early humans and compare them with common causes of death today.

Figure 52-3 **This growth curve shows how large the human population is expected to be by the year 2000. What phase is the human population in currently?**

Info: There are 25 more humans now than there were 10 seconds ago, and by tomorrow there will be over 211,000 more than today.

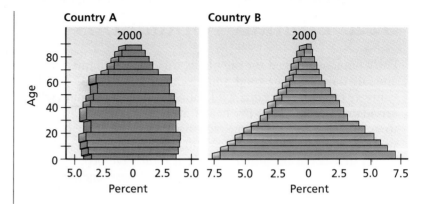

Figure 52-4 These population pyramids, showing distribution of population by age, indicate that Country B has a much greater growth potential than Country A has.

Country A

2000

Country B

2000

In fact, the actual growth patterns for the three villages would differ dramatically. If no density-dependent or density-independent factors limited population growth, you could reasonably make the following assumptions: (1) The first population, a balanced one, could be expected to grow steadily. (2) The second population, consisting of members still too young to reproduce, would not grow at all for a time. Then, once the children reached reproductive maturity, the population would increase rapidly. (3) The third population, which consists entirely of members past their reproductive age, would decline until it reached zero.

Having learned something about the patterns of population growth after analyzing this hypothetical case, you can now make some rough predictions about growth patterns in actual populations. Study Figure 52-4, which shows the population structures of two countries. To which hypothetical village populations can you compare these actual populations? What predictions can you make about changes in the two populations in the course of the next 20 years?

Section Review

1. Define *domesticated,* and list the benefits that people have derived by domesticating plants and animals.
2. To predict accurately the growth pattern of a population, what must you examine in addition to the external limits to growth? Why?
3. In what way would the initial growth pattern of a village consisting solely of children be similar to that of the growth pattern of a village consisting solely of elderly people? In what way would the growth patterns of these two villages then differ?
4. Explain why a population with a pyramidlike structure grows steadily.
5. Although population growth in some nations has slowed to nearly a standstill, the world population is still increasing at an exponential rate. What might account for this rapid population growth?

Answers to Section Review
1. *Domesticated* means adapted to be raised in closer proximity to human beings and to the greater advantage of humans. By domesticating plants and animals people so drastically increased their food supply that they could obtain a surplus of food for storage.
2. You must study the composition of that particular population; the reproductive capacity of the members of that population will affect its growth.
3. It would be similar in that, at first, neither would grow; it would be different in that the population in the village of children would eventually grow in size but the population in the village of elderly would decline.
4. In each succeeding year more individuals are entering the reproductive phase of development.
5. Population in some areas is growing exponentially.

Laboratory

Studying Population Growth

Objective
To study how temperature affects population growth in a protist community

Process Skills
Experimenting, observing

Materials
Tetrahymena culture, test tube rack, pasteur pipette, alcohol lamp, tongs, microscope slides, microscope, medicine dropper, *Tetrahymena* medium in a screw-cap culture tube, graph paper

Method

1 Make an accurate count of a *Tetrahymena* population. Place a tube of *Tetrahymena* culture in a test tube rack. Obtain a pipette. Light an alcohol lamp. Follow your teacher's instructions for using the lamp to sterilize the container and pipette. Use tongs to hold the pipette while it cools. When cooled, use the pipette to withdraw some culture and place a drop of it on a microscope slide.

2 Observe the *Tetrahymena* under low power of a microscope. If there are more than 100 organisms in the population dilute the culture by adding nine drops of water.

3 Use a dropper to place one drop of the diluted mixture on a clean slide, and observe it under the microscope. If less than 10 protozoa are in a drop, count them. If more than 10 organisms are in a drop repeat step 2.

4 Multiply by your dilution factor to estimate the number of *Tetrahymena* in the original drop. For example, if you made one dilution, then for every organism counted, the original drop had 10 organisms. Record your result.

5 Repeat steps 1 through 4 four times. Calculate the average of the five samples. Use this figure as the population of the original drop.

6 Place a culture medium tube in the test tube rack. Following your teacher's instructions for sterilization, remove the tops from the *Tetrahymena* culture and the culture medium tube. Use a sterile pipette to withdraw some culture from the culture container, placing a drop in the culture medium.

7 Your teacher will assign you a temperature. Place your *Tetrahymena* culture in an environment with the assigned temperature.

8 Following the method described, estimate the *Tetrahymena* population each day for the next four days and record your results.

9 Graph your results. Then compare your results with those of students using other temperatures.

Conclusions
1 What is the effect of temperature on the rate of population growth?
2 Is temperature density-dependent or density-independent? How do you know?

Inquiry
Design an experiment to test the results if a competitor, such as *Vorticella,* is grown in the same medium as *Tetrahymena.*

This investigation must be done in five sessions.

Vocabulary

biotic potential (811)
carrying capacity (813)
density-dependent factors (814)

density-independent factors (814)
domesticated (815)
exponential phase (812)

growth curve (812)
J-shaped curve (812)
lag phase (812)
population density (814)

population growth (811)
population growth rate (811)
S-shaped curve (813)

Explain the difference between the terms in each of the following sets.
1. population growth, population growth rate
2. population growth rate, biotic potential
3. lag phase, exponential phase
4. S-shaped curve, carrying capacity
5. population density, density-dependent factors

Review

1. Population growth (a) is a group of individuals of the same species living in the same place (b) is an estimated change in the number of individuals in a given population over a specific period of time (c) will never occur at so great a rate that the members of the population will be unable to exist on the available resources (d) can occur at so great a rate that existing resources cannot support all the members of the population.

2. The biotic potential of a population is its maximum rate of growth (a) if all individuals survive and reproduce at maximum capacity (b) if all individuals survive and reproduce (c) if most individuals survive and some of these reproduce at their maximum capacity (d) if a balance is reached between the number of individuals in the population and the amount of resources available to them.

3. The biotic potential of a population existing in nature is (a) always achieved (b) usually achieved (c) rarely achieved (d) never achieved.

4. The maximum number of individuals in a population located in a stable environment tends to (a) fluctuate near the carrying capacity of the environment (b) fluctuate radically (c) be about 100 (d) be limited by density-independent factors.

5. Availability of space, susceptibility to disease, and number of predators are all (a) abiotic factors (b) density-independent factors (c) factors that provoke greater intraspecific competition (d) density-dependent factors.

6. A predator population usually (a) decreases after the density of the prey population increases (b) increases after the density of the prey population decreases (c) increases after the density of the prey population increases (d) is not at all affected by changes in the density of the prey population.

7. Abiotic factors are (a) never density dependent (b) sometimes density dependent (c) never density independent (d) only density independent.

8. A volcanic eruption is an example of (a) something that would be charted on an S-shaped curve (b) a density-dependent factor (c) something that would be charted on a J-shaped curve (d) a density-independent factor.

9. The following are not domesticated: (a) plants cultivated in plots (b) animals

bred in pens (c) herbs growing in window boxes (d) herbs growing in the forest.

10. In the year 1650, the human population (a) began growing at an exponential rate (b) numbered about 500 million (c) began increasing by approximately 70 million annually (d) numbered about 5 billion.

11. What formula is used to determine the growth rate of a population?

12. What observation helped Charles Darwin to form his theory of evolution by natural selection?

13. Name three common factors that in nature can lead to the decline in the number of individuals in a given population.

14. What is a lag phase, and when does this phase occur?

15. How would you determine the carrying capacity of the ecosystem of a given species?

16. What factors may cause the number of individuals in the Rocky Mountain butterfly population to fluctuate over time?

17. What formula would you use to determine population density?

18. Give one example of an abiotic factor that could be density dependent, and explain how it might be so.

19. What factors made the New Guinea shelf an ideal place to determine whether the number of populations on an island was related to the size of the island?

20. What basic data would you need in order to accurately predict the growth pattern of a given population?

Critical Thinking

1. Humans have tried to control insects with insecticides, such as DDT. Although DDT was effective for the first ten years it was used, eventually the insect population began to increase again. What do you think caused DDT to lose its effectiveness?

2. At right is a map showing the populations of pinnipeds—seals and sea lions— around the world. From examining this map what can you infer about density-dependent factors that affect these populations?

3. Since humans have more power to alter their environment than other animals do,

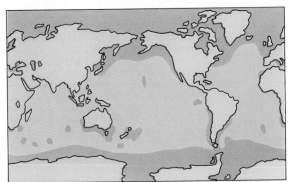

they can affect the carrying capacity of their ecosystems. How do humans increase or decrease the carrying capacity of their local areas?

Extension

1. Read the article by Peter D. Moore entitled "What Makes a Forest Rich?" in *Nature*, September 24, 1987, p. 292. What density-independent factor is discussed in the article? How does this factor affect tree populations?

2. Research and report on how the U.S. census is conducted. Examine the 1980 census for statistics on age, births and deaths, and marriages and divorces.

3. Examine a culture of daphnia, planaria, or rotifers. Separate your culture into ten smaller ones, placing some water from the original culture into separate petri dishes. Watch the cultures under the microscope over a period of time. Make a data sheet to track the growth of each population. Then calculate the rate of reproduction of the separate samples from your original culture.

Chapter 53 | Protecting Life

This chapter discusses renewable and nonrenewable natural resources. Students learn about the effect of pollutants on the environment and human efforts to conserve and protect resources.

Correlation Chart

Section Number	Section Objectives	Pages	Features	Section Review
53.1	1. List three types of renewable resources and three types of nonrenewable resources.	821–824	Biology in Process, pp. 822–823	1, 5
	2. Describe some efforts being made to preserve habitats and wildlife.			2
	3. Describe three methods of increasing soil fertility.			3
	4. List some technological advances that can help conserve fossil fuels and minerals.			4, 5
	5. Compare the advantages and disadvantages of converting waste products to energy.			
53.2	1. List several primary pollutants and their sources.	825–827	Think, p. 826 Biology Lab, p. 827	1, 3
	2. Identify the sources of ozone and acid rain, and describe the effects of these pollutants on the environment.			2, 5
	3. Describe industrial, agricultural, and domestic sources of pollution.			1, 3, 4

Resources*

Print

Abrams, Joanne, ed. *Ecology: Impacts and Implications*. Wayne, N.J.: Avery, 1983.

Myers, Norman. *The Primary Source: Tropical Forests and Our Future*. New York: Norton, 1984.

Repetto, Robert, ed. *The Global Possible: Resources, Development, and the New Century*. New Haven, Conn.: Yale University Press, 1986.

Audiovisual/Software

The Four Corners: A National Sacrifice Area? Film or video, 59 min. Bullfrog, 1983.

The Garbage Explosion. Film, 16 min. Britannica.

The House of Man, I: Our Changing Environment. Film, 17 min. Britannica.

Territory in Conflict. Film or video, 28 min. Centre.

*Additional references are listed in *Modern Biology, Teacher's Program Guide*.

Section Teaching Guides

53.1 Natural Resources

Approach

The main objective of this section is to describe methods of conserving both renewable and nonrenewable resources. The section builds on the concepts of energy flow and nutrient cycling in Chapter 50, and biotic relationships and succession in Chapter 51. In presenting the text, guide students to an appreciation of the importance of protecting natural resources. Focus on the challenges facing humans to find ways of meeting the energy demands of the growing human population without endangering other species or permanently depleting the supply of water, soil, minerals, and fossil fuels.

Concept Framework

Students have most likely heard about endangered species such as the bald eagle and the grizzly bear. They also have probably read about the energy crisis and have heard environmentalists urge people to decrease their consumption of fossil fuels. To prepare students for learning about renewable and nonrenewable resources, ask them why the protection of wildlife, fossil fuels, and other natural resources is of concern to many people. Encourage them as they read the section to consider the long-term consequences of depleting our supply of natural resources.

Emphasis

Make certain that students know the difference between renewable and nonrenewable resources and that they can give examples of efforts to protect both types of resources.

Field Trip

Arrange for the class to visit a facility in your area that uses an alternative energy source. Possibilities include a building heated by solar or geothermal energy. Prior to the trip, prepare a list of questions to answer or points to notice about the facility.

Using the Features/Resources from Wastes

The main technology described in this feature is the trash-to-resource incinerator. This incinerator burns trash such as paper and food wastes, and the energy produced is converted into electricity that heats or air conditions nearby buildings. The incinerator thus addresses two environmental problems: it helps reduce the bulk of trash disposed of by humans and produces inexpensive energy. The feature also describes technologies designed to minimize the emission of hazardous wastes from the incinerator.

In discussing the feature, encourage students to think of advantages and disadvantages of using trash-to-resource incinerators. For example, to avoid polluting the environment, people would probably have to dispose of hazardous wastes separately from burnable wastes such as paper and food.

Section Review Answers

(Continued from p. 824.)

4. Scientists have developed alternative energy sources such as solar and geothermal energy. The public has cut back on its energy consumption and increased its recycling efforts.
5. They are considered nonrenewable because it takes hundreds of thousands of years to form fossil fuels and they are being consumed much more quickly than they are being replaced.

53.2 Pollution

Approach

The main objective of this section is to describe the major types of pollution and identify their sources. The section discusses the effects of pollution on such aspects of the environment as food chains and succession. You might wish to review these topics in Chapters 50 and 51. In discussing causes of pollution, you might also talk about the concepts of population growth in Chapter 52 and note that population growth has contributed to the increasing levels of environmental pollution.

Concept Framework

Students are probably aware that such pollutants as smog and ozone can present health risks to humans. Tap this knowledge to present a conceptual framework for discussion other effects pollution might have on the environment. Elicit the fact that pollutants, especially those that are nonbiodegradable, affect all organisms in an ecosystem and contribute to the depletion of natural resources discussed in Section 53.1.

Emphasis

Students should be able to identify the major sources of primary and secondary pollutants. In addition, they should be able to describe the short-term and long-term effects of pollution on humans and other organisms.

Teaching Pollutants

Have students keep a record of all the pollutants they encounter during the course of one day and the sources of these pollutants. Examples might include automobile exhaust, cigarette smoke, litter, or produce sprayed with pesticides. Compile a chart that lists the students' findings. Then ask the class for suggestions on how to reduce the amount of pollution in their environment.

Section Review Answers

(Continued from p. 826.)
3. Sunlight catalyzes a series of chemical reactions in the primary pollutants in the atmosphere, which forms photochemical smog.
4. Biological magnification occurs when hazardous wastes enter the food chain through streams and soil and become increasingly concentrated in all levels of the ecosystem.
5. They experience the effects of acid rain because pollutants are carried for hundreds of miles by the wind before returning to earth in the form of precipitation.

Laboratory Preparation

Background

In this lab students learn that response to environmental stimuli is included in discussions of the characteristics of living things because organisms generally perceive and respond to environmental cues. As a result, organisms are well-adapted to one set of environmental conditions and ill-adapted to others. Students observe that vagile organisms— organisms that move about freely—perceive changes in their environment and, if possible, move to an area where they can carry out their life functions. Students infer that environmental pollutants, such as thermal or salt pollution, change the flora and fauna of the polluted area because sessile organisms can either tolerate the new environment or die, while vagile organisms can either tolerate the new environment or move.

Preparation

To make 10% NaCl solution, add 10 g table salt to 90 mL water. *Paramecium* and *Ameba* cultures are available from a biological supply house or can be grown in pond water. If pond water is not available for the lab, use water that has been set out overnight in open containers.

Conclusions

1. The amebas tried to move away from the salt solution because the salt interfered with their internal water balance.
2. The paramecia preferred a moderate temperature, as shown by the fact that they congregated near the center of the tubing.
3. Pollutants are substances that cause undesirable changes in an ecosystem. Chemicals present in too great a concentration or at a harmful temperature are therefore pollutants.

Inquiry

Students will think of ways the investigation could have been expanded, such as by testing the ameba's reaction to different concentrations of NaCl in order to determine their tolerance level.

Chapter Review

Vocabulary

1. Contour plowing means plowing across a field rather than up and down; terracing involves building a series of steps down a slope.
2. Deforestation is the clearing of vast regions of forested land; reforestation is the process of nurturing new growths of trees on sites formerly occupied by forests.
3. A primary pollutant is emitted directly into the environment; when it combines with other substances and undergoes a chemical reaction that transforms it, a secondary pollutant is formed.
4. Ozone is a pollutant that forms when nitrogen oxides react with other chemicals and sunlight. It is one component of photochemical smog, another pollutant.

5. An endangered species is currently at risk of extinction; a threatened species is likely to be endangered in the near future.

Review

11. It involves nurturing new trees on sites that have been deforested, preventing soil erosion, and maintaining ground moisture.
12. It defined which species were endangered and established regulations for their protection and conservation.
13. Soil is formed from the breakdown of rocks and the accumulation of dead organic matter.
14. A watershed consists of all the land that drains into a river.
15. Strip mining is an economical method of obtaining minerals. However, it also disrupts the environment when the soil is removed.
16. Geothermal energy is obtained from hot water beneath the earth.
17. The three sources of pollution are industrial, agricultural, and domestic.
18. Inhaled particulates can irritate the lining of the lungs and eventually lead to chronic lung disease.
19. The barrels in which hazardous wastes are stored have corroded and the wastes have leaked out, contaminating water and soil.
20. If pesticides are absorbed into the soil and get into the water supply, they can poison fish and other aquatic animals.

Critical Thinking

1. If enough carbon monoxide binds with the hemoglobin, the hemoglobin cannot carry sufficient oxygen for respiration and the person or animal dies from internal suffocation.
2. Hunting of species can be limited so that the population is not drastically reduced. Specimens of endangered species can be removed from the wild and bred in captivity, and their offspring returned to the species natural habitat to repopulate it. Industrial development can be limited in areas in which it would displace large numbers of indigenous species; if development is unavoidable, the displaced species could be relocated to a similar habitat.
3. Since humus is composed primarily of decomposed plant remains, in deserts where plant life is sparse there is little material available to form humus. Humans could apply fertilizer to supply plants with the nutrients and moisture normally provided by humus in less arid climates.
4. The pesticide is accumulated in lower organisms such as the insects it was used to kill off and is passed on to predators higher in the food chain. For example, birds ate insects that had been contaminated with DDT. The pesticide could have been spread to humans who drank water contaminated with DDT or who ate fish that had consumed plankton tainted with the substance.
5. Filters on smokestacks reduce the amount of sulfur dioxide and nitrogen oxide released into the air, so less acid rain is formed. Catalytic converters on cars change exhaust fumes into harmless products, thereby reducing the amount of pollutants normally expelled from the exhaust system that are transformed into acid rain.

Chapter 53 | Protecting Life

Introduction

As the human population continues to increase, so does its need for natural resources such as soil, timber, water, and fuels for energy. However, as part of our responsibility to future generations, we must both conserve our natural resources and reduce pollution of the environment. In this chapter you will learn about the impact human society has had on such natural resources as wildlife, water, fossil fuels, and minerals. You will also see the efforts people are making to conserve these valuable resources.

Chapter Outline

53.1 Natural Resources
- *Renewable Resources*
- *Nonrenewable Resources*

53.2 Pollution
- *Primary and Secondary Pollutants*
- *Sources of Pollution*

Chapter Concept

As you read, note the ways in which people are using their knowledge of biology in an effort to find ways to protect natural resources.

Giant mirrors of a solar power plant in Daggett, California. Inset: Ponderosa pine seedlings in Klamath Falls nursery, Oregon.

For additional teaching information see pp. T53A–T53D.

53.1 Natural Resources

Natural resources are the raw materials—biotic and abiotic—that support life on earth. Human society, with its increasing population and its industrial way of life, consumes the greatest proportion of available resources. This consumption leads to depletion of resources, which can occur by direct destruction, such as the burning of fuels, by displacement, or by pollution of resources and the environment.

Renewable Resources

Renewable resources are those materials that can be regrown or replenished. However, the increasing human population is consuming renewable resources faster than they can be replaced. An example of a renewable resource is a forest. Although its trees are cut for timber and firewood the forest, if properly managed, can be reforested and reharvested. Wildlife and soil are also potentially renewable resources.

Forests

Forests are biological communities dominated by trees and composed of hundreds of species of organisms. The increasing need for wood to produce paper, fuel, building materials, and furniture has led to the **deforestation,** or clearing, of vast regions of forested land. **Reforestation** is the process of nurturing new growths of trees on sites formerly occupied by forests, thereby preventing soil erosion and maintaining moisture in the ground.

Wildlife

Wildlife refers to all the plants and animals indigenous to a community. As you have learned in Chapter 50, all species in a community interact in a complex system of food chains and food webs. The destruction of any one species can have an impact on many other organisms in the community. Preventing the loss of species through conservation and wildlife management is an important area of scientific concern.

The primary threat to wildlife is habitat destruction. **Endangered species** are those plants and animals that are currently at risk of extinction in their native habitats. **Threatened species** are those plants and animals that are likely to become endangered in the near future.

The U.S. government has passed laws, most importantly the Endangered Species Act of 1973, that define endangered species and establish regulations for their protection and conservation. In addition, nearly 100 nations have signed a treaty—C.I.T.E.S., an acronym for Convention on the International Trade in Endangered Species—that outlaws the trade of endangered species and their products across international borders.

Section Objectives

- *List three types of renewable resources and three types of nonrenewable resources.*

- *Describe some efforts being made to preserve habitats and wildlife.*

- *Describe three methods of increasing soil fertility.*

- *List some technological advances that can help conserve fossil fuels and minerals.*

- *Compare the advantages and disadvantages of converting waste products to energy.*

Figure 53-1 The manatee, *Trichesus* sp., is one animal whose survival as a species is in jeopardy because of human interference.

Soil

Soil is constantly formed as rocks break down and as dead organic matter accumulates. Soil is a mixture of rock fragments, minerals, and animal and plant matter. Soil retains water, which thus supports plant and animal life and controls flooding. **Humus,** the organic component of the soil that helps it retain most of its moisture and nutrients, is composed primarily of plant remains. A single acre of **topsoil,** the uppermost layer of soil, can hold up to 11 tons of worms, insects, bacteria, fungi, and other organisms.

Each year erosion carries away 2.7 billion tons of soil from U.S. farmlands. **Erosion** is the loss of soil resulting from the effects of abiotic factors such as wind and water. This eroded soil enters streams and pollutes the water. To prevent erosion caused by wind, farmers may plant grasses or other vegetation on otherwise bare fields, add plant wastes and manure to depleted soil, put up windbreaks or shelter belts of trees, or build irrigation ditches. To limit erosion caused by water, several techniques are used. **Contour plowing**—plowing that follows the contour of a slope rather than going up and down—and **terracing**—carving a series of steps along a hillside—help to lessen erosion by water.

Tip: Prepare two trays, one with only dirt and the other with a healthy crop of grass, possibly sod. Tip the trays as you pour equal amounts of water over the soil. Collect the runoff water. Let the soil particles settle and compare the amount of soil that has washed away from the trays.

Biotechnology

Resources from Wastes

Americans dispose of millions of pounds of trash each day. Most of this trash is dumped in one of the 15,000 landfills throughout the United States. As landfill space runs out, new ways must be found to dispose of trash.

At the same time the supply of fossil fuels is dwindling, prompting a search for new energy sources. A trash-to-resource incinerator, such as the one in Dade County, Florida, offers a solution to both problems. Such an incinerator produces energy from the burning of trash, a free fuel.

Most of the daily flood of trash consists of paper and other combustible materials. Even when paper recycling programs are instituted, a large amount of burnable trash still remains. A trash-to-resource incinerator reduces the bulk of this burnable trash to ash while producing cheap energy.

Burnable trash, consisting mostly of paper, is shredded and dumped into a funnel-shaped furnace. Heat from the furnace produces steam in an adjacent boiler. The steam, in turn, provides energy to power turbines, which produce electricity. Waste steam is piped to nearby buildings to provide

Nonrenewable Resources

Nonrenewable resources are those that cannot be replaced or replenished by nature or people. As a result, they require special management and attention. Among the most important nonrenewable resources are fresh water, fossil fuels, and minerals.

Water

The biosphere contains a fixed amount of water, of which only 3 percent is fresh water available for human consumption. Because fresh water is unevenly distributed over the earth, some highly populated areas may soon face severe water shortages.

Communities can conserve water by managing its flow in their **watershed**—the area of land that drains into a river. Creating water reservoirs by building dams also conserves water and prevents flooding. However, this practice may endanger species that live behind the dam by inundating their habitat and may harm species downstream by decreasing their supply of fresh water. Finding ways to use less water and to distribute it more efficiently is crucial. The data in Table 53-1 will give you an idea of how much water is required to produce food and other items.

Table 53-1 Water Required to Produce Consumer Goods	
Consumer Good	**Gallons of Water**
One egg	40
One pound of flour	75
Loaf of bread	150
Sunday newspaper	280
One pound of beef	2,500
Automobile	100,000

heat or to power air conditioners. The incinerator uses some of the electricity produced by its generators to run the plant. Any excess electricity is sold. Each day a trash-to-resource plant may produce about 100 million watts from about 4,500 tons of refuse.

The technology is not yet perfect. Paper and food wastes such as apple cores and old spaghetti can be burned efficiently. However, other domestic wastes, including old batteries, cleaning solvents, and pesticides, contain toxic substances. Incinerating these products could release lead, PCBs, or dioxins into the atmosphere.

Air-pollution-control equipment helps minimize the emission of hazardous wastes from the incinerator. A scrubber adds a calcium carbonate solution to the waste steam to help neutralize acid gases. The scrubber also removes particulates and helps reduce dioxin levels. Later the gases pass through other filters that remove any remaining airborne dust or ash particles. In addition, the ash at the bottom of the burner and the material collected by the pollution-control equipment must be safely disposed of.

A plant such as the one in Dade County may cost hundreds of millions of dollars, a massive financial undertaking. However, the expense is offset by the energy produced by the plant and by the many benefits to the environment.

Figure 53-2 Steam rises above the wells and pipes of a geothermal installation in Wairakei, New Zealand.

Info: The use of gasohol, a mixture of a renewable resource (alcohol made through the fermentation of corn) and a nonrenewable resource (gasoline, a fossil fuel), reduces the demand for gasoline.

Fossil Fuels

Fossil fuels are the result of the incomplete decay of prehistoric plants and animals that lived during the Carboniferous period, about 300 million years ago. Fossil fuels include coal, oil, and natural gas. We use fossil fuels for about 95 percent of our energy needs. Fossil fuels are used to heat homes, propel cars, and produce electricity. Researchers have developed ways to use alternative energy sources to conserve the limited supply of fossil fuels. Both private homes and major utility companies now take advantage of solar energy, using photovoltaic cells to capture it. Other utility companies around the world are using **geothermal energy,** which is heat from beneath the earth, to generate electricity, as shown in Figure 53-2.

Minerals

A **mineral** is an inorganic solid formed in the earth. Minerals such as iron, silver, aluminum, zinc, and copper are important components of many materials that humans use. Most minerals are obtained by mining, a process that often disrupts the environment. **Strip mining,** in which layers of earth are removed to gain access to minerals, is more economical than mining ores in tunnels, but it damages the land. Treating and containing runoff, controlling emissions released into the air, and revegetating strip-mined areas can minimize environmental damage.

Figure 53-3 This quarry, from which rock formed from lava is removed, in Oahu, Hawaii, is an example of strip mining.

Answers to Section Review
1. Renewable resources can be replaced; nonrenewable resources cannot be replaced.
2. The government has identified endangered species and established regulations to ensure their protection and conservation.
3. Answers may include: contour plowing, terracing, irrigating, growing plants on bare soil, and creating windbreaks or shelterbelts. (See pp. T53A–T53D.)

Section Review

1. Distinguish between *renewable* and *nonrenewable resources.*
2. What steps have been taken to protect endangered species?
3. Name three farming practices that lessen soil erosion.
4. Summarize efforts being made by scientists and the public to conserve fossil fuels.
5. Since fossil fuels are formed from decaying organic matter, why are these fuels considered a nonrenewable resource?

53.2 Pollution

Pollution is an undesirable change in the physical, chemical, or biological characteristics of an ecosystem. One source of pollution is the accumulation of impurities, called pollutants, in the environment. Some pollutants, such as organic garbage, are **biodegradable**—subject to decay by microorganisms. Those pollutants that cannot be decomposed by microorganisms are **nonbiodegradable.** Pollutants are also classified as either primary or secondary pollutants.

Primary and Secondary Pollutants

Pollutants emitted directly into the atmosphere are called **primary pollutants.** Some primary pollutants are injurious to humans. For example, carbon monoxide (CO), a component of automobile exhaust, combines quickly with hemoglobin and interferes with the ability of the blood to transport oxygen. In addition, **particulates,** the tiny solid particles found in smoke, can irritate the respiratory system and lead to chronic lung diseases. Other primary pollutants include lead, sulfur dioxide, and hydrocarbons.

Secondary pollutants result from some effect acting on primary pollutants. Sunlight catalyzes a series of chemical reactions in the mixture of primary pollutants in the atmosphere. The product of these reactions is **photochemical smog.** Photochemical smog damages the photosynthetic tissues of plants. Nitrogen oxides, primary pollutants produced as by-products of combustion in furnaces and engines, also react with sunlight and chemicals to form **ozone,** another secondary pollutant.

Conditions that exist in the upper atmosphere are the catalyst for a different series of chemical reactions. Under these conditions primary pollutants such as chlorines and fluorines destroy the natural ozone layer that protects the earth from harmful ultraviolet radiation.

1979 1980 1981 1982

1983 1984 1985 1986

Figure 53-4   These computer-generated images show changes in the ozone level over the South Pole from 1979 to 1986. The dark purple indicates a very low ozone level. In 1987 the ozone level decreased again.

Think

The figures show how pesticide levels increase at each level of a food chain. How might such biological magnification affect humans? How might scientists direct their research in order to solve the problem?

Host	Concentration (parts per million)
Water	0.000003
Plankton	0.04
Small fish	0.05
Large fish	2
Fish-eating bird	25

Answer to Think
A diet of contaminated fish or fowl could result in a hazardous level of the pesticide in a human. Ways to reduce the risk include developing pesticides that are not bioaccumulative and developing pest-resistant crops.

Info: An example of biological magnification can be seen in environmental levels of lead. Most lead comes from fuels in the emissions of cars that burn leaded fuel. The lead is washed into lakes, rivers, and streams and then enters the food chain.

Answers to Section Review
1. Biodegradable substances can be broken down by microorganisms; nonbiodegradable substances cannot.
2. The ozone layer protects the earth from solar radiation.
(See pp. T53A–T53D.)

Acid rain is precipitation with below-normal pH, the result of industrial pollution and automobile exhaust emitted into the atmosphere. Sulfur dioxide and nitrogen oxide found in these pollutants combine with water vapor to form acids. The accumulation of acid rain lowers the natural pH of lakes and ponds, often killing the organisms that live there.

Sources of Pollution

To control pollution we must understand its sources. Major national efforts are under way to monitor and control industrial, agricultural, and domestic sources of pollution.

Industrial Sources

Industries cause pollution by creating chemical wastes in factories and processing plants and then either releasing them into the air or water or storing them incorrectly. **Hazardous wastes** are industrially produced materials that are toxic, radioactive, explosive, or otherwise dangerous to humans and the environment. Wastes such as industrial cleaning chemicals and waste oils used to be placed in steel barrels and buried. However, over time the barrels have corroded, allowing the wastes to leak out. Today regulations require the safe disposal of hazardous wastes. Still, hundreds of these former hazardous waste sites must be cleaned up.

Agricultural Sources

Many North American and European farmers rely on manufactured chemicals called **pesticides** to control insects and other pests. However, pesticides can soak into the soil and flow into streams, where they enter the food chain. **Biological magnification** is the process in which hazardous wastes become increasingly concentrated in successive levels of the food chain.

Domestic Sources

The United States accumulates more than 130 metric tons of solid wastes per year. Over 70 percent comes from residences. Sewage is one form of solid waste. To prevent sewage from flowing into rivers and lakes and tainting the water supply, cities and other municipalities have developed sewage treatment facilities to purify and recycle water.

Section Review

1. Explain the difference between *biodegradable* and *nonbiodegradable* substances.
2. What purpose does the ozone layer serve?
3. Describe how photochemical smog is produced.
4. Describe the process of biological magnification.
5. Why do areas far from sites of industrial pollution experience the effects of acid rain?

Laboratory

Effects of Pollution on Living Organisms

Objective
To examine the effects of some common forms of pollution on living organisms

Process Skills
Experimenting, inferring

Materials
Two 400-mL beakers, ice, thermometer, 12 cm of glass tubing bent into a U-shape, corks to fit both ends of the tubing, pond water, 125-mL beaker, *Paramecium* culture, 2 medicine droppers, *Ameba* culture, microscope slide, coverslip, compound microscope, 10% NaCl solution, paper towels, dissecting microscope

Method
1 Fill a 400-mL beaker with half water and half ice. Fill another 400-mL beaker with hot tap water of at least 60°C.
2 Cork one end of a piece of U-shaped glass tubing.
3 Pour 50 mL of pond water and 50 mL of *Paramecium* culture into a 125-mL beaker. Swirl the mixture gently and pour it into the glass tubing, leaving just enough room for a cork. Cork the open end of the tubing.
4 Place the tubing so that one end is in the beaker of ice water and the other end is in the beaker of hot water. What will eventually happen to the temperature inside the tubing? Note the time. Set the apparatus aside while you continue the procedure.
5 Use a dropper to place a drop of *Ameba* culture on a microscope slide. Carefully place a coverslip on top.
6 Check the slide under a microscope set to low power to make sure that there are a number of active amebas on your slide. If there are not, repeat step 5.
7 Use a clean dropper to place a drop of NaCl solution at one edge of the coverslip. Place one edge of a paper towel against the opposite edge of the coverslip. As the towel absorbs water, the NaCl solution will be drawn under the slide.
8 Examine the slide under the microscope's low power. What is the reaction of the amebas to the solution?
9 When at least 15 minutes have elapsed from the time you placed the tubing in water, remove the tubing from the beakers. Examine the tubing under a dissecting microscope. Are the paramecia uniformly distributed throughout the tubing? Feel various areas of the tubing with your fingers. Is there a temperature range?

Conclusions
1 Why did the amebas have an adverse reaction to the salt solution?
2 Was there evidence that the paramecia preferred a certain temperature?
3 Why are NaCl and water considered pollutants in this investigation when both are necessary to sustain life?

Inquiry
In what way might this investigation be expanded?

Vocabulary

acid rain (826)
biodegradable (825)
biological
   magnification
   (826)
contour plowing
   (822)
deforestation (821)
endangered species
   (821)
erosion (822)

fossil fuel (824)
geothermal energy
   (824)
hazardous waste (826)
humus (822)
mineral (824)
natural resource (821)
nonbiodegradable
   (825)
nonrenewable resource
   (823)

ozone (825)
particulate (825)
pesticide (826)
photochemical smog
   (825)
pollution (825)
primary pollutant
   (825)
reforestation (821)
renewable resource
   (821)

secondary pollutant
   (825)
strip mining (824)
terracing (822)
threatened species
   (821)
topsoil (822)
watershed (823)
wildlife (821)

Distinguish between the terms in each of the following pairs.

1. contour plowing, terracing
2. deforestation, reforestation
3. primary pollutant, secondary pollutant
4. ozone, photochemical smog
5. endangered species, threatened species

Review

1. The primary cause of species endangerment is (a) disappearing habitat (b) biological magnification (c) acid rain (d) exposure to hazardous waste.
2. Erosion is the loss of soil that results from (a) recycling (b) reforestation (c) abiotic factors (d) contour plowing.
3. Renewable resources include all of the following except (a) wildlife (b) minerals (c) soil (d) forests.
4. The organic part of the soil that helps it to retain most of its moisture and nutrients is called (a) topsoil (b) organic fertilizers (c) watershed (d) humus.
5. All of these are fossil fuels except (a) oil (b) lead (c) coal (d) natural gas.
6. Energy sources such as solar energy and geothermal energy may help reduce consumption of (a) fossil fuels (b) minerals (c) electricity (d) water.
7. Carbon monoxide restricts the oxygen supply of the body by (a) irritating the lining of the lungs (b) accumulating in blood-forming organs (c) causing lung disease (d) combining with hemoglobin.
8. The major component of photochemical smog is (a) acid rain (b) ozone (c) particulates (d) sulfur dioxide.
9. Acid rain is a (a) primary pollutant (b) secondary pollutant (c) by-product of ozone (d) particulate.
10. The primary source of trash that is burned to produce energy in a trash-to-resource incinerator, such as the one currently being used in Dade County, Florida, is (a) plastic (b) chemicals (c) paper (d) metal.
11. Explain the process of reforestation.
12. What steps were taken in the Endangered Species Act of 1973 to protect species in danger of extinction?
13. How is soil formed?
14. What is a watershed?
15. Describe the advantages and disadvantages of strip mining.
16. What is the source of geothermal energy?
17. Name the three sources of environmental pollution.
18. Describe what effects the inhaling of particulates has on the body.

19. Why has the burial of hazardous wastes not been a satisfactory method of disposing of them?

20. How can pesticides harm organisms other than the pests they are intended to kill?

Critical Thinking

1. Carbon monoxide is an air pollutant that combines with hemoglobin in a person's blood. Why is carbon monoxide such a dangerous pollutant to human beings and animals?

2. Through conservation and wildlife management the United States and other nations are making efforts to prevent the loss of species. What are some ways that this prevention can be achieved?

3. Humus is made up of decomposed organic substances. It is the component of soil that helps it retain moisture and nutrients. Why does desert soil have such a low percentage of humus? How could humans compensate for the lack of humus in desert climates in order to spur plant growth?

4. DDT was effectively used to kill crop-eating insects for a number of years before its use was banned by the United States in 1973. DDT was considered unsafe after it was linked to defects in other species such as birds. Some people were even found to have a low concentration

of DDT in their bodies. How might a pesticide such as DDT be present in species other than insects?

5. The accumulation of acid rain, produced by industrial pollution and automobile exhaust, can damage or even kill off many plants and animals. The photo above shows a situation that contributes to the production of acid rain. What methods could be used to reduce the pollutants that cause acid rain?

Extension

1. Read the article by L. Davis entitled "Biological Diversity: Going . . . going . . . ?" in *Science News*, September 27, 1986, p. 130. In an oral report discuss what prompted the formation of the Club of Earth. Describe the places where more than half the species on the earth live, what percentage of the species in the world is unknown, why the loss of these unknown species is so important, how much of the rain forest is already gone, and who should shoulder the burden of paying for conservation efforts.

2. If you live near a zoo, ask the director if the zoo is participating in the breeding of endangered species of animals. Find out which methods are most successful in producing offspring when the animal is not breeding in its natural habitat.

3. Contact your state department of natural resources to find out what animals and plants in your state are on the endangered species list. Pick one and make a poster showing what it looks like, its location in the state, and what efforts are being made to protect it.

Intra-Science: *How the Sciences Work Together*

The Question in Biology: *How Toxic Are Air Toxics?*

In the 1980s environmental scientists began noticing some disturbing occurrences throughout the Great Lakes region. The incidence of birth defects among farm animals and fish-eating birds had increased. Water samples taken from Lake Superior contained harmful levels of chemicals that could not have originated from anywhere near the Great Lakes. These findings prompted scientists to investigate Lake Siskiwit, located in the United States on Isle Royale, a small unpopulated island at the north end of Lake Superior.

The researchers chose to study Lake Siskiwit because it is far away from farms, factories, or any other source of pollution. Consequently, they were alarmed to discover that mud taken from the bottom of the lake contained an assortment of highly toxic substances, including furans, dioxins, PCBs, and toxaphene. Because the lake is so isolated, the scientists inferred that the chemicals must have been carried there by the wind. Investigators have since discovered these air toxics, as they are called, in air, water, and tissues of animals throughout the earth.

Air toxics differ from the more familiar air pollutants. They are carried through the atmosphere, not by precipitation, but on dust particles or as condensed gas particles. These particles are so small that they are measured in parts per billion or trillion. In comparison, pollutants such as sulfur dioxide are measured in parts per million. Moreover, because air toxics are so minute and because they are fat-soluble, they can accumulate in animal tissues, causing mutations or, in some cases, cancer.

Where do these toxic substances come from? Dioxins and furans, two groups of particularly toxic substances, are the by-products of incomplete combustion. Toxaphene is a pesticide that was sprayed on cotton fields in the southern United States until its use was banned in 1982. Other air toxics include wastes from factories and refineries and combustion products from gasoline and diesel fuel.

The Connection to Earth Science

Efforts to trace the movement of air toxics have relied heavily on the expertise of earth scientists, particularly geologists and meteorologists. For example, analysis of soil samples taken from the bottom of Lake Siskiwit indicates that the accumulation of toxins began after World War II, when industries increased their use of modern chemicals. Samples of mud laid down before 1940 are virtually free of pollutants.

Studies of precipitation and wind patterns have also enabled scientists to trace the path of air toxics around the earth. These studies may help explain why, for example, toxaphene applied to crops in the southern United States in the early 1980s has been found in Lake Siskiwit mud, Antarctic fish, and even reindeer milk in Scandinavia. Additional study of wind patterns and atmospheric conditions may help scientists target other areas at risk of contamination.

The Connection to *Chemistry*

If air toxics have been present in the atmosphere for forty years, why were they only recently discovered? The main reason is that until the 1970s we simply didn't have equipment sensitive enough to detect such minute amounts. Today, however, with the use of the gas chromatograph and the mass spectrometer, scientists can measure the amount of almost any organic pollutant present in air or water.

The gas chromatograph measures the boiling point of carbon-based compounds, which include most combustion products, solvents, and pesticides. It also measures the degree of attraction between molecules of these compounds and the electromagnetic charge of an electric coil. The mass spectrometer splits the molecules of an organic compound into individual atoms and ions and measures the mass and charge of each. With the combined use of these instruments, scientists have detected toxaphene, PCBs, and other substances at levels that had previously gone unnoticed.

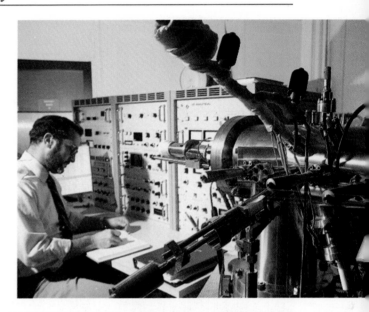

The Connection to *Other Sciences*

The development of instruments that can detect the presence of air toxics required application of the principles of physics as well as of chemistry. In addition, health professionals are closely involved in research on the long-term effects of exposure to air toxics. Years of research will be necessary to determine what health risks, if any, minute levels of these substances pose to humans.

The Connection to *Careers*

The field of environmental protection employs experts from many branches of science, including ecology, chemistry, earth science, physics, and medicine. The Environmental Protection Agency and other government agencies, universities, hospitals, and industries all hire people to conduct research, perform inspections, analyze data, and develop ways to reduce the amounts of toxic substances in the atmosphere. In addition many lawyers and economists are actively engaged in environmental safety jobs. For more information on related careers in the biological sciences, turn to page 842.

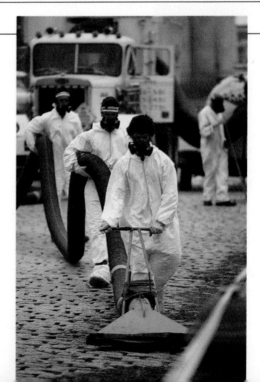

Toward the 21st Century

*T*ake just a moment to flip through Modern Biology. You will spot many pictures and diagrams that recall the biological principles you have studied. The unifying concepts of evolution, genetics, biochemistry, and cell structure that underlie the amazing diversity of organisms; the complex web of ecological relationships that tie living things to one another; the means by which this great diversity of life evolved and is still evolving— these are the concepts we hope this course has taught you.

You have learned many scientific processes, skills, and techniques. Most important, you have learned to think logically, to approach problems in a scientific way. As you face your adulthood in the next century, remember these lessons. You will be better equipped to seek solutions to the problems that confront you.

*T*he biological sciences are in the midst of a golden age of discovery. Each year promises greater insights into virtually every facet of biological study. Some of you may become physical scientists, biologists, and physicians and will share in the discoveries that lie ahead. All of you will be asked to make decisions based on your understanding of the life sciences.

So far genetic engineering has given us drugs to treat heart attacks and diabetes, and genetic engineers are working to create microbes that will improve crop yields and degrade environmental pollutants. Yet many thoughtful people fear the unforeseen consequences of

A Five-Kingdom System for the Classification of Organisms

Kingdom Monera

Unicellular, sometimes forming groups or filaments; prokaryotic (lacking nuclear membrane and organelles with membranes); nutrition mainly by absorption, but some are photosynthetic or chemosynthetic; cell wall a polysaccharide with polypeptide cross-links; reproduction usually by fission or budding, but some may exchange genetic material.

Phylum Archaebacteria: Anaerobic bacteria adapted to harsh environments; differ in terms of biochemistry from all other organisms: methanogens, extreme halophiles, thermoacidophiles.

Phylum Schizophyta: Mostly parasitic or saprophytic organisms; cells lacking an organized nucleus, with nucleoproteins in contact with cytoplasm; reproduction by fission, certain forms producing endospores: bacteria, rickettsias, actinomycetes, spirochetes.

Phylum Cyanophyta: Cells containing chlorophyll and other pigments not localized in plastids; cells lacking an organized nucleus, with nucleoproteins in contact with cytoplasm; reproduction by fission and spores: blue-green algae (*Nostoc, Anabaena, Gloecapsa, Oscillatoria*).

Phylum Prochlorophyta: Unicellular, nonmotile, and spherical; photosynthetic bacteria, resembling chloroplasts of green algae and plants; symbiotic; containing chlorophyll b in addition to a; seven strains, all of the same genus: *Prochloron*.

Kingdom Protista

Unicellular, colonial, and multicellular organisms with eukaryotic cells (cells with an organized nucleus and organelles with membranes); protists lack specialized tissue and include both heterotrophic and autotrophic forms.

Phylum Sarcodina: Move by means of pseudopodia; pellicle lacking; some with shells; reproduction principally by fission; freshwater and marine: *Ameba, Foraminifera, Radiolaria*.

Phylum Ciliophora: Move by means of cilia; pellicle present; many forms with macronucleus and micronucleus: ciliates (*Paramecium, Vorticella, Stentor*).

Phylum Zoomastigina: Move by means of one or more flagella; some parasitic; pellicle usually present; fission longitudinal: zooflagellates (*Trypanosoma, Leishmania, Giardia*).

Phylum Sporozoa: Nonmotile; spore-forming; all parasitic: sporozoans (*Plasmodium, Toxoplasma*).

Phylum Myxomycota: Slime molds with multinucleate amebalike feeding stage (plasmodium); reproductive stage forms sporangia, producing many spores; some spores may act as gametes: plasmodial slime molds (*Physarum*).

Phylum Chytridiomycota: Primarily unicellular organisms whose gametes and zoospores have a single, posterior flagellum; mostly aquatic; parasitic and saprophytic forms; chytrids.

Phylum Acrasiomycota: Slime molds with a one-celled amebalike feeding stage (myxameba) with a single nucleus; forms a mass of discrete cells (pseudoplasmodium) that form fruiting structures producing spores: cellular slime molds (*Dictyostelium*).

Phylum Oomycota: Branched filaments with many nuclei; mostly aquatic; reproduce asexually by means of motile zoospores; also reproduce sexually: water molds (*Saprolegnia*).

Phylum Chlorophyta: Cells containing chlorophyll a and b and other pigments localized in plastids; food stored as starch; unicellular, colonial, and filamentous forms; motile, free-floating, and sessile: green algae (*Spyrogyra, Oedogonium, Ulva, Chlamydomonas, Volvox, Chlorella,* desmids).

Phylum Phaeophyta: Cells containing chlorophyll a and c usually masked by fucoxanthine; food stored as oil and complex carbohydrates; multicellular; nonmotile; plant body usually large, complex, and sessile; mostly marine; motile creating new life-forms and releasing them into our environment. The question of regulating this technology will also confront us in the years ahead. As tomorrow's voters, you will need your knowledge of biology in order to understand such issues.

The world's rain forests are being destroyed at an alarming rate. With their demise comes the potential extinction of thousands—perhaps millions—of species, some not yet known to biologists. Can these irreplaceable ecosystems be saved? This, too, is a complex question, one that pits the economic growth of developing nations against the ecological well-being of the earth.

Finally, on a personal level, we hope you will recall this course as you observe the ebb and flow of life around you. When you view the splendor of a coniferous forest or the grandeur of a tallgrass prairie, when you plant a garden, or perhaps witness the miracle of the birth of your own children, we hope you will reflect on what you have learned. Your knowledge of biology will make your observations meaningful as you survey the beauty and complexity of our world.

zoospores and gametes: brown algae (*Fucus, Sargassum, Laminaria, Postelsia*).

Phylum Rhodophyta: Cells containing chlorophyll a and d usually masked by phycobilins; some unicellular, most filamentous; carbohydrates stored as a type of starch; cell walls with cellulose and polysaccharides outside of walls; some epiphytic or parasitic; no motile cells in life cycle; some with stiff thalli (coralline algae): red algae.

Phylum Crysophyta: Cells containing chlorophyll a and c and other pigments localized in plastids; cells often yellow-green, golden-brown, or brown; food stored as oil and complex carbohydrates; cell walls often with silicon; unicellular, colonial, and filamentous; motile and free-floating: golden-brown algae: diatoms.

Phylum Pyrrophyta: Cells containing chlorophyll a and c and other pigments localized in plastids; cells often yellow-green or golden-brown; food stored as starch or oil; unicellular flagellates with two flagella, one lateral and one longitudinal; mostly marine organisms: dinoflagellates (*Noctiluca, Gonyaulax*).

Phylum Euglenophyta: Unicellular autotrophic or secondarily heterotrophic organisms with chlorophyll a and b in chloroplasts; motile by means of a single flagellum; water content controlled with contractile vacuole; no cell wall; mostly freshwater; asexual reproduction by fission.

Kingdom Fungi

Eukaryotic heterotrophs with nutrition by absorption; most with a continuous filamentous body with many nuclei; cross walls may form at some time in the life cycle; some unicellular forms; cell walls made of chitin; reproduction asexual (by means of spore production, budding, or fragmentation) or sexual.

Division Zygomycota: Terrestrial saprophytes; hyphae lack cross walls: bread molds (*Rhizopus, Pilobolus*).

Division Ascomycota: Terrestrial, marine, and freshwater species; hyphae with perforated cross walls; sexual reproduction by formation of haploid ascospores in an ascus; asexual reproduction by means of conidia or budding; one unicellular form, the yeasts; red bread mold, wilt fungi, powdery mildews, ergot, morels, truffles, yeasts.

Division Basidiomycota: Mycelium with cross walls; sexual reproduction by means of basidiospores borne on basidia following meiosis: mushrooms, bracket fungi, rusts, smuts.

Division Deuteromycota: Fungi Imperfecti (fungi whose sexual phase is not known); asexual reproduction by means of conidia: *Aspergillus*, some *Penicillium*, athlete's foot fungus.

Kingdom Plantae

Multicellular, eukaryotic, autotrophic organisms having tissues and organs, cell walls containing cellulose; chlorophyll a and b present and localized in plastids; food stored as starch; sex organs multicellular.

Division Bryophyta: Multicellular green plants living on land, usually in moist situations; vascular tissues lacking; alternation of generations with the gametophyte the dominant generation; motile sperm.

Class Hepaticae: Gametophyte leafy or thalloid, usually prostrate: liverworts (*Marchantia*).

Class Anthocerotae: Gametophyte thalloid, sporophyte elongated and cylindrical: hornworts (*Anthoceros*).

Class Musci: Gametophyte usually an erect leafy shoot, sporophyte inconspicuous and parasitic on the gametophyte: mosses (*Polytrichium, Sphagnum*).

Division Psilophyta: Vascular seedless fern allies; no roots or leaves; produce sporangia on the ends of short branches; tropical and subtropical: whisk ferns of the genus *Psilotum*.

Division Lycophyta: Vascular plants; sporophyte dominant; trailing or erect with simple leaves; spores produced in a strobilus; gametophyte underground: club mosses (*Lycopodium, Selaginella, Isoetes*).

Division Sphenophyta: Vascular plants; sporophyte dominant; stems containing silica, hollow and jointed; leaves tiny scales, in whorls at stem joints; fertile stems with strobili: horsetails (*Equisetum*).

Division Pterophyta: Vascular plants; sporophyte generation dominant with leafy fronds; creeping rhizome; gametophyte a free-living prothallus; motile sperm: ferns and tree ferns.

Division Cycadophyta: Palmlike plants; male and female cones on different trees; seeds naked: cycads, sago palms.

Division Ginkgophyta: Pollen in conelike structure; seeds naked; one living species: *Ginkgo biloba.*

Division Gnetophyta: Specialized gymnosperms; seeds naked; desert species with xylem like that in angiosperms; few species: *Welwitschia, Ephedra, Gnetum.*

Division Coniferophyta: Gymnosperms with sex organs in cones; leaves as needles or scales; most evergreen; seeds naked: pines, spruces, firs, larches, yews.

Division Anthophyta: Flowering plants; sex organs in whorls in flowers; seeds enclosed in ovary that ripens into a fruit.

Class Monocotyledoneae: Embryo has one cotyledon; flower parts in threes; leaf veins parallel; vascular bundles scattered through stem tissue; grasses, sedges, lilies, irises, palms, grasses.

Class Dicotyledoneae: Embryo has two cotyledons; flower parts in fours or fives; leaves with netlike veins; vascular bundles in orderly arrangement in stems: roses, maples, elms.

Kingdom Animalia

Multicellular, eukaryotic, heterotrophic organisms; nutrition mainly by ingestion; specialized tissues and many have complex organ systems; no cell walls; reproduction mostly by sexual means; tissues often considerably reorganized during embryonic development.

Phylum Porifera: Body wall consists of two cell layers, penetrated by numerous pores; "skeleton" formed by siliceous or calcareous spicules or horny spongin; sessile marine and freshwater animals: sponges.

Class Calcarea: Simple sponges of shallow waters; calcareous spicules: *Grantia.*

Class Hexactinella: Deep-water sponges; siliceous spicules: Venus's flower basket.

Class Demospongiae: Large sponges; spongin or a combination of spongin and siliceous

material; freshwater and marine: bath sponge, finger sponge.

Class Sclerospongiae: Spongin and siliceous and calcareous spicules.

Phylum Cnidaria (also called Coelenterata): Usually free-swimming; baglike body of two cell layers with a noncellular substance between them; gastrovascular cavity with one opening leading to the outside; many with tentacles and all with stinging capsules; solitary or colonial forms; marine and freshwater: hydroids, jellyfish, corals.

Class Hydrozoa: Solitary or colonial; freshwater and marine; reproduction by asexual buds and gametes; alternation of generations in many: *Hydra, Physalia.*

Class Scyphozoa: Exclusively marine; most have mesentaries; polyp stage usually absent: jellyfish.

Class Anthozoa: Marine forms; solitary or colonial; numerous tentacles: sea anemones, corals, sea fans.

Phylum Ctenophora: Marine forms resembling jellyfish; hermaphroditic; definite digestive system with anal pore; biradially symmetrical: comb jellies.

Phylum Platyhelminthes: Body flat and ribbonlike, without true segments; acoelomates; no skeletal, circulatory, or respiratory systems; head provided with sense organs; nervous system composed of two longitudinal nerve cords: flatworms.

Class Turbellaria: Mostly free-living aquatic or terrestrial forms; many with cilia on the epidermis: planarians.

Class Trematoda: Parasitic forms with mouth at anterior end; intestine present; no cilia on adults: human blood flukes (*Schistosoma*), sheep liver fluke.

Class Cestoda: Parasitic forms; body a series of proglottids; intestine lacking; hooked scolex: tapeworms.

Phylum Rotifera: Crown of cilia on anterior end; pseudocoelomates; chitinlike jaws and well-developed digestive system; body usually cylindrical: rotifers.

Phylum Nematoda: Body slender and elongated; pseudocoelomates; unsegmented body wall in three layers; body cavity present; bilaterally symmetrical; free-living and parasitic forms: roundworms (*Ascaris, Trichinella*, pinworm, hookworm, vinegar eel).

Phylum Nematomorpha: Body slender and elongated, resembling a hair; larvae parasitic in insects, adults free-living in fresh water; mouth often lacking in adults: horsehair worms.

Phylum Acanthocephala: Body elongated, digestive tract lacking; anterior proboscis armed with many recurved hooks; parasitic in vertebrates: spiny-headed worms.

Phylum Bryozoa: Microscopic organisms forming branching colonies; row of ciliated tentacles at anterior end; usually marine: bryozoans (sea mosses).

Phylum Brachiopoda: Body enclosed in dorsal and ventral shells; two spirally coiled arms within shell bearing a row of ciliated tentacles; simple circulatory system; marine; mostly fossil forms: brachiopods.

Phylum Mollusca: Coelomates; undergo spiral cleavage in early stages of development; body in three parts; all organ systems present; mantle, secreting a shell in many genera; rasplike radula; trochophore and veliger larvae; terrestrial, freshwater, marine: clams, snails, octopuses.

Class Gastropoda: Stomach-footed, with or without single coiled shell; head, distinct eyes, and tentacles present: snail, slug, whelk.

Class Bivalvia (also called Pelecypoda): Hatchet-footed with bivalve shell; gills in mantle cavity; head, eyes, and tentacles lacking: clam, oyster, scallop.

Class Cephalopoda: Head-footed; foot modified into grasping tentacles; marine animals: squid, octopus, chambered nautilus.

Class Polyplacophora: Elongated body and reduced head, without tentacles; many forms with a shell composed of eight plates: chiton.

Class Scaphopoda: Body elongated and enclosed in a tubular shell, open at both ends; gills lacking; marine: tooth shells.

Phylum Annelida: Segmented coelomate worms; all organ systems present; bristles or setae on most; spiral cleavage in early development stages: segmented worms.

Class Polychaeta: Fleshy outgrowths, or parapodia, extending from segments; marine; many bristles: sandworm.

Class Oligochaeta: Head not well developed; few bristles; terrestrial and freshwater forms: earthworm, *Tubifex*.

Class Hirudinea: Body flattened from top to bottom; no bristles on body; suckers at both ends; mostly freshwater forms, but may occur as terrestrial or marine organisms: leeches.

Phylum Arthropoda: Segmented bodies, the segments bearing jointed appendages; all organ systems, including ventral nervous system and open circulatory system; chitinous exoskeleton; aerial, terrestrial, and aquatic forms: arthropods.

Subphylum Chelicerata: Distinguished by absence of antennae and presence of chelicerae; nearly all are land-dwellers.

Class Arachnida: Head and thorax usually fused into a cephalothorax; antennae lacking; four pairs of legs; book lungs and trachea for respiration: spiders, scorpions, ticks, mites.

Subphylum Crustacea: Antennae, mandibles, and branched appendages.

Class Crustacea: Head and thorax joined in a cephalothorax; usually five pairs of legs; two pairs of antennae; mostly aquatic; gills for respiration; many with calcareous deposits in exoskeleton: crayfish, lobster, crab, shrimp, water flea, sowbug, barnacle.

Subphylum Uniramia: Antennae, mandibles, and unbranched appendages.

Class Chilopoda: Body flattened and consisting of 15 to 170 or more segments; one pair of legs attached to each segment; maxillipeds developed into poison claws: centipedes.

Class Diplopoda: Body more or less cylindrical and composed of 25 to 100 or more segments; most segments bearing two pairs of legs: millipedes.

Class Insecta: Head, thorax, and abdomen separate; three pairs of legs; one pair of antennae; usually two pairs of wings; trachea for respiration: insects.

Order Thysanura: Silverfishes.

Order Ephemeroptera: Mayflies.

Order Odonata: Dragonflies, damsel flies.

Order Orthoptera: Grasshoppers, cockroaches, walking stick, mantis, crickets.

Order Isoptera: Termites.

Order Dermaptera: Earwigs.

Order Mallophaga: Chicken lice.

Order Anoplura: Human body louse.

Order Hemiptera: True bugs, water bug, water strider, water boatman, back swimmer, bedbug, squash bug, stinkbug.

Order Homoptera: Cicada, aphids, leaf hopper, tree hopper, scale insects.

Order Neuroptera: Dobson fly (hellgrammite), aphis lion, lacewing.

Order Coleoptera: Beetles, ladybugs, firefly, boll weevil.

Order Lepidoptera: Butterflies, moths, skippers.

Order Diptera: Housefly, bot fly, blowfly, midge, mosquitoes, crane fly, gall gnat.

Order Siphonaptera: Fleas.

Order Hymenoptera: Bees, ants, wasps, hornets, ichneumon fly.

Phylum Echinodermata: Pentaradially symmetrical; adults covered with spiny or calcareous plates; most forms with tube feet for locomotion; radial cleavage in early development; digestive tract and nervous system; endoskeleton; marine: echinoderms.

Class Crinoidea: Five branched rays and pinnules; tube feet without suckers; most forms with stalk; many fossil forms: sea lily, feather star.

Class Asteroidea: Body usually with five rays and double rows of tube feet in each ray; eyespots: starfish.

Class Ophiuroidea: Usually with five slender arms or rays: brittle stars, basket stars.

Class Echinoidea: Body spherical, oval, or disk-shaped; rays lacking; tube feet with suckers: sea urchin, sand dollar.

Class Holothuroidea: Elongated, thickened body with tentacles around the mouth; no rays or spines: sea cucumber.

Phylum Hemichordata: Wormlike chordates; body in three regions with a proboscis, collar, and trunk: acorn worm (tongue worm).

Phylum Chordata: Notochord present at some time, disappearing early in many forms and replaced by vertebral column; paired gill slits temporary or permanent; dorsal nerve cord.

Subphylum Urochordata: Saclike covering or tunic in adult; marine; free-swimming or attached: sea squirts and other tunicates.

Subphylum Cephalochordata: Fishlike with a permanent notochord: lancelet (*Branchiostoma*).

Subphylum Vertebrata: Chordates in which most of the notochord is replaced by a spinal column composed of vertebrae and encasing the dorsal nerve cord: vertebrates.

Class Ostracodermi: Extinct; jawless primitive vertebrates; marine; protective scales: ostracoderms.

Class Agnatha: Freshwater or marine eellike forms without true jaws, scales, or paired fins; cartilaginous skeleton: lamprey, hagfish.

Class Chondrichthyes: Fishes with true jaws and paired fins; gills present but not free and opening through gill slits; no air bladder; cartilaginous skeleton: sharks, rays, skates.

Class Osteichthyes: Freshwater and marine fishes with gills free and attached to gill arch; one gill opening on each side of body; true jaws and paired fins; bony skeleton: bony fishes.

Class Amphibia: Freshwater or terrestrial forms; gills present at some stage; skin slimy and lacking protective outgrowths; limbs without claws; three-chambered heart in adults; numerous eggs, usually laid in water and externally fertilized; metamorphosis: amphibians.

Order Apoda: Wormlike; tail short or lacking; no limbs or limb girdles: caecilians.

Order Urodela: Body elongated and with a tail throughout life; most forms with two pairs of limbs: salamanders, newts.

Order Anura: Body short; tailless in adult stage; two pairs of limbs, the hind limbs adapted for leaping; gills in larval stage, lungs in adult stage: frogs, toads, tree frogs.

Order Trachystoma: Aquatic, eellike, tiny forelimbs and no hindlimbs; three species: sirens, mud eels.

Class Reptilia: Terrestrial or semiaquatic vertebrates; breathing by lungs at all stages; body covered by scales or plates; toes with claws, amniote eggs provided with a leathery, protective shell; fertilization internal: reptiles.

Order Rhynchocephalia: Tuatara (*Sphenodon*).

Order Chelonia: Turtles and tortoises.

Order Crocodilia: Alligators, crocodiles, gavials, caimans.

Order Squamata: Lizards and snakes.

Class Aves: Body covered with feathers; forelimbs modified into wings; four-chambered heart; warm-blooded; amniote egg: birds.

Order Gaviiformes: Loons: common loon.

Order Pelecaniformes: Tropical birds: white pelican, brown pelican, cormorants.

Order Ciconiiformes: Long-legged wading birds: herons, bitterns, ibises, spoonbill.

Order Anseriformes: Short-legged gooselike birds: ducks, geese, swans.

Order Falconiformes: Large birds of prey: hawks, falcons, eagles, kites, vultures, buzzards.

Order Galliformes: Fowllike birds: pheasant, turkey, quails, partridges, grouse, ptarmigans.

Order Gruiformes: Cranelike birds: cranes, coots, gallinules, rails, limpkin.

Order Charadriiformes: Shore birds: snipes, sandpipers, plovers, gulls, terns, auks, puffins.

Order Columbiformes: Pigeons and doves: mourning dove, white-winged dove.

Order Psittaciformes: Parrots and parrotlike birds: parrots, parakeets, macaws.

Order Cuculiformes: Cuckoos: cuckoo, roadrunner.

Order Strigiformes: Nocturnal birds of prey: owls.

Order Caprimulgiformes: Goatsuckers: whippoorwill, chuck-will's widow, nighthawk.

Order Apodiformes: Swifts: chimney swift, hummingbirds.

Order Coraciiformes: Fishing birds: kingfishers.

Order Piciformes: Woodpeckers: woodpeckers, sapsuckers, flickers.

Order Passeriformes: Perching birds: robin, bluebird, sparrows, warblers, thrushes.

Class Mammalia: Hair on at least part of body; young nourished with milk secreted by mammary glands; warm-blooded; four-chambered heart; lung breathing; many with highly developed cerebrum and cerebellum.

Order Monotremata: Egg-laying mammals: duckbilled platypus, spiny anteater.

Order Marsupialia: Pouched mammals: opossum, kangaroo, koala.

Order Insectivora: Insect-eating mammals: moles, shrews.

Order Chiroptera: Flying or hand-winged mammals: bats.

Order Edentata: Mammals that are toothless or have peglike teeth: armadillo, sloth, great anteater.

Order Pholidota: Anteaterlike mammals that have scalelike skin: pangolins.

Order Rodentia: Gnawing mammals: squirrels, woodchucks, mice, rats, muskrats.

Order Lagomorpha: Harelike mammals: rabbits, hares, pika.

Order Cetacea: Marine mammals: whales, porpoises, dolphins.

Order Sirenia: Aquatic mammals: sea cow.

Order Proboscidea: Trunk-nosed mammals: elephant, fossil mammoth, fossil mastodon.

Order Carnivora: Flesh-eating mammals: bears, weasels, mink, otters, skunks, lions.

Order Pinnipedia: Flesh-eating marine mammals: seals, sea lions, walruses.

Order Perissodactyla: Odd-toed hoofed mammals: tapir, rhinoceroses, horses.

Order Artiodactyla: Even-toed hoofed mammals: hippopotamuses, camels, llamas, deer, giraffes, cattle, sheep, goats.

Order Primates: Erect mammals: monkeys, lemurs, gibbon, orangutan, gorilla, human.

Measurement

All measurements in this book are expressed in metric units. Scientists throughout the world use the metric system, and you will always use metric units when you make measurements in the laboratory. The official name of the measurement system is the Système International d'Unités, or International System of Measurements. It is usually referred to simply as SI.

The metric system is a decimal system—that is, all relationships between units of measurement are based on powers of 10. Most units have a prefix that indicates the relationship of that unit to the base unit. For example, a meter equals 100 centimeters or 1,000 millimeters. The lists below show the most commonly used prefixes as well as the main units used for each type of measurement.

Metric Prefixes

Prefix	Symbol	Factor of Base Unit
giga	G	1,000,000,000
mega	M	1,000,000
kilo	k	1,000
hecto	h	100
deka	da	10
deci	d	0.1
centi	c	0.01
milli	m	0.001
micro	μ	0.000001
nano	n	0.000000001

Temperature

In the metric system temperature is measured on the Celsius (C) scale. On the Celsius scale, 0° is the freezing point of water, and 100° is the boiling point of water. Thus 1° C equals 0.01 of the difference between the freezing point and boiling point of water. You can use the scale shown to convert between the Celsius scale and the Fahrenheit scale, which is used in the United States.

Area

square kilometer (km²)   = 100 hectares
1 hectare (ha)   = 10,000 square meters
1 square meter (m²)   = 10,000 square centimeters
1 square centimeter (cm²)   = 100 square millimeters

Mass

1 kilogram (kg)   = 1,000 grams
1 gram (g)   = derived from kg (base unit of mass)
1 milligram (mg)   = 0.001 gram
1 microgram (μg)   = 0.000001 gram

Liquid Volume

1 kiloliter (kL)   = 1,000 liters
1 liter (L)   = base unit of liquid volume
1 milliliter (mL) = 0.001 liter

Note: When measuring liquid volume in a graduated cylinder, be sure to read the measurement at the bottom of the meniscus, or curve.

Length

1 kilometer (km)   = 1,000 meters
1 meter (m)   = base unit of length
1 centimeter (cm) = 0.01 meter
1 millimeter (mm) = 0.001 meter
1 micrometer (μm) = 0.000001 meter
1 nanometer (nm)   = 0.000000001 meter

Freezing point of water   Boiling point of water

Safety

You will avoid accidents in the biology laboratory by following directions, handling materials carefully, and taking your work seriously. Before you begin working in the laboratory, be sure that you are familiar with safety procedures and that you know the location of firefighting equipment and first-aid supplies. Be aware of the safety of your classmates as well as your own. Never attempt any laboratory procedure without the direction of an instructor, and never work in the laboratory by yourself.

Read the text that follows to familiarize yourself with the safety symbols used in your text and the guidelines that you should follow when you see these symbols.

Humane Treatment

You are both legally and ethically required to treat all animals in as humane a way as possible.

Fire Safety

Keep combustible materials away from sources of fire. Shield any open flame with asbestos-protected screen.

Clothing Protection

Wear a laboratory apron. Roll up or button loose sleeves. Keep work surfaces clean.

Glassware Safety

Use care when handling glassware. Never attempt to clean up broken glassware. Notify your teacher immediately.

Eye Safety

Wear goggles when working near an open flame or handling chemicals. If any chemical gets into your eyes, flush it out thoroughly with water and notify your instructor immediately.

Animal Safety

Animals can injure you or spread disease. If you are not familiar with the correct way to handle an animal, get professional advice before you do so.

Gas Precaution

When using a Bunsen burner, wear protective goggles, gloves, and an apron. If the burner works improperly, turn the gas completely off at the laboratory gas valve.

Plant Safety

Some plants can cause ill effects when touched or eaten. Use a reliable field guide when collecting specimens, and never eat any part of an unknown plant.

Hand Safety

Dissect specimens in dissecting pans—never in your hand. Never use chipped or cracked glassware. Wear gloves when working with an open flame or with caustic chemicals.

Poison

Use poisonous chemicals with extreme caution. Two harmless chemicals may become poisonous when combined. Never mix chemicals without teacher direction. Never put chemicals in your mouth and avoid chemical contact with your eyes or skin.

Caustic Substances

Use extreme care when handling caustic substances such as hydrochloric acid. These substances can injure the skin. If your skin comes in contact with a caustic substance, flush thoroughly with water and notify your teacher immediately.

Electrical Safety

Watch for loose plugs or worn electric cords. Be sure that cords are placed where they will not cause someone to fall. Do not use electrical equipment near water or with wet hands.

Explosion Danger

Many chemicals are explosive when combined, and some will explode when jarred, heated, or exposed to air. When heating chemicals, always point the test tube away from people.

Careers

The following career listing provides the names and addresses of organizations that can supply information on a broad range of careers in biology. The listing presents possible careers in areas related both to the contents of the units and to the subjects discussed in the Intra-Science features. Keep in mind that almost any interest or talent can be developed and used in a career in the biological sciences.

Research and Communication
Science Writer and Editor
National Association of Science Writers and Editors
P.O. Box 294
Greenlawn, NY 11740
Biomedical Researcher
Federation of American Societies for Experimental Biology
9650 Rockville Pike
Bethesda, MD 20814
Biophysicist
Biophysical Society
9650 Rockville Pike
Bethesda, MD 20814
Exobiologist
Extraterrestrial Research Division, Code LX
NASA Ames Research Center
Moffett Field, CA 94035
Scientific Illustrator
Association of Medical Illustrators
2692 Hugeunot Springs Road
Midlothian, VA 23113

Cell Biology
Cellular Physiologist
American Physiological Society
9650 Rockville Pike
Bethesda, MD 20814
Photobiologist
American Society for Photobiology
8000 Westpark Drive
Suite 400
McLean, VA 22101

Genetics
Medical Geneticist
Genetics Society of America
9650 Rockville Pike
Bethesda, MD 20814
Agrigeneticist
American Society of Agronomy
677 South Segoe Road
Madison, WI 53711

Diagnostic Medical Sonographer
Society of Diagnostic Medical Sonographers
Building 1, Suite 276
10300 North Central Expressway
Dallas, TX 75231

Evolutionary Biology
Botanist
Dr. Robert H. Essman
Publications Manager
Botanical Society of America
Department of Genetics
Ohio State University
1735 Neil Avenue
Columbus, OH 43210
Physical Anthropology
American Association of Physical Anthropologists
Department of Anthropology
SUNY/Buffalo
479 Spaulding Quad, Building 4
Buffalo, NY 14261

Microbiology
Virologist
American Society for Microbiology
1913 I Street, N.W.
Washington, DC 20006

Nurse Epidemiologist
Director
Epidemiology Program Office
Center for Disease Control
Atlanta, GA 30333

Agriculture
Farm Manager
American Society of Farm Managers and Rural Appraisers
950 S. Cherry Street
Suite 106
Denver, CO 80222
Seed Analyst
Association of Official Seed Analysts
Illinios Department of Agriculture
Seed Lab
P.O. Box 4906
Springfield, IL 62708
Plant Pathologist
American Phytopathological Society
3340 Bald Knob Road
St. Paul, MN 55121

Invertebrate Zoology
Parasitologist
American Society of Parasitologists
1041 New Hampshire Street
Lawrence, KS 66044

Marine Biologist
American Society of Limnology and Oceanography
Great Lakes Research Division
University of Michigan
Ann Arbor, MI 48109
Entomological Inspector
Entomological Society of America
4603 Calvert Road
College Park, MD 20740
Aquaculturist
U.S. Aquaculture Federation
Box 276
Lacey Spring, VA 22833

Vertebrate Zoology
Zoology Professor
American Society of Zoologists
Box 2739
California Lutheran University
Thousand Oaks, CA 91360
Wildlife Manager
U.S. Department of Interior
Fish and Wildlife Service
Division of Personnel Management
Room 3458
Washington, D.C. 20240

Health and Medical
Biomedical Engineer
Biomedical Engineering Society
P.O. Box 2399
Culver City, CA 90230
Fitness Instructor/Physical Education Teacher
VGM Career Horizons
4255 W. Touhy Avenue
Lincolnwood, IL 60646
Pharmacologist
American Society for Pharmacology and Experimental Therapeutics
9650 Rockville Pike
Bethesda, MD 20814

Environmental Protection
Sanitation Inspector
National Sanitation Foundation
P.O. Box 1468
3475 Plymouth Road
Ann Arbor, MI 48105
Public Health Microbiologist
American Society for Microbiology
1913 I Street, N.W.
Washington, DC 20006

Glossary

The phonetic respellings throughout this text are based on the pronunciations given in *Webster's Ninth New Collegiate Dictionary* and *Webster's Third New International Dictionary*. Primary stress is indicated by capital letters; secondary stress, by small capitals. Vowels are respelled according to the following system:

Webster's Symbol	Respelling	Note
ā	ay	At end of syllable
	a-e	When consonant follows
a	a	
ar	ar	
ä	ah	At end of syllable
	ah	When consonant follows
är	ahr	
au̇	ou	
ē	ee	
e	eh	At end of syllable
	e	When consonant follows
er	er	
a(ə)r	air	
ī	ie	At end of syllable
	i-e	When consonant follows
i	ih	At end of syllable
	i	When consonant follows
ir	ir	
ō	oe	At end of syllable
	o-e	When consonant follows
ȯ	aw	Except when *r* follows
ȯr	or	
ȯi	oi	
ü	oo	
u̇	ooh	
yü	yoo	
ə	uh	At end of syllable
	u	When consonant follows
ər	ur	
ə (schwa)	uh	For indefinite sound at end of syllable
	,	When schwa is elided

A

abdomen a posterior segment of an animal that usually houses the organs of digestion and excretion (483)

abdominal cavity the posterior portion of the human ventral cavity (637)

abiotic factor any nonliving component of an ecosystem (765)

aboral surface the surface opposite the mouth on a sea star (515)

abscisic acid a hormone in plants that plays a role in regulation of both the growth of buds and the germination of seeds (424)

abscission zone the area at the base of a leaf petiole where the leaf breaks off from the stem (428)

absorption in excretion, the movement of wastes from the bloodstream to the excretory system (698)

acetyl-CoA in aerobic respiration, a molecule formed when a two-carbon acetyl group combines with coenzyme A (106)

acid rain precipitation with below-normal pH, often the result of industrial pollution and automobile exhaust (826)

acid solution a solution containing more hydronium ions than hydroxide ions; also called acid (40)

acoelomate an animal with no coelom, or body cavity (453)

acquired immune deficiency syndrome (AIDS) a group of diseases caused by a viral disruption of the immune system (680)

acquired immunity disease protection gained after birth (678)

actin one of the two protein filaments in a muscle cell that function in contraction (647)

actinomycotes members of a class of bacteria characterized as rod-shaped monerans that form branched filaments (291)

action potential a wave of depolarization that moves down through the neural membrane (715)

activation energy the amount of energy required for a chemical reaction to get started and to continue on its own (42)

active transport the movement of any substance across a cell membrane with the use of energy from ATP (84)

adaptation an inherited trait that increases an organism's chance of survival in a particular environment (6, 230)

adaptive radiation an evolutionary pattern in which many species evolve from a single ancestral species (231)

addiction a dependency on the presence of a drug (749)

adenine a nitrogen base that is a component of a nucleotide and ATP (113)

adenosine diphosphate (ADP) the molecule adenosine with two phosphate groups attached (94)

adenosine monophosphate (AMP) the molecule adenosine with one phosphate group attached (94)

adenosine triphosphate (ATP) the molecule adenosine with three phosphate groups attached (94)

adrenal gland an endocrine gland located on the top of a kidney (729)

aerobic respiration the breakdown of pyruvic acid with the use of oxygen (103)

afterbirth the remains of the placenta and the amnion, expelled from the mother's body following birth (744)

agroforestry the planting of trees and crops to maintain a forest, prevent erosion, and shield young crops (355)

AIDS-related complex (ARC) a group of diseases similar to but less virulent than those of AIDS (681)

air sac in birds, a section that stores air, functions in respiration, and reduces density (593)

albumen the proteinaceous portion of an egg; the egg white (594)

alcoholic fermentation the process by which pyruvic acid is converted to ethyl alcohol; the anaerobic action of yeast on the sugars found in fruits and grains (105, 751)

alcoholism the disease of addiction to ethanol (751)

algal bloom a population explosion of algae that colors the environment (327)

alimentary canal the digestive tract (560, 694)

allantois in amniote eggs, the membranous sac that contains many blood vessels; in humans, a membrane surrounding the embryo that becomes the umbilical cord (571, 742)

allele a contrasting form of a gene (152)

allele frequency the percentage of an allele in a gene pool (240)

allergen a usually harmless antigen in the environment (678)

allergy a reaction to an allergen (678)

alternation of generations a sexual life cycle in plants and algae involving two or more phases (368)

altricial referring to birds that, when hatched, are immature and in need of parental care (596)

altruistic behavior the sacrifice of immediate gratification by one individual that results in a benefit for another individual (628)

alveolus one of the air sacs found in the lungs (667)

amebocyte an amebalike cell in a sponge that moves through the body cells, supplying nutrients and removing wastes (442)

ameboid movement movement by means of pseudopodia involving a cytoplasmic gel–sol conversion (311)

Ames test a procedure used to identify carcinogenic substances (164)

amino acid one of 20 monomers that form proteins (52)

ammonification the process in the nitrogen cycle in which ammonia compounds form (790)

amniocentesis a procedure used in fetal diagnosis in which cells are removed from the amniotic fluid (176)

amnion one of four membranes of an amniote egg (571, 742)

amniote egg the four-membrane egg of a terrestrial vertebrate (571)

amotivational syndrome a suppressed activity level caused by habitual use of marijuana (756)

amplexus the mating embrace of frogs (565)

ampulla a bulblike sac in an echinoderm that functions in movement (515)

anaerobic respiration the combined processes of glycolysis and fermentation, which do not require O_2 (103)

anal fin medial fin near the anus of a fish (547)

anal pore an opening where protozoan wastes are eliminated (313)

analyzing data the process of determining whether data are reliable (20)

anaphase a stage of mitosis and meiosis in which the chromosomes separate (133)

anatomy internal morphology (8)

androgen a hormone secreted by the testes that controls secondary sex characteristics (730)

angiosperm a flowering plant (367)

annual rhythm a pattern of activity of organisms that occurs about once a year (801)

annual ring a yearly growth ring in a woody plant (389)

antenna an appendage of some invertebrates specialized for touch and taste (484)

antennule an antennalike appendage with receptors for touch, taste, and equilibrium (484)

anterior the front end of a bilaterally symmetrical organism (437)

anther the microsporangium of an angiosperm in which pollen grains are produced (406)

antheridium in plants and algae, a reproductive structure that produces flagellated sperm by mitosis (330, 401)

anthropoid a subgroup of primates including monkeys, apes, and humans (256)

antibiotic a chemical produced by microorganisms that is capable of inhibiting the growth of some bacteria (304, 678)

antibody a blood protein produced by B cells that destroys antigens (663)

anticodon a region of tRNA consisting of three bases complementary to the codon of mRNA (121)

antigen a substance that stimulates the production of antibodies (174, 663, 676)

antihistamine a drug that counteracts histamines (679)

anus the posterior opening of the alimentary canal (458)

aorta the largest artery in the body (499, 657)

aortic arch in invertebrates, a tube that links the dorsal and ventral blood vessels (473)

aphotic zone the ocean layer that receives no light (774)

apical meristem the growing region at the tips of stems and roots in plants (382)

apical dominance a plant growth pattern in which branches near the shoot tip are shorter than branches farther from the shoot tip (422)

appendage any complex part or organ extending from the body (479)

appendicular skeleton in vertebrates, the bones extending from the axial skeleton (539, 641)

applied genetics the science of manipulating heredity characteristics to improve or create specific traits in offspring (191)

aqueous solution a solution in which water is the solvent (38)

arachnoid layer the middle layer of the meninges (710)

archaebacterium a type of bacterium that has physiological adaptations to extreme environments (296)

archegonium in plants, a reproductive structure that produces a single egg by mitosis (402)

archosaurs a group of reptiles that gave rise to crocodiles, dinosaurs, and those reptiles that evolved into birds (573)

Aristotle's lantern a jawlike feeding mechanism in Echinoidea (514)

arteriole a small artery (657)

artery a vessel that carries blood from the heart (549, 657)

artificial propagation in plants, asexual reproduction manipulated by humans (416)

ascocarp the reproductive body of an ascomycete (339)

ascogonium an egg-producing structure in Ascomycetes (339)

ascospore one of eight haploid cells in an ascus (339)

ascus a sac that forms on the surface of an ascocarp (339)

asexual reproduction the production of offspring that does not involve the union of gametes (7, 138)

association neuron a cell that maintains neural connections within the spinal cord (711)

aster a centriole with its radiating fibers that is active during cell division (132)

atom the fundamental unit of matter (33)

atomic number the number of protons in an atom (34)

ATP synthetase an enzyme that catalyzes ATP synthesis during respiration (95)

atrioventricular node a group of nerves in the heart that functions in establishing the heartbeat (657)

atrium the anterior chamber of the heart (549, 655)

auditory canal the tube through which air enters the ear (720)

auditory nerve the nerve along which impulses from the cochlea travel to the auditory region of the brain (720)

autoimmune disease a disorder in which the immune system produces antibodies against the organism's own cells (679)

autonomic nervous system   a part of the peripheral nervous system that controls involuntary actions (712)

autosome   a chromosome that is not a sex chromosome (165)

autotomy   the ability of an organism to regenerate body parts (579)

autotroph   an organism that synthesizes organic molecules from inorganic substances (9, 93, 301)

auxin   a plant hormone that regulates cell elongation (421)

avoidance behavior   a reaction away from danger (313)

axial skeleton   the backbone, skull, and associated bones of vertebrates (539, 641)

axon   in a neuron, an elongated extension that carries impulses away from the cell body (713)

B

bacteriophage   a virus that infects bacteria (288)

baleen   the bony plates that function to filter food from water in some whales (610)

barb   a branch of a vane in the feather of a bird (589)

barbule   a branch of a barb in the feather of a bird (589)

bark   the protective layer of cells produced by a cambium in woody plants (389)

basal metabolic rate   the energy expended by the body when resting (668)

base solution   a solution containing more hydroxide ions than hydronium ions (40)

basidiocarp   the reproductive body of a basidiomycete (338)

basidiospore   a cell resulting from meiosis that is found at the end of a basidium (338)

basidium   a specialized club-shaped reproductive structure that forms on the gills of mushrooms (338)

Batesian mimicry   a defensive adaptation in which a nonpoisonous organism has the appearance of a poisonous species (502)

B cell   a lymphocyte that produces antibodies in response to antigens (676)

behavior   a set of responses to a stimulus (11)

behavioral variations   differences in responses among closely related organisms (493)

bell curve   a graph in the shape of a bell that usually describes the normal range of a trait (240)

benign tumor   an abnormal but nonthreatening cell mass (186)

benthic zone   the ocean bottom (774)

bicuspid   a tooth with two points used to shred food (604)

bilateral symmetry   a type of symmetry in which the left and right sides mirror each other (437)

bile   a yellowish fluid secreted by the liver that functions as an emulsifier (697)

binary fission   an asexual, mitotic division of unicellular organisms that produces identical offspring (302)

binomial nomenclature   a system of naming organisms that uses the genus name plus a specific epithet (267)

biochemical pathway   the series of complex processes involved in intermediary metabolism (94)

biodegradable   referring to any pollutant that is subject to decay by organisms (825)

bioethics   the application of ethics to biology (10)

biogenesis   the theory that living organisms come only from other living organisms (207)

biogeochemical cycle   the process by which materials necessary for organisms are circulated through the environment (788)

biological clock   an internal timing mechanism that controls endogenous rhythms (800)

biological magnification   the process by which hazardous wastes are concentrated as a result of feeding in a food chain (826)

biological species concept   the idea that organisms can be classified by their ability to breed with one another (238)

biology   the science of life (5)

bioluminescence   the ability of an organism to produce light by means of a chemical reaction (327)

biomass   the dry weight of organisms (787)

biome   a geographic area characterized by specific kinds of plants and animals (767)

biosphere   the area on and around the earth where life exists (765)

biosystematics   the study of reproductive compatibility and gene flow (271)

biotic factor   a living component of an ecosystem (765)

biotic potential   the rate at which a population grows if all members survive and reproduce at maximum capacity (811)

biotic relationships   the interactions among organisms (797)

bipedalism   the ability to walk upright on two legs (256)

bipinnaria   the larva of echinoderms (516)

blade   the broad, flat portion of a typical leaf (324, 394)

blastocyst   a mass of cells that will become the organs of a new human being (740)

blastopore   a depression formed when cells of the blastula move inward (438)

blastula   a hollow ball of cells formed when a zygote undergoes repeated cycles of cell division (184, 438)

blood alcohol concentration (BAC)   a measurement of the amount of alcohol in a person's blood (751)

blood clot   a structure caused by the entrapment of red blood cells in fibrin (663)

bloom   a population explosion of a species (298)

blue-green bacterium   a photosynthetic moneran classified into Phylum Cyanophyta (297)

bolting   the rapid lengthening of internodes caused by gibberellic acid (424)

bolus   a ball of thoroughly chewed and moistened food in the digestive tract (695)

bone marrow   the soft tissue in the center and ends of long bones where blood cells are produced (641)

book lung   a structure in the abdomen of an arachnid where gases are exchanged between air and the blood (487)

bouton   the swollen tip of an axon terminal (716)

Bowman's capsule   a cup-shaped structure of the nephron of a kidney that encloses the glomerulus and where absorption takes place (701)

brain   the anterior end of the central nervous system; in vertebrates, the cerebrum, the cerebellum, and the brain stem (708)

brain stem   the ventral region below the brain that connects the brain to the spinal cord (709)

bristle   a hairlike structure (589)

bronchus   one of the two branches of the trachea that enter the lungs (667)

bronchiole   a small tube that branches from the bronchi within the lungs (667)

brood patch   in birds, a patch of the parent's skin used to cover eggs (595)

bundle sheath   the parenchyma cells around a vein in a leaf (395)

bursa   a fluid-filled sac that aids lubrication in some joints (644)

C

C₄   a form of photosynthesis in which carbon dioxide binds to a three-carbon compound found in leaf cells (102)

calorie   the amount of heat required to raise the temperature of 1 g of water 1° C (687)

Calorie one kilocalorie; the unit used to measure the amount of energy a particular food provides (687)

Calvin cycle the dark reactions of photosynthesis in which glucose is produced (100)

CAM crassulacean acid metabolism, a form of photosynthesis (102)

camouflage the characteristic of blending with the surroundings (502)

cancer the uncontrolled growth of cells (186)

canine a pointed tooth used to grip and tear (604)

cap the flattened, top portion of a mushroom (338)

capillarity the ability of a liquid to move upward against the force of gravity by molecular attraction to a surrounding surface (47)

capillary the smallest vessel in the blood vessel network (549, 657)

capsid the protein covering that encases a virus (283)

capsule in mosses, a sporangium that produces spores; in bacteria, a protective layer of slime around the cell wall (300, 369)

carageenan a substance found in the cell walls of red algae that is used commercially as a smoothing agent (325)

carapace the dorsal part of a turtle shell (577)

carbohydrate an organic compound composed of carbon, hydrogen, and oxygen in which the hydrogen-to-oxygen ratio is two to one (49, 687)

carcinogen a cancer-causing substance (186)

carcinoma a malignant tumor that grows in tissues (186)

cardiac muscle the involuntary muscle of the heart (646)

cardiac sphincter a circular muscle located between the esophagus and the stomach (696)

carnivore a consumer that eats other consumers (574, 783)

carotene an accessory pigment of plants (98)

carotenoid the yellow, orange, and brown plant accessory pigments (98)

carpel the ovule-containing reproductive structure of flowering plants (406)

carrier a person heterozygous for a trait, the recessive gene of which causes a disease (171)

carrier molecule a protein that functions in transport of molecules across a membrane (84)

carrying capacity the maximum number of individuals of a species that an ecosystem is capable of supporting (813)

cartilage a strong, flexible connective tissue (642)

Casparian strip a narrow band of suberin belting each cell of the endodermis in a root (384)

cast a type of fossil formed when sediments fill in the cavity left by a decomposed organism (219)

caudal fin in fishes, a fin extending from the tail that moves from side to side and amplifies the swimming motions (547)

cecum a blind sac usually found in the digestive system (612)

cell a membrane-bound structure that is the basic unit of life (63)

cell culture a population of identical cells grown in the laboratory (27)

cell cycle the events of cell division; includes interphase, mitosis, and cytokinesis (131)

cell differentiation the change in morphology or physiology of a cell in relation to its neighboring cells (184)

cell division the formation of cells from a parent cell (12)

cell enlargement an increase in cell size (12)

cell-mediated response the immune response involving a T-cell attack on an antigen (677)

cell membrane the outer bilipid boundary of a cell (66)

cell plate a membrane that divides the daughter cells following mitosis of a plant (135)

cell theory the theory that all living things are made up of cells, that cells are the basic units of organisms, and that cells come only from existing cells (64)

cell wall an extracellular, strengthening component of a plant cell (72)

central nervous system the brain and the spinal cord (707)

centric diatom a member of Chrysophyta algae with circular valves (326)

centrifugation a technique used to separate cell parts by type by spinning fractionated cells at high speeds (28)

centriole a rod-shaped structure that appears during mitosis and is responsible for spindle formation and movement (132)

centromere the structure formed where chromatids join (129)

cephalization a concentration of nerve tissue and sensory organs at the anterior end of an organism (438)

cephalothorax a body part formed by the fusion of the head with the thorax (483)

cereal any species of the grass family whose grain is eaten (353)

cerebellum a posterior portion of the brain that controls movement and muscle coordination (550, 709)

cerebral cortex the folded outer layer of the cerebrum that controls motor and sensory activities (708)

cerebral hemisphere one of two halves of the cerebrum (708)

cerebrospinal fluid a watery substance that provides a cushion that protects the brain and spinal cord (711)

cerebrum the anterior portion of the brain where higher brain functions occur (550, 708)

chalaza the ropelike strands that support the albumen in a fertilized egg (594)

chelicera a pincerlike mouthpart of some arthropods (482)

cheliped in arthropods, a claw used to capture food and to provide defense (484)

chemical equation an equation showing the products and reactants of a chemical reaction (42)

chemical reaction the process of breaking chemical bonds, forming new bonds, or both (42)

chemiosmosis the synthesis of ATP during the passage of H^+ through a membrane (99)

chemoautotroph an organism that synthesizes inorganic compounds by using chemicals instead of sunlight (301)

chemotropism plant growth in response to a chemical (428)

chitin a carbohydrate that forms part of the arthropod exoskeleton (479)

chloroplast a plastid containing chlorophyll; the site of photosynthesis (74)

cholesterol a type of steroid; a lipid (689)

chorion the outer membrane surrounding an embryo (571, 741)

chorion villi sampling a procedure involving the analysis of the chorion villi to diagnose fetal genotypes (176)

chromatid one of two identical parts of a chromosome (129)

chromatin the DNA and proteins of the nuclear membrane of a nondividing cell (72)

chromatography a method of separating components of a chemical based on their varying rates of movement up a piece of paper or down a column of beads (28)

chromatophore skin pigmentation that creates color patterns; a pigment cell that allows organisms to blend with their surroundings (470, 547)

chromoplast a type of plastid that synthesizes and stores pigments to trap sunlight for energy (74)

chromosome DNA in a coiled, rod-shaped form that occurs during cell division (72)

chromosome map a diagram of allele positions on a particular chromosome (167)

chronic bronchitis an inflammation of the bronchi and bronchioles (749)

chrysalis the outer covering of a butterfly pupa (501)

chyme the mixture formed in the stomach from digested food particles and gastric fluid (696)

cilia a structure made up of microtubules, extending out from a cell, and aiding movement (71, 312)

circadian rhythm a pattern of behavior that is about one day in length (800)

circulatory system the system that distributes oxygen and nutrients to cells in all parts of the body (655)

cirrhosis a condition in which normal tissues are replaced by scar tissue (752)

class in taxonomy, a group of similar orders (267)

classifying the grouping of objects, organisms, or phenomena into a new or established organizational scheme (19)

climax community a community of organisms that will remain stable in an area as long as the area is undisturbed (803)

clitellum a noticeable swelling around the sex organs of an earthworm during reproduction (474)

cloaca a cavity holding undigested wastes (561)

cloacal opening the site where waste materials and eggs leave the body (561)

clone an offspring produced by asexual production (415)

closed circulatory system a system in which blood is contained within vessels (470)

cnidarian a member of the Phylum Cnidaria (444)

cnidocyte the stinging cells of cnidarians (444)

CO₂ fixation the binding of carbon dioxide to ribulose biophosphate during photosynthesis (101)

coacervate a cell-like droplet formed from dissimilar substances (210)

cochlea a coiled tube in the inner ear filled with fluid and lined with hair cells (720)

cocoon the covering of an insect during its pupal stage; a chrysalis (501)

codominance the condition of two alleles of the same gene neither of which totally masks the other (157)

codon a group of three sequential nitrogen bases of an mRNA molecule (120)

coelenterate another name for members of the Phylum Cnidaria (444)

coelom a body cavity formed from the mesoderm (453)

coenocytic referring to filaments without internal cross walls (336)

coevolution the joint change of interacting species (232)

cohesion the bonding of water molecules to each other (48)

colchicine a chemical that prohibits the formation of the cell plate during plant cell division (193)

coleoptile in plants, a cylindrical sheath that protects the shoot during germination (411)

coleorhiza in plants, a sheath that protects the embryonic root (411)

collar cell a cell lining the inside of sponges that circulates water; also called a choanocyte (441)

collecting data the gathering and recording of specific information based on observations (18)

collenchyma plant tissue made up of elongated cells with unevenly thickened, flexible walls (381)

colloid a mixture in which particles do not settle out over time (39)

colon the large intestine (699)

colonial organism a group of unicellular organisms that live together in closely connected groups (76)

columella a bone that transmits sounds from the eardrum to the middle ear (559)

commensalism a form of symbiosis in which one organism benefits and the other neither benefits nor is harmed (799)

common name an everyday name given to an organism (267)

communicating the process of sharing information (22)

community all the populations in one area (765)

compact bone the layer of bone just beneath the periosteum that gives the bone its strength (541)

companion cell a specialized plant cell that helps control the movement of sugars through sieve tubes (381)

competition the use or defense of a resource by an organism that reduces availability of that resource to others (797)

competitive exclusion principle the extinction of all but one species when two or more species occupy the same niche (798)

complete digestive system a system in which food enters via the mouth and passes sequentially through the system to the anus (526)

compound two or more elements that are chemically joined (35)

compound eye an eye with many lenses (485)

compound leaf a type of leaf in which the blade is divided into leaflets (394)

concentration a measurement of the amount of solute dissolved in a fixed amount of solvent (38)

concentration gradient the difference in concentration of a substance across space (81)

condensation reaction a chemical reaction, also called dehydration synthesis, in which water is produced (48)

conductor a specialized nerve cell that transmits messages to the central nervous system (707)

cone a photoreceptor within the retina that can produce sharp images and distinguish colors (719)

conidiophore the specialized branch where conidia form (340)

conidium a spore produced during asexual reproduction in ascomycetes (340)

coniferous forest a biome dominated by evergreens (769)

conjugation a form of reproduction in which genetic information moves from a cell of one filament to a cell of an adjoining filament (302)

connective tissue the most common type of human body tissue, which supports and holds body structures together (638)

consumer a heterotroph that obtains energy from chemical bonds in nutrients it eats (783)

contour feather a feather that provides coloration, insulation, and a streamlined shape to adult birds (589)

contour plowing forming rows for farming perpendicular to a slope (822)

contractile vacuole an organelle in microorganisms that excretes water (83, 312)

controlled breeding the manipulation of heredity by selecting parents with specific phenotype traits (191)

controlled experiment a comparison of a control group or phase with an experimental group or phase (20)

conus arteriosus the vessel in front of a frog's heart (563)

convergent evolution the process by which unrelated species become more similar as they adapt to the same kind of environment (231)

cork a layer of bark that functions for protection (382)

cork cambium the lateral meristem of a plant; produces cork (382)

coronary artery a vessel supplying blood to the heart (660)

coronary circulation the systemic subsystem that supplies blood to the heart (660)

corpus callosum the structure formed by nerve fibers connecting the cerebral hemispheres of the brain (708)

corpus luteum the structure that forms from the ruptured follicle in the ovary at ovulation (738)

cortex in plants, a mature ground tissue located just inside the epidermis; in animals, the outermost portion of an organ (384, 700)

cotyledon a seed leaf (375)

cotylasaurs a group of ancestral reptiles resembling small lizards that lived about 300 million years ago (573)

courtship behavior used to attract a mate (597)

covalent bond a bond that forms when two atoms share one or more pairs of electrons (37)

coxal glands in some spiders, organs that remove wastes and discharge them at the base of the legs (487)

cranial capacity the size of the brain case (256)

cranial cavity the area in which the brain rests (637)

cranial nerves nerves between the major sense organs and the brain (550)

critical length the length of daylight above or below which plant species will flower (427)

critical period the period when imprinting may occur (627)

crop in earthworms, a structure that stores soil; in birds, a structure that stores food (472, 592)

crossing-over the exchange of genes by homologous chromosomes during meiosis (137)

cross-pollination a reproductive process in which pollen from one plant is transferred to the stigma of another plant (148, 408)

cutaneous respiration in frogs, the exchange of gasses through the skin (563)

cuticle in plants, a waxy covering that prevents desiccation; in parasites, a nonliving layer that protects them from the host's digestive and immune systems (365, 454)

cyst a dormant cell with a protective outer wall (312, 456)

cytokinesis the division of the cytoplasm of a parent cell into two daughter cells (131)

cytokinins plant hormones that promote cell division (423)

cytolysis the intake by a cell of so much water that it bursts (83)

cytoplasm the gellike fluid of a cell interior (66)

cytoplasmic streaming the circular motion of the cytoplasm (67, 311)

cytosine a nitrogen base; a pyrimidine of DNA and RNA (113)

cytoskeleton the framework of a cell (70)

D

dark reactions in photosynthesis, the process by which ATP and $NADPH_2 + H^+$ react with CO_2 from the atmosphere and form glucose (96)

day-neutral plant a plant unaffected by the daylight length (427)

deciduous forest a biome characterized by trees that lose their leaves in the autumn (770)

decomposer an organism that uses nutrients from dead plants and animals (784)

deforestation the clearing of forested regions (359, 821)

dehydration the condition resulting from excessive water loss (693)

deletion a mutation in which a segment of DNA breaks off of a chromosome (163)

dendrite an extension of a neuron that receives impulses from other neurons and carries the impulses to the cell body (713)

denitrification a final step in the nitrogen cycle, during which nitrogen gas is returned into the atmosphere (790)

density-dependent factors variables related to the density of a population that affect population size (814)

density-independent factors variables that affect population size regardless of population density (814)

deoxyribonucleic acid (DNA) a double-helix-shaped nucleic acid (7, 54)

deoxyribose a five-carbon sugar molecule found in DNA (113)

dependent variable the responding variable in an experiment (20)

depressant a drug that decreases the activity of the central nervous system and slows many body functions (751)

dermal tissue system one type of plant tissue system, which forms the outer covering of plants (382)

dermis the layer of skin beneath the epidermis (649)

desert an area where rainfall averages less than 25 cm per year (772)

determinate cleavage a process in which each cell develops into a specific part of the gastrula (522)

deuterostome an organism in which the blastopore develops into the anus and the coelom arises by enterocoely and whose embryo has indeterminate cleavage and develops radially (521, 537)

development the process by which an organism grows (7)

diabetes mellitus a disorder of glucose metabolism caused by insufficient insulin (732)

diaphragm a thoracic cavity muscle used in breathing (604, 665)

diastole a phase in the heartbeat when the ventricles relax and the blood flows in from the atria (657)

diatomaceous earth a layer of material formed of the rigid shells of dead diatoms (326)

dichotomous key a written set of choices that leads to the name of an organism (269)

dicot an angiosperm with two cotyledons, net venation, and flower parts in fours or fives (375)

diffusion the process by which molecules move from an area of greater concentration to an area of lesser concentration (81)

digestion breaking down food into usable molecules (694)

digitalis a medicine used in the treatment of heart disease (356)

dihybrid cross a cross between individuals that involves the pairing of contrasting traits (157)

dipeptide a formation from two amino acids bonded together by means of a condensation reaction (53)

diploid a cell that contains both chromosomes of a homologous pair (130)

direct development a process by which a young animal is born in a smaller form of the adult (523)

directional selection a type of natural selection in which the distribution of a trait is shifted toward one of the extremes (245)

disaccharide a double sugar formed from two monosaccharides (50, 687)

disruptive selection natural selection in which individuals with an extreme form of a trait have an advantage (245)

diurnal referring to organisms that are awake and active mainly during the day (800)

divergent evolution the process of two or more related species becoming more and more dissimilar (231)

division a grouping of similar classes of plants (268)

division of labor specialization in form and function in a cell, colony, organism, or society (503)

domesticated adapted for human use (815)

dominant referring to an allele that masks the presence of another allele for the same characteristic (150)

dormancy a state of decreased metabolism (413)

dorsal the top of a bilaterally symmetrical animal (537)

dorsal cavity the cavity containing the organs of the nervous system (637)

dorsal fin the median dorsal fin or fins of a fish (547)

dorsal nerve cord a neural tube dorsal to the notochord (537)

double fertilization in plants, the process in which two types of cell fusion take place in the embryo sac (410)

double helix spiral-shape characteristic of the DNA molecule (114)

down feather soft, fluffy feathers that cover the body of nesting birds (589)

Down syndrome a disorder caused by an extra twenty-third chromosome (175)

drug any chemical taken into the body that alters the normal processes of either the mind or the body (749)

duodenum the first section of the small intestine (561, 697)

dura mater the outer layer of the menenges (710)

E

echolocation a method of navigation similar to sonar in which the sound produced bounces off an object (610)

ecology the study of the relationship between organisms and their environments (10, 765)

ecosystem all the biotic and abiotic components of an environment (765)

ectoderm one of the three germ layers of the gastrula (439)

ectoplasm the region in the cytoplasm located directly inside the cell membrane (311)

ectothermic cold blooded; referring to an animal whose body temperature is determined by the environment (573)

effector peripheral nerve cell that responds to messages from the central nervous system (707)

egg tooth scalelike tooth on a bird embryo used to crack the surrounding egg (595)

electron a particle with a negative electric charge that orbits the nucleus (33)

electron transport chain a sequence of reactions in which an electron is passed from one molecule to another molecule in the chain (98)

electrophoresis a procedure that separates proteins by acting on differences in their electrical charges (284)

element a substance that ordinarily cannot be broken down chemically to form simpler kinds of matter (34)

embryo an organism in its early stages of development (438, 741)

embryo sac in plants, the megagametophyte containing seven cells and eight nuclei (407)

emphysema a degenerative lung disease (749)

endangered species those plants and animals that are currently at risk of extinction (821)

endocrine gland a ductless gland that secretes substances directly into the blood (728)

endocytosis the process by which a cell engulfs and surrounds large substances (87)

endoderm one of the three germ layers; found in the innermost layer of the gastrula (439)

endodermis in plants, a specialized layer of cells that regulates substances entering the center of the root (384)

endoplasm the region in the cytoplasm found in the interior of the cell (311)

endoplasmic reticulum a membrane system in the cytoplasm (68)

endoskeleton an internal skeleton; in humans, the 206 bones that are the body framework (511, 537, 640)

endosperm a tissue in angiosperms that provides food for the developing embryo (409)

endospore a dormant cell enclosed by a tough coating, often found in bacteria (300)

endosymbiosis the theory that prokaryotic intracellular parasites evolved into various cell organelles (214, 309)

endothermic warm blooded; referring to an animal that generates its own body heat (573)

endothermic reaction a chemical reaction that involves a net consumption of energy (42)

endotoxin a substance that causes a poison reaction; a lipopolysaccharide compound that makes up part of the cell wall of Gram-negative bacteria (303)

energy the ability to do work (41)

enhancer a region adjacent to a eukaryotic gene that must be activated if the gene is to be expressed (182)

enterocoely the method of mesoderm formation in deuterostomes in which the coelom forms in folded mesoderm (523)

entomologist a scientist who studies insects (494)

enzyme a protein that acts as a catalyst in intermediary metabolism (53, 688)

epicotyl in plant embryos, the region above the cotyledons (411)

epidermis in plants, the outer layer of cells; in animals, the outer layer of skin that encloses the body in a continuous sheet (382, 649)

epiglottis a flap of tissue that covers the trachea (695)

epiphyseal plate a band of cartilage located between the shaft and the end of a long bone (643)

epiphyte a plant that grows on other plants (799)

epithelial tissue tissue composed of cells that create a solid sheet that covers a body part (638)

equilibrium a state that exists when the concentration of a substance is the same throughout a space (81)

erosion the loss of soil that is caused by wind or water (822)

erythrocyte a red blood corpuscle (662)

esophagus the tube connecting the mouth and the pharynx to the stomach (472, 548, 695)

estivation an annual rhythm characterized by reduced activity during the summer (556, 801)

estuary an aquatic biome found where freshwater streams and rivers flow into the sea (775)

ethanol the type of alcohol found in beer, wine, and liquor (751)

ethics the study of what is morally good and bad (10)

ethologist a scientist who studies social behaviors (628)

ethylene C_2H_2; a gaseous hormone produced by various parts of plants and released into the air (424)

eubacterium a bacterium in the Class Eubacteria (297)

euchromatin the uncoiled form of chromatin (182)

eukaryote a cell that contains a nucleus and membrane-bound organelles (65)

Eustachian tube a canal that connects the middle ear to the mouth cavity (559, 720)

eutrophication the increase of nutrients in an environment (806)

eutrophic lake a lake or pond rich in organic matter (776, 806)

evisceration the ejection of internal organs (514)

evolution the theory that species change over time (6, 219)

excretion the process of removing metabolic wastes (700)

excurrent siphon a hollow tube through which water exits the mantle cavity of a clam (468)

exocrine gland a gland that secretes substances through a duct (728)

exocytosis the process by which wastes are packaged in vesicles and leave the cell (88)

exon a segment of DNA and mRNA that is not excised and that functions in protein synthesis (183)

exoskeleton a hard external covering of some invertebrates that provides protection and support (479)

exothermic reaction a reaction that has a net release of energy (42)

exotoxin a protein secreted by pathogenic bacteria (303)

experimenting the process of testing a hypothesis by gathering data under controlled conditions (20)

expiration the process in which air in forced out of the lungs (666)

exponential phase a period of population growth in which the population keeps doubling in increasingly shorter time periods (812)

external respiration the exchange of gases between the atmosphere and the blood (665)

extinct the state of a species when it no longer exists (249)

extreme halophile an archaebacterium that lives only in areas of high salt concentration (296)

eyespot a localized region of pigment in some invertebrates that detects changes in the quantity and quality of light (310)

F

facilitated diffusion a passive form of carrier transport (85)

facultative anaerobe an organism that can metabolize with or without oxygen (301)

Fallopian tube a tube that carries eggs from the ovary to the uterus (736)

family in taxonomy, a group of similar genera (267)

famine a food shortage that exists over a wide area over a long period (355)

fat a triglyceride that is solid at room temperature (51, 688)

fatty acid a monomer that is part of most lipids (51)

fatty liver a swelling of the liver due to excessive alcohol use (752)

feces waste materials that pass through the anus or cloacal opening (699)

feedback mechanism a mechanism in which the end product of a process controls the first step in the process (727)

fermentation the breakdown of pyruvic acid with the use of oxygen (103)

fertilization the fusion of gametes (739)

fetal alcohol syndrome a condition affecting some of the children born to women who drink alcohol during pregnancy (752)

fetoscopy a visual, observational procedure for diagnosing disorders in a fetus (176)

fetus an unborn child; referring to the embryo from the end of the eighth week until birth (741)

fever an increase in body temperature in response to the presence of pathogens (675)

fiber a type of sclerenchyma cell in which the wall is long and thin (381)

fibrin the long sticky threads that function in blood clotting (663)

fibrinogen a blood protein essential for the formation of blood clots (662)

fibrous root one of many small roots that spread out near the top of the soil (383)

fiddlehead a coiled fern leaf (373)

filament in flowers, the structure that supports the anther; part of a stamen (406)

filoplume the hairlike pinfeathers near the nostrils of birds (589)

filter feeding the method by which a sessile organism screens food out of the water (442)

filtration the process in the kidney in which materials from the blood are forced out of the glomerulus and absorbed by the cells of Bowman's capsule (701)

first filial, or F₁, generation the offspring of cross-pollinated P₁ generation plants (148)

flagellum a hairlike structure made up of microtubules that function for motility (71, 313)

flame cells in platyhelminths, cells that enclose a tuft of cilia and function in excretion (455)

flight feather a specialized contour feather on the wing or tail of a bird (589)

fluid mosaic model a model of cell membrane structure representing the dynamic nature of lipids and proteins (67)

follicle a pit in the skin in which hair or a feather develops (589)

food chain the specific feeding sequence in which organisms obtain energy in an ecosystem (785)

food vacuole the portion of a cell membrane surrounding the food during phagocytosis (312)

food web the interrelated food chains in an ecosystem (786)

foot an organ in a mollusk used for locomotion (466)

fossil the remains or traces of a once-living organism (209)

fossil fuel fuel formed from the incomplete decay of prehistoric plants and animals (824)

fractionation a technique that breaks up cells and releases cell components (28)

frameshift mutation a specific type of point mutation in which the genetic code is read out of sequence (164)

freshwater biome an aquatic biome characterized by low levels of dissolved salts (776)

frond the mature leaf of a fern (373)

fruit a mature plant ovary; a modified flower part that encloses a seed or seeds (355)

Fungi Imperfecti a group of about 25,000 species of fungi whose sexual phase has not been discovered (341)

furculum the keel-shaped sternum unique to avian skeletons (587)

G

gallbladder the saclike organ connected to the liver that produces bile (697)

gamete a sexual reproductive cell (138)

gametophyte in plants and algae, a multicellular structure that produces gametes (330, 368)

ganglion a mass of neurons (455, 707)

gas bladder an organ in fish that provides buoyancy and prevents sinking (549)

gastric cecum one of the pockets in the stomach that secretes enzymes (499)

gastric fluid a liquid secreted by gastric glands (696)

gastric pit the open end of gastric glands through which secretions are released (696)

gastritis an irritation of the stomach lining (752)

gastrovascular cavity the hollow gut of some lower invertebrates (445)

gastrula a cup-shaped embryo formed as the blastula folds inward (184, 438)

gated channel a protein-lined passage making cell membranes permeable to certain large molecules (86)

gel a colloid in a semisolid state (39)

gemmule an asexual reproductive structure of amebocytes surrounded by organic material and spicules (442)

gene a short segment of DNA that contains coding for a polypeptide or protein (7, 123, 152)

gene cloning the process by which a gene is replicated (196)

gene expression the transcription of DNA into RNA and then into proteins (181)

gene flow the movement of genes into or out of a population (242)

gene pool all the genes for all the traits in a population (240)

gene sequencing a process that determines the exact order of bases in a fragment of DNA (196)

gene splicing the process by which a gene from one organism is placed into the DNA of another organism (195)

generative cell in a pollen grain, the cell that forms two sperm (407)

genetic code the system that contains information needed by the cell for proper functioning (120)

genetic counseling the process of informing a couple about genetic problems that have the potential to affect their offspring (176)

genetic drift a shift in allele frequencies in a population due to random events or chance (243)

genetic engineering a form of applied genetics in which scientists directly manipulate genes (194)

genetic equilibrium the condition in which allele frequencies in a population do not change from generation to generation (241)

genetic marker a short section of DNA that indicates the presence of an allele that codes for a trait (173)

genetic recombination the process that results in chromosomes being rearranged in many different ways (138)

genetic screening an examination of a person's genetic makeup (175)

genotype the genetic makeup of an organism (153, 240)

genus in taxonomy, a group of similar species (267)

geographic isolation the physical separation of populations (247)

geothermal energy the heat originating beneath the earth (824)

germ cell mutation a change in the DNA of a sex cell (163)

germ layer a specific layer of cells in an embryo from which specific organ systems are derived (439)

gestation period length of pregnancy (608, 741)

gibberellin a plant hormone that, among other actions, stimulates bolting and cessation of dormancy (423)

gill in mushrooms, the radiating rows on the underside of the cap; in animals, an organ specialized for the exchange of gases with water (338, 466, 540)

gill slits the openings in the throat region of a fish that lead to the gill chambers (537)

gizzard a muscular organ in birds that crushes food; a digestive structure in earthworms in which thick, muscular walls grind soil to release organic matter (472, 592)

glomerulus a cluster of capillaries that receives blood from the renal artery and that is enclosed in Bowman's capsule (701)

glottis a structure at the end of the passage from an organism's throat to its lungs (563)

glycocalyx a covering in some prokaryotes that functions to stick the organism to a surface (300)

glycolysis the breakdown of glucose to pyruvic acid (103)

glycoprotein a molecule of protein with attached sugar chains (285)

Golgi apparatus an organelle of a cell that functions in secretion (68)

gonads the organs that produce gametes and some sex hormones (730)

G_1 phase the first period of interphase, in which the cell doubles in size (132)

G_2 phase the final period of interphase, in which the cell undergoes rapid growth that prepares it for mitosis (132)

grafting a technique in which a portion of one plant is inserted into and grows with the root or shoot of another (416)

Gram-negative bacterium a bacterium that stains pink with Gram stain and that usually has an outer covering on its cell wall (297)

Gram-positive bacterium a bacterium that stains purple with Gram stain and that usually lacks an outer covering on its cell wall (297)

Gram stain a series of dyes that turn either purple or pink according to the chemistry of a bacterial cell wall (297)

granular gland an organ in frogs that secretes foul-tasting or poisonous substances (560)

granum a stack of thylakoids in a chloroplast (96)

grassland a biome dominated by grasses (771)

gravitropism a plant growth response to gravity (425)

green gland an excretory organ in crayfish that removes wastes from the blood and helps retain salts (484)

greenhouse effect a condition of increased heat due to reflection of sunlight back to earth from the atmosphere (791)

Green Revolution a change in worldwide food production implemented by improving productivity (355)

gross primary productivity the total amount of energy produced in an ecosystem (783)

ground system one type of plant tissue system made up of parenchyma, collenchyma, and sclerenchyma (381)

growth curve in populations, a graph of the number of individuals in a population over time (812)

guanine a nitrogen base; one component of a nucleotide (113)

guard cell one of two cells bordering a stoma, each of which has flexible cell walls that allow it to bend (396)

gullet the structure that forms food vacuoles that circulate throughout the cytoplasm of the paramecium (312)

gymnosperm a seed plant whose seeds are not enclosed by an ovary (367)

H

habitat the physical area in which an organism lives (781)

hair follicle a structure in the skin that produces hair (650)

half-life the period of time in which half of a radioactive substance decays (220)

haploid a cell that has only one chromosome of a homologous pair (130)

hard palate the bony roof of the mouth (695)

Hardy–Weinberg principle a theory that states the conditions necessary for gene frequencies to remain stable across generations (241)

Haversian canal a channel containing nerves and blood vessels making up compact bone (641)

hazardous waste industrially produced material that is toxic, radioactive, explosive, or otherwise dangerous (826)

head the anterior end of a cephalized animal (466)

heart a muscular organ that pumps blood (655)

helper cell a T cell that plays a role in the cell-mediated immune response (677)

hemocoel the blood-filled space or body cavity in the tissues of some invertebrates (467)

hemoglobin an iron-containing protein in blood (622)

hemophilia a trait in which the blood lacks a protein that is essential for clotting (175)

hemotoxin a component in snake venom that destroys red blood cells and disrupts the clotting of blood (581)

hepatic portal circulation the system of blood movement through the digestive tract and liver (661)

herbaceous plant an annual plant that is nonwoody (382)

herbivore a consumer that eats primary producers (574, 783)

heredity the transmission of traits from parents to their offspring (147)

hermaphrodite any individual that produces both eggs and sperm (443)

heterochromatin the chromatin that contains inactive genes not currently being transcribed into RNA (182)

heterocyst a cell of cyanophytes that fixes atmospheric nitrogen (298)

heterokaryotic referring to a hypha with nuclei from genetically different mating strains coexisting within it (336)

heterospory a phenomenon in which two types of spores are produced by an individual plant (403)

heterotroph an organism that cannot make its own food (9, 93, 301)

heterozygous referring to a gene pair in which the two alleles do not code for the same trait (153)

hibernation a state of severely reduced physiological activity occurring during the winter (556, 801)

hilum a scar along the edge of a plant seed marking where the seed was attached to the ovary wall (411)

hindgut the colon and the rectum of some invertebrates (499)

histamine a chemical associated with the sneezing and wheezing of allergies (679)

histone a special protein molecule that DNA wraps around during chromosome formation (129)

holdfast the part of an algal thallus that anchors it to a rock (324)

homeostasis the biological balance between a cell or an organism and its environment (81)

homeotic gene a gene that controls the development of a specific adult structure (185)

hominid a subgroup of primates that includes human beings and their immediate ancestors (258)

homokaryotic referring to a hypha with genetically similar nuclei (336)

homologous chromosome one of two members of a pair of morphologically similar chromosomes (130)

homologous structures structures that are embryologically similar (223)

homospory a phenomenon in which all spores look alike and produce similar gametophytes (402)

homozygous referring to a gene pair in which the two alleles code for the same trait (153)

hormone a substance produced in one part of an organism that specifically influences the activity of cells in another part of the organism (421, 727)

host an organism that supports a parasite (454, 799)

humus the organic component of the soil (822)

Huntington disease a human genetic disorder caused by a dominant allele (172)

hybrid the offspring of two dissimilar organisms (192, 238)

hybridization a type of controlled breeding in which two different but related species are crossed (192)

hybrid vigor the quality displayed by hybrids that grow faster and are healthier than their parents (192)

hydrogen bond a chemical bond in which atoms share a hydrogen nucleus (114)

hydrolysis the splitting of a molecule through reaction with water (48, 687)

hydrophilic referring to the molecular attraction to water (51)

hydrophobic referring to the molecular repulsion to water (51)

hydrotropism a plant growth response to water (426)

hyperthyroidism a condition that results when too much thyroxine is secreted by the thyroid gland (728)

hypertonic the condition when the concentration of solute molecules is higher outside than inside a cell (83)

hypha the vegetative filament of a fungus (336)

hypocotyl a stemlike area in plants between the cotyledons and radicle (411)

hypoglycemia a condition caused by a lack of glucose in the blood (732)

hypothalamus the structure located below the thalamus (709)

hypothesis a statement that can be tested experimentally (19)

hypothesizing the process of forming testable statements about observable phenomena (19)

hypothyroidism a condition in which too little thyroxine is secreted by the thyroid gland (728)

hypotonic the condition when the concentration of solute molecules is lower outside than inside a cell (83)

I

icosahedron a polyhedron or a capsid of a virus with 20 triangular faces (284)

ileum the middle portion of the small intestine (561, 698)

imaginal disk the specific tissue area in the pupal stage that leads to a specific structure in the adult stage (185)

immune response a reaction of the body against a foreign substance (675)

immunity a resistance to a specific pathogen (678)

imprint a type of fossil in which a thin film of carbon remains after the organism has decayed (219)

imprinting the behavior process that causes young animals to follow the first moving thing seen after birth (627)

impulse an electrical signal carried by neurons (713)

inbreeding a type of breeding based on the idea that individuals with similar phenotypes also will have similar genotypes (192)

incisor the chisellike tooth used for biting and cutting (604)

incomplete digestive system in invertebrates, a system in which one opening serves as both mouth and anus (526)

incubate the process of warming eggs before they are hatched (595)

incurrent siphon a hollow tube through which water enters the mantle cavity of a clam (468)

independent variable an experimentally manipulated variable (20)

indeterminate cleavage a process in which each embryonic cell will develop into a complete gastrula (522)

indirect development in invertebrates, morphogenesis with an intermediate larval stage (523)

indoleacetic acid (IAA) an auxin produced in actively growing regions of plants (421)

inducer the molecule that binds to repressor molecules and induces gene expression (182)

infectious disease a transmittable disease (673)

inferior vena cava the large vein carrying blood from the lower part of the body (657)

inferring the process of drawing conclusions on the basis of facts or premises instead of direct perception (21)

inflammatory response a response in which white blood cells engulf foreign substances and body temperature rises (675)

inheritance the acquisition of traits by offspring (7, 147)

inorganic compound a compound generally derived from nonliving things (47)

insertion the attachment point of a tendon to a moving bone (648)

insight learning the ability of individuals to apply what they learn in one situation to other situations (627)

inspiration the process of taking air into the lungs (666)

instinct genetically controlled behavior (503)

insulin a hormone that lowers the blood sugar level (732)

integument in plants, one or two outer layers of the ovule; in animals, the outer covering (407, 547)

interferon a protein produced by the body that inhibits the reproduction of viruses (194, 676)

intermediary metabolism a series of chemical reactions in which energy is released slowly (42)

internal fertilization the method of fertilization by which sperm fuses with an egg inside the female's body (571)

internal respiration the exchange of gases between the blood and the cells of the body (665)

interphase a period of cell growth and development that precedes mitosis (131)

interspecific competition competition between organisms of different species (797)

intertidal zone a biome along ocean shorelines that is repeatedly covered and uncovered by seawater (775)

intraspecific competition competition between organisms of the same species (797)

intron a DNA segment producing an RNA segment that is excised before it codes for a protein (183)

inversion a mutation that occurs when a chromosome piece breaks off and reattaches in reverse orientation (163)

invertebrate an animal without a backbone (437)

ion an atom or a compound with a net electrical charge (36)

ionic bond a bond formed by electrical attraction between two oppositely charged ions (36)

iris the structure of the eye that controls the amount of light entering the eye (718)

islets of Langerhans the special cells in the pancreas that secrete two hormones—insulin and glucagon (732)

isomer one of two or more compounds that differ in structure but not in molecular composition (50)

isotonic the condition when the concentrations of solute molecules inside and outside a cell are equal (83)

isotope an atom of an element that has a different number of neutrons than other atoms of that element (34)

J

Jacobson's organ the two pits in the mouth of a snake that contain nerves which are extremely sensitive to odors (581)

jejunum the middle section of the small intestine (698)

joint the place where two bones meet (643)

J-shaped curve a growth curve that describes the lag phase and the exponential phase of population growth (812)

K

karotype a picture of an individual's chromosomes (175)

keratin a protein substance that forms hair, bird feathers, human fingernails, and the horny scales of reptiles (572, 649)

kidney an excretory organ in which wastes are filtered from the blood (700)

kinetic energy the energy of motion (41)

kinetochore fiber a fiber extending from a centriole to a centromere during mitosis (132)

kingdom in taxonomy, a group of similar phyla or divisions (267)

Koch's postulates the steps that Koch formulated for determining the cause of a given disease (673)

Krebs cycle the central process of aerobic respiration (107)

L

labium a liplike structure (498)

labrum a structure forming the roof of the mouth in insects (498)

lac operon the DNA segment on the chromosome of the *E. coli* bacterium that codes for polypeptides that cleave lactose (181)

lacteal a tiny lymph vessel (698)

lactic acid fermentation the process by which pyruvic acid is converted to lactic acid (104)

lag phase a period of little increase in the number of individuals in a population (812)

laminarin a type of polysaccharide with glucose units linked differently from those in starch (324)

large intestine the organ of digestion in which water is reabsorbed (699)

larva an immature, morphologically distinct form of an organism (443)

larynx the voice box that houses the vocal cords (667)

lateral line system a row of sensory structures that run the length of the body of a fish (540)

lateral meristem a growing region that increases the circumference of a plant (382)

leaf the plant organ often specialized for capturing sunlight (381)

legume a member of the pea family (354)

lens a convex crystalline structure in the eye (719)

leucoplast a plastid in plants that stores food (74)

leukemia a disease characterized by the abnormal growth of immature white blood cells (186)

leukocyte a white blood cell (663)

lichen a symbiotic association between fungi and green algae (342)

ligament a connective tissue in a joint (643)

light reactions the phase of photosynthesis in which radiant energy is used to produce molecules of ATP and NADPH + H$^+$ (96)

limbic system a group of neurons in the brain that helps regulate the emotions (710)

linkage group the group of genes, located on the same chromosome, that are usually inherited together (166)

lipid a fat, an oil, a wax, or a steroid made up of carbon, hydrogen, and oxygen (50)

liver a large organ that aids digestion (697)

lobe-finned fish a fish having paddlelike fins with fleshy bases (545)

long-day plant a plant that flowers when exposed to day lengths longer than the plant's critical length (427)

lumen an internal reservoir in a thylakoid (96)

lung the respiratory organ where gas exchange occurs (545, 665)

lungfish a fish having both lungs and gills (545)

lymph the fluid of the lymphatic system (661)

lymphatic system a series of one-way vessels that carries intercellular fluid from tissues to the bloodstream (661)

lymphocyte a type of white blood cell occurring in two primary forms—T cells and B cells (676)

lymphoma a tumor in the tissues that form blood cells (186)

lysis the disintegration of an infected cell (289)

lysogenic cycle the life cycle of a temperate virus (289)

lysosome an organelle containing digestive enzymes, existing primarily in animal cells (70)

lytic cycle the basic reproductive process of viruses (288)

M

macromolecule a large molecule, often a polymer (48)

macronucleus in paramecia, the larger of the two types of nuclei (312)

macronutrient an element used by plant cells in relatively large amounts (386)

macrophage a large white blood cell that can engulf hundreds of bacteria (675)

magnification the increase in an object's apparent size (25)

malignant tumor a cancerous mass of cells (186)

Malpighian tubule the main excretory organ of some invertebrates (487)

mammary gland one of the milk-secreting glands of mammals (604)

mandible a movable mouthpart that usually functions in chewing (482)

mantle the epidermal layer of mollusks (466)

mantle cavity a space between the mantle and the visceral mass in mollusks (466)

map unit a unit in chromosome mapping equal to a 1 percent occurrence of crossing-over (168)

marine biome the earth's oceans (774)

mass the quantity of matter of an object (33)

mass number the sum of the protons and the neutrons in an atom (34)

mass selection the process of choosing a few individuals to act as parents from a larger pool of individuals (191)

mastax a muscular organ that chops up food in rotifers (460)

matter anything that occupies space and has mass (33)

maxilla a structure located behind the mandibles that helps hold and cut food (484)

maxillary teeth the teeth in the upper jaw (560)

maxilliped a specialized appendage used to hold food (484)

measuring the process of determining the characteristics of an object or a group in quantitative units (18)

medulla the innermost portion of an organ (700)

medulla oblongata in fish, the posterior brain lobes that regulate the internal organs; in humans, a part of the brain stem (550, 709)

medusa a bell-shaped stage in the life cycle of a cnidarian (444)

megaspore a spore produced by meiosis in a megasporangium (404)

meiosis the process of nuclear division that reduces the number of chromosomes in a cell by half (136)

meiosis I a division during meiosis in which the homologous chromosomes are separated into separate cells (136)

meiosis II a division during meiosis in which the chromatids are segregated into separate cells (136)

melanin a substance that determines skin color and protects the body from overexposure to sun (650)

memory cell a specialized white blood cell that remains in the blood after an antigen has been destroyed (677)

meninges the three protective layers that surround the brain and spinal cord (710)

menstrual cycle the monthly changes in the female reproductive system (737)

menstruation the release of uterine lining, blood, and unfertilized egg at the end of the menstrual cycle (738)

meristem the growing region in plants (382)

mesentery a fanlike membrane that holds the digestive organs in place (561)

mesoderm a layer of cells in the gastrula that gives rise to muscles and to interior body linings (439)

mesoglea in cnidarians, the jellylike material located between the endoderm and the ectoderm (445)

mesophyll in leaves, the tissue where photosynthesis occurs (395)

messenger RNA (mRNA) the type of RNA that carries genetic information from the nucleus to the ribosomes (118)

metameric unit a repeating part in a segmented body (524)

metamerism segmentation (471)

metamorphosis the changes through which an immature organism passes into adulthood (500, 556)

metaphase the second phase of mitosis, during which all the chromosomes move to the cell's equator (133)

metastasis the breaking away of cells from a malignant tumor to cause new tumors (186)

metazoan a multicellular organism (438)

methanogen a bacterium that produces methane during anaerobic respiration (296)

microfilament a fine protein threadlike organelle (70)

microfossil a microscopic fossil (209)

micronucleus in paramecia, the smaller of the two nuclei, involved in sexual reproduction (312)

micronutrient an element used by plants in relatively small amounts (386)

micropyle in plants, a small pore at one end of the ovule through which the pollen tube passes (407)

microscope an instrument that can form an enlarged image of an object (25)

microsphere a type of microscopic droplet closely resembling a living cell (210)

microspore a spore produced by meiosis in a microsporangium (404)

microtubule a long, slender tube-shaped proteinaceous organelle (70)

microvillus one of the extensions of membranes covering the villi (698)

midbrain the section of the brain stem controlling reflexes (709)

middle lamella the thin, gluelike layer containing pectin that forms in plant cell division where the new cells meet (73)

midgut the stomachlike region of some invertebrates (498)

migration the movement of individual organisms into or out of a population, community, or biome (242, 801)

milt in fish, a fluid containing sperm (550)

mineral an inorganic solid formed in the earth (691, 824)

mitochondrion the organelle that is the site of aerobic respiration in cells (69)

mitosis the division of the cell nucleus (131)

mixed cropping the simultaneous planting of different crops in one field (355)

mixture a combination of two or more substances in which each substance retains its original chemical properties (38)

modeling the construction of a representation of a process or phenomenon (22)

molar a flattened tooth used to grind and crush (604)

mold a type of fossil formed from an impression of the shape or tracks of an organism (219)

molecule a group of atoms held together by covalent bonds (37)

molting the process of shedding an exoskeleton (480)

monocot an angiosperm with a single cotyledon, parallel veins, and flower parts in multiples (375)

monohybrid cross a cross between individuals that involves one pair of contrasting traits (154)

monomer a repeated, single-molecule unit in a polymer (48)

monosaccharide a simple six-carbon sugar such as fructose or glucose (49, 687)

monosomy a condition resulting from nondisjunction during meiosis (175)

morphogenesis the change in form of an organism resulting from cell differentiation (184)

morphological species concept the idea that organisms can be classified by differences in their appearance (237)

morphology the internal and external appearance of an organism (8, 237)

motor neuron a nerve cell in a ventral root that carries impulses from the spinal cord to the effectors (711)

mouth pore the structure at the oral groove into which the paramecium sweeps food (312)

mucus gland an organ that supplies a lubricant to keep the skin moist (560)

Müllerian mimicry the adaptation among several poisonous or bad-tasting species of similar patterns of coloration (502)

multicellular gametangium one of the multicellular structures that form gametes in plants (365)

multicellular organism a living thing that consists of more than one cell (7, 63)

multiple alleles three or more contrasting genes that control a trait (153)

muscle a contractile organ consisting of many cells (645)

muscle fiber a skeletal muscle cell that runs the length of the muscle and contains many nuclei (646)

muscle tissue the body tissue that enables movement (638)

mutagen an environmental factor that damages DNA (164)

mutation a change in DNA (163, 242)

mutualism a form of symbiosis in which both organisms benefit from living together (342, 799)

mycelium a mat of interwoven hyphae (336)

mycorrhiza a symbiotic association between fungi and plant roots (342)

myelin sheath a cellular covering that protects an axon (713)

myofibril a contracting thread in a skeletal muscle (646)

myosin the thick protein filament of a sarcomere (647)

myriapod the common name for centipedes and millipedes (488)

myxameba a uninucleate cell that is the feeding stage of a slime mold (343)

N

narcotic a drug derived from opium (753)

nastic movement a type of plant movement that is independent of the direction of the stimuli (426)

natural immunity an immunity that is present at birth (678)

natural resource a raw material that supports life (821)

natural selection the process by which organisms with favorable variations survive and reproduce at higher rates than those without such variations (230)

necromass dead organic matter in soil (804)

nectar in flowers, a nourishing solution of sugars and amino acids (408)

negative feedback a process in which an end product inhibits the first step of the system (727)

nematocyst a coiled stinger in cnidarians (445)

nephridium a long tubule through which some invertebrates eliminate wastes (473)

nephron the functional unit of the kidney (701)

neritic zone the ocean subarea over the continental shelf (774)

nerve a strand of nerve cells through which impulses travel (707)

nerve ring a nerve surrounding the mouth of a sea star (516)

nervous tissue a type of human body tissue that conducts electrical impulses (638)

net primary productivity the rate at which plants store energy that is not used in plant respiration (783)

neuron a cell that carries electrical signals (707)

neurotoxin a component in snake venom that disrupts the transmission of nerve impulses (581, 673)

neurotransmitter a chemical messenger that carries a nerve impulse from cell to cell (716)

neutron a subatomic particle with no electric charge (33)

neutrophil a small white blood cell with an irregular nucleus (675)

niche the way of life of a species (781)

nicotine the stimulant found in tobacco (749)

nictitating membrane a third membrane that can cover the eyeball (559)

nitrification the process in the nitrogen cycle by which nitrites and nitrates are produced (790)

nitrogen fixation the process by which gaseous nitrogen in the air is converted into nitrates and nitrites (301, 790)

nocturnal referring to organisms that are primarily active by night (800)

nodes of Ranvier gaps in the myelin sheath (713)

nonbiodegradable referring to a pollutant that cannot be decomposed (825)

nondisjunction a mutation caused when a replicated chromosome pair fails to separate during cell division (164)

nonrenewable resource a natural resource that is unreplaceable (823)

nonvascular plant a plant that lacks vascular tissue, and true roots, stems, and leaves (367)

notochord a firm, flexible rod found in the dorsal part of chordates (537)

nuclear envelope the double membrane of lipids and proteins that surrounds the nucleus of a cell (72)

nucleic acid an organic molecule, DNA or RNA, that stores and carries important information for cell function (54)

nucleolus the structure found in most nuclei in which ribosomes are synthesized and partially assembled (72)

nucleosome the histone complex around which DNA wraps itself when coiling into heterochromatin (182)

nucleotide a monomer of DNA and RNA, consisting of a nitrogen base, a sugar, and a phosphate group (54, 113)

nucleus in atoms, the core of protons and neutrons; in eukaryotic cells, the organelle that contains most of the DNA (33, 65)

nutrient a chemical substance an organism needs to grow and function properly (687)

nymph an immature form of some invertebrates (500)

O

objective lens in a microscope, the lens closest to the specimen (25)

obligate aerobe an organism that requires oxygen (301)

obligate anaerobe an organism that lives in the absence of oxygen and dies when exposed to it (301)

obligate intercellular parasite a parasite that requires a host cell in order to reproduce (283)

observing the use of one or more of the five senses to perceive objects or events (18)

oceanic zone one of two subareas of the pelagic zone (775)

ocular lens in a microscope, the lens in the eyepiece (25)

oil a triglyceride that is liquid at room temperature (51)

oil gland a skin gland that secretes sebum (650)

old field succession the replacement of populations in abandoned farm fields (806)

olfactory lobe the brain region that interprets smell (542)

olfactory nerve the nerve that carries impulses from olfactory receptors in the nasal passage to the brain (721)

oligosaccharin a plant hormone that regulates growth and may fight disease (424)

oligotrophic lake a lake that is very low in nutrients (776)

omnivore an animal that eats both plants and animals (611, 784)

oncogene a cancer-causing gene (186)

oogonium a unicellular gametangium that produces eggs (330)

oospore a thick-walled, diploid structure that develops from a zygote (344)

ootid a cell that develops into an egg (138)

open circulatory system a system in which the blood does not circulate entirely within vessels (467)

operator one of three segments of the operon that controls expression of a gene (181)

operculum in fish, a hard plate attached to each side of the head that covers gills and is open at the rear (546)

operon the operator and the structural genes on a DNA segment (181)

opposable thumb a thumb that can be positioned opposite the other fingers (256)

optic lobe one of the paired brain lobes that receives nerve impulses from the eyes (549)

optic nerve the nerve that carries impulses from the retina to the occipital lobe of the cerebrum (719)

oral groove a funnellike structure lined with cilia, used by ciliates for feeding (312)

oral surface the underside of a sea star, opposite the aboral surface (515)

order in taxonomy, a group of similar families (267)

organ several types of tissues that together perform a bodily function (75, 639)

organelle any structure in cells that performs a special function (65)

organic compound a compound that is derived from living things and contains carbon (47)

organism a living thing (5)

organizing data the process of placing observations and measurements in some kind of logical order (18)

organ system a group of organs that interact to perform a set of related tasks (76, 639)

origin the attachment point of a tendon to a stationary bone (648)

osculum an opening at the top of sponges (441)

osmosis the diffusion of water (82)

ossicle a protective spine on a sea star (515)

ossification the process by which cartilage is converted to bone (642)

osteocyte a living bone cell (641)

oval window a membrane that separates the middle ear from the inner ear (720)

ovary in plants, the base of the pistil; an egg-producing organ (406, 446, 730)

over-the-counter drug a drug that can be obtained without a prescription (754)

oviparous referring to organisms that produce eggs that hatch outside the body (582)

ovipositor a pointed organ in some female invertebrates that digs a nest hole and deposits eggs (500)

ovoviviparous referring to organisms that produce eggs that hatch inside the body (582)

ovulation the second stage of the menstrual cycle, in which the follicle ruptures and releases the ripe egg (738)

ovule the megasporangium of a flowering plant (404)

ovum a female gamete or an egg (736)

ozone a secondary pollutant formed when primary pollutants react with sunlight and chemicals (825)

P

paleoanthropologist a scientist who studies human evolution (255)

palisade parenchyma in leaves, a type of mesophyll directly below the upper epidermis, where most photosynthesis occurs (395)

pancreas the organ that lies behind the stomach against the back wall of the abdominal cavity (697)

papilla one of the bumps, mostly on the tongue, between which most of the taste buds are embedded (721)

parapodium one of the fleshy appendages of annelids (471)

parasite an organism that obtains its nutrition at the expense of another (454, 799)

parasitism a symbiotic relationship in which one organism obtains its nutrition at the expense of another (799)

parasympathetic nervous system the part of the autonomic nervous system that controls internal organs (712)

parathyroid gland a gland that secretes hormones that regulate calcium and phosphate ions in the blood (729)

parenchyma the living tissue of the ground system consisting of cells with thin cell walls (381)

parental, or P$_1$, generation a strain obtained through self-pollination (148)

parietal eye a third eye on the top of the head, characteristic of the tuatara (576)

particulate a particle found in smoke (825)

passive transport the movement of any substance across a cell membrane without the use of cell energy (84)

pathogen an organism that causes a disease (303, 673)

pectoral fin a fin located on the side of a fish body just behind the head (543)

pedicellarium in sea stars, a structure surrounding the spine (515)

pedigree a record that shows how a trait is inherited over several generations (171)

pedipalp in arachnids, an appendage that aids in chewing (486)

pelagic zone the open ocean (774)

pellicle a rigid protein covering of some unicellular organisms (312)

pelvic fin in fish, one of the paired fins positioned in the ventral, posterior portion of the body (547)

pennate diatom a type of golden-brown algae with long, rectangular valves (326)

pentaradial symmetry an organism with radii in multiples of five (511)

peptide bond a covalent bond between a nitrogen atom and a carbon atom (53)

pericardium the tough saclike membrane that surrounds the heart (655)

pericycle the outermost cylinder of cells in the central vascular area (384)

periosteum the tough outer membrane of a bone; the point where tendons attach muscles to bones (641)

peripheral nervous system the network of nerves that connects the central nervous system with all parts of the body (712)

peristalsis a series of rhythmic muscular contractions (695)

permafrost a permanently frozen layer of ground (768)

permeable referring to a membrane through which molecules can pass (82)

pesticide a chemical used to control insects (826)

petal the showy part of a flower (405)

petiole in leaves, the structure that attaches blade and stem (394)

phagocyte a white blood cell that engulfs foreign cells (663)

phagocytosis the engulfing of food by a cell (88, 312)

pharyngeal pouch a small outpocket of the anterior gut in a chordate (537)

pharynx the open area at the back of the mouth; in planarians, a muscular tube that extends out of the mouth; in earthworms, a structure through which soil is pumped into the gut (455, 548, 667)

phenotype   the external appearance of an organism (153, 240)

pheromone   a chemical released by an animal that affects the behavior of other animals of the same species (505)

phloem   in plants, the specialized tissue in the vascular system that transports dissolved sugars (371)

phosphorylation   in plants, the transfer of a phosphate group from one molecule to another molecule (95)

photic zone   the layer of the ocean that receives light (774)

photoautotroph   an organism that uses sunlight as an energy source (301)

photochemical smog   the product of sunlight catalyzing a series of chemical reactions among pollutants (825)

photoperiodism   a plant response to changes in day length (427)

photosynthesis   the process of converting the radiant energy of sunlight into the chemical energy of glucose (93)

photosystem   in plants, a unit of hundreds of chlorophyll molecules and associated acceptor molecules (98)

phototropism   plant growth in response to light (425)

pH scale   a scale for measuring the relative concentrations of hydronium ions and hydroxide ions in a solution (40)

phycobilin   an accessory pigment found in red algae and blue-green bacteria (98)

phylogenetic tree   a visual model of the inferred evolutionary relationships among organisms (271)

phylogeny   evolutionary history (6, 269)

phylum   in taxonomy, a group of similar classes (267)

physical addiction   the changes in body chemistry that result in a demand for a steady supply of a drug (755)

physiological variations   differences in the way systems function (493)

phytochrome   a pigment in plants that helps monitor changes in day length (427)

phytoplankton   a photosynthetic microorganism (321)

pia mater   the inner layer of the meninges (711)

pigment   in plants, a light-absorbing compound (97)

pilus   an appendage bacteria use to attach to objects (300)

pinocytosis   the movement into cells of solutes or fluids (88)

pioneer species   the first species to colonize a new habitat (803)

pistil   a plant structure formed of fused carpels (406)

pit   a thin, porous area of a cell wall (381)

pith   the plant tissue located in the center of the stem (388)

pituitary gland   a gland that secretes hormones to control body growth and to regulate the activity of the other endocrine glands (730)

placenta   the organ that nourishes the unborn young of placental mammals (608, 742)

placoid scale   a toothlike spine found on a shark (543)

plant ecology   the study of the interaction of plants and the environment (359)

plant growth regulator   a hormone that affects growth (421)

planula   a free-swimming, ciliated larva (447)

plasma   the liquid component of blood (662)

plasmid   a single ring of DNA in bacteria (195, 300)

plasmodium   the multinucleate cytoplasm of a slime mold surrounded by a membrane that moves as a mass (344)

plasmolysis   the shrinking or wilting of a cell from low turgor pressure (83)

plastid   an organelle in which food or pigments are stored (72)

plastron   the ventral portion of a turtle shell (577)

platelet   a partial cell needed to form blood clots (663)

plumule   in plants, the structure composed of the epicotyl plus any embryonic leaves (411)

point mutation   the change of a single nitrogen base within a codon or larger segments of DNA (164)

polar biome   the areas that surround the earth's poles (767)

polar body   a cell formed during meiosis that receives little or no cytoplasm and eventually disintegrates (138)

polar compound   a compound with one side having a negative charge and the other side a positive charge (47)

polar fiber   a type of spindle fiber that extends across the cell from centriole to centriole (132)

polar nucleus   in plants, a nucleus that migrates to the center of the cell during embryo sac formation (407)

pollen grain   the microgametophyte of seed plants (404)

pollen tube   a structure that grows from a pollen grain, down which the sperm travels to the eggs (404)

pollination   the transfer of pollen from an anther to a stigma of a flower of the same species (148, 408)

pollution   an undesirable change in an ecosystem (825)

polygenic trait   a trait controlled by multiple genes (173)

polymer   a compound consisting of repeated linked monomers (48)

polyp   the vase-shaped stage in the life cycle of a cnidarian (444)

polypeptide   a long chain of several amino acids (53)

polyploidy   a condition in which cells contain multiple, complete sets of chromosomes (193)

polysaccharide   a complex carbohydrate composed of three or more monosaccharides (50, 688)

pons   a part of the brain stem (709)

population   all the members of a species that live in the same area and make up a breeding group (239, 765)

population density   the number of individuals in a population in a given area at a specific time (814)

population growth   an increase in the number of individuals in a population (811)

population growth rate   the change in the number of individuals in a population over time (811)

population sampling   a small number of individuals chosen to represent an entire population (170)

positive feedback   a process in which an end product regulates the first step of the system (727)

posterior   the hind end of an organism (437)

potential energy   stored energy or energy of position (41)

precocial   referring to birds that, when hatched, are physically active and mature (596)

predation   a form of interaction in which one animal kills and eats another (766)

predicting   the process of stating in advance the result that will be obtained from testing a hypothesis (19)

preen gland   the oil-secreting gland of a bird (589)

pregnancy   the phenomenon of carrying an embryo internally (741)

prescription drug   a drug that can be obtained only on written orders from a doctor (753)

pressure–flow hypothesis   an explanation of the movement of sugars in the phloem of plants (389)

primary consumer   any organism that eats a producer (783)

primary germ layer   any of three layers of the inner mass of cells on the inner surface of the blastocyst (740)

primary growth   the tissue produced by apical meristems in plants (382)

primary immune response   a specific defense involving both B cells and T cells (676)

primary mycelium   a heterokaryotic mass of hyphae (338)

primary pollutant   a pollutant emitted directly into the ecosystem (825)

primary producer   an autotrophic organism (782)

primary productivity   rate at which solar or chemical energy is converted into organic compounds in an ecosystem (783)

primary root   the first root to grow out of a seed (383)

primary succession   the process of sequential replacement of populations in an area that has not previously supported life (804)

primate   a member of Primates, the mammalian order that includes humans (255)

principle of dominance and recessiveness   the principle that one factor in a pair may mask the other factor (150)

principle of independent assortment   the principle that each pair of factors is distributed independently (151)

principle of segregation   the principle that each pair of factors separates during gamete formation (151)

prion   a glycoprotein particle implicated in diseases with long incubation periods (287)

probability   the likelihood that a given event will occur by chance (154)

proglottid   one of the square body sections produced by tapeworms (457)

prokaryote   a cell without a nucleus (65)

promoter   one of three segments of an operon that regulates the production of a metabolizing enzyme (181)

prophage   an intracellular bacteriophage that is harmless to the host cell (290)

prophase   the first stage of mitosis (132)

prostaglandin   a lipid that functions as a cell regulator (734)

protein   an organic compound composed of one or more chains of polypeptides, which in turn are formed from amino acids (52, 688)

protein synthesis   the formation of proteins using information coded on DNA and carried out by RNA (120)

proton   a subatomic particle with a positive charge (33)

protonema   an early stage of gametophyte development in mosses that resembles a slender thread (402)

protostome   an organism in which the blastopore develops into the mouth and the coelom arises by schizocoely and whose embryo has determinate cleavage and develops spirally (521)

proventriculus   the first of two chambers in the stomach of a bird (592)

pseudocoelom   a body type characterized by a cavity between the mesoderm and the ectoderm (453)

pseudoplasmodium   a group of cells that act together as a unit to form a sporangium (343)

pseudopodium   a cytoplasmic extension that functions in movement (311)

psychoactive drug   a chemical compound that changes sensory perception and thought processes (753)

psychological addiction   an emotional dependency on a drug (755)

pulmonary artery   the artery that carries deoxygenated blood from the heart to the lungs (659)

pulmonary circulation   the movement of blood between the heart and the lungs (659)

pulmonary respiration   in frogs, a type of respiration involving lungs in which air is inhaled and exhaled by changing the volume and pressure of air in the mouth (563)

pulmonary vein   the vein that carries oxygenated blood from the lungs to the heart (659)

punctuated equilibrium   a theory that speciation may occur during brief periods of rapid genetic change (249)

Punnett square   a method used to establish the probabilities of the results of a genetic cross (154)

pupa   an inactive stage in metamorphosis of some invertebrates (501)

pupil   an opening in the iris that allows light to enter (718)

pure   referring to a strain that always produces offspring with a certain trait (147)

purine   an organic molecule that has a double ring of carbon and nitrogen atoms (113)

pyloric sphincter   the valve that separates the stomach from the small intestine (698)

pyrenoid   an organelle that synthesizes and stores starch (321)

pyrimidine   an organic molecule that has a single ring of carbon and nitrogen atoms (113)

Q

quinine   a drug used to treat malaria (356)

R

radial canal   a canal in a sea star arm, part of the water-vascular system (515)

radial cleavage   in deuterostomes, the division of cells in a parallel position with or at right angles to the polar axis (521)

radial symmetry   an arrangement of body parts in which many planes can divide the organism into similar halves (437)

radicle   in plants, the embryonic root (411)

radioisotope   an isotope with unstable nuclei (35)

radula   a rough, tonguelike structure used in feeding by many mollusks (467)

ray   a tissue that transports food across the xylem (389)

ray-finned fish   a fish with fins supported by bones (545)

reabsorption   a process in the kidney by which materials return to the blood from the nephron (701)

receptacle   the swollen tip of a branch that is the base of a flower (405)

receptor   a nerve cell that receives information from both internal and external stimuli (707)

receptor site   the area of a cell wall where attachment takes place (288)

recessive   referring to an allele that is masked by the presence of another allele for the same characteristic (150)

recombinant DNA   a DNA segment from at least two different organisms (194)

red tide   discoloration of sections of the ocean caused by algal bloom (327)

reflex actions   involuntary movements caused by external stimuli (712)

reflex arc   the pathway of reflex impulses (712)

reforestation   the process of nurturing new growths of trees on sites formerly occupied by forests (821)

refractory period   the time following passage of a nerve impulse, during which the neuron cannot be stimulated (716)

regeneration   the regrowth of missing parts (443)

regulator gene   a gene that controls the production of a genetic repressor molecule (181)

renal circulation   a subsystem of systemic circulation that involves the movement of blood through the kidneys (660)

renal pelvis   the central structure of the kidney (700)

renal tubule   a long tube with permeable walls in a nephron (701)

renewable resource   a natural material that can be replenished (821)

replication   the process of copying a DNA molecule (115)

repressor   a protein that inhibits an operator and thus stops gene expression (182)

reproductive isolation   the inability of formerly interbreeding organisms to produce offspring (248)

resolution   the power of a microscope to show detail (25)

respiration   the processes that release chemical energy for use by the cell (93)

respiratory system the system of organs that functions to transport gases to and from the circulatory system (665)

response a reaction to a stimulus (11)

resting state the state in which a neuron is polarized (714)

restriction enzyme a protein that cuts a DNA molecule into distinct pieces (194)

reticular formation a network of nerve fibers running through the brain stem and the thalamus (710)

retina the light-sensitive inner layer of the eye (718)

retrovirus a virus that contains RNA and reverse transcriptase (286)

reverse transcriptase an enzyme that makes DNA from RNA (286)

Rh factor a type of antigen found in blood that may cause complications during some pregnancies (664)

rhizoid a rootlike structure attaching a gametophyte to soil, rock, or tree bark (337, 369)

rhizome the underground stem of a fern sporophyte (373)

ribonucleic acid (RNA) a single-stranded nucleic acid distinguished from DNA by containing ribose and uracil (54)

ribose the five-carbon sugar in RNA (118)

ribosomal RNA (rRNA) the type of RNA found in a ribosome (118)

ribosome an organelle that functions in the synthesis of proteins (67)

rickettsia a small Gram-negative moneran (297)

ring canal a channel encircling the mouth of a sea star, part of the water-vascular system (515)

rod a photoreceptor that detects light as gray shades (719)

root the organ that anchors a plant, absorbs water and minerals, and stores food made during photosynthesis (381)

root cap a shield covering the apical meristem (384)

root crop a plant grown for the nutrients it stores in its roots or modified underground stems (394)

root hair an extension of the epidermal cells (384)

royal jelly a high-protein substance secreted by worker bees and fed to the queen and the youngest larvae (503)

rumen the storage chamber in the stomach of artiodactyls (611)

S

salicylic acid the chief ingredient of aspirin, originally obtained from willow tree bark (356)

saliva a mixture of water, mucus, and a digestive enzyme called salivary amylase (695)

saprophyte an organism that feeds on dead organic material (301)

sarcoma a tumor growing in bone or muscle tissue (186)

sarcomere a dark band that runs from side to side down each myofibril (646)

scanning electron microscope an electron microscope that scans the surface of an object with a beam of electrons (26)

scavenger a consumer that eats organisms that have recently died (784)

schistosomiasis a disease caused by a parasitic blood fluke of the genus *Schistosoma* (456)

schizocoely the method of mesoderm formation in protostomes (522)

Schwann cell a cell that forms part of the myelin sheath of nerves (713)

scientific method a procedure that is a logical combination of scientific processes used to solve a problem (23)

scientific name the two-word Latin name of a species, always italicized (268)

scientific sampling a technique in which a small sample can be used to represent an entire population (28)

sclerenchyma the tissue of the ground system that supports and strengthens the plant (381)

scolex the head of a tapeworm (457)

scrotum a pouch of skin that encloses the testes (735)

scutes the scales on the belly of a snake (580)

secondary consumer an organism that eats a primary consumer (783)

secondary growth the tissue produced by lateral meristems of plants (382)

secondary immune response the response to a subsequent infection by the same pathogen (677)

secondary mycelium the result when mating strains of homokaryotic septate hyphae fuse (338)

secondary pollutant a chemical that results from some process acting on primary pollutants (825)

secondary succession the sequential replacement of populations in a disrupted habitat (805)

second filial, or F$_2$, generation the offspring of a self-pollinated F$_1$ generation (148)

seed a plant embryo enclosed in a protective coat (373)

seed coat the protective structure of a plant seed (411)

segmentation a division of a body into units (524)

selectively permeable membrane a cell membrane that keeps out some molecules but allows others to pass (66, 82)

self-pollination the transfer of pollen from one flower to another flower on the same plant (147, 408)

semen a fluid formed from sperm and various secretions (735)

semicircular canal a structure in the inner ear that helps maintain balance (720)

seminal receptacle a structure that contains sperm until the sperm is released (474)

sense organ an organ that monitors external stimuli (718)

sensory neuron a spinal nerve cell of a dorsal root that carries impulses from receptors to the spinal cord (711)

sepal a modified leaf that protects a young flower (405)

septum a perforated cross wall dividing hyphae into cells; a thick wall that divides the heart lengthwise (336, 655)

seral community the intermediate community that arises through the process of succession (803)

serology the study of blood serums (284)

sessile referring to an organism that attaches firmly to surfaces and does not move (441)

seta one of the external bristles on annelids (471)

sex chromosome a chromosome that determines sex (165)

sex-influenced trait a trait that is influenced by the presence of male or female sex hormones (175)

sex-linked trait a trait that is determined by alleles on the sex chromosome (166)

sexual reproduction the production of offspring from the exchange of genetic material (7, 138)

sexual selection the preferential choice of a mate based on the presence of a specific trait (246)

shaft the center of a bird feather (589)

shell gland a gland in female birds that secretes a protective calcium carbonate shell to surround an egg (594)

shell membrane the interior of an egg shell (594)

short-day plant a plant that flowers when exposed to day lengths shorter than the plant's critical length (427)

sickle-cell disease a disease resulting from abnormal hemoglobin (171)

sidestream smoke tobacco smoke flowing from cigarettes or exhaled by a smoker (750)

sieve plate a region at the end of a sieve tube member through which compounds move from cell to cell (381, 515)

sieve tube member the conducting cell of angiosperm phloem (381)

simple leaf a leaf with only one blade (394)

sink the place where a plant stores or uses sugars (389)

sinoatrial node the nerve tissue in the right atrium that regulates the contraction rate of the heart (656)

skeletal muscle voluntary muscle, attached to bones by tendons, that moves parts of the body (645)

skeleton the internal framework of the human body (640)

skin gill a hollow tube through which diffusion takes place in sea stars (516)

skull the protective structure for the brain in vertebrates (539)

smooth muscle an involuntary muscle (646)

social behavior the adaptive interactions among members of the same species (627)

sodium–potassium pump an active transport mechanism that moves ions in order to achieve polarization (85, 715)

soft palate a soft area made of folded membranes that separates the mouth cavity from the nasal cavity (695)

sol a colloid in a liquid state (39)

solute a substance dissolved in a solution (38)

solution a mixture in which one or more substances are uniformly dissolved in another substance (38)

solvent the substance into which a solute is dissolved (38)

somatic mutation a mutation that occurs in a body cell (163)

somatic nervous system the part of the peripheral nervous system that controls skeletal muscles (712)

sorus a cluster of sporangia in ferns (403)

source the place where sugars are made in a plant (389)

spawning the egg-laying behavior of some bony fishes (550)

specialization the adaptation of a cell for a particular function (438)

speciation the formation of a new species (247)

species a group of organisms that look alike and are capable of producing fertile offspring in the natural environment (237)

sperm a motile gamete (735)

S phase the second period of interphase (132)

spherical symmetry the arrangement of body parts in the shape of a globe or sphere (437)

spicule a tiny, hard particle shaped like a spike or a star that forms the skeleton of some sponges (442)

spinal cavity the area around the vertebral column (637)

spinal cord a column of nerve tissue running through the vertebral column (711)

spinal nerve in fish, a nerve connection between the spinal cord and an organ or a muscle (550)

spindle fiber one of the microtubules of protein that develop during the middle period of prophase in mitosis and that aid in chromosome movement (70)

spinneret the structure through which fluid from the silk gland of spiders passes (487)

spiracle an opening that functions in respiration in spiders and grasshoppers (487)

spiral cleavage in protostomes, the division of cells in a spiral arrangement (521)

spirochete a spiral-shaped bacterium (297)

spongin the network of protein fibers making up the simple skeleton of some sponges (442)

spongy bone the lacy network of connective tissue in the center of a bone (641)

spongy parenchyma a type of mesophyll that allows oxygen and carbon dioxide to diffuse into and out of a leaf (395)

spontaneous generation the development of living organisms from nonliving material (205)

sporangiophore an upright hypha in some fungi that produces sporangia (337)

sporangium a spore-bearing sac in plants and algae (330, 368)

spore a resistant reproductive cell of some organisms (366)

sporophyte the diploid generation of land plants that produces spores (330, 368)

springwood secondary xylem with cells that are wide and thin walled (389)

S-shaped curve a growth curve that depicts stable, lag, and exponential phases (823)

stabilizing selection a type of natural selection in which the average form of a trait causes an organism to have an advantage in survival and reproduction (244)

stalk a stem or stemlike structure (338)

stamen the structure in the flower of plants that produces pollen (406)

stem an organ of plants with the primary functions of support, transport, and storage (381)

steroid a lipid in which the molecule is composed of four carbon rings (51, 735)

stigma the tip of a style in plants (406)

stimulant a drug that increases the activity of the central nervous system (749)

stipe the stemlike region of a thallus (324)

stolon the horizontal hypha of some fungi (337)

stoma an opening in the epidermis that regulates the passage of gases into and out of a plant (102, 382)

stomach an organ that prepares food for absorption by both physical and chemical activity (695)

stone canal part of the water-vascular system in sea stars (515)

strain the term used to denote all organisms pure for a specific trait (147)

striation a dark band of skeletal muscle that functions in contraction (645)

strip mining a type of mining in which layers of earth are removed to gain access to minerals (824)

strobilus a conelike structure of sporangia-bearing leaves (372)

stroma in plants, a solution surrounding the grana (96)

structural gene one of three segments of an operon that controls the production of an enzyme (181)

structural variation a difference in morphology (493)

style a stalklike structure of a pistil (406)

substrate the reactant in a chemical reaction (54)

succession the predictable, sequential replacement of populations in an ecosystem (803)

summerwood secondary xylem cells that are small and thick walled and that are produced in hot, dry summers (389)

superior vena cava the large vein carrying blood from the upper part of the body (657)

suppressor cell a modified T cell that prevents further antibody production after infection (677)

suspension a mixture in which particles spread through a liquid or gas but settle out over time (38)

sweat gland a skin gland that excretes water (650)

swimmeret an organ in crustaceans that creates water currents and functions in reproduction (484)

symbiosis the relationship between different species living in close association with one another (766)

symmetry an arrangement of body parts around a point or central axis (437)

sympathetic nervous system the part of the autonomic nervous system that controls internal organs during conditions of high stress or increased activity (712)

synapse the space between the axon terminals of one neuron and an organ or the dendrite of another neuron (713)

synapsis the pairing of homologous chromosomes during meiosis (136)

syrinx the vocal organ of birds (593)

systemic circulation the movement of blood between the heart and all parts of the body except the lungs (659)

systole the phase of the heartbeat in which the ventricles contract and force blood into the arteries (657)

T

tadpole the swimming larva of toads or frogs (557)

taproot the organ that results when the primary root grows downward and becomes the largest root (383)

tar a mixture of chemicals and particles from tobacco (749)

target cells the specific cells affected by a hormone (727)

target tissue an area that is affected by hormones, triggering a physiological response (421)

tarsus the end of the leglike appendages of some invertebrates having spines, hooks, and pads (498)

taste bud a structure in which chemoreceptors for taste are clustered (721)

taxonomy the science of grouping organisms according to their presumed relationships (267)

T cell a lymphocyte that attacks antigens (676)

tegument the tough covering that protects some invertebrates (454)

telophase the final stage of mitosis (133)

telson a flat triangular section at the posterior tip of the seventh segment of the abdomen in crayfish (487)

temperate referring to a virus that does not cause disease for some time after it enters an organism (283)

temperate rain forest rain forest biome characterized by moderate temperatures and high humidity (773)

tendon a tough, solid band of connective tissue that supports and connects body parts (645)

tentacle an appendage with which an organism captures and paralyzes prey (444)

terracing a method of plowing that carves a series of steps out of hillsides in order to limit erosion by water (822)

territoriality the establishment of an area by one organism that it defends against others of the species (597)

test a compact, rigid endoskeleton that is found in some invertebrates (514)

testcross the crossing of an individual of unknown genotype with a homozygous recessive individual (156)

testes the gamete-producing organs of a male (446, 730)

tetrad the grouping of two homologous chromosomes during meiosis (136)

thalamus the structure that directs incoming sensory impulses to the proper region of the cerebral cortex (709)

thallus the body of an alga (321)

theory the most probable explanation for a large set of data (23)

therapsid the reptile group from which mammals evolved (603)

thermoacidophile an archaebacterium that lives only in hot, acidic places (296)

thigmotropism plant growth in response to contact with a solid object (425)

thoracic cavity the upper ventral cavity of the human body, containing the heart, lungs, and esophagus (637)

threatened species those plants and animals that are likely to become endangered in the near future (821)

thylakoid a flattened sac of a photosynthetic membrane in a plant chloroplast (96)

thymine a nitrogen base, one component of a nucleotide (113)

thyroid gland the endocrine gland that secretes thyroxine (728)

tidal rhythm a period of activity that corresponds to the rise and fall of ocean tides (802)

tissue in most multicellular organisms, a group of similar cells that carry out a common function (75, 381, 638)

tissue culture the growing of living cells in a controlled medium (283)

tolerance a lessening response to a drug as a result of physical addiction (755)

topsoil the uppermost layer of soil (822)

torpor a state of dormancy (556)

torsion the process in gastropods that allows an animal to protect its head when threatened (467)

total lung capacity the amount of gas that the lungs are able to hold (666)

toxin a chemical substance that is harmful to the normal functioning of cells (303, 673)

trachea in insects, one of a network of air tubes; the tube that connects the pharynx to the lungs (487, 667)

tracheid a long, thick wall with tapering ends, forming part of the xylem tissue of plants (381)

transcription the process in which RNA is made from DNA (118)

transduction the process in which a virus carries genetic material from one cell to another (290)

transfer RNA (tRNA) the type of RNA that carries amino acids in the cytoplasm to the ribosomes (118)

translation the process of converting the DNA code in RNA into the amino acid sequence that makes up a protein (121)

translocation a mutation in which a broken piece of chromosome attaches to a nonhomologous chromosome (163)

transmission electron microscope a microscope that transmits a beam of electrons through a thinly sliced specimen (26)

transpiration the evaporation of water through stomata (386)

transpiration–cohesion theory the theory that water may move up a plant because it is a cohesive fluid (391)

trichinosis the disease caused by the round worm genus *Trichinella* (460)

triglyceride a lipid made of three fatty acid molecules and one glycerol molecule (51, 689)

trimester one of three divisions of pregnancy (741)

trisomy a condition resulting from nondisjunction during meiosis (175)

trochophore a type of larva of mollusks and annelids (465)

trophic level a feeding level in an ecosystem (782)

tropical rain forest a biome near the equator with large amounts of rain and sunlight (772)

tropism plant movement toward or away from an environmental stimulus (425)

truncus arteriosus a vessel in a frog's heart that emerges from the conus arteriosus (563)

trypanosomiasis any of the various forms of sleeping sickness (314)

tube cell the cell of a pollen grain (407)

tube foot a small protrusion on an echinoderm arm that functions in movement (511)

tumor an abnormal mass of cells (186)

tundra a biome of low-growing vegetation that forms a continuous belt across North America, Europe, and Asia (768)

turgor pressure water pressure within a cell (83)

tympanic membrane the eardrum (559, 720)

tympanum a sound-sensing membrane in a grasshopper (499)

U

ultrasound a procedure using sound waves to diagnose genetic disorders in a fetus (176)

umbilical cord the cord that connects the placenta and the body of the unborn mammal (742)

ungulate   a hoofed mammal (611)

unicellular gametangium   a single-celled gamete holder (321)

unicellular organism   a living thing composed of one cell (7, 63)

uracil   a nitrogen base found in RNA (118)

ureter   a narrow tube through which urine flows from the renal pelvis to the urinary bladder (702)

urethra   the tube through which urine flows from the urinary bladder out of the body (702)

urinary bladder   a muscular sac that stores urine (702)

urine   a liquid containing wastes that have been removed from the blood (701)

uropod   an anterior flipper that propels some crustaceans through water (484)

uterus   in flukes, a tube that shelters eggs until they are released; a structure that houses a developing fetus (456, 736)

V

vaccination   inoculation with a pathogen to produce immunity to a disease (678)

vacuole   a fluid-filled cavity found in plant cells (72)

valve   one half of a diatom wall (326)

vane   the projection on either side of a bird's feather (589)

vascular bundle   a strand of vascular tissue formed near the apical meristem that lies embedded in the cortex (388)

vascular cambium   the lateral meristem in a plant that produces additional vascular tissue (382)

vascular cylinder   the innermost part of a root, containing xylem and phloem (384)

vascular plant   a plant that has xylem and phloem (367)

vascular system   the system that includes the xylem and phloem tissues of plants (381)

vascular tissue   the tissue that transports water and food from one part of a plant to another (365)

vegetative propagation   asexual reproduction in plants (415)

vein   a vessel carrying blood to the heart (549, 657)

venation   the arrangement of veins in a leaf (394)

ventral   the bottom surface of a bilaterally symmetrical organism (437)

ventral cavity   the cavity containing the thoracic and abdominal cavities (637)

ventricle   the lower chamber or chambers of the heart (549, 655)

venule   a small vessel in the network of veins (657)

vernalization   the requirement that some seeds or spores be exposed to cold before germination (428)

vertebral column   the backbone (537)

vessel element   an elongated, water-conducting plant cell wall with slanting ends and holes through which water can pass (381)

vestigial organ   a functionless structure that may be homologous to useful structures in other species (223)

villus   one of the small projections lining the small intestine (698)

viroid   a short, single strand of RNA that causes disease (286)

virulent   referring to a disease-causing agent (283)

virus   a biological particle composed of genetic material and protein that is not usually considered to be a living organism (283)

visceral mass   the interior of an organism, containing the organs of digestion, excretion, and reproduction (466)

vital lung capacity   the maximum amount of air a person can inhale or exhale (666)

vitamin   a complex organic compound that serves as a cofactor for enzymes (690)

viviparous   referring to organisms that carry and nourish the young inside the body before birth (604)

vomerine teeth   the teeth in frogs used to hold prey (560)

W

warning coloration   in poisonous or bad-tasting organisms, bold bright color patterns that warn predators of possible danger (502)

watershed   the area of land that drains into a river (823)

water-vascular system   a network of water-filled canals in echinoderms (511)

wax   a type of lipid (51)

weed   a plant that grows quickly in disturbed areas (415)

wildlife   all the organisms indigenous to a natural community (821)

withdrawal   a reaction to the lack of a drug (755)

woody plant   a perennial plant adapted for growth year after year, such as a tree (382)

X

xanthophyll   a common carotenoid found in plants (98)

xylem   the vascular tissue in plants that transfers water and minerals from the roots to the leaves (371)

Y

yolk sac   in amniote eggs, the membrane that encloses the yolk and supplies the embryo with food (571, 742)

Z

zooplankton   unicellular, heterotrophic organisms that drift in water (310)

zoospore   a flagellated spore (328)

zygospore   the protective structure that results when the wall surrounding a zygote thickens (328)

zygote   the cell that results from the fusion of gametes (438)

Index

Page references to illustration are printed in boldface type.

arthroscopic surgery, 642–643
artificial propagation, 416
Artiodactyla, 611–612
Ascaris, 459
ascus, 339
ascocarp, 339
ascogonium, 339
Ascomycota, 339–340, **339**
ascospores, 339
asexual reproduction, 7, 7, 138. *See also* reproduction
    in invertebrates, 528
    in plants, 415–416
Aspergillus, 341
association neurons, 711
Asteroidea, 513
asters, 132, **132**
atomic number, 34, **34**
atoms, 33, **33**, 34–35
ATP, 69, **104**, 210
    in chromosome movement, 133
    conversion of sugars into, 687
    energy storage in, 94, **94**
    hormone action on, 733, **733**
    molecular structure of, 94, **94**
    in muscle contractions, 647–648
    in respiration, 104, 107–108
    in sodium–potassium pump, 85
    thyroxine and, 728
ATP–ADP cycle, 94–95, **95**
ATP synthetase, 95
atrioventricular node, **656**, 657
atrioventricular valves, 655, **656**
atrium, 549, 655
auditory canal, 720
auditory nerve, 720
Australopithecus, 258–259
autoimmune disease, 679
autonomic nervous system, 707, 712
autosomes, 165
autotomy, 579
autotrophs, 9, 93–94, **94**, 301, 342, 782–783
auxins, 421–423
    discovery of, 422–423
    synthetic, 421–422
    and thigmotropism, 425
AVEC, 70–71
avoidance behavior, 313
axial skeleton, 539, 641
axons, 713

B

bacteria. *See also* monerans
    classification of, 295–298
    diseases caused by, 303, 304
    evolution of, 295
    resistance of to antibiotics, 304
bacteriophages, 288
Bakker, Robert T., 574–575
bald eagle, 781
baldness, 175
baleen, 610, **610**
baleen whale, 610, **610**
Banks, Harlan, 366–367
barbs (in bird feathers), 589

barbules, 589
bark, 389
barnacles, 483
basal metabolic rate (BMR), 668
    and thyroid disorders, 728
bases, 39–40
basidiocarp, 338
Basidiomycota, 338–339
basidiospores, 338
basidium, 338
basket stars, 513
bats, 610, **610**
Batesian mimicry, 502
B cells, 676–677, **676**
bears, 611
beavers, 781–782
behavior, 11, **12**
behavioral variation, 493
bell curve, 240, **240**, 244, **245**
benthic zone, 774
beriberi, 692–693
berries, 355, 412
beta-carotene, 98
bilateral symmetry, 437, **437**, 524
bile, 697
binary fission, 302, **302**, 312
Binnig, Gerd, 26
binomial nomenclature, 267–268
biodegradable pollutants, 825
bioethics, 10
biogenesis, 207
biogeochemical cycles, 788–791
biogeography, 812–813
biological clock, 800
biological magnification, 826
biological species concept, 238–239
bioluminescence, 327
biomass, 787, **787**
biomass pyramids, 787, **787**
biomes, 767
    aquatic, 774–776
    coniferous forest, 769
    deciduous forest, 770
    desert, 772
    freshwater, 776
    grassland, 771
    marine, 774–775, **774**
    polar, 767–768
    and remote sensing, 770–771
    terrestrial, 767–773, **767**, 768
    tundra, 768–769
biorhythms, 733
biosphere, 765
biosystematics, 271
biotic factors of ecosystems, 766, 781
biotic potential, 811
biotic relationships, 797–799. *See also* competition; food webs; predation; symbiosis
bipedalism, 256
bipinnaria, 516, 523
birds
    adaptations of, 590, **590**
    behavior of, 596–598
    characteristics of, 589–594
    classification of, 588, **588**

embryo development of, 595, **595**
evolution of, 587–588
migration of, 596–507, 598
reproduction in, 595–596
birth, 744
bison, 611
Bivalvia, 468–469
bladder. *See* urinary bladder
blade, 394
    on thallus, 324
blastocyst, 740, **740**
blastopore, 438
blastula, 184, 438
blastulation, 438
blood, 662–664
blood alcohol concentration (BAC), 751–752
blood clot, 663
blood types, 174, 663–664
blood vessels, 657–658
blue-green bacteria, 98, 212–213, 297–298
bolting (of plants to seed), 424
bone marrow, 641
bones, 641, **641**. *See also* skeletal system
    development of, 642–643
bony fishes, 545–550, **546**, 548
book lungs, 487
Borlaug, Norman, 355
botulism, 303, 304
bouton, 716
Bowman's capsule, 701
Brachiosaurus, 574
Bragg, Lawrence, 116–117
brain (human), 708–711, **708**
    and the senses, 718, **718**
brain stem, 709–710
Branchiostoma, 538, **538**
breathing, 666, **666**. *See also* respiration; respiratory system
    regulation of, 668
bristlecone pines, 374
bristles (in birds), 589
brittle stars, 513, **513**
bronchi, 667
bronchioles, 667
bronchitis, 749
Brontosaurus, 574
brood patch, 595
brown algae, 324
Bryophyta, 367, 369–370. *See also* nonvascular plants
buds, 387
bud scales, 387
Bufonidae, 557
bundle sheath, 395
Burbank, Luther, 191
bursa, 644

C

cactuses, 393, **393**
caecilians, 558, **558**
Caenorhabditis elegans, 458–459
caimans, 577–578, **578**
calcium, and parathyroid glands, 729

hydrophobic molecules, 51
hydrotropism, 426
Hydrozoa, 446–447
hyenas, 611
hyperthyroidism, 728
hypertonic solution, 83, **83**
hypha, 336, **336**, 337
hypocotyl, 411, 413, **414**
hypoglycemia, 732
hypothalamus, 709–710, 730–732, **730**
hypothesizing, 19
hypothyroidism, 728
hypotonic solution, 83

I

ichthyosaurs, 573
ileum, 561, 698
imaginal disks, 185, **185**, 501
immune disorders, 678–679
immune system, 639, 675–679, 732
immunity, 678–679
implantation, 740
imprint fossil, 219
imprinting, 627
inbreeding, 192
incomplete digestive system, 526
incomplete metamorphosis, 500–501, **500**
incurrent siphon, 468
independent assortment, 151
indeterminate cleavage, 422
indirect development, 523, 622
indoleacetic acid (IAA), 421, **421**
inducer, 182
infectious disease, 673–674. *See also specific diseases*
  nonspecific defenses, 675–676
  specific defense, 676–677
  transmission of, 674
inferior vena cava, 657, 660, 661
inferring, 21–22
infertility, 740–741
inflammatory response, 675, **675**
inheritance, 7, 147
"inheritance of acquired characteristics," 225, **225**
innate behavior, 503
inorganic compounds, 47–48
inorganic nutrients, 687
Insecta, 494
Insectivora, 608–609
insectivorous plants, 393, **393**
insects, 493–496
  adaptation in, 493, **493**
  behavior of, 503–506
  characteristics of, 493
  classification of, 493
  communication among, 504–506
  defense systems of, 502
  effects of on humans, 494–495, 496, **496**
  evolution of, 495
  external structure of, 497–498, **497**
  internal structure of, 498–500, **498**, **499**

natural selection in, 495
  orders of, 494–495
  and pollination, 496
  success of, 493–495
insertion (muscle), 648
insight learning, 627
inspiration, 666
instinct, 503
insulin, 52–53, 732
integumentary system, 623, 639, 649–650
integuments, 407, 547
interbreeding, 237
interferon, 194, 676
intermediary metabolism, 42
internal respiration, 665
International Shorebird Survey, 596–597
interphase, 131, **131**, 132, 136
interspecific competition, 797–798
intertidal zone, 775, **775**
intestine, 698, 699
intraspecific competition, 797
introns, 183, **183**
inversion (chromosome mutation), 163
invertebrates, 437–439
  classification of, 439
  coelom formation in, 522–523
  development in, 521–523
  embryo development of, 521–522
  evolution of, 439, **439**, 521
  germ layers in, 522–523
  internal anatomy of, 526–528,
  larva development in, 523
  morphology of, 524, **524**
  phylogeny of, 524, **525**
  structural support of, 524
iodine, and thyroid disorders, 728–729
ionic bonds, 36
iris, 718
islets of Langerhans, 732, **732**
Isoetes, 372, **372**
isomers, 49
isopods, 483
isotonic solution, 83
isotopes, 34

J

Jacob, François, 181
Jacobson's organ, 581
jejunum, 698
jellyfish, 444–445, 447–448, **447**
jimson weed, 360
joints, 643–644, **643**, **644**
jojoba, 357
J-shaped curve, 812, **812**

K

Kanno, Yoshinobu, 86–87
Kaposi's sarcoma, 681
karyotype, 175–176
keratin, 572, 649
Kettlewell, H. B., 228–229
kidneys, 549, 561, 592, 624, 660, 700–702, **700**
kilocalorie, 687

kinetic energy, 41
kinetochore fibers, 132, 133
kingdoms, 272–274
Klinefelter syndrome, 175
Koch, Robert, 673
Koch's postulates, 673, **673**
Komodo dragon, 579
kozo fiber, **356**
Krebs, Hans, 107
Krebs cycle, 105, **106**, 107–108
Kuffler, Stephen W., 86

L

labia (in humans), 736
labium (in insects), 498
labrum, 498
lac operon, 181–182
lactase, 698
lacteals, 698
lactic acid fermentation, 104–105
Lagomorpha, 609
lag phase, 812–813
lakes, 776
Lamarck, Jean Baptiste de, 225
laminarin, 324
lampreys, 541–542
lancelet, 538, **538**
LANDSAT satellites, 770–771
large intestine, 699
larva
  of amphibians, 556
  ancestral and modern, 619
  development in invertebrates, 523
  of flukes, 456
  of mollusks and annelids, 465, **465**
  of sponges, 443
larynx, 667
late blight fungus, 344
lateral line system, 540
lateral meristems, 382, 387, 388
Latimer, Mary, 545
LAV virus. *See* HIV virus
Leakey, Richard, 260
leaves, 381, 393–396
  adaptations in, 393, **393**, 395
  evolution of, 365
  functions of, 395–396
  gas exchange in, 396
  structure of, 394–395, **394**
leeches, 472–473, 474
Leeuwenhoek, Anton van, 63
legumes, 354, 412
Leishmania, 314–315
lemurs, 613
lens, 719, **719**
leucoplasts, 74
leukemia, 186
leukocytes, 663
Lyell, Charles, 226
lichens, 342
life
  appearance of on earth, 208–211
  characteristics of, 11–12
  first forms of, 212–214
  origin of, 204–214

T cells, 676–677, **676**
tegument, 454
telophase, 133–134, 136, 137
telson, 483
temperate rain forest, 773
temperate viruses, 283
tendrils, 393, **393**
terracing, 822
terrestrial biomes, 767–773, **767**
territoriality, 597
test (in endoskeleton), 514
testcross, 156
testes, 730
testosterone, 730
tetanus, 303, 304
tetrad, 136, **136**, **137**
T4 cells, 680–681
thalamus, 709–710
thalloid algae, 321
thallus, 321
thecodonts, 573, **574**, 587
theory of endosymbiosis, 309
therapsids, 603, **603**
thermal vent communities, 775, 775
thermoacidophiles, 296
thiamine, 693
thigmotropism, 425
Thomson, J. J., 34
thoracic cavity, 637, **637**
thorax, 497
threatened species, 821
thrombin, 663
thylakoids, 96, **97**
thymine, 113
thymosins, 732
thymus, 732
thyroid gland, 728–729, **729**
thyroxine, 566, 728
tidal rhythms, 802
tissue culture, 283–284
tissues, 75–76
    of humans, 638, **638**
    of plants, 381–382
toads, 557
toadstools, 338
tobacco, 749–750, **750**
toenails, 650
tomatoes, 355
toothed whale, 610
topsoil, 822
torpor, 556
torsion, 467
tortoises, 576–577
total lung capacity, 666
touch, sense of, 722
toxins, 303, 673
Toxoplasma, 315
toxoplasmosis, 315
T phages, 288, **288**
trachea, 487, 499, 667
tracheids, 381, **381**, 389
Trachystoma, 558
transduction, 290
transfer RNA, 118
translation, 121–123
translocation, 163

transmission electron microscope (TEM), 26, 63
transpiration, 386, 395
transpiration–cohesion theory, 391, **392**
transverse colon, **694**, 699
tree ferns, 372, **372**
trees, evolution of, 366–367
Trematoda, 455–457
Trichinella, 460
trichinosis, 460
triglycerides, 51, 689
Trilobita, 481
triploid, phase in plant reproduction, 409
trisomy, 175
tRNA, 118
trochophore, 465, **465**, 523
trophic levels of ecosystems, 782–783, 785
trophoblast, 740
tropical rain forest, **359**, 772–773, 783, **783**
tropisms, 425–426
Trypanosoma, 314–315
trypanosomiasis, 314
trypsin, 697
tuataras, 576, **576**
tube cell, 407
tube foot, 511
tubers, 387, **387**, 392
tube worms, 302–303, 775
tumor, 186
tundra, 768–769, **769**
tunicates, 538, **538**
Turbellaria, 454–455
turgor pressure, 83, 396, 524
Turner syndrome, 175
turtles, 576–577
    adaptations of, 577
twin studies, 170
2,4-D, 421–422
tympanic membrane, 559, 720
tympanum, 499
typhoid fever, 303
Tyrannosaurus, 574

U

ulcers, 696
ultrasound, 176
Ulva, 323
umbilical cord, 742
ungulates, 611–612
unicellular gametangia, 321, 330
unicellular organisms, 7, **62**, 63, 65
Uniramia, 481–482
unsaturated fats, 689
uracil, 118
ureter, 702
urethra, 702, 735
Urey, Harold C., 210
urinary bladder, 702
urinary ducts (in frogs), 563
urine, 702, 735
Urochordata, 537, 538
Urodela, 558

uropods, 484, **484**
uterus, 456, 736

V

vaccination, 678
vacuoles, 72, 73–74, 83
vagina, 736
vanes (bird feathers), 589
van Leeuwenhoek, Anton, 63
variation, 239–240
    and adaptation, 415
    behavioral, 493
    physiological, 493
    structural, 493
vascular bundles, 388, **388**, 395
vascular cambium, 382, **385**, 389, **389**
vascular cylinder, 384
vascular plants, 367, 371–376
    gametophytes and sporophytes in, 368
    specializations of, 371
vascular system, of plants, 381–382
vascular tissue, 365, 371
    in plants, 382
    in rhizomes, 373
    in roots, 384
vas deferens, 594, 735
vasopressin, 732
Vedder, Edward, 693
vegetative propagation, 415, **415**
veins, 549, 657, **657**
    valves in, 658, **658**
venation, 375–376, 394
ventral cavity, 637
ventricle, 549, 655
venules, 657
vernalization, 428
vertebral column, 537
Vertebrata, 537, 539
vertebrates
    behavioral adaptations of, 627–628
    embryo development of, 621–622
    evolution of, 619, **619**
    morphology of, 621–626
    phylogeny of, 620
vessel elements, 381, 398
vestigial organs, 223
villi, 548, 698, **698**
vipers, 581, **581**
Virchow, Rudolph, 64
viroids, 286, **286**
viruses, 283–284
    classification of, 284–286
    compared with cells, 283
    evolution of, 290
    reproduction of, 288–290, **288**, **289**, **290**
    structure of, 284–285, **284**
    virulence of, 283
visceral mass (in mollusks), 466
vision, 718–719
vital lung capacity, 666
vitamin B_1, 692–693
vitamins, 690–691
    overdoses of, 691
    synthesis of in body, 691

Credits

Illustrations

Ligature 303, 421, 423, 424, 427 **Precision Graphics** 6, 18, 21 (b), 34, 48, 50, 51, 52, 53, 54 (b), 83, 94 (b), 98, 129, 131, 137 (t), 138, 146, 147, 149, 152, 164, 168, 171, 173, 175, 197, 199, 207, 208, 209, 210, 212, 219, 222, 228, 239, 240, 241, 243, 245, 249, 251, 253, 259 (chart), 261 (chart), 262, 263, 274, 293, 368, 376, 391 (t), 423, 429, 439, 451, 482, 493-494, 493 (t), 497, 498, 500, 503, 504, 505, 509, 525, 539, 541, 543, 545, 546, 548, 555, 557, 560, 561, 562, 563, 564, 565, 566, 573, 574, 585, 587, 588, 591, 592, 596, 605, 619, 669 (graphics), 681, 687, 689, 697, 727, 737, 755, 779, 782, 783, 786 (b), 787, 789, 797, 798 (b), 800, 803 (graphics), 812, 813, 815, 816, 819, 823 **Nadine B. Sokol** 13, 21 (t), 43, 154, 233, 422, 705, 723 **George V. Kelvin, Science Graphics** 33, 35, 37, 39, 114, 115, 119, 121, 122, 123, 124, 146, 150, 155, 156, 157, 158, 181, 282, 284-285, 288, 289, 290, 676-677, 679, 682, 683 **Douglas Schneider** 22, 24, 236, 270, 331, 358, 363, 391 (b), 417, 440, 456 (sheep), 461, 551, 793, 804-805 **Cecile Duray-Bito** 2, 60, 67 (br), 144, 202, 235, 257, 261 (skulls), 280, 350, 422, 435, 453, 454, 456, 457, 460, 465, 466, 467, 469, 473, 507, 534, 538, 540, 549, 620, 762, 772, 798 (t), 803 (trees) **Sarah Woodward** 220, 224, 364, 401, 402, 403, 405, 406, 407, 409, 410, 411, 412, 414, 416, 419, 437, 438, 441, 443, 445, 446, 447 **Harry Whitver** 22 (arrows); 25, 27, 28, 93, 94 (t), 95, 96, 99, 101, 102, 103, 104, 106, 107, 108, 790 **Joe LeMonnier** 40, 47, 49, 54 (t), 87, 163, 165, 166, 167, 179, 183, 185, 195, 196, 217, 221, 233, 300, 302, 311, 312, 313, 341, 353, 354, 489, 522, 523, 529, 714-715, 767, 771, 774 **Stansbury, Ronsaville, Wood, Inc.** 66, 67 (t), 68, 69 (t & b), 70, 73, 75, 77, 219, 223, 225, 366, 367, 374, 381, 382, 387, 389, 392, 394, 396, 571, 578, 581, 728 (t), 729, 733, 734, 735, 740 **Cynthia Turner Alexander** 80, 81, 84, 85, 88, 132, 133, 134, 135, 136, 137 (b), 237, 471, 589 **Andrew Grivas** 97, 105, 327, 328, 329, 330, 333, 336, 338, 339, 511, 514, 515, 593, 594, 595, 728 (b), 730, 732, 736, 739, 750 **Howard S. Friedman** 259 (figures), 603, 604, 607, 608-609-611, 785, 786 (t), 791 **Scott Thorn Barrows** 316, 694, 695, 696, 698, 700, 701, 702 **Walter Stuart** 479, 484, 485, 487, 655, 656 (b), 657, 668, 671 **Floyd E. Hosmer** 621, 622, 623, 624, 625, 626, 656 (t), 658, 660, 661, 665, 666 **Keith Kasnot** 637, 638, 640, 641, 644, 645, 646-647, 648, 649 **Robert Margulies** 707, 708, 709, 711, 712, 713, 716, 719, 720, 721, 725

Photographs

AA—Animals, Animals; PR—Photo Researchers
Front cover Charles Krebs; **Back cover** David Muench; **i** Anne Moos, Southern Light; **ii-iii** Jeffrey Rotman; **xii** Ken Brate, PR (b); Peter Turner, Image Bank (t); **xiii** Lou Jones (tr); Jeff Foott, Tom Stack and Assoc. (l); David Muench (br); Art Wolfe, Image Bank (in); **2** Carl Roessler, FPG (b); **3** C. Allan Morgan (l); G.C. Kelley (b); **4** Jeff Gnass, West Stock (t); **5** Bob Gossington, Bruce Coleman; Gerry Ellis, Manhattan Views (l); **7** CNRI Science Library, PR (r); P.A. McTurk, PR (l); **8** Culver Pictures (b); **9** Jeffrey Rotman (r); **9** William Hopkins (l); **10** G. Zysler, Peter Arnold; **11** Gerald Corsi, Tom Stack and Assoc. (r); Kjell Sandved, Bruce Coleman (l); **12** Root Resources, Hillstrom Stock Photography (r); C.B. & D.W. Frith, Bruce Coleman (l); **15** Kevin Schafer, Tom Stack and Assoc. (c); **16** David Scharf (b); Henley and Savage, Click/Chicago (t); **17** Steve Mangold; **18** Richard/Mary Magruder (r); Luiz Claudio Marigo, Peter Arnold (l); **19** Ruth Dixon; **20** David Cavagnaro; **23** Fred Whitehead, AA; **25** Michael Abbey, PR; **26** Jerry Burgess, PR; IBM (c); IBM (t); **29** Dan McCoy, Rainbow; **31** Tony Brain, PR (t); Tony Brain, PR (b); **32** John Shaw, Tom Stack and Assoc. (t); M.I. Walker, PR (b); **34** Culver Pictures; **35** William Ferguson; **36** John Colwell, Grant Heilman; **38** Runk/Schoenberger, Grant Heilman; **39** Biology Media; **40** Gary Milburn, Tom Stack and Assoc. (l); Doug Wilson, The Stock Solution (r); **42** Omikron/Science Source, PR (l); **45** D.R. Specker, Earth Scenes; **46** M.I. Walker, PR; **47** William Hopkins; **50** Fred Whitehead, AA; **52** Arthur Lesk, PR; **55** Runk/Schoenberger, Grant Heilman (t); Runk/Schoenberger, Grant Heilman (br); Runk/Schoenberger, Grant Heilman (tr); Runk/ Schoenberger, Grant Heilman (bl); **57** D. Kirkland, Image Bank; **58** Steve Ogden, Tom Stack and Assoc. (r); Chris Arend, Alaska Photo/Aperture (l); **59** Chris Arend, Alaska Photo/Aperture by Nancy Simmerman, Alaska Photo/Aperture (l); **60** Michael Abbey, PR; **61** The Incredible Machine/ Nat'l. Geo. Soc., Lennart Nilsson (t); Jerome Pickett-Heaps, PR (b); **62** Biophoto Assoc. /Science Source, PR (in); The Incredible Machine/Nat'l. Geo. Soc., Lennart Nilsson (b); **63** Michael Abbey, PR; P. Dayanandan, PR (br); Kitty Kohout, Root Resources (bl); **64** Komura/D.W. Fawcett (l); David Scharf (c); PR (r); **65** Lee Simon, PR (b); Ann Smith/ Science Source, PR (t); **68** Don Fawcett, PR; **69** Bob Bolendern, PR (b); K.R. Porter, PR (t); **71** Gopal Murti, PR (t); Don Fawcett/E. Roth, PR (b); **72** K. Hausmann, Peter Arnold; **74** Biophoto Assoc., PR; **76** Peter Parks, AA; **79** P. Manfred, Peter Arnold (l); Walter Dawn (r); **80** Omikron, PR; **82** Lester Bergman & Assoc. (t); Lester Bergman & Assoc. (b); **86** Robert Knauf, PR; **89** William Hopkins; **91** D. W. Fawcett, PR; **92** Jeremy Burgess, PR (inb); Keith Porter, PR (int); Frederick Atwood; **94** Stephen Krasemann, Alaska Photo/Aperture (t); **97** Jeremy Burgess/Sci. Photo Libr, PR; **100** E.R. Degginger; **101** George Gersier, PR; **109** R. Taylor, PR; **112** Dan McCoy, Rainbow (b); Tom Rooker, Rainbow (t); **116** Science Source, PR; **117** UPI/Bettmann Newsphotos; **118** Omikron, PR; **120** Michael Abbey, PR (t); Jeffrey Rotman (b); **125** Dan McCoy, Rainbow; **128** CNRI/Science Photo Library, PR (t); Dan McCoy, Rainbow (b); **130** Biophoto Assoc., PR (l); **133** Lester Bergman & Assoc. (l); Lester Bergman & Assoc. (c); Lester Bergman & Assoc. (c); Lester Bergman & Assoc. (r); **134** Biophoto Assoc./Science Source, PR; **135** D. Fawcett, PR; **139** Runk/Schoenberger, Grant Heilman (l); Runk/Schoenberger, Grant Heilman (c); Runk/ Schoenberger, Grant Heilman (r); **141** M.I. Walker, PR; **142** UOP Fluid Systems (l); Gianni Torotre, PR (r); **143** Tom Pantages; **144** David Cavagnaro; **145** Biophoto Assoc. /Science Source, PR; **146** Fiona Sunquist, Tom Stack and Assoc.; **150** Biophoto Assoc. /Science Source, PR; **151** Dan McCoy, Rainbow; **153** Karl Maslowski, PR (l); Karl Maslowski, PR (r); **159** Runk/Schoenberger, Grant Heilman; **162** The Incredible Machine/ Nat'l. Geo. Soc., Lennart Nilsson, (t); R. Knanft, PR (b); **169** Dan McCoy, Rainbow (r); **172**

R. Roseman, Custom Medical Stock Photo (t); M. Fisher, Custom Medical Stock Photo (b); **174** L. Rhodes, Click/Chicago (c); L. Rhodes, Click/Chicago (l); L. Rhodes, Click/Chicago (r); **176** Diana Rasche; **177** William Hopkins; **180** Louis Quitt, PR (t); John Gerlack, Tom Stack and Assoc. (b); **182** Don Fawcett, PR; **183** Gerald Corsi, Tom Stack and Assoc. (r); Robert Simpson, Tom Stack and Assoc. (l); E.R. Degginger (in); David Dennis, Tom Stack and Assoc. (int); **184** Science Photo Library, PR; **185** Dan McCoy, Rainbow; **187** E.R. Degginger; **189** David Scharf (b); **190** Gopal Murti, PR (t); Grant Heilman (b); **191** Runk/Schoenberger, Grant Heilman (l); Luther Burbank Home & Gardens (r); Grant Heilman (c); **192** B. Ullman, Taurus Photos; **193** William Ferguson; **194** Hieter P., Mann C., Snyder M., Davis R. © 1985; **195** Hieter P., Mann C., Snyder M., Davis R. © 1985; **200** Scott Camazine, PR (l); Stephen Brown, Life Magazine 1985 Time Inc. (r); **201** Stacy Pick, Stock, Boston (l); David Doody, Tom Stack and Assoc. (r); **202** William Ferguson; **203** Edward Ross (c); Edward Ross (b); **206** Photofile (t); California Institute Tech., Palomar Observatory (b); **211** Michael Abbey, PR (t); S.M. Awramik, University of California, Santa Barbara (b); **213** D.W. Deamer; **214** Lee Simon, PR (b); **215** Omikron, PR (r); Biophoto, PR (l); **216** Richard Kokar, AA (b); Rod Planck, Tom Stack and Assoc. (t); **218** Tui DeRoy, Bruce Coleman; Julia Cameron, The Bettmann Archive (t); **223** Hans Pfletschinger, Peter Arnold (t); Christy Volpe, Marilyn Gartman (b); **226** T. Pritchett, Mary Evans Picture Library; **228** Heather Angel, Biofotos; **230** Mark Kelley, Alaska Photo/ Aperture (b); Breck Kent (t); **231** C. Allan Morgan, Peter Arnold (r); Kevin Schafer, Peter Arnold (l); **232** Mickey Gibson, AA (l); Patti Murray, AA (r); **236** Charles Mauzy, Alaska Photo/Aperture; **238** William Hopkins; **242** D. & J. Heaton, Click/Chicago (t); D. & J. Heaton, Click/Chicago (b); **243** Tui DeRoy, Bruce Coleman; **244** E.R. Degginger; **246** John Chellman, AA; **247** Peter Sanchez, Death Valley National Monument (int); Peter Rowlands, Death Valley National Monument (inb); Peter Rowlands/Bessken, Death Valley National Monument (b); Tim Coonan, Death Valley National Monument (t); **248** Zig Leszczynski, AA (b); Zig Leszczynski, AA (t); **250** Stephen Krasemann, Alaska Photo/ Aperture (t); Tim Clark (r); **254** Enrico Ferorelli, DOT (b); C. Allan Morgan, PR; **255** Michael Diek, AA; **256** Nancy Adams, Tom Stack and Assoc. (b); Co Rentmeester, Image Bank (t); **258** Steve Ward, Anthro-Photo File; **259** John Reader; **260** Alan Walker, National Museum of Kenya (b); E.R. Degginger (t); **265** John Chellman, AA; **266** Patti Murray, AA (c); E.R. Degginger, AA (r); Sea Studios, Peter Arnold (t); Roger Archibald, AA (l); **267** Zig Leszczynski, AA; **268** Dee Wilder; **269** Johnny Johnson, Alaska Photo (t); J. & M. Ibbotson, Alaska Photo/Aperture (t); **271** Science Photo Library, PR; **272** John Edmond, Woods Hole Oceanographic Institute; **273** John Shaw, Bruce Coleman; **275** E.R. Degginger; **277** Eric Grave, PR; **278** NASA (t); NASA (b); **279** Dan McCoy, Rainbow (l); Ray Nelson, Phototake (b); Dan McCoy, Rainbow (r); **280** F.E. Unverhau, AA; **281** CNRI/Science Photo Library, PR; **282** Science Photo-Library, PR; **286** T.O. Diener, United States Department of Agriculture (t); Historical Pictures Service (b); **287** Science Photo Library, PR; **288** Science Photo Library, PR; **289** Lee Simon, PR; **291** Larry Lefever, Grant Heilman; Grant Heilman (in); **294** Jeremy Burgess, PR (t); Science Photo Library, PR (b); **295** Tony Brain, PR (l); David Scharf (r); Science Photo Library, PR (c); **296** William Ferguson; **297** Science Photo Library, PR (c); Michael Abbey, PR (b); Science Photo Library, PR (r); Tony Brain, PR (l); **298** T.E. Adams, Click/Chicago (l); Biophoto, PR (c); T.E. Adams, Click/Chicago (r); **299** CNRI/Science Photo Library, PR; **300** Science Photo Library, PR (b); Woods Hole Oceanographic Institute (t); Woods Hole Oceanographic Institute (c); **305** Runk/ Schoenberger, Grant Heilman; **307** Manfred Kage, Peter Arnold; **308** Michael Abbey, PR (r); Manfred Kage, Peter Arnold (l); **309** Michael Abbey, PR; **310** E. McArdle, Custom Medical Stock Photo (r); Center for Disease Control (l); E.R. Degginger; E.R. Degginger (l); **314** Richard/Mary Magruder; **315** Oxford Scientific, AA; **317** Manfred Kage, Peter Arnold; **319** Eric Grave, PR; **320** Biophoto Assoc., PR (r); Howard Hall, Tom Stack and Assoc. (l); **321** M.I. Walker, PR; **323** Biophoto Assoc., PR (b); M.I. Walker/Science Source, PR (l); Phil Degginger, E.R. Degginger (r); **324** Tom Stack, Tom Stack and Assoc. (t); Tom Smoyer, Harbor Branch Foundation (b); **325** E.R. Degginger (t); Julie Houck, Stock, Boston (c); **326** Biophoto Assoc., PR (l); Biophoto Assoc, PR (r); M.I. Walker, PR (t); **329** Dan McCoy, Rainbow; **334** Ed Reschia, Peter Arnold (t); Michael Fogden, AA (b); **335** C.C. Lockwood, AA; **336** Steve Solum, West Stock; **339** J.L. Lepore, PR; **340** Jeremy Burgess, PR (t); J. Forsdyke (b); **342** G.R. Roberts (t); Biophoto Assoc., PR (r); **343** J. T. Bonner, Princeton University Press; **344** Rod Planck, Tom Stack and Assoc. (t); Kim Taylor, Bruce Coleman (b); Biophoto Assoc., PR (c); **345** E.R. Degginger; **347** Ray Simons, PR; **348** Jeffrey Rotman (t); Norma Morrison (c); **349** Greg Vaughn, Tom Stack and Assoc. (t); Richard Smith, Tom Stack and Assoc. (b); **350** David Muench; **351** Kerry Givens, Tom Stack and Assoc. (t), Chuck Place (b); **352** Dan McCoy, Rainbow (t); Martin Rogers, Click/Chicago (b); **354** Photri, Marilyn Gartman (t); Herta Newton, Click/Chicago (c); **355** Odyssey Productions, Robert Frerck (c); G.R. Roberts (t); **356** Chuck O'Rear, Click/Chicago (b); William Hopkins (t); **357** Dan McCoy, Rainbow; **359** G.R. Roberts (b); Richard LaVal, AA (l); **360** James Carmichael, Nature Photographics (r); E.R. Degginger (r); Marcia Griffen, AA (c); **361** William Hopkins; **364** Chuck Place; **365** Linda Graham; **366** Harlan Banks; **367** Breck Kent, AA; **369** Fred Bavendam; **370** Heather Angel, Biofotos (l); E.R. Degginger; **371** Peter Zika; Heather Angel, Biofotos (l); G.R. Roberts (r); **372** Peter Zika (tc); Heather Angel, Biofotos (l); William Ferguson (tr); **373** William Ferguson (b); William Ferguson (t); **374** Michael Fogden, AA; **375** Matt Brown, West Stock; **377** Runk/Schoenberger, Grant Heilman (l); Runk/Schoenberger, Grant Heilman (b); Barry Runk, Grant Heilman (r); **379** Pamela Zilly, Image Bank; **380** Dan McCoy, Rainbow (b); Chuck Place (t); **383** Walter Chandoha (tl); Dan McCoy, Rainbow (tr); Brian Parker, Tom Stack and Assoc. (tc); Thomas Hovland, Grant Heilman (tr); **384** Runk/Rannels, Grant Heilman (t); Runk/Schoenberger, Grant

Heilman (b); 385 George Whiteley, PR (r); Omikron, PR (t); Manfred Kage, Peter Arnold (l); 387 Walter Chandoha (b); Walter Chandoha (l); Frank Ledermann, Manhattan Views (t); 388 P. Dayanandan, PR (tr); John Shaw, Bruce Coleman (b); E.R. Degginger (tl); 389 Bruce Iverson; 390 Science Photo Library, PR; 393 C. Allan Morgan (c); John Colwell, Grant Heilman (bl); Ruth Dixon (r); 394 Walter Chandoha (t); Runk/Schoenberger, Grant Heilman (b); 395 Runk/Schoenberger, Grant Heilman; 396 Jeremy Burgess, PR; 397 E. J. Cable, Tom Stack and Assoc.; 399 David Muench; 400 Rod Planck, Tom Stack and Assoc. (t); Edward Ross (b); W. Eberhart, Click/Chicago (c); 403 Heather Angel, Biofotos; 404 W. Geiersperger, Click/Chicago; 405 Raymond Barnes, Click/Chicago (b); Wally Eberhart, Click/Chicago (l); 407 David Phillips, Taurus Photos (c); David Phillips, Taurus Photos (t); 408 Heather Angel, Biofotos; 413 Lois Moulton, Click/Chicago; 415 Kitty Kohout, Root Resources; 420 Kenneth Hayden, Click/Chicago (l); Joe Devenney, Image Bank (t); John Sims, Click/Chicago (c); Willard Clay, Click/Chicago (r); 422 Runk/ Schoenberger, Grant Heilman; 425 Terry Domico, AA; 426 Zig Leszczynski, AA (l); Randy Moore, Flora and Fauna Stock Photos (t); 428 Bill Ross, West Light (l); Grant Heilman (r); 431 Phil Degginger, Click/Chicago; 432 Stuart Westmoreland, Tom Stack and Assoc. (l); Richard Quataert, Taurus Photos (r); 433 AB Assoc. (t); Lawrence Migdale (b); 434 Michael Fogden, AA (r); Lynn Funkhouser (l); 436 Zig Leszczynski, AA (l); Brian Parker, Tom Stack and Assoc. (r); 442 Al Grotell (l); David Hall, Image Bank (r); 444 David Hall, Image Bank (b); Carl Roessler (t); 446 Heather Angel, Biofotos; 448 F. Stuart Westmore, Tom Stack and Assoc.; 449 Oxford Scientific, AA; 452 CNRI/Science Library, PR (b); Sea Studios, Peter Arnold (t); 454 Carl Roessler; 455 T.E. Adams, Click/Chicago; 458 Martin Rogers, Click/Chicago; 459 E.R. Degginger (t); Lester Bergman & Assoc. (br); E. McArdle, Custom Medical Stock Photo (bl); 463 E.R. Degginger; 464 Robert Holland, Marilyn Gartman (t); Alex Kerstitch (c); 467 Jeff Foott (t); Don Fawcett, PR (b); 468 Mr./Mrs. M.B. Proudfit; 470 Steinhart Aquarium, PR (c); Carl Roessler, AA (t); 472 Jim Stevenson, PR; 473 Michael Markin, Bruce Coleman; 474 Sea Studios, Peter Arnold; 475 Brian Parker, Tom Stack and Assoc.; 477 W. Gregory Brown, AA; 478 W. Curtsinger, PR (l); Jim Shives, Alaska Photos/Aperture (c); Tom McHugh, PR (t); James Carmichael, Nature Photographics (b); 480 Tom McHugh, PR (b); James Rowan, Marilyn Gartman (t); 481 Bentech (l); William Hopkins (r); 483 Oxford Scientific, AA (l); Bill Curtsinger, PR (c); J.P. Jackson, PR (r); 486 Tom McHugh, PR (l); George Dodge, Tom Stack and Assoc. (c); Jeremy Burgess, PR (r); 487 Rod Planck, PR; 488 C. Perkins, AA (t); Oxford Scientific, AA (c); 491 K. Preston-Mofham, AA; 492 Donald Speaker, AA (t); William Ferguson (c); 493 Science Photo Library, PR (b); Biophoto Assoc., PR (t); 496 D. Wilder, PR (r); David Scharf, PR (l); 498 Richard LaVal, AA; 499 David Scharf; 501 Charles Palek, AA (t); Patti Murray, AA (b); 502 Rod Planck, Tom Stack and Assoc. (b); Patti Murray, AA (t); 504 Edward Ross; 506 Rudolf Freund, PR; 510 L. Rhodes, AA (t); Carl Roessler (b); 512 Carl Roessler, Tom Stack and Assoc. (t); Jeffrey Rotman (b); 513 Carl Roessler (b); Charles Seaborn, Odyssey Productions (c); G. Schatten, PR (r); G. Schatten, PR (l); 514 Steve Lissau, Rainbow (c); Carl Roessler (t); 516 Thomas Kitchin, Tom Stack and Assoc.; 517 Doug Wechsler, AA; 519 Dave Woodward, Tom Stack and Assoc.; 520 Rod Planck, Tom Stack and Assoc. (r); Jeffrey Rotman (b); Carl Roessler (t); 521 Michael Abbey, PR; 524 Jeff Foott (c); Robert Holland (r); 528 David Scharf; 529 Lynn Stone (b); Jeff Foott (t); Carl Roessler, AA (in); 531 E.R. Degginger; 532 Rod Planck, Tom Stack and Assoc. (r); William Ferguson (l); Grant Heilman (t); 533 Grant Heilman (t); Stephanie Ferguson, William Ferguson (b); 534 C. Place, Image Bank; 535 Art Wolfe, Image Bank; 536 Alex Kerstitch (int); Carl Roessler, FPG (inb); Dave Woodward, Tom Stack and Assoc.; 537 Dave Woodward, Tom Stack and Assoc.; 538 G. Bernard, AA; 541 Heather Angel, Biofotos; 542 Gretchen Allen, Mote Marine Foundation; 543 Carl Roessler, AA; 544 Jeffrey Rotman; Carl Roessler, Tom Stack and Assoc. (b); 547 Hans Reinhard, Bruce Coleman; 550 Terry Domico, West Stock; 553 Lynn Funkhouser; 554 Joseph Collins, PR (t); Michael Fogden, AA (b); 556 Ruth Dixon (b); Oxford Scientific, AA (c); 557 Zig Leszczynski, AA (b); Michael Fogden, AA (tl); Zig Leszczynski, AA (b); 559 Stephen Dalton, AA; 566 Oxford Scientific, AA; 567 Zig Leszczynski, AA; 569 George Porter, PR; 570 Breck Kent (t); David Dennis, Tom Stack and Assoc. (b); 572 Zig Leszczynski, AA (c); Henry Aysloos, AA; Zig Leszczynski, AA (t); 575 E.R. Degginger; 576 Heather Angel, Biofotos; 577 Ann Duncan, Tom Stack and Assoc. (l); Breck Kent (r); 579 C. Allan Morgan (l); J. Robinson, AA (c); Carl Roessler (r); 580 Anthony Bannister, AA; 582 Zig Leszczynski, AA; 583 Runk/Schoenberger, Grant Heilman; 586 David Cavagnaro (c); Robert Perron, Nawrocki Stock Photo (b); R.J. Erwin, DPI (t); 590 William Boehm, West Stock (br); Ralph Hunt Williams, West Stock (c); Nicholas Devore, Photographers/Aspen (bl); James Carmichael, Image Bank (r); Stan Osolinski, FPG (l); 595 Bo Jarner, Marilyn Gartman (c); Bo Jarner, Marilyn Gartman (t); Bo Jarner, Marilyn Gartman (l); 597 Jeff Foott, Tom Stack and Assoc. (b); Richard Kolar, AA (t); 598 Anthony Mercieca, Root Resources (tl); Terry Murphy, AA (b); Pete/ Alice Bengeyfield, The Stock Solution; 599 Stephen Dalton, PR (t); David Cavagnaro (b); 601 G.C. Kelley; 602 Zig Leszczynski, AA

(b); Paulo Curto, Image Bank (t); 606 Robert Frerck, Odyssey Productions (t); Hans/Judy Beste, AA; 609 Hans/Judy Beste, AA; 610 Stephen Dalton, AA (t); Bill Byrne (b); 611 Michael Dick, AA (l); Jack Swenson, Tom Stack and Assoc. (r); Art Wolfe, Image Bank (c); 612 Alfred Thomas, AA (t); Jenny Garst, Tom Stack and Assoc. (b); 613 Rick Browne, Stock, Boston; 614 Nicholas Devore, Bruce Coleman (r); K. & K. Ammann, Bruce Coleman (l); 615 Fred Whitehead, AA (r); David Fritts, AA (l); 617 Gary Milburn, Tom Stack and Assoc. (r); Anthony Bannister, AA (l); 629 Bill Frantz, Click/Chicago; 631 Zig Leszczynski, AA; 632 William Boehm, West Stock (r); Jack Elness, Photofile (l); 633 University of Hawaii (l); Thomas Kitchin, Tom Stack and Assoc. (r); Jack Wilburn, AA (t); 634 Gary Gladstone, Image Bank (r); Michael Radigan, The Stock Market (l); 635 Shostal Assoc. (t); NY Hospital, Peter Arnold (in); 636 Michael Abbey, PR (b); Team Russell, Click/Chicago (t); 642 David Madison (b); Martin Rotker, Taurus Photos (t); 643 The Incredible Machine/Nat'l. Geo. Soc., Lennart Nilsson (t); Art Stein, PR (b); 646 Michael Abbey, PR; 647 Terry Kirk, Tom Stack and Assoc.; 650 Sipa-Press, Focus West; 651 Townsend Dickinson, PR; 653 Ed Reschica, Peter Arnold (t); 654 The Incredible Machine/Nat'l. Geo. Soc., Lennart Nilsson (t); Tony Brain, PR (b); 657 The Incredible Machine/Nat'l. Geo. Soc., Lennart Nilsson; 659 North Wind Picture Archives; 662 David Scharf; 663 Manfred Kage, Peter Arnold; 667 Manfred Kage, Peter Arnold; 669 William Hopkins; 672 NCI/Science Source, PR (t); The Incredible Machine/ Nat'l. Geo. Soc., Lennart Nilsson (b); 673 CNRI/Science Photo Library, PR; 678 Jeremy Burgess, PR (c); UPI/Bettmann Newsphotos (b); 680 Luc Montagnier, PR (c); CDC/ Science Source, PR (t); Alfred Pasieka, Taurus Photos; 685 A. Liepins, PR; 686 Manfred Kage, Peter Arnold (b); The Incredible Machine/Nat'l. Geo. Soc., Lennart Nilsson (t); 688 William Hopkins; 693 Manfred Kage, Peter Arnold; 696 Ed Reschke, Peter Arnold; 698 Biophoto Assoc./Science Source, PR; 699 Lester Bergman & Assoc.; 701 Biophoto Assoc., PR; 703 William Hopkins; 706 The Incredible Machine/Nat'l. Geo. Soc., Lennart Nilsson (b); The Incredible Machine/Nat'l. Geo. Soc., Lennart Nilsson (t); 710 Dan McCoy, Rainbow; 711 The Incredible Machine/Nat'l. Geo. Soc., Lennart Nilsson (r); Spencer Grant, Picture Cube (b); 713 Science Photo Library, PR; 717 Don Fawcett, PR; 718 Dan McCoy, Rainbow; 719 Ralph Eagle, PR; 722 David Scharf; 726 A Child Is Born-Dell Pub. Co., Lennart Nilsson (t); Dan McCoy, Rainbow; 729 Biophoto Assoc., PR; 738 David Scharf; 740 Dan McCoy, Rainbow; 741 UPI/Bettmann Newsphotos (r); Hank Morgan, Rainbow (l); 743 The Incredible Machine/Nat'l. Geo. Soc., Lennart Nilsson; A Child Is Born-Dell Pub. Co., Lennart Nilsson (tl); Behold Man-Little Brown & Co., Lennart Nilsson (bl); The Incredible Machine/Nat'l. Geo. Soc., Lennart Nilsson (br); Behold Man-Little Brown & Co., Lennart Nilsson (cl); 745 Runk/Schoenberger, Grant Heilman; 747 A Child Is Born-Dell Pub. Co., Lennart Nilsson; 748 Breck Kent, AA (b); Gary Gladstone, Image Bank (t); 749 Martin Rotker, Taurus Photos (c); Martin Rotker, Taurus Photos (b); 752 F. Skvara, Peter Arnold; 754 Benn Mitchell, Image Bank; 756 Michel Viard, Peter Arnold (r); W.H. Hodge, Peter Arnold (c); W.H. Hodge, Peter Arnold (l); 757 M. Walker, PR; 760 Richard Kolar, AA; 761 Steve Dunwell, Image Bank; 762 Keith Murakaml, Tom Stack and Assoc. (t); NASA (b); 764 Jeff Gnass (t); Ed Robinson, Tom Stack and Assoc. (b) 765 Leonard Rue, Click/Chicago; 766 Harold Simon, Tom Stack and Assoc.; 769 Leonard Rue, AA (b); Nancy Simmerman, Click/Chicago (t); 770 John Shelton (t); Jet Propulsion Laboratory (b); 771 E.R. Degginger; 772 K.G. Preston Mafhamr, AA (r); G.C. Kelley, Tom Stack and Assoc. (l); C. Allan Morgan (c); 773 William Ferguson (br); Co Rentmeester, Image Bank (tr); James Carmichael, Nature Photographics (l); 775 Dudley Foster, Woods Hole Oceanographic Institute (t); Art Wolfe, Alaska Photo/ Aperture (b); 776 Sheryl McNee, Tom Stack and Assoc.; 777 E.R. Degginger, AA; 780 Gary Ladd (t); R. Armstrong, AA (b); 781 Chuck Place; 782 Ron Austing, PR; 783 Anthony Mercierk, Root Resources; 784 Peter Katsaros, PR (l); Zig Leszczynski, AA (r); 788 United States Department of Agriculture; 790 G.R. Roberts; 792 Kenneth Fink, Bruce Coleman; 795 Mickey Gibson; 796 Anthony Mircieca, Hillstrom Stock Photography (t); D.C. Lowe, Alaska Photo/Aperture (b); 799 John Shaw, Tom Stack and Assoc. (r); Jeffrey Rotman (b); Charles Seaborn, Odyssey Productions (c); 801 Tim Sharp, Earth Images; 802 Art Wolfe, Alaska Photo/ Aperture; 806 William Ferguson; 807 E.R. Degginger; 809 Gary Braasch, Alaska Photo/ Aperture; 810 Leonard Nadel, After Image; Russ Kinne, COMSTOCK (b); 812 Leonard Nadel, After Image; 814 Richard Laird, FPG; 817 Eric Grave, PR; 820 Phil Degginger, E.R. Degginger (t); Earl Roberge, PR (b); 821 F. Gohier, PR; 822 Len Kaufman; 824 William Ferguson (t); Greg Vaughn, Tom Stack and Assoc. (c); 825 NASA; 827 Bohdan Hrynewych, Southern Light; 829 Doug Wechsler, AA; 830 K. Fink, Bruce Coleman (r); D. Lowe, Alaska Photo/Aperture (l); 831 Andrew Popper, Phototake (b); Dan McCoy, Rainbow (t); 832 David Scharf (b); Lou Jones (c); Lou Jones (t); 833 David Austen, Click/Chicago (t); David Muench (b); Behold Man—Little Brown & Co., Lennart Nilsson (tr); Herbert Wagner, Phototake (cr).